MODERN GENERAL PSYCHOLOGY

Thoroughly Revised and Expanded

VOLUME-I

About the Author

M. Rajamanickam is the Founder-Director of the Institute of Community Guidance and Research, which is one of the permanent Members of the Indian Council of Social Science Research. Formerly, he was the professor and Head of the Department of Psychology and the Dean of the Faculty of Education and Psychology of the Annamalai University. He was been teaching Psychology to the Post-Graduate students and guiding research for over three decades of the University and produced many Ph.D. and M.Phil. Degrees.

He has been elected to the Syndicate of the University three times. He was mainly responsible for introducing Psychology in the scheme of Distance Education, which has attracted more than 500 students every year all over India for the study of Psychology at the M.A. and B.A. Degree levels. He extended the Psychological services to the Rural people through the Institute of Community Guidance and Research which he founded in 1980.

He has been associating himself with a number of Academic and Professional Organizations and contributed quite large number research papers to the national and international journals. He founded the Journal of Community Guidance and Research in 1984 and has been the Editor of it for more than twenty years, which has benefited many young scientists in India and in other countries. He is a prolific writer, has published more than ten books on Psychology. Some of the important titles are, Contemporary Fields of Psychology and Experiments, Statistical Methods in Psychological and Educational Research, Modern Psychophysical and Scaling Methods and Experimentation, Modern General Psychology, Abnormal Psychology, Experimental Psychology with Advanced Experiments (in 2 Volumes) and other publications.

By the Same Author

1. Contemporary Fields of Psychology and Experiments
2. Statistical Methods in Psychological and Educational Research
3. Modern Psychophysical and Scaling Methods and Experimentation
4. Psychology of Perspective of HIV and AIDS
5. Experimental Psychology with Advanced Experiments (in Two Volumes)

MODERN GENERAL PSYCHOLOGY

Thoroughly Revised and Expanded

VOLUME - I

M. RAJAMANICKAM

M.A., M.Litt., Ph.D.

Formerly Dean, Faculty of Education and Professor and Head of the Department of Psychology

Annamalai University

CONCEPT PUBLISHING COMPANY, NEW DELHI-110 059

Cataloging in Publication Data—DK
Courtesy: D.K. Agencies (P) Ltd. <docinfo@dkagencies.com>

Rajamanickam, M. (Muniappa), 1925-
Modern general psychology./ M. Rajamanickam.— Rev. ed., thoroughly rev. and expanded.
2 v. cm.
Includes bibliograpahical references (p.)
Includes index.
ISBN 13: 9788180694219 (set)
ISBN 10: 8180694216 (set)
ISBN 13: 9788180694226 (v. 1)
ISBN 10: 8180694224 (v. 1)
ISBN 13: 9788180694233 (v. 2)
ISBN 10: 8180694232 (v. 2)

1. Psychology—Textbooks. I. Title.

DDC 150 22

Set	ISBN - 13: 978-81-8069-421-9 (HB)	ISBN - 10: 81-8069-421-6 (HB)
Vol. I	ISBN - 13: 978-81-8069-422-6 (HB)	ISBN - 10: 81-8069-422-4 (HB)
Vol. II	ISBN - 13: 978-81-8069-423-3 (HB)	ISBN - 10: 81-8069-423-2 (HB)

Revised Edition 2008

Published and Printed by
Ashok Kumar Mittal
Concept Publishing Company
A/15-16, Commercial Block, Mohan Garden
NEW DELHI-110059 (India)
Phones : 25351460, 25351794
Fax : 091-11-25357103
E-mail : publishing@conceptpub.com

PREFACE TO THE SECOND EDITION

In bringing out the Second Edition of the book "Modern General Psychology", I was very much profited and inspired by the suggestions made by many of the teachers who are teaching psychology in various Universities and Colleges in India. The increased acceptance of this book at a different levels resulted in bringing out the second edition, not in the original form, but by making great amount of changes in the light of recent development of the discipline. All the chapters are thoroughly revised, expanded and reinterpreted. The structure of each chapter is reshaped and enriched with many new information. The recent experimental results from various disciplines like biology, physiology, anatomy and neuropsychology are related to the subject matters of this book. While revising the chapters special attention was given to some of the chapters particularly to the brain and nervous system, intelligence and personality and motivation and behaviour.

The scope of this book has been increased by adding twelve more chapters in addition to the revision and expansion of the other chapters. These additional chapters include five in general psychology, namely Human Development, Consciousness, Frustration, conflict and adjustment, Nature of intelligence, and Personality assessment. The other chapters are Psychopathology, Psychotherapy, Cognitive and behaviour therapy and Health psychology under the Part of Psychopathology and Mental health. The chapters on Social psychology are Social behaviour and Attitudes and related traits. A chapter on psychological Statistics is also added.

Thus the subject matter of this book may suit to the syllabus of any University in India and I am quite sure that the students would be highly benefited by this book. The second edition is brought out in a quite different manner. The text is divided into eleven parts based on the functions of the psychological processes. Some Textbook authors do not give equal importance to all the psychological processes. They combine two or three themes under one heading and discuss the subject matter briefly. Some Universities include in their syllabus some aspects of Behaviour disorder and Social behaviour. This book would fulfil those requirements and may prepare the students for advanced work in the subject.

Psychologists may find that this book would survey the whole basic psychological processes of the behaviour of an individual and also an individual's place in a group. The student of psychology would have a comprehensive understanding about the role of the brain and nervous system in behaviour. This book is a sincere attempt to present the scientific biological basis of contemporary psychology and to explain the fundamental process of human behaviour. It is a modest attempt to present the origin and development of the human organism and how it functions through various stages of development. I have also attempted to present and maintain the biological emphasis which considered the human individual as a complex organism, existing in a complex environment and gradually adapting it. The student of psychology in this book is exposed, as far as possible, to the diversity of basic psychological research and also to provide him an understanding of the contemporary psychological approach with a scientific spirit and analyse the complex human behaviour.

I gratefully express my indebtedness to Mr. Ashok Kumar Mittal, the Publisher of the Concept Publishing Company for having brought out this book in an exciting manner.

Chennai
2007

M. RAJAMANICKAM

CONTENTS

Volume - I

PART ONE : Scientific Study of Behaviour, Theories and Systems

James, W. (1842-1910)

J.B. Watson (1878-1958)

I.P. Pavlov (1849-1936)

Chapter - 2

P.F. Skinner (1904-)

A. Maslow (1908-1970)

PART TWO : Biological Bases and Developmental Processes

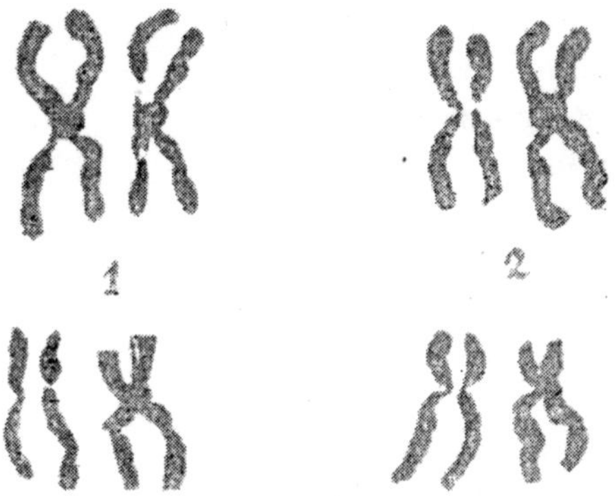

Chromosomes

Fertilized Ovum

Chapter - 5 **109-182**

Structure and Functions of Nervous System and the Brain

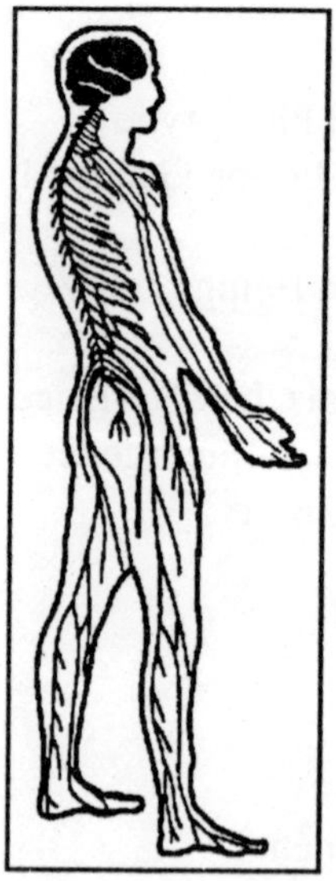

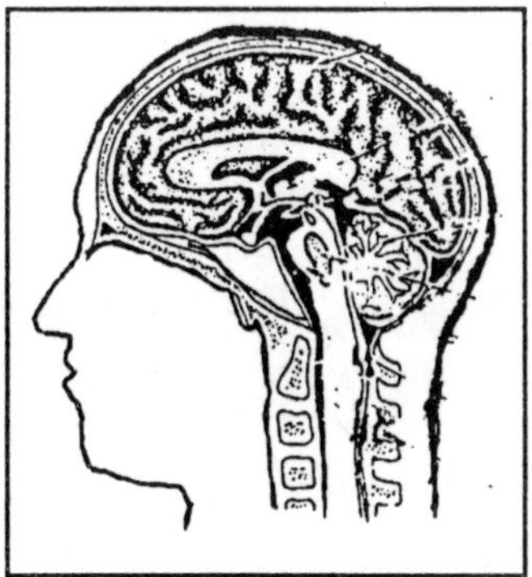

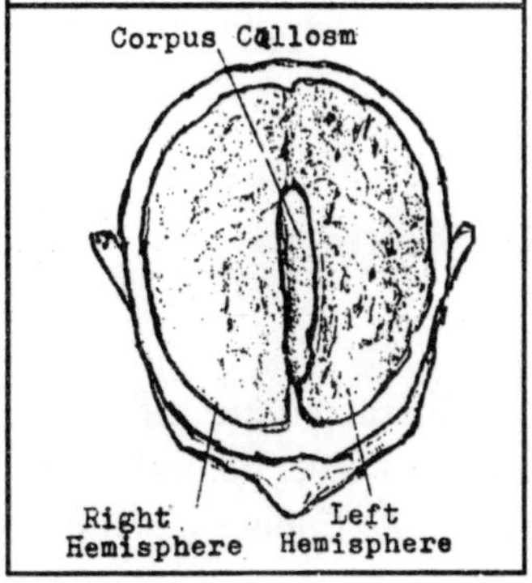
Corpus Callosm
Right Hemisphere
Left Hemisphere

PART THREE : Perception and Awareness

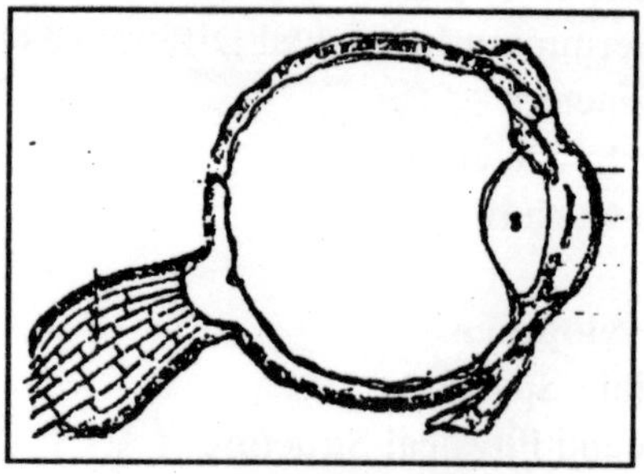

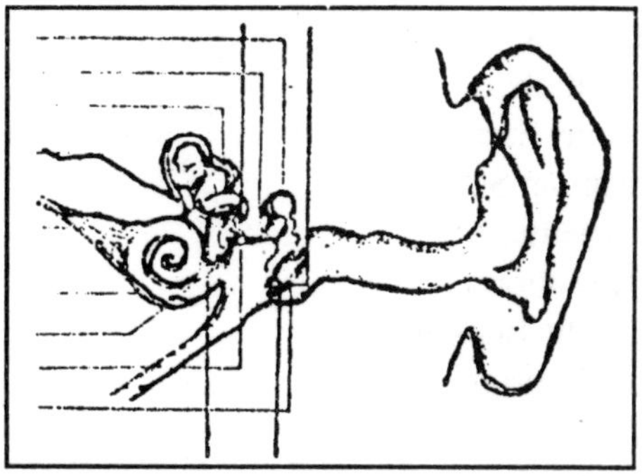

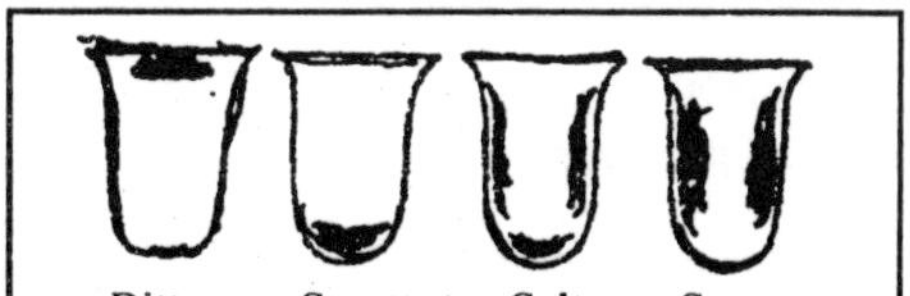

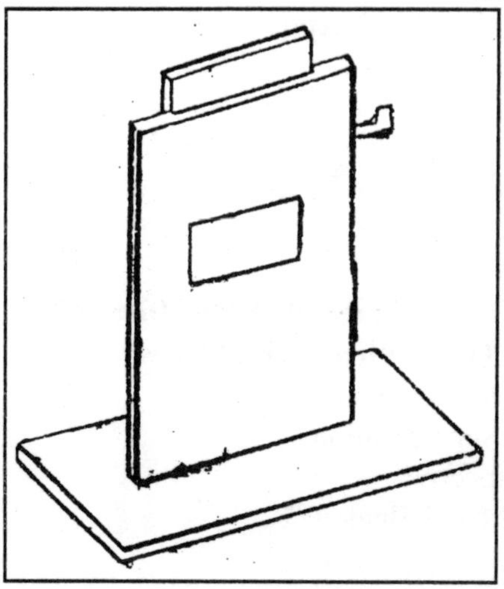

Tacistoscope

Figure - Ground

Meditation

PART FOUR : Learning and Forgetting

Volume - II

PART FIVE : Thinking and Imagination

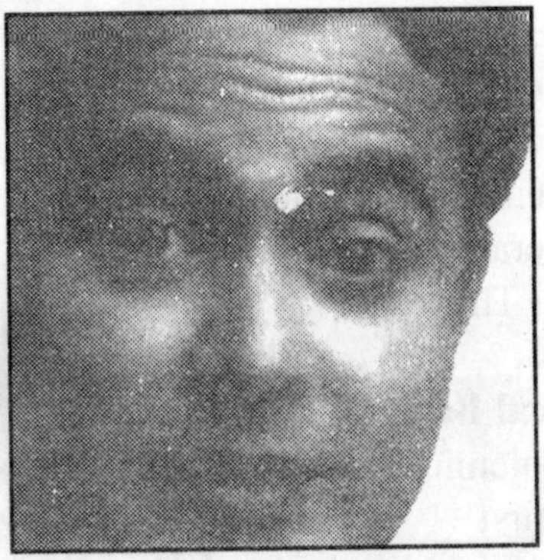

PART SEVEN : Intelligence and Creativity

Binet, A. (1857-1911)

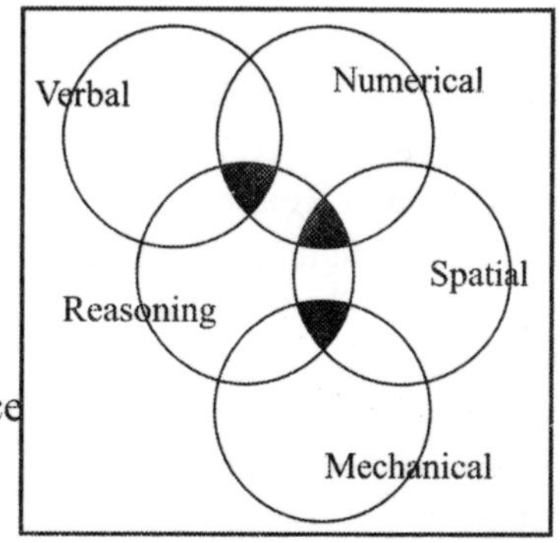

Sir Francis Galton
(1822-1915)

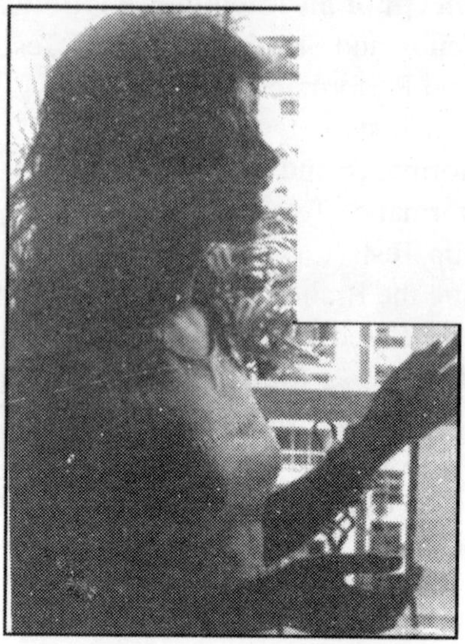

PART EIGHT : Personality and Assessment

PART NINE : Psychopathology and Mental Health

PART TEN : Social Behaviour and Related Traits

PART ELEVEN : Psychological Statistics

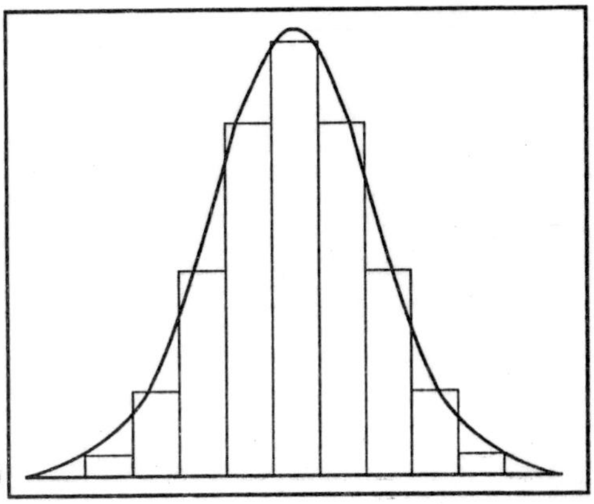

PART ONE

SCIENTIFIC STUDY OF BEHAVIOUR, THEORIES AND SYSTEMS

1. Scientific Study of Human Behaviour
2. Development of Theories and Systems in Psychology

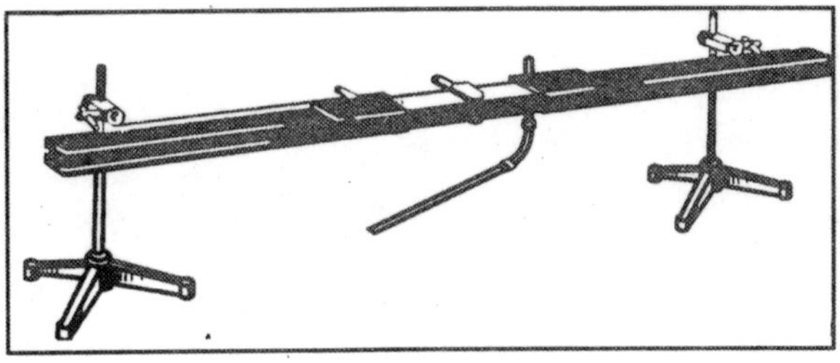

1

SCIENTIFIC STUDY OF HUMAN BEHAVIOUR

Nature and Characteristics

Psychology is the discipline that studies human behaviour scientifically. It studies human behaviour systematically and experimentally, collects reliable information, describes and interprets them in terms of various types of characteristics and behavioural patterns. Science is empirical and it is based on the actual experience of the Scientists. The experimental results are subjected to verification by other scientists.

Psychology was spoken about even in ancient times. The ancient Greeks have discussed everything about human beings. **Aristotle** (384-322 B.C.), the Greek philosopher was the first to write a book on psychology. Aristotle dealt with human behaviour philosophically and rationally. During this period and later on, psychology was considered as the study of human soul, consciousness and mind or mental life. These three terms were invariably used in the study of psychology. But as and when the progressive thoughts developed, the terms were slowly diminishing in the usage of the studies of psychology.

During the medieval period and renaissance the terms, consciousness and mind were not totally neglected. These terms were appearing, disappearing and reappearing now and then in the studies of philosophy and psychology and received importance. **William James** (1842-1910), the American philosopher-psychologist has stressed the importance of the term, consciousness in one his important works, "*Does consciousness exist?*". He states that consciousness stands for a function and not an entity. Similarly, the term mind was applied in most, of the studies. The main reason has been that these are philosophical terms and psychology has been a part, of philosophy for a long time.

Psychology began to function as an independent science only in the last quarter of nineteenth century. Some fix its date as 1879, the year in which **Wilhelm Wundt** (1832-1920) had founded the first psychological laboratory University in Germany. In the beginning, when the psychological was started psychology was mainly concerned with sense, sensory processes, perception, simple learning and memory. At the same period **William James** had started psychological laboratory in America and studied human behaviour from his angle and brought out this famous work, "*The Principles of Psychology*" in two volumes in 1890. This work has received greater importance all over the world.

It should be also known that almost at the same period between 1880 and 1890 **Sir Francis Galton** (1822-1915) an English naturalist and mathematician had developed the first test to measure and developed quite large number of tests in his laboratory. Galton applied genetic principles in his studies. On seeing the vast individual differences in intelligence among the people, he wished to improve the human potentials by the method of *eugenetics*.

Gradually the field of psychology has expanded to the level of covering a very wide area of study, broad enough to include all aspects of human life. As and when this expansion has occurred psychology has developed on two parallel lines, to qualify itself as a science and also making it as a profession to deal with the human life problems. All these developments have contributed to function psychology as an independent discipline.

Contemporary psychologists define psychology as the scientific study of behaviour of an organism. But an organism or a human individual does not reveal any behaviour or an activity on its own spontaneously or automatically, without any purpose. There are two aspects involved in the behaviour of an organism. The organism's behaviour is *a response* to some *stimulus* and also the organism *has a purpose* or motive in responding to some. stimulus. The *Stimulus* is *from outside* or the environment and the *Motive is from inside*, that is within the organism. This is applicable to all living creatures, though this formula is applied to human beings. The scientific study of behaviour is with reference to a specific. organism in which the scientist is interested. But the behaving organism, against which the study is directed, has its inherited structure and functions. Because of this the behaving organism reveals all its potential capacities, the effects and changes produced by its past learning and experience.

The organism's response in this situation is also influenced by the environmental factors from where the organism makes its response. This means that the environmental factors are having their influences in the behaviour of the organism. Scientific study of behaviour includes such information. In the scientific study of behaviour all these and other resultant responses to the stimuli are registered. Though the processes explained here appear to be simple, they are very complex functions that are involved in the behaviour of an organism. In any behaviour an organism exhibits some response to some stimulus. This response (or the behaviour) is a dependent variable. This depends upon the strength of the stimulus (independent variable).

Persons who are familiar with psychology know pretty well that psychology is the study of mental life of human beings. The human mind is expressed in behaviour of various types. Behaviour involves movements and activities. The activities may be conscious activities with some motives" and also unconscious activities without any definite motives, whatever they after they are the expressions of the, mind.

What is mind? Mind not a matter, cannot be seen. But it is it process of function, seen in the behaviour, of the organism, in the activities. Several definitions and descriptions are given, to the mind. It is also said that the mind is the faculty of consciousness and thought. It is the, intellect of a person. A person's mind is "identified with his intellectual faculties, we sometimes used to say that the person was one of the greatest mind of time. This statement only indicates haw brilliantly the person was performing his mental activities which were expressed in all his behaviour patterns.

But modern psychology bases all its subject matters on the experimental results. Psychological studies in the twenty first century are trying to adapt an interdisciplinary approach for collecting facts and figures and information. It does not meant that all the - earlier studies and information are nullified or discorded. Modern psychology taking into confidence all the earlier findings and proceed further for more accurate and reliable information in the study of human behaviour. Modern psychology in studying the human behaviour puts more emphasise on the role of the brain.

Right from the ancient time to the present period the legends, mythology and other literature put more importance on the role of the rain. The individual's identity, personality and talents are mainly the function of the brain. Our thoughts and feelings, dreams and imaginations, our fantasies and memories, beliefs and desires, our fears and courages are all the functions of the brain.

Therefore, the study of psychology is an attempt to explore all the possible ways in which the structure and the functions of the brain that produce mind and behaviour. This is one side of the aim. Apart from this we are also interested in knowing the ways in which that behaviour in turn influences and modifies the structure and actions of the brain. This is because, modern biological psychologists are of the view that there is interaction between the brain and behaviour and this function is highly reciprocal, each influences the other.

The revolution that took place in neuroscience at the last quarter of the twentieth century has reduced the distances between many scientific disciplines, which were interested in the study of human brain in one way or the other. Psychologists are mainly interested in the study of brain and behaviour, the anatomists are interested in the form and structures of living things. The physiologists are interested in the functioning of the living organisms. Anatomy and physiology are complimentary sciences. The biologists are interested in the study of living organisms, including their structure and functioning and their internal relationships. The endocrinologists study the endocrine glands and their secretions and, functions. The geneticists study the heredity and the variations of the organisms. The embryologists study the embryos, the structure and development of the embryos of the organisms. The zoologists study the biological characteristics and the classification of the organisms. The engineers apply the scientific principles and study the relationship between the skills of man and the machine. The chemists are interested in the composition of properties and the combination of substances. All these sciences have made notable contributions for solving the mysteries of the brain. In the present day study of psychology we can observe what amount of shares these sciences are having.

There are also several social sciences that deal with human behaviour in one way or other at some stage. But the dealings of these social sciences will not help them for fulfilling their curiosities fully. For example anthropology deals with antiquities of man, archaeology unearthed the buried cities and forgotten states. Sociology deals with the group life, history studies the stories of dead kings and queens and the chieftains and worriors, economic is interested in the production, distribution and consumption of goods.

Thus we can see that every science is interested in one aspect of the function of the brain from its own angle. But the mysteries of the brain cannot be fully understood in part and parcel. It needs a combined and synthesized approach. Psychology, since it studies the man as a whole right from the time of fertilization in the mother's stomach, after the birth of the child, through various stages of development and till the old age, it combines the contributions made by all these sciences and studies the living man as whole involving each and every movement and the behaviour patterns.

Psychology tries to answer any reasonable question and clarifies genuine doubts. Psychology has also acquired additional and valuable information about the external physical world through physical sciences, like physics, astronomy and chemistry and the internal world through anatomial biological physiological and the emotional world through physiology and medicine and also through social sciences like anthropology, sociology, history and economics for its use. Psychology hopes that it could approach the mysterious of human behaviour with these additional information and finds out solution for the problems. Therefore, psychology is the only science, which has synthasized all the information of the external and internal worlds for its study. But it has never enchroached into the realm of other sciences.

Definitions and Descriptions

One can ask the question, *what psychology is about* and he can also try to examine it. In his attempt to examine it he can find several definitions and explanations. To define psychology in precise terms is a difficult task as the word psychology is having a long history and is also combined with other faculties like religion and philosophy. Further psychology is defined in a variety of ways. Psychologists, for very long time have been searching for a complete and satisfactory definition of psychology. It seems that the search is still going on. But a search for correct knowledge is a never-ending process.

In the definitions and explanations the aspects of behaviour might have, been stated. Originally, psychology was defined as the science of the soul. The word psychology is derived from the Greek words '*Psyche*' and '*Logos*'. Psyche means soul or spirit. Logos means science or department of knowledge or character or action of a person. Therefore, combining these two Greek root words, the Oxford English Dictionary defined psychology as the science of nature, functions and phenomena of a human soul or mind. Various meanings are given in other dictionaries. The word 'Psychology' is the combination of the words, Psycho and logy, which means the *science of the mind.*

In Greek mythology 'Psyche' was personified as the mortal maiden of extraordinarily beautiful, greatly admired by men. Psyche inspired men and distracted them from Venus, the Goddess of beauty who was worshipped by men. Venus became very angry and sent her son Cupid (Eros) to punish the girl, Psyche. But Cupid was not able to execute his mother's order as he found that Psyche was an extraordinarily beautiful girl and he fell in love with her. Cupid made arrangements to meet her during night time every day, and strictly told her not to see his physique. But when psyche's sisters visited her and they saw him and told psyche that Cupid was not handsome, but an ugly creature and hiding his ugliness in darkness. Knowing this by some means, Psyche awakened Cupid from his sleep. Cupid got angry and at once left her and went away as she did not carry out his order. Psyche searched and tried to find out her lover, Cupid, At last she met Venus and wanted that Cupid should join her. Venus put some conditions to psyche, one of which was a journey to the *underworld*, the home of the soul of the dead, to bring back a box of beauty from the queen, Persephone, the wife of Hades and the daughter of Zeus. Psyche could not carry out this task as it was impossible for a mortal like Psyche. There upon, Cupid earnestly requested Zeus (for Romans, it is Jupiter) the Supreme God, to show sympathy in this matter. On that, Zeus conceded immortality to Psyche and granted equal status with Gods. Cupid and Psyche reunited for ever. Thus Psyche became immortal and this was attributed to the human soul by the Greeks. The soul is a disembodied spirit within every individual. The Psyche is an immortal substance, a spirit. Psyche is a Greek word, which also means breath, life, soul, spirit, distinct from the body.

In Greek mythology 'Psyche' was also personified as the Goddess of love, the wife of Cupid and represented as the beautiful maiden having the wings of a butterfly. Psyche symbolized the soul, and the butterfly symbolized the human immortality. The soul is regarded as an entity distinct from the body. *It is the spiritual past of man. It is spoken of after the death of man.* It is purely a religious explanation, and does not fit into the scientific study of human behaviour. Therefore, the religious associations have to be dropped out completely. At this stage the word *psyche* has *acquired the meaning of mind* and it is used in the definitions. Therefore Psychology was defined as the study of the mind. **William James** (1891) says, "Psychology is the science of mental life." It is also stated that psychology is mainly concerned with mental activities.

In the olden days it was referred to, as remembrance, in the middle age it was referred to, as the faculty of memory. Mind may be defined as the entity in an individual, responsible for thought, feelings and speech. It is also the faculty of original and creative thought. It involves recollection or remembrance, memory and imagination. All these are the aspects of the mind. It is also defined as the seat of consciousness in which thinking, feeling and speaking take place.

Some defined, psychology involving the term consciousness. Then psychology is defined as a systematic description and explanation of consciousness. But any definition involving consciousness is subjected to severe criticism, because, it is private and personal in an individual. If the meaning of consciousness is awareness, only the concerned individual knows it. Some associate consciousness with emotion. While defining, emotion, it is said that it is the stirred up state of consciousness. The organism or the human individual as a whole is involved in it.

Consciousness is also often associated with unconsciousness. A small child or baby is largely in a state of unconsciousness of the world in which it lives. A man is unconscious of many of his actions. While he is writing, he knows that he is writing, but he does not know that he is also breathing and his lungs are in action. He is not aware of or conscious of the actions of his kidney or the actions of digestive organs. When a man goes on to sleep, he goes into unconsciousness, but it does not mean he is inactive. Whether it is consciousness or unconsciousness, it is not subjected to a systematic study. We do not know what a man thinks or imagines or seeing a strange object or event, unless he himself tells us. Therefore it is not subjected for objective observation.

The philosophical definition (1910) is that it is the *experience* that includes all consciousness. Experience is the whole field of consciousness or awareness. If one is conscious means that he is experiencing. **Bertrand Russell** (1951) says that consciousness is *an inference* and it is included in perception. According to him perception involves something more than what is found in the scientific instruments. What is that something more? It is the consciousness. Consciousness can be easily understood and it is *all pervasive aspect of life*. It was studied through introspective method. Introspection means looking with in. It is self-observation. It is said that by careful and systematic introspection, many useful facts about the conscious life may be obtained.

It is true that consciousness is *private*. But when we express it in words what we have experienced, it is becoming the *overt activity*. The experience is put into words or in speech, speech is a form of behaviour. In this way, we may state that it is objective and also acceptable. Therefore through introspection we understand what is happening in consciousness. We may state that in a verbal report, consciousness becomes a form of behaviour and all the underlying processess may be inferred. This is, what Bertrand Russell said. The verbal reports are obtained through introspection of what is happening in the consciousness and from the reports we can infer everything. Therefore psychology may be defined as a systematic description of conciousness.

In the foregoing passages we have seen using the concepts, soul, mind and consciousness for defining psychology. Modern psychology gradually dropped all these three phenomena as they are not observable and experimentally verifiable. What is observable and verifiable is the behaviour of the organism. It is the response to a stimulus that is observed and recorded. It is the overt bodily activities in response to some stimuli that are observable. Hence psychology may be defined as the science that studies the activities or responses that a living organism makes to its environment.

But whatever the contemporary psychologists state with regard to the definition and the study of psychology, rejecting the concepts of soul, mind and consciousness, the use of mind in psychological studies could not be totally rejected from the field. Modern psychology has accepted this concept of mind by giving different meaning to it. Whatever the bodily activities revealed are the mental activities. Mind is a functional aspect of the body revealed in various ways. Mind is the coordinating body of all the activities. There cannot be any activity independent of the mental process. The seat of the mental process is the brain. Mind is revealed in various functions, or activities and thus we feel and infer the presence and the actions of the mind, though it is not visually or directly seen. Therefore psychology also may be defined as the science that studies the mental life of the human beings.

All these definitions did not receive the general acceptance from all the psychologists. Further, the term, mind seems to have certain philosophical implications. Therefore, the psychologists who were interested in the scientific study of human behaviour were not ready to accept a philosophical meaning. So long as psychology was considered as the part of philosophy, there is no disagreement with regard to these kinds of definitions.

In the beginning of twentieth century the emergence of behaviouristic movement made radical changes in the definitions and in the methods of study. Modern psychology cut off all its associations with religion and philosophy and it has become an independent faculty. It has its own terminology, meaning, definition and interpretation. Therefore, psychology tries to explain its subject matter from scientific point of view. **Watson** (1925) the founder of the school of behaviourism has objected to any reference to mental processes in the scientific study of psychology. He strongly felt that psychology studies behaviour and not thoughts or ideas or any other subjective events. Contemporary psychologists defined psychology as the science of behaviour. Science studies its subject matter objectively. Psychology too tries to study its subject matter objectively. It studies only that which can be observed. It does not deal with unscientific matters like soul or mind or consciousness which no one can see, touch or examine, but deals with the subject matter which can be examined and directly dealth with.

The subject matter of psychology was looked upon from different angle. Therefore, some stated that psychology may be defined as the science that studies human behaviour systematically. Why is psychology, the study of behaviour? Why not mind or thought? The answer is straight forward. You can study only what you can observe and behaviour is the only aspect of a person that is observable. **Kendler** (1963) defines behaviour as the observable responses of the organism. There are events going on within a person events that can be called thoughts, feelings and mental activities. We can make inferences about these events, but we always make the inferences from the way a person behaves or acts. It is what the person does, writes, says that we can observe and record as the scientists do. Hence it is only behaviour that we can study. What is seen, is only the behaviour. Behaviour is not inert in nature nor it is conceded. Behaviour is revealed in various kinds of activities.

Therefore, psychology may be also defined as the scientific study of the activities or behaviour of the individual in response to the environment. It is greatly interested in the general law of activity. The term activity is used in a broader sense. It includes *cognitive activities* like seeing, hearing, remembering and thinking, *motor activities* like walking and speaking and *emotional activities* like, laughing, weeping and feeling happy or sad.

Therefore, it includes three kinds of activities. All the three kinds of activities may appear in some form in an organism. What is an activity? An activity may be defined as the manifestation of life. It is the lively action or movement. It is the process of doing something. Human life is revealed in various ways. An event of life may be called an activity. Every living creature is involved in some kind of activity. For example, lifting a book, writing a letter, watching a game, listening to a radio, music or television programme, smelling a flower, eating the breakfast, tasting a sweet meat, touching a smooth surface, walking, running, speaking, singing may also be called an activity. An organism of a human being becomes inactive only after his death. Human activity is looked upon by psychology from different angles. But there are also other sciences that study human activities from different angles. Therefore this requires some clarification. Because there are variations in the approach and in the nature of the subject matter.

On analysing the definitions and the subject matter of psychology it is ordinarily concerned with activities such as learning, remembering, thinking, planning, observing, wishing, loving, hating, walking, eating, etc. These are generally grouped under three main headings of knowing, feeling and doing. But any mental activity can be called at the same time a bodily activity. In any bodily activity in which the mind is active, the muscles and sense organs also play an important part. It is the province of physiology to discover how the various organs operate? Physiology analyses the organs to pieces and thus sees what each organ contributes to life. It tries to find out what goes on in the ear while we are listening to a music programme, what goes on in the eyes while looking at an object and what goes on in the speech organs during talking. It tries to see how the muscles work in every part of the body. It further examines how the complicated organ - the brain integrates the activities of the individual and makes him to deal with the complicated environment. How it directs man to solve complicated problems?

Every organ is specialized in a particular activity. But, if the activities of the individual can be analysed into the activities of his organs why should we study the individual as a whole? The answer is, that physiology studies only a part of what we need for our purpose. The individual is a real and whole unit. It is the individual who loves and hates, acts and rests. He has many tasks to perform and problems to solve in his life. He deals with other persons and things happily or sorrowfully. Therefore, there is a vast amount of interaction between the individual as a whole and the world around him. That is why psychology studies the individual's activities in relation to the environment.

A newspaper, for example, may contain many information about a cricket match, but the reader may riot get the information that he wanted. He may be interested in seeing the photograph of particular player, which may be missing in the information. We can get physiological, psychological and sociological pictures of a human activity. Each one may present a valuable information. In this way, psychology is interested in the individual's activities in relation to the environment, because the individual is always responding to the environmental forces in his behaviour patterns. This is revealed in his each and every movement.

What is Behaviour?

The term behaviour is used in several places in the earlier section with some simple definition. But the student of psychology is expected to know it in detail. Hence the term behaviour is described here with all its characteristics as it is the key concept in any discussion on psychology. Every living organism exposes some type of behaviour in the form of movements, actions, reactions and interactions with other living organisms. Behaviour is the response of an organism to some stimulus in the situation. Experimentally behaviour is a dependent variable.

Modern psychologists stress that psychology is the science of human behaviour. Behaviour can be studied in an individual and also in a group of individuals or organisms. An organism is involved in constant movements or actions. An organism keeps on moving or performing something always in its living situation. The organism functions internally in relation to the outside situation. While the organism is involved in some type of action or behaviour some kind of bodily function internally is going on. This internal function in the organism's body is going on even when the organism is not involved in any type of overt behaviour in the situation. The internal behaviour is a continuous process that is going on even when the organism is in a sleeping state, and also the functions like the breathing, dreaming, thinking and when closing the eyes and sitting and doing nothing. In all the internal functions each part of the nervous system in the body is involved in a continuous activity. This internal behavioural mechanism is very complex and psychologists have to categories the behaviour in discriminately responses. This is a very complex type of task, very difficult to study.

There are also other behavioural patterns, like the winking of the eyes, sneezing, becoming aggressive or angry against the other person, feeling sad or weeping are very complex type of behavioural patterns. These are also complex type of responses very difficult to explain relating the internal parts of the body. But, however, psychologists try to study these complex response behavioural patterns by scientific methods. Any response, according to psychologists must be identifiable, observable and also must be meaningful. Further, *any part of the total activity or behaviour* of the organism, if identifiable and observable can be studied by psychologists by scientific methods, using suitable instruments and tools. Any type of response an organism makes in a situation should have some *consistency*. Behaviour refers to an activity or an event occurring naturally or in a planned situation or to a habitual occurrence. An accident on the road or the consequences of a natural calamity is not a qualified event to be called as behaviour.

The important characteristic of behaviour is the *functional aspect*. It should have cognizable aspects or segments, which should *serve some purpose* in the total behaviour of the organism. In an individual's total behaviour or activity there must be responses that could be psychologically identifiable and meaningfully interpreted. There may be also some irrelevant responses, which should be eliminated. Thus behaviour includes only functionally meaningful responses that could be segmented for investigation and analysis.

Classification of Behaviour : What we call behaviour is referring to responses the organism makes in a situation. Behaviour is reaction to a stimulus or to several stimuli. Psychologists classify behaviour into three classes such as overt behaviour, verbal behaviour and physiological behaviour.

The *overt behaviour* refers to observable movements of the body of the organism or an individual person. Every type of response needs different method of study. The first overt behaviour is the general movements of the body, which can be described by careful observation by most of the investigators. But certain types of behavioural patterns need proper training and experience. Some of the actions of the organism cannot be identified with normal perceptual process. Certain special instruments and tools are required for the study of the movements of the bodily parts.

Secondly, the *verbal behaviour* is a part of a general activity of the organism. Though it is the part of the general activity it is important in the case of human beings. Therefore, it is dealt with as a separate class of activity. The verbal behaviour of an individual can be reliably studied. It is the language the human individual speaks. The verbal behaviour may be in vocal form or in written form. The verbal form or the speech may be tape-recorded and analysed.

Thirdly, *the physiological behaviour* refers to the physical action within the organism. This is not easily observable. This involves *electrical, chemical* and *physical activities*. The recently developed psychological techniques are applied to the study of physiological responses. The functions of the nerves are within the organism. The recently developed instruments and techniques are used to study the brain functions in animals. It is very difficult to understand the activities of the brain. In this study the microelectrodes are used for recording the impulses. The instruments are very sensitive which can identify and transmit the *electrical activity* of the nerve cells in the brain. Information about the body fluids in the function of hunger and other physiological drives are also obtained by the method of *chemical analysis*. Every aspect of such studies adds something to the total behavioural pattern of the organism. Thus every fragment of the movement either simple or complex type contributes to the understanding of the behaviour of the organism. The study of behaviour of an organism is not conducted only on easily observable overt activity. Psychologists go deeply into other aspects of behaviour such as verbal and physiological for correct understanding of behaviour.

The neurobiological approach to the study of human behaviour is a very recent development. The neurobiologists are of the view that it is possible to study the internal changes that take place in the nerve cells, brain and other nervous systems. Man is the main source of information for the psychologist to peruse his studies. Man emits behaviour. Behaviour includes actions and reactions of the human organism. These actions and reactions take place internally and externally. In the case of internal, various nerve cells and nerve fibers, muscles, the senses and all other systems are functioning in human body with *infinite number of pathways and interconnections*. These functions that are going on in the human body become the most complex aspect of the human body. The brain and the nervous system regulate everything in a harmonious way. Psychologists adapt different approaches to study the human behaviour.

One of the approaches is *to relate the actions and events* that are taking place inside the body, specifically within the brain and nervous system. The approach that is applied to study all the inner parts of the body is the neurobiological approach. The neurobiologists state that the mental activities like thinking and emotional reactions are neurobiological processes.

Generally, the psychologists who wish to study the learning process, study it biologically by observing the changes that take place in the nervous system. Auditory perception is studied by recording, the activities of the auditory nerve cells in the brain. The ear is exposed to the sounds of various types. The neurobiological studies of recent times state that there is intimate relationship between the activities of the brain and behaviour. These studies ascertain that the neurobiological mechanism can identify even the most complex human actions. Psychologists are trying to incorporate the new developments that have taken place in the allied disciplines for the study of human behaviour.

Psychology as a Science

Is psychology a science? Yes, Psychology is a science because psychology studies its subject matter, the human behaviour, systematically, applying the scientific methods as the other science study their subject matters.

Psychology has been defined as the science that studies the responses or activities or the mental life of a living organism in its environment. If psychology is accepted as a science, it must possess all the characteristics of a science. Then what is a science? What are the characteristics of a science? A science is a body of knowledge, obtained in a systematic manner. It is also defined as the systematic study of the nature and behaviour of the material and the physical world based upon the experiment, observation, measurement and the formulation of laws to describe the facts in general terms and accepted by all. It may be also defined as systematized knowledge derived by controlling the variables in the experiment and predicting the results. The facts or the results that are obtained by systematic means are objectively assessed, and the same results are obtained by repeating the same experimental procedure. The result obtained by the experimental procedures are in quantitative terms.

In these definitions and descriptions of science, we can see that the important characteristics of a science are systematic, objective observation, experimentation, repetition, measurement, quantities, formulation of laws, description of the facts in general terms, control of the variables in the experiments, prediction of the results. In physical and biological sciences, studies are made systematically. They observe the results carefully and record them. They record the results as they see in the events of the material world. Thus there is objectivity. Things may occur in haphazard manner or incidentally. Science does not take into consideration such things. But makes the events to occur as they required and thus makes everything *systematic*. In all physical sciences this is carried out. In physics the variables are controlled to the requirements, i.e. energy. There is prediction in every science. Physicists predict the interaction of matter and energy. The geophysicists predict the effects of weather, wind etc., Experiments are conducted for evolving the expected results, Instruments and tools are used to make the events to occur in matter. The physical sciences, like the physics or chemistry carry out the experiments according to certain laws or principles. The experimental results are subjected to measurement, the results are measured or assessed for accuracy. How accurately the experiment is conducted or how accurately the experimental event has occurred, is tested only by measurements. Whether the result obtained in a experiment is correct or not is tested by repeating the same experiment. This repeating act can be done by the same person or some other person. But the assumption is that the same result should occur irrespective of the number of repetitions. If the same result is obtained by repeating the experiment several times, then the law is formulated. The *formulation of laws or principles* is very important to establish the scientific characteristic of a science.

If psychology is to be called as a science, it must possess all these characteristics. Modern psychologists claim that psychology possess all these s characteristics. Modern psychologists know that human behaviour is so complex that it is influenced by too many variables.

Therefore they control variables in their experiments and predictions are made. For example, to predict the consumer market, psychologists may design the product or goods in an attention getting attractive manner for the industrialists. Psychologists also conduct opinion poll survey and predict the election results.

The modern psychological laboratories are with full of newly invented instruments and tools. Many of the human behavioural aspects are measured accurately by conducting the experiment. There are instruments to measure sensory experiences and bodily charges. In the emotional states of fear and anger, there is an increase of blood pressure, increased heart beat, pulse rate and so on. *The Fechner's laws and the Weber's* laws were formulated only after serious experimental evidences. Further, the stimulus - response, mechanism is observable in the same manner as the cause and effect principle. Pavlov and Watson have formulated the laws with regard to the classical conditioning in animal and human learning activities. The S-R formula can be applied to explain any behavioural event.

The scientific studies when started were only in pure sciences. But as and when the studies advanced in nature, scientists started applying principles to life aspects and developed applied areas. All the machines, tools, industries and the technologies are the results of applying pure scientific principles for manufacturing machines and tools that we are using today. Similarly psychologists applied the psychological principles for various purposes, which resulted in emerging the fields like the clinical, counseling, industrial, educational to deal with the problem in the specific areas. Thus psychology as a science has brought out solutions to human behavioural problems. Having understood how psychology has become a science, it is also important to know the background under which it has become a science.

Development of Scientific Spirit and Approach

Some of the modern scholars in psychology are of the view that scientific approach to the study of human behaviour had its start when **Wilhelm Wundt** founded the psychological laboratory in 1879 in Leipzig as his name is associated with scientific study of human behaviour. It is true that a systematic experimental study of psychology had its start and psychology began to function as an independent discipline only after 1879. But with regard to scientific approach to the study of human behaviour some psychologists express different views. Moskowitz and Orgel (1969) state that the scientific spirit and approach to the study of human behaviour had its start in seventeenth - eighteenth centuries. These psychologists refer the occurrence to the scientific pursuits of the empirical philosophers of the seventeenth-eighteenth centuries.

The empirical philosophers of these periods had free thinking for new pathways for life. They evolved a new pathway, which they called as empiricism. These empiricists include the English philosophers, **John Locke** (1632-1704), **George Berkeley** (1685-1753), **David Hume** (1711-1776), the French philosophers, **August Comte** (1793-1857), the Austrian philosopher, **Ernst Mach** (1838-1914). Each one of these philosophers independently contributed to the development of the school of philosophy, empiricism. Empiricism replaced then existed intellectual background with the direct empirical observation and logical reasoning.

The empiricists conception is that the human mind at birth is a blank slate and man, only through the experience of the senses, becomes aware of the world and develops ideas and intelligence. The ideas of the mind are composed of many different sensations. To explain this process the empiricists formulated the principle of *associationism.* According to this principle events and sensations occur at the same time in space (environment) and time and these are connected to the mind.

Empiricism in general is the emphasis that all opinions and ideas or theories in order to be validated must be subjected to the *test of experience.* All assumptions must be *experimentally verified. It stresses experimentation,* a scientific approach. In the development of empiricism there was some intermixture of some important ideas. It had the influence of the German philosopher, **Immanuel Kant's** (1724-1864) skepticism and Hume's influence of empiricism. These influences let the empiricists Locke and Berkeley to form two related forms of empiricisms, namely *Positivism and Pragmatism.* Thus the outgrowths of modern empiricism are positivism and pragmatism.

The Austrian philosopher, **Mach** and the American philosopher, **Pierce** has had a considerable influence of the flourishing of the empirical movements. The positivists and the pragmatists have made many applications on many disciplines using their empirical principles. There were striking incidents in the sciences like, *psychology*, sociology, *anthropology and education.*

Positivism : The positivists view of scientific method is that scientific procedure admits theories and go beyond what can be tested directly in experience. Science should confine itself either to descriptive generalization or to explanations, which sooner or later can themselves be directly verified. According to the positivists a theory about molecules is scientifically admissible for verification, because molecules are capable of direct observation. Positivists reject anything that is not directly observable and verifiable scientifically.

Pragmatism : Pragmatism is the second outgrowth of empiricism, which is also stated as experimentation. Pragmatism was born in America. It was formulated by the American philosopher, **Pierce** and it was nourished by **William James, John Dewey** (1859-1952) and **Mead** (1863-1932). Pragmatism advocates that everything should be detected by *practical and experimental methods*. According to this system a theory should be subjected to scientifc *verification* any type of object must be experimentally identified and also verified by testing.

According to William James pragmatic method is to interpret each notion by testing its respective practical consequences. James asserts that pragmatism is practicalism, experimentally verifying the truth. Pragmatism by the process of experimentation provides the correct experience and knowledge. An experience when tested experimentally, it becomes the correct knowledge. In this way the seventeenth-eighteenth centuries conception of empiricism had developed the scientific spirit and scientific procedures for the *study of mind and behaviour*. Though the outcomes of these philosophical enquiries appear to be theoretical, they have promoted the scientific spirit and experimental approaches to the study of mind and behaviour.

Similar to the English and American empiricism the German Structuralism also promoted the scientific and experimental methods for the study of human behaviour in various ways. **Hermann von Helmholtz** (1821-1894) the great German scientist has made very notable contributions to the scientific study of human behaviour. This scientific genius of the century discovered the speed of nerve impulses. In his experimental laboratory Holtz brought out many valuable information and formulated the famous theories of *vision and hearing*. In the study of *sensation and perception* his findings remain indelible.

Almost at the same period **Gustav Fechner** (1801-1887), the German physicist, philosopher and psychologist had made notable contributions for the development of scientific psychology. He brought out the *relationship between physiology and psychology* and evolved theories and methods for determining the strength of physical stimulus and the corresponding response in an organism. He formulated mathematical equations for the study of sensory stimulus and the related response. This resulted in inventing the psychophysical theories and methods. This Fechner called as *Psychophysics*. This has received such an important place, in Psychological measurements as to be called as the science of Psychophysics. Psychophysics is applied in all psychological measurements and no behavioural assessment can be made without the knowledge of psychophysics.

In spite of this much of development for the study of human behaviour, the emphasis is still placed on the mental processes. There were some problems faced by the scientists who were more interested in pure scientific approaches even though the idea that the study of mental function can never be directly observed was prevailing. However there were attempts at some level to promote the scientific approaches. The Russian physiologist, **Ivan Pavlov** (1849-1936) made many outstanding contributions in the field of physiology and medicine, for which he was awarded the Nobel Prize in 1904. He mainly concentrated on glandular functions and physiological properties and animal learning. Pavlov's experimental animals were dogs. He made his experimental animal to respond to two types of stimuli, one the *natural stimulus the food* and the other an *artificial stimulus, the bell sound.* Pavlov observed the functions and changes that were taking place in the *internal organs* of the animal and related them to the respective stimulus. Pavlov was able to elicit the functions of the internal organs and the brain. He used the term, *Reflex action* to explain the internal changes. Later on it was named as *Reflexology* by his student **Bechterev** and it has become a part of psychological study in Russia. The term was later on reformulated as conditioned response.

In the first quarter of twentieth century there was a strong belief in America that the study of animal behaviour provides valuable information, which can be adapted to human behaviour. **John B.Watson,** (1878-1958), an American psychologist started studying animal behaviour. The assumption is that animals can be manipulated in whatever manner we wanted. The hypothesis, a group of ps9chologists who were interested in animal behaviour, was that 'the mental characteristics of the, animal can be inferred by analysing the observable behaviour of the animal. But **Watson** raised the issue by stating, 'If mental characteristics can be understood by studying the observable behaviour of the animals, why not we straightaway study the observable behaviour of the animals? Then, it is possible to control and predict the behaviour without going into the mental functions. In this case, psychology becomes a science concerning only the observable behaviour.

By developing this idea, Watson was able to formulate a system of psychology called Behaviourism. Watson's scientific approach to the study of human behaviour has attracted many young psychologists and they began to adapt it in their studies.

The scientific spirit and approaches that had the start in the middle of seventeenth century were going on developing step by step, inspiring many scientists time to time and ended in the first quarter of twentieth century, qualifying psychology as the science of human behaviour and by providing a professional status to deal with human behavioural problems in a scientific manner.

The scientific spirit of the psychologists did not end with the study of the individual's behaviour and dealing with the behvioural problems. They extended their area of study and evolved scientific hypotheses with the view of applying them wherever human behaviour can be studied. Since psychology is the study of human behaviour, it can find its scope wherever human beings act, react and interact with other human beings.

Scope of Psychology

Modern psychologists define psychology as the scientific study of human behaviour. Its subject matter includes the types of activities or the behavioural processes that are observable, to be manipulated, assessed and described. The behavioural processes are the gestures, actions, movements, speeches and all the physiological changes and processes. But in the history of psychology we find that most of the textbooks of psychology till 1930s define psychology as the study of mental activities and consciousness. The definition of psychology varied from time to time. In the definitions and descriptions section we have presented the historical development of the definition of psychology beginning with the ancient period to the present time. But this trend changed after 1930s when Watson's Behaviouristic school has emerged as the revolutionary approach to the study of psycholagy.

Watson, who was originally interested in the study of animal behaviour had a strong view that the information obtained from the experiments on animal can be generalized to human organism.

The present day psychologists are also interested in genetics as the genetic factors are having strong influence in the human developmental processes. They start the study of human behaviour right from the time of fertilization after birth and throughout the span of life. Thus psychology studies the human organism's life before birth, up through infancy, childhood and adolescence, adulthood, old age and through his declining years. psychology compares the child and adult, normal and abnormal and human and animal and also studies the differences between one individual and another. Psychology is very much interested in the study of the *general laws of activity* such as learning, remembering, thinking, problem solving, imagination, emotion and motivation. In this way every psychological process has become a specialized field of study.

During the long course of study the subject matter of psychology is enlarged into several main fields and branches of study. Each field or branch is interested in one aspect of human behaviour. Its scope has grown so vast that no single individual can specialise in all the fields. Therefore, there are specialists in every field who have mastered it. After the world was II, the fields of professional psychology have grown very rapidly. An elaborate description of all the fields cannot be done here. Let us describe some of the major fields or areas like developmental psychology, social psychology, physiological psychology, educational psychology, psychometry. Industrial and Engineering psychology and biological psychology. Apart from these there are some more new fields also have developed very recently.

Developmental Psychology

Developmental psychology is a branch of psychology that studies all the changes that take place in physical, mental and social functioning that occur through out the life span. Developmental psychology is mainly concerned with the behavioural changes at different levels of development. These changes accompany the growth and development of the body of the organism. It is also interested in the factors that shape the human behaviour from the time of birth to old age. Developmental psychology studies the behaviour of the infant, its mental development, physical development, motor development, speech development, emotional development, social development, its relation to religious activities, development of aesthetic sense, its play, interests and personality development. In short it deals with all aspects of growth and development of an organism.

Developmental psychology studies the basic process of development. It is greatly interested in the types of relationship the child has developed towards its parents, siblings, teachers, schoolmates, playmates and strangers. It shows more concern in the child-hood problems and mental health problems. Developmental psychology studies these aspects of the individual from infancy to old age. It is mainly divided into child psychology and the psychology of adolescence. It is a very vastly developed area that it has both the research and applied areas. The applied area at present mainly concentrates on disturbed children or the disadvantaged children. It pays more attention to the developmental processes.

Social Psychology

Man is a social animal. He develops his skills only in the society. The development of the individual is taking place in the midst of other human beings such as parents, siblings, schoolmates, playmates and teachers. Therefore, a larger part of human development is social in nature. Social psychology is interested how the individual's interaction with the group influences the behaviour of other individuals and the group and how he is influenced by the group. Social psychology is mainly concerned with individual's behaviour in the group or society. It studies how the individual is influenced by the structure and function of the society in which he is a member. The individual acts according to the customs, manners and stereotypes of his group. Modern social psychologists give more importance to the phenomena resulting from interaction among individuals and groups. Social psychology is interested in various social phenomena like formation of public opinions, attitudes and interests and how they are changed. It shows more interest in the measurement of opinion and attitudes, the effects of propaganda, market research and mass media of communication like, newspaper, motion pictures, radio and television. The subject matter of social psychology includes riots, revolution, wars, the effects of segregation and racial and communal prejudices. It is more interested in the emergence of leadership, in the groups and in the characteristics and the roles of leaders.

Psychology of Personality

This area of psychology is primarily interested in the individual's case. It studies individual's personality traits or characteristics. The individual's personality develops in relation to mental, physical, social and emotional developments. Personality development is greatly influenced by the physical and social factors. Drugs, medicines, vitamins, minerals, acids and social situations are said to have influence upon the personality development.

Personality psychologists have developed refined methods to study and measure personality traits of an individual. Large number of personality and interest inventories have been used by them to assess the individual's personality. With the help of these instruments they are able to classify human beings into different types of personality. Personality psychologists show interest not only in the study of normal persons, but also in the abnormal. They have evolved several theories with which they can explain the individual's behaviour.

In 1960 a group of psychologists under the leadership of **Abraham Maslow** (1966) have revolted against the existing study of personality and wanted that psychology should study the healthy personality and not abnormal persons and evolved a separate field of study called humanistic psychology. They were of the views that man, the human person and his experience must be the center of study. They called it a third force in psychology which emerged as a protest against the major schools of psychology, psychoanalysis and behaviourism.

Physiological Psychology

Physiological psychology is not an independent science. It has become ' a distinct field of study in psychology, as its subject matter is at the frontier of some of the biological sciences. However, when one is a trained expert in anyone of these sciences he cannot work effectively in physiological psychology. Relevant data from psychology, physiology, anatomy, pharmacology, neurophysichology and chemistry have to be collected and integrated to constitute physiological psychology. Therefore, it is a combination of several sciences. Physiological psychology studies in detail the parts and the functions of the brain and nervous system, the nature and characteristics of the sense organs, the functions of the glandular system and all those things that are closely related to our mental life. Physiological psychologists are of the view that the first step in understanding the human behaviour is to understand how the external world come to be a part of the mind. They stress the point that a man can learn only by what he perceives and what he perceives is only by means of his sensory apparatus. All knowledge and behaviour come about only by perceiving the external world through the senses like vision, audition, taste, smell, touch and kinesthesis. Therefore, the important subject matter of physiological psychology is the physiological and chemical processes and their anatomical bases that come in between the arrival of sensory signals in the central nervous system and the appropriate responses to them.

Psychometry

Psychometry is one of the areas of psychology, which is mainly interested in the measurement of qualities or properties of objects or persons. It is interested in the measurement of duration and intensity of mental states or processes. It is the application of mathematics and measurement to psychology in general and mental testing and analysis of experimental results in particular. The psychometrician is interested in developing tests and methods to measure sensory and mental abilities of human beings. Psychometry provides tools and test materials to applied areas like Industrial Psychology, Vocational Psychology, Clinical and Counselling areas. Psychometry is a modified area of experimental psychology.

Educational Psychology

Psychology tries to explain human adaptation to the environment. Education tries to maximize human adaptability. Psychology deals with the nature of human adaptation. Educational Psychology applies what is known about human adaptation to the actual practice of enhancing adaptability in individuals. To put it in a simple manner, it is the union of these two areas. Therefore psychology is more useful to an educator and education is equally useful to a psychologist. Educational psychology is primarily concerned with the learning processes. Thus the aspects of psychology more important for education are learning, motivation and human development. The other aspects such as social learning, creativity and intelligence are having direct relationship with the instructional or teaching strategies, discipline and assessment. Generally, educational psychology is concerned with the application of psychology in the class room situation.

Educational psychology has emerged from the development of subdivisions within psychology. Educational psychologists included those aspects of psychology which are relevant for understanding of human behaviour of both within the classroom and outside the classroom. Therefore the classroom teacher must be concerned with the causes of behavioural and the symptoms, because what is taking place outside the classroom may pervade inside the classroom. Thus the content of education psychology has increased through years of application of psychological principles.

Educational psychology may be defined as the study of human behaviour in educational settings, primarily in the classroom setting where there is the acquisitions of academic subject matter by the learners. In the classroom setting there are three important elements, the learner (the student), the teacher and the environment. All these three elements are contributing to the outcome of education. What sort of interaction takes place among these three becomes the source of influence. The classroom teacher must be fully convergent with the teaching and learning methods, and also must identify the general and specific skills of the students. He also must provide opportunities for them to demonstrate the specific skills of the students. The classroom teacher has got the responsibility of transmitting knowledge, facts, ideas, skills and creative processes.

The educational psychologist have greater responsibility in schools and colleges. The students may have emotional or behaviour problems, which may manifest often in the early grades. Some schools employ school psychologists and they work with individual children and assess the level of their learning and emotional problems. The educational psychologist is a specialist in the principles of learning and teaching. Psychological researches provide immense assistance to the classroom teachers in dealing with the teaching materials and the problems children.

Clinical and Counselling Psychology

Clinical psychology is the area of psychology that studies the psychological problems. It studies the causes of the problems and applies the psychological principles to diagnose and to treat them. It mainly deals with the emotional and behavioural problems such as all kinds of mental illnesses, criminal behaviour, alcoholism, mental retardation, juvenile delinquency, marital conflict and other adjustment problems. According to Cate's study, there are about twenty-nine percent of the psychologists engaged in work in hospitals, juvenile court, probation office, prisons, community centres and in schools. He may even do private practice. But he must have close contact with psychiatrists. The clinical psychologist's activity may range from administering personality and adjustment inventories to individuals and groups to and apply therapeutic techniques and finding out the effectiveness of various therapies. The work a psychologist who is engaged in counselling and guidance is not completely different from that of a clinical psychologist. He too works, with individuals having mild problems. The counselling psychologist's work is mainly with students in High Schools and Colleges. He deals with individuals who need expert guidance for their emotional and social problems. He guides the students to find out their correct educational and vocational goals.

A group of clinical psychologists and the psychotherapists who were not satisfied with the outcome of the field of mental health have planned to evolve a new field of study, calling at a 'third revolution' brought under the name of community psychology. According to them the community is the cause of all the behavioural and social problems and it is also a potential agent for solving these problems.

Industrial and Engineering Psychology

Modern human life is entirely depending upon the scientific and technological developments. The technological development can be seen in the growth and development of various industries, which make available all the material goods that provide comforts, satisfaction and happiness in our everyday life. Industrial psychology as a separate discipline has passed through different stages during the lost fifty years. It is evolved from social psychology since it deals with group behaviour. Modern Industrial Psychology mainly deals with industrial recruitments, selection and placement, human factors in industrial work, human abilities and their measurements, personnel training, training methods and techniques, interests and attitudes of the employees, motivation, job satisfaction, working conditions such illumination, noise, music, elimination of fatigue and rest pauses.

Industrial psychology also studies about human errors, industrial accidents and reducing accidents, morale of the industrial employees and also the psychological contract between the management and the employees. This is one of the important aspects that industrial psychology deals with. If there is any misunderstanding between the management and the employee, the management may fire the employee and the employee may quit the organization. But at the same time, the management should not pass any damaging remarks about the employee and the employee also should not issue any statement against the management.

The industrial psychologist helps the management by devising techniques for increasing the output. The employee may be highly motivated to increase the production by providing the incentives like the praise well done, 'good', `excellent' and so on, and also the role of honour, bonus etc. The time and study has solved several problems in industries like the fatigue effect, rest pauses, eliminating erroneous methods of moving the instruments and tools, avoiding accidents and injury, reducing the losses due to unavoidable accidents or incidents. There is a saying, that a bad worker often finds fault with his tools' may be true in more than one way. Though the organization may find fault with the worker, in many cases the real source of trouble may be within the executive leadership and the management. The industrial psychologist tries to solve such problems and works as a middle-man.

In cases of low production, inefficiency, accidents and loss, the industrial psychologist makes recommendations to any organisation to provide better training or reorientation to the employees to avoid such problems. Thus the industrial psychologist helps the organization in many ways.

Now-a-days the emphasis is made on man-machine relationship. The instruments and tools the employees use must be within the limitation of human capacity. For example, the hummer must be within the lifting capacity of the worker. If the worker struggles in handling the instrument it may cause accident or the employee get fatigue very soon and he may need rest. Therefore in designing the equipments and tools, they must be easily handled by the employees. This man-machine relationship has developed a new field in industrial psychology. This has been called Human Engineering or Engineering psychology. This mainly deals with design of equipment and tools, bearing in mind the human capacity and the easy way of operational procedure and maximising the production. It is a way of fitting the machines or tools to man instead of fitting the man to the machines and tools.

Though human beings are enjoying comfort and satisfaction modern industrial technologies cause more destruction and problems. There are quite lots of misuse of technologies which result in the pollution of air, water and natural resources. Many of these problems cause health hazards to the people. The industrial and engineering psychologists have become very anxious about these, as these problems may affect the health and behaviour of the people. The quality of life of the people may be affected by the environmental pollutions, over crowding everywhere and other misuses of natural resources. This resulted in the emergence of some new fields of psychology called environmental psychology and population psychology.

Biological Psychology

Biological psychology is a newly developed field of psychology which places before us several hypothses which cannot be answered or tested with the existing knowledge or testing procedures. Some of the important questions raised by the biological psychologists are, can experiences increase the capacity of the human brain? Is human sexual behaviour genetically evolved? Answers to these type of questions cannot be provided with out sufficient background of knowledge about the behavioural and the bodily systems. Biological psychology is the science that tries to deal with these problems.

Biological psychology mainly takes into consideration the bodily bases of human experience and the behaviour exerted by an organism. Biology is the scientific study of living organisms, the features of a particular organism. Psychology is the scientific study of human behaviour governing various activities. Integrating the results of the studies of these areas, it is evolved that the bodily system is having influence in the behaviour and in turn the behaviour is having influence in the bodily system. In other words biological psychology is the systematic study of the bodily characteristics on the behaviour of an organism and the behavioural influence on bodily characteristics.

Biological psychologists state that the behaviour and the behavioural experience of an organism make change in the faculties of the brain. They are of the view that the bodily states and the processes produce and control behaviour and knowledge-getting activities and also the way in which behaviour and knowledge-getting activities cause influence on the bodily system. **Rosenzweig** *et al.* (1996) state that many scientific disciplines have contributed to this phenomenon. The area of study, which relates behaviour to bodily system and process has been given various names. Some of the names attributed are physiology of behaviour, behavioural neuroscience, neuro-psychology behavioural neurology and so on. But of all the names the name, biological psychology is the most suitable one since biology is considered to be the broadest science involving endocrinology, genetics and embryology.

The important aim of the study of biological psychology is to understand behaviour and experience through biological mechanisms. Investigators of health sciences have observed more reciprocal interactions between behaviour and bodily structures and functions. A new area of study called psychoneuroimmunology has emerged which includes in its study, bringing connections between illness, the immune system, the brain and the psychological factors.

The important subject matters incorporated in biological psychology are mostly relating bodily systems and behaviour to the brain, basing behaviour on neuroanatomy, the neural connections, psychopharmacology, hormones, informations processing, mainly sensory processes and experience, control of behavioural state mainly motivation, emotion and mental disorders, psychobiology of emotions, learning, memory and cognition.

Biological psychology is an outcome of the innovations made in various scientific fields. It has made use of all the new information brought out very recently in the fields mostly anatomy, physiology, neurology, endocrinology, embryology, psychopharmacology, psychobiology and immunology. The contributions made by these sciences in their respective areas contributed for the emergence of biological psychology. Biological psychology is very much interested in solving many of the mysteries of mind-body problems.

Methods of Psychology

Psychologists have devised various methods to study human behaviour. Psychology is defined as the science of behaviour. Therefore, psychologists believe that they can study human behaviour in the same manner as the other scientists study their subject matter. Physical and biological sciences base their studies mainly on experiments, though some sciences depend upon objective observations. Modern psychology meets with many difficulties from its experimental point of view. If psychologists wanted to use the methods of science, psychology also must possess all those qualities possessed by other sciences. The aim of psychological study is to provide us with new and useful information. These informations first of all must be obtained empirically. They must be verifiable. Those who are interested in the same study can also make similar observations and obtain the same results. These results must be presented quantitatively.

Secondly, a science is systematic. In scientific studies observations and experimentations themselves do not make meaning. The data must be presented meaningfully. This is done according to certain principles and laws. This process we may call as the system of classification. The biological scientists often classify their subject matter systematically. Thirdly science is measurable. In every science some kind of measurement is made and the data is presented precisely and accurately. Physics is the science that could present its data more precisely and accurately than all the other sciences. But the accuracy of measurement depends upon the nature of science. In the preceding section we have stated that science is empirical, systematic, measurable and also some other important characteristics of a science. Psychologists believe that psychology also possesses all these characteristics and therefore, they also can aim at the same goals as the scientists do. Towards the achievement of their goals they have developed certain scientific methods and techniques. We shall now describe some important methods of psychology.

Introspective Method

Introspective method is one of the oldest methods of psychology. Introspection means looking within. It is the action of looking within one's own thoughts, feeling and actions. In the beginning of psychological study introspection was the main source of getting information about the mind. It is assumed that human beings can have access to mental activities and can observe the facts systematically, collect and interprete them. Human beings can introspect into their various sensations, feelings, elementary motives and cognitive process. It was Wilhem Wundt who founded this method as one of the approaches to psychology. For him, introspection is a method of research that requires subjects to observe and report upon the experiences of their own. In the early days intro-spectionism revealed the idea that mind and consciousness are the same. It is believed that the subjects are able to know and describe their own thoughts. But later on the weaknesses of this method were revealed. In the study of personality, motivation and emotion introspective methods had its own limitations.

(1) One of the weaknesses of this method is that it is not possible for another person to observe and verify whether the information revealed by this method is correct or not. (2) Certain aspects of human behaviour cannot be introspected. For example, our emotions and motives. The moment we start introspecting the emotions like rage and fear they fade away in their intensities and change their nature. (3) We cannot use this method on children and animals, as they are incapable of introspection and reporting their experiences. (4) This method is subjective in nature and private to the observer. We will have to depend upon his statements for everything. Therefore, it is not an objective and scientific method. Modern psychologists reject this method on these grounds. But the method with all its limitations and weaknesses is still used for the important psychological researches.

Experimental Method

The experimental method has been mainly responsible for the great advancement of the world of natural sciences. This method can be used both inside and outside the laboratory. An experiment may be defined as an attempt to discover the relationship that exists between two variables or among a number of variables. An experiment may be also defined as a planned test or investigation to provide evidence for or against a hypothesis. It is an act of conducting such as investigation or test, an attempt at something new or different. It is an effort to be an original in nature. It is also defined in another way. An experiment is a controlled condition in which an independent variable is manipulated and the corresponding dependent variable is measured. A variable is any aspect of a person, event, environment or object that can be changed or held constant in an experiment and can be measured. But experimenting on human nature is a more complicated task than experimenting on matters, hydrogen or animals or frogs. A human being is affected by many kinds of influences whereas matters and animals are not.

The experimental method is a procedure of obtaining data that are directly observed and collected in an experiment conducted in a laboratory situation. An experiment is something which can be repeated in the same situation. An important characteristic of a laboratory is that it is a place where variables can be controlled by the experimenter, measurements can be taken and relationships between variables can be discovered. Anyone with proper equipment can demonstrate that water is made up of oxygen and hydrogen in chemistry. In psychology the subject matter is man himself. Therefore, two groups of men must be brought into the laboratory.

In every experiment we must have two variables, one independent and one dependent. Suppose, we are interested in finding out whether certain condition improved our ability to learn some material. To test this we must have two groups of subjects, one experimental group and the other control group. Both groups must be as identical as for as possible in all respects. The condition that would improve the learning ability may be the effect of a tranquillising drug. Therefore, this drug is given to the experimental group before learning takes place. The drug is not given to the control group. Now both the groups are given the same material to learn. The experimental group learns under the experimental conditions and the control group learns under normal conditions. The average scores of the two groups in learning the material under the two conditions are compared. Whether the experimental condition is having more effect on learning than in the normal condition or not is determined. In physiological psychology, in studying sensory and motor functions, laboratory experiments often involve elaborate procedures and the use of sensitive apparatuses. This method possesses all the characteristics of a scientific method. But psychological experiments must be conducted with all seriousness. Otherwise we should be collecting only wrong information. Therefore, we cannot neglect the possible errors occurring in psychological experiments.

Objective Observation Method

The method of objective observation is also called naturalistic observation. This method is concerned with systematic investigation of phenomena, which occur without any effort or arrangements on the part of the investigator himself. Some of the biological and astronomical or physical events require naturalistic observation just as they occur since they cannot be brought into the laboratory situation. In the same manner some of the psychological phenomena are studied by. This method. In the fields of developmental psychology, social psychology and industrial psychology, particularly this method would be of very much useful to the investigators. In the determination of average age at which walking begins in small children in the study of various stages of speech development and in other developmental processes this method may be of great help. Further, public opinion, the effects of propaganda, animal behaviour, the nesting habits of birds may also be studied by this method. This method is used as a substitute to anedotel method. The faults in observation and the errors due to biases can be reduced in this method. The observer can be trained well to record the events and conditions to make the study more accurate. One difficulty with the method is that the observer has no chance of influencing the event observed by manipulating the independent variable like the experimental method. The observer simply makes systematic study of the conditions and events as he observes them. The difficulties that arise from the introspective method can be avoided here. One limitation of this method is that it is not possible to observe the inner motives and the emotional aspects of the individual. We will have to depend upon only the outward expressions of the organism.

Case History Method

This is one of the methods used in early days of psychological studies. Case history means a biography of a person obtained for scientific purposes. The material for this kind of study is obtained over several years and sometimes by the method of interview. When the biographical data obtained scientifically, we call it 'case history.' This becomes an important source of information for the psychologists to study different types of persons. He has before him a genius, a criminal, a maladjusted worker, a problem child or a fine personality. He beliefs that he can find out the antecedents of a specific form of behaviour. The case history method usually has three basic forms such as (a) day book method (b) the clinical method and (c) the biographical method.

(a) The day book method : This method is most often used in the developmental psychology, particularly to study the child's behaviour. The child's various developmental processes are very carefully observed and recorded in a standard manner. Such records will include the age at which the child makes his first smiles, sits, crawls, stands, walks, recognition of siblings and strangers and so on. This kind of facts become the accumulated records. They provide us with standard information to judge the development of other children.

(b) The clinical method : This method is used to study an individual when he is handicapped in some form. In psychological study the handicaps may include some form of personal or social maladjusments or some kind of deficiency such as mental deficiency or a severe reading disability. This method may be in the form of verbal interviews with the subject. This was first used by the doctors. When a person goes to him with certain problems he uses this method. On the basis of the information collected from the interviewers, past life history and other information collected from various sources the observer may prepare a case history of the individual. The life history of a person is reconstructed. In this way the clinical method represents an elaborate and extensive record of the day-book method. This reconstructed record of the case history may be useful for clinical diagnosis and for systematic treatment. The clinician through proper training could make very accurate reports.

An effective clinical study requires the help of three professionals such as the psychiatrist, the psychologist and the social worker. The psychiatrist or the physician examines the individual's physical and mental illness. The psychologists evaluates his mental abilities and personality characteristics. The social worker examines the home conditions and the social background. A good psychological clinic provides for a follow up to see that the recommendations are observed and reports the progress shown. This method is used on children and adults. The clinical method is important in two ways. It provides us with information concerning the adjustment problems of human beings. Secondly it attempts to modify behaviour in the interest of personal and social welfare.

(c) The Biographical Method : This is an attempt by which we can obtain certain psychological knowledge about the individuals. An understanding about the individual is obtained from an analysis of the records of the lives of men as set down by themselves or by others. Thorndike made an elaborate study by using this method. He made personality analysis of 92 famous men in terms of 48 different personal characteristics such as intelligence, sensitiveness, sociability, agreeableness, liking for beauty and so on. He can understand from this study the characteristics, which underlie the achievement of greatness in various fields.

Interview Method

If we wish to know something about a person, one method to do so is to ask him questions and to evaluate his answers. This kind of enquiry is called interview. It may be structured, controlled and standardised with respect to the questions asked. Psychologists and researchers in other fields believe that interviews can be used to study the personal inner life of the individuals and would yield knowledge that can be used for the prediction of behaviour. The behavioural sciences started with this intention and this was dominant in nineteenth century psychological studies.

What is an interview? An interview is conversation with a definite purpose. There are four purposes in the interview. (1) Obtaining information from the interviewee, (2) Providing information to the interviewee, (3) Finding out the motivations of the interviewee, (4) Assessing behaviour under standardised conversation in a face to face relationship between two people. There are many definitions of an interview. But all definitions include these aspects. One great advantage in interview method is that the interviewee is seen in the face to face situation. There is a tendency to see the person as to how he/she looks like and speaks. This desire is fulfilled in the interview method.

Interviewing is a special kind of technique that requires a special knack or training to deal with the interviewees successfully. Interviewers can be trained to adapt the method in such a way that the interviewee may feel at ease to give truthful responses. Within a short time the interviewer has an opportunity to observe some of the aspects of the person's behaviour. The person's way of looking at the interviewer, manner of speech, facial expression when he answers well and when he fails to answer, tendency to avoid and accept certain things and so on. The interviewer should provide the interviewee with more chances to speak and express his personal experiences, feelings, interests and attitudes. He must establish rapport in such a way that the interviewee speaks out all that he has in his mind. There are two broad types of interviews. They are structured interview and unstructured interview.

Structured Interview

In structured interview, the interviewer, uses standardised questions in a predetermined manner. He may use either a checklist of some general topics or a list of short questions. These questions are in printed form supplied to the interviewer at the time of interview. The structured interview method has certain advantages. If the questions are not patterned, the interviewer may forget or omit important aspects of the study or enquiry. There is definite answer for every question in this method. It is possible to compare the interview conducted by several other interviewers with this kind of information. This method is mainly used in the employment situations particularly in selections and placement. The structured interview is also called patterned interview or standardised interview. This was designed by Murray to measure the personality traits that are wanted for employees.

Unstructure Interview

In this interview method the interviewer has no definite plan as to the questions. He is at the liberty of asking any question that he finds suitable to the situation. During the course of conversation he may find certain straight moments. He may ask certain specific questions at this moment. The interviewer makes observation of all evaluation of the interviewee and develops certain general opinion. The clinical psychologist are the chief advocates of this method of interview. They allow the conversation to wander freely. They say that the causes of the particular type of behaviour vary from person to person. Therefore, questions that are suitable for probing one person are not suitable for another.

The interview method may be used in three important situations. They are (1) personal counselling situation, (2) measurement of personality, (3) vocational guidance situation. In all these situations the individual's various aspects of behaviour are studied and predicted. The chief advantage of this method is that abilities and traits cannot be accurately measured by tests and therefore, interview must be substituted by paper-pencil tests.

On the positive side we see whether the candidate is good looking or not and how he reacts to conversation. One important disadvantage of the interview method is this. The interviewer cannot judge man's personality and character traits such as honesty, loyalty, etc., from his face. A shifty gaze need not be mistaken for dishonesty. Conduct and character do not agree with contours of the face. That is why other methods also used as supplement to interview method.

2

DEVELOPMENT OF THEORIES AND SYSTEMS IN PSYCHOLOGY

The Beginning of Psychology

Psychology is often said to be having a long past, but a short history. This is because, psychological studies were not made systematically and continuously till the end of the 19th century. Strictly speaking psychology had its origin in 4th century B.C. when **Aristotle** (384-322 B.C.) the Greek Philosopher pupil of Plato and teacher of Alexander the Great and author of many sciences, made careful observation and described the nature of man. According to him man has five senses such as sight, hearing, smell, taste and touch. We can state that he was the first scientist and psychologist. But the trend, which Aristotle started was not continued until the period of Renaissance. The great revival of arts and learning in Europe took place during the periods of 14th, 15th and 16th centuries, since *the fall of the 4th Roman Empire in 4th century A.D.* to the period of renaissance the valuable writings of the ancient Greeks were lost. During the period of Renaissance remarkable advancements that took place were in art and science. Several Greek scientists made worthy contributions. **Galileo** (1564-1642), an Italian mathematician, astronomer and physicist discovered the isochronism of the pendulum and demonstrated that falling bodies of different weights descend at the same rate, which became the basic laws of motion of physical bodies. **Copernicus** (1473-1543) opposed the *Ptolemaic system* and established that the earth and other planets go around the sun.

Apart from these developments in physics and astronomy, valuable contributions were made in the field of physiology. **Willam Harvey** (1578-1657) an English Physician discovered experimental method and demonstrated the mechanism of blood circulation. During this period the British philosopher **Francis Bacon** (1561-1626) proposed revolutionary ideas in the experimentation of natural sciences. He was also interested in some aspects human behaviour such as the influence of habit, education, praise and friendship. The 17th century was considered to be the great intellectual excitement in Europe, since the philosophers like **Descartes** (1596 -1650), **Spinoza** (1632-16771, **John Locke** (1632-1704) were showing great interest in the mind and matter relationship. The universities were interested in teaching classics and also there was great enthusiasm on the successful experimentation of science. Between the interest and enthusiasm psychology did not find place till middle of nineteenth century. Thus the whole of eighteenth century, psychology did not make any advancement. It was because that it was believed that the properties of mind are completely different from the properties of the material objects and therefore, the methods of natural sciences could not be applied for the study of mental processes. The highly influential philosophy of Descarteds was strongly against the development of psychology since psychology at this time was considered as the science of the soul. Descartes was of the view that all material things and the human body are subjected to mechanical laws and can be scientifically, studied, but the human soul cannot be brought under this law and therefore cannot be studied scientifically.

Therefore what the psychologists were studying regularly, were based upon Descartes methods of dualism. This influence was prevailing for more than 150 years (from 1700 to 1859) which may be said to be the cause of delay of the scientific psychology. Around 1859, a most influential work. "*On the Origin of Species*" was brought out by the English naturalist, **Charles Darwin** (1809-1882). He formulated the theory of evolution By natural selection and he applied his theory to man and brought out the work *"The Descent of Man"* in 1851. Darwin concluded that man is a by-product of a continuous structural adaptation to the demands of biological survival. Darwin believed that evolution was slow and gradual, shaped by the processes of natural selection and sexual selection. According to him natural selection meant that animals developed skills accidentally that allowed them to exploit new region where animals can naturally live. By sexual selection meant that if other sex found the characteristics of the other sex attractive, when these characteristics are selected for the appearance" and quality of the human sexes.

Darwin's theory emphasised man's functionalism on goals and on the process of adaption to the environment. The primary problem of all living creatures is that of survival or the survival of the fittest. The power, the reason that man has in him makes him fit to survive in the world of nature. Man evolved this form of adoptation through genetic change and natural selection. Man is a biologically matured creature and Darwin puts more emphasis on instincts and drives. This is the contribution made by Darwinism to the development psychology.

During the same time, **E.H. Weber** (1795-1878), a German physiologist and anatomist, introduced the psychological concept of the just noticable difference between stimuli. According to Weber, a small perceptible increment of response correspond, to the perceptible increment in the stimulus. But the stimulus must be changed in certain ratio to produce change in the response. Weber stated this in the equation, $ds = Ks$ and the law is stated as $ds/s = K$. The ds refers to any increment in the corresponding stimulus (s) K is the ratio obtained, which is constant for any given stimulus continuum. This method is called just noticeable difference (JND).

Almost at the same time **Gustav Fechner** (1801-1887) a German Physicist, Philosopher and psychologist brought out his famous work *"Elemenateder psyphophysik"* in 1860, in which he stressed the relationship between physiology and psychology. Fechner was a contemporary to Charles Darwin, had started a systematic empirical study and soon formulated the famous classical psychophysical laws and believed that the relationship between mind and matter can be stated in mathematical equation. Fechner was mainly concentrating on the study of aesthetic valves. His study was an expansion of Weber's study.

Fechner restated Weber's law by applying the method of calculus to state the relationship between the mental events and the corresponding material events. It is R = C log S+a, where R refers to the mental event or response, S is the stimulus magnitude in physical units, and 'C' and 'a' are the constants. Fechner believed that the relationship between any physical event and the corresponding material events can be determined objectively by using this method. He evolved the classical psychophysical method of constant stimuli. Fechner divised several techniques for determining the absolute thershold and just noticeable difference (JND). The other two psychophysical methods are method of average error and the method of minimal changes. All the three classical psychophysical methods were employed in determining absolute limen by conducting experiments on sensory areas. Of the three psychophysical methods, the method constant stimuli, the method of average error and the method of minimal changes, the method constant stimuli is considered to be the most accurate of all the methods. The method of average error is the most fundamental and the oldest of all the methods. The aim of this method is to determine the variable stimulus (SV) equal to that of the standard stimulus (SS) by moving forward and backward the variable stimulus. Because of this, the method is also called method adjustment and method of reproduction. This method is associated with G.E. Muller.

The method of minimal changes is also called the method of limits. It is assumed that the human subject can notice the difference between two stimuli presented with some difference in an experimental situation. From these three methods several methods have emerged as variations. These methods are the method of paired comparison, method of ranking, the rating method and the method of successive categories. Various techniques were used to study the samples of handwriting, written essays, popular songs and also various other stimuli that may vary along a single dimension. Fechner's influence extended to the study of sensation and perception and also continued in the fields of contemporary psychology which became the foundation' upon which all the psychophysical methods and techniques of psychological measurements, the psychometric methods were built. It was because of these that Fechner's methods were considered so important the development of psychology.

Physiological Influence on Psychology

There were many discoveries in other fields in the first half of the nineteenth century, which had direct bearing on psychology. The physiological researches have been very much successful in investigating the neurological mechanism. In the first half of the nineteenth century most of the discoveries about behaviour were. mainly physiological discoveries. **Charles Bell** (1774-1842) (1811), and **Francis Magendie** (1783-1835) (1839), discovered independently the relationship between the spinal cord and the motor sensory nerves. This discovery showed that the nervous system, like the behaviour, has two components, one the sensory component and the other a motor component. **Johannes Muller** (1801-1858) developed several theories, especially the theories about the functioning of the'r8kin and the sensory systems of taste and smell. These theories suggested that the nerve is able to code the incoming information. At the same time another scientist, **Joseph Gall** (1758-1828) perceived the brain as the centre of physiological and behavioural functioning.

Apart from these, a host of scientists in the first half of the nineteenth century applied the nature of electricity discovered by **Volta** (1745-1827), **Galvani** (1737-1798) and **Simon Ohms** (1787-1854) to physiology and behaviour. **Remond** (1818-1896) 'discovered the presence of an electric discharge in nerve by using the Galvan,:)meter on animals. There were also further discoveries made by **Rolando** (1770-1831) the specific components of the brain that controlled sleep. **Flourens** (1794-1867) discovered the specific components of the brain that controlled the mood. These scientists developed experimental methods. This was the beginning of experimental approach to the brain physiology. The same approaches were used by physiologists, and the physiological psychologists in these days.

Further the works of **Fritsch** and **Hitzig** (1870) (1960) indicated the electric stimulation of various parts of the cerebral cortex and also revealed the relationship between the parts of the brain and behaviour. There were also further discoveries in brain physiology which indicated the centres of the brain for eating, sleeping and so on. In the last quarter of nineteenth century there were discoveries of reflex reactions. The works of **Ivan Pavlov (1849-1939)** on conditioning and on learning behaviour were the contributions to the twentieth century psychology from the field of physiology.

There were contributions to psychology from other fields, which had the direct bearing on psychology. **Hermann van Helmholtz** (1821-1894), a German physiologist, physicist and mathematician, founded the theory of conservation of energy and also invented the opthalmoscope in 1850, and investigated the mechanics of sight and sound. He put forth the theories of colour vision making improvement over **Young's** (1773-1829) theory and therefore his theory of hearing is called a *place theory.* Helmholtz proposed that there are three receptors like blue sensitive, green sensitive and red sensitive cones and each would tune to the part of spectrum and each type has separate path to the brain. The theory involved both the scientists and is called Young-Helmholtz's Trichromatic hypothesis. According to place theory the particular region of the basilar membrane respond to sound depending upon the sound frequency. Further, Helmholtz measured the speed of nerve, impulses. **Paul Broca** (1824-1880) a French surgeon and anthropologist discovered the motor speech centre of the brain and did pioneering work in brain surgery and found that brain injury was the cause of language impairment. He located that damage to the left inferior frontal region (left cerebral hemisphere) was the cause of language impairment and the area is named after him as *Broca's area* and the brain injury was called aphasia.

These are the contribution made by Philosphers, mathematicians and physical scientists to the development of phychology. Therefore a systematic academic work in psychology was started only after 1879 when **Wilhalm Wundt** (1832-1920) setup in Leipzing the first psychological laboratory. Several scholars from various fields have made notable contributions before that time. Close to the end of the 191h century, psychological studies-were mostly a matter of mind-matter-problem. Thus the influence of experimental physiology and psychophysics made the mind-matter problems as brain behaviour-problems. The mind-matter issue was discussed long back from the time of the Greek thinkers like Aristotle, Hippocrates, (400 BC) and Herophilus (350 BC) and through ,out the period of renaissance and **Rane Descartes** (1596-1658) and also till the middle of nineteenth century, and till the western scholars accept that the brain as mechanism coordinates and controls the human behaviour. Fechner applied in his experiments the psychophysical methods and determined the stimulus-response relationship. **Sir Francis Galton** (1822-1919) an English, explorer and scientist, a cousin of Charles Darwin was a noted researcher in the areas of intelligence, heredity, meterology and statistics. Galton was an extraordinarily brilliant man, made many contributions to the development of psychology. He was the first to use the questionnaire method in psychological research and invented the correlation coefficient method and developed intelligence tests. Does a bigger brain indicate greater intelligence? Galton investigated this. Galton invented innumerable number of psychological tests which includes the Galton's bar. **Lewis Terman** (1877-1956) described about Galton's extraordinary mental development very colourfully in his work (1917). At the age of 5 Galton's mental age was 10 and his intelligence quotient (IQ) was 200.

A little later **Harmann Ebbinghaus** (1858-1909) a philosophy graduate from Bonn, in Germany conducted experiments on verbal learning and memory. It was Fechner's work, Elements of psychophysics gave the idea to Ebbinghaus to apply the scientific methods for studying the mental processes other than sensations. He conducted experiments on memory, rate of learning, overlearning, spaced learning and also invented the famous- nonsense syllabus.

From the beginning till the end of 3rd quarter fo 19th century, psychologists concentrated on the study of sensory processes, perception, learning, memory and intelligence. Psychology had its beginning as a separate discipline when **Wilhalm Wundt** started the first psychological laboratory at Leipzig University in the year 1879, after four years of his going to Leipzig in 1875. Psychology gradually expanded to the level of covering a very wider range of topics to touch the lives of human beings. At the same time **William James** (1842-1910), an American psychologist founded the psychological laboratory in 1875 and also published his famous works, *"Principles of Psychology"* in two volumes in 1891. Wundt built up his laboratory on the basis of the works of **Weber** and **Fechner** and also formulated an approach known as the introspectionism, which later on was called introspective method. Wundt primarily was interested in philosophy, then in folk psychology and physiology. He was a prolific writer. Wundt's standard works and research papers run more than 550 titles. He also started a Journal, *"Philosophical Studies"* in 1883, exclusively for psychology. Wundt was the first to speak of experimental psychology and psychology as a science. Many American psychologists went to Leizip and worked under Wundt.

Wundt built up his psychological laboratory to verify the psychophysical methods founded by **Weber** and **Fechner** and the same was also- followed in the United States. Psychology Departments were started in many of the Universities in America and experiments were conducted on a large scale. **Mckeen Cattell** (1860-1944) became the professor of Psychology in Pennsylvania University. While the experiments in laboratories were going on different schools of thought were emerging both in Europe and in United States. The German scientists, Weber. and Fechner made physiological approach to study behaviour. Men Who were interested in physiology started to study psychology as they were interested in the problems of human behaviour. The other approach to study behaviour was the English School of Philosophy of Associationism and Empiricism. Modern psychology is said as the blend of these two approaches.

Associationism

The school of associationism was the earliest theory in psychology. It has its origin in Plato's formulation of a theory called association. According to Associationism all mental activities are based on connections between basic mental events such as sensations and feelings. When an animal or a human is sufficiently exposed to two stimuli simultaneously, the earlier of the two alone tends to call out as the response. If the other one is connected to the first one is some way, it is also brought to mind. The first object reminds us of another because the two things are either similar to each other or quite and close to the other. Thus the two factors have similarity and contiquity (or nearness). Later on Aristotle contributed a third factor 'contrast' for the successful association.

Associationism is a mechanical concept. According to Aristotle all mental functions are mechanistic and the experiences stimulate the specific part of the brain. If an idea occurred after sometimes, it brings out all the other ideas related to it. This is called association of ideas. The association of ideas has played a greater role in the British philosophy. But now it appears that it is a consequence of a wider and more primitive principal. It is now association of bodily processes. This bodily process applies not only to human, but also to any organism. Later on **Thomas Hobbes** (1588-1679) an English materialistic philosopher states that all that exist is body, all that occurs is motion. The extend stimuli cause internal motions. This is continued even when there are no stimuli. This continued motion is composed of memory. According to Hobbes there are various reasons for the association of memory elements. One of the reasons for this type of association is that in the original experience all the memory elements are connected together. Ideas or mental elements are connected through the action of association by contiquity in experience (**Levitham**, 1651)

Later on **David Hume** (1711-1776), Scottish empiricist philosopher, had explained association through the principles of cause and effect. Thomas Brown, as early as 1920, a very useful list had provided, Brown has classified the laws of association into primary or qualitative laws, the laws of similarity, contrast, and contiguity and secondary or quantitative laws of associations, frequency, vividness, emotional contiguity and mental set. The primary laws started for the essential general conditions for the formation of association and the secondary laws decide which of the many possible associations is formed or recalled. He was also called as the association psychologist. In the learning processes it was he who spoke of the process of recall.

Associationism mainly evolved from -the mechanistic interpretation and knowledge could be acquired only by mechanism. The classical-empirical philosophers like Thomas Hobbes, John Locke, David Hume, James Mill and John Sturert Mill elaborated and applied to many different cases. Thus there are several varied views opposed for each other.

The mechanistic interpretations of associationism was based upon the idea of the English philosopher, Hobbes. But James Mill did not accept the mechanistic interpreation of associationism. He emphasized the Hedonism for interpreting associationism and advocated utilitarianism. He converted experience into arithmetical sum. All the sensory components of the individual are accounted in terms pleasure. According to hedonistic principle all human actions derive from the desire to achieve pleasure and to avoid pain. Thorndike also followed the pleasure pain principle. He called it as the law of effect.

Therefore modern psychology began within the attempts of synthesising the conflecting views. It gradually developed by elaborating of the Ariskalian scientific points of views, slowly eliminating Plato's views. Rane Descartes' writings of the philosophical approach brought out many changes in the development of psychology. Descartes at this stage was considered as the father of modern philosophy and psychology. His psychological thinking was said to be much modem scientific psychology as he rejected the idea of associating psychology with soul. According. to Descartes, man is consisted of a mind and a body. These two systems are independent, but very much interacting. The body is a physical system. The motions of the body both in humans and animals are automatic. Here Descartes accepts Aristotle's view that the automatic motions or the reflexes are the responses to outside or external stimuli. Man differs from animal as he has mind whit-h is not a physical matter, but functioning as a store-house of ideas. Descartes here accepts Plato's views, that the ideas of the mind are already present at birth.

According to Descartes mind and body are in constant interaction and the mind receives the images from the motions of the body and the body constantly receives commands from the ideas of the mind. The influence of Descartes spread out throughout in the history of psychology. The automatic motion and refl6x, response to external stimuli can never be replaced with any alternative in modern psychology. This has become the foundation for the twentieth century psycho-physicists and behaviouristic schools.

Associationism was also looked upon from physiological point of views. **William James** (1842-1910) states that if two particular areas of the cortex region were stimulated simultaneously, the nerve energy would pass from one area to the other which may reduce the resistance of the particular pathway. After sometime when one experience was revived the other also may be revived in the consciousness.

The theories of associationism were existing only at the hypothetical level. There were no attempts, to provide experimental evidences. Further, the theories were not related to the experimental results. Hermann Ebbinghaus has published his major work on memory in 1885. He realized that materials with meaning have an effect and easily learned and also remembered. He set out to control this by eliminating the meaning in the concepts. For this he used the non-sense syllables, with three letter consonant-vowel-consonant sequences. This is the beginning of new type of thinking in the history of psychology. He introduced the concept of memory span, curve learning and forgetting. The theory of associationism is one of the basic principles in the history of psychology.

Structuralism

While experiments in psychological laboratories were going on both in Europe and America, different types of thinking were emerging for and against the existing systems of psychology, which were greatly influenced by the contented philosphies of rationalism and empiricism. The German psychologists like, Fechner and Wundt were following the structuralism and introspection was their important method of study. This school of thought was originally founded by Wundt in Germany. According to structuralism, psychology is the study of conscious experiences like the sensation, images, feelings and emotions. Consciousness is made up of mental elements. Those .who followed the structuralism emphasised more the structure of the mind than the function. Therefore it is called structuralism.

Structuralism finds it origin in the philosophy of the British empiricism. Experience or conscious is the essential phenomenon for them. To be conscious is the experience. Experience is identified with the quality of sensation or emotion. Suppose, if one says, "I had frightening experience yesterday". Experience in this context is a psychological reaction. For the structuralists introspection is the rigorous method of experimentation. Edward B. Titchner, one of the students of Wundt, carried structuralism to America and started the experimental studies in 1850 in the. Cornell University. Titchner preferred to call it as existentialism rather than structuralism. Existentialism is a modem philosophical movement which stresses the importance of personal experience and responsibility and the demands that they make on the individual who is seen as a free agent in a deterministic and seemingly meaningless universe. Existentialism puts more emphasis on human existence. It is based on personal experience, empirical as opposed to the theoretical, concrete as opposed to abstruct. It is denting to a formula or proposition, asserting the existence at least one object fulfilling a given condition.

Originally, structuralism is an approach to anthropology and other social sciences and to literature that interpretes and analyses its material in terms of opposites, contrasts and hierarchical structures that might reflect the universal mental characteristics. When this approach is applied to psychology, it is called structural psychology. It uses introspection to analyse experience into basic units. The school of structuralism is associated with German, since it was started in Germany. In psychophysical laboratories the German physiologist, physicist and the mathematician, Hermann Van Helmholt one of the great scientist of this century found the theory of conservation of energy and investigated the mechanism of sight and sound. He also measured the speed of nerve impulses. Philosphers and physicists were applying the mathematical equation to the process of the mind.

The English Philosophers have already formulated theories of rationalism, and empiricism. The debates were based on search for knowledge. The rationalistic philosophers stressed reasoning or thought as the fundamental factor of knowledge. But the empiricists stressed on the role of observation or, experience or the senses as the fundamental factor of knowledge. **John Locke** (1632-1704), the English philosopher-was the first to discuss empiricism and whose political thought greatly influenced France and America. He sanctioned the right to the individual to revolt.

The German psychologists were stimulated by the theories of rationalism and empiricism and started investigating the theories in their laboratories. Wundt started the first psychological laboratory in 1879 in Leipzy University. Fechner formulated theories and methods of psychophysics and applied mathematical equations for studying the strength of sensation and the corresponding physical stimulus. **Helmholtz's** psycho-physiological studies of sensory processes of sight and hearing, all these contributed to the development of modern psychology.

The structural psychologists were of the view that mind is a repository of simple and complex ideas, sensations and feelings. Wundt, who founded this school stressed the conscious experience which according to him is of two categories such as the sensations and feelings. Sensations come from outside as a result of our contact with the environment. The feelings are our internal experiences. Wundt studied these in detail and proposed the three dimensional feelings such as pleasant-unpleasant, tension-relaxation and excitements-quiescent. According to Wundt, the qualitative and the quantitative blending of these feelings may produce all types of feelings, which we experience in our everyday life.

Functionalism

Functionalism is American in origin. The school of functionalism is an outcome structuralism. Titchner who worked under Wundt in Germany, carried structuralism to America and started a psychological laboratory at Cornell University and systematic studies on sensations were carried out. He tried to verify Wundt's elements of feelings whether pleasafitness-unpleasantness could exist in consciousness at the same time. Titchner's structuralism in America met with servere criticism. It was Darwin, who introduced the concept functionalism in his theory for the study of human beings. Functionalism emphasised the goals on the process of adaptation for environment. According to Darwin the primary problem of living creatures is that of survival. Functionalism stresses that behaviour is adaptive and very much useful in the process of survival and adjustment to the environment.

The school of functionalism partially is an off shoot of the criticisms levelled against the structuralism and partially by the combination of Darwin's theory of evolution and the pragmatic philosophy founded by **Charles S. Peirce** (18391914), **William James, John Dewey** (1859-1952) and **Mead** (1836-1931). Peirce was the founder of pragmatism and the other three were influenced by Peirce.

Since structuralism studied sensations, feelings and emotions, which are the elements of experience, how it can be considered as the study of the mind, which is dynamic in nature? Analysing the structure of sensations, the feelings and emotions or the experience is only a partial study and not a complete study of the mind. This type of contrasting views against the structuralism let to the birth of the school of psychology called functionalism.

The pragmatists stressed that the hypotheses should be empirically testable and the theories should have practical results. The important problem the functionalists dealt with is the function of the mind. The important function of the mind is that man is superior to the lower organisms. The functionalists were more influenced by the writings of William James. James wrote in his book. The principles of psychology' that the function of the mind is much more dynamic and consciousness and it is not the by product of bodily effect. Consciousness is an incidental or expressing agent in the life of an organism. The functionalist in the beginning were holding the view that psychology is the science of consciousness. This was the reason that they primarily dealt with sense perception, memory, images, and emotion. They also dealt with the various evolutionary mental processes and also increased the animal laboratories in the universities for comparative studies and also' studied the individual behaviour.

During the same time, Darwin expressed his view, that there are variations in the structure and function of any species, which make some to survive better then others, especially in man the mental functions differ and these differences are sufficient facts to show the nature of the functions. Some American fundamentalists went to the extend of dropping the introspective method in psychological studies.

There was fast growth and development of psychology in the United States during the beginning years of 19th century. Psychology has become one of the special fields of study. The traditional teaching methods were strongly opposed and certain evolutionary ideas were introduced in education under the influence of the functionalism.

Psychoanalysis

In the history of psychology, Fechner and Wundt and their followers were time dominating psychological world for a considerable period of through their experimental methods, theories and findings. But subsequently there were also the development of other exciting ideas and methods. Just like the structuralism and functionalism had their origin in Germany and America, a school of thought had emerged in Austria. **Sigmund Freud** (1856-1939), a Viennese physician and neurologist was strongly influenced by the evolutionary ideas, founded a school of thought called psychoanalysis. Freud was a medical scientist, but turned to the field of psychology as he often observed in his methods of treatment to his neurotic and hysterical patients that there was something to do with human nature and mental activities.

Freud was of the view that Wundt's introspective method and the structuralistic approach cannot fully explain the richness and the complexity of human personality. Freud did not take into consideration the vigorous quantitative emphasis of the American behaviouristic concepts. Freud on the other hand thought more on materialistic terms and used freely the concepts of physical sciences. Because, of these psycho-analytic concepts were different from the concepts of the language used by modern psychology. Freud used the concepts like, mind, psychic energy, consciousness, unconscious, libido, eros, thanatos and so on. According to Freud, man is neither rational nor moral.

Freud strongly emphasised that man is not aware of most of his mental activities that were going on in the unconscious level. Since the source of his behaviour lies in the unconscious mind, he is not aware of the reasons for most of his behaviour. Unconscious is a state of unawareness without sensation or thought. It is the part of the personality. Freud felt that in the mind, there is A part called preconscious. It may be made visible through active recall. The memories and the images that can make conscious are from the preconscious mind.

Freud while treating his neurotic or hysterical patients often observed that they were often referring to their dream experiences. Freud used the hypnotic method for treating his patients. Later on he developed the free association method, which helped him to, go into the depth of the unconscious mind of the patients. In free association method the patient was seeking psychological help. The patient expresses everything that comes into his conscious mind, however trivial or embarrassing they might be. By this means Freud was able to bring out the patient's thoughts, feelings, desires and drives or motives, which were suppressed and sent to the unconscious mind.

Freud observed that most of the unconscious materials of the patient were once suppressed and sent to unconscious level as they were not fulfilled in his real life situation and they were of sexual in nature and of emotional problems, which were the results 'of conflicts and frustration, sexual drives, aggression and hostility. While treating the patients, Freud was collecting the patients' dream contents and published them under the heading of "The Interpretation of Dreams" in 1899. This was his pioneering work. All dreams are not vividly remembered after waking up. Only some dreams are vividly remembered.

According to Freud every dream has two aspects, one is the manifest content and the other is the latent content. The manifest content is the dream remembered actually. This is not the true picture of the dream. By interpreting the manifest content of the dream, we can arrive at the latent content. The latent content is the original dream material. When a person develops strong desire for something and if the desire is not fulfilled he suppresses and send it to the unconscious mind. There might have been several such unfulfilled desires in the unconscious. During sleep these desires try to come out. Since the desires are not socially acceptable they could not come out in dreams in their original forms, because the - censor, (super Ego) which is the moral aspect of the personality, pushes the material below and does not permit it to go out. As result of it the libido or the desire tries to come out in a distinguished form which may be socially acceptable. This Freud calls the manifest content. The manifort content is termed as sublimation. What is not socially un-acceptable is converted into socially acceptable form. All dreams are like this. This process is called dream mechanism. Freud says that the dream is a wish fulfillment. Whatever is not fulfilled in the waking life is fulfilled in the dream life.

Freud gradually, developed a theory of personality from his studies. The components of personality are the id, the ego and the super ego. The id is primitive, present at birth and it is a part of the personality. The id fulfils the basic urges, the bodily needs, elimination, warmth, affection and sex. The id consists of two basic instinct the Eros and the thanatos of which the Eros in more important, which is life-related instinct, sexualized and always pleasure seeking. The Eros is called the Libido. The Thanatos, the death instinct, is aggressive and destructive. The id seeks immediate gratification and operates on pleasure principles.

The ego is the second part of personality. It develops out of the id. The ego deals with the reality of the world and operates in between the id and the reality and also the super ego. The ego drives its energy from the id. It is the conscious self. It exercises restrain on the unconscious desires. Then the ego is having two aspects. The unconscious self is having contact directly with the id and the other is the conscious self having touch with the external world, the reality. In the childhood period the ego is undeveloped and functions according to the desires of the id. Later on it slowly becomes to suppress the objectionable desires.

The third part of personality is the super ego. The super ego acts upon the moral principle. It is the conscience. The super ego judges the behaviour as good or bad, moral or immoral and also judges whether the action is socially acceptable or not. According to Freud human behaviour is a complex interplay of three psychic systems, the id, the ego and the super ego. All these striving for the achievement of goals. This interplay of active forces is stated as the psychodynamics.

Freud's associates, **Carl Gustau Jung** (1887-1961) and **Alfred Adler** (1870-1937) had difference of opinion on certain aspects and broke away from Freud and established separate schools of thought. Jung did not accept the sexual nature of the libido and regarded it as, the general biological life energy. Further, Jung, in addition to personal unconscious, also spoke of collective unconscious, which contains the social history of mankind. It is the storehouse of all the experiences people had over centuries. He developed the 'Analytical psychology'. He introduced the concept, self-actualization, a state of fulfillment of a person's desires, giving expression to all aspects of his personality. Similarly, Adler (1924) developed the Individual Psychology'. Later on Adler developed the humanistic psychology in association with Maslow.

Sigmund Freud was a prolific writer and his complete works were published in 25 volumes. Any history of psychology written in the twentieth and twenty-first centuries is not complete and not substantial without making reference to sigmund Freud. His followers had high praise for him and considered Freud as the Apostle of twentieth century.

Behaviourism

In the first quarter of twentieth century the psychological trend was towards the development of comparative psychology and ethology. Experiments on animals were going on rampantly in the United States and in Europe. **Charles Darwin** (1872) was the first to investigate behaviour in man and in animals. Morgan (1894) brought out his book, comparative psychology. William James (1890) described much about animal behaviour in his book 'principles of psychology'. "Thorridike (1890) studied animal intelligence. Following these many scientists brought out their works in comparative psychology. Russian physiologist Pavlov studied the verbal behaviour as the important regulator of human behaviour, which is absent in animal behaviour. The assumption was that once if we reach understanding of animal language, it is easier to comprehend the language of man. **J.B. Watson** (1878-1958) published his book, "Behaviour: An Introduction to Comparative Psychology" in 1914. All these studies created an atmosphere that all those applicable to the study of animal behaviour can be also applied to the study of human behaviour.

Guthrie (1886-1959) was said to be first to state that a science of psychology must be based upon a study which should be overtly observable. This includes physical stimuli, the muscular movements and the glandular secretions and the environmental products. There were difference of opinions among the scientists as to what may be measured and what may be inferred-:-But they rejected the self-observation, the introspective method and they preferred experimentation on animals and infants. They presumed themselves as biologists and very much interested in knowing how the organisms behave in various circumstances.

In the last quarter of ninteenth century **Ivan Petrovich Pavlovf** (1849-1936) a Russian physiologist was working on the circulation of the heart and on the physiology of digestion, for which he was awarded the Nobel Prize in 1904. He was working on conditioned reflexes as psychic secretions and this was later on stated as conditioned reflex. His experimental animals were dogs. Pavlov, while working on animal learning, had evolved the principle called classical conditioning. He observed that animals could learn to salivate to suitable signals proceeding to the supply of food. He noticed that the salivary reflex showed adaptive properties. In the properties of simple reflex Pavlov found a model for complex adaptive mechanisms of the brain. This model was expanded and it was called reflexology for the objective study of human behaviour without making any reference to the mental processes. The reflexology became the foundation for psychological researches. Learning is a response to an external events or to a stimulus to which organisms respond.

Humans and animals become accustomed to it. Psychologically it may be stated as conditioned. Pavlov, in this experimental situation, served food to the dog, his experimental animal, at a fixed time. He observed salivation in dog's mouth, for food which is the natural response. It is the unconditioned response and the food is unconditioned stimulus. In the next stage Pavlov served food of the dog at the fixed time and also made a bell sound. The dog as usual salivated in its mouth. The supply of food and the bell sound were repeated several times till the dog leaned to associate the supply of food and the bell sound. In confirming the dogs learning that the food supply and the bell sound go together, Pavlov made only the bell sound and the food was not served. But the dog made salivation in its mouth even though the food was not served Pavlov called this salivation as the conditioned response and the bell sound as conditioned stimulus. Pavlov called this process as the classical conditioning.

At about the same time, J.B. Watson in America started his career in psychology, studying the animal behaviour. Watson rejected the introspective method as it was not possible to get informations from the animals, but accepted the technique of inferring the mental characteristics from analysing the observable animal behaviour. Watson realized that, if mind could be understood by looking at the behaviour and inferring the future behaviour, why not directly deal with the behaviour itself. Therefore, he started studying directly the observable behaviour and stated that psychology is a science and that psychology can directly deal with observable behaviour. Watson made use of Pavlov's reflexology, and developed behaviourism or behavioural psychology. Watson applied Pavlov's theory to human beings and studied directly the human behaviour without making any reference to mind. He found in his experiments that human beings were good subjects in classical conditioning experiments.

Watson stressed that scientific study of psychology exclusively refers to the study of behaviour and not i.e. as, thought and other subjective phenomena. According to Watson what generally thought of as thinking and imagining are only the verbal behaviour. Thinking is silent speaking and speaking is louder thinking. An emotion is not a feeling but a inction of glands and bodily chemistry. Watson put more emphasis on the importance of learning in human growth and development, and behaviour is the products of experience. Watson was the champion of this view and founded the school of behaviourism in 1913 and published his work, "Behaviourism. - An introduction to comparative psychology" in 1914. Based upon Pavlov's principles of classical conditioning Watson evolved the concept memory and stated that all learning is conditioning.

Watson's behaviourism dominated the American psychological world for more than 50 years. He had a good number of followers. Some of the eminent scientists were **Clark L. Hull** (1884-1952), **Edward C Tolman** (1886-1959) **Skinne** (1904) and **Edwin R-Guthrie** (1886-1959) Hull made some advancement over Watson's behaviourism. which he called 'Systematic Behaviour Theory'. Talman developed a theory of his own called, 'Purposive Behaviourism' (1952), later on it was called "Talman's Sign Learning". Skinner formulated a theory called "Operant conditioning'' in 1950, Guthrie profounded a theory, "Contiguous conditioning". All these theories are mainly learning theories and also explained various aspects of human behaviour and they all co-existed.

Gestalt School

During the first quarter of twentieth Century there were several schools of thought in psychology functioning indepentently in America. They all had their frames of reference in association theory of psychology. The schools which had their framework in associationism were structuralism, functionalism and behaviourism. They were working out of the methodology of the philosophic empiricism. The smooth contentment was interrupted by a new doctrine called Gestelt school. A set of psychologists in Germany evolved this new principle and **Max Werthimer** (1884-1943) was the champion of this venture and made announcement in 1912. The other two founders were, **Wolfgang Kholer (1887-1968),** and **Kurt Koffka**. Kohler's "The Mentality of Apes" appeared in 1925, and Koffka's "The Growth of the Mind" appeared in 1924. These two works were published in English and they attracted the attention of many psychologists. The school of Gestalt emerged in the context of an experimental situation of perception.

Gestalt psychology is a model of rationalist psychology. Gestalt psychologists started it with certain abstract ideas about the nature of perception, thinking and learning. Gestalt doctrine very much influenced the American learning theories and they were severely criticised. Kohler and Koffka went to America with their Gestalt theories and proceeded to interprete the familiar observations in terms of some new concepts and also made striking demonstrations of their theory. The gestalt psychologists were primarily interested in perception and problem solving processes. Kohler strongly emphasised the notion of insightful learning of the organism into the foreground which he stated as an alternative to trial and error learning. Gestatt is German word which means pattern or form or organized whole. According to this theory learning was viewed as a secondary derivative phenomenon of an importance can be attached to it.

According to the Gestalt theory, when we look at a scene, our minds grasp its full significance intuitively or by insight. Gestalt psychologists argue that objects are perceived as whole and not in parts. We see the object as complete, as unitary and not as a elements put together. When we perceive a chair, we see it as a whole as a furniture and not by its parts as four legs, as arms, back or the seat.

Gestalt psychologists explains the principles through the concepts figure-ground relation, perceptual grouping, proximity, similarily, continuity, unitary, insight, organisation and phiphenomenon. The three schools discussed above had their framework in associationism. But the gestalt theory is based upon rationalism. The gestalts were primarily interested in perception, and problems solving and learning was viewed as a secondary matter. We solve many problems with insights. Solutions to problems occur suddenly after many trial and error. The solution comesin a new pattern. The classical experiments were conducted by Kohler during World War I in one of the Canary Islands during the years 1913-1917. The chimpanzees were his experimental animals. Experiments were conducted both inside and outside the cage. Banana and other temptings were placed at some point which could not be reached ordinarily, but only with some aid or a tool. In one experimental situation the banana were suspended at a height in the centre of the ceiling. In one corner of the cage, a box was placed which could be moved and used as a platform and by standing on it the banana could be reached. In another case two boxes should be stacked one on the other to reach the banana. In the third situation, the banana was kept outside the cage and it can be got by using a stick. In the fourth situation, it needs two sticks to be fitted together and made it longer to reach the banana. Kohler observed that some chimpanzees solved all the four problems. This was good evidence to demonstrate the insight in solving the problems. These experiments not only revealed the chimpanzees (insightful learning in solving the problems, but also the chimpanzees were able to perceive the whole situation and organize things in solving the problems. Similarly gestalts believed that the facial expression is not perceived in piecemeal.

Normally we respond to a face or a scene as a whole, to the overall characteristics. Gestalt psychology had immediate influence upon all psychological studies of perception, learning, thinking, memory, motivation, personality, social psychology and human development. This has provided a good foundation to **Kurt Lewin** (1890-1947) for formulating a comprehensive theory for this study of human development. The gestalt psychology had its impact in the holistic theories of **Kurt Goldstein** (1878-1965) and in the studies of Andhras Angyal (1902-1960) who stressed that human organism should not be perceived into psychological and physiological units, but should be seen as a total integrated organism. Another important concept the gestalt used in their studies is the phenomenon. It is apparent movement of the objects. If two lights are put up at a short distance and flashed alternatively in rapid succession, we can see the one light in moving from one position to another. The phiphenomenon is the basic principle on which the motion picture was produced. Still pictures when presented in rapid succession the objects seem to be moving.

Mental Testing

The study of intelligence has a long history. The earliest testing of intelligence took place in the last quarter of nineteenth century.

The assumption about intelligence was that it has physiological, background. Many of those who studied intelligence attributed hereditability to intelligence and some assumed that the speed of neural functioning determine intelligence. The earliest intelligent tests were used to measure sensory accuity, eye-hand coordination, reaction time, and mauscular strength. Francis Galton's approach in the beginning was physiological and he believed that the well tuned nervous systems of the body makes man physically vigorous and intelligent. But this approach to the study of intelligence did not make any further progress.

The school performances of the children were not adequately assessing the hip-her mental development of the children. **Alfred Binet** (1857-1911), a French psychologist was the first who developed widely applied intelligence tests. Binet developed a number intelligence tests in 1900 and administered them first to his own daughters for three years and published these tests under the heading of "The experimental study of Intelligence". This was an exhaustive report of his study. While analysing the test results, Binet observed that girls as they grow older and older perform better on intelligence tests. This has given an idea to Binet that intelligence is age binding. Therefore intelligence tests were constructed age wise and with increasing difficulty. The educational authorities in the city of Paris assigned the task, of developing tests that could objectively indicate the children who are falling behind the children of the same age. Binet in collaboration with **Theodore Simon** in 1905 undertook the task of developing tests to identify the children who were not benefitted from the normal schoolroom education. The tests were in the order of increasing difficulty.

Binet-Simon observed that the older children have answered more items correctly than the younger children. Binet-Simon revised the 1905 tests and the revision was published in 1908. The revision was designed according to each age level, from 3 year to 13 year and it was expected at each-age level 50 to 90 percent of the children should pass the tests. This has resulted in the large scale testing measurement and also it was a more general study of individual difference. Binet-Simon scales underwent further revisions in the United States and it was done by Lewin Terman one of Wundt's students and others. The mental testing was found to be very much useful in World War I and from that time onwards psychological tests were regularly administrated in armed forces, in schools, clinics and industries for guidance, selection and placement of personnel for various occupations.

By the same time in England, **Charles Spearman** (1863-1945), who studied under Wundt, made extensive studies on intelligence and published his work in two volumes under the heading "Psychology Down to Ages'. He formulated the two factors theory of intelligence, According to him, people generally cope with their general abilities in most of the life situations, which he called the general factor. Since people do not equally perform in certain areas like mathematics, music, dance, painting, language etc., Spearman called these as specific abilities or specific factors. Spearman suceeded William McDoughall as the professor of psychology in London University in 1907, and turned out large amount of work on intelligence. His other important works are "The nature of Intelligence" and "The Principles of cognition and the abilities of man". He founded the 'factor school', which is considered to be the highly sophisticated statistical technique applied in psychological researches.

The American Psychologists did not agree with spearman in certain things. **Edward Thorndike**, (1874-1948) was of the view that intelligence consists of larger member of independent abilities. **Louis Thurstone** (18871955) said that intelligence is comprised of specific ability and each one also has some general abilities. He called such abilities as primary mental abilities. Others who worked on intelligence were Henry Goddard (1866-1957), David Wechsler and others.

Henry Goddard redefined intelligence as, the degree of availability of one's experiences for the solution of the immediate problems and the anticipation of future ones. It has been criticised that some of the studies on intelligence testing were of unconscious bias. Bias influenced the 19th century researcher Richard L. Dugdate. In his study he found a high rate of disease and depravity in a poor New York State family which he called the jukes (Wickets). He accused their ailments due to inbreeding. In these families the cousins had married cousins for generations. But it is said that this is an exaggeration of the role of heredity, ignoring environmental factors. The jukes later on linked with the public, the Kallikaks, a New Jersey families. Goddard, a psychological researcher desired to get rid of the undesirability of the population of feeble mindedness and criminality.

Goddard found out two branches of family of single ancestory, who had begot children from two women, one virtuous wife and the other a promiscuous lady. The first wife's dissidents were stated to be pillars of the community, while the other woman's children were morons, prostitutes and, thieves. Goddard's study yielded this type of result because the early studies misinterpreted the IQs tests of these people. The eugenics were urged to sterilize or to lock up the mentally defective and to stop immigration of the people of inferior stock. Very recently this has been revealed that the early studies were not only biased but also big fraudulent. The researcher had misinterpreted the data and crudely described the Kallikaks look idiotic and stupid. Goddard later on changed his views that mild feeble mindedness was not an inherited characteristic.

David Wechsler (1975) had developed several intelligence tests that are used to assess a broader range of abilities and they are also quite suitable for different age groups. The best known tests are the Wechsler adult Scale Intelligence/ (WA14, the Wechsler Intelligence scale for children (WISC), and Wechsler Pre-School and Primary scale of Intelligence (WPPSI). According to Wechsler intelligence is the capacity of an individual to understand the world about him and his, resourcefulness to cope with its challenges.

Instinct Theory of School

Any writing of the history of psychology would be incomplete if mention was not made about the pioneering works done by **William McDougall.** He was the doughty champion of the instinct theory. According to McDougall an instinct is a disposition, which determines the organism to perceive and to pay attention to any object and to experience in its presence a certain emotional excitement and an impulse to act which finds expression in specific mode of behaviour in relation to the object. McDougall postulated 14 instincts and every instinct has an emotional core which makes itself manifest whenever instinctive behaviour is in action. The parental instinct, for McDougall is the brightest and most beautiful gift of nature to living creatures.

The instinct phenomenon has been discussed from the beginning of 20th century when McDougall (1908) put forth his theory of instinct in his pioneering work *"Introduction to Social-psychology"*. According to him an instinct is an unlearned behaviour pattern found in all living creatures. It is also defined as the innate capacity of our organism for responding to a given stimulus in a relatively fixed manner. It is an in born intuitive power. It is a natural and apparently innate aptitude. An instinct according to McDougall is not modificable. Instinct shows some degree of adaptability under certain conditions. For example in a dangerous situation the organism takes a flight to escape from the danger out of fear. It is a common pattern of behaviour in all organisms. Instinct is quite servicable at the very first appearance and that no initial learning is necessary for the appropriate response.

The instinctive behaviour pattern was tested to see whether it is really innate or learned. Marais brought up four generations of the South African weaver bird (a kind of song bird) under with carnaries (an yellow song bird) without giving them any chance for building their nests. The fifth generation was set free. The bird built nest exactly as the forest weaver bird built. This shows that there was some innate pattern of behaviour in these birds to built up their nests. Therefore we can assume from this that all birds, and animals including human have the innate pattern of behaviour to face the situation. McDougall has done a great amount of work on instincts. He evolved a list of fourteen instincts and each instinct arouses a corresponding emotion. These are (1) Escape - Fear, (2) Combet - Anger, (3) Repulsion - Disgust, (4) Appeal Distress, (5) Sex or pairing-Lust, (6) Curiosity - Wonder, (7) Submission-Negative-Self feeling, (8) Parental - Tendder emotion, (9) Assertion - Elation, (10) Gregariousness - Loneliness, (11) Food Seeking - Appetite or Craving, (12) Acquisition-Ownership, (13) Constructive - Creativeness or making, (14) Laughter - Amusement. Bernard (1924) has broadly defined instinct to include any type of behaviour that appear to be generally human. Bernard gave a list of 6000 human instincts.

There were also studies on instincts in the second a quarter of 20th century, specifically after 1930s. Ethology is a new science which occupies a place in between zoology and psychology. Some of the ethologists like, Konrad Lorenz (1936), Tinbergen (1952) Eckhard Hess (1958), Gottliah (1973) and others have made experimental studies on animals and birds.

Later on this resulted in the development of animal psychology. They observed the innate pattern of behaviour in goats, geese, ducks, and in other animals, and also in fish, monkeys and birds, deers, zebras, mouse, buffalo and seals. They called the behaviour pattern of instinct as imprinting. Lorenz (1952) used this term. Later on this resulted in the development of animal psychology. For the ethologists instinct means a complex pattern of serially organized unlearned behaviour, which is the typical quality of the species. Kohler, in his experiments on chimpanzees observed that the monkeys exhibited the instinct of cleanliness and the instinct of aesthetics. Freud spoke of two instincts, the eros and *thanatos*. The first is sexualized, pleasure oriented, life instinct, while the second is aggressive, destruction oriented, death instinct.

The instinct theory has been discussed by a number of scholars and it was severely criticised by some because it offends the hopeful views of human potentialities, and the evidences are inconclusive. But whatever criticism levelled against the instinct theory of school, it cannot be thoroughly rejected. in psychological studies. It is true that certain human behaviour patterns cannot be explained with the available experimental results. Thomas Hobbes an English political philosopher in his' work Leviathan published in 1651 strongly asserted that man by nature is short, brutish, nasty and aggressive and supreme (sovereign) power is a necessity to control human beings. Man is always at war with another, because he is instinctively aggressive and if there is no institutional restrictions on him he would be at war With his fellow beings always. Freud's interpretation of instinct theory endorsed Hobbe's view. Since nations in the world were often fighting wars, the League of Nations requested the scientist, Albert Einstein (1933) to find out ways and means of saving the mankind from the menace of war. Einstein referred the matter to Freud. Freud (1937) replied by saying that war cannot be stopped, it has to be fought, because man has an active instinct for hatred and destruction. Einstein (1937) accepted Freud's reply and endorsed the same view stating that, a man is forced by an innate lust for hatred and destruction.

The contemporary writers like Ardrey (1966), Lorenz (1966), Storv (1968), Morris (1968) and Klineberg (1954) have discussed human aggression in terms of instinct, because they do, not find any other alternative concept through which they can explain the aggressive behaviour pattern. The critics of the instinct theory cannot abandon and reject the instinct theory totally as there is no other single hypothesis that can replace instinct. On one side the contemporary critics reject the instinct hypothesis, but on the other side they use it to explain or interprete certain behaviour pattern in terms of the instinct theory. This is like sending the man out through the front door and inviting him through the back door. Therefore, what Hobbes, Freud, McDougall, LeBon have said about behaviour pattern on the basis of the active instinct cannot be ruled out from the psychological studies.

Development of Social Psychology

The present development of psychology is the continuity of human knowledge about behaviour right from the time of Aristotle through Thomas Hobbes and to the present time. Thomas Habbes, discovered the roots of Government and the institution in human nature, the nature that man is instinctively aggressive and because of this nature every man is at war with every other man. Similar views were also expressed by the scientists of the twentieth century. Their contributions resulted in the origin and development of social psychology.

Though there were some instances in the earlier periods to quote some date of the origin of social psychology, the year 1908 stands as singular land when William McDougall brought out-his classical work, *"An introduction to social psychology"*. This has awakened the social scientists to show the intimate relationship between man's innate tendency or the instinct and the society. Again in 1920 McDougall published his "*The group mind*", based upon his instinctive theory in which he advocated the Fherd instinct subsequently, **F.H. Allport** (1924) publishg4his social psychology, and LeBon brought out as his *'The Crowd'*. Fellowing these works quite number of works on social behaviour were published.

Though the social scientists disagreed with the instinct theory and many of McDougall's ideas, the influence of the instinct theory still preserved and the fundamental relationship is carefully maintained in the vital part of psychological studies. The interpretation which McDougall given to gregarious (herd) instinct in his book "*The Group Mind*" cannot be fully rejected. Therefore, we can state that William McDougall was the founder of social psychology and the year may be stated as 1908.

Following McDougall, those who were greatly responsible for the development of social psychology. in the first quarter of 20th Century were, F.H.Allport, Sigmund Freud, G.H. 'Mead and Kurt Lawin. Students of social psychology may quote Allport's theme of emphasising the necessity of observable variables and the controlled experimentation, and bear in mind what 4Sigmund Freud said about man's unconscious Strivings and I their management. Freud strongly assessed that there is no difference between individual psychology and social psychology. The human interaction and the socialized self and the observation made by Mead would certainly have impact in the studies of social psychology. The student of social psychology in no way can reject the stress made by Kurt Lewin in his field theory and experiment in social psychology that the child from his first day of life is objectively a part of a social setting and would die within few days if he were to be withdrawn from the social settings. Human's early life-space is greatly influenced by social facts and social relations than any one would have expected. The child reacts to another person's smile and voice in specific ways. Thus the child's behaviour in molded in every respect by his social situation and interaction.

Physiological Psychology

The development of physiological psychology may be said as a branch of biological sciences. It is not a separate field of psychology in the real sense. The subject matter of physiological, psychology is drawn from various sciences like psychology, physiology, anotomy, pharmacology and chemistry. Researches conducted in these areas contributed to the development of physiological psychology. Since no one can master all these areas, the relevant research papers which interpreted human behaviour were collected and synthesised into a discipline, calling it as the physiological psychology. Apart from the areas mentioned above information was also collected from neuro-physiology and neuroanatomy. In the formulation physiological psychology the nuroanatomical, physiological and chemical techniques and the concepts to the study of psychological processes were used. The rapid development of some of the areas of physiological fields makes it difficult to fix the areas of physiological psychology. For example the field of psychopharmacology is rapidly expanding and all recent discoveries in this area should be included in it.

Most of the psychologists would accept that the field of physiological psychology has grown very rapidly within few years. But the first physiological psychology book was written by Wilhelm Wundt under the title *Principles of Physiological Psychology* in 1902. Many influential physiologists of nineteenth century were publishing research papers on various aspects of physiology. But a systematically written physiological psychology book after wundt appeared after very long time. Many new concepts were used in physiological psychology as it expanded and all such concepts have spread out to many areas of psychology. These concepts were commonly used previously. Physiological discoveries in addition to the classical concepts such as sensation, emotion, motivation and learning have our better understanding of perception, intelligence, psychosomatic disorders, stress, neurosis, functional psychoses, personality and psychotherapy.

Physiological psychology for sometimes was studied with animal psychology, because many new information were drawn from the researches in the field of animal psychology. Lashley was maintaining the tradition of physiological psychology during 1920s, 30s, and 40s. Lashley was a student of Watson and taught physiological psychology to many Americans and some of them became outstanding in physiological psychology during and after, the Second World War. Hebb a student of Lashley set up a stage for physiological psychology in 1949 and published a book, "*The organization of Behaviour*". Since physiological psychology was combined with comparative psychology, it is very difficult to isolate the subject matter of physiological psychology. However some outstanding books were published in the field of psychology. Some of them are *Physiological Psychology* (1956) written by Wenger and Jones, a text book of *Physiological Psychology* (1967) written by Grossman, *Introduction to Physiological Psychology* (1968) by Francis Leukel, *Physiological Psychology* (1965) by Morgan, *Physiological Psychology* (1948) by Freeman, *Physiological Psychology* (1960) by Deutsch and Deutsch, and *Handbook of Physiological Psychology* (1951) by Stevens. These books were written on the basis of the information provided by innumerable number of research papers, published on various aspects of physiological psychology in various research Journals. These research papers contributed to the development of physiological psychology.

Comparative Psychology

The field of comparative psychology mainly deals with animal behaviour. Comparative psychology, unlike other areas, is a very difficult area, because it is multifacted discipline. Nature has yielded to evolve innumerable number of creatures in this world. Out of which many are vertebrates, having backbone and spinal column. Man is classified as one of the vertebrates. Man has certain unique features like developing languages and more complex communication networks, construction of huge buildings and dams, invention of television, the satellite and capable moving in the space, which are not possible for any other creatures.

But with all these man must know that he too is an animal like other creatures as a part of nature. Just like all other vertebrates man moves, eats, drinks, urinates, executes, copulates and brings out young ones, as other verterrates do and also learns to live in harmony without any antegonism with the nature. The aim of comparative psychology is to understand the behaviour of a broad Rrea of animal species as a part of our liberal education which may help us to understand man'-four and his place in nature. In his way comparative, psychology is a basic science. It helps us to acquire basic knowledge, which is applicable in relevant situations. The knowledge, man acquired from the comparative study of instrumental learning, is applied for training animals in Zoo, circus and at home. Comparative psychology may be defined as the systematic study of the behaviour of every creature. It is the study of what every creature or species does and also what the creature or species is capable of doing. How did the comparative psychology came into being what we know to-day?

Comparative psychology has not emerged suddenly today. It has also history of its own. Right from the nineteenth century, the physiological subject matter of study was animal behaviour. The antecedent to comparative psychology was philosophy and natural history. Beach (1955) stated that according to the Greek Philosopher, Heraclitus, there was two types creations in this world. One type was man and gods. They were Rational creation with souls. The other type was brutes which were irrational and separate category without souls. If the beast was found in possession of reason and soul it was possessed by man or God. This view was held for a long period till the twentieth century.

There were divergence of views between the traditional comparative psychology and the classical ethology. In the midst of the diverse views, the first systematic comparative psychology book was published in 1864 by Pierre Flourens under the title of '*Psychologic comparee*'. During the same time Darwin's theory of evolution brought out many new findings about animals and humans. This made the study of animal behaviour more important. Further Darwin published a book under the title of "*Expression of the Emotions in man and animals*" in 1873. This book was mainly dealing with the comparative study of animal behaviour. The behaviour patterns of different species' were clearly dealt with in this book. George John Romanes, a friend of Darwin brought out a book 'Animal Intelligence' in 1882 which provided lots of information and served as the source book in comparative psychology. Romanes was of the view that mental continuity existed in species. He used mainly the anecdotal method.

At the close of nineteenth century and in the beginning 20th century there were some eminent scientists whose impact was felt greatly in the development of comparative psychology. Among them an English naturalist Lloyd Morgan (1894) brought out important findings and formulated the laws of 'Occam's rezor', the law of parsimony and another law called, 'Lloyd Morgan's Canon'. He was of the view that any behaviour pattern that stands lower in the psychological scale cannot be interpreted its action at the higher psychical level. Morgan's theory is applicable even today. Because of these contributions Robert woodsworth (1952) states that Lloyd Morgan is one of the founders of animal psychology. Following this Jacque Loeb proposed a theory called tropism in 1890 which is an extension of Morgan's theory. William James (1890) in his principle of psychology said more about animal behaviour. Douglas Spalding, Henri Fabre, Augusta Forel, Wheeler and John Lubbock have done more work on animal behaviour. The work of Donglas Spalding Led Gray (1922) to call him as first experimental behaviourist.

Later on in the first quarter of twentieth century, Thorndike (1898) worked on "Animal Intelligence" and formulated the 'Law of effect' in trial and error learning. He has introduced his theories in education and became one of the leading educational psychologist. Small studied white rats and maze learning problem. John Watson brought out a book, *"Behaviourism : An Introduction to Comparative Psychology"*. This line of research reached its peek in 1930s and 1940s with the theories of Hull, Tolman, Skinner, Yerkes and others. The present comparative psychology is the outcome of the intraction of American Animal psychology and the European ethology. Ethology is the science that studies the animal behaviour in their normal environmental conditions. Ethology was developed by the European zoologists. This development took place after World War 11.

Development of Psychology : Psychophysics after World War II

The validities of the psychophysical methods that were founded by the early psychophysicists, Fechner and Weber were not questioned till 1950s. After World Q War II there were several developments in science and technology. Therefore some. psychologists re-examined the psychophysical methods in the light of new developments. Psychophysical research has become very complex. But it may be simply stated as the problem of measurement. In classical psychophysics the determination absolute they thresholds and difference thresholds were stressed and they were fixed. But the present day psychologists realizes that these two are not fixed quantities, but vary systematically and they evolved different methods.

One of the modern psychophysicists, Helson (1947) made some reformation in Fechner's law. He evolved the method called adaptation level. This was in connection with the judgements of visual qualities. The adaptation level is not limited to visual or other sensory processes. It is also applied to complex perceptual processes and to learning and conditioning processes. The social psychologists made use of the adaptation level for describing the frames of reference. The classical psychophysicists tried to avoid the sources of errors in the judgements, like the error of expectation and the error of habituation. Therefore Helson designed a procedure on the basis of adaptation level theory.

Helson (1947-48) had designed an experiment for verifying the relative influence of attributes of colours such as colour constancy, colour contrast, colour conversion and colour adaptation. Previously, these four were regarded as unrelated extents and shifts from one to another with very little change. Helson states that colour has three variables such as colour object, its background, and the intensity of illumination. By varying these three variables in many combinations and varying the background colours in three steps like white, grey and black, there were four varied colours like red, green, yellow and blue for every combination of background. Helson observed in this experiment that judgement of the colour of the object depended upon other variables (background colour). The level the subject adapted to illumination determines the judgement of colour only partially. From this Helson evolved the general conclusion that colour constancy event, for white light stimulus, occurs when the object seen was at the adaptation level of the moment. Thus adaption level becomes a function of many stimulus condition. Therefore, according to Helson (1938-64) adaptation level is a pooled effect of all the stimuli on the organism. The pooled effect for Helson is from within, without and some of the past experience. Helson generalized these findings to other experiments.

Steven's Power Function

Fechner provided the psychophysical methods to the psychologists for investigating the sensory phenomena. For more than hundred years Fechner's psychophysical methods occupied important place in textbooks, in laboratory experiments, and in classrooms discussion all over the world. Criticisms appeared often about the limitations of the psychophysical methods. But there was no universally accepted national principle to substitute Fechner's law till 1959. At this time Stevens (1953) from Harward University came out with evidences for the long expected efforts. Stevens (1975) asserted that his prothetic (quantitative) continuum can replace to Fechner's logarithmic law. He found that his power function expressed the relationship between reports of stimulus intensities and actual stimulus magnitudes better than Fechner's logarithmic function. Steven's power law was applied to large number of modelities and physical stimulus was related to sensation (Psychological effect). The formula of his power law is:

$$R = KS^n$$

Where 'R' is the report of apparant magnitude, 'S' the stimulus magnitude measured in physical units and the 'K' and 'N' are experimentally determined constants. When the obtained data by Steven's method were plotted on a graph, the curve was of complex type, in a bizarre way. But when the data were plotted on log-log coordinates, it was a simple straight line. Steven's investigations led to the formulation of several new psychophysical scales. Stevens had conducted most of his experiments in the sensation of sound, one of the most important, (scale is the scale of loudness, which is based on recently defined unit, the one. Since the acoustical engineers found that the decibal scale was very unsatisfactory, this scale was developed. They wanted a scale, which should reflect accurately the loudness of sound to the subjects. Stevens (1957) stated that there are two kinds of continua in the same sensation. In the case of sound, there are two kinds of perceptual continua, the loudness and the pitch. Loudness is an aspect of sound and it is described in degrees of magnitude or amount of quantitative, whereas pitch is the highness of sound due to vibrations of sound waves.

From the perceptual points of view, pitch is a qualitative continuum. Stevens called the loudness a prothetic (quantitative) continuum and the pitch as a matathetic (qualitative) continuum. According to Stevens these two continua are the physiological processes, that underlie sensory continua. Stevens put more emphasis on prothetic continua as it provided him with more satisfactory results for the measurement of magnitude. Stevens (1953) conducted more than 36 experiments on the sensation of sound and had the same results in prothetic continua. Stevens in 1953 had demonstrated a simple quantitative law under many diverse circumstances, applying the power law to both light and sound. From that time onwards experiments were conducted in the psychophysical laboratory of Harvard University and the power function was demonstrated for taste and smell, warmth and cold, vibration and shock and a power law as evolved for every sensory continuum. Stevens claimed that at this stage his psychophysical power law has substituted Fechner's logarithmic law.

Galanter's Decision Theory

While Stevens was evolving his psychophysical power law at Harvard University, Galanter (1962) was also working on different psychophysical methods in a now approach. He called his psychophysical theory as 'Decision theory'. Galanter's decision theory'is an interesting account of a new approach to psychophysics. He has presented it in a very vivid manner. According to this theory the sensory threshold is a statistical approximation. The threshold is a statistical device and the threshold changes sometimes. The theory is named after Galanter and it is also called signal detection theory.

Galanter based his theory on a simple assumption that psychophysical measurement may be considered as the measurement of detection and recognition of the thresholds. A decision is made when a person is able to say that a stimulus is present. Then the person is able to say that the stimulus has certain characteristics. A person is coming in the front at a distance. This is also meant that a person is coming, but it is not known whether the person is a man or woman, known or unknown. When the person approached quite nearer, then the person is able to say that the person is a young man. At this stage the person is able to recognize and the recognition threshold is determined. This theory stresses the need on the part of the subject to make decisions about the stimuli and ascertain the presence of the stimuli indicating the character. Therefore Galanter called this as decision theory.

Galanter states that even in classical psychophysical theory the subject makes his own decision by responding, 'Yes' or 'No'. In the procedures of classical psychophysics a threshold is perceived 50 per cent of the times. Detection takes place 50 per cent of the times and hence it may be called 'Detection threshold' instead of stimulus thereshold. Galanter questions even the fundamental concept, the thereshold. He says, even if we accept it, there is no one threshold, but there are three, one below what is not detected, one the signal detected and one that is fully and always detected. Thus modem psychophysicists think from different angles, the problems of discrimination and scaling. The subject in the experimental situation is sensitive to the stimulus energy at the detection threshold, but he/she is also sensitive to many other factors. As a result of this the determination of threshold is also unstable. Therefore Galanter is of the view that the psychophysical scaling methods should be revised.

Information Theory

The information theory of psychophysics seems to be the contribution of Galanter's signal detection theory. If we combine the concepts, decision and detection, a third concept also seems to be immanent, that is recognition. This is with regard to the amount of information the stimulus contains and the amount of information the participant can process and registered at one look. Psychophysicist are of the view that the information theory may provide answer for this problem. The function of speech is to communicate information. But all speeches do not provide the same amount of information. Information theory could tell the experimenter how much information is contained in a stimulus presentation.

The psychophysicist may be controlling the stimulus displays in his experimentation. Therefore it is possible for the experimenter to know how much minimum information his participant can deal with and the accurate judgements (responses) he can produce. According to information theory the participants can learn to recognize perfectly only about seven alternatives with more stimuli on a single sensory dimension such as pitch of a tone or brightness of light. The information recognized is little less than three bits of information. Shannon (1948) states that the amount of information depends upon the number of alternatives. Therefore the amount of information transmitted by an occurrence may be stated as the natural logarithm of the number of alternatives. As the number of alternatives increase the amount of information also increases. Galanter (1962) states that recognition is easier when complex stimuli are used instead of simple stimuli -and the number of bits controlled are relatively higher. If the stimuli were more complex a process of diminishing returns occur and the absolute limit for any complex stimuli will not be more than 10 or 11 bits. But some people in real life manage many more alternatives then this.

Psychophysical methods have been developed by Lane (1961) and others for the measurement of speech perception. It is better to know the number of alternative stimuli involved in this for assessing the amount of information conveyed by the stimulus. For example, an 'Yes' answer to 15 questions may elicit little information from the respondents. But for answers 'Yes' and 'No' there may be more information. But if the number of alternatives are increased like 'May be' 'some what' still more information are elicited. Therefore if the number of alternatives are increased more information can be elicited from a single answer. In this way psychophysics has gone long way from Weber and Fechner, that is from the determination of sensory thresholds to the measurement of information processing.

Development of Psychology from 1960 to 2000 AD

The beginning years and the first quarter of twentieth century were the remarkable periods in the history of psychology for more than one reason. These were the periods in which the experimental methods under the title of psychophysical methods have emerged for the study of human behaviour, and also different schools of thought were emerged almost at the same time. Though each one had its own way of studying human behaviour, they all coexisted somehow for a considerable period of time till the end of the second quarter of the twentieth century. After 1950s - the validities of these schools, particularly, the dominant schools, the psychoanalysis and behaviourism were questioned by some critics and radical thinkers in the field of psychology. Evidences were collected for and against the services and the contributions made by those schools for more than fifty years. This resulted in the emergence of a third revolution in the mental health field and a third force in the field of experimental psychology.

The neuroscientists have shown extreme interest in studying the human behaviour through the functions of the brain and nervous system by using new techniques for the development 6f biological psychology.

Humanistic Psychology

The main argument of the revolutionists is that psychoanalysis studies only abnormal persons and behaviourism studies animal behaviour and attributes animal characteristics to human beings. They come out with a strong view that psychology should mainly deal with mentally healthy persons and the study of human personality should be holistic, the person as a single whole. These views were gaining more importance in the United States. At last a group of psychologists under in leadership of Prof. Abraham Maslow (1908-1970) came out as humanistic psychologist in the year 1962 which is considered to be birth date of humanistic psychology. The other psychologists who were equally responsible for starting the humanistic psychology, were Gardner Murphy (1895) and Carl Rogers (1902).

Human society is rapidly multiplying and also becoming more complex. Psychological studies began to touch almost every aspect of human life. Problems of various types cause people more stressful. The mental health problems of the people were, receiving serious attention for a systematic study. Therefore it has become necessary to study human behaviour and mental health processes scientifically. The existing schools were studying different aspect of behaviour and mental health processes. The dominant force in American psychological world was behaviourism founded by J.B. Watson. In Watson's behaviourism, the key concepts were conditioning, conditioned stimulus, conditioned response and stimulus-response inechanism. There were strong expressions of dissatisfactions about the services of the schools of psychoanalysis and behaviourism. The complete nature of man, his potentialities and resources were neglected. Man should be studied as a whole person and not in fragments. This type of thinking resulted in the emergence of humanistic psychology.

Humanistic psychology deals with the human person in a deeper and broader perspective combining science and philosophy. Humanistic psychology tries to answer the questions, what is human life about? and what human life can be? Is life lived positively. Humanistic psychology tries to explore the behaviourial characteristics and emotional dynamics of full and healthy human living. It tries to understood the person as a whole. According to humanistic psychology the healthy individual's need is increasing the self actualization which is the basic motive and goal of life. It puts more faith in a healthy person for its study. The individual person is basically whole, sound, healthy and sole with innate drives for growth and self actualization. Human sufferings come to him when these basic resources are denied. Humanistic psychology is greatly influenced by Smut's (1870-1950) holism. According to this theory personality is the highest form of integration. In a short period, quite number of books an"' research papers on Humanistic psychology have come up to enrich the field.

Community Psychology

One of the important problems the psychologists wanted to study in recent times is the problem generated by the inter-personal relationships of the individuals in the communities. Rappaport (1977) states that community psychology is the history of the efforts taken by the mental health personnel to deal with the margional people in the community who reject the society and those who are rejected by the society. The antecedent of community psychology is the history of mental health tradition goes back to the first mental health revolution which took place during the period of phillippe pinal (1745-1826), a French Physician and Benjamin Rush (1745-1813), an Arnerican physician and Devothea Lynde Dix (1802-1882) an American teacher. Community psychology is a direct product of the attempts of a section psychologists to deal with the behaviour of the individuals who do not fit into the society. Rappaport points out that the roots of community psychology are in clinical psychology, in psychiatry, in the academic study of human behaviour and in the study of individual differences. But it puts more emphasis on the abnormal-the pathological and on the deviant.

Many psychologists who were working in community mental health centres identified themselves with the goals of the 1965 Boston -conference and set a division of community psychology with the American Psychological Association, *The Community Mental Health Journal* was started in 1965. At the same time the *American Journal of Community Psychology* was published under the editor ship of Charles D. Spielberger. Cowen (1967, 1968, 1969, 1971, 1975, 1970, 1966, 1973) made many contributions. Zax and spector have brought outs a textbook in community psychology. Juliam Rappaport published a community psychology book on values and research, Micheal T Nietzel and others brought out a book under the title of "Behavioural approaches to community psychology".

Therefore, there is no one single person who can be named as the founder of community psychology, but, we can name a few of the psychologists who have made notable contributions for the emergence of commonlogy psychology. Some of them are Gerald Caplan (1964) Cown, E.I. (1976), Hersch (1969), Zax and speqter, (1974) Rappaport (1977) Charles D Spielberger (1970) Iscoe (1970). Community psychologists are of the view that community is cause of all mental health problems and community is also serving as a the repuretic agent. Community psychology is an offshoot of clinical psychology and psychoanalysis. It emerged as a passionate protest against the limited services performed by the existing schools of psychology, the psychoanalysis and clinical psychology in the year 1965. Further the community mental health discipline has redefined its scope involving the individual's mental health problem and also the society as a whole putting the responsibility mainly on the community. Community psychology is interested in the comprehensive community based services. Community psychology involves the community and the individual persons. According to community psychologists it is the transectional process between the individuals and the environment that causes and develops the human behaviour. The transection may result in an unhealthy or healthy well being of' the person.

Generally in psychological studies there may be more marginal people, people who are mentally retarded, or mentally ill or devient. There may be behaviour devients, social devients and sexual devients The social deviance is viewed from a person's biological make up and sometimes from current environmental point of view. Community psychology is not having any faith in the traditional methods of dealing with human mental health problems. Community psychology wanted to carryout the services directly to the community where the clients are, instead of the clients going in search of clinics and hospitals and waiting for hours and getting and expensive treatment. It also tries to make immediate changes in the existing social system of the community, so that greater disaster that is likely to take place can be averted. It has formulated action oriented programmes.

The cordinal characteristic of community psychology and community mental health has been the emphasis on interventions. There are three types of interventions such as primary prevention, secondary prevention and tertiary prevention. Anyone of the preventive programme may be applied according to the nature and severity of the problems. This field is very fastly deve ing all over the world. Large number of books and research papers are lopi published and universities teach the subject at post-graduates level and students register for M.Phil and Ph.D., Degrees by research.

Environmental Psychology

Following these, human behaviour problems were conceived-also from other angles like ecological point of view involving the broad spectrum of environment and its influence on human and the reaction of human to it and evolved a field of psychology called environmental psychology in the year 1968.

The behaviour of human beings is highly influenced by the environmental factors and in return human behaviour changes the environmental atmosphere. Though environment provides all facilities for the survival of the organisms, human behaviour seems to be very cruel and unsympathetic towards environment. Man cuts and destroys the green plantations and builds industries, which cause pollutions of various types by throwing the wastages in the environment, and pollute river water by canalizing the unwanted chemical substances into the fresh water and also kill the creatures. Man pollutes the pleasant environmental conditions. The smokes, dusts, gases from the industries and the from the moving transport pollute the air, water, residential areas which spoil the health of the people.

The environment includes the land, under the water, in the space and in the unusual environments. Human behaviour varies in all these areas. Environmental psychology is interested in the natural environment as it influences human behaviour. Some of the psychologists use the term ecological psychology. According to them, ecological psychology is the study of the relationship between living organisms and their environment. When it is applied to human, it is stated as human ecology, which is the study of human groups and their physical environment. But, basically there is no difference between the terms ecology and environment from the environmentalists of point of view. Environmental psychology is concerned with practical world problems, the field of it is not laboratory. Its approach is holistic and synthesises other disciplines like work and landscape, architectural, urban planning, designing and building, pesticide and other chernical areas, geography, sociology and anthropology. One of the important subject matter is the environmental pollutions. Environmental pollution means making the environments unclean and impure and contaminating the environment with poisonous or harmful substances, Now all over the world there are environmental pollution control boards to look into this aspect as the health of the people is greatly effected by various pollutions. It also deals with the effects of noise, effects of seasons, temperature personal space, culture and the inter personal contacts, territorial behaviour, population density, consequences of overcrowding. Thus the scope of environmental psychology is very wider with varied subject matter.

Population Psychology

A fourth group of psychologists, who paid more attention on the miseries and the vices of humans strongly viewed that humans were loosing their happiness of life periodically due to the serious population problems and found strong evidences for these in the history of human kind and evolved a field of psychology called population psychology. It has its origin in 1963.

The American population council has started a separate division for studying the population problems from the psychological perspective. The human crises miseries and vices affect the harmonious life of the people very much and people are getting disgusted with the unwholesome human crowds everywhere; which are causing adjustment problems in their interaction with others. Therefore it has been realized whether these human problems could be brought under the fold of psychological perspective and remedies can be find out. Though these problems involve the services of several other social sciences, effective strategies can the evolved from psychological perspective and the problems may be dealt with. This has awakened many social scientists all over the world and a separate field of study was felt which resulted in the birth or population psychology.

There is a great deal of imbalance behaviour between the birth rate and the death rate. It is observed that more than 80 percent of this growth has taken place in the developing nations. These nations face several problems and because of this the most serious problems is the high proportion of population in the economically dependent age level. The population pressure damages the qualitative aspects of human life. According to Julian Huxley (1963) the present (generation of man is, laying a big burden on his own future. He is putting his, childrens' in great misery. He is making the world difficult to be free from frustration and greed. The rapid increase of population is seriously affecting the behaviour of people in many ways. The main problems faced by the population explosion are severe unemployment, shortage of food, place to live in, migration, immigration, increase of antisocial problems, like, theft, dacoity, house breaking, way laying, murder, all sorts malpractices, increased sex crinies, and so on. Thus the behaviour of people is much affected by the increased population and psychologists wanted to find out some remedy for all these.

Biological Psychology

Studies of human bodily parts and functions are the continuous processes and they are non-stop businesses of biologists, anatomists, and physiologists for centuries. Since the brain and the nervous systems are the important human parts that are having control over the behaviour, the importance of their functions can never be concluded. Though the brain has long been studied, scientists have recognised the central role of the brain only recently. The central role of the brain is the controlling of behaviour. From the time the scientists started studying the brain, diverse information about the brain were released in the human history. But only recently the knowledge about the brain universally acceptable was brought out, that, it is the brain that mediates and controls the behaviour. But a science that explores the bodily bases of our experience and behaviour, the ways in which bodily states and the processes exercise and control behaviour and cognition, and the way in which behaviour and cognition in turn influence the bodily were not in yoke. It is because not one, but many scientific disciplines contribute to this whole function. Therefore we have to depend upon the research findings from the fields of psychology) anatomy, genetics, physiology, bio-chemistry, endocrinology, neurology, psychopharmocology, zoology and psychoneur-immunology. It is not limited specifically to any one of these disciplines. The finest aspects of these disciplines put together in full-fledged manner, which can fully deal with the bodily bases of' our experience and behaviour constituted a new scientific field known as biological psychology. Though the idea about biological psychology was in air for a long time, it was inculated in 1980s and the first book which is systematically dealt with the subject matters of the recent developments under the heading of 'biological psychology' was brought out only in 1996 by Rosenzweing and his coworkers of the university of California, Berkeley. There books on physiological psychology, but not on biological psychology.

Biological psychology is now rapidly developing field. Biological psychology mainly deals with the organization of the brain and the-functions through the processes of evolution and behaviour. Biological psychology has acquired several names in the past few years and when more and more research papers were published. The biological psychologists had some difficulty in giving the correct name when many of the scientists from other areas naming it in various ways. Biology is a broader life science, encompassing several sciences like physiology, anatomy, genetics, endocrinology, neurology, zoology and neurosciences. Some scientists called it as physiology of behaviour, some as behaviour of neuroscience, some as clinical neurology and some as neuropsychology. But however, the founders Mark R. Rosenzweig, Arnold L. Seirnan, and S. Marc Breedlove named it as 'Biological psychology'. The aim of the biological psychology is to understand behaviour and experience in terms of their biological mechanisms.

The American Psychological Association in 1989, while formulating the area of study for the professional psychologists specified that each student should study four substantial areas, of which the first is being the biological psychology, and then comperative psychology, neuro-psychology and psychopharmacology. Those who study and teach neuroscience in many departments like psychology, physiology, and psychopharmacology are required to be trained in behavioural and biological sciences, because biological psychology is devoted for improving human bodily conditions and behaviour. Biological psychology finds its roots in the history of research on brain and behaviour.

There are references about brain research in ancient Greeks' works. Hippocretes (400 BC) the ancient Greek physician considered that the brain was the of tellect and the part that controls behaviour. Aristotle (350 BC), the Greek philosopher considered that the brain is the seat of mental capacities and controls the bodily temperature through heart.

Later on the period of renaissance is considered as the period of rebirth, the great revival of art, science and learning in Europe during the 14th, 15th and 16th centuries The great Italian painter, Leonard Davinci (1452-1519) portrayed the functioning of the human body in his paintings. He has painted the nerves and the shape and the appearance of the external surface of the brain. Rane, Descartes the French philosopher and mathematician, explained in his works the behaviour of animals and compared it to human. He also dealt with the spinal reflexes and the neural pathways.

For about two hundred years there were no substantial work on brain or human behaviour Around middle of nineteenth century in Europe it has been accepted than the brain is the mechanism that coordinates and controls human behaviour Phrenology which is considered as the brain science is concerned with localization and the function of the human brain. It is believed that the cerebral cortex has separate functional areas, which are responsible for behavioural faculties.

But this idea was not accepted by another group of scientists. They were of the view that the brain like the mind functions as a whole and there are no separate faculties in the brain. But around third quarter of nineteenth century, the French surgeon Paul Broca was able to establish that the language ability is not the function of the entire brain, but it functions in a restricted area of the brain. Broca by an operation was able to demonstrate the region of language within frontal lobe of the brain, in 1861, which is the master area responsible for speech, Patients with injury in BrocWs region are aphesia. Apasta is the general term to describe the inability to understand speech Thus the nineteenth century backgrounds of brain studies were locating the different regions of behaviour in different parts of the brain which were the major contributions to the development of biological psychology.

There was also an additional information emerged from these studies, This was relating the size of the brain to ability or intelligence. One remarkable feature of the nineteenth century anatomical study was the use of microscopic observation of the regions of the brain, which revealed something about the shapes, sizes and identity of the nerve cells of the brain. At the end of nineteenth century the American psychologist William James explained in his book, *'The Principles of Psychology'* (1980) the psychological ideas and the properties of nervous system which may be stated as the beginning of biological psychology. Other than these there were also many important developments in the subject matter relating the to biological psychology. Hermann Ebbinghaus a German psychologist was the first to use non-sense syllables in experiments to measure learning and memory in humans. At the same time American psychologist Rebert L. Thorndike broughtout his classical work (Ph.D Thesis) in 1898, out of the experiments conducted on animal learning and memory.

In the first quarter of twentieth century many research papers on animal learning were published. Russian physiologist, Ivan V. Pavolov had published his research findings, obtained from the experiments conducted on dogs (1927, 1928a, 1928b, 1932). His main hypothesis was that animals could be conditioned in learning and his famous classical conditioning can be applied on any animal. He also demonstrated how neurotic behaviour could be produced by conditioned stimulus in dogs.

A search was made to find out the location of memory functioning in the brain by Karl S. Lashley (1890-1958). In hi,s long career of research work he had published many papers on the behavioural effects of brain damage (1920, 1922, 1929, 1933). After Lashley, his student Donald Hebb, a Canadian Psychologist continued in the same field by studying the nervous system and the brain. Hebb (1904-1985) explained the principle, how complex cognitive behaviour could be carried out successfully by the networks of active neurons in his book *"The Organisation of Behaviour"* (1949). According to Hebb there are random connections between brain cells in the beginning, that could be organised by sensory input and stimulation and become strongly interconnected groups. Hebb called this as 'cell assemblies'. Cell assembly for Hebb means that a large group of' cells that tend to be active at the same time because they have been activated simultaneously. How neurons strengthen their connections by use, was the hypothesis for Hebb. This is known as 'Hebbian Synapse', a subject matter which the present neuroscientists study very much in these days.

There are also some new techniques, evolved that would help to identify the genetic mechanisms and nervous systems and the mysteries of the brain. All these research findings about the behavioural functions or nervous systems and the brain, though in piece-meal form have all contribute to the development a full-fledged *biological psychology,* a new field of psychology which fully deals with the knowledge about the behavioural and bodily systems of the brains and nerves system. With in a short period of time enromous amount of materials have been amassed ior the field of biological p Sychology by the neuroscientists all over the world.

REHABILITATION PSYCHOLOGY

Introduction

Human mental health problem is as old as man himself. The body suffers from physical illness, similarly the mind suffers from mental illness which is revealed in various behaviour patterns. Man does not bother about his physical illness so much, as he is often gets some illness and gets it recovered by some treatments. But the mental illness whether it is of some mild type or seven type causes worry and anxieties and the person gets himself involved in some kind of trouble from which he has to struggle to get rid of it. If he succeeds, it is well and good, but if he succumbes to its sever-hold, then it becomes not only an individual problem, but also a social problem or a community problem. Human kind has been facing severe mental health problems of various types for a long time and professional workers who are engaged in dealing with such problems have been trying to find out remedical measures, medicines and methods of treatment.

The treatment techniques, medicines, tools and instruments that we find today in various medical fields are the consequences of continuous hardworks of the mental health scientists for generations. There has been continuity from one stage to the next stage without any break in dealing with the mental health problems, resulting in remarkable improvements. The mental health revaluation started by **Phillippe Pinal** (1745-1826), **Banjamin Rush** (1745-1813) and **Devothea Lynde Dix** (1802-1888) did not end with it. The efforts were continued with some improvement which was called as the second mental health revolution made by **Sigmund Feud** (1856-1939). Since the human mental health problems were varied time to time different type of approaches were needed. The conventional methods of treatments were found to be expensive and time consuming. The mental health scientists felt that a different type of approach was inevitable. This dissatisfaction resulted in the emergence of a *third revolution*. The outcome is the development of a separate field called *community psychology* in the year 1965. The existing systems of psychoanalysis and the clinical psychology were combined and the objectives were evolved. They approached the mental health problems from different dimension. According to the community psychologists, the community is the causative factor of mental health problems and also it serves as a therapeutic agent for alleviating the problems. They formulated the *intervention programmes* to deal with the mental health problems prevailed in the community.

Community psychology is a product of the attempts of a section of psychologists to deal with the behaviour of the individuals who do not fit into the community. It puts more emphasise on the pathological individuals and on the deviants. In psychological studies there may be more marginal people, persons who arc mentally retarded or mentally deviants. There may be behaviour deviants, social deviants, and sexual deviants. Community psychology carries out services directly to the community where the clients are not made to go in search of clinics and hospitals for treatment. It has formulated *action-oriented* programmes.

There are three types of intervention programmes such *primary prevention, secondary prevention* and *tertiary prevention*. The main objectives Of these preventive programmes are preventing the incidents of the mental health problems in the community by removing the causative factors from the society and keep the people away the attacks of any such mental health problems. Any one of the preventive programmes may be applied according to the nature and severity of the problems.

This fields is very fastly developing all over the world. Large amount of study materials have been produced. Researches are going on enormously everywhere. Intervention programmes are widening. This includes not only the alleviating the grieved, but also providing the needed rehabilitation facilities and make the disabled and the deviants to live on their own. Since the programmes are for the community as whole which includes socially deviants, mentally deviants, and sexual, deviants a-rid physically disabled, mentally retarded or mental deficiants and all other mentally impaired persons. Therefore the services of the mental health professionals are widely distributed.

Emergence of Rehabilitation Psychology

A section of the mental health professionals have felt that a considerable percentage of the mentally retarded category of persons are not getting concentrated attention, care and treatment. A group of psychologists who were working in the action-oriented intervention programmes were not satisfied with the services rendered to the mentally retarded persons. Any rehabilitation programme should make its basic principle for the study of the problem person from various dimensions, and needed form of rehabilitation programme should be formulated. Therefore it was felt by a section of psychologists to have a separate field of study with deep concentration on mentally retarded alone can serve the purpose. This revolutionary view resulted in giving birth to a field of psychology called *Rehabilitation psychology*. In other words we may state that community psychology begot the rehabilitation psychology as a branch of its field.

Rahabilitation psychnlogy, therefore is a recently emerged branch of psychology. The main aim of this field of psychology is action- oriented services to the individuals who axe mentally, physically, socially and environmentally handicapped and to help them to achieve the maximum possible benefits mentally and physically. These individuals, because of their disabilities, they could not participate in any legitimate activities in the community. Rehabilitation psychologists have become active in putting themselves in action and taking care of the disabled persons in various ways.

Rehabilitation psychologists are seriously engaged in identifying the mentally retarded, physically disabled, persons of sensory deprivation, emotionally disturbed and developmental defects and analyse their problems individually and bring them under clinical observation wherever necessary, conduct counselling for identifying the specific problem and formulate suitable methods of alleviating their sufferings. The individual person is studied as a whole and the existing assets are identified and accordingly the rehabilitation psychologists designed the programmes by which they hope that the individual person must get the maximum benefits.

The rehabilitation system is a multifaceted and dynamic in characteristic, consists of both the professional of various categories and the deviants of different problems. An individual person's disability of any type is an obstruction for normal and easy way of habituation in such cases special provisions are made for facilitating for the normal functions. It is by way of suitable training or exercise or education the impairment is overcome to some extent.

Rehabilitation psychology also pays special attention on exceptional children. The concept exceptional covers a very wider area of handicapped state of any type among the children. Generally, they are called deviants. In any community we can find certain percentage of children deviating from the normal level. This deviation may be mental, physical, social and emotional. There may be also variations in the level of deviation.

It has been observed that of the disabilities the mental retardation is the most difficult to deal with. Mental deficiency and mental retardation are synonymously used. Generally mental retardation is defined in various ways. In order to avoid confusion and to have a clear understanding, the American Psychiatric Association (1968) has formulated its own definition. According to them, mental retardation is a subnormal general intellectual functioning which originates during the developmental period and is associated with impairment of either learning and social adjustment or maturation or both. According to the statistics of the Association as given by **Albee** (1968) the incidence of mental retardation in the United States is about 6.5 million persons which is 3 per cent of the total population in 1968. But the American Association of Mental Health made a classification of various types of mental deficiencies and used a suitable concept, called *mental retardation.*

According to this classification, mental retardation is defined as the state, significantly sub-average general intellectual functioning existing concurrently with deficits in adaptive behaviour and manifested during the developmental period. Mental retardation may be also. defined as the condition of low intellectual development requiring special education, training and employment. The concept mental retardation has replaced the concepts, mental handicap, mental abnormality, mental deficiency and feeble mindedness.

Assessing Mental Retardation

The level of mental retardation is assessed by intelligence tests. A person is rated as highly intelligent when he successfully handles complex situations or performs more difficult task. Some of the mental abilities included in the intelligence tests are mainly verbal abilities, vocabulary, arithmatic abilities, logical reasoning, special relation and visual hension of geometrical forms. If a person is found to be deficient in all or in any of these mental abilities he is said to be mentally deficient or defective in his attempts to solve his problems of life.

Intelligence of an individual is measured in terms of Mental Age (M.A.) and Intelligence Quotient (IQ). A person's mental age is the level of his mental development. This should be revealed as equivalent to his actual age which is the chronological Age (CA). A person's Intelligence Quotient (IQ) is the ratio between Mental Age (MA) and the Chronological Age (CA). The Intelligence Quotient (IQ) is hypothetically determined that it may range from 0 (Zero) to 200 and the average is 100. The ratio is multiplied by the average (100). Thus the formula for determining the Intelligence Quotient (IQ) is

$$IQ = \frac{MA}{CA} \times 100$$

The mental age is always determined by intelligence test scores.

Classification of Mental Retardation

The rehabilitation psychologists' primary work is to assess the mental level by using intelligence test and classify the persons under various categories. According to the norm, those whose IQs are above 110 are very bright or super-lor. Those whose IQs fall between 90 and 110 are average. Those whose IQs fall below 90 are said to be mentally retarded. The rehabilitation psychologists are more concerned with these people and their behaviour.

According to the American Association of Mental. Health classification, 25 percent of the people fall below 90 IQs. These 25 percent of the people vary in their IQs considerably. Among the 25 percent of the people those who obtain the IQs between 70 and 80 are said to be borderline category and those whose IQs fall below 70 are classified as feeble minded. These feeble minded persons are about three percent of the population. These feeble minded persons fall under three categories as idiots, whose IQs fall below 25. Those IQs fall between 25 and 50 are imbeciles and those whose IQs fall between 50 and 70 are morons. This system of classification was continued for a long time. But, since the mental health workers felt that this system of classification is derogatory, they dropped this system and a different type of classification was attempted. They used the concept mental retardation in place of feeble mindedness.

Several studies on mental retardation were made all over the world. On the basis of these studies mentally deficit persons were classified into five categories. According to the correct diagnostic and statistical manual of the American Psychiatric Association DSM-III four levels of mental deficiencies were recognised. They were : (1) Mild Mental Retardation, (2) Moderate Mental Retardation, (3) Severe Mental Retardation and (4) Profound Mental Retardation.

According to this classification the mildly retarded people constitute 85 percent of the total mentally retarded whose IQs range from 50-55 to 70. As they are, there is no such difference between them and the normal children in their behaviour. But these persons always need help in social and financial matters.

The Moderate Mental Retarded persons constitute ten percent of those whose IQs fall below 70. The IQs of these persons range from 35-40 to 50-55. These persons often subjected to brain damage and other pathological conditions. These persons have also physical defects and nervous disorders and also their motor skills are affected.

The severely mental retarded persons constitute about four percent of those whose IQs fall below 70. The IQs of these persons range from 20-25 to 35-40. These individuals have physical abnormality by birth. The motor skills of these individuals are limited. Most of them need constant help and supervision from others. They have severe brain damage which makes them very much possessive.

The profoundly mental retarded individuals constitute two percent of those whose IQs fall below 70. The IQs of these persons range between 20 and 25. These individuals need close supervision and care all the times. These persons have physical defects and neurological impairment. The death rate of these retarded individuals is very high in the childhood period.

Other Classifications of the Mentally Retarded Persons

Mentally retarded problems were also studied by-educational psychologists from educational point of view. This classification is quite similar to the classification made by DSM III-R.

These studies were made with aiding instruments and tools and with institutional programmes for the mentally retarded. They made the classification of the mentally retarded as educable and trainable individuals. The educables are mildly retarded and the trainable are moderately retarded.

There is also another classification of the mentally retarded made by the American Psychological Association. This 1968 classification identifies five levels of mental retardation. They were : (1) borderline mental retardation, (2) mild mental retardation, (3) moderate mental retardation, (4) severe mental retardation, and (5) profound mental retardation.

This classification is made first like all other classifications based upon the IQ levels. But this classification includes other factors like the individual's early development, academic and vocational achievements, motor skills, social and emotional maturity apart from the IQs.

The Causes of Mental Retardation

Mental retardation is mainly a developmental problem and therefore it is to be determined genetically. These are prenatal stage, perinatal stage and the postnatal stage. The prenatal stage is the stage, present before birth during the period of pregnancy. The chromosomal irregularities of impairment might have had effect upon the growing fetus, which might have resulted in mental retardation when the child was born and grown. The chromosomal abnormalities may also occur due to the increased number of chromosomes and the child is likely to be born with XYY or with XXY chromosomes. There are also several other reasons like the mother taking too much of *medicines* during pregnancy, the *thyroid* over or under secretion, exposing often to *x-rays* or *scanning*, all may cause mental retardation. The child may also have physical deformities.

The second stage is the perinatal at which is relating, to or occuring in the period, immediately before or after the birth of the child. This mainly refers to the premature birth. The child is born with less weight. The child might have experienced oxygen deficiency, developing in an incorrect position in the stomach, abnormal umbilical coiling, all contribute to the development of mental retardation.

The postnatal stage refers to the period of after birth of the child. After the birth if the child had malnutrition during the period of first-two years there is every liklihood of causing mental retardation. The first two year period is a crucial period and infectious diseases are quite common. Brain injuries also may occur, all may pre-dispose to mental retardation.

Rehabilitation

Rehabilitation psychologists are very curious about the rehabilitation programmes for the mentally retarded persons. They apply all the methods, procedures, techniques adapted by the community mental health professional. In addition, they formulate and design certain specific programmes to meet the needs of the mentally retarded children, adolescents and adults in the community. Though they accept the institutional programmes in principle, they strongly advocate individual attention as there arc serious individual difficulties in the problems of mental retardation Therefore, they first of all identify the individual problems and the levels of mental abilities by administering suitable tests.

They are of the view that the mental abilities and the educational interests of the mentally retarded considerably vary from those of the regular class of the normal children. Therefore, there should be special classes for the mentally retarded children on the basis of the levels of the mental abilities or I.Qs. The rehabilitation psychologists first test the mentally retarded persons as to what extent each one person is educable and trainable. On the basis of their characteristics they design the rehabilitation programmes.

The rehabilitation professionals first of all take into consideration the causative factors of mental retardation, the nature and the characteristics of the problems, then the preventive and curative techniques. Then they formulate intervention and training programmes. The rehabilitation professionals are trying to integrate the problem individuals into the larger community. The professionals also concentrate on enlightening the people of the community the genetically oriented causes of mental retardation and 'i' al make them aware of the consequences of the prenatal, perintal and the postnatal occurrences. It is believed that if people understand the seriousness of their negligence of their responsibilities during the period of pregnancy and after the birth of the child it is possible to avoid or reduce the occurrences of the mental retardation in the community. Since the *main cause is rested in the chromosomal impairment or abnormalities, the mother is mainly responsible for the cause of mental retardation.* This is one of tile important rehabilitation strategies of the rehabilitation professionals.

The next stage of the rehabilitation programme is making some improvements in the skills of the mentally retarded individuals. As it is genetically rooted and no change can be made in the brain structures, the only way of dealing with the problem is to make some improvements through specially designed educational programme, training the individual in certain specific muscular and motor coordinating tasks and in some semi- skilled, and in some simple motor skilled tasks and making the person as far as possible to stand on his own,

If the mentally retarded individuals need super-vision and care they are provided till they carry out the task without care. Rehabilitation programmes are designed to suit the individual person's mental abilities and also on the basis of the physical capabilities, speech and language development and also emotional and social adjustment the individuals are promoting in various situations.

PART TWO

BIOLOGICAL BASES AND DEVELOPMENTAL PROCESSES

3. Heredity
4. Environment
5. Structure and Functions of Nervous System and the Brain
6. Endocrine System
7. Human Development

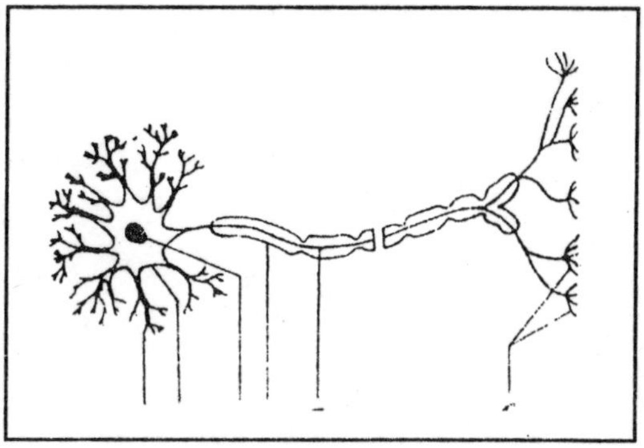

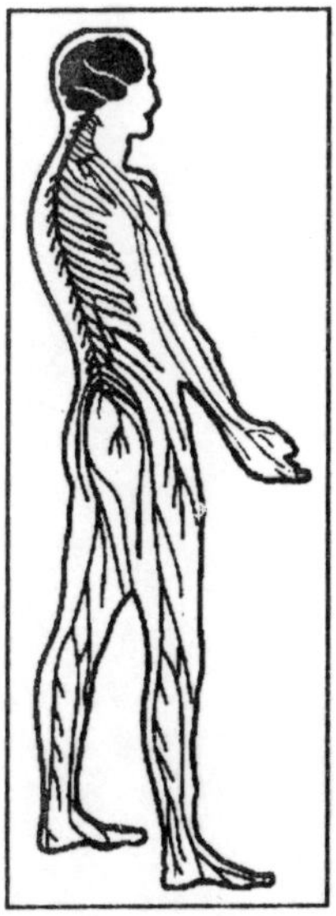

3

HEREDITY

The Problem of Variability in Human Behavioural Achievement

Human behaviour is revealed in various kinds of activities. They may be physical activities or mental activities. Man's achievements are made both in physical activities and mental activities. In performing either physical or mental activity individual persons do not make the same level of achievements. Some perform certain physical or mental activities very successfully without much strain. Some perform the same activities either physical or mental with some difficulties or with some strain. But some perform the same activities with much difficulties or even fail to complete the activities successfully. It may be simple activity like arranging the books according to the code number or complex activity like solving some arithmetic problem.

In the case of some difficult tasks or memorising and reciting some passages of some poems, few students recite them without any mistake, while some do not remember anything even after learning the passage reveal times. The teacher with all his efforts could not bring him to the average level. What must be reason? Some carry out the instructions correctly while some fail to do the same. The reason generally attributed is *inborn quality*. The inborn quality of the mind is stated as *heredity*. In the classroom, some students are bright and some are dull and few very dull. Are these differences really due to heredity? This question has been raised for decades and centuries. Or is it due to circumstances or environmental conditions? This problem has been a riddle and many experts tried to find out the correct answer for it.

The parents and the ancestors of **Thomas Alva Edison** were not scientists. The parents of **Descartes** were not philosophers. If heredity alone is responsible for producing bright students, intelligent men and women, experts and geniuses, so much scientific and technological developments would not have taken place in the world. Bright and intelligent people are found in every racial stock and culture. But what is needed for achievement is hard work. It is by burning the mid night candle. Many experts accept this view that hard work alone can bring achievements.

The strong environmentalists emphasise that the environmental conditions are mainly responsible for great achievements. They state that in the environmental condition we can bring up a child as a genius or as a dull and stupid person. On the other hand the anatomists and the biologists emphasise the genetic factors alone can decide whether a person could be an extraordinarily intelligent or genius or a stupid. In determining the level of intelligence these two extreme views are always prevailing on one another. It is true that we cannot totally reject the hereditary factors nor can totally accept the environmental factors.

Hereditary factors need not be attributed only to the development of intelligence. A person's eye colour, hair colour, skin colour, formation nose, face, head and other bodily features may be similar to that of a person's parents and ancesters. Also his strength, muscle and skeletal structure and formation may resemble to those of his parents and grandparents. But the language he speaks is learned in the culture and society in which he is born and reared up. The diet, climate and the habits of bringing up, all may have influence upon the skill and talent of the individual.

Every human being or animal is born with some skills and talents which are based upon the muscular and skeletal and other bodily structures with the source of movements. These are inherited, from the forefathers. These native resources would be of very little use unless they are shaped and moulded by learning,, training, practice and experience. If behaviour were arranged on the scale of rank order or complexity from simple reflex action to most complicated or abstract thinking or problem solving activities and scientific achievements, there would be no chance of totally leaving off heredity and taking into account only the environment. In any given act of behaviour whether it is simple or complex there would be the constitution of both the native and the environmental influences functioning together in certain proportion.

Therefore, asking the question whether heredity is more important or environment for successful performance or behaviour becomes meaningless. But a legitimate doubt we can raise is, whether the observed differences in behaviour between two individuals are depending more upon the hereditary differences than on the differences in environmental conditions? To what extent environment can bring changes on hereditary factors? Can environmental influences remove the differences in hereditary traits? Nor we can strongly assert that a strong favourable environmental condition influence and correct fully the hereditary defeats. But we are to accept the development and *contribution made by the nature and nurture* for various types of physical traits and mental abilities, the development of conduct and character in the culture and society and the development of personality characteristics. At this stage it is also very essential to know the evolutionary processes that were going on for millions of years resulting in the production of innumerable number varied species (humans, animals and plants) with the characteristics of *variability* and *complexity*.

Evolution and Genetic Basis of Behaviour

Nature is endowed with enormous amount of resources, with which it could produce innumerable number of species with varied potentialities. Species whether human or other vertebrates, because of the limited physical capacity, are not able to develop beyond the level what they are assign with. There appears to be enormous amount of wastefulness in the species and also constant tentative beginnings in new directions one after another. This is the information we find in the biological background from the evolutionary processes and the emergence of humanity is one of the species as the result of the constant beginnings in the new directions of the evolutionary processes.

Through millions of years of evolution there has been heavy deposits on those living organisms which were capable of maintaining rich and healthy foundation and also capable of producing improved variations in the species. These types of living organisms were constantly producing offsprings. Among these offsprings few of them were capable of making better, some moderate, some become assets and some vanish away. Life of some type always takes new directions and these new directions appear in the form of young ones and these produce generations of young ones. In this way enormous amount of species of different types come into being. All these different types of the existing species are the results of *divergent evolution.*

The enormous amount of groups of the living species, by natural causes, get isolated. These isolated groups get more and more isolated geographically from one another. These isolated groups which are physically separated, are likely to evolve in varied directions. After very long period of time the varied isolated groups gradually and slowly change in characteristics and show no resemblance to other groups of the same species, isolatedly living elsewhere. This varied appearance in physical characteristics of the same species, would not have taken place in a short period. It would have taken thousands of years, because, evolution is a slow process and changes in species take place gradually and slowly. Thereis always a place in nature for the emergence of new species which may go on changing in the evolutionary process.

Any type of development in any species is a *modification of the existing conditions* or characteristics. There may be some type of similarities in the characteristics of the species of some groups and also some dissimilarities in other characteristics in the same species which might have occurred due to geographical variations. In the evolutionary process the species take different directions which generally results in *modification of biological and psychological characteristics*. The study of the stimulus-response mechanism and the behavioural patterns of the primitive organisms provide us information for understanding the basic characteristics upon which the complex human behaviour is organised.

While going through the successive evolutionary stages we can understand and get into the background of the development of the psychological processes and can have a clear idea of the manner in which they function It is to be known that in recent times those who are interested in the *study of genetics* with the view of interpreting inheritance and behaviour have made it a point to study genetics and evolution. By this means more information could be brought out about the *laws governing inheritance* of physical characteristics and also they have given more importance to behaviour genetics which is the study of inheritance of *behaviour traits*.

In recent times biologists and psychologists have studied some species like the mouse and fruit flies and observed in them certain psychological characteristics like the simple responses to stimuli, mating patterns and the preference for temperature. These are all inherited characteristics according to the *law of gene domination and chromosome transmission.* Animal breeders for a long time have been manipulating temperament and behaviour and also the physical appearance in animals, mostly in dogs, horses and in other domestic animals.

Modern psychologists on the basis of the genetic laws are trying to discover those aspects of *human behaviour in individuals' genetic makeup.* In these studies they try to eliminate the mental retardation which is supposed to be an accidental occurrence caused internally due to human errors. This defect is caused in the developmental process and should not be viewed as inherited characteristics.

Generally the process of evolution depends upon *mutations* because they provide the major source of *variability*. The concept of evolution is an important source for modern biology and for psychology. The theory of evolution is associated with **Charles Darwin**, who formulated this theory. He provided a scientific recognition to his theory by basing it on two principles, namely, *natural variation (selection) and survival of the fittest.* According to him, natural variation refers to *observable and demonstrable facts* that the variations exist in each species and in each species there exist wide variations in structure and behaviour. These variations provide potentialities to some species and they are better adapted to the given environment than others.

Those organisms that are *better adapted* to the environment are said to be the survival of *the fittest* and they produce offsprings. These adaptive structures and functions are determined to the extent that they are transmitted to the future generations. But the isolated and poorly adapted species are not likely to have any great evolutionary development. But in some species a major evolutionary development is likely to occur under a particular environment than in other species In certain areas like the coastal and mountainous regions some type of *rare species* are found. These species might have been isolated long-long ago and existing independent of other species. This enormous number of variety of species that exist in different environments might have come into being due to divergent evolutionary processes.

According to the theory of evolution, wherever the existing large groups of species are geographically or environmentally isolated long ago, rare groups of species are likely to evolve in different directions These groups after a long time show *little resemblance to the descendants*. This means that each development is a *modification of the existing condition.* We have seen above that in the evolutionary process there are two important aspects, the *complexity* and the *variability* and there is always a place for nature for the occurrence of new species and a new life begins. As and when the species develop new way of life they also get ready to perform complex activities. The complex activities of the species bring new achievements for them. This means that the species gradually develop from simple to complex behavioural patterns.

This is the result of the development of *complex nervous system*, which takes the species to a *higher level of intelligent state*. This state makes the species to *solve problems*. Species at lower level make only blind attempts and fall to solve problems. The nervous system makes the species to have communication among various parts of the body and also provide ability to make complex responses and also *the ability to learn.*

What is Heredity?

Life at present is not the same as it was some years ago. It has been taking always a new direction which may be beneficial to human beings, but sometimes it causes troubles and disasters. Generation to generation man's physical and mental capacities reveal enormously more than can be expected. Where from all these come? Though we can attribute this to the evolutionary process which brings perfection in the existing things, a correct answer yet to be discovered. It is also convincing that it is the result of man's unending and relentless efforts to see something new in everything. But can all men do this? Thorrias Alva Edison remarked once, that "Genius is a matter of perspiration and not inspiration".

A classroom may consist of many students, drawn from various cultures. They may range from twenty to one hundred students. Look at them all very carefully. They do not look alike. Each one differs from the other in one way or other. All of them may be of the same age, may come from the same home town, live in the same hostel, eat the same type of food and sleep for the same number of hours. Physically they are not equal. Few may be very tall, few may be very short and most of them may of moderate height. When arithmetic exercises are given to them some of them do the exercises excellently, some do with some difficulty and few do not understand them. When they go to play grounds, few master all the games, some do only some games, some afraid of going near the pit for jumping and some do everything with pleasure. Thus students may differ from each other physically, mentally and emotionally. Why these differences in individuals? Are there not any causes behind all these differences? It is impossible to give complete and perfect answer to all these questions. Many causes have been attributed for these differences.

Scientists believe that the differences are due to biological foundations. Every individual is having different biological foundation, Therefore, if we want to know more about the biological foundation, we must know something about the influence of heredity. The course of human development is influenced by the hereditary endowment of his species. All human beings as a homosapiens share certain characteristics. They walk upright on two legs, follow the same method of getting their needs. In this task they are able to handle their environment with a high level of manual accuracy. They all possess a large size and complicated material, the brain. All those and some other characteristics are present at the birth and the other characteristics develop later on. Even though all human beings share many attributes that are visible in the course of development, each individual also possesses certain unique characteristics. But many of them appear to be inherited. Thus the inherited internal differences determine all our behaviour which each one of us exhibit.

Therefore, heredity may be defined as the transmission of characteristics from one generation to another of the genetic factors that determine individual characteristics, responsible for the resemblance between parents and offsprings. It is the sum total of the inherited factors or their characteristics in an organism. The science that studies the hereditary characteristics is called Genetics.

Functioning of Inheritance

Heredity functions generally on two main principles. One is that "likes beget likes" The other is, "likes beget nearly likes". We all know that children normally resemble their parents or grand parents and their relatives more closely than they resemble the unrelated persons. But in some families the child is not exactly like either of his/her parents. A boy may resemble his father in his eye, hair the skin colour and height or his father's relatives. But in other characteristics like physical structure, emotion, he may resemble like his mother or mother's relatives. The boy may duplicate his father' physique and duplicate mother's skin colour in some families. The girl may duplicate in all respects her mother or mother's relatives.

But in some families the boy or the girl neither duplicate his/her father's characteristics or mother's characteristics. When the boy or girl in a family does not duplicate exactly his/her parents and their relatives variations take place. In such cases the second principle of variability is applied. This means that "likes beget almost or nearly likes". Thus variability in inheritance occurs. Generally inheritance means deriving something from the past to the present. Inheritance may be defined as the characteristics that the earlier generation had transmitted to the later generation through the process of heredity. The transmitted and the derived characteristics may be anything that were found in the persons of the previous generations. When this occurred in the family the members of the family resemble each other to some degree. But they also differ sometimes to some degree, from one another, in physical or mental characteristics. To know the resources of the variability we must go into the details of the functions of genetics.

Genetics

What is genetics? Genetics is a branch of biological science that studies the hereditary characteristics and the transmission of the characteristics from one generation to another generation. It is the study of heredity and the variations in organisms. According to genetic studies, traits are passed on from one generation to the next in people, plants and in animals. We often hear the words "heredity runs in our family". We all have inherited certain traits from our parents, grand parents and ancestors.

We have certain conviction that children usually resemble their parents and some relatives more closely than unrelated persons, but strictly speaking the child is not exactly like his/her parents. In certain physical characteristics like height or body built the child may be resembling his father or grandfather and in certain characteristics like temperament or mood he may be resembling his mother. He may very rarely duplicate his father's physique. Therefore, there is variability in inheritance. Though a group of members of a family resemble one another to some degree they also differ considerably from one another to some degree.

Gregar Mendel (1822 - 1884), (1959) a Austrian Monk and a botonist founded the science of genetics. He developed his theory of organic inheritance from his experiments on the hybridization of green peas. His findings were published in 1865, but it was recognized only in 1900. Mendel's first law is the *law of segregation.* According to this law, genes occur *in pairs* and therefore one member of the pair is contributed by each parent to the offsprings. According to modern genetics it is believed that every trait. was controlled by some inner elements which were transmitted without any change from one generation to the next. These elements are called genes. We often say that *genes are the careers of heredity.* During the course of fertilization, the.genes are separated and recombined in the ovum. Thus the child inherits one trait from his father and one from his mother. The inherited traits may have both similarities and differences. Human behaviour depends upon heredity and environment. The inherited traits may be changed due to different interactions of the environmental factors.

Behaviour Genetics

The main objective of psychology is to study behaviour. How behaviour originates? In studying the origin of behaviour we are to examine the factors which influence the development of behaviour in individuals or in organisms. It is in studying the origins of behaviour, we see the manner in which the genotype, and the environment interact to produce behaviour. All beta ur are depending upon both the genotype and the environment of the organisms Genotype means the genetic constitution of an organism. Behaviour genetic is the study of individual differences in behaviour that is attributed in part of the genetic make up. But all organisms do not behave in identical ways. The reasons for their behaviour differences can be found in the history of environment and in the genotype or in both. Some of the behavioural differences between normal healthy individuals appear to develop essentially of independent of specific learning. The ways in which the genetic and learned differences, between organisms produce the behavioural differences have drawn the attention of many disciplines.

During the first quarter of twentieth century the environmentalists denied the growth of behaviour genetics. **Watson** in 1924 has thrown a challenge that he can mould a person of any genotype into whatever manner I you want' by providing such environmental conditions. But today the important role played by the genetics in determining behaviour in human and in non-human is recognized by all scientists. The genetic determinants of behaviour has implied to all areas of behaviour. Behaviour genetic variables have become important to all psychologists, Genetic variables cause substantial individual differences in behaviour. These differences in behaviour are frequently genetic in origin. Behaviour genetic is supposed to be a recently emerged division within genetics.

Nature has produced a tremendously effective kinds of nervous systems, sense organs and effectors. Behaviour genetic is also interested in the distribution and causes of individual differences within populations. Among the humans how different they are, in socially important traits, such as intelligence, aggressiveness, response to stress and the like (Wechsler, 1952). In some situations the relative contributions of heredity and environment become immediate practical importance.

The roots of behaviour genetics go back to **Darwin's** (1872) study of sexual selection as an evolutionary force and to his studies of domestication or expression of emotions.

Almost at the same time **Francis Galton** (1883) reported the resemblance between relatives in talent and ability. Since **Mendel's** (1865) works on genetic were recognised, there was renewed interest in the inheritance of behavioural characteristics in humans particularly in intelligence.

The important feature of genetics is the use of biological relationship as an independent variable. The genetic approach to biology encircles a broad range of phenomena as known descriptively such as molecular genetics, physiological genetics population genetics and behaviour genetics. We are not going into the details of all these, but only behaviour genetics.

Mechanism of Heredity

If an organism is matured, it produces germ cells. These cells are sperm (spermatozoon) or ova. These cells arc 'reproductive cells. Every organism, whether human, or animal or plant begins its life as a single cell. Human life begins as a single cell in a fertilized ovum. Like all cells of the human body the reproductive cells contain chromosomes and each chromosome has thousands of genes. The genes are the segments of deoxyribonucleic acid (DNA). These genes bear the genetic information which is passed on from one generation to the next generation. The male cell is called sperm (spermatozoon) cell and the female cell is called ovum or egg cell. The sperm cell of the male penetrates into the egg cell of the female during the sexual intercourse and fertilizes it. The fertilized ovum or egg cell is called zygote. The fertilized ovum grows and divides into two cells. These two cells divide into four, four into eight, and eight into sixteen and into incalculable number of divisions that go into the formation of the organism. This means that they multiply into millions. The cell divisions become various parts of the organisms. Some become nerve cells, some muscle cells and some gland cells.

Chromosomes : Every normal human cell contains a nucleus which. is surrounded by cytoplasm. In the nucleus there are very small thread-like bodies called chromosomes (Fig. 3.1). These are involved in genetic transmission. This can be seen under microscope.

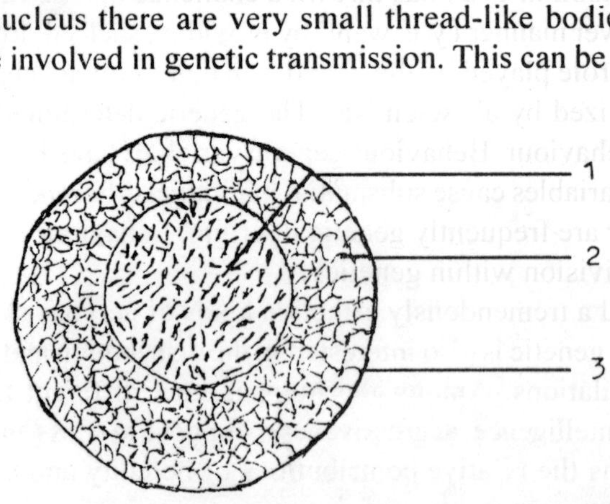

Fig. 3.1 : Cell Plan
1. Nucleus, 2. Cytoplasm, 3. Cell Membrance

The number of chromosomes are distinct and characteristic for each species with numbers ranging from two to as many as 127 pairs. (Sinnott *et al.*, 1958). There are 46 chromosomes in the nucleus of the cell. They are always found in pairs and there arc 23 pairs of chromosomes. They vary in size and shape, Some of them are large, some shorter, some with curves and some straight. The term chromosomes refers to the combination of two things, 'Chromo' and 'Soma'. Chromo means colour and soma means body. Therefore chromosomes are bodies consisted of colour. The number of chromosomes vary in animals, birds and plants. But the same number of pairs are found in every species. During the course of fertilization the 23 pairs of chromosomes from male's sperm and the 23 pairs of chromosomes from female's ovum are combined into a single cell called zygote. They do not make into 46 pairs. When the reproductive cells called gametes are formed, the chromosomes from sperm or ovum do not divide. Therefore, there are only 46 chromosomes, 23 chromosomes from sperm cell and 23 from ovum (Fig. 3.2).

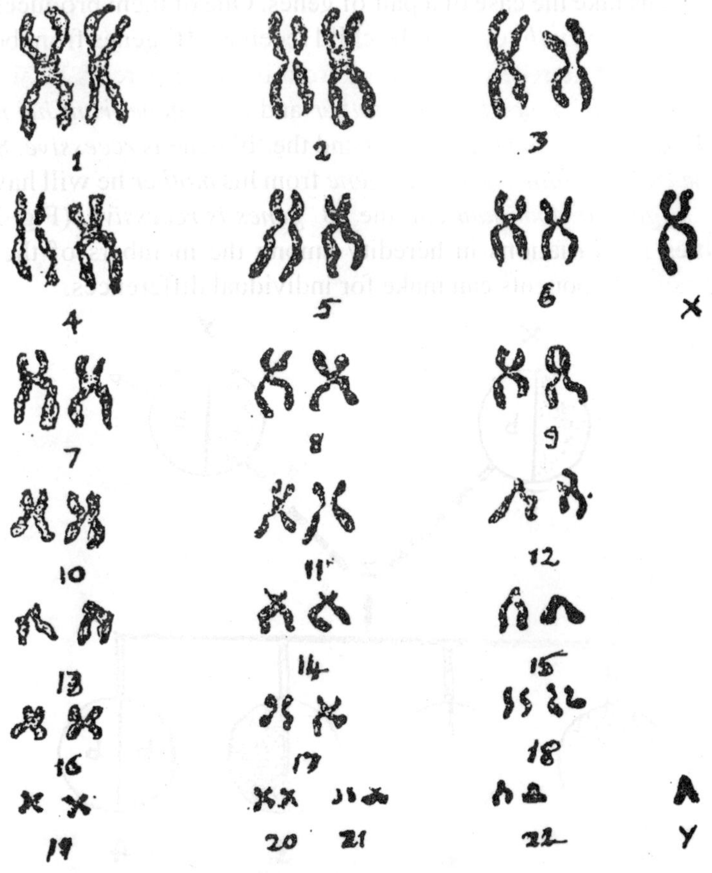

Fig. 3.2 : Chromosomes

Every normal human cell contains a nucleus which is surrounded by cytoplasm. In every nucleus there are thread-like bodies called chromosomes. This can be seen only under microscope. There are 46 chromosomes in the nucleus of the cell. They are always found in pairs and there are 23 pairs of chromosomes. The vary in size and shape.

Genes : The genes, that we have discussed earlier under the heading of genetics, are the primary elements in heredity. They are found in chromosomes. If a chromosome is examined under a very powerful microscope a string of beads like material is seen. It is actually a string of genes arranged in an order. They are found in thousands. The genes are equally distributed in chromosomes. They also come in pairs like the chromosomes, Usually the genes do not stay in their own chromosomes. The genes from one chromosome will break off and go into another chromosome and exchange the chromosomes. This process is known as *crossing over*.

The genes of any given pair are alike and produce resemblances between parents and children. In certain pairs they may produce differences between parents and children. But there would be more resemblances between parents and children then differences. The genes may differ in certain pairs. Since the genes are in thousand with different traits it is not likely that two human beings would haev the same characteristics though they were born to the same parents, except in the case of identical twins.

For example, let us take the case of a pair of genes. One of them produces *blue (B)* eyes and the other produces *brown (b)* eyes. If the child receives 'B' genes from both the parents he will have *blue eyes*. If he receives V genes from both the parents he will have *brown eyes*. But if he receives *'B' genes from his father* and a *'b' gene from his mother* he will have *blue eyes*. Here the *'B' gene is dominant* and the 'b' gene is *recessive*. Similarly, if he receives a *'b' gene* from his *father* and a *'B' gene* from his *mother* he will have *brown eyes*. This means, the *'b' gene is dominant* and the *'B' genes is recessive*. (Fig. 3.3) There are several possibilities for variations in heredity among the members of the same family. Therefore, hereditary components can make for individual differences.

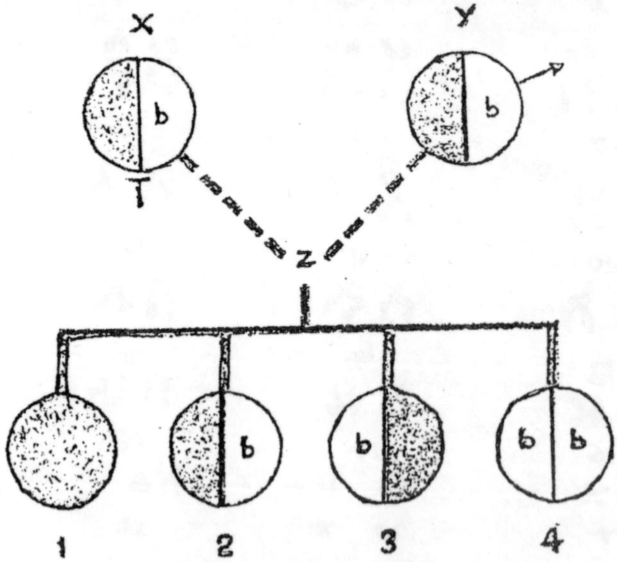

Fig. 3.3 : Transmission of eye-colour.
X = Brown eyed mother. Y = Brown eyed father.
Z = Offspring. 1, 2, 3 = Brown eye. 4 = Blue eye
'B' genes from father and 'b' genes from mother.

Sex-Linked Genes : There are 23 pairs of chromosomes in human cell. Of these 23 pairs, 22 pairs are having similar and equal characteristics and contain the same genetic information. These characteristics are common to the chromosomes of both the sexes. The twenty third pair alone differs in characteristics and information. We may call this pair as sex chromosome, because, it is this pair that determines the sex of a person.

The sperm cell and the egg cell have each 23 pairs of chromosomes. During the process of fertilization, the 23 pairs of the sperm cell join with each of the 23 pairs of the egg cell completely. This process is very much needed for proper cell functioning.

Females have two X chromosomes (XX) and males have one X and one Y chromosomes (XY). These two chromosomes are named as X and Y because they look like X and Y letters. If both the parents (father and mother) contribute an X chromosome the child would be a female child. But if the father contributes Y chromosome then the mother has to contribute only the X chromosome. In this case the child would be a male child. So the male inherits an X chromosome from his mother and the Y chromosome from his father. Therefore, the Y chromosome of father has the fundamental biological characteristics to determine the sex. The Y chromosome contains genetic source which is the cause of the development of testes and male c ha racteri s tics. All other sex developments are due to environmental factors (Fig. 3.4).

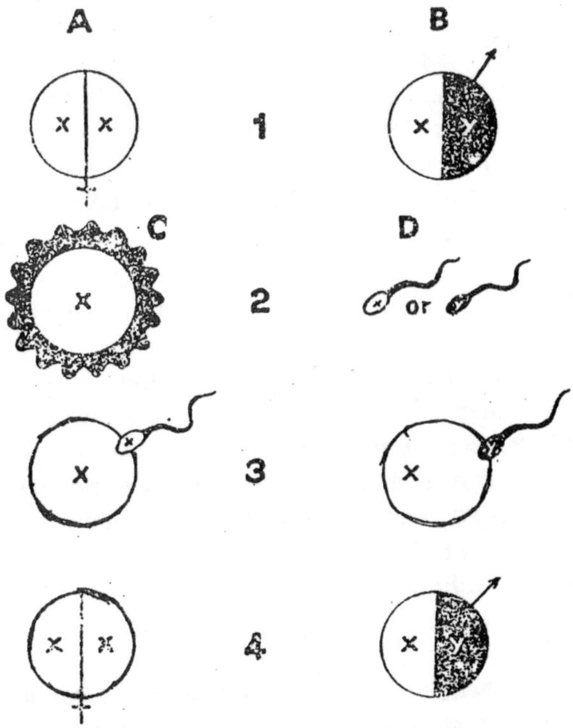

Fig. 3.4 : Determination of Sex.
A = Female. B = Male. C = Egg. D = Sperm.
1. Parent. 2. Gene Cell. 3. Conception. 4. Offspring.

Genetic Abnormalities : When a cell is fertilized, first it divides -into two divisions and then they go on multiplying in number. Sometimes, during this process part of a chromosome may be lost or the female may possess only 45 chromosomes. One is lost. This may result in the death of the organism. This means that the fertilization is not complete and perfect. But if the organism survives a female child may be born with one X chromosome (XO) and not with two chromosomes (XX). This kind of abnormality is called *Turner's syndrome*. Females in this condition are short, lack verbal abilities and spatial abilities. They may develop webbed neck and fail to develop sex characteristics at puberty. They may be treated with doses of the hormones estrogen which induce the development of secondary characteristics like public hair and breast. Estrogen therapy in adolescence can help women with the Turner syndrome which can produce more normal appearance to have normal life.

Another abnormality due to sex chromosomes may develop in males. It causes breast development in males. The males may fail to produce sperm. It may also cause mental retardation. These men are tall, thin, have long arms and legs, their testicles are small and they are usually incapable of producing offsprings. They have low intelligence quotinet (IQ), show poor social adjustment, The genetic make up of such patients is two -X chromosomes and one Y chromosome (XXY). This abnormality is called **Klinefelter's syndrome**. There is another type of chromosome abnormality. Men will have the extra Y chromosome (XYY). They are called XYY males. Normally they are taller than the average men and are extremely aggressive and with low (IQ). The extra Y chromosome causes the adrenal glands to secrete large amount of male hormones. These persons reach sexual maturity earlier than the normal males and have abnormal sex drive. Further, they may be violent and have high criminal tendencies. The cause of their aggressive behaviour is social and not physiological. They are criminals with XYY *Keryotypes*. In the countries like Australia and France they were defended on the ground that they are helpless victims of their genetic inheritance and hence they could not be held responsible for their offences.

Polygenic Inheritance : The interaction of many separate genes determine most of our traits. Sometimes a single gene may have influence in many of our traits. In most of the cases/the interaction of several genes determine a single trait. This process is called *polygenic inheritance*. The genes that determine a particular trait may be found in one chromosome or may be equally distributed in several chromosomes. For example, in most people there is no absence of intelligence. It is distributed over wider range of population. The same is true in the case of height. The height of most people fall in the middle of the distribution. In most of the cases a person's intelligence is determined by the interaction of many genes that influence different abilities. In the development of a particular' trait the influence of the environment also must be taken into account.

Twins : In genetics we study mainly the similarities and the differences in human behaviour. Family inheritance shows that there are certain traits in families. The differences among them, may be due to environmental factors. There may be certain talents in families. For example, a man may become expert in certain fine arts like music, dance or painting. His son also may develop the same taste. Is it due to genetic potentialities or training? Was Beathovan's father a musician? Was Shakespeare's father a poet-play-wright? Have these two persons inherited the trait genes from their parents? They need not inherit. Here heredity means individual differences, except in the case of identical twins.

Sometimes a mother may give birth to two children at the same time one by one. These two children are called twins. Twins are of two kinds. One is identical twins and the other is *fraternal twins*. The identical twins have more psychological importance than the fraternal twins. Because, they are genetically identical and resemble each other in many respects.

The fertilized ovum first divides into two divisions. These two are exactly the same. The two divisions divide into four, four into eight and go on multiplying into millions and developing into an organism. This process is called cell division or *mitosis*. The cell divisions develop first into muscle cells, nerve cells, bone cells, gland cells and skill cells, etc., and later develop into various parts of the body, like head, eyes, limbs and trunk. This process is taking place in the normal course of cell division and development. This matter we have stated elsewhere in different context in precise form. But here it is important to state the same matter little more in detail.

But sometimes the fertilized ovum after making the first division into two, may start functioning *like two separate fertilized cells*. The two divisions do not mix with each other in further divisions and develop independently. As a result of it two organisms grow to develop into two individuals and two children are born. They are called *identical twins*. They have been developed from the same zygote called *monozygote*.

The other kind of twins, *fraternal twins*, are born from two different egg cells. Sometimes the mother may produce two egg cells or ova and both may be fertilized by two different sperm cells from father. These two fertilized ova also will undergo the same process of cell division independently and develop as two separate fertilised ova and they develop as two separate embryos. Therefore it is called *dizygote*. The fraternal twins are genetically separate individuals. They may have certain similarities like any other siblings in the family but not like identical twins. Fraternal twins sometimes may vary in sex, one may be male and the other may be a female. But in most of the cases they are of the same sex, either both are males or females. There is every likelihood that giving birth to twins is also determined by heredity. Generally aged husbands are likely to conceive twins irrespective of the age of the wives.

Therefore it is observed in the study of twins that most of the twins are fraternal or two egg twins. They are twins because two separate egg cells are fertilized, each one by a different sperm cell, practically at the same time. This type twins have nothing more in common than other siblings. But identical twins are necessarily of the *same sex*, either both are male or both are female. It is determined by the fertilized ovum as either XX or XY. While testing the mental abilities, the identical twins were found to be much more alike in intelligence than the: fraternal twins and also the other brothers and sisters. Identical twins normally do not differ in intelligence except on some special cases like brains injury or such things. But the fraternal twins are like other siblings may vary from each other as they have nothing in common genetically. They have more resemblance in common than the other siblings, who were born before them or after them. They have more common environment than the other siblings such as same living place, same school, some games and the same treatments.

Though the genetic factors determine mostly the similarities in identical twins and in fraternal twins, the environmentalists claim that it is the common environmental factors that brought out the similarities in these children. **Woodworth** (1952) states that we cannot reject either of the two theories as there is some justification in both.

Genotype and Phenotype

A person's physical characteristics and social disposition, are passed on through blood from one generation to another, is an old idea. Recent researches of genetic transmission were investigated and the facts were well established. The distinction between *genotype* and *phenotype* can be made. Genotype is concerned with the *entire genetic endowment*, while Phenotype is concerned with the *outward manifestations* of the genes.

The total genetic make up of an individual, which is consisting of inherited genes, is referred to as genotype. Generally genotype refers to an individual's unobservable, physiological genetic constitution in contrast to phenotype. In behaviour genetics, genotype refers to the whole of the genetic material of a cell or organism. Genotypes are named for their obvious phenotypic effects and the phenotype is often diagnostic of its genotype.

The individual's *total observable characteristics* are referred to as phenotype. A phenotype is a characteristic assigned to living things of different groups on the basis of some observable physical characteristics which they possess such as colour, size, weight, etc. In behaviour genetics phenotype includes behavioural processes such as nest cleaning, learning ability and alcohol preference. For many human traits, the phenotype is constantly changing, depending upon the nature of the interaction between genotype and the environment. For example, a man's physique is strongly determined by his genes, and nutrition and environmental factors also have influence in shaping the physique. The phenotypes are divided into two broad, categories such as *psychophene* which is mainly attributed to behavioural processes and the *somotophene* attributed to mainly anatomical and chemical characteristics.

The psychophenic differences have physical bases and the genes modify psychophenes through intermediary somotyphenes. *The genotype* is fixed at birth. But the phenotype changes and generally is thought of as the product of an interaction between the genotype and experience. Suppose, for example, an individual is born with the capacity for high intellectual achievement. Whether, the person develops this capacity genetically endowed potential, is depending upon the environmental factors such as rearing practices and education. If any measurement of intelligence yields a higher intelligence quotient (IQ) then it may be stated as the indication of phenotype. It has been observed that there are various clinical syndromes. They are of phenotype disorders. It is only the genotypes inherit the disorders. But will it also harm and become phenotypic behaviour disorder? This depends upon the environments and experience. A susceptibility may be inherited and not the behaviour disorders itself.

The genotypic and the phenotypic behaviour genetics are used to compare the members of a family and persons of twins. By comparing the members of a family we can determine on the average how many genes are shared by the two blood relatives. For example, children in the family receive half of their genes from one parent and the other half from another parent. Other brothers and sisters (siblings) will receive on the average fifty per cent of their genetic backgrounds. But relatives, in contrast, not so closely related, receive few genes. Generally, for example, nephews, and nieces may receive twenty-five per cent of the genetic makes up from an uncle. But in the case of susceptibility for a mental disorder, it is inherited. If a study of the family is made it may reveal a correlation between the members the genes shared and also the incidence of the disorder occurrences among the relatives. The study of investigation may be started by collecting sample of individuals who exhibit the symptom of the disease inherited. These are indicated as probands or index cases. Proband means the person on whom a genetic investigation is conducted as the symptom is found in him.

But twins also may be compared as they are supposed to be genetically identical Twins may be monozygotic and dizygotic pairs. The monozygotic twins are genetically identical sibling who were developed from a single fertilized ovum (egg). The dizygotic twins are called fraternal twins. The dizygotic twins, are the birth partners, born at the same time, who' have developed from two separate fertilized ovum (eggs) and who are only fifty percent alike genetically and *not born from different pregnancies.* The monozygotic twins are always of the same sex, but the dizygotic twins are not of the same sex, but rarely may be of opposite sex. If the twins are similar diagonistically, they are said to be *concordant behaviour* genetically. The susceptibility to mental disorder may be inherited but concordance for the disorder may be greater in the case of monozygotic twins than in dizygotic twins.

However clear cut information is received from the studies of family and twins, it is very difficult to interprete whether the children inherit the same amount of disorder symptoms or less or more. The child caring practices and exposure of the children to neutotic adults models may reflect greater number of anxiety disorders.

Experimental Studies on Genetic Influence

The human development is no doubt very much influenced by genetic factors, But to what extend the physical structures, temperament and intelligence might be genetically determined? When a child exhibited very similar intelligence and sociability to both or either of the biological parents, it is quite but natural to infer that the child has inherited these characteristics from his parents.

A young man born to a very bright parents may appear very brilliant, I because his parents have provided him with rich environment apart from the genetic contributions they have endowed in him. Can we judge and discriminate between these two factors as to which one had greater degree of influence upon the young man's brilliances? Unless a systematic research is conducted, in which, either the heredity is kept constant and the environment is varied or vice-versa, we cannot sort out the effects of genetic factors and the effects of environmental factors on the behaviour of the person. Researches have been conducted to this effect on animals and on human subjects.

Studies on Animals

Since it is easy to conduct researches of this type on animals, psychologists first try to test their hypothesis on animal behaviour genetically. One of the approaches to behaviour genetics involves testing and comparing the behaviour of pure stock or ancestory genetically. If the behaviour of the stock (animal) differ significantly from that of another, than some genetic influence is exhibited. But a common method is to breed animals selectively and conduct experiments.

Men have been breeding animals for thousands of years, for physical and temperamental traits. We know now-a-days, the dog breeders know much about the temperaments of these animals. There were also research centres like Jackson Laboratories at Bar Harbor, were known for long time for breeding, mice, rats, dogs, cats, and other animals to produce in them certain specific attributions. It was found that it was possible by long series of in-breedings to produce animals, which are uniform in terms of heredity. It was also observed that every gene in one animal or individual corresponds accurately with a gene in every other animal or individual in the whole stock. There were hereditary differences in temperament between one animal and another in the pure stock. There were also wild and docile animals. The ancestral temperamental differences were produced in the animals by the environmental conditions.

Attempts have been made over years 'for determining the hereditary factors in the ability to learn to run a maze among the rats. **Calvin Hall** (1941) conducted this experiment on rats. He selected bolder rats and timid rats and breeded bold rats with only bold and timid rats only with timid with the view of producing a well defined bold rats and well defined timid rats in isolation and identical environments. The animals had no chance to learn from others. The animals showed the boldness and timidity at the appropriate ages as their ancestors. In this situation also the environment can be altered and by handling by some means in the individual animal bold or timid. In this way the interaction of hereditary and environmental factors can be affected. These were the original characteristics in the behaviour of the animals of different stocks. But at the same time the animals reflected the behaviour how they were reared. These behaviour patterns of the animals showed that they were subjected to the influences of hereditary and environmental factors.

One of the experiments, **Thomson** (1954) had conducted, was on rats. He interbred maze bright animals and maze dull animals separately for six succeeding generations. Their learning abilities were tested. It was observed that the learning ability between the two groups was increased steadily as the selected breeding went on. Finally, in the sixth generation, the maze dull animals made 100 percent more errors in running the maze than the bright animals. But animal studies of this type do not provide us sufficient evidence genetically for human attributes. We can conduct this type of experiments only on animals. We cannot conduct such investigation on humans with selective breeding and with tight environmental controls. Therefore the models or approaches applied to animals cannot be applied to humans to test the influence of hereditary factors.

Studies on Human Beings

In a study of ancestral analysis, the investigator, Dugdale identified a representative of the characteristics. He then studied the family history, examining each relative of the index case, including both lineal or direct descendents and collecteral relatives such as siblings, aunts, and uncles. **Richard Dugdale** (1877) a researcher of 19th century, who found a high rate of disease and deprivity in a poor jukes. He blamed their ailments on inbreeding, that cousins had married cousins for generations. Jukes later on became to link with public with the Kallikaks, a new Jersey family which was studied by **Gaddard** (1913) who was interested in the population of feeble mindedness and criminality. Gaddard was concerned with two distinct family lines of Martin Kallikak. Kallikak, a revolutionary soldier reputedly had an fair with a feeble minded tavern girl which eventually resulted in 480 descendents. Another line of 496 descendents was produced as the consequence of his later marriage with a better stock. Gaddard traced the two branches of the family to a single ancestor who had fathered children by two women, one a virtuous mistress. The virtuous wife descendents were considered to be pillars of the community. But the other woman's descendents were morans, prostitutes and thieves. The differences were attributed to hereditary factors. Gaddard conducted his study by stating that he had separated the effects of nature and nurture (McClearn, 1963). The study was criticised by many researchers stating that it has many weaknesses. One ·important point is that the two family lines had experienced different environments, this was not considered seriously.

Analyses of closely related ancestory are called *consanguinity studies*. In such cases information must be obtained about the degree to which a particular characteristic is present in the relatives of an index case and in the general population. The characteristic must be genetic in origin and must occur more frequently among his first degree relatives such as parents, siblings and children than among the second degree relatives such as grand parents, aunts and nephews. On the basis of this assumption, **Lennox, Gibbs and Gibles** (1940) observed that only 10 per cent of the population showed the specific characteristic occasionally, only in some abnormalities, in brain wave patterns (EEG).

Studies of Twins

Investigators call the birth of twins as multiple birth. **Lerner** (1968) in his study, 'Heredity, Evolution and society', has indicated that about 1.2 percent (A all births are multiple, by producing two children. According to him about two-thirds of' the twin pairs are dizygotic and these members are more like each other genetically then are the siblings born after this. The monozygotic twins who develop from the same union of ovum and sperm are genetically identical. It is the influence of genetic endowment that determines the degrees to which the monozygotic twins are more alike on a characteristic than the dizygotic twins or siblings. Generally, comparisons are made only among the children who have been brought up together since the enviroment is fairly constant.

In studies of similarity or concordance of pairs of individuals usually the measures of intelligence quotient and personality trait assessments are made. There may be instances of dissimilarities between twins in these measures. Here we come across a problem. That is, is there more concordance among the monozygotic twins on a particular measure (I.Q or personality) than among the same sexed dizygotic twins brought up together? The answer could be only 'yes'.

It is because the monozygotic twins are physically more identical than the dizygotic twins. But with all these, **Sydow and Rinne** (1958) from their Students present an interesting example of monozygotic twins who were found to be very unequal in blood values and size at birth. The surprising result was that one twin was dramatically more than twice the birth weight than the other. This example brings to light against the conventional results that monozy twins are always physically identical. Therefore in twi studies adequate discrimination should be made between the same sexed monozygotic twins and the dizygotic twins with regard to physical appearance. If the twins are of different sex, we must know that they are only dizygotic twins of different sex. Sydow and Rinne state that as it appears the task is not easy. There are quite number of instances in the earlier studies of dizygotic twins indicated as monozygotic as their appearance was very much similar. But now the discrimination is made easier as correct information about blood group, serum protein and other physical measures are obtained.

In one of the studies, **Gottesman** (1963) had quoted that observations made on blood typing of nine blood groups, finger print ridge counts, height and weight and the judgement made by geneticists, psychologists, and artists, were taken into confidence. In this method, we are quite sure- that the monozygotic and dizygotic twins, can be distinguished on all these characteristics. If the twins were dizygotic, it was one in two hundred. **Lerner** (1968) states that the skin grafting although is more difficult to use to identify the twins, it is best method to test the relationship between twins. It is because only the monozygotic twins accept grafting from each other. Further, the environments of other kinds of twins did not cause any substantial difference when the twins brought up together. **Smith** (1965) states that monozygotic twins had more common experiences than the dizygotic twins. The fraternal twins only differ on sex. The members of different sexed pairs, no doubt are treated differently from that of the same sexed pairs. The monozygotic twins are more likely to have the same friends, and also wanted to have the same type of dress, share in all such activities like studying, attending sports events than the same sexed dizygotic twins.

Inheriting Physical Characteristics

The physical appearance is one of the most clear way in which children resemble their biological parents. Actually the similarities between parents and children in some physical characteristics are inclined to be greater mainly during the childhood and adolescence than during the first year of life. **Bayley** (1954) studied this aspect and reported that there is correlation between the size of parents and their children at six months of age and also at different later times. The height of the child is related to that of his parents and this relationship is inclined to be stronger for the same sexed than the opposite sexed comparison. This means that the male children -esembled more closely their fathers and the female children more clearly their mothers.

It is also observed that where twins were compared, there were impressive relationships. Studies were made by **Newman, Freeman, and Halzinger** (1937) and found that there was a correlations of 0.94 between the heights of monozygotic twins. But in the case of dizygotic twins the correlations were 0.58 and 0.50 respectively and also for the siblings of different ages. Studies were also made comparing the abilities and social behaviour of the two kinds of twins to test whether the individual's genetic endowment had any influence.

Heredity and Human Intelligence

Relating hereditary factors to the development of intelligence and to give a correct answer to the question, what amount of heredity functions in intelligent performance of an individual, is extremely difficult. Members of the present society may hesitate to accept that human intelligence is genetically determined either in part or fully, They have strong faith in the statements that all men are created equal. But great deal of researches involving humans and animals show that there is considerable amount of variations among the individuals which can be traced in genetic endowment. Greatly influenced by the workers of **Charles Dar-win, Sir Francis Galton** (1822-1911) argued that eminent men tend to be related to one another. Galton was the first to study the factors influencing intelligence. He stated that there is possibility that intelligence might be inherited. He declared this only after making extensive studies among the families of prominent men in the fields of art, science, law and military. The results of this study indicated that greatness ran in certain families.

Galton's proposition was tested by many investigators repeatedly and the pedigree analysis did not yield satisfactory result to support the effects of heredity. Investigators, administered mental ability tests to individuals who differ in familiar relationship. Then correlations were worked out between different degrees of kinship. Some of the important workers in this field were **Erlenmeyer-Kimling and Jarvik** (1963), **Eurt** (1966) and **Jenson** (1969).

Jenson has summarised more than one hundred studies drawn from eight countries which provided clear evidence that more of the variability in the test scores between groups was attributed to heredity. There was a substantial correlation between the Intelligence Quotient (IQ) of biological parents and their children irrespective of other factors. It was also observed that the correlation between the children and their faster parents was lower. The correlation of scores for dizygotic twins was found to be substantial. They had the same level as their siblings. The monozygotic twins have shown more similar test scores and when they were reared together there was greatest similarity.

Evidences were provided by three major studies involving identical twins to the existence of genetic determinants of Intelligence Quotient (IQ). These identical twins were separated at early in life and brought up in different homes. **Newman** *et al.* (1937) in their study, administered the Standard Binet Scale to nineteen pairs of separated monozygotic twins. The correlation between the scores of each members of the pair was found to be an average of 0.77. In **Burt's** (1966) study a correlation of 0.86 was obtained between the 10 scores of two members of a pair involving fifty three pairs of monozygotic twins who were also separated and brought up. Shields' (1962) study seems, to support these results, but with different test of intelligence, there was high degree of similarity between the I.Q. scores of the twins of identical genetic constitutions, with different environments. All these support the idea that inheritance plays important role in determining the intelligence quotient at least in part.

Heredity and Creativity

We have been discussing in the preceding sections that heredity plays important role in determining the levels of intelligence. But, is there something else also beyond the intellective skills, as usually measured, which contribute to the production of great idea and significant accomplishments, the creativity? The question raised is whether creativity can be identified as a characteristic in a person and who can be trained for creativity? Evidences are rare in this area. Creativity is a capacity to produce through thought and imagination something in original. There were many creators lived in the world. They created wonderful things. Were their descendents continued in the same line and created something new? Or is creativity inheritable?

We can see in the world that creativity is appearing in generation after generation in some families. It is said that Johann Sebastian Bach was born in the family that has produced musicians for the previous hundred years and three of his sons had become composers. And also the Wyeth family has had three generations of artists. But such specific creative talents are rare in the families. There is no evidence for some sort of 'creative gene'. It may be that there may be a general kind of creative temperament running in families, whose members show creativity in diverse fields like, business, politics, art, painting, and so on. In any case heredity may not be the only factor that can predispose an individual to creative talent. There may be equally important influence on an individual, that is the influence of his stimulating family environments.

4

ENVIRONMENT

What is Environment?

Environment, generally may be defined as the surroundings or conditions in which a person, animal or a plant lives, grows, develops and operates. Air environment is the natural world which may influence in some the growth and development and the total activities and behaviour of the organism. An environment may be also defined as the external conditions or surroundings especially those of which human beings live and work. The environment generally tends to influence the development and behaviour of an organism. it is the state of' being encirclement and an organism cannot survive outside of as the environment provides the organism's livelihood.

Organisms, whether humans, animals or plants can flourish only in suitable environments where they can find nourishments for their survival. For example, people who live in plains find it difficult to cope with hilly or tropical regions. Similarly, animals in artic regions will not survive in hot surroundings Tea or Coffee plantations grow only in mountain regions. Thus the surrounding conditions determine the existence of living beings and plants.

The behaviour of human beings is highly influenced by the environmental conditions. Hypotheses have been formulated and tested in various situations. One of the hypotheses tested by the Berkeley group was, whether the environmental stimulation could affect the brain chemistry, and the anatomy of the brain. Conducting experiments on rats tested the hypothesis. It was observed in the experimental results that the experimental animals had been very much influenced by the environmental stimulation. The experimental animals had thicker and heavier cortical tissues, better blood supply to the brain, and biochemical changes than in the controlled animals. There was also more brain activity in the experimental animals than in the controlled animals. Therefore, environmental factors are having definite influence on behaviour. These results are also applicable to human beings.

The human individual right from the time of birth till his death he has to live in the midst of other human being and he has to eat the same food it others eat for his growth and development. He has to be Just like all other human beings in all respects if he has to survive in this world. The place where he lives, into which lit moves and withother humans, and the situations where lie lives have tremendous influence upon him, in determining his growth and development physically, mentally, socially, or culturally and behaviourily The growth and developmental processes are fully depending upon adequate supply of food, water, oxygen, favourable environmental temperature, fresh air, humidity, healthy climatic condition and so on. The individual person has to develop certain habits, customs, environmental conditions, and the extend to which he is permitted to practice his skills and talents.

Man is a social animal. He can be happy and carry out his activities for getting his requirements only in the company of other human beings. His movements with other humans in most of the situations qualify him that he is socially a desirable person. His participation in the every day activities makes him to learn many things consciously and unconsciously. Right from the time of' his birth the environment is having continuous influence upon him. As a child he has to develop into an adult human being. Therefore, the growth processes are very much dependent upon many requirements, not only food and water, but also the company of other human beings and also the effective environment.

The environments which are having more influence in the growth and development of an individual are mainly of three types. One is the biological and the other is the physical environment and the third is the social environment.

Biological Environment

The biological environment is nourishing the cell. The genetic properties are protected by the cytoplasm. The genetic characteristics that a man inherits through genes are passed on from one generation to the next without any change. Since we associate our traits with the genes, we say that they do not change. But genes change by chemical processes. Therefore there are conditions in which the genes and the traits may be made to change. The fertilized ovum at every developmental stage is nourished by some internal environment - the uterus. This shapes the fetus and also the behaviour of the organism. But there may be differences in the environment after the birth of the child. **Muller** (1844) says that *X-rays could alter the genes permanently.* This kind of change may take place even before the birth.

The genes are in the chromosomes and the chromosomes are surrounded cytoplasm. Events may occur in chromosome, in the cytoplasm or in the cell and even outside the cell and can have change in the genes. Therefore the cell environment itself can influence the genes. It may take place even earlier to the process of fertilisation. The genes may be identical, but the may be different. This will influence the genes. Therefore, the makeup of the embryo may be influenced by this kind of environment. For example, certain kind of fish in a chemical solution develops one eye only and not two. It is also observed that certain rabbits their skin colour according to the climatic conditions. It is the that changes the expression of the genes. This shows that genes could retain the characteristics of the organism only to certain extent and not permanently. There are acids and sugars that may have influence upon the intellectual abilities. **Zimmerman and others** (1946) found that *glutamic acid* accelerates mental functioning in human beings. We cannot neglect the *effects of food, vitamins and minerals* and health and diseases upon the genes.

Therefore, it is evident that the genes themselves are influenced by various kinds of environmental conditions. The term environment in psychological studies is referred to all sorts of chemical products, the physical structures and forces that surround and impinge upon the organism. It also means that the surrounding conditions of the organism influence and modify its behaviour. Therefore, the term environment is referred to the conditions of physiological structure, physical forces and the conditions of cultural and social factors.

Physical and Social Environment

The Physical environments provide all the physical material or instrumental type of things which are essential for the human individual for developing his physique, skills and talents, strength and energy, which are required individually for the physical work. The individual gets all these requirements only from the environment. The individual's personality development is fully depends upon the environment in which he is placed.

The effects of drugs, blood substances, diet, are acquired only in physical environment. He has to keep himself free from diseases an health and keep himself healthy. A man can be healthy only in a healthy environment which is free from infections and epidemic diseases. The living should be free from litters, rubbish things, garbages and should be clean. Many things may be found in the living environment. But all of them do not have much effect on an individual. Because they do not stimulate the person for any action or behaviour. He has no interest in many things. **Woodworth** (1952) says that a field of luscious green grass would be a barren desert to a hungry dog and a field full of rabbits would have nothing to do with a cow. Therefore any environment may be effective as far as it is essential for some needs or provides interest to an individual and stimulates him to react in some manners. The nature of the effect depends upon the individual.

The materials available in an environment may be essential for some and may stimulate them. It also depends upon one's age, experience and the purpose. For example, a room may be full of adult persons and they are all seriously engaged in some political or philosophical discussions. This environment may not be stimulating a little boy aged eight or ten years old for an intellectual discussion. A child might have been left out in a well furnished house and we may state that he is in a good home environment, simply by seeing the home and the members of the house. But what is important for the child? How the little boy is reacting to the home environment. The home environmental situation may not be stimulating the boy to his purposes. If there were two boys in the home environment, the same home situation does not produce the same feeling and reaction in the boys. They may differ from each other very much in their choice of environments

However there are individual differences, the environment can eliminate the individual differences. **Woodworth** (1952) says that individuals of different heredity when exposed to the same environment are not made alike. They differ in their abilities. In order to a make them equal in ability compensating measures must be taken by providing suitable stimulation and raining. By this means the environment compensates the differences caused by heredity. In certain schools when some students fall back in grade, the teachers take pains and provide them special attention and bring them up to the level of the bright students. By this means at the end of the year or course of study all the students in the class achieve almost level. In certain schools the individual differences come out strongly they encourage each child to come up or advance at their own. In some schools there are unlimited opportunities which favour individual differences. If the hereditary factors, are unequal in abilities or in school perform by providing varied unequal environmental pressures the hereditary differences can be iron out and normal or equal abilities and performance can be brought out. Rigid training in certain areas compensate the heredity deficiencies. Environment may provide opportunity and stiinulation, but environment will not force an individual and mould if the individual is passive in the situation. If the individual I,, riot active or does not react to the stimulating environment, the environment will not be held responsible for the individual development.

Studies have been conducted to test whether personality and intelligence can be improved by making improvements in the environment, One method is to keep the heredity constant and exposing the individual first to an unfavourable environment. The unfavourable environment is very unstimulating and unattractive. The same individual then is exposed to a favourable environmental condition, which is highly stimulating and attractive. The favourable and attractive environment is likely to raise the Intelligence Quotient (IQ).

Suppose, for example, if a group of school children were shifted from a dull environment to a more stimulating environment, we expect that their IQ may be raised to certain extent. How much could be raised or improved? This is very difficult to assess as young children do not always express their hereditary traits either in personality or in intelligence. Certain abilities are dependent upon the use of language, and since young children have not learned the language to the extend of using it for some purpose all their assets in intelligence cannot be assessed Though the same children carry the same hereditary factors throughout their life all the genetic factors do not come into action in these years. But when they become older children, it is possible to assess the level of intelligence by using language. However there are evidences that by improving the environment it is possible to make improvement in children's personality and intelligence.

In the experiments conducted by **Carpenter and Eisenhery** (1938), **Jack** (1934), **Page** (1936) and **Kerster** (1938) it was observed that by teaching the school children mastery of games and skills which were their group values, the personality factors like ascendency in children can be fostered. Trained mental hygienists have undertaken the responsibility of training the children. They found that a change in environment was very much helpful for them Since personality is the product of interaction, the development of the individual was very much influenced by social and environmental factors.

The effect of the environment was observed in many animal studies. How differences in environment interact with genotype differences in determining behaviours? The effect of early environment was studied by **Fuller** (1967) in maze learning of two breeds of dogs. Some pups of each breed were raised normally and others were bred in isolation. The effects varied in pups reared in isolation, but the learning ability of dogs reared in isolation was generally poor. Fuller found that these dogs were kept in back more in emotionality than in intelligence and maze learning. When the emotionality of the dogs overcome there was improvement in intelligence and in maze learning. In one of the earlier studies also dogs reared in isolation had shown poor performance in maze learning.

Psychologists believe that genetic and environmental variability operate in producing behaviour and therefore the influence of each one can be separated out in behaviour analysis. **Moskowitz and Orgel** (1969) state that this leads to the idea that one or the other is having more importance in producing intelligence. Actually, the interaction of genetic and environmental factors are such that they are essentially inseparable, except on a very abstract level.

Internal Environmental Influence

In the developmental process of an organism there are two important factors operating constantly. Though there are some controversies as to which is the most important factor in this process whether *heredity* or *environment*, we may treat that both are equally important. In the case of heredity it is the genes that determine the specific characteristic of an organism. The internal environment is having the influence on the development and on the behaviour of the organism. The psychologists are generally interested in *two important sources* of environmental influences These are *internal environment* which includes the stimuli acting within the organism and also the stimuli that are striking the organism from the *external environment*. The internal environment consists of *two varied subsystems* such as the *intracellular system* which includes the physical and the chemical substances within the cell. These substances influence the genetic particles of the nucleus. The other system is the *intercellular system*. This system includes the fluids around the cell which influence the growth and development of the cells.

The other is the external environment, which also has *two phases* such as the prenatal *environment* and the *postnatal environment*. The prenatal environment consists of the *amniotic fluid*, which is located around the fetus. The stimulation is provided by its position in the mother's uterus. The postnatal environment consists of complex series of stimulations that confront the child after its birth. Taking the whole situation together it may be considered that these various sources of the *environmental influences* functionally seem to be the *psychological environment* of the developing organism. But this should not be considered as the physical environment as to the nature of the influence in this case is different from that of the other physical influences.

It is because that the effects are more of psychological than physical. The psychologists always consider that the physical environment consists of those aspects of the *physical atmospheres* of the organism that could influence the organism's structure and the behaviour.

Interaction between Structure and Environment

The organism's behaviour mainly depends upon the development of the structure. The organism's behaviour has *three important* characteristics such as the capacity to get provoked by some stimulus, the capacity to react to the stimulus and the capacity to assimilate and make it as a whole or to coordinate the responses received by various means. These three essential characteristics are common to all organisms. In the case of higher species like human beings, these three components of behaviour are put in some *specialized structures* that make the organism to respond *to its environment* quite suitably. These specialized structures in higher organisms are the *receptors* or sense organs, the *effectors* or the muscles and glands and the *integrators* or the nervous system and the endocrine glands.

As the organism grows and develops the structures also get developed into complex systems and the organism behaves in a complex manner in any environmental situation.

Therefore, it is stressed here that an organism's responses are aroused through the interaction of its structures (senses, glands nervous systems) with its environment. Hence it can be viewed that the final products of the complex interaction between the *structure and the environment* becomes an essential function.

The Effects of Environment

The effects of environment on the behaviour of the organism have been examined by several investigators. In their efforts various kinds of results have been exposed. In one of the genetic studies an important principle was evolved. One assumption the investigator had for his study was, that the same environment produces different effects on different genotype organisms. The principle of *Nature-Nature interaction or Gene -Environment interaction* was explained by **J.B.S. Haldane** (1946). He has explained this principle citing some theoretical norms of interaction. According to him individuals with genotype 'A' always score higher than those individuals with genotype W. But when both types of individuals are exposed to the environment 'Y' both the genotype individuals have shown improved scores considerably, and not in the environment X. They did not score high in the environment X. **Haldane** states further, that better nutrition increases the height for both men and women.

There was also another study on gene-environment interaction. In this study, **Haldane** (1946) had observed that the *genotype 'A' was not found superior to the genotype 'B'*. It was found that the environment X was superior to the environment Y. It was observed in this study that these two genotype groups reacted differently to different environments.

People from one cultural environment migrate to the environment of another culture. These people react differently from the people of the native culture. In this case two cultural groups react differently in the same environment. The reason is that people from different cultures do not take up the ways of life of the new environment. It is not easy for them to react in the manner what the cultural environment demands.

It will take longer time for them to get accustomed to the new environmental conditions. The food habits, dressing habits, earning means, personal contacts and interaction with others all vary considerably and it will not be possible for the people of an alien culture to adapt the way of life in the new environment. In India there are hundreds of species of tribal groups living in various parts of the country. Many tribal groups live in forest areas, hills and in tropical areas, having totally different ways of life. These people go to the nearby villages, suburban and urban areas for buying food materials and immediately go away to their living places. They live totally in different environmental situations. By some means if the young men or boys of these tribal groups employed in some type of works, after few days they runaway. The tribal groups in India live totally in different type of environment and they do not react to different cultural situations. India is a live laboratory for studying the effects of environmental conditions as it is a nation of multicultural, multicommunal and multitribal people with varied norms.

Psychologists have made several studies to understand the effects environmental conditions. **Vandenberg** (1965) has reported that environmental differences did not cause any effect on genotype persons, but it had considerable effect on another type of persons. This means that environmental effect varies from group to group. **Naeslund** (1965), in his study on the effect of environment, used two methods of teaching reading on eighteen pairs of twins. They were of average intelligence and some were superior. The assumption **Naeslund** had was that there will not be any varied effect in these children in the environment. The twins of each pair were separated and assigned to different classrooms. Reading was taught by *phonic method* in one room and the *sight method* in another room. In both rooms the same type of environmental treatment was given to the children of the two genotypes who were average and superior in intelligence. It was found that the children of the average intelligence have learned to *read better* with the *phonic type* of method of teaching *than* with the *sight method*. But *the superior* children *did well in both methods* equally. In this study it was found that the two different environments had the same effect in one genotype, the superior children. But in the case of the genotype of average children it had different effect.

The assumption the investigator had in mind was that the *environment can act on the genotype*, but the genotype will not act on the environment. The genotype was considered to be having much of potentialities, but mostly passive. But, however the environment is an agent operates on the genotype and indicates on whom it can have its effect. In this experiment the assumption was not correct. Because the geonotype individuals did not react in a simple act of reflex to the inputs of the fixed environment. It was observed that the individuals were actively participating and manipulating the environment meaningfully.

This experiment indicates that each genotype person has a range of behaviour which is expressed in some pattern. The environmental impact cannot produce any trait, but the environment can influence where the trait falls to perform within the range of the environment. Some environments may affect some particular traits differently. In this way a gene-environment or the nature-nurture principle is evolved.

The effects of environment on human development and behaviour have been studied by the psychologists from different angles. One of the areas from where we get information for this is the *cultural anthropology*. This discipline is also interested in groups which have not developed the written language. These groups speak something to convey that they needed. The effects of environment in these groups function mostly in their life. The people of these groups use only verbal expressions, signs and symbols for communicating their ideas and thoughts.

Another discipline that is interested in the study of groups is *sociology*. Sociology studies group life in more complex situations. Groups of different cultures live in varied environmental conditions and the studies of their life brought out lots of interesting information about the *effects of environment on their temperament and the behavioural patterns*. The information obtained from the studies tell us about the way in which the child is reared through the customs and traditions of the groups and how the child learns not only in the *ways required by the habits* they have been adapting but also to *learn the mental habits* which symbolize the group's way of life. The child's *intellectual developments are moulded* by the *social environment* of the groups. **Otto Klienberg** (1935) has studied the social effects in the development of intellectual abilities of the *Negro children* in America. The performances of the children in *rural areas were found to be poor. But when the Negno Children brought to Urban areas in the early years* of life, after few years their school performances were found to be *much improved*. Similarly, in the case of white children, their school performances were *found to be inferior* to that of the white children *in urban areas*. The results of these studies indicate that the social environment is exerting profound effect upon the performances of both the Negro and the white children's school activities. The effects of the social environment were also tested in various cultural and environmental situations and found to be having profound effect on the behaviour of the people.

There are also interesting studies to show how the social environments in the early years of life of human beings had the profound influence in the developmental processes and behaviour. One of the dramatic studies was the 'wolf child and the human child' reported by **Cesell, A.** (1941). It was with regard to two little girls discovered by an Indian missionary from a wolf's den. These two little girls were living in wolf's den, eating berries. They were brought to human social environment. The children were displaying howling, barring their teeth and were moving on all fours like the wolves.

These children, in a number of ways have shown the wolves's affinity and did not show human manners. They were brought up in the missionary's family and were treated like the human children. One little girl because of the poor physical condition did not survive. The other girl was brought up like the human for about twelve years with great difficulties. But things were going on hard. She has shown the original bodily posture in the manner of a wolf. Since she has been always moving on all fours her kness were calloused heavily. Any mistake she had committed was jealously guarded from others who were nearby. When she was brought into human environment she could not adjust to the situation to the discipline and the language acquisition of speech. But however she made some progress gradually during the period of twelve years. This shows that it is clear, to get into the human habits, it needs time and efforts. It may take very long time to make any changes in the behavioural patterns of the girl. Any environmental effect in the very early years of life may have strong adaptation in the life of an organism and it will be very difficult to bring into a different environmental habits of life.

Interaction of Heredity and Environment

Any or ism for its development requires the interaction of heredity and environment.

Either the heredity or the environment can never independently shape an organism. Both the factors are equally responsible for the development of an organism, which is important? Heredity or environment. Both are important. But to what extent heredity determines the trait and to what extent environment does it? Several studies have been made to find out answers for these questions (Fig. 4.1).

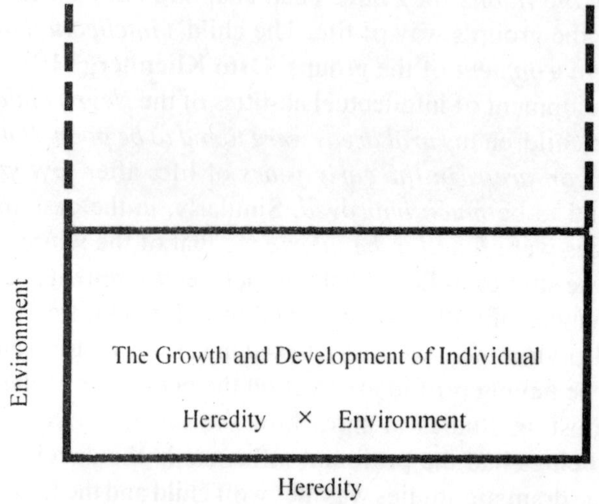

Fig. 4.1 : Action of Heredity and Environment.
The individual is a product of heredity and environment. any increase in the environment also incieases the stimulation. The area of environment may be increased, but the heredity is the same (the width).

Experiments were conducted both. on animals and on human beings. The rats found to be the most suitable animals for psychological experiments In laboratories. **Tryon** (1940) has selected a set of bright and dull rats They were made to run the maze. The number of trails were nineteen. It was found that the brightest rats made only 14 errors in learning the maze where as the dullest made 174 errors. The bright females then mated with bright males and the dull females with dull males. This was done on the basis of their ability to run the maze. This kind of selected breeding experiment was carried out for 18 generations. But the performances of these rats were tested in the 3rd, 7th and 9th generations. After the 9th generation, it was found that the dullest of the brightest of rats were brighter than the brightest of the dull set of rats. At this stage changes were made in the environmental conditions. All the rats were given the same care. The pups of the bright mothers were cared by the dull mothers and the pups of the dull mothers were cared by the bright mothers.

After few years, **Searle** (1949) in Tryon's laboratory tested the rats" learning abilities. He found that the bright rats were not bright in everything and the dull rats were not dull in everything. There were different patterns of abilities in the rats. The changed environmental conditions have altered the learning, abilities of the rats.

In studying the learning abilities in rats Tryon found that the capacity to learn this particular maze problems was not linked with the capacity to solve other problems. Further, the characteristics of brightness and dullness of the rats were applied to the particular task involved. Therefore, the dull rats were lacking in general intelligence. Tryon's experiments on rats were further continued by **Cooper** and **Zubeak** (1958). They have raised one group of bright and dull mixed rats in plain surroundings and another group of the same type in a stimulating environmental conditions. When the rats of these two groups were grown they were made to run the maze. Actually there was no difference between the rats that were bright and dull formerly, This indicates that the inherited abilities of the bright rats had not shown any difference. Psychologists believe that the same results could be observed in human beings.

In studying the contributions made on heredity and environment, the behaviour of the identical twins receive our attention. Identical twins are genetically the same. Therefore if there is any difference in their learning abilities it is due to environmental factors. In one of the studies made by **Newman** (1937) and others 19 pairs of identical twins were separated at different age levels. They were all young and brought up separately in different homes. Different educational opportunities were provided to the twins. The educational performances and the IQs were tested. It was found that there was significant difference between the twins.

Theinvestigation made by **H.M. Skeels** (1942) among the orphan children reveal the fact that stimulating social environment was the cause of increase in the children's IQs. There were only four attendants for 35 orphan children. They were responsible for washing, dressing, feeding all the 35 children. The attendants did not find time to talk to the children. So there was no play, no learning and no story for them. There were many subnormals and were sent to the institution of mentally retarded. Skeels was interested in studying two children. They were sent to a ward of adult Skeels tested the IQ of these girls before going into this ward and after spending a year there. He found that they were mentally retarded in the beginning and became normal after one year stay. Skeels repeated his study on the other children also. He was greatly surprised to see that the IQs of some 13 children had risen from 64.3 to 93 when they were made to stay in adult wards. They had all the facilities there to play with, to read something and to learn. They received all the encouragements from adult members. But the IQs of the children who were left in the orphanages reduced from 86 to 61. If intelligence were stable and hereditary characteristics were unchanged, these changes could not have occurred in these children.

Some psychologists suggest that the socio-economic status and the family size determine the IQ level in children. Usually middle class mothers play with their children and encourage them. Lower class mothers and mothers with more number of children do not play with their children. They do not provide their children with stimulating environmental conditions. This lowers the IQ levels and destroys the curiosities of poor children.

It is true that some general abilities are inherited. But, unless the child is placed III stimulating Situation and activated to work there is no chance for the development of his intelligence. Therefore neither the heredity nor the environment is individually responsible for the development of an organism. Both are jointly shaping the individual's abilities and characteristics. The genetic potentialities must be nourished by the environment. For example, a good seed will not germinate in a bad soil, nor a bad seed will germinate in a good soil. Therefore the organism is the product of both heredity and environment. **Woodworth** (1952) says that the individual does not equal environment, but does equal heredity × environment. According to him, it is like the rectangle having the base and altitude. The heredity is the base and the environment is the altitude and the individual is the area of the rectangle. Therefore, we cannot conclude whether an organism is having more of heredity or an environment. Both are equally important. Therefore, the student of psychology has to study the interaction of heredity and environment (Fig. 4.2).

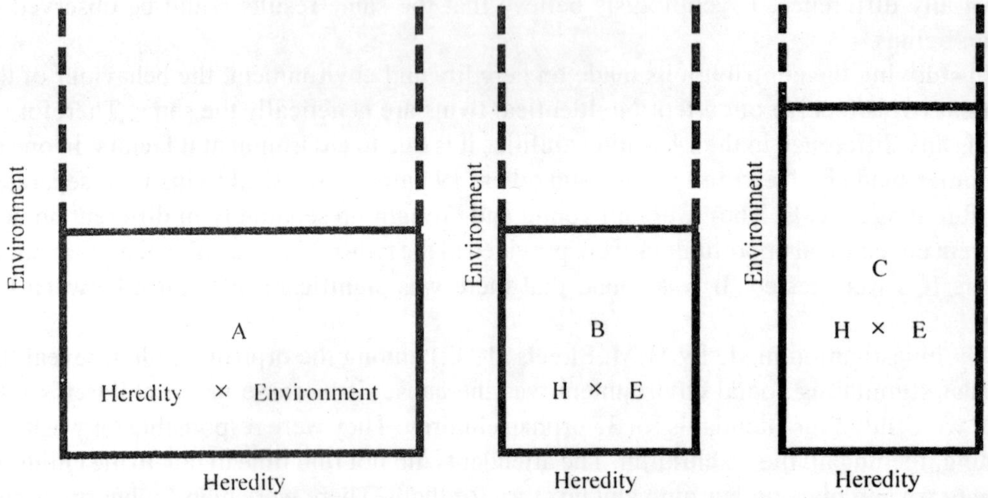

Fig. 4.2 : Individuals may differ because of Heredity and Environment.
Subjects A and B receive the same amount of environment stimulation, but vary because of varied theredity. Subjects B and C have the same heredity, but differ in environmental stimulation. Hence, they may differ in behaviour.

Environment and Behaviour

In the preceeding passages we have been discussing mostly the interaction between heredity and environmental factorsin human development and the development of behaviour. The behaviourists have put up a might in protesting the influence of hereditary factors in support of the environmental factors in human development and behaviour. Since psychology studies human behaviour objectively observable which is the result of the response to a stimulus and anything which is inherent and not objectively observable cannot be accepted as the casuative factor of human behaviour. Experimental evidences were produced for and against this view.

But the emergence of the environmental psychology around 1960s has caused heightened excitements among the psychologists as it has explained several intricacies in human development and behaviour and also resolved certain conflicting views with regard to the environmental factors in human behaviour.

Human development and behaviour are influenced by several environmental factors. The studies made by the experimental scientists mainly stressed the unpalatable, unpleasant and unhygienic environmental factors interact with other pleasant factors and cause developmental and behavioural discrepancies in human beings and in other living creatures.

The individual is dealing with the environment by responding to the stimulus in the environment. The phenomenon can be explained in this way. The individual is aroused by the stimulus from the environment and the stimulus. Therefore, the formula put forth by **Woodworth** (1952), W-S-O-R-W is rightly applicable to the view held by the environmental psychologists. The basic principle on which the environmentalists base their idea is that the with the environment through the stimuli which are constantly influencing the individual in his behaviour.

There is close relationship between human behaviour and man's *physical environment.* **Proshanasky** (1976) states that there is a close relationship between man's behaviour and experience and his built environment. **Kantowitz** (1977) is of the view that psychology is the study of man in the physical setting. **Holahan** (1976) relates physical world with human behaviour. Based upon these views we may state that human behaviour is highly influenced by the effects of built, or non-built, or physical and natural environment and in turn human beings put Lip their behavioural pressures on the built, physical and natural environments.

The relationships between human behaviour and the environment are built upon some extensive information obtained from many experimental results. Based upon these results several theories and mediators were evolved. These theories try to explain the relationship between human behaviour and the environment. Each theory uses one or more mediators to the relationship or the inforaction between the individual and his behaviour. The mediators are stated as the antecedent causes and the consequences are the coping responses or the protective devices. Thus the cause and effect theory is applied to explain the relationship between the environment and behaviour. There is a cause behind every act of the individual.

An event when occurred, certainly there is another event followed by it. Therefore there is a cause behind every effect. This means that something determines the effect. Thus psychologists accept the idea of determination and something determines the human behaviour in the environment. Therefore, we may state that the fundamental assumption on which the psychological studies are made, is that human behaviour is determined by the environmental events and environmental events may determine human behaviour cognitively, conatively and emotionally. Therefore, psychological studies are directed in finding out the antecedent causes of various behavioural events or activities.

There are certain theoretical systems which can be used for the prediction and generalization of the effects of environmental components on human behaviour. Generally in the environmental situations in which the individual organism is put into constant stress, arousal experiences, over stimulation or under stimulation These are some of' the behavioural components with which the individual organism deals with.

Stress

The concept stress is generally used in the common sense world and in the psychological studies. Stress may be physical stress or mental stress. Stress generally refers to the physical or mental tension a force that causes strain physically or mentally. To put it in a psychological sense, stresses a response to an external force threat that involves fear, anxiety, anger or irritation. In physiological sense, stress refers to severe physical discomfort, or physical strain or some physical impairment. If the environmental component is strong or severe it causes physical stress and strain. Man experiences stress in the environmental situation when there are heavy demands on him, which may impair his adjustive mechanism or resources.

Man's adjustive resources in the environment are limited. When the environmental pressures are severe on him or exceeds he is put into severe stress, which is a kind of disorganization of the body of the individual. They cause harm to the normal integrated functions of the bodily tissues.

Man in the environmental conditions face several physical stressors such as heat, cold, humidity, natural disasters like floods, earthquakes, storms, depression of the climatic conditions and so on. Further there are various types of environmental pollutions caused by dumping of toxic wastages and garbages in the public places and polluting the air, water and the living areas result in producing virus or microorganisms which may cause human and affect the behaviour of the people. All these cause stress to human beings and these are stated as the environmental stressores. These are having influence in the behaviour of the human beings.

Another environmental component is the noise. Noise causes much stress and strain in its extreme level. Stress is a reaction to the environmental stressors. The environmental stressors serve as stimuli in an environmental condition and make the individual organism to respond with stress to the condition. Every environmental component in its extreme level whether it is high or low causes stress to the organism. Thus in a stimulus response situation the stressors are all environmental components.

It is also observed that certain social environmental conditions may cause damage to the organism. For example, over crowding is a social environmental condition which causes much stress and strain and influence social environmental condition may result in aggression, hostility and violence in a situation. The over crowding social environmental condition may cause stampeding and death and also all sorts of criminal activities. Though the stress originates in an environmental condition it is not always there. The stress also depends upon the weakness of the individuals and also the strength of the System of defense mechanisms built by the individuals physiological and psychological factor. Generally, mail learns to meet with certain stressful environmental situations.

In all the stressful environmental situations man's behaviour IS very much affected inspite of his learning to face them. Some of the stressful environmental situations, which influence and interfere in human behaviour are the natural disasters like the heavy rains which often wash away the lives and properties of the people, stroms, earthquakes, fires, explosions, military operations, imprisonment, capativity, disabling injuries, incapacitating illness, fatal diseases, like the cancer, tuberculosis, the AIDS, loss of loves ones. All these stressors depend upon the defence mechanism built by the individuals.

Individuals react to the stressors in varied ways. The individual's reaction to the environmental conditions varies according to the severity of the environmental components. For example, the heavy rains and floods, wash away everything on the route. In this environmental event those who have lost their properties and dear ones get disorganised by the stress while some consider it as an adventurous event which they rarely come across in their life. Thus every environmental component in its extreme level interfere with human behaviour and cause stress and the stress is the reacting behaviour stressors.

These environmental stressors cause harmful reactions in the individuals. There is increased heart rate, blood pressure, and glandular secretions. The other state of reaction to the stressors is resistance. But this may take place only to certain extent. For example in the case of the environmental component, temperature, it could be tolerated only to certain level. If the environmental temperature is increased above the normal level, then there is a severe sweating all over the body and if the same condition continues for a longer period then the body get dehydrated and it results in disorganisation and death.

It Is often observed that some of the environmental stressors pollute the air and water and cause impairment in human behaviour. Experiments have been conducted on certain environmental stressors to find out the effect of them on human behaviour.

Arousal

Arousal is one of the reactions of an organism to the environmental hen an organism is exposed to environmental stressors the organism is made awaken swiftly and vigorously. This we may call as arousal. Thus arousal Is an increased reaction effect to the environmental stressors. The evidence for this reaction can be seen in the increased physiological changes mainly in the autonomic nervous system. There are several autonomic changes if the organism is aroused by the environmental stressors. Some of the autonomic changes are the increased heart rate, blood pressure, increased respiration rate, constriction of blood vessels, glavanic skin response, palm sweating, urin secretion, muscular tension, adrenal secretion and so on. Generally, arousal is used to explain human behaviour. Arousal may have varied effect on behaviour, depending upon the level of arousal in the environmental situation. For example, a loud noise at distance may have very lowered arousal on behaviour. But some loud noise may have startled and hightened arousal on behaviour to the extent of even becoming a severe a stress. An arousal of both the extremes, either too low or too high will have disadvantageous effect on behaviour. Every environmental component is a stimulation in certain situation if the organism is to react to it.

Human beings need optimum or intermediate level of stimulation for their normal behaviour. **Hebb** (1972) and **Kahneman** (1973) have observed in their studies of environment- behaviour relationship that when the environmental stimulations from various sources like noise, temperature, air pollution and overcrowding were increased the arousal, the behavioural performance or the tasks were also increased or decreased. It was found that this type of behaviour depends upon the level of response of the individual in a particular task. The response in such cases might have been below or above the optimum arousal level for performing the task. These authors have observed that low arousal was found to be facilitating the maximum performance of the task and extremely heightened arousal was preventing from paying more attention on the performance of the task. Therefore, it is asserted that arousal serves as an intermediate variable between stimulus and behaviour.

Environmental Stimulation and Behaviour

The environmental stimulation may be overstimulation or understimulation. Since human capacity is limited man cannot put sufficient efforts to respond to all the stimulations that are attacking him in a particular environmental situation and respond to them. Studies were made by **Cohan** (1977) on this aspect of overstimulation in environment-behaviour relationship. In a complex environmental situation there are many stimulations In which a person with his limited capacities cannot process them and make out meaning out of them. Therefore, a different type approach should be made. **Cohan and Milgram** (1970) have proposed an environmental overload approach to explain the environment behaviour relationship.

Environmental Overload

The environmental stimulations determine man's behaviour. There are many environmental conditions or components Operating in an environment. In this way it becomes a complex environmental situation and too many types of stimulations from too many sources are attacking the man and man cannot respond to all the stimulations at a time and make them meaningful stimulations. Thus the amount of stimulations from the environmental sources exceed the man's capacity to receive the relevant stimulations and to process them meaningfully. This process may be called as environmental overload or information overload. When there, are too many stimulations from several environmental sources or when there is environmental overload how an individual human being can deal with them?

One way of dealing with such situation is to respond to relevant stimuli as far as possible and to ignore which are not relevant for the immediate purpose. This is what human beings do in their behaviour. In this type of a situations an individual may be positively reaching to certain stimulations and negatively reacting to certain stimulations. Therefore man's coping, up plan is negative behavioural effect to excess or over environmental stimulations. Therefore in the environmental overstimulation situation the behaviour environmental relationship is negative.

It is also observed that sometimes the less important stimuli when ignored may interfere with the main behavioural task. The ignored stimuli may facilitate and improve the main task. But if the ignored stimuli interfere with strong force then it may impair the efficiency of the task. The ignored become the strong distracting forces. In such cases the only way of overcoming the distractor is to convert it as an advantageous stimulus., which may even increase the efficiency. For example, the piano player while playing the piano with his right hand often experiences that his left hand in a constant source of distraction and spoiling the efficient performance of the piano. This can be overcome by putting both the hands in playing the piano, which increased the efficiency ill performance. Thus, the environmental Overload of the distracting stimulus is used as one of the positive stimuli and the overload is avoided. In many situations human beings overcome the distracting environmental forces in these ways.

The environmental overload or the environmental distracting forces can be overcome in four ways. These environmental stimuli are impairing human performance and turn the attention from the main task. (1) These environmental distracting forces can be overcome by putting more attention and effort on the main task; (2) The environmental overload or distracting forces can be overcome by putting more energy in the work and ignoring the overload; (3) When the distracting environmental forces could not be avoided or ignored by any means the distracting forces themselves should be taken as the source of interest for the work and may become a source of inspiration for the same task; (4) But one best way of overcoming or avoiding the overload of the environmental distracting force is to get rid of it physically by removing it from the place or changing the very place of work.

In the environmental overload situation if the work is performed for longer period of time the workers' energy and capacity get exhausted by putting more effort in the work. As result of it even very small need for work is considered as environmental overload. Further the environmental overload may also cause behavioural negative after effect such as frustration, more errors in the work, less interest in the welfare of others and also accidents. Thus the normal behaviour would be very much affected by the environmental overload as it may incur much energy in the work.

Environmental Understimulation

Human capacity for work is limited only to certain extent. Man tries to make use of it to meet all his demands in .1, short period of time. When this is not possible he gets exhausted, vexed and frustrated. It involves also emotional disturbance. One extreme is the overstimulation. The other extreme is under stimulations or inadequate source of stimulations. Human behaviour is also related to understimulation. What we call understimulation is mainly inaction or non-response on the part of an individual person when the environmental stimulation is weak or feable which fails to arouse response in the individual. Man needs an optimum stimulation to respond or for action.

If a normal human being with normal sensory apparatuses fails to respond to a stimulus either by seeing, hearing or smelling or in any other form the stimulus is said to be a weak stimulus and there is too small or too Jade stimulation. Understimulation also may mean opposite of over stimulation which is referring to the environmental conditions causing excitements with fear stimulations.

There are social environmental conditions, which fail to provide the optimum stimulation for response. Man in these days feels that he is shut-up in all directions without having any chance for free expression of sensory aspirations These situations cause no, interest in life. There is a feeling of disinterestness or boredom in life aspects which may result in some kind of illness Adolescent boys and young adults get disgusted in life and -turn to delinquency criminal and all kinds of antisocial activities and damage the properties. When the social environment falls to provide sufficient stimulations for the expression of sensory experiences people go in search of natural physical environment.

Thus according to urbanologists understimulation means the environment which is providing sufficient stimulation for happy living and to have exciting experience in life. Thus the environmental under- stimulation is having relationship with human behaviour in these ways. Here environmental is not defined in terms of the of stimulation.

Optimum Environmental Stimulation

We have a seen that both the stimulation and under stimulation are having harmful effects on human behaviour. Therefore any stimulation of' intermediate or medium type may be the most suitable for human behaviour. **Nelson** (1964) formulated a theory called *adaptation level theory* to explain the environment behaviour relationship Man is reacting to the environmental situations through his sensory responses, social responses and his movements. In these functions the optimum level of stimulation is most desirable.

In the optimum environmental situation man is not meeting with any problem. Man's reactions and behaviour are found to be normal. Human beings in the environmental would is always exposed to all these three types of stimulation such as over, under and optimum levels. But the optimum level of stimulation is not always possible at all situations. These levels may be going on changing from time to time. If the stimulation in an environmental situation is overstimulation and if it is not possible to avoid it, man learns to change the situation himself to suit to his comforts. This is the way man learns to adapt himself to the overstimulation or understimulation. This is how man adjust himself to the situation.

There may be also different levels of intensity in the environmental components For example, the physical environmental components noise or sound, temperature, low level of oxygen concentration and wind may be having varying intensities from place to place. In a place where there is loud noise continuously, is a source of disturbance. But after sometime ,h(- lear adapt himself to the situation. Similarly, there may be a place where there is absolute silence for a long time which is also a disturbing environmental condition. In the same way, high and low temperatures may be very comfortable. But after sometime man learns to adapt to the situation.

There may be varied environmental diversities, specifically in the sensory areas,, like seeing, hearing, smalling and touching but too much and too low diversities also are not desirable. People may like certain diversities very much. For example, too hot, too cold, too Sweet, salty and sour may cause feelings. In the sensory diversities there may be individual difference. Thus the environment -behaviour relationship can be explained.

5

STRUCTURE AND FUNCTIONS OF NERVOUS SYSTEM AND THE BRAIN

BASIC IDEAS AND BACKGROUNDS

The study of human nature and behaviour dates back to the period of the ancient Greek Physician, **Hippocrates** (460-370 BC) and the Greek Philosopher, **Aristotle** (384-322, BC) and other Philosphers and Anatomists. Aristotle was the first to write a book on psychology. But till the middle of the nineteenth century there were no systematic studies on human behaviour, although Philosophers like **Descartes** (1596-1650, AD), **Thomas Hobbes** (1588-1679 AD), and **John Locke** (1632-1704 AD) have shown tremendous interest in the study of relationship between mind and matter and interpreted human behaviour philosophically.

It was **Weber** (1795-1878, AD) an Anatomist and Physiologist, who had first introduced the psychological concepts in his studies and subsequently **Fechner** (1801-1887 AD), a German Physilogist, Philosopher and Psychologist started scientific study of human behaviour and also started a Psychological Laboratory. During the same period many Anatomists, Physiologists and Biologists made many interesting innovations in the study of human brain and its functions and behaviour.

All these Developments have inspired the psychologists to make use of the new information and they started rapid scientific studies on human brain and behaviour. This resulted in the emergence of several fields of psychology in which the study of nervous system and the brain received more importance. One of the contemporaries of this period was the famous Biologist, Charles Darwin.

The English naturalist, **Robert Charles Darwin** (1809-1882), who formulated the theory of evolution by natural selection, expanded it in his work, *On the Origin of Species* (1859) and also applied it to man in his work, *The Descent of Man* (1871). According to Darwin the variation that occurs in any species leaves some animals better adapted to their environment than others. The evolutionary thought began as a mighty challenge to many of the existing ideas. Darwin's comparative study of animal behaviour was received with a great boost. It has brought out several changes in the biological and scientific world which has also its influence on psychological world. Modern biology and psychology are very much benefited by the evolutionary theories. This perspective provided rich insight into many kinds of behaviour and mechanisms. There are different perspectives and each one yields different information.

We may describe behaviour in terms of detailed acts or processes or in terms of results or functions or responses. A functional behavioural description may be whether the limb is being used in walking, running, hopping and swimming. The evolution has resulted in many features that are common to all animals, but each species has its own features in a unique way. Bodily or behavioural inventions are maintained without break for millions of years. There was no change in the nerve impulses of the humans all these million years, though the behavioural patterns were changing time to time according to situations. Psychologists stress that the fundamental task of psychology is to explain human behaviour. What is meant by behaviour?

We think of behaviour in terms of response. Behaviour is a response or a series of responses that the individual makes in an environment. Responses can be seen in various forms. Raising the hands to catch the approaching cricket ball, swinging the tennis rocket to hit the ball, kicking the football to the goal, running and getting into the moving train, eating the breakfast, writing a letter, reciting a poem, recalling a pleasant event, feeling happy on winning the match, feeling sorry for a painful event, all refer to behaviour.

There are also other kinds of behaviour which may be called responses. When you are asked to describe a story you read yesterday, you do so. This is a response. What are these responses? These responses are movements of bodily muscles. They do not occur themselves. They are aroused by some stimuli or by stimulus situations.

Stimulus-Response mechanism

Stimulus

What is a stimulus? Stimulus means a goad or sharp stick. It is used for driving animals. The bullock cart man or the horse man uses the goad or stick to drive his animal and also supplements with sound. Similarly, the stimulus is used as goad to the sense organs or receptors. For example, light is a stimulus to the eye. But it is a stimulus to the whole organism. The whole organism reacts to this stimulus even though it attacks only the eyes. Within the organism a chain like activity is going on.

Therefore, a stimulus may be defined as an energy change in the physical environments that excites the sense organs. The change in energy may lead to seeing, hearing, tasting, smelling. The sense organs stimulate the sensory nerves and they in turn stimulate the brain or the spinal cord. From the spinal cord or brain the motor nerves receive stimulation and they move the muscles. So the seeing of an object and responding to it may be explained in this way.

Object ⟶ sense organ ⟶ sensory nerves ⟶ brain or spinal cord

moving the object ⟵ muscles ⟵ motor nerves ⟵

These are the functions taking place inside the organism when an individual sees an object and moves it.

When your lecturer calls your name from the attendance register, you immediately stand up and say "yes, Sir." The lecturer's voice of calling your name is a stimulus and your giving attendance as "yes, Sir," is the response to it. This is in the classroom situation. Someone, your father, mother or friend may also call your name and your response is not the same. It varies because the environment is different. Therefore a specific stimulus produces a specific response. Krech and Crutchfield say that a stimulus is a physical energy that can excite a receptor (sense organ) and thus produces an effect, it is not properly called a stimulus. According to **Woodworth**, (1952) any force acting on a receptor and making it to activate, is called a stimulus.

Response

It is through stimulus that we speak of response. We believe that every stimulus is supposed to arouse a response in an organism. But all physical events do not become stimuli, because some events fail to arouse responses. Only those events that can arouse responses may be called stimuli. Therefore, a response refers to an activity behaviour in an organism that is instigated by a stimulus. On this ground, therefore, we are compelled to speak about these two terms together as they are interdependent. Generally any behaviour of an organism is called a response. A response may take place in some part of the organism. But the organism functions as a whole. Therefore, Woodworth says that we need not limit a response to a muscular activity. According to him any kind of activity instigated by a stimulus can be called a response.

There may be also indirect responses. When you attend to some musical performance and enjoy it, you nod your head, move your finger, recite the song within yourself. These are responses to the performance. This is a sensory response to sound — the music. After some days you hear the same song in a radio programme and you remember it. This is another activity aroused by the sound — the music. You have recalled all those things of the past. Therefore, it is an indirect response to sound — the music. A related external stimulus may arouse series of responses in an organism even though that was not present there.

Stimulus-Response Formula

In evolving a stimulus response formula we must take into account the definitions of both the terms given above. A stimulus elicites a behavioural response in an organism. The letter 'S' stands for stimulus and 'R' for response.

$$S \longrightarrow R$$

The formula means a stimulus arouses a response or a response is aroused by a stimulus. According to this formula a response depends upon a stimulus. But at the same time we cannot conclude that a particular stimulus will arouse the same response in all the organisms. There may be individual differences in behaviour and the individual differences may be based upon the auditory sensory apparatus. Therefore, when a stimulus (a) is presented to an observer (0) it arouses a response (R) in him. Thus response depends upon two factors, first on the stimulus and second on the organism. The formula is:

$$S \longrightarrow O \longrightarrow R$$

The stimulus acts upon the individual and makes him to respond to it. The type of response to the organism has made, depends upon the nature of the stimulus properties and the nature of the organism. There are certain characteristics attached to the stimulus which may be called stimulus factors, S-factors and also to the organism which may be called O-factors. The stimulus factors may be the strength of the stimulus and the relevance of its arousal. The O-factors may be the strength of the sensory apparatus, the state of the organism.

The strength of the stimulus refers to the intensity and size of the stimulus and the relevance refers to the cause of the stimulus. If an organism is to respond to a bell sound any bell sound will not arouse a response in it. For example, there are several kinds of bells such as the temple bells, church bells, college bells, railway station bells. The church bells will not arouse response in a Hindu or Muslim devotees. Even among the church bells, it may be for the usual prayer or for a special communion or for funeral ceremony. The differences make the organism to respond relevantly.

With regard to O-factors, the sensory apparatus or sense organ must be in sound condition so as to receive various kinds of stimuli and differentiate the meanings. Two persons must attend to the same function. One attends and the other is not. He says that he did not hear the bell. The absence is due to the difference in the sense organs. The structure of the organism and 'the sensory apparatus have certain permanent characteristics in the organism. **Woodworth** (1952) says that the genes that shape the characteristics of the whole organism is responsible for the non-response of the organism.

Secondly, the organism may not be in a state of awareness to respond to the stimulus. It also depends upon the readiness of the organism. A state of drowsiness, fatigue, boredom, depression, worry, extreme excitement, sickness and unpleasant news all have effect in the type of response the organism will make to a given stimulus.

Thirdly, how is important the stimulus to the organism at the particular moment? The individual is already engaged in some kind of activity in response to some stimulus. Another stimulus at this moment may put him in a fix. He is to respond to it or not. It depends upon the nature of the stimulus. If it is going to put him in a better position, he would respond to it immediately. Otherwise he may ignore it or resist it. Therefore, in evolving a stimulus-response formula all these factors must be considered.

Basic Units of the Nervous System

We have been, so far, looking into the stimulus - Response mechanisms based upon the behaviour of the organism. But what is going on inside the organism? If we want to know more about what is going on inside the organism, we must know more about the nervous system. The nervous system is a complex pattern. It controls and regulates the individual behaviour. In all the sensory activities the individual is regulated by the nervous system. Physiologists, anatomists and chemists studied nervous system with a hope of getting more information about the behaviour. Psychologists and physiologists have many things in common in human behaviour. Physiological psychology studies the relationship between physical structure, function and biochemistry on one hand and behaviour and experience on the other hand. In a broader sense it is called Bio-psychology.

Reflex Action

Bio-psychology is a branch of science that studies mainly the role of the nervous system in behaviour. The bio-psychological basis of behaviour is divided into three functions. They are the receptors, nerve cells and effectors. Each function is having its basis in different bodily structure (mechanism). They are doing the functions of receiving, connecting and reacting mechanisms respectively. These three kinds of mechanisms cooperate and produce (simple sensory-motor) reflex actions. We may call this function as sensory-motor-reflex action (Fig. 5.1)

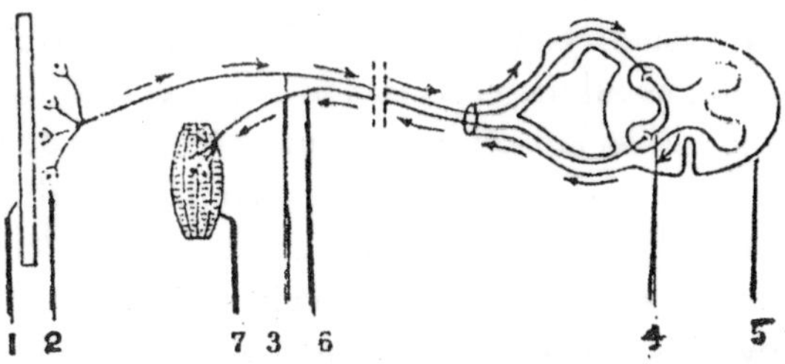

Fig. 5.1 : Simple Reflex Arc.
1. Skin. 2. Receptors. 3. Afferent Nerve Fibres. 4. Association Neurons. 5. Spinal Cord. 6. Efferent Nerve Fibres. 7. Muscle Cells (effectors).

Reflex actions are simple involuntary actions. They are innate in human beings. The actions are profound and aroused automatically and unconsciously in response to certain stimuli. There are several reflex actions. If there is a pin prick suddenly in your finger you immediately give a jerk and jump and try to withdraw from it. Similarly, we withdraw our hand from a burn, mild electric shock. An unexpected sudden loud noise behind you would make you to twist your body. An irritation in your nose may cause you sneezing. We all blink at a very bright light. When somebody tickles your body you jump instantly. These are all some of the reflex actions. In reflex action generally there is a tendency to withdraw our body from the source of painful stimulation.

When some one pricks your skin with a pin you experience a painful feeling and at once you withdraw from it. It is called protective reflex. When the pin enters your skin it activates some tiny sensory nerves or receptors. The message is immediately transmitted through connecting mechanism to the spinal cord. It is not necessary that it should go to the brain. A message from the spinal cord is sent through the motor nerves and the contraction of the muscles takes place.

Skin (pinprick) → Sensory nerves → Spinal cord

Muscles ← Motor nerves ← (Spinal cord)

The communication process from a sense organ through the nerve centre to the muscle is called a *reflex arc*.

Neurons

The nervous system is an important coordinating and integrating mechanism of the body. The nervous system is involved in the extraordinarily intricate activities of the body. The parts of the nervous system are many and complex. But it is easy for us to understand it if we approach it though its various parts. The smallest part is the individual neuron.

The nerve cell or neuron is the structural and functional unit of the nervous system. The structure of the neuron is the branching filament or minute fibre with a cell body in it. It has been estimated that there are twelve billion neurons or nerve cells in human body and it is they that make up the nervous system. Neurons are of different sizes and shapes.

A neuron is a living cell. It has all the parts that all other cells have. For example, the *nucleus*, the *cytoplasm* and the cell *membrane* are common to all the cells. What makes the neuron to differ from other cells? They are the specialised parts of the neuron. There are many short fibres extending from the cell body. They are called *dendrites*.

The dendrites receive messages from surrounding cells and carry them to some distance and pass them on to the nearby cells. There is a single long fibre extending from the other side of the cell body. It is called axon. The fibre of an axon is very thin and longer than the dendrites. In adult human being it is about three feet long but most of them are short. The axon carry messages to outside either to the next neuron or to the muscles and glands. Therefore, the neurons do the work of coordination from one part of the body to another (Fig 5.2).

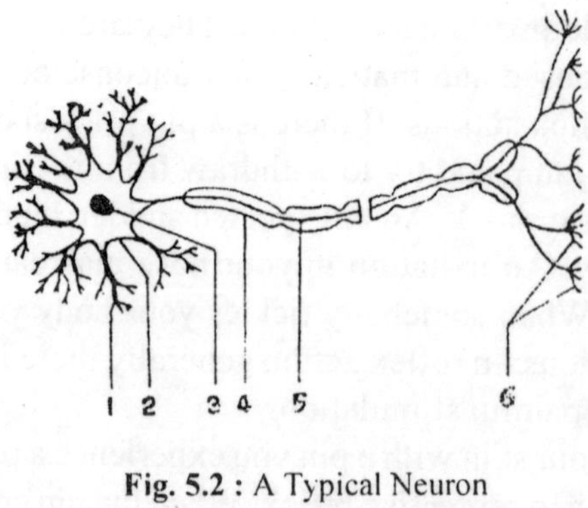

Fig. 5.2 : A Typical Neuron

1. Dendrites. 2. Cell Body. 3. Cell Nucleus. 4. Myelin Sheath. 5. Axon. 6. Terminal Branches.

When we speak about a nerve, it is a group of axon bundled together. In a nerve bundle there may be thousands of neurons. The axon is covered by a fatty substance called *myelin* sheath. The sheath is like an insulator. The neurons are of three kinds. They are *afferent* or sensory neurons, *efferent* or motor neurons and *association* or inter neurons. The afferent neurons collect messages or impulses from outside and inside of the body and carry them to the spinal cord or brain. Secondly, the efferent neurons carry messages from the spinal cord or brain to the muscles and glands. Thirdly, the association neurons that are smaller in size connect the afferent and efferent neurons. They are found in the spinal cord and brain. Figure 5.2 illustrates that the function begins at I and ends at 6.

The Nerve Impulse

This refers to the action of the nerve fibres. A Nerve impulse is an electrochemical discharge of a nerve cell. Nerve impulses are conducted through axon or dentrites. A single neuron functions in the same manner as the fuse of a fire cracker. If it is activated once a small electrochemical action (nerve impulse) runs forcibly along the length of the nerve fibre. It is the basic message unit. Every firing neuron functions in the same manner. The same amount of energy is spent and transmitted each time. The main point is that if a neuron fires at all, it fires completely, otherwise it is not. It obeys all-or-none law. Once a neuron is fired, it cannot be fired again until its electrochemical conditions are restored. But the amount of time taken for this process of restoration is one-thousandth of a second (.001). The nerve impulses are weak and have little energy, but able to arouse action in muscles and nerve centre. In these processes there are two kinds of transactions. One is that messages through afferent neurons are taken along the membrane of the body cell and fibre to the spinal cord and brain and the other is sent from spinal cord and brain to the efferent neurons or muscles. The messages (impulses) are taken on either side. This process is called nerve impulse. The speed of the nerve impulse depends upon the size of the fibre. The speed is slow in small fibres and fast in fibres of large diameter.

The Synapse

We have been, so far, considering the function of a single neuron. The neurons do the work of co-ordination of the body activities. How does it take place. Each neuron remains separate and there is no physical connection between two neurons. A nerve impulse begins when one of the dendrites of a neuron receives a message. There may be several dendrites in a neuron. These dendrites carry the message to cell body and from there to the axon. The axon terminates at a place where there are tiny branches known as axon terminal. The axon terminal of a neuron is put close to the dendrite of another neuron. There is a very small gap between the two. This gap is called synapse. This is a junction or area where an axon of one nerve cell meets a dendrite of another cell. The nerve impulses somehow travel through gap. An important characteristic of the synapse is that it allows impulses to pass in one direction only, that is from axon to dendrite. The axon is the conductor and the dendrite is the receiver.

The idea held for 'long time was that nerve impulses run so fast that they jump across the gaps. But a recent discovery in this field says that the transfer is chemical and not electrical. According to this theory some axon terminals have tiny oval sacs called synaptic vesicles. The synaptic vesicles contain some chemical transmitter substance. When the nerve impulses go near the end of the axon, it makes these vesicles to burst and releases the chemical substance. This substance actually travels across the gap and makes the second neuron to fire. There are several chemical transmitter substances. But all of them do not make the second neuron to fire.

There are certain complicated functions in this process. Each neuron can effect the activity of thousands of other neurons and in turn may be affected by them. Therefore each neuron is indirectly connected with every other neuron in the whole nervous system.

STRUCTURE AND FUNCTIONS OF NERVOUS SYSTEM

The human nervous system is a very complicated one. If we are interested in knowing more about the role played by the nervous system in human behaviour, we must know something about its anatomy. The nervous system contains many parts. But all of them can be brought under two broad divisions as the Central Nervous System (CNS) and the peripheral nervous system.

Divisions of the Nervous System

The central nervous system consists of the brain and spinal cord. These two are distinct structures. The brain lies within the skull, a bony case. The brain is a large soft mass of nerves and supporting tissues contained within the skull. The brain is the most important part of the body. It is the seat of awareness and reason. All the nerves are having their origin only in the brain and spinal cord. The nerves run from the brain and spinal cord to the effectors, the muscles and glands and to the receptors, the sense organs, and also run toward the brain and spinal cord from the receptors. The brain and the spinal cord constitute the central nervous system. (Fig. 5.3)

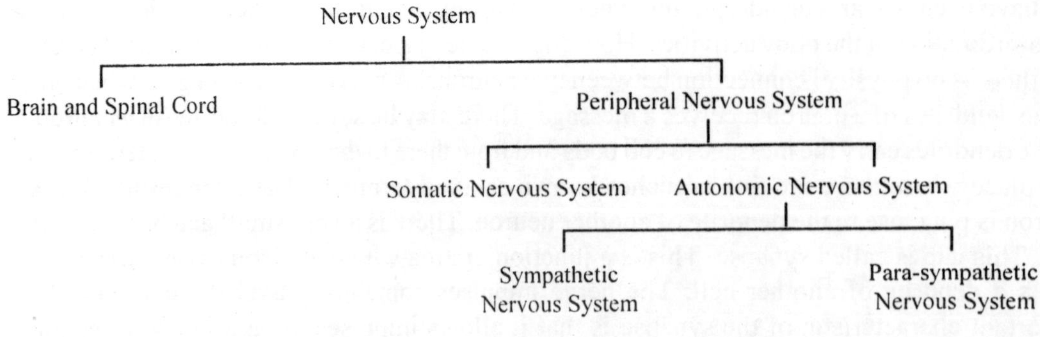

Fig. 5.3 : The Structure of the Nervous System

The Peripheral Nervous System

The peripheral nervous system consists of nerves that lie outside the brain and spinal card. It is the portion of the nervous system that includes all the nerves and neurons or the parts that are outside the brain and spinal cord. The peripheral nervous system is the collection of axons (extension part of the cell body) bundled together and these are extended throughout the body. The peripheral nervous system connects the spinal card with receptors and affections. The peripheral nervous system carries information from the central nervous system to the muscles and glands through the motor nerves. The sensory nerves carry impulses or excitements from the receptors, the sense organs, to the central nervous system. The peripheral nervous system includes the pathways that transmit the information in both directions through separate axons between the central nervous system and various internal organs, the heart etc. Then the sensory nerves carry the excitement (impulses) to the central nervous system from the sensory apparatuses, and the motor nerves carry the commands of the central nervous system to the internal organs, the glands and muscles.

All the nerves originate only in the brain and the spinal card. Near their origin, all the nerves are mixed (sensory and motor) in bundled form and contain both the sensory and motor fibres. The sensory and motor nerves, which are in bundled and in mixed form near their origin, the brain and spinalcard, branch out repeatedly at every stage as they reach the receptors and effectors over the entire body. Their final branching out ends only in sensory fibres and motor fibres and this can be also called, sensory and motor nerves.

The peripheral nervous system is bundled in mixed form. It consists of four nerve fibres and they are branching out for every part of the body. They are:

(1) The Somatic Sensory Nerve Fibres

These nerve fibres carry information or impulses from the receptors (sense organs) — the eyes, ears, nose, skin and other receptors to the central nervous system from one region to the regions of interest. These nerve fibres are called *afferent neurons* or afferent fibres. It is in *contrast to efferent neurons.*

(2) The Somatic Motor Nerve Fibres

These nerve fibres carry commands from the central nervous system to the skeletal muscles attached to the bones, ordering them to move the body, These nerve fibres are called efferent neurons or efferent fibres. This is in *contrast to afferent fibres.*

(3) The Visceral Sensory Nerve Fibres

These nerve fibres carry information about the internal state of the body and internal organs to the central nervous system. These fibres are called *afferent neurons*.

(4) The Visceral Motor Nerve Fibres

These nerve fibres carry commands of the central nervous system to the *smooth muscles* of the blood vessels and internal organs, to *heart muscles* and *glands*. These nerve fibres are called *efferent neurons*.

Thus the two types of sensory nerve fibres (1&3) namely afferent nerves carry message (impulses) toward central nervous system. The other two types of motor nerve fibres (2 and 4) namely, afferent nerves carry message away from the central nervous system.

There are three components which make up the peripheral nervous system. They are (1) the cranial nerves (2) the spinal nerves and (3) the autonomic nervous system. The autonomic nervous system originates from both the brain and the spinal cord. All these three components carry information to the central nervous system from the environment and also transmit the commands from the central nervous system to the bodily parts.

The Cranial Nerves

The cranial nerves are originating from nuclei in the brainstem and these nerves are directly connected to the brain. There are 12 pairs of cranial nerves in the human brain. The cervical region of the spinal cord is continuously upward into the brain stem where the cranial nerves originate.

They are mainly concerned with sensory and motor system associated with the brain. The cranial nerves carry information from the specialized sensory system in the brain and control the special movements of the muscle system in the brain. These nerves pass through small opening in the skull and enter in the brain and also leave the brain. Some cranial nerves are specialized only for sensory pathways to the brain and some are specialized only for motor pathways to the brain. The cell bodies within the brain, send axons out to form 12 pairs of cranial nerves. These axons are long enough to reach the respective parts of the body such as nose, eyes, face and teeth, tongue, heart, lungs and so on. For example the vagus nerve extends to the heart, liver and intestine far from the brain. The vagus and some other cranial nerves are also parts of the parasympathetic nervous system. Many of the cranial nerves are mixed. Two of the cranial nerves, the olfactory and optic are not really peripheral nerves, but included in the division from functional point of view.

The Spinal Nerves

The spinal nerves arise from the spinal cord. The spinal nerves connect the spinal cord with body. The spinal nerves consists of 31 pairs. They are lying along the spinal cord, each one of the pair of the nerves runs to each side of the body. These nerves are connected to the spinal cord at regular space interval. Each spinal nerve consists of blending of two distinct roots such as drosal root and ventral root but they are functionally different, One is the *dorsal root* (afferent nerves) - carries sensory information from peripheral nervous system to the spinal cord. Each of this spinal nerve consists of *sensory pathways* from the body to the spinal cord. The other one is the *ventral root* (efferent nerves) consists of *motor pathways* from the spinal cord to the muscles. This ventral root carries motor messages from the spinal cord to the peripheral nervous system. The name of a spinal nerve carries the same name of the parts of the spinal cord with which it is connected, for example, the cervical nerve is for neck, the thoracic nerve is for trunk, the lumber nerve is for the small of back, and the sacral nerves is for the bottom of the spinal column. Nerve fibres from different spinal nerves join and form the peripheral nerves. This takes place at some distance away from the spinal cord.

Autonomic Nervous System

The central nervous system and the peripheral nervous system are concerned with the regulations of the functions of the internal organs, the viscera. The somatic nervous system is concerned with the sensory inputs from exteroceptors and control of the skeletal (striped) muscles. The autonomic nervous system is closely mixed with the somatic aspects of the brain and spinal cord, peripherally the autonomic nervous system anatomically, physiologically is different from the somatic nervous system.

The autonomic nervous system is a system of neural structures lying out side of the bony covering, the central nervous system. It is a series of bunches of nerve cells, arranged segment by segment parallel to the spinal cord. The autonomic nervous system consists of nerves making up of motor pathway to the heart, stomach, intestine and to other internal organs. These nerves are not under voluntary control. This consists of collection of neurons outside the central nervous system which is generally called *ganglia*. These neuron aggregates, since they are outside the central nervous system, are named autonomic ganglia. They are said to be independent (autonomic). Now it is found according to **Rosenzweig** *et al.* (1996) that the autonomic ganglia are controlled by neurons in the central nervous system and the autonomic nervous system stretches to the central and peripheral nervous systems. The autonomic neurons within the brain and spinal cord send their axons out to stimulate neurons in ganglia. These neurons have their cell bodies within the ganglia, in turn send their axons out to stimulate all major organs (Fig. 5.4)

The nerve axons of the autonomic nervous system leave the spinal cord and brain and are mixed with other nerve axons in the cranial and spinal nerves. The autonomic nervous' system is divided into *two divisions*, as the sympathetic nervous system and parasympathetic nervous system.

The sympathetic nervous system is also called *thoracolumbar division* and the parasympathetic is also called craniosacral division. The autonomic nervous system functions through three main parts. The middle part is sympathetic division, which plays large part in regulating the body in times of emergency. The parasympathetic system falls at the top and bottom of the body parts and functions continuously in vegetative activities such as growth, nutrition and self repair. We shall discuss in detail about these two systems.

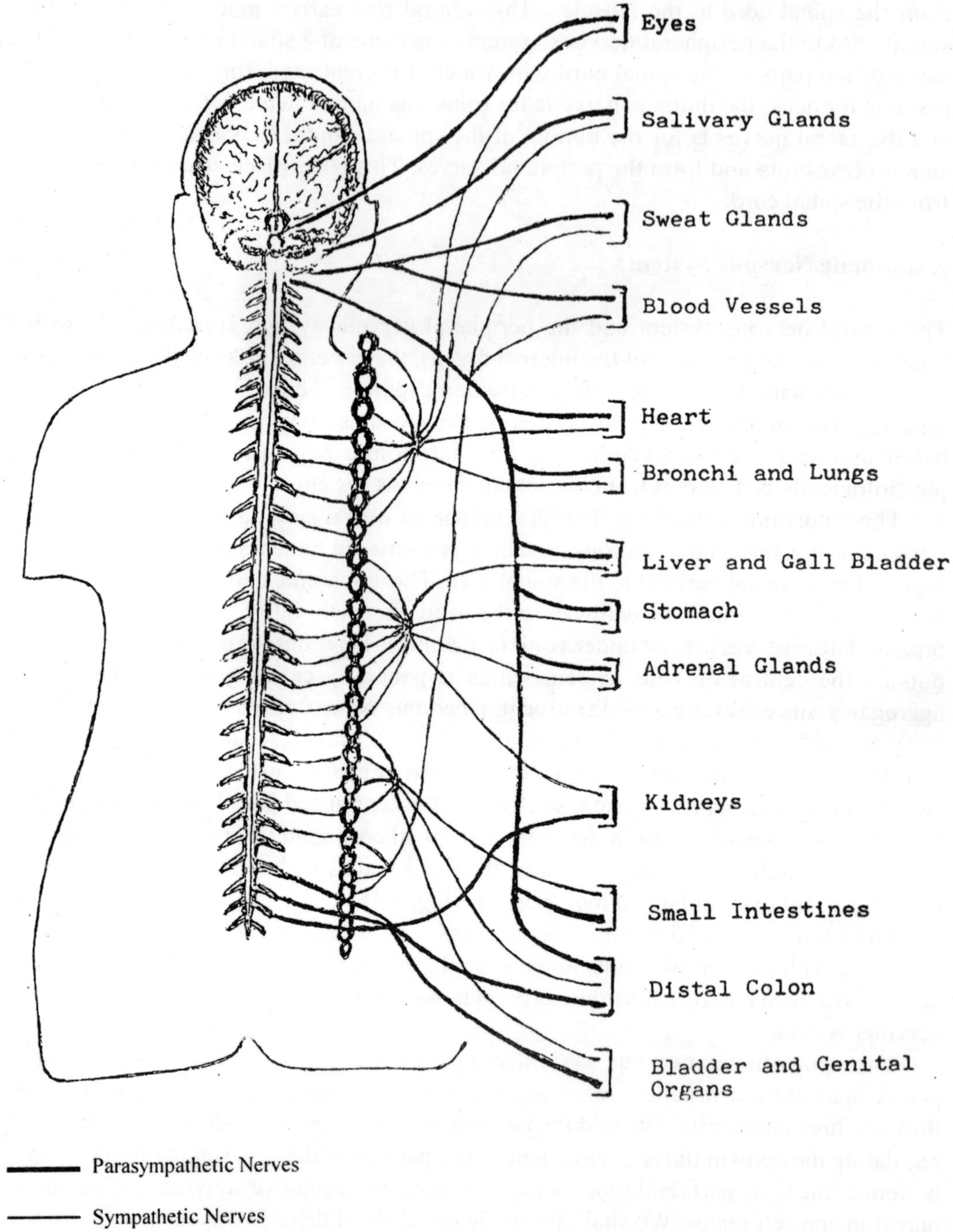

Fig. 5.4 : Autonomic Nervous System and Major Pathways of the Sympathetic and Parasympathetic Nerves.

The Sympathetic Nervous System

The sympathetic nerves leave the middle regions of the spinal cord at the level of chest. The parasympathetic nerves leave the central nervous system from the upper regions where they travel in cranial nerves, and from the bottom regions of the spinal cord where they travel in spinal nerves. Most of the internal organs of the body are supplied with, both sympathetic and parasympathetic nerve fibres. These two nerve fibres work in opposite ways. For example, impulses going along the sympathetic nerve fibres toward heart increase the heart beat rate, whereas impulses going along the parasympathetic nerves leading to the heart decrease the heart beat rate.

When impulses travelling along autonomic nerves fibres reach the end of the line, that is the contact point with some sense organ, the nerves put into action by releasing a chemical transmitter. The sympathetic nerve fibres release a transmitter called *nonadrenaline* and the parasympathetic nerve fibres release *acetylcholine*.

But the functions of some of the nerve fibres of the sympathetic nervous system are not taking place according to the general rule of ending at the contact point with same organs. But they go directly to the adrenal gland and release there the chemical transmitter acetylcholine which stimulates the *adrenal gland*. This results in secreting a mixture of adrenaline and nonadrenaline substances. Then this mixture of these two substances is carried by the blood streams to all parts of the body and functions on various organs. Thus the adrenaline and the non-adrenaline chemical substances have become hormones and the actions of these are quite similar. The non-adrenaline is a transmitter of sympathetic nervous system. The secretion of the adrenal glands produces the same psychological results of much stronger stimulation of the sympathetic nervous system. Because of these, the heart beat becomes stronger when the sympathetic nerves are stimulated and when the non-adrenaline is also released at the sympathetic nerve endings. The same result is expected also when the adrenal gland secretes the hormones and the non-adrenaline and adrenaline are carried on to the heart by the blood streams. In these two instances the heart is reacting in the same manner to the substances.

On examining the effect of sympathetic stimulations of various organs, we find that a number of events are taking place. The cells of the sympathetic chain stimulate the smooth muscles in the organs and in the walls of blood vessels. The sympathetic stimulation prepares the body for action. It causes widening of the air passage leading to lungs, constriction or narrowing of blood vessels in the skin and in the intestinal track, blood being shifted from the digestive organ and skin to the muscles, a slowing down of movements in the intestinal track, an increase of force in the rate of heart beat, release of the blood sugar from the liver, widening of the pupils in the eyes, all these functions make the organism ready to face any emergency situation such as running away from the dangers or fighting with enemy.

The widening of the air passage in lungs makes it easy for an animal or organism to breath faster and get more oxygen. Blood is made to flow from regions, where it will not be needed during the emergency, to the skeletal and heart muscles which require more oxygen and blood sugar. The heart beats faster, so that the circulation takes place through the muscles at greater rate.

The Parasympathetic Nervous System

The parasympathetic nervous system is named like this because the neurons of the central nervous system are found above and below the sympathetic system, the brain is above this and the sacral and the spinal cord are below this. This is one of the two systems that composes the autonomic nervous system. The parasympathetic division arises from both the cranial nerves and the sacral spinal cord. It is in contrast with the sympathetic nervous system. The parasympathetic nerves are dispersed throughout the body, near the organs. The parasympathetic functions are slowing down the heart rates accelerated by the sympathetic system. It relaxes the walls of the blood vessel and smoothen all the functions raised by the sympathetic division.

Certain comparative analysis between these two systems are often made. The sympathetic and the parasympathetic systems are performing antagonistic functions on certain organs. The sympathetic system increases the heart beat rate and the parasympathetic system decreases the heart beat rate. One slows down certain actions, while the other speeds up the functioning. One conserve the energy while the other mobilizes the energy. The sympathetic system has the function of the expenditure of energy, while the parasympathetic has the functions of conservation of energy. Particularly in the stressful situations the sympathetic system is dominant. At such times there is increase of blood pressure, and respiration rates, and the parasympathetic hinders the digestive action.

Further the autonomic nervous system plays important roles in emotional behaviour of the organism. Theorists suggest that the levels of pleasantness and unpleasantness of emotional states are correlated with the stimulation of the parasympathetic and sympathetic nervous systems. The parasympathetic activities are involved in salvation, sacretion of gastric juices, thermoregulation and in certain muscular responses involved in sexual orgasm. Therefore, the parasympathetic activity is correlated with the sensation of emotional pleasantness.

THE CENTRAL NERVOUS SYSTEM

The nervous system consists of two major divisions such as central nervous system and the peripheral nervous system. The central nervous system is the portion of the nervous system that includes the spinal cord and the numerous structures and divisions of the brain. The brain is the mass of nerve tissues that controls and co-ordinates the activities of organisms. In vertebrates it consists of the brain and the spinal cord. The neurons in the central nervous system are grouped into structures. The brain alone contains about ninety percent of the neurons of the nervous system. But the basic function of the neuron is the same in the whole body. The central nervous system basically serves two important functions. First of all it interconnects the sensory receptors and the motor effectors, which are located in different parts of the body. Secondly, it allows the integration of the various sensory inputs and making the motor different to ascertain that the organism is functioning in only one direction at the same time. In higher organism, specifically in human the integration function is very complex.

The whole design of the nervous system reflects these basic functions. The sensory nerve pathways bring information into the central nervous system and it distributes to various parts. The motor pathways bring detailed information from various parts of the brain and the spinal cord to the muscles and glands. In between these, a complex network is taking place. The central nervous system on receiving the information does the function of organization and integration. As no section of the central nervous system operates independently these functions are taking place in a network system. Each part adds its activity to the total function of the body.

The nerve structures of the central nervous system are completely different from that of the peripheral nervous system in several ways. The peripheral nervous system takes the information into the central nervous system and also away from it and the central nervous system on receiving the information carries out the function of co-ordination and integration.

The central nervous system consists of the *brain* and the *spinal cord*. The brain is placed inside the head with all its divisions and sub-divisions. But the spinal cord starts from the lower end of the brain and goes into the body through the spinal column in the backbone. Though these two parts, the brain and spinal cord are located in different parts of the body and function independently these two parts constitute the central nervous system. We shall first discuss the spinal cord and go into the brain and other divisions of the brain (Fig. 5.5).

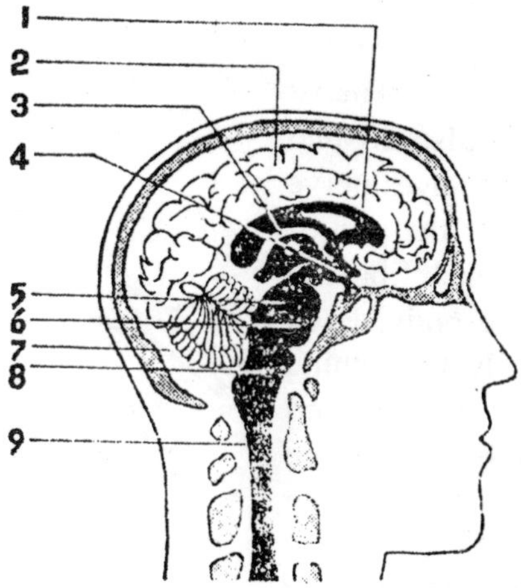

Fig. 5.5 : The Cross-section of the Human Brain
1. Corpus Callosum. 2. Cerebral hemisphere. 3. Thalamus. 4. Hypothalamus. 5. Bid Brain. 6. Pons. 7. Cerebellum. 8. Medulla. 9. Spinal Cord.

Spinal Cord

The spinal cord is a part of the nervous system enclosed *within the backbone*. It is the bundle of complex nerve fibres going through the spinal column. The spinal column runs from the base of the brain to the sacrum (large wedge shaped bone at the longer backbone). A pair of nerves emerges from each level. There are four levels. The spinal cord is surrounded by (bony structure) vertibrac and enclosed in a membrane called the dura matter. Dura matter is the outermost part of the three coverings that embrace the brain and the spinal cord. Each vertebra has an operning on each side through which the spinal nerves pass.

The spinal cord has four levels, such as cervical level, thoracic level, lumber level and sacral level.

The spinal cord connects the brain to the rest of the nervous system. The spinal cord is made up of long and cylindrical nerve fibres. There are two distinct components in the spinal cord. (1) The gray matter and (2) the white matter which surrounds the spinal cord. In the gray matter there are interneurons and the motoneurons that send axons to the muscles. The white matter consists of the myelin sheath of the axons covered by fat substance which makes the matter white and run up and down the spinal column. The stain (discolouring) turns the fatty myelin sheaths of the axons into dark. Thus the white matter turns into (dark) gray in this function. The phenomenon of gray matter and white matter originates in the brain and is extended throughout the spinal cord.

The spinal cord consists of cell bodies of neurons. The white matter contains conducting paths. The nerve fibres in the white matter carry information to the brain and back to the spinal cord. Therefore the spinal cord is primarily dealing with the transmission of impulses between the brain and the peripheral nervous system.

The basic structure of the spinal cord is divided into segments. There is a tube of nerve cells within the body, which is also divided into segments. Each segment receives nerve fibres from sensory receptors of the part of the body close to it and sends back fibres to the muscles of the part of the body. Each segment functions independently. Everything is interconnected through fibres and coordinate their activities. This basic structure is common to all vertebrates and to human body.

Each body segment of the spinal cord is called *dermatome*, which means dealing with skin. The dermatome refers to the area of the skin, which is supplied with afferent nerve fibres by a single spinal dorsal root. This surrounds the body in a ring like manner. The human body has an upright posture. Because of this the ring formation of the dermatome is twisted from the shape into the pattern (Fig. 5.6)

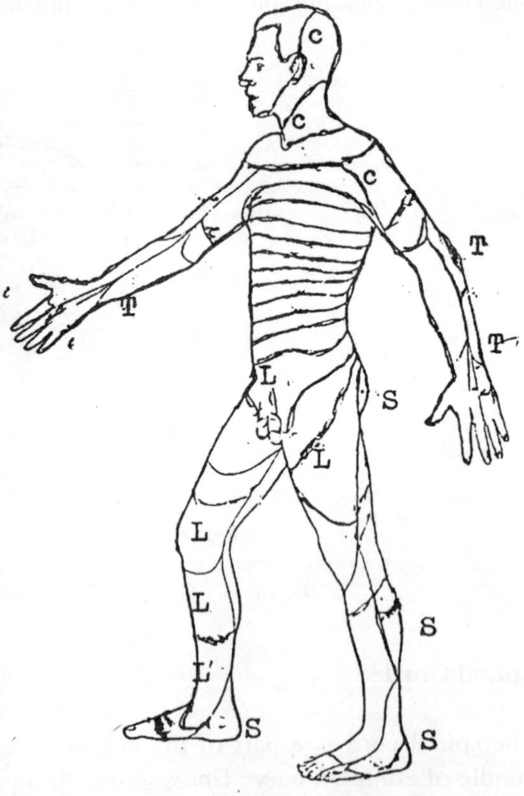

Fig. 5.6 : Functions of Spinal Cord : Human Dermatomes
C = Cervical, T = Thoracic, L = Lumbar, S = Sacral

In animals there are six levels of segments. In humans there are four levels of segments. Each segment is connected with the organs and musculature of a specific body segment. There are thirty spinal cord segments at four levels. But some authors like **Grossman** (1967) state that there are thirty-one pairs of segments. In (1) the *cervical* (*C*) level there are eight segments, in (2) the *thoracic* (*T*) level there are twelve segments, in (3) the *lumber* (*L*) level five segments and in (4) the *sacral (S) level* five segments. The nerve fibres going into the dorsal area of the spinal cord carry information from the sensory receptors of the body (Fig. 5.7 and 5.7(a)).

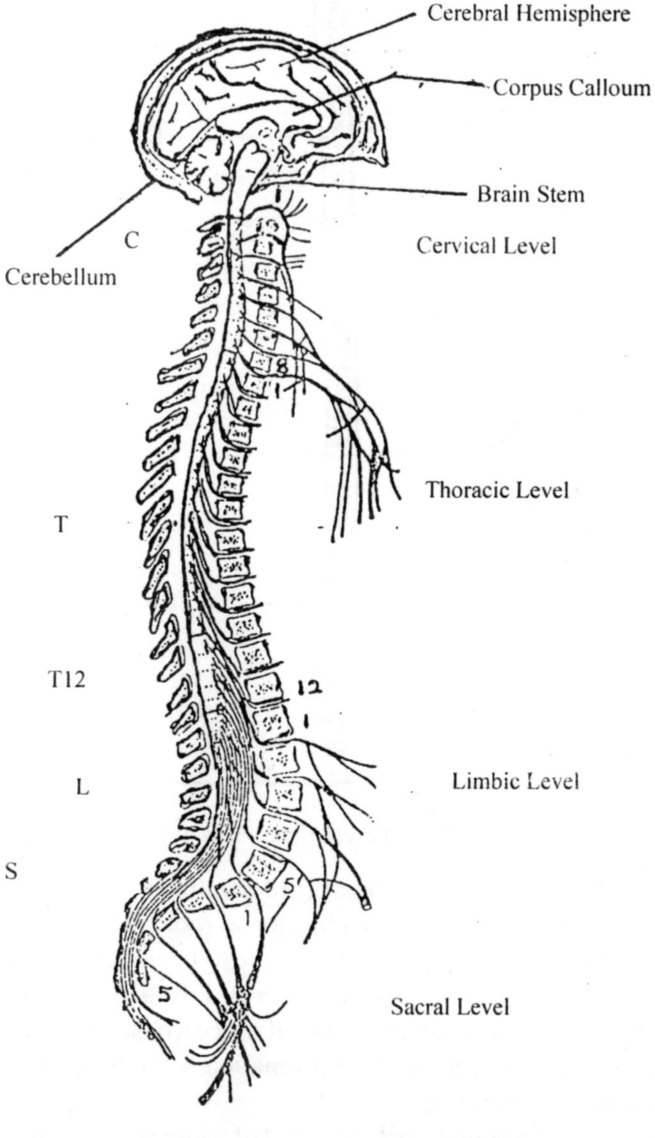

Fig. 5.7 : Spinal and Cord and Levels of Segments.

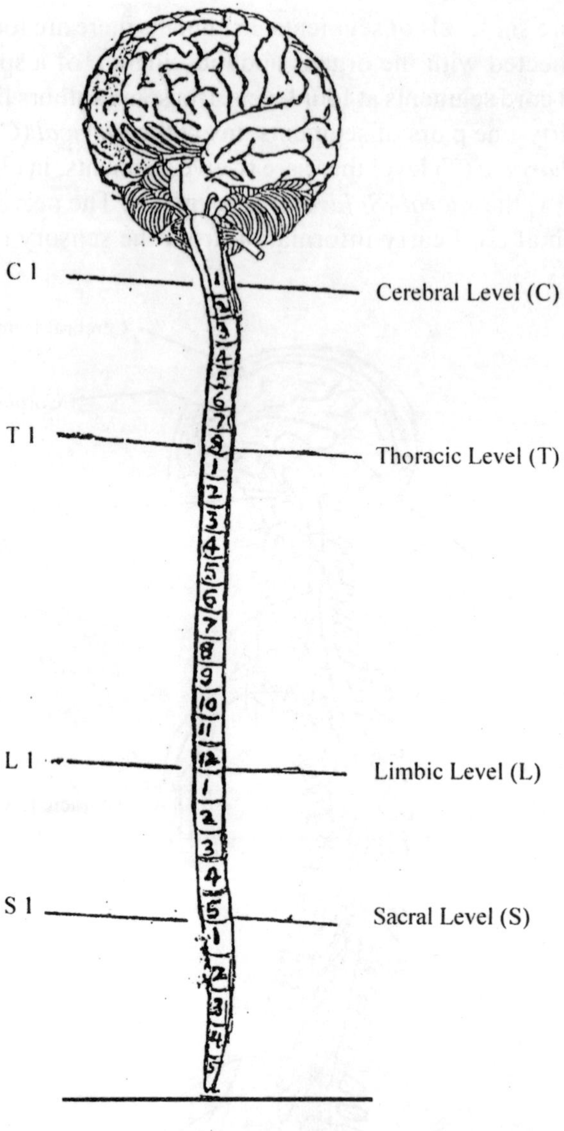

Fig. 5.7 (a) : Levels of Segments of the Spinal Cord

The nerve fibres *entering* the spinal cord are collected together and this collection of the fibres is stated as the *dorsal root*. This dorsal root carry sensory information into each segment of the dorsal portion of the spinal cord. The collection of nerve fibres which are *leaving* the ventral part of the spinal cord carry information from the spinal cord to the muscles is called ventral root. Thus the dorsal root carry sensory information into the spinal cord, and the ventral root carry motor information from the spinal cord to the muscles for action. It is for a motor function.

The outer part of the spinal cord consisted of *white matter* or tracts that are so organized in such a way that with *few exceptions* the dorsal root track functions as *motor nerves* and the ventral root functions as sensory nerves. The inner portion of the spinal cord is a grey matter, which is mainly composed of cell bodies (Fig. 5.8).

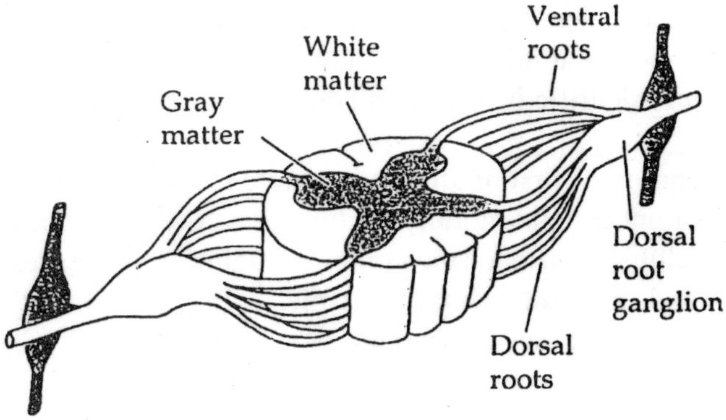

Fig. 5.8 : Cross cut of the Segment showing the Cell Bodies of White Matter and Gray Matter

Spinal Reflexes

The sensory impulses sweep through the ascending neural-pathways towards the spinal cord and the brain. The efferent or motor impulses violently go through the descending pathways to the muscles and glands. The two neural ascending and descending pathways to the brain and from the brain go only through the spinal cord. But sometimes certain sensory stimuli evoke sudden motor responses at the spinal cord. This is called *spinal cord reflex.* The brain is not involved in this reflex action. For example, the pinprick to the skin damages the skin and stimulates pain receptors in the skin. This stimulation activates a pain neuron, which sends the information to the spinal cord. In the spinal cord the pain fibres activates interneurons which in turn activate the motor neurones which contact with muscles and the muscles contract and twist the limb. This is called *withdrawal reflex*, which removes the limb from the point of painful feeling. This reflex depends upon the nature of the stimulus, which may be mild or stronger. For example, the knee-jerk reflex is produced by sharp tap to the knee. The tap to the knee is the stimulus and it produces impulses. These are transmitted through peripheral sensory nerve to the spinal cord. The spinal cord connects the impulses to the peripheral motor nerve and activates the leg muscles. These kinds of simple reflex activities may take place at the spinal cord level. The spinal cord reflex actions are protective.

THE BRAIN

A correct understanding of the Brain-Behaviour relations has been a very big riddle right from the time of the ancient Greek and Roman civilization. It is important that certain significant background information must be known before any conclusions to be drawn out on this issue. It stands always an unanswered riddle. The riddle posed by Sphinx to the travelers who entered the city of Thebes and killing them when they failed to answer it.

Sphinx, a monster with a woman's head and alion's body lay at the entrance of the city of Thebes. Oedipus, the celebrated hero in Greek mythology, answered the riddle, then Sphinx killed herself. It was about *Human Nature : But what are humans? What is human nature?* A deeper meaning is needed. The deeper meaning of the riddle of human nature is still at large. Experts are of the view that it is possible to get the deeper meaning of human nature to some extent by certain scientific investigation on the nature and functions of the human brain.

The student of psychology,' first of all must be quite familiar with the basic principles of the brain functions for understanding fully the human nature. The student of psychology should know the psychological basis of the complex cognitive processes, specifically the evolution of the brain and the fundamental principles of its functions, the chemistry, the principles of the operation of the sensory and motor systems.

The Human Brain

Researches on the functions of the brain were going on from time immemorial and the final acceptance of the correct functions yet to come out. The reason is that it is composed of very complex system and it is accurately inaccessible for experimental analysis of its functions. Since every small part of the brain has certain specific function and the degrees of functions vary from person to person, a generalized idea cannot be accurately achieved.

The brain is a soft convoluted mass of nervous tissues carefully packed within the skull of the head. The brain is the controlling and coordinating centre of the nervous system and the seat of thought, memory and emotion. It includes the major divisions of *cerebrum, brainstem* and the *cerebellum*. The Brain may be also stated as an organ of soft nerve tissue contains in the skull, functioning as the co-ordinating centre of sensation, intellectual and nervous activity. It is the main intellectual, sensaory and motor capacity of the human beings. Right from the time of **Hippocrates** (460-370 BC), the Greek physician, the brain has been the main subject matter of study. Every time some new information about the brain is being released by the experts in the biological, anatomical and medical fields. The nomenclature of the parts vary time to time and experts persent varied classifications of the parts.

New machines and tools for testing the brain parts and for treating the patients who suffer from some kind of brain disease have been brought out. There are new scanners making it possible to explore the living brain, actually to see its functions, like the news and information on Television. These types of wonderful probes have greatly extended our knowledge about the human brain. The deeper we go into the intricacies of the working conditions, the more magnificently the brain appears. Every human being is carrying with him an actual universe, the brain.

In certain ways, which involve the brain, some of the animals and birds are found to be clearly superior to human beings. Some kind of birds are involved in wonderful facts of navigation in making long distance migrations. Dogs and cats of some species have senses of smell and hearing that are much more than humans. Behind these super human talents, there is an area in the animal's brain, which is enlarged in some manner that could make the most possible special skills.

Size of the Brain

But what makes the human brain to distinguish it from the above said more talented other brains, is the relative size of the cerebral cortex. The human *cerebral cortex* is about eight millimeter thickness. The whole area is covered by the *gray matter* on the lobes and the hemispheres of the cerebrum. The cerebral cortex is so large only in human beings in relation to the body size. The human cerebral cortex is further distinguished from other animals' brain by its greater quantity of foldings, convolutions or gyri (twisting and coiling), valleys (fissures and suli) and ridges. All these increase the surface area of the cortex and make the maximum amount of gray matter to be packed within the boundaries of the skull. But the lower animals or mammals have relatively smoother cortexes, have smaller surface areas and less *gray matter*.

The brain size is not depending upon the size of the physical body built. The size of the brain is a fascinating problem. Is the size of the brain related to the abilities of the organism? The most exciting characteristic of the human brain is that it is larger than the brains of the most other animals. According to **Kolb and Whishaw** (1996) the size of the human brain ranges from 1000 to 2000 grams with an average size between 1300 and 1400 grams. The brain size of the women is little smaller than that of men. The sex difference in the size of the brain is reflected in the body size and in some of the functional differences caused by the hormones.

But how large is the human brain compared to other animals? The human brain is larger than the brains of the other animals, except the brain of the elephant which is three times larger than the human brain, because of the huge body. Neuroanatomists have been trying for a long time to evolve a formula for assessing the relationship between the brain size and the body size. **Blinkov and Glesner** (1968) were able to present the results of their studies on brain and body size ratios in tables and figures by applying the Jerison's formula. Jerison has evolved a formula called Encephalization Quotient (EQ). It is just like the Intelligence Quotient (IQ) and the Emotional Quotient(EQ).

The encephalization quotient is a ratio of actual brain size to the expected brain size for a typical mammal of that body size. The expected brain size is a kind of average for the living mammals that takes the body size into account. The formula Jerison evolved is

$$EQ = \left(\frac{Ei}{2 \times Pi\left(\frac{2}{3}\right)}\right)$$

Where Ei refers to the actual brain size of the species.

Pi refers to the body size of the species.

This is applied in most of the anatomical studies in recent times.

Another similar ratio in the brain studies was also evolved by Stephen and his co-workers (1967). They have calculated a Cortical Quotient (CQ). It is the ratio between the actual neocortex and the expected neocortex of a typical mammal. The cortical quotient shows that the volume of the human neocortex is 3.2 times greater than that is predicted for a chimpanzee of the same body weight. These result infer the idea that the differences in behaviour between individuals are much more likely to be related to the differences in the brain size. There are a good deal of discussions and debates on this issue. It is about both the contemporary human brain size and intelligence. But relating the brain size of the present generation of human beings and their intelligence has not brought out the expected results, because it is anatomically found that intelligence depends upon the *functional cortex area of the brain* and not to the total size of the brain.

Experts in this field of study agree that the superior functions of the human brain are directly related to the *cerebral cortex*. Man is significantly distinguished from other mammals or animals by the powers of speech and written languages. These abilities are stored in this area and these separate us from the animal world. These powers of abilities also go hand in hand with other functions like observing, perceiving, thinking, analysing coordinating and integrating experiences, in solving problems, imagining, planning and with all future happenings.

While discussing about the functions of the brain, Koitband Whishaw have reviewed the recent anatomical studies made by **Balikov** and **Glesner** and **Stephen** and his co-worker on brain. They commended that it is not only the weight and volume of the brain have influence on the faculties, but also *certain areas* of the cerebral cortex may have the determining effect on certain faculties which may be *genetically rooted.* In the case of sex differences the *'Y' chromosome* may be *having influence* on certain faculties like, *self-realization* and *intuition* which women lack. Sri Ramakrishna has observed this view in his Gospels. But this hypothesis needs to be tested on a larger population of the present generation. Self-realization is a potential ability for probing the past and the future. Intuition is a sudden awareness of happening something in the situation and revealing it (Fig. 5.9).

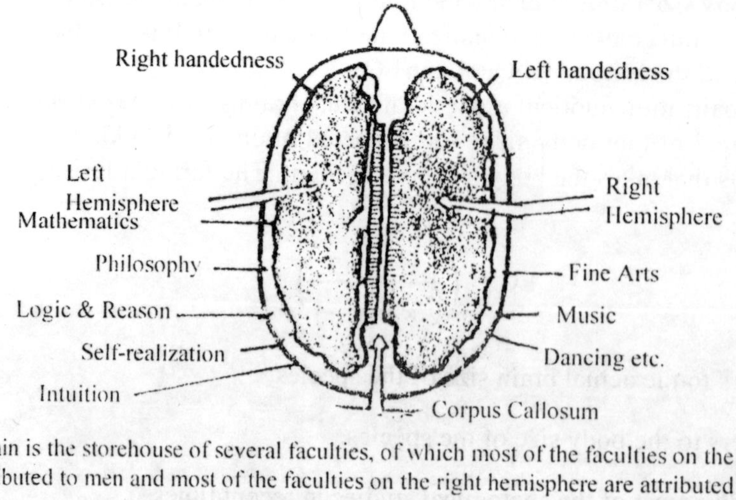

Fig. 5.9 : The brain is the storehouse of several faculties, of which most of the faculties on the left hemisphere are attributed to men and most of the faculties on the right hemisphere are attributed to women.

Earlier Divisions and Functions of the Human Brain

The human brain is the most important part of the body and it is strongly associated with the study of human behaviour. It is involved in all complex activities like *perception, learning, thinking, problem solving* and so on. In order to have a clear understanding about all these complex activities We must go into the details of the *structure, the divisions and their functions. The adult human brain weights about 1400 grams*, which is two percent of the body weight. The brain is a soft mass of nerves and supporting tissues, which are safely packed inside the skull. The *newly assigned name for the brain is encephalon.* Anatomical experimental Fescarches often yield new divisions and names for the bodily parts. We are applying here several new terms for the parts of the brain, which were not used earlier.

Origin and Development of the Human Brain

The human *central nervous system* of an adult person is *the brain* which is an extremely complex structure of all other structures. But *its origin begins in a simple manner.* It has its origin and the beginning as a simple *line of cells* in the back of the *developing embryo.* The embryo is a developing organism from conception to three months of prenatal (before birth) development.

The cells that formed line begin to divide and then form *into a groove* which is called as the *neural groove.* This deepens and then pinched off from the surface to *form a tube beneath the surface layer of the cells.* This tube thus formed is called the neural tube. It is the *neural groove* that is developed into the neural tube, which is a *hollow tube.* It is from the neural tube the brain and the spinal cord develop (Fig. 5.10).

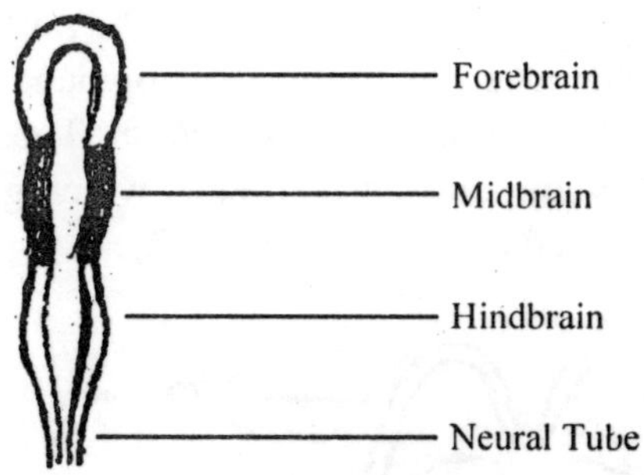

Fig. 5.10 : Showing the Neural Tube with three bulging shapes which indicate the locations of the Forebrain, Midbrain and the Hindbrain.

The cells that ah7eady connected the neural tube to the *ectoderm* migrate to either side of the tube for forming the *dorsal root ganglia.* The ectoderm is the outermost layer of the cells in the embryo- from which the skin, hair, teeth and nervous system develop. The dorsal root is the nerve composed of fibres carrying sensory information that enters each segment of the dorsal (posterior) portion of the spinal cord. They send fibres into the neural tube to form the dorsal (tail end) roots of the spinal cord and other fibres to form the sensory nerves.

Thus the neural tube develops into the *spinal cord*. Cells in the ventral (front) part send fibres to form ventral roots of the cord and the *motor nerves*. The cells of the anterior (head end) part of the neural tube begin to develop and get multiplied rapidly than the cells at the posterior (tall end) part of the tube. The rapid development of the cells in the anterior (head end) part of the neural tube form the brain and the other posterior (trail end) part of the neural tube develop into *spinal cord* (Fig. 5.11).

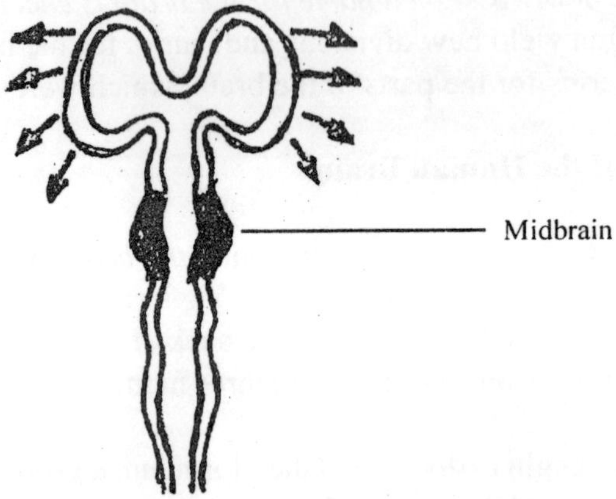

Fig. 5.11 : Showing how the Forebrain is rapidly enlarging to become the largest part of the brain, the *Cerebrum*

At the anterior (head end) part of the embryo or the neural tube the brain develops from *three bulging structures*. These three bulging structures when developed become, what we call as the *forebrain, middlebrain* and *hindbrain*. The portions of *forebrain* and *hindbrain* greatly increase in size and make way for the development of two very prominent structures of the adult brain, the *cerebrum* and the *cereballum*. These two structures, cerebrum and cerebellum are covered by an outer layer of cells which give them a look of *gray colour*. This outer *gray layer* is called the *cortex* (Fig.5.12).

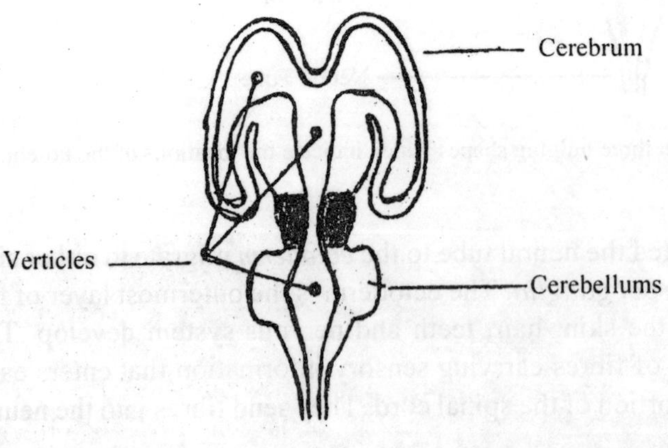

Fig. 5.12 : Showing the portions of the Forebrain and the Hindbrain increasing in size and they become the structures of the adult brain called as Cerebrum and Cerebellum. These two parts are covered by an outer layer of cells which become gray in colour and it is called the *Cortex.* In the adult brain the hollow interior part form into chambers called ventricles which contain the Cerebrospinal fluid.

Thus the human brain develops from a *neural tube*, which is *hollow tubular structure*. In the adult human brain the hollow interior functions in the form of an intricate system of ventricles (chambers), which consist of the *cerebrospinal fluid* (CSF). The cerebrospinal fluid is the *tissue fluid* surrounding the brain and the spinal cord and filling the cerebral ventricles (Chambers).

The cerebrospinal fluid is the clear solution of *sodium chloride* and other *salts* that fill the ventricles (the chambers) inside the brain and circulate around the brain beneath the *archnoid layer* (thin delicate connective tissue) that follows the outline of the brain in the subarchnoid space.

Earlier Divisions of the Human Brain (Primitive Brain Divisions)

The human brain till the end of the third quarter of the nineteenth century was said to be composed of *three main divisions*. They are (1) the forebrain, (2) the midbrain and (3) the hindbrain. All the parts of the brain *except the spinal cord* are brought under these three divisions and discussed. From the beginning of the twentieth century the rapid development of science and technology started influencing every science that studied human and other organisms altogether from different angles.

The human sciences such as the anatomy, biology, physiology genetics, embryology and psychology also are very much benefited by the recent developments. The theories of the subject matters in these sciences were very rigorously and drastically tested through experiments. The human brain was examined from different angles. As result of these developments the subject matters of the brain were *revised with new divisions, new names* and descriptions. But, however, the central theme of the knowledge of the human brain is retaining its status in most of the cases.

The study of human behaviour is very much enlarged with new concepts, knowledge and experiments. We shall discuss first the *earlier divisions* and functions of the brain briefly and then go in detail to the *recent developments*. Our approach to the recent development of the study of the brain is only very modest to the extent of providing the most important aspects precisely. But, however, to avoid some gaps or confusions in understanding the structure and functions of the brain parts we have also incorporated some new information in the origin, development and the functions of the parts.

The Forebrain

The forebrain is the most important of all the three divisions. In the evolutionary process the forebrain was evolved lastly. It is the anterior (at the front) part of the three primitive enlargements of the developing brain, the embryo. It occupies the entire upper portion of the skull and widely bulged out *above the cerebellum*. The shape of the central nervous system is *like a stick* on the roughly curved handle of a *walking-stick*. The stick is the spinal cord, the handle is the brain stem and cerebellum and the overlapping hand is the forebrain (Fig. 5.13).

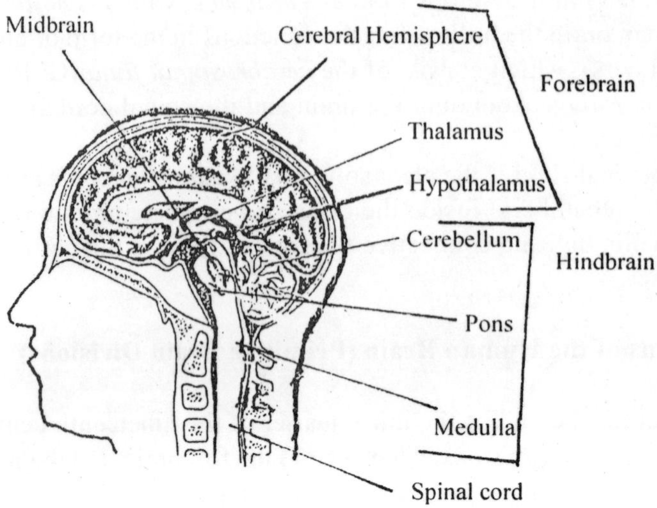

Fig. 5.13 : Fully developed human brain and its parts : Earlier Devisions.

The forebrain is a mass like substance and greater than the other two parts of the brain, the midbrain and the hindbrain. The outer part of the forebrain looks like two great circular hemispheres. The forebrain then divides into two main parts such as (1) anterior part or the endbrain and the (2) posterior part as second brain.

From the anterior part of the forebrain, the cerebral hemispheres, the corpus stratum, subcortial centres within cerebral hemispheres are developed. From the posterior part of the forebrain, thalamus, hypothalamus are developed. The end-brain is the most anterior (head side, most front) part of the forebrain. This consists of two major structures the (interior) (1) *basal ganglia* and (2) *the large surface cerebral cortex.* The basal ganglia contains subcortrical nuclei of the endbrain, the putamen and the globus pellidus.

The basal ganglia and the corpus straturn are the centres that develop from the inner walls of the cerebral hemispheres. The cerebral hemispheres develop gray matter on the surface. The surface gray matter is called the cerebral cortex, cortex means tough outer cover. The cortex covers both the *cerebrum* and the *cerebellum.* The two left and right, cerebral Lemispheres are connected by a thick band of *white matter* called the *corpus callosum.* The forebrain contains three important areas such as (1) the cerebrum, cerebral cortex, (2) thalamus and the hypothalamus and (3) the limbic system.

The limbic system is a ring like structure of inter-connecting pathways and centres that includes the septum, cingulate gyrus, hippocempus, entorhinal area, amygdala and the anterior thalamus.

The cerebrum is the most highly evolved and most complex part of the forebrain. It is formed by millions of neurons and divided into two connected halves, as the right and left cerebral hemispheres, Therefore the functions of cerebrum are always discussed as cerebral hemispheres. The cerebrum is folly involved in these hemispheres. We are discussing this in the next section.

The Midbrain

What we call as midbrain is the middle of the three primitive enlargements of the developing brain in the embryo. The midbrain is located above the brainstem, the *pons* and the *medulla* and lies between the forebrain and the hindbrain. It is a complex centre, but relatively, is very simple. It evolves four pea-shaped enlargements on the dorsal side (backside). The ventral part (front part) of the midbrain is made up of tracts, *bundle of axons* including the cerebral fibres going from cerebral hemispheres to the lower centres.

Above the cerebral equeduct (tubular passage inside the midbrain) on the roof of the midbrain are the pea-shaped colliculi. These are four *visual and auditory reflex centres.* Below the cerebral equeduct there are fibres connecting the midbrain and the spinal cord.

The midbrain consists of major synapses for the visual and adutory nerves. It controls the muscles of the eyes, which are involved in *visual focus*. It also contains centres for controlling of *equilibrium of the body*. It includes complex network of fibres called the *reticular formation*, which is the general arousal centre. The reticular system alerts the higher centres of the brain when activated. It also regulates the sympathetic nervous system. The midbrain contains the structures like the *tecturn* and *tegmentum*.

The functioning system of the midbrain is much like the hindbrain. Tracts of the motor system orginating in the forebrain go down through the midbrain in their way to lower motor centres and the tracts of the sensory system from the spinal cord and the hindbrain *stream upward* into the forebrain. The midbrain has some important integrative functions. (Fig. 5.13).

The Hindbrain

The hindbrain is the earliest to evolve in the neural tube. This is organized in the same manner as the *midbrain*. The hindbrain is the posterior (end part), one of the three primitive enlargements in the developing brain of the embryo. The hindbrain develops the *cerebellum* and the *pons* in its *anterior* (end part) portion and the *medulla* in the posterior part. Thus the region of the hindbrain consists of mainly of the cerebellum, the pons, the *medulla oblongata* and the *forth ventricle*. The medulla oblongata is the lowest part of the hindbrain and its primary functions are controlling breathing, heart rate, reflexes and body posture. The pons are located above the medulla which consists of nerve fibres for connecting the cortex to the hindbrain. (Fig. 5.13).

The major part of the cerebellum receives impulses from the cortex and controls the skilled movements. The recent observation is that the hindbrain includes almost all the parts of the brain except the cerebral hemispheres and cerebral hemispheres.

The hindbrain is surrounded by several structures of the brain, the cerebellum is a two hemispheric' structure with a small wormlike central lobe between the two hemispheres and it is behind the brainstern and connected to the pons through three large pairs of nerve fibres. The hindbrain is next to the forebrain, the second largest in size in human brain. The *gray matter* of the cortex is covering this like the cerebrum.

The brainstem part of the hindbrain plays an important role in the functions of the nervous system. The medulla oblongata is *a narrow structure* lies immediately above the spinal cord. It contains many centres for the vital reflexes of the body which regulates the heart rate and breathing. Nerve tracts originating in the spinal cord go through the medulla to the higher brain centres. The nerve tracts originating in the brain also go through the medulla to the spinal cord and to the lower centres.

The pons is a part of the hindbrain, which is mostly composed of motor fibres going to the areas of cerebellum and spinal tract. The pons located in the anterior surface of the brainstem. The pons and the medulla also receive sensory nerve fibres from the ears, skin, mouth and muscles of the head.

BRAIN DIVISIONS AND FUNCTIONS BASED ON RECENT DEVELOPMENT (MAMMALIAN BRAIN DIVISIONS)

Development of Brain and Divisions

The human brain consists of several complex system. It is a difficult task for a person who wants to understand it clearly without going into the origin and development and the nomenclatures of the brain. Anatomists, biologists and the embriologists often bring out new information about the brain and its parts and their functions. According to them the *weight and the volume of the human brain* are going on *changing from the time of birth to the old age*. The changes in these two areas make changes also in other areas like the developmental processes, functions and the behaviour of the organism. Apart from the psychological point of view, even from the anatomical and biological point of view it causes surprise and flux. The adult human brain weighs about 1400 grams, which is two percent of the total body weight.

Cephalon

The human brain emerges from the neural tube. At the anterior part (frontal part) of the neural tube, three sub-divisions are formed. This neural tube is an embryonic structure with three sub-divisions that correspond to the origin and development of the forebrain, the misbrain and the hindbrain. There is cavity (hollow space) in the neural tube where there are the cerebral ventricles (chambers) and the passages that connect the three divisions. These three divisions of the developing brain, which are embryonic are labelled as the *prosencephalon* (forebrain), the *mesencephalon* (midbrain) and the *rhombencephalon* (hindbrain).

As they are, the prosencephalon is involved in olfaction, the mesencephalon is involved in the functioning of vision and hearing, and the rhombencephalon is involved in maintaining the balance of the body. These three divisions are considered as the *primitive* divisions of - the brain as they are in the embryonic stage. These divisions are also making further development when the organism grows and develop *into the adulthood*.

What are these cephalons? Cephalon means brain. The prosencephalon is the front brain. The term prosen is used to indicate the most anterior part of the embryonic brain. When the organism grows and develops the prosencephalon brings out an important structure called the cerebral hemispheres. The prosencephalon subsequently evolved two important divisions of the brain called *telencephalon* and the *diencephalon.*

The telencephalon is also known as the endbrain which includes, apart from the cerebral hemispheres, the cortex, *basal ganglia, limbic system* and the *olfactory bulbs.*

The diencephalon which is evolved from the prosencephalon is also called as the between brain. It is the second division of the prosencephalon which includes the *thalamus*, *hypothalamus*, and the *epithalamus.*

The *rhombencephalon* is the hindbrain, the backside of the brain. This develops further as the organism grows and develops, and becomes the lower cerebellum. This part of the brain further develops into metencephalon which is also known as the cross-brain and myelencephalon which is also known as the spinal brain. (Fig. 5.14).

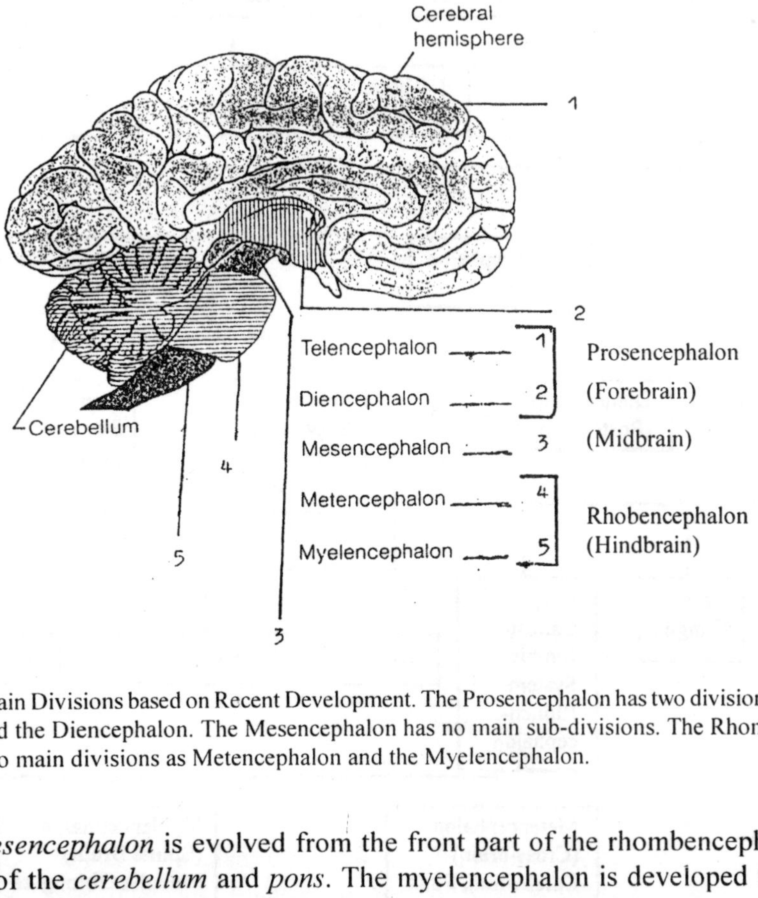

Fig. 5.14 : Brain Divisions based on Recent Development. The Prosencephalon has two divisions as Telecephalon and the Diencephalon. The Mesencephalon has no main sub-divisions. The Rhombencephalon has two main divisions as Metencephalon and the Myelencephalon.

The *mesencephalon* is evolved from the front part of the rhombencephalon which is composed of the *cerebellum* and *pons*. The myelencephalon is developed from posterior (back) portion of the rhombencephalon which includes the *medulla oblongata* and the fourth ventricle. But it must be known that the brain and the divisions at-this stage are hollow and the hollows in each portion are referred to as ventriclesbladders. These ventricles are the cavities of the brain: The ventricles or hollows are filled with the fluid called *cerebrospinal fluid* (CSF). These ventricles are four in number and referred to as I to IV (Fig. 5.15).

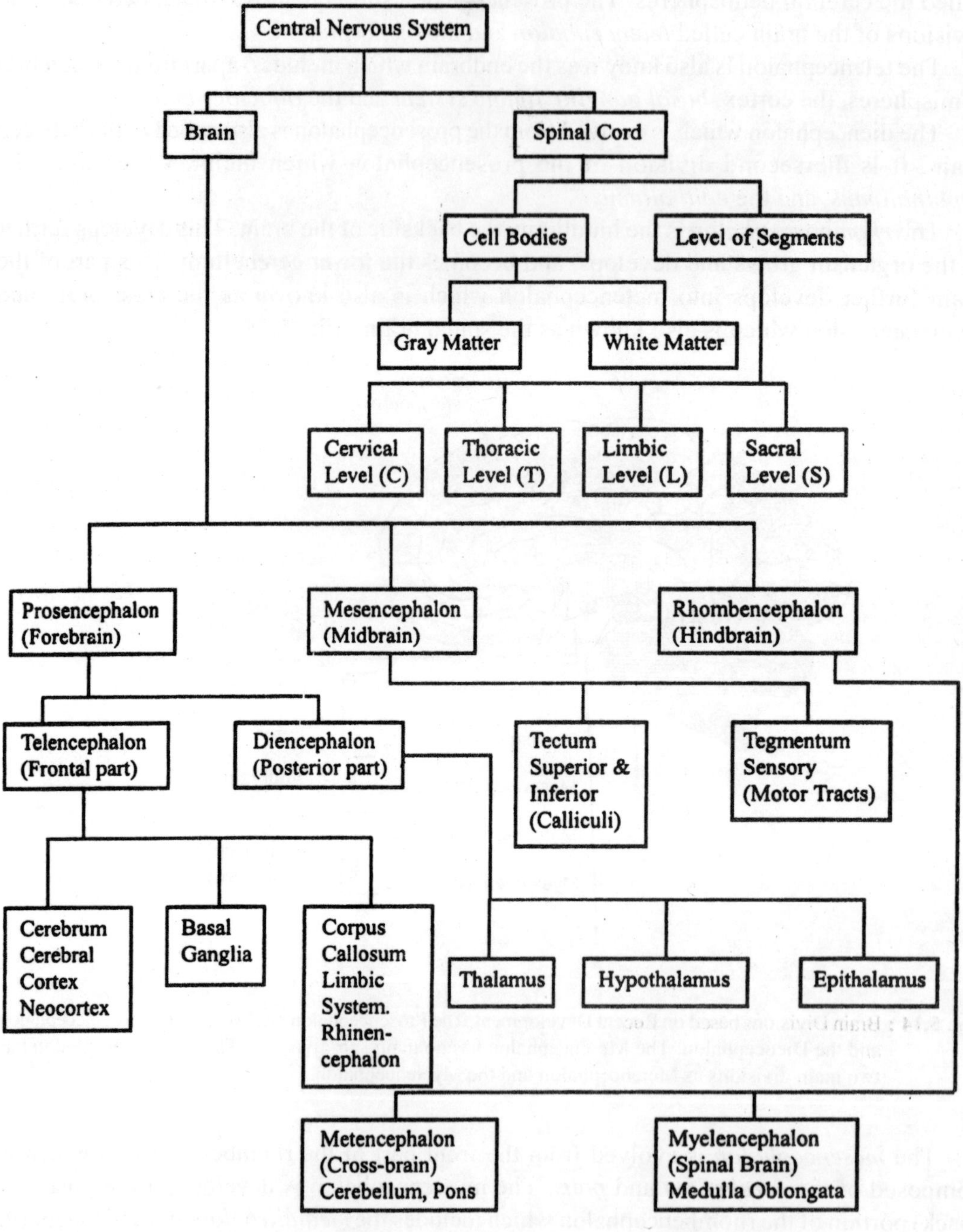

Fig. 5.15 : Brain Divisions based on Recent Development

The cerebrosphinal fluid (CSF) that fills the cerebral cavity is a solution of sodium chloride and other salts fills the ventricles inside the brain and circulates around the brain. This fluid (CSF) is produced by the choride plexus, which is a specialized glial cells found within the cavity. The cavity is bulged in some parts of its length than in other portions. These buldged or enlarged parts are called ventricles. In human brain there are four ventricles, I to IV. The fluid is a colourless solution produced by the plexus of cells continually and made to flow from the ventricles for circulation in the brain. The first two ventricles I and II are lateral ventricles, the third III ventricle is in diencephalon and ventricle fourth IV is in metencephalon and myelencephalon.

Authors who deal with the brain divisions and functions of the recent development differ with regard to certain divisions. On the basis of the newly evolved functional information, **Kolb** and **Whishaw** (1996) used seven new names of the sub-divisions of the brain such as *prosencephalon* (forebrain), which is divided into *telencephalon* (frontal part), and the *diencephalon* (between brain), the *mesencephalon* (midbrain), which has no subdivisions. Then the third main division is the *rhombencephalon* (hindbrain). According to them it is divided into two divisions such as the *metencephalon* (crossbrain), and the *myelencephalon* (spinal brain).

But another group of authors, **Rosenzweig** *et al.* (1996) take into consideration only the five mammalian brain divisions as the basic divisions of the brain. These are the *telencephalon* (frontal part), the diencephalon (between brain), the *metencephalon* (cross brain), and the *myelencephalon* (spinal brain), the mesencephalon (midbrain) is taken by Kolb and Whishaw as the single sub-division. But **Rosenzweig** *et al.* treat the mesencephalon is part of brainstem and not as a separate division. (Fig. 5.16).

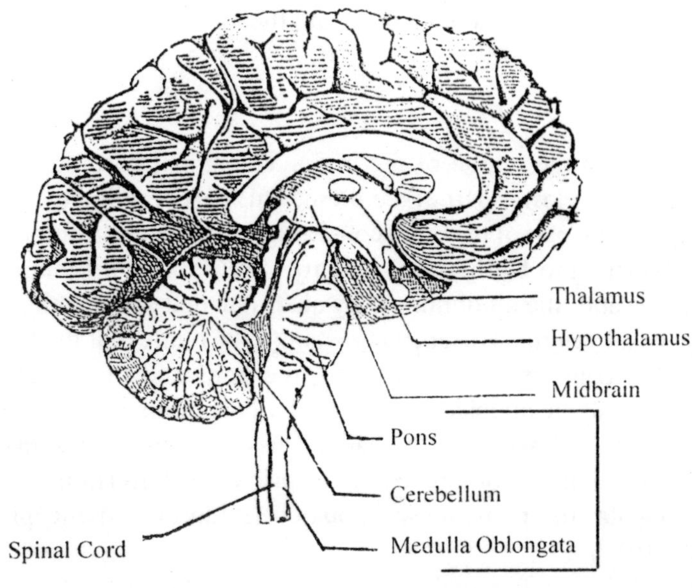

Fig. 5.16 : Brain Stem and its parts

Table 5.1 : Schematic Presentation of the Recent Development of the Brain

Earlier Division (Primitive)	*Recently evolved Divisions*	*Functioning Parts (Fully Developed portions of the Human Brain)*
Forebrain	1. Prosencephalon (a) Telencephalon (Frontal Part) or (End brain)	Cerebral Hemispheres, Cerebral Cortex, Cerebral Lobes, Basal Ganglia, Olfactory Bulb and Limbic System
	(b) Diencephalon (Between brain)	Thalamus, Hypothalamus, Epithalamus
Midbrain	2. Mesencephalon (Midbrain)	Tectum, Tegmentum
Hindbrain	3. Rhombencephalon (a) Metencephalon (Cross Brain)	Cerebellum
	(b) Myelencephalon (Spinal Brain)	Pons, Medulla Oblongata.
Brainstem and Spinal Cord		

All the twelve pairs of the cranial nerves originate in the brainstem area and are projected to various structures of the forebrain, midbrain and hindbrain and activate the brain parts. The functions of the cranial nerves are reciprocal in nature and therefore the whole brain parts are influenced by the cranial nerves. In view of this it is better that we should first know about the brainstem and the cranial nerves and then go to the brain divisions.

Brainstem

The brainstem is located in the region of the brain. The spinal cord, at its upper end is enlarged and merged into the lower part of the brain. This part is called the brainstem. The brainstern is the stalk or stem part of the brain consisting of the medulla ablongata, the pons, and the midbrain. According to **Kolb** and **Whishaw** (1996) the brainstem consists of the midbrain, hypothalamus and 'the hindbrain. They also state that some authors include also the thalamus and the basal ganglia. **Rosenzweig** *et al.* (1996) state that the brainstern is the region of the brain that consists of the midbrain, the pons and the medulla ablongata. According to **Leukal** (1968) the brainstem includes all of the brain parts except the cerebral and cerebellum hemispheres. **Easton** (1978) had discussed about the organization and the pathways of the brain in detail. According to him the primary subdivisions of the brainstem are the medulla ablongata, the pons, mesencephalon (midbrain), diencephalon and the telencephalon (Fig. 5.16).

All these views of the authors indicate that the brainstern is a very important major division of the brain. The midbrain, the hindbrain and some of the parts of the forebrain, all included in the brainstem. Further all these divisions are involved in the brainstem, because of its vital structure, pathways and the functions. The brainstem integrates, regulates and controls the involuntary activities of the body. It can respond to the most of the sensitive features of the environment and also it integrates the eating, drinking, sleeping, walking, body temperature and the sexual behaviour.

Authors who dealt with the brainstem long ago referred the brainstem to the midbrain, pons and the medulla ablongata combined. But the present day authors include all the brain parts except the cerebral and cerebellar hemispheres in the brainstem. The brainstem is located above the spinal cord and beneath the limbic system, where the areas regulate the basic body functions such as heart rate and breathing.

The brainstem is the main axis of the brain which is supposed to be holding the brain and its divisions the cranial nerves which supply nerve fibres to various parts of the brain are mainly functioning from the brainstem, the midbrain, hind6rain and the forebrain.

It is also stated that the brainstem contains the oldest structure of the brain. The brainstem functions even when the consciousness is lost. Lack of function of the brainstem actually indicates the brain death. Generally changes in the brainstem cause sleep paralysis, which prevents us from acting out our dreams. Within the brainstem, the reticular activating system (RAS) controls attention and alertness by sending signals into the limbic system and cortex. It is observed that in soma state the patient loses the function of the cerebral cortex, the part of the brain that controls the higher brain activities, like perception and conscious thought. Even at this stage the brainstem continues to function, maintaining activities such as breathing and heartbeat. The patient may laugh out or make spontaneous movements and the eyes may be able to trace out a moving object.

The brainstem is located beneath the limbic system, where the areas regulate basic bodily functions such as heart rate and breathing. These parts of the brainstem also stimulate emotion and consciousness. The cerebellum is attached to the brainstem, which is supposed to be the miniature version of the whole brain which controls mainly movement and coordination.

The cervical part of the spinal cord anatomically is continuously moving upward and get into the brainstem where the cranial nerves originate. Though the basic functions are mainly movements and coordination, it is also concerned with reflex centres of the muscles of the head and face, like the spinal cord. The brainstem in addition to the association with the cranial nerves, it contains great fibre tracts which carry impulses between spinal cord and to other parts of the brain. The important group of nerve cells in the brainstem exert control over the functioning of the circulalory, digestive, respiratory and other systems of the body. Super imposed on the brainstem are the great cerebral lobes or hemispheres. The simple reflexes of the spinal cord and brainstem are organized into more complicated pattern of behaviour by means of the impulses arrived at the motor neurons from the higher centres of the brain.

The spinal cord or the brainstem cannot produce these functions alone. Since the *cranial nerves originate in the brainstem* and function, these may be also treated as the parts of the *brainstem*. Thus the brainstem has five broad subdivisions. These are : (1) The medulla oblongata together with pons, (2) The cranial nerves, (3) The mesencephalon or the midbrain, (4) The telencephalon first part of the forebrain, and (5) The diencephalon, the second part of the forebrain (The great cerebral lobes and the cerebellum are just above the brainstem. But the brainstem is not a part of them nor they are part of the brainstem). We shall discuss (3) the mesencephalon, (4) the telencephalon and (5) the diencephalon in the respective divisions.

Brainstem and the Cranial Nerves

We have already stated something about the cranial nerves elsewhere in this chapter. But here we are concerned with the cranial nerves for different purposes. Cranial means some thing relating to the skull, that is, the matter packed in the skull. This matter is the brain and its related divisions. The cranial nerves are thus function in the brain regions. There are twelve pairs of cranial nerves in the human brain regions. They are mainly concerned with sensory and motor functions associated with the head, and its parts.

Most of the cranial nerves find their origin from the nuclei of the brainstem. Nuclei are the groups of nerve cell bodies. Some of the cranial nerves are specialized fibre system which stand for special sometic or visceral afferent (sensory) nerve fibres, (Grossman, 1967). These cranial nerve fibres carry information conveyed by the sensory nerves (afferent) fibres to the brain. The other nerves function as specialized visceral efferent nerve fibres from the brain.

The organisation and the pathways and the functions of the brainstem are mostly associated with the cranial nerves The cervical (Neck) region of the spinal cord is çontinued upward and joins the brainstem. This is the region where the cranial nerves originate. **Easton** (1978) states that this part of the brain is basically organized like the spinal cord. A group of nerve cell bodies (nuclei) are associated with the cranial nerves and the brainstem. The brainstem contains great fibre tracts which transmit nerve impulses between the cranial nuclei and other parts of the brain.

Thus it is observed that all the cranial nerves are originated in the head, mostly in the brainstem, midbrain, forebrain and in other parts of the brain. The cranial nerves are made known by using the Roman numerals. Some of the cranial nerves are functioning only in the *sensory pathways* to the brain. These are the olfactory nerve, the first (1) nerve, the second nerve (11) thę-f)ptic neT-vc and the eighth nerve (VIII) the auditory vestibular, the statocaustic nerve.

Some of the other cranial nerves are functioning only as *motor pathways* from the brain. These cranial nerves are the fifth nerve (V) the trigeminal nerve, the nineth nerve (IX) glossopharyngeal, the tenth nerve (X) vogus and the eleventh nerve (XI) spinal ascessory.

A third category of cranial nerves are functioning for both sensory and motor pathways. These nerves are the third nerve (III) oculomotor nerve, the fourth nerve (IV) trochlear nerve, the sixth nerve (VI) abducens nerve, the seventh nerve (VII) facial nerve and the twelfth nerve (XII) hypoglobal nerve (Fig. 5.17).

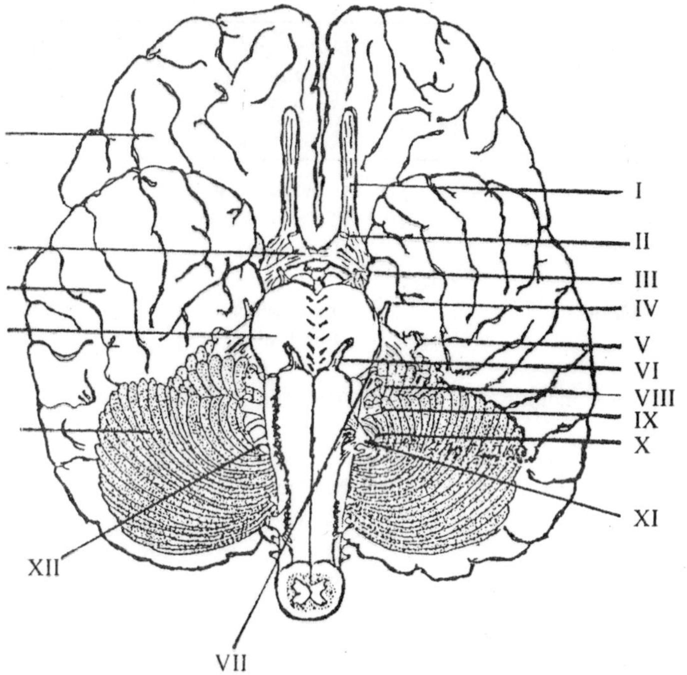

Fig. 5.17 : Origin and Location of Cranial Nerves
I. Olfactory, II. Optic, III. Oculomotor, IV. Trochlear, V. Trigeminal, VI. Abducent, VII. Facial. VIII. Statoacoustic, IX. Glossopharyngeal, X. Vagus, XI. Spinal accessory, XII. Hypoglossal
(A Ventral view of the brain. The Human Brain as seen from below)

The set of twelve pairs of the cranial nerves receive information from the peripheral nervous system and communicate to the brain centres, through the sensory pathways and bring back the orders from the brain through the motor pathways and send them back to the peripheral nerves. All the sensations such as seeing, hearing, tasting, smelling and feeling pain and pleasure are experienced by the functions of the cranial nerves. The cranial nerves play important role in the behaviour of human beings. Therefore, it is important to know the essential characteristics of the cranial nerves and to relate them to human behaviour.

Table 5.2 is the schematic presentation of the twelve cranial nerves, their origins and their functions.

Almost all the cranial nerves are within the brainstem, spinal cord, hindbrain, midbrain and the forebrain. The cranial nerves are going continuously with spinal nerves. Some of the cranial nerves are specialized sense organs. These nerves are centrally connected for specialization. In the spinal cord region the dorsal and ventral roots are equally developed. In the brainstem region if some nerve root is lacking in any way, there is greater development in other root for compensating the deficiency.

Table 5.2 : Showing the Names, Origin and Functions of the Cranial Nerves

Numbers assigned to the Nerves Roman No.	*Name of the Nerves*	*Origin and Location of the Nerves*	*Nature of Functions and the Organs*
I	Olfactory	Olfactory bulbs (Telencephalon) (Forebrain)	Sensor, smell
II	Optic	Diencephalon (Forebrain)	Sensory, Visual
III	Oculomotor	Midbrain	Sensory and Motor to both eye muscles.
IV	Trochlear	Midbrain	Sensory and Motor to one eye muscle
V	Trigeminal	Pons	Sensory, skin
VI	Abducent	Pons	Sensory and Motor one eye muscle
VII	Facial	Medulla Oblongata	Sensory, Taste buds and Tongue (Anterior part) Motor to face muscles and salivery glands.
VIII	(Auditory vestibular) statoacoustic nerve	Medulla Oblongata	Sensory from the inner ear, hearing and balance
IX	Glossopharyngeal	Medulla Oblongata	Sensory from throat, rear tongue and taste buds in the posterior part of tongue
X	Vagus	Medulla Oblongata	Sensory from throat viscera and larynx Motor to viscera and glands, control of smooth muscles
XI	Spinal Acessory	Medulla Oblongata	Motor to viscera, throat neck, shoulder muscles
XII	Hypoglossal	Medulla Oblongata	Sensory and Motor to tongue Muscles

With regards to the origin and location of the centrally connected cranial nerves, VII, VIII, X, XI and XII are in the medulla oblongata region, which is a portion of the hindbrain. These nerves supply nerve fibres for stimulating the muscles of the tongue (XII) and the neck, shoulder and throat (XI). The cranial nerve ten (X) at the *vogus* is the most important of the visceral nervous system, which has both afferent (sensory) and efferent (motor) nerve fibres which controls the smooth muscles at the glands. The cranial nerve nine (IX) is one of the Miost important parts of the visceral nervous system. It carries sensory (afferent) nerve fibres from taste receptors of the tongue and carries motor (efferent) nerve fibres to the salivary glands.

The cranial nerve eight (VIII) is mainly auditory in its function, which is located in the inner ear. It also maintains the bodily balance. This is found in medulla oblongata. The cranial nerve seven (VII) supplies nerve fibres to stimulate the taste buds in the tongue and also to the salivary glands. The cranial nerve six (VI) is the abducens which is located in the pons region. It has both the afferent and efferent nerve fibres and carries the nerve fibres to the stretch receptors and also moves the eye muscles by efferent fibres. The cranial nerve five (V) is the trigeminal, which is also located in the pons. It has remarkable sensory function. It supplies nerve fibres to the skin sense and head. The cranial nerve is centrally connected and moves from medulla oblongata to the midbrain through the pons.

The cranial nerve four (IV) is the trochlear, has its location in the *midbrain*. It supplies nerve fibres to the voluntary *muscles of the eye*. It has both sensory and somato motor nerves. The cranial nerve three (III) is the oculomotor, is located in the *midbrain*. The oculomotor nerve has both sensory and somato motor nerve fibres. It is a part cf the visceral nervous system and it controls the eye muscles. The cranial nerve two (II) is the optic nerve which is located in the *diencephalon*, the *forebrain*.. This nerve is purely a sensory nerve and it sends information to the midbrain. The cranial nerve one (I) is also purely a sensory nerve. It has its origin in the olfactory bulb. It is in the region of the *telencephalon*, the forebrain. The important function of this nerve is the sensation of smell (Fig. 5.17).

1. The Prosencephalon : The Forebrain

The human brain emerges from the neural tube, which consists of three bulges from which the three divisions of the brain, the forebrain, the midbrain and the hindbrain are evolved. The fully developed mammalian brain divisions are named as *prosencephalon*, the *mesencephalon* and the *rhombencephalon*.

The prosencephalon is emerged from the top portion of the bulging in the neural tube and the mesencephalon is emerging from the middle bulging of the tube and the rhombencephalon is emerging from the lowest portion of the tube, the third bulging.

The prosencephalon, when fully developed, is the front brain, which is referred to as the most anterior part (the end brain) of the embryonic brain. This prosencephalon subsequently developes into two divisions as *telencephalon* and *diencephalon*.

It is the prosencephalon that develops to form the *cerebral hemispheres*, *cerebral cortex* and other structures, which are known as *telencephalon* (the end brain), and the *diencephalon* from the remaining portion of the prosencephalon. The diencephalon is said to be the between brain.

Experiments and researches in the areas of anatomy, biology, physiology and psychology on brain and behaviour bring out many new information every time. During the course of four decades enormous amount of facts and figures have been added to the existing subject matter by the anatomists, biologists, physiologists and psychologists. As result of this the study of the human brain and behaviour is becoming more and more complicated and also interesting. We have to inhale and impart all the available information in this area.

Ordinarily, the human brain is highly inaccessible accurately. In explaining the structures and the functions of the brain and its various divisions and to relate them to behaviour needs deep understanding about the structures, parts and their specific function.

The nomenclature of the parts of the brain and the correct placement of the parts and divisions have become highly controversial in some cases. For example, the structure like, the brainstem, spinal cord, the pons, the medulla oblongata and the cerebellum do not have any logical sequence in the placement. **Leukal** (1967) states that the brainstem includes all the parts of the brain except the cerebral and cerebellum hemispheres. **Easton** (1978) includes hindbrain, parts of midbrain and forebrain in brainstem. Therefore exponents in this area should clarify such confusions in the study of the brain.

However, we are presenting the subject matter with all the latest available new information here, so as to make our knowledge up-to-date in this area.

(a) Telencephalon : Frontal Division of the Prosencephalon (Forebrain)

The telencephalon is evolved from the prosencephalon, the forebrain. It is the frontal portion of prosencephalon, which consists of several important structures of the cephalon (brain). The brain is the most fascinating part of the body and it is important for the study of human behaviour. It is involved in all complex activities like *perception, learning, thinking, problem solving* and other higher cognitive functions (Fig. 5.14).

In order to understand all these complex activities we must go into the details of the structure, divisions and their functions. The telencephalon consists of the *cerebral hemispheres, cortex,* the *basal ganglia, olfactory bulbs* and the *limbic system.*

Cerebral Hemispheres

The human brain is dominated by the cerebral hemispheres. It is very recently named as telencephalon. The cerebral hemispheres are at the top and surround the brainstem, which is also connected with the spinal cord. The prosencephalon is the forebrain, is divided into two halves by the corpus callosum. These two are the right and left halves of the prosencephalon (forebrain) which are called as the cerebral hemispheres.

The brain is a soft mass of nervous and supporting tissues which are containing within the skull. It is because of the elaborate foldings together of the tissues the shape of the two hemispheres have the wriggled shape with ridges. The ridges of the tissues are called gyri (gyrus). These are separated from each other by furrows, which are called sulci (sulcus). These are surface foldings. These foldings enormously increase the cerebral surface area. The folding pattern has many features which are common to all human beings. These features help us to understand the divisions of the cerebral hemispheres. The major divisions of the cerebral hemispheres are the *frontal, parietal, temporal* and *occipital regions*. They are also called lobes.

The cerebrum and the cerebral cortex are in the same structure and they are inseparable. Therefore these two parts of the brain are always dealt with together. The cerebrum is highly evolved complex part of the brain. This appears to have two hemispheres. The inner part of the cerebrum is a big mass like substance white in colour, called *corpus callosum*. This corpus callosum is a band of nerve fibres connecting the two hemispheres and other parts of the nervous system. It is like a roof over the thalamus which is another inner part (Fig. 5.18)

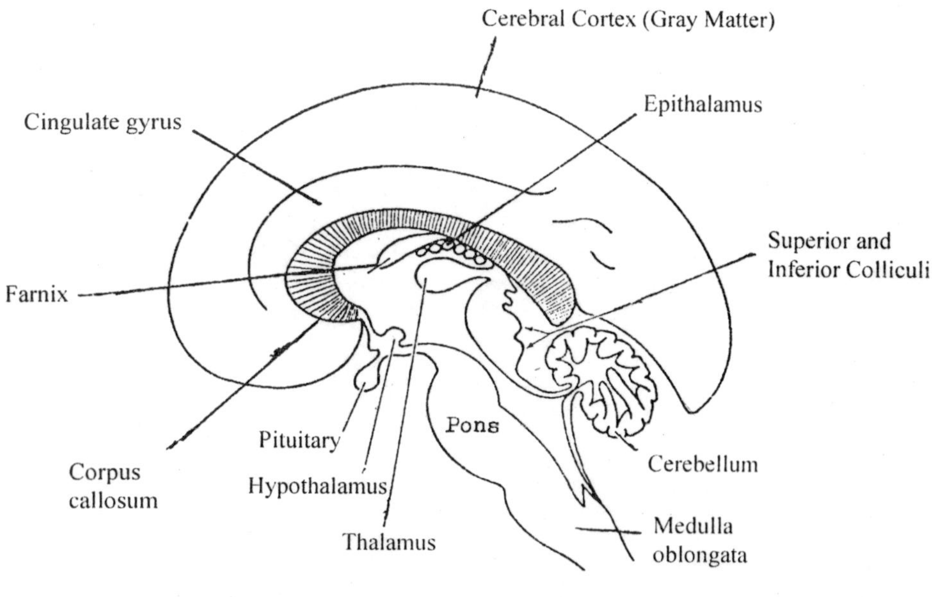

Fig. 5.18 : Medial plane of the Human Brain. Location of the Thalamuses and other parts of the brain

Cerebral Cortex

The outer laver of the cerebrum is made of gray matter which is called the cerebral cortex. The cortex is a covering which encloses the whole of cerebral hemispheres including the forebrain (diencephalon and telencepalon) and the midbrain (mesencephalon). The cortex that covers the cerebrum is called the cerebral cortex. The cerebral cortex mainly consists of nerve cell bodies and it is the outer cover of the cerebral hemispheres. The area of the cerebral cortex is about six square feet. This area varies from brain to brain. This varies in individuals and because of this there may be variations in the behavioural characteristics (Fig.5.18).

The surface area of the cerebral cortex is having many ridges or convolutions. *These convolutions vary in every brain* and from individual to individual. No two brains will have the same convolutions. These convolutions are also called sulcus or fissures. The fissures are of different sizes and the large fissures are used to divide the cerebral cortex into different. areas.

In the middle of the forebrain runs a deep fissure or crack which divides the cerebral cortex into halves. The divided halves of the cerebral cortex are called cerebral hemispheres. The dividing deep midline is called the longitudinal fissure. The fissure divides the cerebrum into right and left as two cerebral hemispheres. There are also other two fissures with prominant landmarks in the cerebral cortex. They are the central sulcus or the fissure of Rolando and the lateral fissure or the fissure of sylvius (Fig. 5.19(a) & 5.19(b))

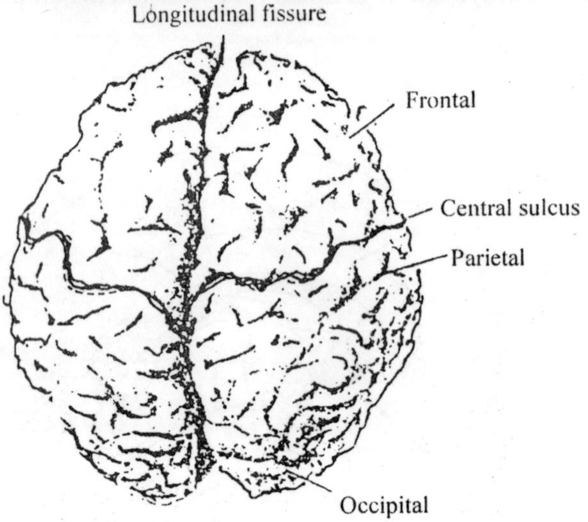

Fig. 5.19 (a) : Dorsal Surface : Longitudinal Fissure runs deep in the middle of Forebrain which divides the Cerebrum into right and left as two Cerebral Hemispheres.

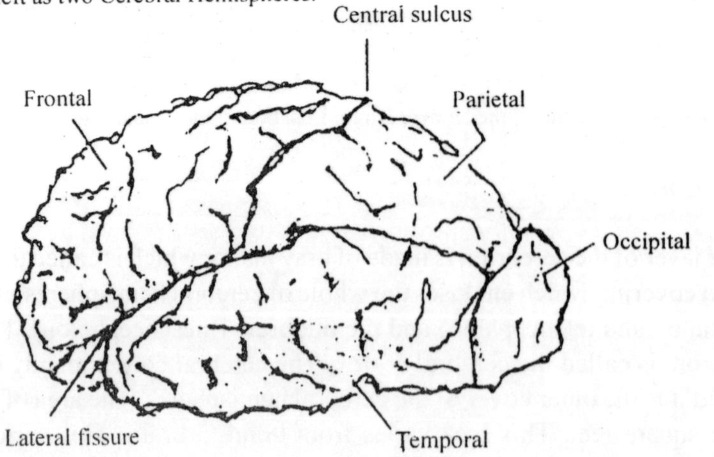

Fig. 5.19(b) : Lateral Surface : Showing the other two Fissure, the Central Sulcus (Fissure of Rolando) and the Lateral Fissure (Fissure of Sylvius) and the location of the four lobes.

The cerebral cortex is a complex structure of three to six layers of a variety of cell types. These are functionally complex synaptic structures. On the basis of the structure, the area of the cortex can be divided into three divisions such as the old cortex which involves paleocortex or allocortex, the transitional cortex and the neocortex. These three divisions of the cortex vary in their anatomical structures. These are inter-connected. Therefore this classification is made functionally. The paleocortex or allocortex has only three layers of cells. But the neocortex has six layers of cells. The allocortex is related to the subcortical nuclei and concerned with smell and put up with these parts of the brain, the rhinencephalon where the parts of the transitional cortex are associated. The neocortex is closely related to the thalamus and basal ganglia. The transitional cortex is interraelated with these two sets of cortex centres.

The allocortex was originally devoted to smell. The other parts such as hypocampus are involved in complex reacting of emotion and motivation. The transitional cortex is a complex structure which was intermediate in the period of development. This cortex includes the cigulate gyrus which lies deep in the longitudinal fissure between the two hemispheres and above the corpus callosum.

Transitional Cortex

The transitional cortex refers to the areas of the cerebral cortex in the longitudinal fissure and elsewhere that are phylogenitically younger than the allocortex and older than the neocortex. This is a complex structure in the longitudinal fissure. This is having connections with pyriform lobe and hippocampus. It is also having connections with that part of thalamus that supplies fibres to the frontal region of the neocortex. In evolutionary terms, the hippocampus, pyriform and the olfactory bulb are said to be the oldest (paleocortex).

The surgery performed by **Kluver-Bucy** in the lateral temporal lobe ablations in Rhesus Monkeys brought out several new informations about the brain structure. The surgery results have become important for various aspects of the syndrome, The hypersexuality, for example, can be produced by the pyriform alone. It can be reduced or stopped by castration in other animals also. The animals can be tamed by removing the parts of amygdala and pyriform cortex.

Neocortex

The neocortex is a very recently evolved *new portion* of the cerebral cortex and it is the newest part in the brain. In more advanced mammals the neocortex accounts for more than half of the volume of the brain. The neocortex covers mostly all the dorsal (back) and lateral (sides) aspects of the cerebral hemispheres. The neocortex contains six layers of cells which give the cortex a laminated appearance. The neocortex includes most of the frontal, parietal, temporal and occipital lobes.

The neocortex also includes the sensory and motor projection areas. The sensory projection areas are having connections in two ways with sensory nuclei of the thalamus.

The motor projection areas are' also having a two-way connection with posterior thalamic nuclei.

The Cerebral Lobes

The two fissures, fissure of Rolando and the fissure of Sylvius divide the two hemispheres into four lobes. The central sulcus is a groove in the cerebral cortex at the top. It divides the frontal lobe from the parietal lobe. The first hemisphere consists of mostly the frontal lobe, which lies in the upper or forward half of the hemisphere. The lateral fissure is a deep cervix in the cerebral cortex at the lower level. This lateral fissure divides the *temporal lobe* from the frontal and *parietal lobes*. The occipital lobes form the posterior pole of the cerebral hemispheres, and it is lying under the occipital bone. The occipital lobe is differentiated from the parietal lobe by the parietal- occipital sulcus (Fig. 5.20).

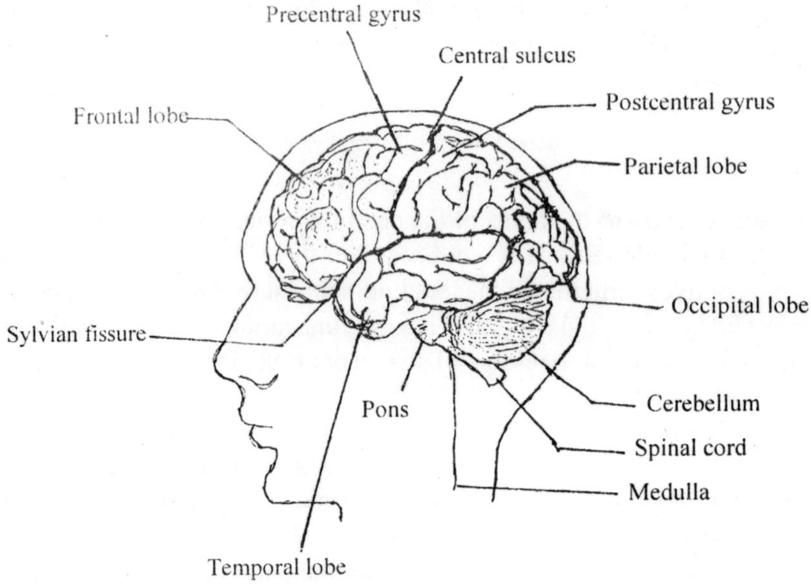

Fig. 5.20 : The Human Brain : Anterior and Posterior Surfaces : Location of Central sulcus and Sylvian fissure which divide the Cerebral cortex into lobes and the location of the lobes.

There are no clear cut indication marks for separating the occipital cortex from temporal or parietal cortex on the lateral surface of the hemisphere. Actually the lateral fissure divides the second hemisphere into three lobes such as the temporal lobe, parietal lobe and the occiptial lobe. These three lobes are generally called the projection areas. Therefore we may state that the first (right) hemisphere consists of the frontal lobe and the second (left) hemisphere consists of the other three lobes, temporal, parietal and occipital lobes. According to **Kolb** and **Whishaw** (1996) the right and left hemispheres are divided by the medial longitudinal fissure and each hemisphere is divided into four lobes, which constitute *eight lobes in both the hemispheres*.

The Frontal Lobes

The frontal lobe is the biggest one of all the lobes in the cerebral area. It contains the motor projection areas and the speech controlling areas. This is responsible for reasoning and problem solving activities. The forward position of the frontal lobe is involved in concentration and also in the ability to plan and to carry out the sequences of activities, It is the expressing part of the brain, generally the frontal lobes are related to the higher order aspects of cognitive and emotional functions.

The human prefrontal cortex is comparatively large in size. It is because of its size that the frontal region has been regarded as the seat of intelligence and abstract thinking. The functions of the frontal lobe becomes unusual due to surgical or accidental damages. The areas of the frontal lobes are not clearly defined. But the human frontal cortex area is almost one-third of the total cerebral cortical surface. At the posterior part of the frontal cortex there are motor and premotor regions.

The clinical assessment states that any injury to the frontal lobe leads to behavioural changes in the cognitive, emotional and motor areas. It is very difficult to point out correctly what type of changes may take *place in the cognitive area* if any damage is caused to the frontal lobe. Some of the most remarkable changes are the loss of divergent thinking, reduced spontaneity and strategy formation. In the case of emotional function, the damage may cause the person to some persistant apathy and exalted sense of well being and the person becomes very much impulsive and has marked changes in social and sexual behaviour and in personality characteristics. The *motor activities* become very much irregular in the case of damage to the frontal lobe. The overall level of motor activities, like the ordinary spontaneous movements are mostly diminished.

The functions of the frontal lobes have become very extreme. Because the frontal lobes are thought to be the seat of the highest intellectual activities. The functions are varied in abstract behaviour like, foresight, intellectual synthesis, ethical behaviour and self-awareness. It also results in Korsa Koff's syndrome, and in the impairment of memory. **Kolb** and **Whishaw** (1996) and **Rosenzweig** (1996) have reviewed the results of the damage caused in cortical regions.

The Temporal Lobes

The temporal lobe is an area of the cortex, which is a major sector of the cerebral hemispheres. It lies below and on the sides of the sylvian (lateral) fissure and anterior to the occipital and parietal cortex. It has the auditory projection areas. Recent anatomical studies have shown more areas in the human brain and also divided the areas into a large number of areas. Most of them are association areas. Several autonomic functions like respiration, change in blood pressure are also affected by the electrical stimulation of some portions of the temporal lobe.

The temporal lobes are having rich internal connections. The afferent projections from the sensory system and the afferent projections from the motor system are made to the parletal and frontal association areas and to the limbic system and the basal ganglia. The temporal lobe consists of the primary auditory cortex, the secondary auditory cortex and the visual cortex, the limbic cortex and the amygdala. Because of these, the temporal lobe is not having a unitary function. We can identity three basic functions of the temporal cortex. The first function, primarily is concerned with the process of auditory inputs. The second is specialized for visual object recognition. The third is associated with long-term storage of sensory inputs. Further, the amygdala functions as the affective tone, (that is in ornotional reactions) and as the sensory inputs and in memories.

The temporal lobes are sensitive to epilepticfon-n of abnomalities and surgical removal of the abnormal temporal lobes. It has been observed by some investigators that specific memory defects differ according to the side the lesion has occurred. Damage to the right side results in deficits in verbal memory, damage to the left results in deficits of nonverbal memory. It also results in the deficits of speech sounds and in processing music. Damage to the temporal cortex results in the deficits of identifying and categorising stimuli.

We may also state the major symptoms of damage in the temporal lobes. Some of the major symptoms are the disturbance of auditory sensation and preception, disturbance in selecting attention of auditory and visual inputs, disorder in visual perception, impairment in the organization of verbal materials, disturbance in the language comprehension, impairment of long-term memory, altered personality and affective behaviour, and altered sexual behaviour.

Damage in the temporal lobes causes auditory disorders. But there are also incidents which cause disorders in visual perception. **Milner** (1958) had observed in her studies that there were deficits in the visual perception of the patient who had impairment in the temporal lobes. Patients with temporal lobe damage performed poorly in the *Moonay Closure Test.* One of the most remarkable visual perceptual deficits is in facial perception and recognition observed by to **Kolf** (1983) and his associates in their experiments.

The Parietal Lobes

The parietal lobe is that part of each of the cerebral hemispheres which lies between the occipital lobe and central- sulcus or fissure of Rolando. It is the major sector of the cerebral hemispheres. It is the general region of the brain lying beneath the parietal bone and the central sulcus. The parietal lobe is primarily concerned with incoming sensory messages. It contains somesthetic sensory projection areas. ihe parietal lobe is functionally related to occipital cortex. This is mainly responsible for the sense of touch, pain and pleasure.

The parietal lobe can be divided into two functional zones such as an *anterior zone* and a *posterior zone*. The anterior zone is the somatosensory cortex and the posterior zone is the posterior parietal cortex. The parietal lobes mostly in the inferior parietal region, have made major expansion in the evolutionary process. Since the anterior (somotosensory) and the posterior zones are functioning separately, two independent functions of the parietal lobes can be recognized. The anterior zone is primarily concerned with the semantic sensations and perceptions. The posterior zone is mainly specialized for integrating sensory inputs from the semantic and visual areas and to a smaller extent from other sensory regions. This mostly controls the movements.

Studies made by **Semmes** *et al.* (1977) revealed important results with regards to the damage to this post central gyrus. Damage in the post central gyrus caused abnormally high sensory impairment. Recent studies indicate that the patients with gross damage in the parietal lobe show curious pattern of symptoms. Some of the symptoms are failure to respond to any stimulation, loss oil ability to do arithmetics, an inability to understand maps and difficulty in reaching for objects. The parietal lobes in the human brain have developed to a much larger size. This means that the human beings may show some symptoms which are not found in monkeys and in other animals.

Further, impairment in the parietal lobe causes deficits in the visual perception. **Balint** (1909) made several studies and found that a patient who had a parietal lobe exposed peculiar visual *symptoms*. This type syndromark seems to be rare. But it is observed in the studies of **Damasio and Benton** (1979) that optic ataxia is a common symptom of posterior parietal lesions. Thus any lesion in the parietal region not only impair somatosensory functions, but also the visual functions.

The Occipital Lobes

The occipital lobe is the most posterior lobe of the cerebral hemispheres. It is that part of each of the cerebral hemispheres, posterior to the parietal lobe. The occipital cortex lies right below the parietal lobe. There is a visual projection area in this lobe and it contains the major centres of vision. Nerve impulses going from the eyes through the optic nerves terminate at the occipital lobe. On the medial (middle) surface of the hemisphere the occipital lobe is differentiated from the parietal lobe by the parietal-occipital sulcus. The three lobes in the lateral surface of the hemisphere are so close that no clear cut line of demarcation can be made to separate the occipital cortex from the temporal and parietal cortexes. Further, the occipital tissues merge with the other regions. Therefore it becoems very difficult to define the extent of the occipital areas clearly.

The anatomists, **Brodmann** (1914) had divided the occipital lobe into three anatomical regions. These regions (17, 18, 19) were considered as the visual regions of the cortex. But recent anatomical studies made still finer divisions. According to these studies the occipital region is divided into six different occipital areas. All the areas play some roles in the visual functions. Each visual area provides some kind of elaboration on the processing of the previous area. This shows that the visual cortex is organized hierarchically on the basis of visual information proceedings from different areas.

Several studies were made in the visual areas and every study comes out with some new information, new division and function which all make—occipital cortex more and more complex and confusion (**Kolb** and **Whishaw**, 1996).

The occipital lobe contains the primary visual projections area, 17, and the visual association area, 18 and the posterior area, 19, The area 17, is the striate cortex, projects to the visual association 18, and also sends some fibres to the area 19. The somatosensory area of the cerebral cortex plays important role in the bodily sensations. Sensations from the upper part and face are evoked in the lower part of the cortex. Sensations from the whole body are projected to the cerebral cortex, to the somesthetic portion, special zone, called projection area.

The person reports flashes of light and other visual effect if the occipital lobe of the brain is stimulated. This lobe is devoted to the retina area of the eye and is designated as the visual projection area. Electrical potentials are recorded here when light flashes to the eye. Blindness may result following extensive injury to this area.

There are various syndromes that may occur from damage at different levels of the visual pathways from the retina to the cortex. Damage to the retina or optic nerve of any of the eyes causes monocular blindness. There will be loss of sight in that eye. Lesion in the medial region of the optic chiasom (connecting fibre) breaks the fibres producing *bitemporal hemianopia* resulting in loss of vision in both temporal fields. Small lesions in the occipital lobe causes *scotomas*, small blind spots, in the visual fields (Fig. 5.20).

Basal Ganglia

The basal ganglia is one the structures in the *telencephalon* division of the forebrain (prosencephalon). It is located in the inner walls of the telencephalon. That is, in the end portion of the brainstem. The telencephalon area consists of a group of large nuclei which are *collectively called as the basal ganglia*, found deep within the cerebral hemispheres. The basal ganglia in the form of nuclei are seen as a mass of gray matter, closely related to thalamus. The basal ganglia are a collection of nuclei (Fig. 5.21)

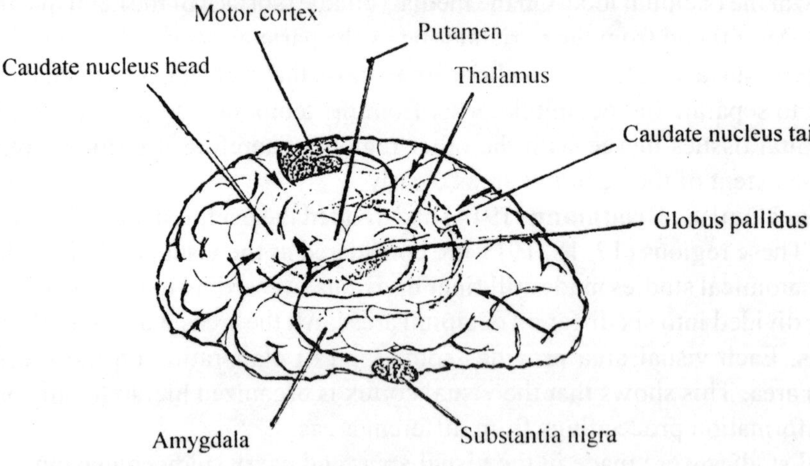

Fig. 5.21 : *Basal Ganglia* structures (Gray matter at the centre) a collection of nuclei including the Caudate nucleus with Head and Tail and other parts in relation to the Cerebral Cortex.

The basal ganglia includes the, *putamen,* (shell), the *globus pallidus* (pale globe), the *caudate nuclei* (tailed nucleus) and the *amygdala* (almond). These structures form the basal ganglia and they form a circuit with the cortex.

Some include the *claustrum* in the basal ganglia strucutre. But the functions of this part are mainly transmission and very little to do with the behavioural aspects. The *amygdala* structure is so complex that it is involved in several structures of the forebrain like the *olfactory bulb* (rhinencephalon) and the *limbic system*. This is discussed in olfactory bulb.

The caudate nuclei receive projections from all the areas of the neocortex and send projects through the putamen and the globus paUidus and to the thalamus and from there to the motor area of the cortex. The basal ganglia makes also reciprocal connections with the mesencephalon, the midbrain. This projection provides chemical dopamine to basal ganglia. The dopamine is a substance produced naturally in the body and used to treat the Parkinson's disease.

When dopamine is lost, a motor disorder called Parkinson's disease is caused. Dopamine is a synaptic (connecting) transmitter produced mainly in the basal forebrain and in diencephalon that is active in basal ganglia and in olfactory system.

The basal ganglia is said to be having three functions. If there is any damage in different portions of the basal ganglia it may result in changes in the bodily postures, increasing or decreasing in the muscle tone and some abnormal movements in the body like jerks and tremours and thus the basal ganglia are found to be involved in motor functions.

Secondly, if any deficits occur, some type changes in specking take place. Thirdly the basal ganglia also facilitate stimulus-response and habit learning. The basal ganglia are very important part of the brain in motor control.

The structures of basal ganglia receive inputs from wide areas of the cerebral cortex and send much of their outputs right back to the cerebral cortex, and this function is stated as cortex-basal-cortex loop. Any lesions in these are in human beings cause movement impairments.

The basal ganglia receive inputs from a large region of the cerebral cortex and from thalamus. The sources of inputs reveal a subdivision of the basal ganglia into two main systems. One of these systems is the caudate nucleus. Inputs from this division are received from frontal association cortex.

The second system is the putamen, which receives inputs from sensory motor cortex areas. Any lesions in these areas result in different effects.

Any lesions in the caudate region impair complex behaviour in dealing with the spatial aspects of behaviour. There will be a confusion as to where to respond. But the effects of lesions in putamen are more of motor type. It affects the nature of responses and the efficiency of the reaction.

Olfactory Bulb : The Rhinencephalon (Smell Brain)

The rhinencephalon is the part of the forebrain that coriiains the olfactory bulb and the tract and it is concerned with the sense of smell. In the earlier studies it was thought as the integrating centre of smell. The anatomists in the beginning were very much impressed with the connections between the olfactory system and the limbic lobe. Having this as an evidence it was suggested that the limbic structure was elaborated to deal with the olfactory sinformation and therefore taking together with the structure they called it as the rhinencephalon or the smell brain.

The rhinencephalon is a recent discovery in the study of brain research. Some authors treat this as a separate part of the brain, giving more importance to certain functions. It is also considered that it is an *alternative to limbic system*. Therefore, some of the structures included in the limbic system in the earlier studies were incorporated in the rhinencephalon structure.

This idea has aroused the curiosity of many researchers and tried to verify this phenomenon. The results of these researches revealed that the limbic system had only very little olfactory function and they felt that the major effect must be from somewhere else. Thus the results ascertained that the rhinencephalon is the main source of the olfactory function.

Several structures of the telencephalon are associated with the rhinencephalon of which the allocortex is one of them. The allocortex is believed to have three observable layers of cells. But the neocortex has six layers of cells. The allocortex is related to the subcortical nuclei which were once considered as the structure that deals with smell and often lumped with these parts of the brain as rhinencephalon. Rhine means smell and encephalon means brain. Therefore the rhinencephalon is said the smell brain, which is mainly dealing with the olfactory functions. Further, certain parts of the transitional cortex are also included in this structure (Fig. 5.22).

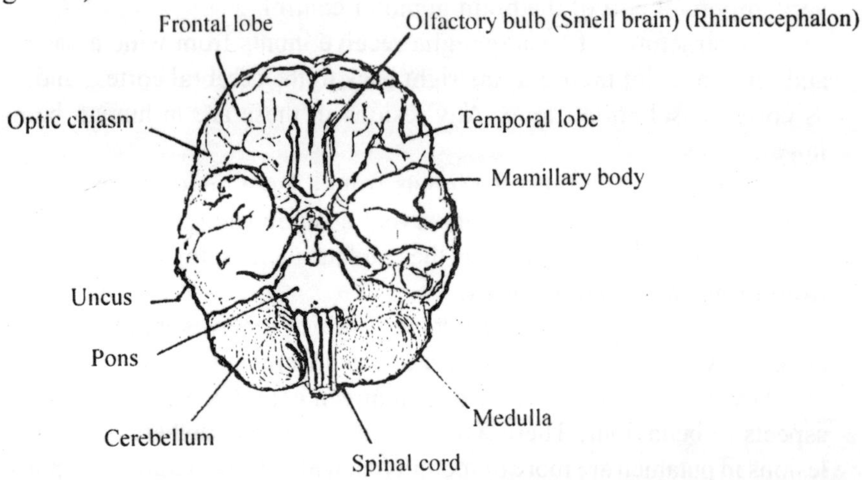

Fig. 5.22 : Brain viewed from below showing the parts of the functions of olfactory bulb and optic chiasm. According to the recent classification the olfactory bulb is considered as the *Rhinencepholon* or the *smell brain* and it is considered to be the alternative to Limbic System.

Olfactory Functions

A better and clear anatomical description of the rhinencephalon mainly includes the olfactory nerves, olfactory bulb and the olfactory stalk, the anterior olfactory nucleus, the anterior perforated area, the paraolfactory area of Broca, the diagonal band of Broca and its nucleus, the hippocampus, the parapyriform cortex and the parts of the hippocampal gyrus.

The primary fibres of the olfactory system projects very thin and poorly mylinated and very slow conducting axons for the brain. These axons enter the olfactory bulb in a sector like form. The olfactory stalk is divided into two as it approaches the surface of the brain. The olfactory bulb and the stalk are made up of primitive cortex.

The olfactory stalk as it enters the cerebral hemisphere divides into a lateral olfactory tract and a medial olfactory tract. These are received by the perforated anterior substance, the terminal fibres from the medial and lateral olfactory tracts and discharges to hypothalmus and epithalamus.

A large part of the visceral brain is composed of nuclei and pathways, related to the olfactory system. Therefore, the term rhinencephelon is also used to indicate the function of smell. Particularly the olfactory stimuli evoke autonomic responses and the rhinencephelon provides a special emphasis to the co-ordination of visceral somatic and sexual functions. The stimuli that have much influence on the entire body are those that are related to the securing food (for survival) and those that involve powerful activities of both visceral and somatic effectors and primarily the close relation to olfaction. It results also in mobilizing the whole body resources when there is pursuit of food or sex involving attempts (Fig. 5.23).

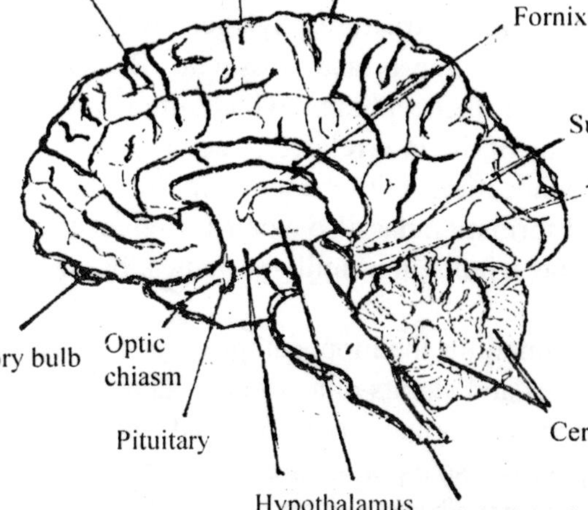

Fig. 5.23 : The brain as seen from the midline showing the various parts which are important in the functions of olfaction and the optic areas. This figure is an illustration of olfactory function of the *Rhinencephalon*. The various parts of the division are involved in the olfactory function.

The visceral nervous system takes active part in actual phN sical compact or in emotional crisis if involved in conflicts. On the basis of the afferent information which is related to the basic drives of food and sex, patterns of impulses leave the visceral brain and transversal to the centres of both somatic and visceral motor control. The somatic and visceral pattern of behaviour thus occur. There are also electric stimulations of certain areas of the rhinencephalon, take place in facial and bodily movements. Some specific facial expressions and the position of the limbs, follow from the impulses which travel over the extra-pyramital system. The visceral effects are taking place through the hypothalamus with which the brain is having strong connections. The hypothalamus is surrounded by the third ventricle in the dinencephalon. It is a co-ordinating centre, has several nuclei that receive olfactory impulses.

The subdivisions and the functions of the rhinencephalon and the limbic system are quite similar to some extent. But still these two structures are given separate treatment taking into consideration of certain specific functions. Particularly in the olfactory functions, the hypothalamus, epithalamus, autonomic functions and emotional expressions, the rhinencephalon and the limbic system have certain common features.

The rhinencephalon according to tlhe recent anatomical studies includes the hippocampus, amygdala, cingulate gyrus, and the pyriform areas. These brain structures once thought to be the integrating centres of smell. Some authors treat this as a separate part of the brain, giving more importance to certain functions.

1. *Hippocampus*

The hippocampus is one of the parts of the cerebral hemispheres found curled in the basal medial part of the temporal lobe that is thought to be important for learning and memory. It is also considered as a primitive structure, phylogenetically, the old name of this is allocortex. It is closely connected with the olfactory system. Because of this, the hippocampus is considered as the part of rhinencephalon, which often called as nose brain or smell brain. The hippocampus is one of the interrelated structures of the hippocampal formation. This has more reciprocal connections with the temporal lobe.

The hippocampal formation has also extensive connections to a number of subcortical structures including the septum, marnmillary bodies, basal forebrain and thalamus and to its most anterior end.

Apart from other functions the hippocampus lays down memories. It stores permanent and important memories and shunts others to permanent storage places in the cortex. It is also observed that all conscious memories involve the hippocampus. This is like the arching, sea-horse shaped organ, is also a part of the limbic system, the primitive unconscious area of the brain, lying beneath the brain's outer layer, the cortex. When we are paying attention to objects and events that are happening around us the informatoin we take in through our senses is sent to the hippocampus along the nerve pathways from the cortex. Some of the information disappear, but memorable events create changes that cause their neural patterns to be retained in the network of hippocampus connections.

Further, the hippocampus is the storage site for spatial memory which is referring to knowledge of routes and places. This organ has a major role in consolidating episodic and semantic memories in the cortex. The hippocampus plays a crucial role in consolidating the long-term memories and store them permanently in the cortex for two to three years.

2. Amygdala

Amygdala is a group of nuclei in the medial anterior part of the *temporal lobe*. It is also considered as a part of the limbic system. It is important in the *organization of emotion* by the limbic system. It is also called anterior thalamus. Amygdala is a Latin word which means almond. The amygdala is having the shape of almond. The four nuclei send axons back and forth to innervate one another forming a neural system. It deals with tasks of incoming sensory information of threatening content, *sending fight or flight messages* to the body and alerting the areas of the cortex connected with fear, anger and sadness. The emotion of fear stimulates amygdala in the brain's limbic system which triggers the neighbouring hypothalamus to send signals to endocrine glands all over the body.

The glands release chemical that activate the nervous system which in turn speeds up the heartbeat, opens the sweat glands and constricts some small blood vessels, draining the face of colour. Information about these events are then fed back to the brain, producing a cycle that may continue, with the sensation of fear, causing each time. The situation is resolved either by action or by thinking parts of the brain sending signals to the amygdala to question its activity.

The amygdala has also something to do with pain. Pain is considered as one of the most useful survival mechanisms. What we feel as pain does not match up with what is happening to our body, because pain is like all other experience is felt in the mind. Once pain signals arrive in the brain they are passed on to amygdala which acts as a relay station, directing pain signals to other areas. The amygdala is the brain's primitive emotional centre which is fully functional at birth and may be involved in the storage of *early memories*.

The amygdala also serves as a *store house of memory*. It is true that most of us cannot remember much of our two or three years of life experiences. It is because of the hippocampus, where the brain stores long term memories. However, some emotional experiences from early childhood may be stored in the amygdala which may be functioning shortly after birth, although these will only be recalled as emotional states rather than explict memories. The fact is that our earliest conscious memories generally date from the age of two or three years, may be connected to the development of language and the emergence of self-awareness which occurs from eighteen months of age onwards. Generally the amygdala is called the brains emotional centre and storage site of powerful emotional memories such as fear and phobias and also plays the role of facial recognition.

The amygdala also has something with the unconscious. Recent research findings provide some information on this. Sometimes we might have experienced some automatic reactions which do not have anything with thinking and judging part of the brain, but by some unconscious processing unit, a subsystem produces automatic reactions.

The human ears detect a signal of danger, for example, the appearance of a snake on our way quite close to us. A signal at once is sent via the brain's auditory system directly to the amygdala, a structure that controls fear response. This signal bypasses the cortex. The danger signal causes us to move much faster in milliseconds time and saves our life from danger. Thus the amygdala's function is like the unconscious awareness to save our life.

There are countless other subsystems in the brain. Every moment these countless systems serve as our unconscious helpers. They deal with all sorts of complicated processing of which we are not consciously aware of them.

3. Cingulate Gyrus

The cingulate cortex is also called as congulate gyrus. It is the strip of limbic cortex lying just above the corpus callosum along the medial walls of the cerebral hemispheres. The cingulate gyrus, the hippocampus and the speturn are interrelated. The cingulate gyrus is the surface area between two sulci (sulcus) in the cerebral cortex. Sulcus is the surface folds that are found in the cerebral cortex, The cingulate gyrus is considered as the part of the transitional cortex. The cingulate gyrus lies between the two hemispheres deep in the longitudinal fissure and above the corpus callosum. Earlier studies state that it is strip of cortex, associated with the limbic system.

The cingulate gyrus area is also connected to reward areas along with other areas like the sieptal areas, dorsal thalamus, tectum and anterior hypothalamus. The papez-mcLean theory evolved a set of notions about a system of nervous centres that control, rather than organize emotional behaviour. The centres and pathways involved are associated with the limbic system. The septal area of the cortex, the cortical cingulate, the hyppocampus and the amygdaloid nuclei are also involved in it. Papez observed a circuit from the *entrohinal cortex* to hippocampus and to hypothalamus through the fornix and from here to the anterior thalamus and finally to the *cingulate gyrus*. Most of these structures had been associated with smell input, but papez states that they are involved in emotional experiences.

It has been observed that animals with bilateral cingulate lesions are slow to learn active avoidance responses, but the bilateral septal lesions animals are slow to avoid passive avoidance responses. The cingulate lesion animals learn passive avoidance readily and the septal animals are superior in learning active avoidance responses. Generally stimulation at cingulate gyrus facilitates autonomic arousal and behavioural activity. But the removal of this area would make an animal less arousal in responses to punishment or less frightened and because of this the animal is slow to learn active avoidance responses.

If the cingulate gyrus is damaged the animal fails to respond to the warning signals to avoid danger. This means that the cingulate gyrus damaged animal is less frightened. The cingulate gyrus lesions impair the active avoidance responses in animals and the septal lesions impair the passive avoidance. The papez-Mclean theory explains these functions in the limbic circuit in emotional responses. The cingulate gyrus and the septal nuclei are the parts of the limbic system which is an alternative to rhinencephalon.

In the experiments it was observed that cutting the pathways between the hypothalamus and that part of the thalamus that sends fibres to cingulate gurus interferes with active avoidance responses similar to the effect of the damage to the cingulate gyrus. Any stimulation to the cingulate gyrus or the amygdala arouses flight response behaviour. The functions of these parts are involved in an interrelated behaviours that are assigned roughly to motivational, emotional and learning behavioural responses.

4. Pyriform Lobe

It is a cerebral cortex area, a projection for smell, located on the ventral surface of the temporal lobe of the cerebral cortex. It is considered to be an old cortex, subserves the olfactory functions. The pyriform cortex or lobs is one of the three layers of the paralimbic cortex. Phylogenetically the paralimbic cortex is older cortex, is adjacent to limbic structure and is composed of roughly three layers. It lays on the medial surface of the temporal lobe where it is known as the pyriform cortex, entrohinal cortex and parahippocampal cortex. It falls just above the corpus callosum where it is stated as *cingulate cortex*. The *paralimbic* cortex has an important role in the functions of memory.

The cerebrum was originally considered as the development of the centre for smell input and the allocortex includes the parts originally developed to smell. Some parts of this cortex area still devoted for smell in higher mammals which includes the of olfactory bulbs and tracts and the prepyriform and the hippocampus have developed into the cortical centres. These are involved in complex reactions of emotion and motivation. The pyriform cortex stands visible on the ventral (frontal) surface of the brain, but the hippocampus get rolled up into the interior of the cerebrum.

The Limbic System

The Limbic System is a widespread group of brain nuclei that innervate (supply with nerves) to each other to form a set of work involved in the mechanism of *emotion and learning*. At the central part of the brain, in the centre of the cerebral hemispheres, there are a number structures 'lying along the inner most edge of the cerebral hemispheres which are grouped together are called limbic system. It is a ring-like structure, which has interconnected path ways and centres. It is a part of the forebrain in the telencephalon.

From the evaluationary point of view, it is a recent discovery like the rhinencephalon. This system is closely interconnected with hypothalamus and seems to control the sequential activities necessary to fulfil some of the basic motivational and emotional needs regulated by the hypothalamus. The limbic system also regulates the instinctive activities of the organism such as feeding, competing, fleeing from danger and mating behaviours. In otherwords, it mainly concered with *hunger, sex* and *emotion* and also higher mental functions just like the cortex.

It is from the limbic lobe the limbic system was evolved. Limbic lobe means border and this was coined by **Broca** in 1876. **Papez** (1937) suggested that emotion is a product of the limbic lobe and not the product of autonomic system as thought earlier (Fig. 5.24)

The limbic system is an abnormally small area enclosed in the high forehead and in a spacious skull, bordering on the corpus callosum. It is considered as the source of superior reasoning. The limbic system is relatively larger in animals like the cats and dogs whose cerebrums are very small. These animals' super smelling powers stem from the giant olfactory connections in the limbic area. The structures of the limbic areas have been called as the minibrain which are controlling the human internal environment automatically.

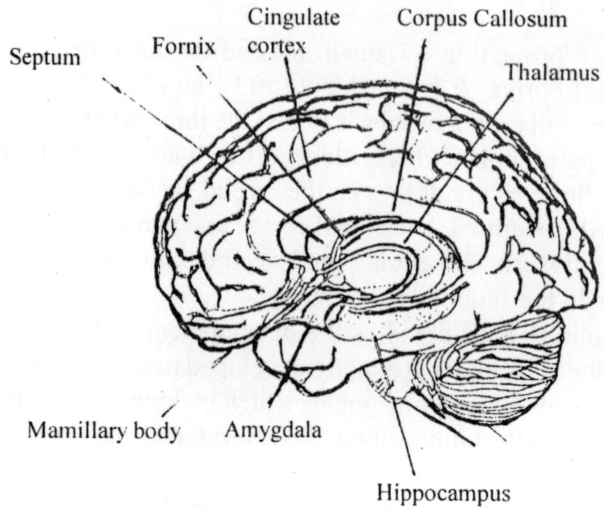

Fig. 5.24 : Showing the Limbic system and the parts

The limbic system includes septum, cignulate gyrus, hippocampus, amygdala and hypothalamus, This structure is generally called as the limbic system or the visceral brain. The other two parts of the brain connecting the limbic system are the fornix and the medial forebrain bundle (MFB).

We have discussed the cingulate gyrus, hippocampus and amygdala in the preceding section under rhinencephalon. The hypothalamus we have discussed under diencephalon in detail. Therefore we shall discuss here the other structures septum, fornix and the medial *forebrain bundle*.

Septum

The septum is also called septal area. The septum is the anterior part of the cingulate gyr-us that lies on the medial surface of each cerebral hemisphere in the longitudinal fissure. It is considered to be the nucleus in the limbic system that when lesioned in rats it produces rage in them and abolishes the rats' EEG wave form. The limbic system consists of a number of structures including the hippocampus (sea horse), septum and the cingulate gyrus or cortex. These are all interrelated. The septum separates the fornix and the olfactory bulb.

The septal area (septum) is a portion of the brain laying deep in the central part between the lateral ventricles (fluid fitted cavities) which when stimulated electrically (in the rat or human) ap:Dears to yeild a state of *skin pleasure*. Generally septum means a partition, separating two chambers. The septum in other words called as *septal complex* which provides subscrotial input to hippocompal formation. The neurons of the septal are located in the medial septal necleus and the ventrical limb of the diagonal band. Many of these neurons use *acetylcholine* as their transmitter. Acetylcholine and adrenalin are often called as harmonal mediators of nervous excitation. They are formed at the parasympathetic and sympathetic endings in negligiable quantities, only about thousand millionth fraction of a gram. But these substances are extra-ordinarily active. A rapid injuction of (2y a very small dose) acety-leholine into the artery produce a tetanic (disease causmig) muscle contraction. A very small dosages are required to excite the muscle fibres.

Several experiments were conducted to assess the septal area's influence in the activity of rats of young and old. **Gallagher** *et al.* (1995) compared brain measures in two groups of aged rats with young rats in learning and memory activities. There were three groups of rats, young, middle and aged rats.

The investigators assessed the activity of the enzyme-choline acetyltransferase (CHAT) in the septal region. The enzyme catalyzes the synthesis of acetylcholine. Any rat revealed pathology in the gross anatomy of the nervous system were eliminated from the study. The CHAT activity in the septal region was high in aged rats that performed better but the aged rats that performed poorly showed lower CHAT activity than the unimpaired aged rats or the young rats.

Fornix

We have discussed in the preceding section that the hypothalamus controls the autonomic nervous system in emotional experiences and the two most important tracts, connecting it with, other parts of the brain are found in the limic system. These two important parts are the fornix and the medial forebrain bundle. The fornix connects the hippocampus with the mammilary bodies.

The fornix is a tract, leading from the hippocampus in each part of the hemispheres to the mammilary bodies of the hypothalamus. The mammilary bodies are paired nuclei that protrude on the ventral side of the brainstem in the posterior hypothalamus. The hippocampus which is buried in the cortical focus, excites the hypothalamus via the fornix. The fornix mainly serves as a communicating structure.

The Medial Forebrain Bundle (MFB)

The hypothalamus is, divided into two main areas as the medial hypothalamus and the lateral hypothalamus. The lateral hypothalamus is paired. From this lateral hypothalamus one main tract is running through each of the pair. This main tract is the medial forebrain bundle (MFB). Thus the medial forebrain bundle is an important tract running through both the lateral hypothalamus zones to interconnect it with the septum and the rhinencephalon. The descending fibres go through the lateral hypothalamus only via the medial forebrain bundle. But some of the fibres continue through lateral hypothalamus through ascending retidular activiting system (ARAS). The medial forbrain bundle (MFB) carries ascending fibres from ARAS and most of them end in the lateral hypothalamus and some go up to rhinencephalon.

The hypothalamus, thus is one of the components of the limbic system and goes through the fornix from the hippocampus and the medial forebrain bundle and from the brainstem reticular formation (BSRF) and the rhinencephalon.

(b) Diencephalon : End Part of the Forebrain

The diencephalon is the end part of the forebrain which includes the three thalamic structures such as the (1) the thalamus, (2) the hypothalamus, and (3) the epithalamus. Among the central connections of the twleve pairs of crainal nerves in the brainstem, half of the nerve fibres of the optic nerves pass below the diencephalon.

1. The Thalamus

The thalamus is a group of nuclei in the diencephalon. The thalamus is the largest part in the diencephalon. The thalamus is divided into two areas like the *ventral area* and the *dorsal area*. The ventral part of the thalamus provides a general input into the neocortex which regulates the activities of the neocortex. The dorsal part of the thalamus (thalamus proper) is composed of a number of nuclei, each one of them projects to a specific area of the cortex. These nuclei send information to the cortex from three sources. One group of nuclei conveys information from sensory system to the respective targets. For example, the lateral body receive visual projections, the medial body receives auditory projection, the ventral posterior lateral nuclei receive touch, pleasure, pain and temperature projections from the body. Another group of nuclei convey information between cortical areas. The third source is the limbic cortex, basal ganglia and the cerebellum which convey their projections to the cortex through thalamus.

The thalamus is the top most part of the diencephalon. and one of the most influential areas of the forebrain and the cerebral cortex. It is a complex nuclei, fuctions between cerebral cortexes.

It is just above the pons and the midbrain. It is said that the thalamus is a complex relay and a control centre of the incoming information. Even other information that travel from one part of the cortex to another part pass through the thalamus. Some of the neurons in thalamus regulate electrical activity of the cortical region and some control the actions of the autonomic nervous system. The thalamus also functions like the pons. Almost all the sensory information enter the thalamus and the neurons send these information to the cortex area (Fig. 5.25).

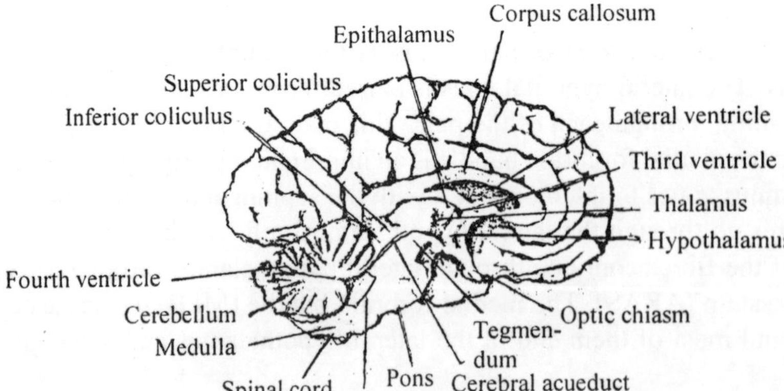

Fig. 5.25 : The brain structure showing the brainstem and the *end parts* of the forebrain. According to the recent classification the *Diencephalon* is the end part of the forebrain. This part consists of *three thalamic structures* such as *thalamus*, the *hypothalamus* and *epithalamus* and other *related parts*. Among the central connections of the twelve pairs of the *cranial nerves* in the brainstem, half of the nerve fibres of the optic nerves pass below the diencephalon.

2. The Hypothalamus

The hypothalamus lies immediately below the thalamus and above the midbrain in the diencephalon. The hypothalamus is relatively small structure in the diencephalon and it is packed with a group of nuclei with vital functions. It exerts lots of influence on several kinds of motivation. The hypotnalamus consists of centres which control many of the visceral functions. It is involved in controlling all behavioural activities like eating, drinking, sleeping, sexual behaviour and controls the temperature of the body. It is also having direct control over all kinds of emotional behaviour like, rage, fear, pleasure and pain. It directly controls the autonomic nervous system and the pituitary glands. It also controls the hormone secretions.

The hypothalamus anatomically is divided into three zones along its axis. These zones are periventricular zone, the medial zone, and the lateral zone. The periventricular zone surrounds the third ventricle. It consists of an extension of periventricular gray of the midbrain. It contains small fibres and cells which are widely disbursed. Here the cells form clear landmark -for this portion of the brainstem.

The medial hypothalamus zone consists of two mammillary bodies. They contain the medial and lateral mammillary nuclei. The lateral mammillary nucleus receives afferent nerve fibres from the hypothalamus and dentate gyrus through the fornix (arched band of white nerve fibres of the brain). The medial mammillary nuclei projects the mammillothalamic tracts to the neucleus of the thalamus. Thus, the communication is conveyed to the thalamus area from the medial hypothalamus zone.

The lateral hypothalamus zone is a mass of undifferentiated cells and fibres which are conveyed to the lateral border of the hypothalamus from the lateral area of the midbrain. The lateral hypothalamus zone consists of several elangated cell connections.

The hypothalamus receives afferent nerve fibres from the cerebral cortex, the thalamus, spinal cord, brain stem, cranial nerve nuclei and from the visceral system. The most important afferent fibre bundle received by the hypothalamus is the fornix. This fornix arises from the cells of the hypothalamus and gyrus and ends in the posterior hypothalamus, which is the marnmillary nuclei. These are the important nerve fibre structures and connections of the hypothalamus,

The hypothalamus exerts too much of influence on Several kinds of motivation. The hypothalamus consists of centres which control the activities like, eating, drinking, sleeping, sexual behaviour and the control of temperature. It is also having direct control over all kinds of emotional behavoiur and expressions like rage, fear, pleasure and pain. It controls directly the autonomus nervous system and the pituitary glands.

3. The Epithalamus

The epithalamus is the collection of nuclei and it is the most primitive portion of the thalamus. It includes the pineal body and the staira medullairs. The pineal body seems to regulate the body rh%thms. The pineal gland develops from the posterior roof of the diencephalon and projects it outward from the dorsal part of the brainstem. The functions of the pineal gland are not yet clearly brought out. They seem to be related to the regulating system which controls many of the biological functions.

Apart from these three divisions of the diencephalon, some authors like **Grossman** (1967) added two more divisions. They are a) the ventral thalamus and b) the metathalamus.

(a) The Ventral Thalamus

This is also stated as subthalamus which is a continuation of the midbrain tegmenturn (inner part). It contains the portions of red nucleus and nigral (a kind of fibre). This part is located between hypothalamus and thalamus.

(b) The Metathalamus

The metathalamus is an area between the mesencephalon and the candal portions of the thalamus. It contains undifferenciated gray matter and it is also referred to as the large-cell-nucleus of the optic tract. It receives nerve fibres from the optic tract and also from some interior portions. It discharges through auditory radiation to the homolateral auditory areas of the temporal lobe (Fig. 5.25).

Higher Functions of the Cerebral Hemispheres—The Forebrain

We shall first discuss the general functions of the cerebral hemispheres and then go for the specific functions of the parts. Though the descriptions of the functions of the parts of the cerebral hemispheres and the general functions of the cerebral hemispheres seem to be almost th-e s-ame there are certain things specific to each area which should be dealt with for making the subject matter clear. In the preceding sections we have stated several new and recent information about the cerebral hemispheres. The two hemispheres are symmatrical, one on the right side and the other on the left side. There is a deep division between them, running from the front side to the back side. Thus our first observation is the division of the cerebral hemisphere into right and left hemispheres.

It is to be noted that for the most part, the functions of the right side of the body are controlled by the left hemisphere and the functions of the left side of the body are controlled by the right hemisphere. Each hemisphere is divided into four lobes such as frontal, parietal, temporal and occipirtal. These four lobes are discussed in detail under cerebral lobes. The functions of these four lobes in both spheres are located equally. But, there are certain functions not broughtout directly under the control of any specific area of lobes. One subject matter is the language.

Language and Speech Area

It has been located that the language and speech areas are in the left hemisphere. It was **Paul Broca** who examined the brain of the patient with loss of speech, found the damage in an area of the left hemisphere in 1861 and the area was named as **Broca's area**. This area is involved in the production of speech sounds. The area controls the tongue and jaws in speaking of the human beings. It was also observed that any damage or destruction of the same type of region in the right hemisphere did not cause any speech impairment. But it was also found that a nearby region of the left hemisphere was concerned with speech sounds.

Further, there are also other areas in the frontal lobes area concerned with the ability to write and to comprehend written words. Although the psychological functions of the linguistic abilities are not correctly located in any specific regions of the brain, the region of the brain in most cases concerned with spoken and written language are found only in the left hemisphere (Fig. 5.26)

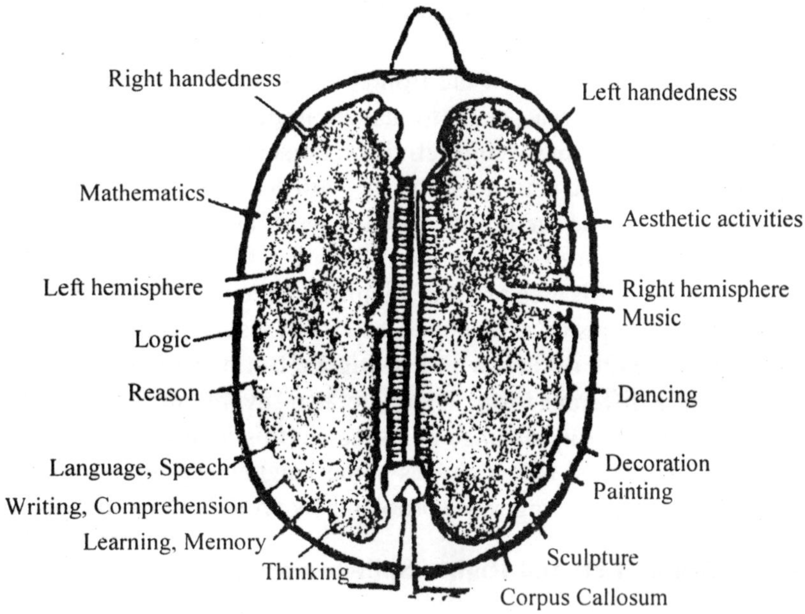

Fig. 5.26 : *Higher Functions* of the Cerebral hemispheres of the brain. There are considerable variations in the higher mental functions between the left and right hemispheres. The functions of the right side of the body are controlled by the left hemisphere and the functions of the left side of the body are controlled by the right hemisphere. All the functions are under the control of the four lobes. There are certain functions do not come under the control of the lobes. The linguistic functions specifically the language is one of the subject matters, which includes speech, writing, memory and thinking.

In the cerebral hemisphere area there is an area called projection area which is supposed to integrate the functions which are called Association functions. These are not concerned with sensory or motor processes.

Therefore, it can also be ascertained that a person who had some damage due to some strokes in the left hemisphere is more possibly to show language imperiment than a person who had the damage in the right hemisphere. Generally this is the case with the right handed persons as the left hemisphere is dominant always in these persons. This is because the right side of the bodv is controlled by the left hemisphere. But most of tlic left handed persons have their normal speech centre in the left hemisphere.

Association Areas

The association areas are the portions of the cerebral hemispheres which are different from the projection areas, Because the functions of these portions are not clearly and specifically located in the cerebral hemispheres. But an assumption is made that these areas have some kind of connecting or integrating functions xvhich are called association functions. Thus the association arcas, are those parts of the cerebral cortex that are not directly concerned with the sensory motor processes. These areas are distinguished from the primary projection areas with that they integrate inputs from more than one sensory channel and are functioning in learning, memory and thinking. We have very little information about the association areas. But the recent researches provide some ideas that these areas perform several different functions.

It is found that some portions of the frontal lobes which are called frontal association areas, anterior to the motor areas, seem to play an important part in the thought processes which are required for problem solving activities. Persons whose frontal lobes are damaged can perform some intellectual activities. But they ar6 deficit of in performing the task when there was a change from one method to another and if any delay is made. In the cerebral structures the lower portion of the temporal lobe is related to the visual perception and therefore any damages in this area produce inability to recognize and discriminate different forms. But the damage or lesion does not cause loss of visual acurity. The person could see the forms and can trace the outline with his fingers. But he cannot identify the shape and differentiate it from a different form. There are also several areas near the temporal lobe concerned with language. Any damage of tissues in this area does not interfere -with the motor production of speech. But it causes some language problem called *aphasia*. The individual in such a state is facing some problems like recalling words, objects and names.

Division of the Brain : Left and Right : Split-brain

The two cerebral hemispheres are quite symmetrical in all respects. Both the hemispheres have the motor, sensory, visual and auditory areas. It is also presumed that any lesion in any one of the areas in any one hemisphere, the functions are not affected. It is because if an area in left hemisphere is damaged the same area from the right hemisphere takes the risk of functioning for the damaged one. Therefore functions in both hemisphere are not affected due to any damage or lesion in one area. Since both the cerebral hemispheres are strongly connected by the corpus callosum, which is a huge band of nerve fibres of white matter,, the mutual communication between them is not affected. But very recently conducted researches brought out some drastic differences between the two cerebral hemispheres in certain psychological functions.

Split-brain

The brain is divided into two halves as left hemisphere and the right hemisphere. They are also called left brain and right brain. When the corpus callosum which connects these two halves is cut off, then the connection between them is severed. This stage is called split brain, thus there are two minds in one head. **Finger** (1994) states that it was firmly established that the cerebral hemispheres are not equivalent in meditating language functions. The left hemisphere controls this function and generally described as the dominent hemisphere. However, when the left side of the brain is injured the right hemisphere does not simply keep quite. But some of the researchers went away from the idea of the cerebral dominance. This idea implies that some functional systems are connected more to one side of the brain than the other. That is, the functions become lateralized. This means that each hemisphere is specialized for specific way of functioning.

But, however at the level of the brain processing in normal individuals the interconnections of the hemispheres generally conceal the evidence of the specialization of the hemispheres. In the study of patients in whom the connections between the right and left hemispheres have been cut, also are stated as split-brain individuals. Research works carried out in 1930s did not reveal clear difference between the functions of the two hemispheres. It was because of lack of suitable methods of the assessment of behaviour. **Roger Sperry** (1974) and his co-worker **Gazzauiga** (1992) applied their experience gained in animal research on this issue of human brain. In their experimental results they found remarkable differences. The results prompted Sperry to speak of separate form of consciousness in the two hemispheres of the human brain. They found that the patients seemed literally to be of having two minds. Sperry observed that one of his patients was seen to button his shirt with one hand and tried to unbutton with the other hand.

Further experimental results indicated similar subtle differences in cognitive style of the two hemispheres even in normal human beings. The left hemisphere is found to be analytical and verbal, but the right hemisphere has been found to be spatial and holistic. The differences in the functions of the cerebral hemispheres suggested that the educational needs and the capabilities of the cerebral hemispheres differ (Fig. 5.27).

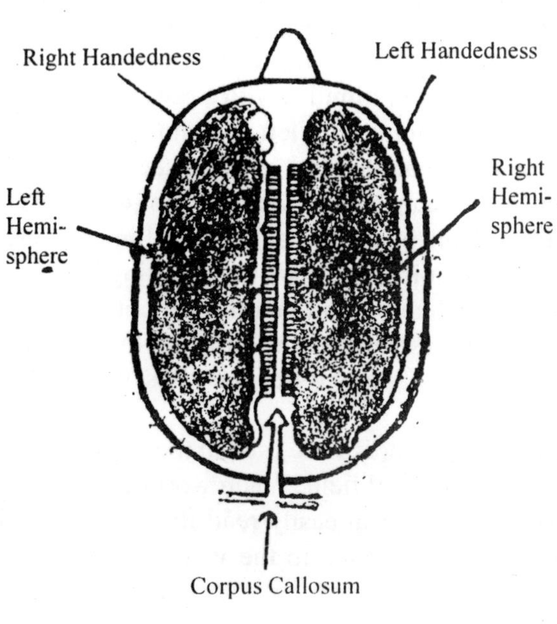

Fig. 5.27 : *Split-brain :* The brain is divided into two halves as left hemisphere and the right hemisphere. They are also called as Left-Brain. The Corpus callosum connects there are two brains and it is cut off as two. The connection between them is severed. This stage is called as Split-Brain. Thus there are *two minds or brains in one head.* It is ascertained that the cerebral hemispheres are not equivalent in mediating language functions. The left hemisphere controls this function and it is stated as the dominant hemisphere. But, however, when the left hemisphere is injured the right hemisphere does not keep quite. Some functional systems are connected more to one side of the brain than the other. Each hemisphere is specialized for specific way of functionining. In the normal individuals if the left and right hemispheres are cut off for experimental purposes the individuals are called *Split-brain individuals.* Roger Sperry and others (1974) found in their experimental studies remarkable results. They have observed *separate form consciousness* in the two spheres of the human brain.

The experiments on this problem involved a small group of human patients who were frequently having disabling epileptic seizures. They underwent a surgical procedure designed to relief from the seizures. It was observed in these patients that the epileptic activity that started in one hemisphere spread to the other sphere via the corpus callosum that connects the two hemispheres. The surgical cutting of the corpus callosum reduced the frequency and the severity of the seizures.

Further studies on this problem had shown that this remedy for the epileptic seizures had not followed by any changes in brain functions, when the behaviour was assessed by general behavioural tests such as the intelligence quotient tests. **Roger Sperry** for his exordinary research work on the human cerebral hemispheres was awarded the Nobel Prize in 1981. However, some investigators emphasize the integrative functioning of the cerebral hemispheres in human *cognitive activities*.

Corpus Callosum

The human corpus callosum is a huge bundle of more than one million axons. It provides the principle connections between the two cerebral hemispheres. This could be cut without producing of any observable changes in behaviour. **Karl Lashley** (1966) a prominent physiological psychologist made a remark on this that the only function of the corpus callosum is to keep the two cerebral hemispheres connected with the bundle of axons. It appears that he disagrees with the findings of Sperry. But, further researches on animals revealed deficits in behaviour as the result of disconnecting the two hemispheres, by cutting the corpus callosum.

Experiments were conducted further to verify the hypothesis on animals (**Sperry, Stamm and Miner**, 1956). In one experiment the cats were used and their corpus callosurn and the optiç chiasm were servered and each eye was connected to only to the respective hemisphere, that is, left eye to left hemisphere and right to the right hemisphere. Thus the cats were made to respond to the particular stimulus assigned to the heii-iisphere. A vertical stimulus with a reward was placed for the left eye and an inverted stimulus with a reward was placed for the right eve. The cats reacted to the vertical stimulus with left eye and to the inverted stimulus with right eye. Each eye was ignorant of what the other eye had seen (Fig. 5.28).

Similar experiments were carried out on human beings. **Joseph Bogen** (1988), a neurosurgeon suggested to split the brain of some epileptic patients for controlling the spread of the seizure from one hemisphere to the other hemisphere. This is by cutting the corpus callosum. He conducted the operation successfully and found that the patients were very much benefited. Several of them were studied at both stages, before and after the operations extensively. A series of psychological tests designed by Sperry were used to test the capability of each hemisphere.

In some of the experiments conducted by Sperry words were used in the testing. Words were presented to the left or to the right side of the visual field. When words are projected to the left hemisphere of the split-brain subjects they can easily read them and verbally communicate the words. But when the words are projected to the visual fields of right

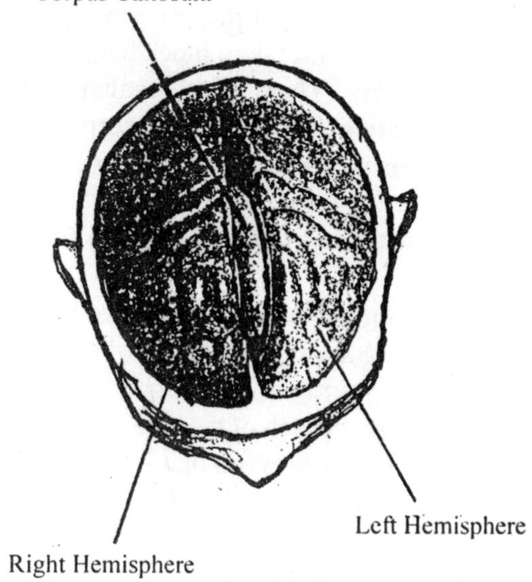

Fig. 5.28 : The Corpus Callosum is made up of huge bundle of more than one million axons. It is located in-between the two hemispheres and it is the main connecting link between the left and right hemispheres with the bundle of axons. One of the main functions of the corups callosum is to pass on the information or stimulus from one hemispheres to another through the axons (or impulses). The corpus callosum divides the brain into two halves as left or major brain and as right or minor brain. It is the great bridge between the two brains of hemispheres.

hemisphere there were no such linguistic capabilities observed. The experiments demonstrated that the right hemisphere has only a small amount of linguistic ability. For example, the right side can recognize only simple words. Further, in general the vocabulary and grammar capabilities of the right hemisphere are much more less compared to the left hemisphere.

The findings of Sperry confirmed the results obtained by him in the study of animals in the earlier periods. In humans the results became very much dramatic as his works later on showed the nature of the processes taking place in the left hemisphere. It is the left hemisphere that processes language and speech mechanism in most people.

There were also criticisms against the split-brain patients behaviour as the operations (cutting the corpus callosum) on the patients were not genuine and fresh. These patients were already suffering from epileptic seizures for many years before the surgery. The epileptic seizures already made some changes in the brains of these patients. Therefore the apparent changes observed after the cutting of the corpus callosum might have aroused from the epileptic brains.

The researchers of the split-brain have weakened this argument by producing further evidences on this problem by demonstrating the changes in non-epileptic person. **Damasio** and his collegues (1980) have reported a posterior callosal section in a sixteen year old boy. This boy had a turnour just below the posterior part of the corpus callosum. For removing the tumour it is required to cut a portion of the corpus callosum. After the surgery the boy showed some signs of failure that was seen in the epileptic split-brain patients. This means that the inter-hemispheric transfer has not taken place because of the cutting of the corpus callosum in the non-epileptic patient. Thus, what results Sperry and his collegues have found in the split-brain epileptic patients by cutting the corpus callosum, that there is failure in the inter-hemispheric transfer, is well established.

Different Processing of Information of the two Human Cerebral Hemispheres

The general functions of the two human hemispheres have been the centre of attention and research for the psychologists for a long time. That is, whether the functions of the left and right hemispheres are taking place jointly or interrelated or separately, particularly on the two modalities of *hearing and seeing*. In each of these two psychologists detected differences in human behavioural responses. They observed that in most people, language is processed by the left cerebral hemisphere.

Advantage of the Right-Ear in Processing Language

Recent researches indicate that in certains functions left and right ears vary to sonic exte~rit. The sensivity and the functions of both ears were tested by using ear-phones. It is by presenting different sounds to each ear at the same time. This testing process is called the dichotic listening technique. The subject is made to hear a particular speech sound in one ear and at the some time different word is presented to the other ear. Then the subject is asked to recall the sounds. The experimental results showed that the right handed persons called the verbal stimuli presented to the right ear more accurately than the stimuli presented simultaneously to the left ear. The result of this experiment is described as the right ear advantage for verbal information.

In further experimental studies **Doreen Kimure** (1973) had observed that auditory information causes stronger neural effects on the opposite ride side than on the same side. The auditory stimuli to the right ear cause stronger effect on the left auditory cortex than on the right auditory cortex. Therefore sounds presented to the right ear cause stronger control over language mechanisms in the left hemisphere and speech sound presented to the left results in less activating language processing in the left hemisphere (Fig. 5.29)

Vision and Speech Stimuli

Several studies were conducted to test the effects of visual stimuli on right and left visual fields. The results show that the verbal stimuli of words and letters when presented to the right visual field which go to the left hemisphere are better recognized than the same stimuli presented to the left visual field which go to right hemisphere. But, at the same time non-verbal stimuli like faces when presented to the left visual field are better recognized than the same stimuli presented to the right visual field. It was also observed that simple visual

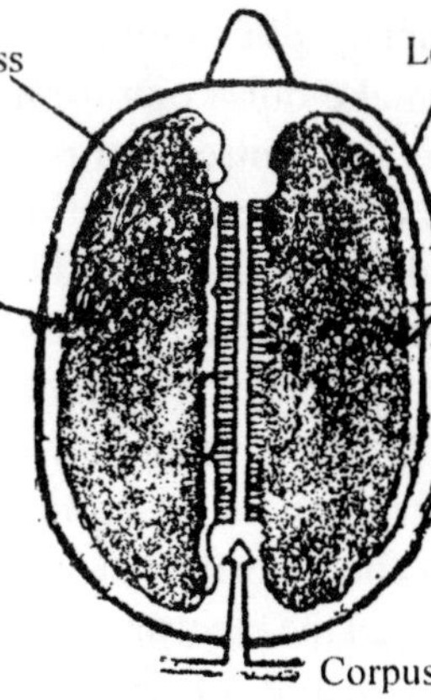

Fig. 5.29 : General functions of the two hemispheres. The functions in both the hemispheres vary considerably. Are the processing of information in both hemispheres taking place jointly or independently? In the case of *Hearing and Seeing* psychologists found difference in human behavioural responses. Specially language is processed by the left cerebral hemisphere. Verbal information is processed by the right ear and it has stronger control over language mechanism. Non-verbal stimuli like faces are better recognized by the left visual field than the right.

processing such as seeing light, hue takes place equally in both the hemispheres. But more complex materials in vision and in audition and certain verbal stimuli are processed better in the left hemisphere in most of the individuals, Thus the left hemisphere processes verbal and visual stimuli much better than the right hemisphere.

The Specific Functions of Major (left) and Minor (right) Hemispheres

The functions of the left and right cerebral hemispheres, we have discussed above with special reference to auditory and visual functions. In most of the functions, the left hemisphere seems to be functioning more actively than the right hemisphere. This has been ascertained strongly by the functions of the split-brain. The brain is split into two hemispheres by cutting the corpus callosum which connects both the hemispheres. Some researchers call this state as two brains in one head. These two brains are also stated as major and minor hemispheres.

The major (left) hemisphere controls our ability in the expression of language. This hemisphere performs many complicated sequential and analytic activities and also performs higher mental activities like mathematical computation. The minor, right hemisphere can deal with very simple verbal tasks. This can respond to simple words and letters and by selecting and responding to objects, pencil, paper, nuts and bolts. It cannot comprehend more abstract linguistic components. The right or the minor hemisphere can add simple two digit numbers, and cannot do anything little beyond this in calculations.

The minor (right) hemisphere even though named as minor hemisphere, it has certain special abilities to perform of its own. The minor hemisphere is having a highly developed spatial and pattern abilities. The left or major hemisphere is capable of making huge constructions, draw difficult and complex designs much better than the right (minor) hemisphere. It was observed that when a split-brain person is asked to use his right hand to assemble the blocks of a design to make out a picture he commits many mistakes. It was observed that the right hemisphere is the centre of musical ability.

Studies were made to locate, the varied specializations of the left and right hemisphere. When some information is presented for a brief period to the right hemisphere for which a verbal or non-verbal response is needed, it is observed that the non-verbal response takes place more quickly than the verbal response. Generally the verbal response is required that the information first to be sent to the left hemisphere through corpus callosum and this takes few seconds (Gazzaniga, 1992), whereas in the case of nonverbal information this does not take place, because the right hemisphere itself processes the information and sends the response. According to **Ornstein** (1972) the electrical impulses as revealed, in EEG indicate that a verbal task activity increases in the left *hemisphere.* But in the case of spatical information the processing activity increases in the right hemisphere (Fig.5.30).

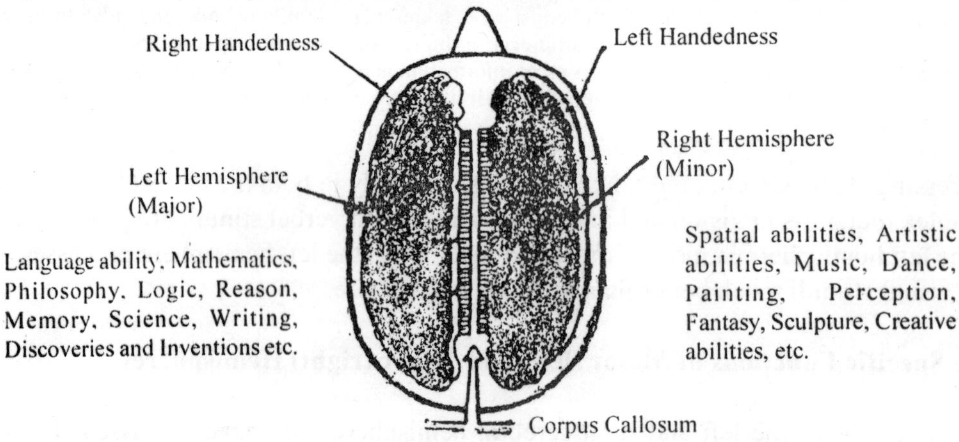

Fig. 5.30 : Specific functions of the Hemispheres : The major (left) and minor (right) function more actively in certain areas. Experiments on split-brain ascertain that the major (left) functions more effectively than the minor (right). It is tested by cutting the Corpus Callosum. Since great differences were not found between these two hemispheres, some called the states as two brains as major and minor hemispheres. The major part has control over language, deals with high mental processes like mathematics. The minor part deals with simple calculations, simple verbal tasks and has highly developed spatial pattern abilities. The left can make huge constructions, draw complex designs, processes the linguistic memories of the past experiences. The right part cannot comprehend more abstract materials. But the right part has certain special abilities. It is the centre of musical abilities. The human brain anatomically and physiologically is asymmetrical. The cerebral cortex structure is larger in the left hemispheres.

Psychologists who studied the split-brain problem are of the view that the linguistic memories of the past are found in the *left hemisphere. The* right hemisphere also is aware of the linguistic information but it receives the information from the left hemisphere through corpus callosum. It is also observed that experiences that take place in the right hemisphere are fastly transmitted to the left hemisphere and processed. When the connection between the two hemispheres is broken by cutting the corpus callosum the right hemisphere becomes silent and inactive and its experiences do not reach the left hemisphere and not processed.

We have discussed that each hemisphere is specializing in certain functions. It is to be known that the specializations of the two hemispheres seem to develop in humans along with language development. But if the left hemisphere in a young child is impaired the right hemisphere takes over the language functions without any difficulty, since it is the organism's developmental stage. But in the case of an adult person any impairment in the left hemisphere it results in producing language disability. Thus the special functions of the two hemispheres vary on the basis of the nature of' the furictions such as motor, sensory and cognitive activities.

Thus, from the proceeding passages we could see that the functions in the left and right hemispheres and in the specializations differ conspicuously. This is because, the human brain is anatomically and physiologically asymmetrical. When we compare the left and right sides of our body critically, we may note and find that they are not perfectly half. The human brain also is anatomically asymmetrical. **Geschwind** and **Levrtsky** (1968) have found in sixty five percent of human beings the brain structures ivere asymmetry. In human brain in the region of the cerebral cortex a structure known as plannum temporale as found larger in the left hemisphere than in the right. In eleven percent of the people the right side is larger. This plannum temporale region contains several auditory association centres and speech area. Any lesions in this area usually produce aphasia. Therefore this region contains important components of language networks. This area may vary in size which reflects the specializations of our cerebral, hemisphere for language.

2. Mesencephalon (Midbrain)

The mesencephalon was formerly called as the midbrain. This part of the brain is developed from the middle part of the neural tube of the developing embryo. It consists of two main sub-divisions. They are the tectum and tegmentum.

Tectum

The tectum is the roof or the dorsal position of the midbrain. The tectum is lying above the cerebral aqueduct which is a part of the ventricular system. The tectum consists of primarily the sets of symmetrical nuclei. They are the two superior and the inferior nuclei (or collicull).

The superior nuclei is the nucleus of the tectur-i of the mesencephalon (midbrain), is also called as the lower hills, which are stated as the posterior pair. The inferior collicull receive auditory projections from the ears. This part of the tectum is involved in the whole body orientation to auditory stimuli. The inferior nuclei are involved in mediating many auditory related behaviours (Fig. 5.30)

The Inferior Colliculi is the nucleus of the tectum of the mesencephalon (midbrain), is also called as the lower hills, which are stated as the posterior pair. The inferior collicull receive auditory projections from the ears. This part of the tectum is involved in the whole body orientation to auditory stimuli. The inferior nuclei are involved in mediating many auditory related behaviours.

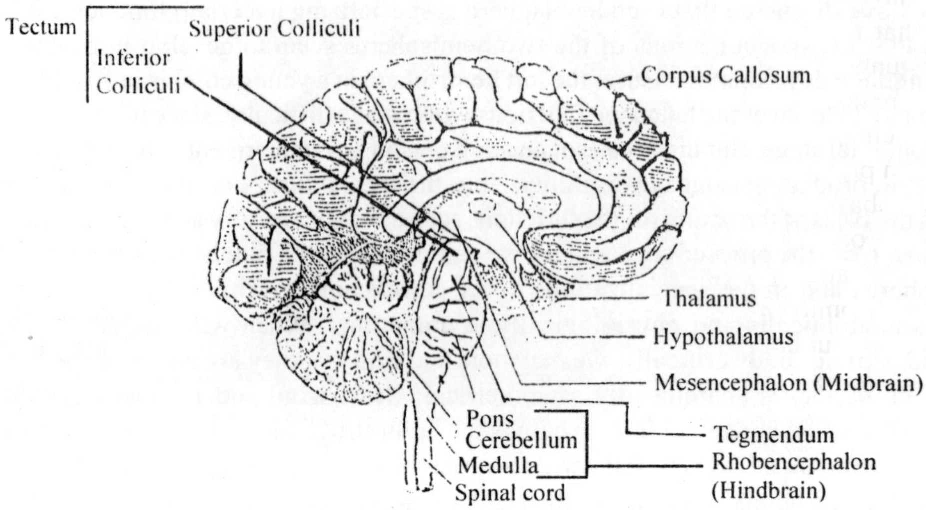

Fig. 5.31 : The *Mesencephalon* is the midbrain located just above the cerebellum. This is developed from the middle part of the neural tube of the developing embryo. The *tectum* is the roof of this part. The tectum consits of sets of symmetrical nuclei called *Superior and Inferior colliculi.* The *Tegmentum* falls at the floor of the mesencephaion which conists of some cranial nerves. The mesencephalon functions also as the sensory area. Its ventral part is related to motor behaviour.

Tegmentum

The tegmentum is also stated as the floor of the mesencephalon (midbrain). This II(-s below the cerebral aqueduct. The tegmentum consists of nuclei for some of the cranial nerves. and a number of motor nuclei intermingle with them. This consists of sensory and motor tracts. Thus it is observed that the mesencephalon (midbrain), following the plan of the spinal cord, the dorsal portion of the mesencephalon becomes the sensory area. The ventral part is related to motor behaviour.

In the study of human brain (cephalon) and behaviour the earlier main divisions were named as the forebrain, the midbrain and the hindbrain. All the structures and parts were included in these three divisions. But in the recent divisions the cephalon (brain) is divided into several divisions and structures and they are given new names and the functions are provided with many new information.

As we understand about the mesencephalon, the functions are centred visual, auditory and motor behaviour of the organism and the main parts are tecturn and tegmentum.

The mesencephalon is a complex centre and it aids in regulating the body for auditory stimulations. In the mesencephalon there is a system called the Reticular Activating System (RAS) which goes through the parts of rhombencephalon (hindbrain) and prosencephalon (fore brain) and performs certain functions. It also regulates the sympathetic nervous system.

Regarding the central connections of the cranial nerves, the cerebrospinal canal in the mesencephalon is very narrow, but surrounding walls are thick and the roof is heavy. In this area of mesencephalon (midbrain) there are reflex centres for the eye muscles.

3. The Rhombencephalon (Hind brain)

The human brain is a very complex system and any one portion of the brain cannot be understood without going into the details of the other portions. The hindbrain or the rhombencephalon is one of the parts of the brain that is very important in the study of human behaviour-, The recent anatomical developments in the study of brain yield several new findings and terms. The central nervous system appears like a tube. It is a neural tube and the walls of this neural tube in the evolutionary process develop three seperate swollen parts at the end of the head. The anatomists and the embryologists have given cumbersome names to the developing regions. These three swellen portions develop into forebrain or prosencephalon, the midbrain or mesencephalon and the hindbrain or the rhombencephalon. We have discussed the prosencephalon and the mesencephalon in the preceding pages.

The rhombencephalon (hindbrain) is located behind the midbrain and also in the region of the brain that consists of the cerebellum, medulla oblongata, the pons and the forth ventricle. The hindbrain is the earliest to evolve into the brain. Anatomically the hindbrain is the rear division of the brain. In the well developed or matured vertebrate the hindbrain contains the cerebellum, pons and medulla oblongata.

The rhombencephalon or the hindbrain is involved in maintaining the balance of the organism. It has two divisions such as (a) the metencephalon and (b) myelencephalon. The metencephalon develops into the cerebellum and the myelencephalon develops into pons and medulla oblongata. The brain even in its developed adult form is a fluid.-filled tube, but takes a complicated shape with folds and ridges. All these five main divisions, telencephalon, diencephalon, mesencephalon, metencephalon and the myelencephalon are also subdivided (Fig. 5.32)

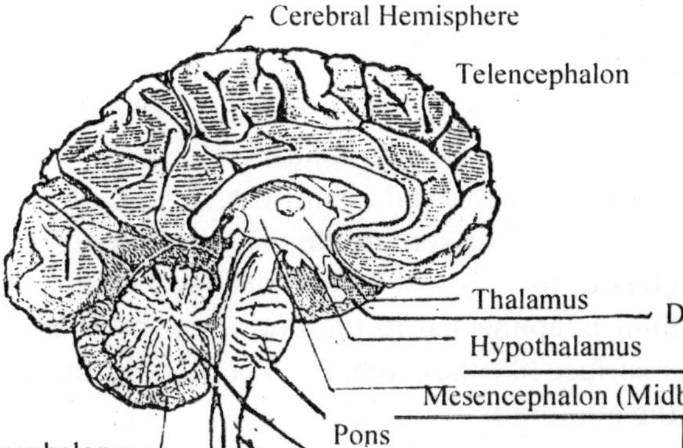

Fig. 5.32 : *The Rhombencephalon* is the hindbrain. In the neural tube of the developing embryo there are three swellings. From the last one the rhombencephalon (hindbrain) is developed. It is located below the midbrain and this region of the brain consists of the cerebellum, pons, medulla oblongata and the fourth ventricle. This is rear division of teh brain. The Rombencephalon is involved in maintaining the balance of the organism. It has two sub-divisions such as *Metencephalon* and *Myelencephalon*. The *Metencephalon* is developed into the *cerebellum* and the Myelencephalon developed into the *Pons* and *Medulla oblongata*. All the five divisions, the Prosencephalon, Telencephalon, Diencephalon, Mesencephalon, Rhobencephalon, Metencephalon and the Myelencephalon also.have the sub-divisions.

(a) Metencephalon

The metencephalon (across-brain) is the anterior portion of the rhombencephalon, which is composed of the cerebellum. The myelencephalon (spinal brain) is the posterior portion of the rhombencephalon which is composed of the pons (medulla oblongata and the forth ventrical.

The medulla oblongata is a narrow structure and it has centres to regulate the primary physiological mechanism of the body. It regulates the breathing, heart beat, temperature of the body and also control several reflexes and maintains the upright posture of the body. It is the point at which the nerves from the higher parts of the brain are crossing to the left and right. The Pons are just above the medulla. The Pons consists of nerve fibres that connect the brain cortex to the top part of the hindbrain and to medulla. The Pons control the sensory and motor functions of mouth and face. It is an integrating system for ascending and descending impulses.

The hindbrain is organized in the same way as the midbrain. The sensory nuclei of the vestibular system overlie the fourth ventrical and beneath this ventricle there are more nuclei of the cranial nerves. The vestibular sensory system is located in the inner ear that controls the balance of the organism. Overlying these cranial nerves is the cerebellum.

The hindbrain consists of part of the cerebral hemispheres, which contains two major sets of centres, the interior basal ganglia and the surface cerebral cortex. The basal ganglia, or the corpus striturn (striped bodies) are the centres that develop in the inner walls of the cerebral hemispheres. They look like striped, as the white matter, of alternate with gray matter in the centres. The other area of the cerebral hemispheres develop gray matter on the surface of the cerebral cortex. It is because of this, the surface of the cerebral cortex is gray. Generally it is referred as cortex. The cerebellum also has -a- cortex.

The surface cerebral cortex is connected to the lower centres by the tracts of the interior white matter. The two cerebral hemispheres are connected by a band of white matter which is called the corpus callosum. The thalami (three in number, some authors state five) develop in the walls of the diencephalon on the sides of the third ventricle. By the side of it there is hypothalamus with scattered nuclei. On the roof of the midbrain there are colliculi, which are four visual and auditory reflex centres, below this are the tracts connecting the brain and the spinal cord. The cerebellum is connected to the brainstem by a group of tract. The cerebellum has the gray matter surface like the cerebrum. The medulla maintains the balance between the spinal cord and the hindbrain. (Fig. 5.32)

The Cerebellum

The cerebellum is a major structure of the hindbrain which is specialized in motor co-ordination. It is a large structure in the brainstem that is involved mainly in movement. It is the central regulation of the movement in which the cerebellum is involved. It lies behind the brainstem. It is made up of circular portions. The outer surface of it is the cortex. it is a gray matter. The cerebellum is responsible for a number of co-ordination activities like the motor co-ordination and sequences in movements. Any injury in this part would destroy the sequences of movements.

The cerebellum is attached to the pons. The cerebellum has a folded surface like the cerebral cortex. But in the cerebellum the folds are very close together. The arrangement of the folds are very close together. The arrangement of the cells within the folded sheet of the cortex is very much simpler than those in the cerebral cortex. In the middle there is a single roVv of very large multipolar neurons called *purkinje cells*. This is after the name of the anatomist ,vho first discovered the fanshaped dendrite patterns. Along the surface of the dendrites there are a number of dendrite spines. Beneath the purkinje cells there is a huge collection of very small granule cells. The axons of these cells form parallel fibres on the surface. The cerebellum is involved in motor co-ordination and in some kinds of learning.

The functions of the cerebellum vary from one part of the structure to another part. This depends on the connections with the rest of the other nervous system. The parts that receive most of their impulses from the vestibular svstem and function for maintaining the bodv's equilibrium. Oil the other hand, the parts that receive impulses mainly from the body senses are the functions with the postural reflexes, and functionally co-ordinate with related muscles. The major part of the cerebellum receives impulsės from the neocortex and controls the skilled movements. The surface of the cerebellum is covered by narrow folds, beneath of it is thick cortex of gray matter which covers a large mass of white nuclei matter. There are many nuclei within this white matter.

The cerebellum is closely attached to the brainstern by three major nerve fibre pathways in which all afferent (sensory) nerve fibres pass to the cortex. All the efferent (motor) nerve fibres originate in the underlying cerebellar nuclei and then they pass on to the other brain structures. Any damage or injury to the cerebellum may cause impairment in the skilled motor activity. Injury or disease in cerebellum also may affect smooth movements, affect the ability to perform rapidly alternating movements and any planned movements may overshoot or undershoot their ranges. Sometimes the muscle tones may turn abnormal which may cause the movements difficult to start with.

The cerebellum is considered as an important part of the motor system. It is not directly sending the projections to the motor neurons. It only projects to a number of nuclei that join the lateral or ventromedial system of the spinalcord directly or indirectly.

The cerebellum receives three gets of inputs and three sets of outputs from the cerebellar pathways. All these inputs and outputs interconnect through the neurons that make marks of cross liner in the cortex of the cerebellum in pathways. Because of this, all inputs to the cerebellum affect all the outputs in the cerebellum. This means one input and one output form a negative feedback loop. This feedback loop interconnect the new part of the cerebrum with the new part of the cerebellum, the neocerebellum. A second input from the stretch receptors, from the extensor muscles stimulates the older part of the cerebellum the paleo cerebellum. This part of the cerebellum regulates the extensor tone by its output to the brainstem, reticular formation (BSRF) and other nuclei of the brainstem. The oldest part of the cerebellum, the *archicerebellum* receives input from the inner ear. Its output regulates the extensor tone. The output of the archicerebellum feeds back the vestibular nuclei. It regulates the excitary output of the vestibular nuclei to the extensor motoneurons. The archicerebellum helps to regulate the adjustments of the other two lobes of the cerebellum.

(1) The Archicerebellum : It anatomically consists of two small-paired structures that hang over the brainstem, beneath the anterior cerebellum. The other is a medically placed structure that is placed down in the nodule which is centrally located cerebellar structure that receives input from the vestibular senses.

(2) The Paleo-cerebellum : This includes the anterior and the posterior lobes of the cerebellum. This paleo-cerebellum was developed when the sea animals left the sea and started living on land. On integration of the activities of the other lobes of the cerebellum the paleo-cerebellum acts on the brainstern and on the reticula- formation inhibitory areas. The paleocerebellum responds to the muscle stratch and other inputs by regulating stretch reflex.

(3) The Neocerebellum : It is supposed to be the most recently developed part of the cerebellum in the evalutionary process. It is considered to be equal to the development of the motor areas of the cerebral cortex. The great cerebral hemispheres cover most of the brain. When the paired cerebellar hemispheres fell on the older areas of the cerebellum the neocerebellum was formed. As and when the motor and premotor cortex started integrating the complex movements, the activities of the paleocerebellum and the archicerebellum have become very essential. The neocerebellum has developed to provide this integration and to have a two way connections between the cortical motor areas and the neocerebellum.

We have discussed briefly the functions of the three major divisions of the cerebellum. With this understanding it is possible to predict the consequences of damage to various parts of the cerebellum. The archicerebellum regulates the postural extensor tone reacting to the vestibular inputs that are controlled by head position and movement. Any damage to the nudole (cerebellar structure) causes disturbances -in the balance. The paleocerebellum when damaged, its functions, are impaired. In this damage the cerebellar lobes (anterior or posterior) are involved and there are impairments in the postual adjustments and also there is lack of co-ordination with the extensor muscle tone. In the case of neo-cerecellar damage to the hemisphere, *intention, tremour,* trembling of the limb is caused. In grasping activity there may be under or over-reaching of the arm to the object. Because of te tremour the person is not able to coordinate the position of the body in space with motion.

The cerebellum is a large paired suprasegmental structure of the hindbrain. The main function is to regulate the postural adjustments of the extrapyramidal system and it controls certain interactions between extrapyremidal and the pyramidal systems. The cerebellum plays an integral role in motor organization. The cerebellur contribution to motor organization is all purprioceptive and cortical inputs to the cerebellum and all cerebellar outputs to cortical and subcortical centres which are involved in movements (Fig. 5.32).

(b) The Myelencephalon

The third or the bottom bulging of the neural tube of the developing embryonic structure evolved the rhombencephalon, the hindbrain. The rhombencephalon during the course of development has evolved two divisions which are called as the metencephalon, which is the anterior portion of the rhombencephalon. From the posterior portion of the rhombencephalon another division has evolved which is called the myelencephalon.

The myelencephalon is the spinal brain. It is viewed by some investigators that it is still hollow and the hollows in each portion are referred to as ventricles (bladders). The ventricles are filled with the fluid called cerebrospinal fluid (CSF). These are numbered from I to IV. The myelencephalon includes the pons, The medulla oblongata and the fourth ventricle.

The Pons

The pons consists of mostly nerve fibres, passing up and down between the medulla oblongata and the midbrain. The pons is located just above the medulla. These two structures are very closely located and also function together in some cases. The pons and medulla constitute the lower part of the brainstem. The pons is a bulge in the anterior surface of the brainstem.

The pons form a bridge of connecting tissues composed of mostly motor nerves going to the areas of the cerebellum and the spinal cord. Immediately below the cerebellum the contributing nerve fibres to the brainstem is the pons. The pons and the medulla receive sensory ner-ve fibres from the ear, skin, mouth and muscles of the head. They send motor nerve fibres to face, throat, neck and to some visceral muscles and glands.

Information from the ear first enters the brain only through the pons. The pons also contains several groups of neurons which send their axons down to the spinal cord for regulating the bodily movements.

The pons supply nerves to the taste buds and to the salivery glands. The cranial nerve seven (VII) also deals with the muscles of the face. The structural end in medulla oblongata is very much enlarged by the mass fibres that connect the two sides of the cerebellum by some nuclei in this area. Because of this structure the fibres appear like a bridge between the two sides of the brain at this area, which is called the pons.

The cranial nerve six (VI) is located in the pons. It contains different nerve fibres from the stretch receptors of some muscles which can serve the eve muscles. The cranial nerve five (V) contains sensory nerve fibres. It is concerned with the skin senses of the face and head. The cranial connection of this nerve goes from the medulla oblongata through the pons into the midbrain (mesencephalon) (Fig. 5.32).

The Medulla Oblongata

The medulla oblongata is more like the spinal cord and it continues with it. The narrow part of the spinal canal opens at the spinal cord into the medulla oblongata where it becomes a broad chamber. This is generally called the fourth ventricle of the brain. The medulla oblongata is immediately above the spinal cord. It contains the controlling centre of the heart and lungs.

The medulla oblongata forms the bottom of the brainstem and marks the transition from the brain to spinal cord. There are twelve pairs of the cranial nerve nuclei-cell bodies of neurons controlling the neck and the tongue muscles, The reticular formation system goes through the pons and ends in the medulla oblongata. The medulla makes contribution to the regulation of breathing and heart rate. Any damage of the tissues (lesions) in medulla oblongata may become fatal. Axons from other parts of the brain that affect the spinal cord also pass through the medulla oblongata (Fig. 5.31).

The medulla oblongata is the hindmost of the brainstem, which contains major ascending and descending tracts between the spinal cord and the higher centres and it also maintains the blood pressure, digestion and other bodily functions.

6

ENDOCRINE SYSTEM

Introduction

Psychology is the study of human behaviour. It attempts to explain and predict behaviour of the individual. The human body is a single physiological structure as it appears with its parts, the head, the hands and legs and the trunk. Psychology makes use of the physiological knowledge of anatomy which deals with the structural relationships of the parts of the body. Psychologists of the present generation are not satisfied with simply observing the various types of activities of the organism, but also interested in knowing how the internal organs and systems of the body function to predict the behaviour. The psychologist at present uses many of the methods of the anatomist, biologist and the physiologist and the psychological techniques. The internal environment of the human body plays important role in the revealed human behaviour. Therefore, it becomes essential in the study of human behaviour to know something about the function of the internal environment of the body.

Biological scientists state that life begins in a single cell. A cell can be an independent living organism and also can be a specialized unit of a complex many celled organism. All living creatures consist of *Protoplasm*, a primitive organic material made up of cells, or a unit of life. The simplest form of life consists of only a single cell. Bat more complex forms of life like the human being are constituted of specialized cells. Man is a many celled and complex organism. The specialized complex cells are organized and integrated into tissues, organs and systems. The specialized cells carry on one single function efficiently that is essential to life. But the cells of a complex organism become dependent on each other as they are also interdependent for various functions. The specialized cells in adapting to the changing environment get reduced in their abilities. By this means the internal environment is kept more constant than the external environment. All the processes of the body and also the behaviour are modified to this effect.

A series of chemical functions are going on in the life of all cells. This chemical function or reaction is called *metabolism*. Metabolism generally refers to the chemical reactions of life which change the foodstuff into energy and waste. The reactions of metabolism require energy and to release energy in a balanced way which is maintained automatically in the body and by the external behaviour of the organism. These chemical reactions that go inside the body are *called cell metabolism.* The reactions that occur around the cells in the internal environment are called *intermediary metabolism.* This mainly refers to the events that take place between the events in the cells and outside events. Thus there is a function of energy exchange between the specialized cells and the outside world. All these processes of function form the basis for all kinds of behaviour. The cells and the tissues of the body are specialized and varied in functions. There is a need for integrating their varied functions for maintaining a constant internal environment. This is taking place in the body by some *automatic mechanisms.*

The human body is constituted of several complex systems of which the endocrine system is one of them. These systems are responding to changes in the environment, not only by initiating and inhibiting muscular activities, but also by secreting various kinds of chemical substances called hormones. These hormones are very essential factors in many metabolic processes. Though the hormonal secretions are very essential to the bodily internal functions the psychophysiologists have for sometimes neglected the study of them for reason that the activity of the specific secretary organ has become very difficult to assess and measure. They should not have neglected this. Becuase the hormones influence and control many metabolic functions that determine the response of the organism to the environmental events.

The sex hormones determine the sexual functions of the organism and represent an important aspect of sexual motivation. There are also many other hormonal influences that affect the motivational mechanism and determine directly or indirectly the organism's response to its environment. For example, the pituitary hormones regulate water loss and carbohydrate metabolism. Many other hormonal mechanisms also affect the organism's response to specific environmental changes.

Nature and Characteristics of Endocrine System

The human body is composed of complicated chemical components. The body is made up of cells and tissues. They are of varied types and also are specialized. There are differentiations in chemical activity in various parts of the body. Each one is specialized in one activity. One is specialized in making use of food, another is specialized in the contraction of some of the tissues and make the organism to move. The third is specialized in supplying food and oxygen through blood. Therefore there is a need for integrating their varied activities for constantly maintaining the internal environment.

The body is integrated through these three specialized systems of cells and tissues. *The first* is the system of organisation and specialization of cells into tissues and organs and systems. *The second* is the system that integrates the tissues, organs and activities by the nervous system. *The third* is the system that integrates the activities of the endocrine glands. The chemical secretions of the endocrine glands play important role in the nutrition and growth of the parts of the body. As stated, the glands are also a group of cells specialized in manufacturing and supplying chemical substances to various parts of the body.

The chemical substances secreted by the endocrine glands are called hormones. The hormones are carried out by the blood-streams to all parts of the body. Some of the glands are under the control of the nervous system.

The glands are mainly responsible for the maintenance of the internal environment of the body. They protect the individual from extreme heat and cold and are important for the behaviour of the individual directly or indirectly. The glands are mainly responsible for the individual differences in growth of the body, muscle and nerve excitability, emotional adjustment, metabolism and sexual development and the behaviour of the individual. Every cell of the body has a function of secretion, secreting some chemical juice. The glands are internal organs made up of tissues. They secreate substances that are required for the body. But the glands are the specialized organs in secreting the chemical substances.

There are glands which are secreting the substances inside and outside. They are called exocrine glands. The exocrine glands are duct glands. They secrete the hormones into a duct (tube) which -carries the hormones to digestive tract, just like the salivary glands and the glands in the intestines. Some of the hormones are sent to the surface of the body. For example, the oil producing glands, secrete the hormones of the skin. Since these glands do not secrete the hormones into the blood they are called exocrine glands.

But the endocrine glands are secreting the hormones directly into the blood vessels. They do not have ducts. It is because of this they are called ductless glands. The hormones secreted by the endocrine glands are carried out through out the body by the circulatory system. Any over-secretion or under-secretion of the hormones may affect the functions of the cells, organs and tissues.

The effect of the endocrine glands are not local but general. They bath the whole nervous system including the brain and the entire body. The hormones of the endocrine glands also influence all those cells and tissues that are determining emotions, actions and thought processes. Human beings differ in their attitudes and emotions, because of the different chemical juices in them. A man may be angry in the morning and quarrelling with his wife and when he goes to his office and returns home in the evening his anger is subsided and he is normal in his moods. The effect of different levels of hormone in the blood may cause different moods and temperament in man. Every gland in its function is having influence over other glands through blood streams. Therefore it is not one gland that is working, but the whole endocrine system. **Wolf** (1939) states that there is great variation in the range and functions of the endocrine glands and the range of size also vary greatly.

The endocrine glands are indentified in the infants and they are mainly responsible in the process of growth and development before and after birth of the child. The growth of the glands depends upon the chemical effect and the size of each gland. The growth of the glands also depends upon the activity of the glands at a given time. Therefore all the glands do not grow uniformly at all levels. As result of this there is shift in the endocrine balance and there are different levels of endocrine balance from infancy, through childhood, adolescence, and old age. This variation in the growth of endocrine balance causes the individual differences. Hormones from one endocrine gland may excite or inhibit the production of hormone in another endocrine gland. Thereby interaction between and among the endocrine glands is taking place. It is also becoming very difficult to study the functions of endocrine glands since each endocrine may produce several hormones (Leukel, 1968). These glands are having in7fluence in the behaviour of human beings and also in the development of personality.

The Activating Power behind the Endocrine System

The endocrine system, apart from functioning under the influence of the autonomic nervous system, is also under the strong influence of the hypothalamus. Hypothalamus is an area of the brain in the walls and floor of the third venricle that controls the reactions of the hypophysis and the autonomic nervous system and it is also sensitive to internal changes in the body.

The hypothalamus weighs less than an ounce and it is to the size of a pea. But is having control over the entire endocrine system. When the blood sugar level is reduced the hypothalamus comes into action and it creates the desire in man to take sugar to meet the deficiency. If the weather condition suddenly becomes warm or hot the hypothalamus reduces the body's internal thermostatic position. Thus the hypothalamus is having control over the endocrine system and regulates the system.

The hypothalamus thus governs the endocrine system through the pituitary glands which serve as an intermediatary. Some of the most complicated bodily functions are the breathing, digestion, circulation, growth, reproduction and repair. These are generally involuntery. These functions are mainly under the control of two different systems which are functioning cooperatively. These are the nervous system and the endocrine system. Here the nervous system is mainly the autonomic nervous system.

The autonomic nervous system sends short information signals very fastly in a fraction of a second. This is by means of electrical impulses for producing rapid responses. These signals pass through nerve cells as fastly as one thousand feet per second. Similary the endocrines system sends all the signals or messages relatively slowly by means of hormones to produce more long lasting effects.

The hormonal signals or messages pass through the blood streams which may take many hours. It appears that certain cells have receptors for one or more hormones. Generally receptors are the energy receiving cell parts. Here a receptor is like a censor and only the right hormones have the entry to a particular kind of cell. When the hormone joins a cell the hormone alters such cell functions in the production of protein and in storing and releasing of the glucose (the carbohydrate, fuel that are used for many bodily activities) and the production of other hormones (Fig. 6.1).

THE WORKS OF THE ENDOCRINE SYSTEM

The endocrine system is a group of ductless glands located in various parts of the body. All the ductless glands secrete hormones of various types directly into the blood streams. The important work of these glands is to keep things set in motion. The motion set in by these glands is slow but make the body to function efficiently. The main work of them is to

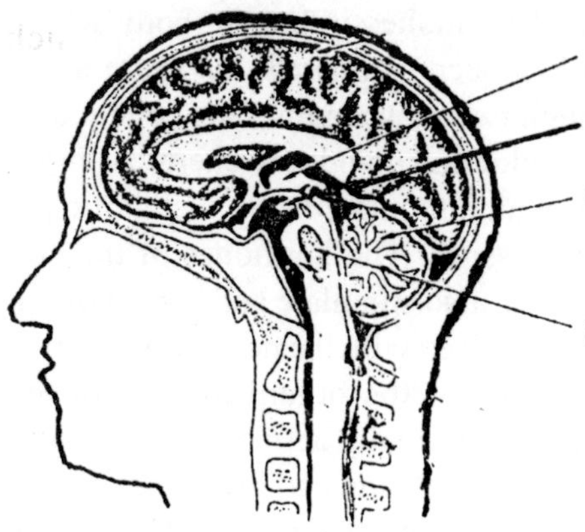

Fig. 6.1 : The Power behind the Endocrime System : The Hypothalamus

control the internal environment which includes the cells, organs and also the whole body. The endocrine glands are fully under the command of the hypothalamus, which is supposed to be the centre from where commands are made. The hypothalamus is operating and controlling this system through the pituitary gland which is located under the hypothalamus. The pituitary on the order of the hypothalamus directs the works of all other glands.

The hypothalamus is very watchful of the functions of the hormones of the endocrine glands. For example, if the hypothalamus finds out any deficiency of the thyroid hormone in the blood, it releases a hormone to the pituitary and orders the pituitary to send the hormone which should stimulate the thyroid gland through the blood streams. When the thyroid gland received, this chemical signal or message or substance the gland releases the thyroid hormone, the thyroxine. By this means the metabolic functions are expedited. Since a deficiency of thyroxine in the blood may cause severe fatigue, it is avoided by this function.

HORMONES

The concept, hormone means to set in motion. The hormones activate all the biological functions of the body. Hormones are the substances manufactured by the ductless glands and they are secreted directly into the blood streams and they are carried by the circulation system to effect metabolic reactions in the selective or respective target tissues all over the body. The effect of the hormone is selective. They act only on certain metabolic reactions in selected tissues. But there are some hormones which are enzymes serve as inhibitors. They block certain metabolic reactions and allow only competing metabolic reactions to proceed on. The other hormones act on the DNA (deoxyribonucleic acid) or RNA (Ribonucleic acid) molcules of the cells for altering the way of their cell function. By these means hormones control growth, general metabolic rate, sexual and pregnancy processes and also the metabolism of many chemicals which are important for normal metabolism and behaviour.

There are quite large number of hormones in human body which are controlling the internal environment. Roughly it has been estimated that there are about 200 hormones produced in the whole body of which twenty hormones are produced in the brain alone.

The hormones secreted by the endocrine glands are chemical substances. These glands function in a reciprocating system which regulate many metabolic processes and also the development and function of many tissues. The hormones of the endoorine glands apart from effecting the metabolic processes, also stimulate and inhibit other glands. The activity of the endocrine system is closely connected with the nervous system in various ways in the process of development and functions of the organism. Certain endocrine glands are functioning independently, but they are normally activated by the neural mechanism.

It is very difficult to state how many glands and hormones are functioning in human organism. Hormones may be defined as a kind of chemical substances that are manufactured and secreted by certain specific organs. They are discharged in the general circulation system, that is in the blood streams directly and transporated to distant sites for action.

Hormones exert their effects by inhibiting and exciting the enzyme systems. An enzyme is an organic catalyst that increases the rate of specific chemical reactions in the cell. An enzyme system is an enzyme made up of an apoenzyme and a'coenzyme. This means that hormones cannot initiate new processes, but only mediate the existing functions.

The production and secretion of certain specific hormones are affected by other hormones and by the neural influences. It is also affected by the general nutritional and metabolic conditions in the organism. Any change or irregularity in secretion activity of the endocrine glands would affect the activity of the other glands of the organism. This means that there is a process of interaction among the functions of the glands. The interaction activity of the glands is more complicated and not easily comprehended. One important point to be emphasized is that the endocrine system tends to function in a homeostatic manner.

The Endocrine Glands

It is observed that some of the endocrine glands are having single structure and some of them are having paired parts of a single structure. The glands vary from single celled glands like the mucus gland of the nasal passages to a very complex glands, the pituitary glands. Therefore, each gland may be referred to a single unit, whether one, or a pair, or of many combined structures.

The study of endocrine glands in various fields give importance only to some glands of the system. They make out the list of the endocrine glands on the basis of their requirements. The importance of the major glands was felt by those who make study and for what purpose. The anterior pituitary gland partly controls the thyroid, the adrenal cortex and the gonad glands. The adrenal medulla, pancreas, posterior pituitary and the pineal glands are under the control of the nervous and the endocrine systems. It is because of this they react to the external and internal events.

There may be also glandular and hormonal abnormalities, which may help us to understand the role of the glands in normal functions of the body and the behaviour of the organism.

The number of glands that are functioning in hi-man body approximately stated as 200, of which most of them are biologically, physiologically, anatomically and medically considered as important. It has been listed ten major endocrine glands psychologically important as, they are mainly involved in human behaviour and also determine human reactions externally and internally.

Therefore, we shall include only those glands which are mainly involved in human behaviour in this study. These are (1) the pineal glands, (2) the Posterior pituitary glands, (3) the anterior pituitary glands, (4) the thyroid glands, (5) the parathyroid glands, (6) the thymus glands, (7) the adrenal cortex, (8) the adrenal medulla, (9) the pancreas, and (10) the gonads (Fig. 6.2).

All these glands are not single. They have all subdivisions in every gland. Therefore, they are always stated as glands. All the glands are internal organs and they are rrade of a variety of tissues. These tissues aid the glands in secretion of the chemical substances called hormones, that are essential for the bodly functions.

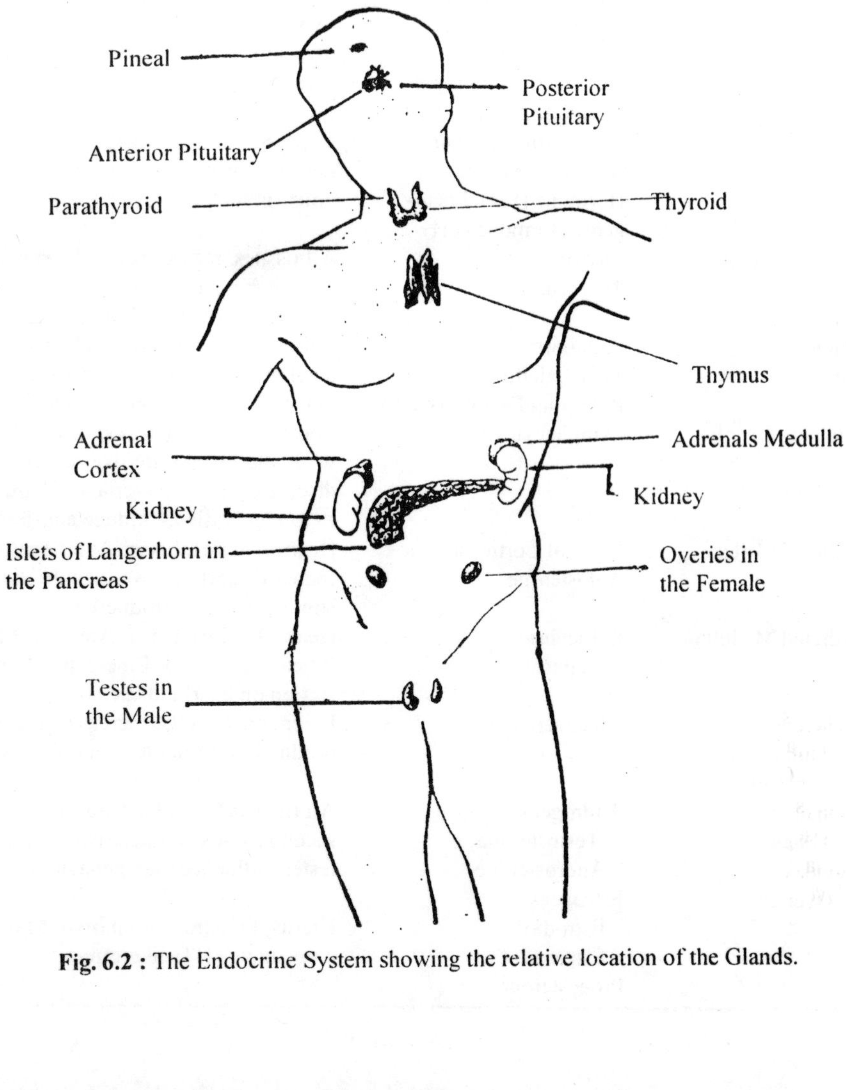

Fig. 6.2 : The Endocrine System showing the relative location of the Glands.

Generally all the cells in the body have secretary functions. But the endocrine glands have specialized cells for secretion. The endocrine glands have no ducts and hence they are called ductless glands. The exocrine glands have ducts (tube or pipelines) and they carry the secretions to the respective destinations (Table 6.1).

Table 6.1 : Showing the Glands, Hormones, Target Tissues and Major Functions

S. No.	*Glands*	*Hormones*	*Target Tissues and Major Functions*
1.	Pineal Gland	Melatonin	Gonads. Supresses the hormone output by Gonads to facilitate sexual behaviour.
2.	Posterior Pituitary Glands (Neuro-hypopysis)	Vasopressin	Arteries, Kidney, Uterus, Mammary Glands, Raisesblood pressure, Stimulates water reabsorpation, Stimulates water reabsorpation, Stimulates contractions, Stimulates milk secretion.
3.	Anterior Pituitatry Glands (Adeno-hypopysis)	Somatotropic Hormonre (STH) Thyroptropic Hormone (TH), Adrenal Corticotropic Hormone (ACTH) Fallicle Stimulating Hormone (FSH) Lutenizing Hormone (LH), Lacto-genic Hormone (LH).	Bone, Thyroid Glands, Adrenal Cortex, Fat tissues, Stimulates growth, Stimulates thyroxin secretion, Stimulates steroid secretion by Adrenal Cortex, breaks down fat tissues, Development of Ova/Serm Cells, Stimulates estrogen and androgen output, Corpus bteum development and milk output.
4.	Thyroid Glands	Thyroxin Thyrocalcitrum	All tissues, raises, metabolic state, increases protein synthesis, Carbohydrate synthesis and transportslower blood calcium.
5.	Parathyroid Glands	Parathormone Thyrocalcitron Releasing Factor (TCRF)	Raises blood calcium level, lowers posphate level, increases phosphate secretion by kidney (by the action of Vitamin-D)
6.	Thymus Glands	Lymphocytes	Located in heart between breast bones. The white blood cells are like security, fight disease causing Bacteria and Virus, Reaction to foreignbodies, Produce anti-bodies.
7.	Adrenal Cortex	Cortisol Corticosterone Aldosterone	Heart, Arteries, Liver, Anterior Pituitary, Increase heart rate, Vasocontriction stimulates RNA production.
8.	Adrenal Medulla	Epinephrine Norephrine	Heart, Arteries, Liver, Anterior Pituitary, Release of ACTH, Emergency Functions, Action on heart muscles.
9.	Pancreas Alpha Cells Beta Cells	Glucogen Insulin	Liver and all cells. Increase Glucose production, Promotes amino acids.
10.	Gonads : Testes	Androgens Testosterone Androsterone	All tissues, Sexual arousal, Primary and secondary sex characteristics, stimulates testes, influences sex behaviour.
	Gondas : Overies	Estrogens Estrodiol Esterone Progesterone	Uterus, Prepares for embryo, Maintains pregnancy, GTH-like-effect.

1. The Pineal Glands

The pineal gland is located in the head and it is found on the top of the posterior of the third vertricle. The third vertricle is the tall narrow hollow space inside the *diencephalon.* The pineal gland is having structure of a pea. In lower animals like the frog it is a primitive visual receptor and serves as a third eye. The hormone secreted by this gland is called *melatonin.* It functions to bleak the skin pigment cells. The functions of the pineal gland are changed drastically in mammals in the eveolutionary process. As result of this its parts in human physiology and in behaviour were not clearly brought out till the present.

The present day researchers were able to identify the functions of this gland. For a long time the functions of the pineal gland were not taken into account and not related to the function of the internal organs of the human body and behaviour. The important activity in which the pineal gland involved is the reproduction. The activity of the reproductive system mainly functions in response to the light changes, the light-dark cycle of the every day changes, especially the day-light with seasonal changes. But the pineal gland in human does not contain the visual receptors like the lower animals. But the gland is stimulated by excitement of the light that comes from the eye. The excitement reaches the pineal gland through the sympathetic nervous system.

Of all the glands only five glands respond to nerve stimulations and the pineal gland is the fifth gland. The other glands are the post piturity gland, the anterior pituitary gland, the adrenal medula, and the pancreas, All these repond to the stimulations of the nerve cells of the posterior pituitary and the hypothalamus.

The pineal gland secretes the hormone, the melotonin. This hormone suppresses the activity of the gonads and also the sexual behaviour in humans. But in contrast, light suppresses the secretion of the hormone, the melatonin of the pineal gland. This results in the release of the gonadal activity by the light. By this means the sexual activity is increased by light. The increased sinlight for longer duration in summer in India and in spring in other cold countries speeds up the annul growth of the gonads and increases sexual activity. Generally, light stimulates the retina area of the eye which reaches the pineal gland in the form of excitation by the way of the fibres of the sympathtic nervous system. These fibres of the sympathetic nervos system secrete *nonadreline*. This nonadreline inhibits the activity of an enzyme which is required by the gland to produce its hormone, *melatonin.*

By this means, the production of the hormone, melatonin is reduced. Normally melatomin suppresses the gonads. Since melatonin production is reduced, the gonads are released form the inhibition and the gonads are free to function and secrete more sex hormones. In women the estrus cycle is accelerated. The germ cell production in both sexes is increased and the sexual behaviour is highly stmulated. This is the reason that in the tropical countries the birth rates and the sex crimes increase rapidly.

The pineal gland produces the hormone, melatonin from *serotonin.* Serotonin supply is made from the brain. This depends upon the light-dark cycle. There are two control mechanisms in this function. The malfunction of the pineal gland in man is rare. Any malfunction in the pineal gland is caused by tumours in the gland or in the tissues and in the surrounding area. Any tumour in the gland results in multiplying of cells and more hormone of the melatonin is secreted. This mainly affects the children because of the *hypersecretion*, and the sexual development in children is delayed.

If the tumour development takes place in the surrounding tissues, it inhibits the function of the pineal gland. This state causes *hyposecration* of the *melatonin* (hormone). This state causes early maturity in children.

Pineal gland thus plays an important role in the sexual behaviour of the organism. One important function governed by the pineal gland is the onset of puberty in human in the early years. Another important function caused by the pineal gland is the sexual maturation occurring earlier in tropical countries where there is more sunlight most of the days. The pineal gland is having influence in the sexual behaviour of human and animals due to the occurrence of light-dark cycle everyday. It also regulates the menstrual cycle in human and in animals. The pineal gland also regulates and governs the sexual behaviour of some animals which usually occurs seasonally.

Recent research findings throw some new light upon the link between brain chemistry and depression. Many severely depressed people feel that their misery is caused by their own favilts. Scientific evidences ignore this view. One recent scientific evidence focussed on this problem is the *brain's neutrotransmitters*, the chemicals that act as messengers between neurons in the brain. In very depressed people there are three such chemicals appear to stop functioning strongly. They are the *dopamine,* the *norepinephrine* and the *serotonin.* The serotonirt is associated with the pineal gland.

We may find that people of some type suffer from seasonal affective disorders (SAD). In India people suffer from seasonal affective disorders severely and die both in winter due to extreme cold and in summer due to extreme heat. What type of mechanism works in our brain? We have no correct answer for this. But in the early 1980s researchers discovered that the hormone, melatonin secreted by the brain's pineal gland is found to play important role in the seasonal affective disorder. The pineal gland produces more *melatonin* hormone when there is *less light*. It means that the depression of hormone caused when there is more light. Therefore the seasonal affective disorder is caused by the higher level of the hormone, melatonin in the blood of the person during the seasons. But it was observed that the level of melatonin is high in some people and not in all people. The low level of melatonin hormone in the brain of some people may be due to environmental conditions. These people may alter their environmental conditions in such a way that the production of melatonin in the pineal giand is suppressed and reduced. This state may normalize the situation without affective disorders.

2. The Pituitary Glands

Among the endocrine glands the hypophysis or the pituitary glands are considered to be the most important glands as these glands influence all other glands in some respect. According to embryological studies these glands are formed into two kinds of tissues. They are the posterior glandular parts and the anterior glandular parts. The posterior is a neural part, originates in neurectoderm, which is from the floor of the third ventricle. This is located beneath the central surface of the diencephalon.

The anterior glandular part is derived from the epithalial tissues, called the Rathke's pouch. The posterior and the anterior parts are also called as *neurohypophysis* and *adenohypophysis*. The neurohypophysis is divided into three parts such as, the median tubler cinereum, the infundular stem and the neural lobe. The adenohypophysis is also divided into three pars as distablis (anterior lobe), the pars tuberalis and the pars intermedia. The main secreting part of the adenohypophysis is the pars distalis.

In all psychological studies more emphasis is placed on the pituitary glands than in other glands. This gland is also called as master gland, because it produces the largest number of hormones and is having control over other glands. The hormones secreted by this gland are carried out throughout the body and they regulate the activities of other glands. It is located in the skull beneath the brain and is connected to the hypothalamus. it is just about the size of a corn.

The malfunction of the pituitary gland causes tremendous physiological changes in the development. The person may grow into a four feet midget or an eight feet giant. The pituitary gland falls under two main glands as posterior pituitary gland and the anterior pituitary gland (Fig. 6.3).

(a) Posterior pituitary gland

The posterior part of the pituitary gland is also known as neurohypophysis. It is named like this because it receives nerve fibres from hypothalamus part of the brain. It appears that this is under the control of the nervous system. This gland secretes two kinds of horniones called *vasopressin* and *oxytocin*. These two hormones have four overlapping effects. When there is a raise of blood pressure in smaller arteries these hormones constrict the blood vessels, they stimulate for contraction of smooth muscles, they stimulate mummary glands for milk production and also they stimulate kidney for reabsorbing water from the urine. The process of stimulating kidney for reabsorbing water from the urine is called *antidiuratic effect*. This is caused by the vasopressin. At the same time the other hormone oxytocin is more to do with the contraction of the uterus and stimulates milk production. The vasopressin and oxytocin function in combiation to aid child-birth. Further the antidiuratic effect of posterior pituitary gland plays an important role of regulating water balance of the body. If there is short of water in the nerve cells in hypothalamus it stimulates the posterior pituitary through the nerves from hypothalamus. Then the posterior pituitary secretes its hormones. These hormones in turn stimulate the kidney for reabsorbing of water. This results in preservation of water in hypothalamus cells.

Any deficit of antidiuratic hormone or undersecretion from the posterior pituitary gland causes water loss and excess of urination. This process is also called hyposecretion of the posterior pituitary. This state is called diabetes insipidus. There is excretion of large afhount of water followed by a strong thirst. The immediate need is large amount of water intake for maintaining the water balance of the body and get rid of dehydration and death. The posterior

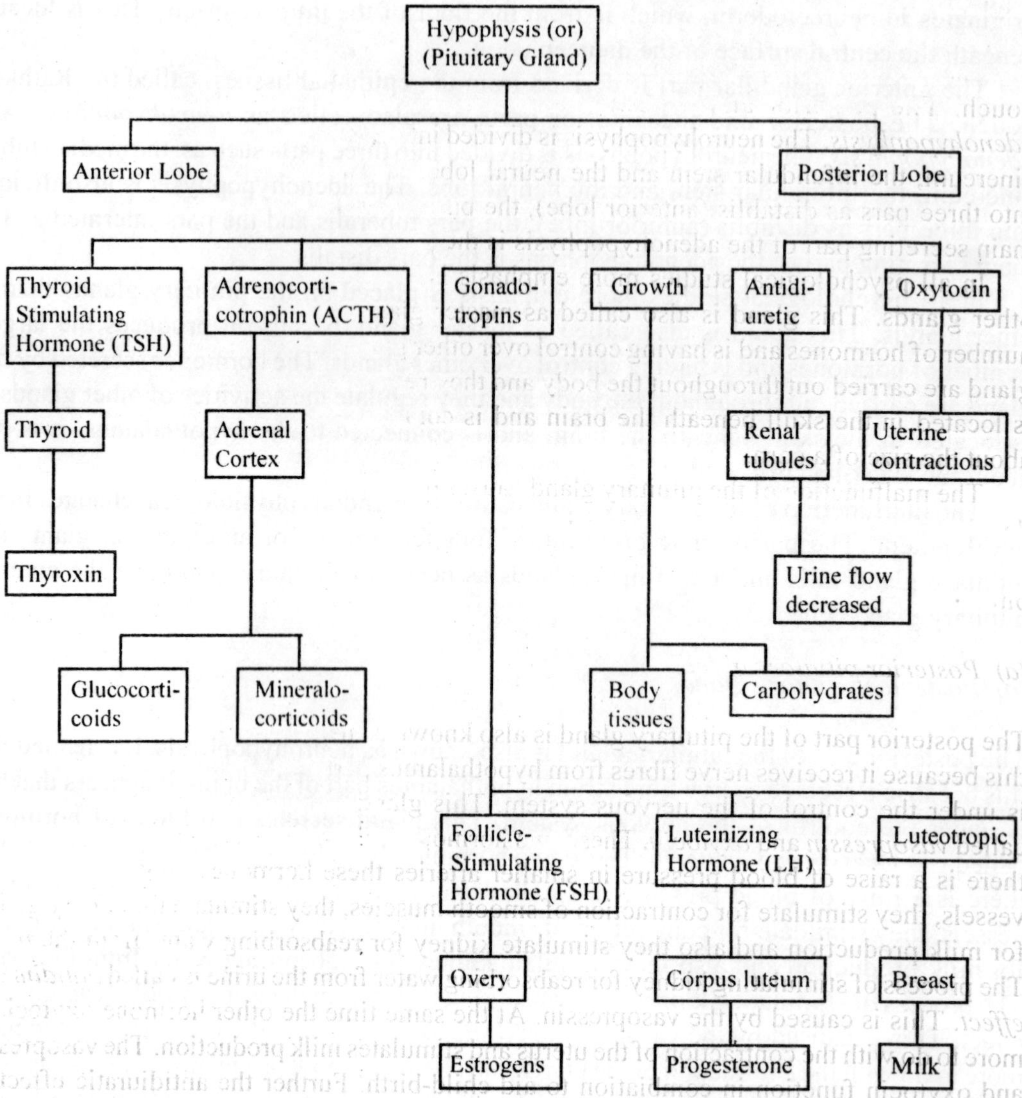

Fig. 6.3 : The Divisions and Functions of the Pituitary Glands (Hypophysis)

pituitary extract can cure this condition. The posterior pituitary or the neurohypophysis is important from psychological point of view as it mainly deals with maternal behaviour and thirst (Fig. 6.4).

(b) The Anterior Pituitary Gland

The anterior pituitary is also called adenohypophysis. This is the exact master gland. The anterior pituitary gland controls the output of adrenal cortex, the thyroid glands and the

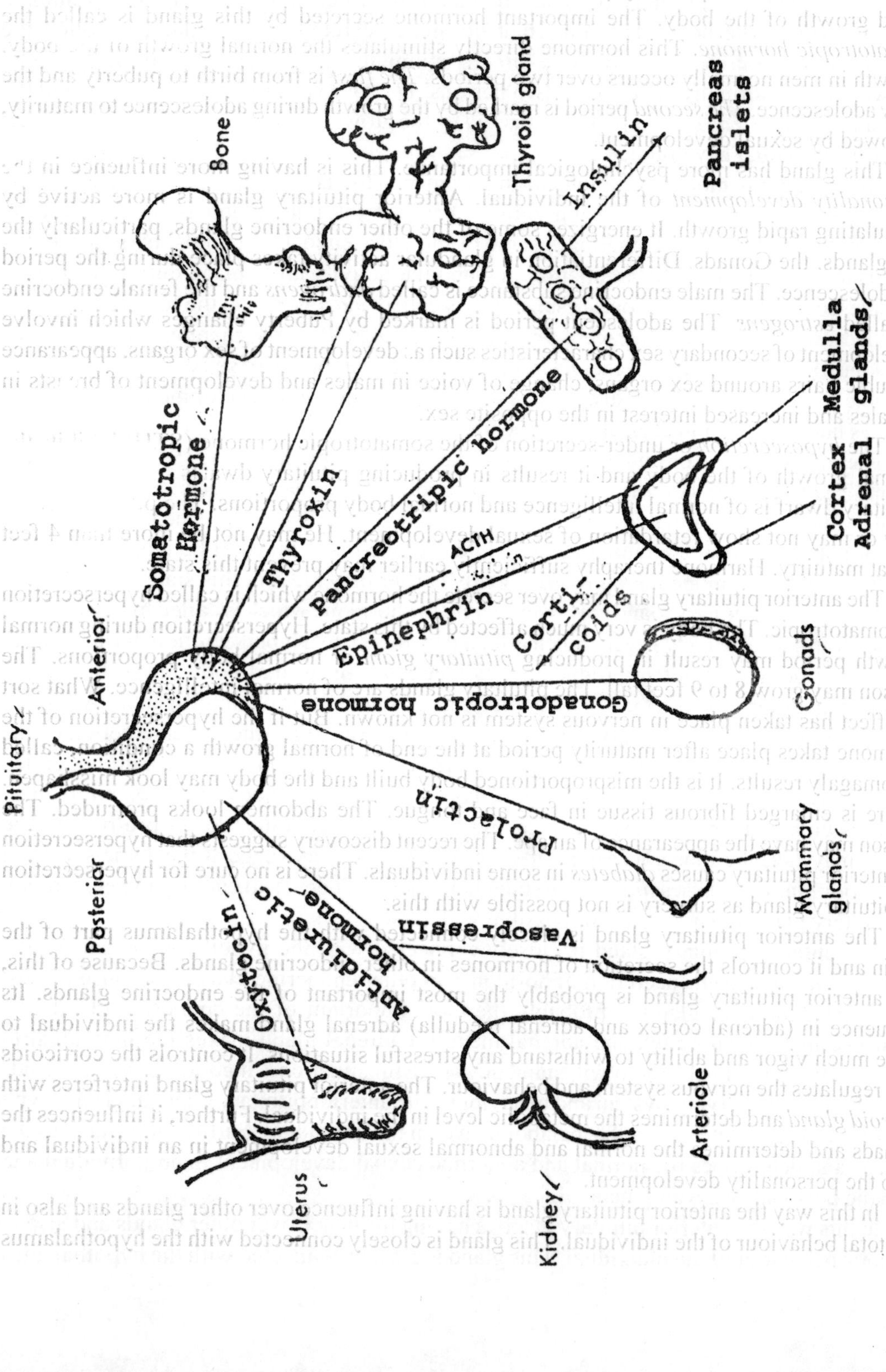

Fig. 6.4 : Glands and Hormones and their Target Tissues of the Anterior and Posterior Pituitary Glands.

gonads. The anterior pituitary produces one additional hormone that serves to stimulate rapid growth of the body. The important hormone secreted by this gland is called the *somatotropic hormone*. This hormone directly stimulates the normal growth of the body. Growth in men normally occurs over two periods. *The first* is from birth to puberty and the early adolescence. *The second* period is marked by the growth during adolescence to maturity, followed by sexual development.

This gland has more psychological importance. This is having more influence in the *personality development* of the individual. Anterior pituitary gland is more active by stimulating rapid growth. It energizes some of the other endocrine glands, particularly the sex glands, the Gonads. Differentiation in glandular activity takes place during the period of adolescence. The male endocrine substance is called *androgens* and the female endocrine is called *estrogens*. The adolescent period is marked by Puberty changes which involve development of secondary sex characteristics such as development of sex organs, appearance of pubic hairs around sex organs, change of voice in males and development of breasts in females and increased interest in the opposite sex.

The *hyposecretion* or under-secretion of the somatotropic hormone (STH) inhibits the normal growth of the body and it results in producing pituitary dwarfs or midgets. The pituitary dwarf is of normal intelligence and normal body proportions. The pituitary dwarf may or may not show retardation of sexual development. He may not be more than 4 feet tall at matuirty. Harmone theraphy sufficiently earlier may prevent this state.

The anterior pituitary gland may over secrete the hormone which is called hypersecretion of somatotropic. The body is very much affected by this state. Hypersecretion during normal growth period may result in producing *pituitary giant* or normal body proportions. The person may grow 8 to 9 feet tall. The pituitary glands are of normal intelligence. What sort of effect has taken place in nervous system is not known. But if the hypersecretion of the hormone takes place after maturity period at the end of normal growth a condition, called acromagaly results. It is the misproportioned body built and the body may look misshapen. There is enlarged fibrous tissue in face and tongue. The abdomen looks protruded. The person may have the appearance of an ape. The recent discovery suggests that hypersecretion in anterior pituitary causes *diabetes* in some individuals. There is no cure for hypersecretion of pituitary gland as surgery is not possible with this.

The anterior pituitary gland is closely connected with the hypothalamus part of the brain and it controls the secretion of hormones in other endocrine glands. Because of this, the anterior pituitary gland is probably the most important of the endocrine glands. Its influence in (adrenal cortex and adrenal medulla) adrenal gland makes the individual to have much vigor and ability to withstand any stressful situations. It controls the corticoids and regulates the nervous system and behaviour. The anterior pituitary gland interferes with *thyroid gland* and determines the metabolic level in the individual. Further, it influences the gonads and determines the normal and abnormal sexual development in an individual and also the personality development.

In this way the anterior pituitary gland is having influence over other glands and also in the total behaviour of the individual. This gland is closely connected with the hypothalamus which governs several drives of the body like hunger, thirst and so on. Through the posterior pituitary and the portal system of the blood vessels and thus regulates the behaviour of the organism.

3. The Thyroid Glands

The thyroid gland is a large mass weighs less than an ounce located on either side of the *windpipe* just below the voice box (larynx). It produces a hormone called *thyroxin* which effects all cells of the body. This hormone regulates the body's metabolic rates. It also determines the rate of transformations of the food we eat into *energy form*. The amount of thyroxin produced by this gland causes difference in metabolic rates. Secretion of too small amount of thyroxin makes a person to sleep too much. His body fails to maintain normal temperature. If this occurs in childhood it leads to a condition called cretinism. This is also called hypothyroidism. It results in inadequate bone growth and related handicapps, puffy face, pot belly and protruding tongue. The central nervous system is not adequately developed. He becomes mentally defective. The intellectual level in him will remain in the level of a small boy. It is possible to remove this defect if sufficient care is taken at the early period. Thyroid deficiency in the adult causes *myxedema*. The person becomes stupid, slim, forgetful and unable to concentrate in thinking and acting. Secretion of too much of thyroxin increases metabolic rates. The persen is *hyperactive*. The person eats too much, but still he is feeling hungry and cannot gain weight. We are calling such people as gluttons. This is called *hyperthyrodism*. This can be cured by removing the part of the gland.

In the thyroxin deficiency state the body's metabolism rate sinks down. There is less oxygen consumption and giving of low level of carbon dioxide. When there is an excess of thyroxin the metabolism rate goes up above the normal level.

Some may be suffering from goiter, which is an enlargement of the thyroid gland. This is indicated by a swelling in the neck area. It is due to *hyperthyroidism*. Simple goiter often caused due to *lack of iodine* in the diet for manufacturing thyroxin. People who eat food grown from iodine deficiency soils often subject to goiter. Iodine can be taken into the body in drinking water and in food eaten. The other division of this gland is called *parathyroid gland* (Fig. 6.5).

4. The Parathyroid Glands

There is a little gland, called parathyroid lying just above the thyroid gland. It consists of four small organs. These are attached to the thyroid gland. The parathyroids secrete- a hormone called *parathormone*. This is very essential for controlling the calcium level and calcium phosphate ratio in the blood. The body uses the calcium taken in with food for nourishing the bones and teeth which are related to the maintenance of normal emotional stability. When the parathyroid is damaged then there is *emotional instability*. Though there are certain other factors for emotional instability when there is no coordination between thyroid and parathyroid emotlonal instability may occur. Both, an excess of thyroid secretion

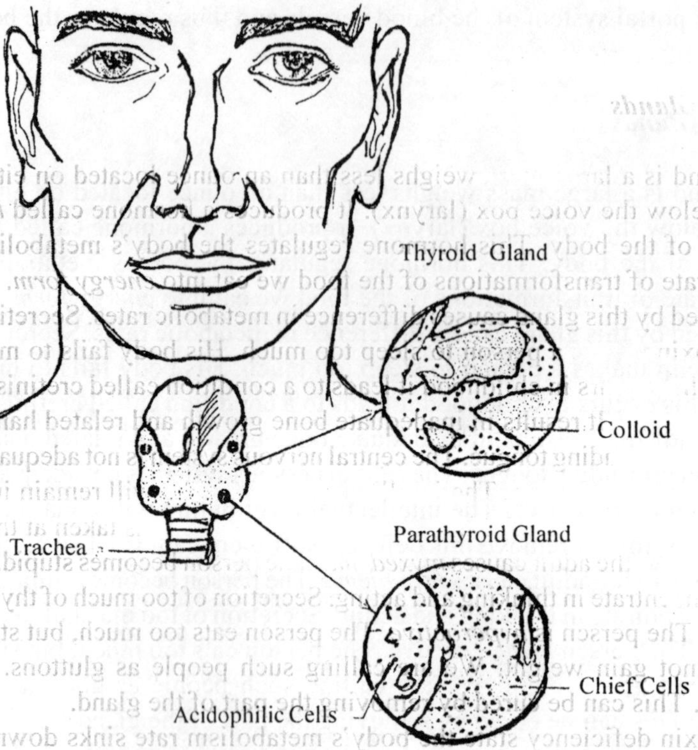

Fig. 6.5 : Thyroid Glands and their Taget tissues.

and a defective (too little) parathyroid may cause *restlessness*. Too much of thyroid secretion and too little secretion of parathyroid go together. This means that these two glands act in oppose to each other in some cases. Therefore the health of the organism is depending upon the balanced function of these two glands.

For adequate regulation of the calcium and phosphate levels the parathyroid gland requires the vitamin-D. The vitamin prevents the loss of calcium in feces and phosphate in urine. Vitamin-D compensates the lack of hormone deficiency (parathormone) in maintaining blood calcium. If there is adequate supply of Vitamin-D to parathormone the body retains the calcium level and the parathormone raises the blood calcium level. In this the parathormone has two functions. One is rapid and limited raise of blood calcium. By this means there is incr6ased absorption of calcium from intestine which prevents the loss of calcium in the urine. The other function causes greater change in calcium level by acting slowly on bone by taking more time. This process is called reserve supply of calcium. That is, the parathormone breaks down the bone into a form of calcium for supply of more calcium and phosphate to the body. This process takes place because bone is made up of *Calcium* and *Phosphate* in order to meet the Parathormone deficiency and maintain the balance.

5. The Thymus Glands

The thymus gland is also classified as one of the endocrine glands that is functioning in the bodily parts internally. It has been viewed by some investigators that the thymus gland is not very important from behavioural point of view and therefore, it need not be classified as an endocrine gland. But from the internal environment point of view thymus gland plays some important role which mainly deals with foreign bodies which are harmful for the organim's efficient function.

The thymus gland is located in the chest between the breast bone (sternum) and the heart. It develops as a large gland of two equal in size of lobes in women. For the children or young calves it serves as a food delivery organ outside. The size of the thymus enlarges gradually with general growth of the organism. In the case of human being it begins to develop slowly until ten years. Then it slowly stops its growth and not functioning in the adult stage.

Some of the authors are of the view that the thymus gland is not an endocrine gland as it is not producing any hormone. But actually thymus gland manufactures the hormone called *lymphocytes*, the white blood cells. It is under the direct control of pituitary gland. The thymus gland is the main source of the hormone of the lymphocytes. This has been discovered very recently. The white blood cells are like the security, fight diseaft causing bacteria and the viruses and prevent the entry of foreign proteins, mostly the infectious germs.

The thymus gland releases the lymphocytes directly into the blood sterms and this is stored by *lymphglands* and the *spleen* and it may go into action whenever necessity arises. The thymus gland produces the lymphocytes to a maximum level during the period of pregnancy in women just before birth and few dayg after the birth of the child. The lymphocytes seem to have two important functions which are of mainly reaction to foreign bodies.

Firstly, the lymphocytes are released by the thymus gland and produce antibodies that neutralize the disease causing bacteria and the viruses. Secondly, they recognize and react against the foriegn bodies. For example, any tissue transplanted from another body under the skin needs to be neutralized. In treating some patients by surgery some part of the body from another human being or animal is transplanted into the part of body where the tissues are damaged. This foreign body must fit into the body of another person. The lymphocytes react to this suitably.

The thymus gland is found to produce the original hormone, the lumphocytes and supply them to the proteins of the body and destroy the foreign bodies that resist. Then they release or send the remaining to the spleen and to the lymph glands for storing for the future when needed. During the storage period the lymphocytes are going on multiplying for future use.

Any failure in the function or mechanism of the thymus gland causes serious impairments. The *autoimmune diseases* are commonly caused in this situation. During these disease periods the body reacts against its own proteins, just like in the diseases like the *leukemia, myasthenia gravis* and so on. The thymus glands functions are mainly contributing in nature for self-repair and for building up the ability of individual to face the diseases. This gland also contributes ability to the individual to withstand the stain, stress, injury and diseases. These contributions of the thymus gland seem to maintain the individual body for healty functioning.

The early investigators and also some of the modern researchers have given the least importance to the functions of this gland. But the recent research findings produce evidences that the thymus gland is functioning to protect the body from autoimmune diseases, and Ikeeps and maintains internal functions efficiently for normal behaviour of the organism which no other gland does. The importance and the values of this gland are realized very recently.

6. Adrenal Glands

The adrenal glands join with thyroid glands in regulating metabolism and also in increasing the pulse rate. The adrenals also influence the emotional excitation of the organism. These are two little glands, located close to the kidney. This is named as adrenal because it is close to the kidneys. The other name for kidney is *renal*. These gl ands are not directly related to the functions of the kidneys. The adrenals consists of two parts such as *adrenal cortex* and *adrenal medulla*. The first is the outer cover and the second is the inner cover. These two differ in their structure and function. These glands contribute to the functioning of nerves and muscles and help the whole body to meet with stress (Fig. 6.6).

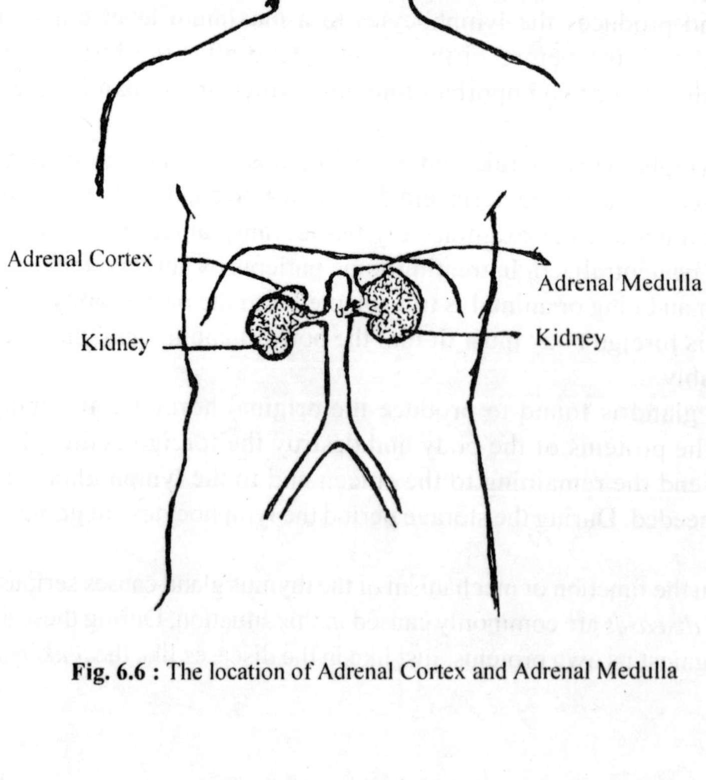

Fig. 6.6 : The location of Adrenal Cortex and Adrenal Medulla

(a) *The Adrenal Cortex*

This gland is under the control of the pituitary glands. The adrenals produce a number of chemical substances (hormones). The most important hormones secreted by the adrenal cortex are *steroids*. These hormones are called *corticoids*. The hormone is also called cortex. These hormones regulate sodium and potassium balance of the body. The cortex has functional relationship with glands for the development of secondary sex characteristics and to maintain the vigor. It also helps to control the sexual functioning. The corticoids stimulate the storage of blood sugar in the liver. The adrenal cortex produces a male sex hormone called *androgen*. The adrenal cortex is directly under the control of anterior pituitary gland and the hormone of this gland stimulates the adrenal cortex to produce the corticoids. The corticoids are produced by the stimulating hormones of the anterior pituitary gland.

The under secretion of adrenal corticoids is called hyposecretion. The hyposecretion causes the *Addison's disease*. In this process there is too much of elimination of sodium and too much of retention of potassium by the kidneys. The excitability of nerves and muscles is reduced. There is loss of sodium. Constant maintenance of salt solution is affected. There is loss of body fluids. There is change in the carbohydrate metabolism. This results in less storage of blood sugar and the reduced utilisation of carbohydrate. This state caused lack of vigor, muscles weakness, lower body temperature and loss of weight, loss of sex interest, low resistance to any infectious diseases, darkening of skin and so on. The administration of sufficient salt and corticoids may remove these symptoms.

The hypersecretion of adrenal cortex may cause sexual prematurity in children and masulinity in women. In women this may cause loss of feminine look, deepening of voice, appearance of beards. But hypersecretion of adrenal cortex or corticoids is rare.

The adrenal cortex plays an important role in the normal functioning of the nerve and muscle excitability. The adrenal cortex is also interacting with the adrenal medulla and pituitary glands and determine the resistance against the environmental stresses.

(b) *The Adrenal Medulla*

This is the inner part of the adrenal gland. The hormone secreted by this gland is called *adrnin*. Adrnin is a powerful hormone produced by medulla. A very small quantity of this hormone is sufficient to cause many changes in the body. This is stimulated by the sympathetic nerves of the autonomic nervous system (ANS). This arouses the body to face stressful situations. The effects of this hormone are rapid heartbeat, high blood pressure, forcing blood through muscles and brain, suspension of activity in the stomach and intestine, release of stored sugar from the liver, dilation of the pupil of the eye. The effect of adrnin is very high in the state's of fear and anger.

The sympathetic nerves cause these effects rapidly for a short period of time. The adrenal medulla is very close to the sympathetic nerves. **Cannon** (1929) says that all these bodily changes occur in animal in an emotional situation and this works out to mobilize the whole energy of the organism to meet the emergency situation. For example, at the time of intense fear, anger or excitement this function takes place as the sympathetic nervous system stimulates the adrenal medulla to secrete more hormone into blood and the organism mobilizes the energy for the specific purpose of meeting the situation. If fear the extra energy facilitates the organism to escape from danger, if anger it is used to attack the object of the anger. But in meeting the stressful and alarming situations both the adrenal cortex and the adrenal medulla are involved. The sympathetic system stimulates the adrenal medulla for secreting more hormones. This in return stimulates the anterior pituitary gland to secrete the hormone called *adrenocorticotripic*. This in turn stimulates the adrenal cortex to release the hormone, corticoids. This increases excitability and basal metabolic rate.

If the organism is put in prolonged stressful situation there is adrenal enlargement. In this both the adrenal cortex and medulla are involved to meet the stress. This makes the organism to mobilize increased energy to meet the stressful or emergency situation and facilitates the organism for survival either by winning the enemy or by escaping from the danger.

7. The Pancreas

The pancreas is a small gland that lies in between stomach and the small intestine. The pancreas has both exocrine and endocrine functions. It is a digestive gland that discharges hormone into the stqAll intestine. There are two kinds of endocrine cells called *alpha cells* and *beta cells*. These are strongly connected to the pancreas. The alpha cells are only 25 percent of the total endocrine cells. These cells produce a hormone called *glucogen* that stimulates the storage of blood sugar in the liver and muscles to release storage blood sugar to body cells. Glucogen is secreted for a short period by the *alpha cells* if there is low blood sugar. This function of the alpha cells is in opposite to the action of the beta cells (Leukel, 1968).

The *beta cells* secrete the hormone called *insulin* that inhibits blood sugar released by the liver and muscles and stimulates blood sugar utilization by the cells of the body. The insulin makes these cells to get more blood sugar thereby it becomes essential for blood sugar utilization. The beta cells secrete insulin in response to high level of sugar in the blood. Their secretion is inhibited when the blood sugar drops. Any under secretion of insulin by the beta cells of the pancreas causes a common disease called *diabetes*.

If sufficient amount of insulin is not secreted the muscles and tissues will not make use of the blood sugar and sugar is not stored by the liver. This results in the *increase of Glucose* (blood sugar) in the blood. Excess of sugar is secreted in the urine. The kidney tries to get rid of the excess sugar by excreting more water than the usual from the body. Then the tissues get *dehydrated*. The blood get accumulated by the poisonous waste products. The person becomes a *diabetic patient*. He may even go into *coma*. It may cause even death. The person needs insulin and sugar free diet to keep the blood sugar at normal level.

The over secretion or hypersecretiQn of insulin is taking place rarely in man. In the case of diabetic it results in low blood sugar level as all stored blood sugar is utilized. But too much of insulin causes too little sugar in the blood and the person suffers from chronic fatigue.

There were many studies made recently on pancreas and these studies brought out valuable results. One of the recent findings is that the pancreas secretes only a single hormonal substance called *insulin.* But this substance is found to be very much essential for the metabolic processes (William, 1960). It is because of its many sided processes. This has been studied very extensively by many investigators. When there is lack of sufficient secretion for metabolic functions it results in several disorders like, *polyuria, glycosuria, ketonuria* marked by *bodily disorders* and death in a *comatose state.*

These are also the symptoms of diabetes which is a common disorder caused by *pancreatic hypofunction.* This condition mostly is fatal unless pancreatic hormone is administered.

8. The Gonads

The gonads are the sex glands which are mainly functioning in the process of reproduction. The gonads are the reproductive organs in males and females. These glands have two roles to play. The first role is the maturation of the reproductive organs and the development of the secondary sexual characteristics. They are mainly the development of the pubic hairs, depending of the voice, enlargement of the sex organs and so on. The second functional role is the growth and development of the germ cells in both sexes and the reproductive events in females. The structural and functional developments are closely related to hormonal changes.

(a) The Reproductive System in Males

The sexual development and growth must take place prior to maturity in males and in females. But the sexual reproductiveness in a male persists mostly without any change throughout a man's adult life and it decreases only gradually as and when the general physical condition of the man deteriorates as he becomes old. Man's sexual motivation with all these is influenced by the endocrine glands. The sexual hormones of a male are collectively called as *androgens.* A variety of anrtogenic substances are obtained from natural sources particularly from the testicular tissue (Fig. 6.7).

The reproductive system of the male is based on a paired testes which are containing in a sac or *serotum.* Their location is outside the body cavity. This condition lowers their temperature which facilitates the *growth of germ cells in the seminiferous tubercles.* The tubular tissue of the test give rise to the *seminal duct,* which goes up into the abdomen to join with the duct that comes from the other testicle and it is connected here by the *prostate gland.* The secretions of this gland and the sperm form the *semen.* This is collected in the *seminal vesicle.* At the time of copulation the vesicles and the outgoing urethra of the penis come to contact for ejaculating the semen which contains the *germ cells.*

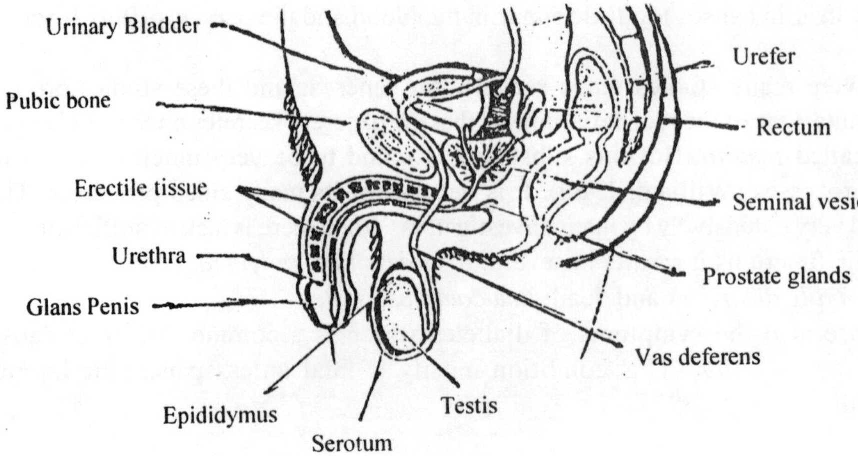

Fig. 6.7 : Gonads : Testes of the Male and the parts. The Reproductive System in Male.

The most active hormones are the *testosterone* and the *androsterone*. Further a group of compounds collectively about seventeen known as *ketosteroids* are closely related to the *androsterone* which are from the human urine. These are biologically active as *androgens*. It is generally believed that *testosterone* is the principal male sex hormone. Testosterone is produced by the gland cells in the intertubular connective tissues of the testes. The growth and function of the entire genital apparatus in man are directly controlled by androgenic hormones. If the androgen level falls below the required level during the period of puberty the scortum (the testes) remains small, the penis remains infentile and the accessary glands of the reproduction (seminal vesicle prostrIte) fail to develop to the normal level.

The hormone androgens also control the development of the secondary sex characteristics. These characteristics are mainly the pattern of body hairs, tonal quality of voice, muscular development, fat distribution and skeletal structures, caused by the fall of androgens hormones. It is also observed that the androgenic hormones affect the physiological functions not related to sexual behaviour. They mainly influence in the general metabolism by facilitating the retention of nitrogen, sodium, potassium, inorganic phosphorus and chloride.

A hypothetical view is that the pituitary glands are having influence in this process and stimulate the production of androgens by the testes and the androgens induce the *spermatogenesis*. It has been estimated that the number of sparmetozoa in a single sexual intercouse are about three hundred to five hundred millions and the mature spermatozoa are incapable of fertilization.

Generally sexual motivation and behaviour are aroused and occur only in special environmental situations which provide some specific type of sensory stimulation. But, however there will be arousal of sexual motivation and behaviour only when the organism is physically fit and ready for mating. This physiological readiness to respond to sexual stimuli is provided to the organism only by the hormones. This affects the neural mechanism of the body.

The sexual hormones of the hypophysical (pituitary) and gonadal influence the structural and functional readiness of the organism for appropriate sexual stimulation and behaviour, Experiments have been conducted by seasonal animal breeders, **Hammond** (1938) and **Day** (1940) on animals like mares, opossums relating hormonal conditions and sexual arousal and demonstrated involving the hormone administration. The result was the arousal of sexual activity. It is believed that. this was caused by the release of the hormones, estragens in the stimulated overies since the administration of estrogens were found to have significant effects. It was found that the animals showed cyclical estrous behaviour response to the hormones administered.

The pituitary and the gonadal hormones are having effect in the human reproductive cycle and also influence the sexual development in adolescence. There is a decline of sexual motivation with age which is caused by hon-none deficiencies. **Runge** (1943) had observed in his studies on animals that the sexual vigfor declined with age on bulls, dogs etc. But according to him this can be restured by the method of transplanting of sexual glands from young animals. There are clinical reports from **Miller** (1938) and **Peczenik** (1936) that aged men respond to continued gonadal hormone treatment for increased 3exual interest and strength. There are many clinical disorders of genital function like retention of testes in abdomen and *hypogenitalism* which are accompanied by complete lack of sexual motivation. Sexual drive can be restored in these cases by administering hormones. It was also tested in farm and domestic animals that hormone injections were found to be successf-W In~cuidng impotence.

In the clinical situations necessities may arise to remove the gonadal glands oy testes. This process is called *castration.* The removal of the gonadal glands does not itself make the organism physically impossible for sexual behaviour. As result of this there is only hormonal deficiency. There is also anatomical and functional disturbances in the reproductive apparatus and also there is significant inhibition of sexual motivation. The severity of postcastrational behaviour is the deficit of normal function and the age at which the cestration is made. The decline of sexual motivation is gradual. In human beings, the castrated male may prolong his androgen treatment. This is generally effective in restoring the sexual motivation and also his penile erection. But his ability to ejaculate in sexual activity does not take place efficiently.

(b) Endocrine Function in Females

The female reproductive system consists of a paired overies in abdomen, each of which has an ovum or egg cell. Generally, in human beings the hormonal and the neural mechanisms are contributing to the regulation of sexual behaviour. The sexual motivation of the female is determined partly by the rhythmic structural and functional changes of the overies. The open ends of the *fallopian tubes* lie near the overies. The follopian tubes are the ducts (tubes). These follopian ducts (tubes) are leading from the *overeies* to *uterus*. The ova (egg cells) are produced by overies, reach the uteries via these tubes. The overies are the female glands or the reproductive organs. The *ova* are the female reproductive cells. The *uterus* is that part of the female reproductive organ in which the embryo develops.

The follopian tubes lie near the overies and when the *ovulation* produces a mature ovum, the follopian tubes carry it to the uterus. Ovulation is the process of producing and releasing of an ovum by the *follicles* of the overies. Follicles are the tissue formation in the overies that produce ova (egg cells) and secrete estrogens, the female sex hormones. If th-e ovum is fertilized by a sperm cell it develops into an embryo in the wall of the uterus. At the time of birth the fetus (developed embryo) is discharged via to cervix and vagina. Cervix is the opening between the uterus and the vagina (Fig. 6.8).

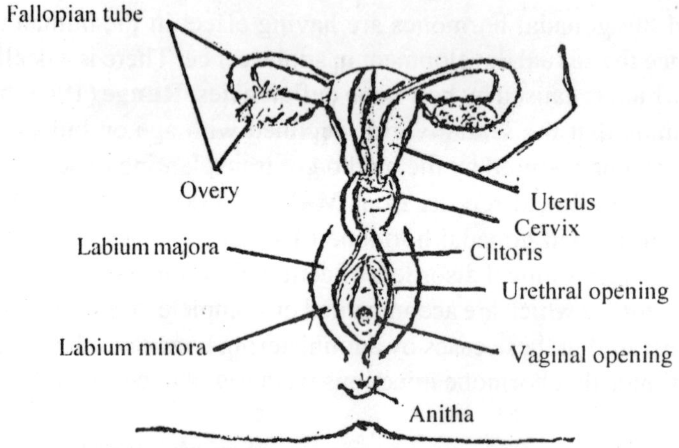

Fig. 6.8 : Gonads : Overies of the Female and the Parts. The Reproductive System in Female.

In most of the species there are three cycles of overian activity identified. The first is the basic cycle limits the life of the organism and proceeds from the beginning stage of development of the overies to the attainment of functional maturity and finally growing old (senescence). Sexual behaviour takes place only during the second stage of this cycle. During this stage the overies undergo a secondary rhythmic cycle because there is a, discontinuous process. This is called *menstrual cycles.* The menstrual cycle is the interval between the onset of one period of utrine bleeding to the onset of the next one.

In human female this meanstrual cycle may range from 20 to 35 days with an average of approximately 28 days. The period of menstruation normally Continues for five days. At the end of menstruation there is an increased out put of *Follicble Stimulating Hormone* (FSH) by the avnterior pituitary gland. This hormone increases the growth of a follicle in the overy. The growing follicle also secretes the estrogen hormone which increases the growth of the vascular and the tissues in the wall of the uterus making ready to receive, the ovum.

The estrogens, further, stimulate the *Luterinizing Hormone* (LH) to secrete through the anterior pituitary gland. The anterior pituitary influences both the Luterinizing Hormone (LH) and the Fallicle Stimulating Hormone (FSH) and stimulates follicle to move to the surface of the overy and release the ovum after forteen days. The ovum is taken by the follopian tubes of the uterus and put into the walls of the uterus after seventeen days. These three days are supposed to be the longest fertiality period. After this period several processes are taking place. The estrogen functions to facilitate the development of the ovum in the wall of the uterus. Mammary glands are also stimulated by the prolaction for the production of milk.

If fertilization is successful, pregnancy takes place. Then a placenta begins to grow in the wall of the uterus for the nurishment of the embryo. The placenta also secretes that of progesterone and chorionic hormones. But these hormones inhibit the follicle stimulatring hormone (FSH) production and the process is made to continue in pregnancy. Now the pregnancy is under the influence of progesterone and choronic hormones. At this stage there is no chance for further fertilization and the fertilized ovum may grow into an embryo, eventually becomes a fetus. Thus the endocrine functions of the overy are regulated by the gonads and other glands, but under the direct control of the pituitary glands. There is also a reciprocating influence of overian hormones on the pituitary secretions. Though the pituitary inhibits and controls the overian function at puberty it does not inhibit the gonad altropic activity.

In summarising the gonadal functions, we may state that when ovum is fertilized by a sperm cell (male reproductive cell), it develops into an embryo in the wall of the uterus. The female reproductive cycle has an average of 28 days beginning with the end of menstruation by the function of the anterior pituitary gland. The hormone facilitates the growth of follibes in the overy. The follicles are the tissues formed in the overies that produce egg cells. The growing follicles secrete estrogens. The estrogens also stimulate the anterior pituitary gland for secreting hormone. Thus the anterior pituitary gland is having direct control over the gonads and in return receives stimulation from the gonads.

The important functions of the endocrine system rest not only in the regulation of the general metabolism of Che body, but also in the normal physiological, emotional and personality developments and the total behaviour of the individual organism. Any severe malfunction of the endocrine glands may result in some physiological and behavioural disorders in the absence of hormonal treatments.

7

HUMAN DEVELOPMENT
(Developmental Psychology)

Introduction

Psychology is the scientific study of human behaviour. Certain behaviour patters are observable and certain not observable. But at what stage behaviour occurs in humans and when the psychologists begin to study the behaviour? This is the question often raised in the common sense world.

Physiogentically and psychologically psychologists are interested in assessing behaviour right from the time of the conception and the development of the human embryo in the early stages in the uterus and the organism's development till it is - born. Then the infant's growth and development physically and also throughout the course of the infants development in the early years, the childhood period, adolescent period, early adulthood period, later adulthood and old age periods. Psychologists show interest in studying the behaviour through all these stages.

The idea behind the interest shown by the psychologists through all these stages is to discover the *potentialities* and the *repositories of traits* with which he begins his life and the ways in which his behaviour changes or modifies and matures as and when he grows and matures with age.

Origin and Development of the Human Organism

Generally, human life begins in a *fertilized ovum.* For example, the physical development begins in the fertilized ovum nine months before the birth. The union of the male and female reproductive cells create a new human organism with same specific growth potential and developmental design. This single cell organism becomes at once active in the process of the first cell division. During the period of first month of existence the tiny organism has already grown to ten thousand times of its original -size (the gene). It is grown and also matured as required for the period.

The total developmental journey of the fetus from a *small spot of watery material* to a weight of six or seven pounds and ninteen inches long human being takes about 200 days and is a sign of wonderful refinement. But in the beginning the process is mainly of growth and development (Fig. 7.1).

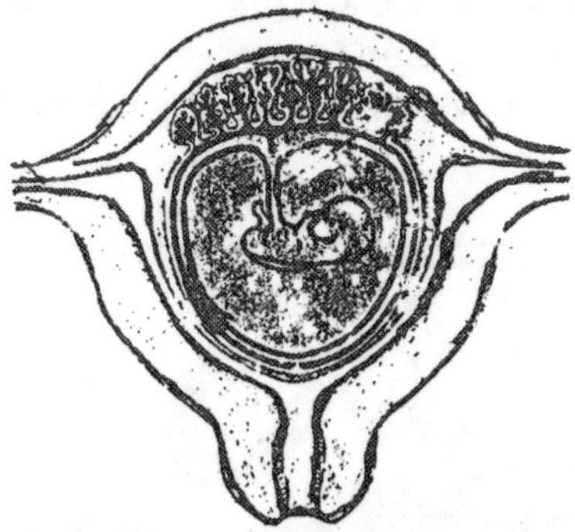

Fig. 7.1 : The human embryo in progressive stages of development.

During the period of first week of conception the tiny cell flots feely and passively in the uterine fluid. At -this time it contacts the wall of the uterus and a dramatic change takes place in its behaviour. It becomes active and aggressive and in a short time it has totally imbedded itself in the wall of the uterus. It then intermingled with blood rich tissues of the uterine wall and getting absorbed with the oxygen, corbohydrates, protines and minerals necessary for servival and further growth and development. The new organism has become a parasite. By the establishment of oxygen resources and the supply of food the growth development process becomes rapid (Fig. 7.2).

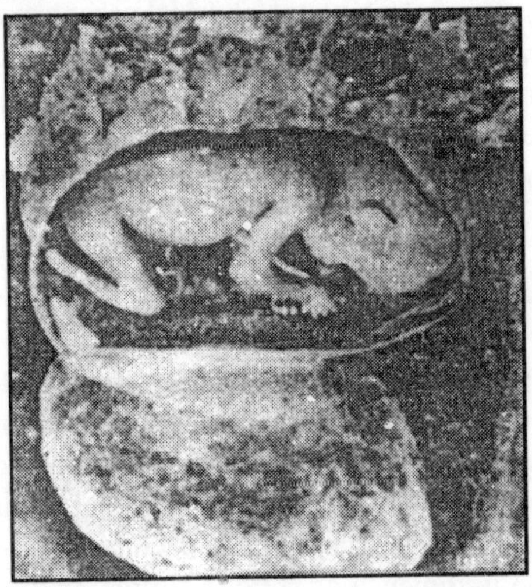

Fig. 7.2 : The infant is lying within the fetal membranes just before birth.

Then the cell layers get, differentiated. One end of the embryo gets the shape of head. At the beginning of third month it grows little longer and the bodily parts are also differentiated and formed. The human organism gets vital qualitative changes by the chemical substances it finds in the mother's body. This period is called *organogentic period,* in which there is rapid growth and maturation. At this stage the grown and matured organism is called *fetus,* which is going to be a baby. The fetus *gets its own* independent *nervous system* and also *circulatory system.* The fetal blood supply and the mother's blood stream supply exchange materials within the organ which is shared by the mother and the fetus through the *placenta.* (Fig. 7.3).

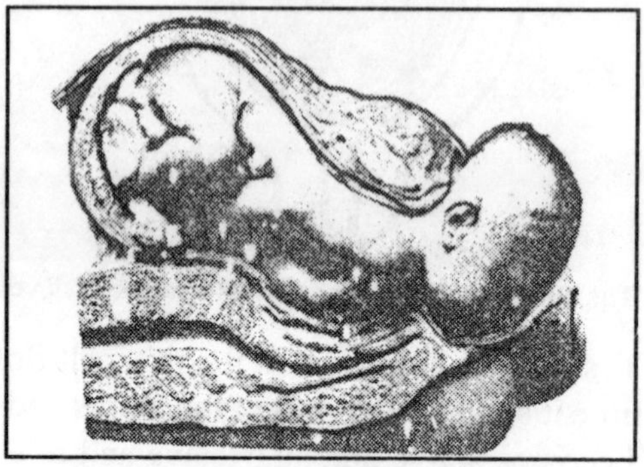

Fig. 7.3 : Shows the infant propelled through the birth canal.

Generally psychologists who deal with human development main4y define development in a broader sense. Development according to them means change. It is any and all changes in a living organism with the passage of time. For them changes include that are on going and generally irreversible, changes in both the physical organism and in its functions dealing with the processing of living and functioning. For the psycholcgists the development of psychological functioning in an organism is the behavioural changes. It is mainly concerned with cognitive developments. This includes developmental changes in emotional patterns with verbal behaviour. These are related to the process of thinking. All these things change with age.

Physical Developments—Growth and Maturation

One of the most remarkable things we must consider about children is their rapid growth and development. We can see that the new-born child with all his disabilities lying on the bed helplessly. But he has the capacity to grow and develop and carry out his activity when he grows. He is capable of growing to become a successful human being.

Everybody is aware of the child's capacity to become a full-fledged man. But it needs time. A continuous change in dimensions is not a simple matter as it appears. Along with other aspects of developmental changes there are tremendous implications in the life of the developing individuals. In assessing the child's growth stages in his various dimensions we can observe that his behaviour is more comprehensive and meaningful. The growth period of a child is often subjected to comparison of his various movements and activities, whether the child is capable of judging the expected level at this stage.

Parents and others may raise the question, 'Is the child developing at a satisfactory rate?' In which an aspect he reached the expected level? In what aspect he is falling behind? Whether his physical growth is normal? All these curious questions the elders may raise and find out answers for them.

Psychologists make the individual studies more significantly. They- also compare children of the same age in various aspects and also prepare the developmental profiles. In comparing the growth and development of some children through their developmental profiles not two developmental profiles look alike in the total pattern. It can be observed that each child may have his peculiarities. The assessment is whether the child's growth and development is normal or he is differing from the average level of development? Whether the difference is above the average or below the average?

The developmental process is a continuous change and a positive progress is always expected in a child. Life is generally a dynamic process of change and we are interested in knowing intensively what is taking place at a particular period of time. But from the time of conception the conditions prevailing around the small organism are of very crucial importance to his development. The interaction of heredity and environment facilitate the child at every state for the magnitude of change and development. Thus the organism changes from the stage of protoplasm to a complex functioning human being.

The change occurs in the organism not only in height and weight, but in various parts and segments as a whole. The child becomes taller, his head and trunk lengthen, his shoulder gets broader, his arms and legs get longer and the dimensions change in each part at his own individual rate at different periods of time. All these changes indicate the significant features of growth and development.

The extremely important development is the relationship between the physical growth and other aspects of development proportionately. There are also many possible connections between the variations in structural development and psychological functioning. Generally growth refers to increase in size. But size is not a full dimension. Along with overall vertical dimension (height), the width on the horizontal dimensions of the different body segments would be involved in a complete assessment of the body size. But the two overall indexes of the most commonly used are the *total height and the total body weight.* These two dimensions in relation to each other would tell us much about the body type which may be useful in approaching of the individual child's growth stage in the judgement of general expectations.

Growth and Maturation in the Human Infant

Of all the living organisms only human beings take longer time to get acquainted with the whole environmental conditions. At every stage the human infant needs care and protection with all safety measures. The growth and maturation in human infant is very slow and at every stage there are instances, which may put restrictions or handicaps which may delay certain development in the organism. Animals of all kinds begin to move and walk soon after the birth. The chicks peak the small grains as soon as they came out of the eggs. But man alone takes years to stand on his own. Therefore growth and maturation of the human infant is not only risky process but also it is time consuming. The child needs a close care and attention continuously for atleast four years. Any deficiency in care or sickness may impair the growth and maturity. If the infant develops sufficient physical strength and cognitive ability to certain level then we can understand that the infant gets ready to learn certain things very essential.

Cognitive Readiness

Growth and maturation in human infant takes place step by step. It takes place from too general to specific. In the beginning the infant proceeds from one of mass activity to one of more specific directed activity or response. In slow movement the child slowly gets up, stands and falls down several times and then stands up steadily and makes himself ready to walk. This development will not take place suddenly, but in a step by step stages. In the beginning he tries to walk by holding the hands of the elders and then he tries by himself. If he could do this he expresses his pleasure by smiling and laughing and by making some noise and by clapping his hands. This may take place in fifteen months in some infants and more in the case of some infants.

Walking alone without the help of elders is an important development in the life of a child. It indicates the child's bodily and muscular control. But if there is practice in the activity by way of encouragement and stimulation by the elders it may speed up the activity. The child in the beginning may raise some sounds when the elders speak to him. This means, he is also trying to speak something with some mono syllables. Therefore, talking, scribbling, writing, drawing are all results of learning activities.

The new-born child may make some reflexes and response patterns. But these reflexes and responses are rarely specific. He makes the same reflexes or response patterns to all kinds of objects, humans, animals and birds. Growth and maturation involve also changes in the nervous system and muscular parts of the body. The increase in body size also should go with growth and maturation. Simple and essential behaviour patterns will take place only under appropriate levels of maturation with some learning or no learning. Providing training may have some effect as the hereditary factors may have some influence in maturation and also may limit it. It is very difficult to overcome the hereditary limitation on learning and training.

As the individual human organism going on increasing in his age, his body structure is also going on changing, in other words, enlarging the body in all aspects. It is not a sudden change in the body building, but gradual, depending upon food, nutrition, heredity, environmental conditions physical exercises and so on. The body changes take place in this way.

Learning to walk, like animals, does not take place soon after the birth. In the first month with great difficulty the infant raises his chin up. In the second month or third month he raises his chest up. In fourth month he could stretch his arm and try to reach some object, but misses. At fourth month he could grasp something. In fifth month he could sit up with some support. At sixth month he could sit on the lap of the mother or father, and see groups of objects. At sixth or seventh month he could sit on chair and hold objects, throw and catch. He crawls and moves on fours (legs and hands). At tight or nineth month the child tries to stand up, but falls down, he also tries again and again till he succeed to stand correctly. At this stage his legs are not straight, but pending and could stand with help. Mostly at this stage he moves on fours fastly. At eleventh or twelfth month he could walk when led and also tries to get up by holding something. At thirteenth and fourteenth month he could claim over the steps by crawling and stand erectly without any help. After completing the fifteenth month he walks alone without any fear. (Fig. 7.4).

Fig. 7.4 : The child is learning to walk. In the beginning he is guided and helped to make his steps and later on when he gains strength he is able to walk by himself.

Physical Growth and Development

The physical growth and development is also stated as the human motor development. The heredity and environment, which we have discussed in the preceding sections are also involved in the study of maturation, the behavioural changes that result from physiological growth. In this context we wish to know the extent to which age-changes the behaviour as the result of maturation.

The body changes take place proportionately right from the time of birth of the infant through the period of childhood, adolescent period and early adulthood, the matural individual state. In this prenatal state the head and the upper parts of the body may become small proportionately and the other parts become larger and stronger. In the development of human organism many dimensional changes take place, for example, the total weight and the total height increase with time. But in this type of development wide individual variations are taking place. It was observed by **Krogman** (1956) in the measurement of various component structures -in relating to different body types.

Some extremely important relations between the physical and other aspects of development were obtained in various studies and were appraised. But certain aspects between physical structural development and psychological functionings are yet to be studied. **Gesell** (1925) made extensive studies about child's physical developmental sequences. According to this author, the developmental sequences differ on the basis of th&' type of home in which the child is born and brought up and the culture and society.

At fourth week, the child is able to control the eye movements, can look at the objects visually. At the age of sixteenth week, that is, when he becomes four months old he can control his head and maintain balance. When the child becomes seven months old he can group and manipulate objects with his hands. When the child completes ten months of his age, he cou!d control his body, can crawl and also could sit, when the child completes one year and two months, he could slowly learn to walk with some help (Fig. 7.5).

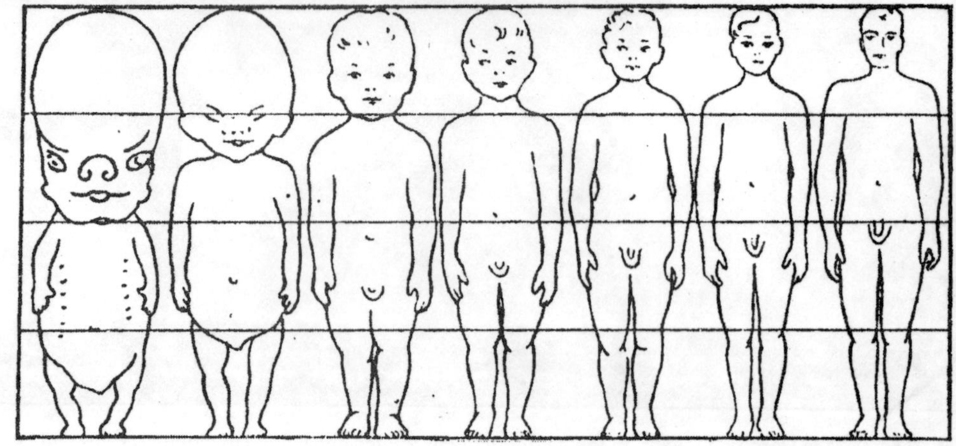

Fig. 7.5 : Physical growth and development right from the time of birth to adulthood. The physical growth is very rapid during adolescent period and then it takes place slowly till the age of twenty-five and in some cases even after the later twenties.

When the child reaches his sixth year of his age he is supposed to have his brain developed ninety percent of the adult size. This is supposed to be the normal state. If this takes place there is no chance for the development of mental retardation after words. The body of the child grows fastly upto the age of five or six years and thereafter the growth is little slow till the age of twelve. Then the period of puberty begins at the age of thirteen and there is rapid growth and maturation. There is sudden growth in physical appearance and the development of sex organs and the bodily growth continues till the age of twenty. The adolescent period is marked by various types of development. The height, weight, body build, strengthening of muscles in various parts of the body, like arm, hand, legs, chest, neck and so on.

As and when the individual person grows and matures, he gains more physical strength and gets ready to perform all types of activities physically. Thus his performances are going on changing from simple physical activities to more difficult and enduring activities. He takes part in various types of games and sports activities and also widens his areas of physical activities. The change of performances with increasing age are mainly in certain physiological and motor capacities. The individual as he grows and matures is very much interested in exhibiting his physical skills and talents in muscular and in motor control. He may show interest in walking, running, swimming, climbing, writing, playing games, taking regular exercises in competitive sports and so on. All these are parts of exercise, development of skills and talents and maturation. The individual's nervous system and muscular system are supposed to have reached an appropriate stage of development. The person has become ready to perform any type of physical and physiological performances.

Growth and Maturity and Sex Differences

Generally the growth in height is very fast in *adolescent period* and it may complete at the age of 20-22 years. In some cases the growth of height may take place a little late at the age 22-23 years. Till the age of 12-14 years the height between the boys and girls may have no difference. But after 14 years the difference in height between the boys and girls differ considerably and the boys grow taller than the girls. Girls attain maturity earlier at the age of 13-15, earlier than the boys. The girls during the maturity period grow taller and are more mature. After puberty the boys get ahead of the girls and grow taller than the girls on the average level.

At the puberty period on the average the girls are heavier than the boys. A girl at the age of ten may show some difference or deviation from the normal height and weight for her age, may be taller and heavier than the other girls of her age. Generally children who are slim body build have small bodies are found to be taller and lighter than the average children. Some children have their body build with large bones grow shorter and heavier. Therefore we can observe in d population children of the same age may very considerably in heights and weights and body builds. The cause may be rooted in genetic and environmental factors, quality of food and bringing up and exercise and so on.

Psychologically girls get maturity earlier than the boys almost in all respects. Apart from physical growth, also in muscular viceral areas. They get mental and social maturity earlier than the boys. Psychologists have observed that at the age of six years the boys fall behind the girls about six months of the same age. At the puberty period the average boys of the same age fall behind the girls by two years. For example, at the age of fifteen, the boy's maturity level is found to be thirteen, that is two years falling behind the girls. It was also observed that at the elementary school level at certain age level girls show much better muscular control than the boys. The girls show much better adaptation with their hands, better muscular control and manipulation than the boys. The muscular control is seen in writing and in handling materials. Generally, it was observed that at the elementary school level and at the early high school level in comparing the performances of boys and girls of the same age the girls are more matured physiologically, emotionally and intellectually than the boys of the same age.

Development of Cognitive Function

In the preceding pages we have dealt with primarily the development of the child's physical and physiological functions and the bodily functions in the environment. This we dealt with from prenatal stage through infancy and to smooth and efficient performance in the later period of adolescence and the manner in which the child acquires the abilities to move and act independently.

We shall now deal with a higher level of functions in relation to the environment in which the central nervous system and the brain play a move important role. The *higher level of activity,* generally, is stated as the *cognitive activity.* It is the activity of the mind. Actually in making a distinction between the physical or motor and mental activities mainly one of the degrees of evert neuromuscular activity is involved. Any physical or motor activity is centrally mediated and controlled by cognition and some of the motor activity is generally the cognitive activity. In the newborn child the mental activity is gradually emerging. The child as he grows generally begins to understand and cope with his environment through his motor responses to the environmental stimulation. There is change in his eyes, responds to sounds of other persons' steps and voices. He reacts to the objects within his reach. He threashes, kicks with his arms and legs. The child grasps and holds objects, which he could reach, mostly mouths the objects. These sensory motor activities involve awareness of the objects of the child to the stimulations. (Fig. 7.6 and 7.7).

Fig. 7.6 : The child grows and develops rapidly with experience and becomes mature. His motor behaviour is mostly associated with his experience and cognitive awareness of objects and persons. A cognitive test was given to four years old child. There were two clay balls of the same size. The boy said they are of the same size. One ball's shape is changed into longer and the boy said that it has more clay.

Fig. 7.7 : A cognitive test was given to some female children. The test consisted of two rows of checkers and they were evenly spaced. Most children said that they were of the same number. The materials of one row were collected in a small space. The children of six and seven years old said that the original row had more number of materials.

The first test was with regard to shape and size. The second test was with regard to shape and number. How these two groups of children have perceived the objects and what type experience they have gained were revealed in these tests.

The child grows and develops rapidly with experience and becomes mature. His motor behaviour is mostly associated with his experience and cognitive awareness of objects and persons. His reaching to the situation becomes meaningful and purpose. From these experiences the child slowly form the concepts of the objects, situations gradually. The child's ability to differentiate between the motor activity and the cognitive activity involves to a greater extent the functioning of the central nervous system. The implicit behaviour of this type is generally referred to mental or cognitive behaviour. The infant reveals his mental ability by means of his motor activities.

The child as he grows and develops and matures his cognitive functions continue expanding in various ways and when he interacts with his environment successfully the efficiency of his behaviour improves. The total controlling functions that take place in the child with his personal guidance and action may be attributed to the development of his intelligence. In every successful motor or physical or physiological performance the level of the development of intelligence is revealed, The term intelligence is used by almost all the people in varied situations irrespective of its correct meaning. The common sense world meaning generally refers to the successful performance of an activity by a person mentally or physically. But we cannot totally reject the understanding of the common sense world about intelligence. There may be some difference in the degree of its description or application.

On observing the behaviour of a person in a private or oraganizational situation we often hear people speaking that he is very clever, very bright, talented, very much thoughtful, did wonders, very stupid, dull, incapable, no brains and so on. All these indicate that they are speaking about the person's mental abilities or intelligence. This means people of the commonsense world think that a person who performs a task or solves a problem successfully or satisfactorily is an intelligent person. Then who is an intelligent? The one who performs the given task successfully or who solves a difficult problem successfully without much strain or a person who deals with a situation amicably or makes achievements in some area, which may be useful for many people is an intelligent person. These explanations can be accepted not only in the commonsense world, but also in the psychological world. Because these explanations apply generally to all people.

Generally, in performing some task, solving some problem, winning a competition, discovering something new, creating a new machine or instrument, evolving some new idea or philosophy are all considered to be cognitive skills and developments. It is an accepted idea that through cognitive development a person becomes a more adequate individual, capable of making greater number of *intellectual achievements*. Man can make achievements when he becomes an adult if any of his mental developments mature in much the same manner as the physical and muscular developments take place. The mental development should always go with the physical and muscular developments proportionately. Otherwise the individual may fall back which may result in mental retardation. They should also develop in making use of wider vocabulary,

Studies show that at the completion of one year, the child may be able to voice two or three words. On completing two years, a simple sentence or use of words in combined manner may be expressed. The child first learns only nouns or learns one-word verbs like come, go, give, put and so on. While using these monosyllables,' either nouns or verbs the child does not use pronouns. To pronounce the pronouns is a very difficult task for the child. If at all, if he uses, it may be referring to something also. For example, he may indicate the cat or dog which he often sees as 'he' or 'she', instead of using the pronoun 'it'.

The mental development and the child's ability to speak or the developments of cognitive abilities depend upon several factors. First of all the bringing up the child, the type of home or family environment, the type of food with which the child is fed all may have some influence upon the mental and cognitive developments. The social environment may be unstimulated and impoverished. The culture in which the child is born may be educationally backward. This may result slow and late understanding of things and events. Psychological studies state the level of the child's vocabulary for different age levels. That is, what a child's vocabulary at the age of eight? It varies from culture to culture. Whatever it is, at the age of eight years he should acquire roughly about twenty five percent of it and at the age of twelve he must acquire roughly about fifty percent of it. The level of vocabulary acquisition mainly depends upon the mental development of the child.

Generally, when a child begins to speak combining three or four words, he has reached a stage in which he can use more words as he progresses. The mental ability of the child is revealed in his usage of words and phrases. In the usage of words and phrases he may commit errors. But the parents and other elders correct it and help him to speak without errors.

In the next stage, he tries to read and write. Generally the child's learning task of reading takes place first and secondly he begins to write. He may commit errors both in reading and writing which is the indication of the developmental stage. Some children in a culture begin to learn at the age of four, in some culture at the age of five, the child is taught to read and write. In some culture the child is taught to read and wrote at his age of six. If the child's mental development is normal than the reading and writing can go together. When the learning takes place with reading and writing the words of a language, there may be some confusion of words. The child may look at the words differently and also may write the words in the reverse order.

For example, the child may get confused between the letters of g and q, h and b, c and e. Similarly in the case of words there may be confusion between book and look, cat and rat. These are common type of errors most of the children may commit in the beginning. The child must be taught frequently the words, which he does not know or fails to recognize. The hard or difficult words, with which the child struggles, may come across at the age of seven or eight) needs to be helped. At the age of eight, the child may go to the stage of guessing or thinking, if he comes across some unknown words. At the age of eight or nine, the child may use the correct pronouns like he, she, it, which indicates that he has understood the correct meaning of the words which he had failed to do ht the age of six or seven.

Generally it is expected that in a normal child the general mental ability develops rapidly from birth to five or six years of age. The rapid development of mental ability should be parallel to the development of the human brain and the spinal cord and the general growth of the body as a whole.

In normal child, who is free from any impairment physically and mentally from the year of five or six years on to fifteen or sixteen, the adolescent stage, the mental ability develops at a slower rate. In a class-room group, all children do not perform the classroom assignments at the same level at the same time. Some complete the task rapidly in a short time while some complete he same task slowly and take more time. This means that their levels of intelligence are varying. Some are bright and some are dull. The bright children at the age of twelve complete the twelve years level tasks successfully and also may go up to even thirteen or fourteen year old level tasks. But the dull children could perform at the age of twelve only ten or eleven year level tasks successfully. They are two years behind the normal level. The bright children are above the normal level and the dull children are below the normal level. The children's learning ability goes with the development of the mental ability, in other words intelligence. This difference in the development of mental ability continues the whole life of an individual. But experiences at certain level may compensate it.

Factors Determining Individual Differences in the Development of Intelligence

The mental activities of the individuals vary considerably in every culture and society. What a child can successfully complete a mental task at the age eight years, another child completes the same task at the age of ten years from a rural or an unstimulated area. But the individual differences in the same culture or society is an important subject matter and problem for psychologists. Man is always interested in his own inner life affairs. Whether a man is normal and average or intelligence or moderate is very curious about his own capacities and abilities and also feels how he has come up with his abilities to this stage. Every man is observing wide differences in the mental abilities and physical skills among the individuals. He finds functional evidences in everyday life. Society finds serious problems of what to do with and how to take care of the *extremely deficients*. The approach to the defective infants in olden days was very cruel and unsympathetic. They were treated very badly by brandishing them as devil haunted or mad. The *mentally deficit* were *punished* and *tortured*.

The development of scientific approach to deal with the mentally deficits emerged after very long time. The scientific era of modern times thoroughly changed the situation and efforts were made to understand the nature of intellectual deficiency and improved many of the mentally defectives through education, training and treatment. In dealing with the extreme deviates the assumption has been that these extreme deviates were born like that and society with its customs and traditions, modes of social interaction could not be responsible in anyway for such complicating problems. Some successful procedures and methods were evolved to deal with such problems only after inventing the intelligence tests to measure the mental abilities by Alfred Binet, Cattell and others.

Biological Factors

Several investigators made valuable contributions- to the development of techniques for the measurement of individual differences. Associating heredity to the individual differences in mental abilities was widely prevalent at this time, in the last quarter of nineteenth century. The strongest reinforcement and the greatest advancement of knowledge came out from the British Scientist, **Sir Francis Galton.** His first work in this problem was the *Hereditary* Genius, originally published in 1869. Galton has strongly implanted the idea that intelligence is biologically inherited. Galton strongly believed that people differ from each other in abilities by measurable amounts and they cannot be grouped as separate types. The prevalent belief at that time was that people can be grouped under the heads of idiots and genius and this belief was demolished by Galton. The strong emphasize that Galton put forth was upon biological heredity which was the only factor that determines the individual differences in intelligence.

This view stressed that children should not be pressurised in anyway. Parents and teachers should take care of children, but should not interfere with the fundamental processes of natural development of children. Children should be protected from any undesirable environmental conditions. **Arnold Gesell** (1925) was of the view that maturation is biologically controlled, which is the unfoldment of an inherent design in the individual development. The idea that the general intelligence was the basic assumption from which intelligence tests were devised and from such results interpretations were made.

The idea stressed is that the inborn general intelligence increases with age. But intelligence was conceived of by some differently and superficially defined as the inherited potentiality for the mental development and was assumed that the growth of the amount of intelligence as the child grows toward adulthood was always within the limit of the inherited potentiality.

But until 1920s the hypothesis that intelligence is an inherited potentiality was not challenged. In 1920s the emergence of behaviourism under the leadership of **J.B. Watson** threw a strong protest and challenged the validity of the assumption that intelligence is an inherited faculty. The Child Welfare Research Station at Iowa University carried out several research programmes about children, their nature and how they develop. They have also studied the potential effects of Nursery School upon the mental development of children. Several investigators carried out research works on this problem. The results of these studies firmly established the assumption of biological determination of intelligence.

Environmental Determination

Research investigations were conducted on human development in various areas regarding the nature of intelligence and the factor that determine the levels of intelligence. The environmentalists emerphsized the importance of the Cultural and Social influences and other factors of environment that determine mental abilities and other characteristics. The child psychologists and the educationists at this time started emphasizing the influence of environmental factors. This led to the stage of almost neglecting the factors of genetic endowments. They even went to the state of developing hopes that by placing the culturally deprived children in intellectually stimulating environment the mental deficiency generally found in such children can be overcome.

The educationist at this time were of the view that the functional intelligence in children can be made to generate intelligence by rich stimulating environmental conditions. They were of the view that it is possible to create intelligence by controlling and manipulating the environment suitably to the needs. But these researchers do not also deny or reject the basic reality and the importance of the genetic endowment.

In psychological studies and in the study of human development the interaction between the biological and environmental factors has been accepted by those investigator, who made intensive studies of basic developmental processes in the fields such as biochemistry, genetics, developmental physiology and psychology. According to this view, the biological factors and the heredity and the rich environmental stimulation are very essential to life and to the physical and mental development. It is also stressed that the factors of these two areas become the cause of the development of all aspects including the physiological and mental faculties. But, however, when one observes differences between individuals in the society, specifically in the levels of development in the structure and function, generally, difference are attributed entirely to the differences in heredity or to the environment. Apart from these factors there may be also other factors underlying the individual differences in a culture with different socio-economic status.

The Nature-Nurture Issue

Researchers of various disciplines wished to investigate the issue of nature-nurture as to whether a particular human trait in the case of intelligence can be entirely due to the result of heredity or environment, **Boyer** and **Walch** (1968) in their research work raised the question, "Are children born unequal?". According to them the extensities of individual developmental potentiality are genetically determined at conceptions.

It is also viewed by some that there is no scientific proof that individual's intellectual capacity is innately unequal. But at the same time there is evidence in the hands of many that a wide range of congenial inequality exists among individuals in the same cultural, social and racial groups of all sorts. It is unexplainable in terms of environmental influences alone. It is very conspicuous that children are not born equally endowed with every trait. We have to accept certain things that we see in the everyday instances. It is true that no amount of love, affection, nurture, no amount of discipline and pressure, or training or tutoring nor any amount of rich and stimulating environment can close the gap of differences between mentally deficient and the more richly talented. Once the gap between the two extremes occur there is no way of bringing them close together with the same level of intellectual performances.

The investigators **Boyer** and **Walch** (1968) stress that it must be realized that a person's genetically inherited potentiality and the potentiality with which the child is born can be very much different. There are also many possibilities for environmental accidents and influences, both prenatally and parentally, which can reduce the developmental potential to varying degrees below that which was set genetically at conception. It is generally accepted at certain levels that individual differences in cogenital-mental-potential seem to be beyond any correct scientific investigation.

It is also well established that children from different cultural socioeconomic status and racial groups differ in average mental capacity on going into the elementary schools. There are also group differences in the level of mental capacity to profit immediately from school performances and experiences.

However, the doubt arises as to what extent, if any, these average group differences are having the basis in genetic mental endowment is one for which we must get the correct answer with evidence. This issue is prolonging for a long time. But there is also an intention indirectly prevailing not to clearly differentiate, the social and cultural groups and to bring out the individual differences within the group members. Because, if the individuals or group members come to know that they are intellectually backward some psychological problems may arise which may result in humiliations and depressions. In view of these the current trends of the educational theorists are in the direction of environmental determinism in relation to social and cultural group differences. Therefore the nature - nurture issue is a never ending one, needs more scientific studies.

Speech Development

All living creatures in one way or other make contact with one another and with the groups by way of communication. This communication with one another may be in the form of behaviour with sounds, gestures and with other motor patterns. Sub-human species generally interact with one another and influence one another's behaviour. The sub-human species influence one another with their gestures and can being about community of action in the group and also with individual behaviour. The ways of communication in subhuman species is limited. But the human communication takes place always with a purpose and with deliberate intention for which human uses a medium, called language. Language is a medium of communication in spoken and written forms. But unlike sub-human species man must learn to make use of language if communication is to take place.

We are now interested in knowing the development of speech in the individual human being, the ability with which man is endowed with for learning to speak a language and to communicate with others. Man must develop his ability to transmit and to receive meaning and information in verbal or speech interaction with others. The child has to learn the vocabulary of a language for speaking and writing. This ability to learn makes the child to transmit and to receive communication with specialized meaning. This is an important experiment for further acquisition of learning difficult and complicated materials. A language in this sense is a system of specialized verbal symbols generally accepted and used in the transmission of ideas with meanings.

The Infant Speech

The human infant is expected to have born with structurally and physically to produce sounds which later on develop into words in speech. The sounds the new born infant makes are uttered in an indifferentiated cry. The infant's first prenatal act is the production of a vocal sound. This sound naturally is a monosyllabical crying sound. This type of sounds constitute about ninety per cent of the infant's total vocal sounds during the first few days of his life. But it can be also observed that some babies going on crying most of the time when he is awaken, except during the time of sleep. Some babies do not cry much. They may be making some laughing sounds.

A large number of investigators have studied the crying sounds of many babies. They have observed that during the early infancy period there was usually very frequent crying. Most of the crying of their studies was emergency respiration. Their observations reveal that most of the crying of the babies was of defense. This is for making the elders to know about his hunger on discomfort. **Brown** (1958) has stressed that the crying mode of communication of the infants is very imperfect. It is because, the infant's crying sounds have the same quality on all situations, whether it is discomfort, or hunger or painful feelings, the nature of the cries are not expressing any specific cause. But after few months the baby learns to use crying in a more efficient manner as a means of getting the attention of the elders apart from any bodily discomfort or needs to be fed. During the early period the infant's crying sounds generally consists of eight different type of sounds which are distinguishable. These eight types of sounds mainly refer to five vowels and three consonants (Irwin, 1949).

Human beings use language for communicating their ideas to others. Language in a broader sense is anything whatsoever is a communicating instrument. Human beings generally communicate or transmit their ideas and feelings to their fellow beings. The language that is used in speech for communication is a highly evolved system of articulating of vocal sounds to express the idea and feelings meaningfully. Psychotogists generally call the speech development in children as verbal behaviour. Verbal communication is the behavioural manifestation and expression of mental functioning of a higher order. Therefore the speech development may be considered as behavioural and cognitive aspects.

Speech and Physical Structure

The main component of speech is the sound emitted by an organism. The sound is produced by an organism by making movements in internal organs. In this, the physical structures are involved in speech are the lungs, the windpipe, at the top the vocal folds and the glottis, the cavities of the throat, the nose, the mouth with its intricate systems of muscles, the soft and hard plates, the gums, the teeth and the lips. All these parts of the body are involved in a speech. The important aspect of speech is based upon coordination and regulation of these various parts of the body. As they all operate in a systematic manner varieties of speech sounds are produced. Any defect or damage in any part of these parts would affect speech sounds. For example, apart from the internal structures, the formation of teeth and lips may cause some change in the normal speech sounds. The teeth may be protruding if care is not taken in child rearing at the time of the appearance of teeth, the lips formation may be thick and projecting. These defects may change the normal voice and expression of words may be very unclear.

The human infant is born with the capacity of producing the components of all possible sound categories (phones). This aspect also functions like all other developments through exercise or learning.

Growing Vocabulary

As the child learns to speak, there may be increasing of new words. He may learn everyday some new words. Studies were made on speech development of child as he goes on increasing in his age. During the early days of speech learning the child's comprehension of vocabulary is much more than the speech vocabulary. At the age of two and half years, the child could speak about 250-300 words and going on increasing day by day. His level of concepts widens and becomes more precise, and specific. The child at first learns mainly generalized words and then he learns the specific words referring to his mother, father and others. By the end of three years of his age the child speaks more meaningful words.

However, the speech development varies from culture to culture. But the normal development and th6 accuracy of words must go with the age of the child. The appearance of the mother often in his presence makes the child to understand and demand many things by pronouncing the words in a clearly undifferentiated manner. The child learns to associate his usual, auditory, olfactory, and gustatory experiences of the objects and learns such words in an incomplete manner and slowly as time passes, he correctly, pronounce the correct words of the objects and understands the meaning of the words. The use of singular and plural words and the relative pronouns he learns by trial and error methods. When he commits mistake the parents correct it and help to use the correct words. As he grows in age his speech development becomes much perfect and never feels shy to use any words of his language.

Factors Affecting Speech Development

The speech development in children is one of the most striking factors observed by the researchers in the field of child psychology. There is very wide range of individual variations in the level of achievements. Speech development in children is a very complicated process as learning is involved in it. Any of the factors that affect the learning process in the child also affect the speech development. The child has to learn to speak.

There are strong evidences that the level of progress in mental development is the basic factor in speech development. It is the mental development that facilities the learning process in children. It was observed that children who are found to be bright mentally show linguistic superiority and learn to speak the language. **Johnson** *et al.* (1950) have observed in their studies that children who are lower in the intellectual development the poorer in speech development.

As in the case of any other life aspects which require favoural situation the environmental conditions also influence the child's speech development. The environment should have the stimulating effect in children for developing speech. Elders at homes if move about without providing any resources for the child's speech development, the child may take longer duration to learn to speak. Children of lower socio-economic status lack facilities for the development of speech. If the parents are going to work place in the morning and returning in the evening the children are let out without any chance for learning to speak a language fluently. These children take longer time for learning to speak.

Emotional Development

Human life would be an empty vessel without feelings and emotions. We cannot think of human life without feelings and emotions. It is said that emotions are the species for human life. It is very difficult to imagine our life without ever experiencing joy or pleasure or displeasure, delight or distress, election or depression, love or anger, mirth or sorrow. Human emotions serve as motive and power to live and to strive. It is the feeling of well-being the joy of freedom of activity and the gratifications that make our live happy. Emotion is a behavioural area which man has in common with subhuman species. Man's cognitive processes are mainly involved in emotions. His emotional behaviour are mainly mediated by lower centres of the brain. These centres and other bodily structures involved in emotional functioning are intact and in a *state of readiness at birth.*

Nature of Emotions

The human being is born well equipped for experiencing feelings of pleasure and displeasure and for displaying emotional behaviour. The infant's emotions are not easily assessed. **Watson** (1924) tried to define the infants' emotional behaviour. According to him fear, anger, and love are the three human emotions which can be studied-by their overt expressions. The infants respond to relatively to few stimuli) revealing to these three emotions. The infants' response to some sudden loud noise or sudden loss of support is stated as *fear*. The infants' reaction to restrictions is labelled as *anger*. The infants' response to *foudling treatment* is stated as *love*. Thus in the case of infants' the emotional responses are either, fear, or anger or eve to certain specific stimuli. Thus emotions may be defined in the case of infants in terms of response to certain stimuli favourably or unfavourably, expressing pleasant or unpleasant experiences. There is also approaching and avoidance reaction in the infants' behaviour.

Quite similar to Watson's view, another objective approach to the study of emotions is in terms of observable reactions or assessable bodily changes that constitute emotional behaviour. It is both internal and external. External behaviour is quite readily seen or observed. If the behaviour is violent and destructive or assertive in nature, it is stated as anger or rage or if the behaviour is not violent or positive it is the expression of affection and it is called as love or lust or elation. But if the behaviour reveals retreat or withdrawal or flight it is named as fear. But it is very difficult to assess the nature of an emotion if it is without any observable response in an organism. If an infant does not convey any information about his reaction to some stimulus it is very difficult to know what emotion functions in an infant's behaviour.

Emotion as a functional aspect in human is a very difficult psychological process to make clear study. Emotion as it is a very important aspect of our life. It is a human behavioural area. Man's affective experiences and his emotional behaviour are mostly mediated by the lower centres of the brain. These brain centres and the bodily structures are involved in our emotional functioning. These are intact and in a state of readiness at birth. The new-born child is well equipped for experiencing the feelings and emotions of pleasure and displeasure and he displays emotional behaviour. In this particular aspect of emotional behaviour there is not much difference between the developing child and adult man in their maturational changes. Any type of emotion may be experienced by human beings right from, the childhood through adolescence and adulthood. The type of behaviour may vary but the nature of emotional reaction would be the same at all stages.

The functions and expressions of emotional behaviour involve the whole organism. It is a response to a stimulus, internally or externally. The process of cognition is always involved. There is no emotion without perception of the provoking situation. A situation as it is does not provoke genuine fear or anger, joy or jealousy unless it is experienced either by seeing or hearing, smelling, tasting or touching or thinking of something about which one is associated. Right from the childhood period a person throughout his life acquires sensory experiences, which may be pleasant or unpleasant. Whether it is pleasant or unpleasant a corresponding emotion or feeling is attached to the experience of the object. If it is pleasant there is joy and acceptance and if it is unpleasant there is anger or disgust and rejection.

Studies on emotions are made from childhood to adulthood. The first expression of a newborn child appears to be an outward expression of an emotional disturbance. It is a protest against the discomfort conditions he is experiencing. In the prenatal period the child shows signs of emotional excitements when the external conditions cause bodily discomfort or sensory pleasure. When he feels a pain from some physical disorder or external condition and when he is hungry or when he is wet or comfortable he expresses it at once with signs of distress. When 'ie experiences discomfort he reacts at once. His demands are immediate relief and gratification of his needs.

We can observe that much greater insight into the nature of emotional behaviour come from the clinical studies. In the clinical field the individual's real emotional problems with widely varied backgrounds, and experiences are revealed. The clinician in each case understands and deals with a total, unique emotional situations. The individual's emotional behaviour is fully analysed in the clinical studies.

It is true that each child has his own inner strivings and frustrations, his own self image, his attudes toward himself in relation to others. Whatever type of emotional reaction the child develops during his childhood period and thereafter all depends upon the home situation in which he is born and brought up. What type of emotional expressions he finds from others and in what manner he reacts to them are all experienced by the child either in a pleasant manner or unpleasant manner. The child's later life is almost based upon these emotional experiences. Clinical studies made in the individual's later life reveal how the child was emotionally reacting to the home, school, playground and to the peer groups. Then the emotional development of the child in the early years play a crucial. In this way clinical studies bring out the emotional behaviour of the problem persons in a more realistic way.

Learning in Emotional Development

The child slowly learns to express his pleasant and unpleasant emotional experiences that he acquires in the home situation first. Thus his emotional expressions begin to vary with the situation. The particular source of discomfort he starts to express differentiated expressive signs of pleasure when he gets gratified his demands and also expresses distress from painful experience or discomfort.

As the child's development continues with experience, he learns to become responsive emotionally to wider range of stimuli and situations. Therefore the direction of the child's emotional developments and the quality of his emotional expressions are very much influenced by his environment. The individual person as he grows and develops physically from childhood period to adulthood period learns in the environmental situations with what type of emotional expressions he should react to the varied type of people. He learns where, when and with whom he should be friendly, cordial, pleasing, amicable, polite, silent, disgust, aggressive, docile, dominant, submissive and so on, Thus we may state that emotional development is a product of *organism-environment* (other human beings and events) *interaction.*

The child's responsiveness to the members of his family and other people becomes pronounced. The accepted view is that the quality of attention and the type of care and affection he receives during the early period of his life has very much to do with the development of his later temperament to accept love and affection from others and to develop love and affection towards others. We get evidences for these types of information about emotional developments come mostly from the clinical observations.

Process of Interaction

Emotion is an active aspect of human life. It functions in various forms throughout the life of human beings. We are not interested here the physiological origin of emotion. But we are interested in behavioural aspects and the consequences of its expression. The functions of emotion is mainly the welfare of the organism. The organism reacts in relation to an outside stimulus which may be another hum4n being or an object or other living creature which may arouse some stirred up state. This may result in the organism to react swiftly. It may be a flight if it causes danger or damage to the organism or to *flight and die* if there is no chance for *flight* if it results in need satisfying the organism expresses soft and agreeable feelings which we 'may also call as love. Thus there are two directional nature of emotionality.

The action of the organism is individualistic and alienating and separating. The individual becomes also angry, fearful and withdrawing from the situation or from another person. Thus there is approach avoidance reaction. The approach behaviour is characterized by the positive type of emotions of the individual organism and the avoidance approach is characterized by the negative type of emotions.

Emotional Deprivation

The child does not carryout the emotional adjustment all the times during the course of his developmental stages. The child sometimes lacks in his positive reciprocal relationship with his family members and others. The child may experience the loss of the reciprocal relationship even with his nurturing persons. The child has regularly gratified his needs and warm loving care, but now he is missing all of them. Either he might have been denied right from the beginning or he may be missing now. Both states are disadvantageous to the child. In both the cases the child experienced the deprivation. This deprivation effects are very profound on the developing child. Results of the investigations on the deprivation effects on children revealed very painful feelings.

The baby spends much of his time in sleeping. When he is awake he is very much responding to the 'stimuli of both external and internal. He cries immediately when he experiences any bodily discomfort and when comforts are provided immediately he smiles and laughs. He finds pleasure and expresses it by clapping his hands and moving his legs. He may express his feelings of excitements on seeing persons moving around him. He finds also excitement on hearing the voices familiar to him. Prolonged deprivation of comforts and cares make the child helpless and he develops the feelings of insecurity which in the long run may create in the child the *feelings* of *destitute*.

Social Development

Man is a social animal. He wishes to be always in the company of others to fulfill his needs, ambitions, aspirations and to make achievements. It is often said that 'the child is the father of the man'. It is because the child grows and develops into an adult person and get married and give birth to children. Thus the explanation goes. But the kernal of the subject matter is that the child gets socialized in the society as he grows, develops and matures. The child slowly learns to identify himself with other human beings and learns to live following the social norms, mores and folkways.

One important characteristic of the human infant is, his desire to be in the company of others, and also to be close with other human beings. Generally, human beings enjoy physical contact. The child particularly enjoys the physical contacts and great care and attention from other people and becomes unhappy when they are not near him and taking care of him. The type of behaviour of this is called *physical attachment*. Generally, there is no process, which is more important in the later development than the child's attachment to his primary caretakers, in any culture and society, particularly to his mother, father and other family members.

Development of Attachments

Attachment is having close contact with other persons, places and objects. But generally *attachment is stated in terms of emotion*. Therefore we may state that attachment is the emotional tie of an infant or a child to mother or father or other caretakers. According to learning theory a child becomes emotionally attached to his mother because she is the source of relief from pain or discomfort. In the beginning the mother is the main source of attachment for the child. When the child begins to enjoy or find pleasure in more pleasurable incidents like feedings, being with attractive stimuli, the mother becomes a secondary object of attachment. According to the British psychologist **Bowlby** (1969) attachment is an innate response which serves to protect the young child from the feelings of insecurity. Bowlby states that children seek close contact with their mothers specially when they are frightened. They always are in need of protective response from their mothers. Another important point Bowlby states is that punishing behavioural events do not extingunish the attachment behaviour. In fact it intensifies the attachment behaviour (Fig. 7.8).

Fig. 7.8 : *Development of emotional attachment :* Attachment is having close contact with other persons. But generally attachment is stated in terms of emotion. Therefore, attachment is the emotional tie of an infant or a child to his mother or father or other caretakes. A child becomes emotionally attached to his mother, because she is the source of relief from pain or discomfort. In the beginning the mother is the source of strong attachment for the child. But when the child begins enjoy or find pleasure in more pleasurable incidents like feeding, being with attractive stimuli the mother becomes a secondary object of attachment.

Several investigator's made intensive studies on young children's attachment behaviour. **Schaffer** and **Emerson** (1964) state that the young children always have an innate need to be close with other people. They also found significant individual differences in a group of young children. They found that some babies had developed attachment to one person and some had the attachment to several people. It was also observed that the mother's love and affection or quick responsiveness to the child's crying and the amount or the interaction she extended with the child were found to be related to the child's attachments.

The development of attachment is not based upon a single factor. It appears that many patterns are prevailing in it. This differs from child to child and from one developmental stage to the next. **Ainsworth** and **Bell** (1973) have presented three different kinds of attachment that children can have with their mothers. When the young children meet their mothers after short period of separation most of the children rush into their mothers arms and wish to be with them. The investigators explain this sort of behaviour of attachment the child had is the secure attachment to his mother. They found another group in their study to be ambivalent in their attachment. These children sought contact, but at the same time resisted it. The third group of children in their studies were found to be avoidant. Sometimes they did not seek contact with their mothers on their return. But if they did, they were silent and looked away from their mothers. It was observed that the third type of response of the children was the result of long separation of their mothers from them. It may be generally observed that the last reaction is the common type after a long periods of separation. Depending upon the duration of separation children may show any of these three types of responses when they are reunited with their mothers. This reveals the problem involved in the mother's attachment to her child.

Importance of Attachment

The importance of attachment between mothers or parents and children has been realized by the parents when they find their children *developing the feelings of insecurity* and the *feelings of destitute*. Therefore it is felt that the attachment between mothers and children is as vital to the child's survival as food and protection to their life and body. They always need a secure relationship with someone who can provide it. **Spitz** (1972) studied the infents put up in institutions. He found that those who received good physical care, did not get the sustained *personal care and contact and perceptional stimulations*. As result of this they *suffered severe ill effects* such as *depression withdrawal*, suceptibility to *illness* and *retarded development*.

The *lack of attachment in the infancy period for a longer time* may result in the *impairment of social bonds in the later life*. The process of attachment has important implications also for *detachment and hatred for life*. The child, in his later life, misses the chance of becoming an independent human being, has lost. The child, when becomes an adult with these experiences develops *deep depression and strong feelings insecurity*. The feelings of destitute he develops in later life make him to *disown everything and cry by saying who is there for me in this world?*

Thus infants who are unable to develop cordial relationship may suffer from severe *developmental disturbances* even when their physical care is sound. A confident relationship with a caretaker may provide the *emotional security* a child needs to explore the situation and to become an independent adult individual in the later life. Children should not be allowed to develop the feeling of detachment, a process that may cause in the later life the *feeling of insecurity*, dependency, fear and the feelings or destitute. (Fig. 7.9).

The term social development involves the observable behaviours in the contact with others successfully. As the child grows and develops in physical size and in the mental ability he has to develop some emotional control in meeting and in contacting other human beings. The contact may be with the children of his own age of preschool, school going age children or even adult persons. Once a child learns to control his emotions in the presence of others we may state that he is socially fit to meet others. In this type of behaviour there may be increasing source of responsibility, development of self-confidence and expanding interest in the adult world.

Fig. 7.9 : *Children are moving from childhood period to the period of adolescence stage on reaching the teenage period.* During this time the children had the required physical growth and development and mentally and socially matured and ready to take up the activities of the next stage. Generally taking part in play with other child may take place in the life of the child only after the age five or six years. Once the child crosses this stage and goes to school then there may be different types of social behaviour in the lift of the child. He is with the peer groups taking part in the organized plays and games which become the common occurrence for the boy and the girl. During this stage and afterwards the boy or the girl comes into contact with various types of persons. As and when the boy or the girl continues his or her contacts with the classmates, schoolmates, playmetes and the teacher there is every chance of getting the leadership activity in these situations. Then he or she moves into the adolescent stage.

Social development involves *social maturity*. It is the identification with adulthood in boys and self-possession in girls. It reveals in the wider adult social relations. These social developments generally may also mean, development of desirable and pleasing *behavioural traits, character and personality characteristics.*

Taking part in the play along with other child may take place or occur in the life of a child only after four or five years. Once the child crosses this stage and goes to school then there may be much different type of social behaviour in the life of the child. He is with the peer groups, taking part in the organized play may become a common occurrence for the boy. During this stage and afterwards the boy or girl comes into contact with various types of people, mainly the teacher in the school, parents at home, play mates and others. As and when the boy or girl continuous in his or her contacts with classmates, school mates and play mates there is likelihood of leadership activity in these situations.

Adolescent Period

Leadership among the school going boys and girls is characterized by instinctive, organizing ability, intelligence to handle others in an agreeable manner, willingness to accept wishes and demands of others. Generally the boys and girls in schools or in playground or in excursion who insist upon exercising their own wishes and ways at all times miss their chance of heading others. Such incidences indicate the nature of social development. A pleasant or cheerful approach in difficult situations may create a successful social approach. By this means one can obtain the help and cooperation and the willingness of others to stand by his side.

Generally, boys at the age of ten or twelve develop *boastfulness*. At the age of twelve the boastfulness among the boys may reach the peak level and it may gradually decline thereafter. At this stage they may try to take the adults as models for their life. They develop knowledge about the importance of *social conventions and norms*. Their social perception is becoming more important aspect of their life. The patterns of social relationships change rapidly. The boy in his later adolescent period reflect changes in his social attitudes. But this change of social attitudes depends upon mainly the culture in which one is born. *Cultural variations* always go with *social development.*

In some culture the adolescent is totally *deficit of social norms and conventions*. But in some culture the adolescent boys and girls are put in greater responsibilities of running the family and going out with others to work places and earn their livelihood. Physically though adolescents come out with the onset of puberty at the age of 16 or 17 they become highly imaginative and involved in fantasy life. *Day dreaming and fantasy* become common and sometimes the adolescents build up their own world of imagination and go much into the distorted way of life, which may affect even their normal way of life. *Adolescents' fantasies* often reflect their feelings of *insecurity and unhappiness*. Since they often go into the world of imagination they do not apply the adult standards life and experience, truth and falsehood to their life. It was also 'observed among the later adolescents that their behavioural activities are mainly antisocial (Fig. 7.10).

Social development among the adolescent boys and girls mainly depends upon the family, school, playground, peer group company situations. If these situations are not socially healthy adolescent boys and girls go astray and become delinquents and antisocial elements. It is the improper social development that makes the adolescents to take up a different way of thinking behaving away from the social norms, mores and folkways. Their type of unhealthy social development is prevailing in every culture among the later adolescents boys and girls and also among the men of early adulthood.

Culture and Social Developments

Social development in Indian culture varies from region to region. People living in hilly areas, forest and tropical areas treat their younger ones with more discipline and with severe social conventions and norms. Particularly the tribal people do not mingle with others closely.

They keep themselves in isolated areas and make their young ones and adolescents within their control. Adolescents and young adults of these communities very rarely rebel against their social norms and customs and traditions. If any such incidents happen such persons are socially isolated and severely punished. Accepting social norms and conventions in such communities becomes a compulsory aspect of life. Even in small villages in India communities enforce the importance of social development.

Fig. 7.10 : *Adolescents moving into adulthood.* It is generally stated that the teenage period is the adolescent period. In such case it falls into two categories, one from the age of 13 to 15 and from the age of 16 to 19 years. Physically, though adolescents come out with onset of puberty at the age of 16 or 17 they become highly imaginative and involved in fantasy of life. Day dreaming and fantasy become common and sometimes the adolescents build up their own world of imagination and go much into the distorted way of life which may affect even their normal way of life. Adolescents' fantasies often reflect their feelings of insecurity and unhappiness. Since they often go into the world of imagination they do not apply the adult standards of life and experience, truth and falsehood to their life. The later adolescents' behavioural activities turn to be anti-social.

Adolescents' Deviations : Adolescents generally in all cultural groups invariably speak lies. Being dishonest and cheating are common behavioural patterns. Cheating and stealing, the adolescents consider as adventurous act and they speak about these things as achievements. But this type of socially unacceptable behaviour patterns do not continue for long time. Gradually they give up such anti-social behaviour patterns and turn to socially acceptable norms.

The adolescents may rebel often vigorously against the principles and disciplines which are honoured customs. It has become customary or fashion for the adolescents to revolt against any changes in the socially oriented activities. They even go to the extent of strikes, demonstrations and even go into violent activities. This type of socially unacceptable behaviour pattern is quite common in these days in schools and colleges. In some communities in India the adolescents' and the young persons' movements and behaviour were found to be highly antisocial and they often become the object of criticism and ridicule. Some adolescents and young men who are sensitive to these do not withstand the remarks on them. They get frustrated and depressed and run away from homes finding some type of life, which may be easy going without any responsibility. Instances are often taking place that boys and girls runaway from homes quarrelling with their parents and others. These informations often appear in Newspapers and in Television programmes. Who are those persons? They are the persons who do not stick to the social standards of life, which is the result of improper social developments.

It is quite but nature that adolescents and young men who are directly, under the care and supervision of their parents and if the parents are very kind and pleasing the adolescents generally do not revolt against anything even such things interfere with their freedom of action and movement. Such adolescents accept whatever their fathers and mothers wanted to do. But in every culture and community we are observing that the adolescents enjoy unlimited freedom in action and movements as the parents are very liberal minded and do not question whatever their children do. This type of parents' attitudes become one of the main cause of adolescents' reckless and indecent behaviour. They come home very late from school or college, do not carryout the assignments given by their teachers, go to school or college late or roam about the whole day elsewhere and return home late. In Indian culture it has become a common occurrence that the adolescent boys and girls make their demands beyond the financial positions of their parents. Some adolescents even go to the extent of threatening their parents that if they do not meet their demands they run away from homes. It is because of their social contact with their peer groups.

The characteristics of adolescence may come to some end when the young person becomes psychologically and socially matured. This happy condition may occur when the young man realise his *responsibility* and begins to be independent, *more self-reliant* and *more resourceful* than what he was earlier. Maturation is a state of the individual's development intellectually and socially responsible to deal with the situation and the way of conducting himself pleasantly in a social situation. By maturity there is a shift of behaviour socially and also the person understands the values of life and behaviour. There is a shift from simple type of roaming form of play to a more organized and principled play and from the glamorous type of occupations to a more realistic, from adventurous type of life to more serious and creative type of life. Which are socially acceptable and appreciable. Some may try effective methods and depend upon activities, which bring the young person into greater social contact and recognition. (Fig. 7.11).

Fig. 7.11 : *Social development and attachment.* Human beings always wish to be in the company of other persons. They also wanted to be close to each other. Generally human beings enjoy the physical contacts and great care and attention from other people and become unhappy when the company of others is not available for them easily. The type of this behaviour is called *physical attachment.* Generally there is no process which is more important in the later development than the child's attachment to this primary caretakers, in any culture and society. Social development involves the observable behaviours in the contact with other perople successfully. As the person grows and develops in physical size and in the mental ability he has to develop some emotional control in meeting and contacting other persons. The contact and the reaction may with all types of persons of the same age group and even with adults. The social development involves *social maturity.* The social developments may also mean the development of the desirable and pleasing *behavioural traits, character and personality characteristics.*

The school or college situation with their group activities should induce in the young person to be more responsible, social being. The young person's interests, attitudes and motives bring greater social admiration and rewards for his achievements by the exercise of his matured way of life. A maturing young man normally show his adult maturity in the early twenties. But it may vary from culture to culture. It may also vary on the basis of socio-economic status and the type of life the family in which he is born, Generally *bright young men* in every type of family mature psychologically faster than the normal persons.

Mental and Physical Development and Maturity

Studies have been made relating the physical and mental growth to maturity. Persons who had normal physical growth and normal mental development get matured earlier without facing any problem, A standard has been assigned for this, which is called *developmental age.* According to this norm developmental age denotes *degree of social participation* as matured way of life. The person's activities, interests and attitudes indicate the matured characteristics. In these studies correlations were made between *physical growth and maturity to the mental development*. It was found that the correlation did not yield significant level. It is meant that a bright person or the child might have been *weak in his social behaviour*. But at the same time a physically small stature child, but well developed revealed socially more mature behaviour than a child who was physically large child of the same age. This shows that children who are good in school works do not come into the classification of athletic, play or social activities. Therefore, we may state that *physical and mental development do not go together with social development* in all the muturing individuals. Individuals differ in their social development eventhough they are physically or mentally proficient.

Social Maturity among the Mentally Backword and Bright Children

Social development in the life of human beings takes place right from the childhood period. The child at the age of three or four months smiles on seeing his mother who used to feed him with milk. This is the *first sign of social behaviour*. Then the behaviour continues and also going on developing to the extent of getting into the company of other children in school, playground, in the street and so on. Thus social development is a continuous process and also it becomes *an aspect human life*. But this social behaviour varies from person to person to some ex tent, in some persons it varies greatly. Mainly it depends upon the mental and physical development. Mental development in some child takes place normally, in some it may be very high and in some it may be very low. For example, at the age of twelve a boy's mental age may be twelve. He is *normal*. But at the some age another boy's mental age is fourteen. This boy is *bright*. But at the same age a third boy's the mental age is eight. This boy's mental development is low. He is *mentally backward*. His social contact, behaviour also may be very backward. These three types of boys and girls and also young men and women can be found in every culture and society.

Children who are backward in social development become problem not only for the parents, but also for the society. These boys and girls have social difficulties more often than the normal or the average children. The *lower the social development the greater the probability of maladjustment*. These mentally backward boys and girls often fail to understand directions and are slow in reacting to play and in other behavioural situations. This type of slow or nil reaction results in neglecting them by other boys and girls who are reacting swiftly in play activities. These backward boys and girls in social situations who are neglected by others develop strong feelings of inferiority, which in due course may make them unhappy and cause them to develop feelings of helplessness. The slow learning over-aged boys and girls become the real problem in the classroom. Further such boys and girls fail to carry out the instructions, show indifference to carry out the assignments and cause disruptive behaviour. These boys and girls need special attention in coaching (Fig. 7.12)

But the bright boys and girls behave in a more trustworthy manner, carry out the directions sincerely, emotionally well balanced and better socialized than the average and backward boys and girls. Boys and girls of very bright in their mental development carry out the instructions, tend to select better companions older than their age groups because of the mental maturity.

Fig. 7.12 : *Adolescent Deviants : Socially and Mentally Maladjusted*

In the process of human development the adolescent period is the most crucial one, causing problems to the parents and to the society. It is a very sensitive period, specifically the early adolescent stage. he boys and girls inthis stage do not wish to face any responsibility of their life. They often fail to carry out the school assignments and remain indifferent to parents' advices. Generally in Indian culture, the boys and girls around 13-15 years of age from rural, sub-urban and slum areas become more succumb to the evil influence of peer groups and find pleasure in their company and easily become deviants. When their parents and the teachers question about their poor school performance and enforce some disciplines they do not withstand them and try to escape from such controls. The only easy way of escaping from disciplines and punishments from the parents and teachers is to run away from homes and roam about in the streets. These boys and girls get caught by the Police and tryto handover them to their relatives. When they could no do it, they put these deviants in destitute homes. Incidences of this type often reported in mass media of communication and in televisions. These boys and girls become delinquents and are involved in anti-social activities. Generally, they are socially and mentally backward and the boys go into thievery and the girls go into prostitute.

Generally, bright boys and girls from the age of eight to fifteen years show very wider range of study habits and also have a tendency to seek butter reading materials and also reveal some creative talents by manipulating materials available within their control. As the boys and girls grow older there is shift in their nature of play and games. They develop increased interest in the adult worldly affairs.

During the later adolescent 'Period there will be sex differences in interests and physical strain of recreational activities. The adolescent boys show more interest in *competitive games and sports* and other physical activities. At the same time the adolescents girls show more interest in *fine art activities* like, music, dancing, painting and other similar activities like decorations and fashion designing and so on. But in all these later adolescent activities there is always close *social contact* with the same age group members.

Moral Development

In the human development process one of the most important aspects is *learning the right and wrong, good and bad, moral and immoral* behavioural patterns. Moral development in human involves mainly knowing what is right and carrying it with all sincerity. It also involves mainly the moral activities, which are to be accepted and the immoral activities, which are to be rejected. In every culture children are taught moral principle. That is what is right and what is wrong. What is good and what is bad and parents expect that their children should carry out this principle. But knowing the moral principles is not the all required, but the practice of it. It is the moral behaviour that is needed and to be practised.

With regard to moral development the *moral behaviour* is the most important. It is observed that in a society knowing moral principles and practising moral behaviour do not go together. Cheating, stealing, burglary, speaking lies are all not moral behaviours. Everybody knows this. But it is taking place in the society. A man may be tempted to cheat another person or steal the property from a house if he comes to know that there is least chance of getting caught and punished, inspite of knowing that this behaviour is not moral.

Understanding moral principles and moral behaviour though considered to be the philosophical issue, psychologists also have shown considerable amount of interest in moral behaviour and moral development. The cognitive developmental theorists are of the view that moral reasoning goes through a series of stages. It is not suddenly emerging in an individual. There were several studies on moral behaviour. **Aronfreed's** (1968) approach to moral behaviour was based upon *social learning theory*. He observed in his study that children learn the forms of behaviour that are approved and disapproved in the same manner as they learn other principles of behaviour. Children believe morally everything because they are *given rewards for moral acts and punishment for behaving otherwise*. **Hoffman's** (1971) study of the development of moral behaviour was focussed on the type of discipline, which parents expected from their children to exercise and adapt. He applied two techniques in his studies such as *punitive techniques and inductive techniques*. He found that the inductive techniques had better effect on children then to punitive techniques. Moral development in children takes place step by step in varied stages.

Development of Adulthood Stages

The physical maturity attainment is not the end of human development. Changes in the bodily parts are going throughout the life. These changes generally affect the individual's efficiency of physical activities, social and personal attitudes, cognitive processes and even the total behavioural patterns. As and when man advances in his age particularly after the late adulthood period physical deterioration is common occurrence. It is revealed in various aspects of life. Several studies have been conducted on aging persons. They wanted to see whether some of the psychological changes that are occurring in elderly persons are the results of the physical changes or due to the limited functions in the everyday life activities. It is true that people when reach certain advancing age level voluntarily limit their physical and cognitive activities.

Though occupational commitments become limited in old age, many people keep themselves busy and would be doing something always. In the old age social contact might have been reduced. As human beings get older and older their activities naturally not only get limited but also vary in nature. But the old age life, patterns vary from culture to culture. In the later middle age, which is considered to be between 45 and 65 years bring new problems. Vocational status at this stage is well established and the persons enjoy a status. Certain changes in his vocation may take place, which may be pleasant or unpleasant. His children reach the state of leaving his home. The mother of these children who has dedicated her life to the welfare of her children may find that she is left alone and feels very much depressed. Sometimes this may cause serious problems. It has been observed in various cultural societies that the rate of suicide cases increase in the age group of 50 years, which reflect some of these problems of separation and depression.

The next adulthood stage falls between the sixty five years and above. At this stage, these different type of problems arise. The man at this stage gets, retirement and sits at home unless he finds some other outlet for his time. Generally the retired person is disengaged and has to adjust to his extra time. Several Studies were conducted among the aged persons and varied views were broughtout. **Rosow** (1967) had examined the relation between the living circumstances and the friendship pattern among twelve hundred aged retired residents. He wanted to see the effects on those who are over sixty-five years, having age peers living close proximity. He found the greater the concentration of the old people, the more friends each resient made. The density of senior citizens living close to the area was important. They were working class individuals, But the higher socio-economic status people were likely to have friends outside the neighbourhood area. He had observed that senior citizens wished to have contact and friendship with the same socioeconomic status citizens. It is a matter of adjustment among the aged people. Whatever type of social development taken place during the early years of life, the later adulthood period after sixty five years of age contact and friendship with young people became a problem.

Though Human development is a continuous process, it is taking place through different stages like infants and childhood period, later childhood, early adolescent and later adolescent periods, early adulthood and later adulthood periods and old age period. Psychologists and physiologists are of the view that development of some type has been taking place at all these stages. The physical development and maturity may be completed at the early adulthood period. This may fully take place before a man completes his thirtieth year of his life.

Though the cognitive, including the mental development and social developments have already taken place to the level of one's potentialities in any society these developments are considered to be continuous processes. It is viewed by many psychologists that the adult years of life are obviously a continuation of much of the functions of what has taken place before. But it is inevitable that the adult years bring physical, social and psychological changes in the life of a man. Many investigators have conceptualized the adult years of human life in terms of some kind of developmental changes and tasks. Freud's last days message to the developmentalists was his insistence on the importance of early experiences that have laid down the patterns of life that will endure through the entire life span.

Erik Erikson (1963), one of the developmentalists who also revised the classical psychoanalytic theory had provided a bridge from which we can understand the psychological characteristics of the adolescence, middle age and old age from social point of view. Erikson discusses the adulthood in *three stages*. Each stage involves some psychological crisis. The *young adulthood* is characterized by intimacy versus isolation. The middle adulthood is characterized by generativity versus stagnation and the *later adulthood* is characterized by ego integrity *versus* despair.

The crisis faced by the young adulthood according Erikson is to establish the psychological process of sharing and earning relationships with others. This mainly involves personal commitments. It is the time generally in which marriage for the young adult person takes place and the person sets up a family life. There may be also failure in this type of life, which may lead to separation and isolation.

In the *middle adulthood* of his life the individual has to extent his commitments to a wider range of people. He is involved in the responsibility of guiding and nurturing of the younger people. Failures may take place in this type of functioning when he is not successful in his efforts. His efforts may come to a stand still and therefore a state of stagnation- may take place. This is characterized by excessive self-indulgence. The individual at this stage begins to look into his own way of living.

The third stage is the *later adulthood*, which is characterized by integration of life's experiences. The individual person begins to like his own past experiences. He accepts what has taken place and what is going to take place and begins to idealize the cardinal vertues of life and death. In case if such integration fails to take place the individual *suffers from despair*. The person finally looses all hopes of life.

Another investigator **Havighurst** (1953) states the human development in terms of different types of tasks that should be carried out at the specific period of time. Successful performance of the tasks in specific period of time depends upon the successful performance at the early periods. The assigned tasks at each stage may be physical, mental, emotional and social. They all tend to be categorized as psychosocial in the adult years. These tasks may be social responsibility, setting up a family and adjusting to the death of loved persons.

Physical and Health Changes : Generally, in human beings there is a stage in which the biological changes are taking place. The change is declining in the bodily aspects after the 30-35 years of age. But this declining is very much gradual which is not noticed by the individual person. It varies generally in the rate and degree of decrements specific to the particular bodily system. For example, the metabolic rate diminishes, breathing capacity reduces and different levels of decrement takes place as and when one advances in age. Marked changes and declining may occur in various sensory areas. The auditory sensitivity begins to decline relatively earlier in humans. It is quite often after fifty years of age. But the visual acuracy and the sense of taste constantly maintained until the age of fifty and then begin to decline. But there may be individual differences.

As and when man advanceg in age is subjected to increasing number of long-term chronic illnesses. By the age of *sixty-five, more than one half* of people in every culture have some common form of heart, kidney and urinary troubles. The rate of acute and short term diseases were at lower level for those over sixty five years than for the persons who are of any other age group. During the course of living time people build up large number of immunities that lower their susceptibility to any disease. However, when old people succumb to an illness their activities are restricted for greater number of days. With all that older people are more optimistic about their health.

In one of the studies **Shanas** (1968) had interviewed two thousand five hundred senior citizens with regard to their own estimation of their health. Among them fifty two percent stated that they were in good health. About thirty percent of them said that their health was fair, and only eighteen percent said that they had poor health. In this study on the whole seventy five percent of them said that their health was better than their contemporaries.

Changes in mental activities : It is quite but natural to assume that people's intellectual activities as measured by Intelligence Quotient (IQ) scores begin to decline as and when people advance in their age. Generally it is believed that intellectual functions decline in adulthood. In a cross-sectional study, it was observed that the IQ scores were lower for the adult persons than in a sample of younger individuals. But longitudinal studies produced different results. According to the study, the intellectual performances were generally maintained in adulthood and also even increased. What must be the reason for this? It was observed that it was due to the nature of different researches that caused the differences.

Several investigations were conducted relating intellectual performances and adulthood. There were different types of information. **Green** (1969) conducted a cross-sectional study on the change of IQ scores among Rican citizens between the ages of twenty five and sixty four and obtained the results of the previous studies that after the adolescent period there was a study decrease in intelligence. This was the result obtained in an earlier study made by **Bromley** (1966). However, **Green** had observed that the older persons in the average had received a fewer years of education and therefore, the reduced educational experience rather than age was responsible for this type of result. **Blum** *et al.* (1970) have made a longitudinal study among the individuals between the ages of sixty-five and eighty five years. They have observed a small decline in the IQ scores between sixty-five and seventy three years and the decline was much deeper between the individuals of seventy three and eighty-five years. They have observed the rate of decline of intelligence was not regular and constant. But there was a general decline in the intellectual performance of the aged persons.

Change in Status and Age : In any society we can observe, that people of certain age level hold considerable amount of power and prestige. This mostly rests with middle aged adults. The elderly persons, because of their old age have lost the power and prestige once they were holding. But in some cultures or societies persons are holding power and prestige even in old age. Age status also interact with other variables. In fact the occupational role and economic status always hold power and prestige in any culture. In the case of occupational status retirement- has become a common incidence in the life of persons. Governments and organizations have fixed certain age level for retirement. This retirement age varies considerably in governments and organizations. In India the retirement age is sixty years. The assumption is that people after sixty years of age become incapable of performing the tasks assigned to them. Therefore persons who are younger than those persons come and occupy their seats.

Generally from physiological point of view there may be decline in the efficiency of people to perform some task. There may be decline in perceptual processes and in cognitive functions. Psychologically persons after sixty years of age feel to retire from the occupation and do some leisurely works. The retired person makes use of his knowledge, experience and skills in some other work and may find pleasure. Retirement from the occupation doelt not mean total withdrawal from useful and profitable tasks. The retired person is retrading his skills for better prospectus, but change in status after the retirement does not impair one's intellectual performance and knowledge he has gained in some area. But retirement and further occupation is a continuous lifestyle in the case of many. For example, statesmen or politicians do not retire from their status and power. Age is not having a different effect on them. They hold some position and power in some area. They spend the whole of their adulthood in some task and enjoy the status.

Later Adulthood Period : The later adulthood period is marked by disengagement in which the aged persons generally make mutual withdrawal. Though there were several studies of this period the study made by **Cumming** and **Henry** (1961) is considered to be highly informative. The study was related to the event of retirement which is based on the study of healthy, economically secured fifty to ninety year old persons. These researchers tried to determine the nature of the sample of the old people were like and they tested this hypothesis by conducting several interviews for few years. They tested the hypothesis that aging is an inevitable mutual withdrawal or disengagement, resulting in decreased interaction between the aging persons and other persons in the social system in which they live. They observed in their subjects a decline in interaction in such aspects as the number of social roles held by the elderly persons, the amount of time spent with other people and the reactions of the aged to their own lives.

One of the important factors observed in their study is the occurrence of preoccupation with the self and change in perception of time. The individual begins to perceive his life *in terms of time from death* instead of the time from birth and often feels that the *years too short*. Many activities once considered as important are put aside. The disengaged persons are calm and quite. They gradually disown their ties with worldly matters. But these events are considered as universal happenings that probably help them to reduce their tensions. Another view prevailed against the view of these investigators is that withdrawal from the worldly affair is not the individual's desire, but the society pushes the elderly persons into inactivity.

Generally retirement puts the man into non-active situation which gradually leads to depression as the individual person has lost the chance of homemaker's role. But the health of aged person is an important aspect irrespective of activity or non-activity. Maintaining a healthy personality is also An important aspect of the disengaged aged persons. **Tobin** (1968) in his study had observed and described that several types of people among the disengagements seemed to be successful with their retirement state. They have high self-concept, self-directed and are interested in the worldly affairs but they are not interested in the interpersonal aspects. Persons who are dependent and passive are also Round to satisfying with disengagement.

Developmental psychologists have made very entensive studies on adulthood period and described the various stage in adulthood. In every stage the human developmental processes were discussed in detail. But they have not given a clear out distinction about the *age-range of the three stages* of the adulthood such as early adulthood, middle adulthood and later adulthood. The last category seems to be having very wider change.

Elderly people have shown a wider range of social life in which they are involved. **Rosow** (1967) was able to identify *four basic themes* that appeared common to fairly large groups within the population. The first theme is labeled as *cosmopolitan*, but thirty percent who had little contact with their neighbours content with what they had. About twenty percent who visit infrequently the desired, had more interpersonal contact are labelled as *isolated.* About twenty three percent had high degree of social contact are labelled as sociable and those who had very little, ten percent of high level of social contacts were labelled as insatiable.

The End of Life : Psychologists are of the view that human development is a continuous process. At every stage man learns something, perceiving something new and experiencing something strange. But man's life is not an endless process. It has an end. But man with all his scientific discoveries and inventions can only event his life. Thus the life span of an average individual has increased considerably during twentieth century. The *birth rate increased* and *death rate decreased* unless there are natural calamities. The central fact of adulthood is that with each year the death rate of aged persons is decreased. It is because of food and medicine, care and happiness in life, reduced physical and psychological problems all increased the span of life and decreased death rate (Fig. 7.13).

But the active functions of the aged people come to an end at certain stage. It is only due to retirement or disengagement, but also due to social pressures. Society slowly avoids the aged people in all their activities. People in this society generally feel that it is a strain for them to pay continuous attention on them and to take care of them. They must provide the aged people at home some special type of food and medical care. Sometimes the young people at home do not take care of them. This causes depression for the aged people. Further physically they may feel weak and could not fulfill their needs and demands by themselves.

They need somebody's help, which is not available for them immediately. Thus the last stage of adulthood is full of several psychological problems.

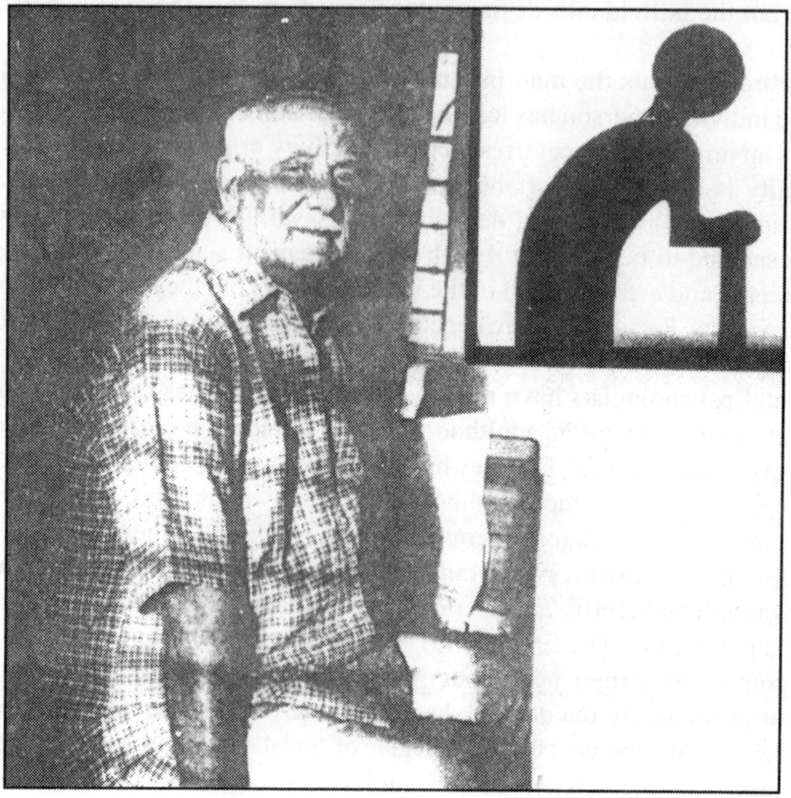

Fig. 7.13 : *The later adulthood period.* This period of human life is said to be the end of life. Psychologists are of the view that human development is a continuous process and man learns something new, perceiving and experiencing something strange for his purpose. But man's life is not an endless process. It has an end. Science and systematic lifestyle increased in life span. But the active function of the aged people come to an end at certain stage. Aged people face several psychological problems. Society also slowly avoids aged people.

In India the problems of the aged are incelesing day by day. The young people at home enjoy the properity earned by the aged people, but they do not take care of them. They develop the feelings helplessness and get much depressed and leave the home and disappear. Such instances are often reported in mass media and in Television with photographs of the elderly persons. They become destitute and some organizations doing charity works. They collect such aged persons and provide them shelter and food. In India such destitute homes are increasing as a consequence of the increased number of aged destitutes.

The Helpage India : A research organization has published a directory of *Old age Homes* in India in 1995. According to this Directory there are two hundred and fifty six old age homes in India. Out which one hundred and sixty two oldage homes are free homes. Thirty of them are pay and stay. There are fifty two oldage homes for women, sixty four homes for both. There are several oldage homes for sick. Aged people who have money by their pensions go into these homes and those who have no money are provided with free shelter and food and care.

The *aims and objective* of these homes are to foster the welfare of the aged in India, especially the needy aged. To raise funds for the projects which assist the elderly irrespective of caste or creed. To create in younger generation and in society a social awareness about the problems of the elderly persons in India to-day.

These facts and figures indicate how the oldage problems have become so serious that the Government of India have taken serious steps to foster the welfare of the aged people. The main cause for the aged people to seek the help of the oldage homes is the negligence of the younger people at home in taking care of them. This is becoming a serious psychological problem in India. These oldage homes also provide medical care to the inmates and in the case of sick and death the organizations make all arrangements for the funeral of the dead. (Fig. 7.14).

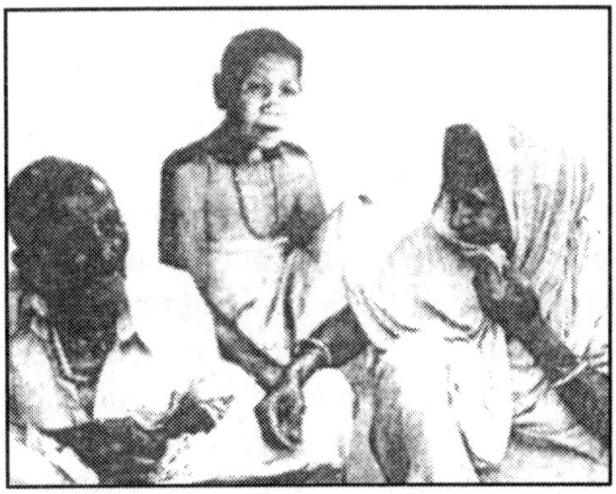

Fig. 7.14 : *The old age problem in India* has become so serious that the Government of India have taken serious steps to foster the welfare of the aged people. The aged people seek the help of the Old age Homes, because of the negligence of the younger people at home. This becomes a serios psychological problem in India. There are 256 old age homes in INdia for both men and women and these homes take care of the inmates with food shelter and medical cares. In the case of death the homes arrange for the funeral of the dead.

Advent of Death : Human development is a continuous process, which is taking place right from the time of birth till the end of life. During the period of time when man is alive man he ft puts himself into continuous activities of some type. But the incidence of death pulkan end to all his functions. Death is inherent in man. He does not know when he is going to die. But he has to die one day. What is death? Death may be defined as the permanent end of all functions of life of an organism or some of its cellular components. But in the case of a human being death puts an end to his very existence and he is no more in the world physically. Death has been studied by several investigations. The event of death is not confined to oldage. It may take place at any age in the life of a man. The newborn baby dies, the child dies, the adolescent boy dies, the adult person dies. Thus death is imminent. It may take place at any age at any time. It depends upon the cause. **Goldman** (1968) had categorized the causes of death as extrinsic and intrinsic. But the categorization does not explain the complete extensity of the causes of death.

Generally in these days we could visualize the causes of death into three categories such as *personal causes, environmental and situational causes* and the *causes of natural calamities*. The personal causes may include mainly sickness, malnutrition, vitamin deficiency, starvation and committing suicide.

Human beings fall sick. It is bodily, the body when lost its immunity for the attack of the germs and bacteria that cause disease and when they kill all the bodily resistances the body dies down. The body may be subjected to the attacks of cancer, heart attack bronchial problems, pneumonia and tuberculosis and brain tumor, all may cause death of an organism. These causes are within the persons. (Fig. 7.15).

Fig. 7.15 : *Advent of death.* Death puts an end to all the functions of a human being. Death is the permanent end of all the functions of the life of an organism. Every man has to face this situation at one time.

The other causes are environmental and situational, which include accidents, water poisoning, pollution of air and epidemic diseases. We are made to know by some 'means hundreds of people die everyday by some accidents occurring in places, where people move about, on the roads. Mainly the deaths caused by motor transports, railways accidents, plane crashes, ship wrecking accidents all cause death of many human beings everyday. They may be young and old and also children. These deaths are unavoidable. Accident means unexpected incident causing lose of life and property.

The third causes are the natural calamities, which cause death of human beings in large number irrespective age, animals and destruction of property. These natural calamities are occurring periodically all over the world and man is helpless against these forces but to die. These are the instance of floods due to heavy rains, storms, earthquakes, tempest and tsunami which all cause the death of millions of people periodically. These natural calamities do not differentiate between the child young and old. All put to death when such calamities take place. Therefore it is very difficult to say how death comes to man. But the aged people at some stage are anticipating their death as they are advancing. (Fig. 7.16).

The study made by **Kubler-Ross** (1969) bring out some information about terminally ill patients. These patients progress through all stages and some were continually denying that they will soon die. The author on studying these patients suggests that many of the terminally ill experience *some psychological stages*. These terminally ill patients varied in their ages. According to her observation, the old people were more aware of death than the young. They report about their thoughts about their death often. But such thoughts did not reduce their life's enjoyment. A number of variables were correlated with anxiety about death. The fear of death was great in both old and young people. The old parents in someway communicate their fears of death to their children and influence them.

Fig. 7.16 : *The Victims of Tsunami expressing their anguish for the sudden death of their family members.* Tsunami is a Japanese term which refers to a huge wave caused by an under-sea earthquake, a sea volconic eruption. It is a a grave natural calamity rarely caused resulting in disastrous consequences on land. This Tsunami eruption has tkane place on 26-12-2004 in the coastal areas of Tamil Nadu, Andhra Pradesh and in Sri Lanka and washed away the people and their properties causing death to thousands of people who lived in the coastal areas. The figure indicates the way in which the victims of the Tsunami expressing their physical and mental agony for having lost their family members. Thus the natural calamities suddenly cause death to human beings.

Therefore, retirement is the question of the optimal condition for the elderly people. The elderly persons voluntarily withdraw from the social system, being pre-occupaied with themselves and feelings, content with this situations. They also reflect the fact of withdrawal from active services and argue that it is *forced upon them and is not optional*. Although many people are living longer to-day than the people lived previously, death in adulthood is increasingly probable after thirty years due several factors, mainly intrinsic and extrinsic factors. As **Kubler-Ross** has stated that patients who are seriously ill and who are at the stage of death, experience death in five stages such as denial, anger, bargaining, depression and acceptance. But healthy aged people think more about death. Their reactions reflect stable attitudes. Death is an enemy, a severe punishment, an unknown and isolating event. In each, there is some commitment of emotions.

But, whatever type of comfortable and happy status one is enjoying, if one is subjected to severe painful suffering due to some ailments, he normally feels that death is better than this type suffering and also expresses his wish to die by some means. In the case of some people death solves several problems indirectly.

PART THREE

PERCEPTION AND AWARENESS

8. Sensory Processes and Sense Organs
9. Attention and Distraction
10. Observing and Perceiving
11. The Psychology of Consciousness

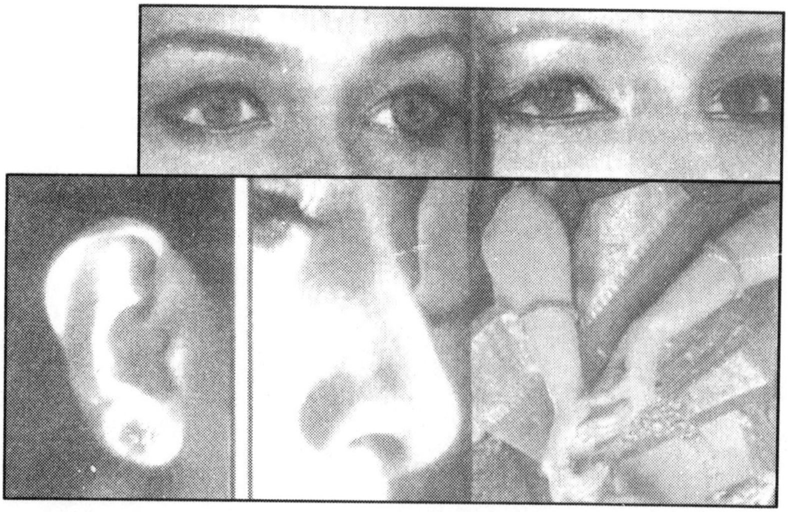

Fig. : Human Perceptual Apparatuses : Eyes, Ears, Nose, Hands, Mouth and Tongue.

8

SENSORY PROCESSES AND SENSE ORGANS

Basic Ideas and Functions

All the events that are taking place around us in the universe produce energy of some kind which the physicists classify them as light, sound, temperature, mechanical events, chemical events and electrical events. A great number of these events affect human beings, but human beings can sense only a few of them and not all. We can see all around us there are many different types of energy affecting us in various ways. Some molecules are passing through the air making us to note down certain particular odours and we smell them. We could detect/waves of compression and expansion of air as sounds. Man cannot see the ultraviolet rays or light, but he can feel and experience it. If air pressure changes, his hearing is damaged and he cannot hear the sound. Certain chemical substances he cannot taste, if he is poisoned.

The events that man can sense are those events that are energy changes which are affecting his receptors and stimulate sensory nerve cells. The nerve cells stimulate the central nervous system to produce sensations which, in turn, are changed and converted by the central nervous system into perceptions. The physical events affect and stimulate the sensory nerve cells in the receptors and the sense organs. The sensory nerve cells transmit the impulses, received from the events to central nervous system with the information that events have occurred outside. Then the input, the sensation is discriminated on the basis of the stimulus whether it is light or sound or odour, The central nervous system reacts to the inputs in an integrated manner.

Some of the integrated activities of the central nervous system that resulted from the realization of sensation by man are reported as light, sound, feeling of hunger and thirst. These reports are said to be perception. It is the conscious awareness of the objects and events. The human knowledge of physiology of sense organs and the central nervous system speak much about the sensation. But not much about the complex system of the central nervous system and perception.

Sensation is an experience that the human organism relaises when some physical event affect and stimulate the sense organs and the receptors and communicating the same to the brain through the sensory nerves. Sensory perception occurs only when the organism experiences the sensation. Perception is the conscious awareness of an object or an event initiated by sensation. Perception of the physical world is unthinkable without sensation. Sensation interact with learning and memory of the past experience in forming perceptions.

Sensation is initiated by the energy of the physical event and physical energy becomes the basis of perception. But a very clear cut distinction between sensation and perception is very difficult. It becomes highly philosophical if this issue is to the discussed in detail.

Therefore while we try to understand senses and sensory organs it is important that we must know the basic ideas and functions that underline them. A sense is anyone of the five faculties of vision, hearing , smell, taste and touch through which humans and animals perceive the external world. These faculties collectively may be stated an senses. Sensation is the power of perceiving through the senses. Sensation generally refers to the experience perceived by the organism through the senses. It is an experience resulting from the stimulation of one of the sense organs, for example the sensation of warmth. Sensation is a general feeling of experience, or awareness. Thus there is a relationship between physical energy that comes from outside and the sensation in man.

One of the earliest ideas in psychology is that mental life (which psychology studies) or consciousness has its origin in sensory experience. The British philospher **John Locke** (1632-1704) brought out his philosophical tretise, "*An Essay Concerning Human Understanding*" in the year 1690. According to Locke the mind of the child is a clean state, 'a tabularasa' and what the child experiences is written in it. Locke was of the view that there are no innate ideas. All knowledge come from experience and experience is gained through senses. The senses give us experience and experience is gained through senses. The senses give us the raw material of sensation and from this knowledge is evolved.

Psychologists started experimental studies. They wanted to investigate the sensory processes experimentally. The sensory processes are closely related to physiology of the nervous system and ultimately to the brain. Psychologist attempted to investigate sensory processes and measure them quantitatively. This led to the emergence of a new science called psychophysics. Fechner was the founder this branch of science. Psychophysics is the science that studies the relationship between changes in the physical energy and the corresponding changes in the attributes of sensation or psychological continuum. Studies in this area brought out lots of information about senses and sensory experiences. All sensory systems have a basic function to begin with at the place where the physical energies are striking on the body, which are converted into a form that activates the nerve cells. All animals and human beings have the specilized bodily parts that are specifically sensitive to some kind of energy. These bodily parts are called sensory organs and sensory receptor organs. These sensory organs act as screening instruments or filters. These sensory organs detuct and respond to the respective stimuli and exclude the other stimuli.

Generally, energy changes are perceived as changes in quality, intensity, extensity, and duration respectively. Man can report these changes in sensation that follow stimulus changes. The physical energy can be compared to the sensation reported by the subject. Generally, the relationship between the energy change and sensation change is not always linear. The physical energy and the corresponding sensation that occurs is measured and compared in psychophysics. With these background we shall discuss the senses and the functions, the nature and characteristics of sense organs.

The world in which we live is full of energies and these energies attack the organisms in various ways. Physicists say that we are constantly assiled by cosmic rays, light rays, sound waves and heat waves. If we want to survive in this world we must understand the information these forces bring to us. Therefore, our bodily parts come into contact with these information. Since our sense organs are specialized in a particular area they receive the respective information from outside in the form of stimuli and convert them into sensations. These sensations become the basis for all our behaviour. Sensation is often associated with perception. Sensations are studied through the subject's response to perception. But the destination between a sensation and a perception is difficult. A sensation is transformed into a more complex meaningful perception. Perception is a conscious event initiated by sensation and integration of the input from several sense organs.

The sun sends energy to the earth in the form of light waves and this becomes a stimulus to the eye. The conversion of this stimulus into sensation, experience and utlimately into knowledge is an elaborate process involving four important stages. They are (1) Sensory stage, (2) Attentive stage, (3) Perceptual stage and (4) Conceptual stage. The.refore, in the knowledge getting process sensation is only the first step. For a sensation to be meaningful it should be attended and then understood. Finally, a general idea has to be drawn by organising several particular ideas about it.

The Nature and Characteristics of Sensation

Sensation is a kind of awareness to sensory stimulation. It is aroused in an organism by a stimulus. A stimulus may be internal or external. The sensation is the act of the organism. Sensation means the activity of the sense organ of the sensory nerves and of certain parts of the brain called the sensory centres. Sensation is the first response of the brain to the stimulus which impinges on the organism.

According to popular belief, man is supposed to possess five senses. But recent physiological research has made it clear that there are eight sensory receptors. So man has got eight senses. They are vision, audition, gustation, olfactory and tactual. The other three are internal such as organic, kinesthesis and vestibular. Further each sense organ is specialized in one particular type of sesation. The ears cannot see nor do the eyes hear. Suppose a light stimulus falls on the ear, the sense organ is not activated. It should be remembered that even when a sense-organ is capable of receiving an appropriate stimulus it responds effectively only according to certain law. There is no sense of hearing to both — a very loud noise and very feable one.

Charateristics of Sensation

The events that produce sesation in an organism should occur in a sequential manner. Some form of energy stimulates the receptor cells in the sense organ such as the eye or ear. The receptor cells are specialized in responding to certain forms of energy called light waves of certain lengths, air pressure, the energy from chemical activity and thermal energy. There are several kinds of receptor cells. Some of the receptors in the eye and nose are specialized neurons. Others in the ear and tongue only activate the neurons that are near to them or enclosed within them.

Table 8.1 : Sense Organs and Corresponding Sensations

S. No.	*Sense-organs*	*Stimulus*	*Sensations*
1.	The eyes	Ether vibrations	Visual (seeing)
2.	The ears	Air vibrations	Auditory (hearing)
3.	The tongue	Chemical substances dissolved in saliva	Gustatory (tasting)
4.	The nose	Chemical substances in gaseous form	Olfactory (smelling)
5.	The skin into contact with the skin	Changes in temperature or objects coming (touching)	Tactual or Cutaneous
6.	Stomach and back of the throat	Churning of the walls of the stomach and certain chemical changes in the back of the throat	Organic (internal)
7.	Muscle spindles and tendons in the joints	Pressure on the muscle spindles and the tendons near the joints	Kinesthetic (muscle)
8.	The semicircular canals in the ear	Disturbances in the fluid of the semicircular canals	Bodily equilibrium (balance)

One of the common characteristics of all the receptors is the ability to convert the energy they receive into a form that the nervous system can understand. The process of change is called electrochemical neural impulse. The conversion of stimulus energy from one form to another form is called transduction. The sensory nerves carry the neural impulse to the central nervous system,the brain and the spinal cord. Then the impulse ends in the appropriate sensory projection area of the cerebral cortex. The sequence continues without any break.

Every receptor cell can respond to the energy that falls within its range. If the energy falls away from the receptor it will not be communicated and sensed. Even the stimulus that falls within the range must have certain amount of intensity or strength. Otherwise the receptor cell will not respond to it. For example, some voice may be heard from the next room, but the voice is not strong enough for our ear to make out a meaning from it. If the duration of the presence of stimulus is increased the stimulation in the senses also increased for a longer period.

Mechanism of Sensation

In psychological studies we are interested in all these eight senses and sensations. Our concern is, how these sensations are caused. How the various forms of physical energy such as cosmic rays, light waves, heat waves and sound waves affect the receptor cells within the sense organ~? How the cells convert these energies into information and send them to the central nervous system? Therefore it is important that we must know something about the mechanism that takes place in the various processes of seeing, hearing tasting, smelling and touching and other things.

The mechanism of sensation is more complex than what we all assume it to be. For a sensation to occur in the human organism stimuli from the external world should come into contact with a sense-organ. This contact will result in the release of some kind of nervous energy which is transmitted by the sensory nerves to an appropriate region in the cortical area of the brain. When this has happened, there will be an awareness in the organism of experiencing a sensation.

Sometimes a question is asked as to which is the most important of the senses. It is very difficult to say which is important. Because the relative importance of the senses actually differs from person to person, depending upon the utility of the particular sense. However, eye receives the first importance, since it is supposed to be the most precious of all our senseorgans. A very valuable object, as we all know, is often compared to our eye. The next important sense organ is the ear. But we are not giving much importance to one sensation. It is often negalected. It is the sense of smell. Though the sensation of smell is valuable from an aesthetic point of view we give very little importance to the sense of smell in our everyday life. As a matter of fact, we do like to smell anything when we get into a bad cold. On the whole, there is great deal of individual difference, in giving importance to the value of our senses. For instance, a musician would rather be ready to sacrifice his sense of smell or taste in preference to his sense of hearing, but not a producer of perfumes or a skillful cook.

VISION : VISUAL PERCEPTION

All living creatures including humans perceive the world only through their senses. Stimuli from outside attack the sense organs and the sense organs communicate them to the brain through the sensory nerves and they are converted into experience and knowledge. In this process of perception of the world, the vision, the experience of the eyes, plays very important role in the lives of the animals including humans. Light comes from the sun, warms up the body skin and stimulates our eyes. All living animals need infori-nation from the light receptors. But some animal have very big eyeballs to their body size. They have large pupils which facilitate them to see objects even in very dim light. Some creatures mostly invertibrates (back boneless) which badly need light, have multiple light receptors spread out all over their bodies. Some of the amphibiaons (animals live in land and water) have their brains with photo receptors.

Thus every animal depends upon how the brain processes the information they receive through the eyes. The visual perceptions of the animals including humans are having spatical distribution with various wavelengths of light. Different kinds of processing are going on the stimulus input in such as way that we could see the objects' form,colour, size, position and the distance of the objects. Visual information processing has become one of the most active part of research in recent times in various scientific fields, including psychology. The reason attributed for this is uiat one third of the human brain is functioning in visual perception and analysis.

The visual experience provides information regarding the objects or the stimulus form, size, colour, distance or locaton and its movement7s. The visual perception has certain limitation. It has a field. The area that we can see the object clearly without moving our head or eyes is called the visual field. We can see the object accurately in the centre of our *visual field.* As we fix our eyes on a book and go on reading the lines we are moving our eyes to the left and right. But if we fix our eyes on a single word at the centre of a line and try to read the word in the upper line or lower line we find it difficult. It is because, our eyes are fixed at the visual field and any attempt to read anything on the left or right or on the up or down results in the fall off the *visual acuity.* The sharpness of the vision falls down. Visual activity falls off from the centre of the visual field in the peripheral area. Visual acuity varies in the visual fields. If this takes place the vision of the individual moves from place to place when we read a passage.

For any sudden appearance of a stimulus away from the visual field we usually shift our vision on the new stimulus placing the visual field on the object and we can see it clearly. By examining the visual field of each eye separately it is possible to find out the blind spot in the eye.

The Eye

The sense organ for the visual sensation is the eye. The mechanical structure and function of the eye can be roughly compared to a camera. The eye ball consists of three sensitive layers. The inner most part is the *retina* which is the *third layer* of the eye, is a sensitive plate. In the case of camera, the photographic plate has to be replaced after every exposure. But the retina recovers immediately after an exposure and as such spontaneously gets ready for another. The eye ball corresponds to the box. The *first layers* is the outer tough coat of the eyeball is known as the *Sclerotic coat*. This ivhe metal or the wood of which the cam-era box is built. The inside of the camera box is painted black. Similarly, the *inner second layer choroid coat* is darkly pigmented in colour. This pigmentation serves the useful purpose of preventing the incoming rays of light from being reflected back (Fig. 8.1).

At the front of the eye, light is admitted. This front portion of the sclerotic is transparent and it is called the *cornea.* The coloured part of the eye which lies behind the cornea is an extension of the *choroid* and it is known as the *Iris.* In the centre of tne iris there is a hole, which is known as the *pupil* of the eye. The Iris looks like the diaphragm of the camera. The lens of the eye is just behind the pupil. Three pairs of muscles regulate the movement of the eye ball in its socket. The lens of the eye is adjustable by the working of the *ciliary muscle.* This is something like the focussing mechanism of the camera. The cornea also functions as a lens but it is not adjustable. The jelly-like substance—aqueous humor and vitreous humor—fill in the eye ball and put it in shape. They are transparent and as such allow the light to pass through them on the way to the retina (Fig. 8.2).

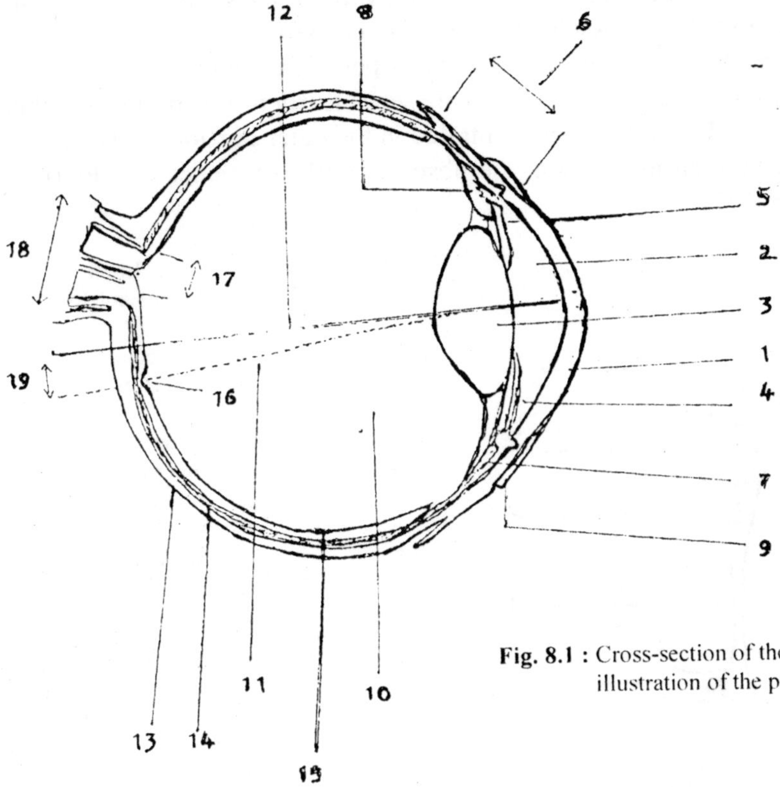

1. Cornea
2. Anterior Chamber
3. Lens
4. Iris
5. Posterior Chamber
6. Ciliary Body
7. Ciliary Muscle
8. Ciliary Processes
9. Conjunctiva
10. Vitreous Body (Fundus)
11. Visual Axis
12. Optic Axis
13. Sclera
14. Choroid Coat
15. Retina
16. Fovea
17. Optic Disc
18. Optic Nerve
19. Yellow Spot

Fig. 8.1 : Cross-section of the right eyeball : Diagrammatic illustration of the parts.

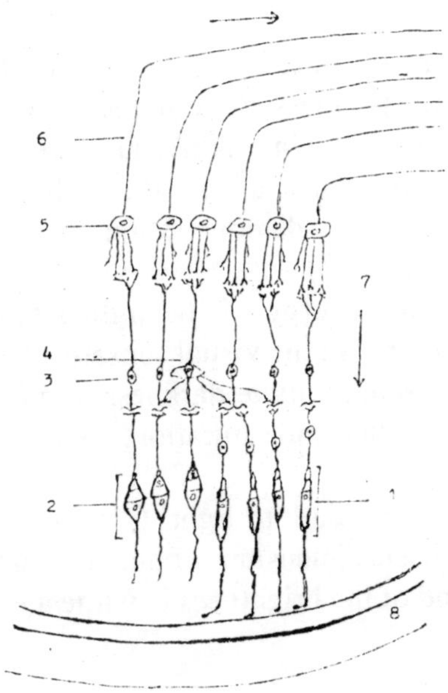

1. Rods
2. Cones
3. Bipolar Cells
4. Horizontal cells connecting Rods and Cones
5. Optic Nerve Cells (Ganglion)
6. Axons leading to optic Nerve and Interbrain
7. Light
8. Choriod coat

Fig. 8.2 : Structure of the Retina and various parts.

At the very back of the eye ball is the sensitive film of the eye and the third layer known as *retina* which lies below the choroid. The nerve cells of the retina are composed by *cones* and *rods*. The cones are meant for normal daylight vision while rods are used in twilight vision. Cones are receptive to colour vision but the rods can receive only light and shade. In the central part of the retina there are more cones and this portion is known as the *Fovea.* There are millions of cones in this area. The colours are seen clearly in this region and this is the point of clearest vision. The rods are concentrated in the outer portion of the retina. Suppose a coloured light falls on this region, the observer will experience a colourless sensation (Fig. 8.3).

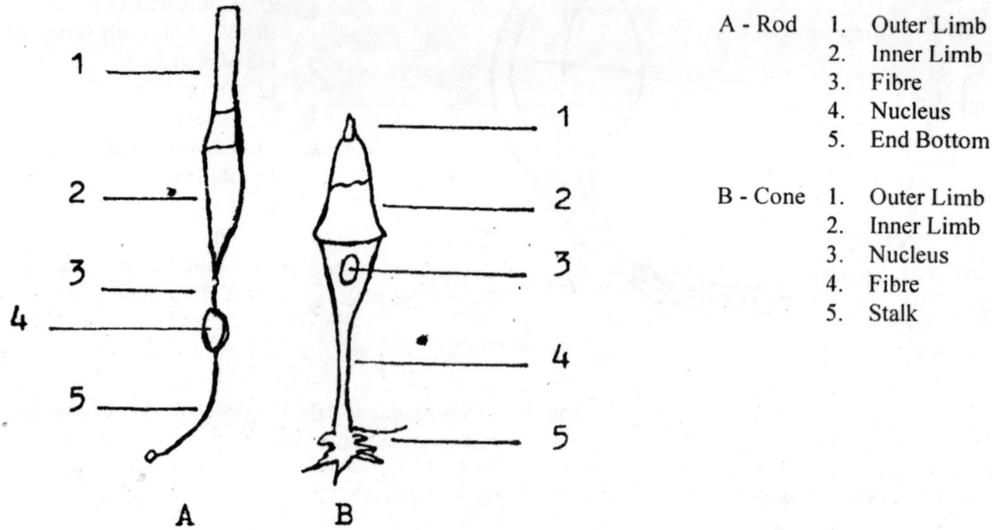

Fig. 8.3 : Rod and Cone.

The optic nerve is that which is formed by the coming together of the nerve fibres from the ganglion cells of the retina. This optic nerve transmits the message from the eyes to the occipital lobe of the cerebral cortex. Not far from the fovea is an insensitive area known as the blind spot where there is no vision at all. We are not ordinarily aware of the blind spot at all, unless we take special care to locate it through laboratory demonstrations.

Dimensions of Visual Sensation : The intensity or the energy of the light stimulus corresponds to the brightness or brilliance of the sensations. In the visual sensation of the light perception there are three dimensions. They vary from absolute darkness to glare or very bright light. The three dimensions are the brightness, hue and saturation.

Brightness : The brightness of light has varied shades from dark to light. It is the visual sensation that results from light intensity. It is one of the basic dimensions of light perception which varies from darkness to very bright light. One end of the brightness is white and the other end is black.

Hue : Hue is the response of the eye to the wavelength of light. The light that is received by the eye from other objects undergoes changes in intensity and in component wavelengths. Therefore, it is changing in brightness. Hue is one of the basic dimensions of light perception. Hue also varies around the colour circle from red through yellow and other colours of the spectrum to violet and then purples and back to red again.

Saturation **:** Saturation is the degree of the response of the eye to a colour, which lacks of gray colour. If there is more mixture of colours, th-en there will be less saturation. It is also one of the basic dimensions of light perception. Saturation varies from rich to pale, from red to pink and from pink to gray in the colour solid. A saturated colour will not be very bright and a bright colour does not require any saturation. For example, pink and reddish-brown are both unsaturated red. These are the general relationships between the physical light and the visual sensation. But these relationships are very complex.

PERCEPTION OF COLOUR AND COLOUR BLINDNESS

Colour Vision

Colour exists only in our head. Without the ability of the brain to interprete the light, every object in nature and in the world would be entirely colourless. Objects themselves have no colour. What we see as colour is actually the result of the way in which different wavelengths of light stimulate certain parts of our brain's visual system.

There are seven colours in the spectrum and they are visible to the naked eye. The seven colours in descending order of wavelength are : (1) Red, (2) Orange, (3) Yellow, (4) Green, (5) Blue, (6) Indigo, and (7) Violet. They are popularly called 'VIBGYOR'. Beyond violet, we have the ultraviolet rays which cannot be seen. It is used for the therapeutic purposes. Beyond red we have infrared rays which are at the other end of the spectrum and also it cannot be seen. It is used in long distance photography.

Colour Mixing

Every colour is determined by the wavelength of the light stimulus. Similarly the hue depends upon the wavelength of the stimulus and each hue corresponds to a particular wavelength. But it is also true that any hue can be obtained without the particular wavelength. It is by mixing the wavelengths lying on both sides.

We can obtain series of deep grays by mixing black and white in different proportions. We can produce less saturated colour by mixing white or gray with any saturated colour. The brilliance of a colour can be lowered by mixing black with any colour. By mixing red and yellow in different proportions we can get all the hues intermixing between red and yellow and also all shades of oranges. By mixing green and yellow we can produce all the greenish-yellow and by mixing green and blue we can have all the greenish-blues. By mixing blue and red we can obtain violet and purples.

It is also observed that purple colour is not seen in the spectrum as it is a sensation that cannot be aroused by any single wavelength. It can be produced by mixing of long and short waves of light stimuli. But what would take place if we mix yellow and blue? Painters, artists and dyers mix different colours for their business works. They state that blue and yellow colours will give green colour. But by mixing blue and yellow light waves we have different effect, not the same as the painters and artists have. If we add blue and yellow we get white or gray colour. When two wavelengths act on the retina area it gives white or gray sensation and not blue or yellow. This is called complimentary. Thus yellow and blue are *complimentary colours*. The complimentary of red is bluish-green and the complimentary of green is purple.

Experimentaly it is observed that red and green are not complimentary colours like blue and yellow. Therefore they are called *disappearing colour pair*. When blue and yellow are mixed both the colours disappear and we get only white sensation and there may be a trace of blue or yellow colour. If red and green are mixed both colours disappear and the sensation of yellow is realized.

It is also observed that certain pairs of colours when mixed or combined may yield a colourless sensation. The process is known as complimentaries. There are two such pairs of complimentary colours, discussed above, the red and green, yellow and blue. If a ray of red light and a ray of yellow are made to fall on the same cone at the same time the resulting colour is orange. Suppose, if red, yellow and blue rays are made to fall at the same time the resulting process is a neutralized gray. In the colour vision we also face the *simultaneous contrast*. It occurs when yellow and blue colours are seen side by side and when looked separately yellow appears as yellow and blue appears as blue. This process is called *simultaneous contrast*.

Colour Perception

The light receptors are the cones, which are located in the retina area of the eyes. The cones are mainly responsible for our colour vision. The *cones are of three types*. Each type is sensitive to certain range of light wavelength by the virtue of a chemical pigment they contain. One type of cone recognizes certain wavelength as the aspects of the colour, *blue*, the second type sees only red and the third type sees only *green*. These three colours, blue, red and green are, in a way, the eyes' primary colours and the mixing of them in the eyes 'palette' forms all the other colours. Colour signals from the cones pass further into the system where neurons on the visual path mix and match them until we can see the full spectrum of about five million colours that make up the world as we see it.

People with normal colour vision see the same colours, but how intense certain colours appear and how subtle the difference between two colours may vary from individual to individual.

It is often believed that, colour affects the human moods and temperments. We cannot deny the effects that colour produces in our moods. But this cannot be scientifically proved. Science cannot explain this effect. The reason is that perception of colour is extremely subjective which depends upon the personal and cultural experiences. For example, a colour that could calm the emotion of one person may cause aggression in another person. The studies conducted among children showed that a majority of them stated that colouns of *yellow depicted happiness* and colours *of brown depicted sadness.* Yet, this issue may very in different cultures. The results of the studies made among the infants were that they prefer toys of pure colours such as red, blue and green. It was also observed that in work places the colour paints on the walls and on the instruments determine the levels of productivity.

Colour perception is an important aspect of vision, which determines the likes and dislikes of materials. Advertisers use different colours in their display of materials, advertised. The colours are chosen and presented in attention getting attractive manner. In these days advertisers use florescent colours in their materials to get quick attention of the people who happend to see them. Printers in these days use attractive colours in their printing works for drawing the attention of the people. Newspapers use invariably attention getting colours such as green, red and blue in different shades in their weekly magazine sections. Colours are applied in fabrics and clothings for promoting sales of the goods produced in mills and factories. In all these areas individuals' feeling aspects are involved.

Colour Blindness

Colour blindness may be defined as the inability of some people to distinguish between colours. It is not a disease and so cannot be cured. It is a natural condition of the eye due to the lack of development of certain zones of the retina. There are two types of *colour blindness — total and partial.* Suppose the entire retinal field is filled by rods alone, then the person will suffer from total colour blindness. But total colour blindness is a rare phenomenon. In the total colour blindness the person looks at all colours as mere variations of light and shade. Partial colour blindness is more common than the other. It is generally red, green blindness. This is very rare among women. But 3 to 4 percent of men suffer because of this. Here the persons have sensations of yellow and blue but not of red and green. What appears to the normal eye as red, orange, green appears to him as more or less yellow. When there is violet and purple, he sees themas blue invariably.

From the practical life point of view people who suffer from colour blindness either partial or full need not get discouraged. In the matter of employment they can go to such of those professions or jobs which do not demand knowledge about colour discrimination. Similarly in the matter of dress selection they can seek the help of their companions or relatives who are free from colour blindness to offer better guidance and direction. Colour blindness is not a common defect in men. and women. Physiologically colour blindness is not a total failure of seeing colours. It is correctly stated as *colour confusion* and not colour blindness. They also see colours, but mistaking one colour for another colour. Colour blindness is an inherited condition, generally it is determined by the cones in. the retina area. When we are outside in the bright light, the cones and rods in the eyes are operating chemically to face the bright light. In a darkroom our sight is greatly reduced to the light. It takes sometime for the rods and for the cones for adapting their nature to the situation by means of chemical reactions. The cones adapt first and the rods take some times for adapting to the condition of dim light. The light receptors of the cones, which are located in the retina area of the eyes are mainly responsible for our perception of colours.

The cones are of three types, each is sensitive to certain range of wave length by its virtue of chemical pigment. This characteristic of the cone makes every type of cone to respond to certain wavelength aspect of a specific colour. The condition of colout blindness occurs due to the lack of cones of the given type in the retina area or reduced number of cones of the specific type. This results only in seeing the objects as black and white.

Studies on colour vision showed that there is gender variation genetically and mostly men are affected. In the genetic make up the chromosome is linked with colour vision. Since men have only one 'X' chromosome and women have two, men suffer from colour blindness, whereas women are not. But women with two 'X' chromosomes becoming colour blind are very rare unless the trait is inherited in the family. In such case the women may produce colour blind sons and daughters. This is how colour blindness found in persons and families. Most of us have several distinguishable hues or shades of colours like blue, red, green, yellow and their complimentations. Rosenzweig (1996) states that about eight percent of men and about half percent of women suffer colour blindness. It is the impairment of colour perception. But most people with impaired colour vision are able to differentiate some shades of colours. Complete colour blindness in humans is very rare, unless it is caused by brain injury or lesions.

After-Image : After-image is something we see even after the perceived object is disappeared. Every one of us in certain situation might have experienced this after-image phenomenon. Suppose, we see an object, while we see, it the light ray falls on the retina area and acts on it for a certain length of time, and its effect persists for a short period, even after the external stimulus has disappeared. This kind of sensory experience is called an *after-image*. **Woodworth** (1952) states that a better name for this kind of visual experience is *after-sensation* and not after-image.

There are two types of aft6r-images. They are : (1) Positive after-image, and (2) Negative after-image. When both the original visual experience or impression and the after-image are of the *same quality*, then that experience is called *positive after-image*. For example, a person looks at a flame of a candle in a dark room for some time and then closes his eyes, but still he will be able to see the candle flame in the dark room for a few seconds. Suppose the quality of the after-image is different from the original experience or impression, then it is called *a negative-image*. The negative after image is also called *successive contrast*. Here every black changes to white and every white changes to black and every colour changes to its complimentary colour.

Theories of Colour Vision

It is the cones which mediate colour vision but it is not known how they do it. There are many theories of colour vision. They all endeavour to explain colour phenomena, but no one of them is satisfactory. A good theory should explain the following matters (a) the number of elementary colours, (b) colour mixture, (c) colour contrasts, (d) after images, (e) colour-blindness.

The Young-Helmholtz Theory

According to this theory there are only three elementary colours — red, green and blue. This three-componefit theory of colour vision was propounded by **Thomas Young** in the year 1801. The theory of Thomas Young was revised and elaborated by Hermann von Helmholtz later on in 1860 and hence it is known as Young-Helmholtz theory. It explains the mixture of wave lengths and negative after images. There are three types of cones for the three colours. Yellow is generally recognized as the fourth elementary colour. But this theory looks upon yellow colour as a blend of red green. Similarly white is a compound of the three primary colours—red, green and blue.

Scientific research has not proved the existence of three such substances in the cones as it is claimed by the theory. Secondly yellow is not a compound of two colours. Physicists and Physiologists treat yellow as a separate colour. Lastly the theory does not give any explanation of colour-blindness.

Hering's Theory

Another theory of colour vision was proposed by **Ewald Hering** in 1874 and it has become an alternative theory for the Helmholtz theory. Hering theory is based on the introspective observation that yellow, white and black look like primary colours rather than the combinations of different hues. According to Hering there are three substances in the retina tissue and they overlap to some degree. These three substances according to Hering are three pairs of colours. They are, red-green, yellow-blue, and black and white.

The two colours in a single substance stand in opposition to each other. Therefore if one is stimulated the other also is thrown into action. Suppose, if both the colours in a substance are stimulated at the same time, we get white colour since they neutralise with each other.

According to this theory light is composed of all the wavelengths and stimulate all the three types of substances equally and produces the sensation of only white. There is also an assumption that this theory deals effectively with the facts of negative after-images, red-green colour blindness and simultaneous colour contrast. But it does not explain the existence of retinal colour zones and certain facts of colour visions. For a long time this theory has been widely accepted and also has been very popular.

One important observation made in this theory is, the treatment of the negative after-images. The assumption is that if one process of a given receptor gets fatigued by the strong stimulation of the appropriate wavelength the second process takes over the action. For example, if the red process of the red-green cones get fatigued the other process green takes over. This theory is an improvement over the previous one. It has accounted for colour mixture and for colour contrast. One major objection to this theory is that it fails to explain the mechanism as to how a single receptor can transmit two different qualities over the same neural pathways.

Ladd-Franklin Theory

The most satisfactory theory of colour vision is formulated by *Christine Ladd Franklin.* Her theory is a genetic theory. The theory makes an evoluationary approach to the problem. It believes that our present colour vision is a development from a primitive stage of colourless vision.

Ladd Franklin has clearly pointed out in her theory that there are three zones in the human retina corresponding to the three main stages of evolution. In a primitive stage of vision the retina was sensitive to only light and shade (white and black). Then the retina at a certain stage developed a second zone which responded to yellow and blue wave-lengths. Afterwards, a third inner zone was developed in the centre of the retina which was responsive to all the four primary colours. Also it is hoped that perhaps man at a future time may develop a fourth inner zone and as such he can see ultraviolet and infra-red rays which are at the moment not visible to the naked eye.

In the totally or completely colour-blind person the whole of the retina is in the first stage of development containing full of rods and no cones at all. But in the case of partially colour-blind person (red-green blindness) the third inner zone is in the second stage of development. So to a partially colour blind person there are really two zones only, instead of three zones. Ladd Franklin has remarked that the whole process of appreciating differences in wavelengths is chemical in nature.

In general, the Ladd Franklin theory has given us a very satisfactory account of the most difficult item like colour blindness. However, her remark that man was originally colour blind is not correct. This is because apes have colour vision and also primitive-cave-dwellers had left behind them bright coloured drawings.

HEARING : AUDITORY SENSATION

When some sound is produced or occurred elsewhere outside we atonce turn our face in that direction and begin to listen what has happened. Our ears receive this sound stimulus and it is communicated to the brain through the respective nerves and the sound is converted into the sensation of hearing. Hearing is an important aspect of the adaptive behaviour of many animals including human beings. For the human beings the sounds of speech form the basic elements of our languages which build up the social, cultural and personal relation among the people. If a man loses his sense of hearing he is deprived of contact with people. The human vocal system produces enormous possibilities of sounds. The successful adaptive behaviour of many animals is linked with their abilities to produce sounds and to preceive the meanings of the sounds and the environments.

Animals produce sounds, understand the meaning of the sounds and react to them. This means some kind of auditory information processing discriminates the acoustic signals accurately and the animals are guided by this in the environment. The human auditory system is capable of detecting the changes of sound intensities. The human ear is as sensitive as possible even to weak sound frequencies. It is surprising, how the small vibrations of the air particles become the speech, music and other sounds we hear. The auditory sense organs play very sensitive role in this process.

The main auditory sense organ is the ear. The auditory or hearing centres of the brain are located in the temporal lobes. It is in this area that the brain hears or receives sounds and registers them as loud or soft or low. The loudness of the sound corresponds to the strength of the sound waves received by the ear. The sound has the characteristics of pitch, timbre. Pitch relates to frequency of the waves, and timbre is the blending of sound waves every voice produces. The frequency is the number of waves per second.

How do we Perceive Sound?

We perceive sound in the left and the right sides of the auditory lobes (cortex) of the brain. Each side receives the sound impulses from both ears. The language processing centres of the left hemisphere convert the sound impulses into meaningful material. This is how sound becomes meaningful. The ears and the sensory apparatuses which receive the sounds and transmit them of the respective parts to the ear and through them to the brain.

Auditory Sensation

The Ear

The ear is the specialized organ in the sensation by hearing. The human ear is a complex mechanism. It is divided into three parts — the *outer* ear, the *middle* ear and *inner* ear. The outer ear is made up of the *pianna* (this is what we call the ear) and the tube which runs inside. It catches the sound waves and carry them to the *ear drum or tympanic membrane* of the middle ear. The soundwaves push this membrane to vibrate. The area of the middle ear begins from the tympanic membrane to the oval and round windows of the inner ear. The middle ear is an irregularly shaped cavity in the skull. And across this cavity there is a chain of three delicate bones the *Hammer* (Malleus), the *Anvil* (Incus) and the *Stirrup* (Stapes). The bones are known by their shape. The middle ear is connected with the throat by means of the eustachian tube. This tube supplies air to the middle ear constantly. The inner ear spreads right back into the skull. It is made of three parts like the semicircular canals, the vestibule and the cochlea. The semicircular canals are responsible for the sense of balance or equilibrium which we shall discuss later. The vestibule is just a receiving and transmitting centre. The cochlea is the real organ of hearing and it is the main part of the inner ear. It is shaped like the shell of a snail. It has wound itself round two and a half times. This kind of anatomical device makes an adequate saving of space. The Basilar membrane is another important part which is running parallel similar to the structure of cochlea. It consists of thousands of fibres which vibrate like the piano strings. The inner ear is water tight and it is filled with a saline liquid known as the perilymph (Fig. 8.4).

Hearing is one of the most important senses in our communication with the outside world. The physical stimulus for the sense of hearing is sound wave. The process of hearing begins when sound waves attack the ear drums and cause it to vibrate.

Anything causing sound vibrates and the air surrounding it is thrown into waves. These waves when they reach the ear we have the sensation of sound. Sound waves) gathered by the outer ear, pass along the auditory canal. They strike the tympanic membrane which vibrates. As a result, when this membrane is pushed in, the three bones move and the stirrup closes the oval window. Now the air current passes upto the auditory area through cochlea and we hear the sound.

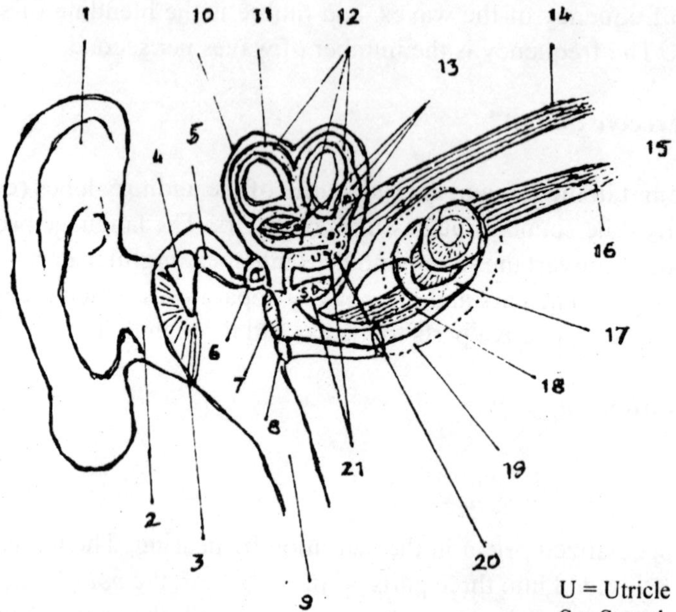

Fig. 8.4 : Showing the various parts of the outer, middle and inner ear.

1. Outer ear (Pinna)
2. Ear canal
3. Tympanic Membrane (Ear drum)
4. Malleus (Hammer)
5. Incus (Anvil)
6. Stapes (Stirrup)
7. Ovel window
8. Round window
9. Eustachian tube
10. Perilymph
11. Endolymph
12. Semicircular canals
13. Cupula in 3 Ampulas (in the semicircular canals)
14. Vestibule nerves
15. Auditory nerves
16. Cochlear nerves
17. Cochlea
18. Scala vestibule
19. Scala tympani
20. Vestibule
21. Utricle and Saccule

The original air-vibration is transformed into a liquid vibration in the cochlea before it is made into a sensation of sound.

An Analysis of Sound

The stimulus which excites nerve cells and causes sound is longitudinal air vibration. When a vibrating body strikes the air, the particles about it vibrate backward and forward and successive waves are produced. These waves move in all directions until they come into contact with the ear or the receiver in a wireless set.

Sound is physically a wave motion and it travels at a speed of 1100 feet a second. Air waves which produce sound are different in three ways:

(a) in the rate of vibration,
(b) in amplitude and
(c) in A form or complexity.

(a) The Rate of Vibration

This is what is called pitch or tone. A tone which is a sequence of uniform vibrations may be high or low. A tone should be differentiated from noise, which is a combination of different types of vibrations. In short, a noise is composed of irregular sound waves. Pitch refers to the lowness or highness of a sound. Pitches range from the lowest notes produced by the longest waves to the highest, produced by the shortest waves. Thus difference of wavelength of sound waves produces differences in the pitch.

(b) Amplitude

This is the distance through which the particles vibrate. This gives to a tone its loudness or intensity.

(c) Form of Complexity

This is called Timbre. Difference of purity is caused by the mixture of different wave lengths in sound. Suppose the sensation is produced by a stimulus consisting of waves of the same length, we get a 'pure tone'. The difference between the different instruments is a difference in quality or purity of the tone. The characteristic quality of different instruments is called the Timbre. Suppose the same tune is played on a Veena and a Violin we are able to notice a difference and this is because of Timbre.

Theories of Hearing

A theory of hearing should try to explain as to how exactly do the fibres in the basilar membrane catch the vibrations in the liquid around it.

(a) Helmholtz Theory

This theory is also called piano string theory or place theory. In explaining it a parallel is drawn between the strings of - a piano and the fibres of the basilar membrane. The wires in a piano resonate (vibrate) to external frequencies. Similarly the fibres of the basilar membrane resonate to external frequencies. The theory put forward by Helmholtz, makes a claim that each region of the basilar membrane is particularly attuned to a certain frequency of vibration. When a tune or note is sung into the piano, a string in the piano can be thrown into sympathetic vibration. In a similar manner a fibre in the basilar membrane will be thrown into sympathetic vibration when waves reach it through outer and middle ear.

There are nearly 20,000 fibres or strings in the basilar membrane. The ordinary human ear can easily distinguish about 11,000 tones. So this is a possible theory and also a satisfactory one.

(b) Rutherford's Frequency Theory

The human ear functions like a telephone receiver according to this theory. Each frequency is carried to the brain. The pitch is determined by the frequency of the nerve impulses reaching the brain. But loudness of a sound wave depends upon the number of auditory fibres activated.

(c) The Volley Theory

This is a modified frequency theory. By themselves the two theories are not exhaustive. However when the two are combined we may get a satisfactory explanation.

THE CHEMICAL SENSES

Smell and taste are called chemical senses. These two are activated by chemical substances. Many studies were not made on smell and taste. Therefore we don't get much information about these senses as seeing and hearing. The reason for non-availability of much information on these two is that it is very difficult to conduct experiments on smell and taste.

Taste

Taste is known as gustatory sensation. Generally we think that there are numerous tastes. Every kind of food is having a unique taste. Apart from tongue, the smell and touch sensations make us think like that. Scientific Research refers to only four elementary tastes like sweet, sour, bitter and salt. The other tastes are only compounds. The different sensations cannot be connected with different parts of the tongue easily. But roughly speaking sweetness is perceived on the tip, sourness on the sides, bitterness on the back and saltness on both the tip and the sides of the tongue (Fig. 8.5).

The real organs of taste are the *taste buds* which lie underneath the surface of the tongue. So the taste-buds of the tongue are the receptors for the sense of taste. The tongue is full of countless pits. The tips of the sense cells run into the sides of these pits. We can taste liquids quickly. But solid food should be well dissolved in saliva to have taste sensation (Fig. 8.6).

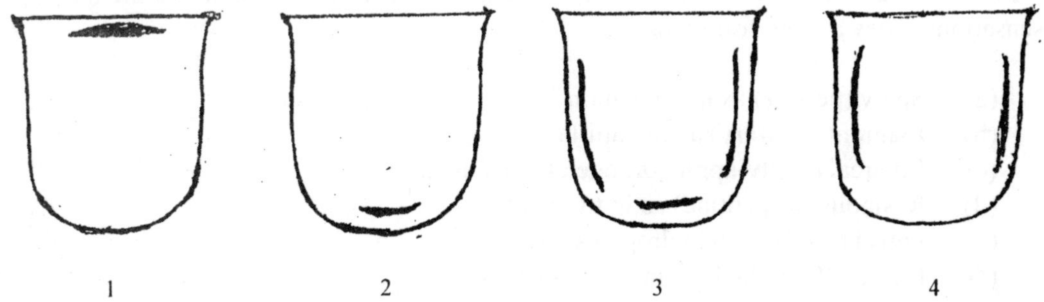

Fig. 8.5 : Location of the four basic Tastes on the Tongue

1. Bitterness 2. Sweetness 3. Salt 4. Sour

Taste is known as gustatory sensation. Different sensations cannot be correctly connected with different parts of the tongue. But roughly speaking sweetness is perceived on the trip, sourness on the sides, bitterness on the back and saltiness on both the tip and the sides of the tongue.

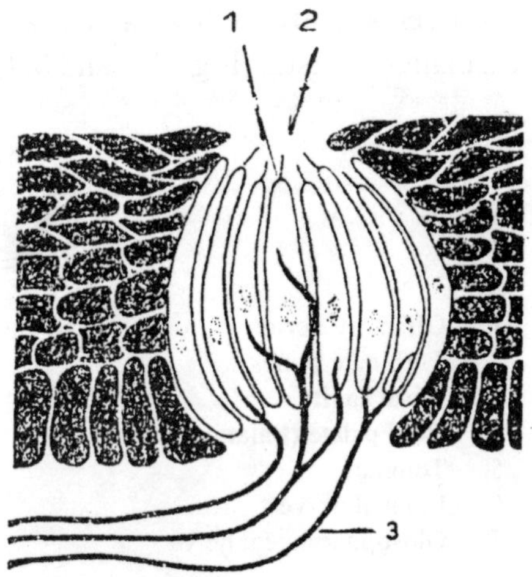

Fig. 8.6 : Showing the description of a Taste Bud.

1. Taste Cell 2. Taste pore 3. Taste neuron

The real taste organs are the Taste buds which lie underneath the surface of the tongue. Therefore, the taste-buds of the tongue are the receptors for the sense of taste. The tongue is full of countless pits. The tips of the sense cells run into the sides of these pits. We can taste liquids quickly. But solid foods should be well dissolved in saliva to have the taste sensation.

The different tastes which we commonly refer to are only compounds of taste smell, touch, temperature and muscle sensations. The flavour of coffee depends on smell. Temperature seems to modify taste considerably. As we all know that ice-cold Idli tastell, different from hot-Idli. Further taste differs according to the roughness or smoothness of the object. Sugar pulverized and unpulverized make a lot of difference in taste. Smell has more influence on taste than that of touch. Much of our delicate tastes are really smells which reach the nose through the air passages that link it with the mouth. A strong cold, as we all know, destroys almost all of our taste sensations.

Smell

The sense of smell is also known as the olfactory sensation. What we know about this sensation is very little. The organ of smell is the olfactory membrane which is situated in the upper nasal cavity. According to experimental evidence there are six elementary smell sensations. They are the following

(a) Spicy (pepper) as in cinnamon
(b) Fragrant (flowery) as in vanilla
(c) Ethereal (fruity-apple; orange etc.) as in ether
(d) Resinous (turpentine) as in turepentine.
(e) Putrid (Foul) as in hydrogen sulphide
(f) Burned (Scorched) as in tarry substances.

It is possible that different parts of the nasal membrane are connected with different types of smell. A person may lose his sensitivity to particular smells alone because of certain diseased conditions. Even persons who have normal sense of smell get tired (fatigued) by certain smell, especially when they continue to smell it for a long time. As a result a period of insensitivity to that smell intervenes. For instance, after smelling camphor for a considerable length of time, alcohol will not be noticed at all. Just like taste sensations, the smell sensations usually get themselves mixed up with other sensations like tactual and taste. The sweet odour of chloroform is to a large degree a matter of taste (Figs. 8.7 and 8.8).

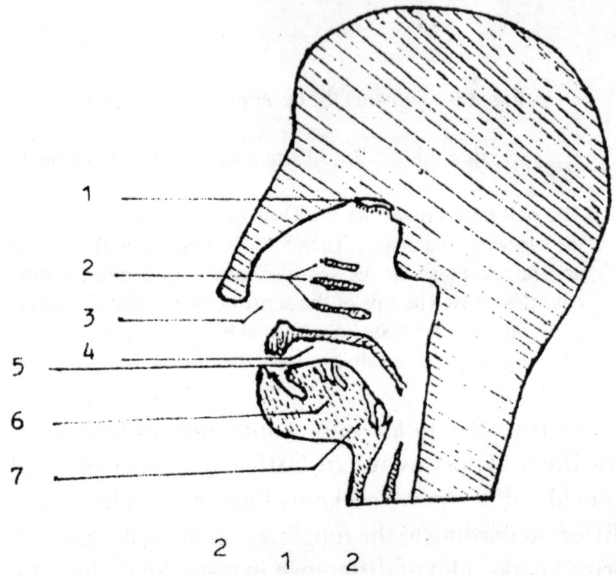

1. Olfactory nerve
2. Turbinate bones
3. Nasal cavity
4. Hard palate (Palatine bones)
5. Tongue
6. Lingual nerve
7. Glosopharyngeal nerve

Fig. 8.7 : Showing the parts of Olfaction

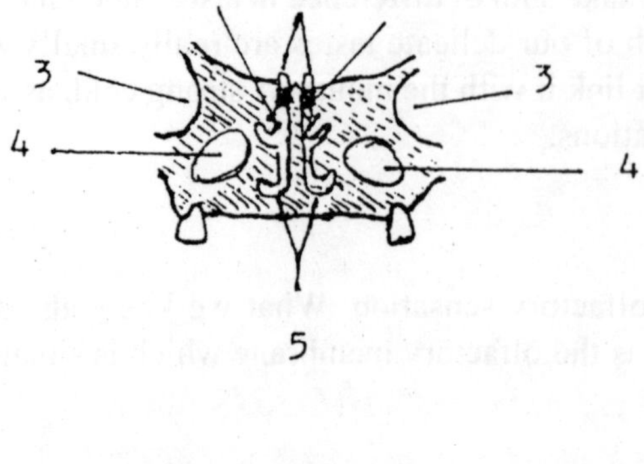

1. Olfactory bulb
2. Olfactory epithelian
3. Orbit
4. Maxillary sinus
5. Nasal passages

Fig. 8.8 : Showing the Frontal section of skull in the region of Olfactory Mucous.

Man's sense of smell is very poor compared to that of the dog. There has been a considerable deterioration of the olfactory capacity with the progress of evolution. The sense of smell is useful for warning us against bad air and food.

The Skin Senses

A touch is actually made up of several different sensations. These are called the cutaneous sensations. This is usually connected with touch or tactuai sensations. But in reality, skin as a sensory receptor has four senses each with a separate quality and different sense-organ. The four sensations of the skin are (A) Cold, (B) Warmth, (C) Pressure (touch) and (D) Pain. The sense-organs associated with these sensations are called spots. They are located below the outer surface of the skin. Any specified area of the skin contains all the four spots in different proportions. Suppose the skin is explored with suitable stimuli, they can all be located.

The appropriate stimulus for locating cold spots is a fairly cold needle. By passing over the marked area of the skin and which is also made hairless, point by point with a fairly cold needle, there will be felt an unmistakable sensation of cold. These points denote the cold spots. At other places there will be no sensation at all. That means our stimulus has been contacting some other kind of spot and as such no response at all. Cold spots are easiest to be traced. To locate warmth spots is a little bit difficult since they lie deeper in the skin then the cold spots. The suitable stimulus for the warmth spots is a fairly hot needle.

The temperature of the skin varies at various places. The exposed parts of the body will be colder than the unexposed on a cold day. The temperature of a given portion of the skin at a given time, is called the *physiological zero point.* The temperature of the warmth stimulus should be above the physiological (zero point in order to be effective).

Generally, a cold spot is responsive to only cold stimuli and similarly a warmth spot to hot stimuli. But sometimes a cold spot is put into activity by a warm or hot stimulus. When this happens, the resulting sensation is cold but not warmth. This is called a *paradoxical cold sensation.* When a cold stimulus arouses a warmth spot, the person will experience a warmth sensation. This is called a *paradoxical warmth sensation.*

Pressure spots are more in number than cold or warmth spots. But they afe less than pain spots. To locate them, the skin should be explored by a horse-hair or bristle. Pressure spots are distributed over the whole surface of the body. The pressure exerted on the skin when it become too much the resulting sensation will be pain but not pressure.

The pain spots lie deep down in the skin. If a pain spot is to be thrown into activity, the appropriate stimulus-a sharp pin should go deep into the skin. This is a protective device. Suppose the pain spots are located in the outer surface of the skin,we will be perpetually alarmed by the danger of pain. Pain spots are very sensitive and they give us a warning of dangers to the body.

The Body Senses

All the internal senses are called body senses. They are the organic senses, kinesthesis and vestibular sense. They mainly help us to maintain inner homeostasis and balance as we walk and stand.

Organic Sensations

Sensations from the internal organs, like hunger and thirst are called organis sensations. Hunger originates in the membranes of the stomach while thirst in the back of the mouth, very near the throat. Hunger is a sensation aroused by the contraction of the stomach walls. The low water position in the human organism and the consequent dryness of the throat and mouth give rise to the sensation of thirst. There may be special organs to inform us about disturbances in the respiratory and circulatory systems but at the moment we have very little knowledge about them.

Kinesthesis

This sense gives us information about the position and movement of our limbs. Further, it enables us to know the difference in the weight of several objects by lifting them with our hands. This sensation is different from a skin sensation. Kinesthetic sensation has been found in people who have lost the skin sensation and so it is more than a skin sensation. Suppose the judging of weight is purely a matter of pressure, then we should be able to remark, immediately we lift an object, how heavy it is. But in reality we are not able to do so unless we resort to move our hand up and down several times. So it is reasonable to suppose that we have a special equipment to find out the movement and weight over and above the help offered by touch sensation. This special equipment is known as the kinesthetic sense. This sensation originates in the muscle spindles embeded in the muscles of the body, the tendons and the joints in general. The sensation of kinesthetic type can be well illustrated by the following example. Suppose, the arm is bent double there is a contraction of one set of muscles and a relaxation of another set of muscles. The muscle spindles in one set of muscles will be actually contracted, but in another set they will be just relaxed. As a result, sensations coming from these muscle spindles enable us to know the exact position of our limbs, even if our eyes are closed or blind folded. Kinesthesis is mainly concerned with mechanical disturbances in the muscles and joints. This system has two sub-components such as the static limb position and the sensation of limb movements.

Kinesthesis is an internal sensation. The basic receptors for the kinesthetic system are specialized endings found in the muscles, tendons and joints. They provide stimulation and also pain and temperature. We feel stomachache, the hot food and pain from distension of the viscera because of the gas. The system that provides sensations for functioning of the viscera is the organic sensibility. The organic sensibility is closely related to two sensations, such as hunger and thirst. These are basic psychological drives and they are responsible for motivating for a portion of behaviour.

The muscles are responsive for the internal stimulation and also they provide information about the external world. For example, the ciliary muscles provide us with important cues about the place of the lens of the eye, giving information of depth perception. The important characteristics of kinesthesis is that the sensory system tell us when the movement is taking place and also tell us the location of the organs being moved. In this way the kinesthetic sensitivity is concerned with maintaining balance, posture and precision and also the motion.

The Vestibular Senses

The vestibular senses are sensory structures in the labrinth of the inner ear that respond to head position and movement. It is a system (Vestibular system), provides information about the force of gravity on our body and the acceleration of our head. When we go up in the elevator, we feel the acceleration. When we turn our head or ride in a car going around tight curve we feel the change of direction. The vestibular sense is very important in birds and fishes for orientation. They relay on it to tell them which way is up and in what direction they are going while flying and swimming. In man this system supplies information about the body position and movement. The vestibular system has functional relationship with both audition and kinesthesis. The same nerve is involved in hearing and in balance. Events in Cochlea, (a part of the inner ear) will affect the vestibular sense.

The vestibular senses respond to head position and movements and sets off many reflexes that enable us to maintain an upright posture. The receptors (sense organs) are located in the non-auditory part of the labyrinth of the innear ear. The labyrinths are located in the temporal bones on either side of the skull, internal to piano the outer ear. Each labyrinth is surrounded by the fluid *perilymph and* the auditory parts of the structure are filled with the fluid called *endolymph.* Labyrinths are soft sensory structures of both audition and propriception in the inner ear. Propriception refers to sensations of position and movement including kinesthesis and the vestibular senses.

The non-auditory part of this structure is classified into two sets of receptors structures called 1) the three semicircular canals which respond primarily to rotation of the head and 2) the secular and utriculus (sacs) which respond primarily to head position and linear motion of the head. (The seculus and the utriculus are sensory apparatuses of the inner ear). (Fig. 8.9).

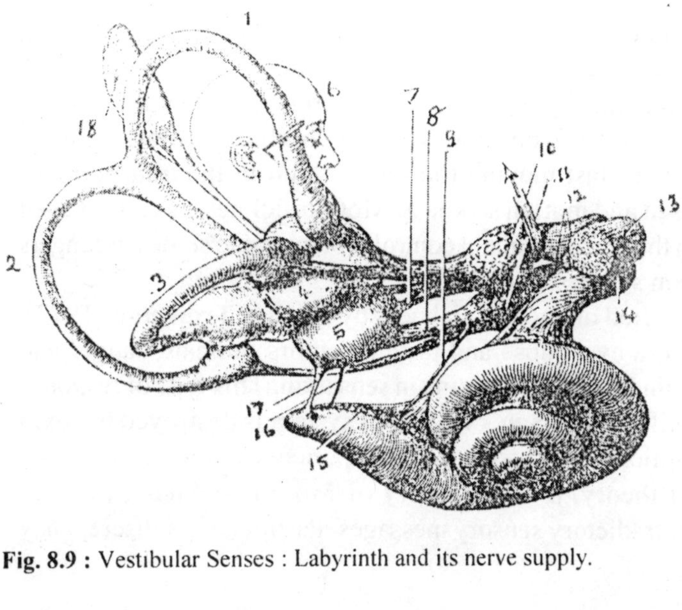

Fig. 8.9 : Vestibular Senses : Labyrinth and its nerve supply.

1. Superior Vertical Semicircular Canal
2. Posterior Verticle Semicircular Canal
3. Horizontal (Lateral) Semicircular Canal
4. Utriculus
5. Sacculus
6. Direction View
7. Rotatory Sac Superior (voit).
8. Non-auditory Sac major
9. Rotatory Vestibular Cochlea
10. Sup
11. Inf (Ganglion Scarpae)
12. Non-auditory Vestibular
13. Non-auditory sac
14. Non-auditory Cochlea
15. Ganglion Spiral Cochlea
16. Rotatory Cochlea Sac
17. Can reuniens
18. Sac end

In humans one portion of the vestibular system consists of three semicircular canals in the inner ear. These canals lie in three planes and right angle to each other. They are the horizontal canal, anterior vertical canal and posterion vertical canal. The horizontal canal inclines at about 30 degrees, which reacts to rotary head movements. The anterior vertical canal and the posterior vertical canal respond to head movements and are at right angle to the first and to each other. These three canals are connected at their ends to a saclike structure called the utricle. There is another small fluid filled secular lying below this. In each semicircular canal, the hair cells are in an enlarged region to ampulla, a bulb like swelling on the end of each semicircular canal, where it contacts the utriculus. The ampulla contains crista, a receptor, responds to head movement.

The three semi-circular canals in the inner ear are responsible for this sensation of equilibrium or balance. These canals are filled with liquid fluid. Every movement of the head or body has a corresponding effect on this fluid in the canals. Hairs in the canals project in the liquid. If the liquid is disturbed the hairs are moved. As a result, they set in motion the impulses which are carried by the nerves, connected with them to the motor area of the brain that controls the movements of the body. Later, suitable adjustments are made to restore the equilibrium of the body. When a person sits in marry-go-round or involved in some swift rotation the semicircular canals are affected and the person complains of motion sickness and giddiness. Suppose the semicircular canals are injured, a person automatically loses his sense of balance and proper motor adjustment becomes impossible.

Generally conscious sensations do not result directly from the stimulation of the non-auditory labyrinth by head portion and movement. Most of the strong sensations that result from changing head position are caused by reflex responses to change. For example, the sensation of dizziness from spinning around and around as the children often do come from reflex responses. The eyes move flexly back and forth, so that the visual field seems to swim by reflex change in extensor muscle tone causes the subject to feel that he is feeling down.

Certain types of body acceleration such as those we experience as a passenger in boat, in an aeroplane, in a car, or in an amusement park ride can produce distress known as motion sickness. Some village people, mostly women avoid bus travel as they feel nausea and get vomiting. There are also some village women who are not familiar to railway journey feel nausea. Though these are not very fast moving transports, even in the transports, of average speed they experience nausea and motion sickness. Motion sickness is caused even by low frequency movements which the persons cannot control. For example some passengers travelling in bus or car suffer motion sickness, but the driver does not suffer.

Several experiments were conducted on motion sickness by **Michal Treisman** (1977). He observed that sensory inputs from eyes, nose and taste receptors indicate danger and cause dizziness and vomiting to eliminate potential toxina in sensation. This type of response has the significance for preserving life. If a person's vestibular system is destroyed by toxin drugs or by the side effects of antibiotics, he will not have the experience of motion sickness.

According to sensory conflict theory, a major theory of Motion sickness, evolved recently, distress arises from the contradictory sensory messages. Particularly a discrepancy between vestibular sense and visual information causes motion sickness. For example, if an aeroplane suddenly starts up, there is excitement in the vestibular system, but the eyes see the planes interior constant. This results in conflict between the vestibular sense and the sensation of sight. The consequents to disorientation is the distress of motion sickness. Investigators like, **Beson** (1990) pointed out that there is also likelyhood of conflict of information within the vestibular system. Moving fastly, for example, 'around a sharp curve may produce conflict between information about acceleration and the forces associated with gravity. **Candland** (1968) states that the vestibular apparetuses are directly responsible for all kinds of motion sickness. A number of responses are involved in motion sickness. A disorder of tinnitus (a buzzing sound) in the ear is one of the reason for motion sickness. There are persons prove to motion sickness, but this varies with age. Any type of vertical motion like droplin of an aeroplane also may cause motion sickness. Medicines are available for inhibiting the symptoms of motion sickness. The vestibular system received importance in selecting men for aviation and to some military services.

9

ATTENTION AND DISTRACTION

Introduction

In the everyday life situation people often come into contact with many things in the world. After few hours or days some of the events they witnessed they remember and some of them they do not remember, If two persons witnessed the same event or a meeting independently and when they meet on the next day, they might have had a short conversation on the event. In the course of their conversation, one might have asked the other person "were you listening to the speech made by the main speaker of the meeting?" The other person might have replied, "I did not observe to the whole thing fully, but, were, you fully *aware* of the whole thing?" "Yes, I did, but I could not pay any *attention* to the matter the speaker has referred to at the closing time as I was *watching* and *looking* out for the bus I wanted to catch".

In this short passage the important concepts are *listening, observe, aware of, attention, watching* and *looking.* All these concepts more or less may convey the same meaning. The process of function involved in all these concepts is the same thing. But the kernel meaning involved in all these concepts is *attention.* The concept attention is invariably used in many situations both in the common sense world and in the scientific world.

In military, while giving the order of command to the soldiers, the most often used concepts are "attention" and "stand at ease". What attention means here? This is a caution word, and the soldiers should respond to the next order of command. They are made to be alert and listen to the next order. Thus the concept, attention made the soldiers to be alert, and to react smartly. Here for the soldiers the word, attention is a word of command which makes them to react switly. It causes a quick response in them.

In the class-room situation, the teacher often uses the word attention, as *'attention please'*. It is making the students to listen to what the teacher is going to say, In hospitals the doctors and nurses pay careful *attention* on the patients. When some aggrieved persons were waiting to see the concerned officer, the officer asks his personal assistant, go and attend on them. Thus the concept, attention in the commonsense world is occupying an important place.

In the scientific world, specifically, in the field of psychology attention has been used to explain an Important process of function from the time of Aristotle to the present time. The psychologists have been using the term attention in the sense of conscious awareness of the individual person or a group of people. They were referring it to the cognitive process. Many of the everyday works require close super-vision and attention.

There are certain methods we all can apply to maintain keen attention on ourselves and on others. Attention is an organism's ability to focus and to concentrate on something whereas alertness is about how awake or sleepy we are. It is true that attention is a very familiar concept, but in the past it has been surprisingly neglected and has become very difficult to define it and to carry out research.

But the phenomenon that requires explanation is how the human beings select information out of many and act upon one. Another problem is how humans select behaviour? Insects, and other lower creatures which have limited sensory capacity, have equally limited stock of behaviour. Animals of some kind have much more sensory capacity and also have corresponding increased behavioural patterns. Human beings, among the primates, have even much more capacity in certain sensory areas. In this way the sensory and motor capacities increase as the brain and nervous systems expand. The learning capacity and the memory increase which provide better response and selection. Thus as and when sensory channels increase in capacity there is a need to relate different inputs to evolve a better response.

In these evolutionary changes there is expansion of brain which results in the increase of sensory and motor capacities and some other processes arc also involved in the sensory and motor areas. One such important process is known as attention. As and when the sensory and motor capacities expand, so also there is expansion in the process of attention and consciousness. Consciousness increase gradually along the evolutionary process relatively with the sensory and motor capacities. The recent conception of attention is that it is a conscious awareness.

Revaluation and Renewed View of Attention

The concept of attention has been treated by psychologists in the same manner as the other psychological concepts. They have been giving importance to attention sometimes and also have been neglecting in other times. At the beginning of twentieth century it wa; an important topic of study, The early authors of psychology have dealt with attention in books and later on after 1920 they have neglected studying the processes like attention (**Kolb** and **Whishaw**, 1956).

Titchner (1902) declared that the doctrine of attention is the nerve of the whole psychological system. But the way in which men judge its value and determine the importance of it is found to be diminishing. One important reason has been the development of *Behaviourism* and the Gestalt Schools of psychology in the first quarter of 20th century. These schools rejected the importance of studying the process of attention. The Behaviourists were of the view that a *full account of behaviour is possible only in physiological* terms and there cannot be any reference to cognitive concepts like attention and consciousness.

According to **Watson** (1878-1948), the founder of Behaviouristic school whatever is observable and measureable alone must be the subject matter of psychology. **William James** (1842-1910), (1890) advocated that every one knows what attention is. Attention is taking possession by the mind in a clear and vivid form, one out of many objects when simultaneously presented. It is selecting one out of train of thoughts. The emergence of cognitive psychology has revaluated the importance and the perspective of the concept attention. Further, the investigators of neuroscience have returned to the position espoused by William James. Thus there was a development of renewed interest in the use of concepts like attention. This has established a distinct group of investigators among the students of attention. A division of investigators, are interested in the automatic processes involved in attention. The others are those interested in the conscious selection of sensory information in attention. Therefore the question arises, whether attention is an automatic process or a conscious selection? One of the views in cognitive psychology is that certain behaviours may be performed with little attention, while the others view 'that it is highly sensitive to the specific object.

Observation and Attention : Defined

The sense organs receive information from the external world in the form of stimuli and convert them into meaningful materials. For example, seeing, hearing, tasting, smelling, touching of the objects, persons, places and other events of life provide essential information. An individual must deal with his environment successfully for which our senses are the guiding elements. **Woodworth** (1952), the famous American Psychologist considers that *observation* is the single concept that would explain the whole processes. According to him observation is a topic of central importance in the whole field of psychology. Observation, he says, has two steps, one is *attention* and the other is *perception.* Attention is the preliminary step for observation and perception is the final step. Here we are mainly concerned with attention.

Attention may mean the action, fact. or state of giving need. It is the earnest direction of the mind. It is the state of mind which prepares us to receive impressions. Attention is a cautionary word used as a preparatory to any particular exercise. It is giving a watchful need to the wishes of others or to something outside of us. When the physical director says, "attention", he means that you are expected to be ready for some task which he is going to give next. Your lecturer often may say "Attention please". He means, "please listen to what I am going to speak to you," Attention is the first step in the perceptual process.

Attention is defined in several ways. It is defined as a set or readiness to respond in certain way. According to **Murphy**, it is narrowing of the range of objects to which the organism is responding. According to **Krech** and **Crutchfield** *(1958)* attention is the focussing of perception involving a hightened awareness of a limited part of the perceptural field.

Titchener (1902) says that an alert organism attends to some stimuli but not to others. The state of attcnflon irnplies not only activation of-Some parts of the behaviour but, inhibition of otlier!;- Tliis process is callcd tuning. In the process of attention an important characteristics of an organism is its bandwidth or narrowness of the range of stimuli to which it will respond. The narrow attentional. bandwidth increases sensitivity and selectivity of additional processes. The organism pays closer attention to a narrow stimulus field. Therefore, we can define attention as the process of narrow downing the perceptual area by selecting anyone of the stimuli at a given moment and ignoring others.

Physiological Mechanism

Attention is a process of listening to something occuring in the presence of an organism. It is also a response to a stimulus or a caution. The organism is made to receive certain information. Attention is not a passive response. It is an active response to a stimulus. In the process of paying attention to stimuli the whole human organism is involved and the organism, is getting ready for action. The organism reacts not to all the stimuli that are appearing in its presence. It pays attention only to a specific stimulus. Though in this respect it is psychological, the process of attention is a physiological aspect, because the organism's sensory apparatuses are put in action. If the stimulus is of visual type, the organism is made to see. If it is of auditory type, the organism is made to hear the sound through the sensory apparatus, the car and respond to it. Unless the visual stimulus is seen there cannot be any response, so also if the auditory stimulus is not heard by the organism there is no response or attention from the organism.

The organism out of several stimuli selects the relevant stimulus from the environment and attends to it. The information received by the organism is utilized approprietely. In this process what is the psychological mechanism that takes place in this process of selection and response -and utilizing the information. Physiologist say that the central nervous system is involved in this mechanism. The role of reticuler system of the brain is responsible for this function.

It is observed that the nerve fibres from spinal cord pass through the brain stem on their way to the forebrain and the fibres from the forebrain also connect the brain stem and pass through it on their way to the spinal card. This mixture of the cells and the fibres in the brain stem is known as the *reticular formation.* This reticular formation is generally known as the reticular activating system. According to **Moruzzi** and **Magoun** *(1949)* the function of reticular formation is to control sleeping and waking states. But the general function of this system is to maintain the general arousal or consciousness.

Fields of Attention

The term attention is used to indicate a variety of situations and its manifestations which are taking place in different forms. An unusual object may attract our attention or we may concentrate on a particular object or an event may be disregarding all other things around it. Attention may also indicate mental set which means we are anxiously expecting to receive a particular stimulus or an object. Sometimes attention may be also guided by a motor adjustment just like catching a city bus.

Quite large number of studies have been conducted in the recent past on attention and on the manifestations of different fields. A variety of situations use the concept, attention in different connections. According to **Moray** (1969) attention has many facts. They are: (1) degree of attention, (2) mental concentration, (3) vigilance, (4) set and (5) selective attention.

The *degree of attention* refers to the state of arousal. For example, a sleeping person will not pay any attention to our environmental stimulation, unless the person is in an awaken state. There is no chance for him to pay attention on anything.

Mental concentration refers to the process of attention on a particular object. In this task the person ignores all other incoming stimuli as they may interfere with the main performance of the task. It is the ability to concentrate on a specific task. This ability is found to be independent of intelligence according to **Wittenborn** (1943).

The process of *vigilance* is just keeping oneself pin down on the task. This is characterised by watching or searching behaviour of an observer. The person is waiting for the signal to react.

The function of *set* is a pre-disposition to perform a task. That is, the person has already decided to perform some task and waiting for the stimulus presentation. The set is a mental set.

Attention is not general. It is specific in character. Therefore it is *selective*. Attention refers to the ability of an organism to process only a part of the information received and to ignore all other aspects.

Thus attention may manifest in these different ways. These variables are the fields of attention seem to have no common mechanism in their functions. But, however, they have certain internal relationship among them. This has been tested by **Kendler** and **Kendler** (1966).

Attention has been studied from different angles. **Trabasso** and **Bower** (1968) have categorized the *action construct* into *three categories* such as *discrimination learning, neurophysiological research* and *information processing* with special reference to human beings.

The study was conducted by them for a long period of time, The objects of their study were both the physiological and psychological aspects of attention. **Travers** (1970) discussed about the physiological and psychological aspects of attention and identified the concept in a number of phenomena. According to him attention is involved in the adjustment of the sensory systems which facilitate the reception of certain type information.

Further, attention has a reference to the process of inhibition of certain class of information in preference to some other information. **Boring** (1970) states that one cannot discuss attention without making any reference to inattention, facilitation and inhibition. Another factor associated with attention is the level of arousal in the organism. In the reception of information certain environmental factors also are involved in the study of attention.

There is some distinction between the selective aspects of attention and the intensive aspects of attention. According to **Kahneman** (1973) the selective aspect of attention has the usual reference to the organism's attention selectively to some stimuli in preference of others. But the intensive aspect of attention has a reference to the level of arousal. There are also some factors called collective properties involved in the process of attention and regulate the arousal levels. According to **Berlyne** (1910) these collective properties are *novelty, complexity* and *incongurity*.

The collective properties are found to cause and interfere with some stimuli which are found to be arousing intensely than other stimuli. Authors are of the view that intensive aspects of attention are considered to be in relation to voluntary and involuntary attention and for this purpose they have proposed a more inclusive term, '*mental effort*' instead of mere arousal. At this stage we will have to consider also about the incident of poor attention.

Observations were also made on attention associating mental retardation. Persons with mental retardation were found to be having poor power of *attention* and concentration. According to **Ellis** (1963) and others distractibility, poor attention and *lack to concentration* are associated with mental retardation. But, **Das** (1975) has observed in his studies that there is not much difference between the mentally retarded persons and the normal persons in some of the aspects of attention. The difference between these two groups is limited only to the passive states of attention, such as involuntary attention or reflex state and the level of arousal. It is found that the voluntary relative attention, which is a cognitive function, is found to be vulnerable for the mentally retarded persons.

According to the studies made by **Posner** and **Boise** *(1971)* it was observed that there are three components functioning in the process of attention. These are (1) alertness, (2) selectivity, and (3) processing capacity. *Alertness* is the same as vigilance and functioning in the same manner. It is watching carefully what is going to take place next. It is the period just before the reaction performance of the organism to the stimuli. The person prepares himself for action.

The second component of attention is *selectivity* which is concerned with the ability to select the information from one source and ignore the others. The selectivity or the selective aspect of attention has become one of the important cognitive factors in psychological studies. The third component of attention is the *processing capacity* of the organism. These three components are functioning in attention in varied levels in an organism.

Selective Attention

The selective component in the field of attention has been recognized and accepted by all the scholars who have studied attention as the important factor functioning in an organism. Selective attention means, the organism in its reaction to the stimuli chooses one stimuli-is that is very important for its own purpose in the field of function. The organism only selectively attends to some of the information in preference to others. Selective attention enables an individual to function in relevant information. Thus selective attention has reference to the ability of an organism to process only to the part of the information registered in the sensory systems and to eliminate the rest. Selective attention has also reference to the control of information, processing, so that the limited capacity channel is not overloaded and the efficiency of the organism for the selected stimulus is not impaired.

Several investigations were conducted on selective attention. In one of the studies made by **Kahneman** (1973) has observed that there were many varients of selective attention.

It was found that situations are required for the organism to select stimuli from a particular source and the targets of a. particular type of objects with certain specific qualities and the response occurred in a particular manner.

An interesting investigation was conducted by **Cherry** (1955) in which he studied the problem of selective attention. This study has brought out interesting information. For his study, Cherry has developed a new technique, *shadowing* in which the listner was asked to repeat one of the two Continuous messages presented to him over the two earphones when he received them and ignore one of them. It was a kind of verbal communication through which information were conveyed. This shadowing technique for the study of a selective attention had become very popular in the psychological studies. Experiments were conducted on group of persons with different passages of prose, one for reading and the other for listening simultaneously.

The respondents were tested by asking questions from both the materials to find out. how much information they were able to receive from both the passages presented to them in two different ways. On analysing the results it was found that the respondents have done well with easy passages on both the materials. But with difficult passages the efficiency was found to be poor.

Some of the later investigators like **Moray** and **O'Brain** (1967) conducted studies with auditory attention. They applied the monitoring technique instead of shadowing technique. According to monitoring method the person is exposed to a continuous message and he has to respond only to occasional targets specified for a particular situation. The difference between these two techniques is that in shadowing the person is required to provide continuous overt response, but in monitoring only occasional response is required. But it was felt that the shadowing method was much letter than the monitoring method.

Focus of Attention

Many events may be taking place around us in the outside world. But it is not possible for us to give equal importance to all of them and pay our attention on them. We give importance only to a single stimulus or event out of many. What we are doing is, that we select one object or event and attend to it. This process is called focussing. When we specifically look at an object, it is said that we are focussing our attention on the object. **Posner** (1990) said that when we search for an object around us in the world, the focus of an attention shifts from one place to another. In such case our attention must suspend from the present object of interest and move to a new object and engage in it. In an environment where objects are visually scattered, another functioning system of inhibiting or filtering of the irrelevant object or information may be required. It may be taking place automatically in this situation.

We are said to be attentive where our sense organ is actively focused upon some determined stimulus. The focussed attention may shift to something else, because it may be tedious or dull and the object or information does not hold our interest for long time. In our every day life many events are taking place in our presence, but we do focus our attention sometimes unconsciously upon them, but soon shift from it and we may not be having any idea about them.

You might have heard or used the concept 'focus' when you sit for a photograph or use a camera. What the camera man does, is focussing the lens on the person or the group. Similarly, in drama or in a play house you might have observed that the lights or colour lights are often turned to the direction where the hero or heroine is playing his/her part, or an important event. This we call, focussing the light. In the same manner we focus our attention on a particular object or act. Sometimes you may he watching, several interesting events and may be desiring to see all of them or more than two, But your power of attention is limited and you are forced to see only one at a time.

FACTORS OF ATTENTION

Attention is based upon the nature of properties of the stimulus. In the applied areas, advertisingmen make use of this technique to attract the consumer's attention and arouse their desires to buy the particular brand of article. There are two kinds of factors that are accounted as determinants, They are objective factors and subjective factors. If the advertising technique is to have any effect upon the consumers both the factors must be involved.

Objective Factors

The important objective factors are: (1) Size, (2) Intensity, (3) Motion or change, (4) Novelty, (5) Repetition, (6) Duration and (7) Contrast or difference.

(1) ***Size*** **:** Stimuli in the perceptual field may be having different sizes. Our attention depends upon the sizes of the stimuli. Therefore, a stimulus which is larger in size is more likely to attract our attention than a smaller one. This is what we find even in society. A tall and stout man can easily take the advantage of his size in commanding others. A big capital letter, a full size advertisement, a big size portrait can catch the attention of the people much better than smaller ones. Further the respondent will have no difficulty, in perceiving the biggest objects, whereas he has to take pains to smaller objects. A mountain is seen even from miles far off but a man in the same mountain is not seen. Therefore, the size of the stimulus field is a strong determining factor of our attention.

(2) ***Intensity*** **:** This refers to the quality of the object. The stimulus has certain strength to attract the organism, A strong stimulus is better noticed than a weak one and receives better attention. For example, a bright light draws our attention at once, whereas a dim light fails to do it. Similarly, a loud noise, a strong smell would attract our attention much better than weaker ones. A much painful incident is more seriously attended to than a painless incident. The phenomenon of intensity of the stimulus is directly related to the various degrees of sensory reaction.

(3) ***Motion or Change*** **:** Motion is an effective psychological factor used by advertisers. A moving object in our field of vision will catch our attention much better than a static one. Even when our attention is not focussed in a direction a rapidly moving object would draw our attention in that direction. For example, a fastly driven motor car Is seen at once by us whereas a parked one is ignored. Many advertising sin boards make the letters and objects keep on moving constantly to attract the consumers. Some of them are devised with blinking lights. In big cities while walking through the shopping centres we could not avoid such advertisements.

(4) ***Novelty*** : It is also called striking quality. The degree of novelty of an object is having a definite effect upon the organism like motion or change. An object in a familiar place will not draw our attention. But an object in an unfamiliar place or a man in unfamiliar seat is in a better position to attract us. For example, a police man in black uniform or a student in police uniform in the class room is not likely to be overlooked. In the fancy dress competition one who outfits others in unfamiliarity gets the first prize. An unusual colour combination, an unconventional picture, an unfamilar sound are some of the objects compelling our attention.

(5) ***Repetition*** : Stimuli that are presented intermittently are having more effect than those presented continuously. This principle is applied in several situations. When caution is required or warming is given repetition of the stimulus is made by electric display signs in many places. The display signs flash on and off. If the danger signal is presented once or in a non-stopping manner people may not look at it. Therefore, the stimulus is repeatedly presented. The repeated stimulus increases our power of sensitivity or awareness.

(6) ***Duration*** : A stimulus must have a minimum duration of time. Otherwise it will not have any effect. The duration depends upon the intensity of the stimulus. If the intensity is stronger the duration may be shorter. A weak stimulus requires longer duration to arouse our attention. For example, a very short time, that is a fraction of the second is sufficient for a bright light to arouse our attention but a faint light requires more time to do the same. The **Bunsen-Roscoe law** explains the relationship between intensity of the stimulus and duration in mathematical equation. But if the stimulus is having very long duration it loses its effect because it is present continuously. This principle is applied in the harbour light-houses.

(7) ***Contrast or Difference*** : A stimulus that is contrast or different to the background is very effective. One aspect of the environment must stand out completely from the reminder. Architects, modern decorators use this idea in building, painting, decoration and furnishings to beautify the area. They say that there must be contrast or striking difference between objects, qualities and colours to show their relative excellence. In an area if all the buildings are skyscrapers and if one is very short, it is a contrast to others and it draws our curiosity and attention. In military the soldiers camoufalge everything that is in contrast to the background. That is, they conceal an object by reducing the contrasting quality between it and the background. Otherwise the enemy planes will easily find out the location where to bomb and destroy.

Subjective Factors

There are certain factors in attention which are based- upon the individual characteristics. They are called subjective factors. These factors originate in the previous learning and experience and in the individual's motivation and interest. They are : (1) motivation, (2) previous experience, and (3) perceptual set.

(1) *Motivation* : We are having several needs. These needs are to be fulfilled, towards which our attention is directed. But all our needs are not having the same amount of strength. Time makes us to forget several things. At times certain needs or motives are becoming strong and when we satisfy the needs they become weak and disappear. When we are hungry our needs are strong for some kind of food. Such physiological needs are strong for us at times and when they are fulfilled we ignore them. Therefore, motivation generally refers to the behaviour instigated by needs and directed towards some goals. It is something that propels action originating from within the person who acts.

If the motivation is weak we tend to respond to variety of stimuli, and we don't see the relevance. If the motive is strong we respond to certain particular stimulus and not going into the details that are not connected to the motive. Advertisers often make use of sex drivers in their advertisements to draw the attention of the consumers. They present in their advertisements something related for the consumer's sex drive which makes them to be attentive. Modern advertisements, whether they are related to men's wear or, women's wear, a soap or hair oil are associated with sex, because sex could draw the attention of the people very easily.

Similarly, if two persons are walking along the country road side one who is a botanist, is interested in plants and the other who is a zoologist, is interested in animals and birds. Every one is interested in different matters. H' you ask the botanist anything about the characteristics of species of animals and birds he would say that he does not know them. So also the zoologist may not be knowing about the plant life. The motives of these two persons are strong because they tend to respond to a restricted area and not interested in other parts of the environment that are not related to their motivation. A strong motivated person is attracted to more information, in a narrow area whereas a weakly motivated person is attracted by many things around him but know little about a particular object.

(2) *Previous Experience* : Every one of us is having some previous experience in something. In certain things our experience is very wider and in certain things it is very narrow. This experience determines our selection of stimuli from an environment. The accuracy of our perceptual process is, therefore, based upon the nature of experience we have acquired in the particular object or event. For example, a deep blue sea is looked upon differently by different people. An oceanographer is interested in finding out the depth of the sea, while a saler looks upon it as wild sea to cross it over. A fisherman may look upon it as the source of big catches. A marine biologist may search for a new species in it. The deep blue sea attracks the attention of all these people, but everyone sees it according to his past experience. The process of selection is in operation here.

The selecting tendency of the individual filters out certain aspects of the environment from his past experience. But if certain aspects are relevant to guide his behaviour and if they are filtered out it may impair his learning ability. This selecting tendency would Improve the performance of the individual in a situation if irrelevant aspect of the stimuli are neglected and relevant things are responded. We are always trying to project our past experience in every perceptual field. Sometimes it may result in certain disadvantages. The personal biases may interfere at certain stage of our past experience and we may even change our perceptual field. This selecting tendency is called by some as personal habitual interest.

(3) *Perceptual Set* : Set is playing an important role in the selection of our perceptual stimulus. Set is also called expectancy and momentary interest. Set is closely related to previous experience. According to **Morgan** (1971) a set is a readness to react in a certain way when confronted with a problem or stimulus situation. Therefore, set means a tendency to respond in a certain way. It is established by events.

The effects of set arc short lived. In set we make ourselves ready to respond to certain stimulus that we have been expecting. The set is purely individual or personal in nature. What is a set to one person is not a set to another person. For example, you want to wake up at 4 O' clock in the morning and you set your alarm clock at 4 A.M. When the alarm clock rings at 4 A.M. you immediately wake up. Whereas your friend who is sleeping in the same room is not waking up. Even if he is not sleeping lie will not wake up. Because you set yourself to wake up at 4 A.M. and the alarm is for you and not for your friend.

A sprinter, who is taking part in the race is ready to start at the crack of the gun shot, The sleeping mother wakes up immediately at the cry of her child, whereas her husband who is accustomed to telephone calls wakes up when the phone rings. All these people have set themselves to respond to some particular stimuli. Therefore, the term set is referred to the readiness of an organism to pay attention to certain perceptual object or stimulus pattern. A set can be created in an organism by training. An organism is allowed to eat a particular kind of food at a particular time for few days. Naturally he learns to respond in the specific pattern to satisfy his needs. Here a set is created in the organism. Pavlov has created this kind of set in his experimental animals.

Some scholars arc of the view that set leads to increase the efficiency in behaviour. But it is not so. It is not increasing the general efficiency in behaviour, The increase of efficiency in behaviour is limited to the particular stimulus situation to which we have set ourselves. We could observe this in certain situations. When we are set to one aspect of the situation we are not prepared to respond to another aspect of the same situation even if it is more important. For example, when you are searching for a lost silver coin, a golden coin on the same ground is not noticed by you. Because you have set yourself to find out only the silver coin and not the golden coin. Previous experience and set may cause the same result even though they are not the same. Previous experience requires more time to establish whereas set is short lived, occurs for a momentary act and may also be shifted.

Division of Attention

Our attention is normally directed to one object at a time. Sometimes it so happens that we try to attend to more than one or two things at the same time. But our attention will have only one form at a given time. Due to certain necessity or inattentiveness to one object we often divide our attention and distribute it among different objects or attend to different activities. When our attention is distributed over the whole field of consciousness there may be greater concentration at one object or activity and less concentration on the rest. Our attention, therefore, is diffused and no particular object or activity is dominent. Therefore, division of attention means, trying to attend to two or more objects or simultaneously. The Indian mystics often speak about *astavanthanam* which means, attending to eight activities simultaneously. It is said that Nepoleon was capable of writing two different matters with two hands simultaneously. But all these things were not experimentally verified. **Woodworth** (1952) says that division of attention means a simultaneous focussing upon two separate activities. According to him if one of these is automatic, it goes well without conscious control. Then there is no division of attention.

There is no doubt that we are all engaged in more than one activity, simultaneously. We are walking fast on the road and speaking with our friends, seeing things around us, rolling the walking stick and smoking the cigar. Are these not simultaneous activities? But are these activities independent or interdependent. These activities are interdependent and therefore, do not satisfy the conditions of the definitions.

Psychologists have been trying to study this aspect of attention experimentally. **Paulhan** was able to do two separate activities simultaneously. While reciting a famous poem orally he was also writing another poem. Further he was able to do a simple multiplication and also recite a poem. Both were done without any interruption. **Binet** as early as 1890 has conducted experiments on division of attention. In psychological laboratories there are several kinds of experiments designed for this. One familiar apparatus is the Division of attention board (Fig. 9.1).

The apparatus is a board, having two different patterns and electrically operated. The two patterns are connected to two electric counters and two styluses. The subject is asked to run the pattern with the stylus as fast as possible with right hand and then with left hand and thirdly with both hands simultaneously. The counter records the number of rotations the subject has made at each occasion. It is assumed that the number of rotations made by both hands must be less than the number of rotations made separately by right and left hands. In this way the percentage of loss of efficiency in each hand can be determined. This loss is due to the *divided attention.*

Different pattern of experiments can also be designed. Two mental tasks simultaneously may be assigned or one mental task and one physical task or two physical tasks may be assigned. The efficiency of double performances is determined by finding out the percentages of the performance of the two tasks.

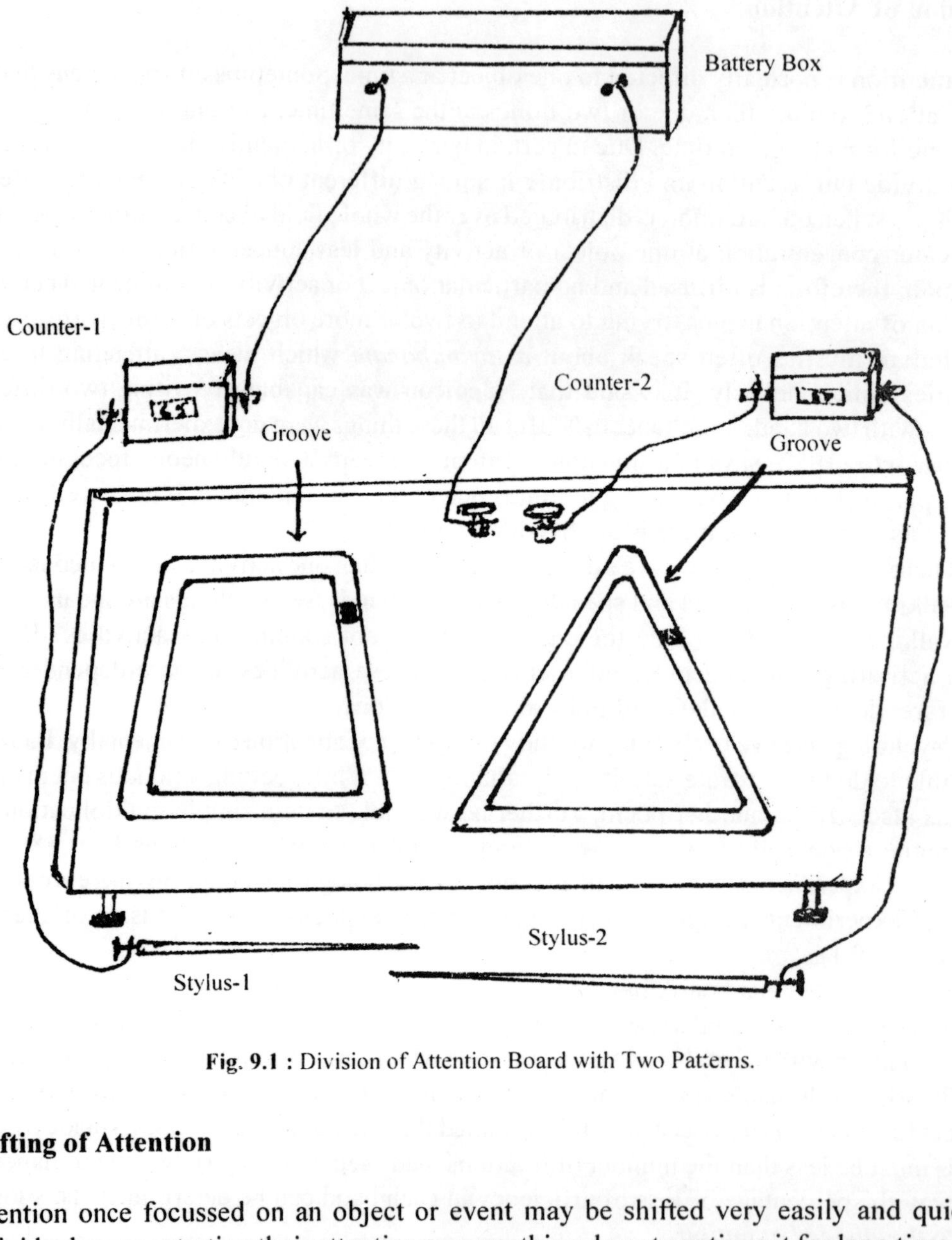

Fig. 9.1 : Division of Attention Board with Two Patterns.

Shifting of Attention

Attention once focussed on an object or event may be shifted very easily and quickly. Individuals concentrating their attention on something do not continue it for long time. Our attention often moves like a movie or television camera. It centres first on the object or event and then moves to another. Our attention is always directed to objects or events that have strong attractions to our sense organs. This process is called *shiffing of attention.*

Shifting of Attention may occur very easily while looking at ambiguous figures. When you look at a person, what is that you are looking at him? First you look at his face, then go on shifting from one place to another. The face as whole, then eyes, nose, mouth etc. After few seconds your attention replaces the objects as a relief. You do not see a person's eyes all the times. You also look at other parts of his body. Then you may look through the window. When you fix your attention on a picture or person or an object your eyes may be going on shifting from one view to another in accordance with the intensity of the stimulus. There is change of position or direction in your attention. The change or shift may be voluntary or involuntary.

The shift of visual attention may be measured objectively. We can record a person's successive eye movements when he is looking at an object. We shift our attention for several reasons. In attention there is form of satisfaction. This makes us to inhibit the continuance of our attention in a place or direction. In the total perceptual process the shifting of attention is an essential process.

Shift and Fluctuation

These two process appear to be the same. But they are two processes. They can be differentiated and related. Our attention has two kinds of processes. It shifts from one response to another and secondly it fluctuates in the efficiency of response. Our attention often fluctuates from its original position. Just like the waves rise and fall our attention also makes repeated variations in its intensity. We do not pay the same amount of interest at an object we see.

Fluctuation is lapsing from high level of efficiency during continuous work. The responses change but the stimulus remains the same. For example, when we look at certain event or object we do not put the same amount of attention on it all the times, In the beginning, it is, very strong, we have more amount of curiosity. Later on it slowly decreases and dimishes. We may even withdraw from it for few seconds and go back to it. If the motivation is strong in attention there would be no fluctuation. But attending to an object with same amount of curiosity is not possible. There may be vacillation in attention.

Modern psychologists are of the view that there is no difference between shifting of attention and fluctuation of attention. Therefore, they do not give different treatments to these two processes. But it is possible to differentiatc these two to some extent. In shifting of attention we withdraw our attention after a few seconds from an object and move to another object. But in fluctuation of attention, we do not withdraw our attention from the object. We fail to attend to it with the same amount of strength. It is strengthening and weakening the process of attention.

The Span of Attention

There is certain amount of limitation in our process of attention. We cannot see everything and anything at one glance or look. Therefore there is a *limit for our attention.* It is called *span of attention.* It is just like an individual clearing a pole-volt or high jump or long jump. He has capacity to clear certain height or length. For example, an athelete can run 100 meters 9 seconds. Similarly there is a limit our attention to see a number of objects in a single act of perception.

Psychologists have devised certain apparatus to measure our span of attention. *Tachistoscope* is the apparatus generally used for this purpose. **Cattell** has invented this. Now-a-days we can see various patterns of tachistoscope in psychological laboratories. It is a board held vertically with a window at the centre. There are cards designed for this purpose with dots varying in number ranging from 5 to 8.

The card is kept in the pocket of the shutter at the backside of the board and the shutter is held by a lever at the top of the board. The -subject is asked to watch at the window. When the shutter is released from the lever by the experimenter the shutter falls through the window exposing the dots in the card for a fraction of a second. The subject is expected to say how many dots he saw in the card. The cards are presented at a random order. Every time when the card is presented to the subject through the window the number of dots may be varying. The maximum number of dots, the subject could discriminate clearly in the card, is his span of attention. The span of attention of a subject can be five, six or seven. It varies from person to person depending upon the person's power discrimination of the dots in a fraction of a second (Fig. 9.2).

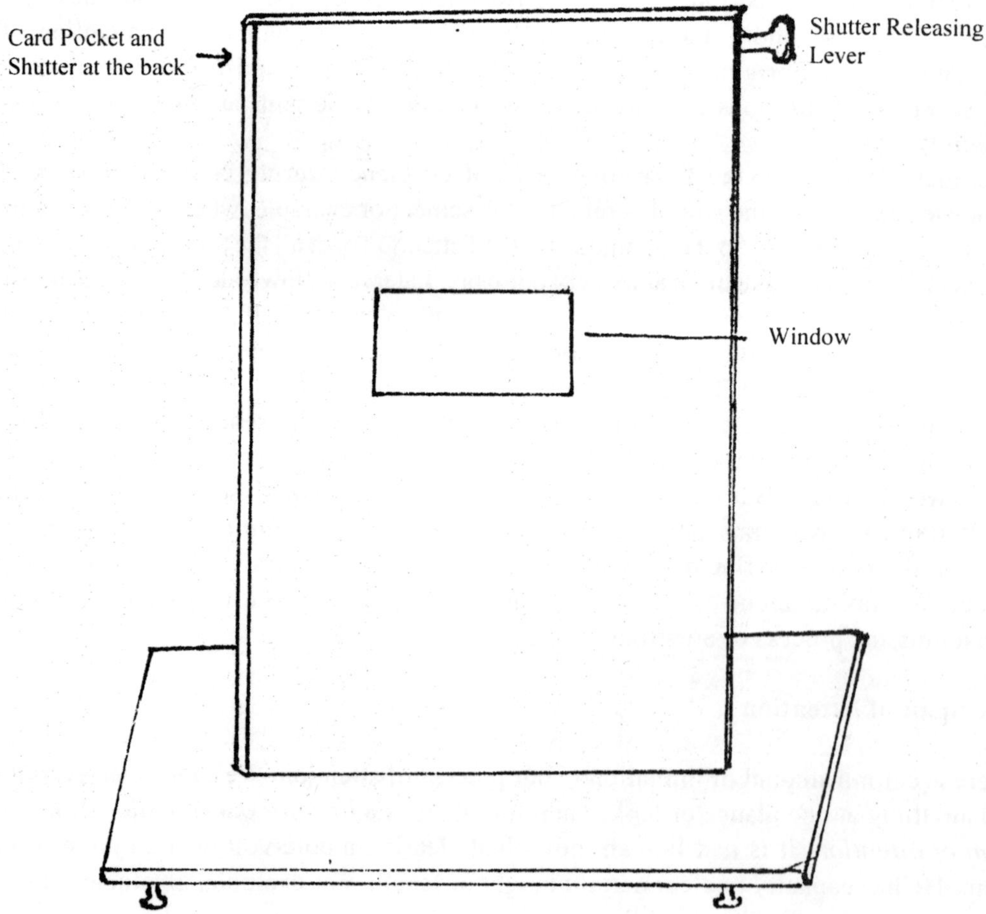

Fig. 9.2 : Tachistoscope : Span of Attention Test

It has been concluded from experiments that the span attention for a normal human being is three letters and four numbers. Anything beyond this limit will result in errors. If the letters and numbers are more than this in any situation, there will be more error responses than correct responses. This psychological idea is applied mostly in giving numbers to vehicles. You might have seen a vehicle either a motor car or bus or lorry having only three letters and four numbers. This is the normal human span of attention. The letters and numbers are presented like this: TMX 8116, TMY 7812, TMZ 9999. The three letters are not non-sense syllable. They indicate the State and the District and the numbers indicate the serial registration number of the vehicle. There are two groups of stimuli here, one the letters and the other numbers. It is easy for our mental faculty to organise the perceptual field into units of letters and numbers.

There may be also arrangements to have more than two groups as three or four or five. For example, we may have three groups and four- units like this with twelve dots.

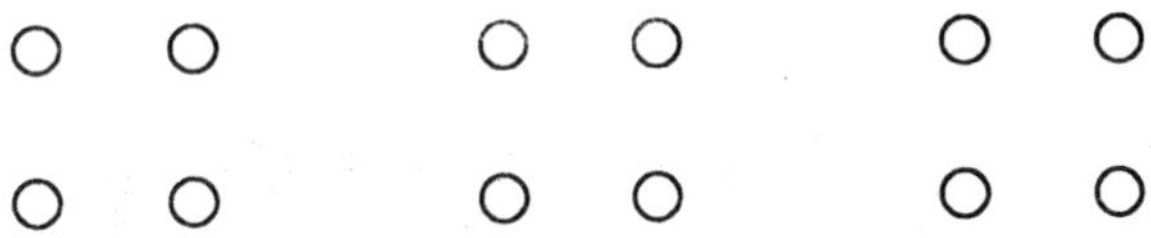

When we have more than one group with four units or five units the groups must be presented in the card in some order. Otherwise it will not be possible for the subject to distinguish the number of groups and to organize the perceptual field into units. It is important that all the groups must have the same number of units. Whether a group consists of four or five units (dots) the same principle must be followed. The groups must be placed with sufficient interval of space. When this kind of complex units are presented it is better to have a wider window in the board. The span of attention in this type of units is only four and not twelve. The number twelve is given by calculation. This kind of experiment is conducted with visual stimuli.

In the same manner we can also design experiments to measure the span of attention with auditory stimuli. Sound is caused by rapid tapping and the subject has to tell the number of taps he heard and this becomes his span of attention.

DISTRACTION OF ATTENTION

The distraction of attention generally refers to the failure of concentration of attention. Every body is interested in focussing his attention on certain object. Particularly in the field of education the failure of concentration causes much anxiety to the teachers and parents. Why the student himself is much worried about it at the time of examination. Our attention is not pinned up to the object or activity as we desire to be. Because there are some forces that are impairing our performances and turn our attention from the perceptual field. This process is called distraction.

The forces that are impairing our activity are called irrelevant stimuli or distractors. It is possible to verify whether there are any forces called distractors in our attention. A *distractor* is one that interferes in the efficiency of the performance. This does not mean that performance itself is affected. We can experience, and find out the interferences in our various performances. In mental and physical works we ordinarily experience interferences.

Students often complaint that they could not concentrate their attention in their studies owing to the constant disturbances caused by the noises outside. Some say that they could not study because there are four or five students in the room and they are always engaged in some kind of talks. Some might say that the light is very dim and few may have the complaint of constant stomach pain. So there are several sources of distractions in our attention. The sources of distraction may be external like sounds or noises, moving visual objects, insufficient light, internal sources like stomach pains, headaches and mental sources like interferences of ideas.

Thus attention is often distracted due to various reasons. Generally the environmental conditions are easily distracting our attention. Background loud noises and activities can be highly detrimental to human abilities to keep the attention focused at some work. While some workers thrive on the buzz of frequent interactions, others suffer because they find it very difficult to concentrate on the work.

When there are loud noises outside how can we concentrate on our studies? Noise is an irritating and unwanted force. Noise is an irregular sound. We don't like it. If noise is regular it becomes music and we like it. But we don't want it at the time of our study. In some kind of works noise is required because it increases efficiency. For example, in spinning mills it is required. The typist who is accustomed to the clattering noise feels that the absence of noise is a source of distraction for him.

If there is any moving of objects or persons constantly we may feel irritated at the time of study. Similarly when the student is having severe stomach pains how can he concentrate his attention on his lessons. Several ideas may be appearing in a chain like manner in our mind. You may stop studying your lessons and go into the world of imagination. Therefore, distractions are functioning in these manners to destroy our concentration of attention. **Woodworth** (1952) says that anger is the worst distracting force than noise.

Experimental results suggest that the effects of distraction are not so severe as they are thought to be. An experiment has been designed by **Hovey** (1928) to study the effect of distractions. One form of the Army Alpha intelligence test was administrated on a group of college students. Then the group divided into two on the basis of the scores. One was the control group and the other experimental group. Six, weeks later another form of the Army Alpha test was administered on the control group under normal conditions and on the experimental group under conditions of distractions. There were lot of distracting forces like loud noises of different types and bright lights etc., on the experimental group. But the scores of the second test were not much affected.

In first test both the groups had equal scores. In the second test through the control group scored little higher than the experimental group the difference between them was only 3.7. The distracting forces did not cause appreciable difference. Therefore, it is inferred that the type of attitude the individual has developed towards the forces of distraction determines the results. If one believes that music is accelerating the efficiency in performance he is sure to make gains and loss if he holds the opposite view.

DISTRACTION

The concept of distraction is always associated with attention, reading, studying and performing some mental and physical works. Distraction causes negative effect in mental and physical works. It has destructing consequences. Because of this human beings mostly try to a Avoid it and also dislike it as it interferes in the efficiency of their works. Due to severe distractions to the industrial workers production is heavily affected, Therefore it is always considered as destructing force and should be eliminated.

The process of distraction has become an important problem in psychological studies when psychologists started dealing with the phenomenon of attention. But the influence of distraction has been felt long before the study of attention as it is a common disturbing force. Therefore psychologists felt that the phenomenon of attention cannot be discussed without making any reference to the problem of distraction. Authors like **Kulpe** (1903), **Geissler** (1909) and **Wundt** have dealt with the problem of distraction in the early years of twentieth century. Experimental studies were made on behaviour by using the method of distraction. First they were interested in assessing the level of the influence of the distracting stimuli on the tasks performed by the workers. Then the method was extended to the study of attention.

The experimental studies on distraction were made first on four categories of subjects such as animals, psychiatric, patients, mentally retarded persons and on normal human beings. **McDowell** and others (1960) have conducted the experiments on monkeys. They determined the effects of some external auditory stimuli on the performance of monkeys. The auditory stimuli had definite influence on the performance of the monkeys. Quite large number of authors like **Cameron** and **Megarat** (1951) have conducted experiments on schizophrunic patients. They determined the effects of distraction on tasks they performed. It was also observed that their thought processes were also affected by the distracting forces. Some investigators have observed in their experiments on schizophrenic patients that the results of their experiments were based upon the nature of the functions of the experimental conditions and the tasks involved in the performance.

Large number of studies were made on the effects of distraction among the mentally retarded persons, brain injured individuals and other abnormal persons. Various types of experiments were designed and conducted using extraneous stimulations in the experiments. There were visual, auditory and other sensory type of stimulations in the experimental rooms. Brain injured and mentally retarded familial children were tested in **Gallagher's** (1957) study. It was found that the brain injured children were more distractible and responsive than the mentally retarded children.

Studies were conducted by **Brown** (1964) and others among the institutionalized mentally retarded and the familial mentally retarded children using the variables that could cause effects of distractability. He found that the age of the children who go into the institution and the length of the period of institutionalization had the determining effect in the problem of destraction and the brain damaged subjects were found to be more distractible than the familiar type of mentally retarded children.

Experimental studies have been made on normal persons for assessing the nature of distracting effects by using various sense modelities. The effect of noise has been tested by **Dockeray** (1942) in his studies found that auditory stimuli had negative effect in the communication process. Sensory stimuli like noise and light were used as distractors in an experiment conducted by **Geissler** (1909) in a visual discrimination task. The stimuli were presented in varied combinations. The experiment was carried on for few days. It was found that even the most complex combination of distractors did not induce great amount of variations in the process of attention. The surprising result in this experimental study was that towards the end of the whole study most of the normal controlled subjects had performed the task at a slower rate than the normal experimental subjects. Geissler also observed that the level of concentration was not so much depended upon the nature and number of distracting stimuli. But it *was* depended upon the nature and complexity of the tasks.

Though is an unwanted auditory sensation, it is inevitable. In the last quarter of nineteenth century, most of the psycho-physical experiments were conducted only by using noise as a sensory modality. Experimental results on noise were not rejecting noise in the environmental situations. **Broadbent** (1957) on reviewing the studies on the effects of distraction, has stated that the human sense organs and the muscles of human beings were not completely disorganised when human beings were put into the noisy situations. There were only tentative effects which might have been due to experimental conditions.

It was also observed that any decrement in the efficiency in the task was due to more complicated task and not due to noise. There were instances in which the phenomenon of distraction facilitated the performance of simple tasks, but at the same time impairment of the performance in complex tasks. There are also many Indian studies on the effects of noise as a distracting force and interferes with the efficient performance of the workers in industrial situations (Ganguli and Rao, 1952). They reviewed the effects of continuous exposure to high factory noise on worker's auditory acuity and productivity. They were of the view that the term noise indicates not only disturbing sound consisting, of non-periodic air vibrations but also certain musical sounds. It is under certain particular circumstances that noise may be distracting the workers from the performance. It was also observed that intermittent noise stimulus was more distracting than the regular noise stimulus.

The distraction effects were also tested by using group Intelligence Tests. **Super** *et al.* (1947), **Smith** (1951) and others tested their hypotheses by using group mental ability tests. The results of these studies showed that distraction did not cause any substantial damage in the performance of group intelligence tests. These authors have also observed that there was no difference in the effects of distraction on the basis of sex and age of the subjects.

The effects of distraction were also tested on the tasks of higher mental processs. It was found that higher cognitive tasks were more affected under conditions of distraction. **Fenrick** (1957) had observed that the effects of distraction in phonagraph recording were very high where the subjects were functioning at a higher intellectual tasks. Similar higher level of distracting effects on higher cognitive tasks. Were observed by **Hassol** *et al.* (1952) and **Fisher** (1973) in their experimental studies. Since Higher mental tasks generally need better concentration than the simple mental tasks distracting effects cannot be ignored or avoided.

Overcoming Distraction

Since distraction is an undesirable experience we are trying to avoid it. Therefore, we havebeen constantly searching for suitable methods to overcome it. Psychologists were able to suggest some methods:

(1) The easiest way of overcoming this distracting force is to put *more energy* into the performance. You can put your face into the book and start reading the passage loudly. On one of the experiments conducted on a typist it was found that noise had little effect in the performance in the beginning, but later on the effect disappeared when the typist put more effort in his work. In overcoming the distraction the subject put increased muscular energy. It is instinctive to put some additional energy into the activity if there is instinctive to put some additional energy into the activity if there is any distraction.
(2) If work is continued for many days with some distractions there will be greater output in the work without putting extra energy into the work. The individual has learned to do the work inspite of the distraction. It is called negative adaption. He has learned to ignore it. This is supposed to be another method of overcoming the distractions.
(3) There are certain distractions that we could not avoid or overcome them. Further, some distractions have disturbing backgrounds in various degrees. Sometimes the distractors themselves become centre of and the worker *forgets the disturbance.* **Woodworth** (1952) says that the background of music which was a source of disturbance for the work, later on becomes a source of facilitation for the same work. This process takes place when we could not overcome the distractions.
(4) The best way of overcoming the distracting forces is *to avoid it physically*. It should be removed from the place.

Experimental psychologists have conducted several studies on attention and on distractions. Some of the results of these studies indicated that the effects of distractions, are complex. Some studies were carried out for determining the effects of auditory distraction on certain. visual and motor tasks. **Baker** (1965) in his studies had observed that by and large the factors, that have been found to influence the results in distraction experiments are: (a) the nature of distracting stimulus—its intensity, (b) the nature of task—its relationship with distracting stimulus, (c) temporal relationship between the task and the distracting stimulus, (d) the individual differences, (e) the effects of instruction, (f) the length of the task. These factors were found to be operating in the process of distraction in the subjects. All manners of complexities are involved in determining the effects of distraction. But the effects of distraction are not constant. It tends to loose its value in the long run.

PHYSIOLOGICAL EFFECT OF DISTRACTION

We have discussed the effects of distraction on physical and mental tasks performed by different type of subjects. One critical idea we can evolve from the results of various types of experiments is that, whether it is simple task or complex task or whether it is physical or mental task the distracting effect on performance is present to some extent. The distraction effect is not totally absent or nil. The physiological effect of distraction is there in some form. Several Scholars have observed this effect in their studies. **Morgan** (1916) had observed and recognised that distraction results in greater expenditures of energy on the part of the person who performs the task under distraction situation. He was able to demonstrate this phenomenon that the subjects were affected by distraction, physically or mentally by the task they performed tinder distraction, but were able to adjust themselves quickly to the situation.

But the subjects were able to achieve this status only at the cost of their increased expenditure efforts or energy. Under noisy situation the subjects have increased their vocal activities. It was observed that under continued periods of work, under distracting condition the increased efforts or energy in the work was reduced to the normal level. Several studies were conducted an this problem and they all seemed to confirm the result obtained by **Morgan**. Under the condition of distraction, **Harmon** (1933) had observed in his studies that there was an increased expenditure of energy for about sixty percent in the beginning of the task. But after few clays the subjects learned to adapt to the situation and the extra energy expenditure was not noticed.

Noise as a distracting, force has been studied by several investigators. The *physiological affect* of this factor is *annoyance.* But it is not having any damaging effect on behaviour or health of the subjects. There may be individual differences. For some, even some feeble noise does not interfere with their performances. These physiological effects of distraction may vary from person to person.

THEORIES OF ATTENTION

Generally, theories are evolved from the experimental results. In the area of attention enormous amount of experimental studies were made in the past one hundred years. In the process of attention one of the important aspects is the *selectivity*. According to behaviourists, attention is selectively projecting our senses on the stimulus. In this activity we ignore other stimuli (objects) even though they are also present. The development of any theory needs some explanation. There are several theories in every field and every theory should be convincingly explained.

The first theory of attention was evolved by **Broadbent** (1958). He called his theory as *filter theory* of attention. Since the important aspect of attention is selectivity and the process of selectivity is like filtering the matter, it is named as filter theory. It allows only one channel to function and rejects others. According to Broadbent the cognitive mechanism has a limited capacity in functioning and therefore, a filter mechanism is needed to avoid much strain on the capacity. The peripheral nervous system consisted of a number of different input channels like the vision, hearing and somaesthesis and each one has an input line. The central information channel has a limited capacity compared to the total capacity of the peripheral receptors. Since enormous amount of information are flowing into the body surface, the central information processing mechanism frequently gets overloaded. Therefore, the only possible means of dealing with the information processing is to limit the input channels. The sense organs are incapable of analyzing all the information simultaneously. As result of this a selective function is taking place in all the input channels. Therefore, only a limited amount of information can be analysed and admitted in the human sensory inputs and rejecting all others. This process **Broadbent** named it as *filter mechanism* the theory he called as *filter theory*.

According to Broadbent there are three main principles in the filter theory. They are thought to be the sequence of three systems. They are (1) a short term store system or *S-system*, (2) a selective *filter system* and (3) a limited capacity channel system, or *P-system*. The S-system has the capacity of receiving a number of information that enter the system in parallel form and they are analysed on the basis of their physical characteristics. The second system, the selective filter system allows information from only one channel which is desired and this goes into the P-system. A channel, here is defined as any physical characteristic relating to a particular sense. For example, sound, pitch are referred to auditory channel and form, colour are referred to the channel of vision. But some authors attribute the term channel to auditory attention. The third system, P-system is involved in processing more details analysis on the specific related channel. On clearing the process of one information from the channel, the filter mechanism permits another new information to enter into the system and the processing is made successively one after the other.

The information are temporarily stored in the S system and presented one by one to the P-system through the filter system and the P-system deals with the information serially. The time spent on each of the information or stimulus depends upon the nature of information provided. Generally, the filter system selects and conveys more suitable and relevant information that is stored. This is the function of the filter theory.

When the filter theory of attention was published there were both appreciations and criticism for the theory. **Gray** and **Wedderburn** (1960) conducted experiments to verify Broadbent's theory. According to the result obtained by them is that attention followed meaningful sequences and not based upon the physical characterstics of the input. They questioned the basic idea of Braodbent that attention is based upon the physical characteristics of sensory channel and they presented an alternative idea that psychological attributes play important role in attention.

Following Brodbent's Filter theory of attention several theories of attention have been evolved by conducting experiments on attention.

But all the theories are more or less based upon Broadbent's Filter theory by making either some minor modification or adding something to it. But one theory called *Kahneman's theory of 'Mental Effort'* varies from the filter theory of some extent by putting greater emphasise on mental capacity. According to this theory, the spare mental capacity is mostly associated with the processing of perceptual units. The organism focuses attention to a task and allocates only the spare mental capacity to the continuous processing of other irrelevant information.

If the primary task needs greater effort the unwanted message is rejected and there by more capacity is available for unit formation which performs some functions. This is assumed to be performed by the pre-attentive mechanism. In figural emphasis some of the units are selected for detailed processing. The emphasis on selecting message determines the amount of spare mental capacity that is available for processing of other irrelevant information. The selective effectiveness depends upon the ease with which the relevant stimuli can be identified and the processing of information is made.

ATTENTION DEFICIT DISORDERS

We may observe on certain occasions that some people often find it difficult to pay attention for a longer periods. Researches were conducted recently on the focus of attention and found that there were extreme cases of the problem of attention deficit. These problems have been classified as *attention deficit disorder* (ADD) or attention deficit and *hyperactivity disorder* (ADHD). Normally children are diagnosed with this type of disorder. But it is found that adults may also suffer from this kind of disorder.

The symptoms of attention deficit and hyperactivity disorder are found to be many. The attention deficit and hyperactivity disorder person is found to be inattentive and of impulsive behaviour. The person who is affected by this disorder is often found to be not listening to what others have spoken to him directly. The person often fails to pay close attention to the details of the information conveyed to him and also makes careless mistakes. The person often finds difficulty in retaining his attention in the activities or in any work in which he is engaged.

The person who suffers from this disorder gets easily distracted by other outside stimuli. He provides unclear and confused answers to the questions and never completes his answers. He often shows signs of restlessness. He finds it very difficult to wait for his turn and intrudes upon others. The person easily interferes in the conversation of other persons and in the other persons' play or games. The cause of this type of disorder is tentatively stated as both biological and psychological The dietary habit also is taking place on the same line.

10

OBSERVING AND PERCEIVING

Introduction

The study of perception refers to the problem of how we sense and know the rich, varied and the complicated world of things and people that are surrounding us. This enquiry is one of the oldest one that human beings are often raising and speculating over it from time immemorial. As such it has a long history with adequately unanswered problems and issues. Scientific questions were raised and answered by the results of factual research. In fact the study of perception was started long before the science of psychology was started.

A great deal of research works were carried out in this area much earlier than in the other areas. The works were mainly of the physicists and physiologists and many important contributions were made by men who were not usually thought of as psychologists. Even now it is found true, because tl ie problems of perception cut across other sciences. We cannot understand man's perceptual problems of the world unless we know something about the world in terms of the physical events and also man as a physiological structure. In this way the two important fields involved in the study of perception are physics and physiology and through the contributions of these sciences the *processes of psychological perception* of the world were evolved.

All the *psychophysical theories* and methods were designed and developed first only by the physical scientists. On understanding the values of these theories and methods psychologists instantly started studying various aspects of perception by using the psychophysical methods. Measurements on various psychological processes were designed, of which t-he study of perception occupied an important place.

Some of the early psychologists who dealt with perception did not deal with the process of perception directly. They brought the process of perception under *'observation'*. For example, **Woodworth** (1952) deals with perception under "Observation" and **Murphy** (1964) deals with the process of perception under "Sensing and Perceiving".

They are of the view that there is no perception without *seeing* or *sensing* the object in the environment with our sense organs. What they said about perception is true. Without seeing and sensing the objects, there is no perception. Since the individual deals with the environment all the times, it is important for him to know the environment accurately and master it for his needs and purposes. He gets to know the world around him only by using his senses and he acts on it and masters it to some extent by using his physical parts.

But how the human individual comes to know and masters the environment? It is a difficult question to provide a correct answer. This is one on which psychologists have done enormous amount of studies. The studies arid explanations and interpretations went on into the areas of senses, learning and thinking. The fundamental question raised is, how the use of the senses enables a person to get familiar with the environment. A detail answer to this question can be had only by going into the technical investigation. But a simple and clear answer can be obtained by careful and vivid observation. **Woodworth** (1952) states that observation is the subject matter of central importance in the study of psychology. The process of observation involves several important factors. *Observation* is made through sense organs. The experience is called *sensation.* Observation is always having two important steps. The first step is *attention* which is the preparatory step with which the organism responds to the object. The organism by this means is getting ready to observe the object.

The second step in observation is *Perception*, by which the *actual process of observation* on some object is taking place. Therefore an actual and clear process of observation may be stated as perception. Individual persons may vary in their power of observation. Some persons may have clear and vivid observation or perception and some persons may have poor and unclear observation of objects.

Thus the process of perception is conceived from varied angles. Of late, psychologists associate perception mainly with *Vision.* The latest development in the field of psychology is the *Biological Psychology,* which tries to explain every psychological process through *Biophysiological processes,* disregarding the cognitive interpretations.

In the preceding chapters, while discussing about *sensation* we have stated that stimuli from the external world are attacking the organism in various forms and the organism is responding to them with its specialized senses. In *attention* we have stated that there are several stimuli in the external world. We select and attend to a particular stimulus. But here we are coming across a complex process which is mainly a synthesis of these two and several other processes. This process is called *perception.* A man must know that there is an object outside of him at a given distance in a particular direction with certain characteristics which are important for him. Varieties of stimuli come in complex form and stimulate us. However, complex they are, we are able to *organize them* in some way automatically and we are able to describe what they are. How this process of organization is taking place within us to give the correct meaning to the objects? This is the subject matter of perception. We shall discuss this matter here.

What is Perception?

All living organisms in the world are endowed with the wonderful sensory apparatus called the eyes, with which they see the world out of them, including the human beings.

We, the human beings see everything outside with our eyes. All the people have much the same sort of sensory equipments, their eyes, ears and other senses, work according to the same physical principles. If two individuals witness the same event, the raw information that they receive will be more or less the same. But at the same time no two people will interprete the information in exactly the same way. Those who belong to the same cultural background, socio-economic status and education may most likely have similar conclusions. But people who are belonging to different cultural background may form wide variant pictures of them. Perceptions differ on a number of levels from purely *sensory perception* to the complex *psychological* and *philosophical perceptions.* Perception is something more than just getting the raw information by seeing.

Scholars have observed that the culture to which one belongs to and Ole kind of group in which one is a member pervade in an individual's perception. The individual in his perceptual process may project his own personal experience and knowledge. Thereby every person's perceptions in various events are private and personal. Whatever one sees outside in 1.fic world is his personal observation and experience. But whatever we see or perceive must be communicatable and others must understand it and agree with us. If this is not taking place, then there will be confusion and we do not understand anything correctly and meaningfully. For this purpose we must have some idea of the world and things that other people form and construct in their minds. In this process people formulate common terminology for easy understanding of what other people have perceived and understood.

The language is the medium through which everyone communicates what lie has seen and perceived. Not only the language, there are quite large number of sophisticated mechanisms humans have evolved for knowing how other people are seeing things. One such mechanism actually organized and programmed is the human brain. Psychologists view this as the function of the mind. One of the wonderful functions of the human mind is that it functions in a specific activity of seeing or perceiving in the same manner in every normal human being and communicates the same thing to others. It is, because, the brain responds to the visual or auditory or any other sensory information by constructing the visual or auditory images.

Recent studies on perceptions bring out many surprising informations. The main idea revealed by these studies is that *perception is altered by mood.* Change of moods have profound influence upon human perception. We have sufficient metaphysically presented information in philosophical studies and empirically obtained information from psychological studies. The altered state of perception by the change of mood need not be mistaken for pathological perception. The recent theory of the mind is that it has a physical basis in the brain cells. These brain cells are known as *mirror neurons.* When these cells are activated, they produce a very wide range of perceptions, emotions and compulsions to act. This process of function takes place just like in all other brain cells. But the important matter involved in this is the manner in which these cell neurons are activated. All other neurons respond to a wide range of stimuli. But the mirror neuron cells become active and produce response only when it is observed in another person. For example, when a person sees another person and expresses the feeling of disgust, the disgust mirror neurons become active and produce similar feeling of disgust in another person. The mirror neurons in other words can be stated as the built in empathy. It is an automatic instinctive feeling or sense what the other person is experiencing. It is a feeling into-a thing. This aspect of perception is the power of understanding and imaginatively entering into another person's feelings.

Altered state of perception is caused by various reason. Apart from moods and emotions the mind altering drugs cause perception in various altered states. Many writers, artists, musicians and painters have reported experiences of altered states of perceptions. This experience is called *synaesthesia.* The Russian painter **Wasbily Kndinsky** made use of this experience of perception in his work *"colour"*. Mind altering drugs have played important role in the works of many writers, poets and artists. By means of these they achieved their creative insights and perceived everything in fantastic manners.

People generally assume that our sense organs must function accurately in order to have correct perception. But people usually believe that the sense organs furnish the brain with the naked information of the external world. The sense organs present to the brain only the copies of the objects what the eyes see. Though apparently it appears to be correct, it is not so. What is taking place, is something different? Every object presented to the brain is processed into meaningful object through the past experience and knowledge.

Therefore perception is thought of as the interpretation of sensory information as a constructive, creative process which endows sensory experience with meaning. But what is the basic rule to make this type interpretation of perception. The painters make as an acceptable representation of the world. It is said that they have an intuitive grasping or Understanding of the worldly objects. Though everybody is said to have this intuitive grasping to some extent the use of this ability varies considerably iron person to person. The painters and some fantastic writers develop this ability to a greater extent and they perceive every minute thing in objects May observe in the world.

Some of the modern scholars explain and interprete perception in terms of sensations. According to them perception is just more than a complex sensation and therefore, they interprete the phenomena of perceptions as visual perception, auditory perception and so on. The meaning of perception in the common sense world is the act or the effect of perceiving. It is also meant as the insight or intuition gained by perceiving or observing by the individuals. It is also stated as the ability or capacity to perceive. It is the process by which an organism detects and interpretes information received from the external world by means of the sensory receptors. Based upon the gist of the various interpretations recent authors combine the psychological processes of sensation and observation and discuss them as sensing and perceiving. But the old interpretations cannot be thoroughly neglected from the psychological studies.

Perception is the process of integration and interpretation of what the mind receives. Perception involves responding to the world through the senses. It is not a total response to everything that we see the outside world with our senses. Some of the responses are specific and serve some purpose on a particular occasion. Therefore, our response is selective, vivid, purposive and relevant to our needs. We select and organize only those things which are needed for our purpose and leave the rest of the objects at the background of our perceptual field.

According to **Moskowitz** and **Orgel** (1969) perception is a wide range of response to a stimulus or set of stimuli which integrates the information beyond that what the stimulus contains. The meanings of other stimuli obtained from past experiences may also be added to this.

Wickens and **Meyer** (1961) defined perception as a process which gives significance to the stimuli making the perceiver to react to the event in an adaptive manner.

Murphy (1964) says that perception is a way of coming in terms with the environment playing certain features up and down. He emphasises four different aspects in perception such as sensitivity, quality, past experience and set. According to **Stagner** and **Karwoski** (1952) perception is the process of obtaining knowledge of external objects and events by means of senses.

Man makes use of his sensations and arrives at certain conclusions about the objects and events of the world. If the objects and events are real and well established he does not doubt his knowledge too much. But still while dealing with physical objects he resorts to probability for accurate knowledge. The physical objects and events may be new and unfamiliar and also not well established and they are further away from him. In such cases his knowledge is subjected to verification as chances of error are possible in it.

Let us assume that a man from a remote village, where there were no trains, motor cars, buses, electricity, industries, big buildings, sea. etc., was brought suddenly into a big city like Chennai. Under such new situation what he would look at? He would be in a stage of buzzing. There are innumerable number of colours, contours, shapes, brightness. The person is visually stimulated with many new objects. He is not able to identify objects. Even though he could make use of them, and knows the language, he is not in a position to answer any questions relating to the objects. We must see, LAse and handle the objects and learn to answer any questions related to the Objects. It is easy for such persons and children to name colours. In case of size, distance and motion they must acquire knowledge. But it requires longer time for them to identify shapes.

William James (1890) said that the child's world is a big booming, buzzing and confusion. Perception takes place when a part of the stimulus is used. to predict the presence of the total stimulus situation.. In other words, it is like the prediction of the object flower or dinner from the smell without seeing it.

In our every-day life, we are making several perceptual reactions to things and events. When dark clouds are collected in the sky and cold wind blows we understand that there is going to be rains and immediately run for shelter. You are waiting for a train and the railway bell is ringing, you make all the preparations to get into the compartment thinking that the train is arriving at a distance. In all these incidents we are combining several things in our perceptual process. Perception reduces chances of our failure to achieve something and simplifies our life process and makes our adjustment more easy and effective. Therefore, perception may also be defined as the process of knowing the objects and events in the external world through our senses and making them a meaningful whole with the help of past experience and learning.

Perception According to Gestalts

Any discussion on perception is incomplete and inadequate in the study of psychology without making any reference to the Gestalt Psychologists' contribution to perception. According to Gestalt psychology perception is the most important process of function in understanding the objects in the world. Gestalt psychology emphasises the wholeness of behaviour, that is, the perception. The whole determines and gives meaning to the parts. According to Gestalt the laws of organization form the heart of Gestalt theory.

On seeing an object in the outside world, for example, a chair, what we actually see in it. Do we see the seat, or the arm, or the back, or the legs, or the colour or the finish? We do not see anything separately and call it by its name. We see the chair as a whole and describe it. We see the chair putting all the parts together. This process is called organization. We perceive the object as a whole and interprete the object. Any single part of an object does not convey the correct meaning. The adding up of the parts and seeing them cannot be stated as the perception of the object, unless the process of organization takes place as a whole.

The Gestalt Psychologists, **Max Wertheimer, Kurt Koffka** and **Wolfgang Kohler** had conducted their experiments first on perception of *apparent movements* and later on they named it as *phiphenomenon.*

Their experiments on chimpanzees brought out very interesting results. The chimpanzees were able to solve the problems by perceiving the whole situation. The same principle, according to them can be also applied by human beings.

Another important result they brought out was the phiphenomenon. It is referring to the apparent movement of the objects. If two lights are put up at a short distance and flashed alternatively in quick succession we can perceive that one light is moving from its position to the side of another. It is not that the light is moving. It is the short distance and the rapidly flashing of the lights alternatively make us to perceive that they are moving.

The Gestalt psychologists' phiphenomenon is the basic principle on which the motion picture was produced. The principle is that when the objects are presented in quick succession one by one we perceive the objects seem to be moving which provide continuity in the events of the story presented in the motion picture.

Advertisers and electricians are making use of this principle in their professional activities. They design advertising banners in different colours in such a way that the figures are moving and the indicating arrows are made to point out the object they wanted display. Thus the Gestalts' perception is referring to the whole situation and organizing the relevant parts into a meaningful whole.

Perception and Sensation

While explaining perception we have been using the concepts *senses and sensation.* This naturally infers that there is relationship between *perception* and *sensation.* Sensation is a response to specific stimulus such as visual, auditory and the like. Perception is a response not only to the specific stimulus, but to the whole situation. Perception relates our behaviour to a complex pattern of stimuli. Sensation refers to the function of sense organs with their nerves and nerve centres. Perception refers to the objects of the Id known through the sense organs.

For example, in an experiment on sensation we present a simple stimulus to an observer and try to measure what reaction the stimulus has aroused in the observer. In an experiment on perception we present objects to the observer and find out how well they are observed by the observer. The observer is attentive to what is presented in both sensation and perception and the responses in both cases are also verbal. In sensation we are interested in relating the verbal report to the stimulus whereas in perception we relate the report with some facts.

Attention and Perception

In the perceptual process both attention and perception are taking place in an organism. Attention takes place first and then perception occurs. Attention makes the organism to be aware of the occurrence of the event for perception. Therefore attention is active, quick and rapid. Since attention shifts rapidly it exists only for a short period. An event may exist only for a short duration or appear and disappear, because of this our sense organs must be very attentive to the events or movements or appearance of the objects. For example, in a race, in a sports event we watch very attentively to know who comes first and who comes second. Our attention is fixed at a particular place and we do not turn our attention even for a second. After everything is over we provide the complete note of everything what has happened there which may be called perception.

Therefore attention may be stated as the part of perception and it is attention that provides information for perception. The attentive processes arc mediated in perception. Attention is always selective in the perceptual field. Sometimes in a situation we may be in the midst of many noises, but may be attentive to one particular voice in which we are Interested neglecting all others. There are several attention getting characteristics in the stimulus. They are mainly intensity, contrast, change, size, novelty, repetition and duration. These are all, also contributing factors of perception as antecedent causes. Sensation, attention and perception are all having many common characteristics. Therefore, it is very difficult to accurately differentiate attention from perception and attention from sensation and also sensation from perception.

Organisation of Perception or Perception of Form

In perceptual process the organism is reacting to certain stimuli and not to all in a complex environmental situation. The stimuli are of various kinds. They may be having different shapes, forms, sizes and colours. They may be square, curved and angle. The stimuli may sound with high or low intensity and the sound may be intermitten or continuous. There may be also many stimuli of various kinds reaching our senses at the same time. All these types of stimuli are responded by the organism in the appropriate manner. There is no confusion or buzzing in our sensory process. In spite of innumerable number of stimuli and welter of stimulations, things emerge with organized pattern. This *organized pattern* becomes basis for our appropriate reaction to the situation.

The gestalt psychologists were very much interested in the study of perception. They tried to determine how sensory impressions are organized. They have established that the whole is more than the sum of its parts. That is, a perception is more than the sum of a number of sensations. There is a tendency in the individual to follow certain basic principle to organize the sensations.

Figure-ground : The figure-ground perception is the fundamental principle in the organization of sensations. It is the most primitive process in perception. The separation of figure from the ground is basic in every perceptual process. This perceptual process is a development in the-organising independent of practice or training. It is found in human beings and also in animals. **Munn's** (1962) the famous ambiguous figure is a good example to illustrate figure-ground perception. Persons who were blind throughout their lives organized their perception into figure-ground relationship without any experience. They did the organization spontaneously (Fig. 10.1).

Fig. 10.1 : Figure-ground

The figure-ground perception is also fundamental in another way. When we look at a stimulus pattern the figure-ground relationship emerges first before all other aspects are perceived. This has been experimentally tested by a projection tachistoscope. In visual perception a figure is often differentiated from the background by its quality such as shape, size, colour etc. The object appears closer and the ground recedes to a distance. A colourfully printed book on a table stands out as a distinct figure against the black sun glassed table top. The figure-ground relationship is not limited to visual perception. In the auditory sensory area the sound of a violin may stand against that of a symphony.

Closure : This is a tendency in our perceptual process to see things as a whole. When we receive sensations of an object in an incomplete form we try to see it complete. The incompleteness of the object is overlooked and the 1-1gure or object is perceived as a completed one. The term closure refers to the tendency to fill in the gap in the stimulus presented to us (Fig. 10.2).

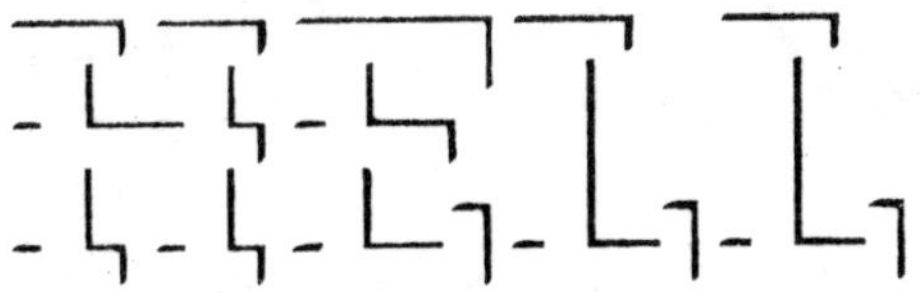

Fig. 10.2 : Closure

***Continuity* :** We like to see that elements should go with other elements in such a way that they have continuity of a line, curve or a movement in the same direction. But the principle of continuity sometimes may lead to inaccurate perception.

***Proximity* :** The perceptual organization occurs when objects are put close to each together. We don't perceive them as separate objects. The 1,1dividuals items may vary to some extent. Still we group them in a single pattern. Proximity may be temporal or spatial, When a cracker is Fired in a dark place we see bright light and loud noise. We tend to perceive them as parts of the same element. But if the light is seen after few minutes or sound is heard after few minutes we treat them as separate elements.

***Similiarity* :** Stimuli that are more similar to each other will have greater to be grouped. Similarity means that the object may resemble each other in several respects such as physical attitudes, size, shape, colour, intensity, weight and odour and the like.

Perception of Space, Depth or Distance

The world in which we live and which we perceive is spatial in character. It is a three dimensional space and we cannot imagine or think of a world without a three dimension. When we look around us we see objects in three dimensions. The three dimensions are height, width and depth. We perceive an object, the three dimensional room. The room has a floor, the walls of the sides and the space in between the *side walls.* Similarly an object has three dimensions like length, width and thickness. Man's biological make up is such that the retina of his eyes has two dimensions, height and width and no depth.

A person is facing certain situations constantly, in which he has to judge the distance between himself and the objects. When we walk through the streets filled with people the perception of distance helps us to avoid dashing against other people or carts or motor cars or other fast moving vehicles (Fig. 10.3).

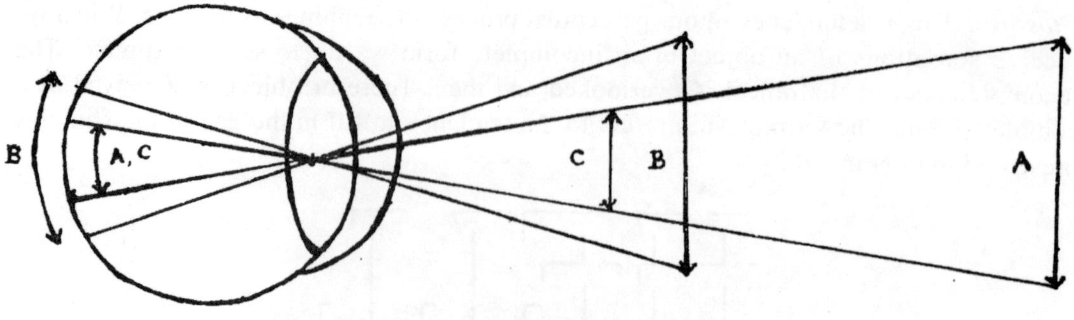

Fig. 10.3 : Retinal Image and Distance.

If we want to pick up a *coin from* the floor we judge the distance automatically and stretch or extend our arm. Running trains stop at railway station platforms without any errors. Aeroplanes from sky land in airports without any deviation, buses halt at bus stops without any failure, the paper boy adjusts his cycling in such a way that he never fails to drop the paper al our door-step. In all these instances persons involved make judgment of distances accurately. Every day we may be doing some judgments regarding, the depth of the objects. How deep the well is? How much the barrel can hold? Can we explain the processes involved in all these?

When we look at two objects at different distances, the images of the objects do not fall on the retina one behind the other. Because the retina t'S flat the image of the farther object can be either on the left or right below or above of another object. Our visual images are having only two dimensions. Therefore, a three dimensional world is a puzzle for us. How it occurs? The two dimensional images that fall on the retina area are converted into three dimensional perceptions. The two dimensions provide stimulus informations to evoke a third dimension. This is depth perception or distance. The information fall under *two sets.* One is related to the *environments* and the other is related to the *organism.*

1. The Environmental Cues

The environmental cues are monocular cues, superposition, perspective, light and shadow, movement and texture gradient.

Monocular cues : The information received from the objects by using *one eye* are called *monocular cues.* We can make accurate judgments about distance and depth with two eyes. But our monocular cues often help us to judge distance and depth successfully. We cannot aim certain things accurately with our both eyes. For example, *in shooting, we aim with one eye only.*

Superposition : If two objects are on the same line of our vision and one is superimposed on the second, it partially blocks the distant One. The first ot-ie is nearer to us and the second is farther away. We could also see that the nearer objects are continuous and unbroken and distant objects appear broken by the nearer objects.

Perspective : It is a technical word used in painting and drawing in the estimation of distance and depth. Artists have found out that they can indicate the distant object and nearer object in their paintings. The nearer object is painted as larger and the distant one smaller. This is the way the *artists evoke the depth perception* in the perceiver.

Two lines that run parallel to each other may appear to come together at a point on the horizon. These converging lines are providing informations important to the aeroplane pilot. The pilot could understand the length of the runway which he is going to take. These cues to distance and depth are called *linear perspective.*

Distant buildings, objects appear hazier and less clear, whereas nearby objects look clear and bright. This is called aerial perspective.

Light and Shadow : Shadow is an important cue in depth perception. If the shadow appears on the parts of the object it is farther away. A two dimensional drawing may have a three dimensional look if lights and shadows are distributed appropriately. By this method a geometrical figure may give an appearance of a solid matter. A circle drawn on a paper may look like a flat disk. But if you give appropriate shadows on the edges it would look like a ball. The shadow gives a different look to an object.

Texture Gradient : The texture of the object provides certain cues to our perceptual process. The nearer objects appear rough, irregular -and full of dusts while the distant objects appear smooth, regular and clear. For example roads, houses and lawns have these characteristics.

Movement : The speed and direction of the object as it moves in the space works as a cue to perceive distance. While you are travelling in trains and buses you might have often observed through the window that objects like trees and telegraphic posts close to the railway lines and roadside are running fast in the opposite, direction and the distant objects are slowly moving with you.

2. *Organismic Cues*

This is with reference to the eyes, cues provided by the eyes. The organismic cues are binocular cues, accommodation, retinal disparity and convergence.

Binocular cues : In the perceptual proccss two eyes may have certain advantages. *Two eyes* provide *wider visual area* and we can *perceive depth and distance* more accurately that with one eye.

Accommodation : Accommodation in perceptual process refers to the change of the eye lens in its curvature. This function is needed *to focus* different objects on the retinal area. If the *object is quite close,* the *lens becomes round* and if the object is *farther away the lens becomes flat.* Kinesthetic sensations that affect these changes also provide some information regarding the position of the object.

Retinal disparity : Our two eyes receive slightly different views of things. The *difference between the images of the two eyes* is called *retinal disparity.* The right eye could collect more information from the right side and the left eye from left side, and we could see that these two informations, are different to some extent. Hold your pointing finger three inches away against your nose. Close your left eye and see the finger with your right eye and then close your right eye and see the finger with your left eye. Do it rapidly for several times. What do you see? Your finger is moving fast from left to right and from right to left. In reality it is not moving. This is due to retinal disparity. But when you see the finger with your both eyes the two different images merge into one and you see the finger is not moving. This is *stereoscopic vision,* makes the depth perception more accurate. Our two eyes may have overlapping perceptual fields or slightly different pictures. The binocular perception removes such differences.

Convergence : When we look at *objects* that are *farther away* from us the lines of our *vision* in both eyes are *parallel.* But if the objects are moved towards us at a close distance our eyes tend to converge. They turn slightly inward towards each other. The Kinesthetic sensations control this movement. In this way our eyes provide additional information.

Perception of Motion

This is perceiving the objects which are moving. Our ability to perceive the moving objects is more important than perceiving the objects which are stationary. We detect an object in motion when a succession of neighbouring rods and cones in the retina area of our eyes are stimulated. When we are looking at the garden plants outside, we see the bugs on the flowers moving rapidly and when there is some wind, the plants move and the bugs fly away. We also move our eyes in the direction of the bugs till they disappear. At this time the images of the bugs fall consistantly on the fovea region and objects in the background also stimulate rapidly the rods and cones. We see only the movement of the bugs and not the background. This is the important characteristic of perceiving the objects in motion.

In the preceding section we have discussed perception of form and perception of distance and depth in space. In these, the subject matter is concerned with static objects and not with moving objects. We keep ourselves constantly moving place to place. The space in which we move is filled with variety of objects moving most of the times. People walk on the street, some fast while others slowly, cars run fast making sounds along the high way, trains run on the rails, birds fly in the sky, aeroplanes fly in the air, ships move in the seas and so on. In all these, persons, organisms and objects keep them in motion. The motion or movement of persons or objects in space is a part of perception of objects in space. The perception of motion is determined by certain factors. **Postman** and **Egan** (1949) say that the conditions that determine the perception of stationary objects also seem to be operating in the perception of motion. Therefore it becomes necessary for us to know the various types of motions that could be perceived and the factors that determine the perception of motion in space.

One of the important characteristics of our eye is that it is able to perceive motion and provides us with information regarding the destination and the movement of people and vehicles towards us. We are also able to see the speed of the on coming motor cars, so that any untowards can be avoided. We see the motion pictures and television scenes without any difficulty and could make out the correct meanings of them. The perception of motion is a basic human problem, but we rarely give importance for the study of it. The opposite of motion is stability and we study both. But the ability to see and detect the movement of the object is more important than seeing and detecting the stationary objects. Our eyes are in constant movement. As the eyes are analysing the world, images of objects move backward and forward over the retinas of the eyes. In this process the rods and cones are stimulated in the retinal areas. Thus al! the proximal stimuli art moving and shifting while the eyes are watching the objects in motion. When an object moves across the field of vision, the rods and cones are stimulated in quick succession and motion is perceived. At the same time the distant objects like the buildings, trees, lamp posts appear motionless (Stationary). But when the eye, moves the images oil the stationary objects stimulate the rods and cones in succession. Thus a complex process, perceiving moving and static objects, is taking place and it takes place automatically, because of the system of *compensation.* That is, the nervous n coordinates the movements of the eyes and head and changes in the retinal pattern also take place. The rods and cones are stimulated and there is movement of the retinal image, which is in agreement with the movement of the eyes and the object is perceived as stable. But if the eye and head movement is abnormal then the compensation is affected and stability disappears.

For example if you sit in a revolving chair and rotate it for few minutes you will see the world is going round and not stable. But still we perceive the world as stable. This ability to compensate the iliusion caused by the eye movement is learned, learned when we were infants. The perception of motion is relative. We state that an object is moving because another object is stable in the same field. This means that a moving object is producing a moving image in the retinal area in contrast to the stable retinal image in the stationary field. The perception of real motion depends upon the relationship between objects within the visual field. The accurate perception of what is moving and what is stationary depends greatly upon the frame of reference in which it is perceived.

The only information we have for accurate perception of motion is the visual field. In the absence of frame of reference illusory perception of motion occurs. One assumption we often make in the absence of frame of references is that large objects are stationary and small objects are moving. For example a person sees a. small spot of light on a large dark screen background and the large screen is slowly moved while the spot of light is stationary. The person sees that the spot of light is moving. An illusion of movement called *autokinesthetic effect* has taken place. This is induced movement of the object. Another illusion of motion that is often experienced is the moon-cloud perception. On a cloudy evening when the moon is visible with a thick cover of moving clouds, the moon appears to be fastly gliding in and out of the clouds in the opposite direction. In spite of our knowledge that it is the clouds which are in fast motion rather than the stationary moon we assume that the moon is in motion. **Wallach** (1959) states that the clouds are in the field of vision and they become the frames of reference. In reality the frames of reference are stationary and hence we are inclined to see the clouds a stationary and the moon appears to be moving.

In the perception of motion, the motion receptors which are located in the semicircular canals of the inner ear are involved. These receptors are stimulated whenever the body is involved in motion. It is this stimulation that provides us the feeling that we are in motion. Any irregular or violent motion may produce motion sickness or a vomitting sensation (nausea). The motion receptors feedback the sensory system. This feedback system contributes to perception of illusions of motion which is a tendency to retain the (illusory) motion even after the motion is ceased. Because of feedback we are conditioned to also other forms of stimulation. Hence the train even after it has come to complete halt the illusion of motion is continuing. Further we are also not sure whether it is our train is moving or the train on the next track is moving.

There is another kind of illusion of motion or apparent motion called *stroboscopic* motion. This forms the basis mainly for films. This illusion of Motion can be produced by presenting separate stimuli in quick succession. This illusion of motion is adopted mostly in motion pictures, advertisements and televisions. In all these, separate stills of pictures, each one slightly different from the preceding ones is used. They are presented so rapidly that the pictures blend with one another and maintain a sequence of the events.

The simple form of stroboscopic motion known as the *phiphenomenon,* an illusion studied extensively and demonstrated in the laboratory situations. The flicker-fusion apparatus experiment is a good example to demonstrate the phiphenomenon. This is a very sophisticated instrument for this purpose. All other instruments are crude. Light is flashed in a fraction of a second and the observer will have to judge whether it is flickering or fusion by seeing through the viewing chamber.

The instrument is designed in such a way that the light is made to flash only for a fraction of a second. It appears to the observer as if the light is burning continuously without any break. But actually there is break between two flashes. The time interval between flashes is the crucial factor in this illusion.

In motion picture and television only stills of different pictures are presented in quick succession to produce the illusion of motion. The gap between one still picture and another still picture is not observed and we see as if there is a continuous motion of the pictures and we do not see several still pictures appearing on the scene one by one. How this illusion is caused? Several studies have been made on this phenomenon. **Gibson** (1954) says that the *temporal-spatial relationships* produce the illusion of motion. According to him there is a small real difference between real movement and the illusion of motion in the temporal patterns of retinal stimulation.

We see in certain situations that objects standing still are seen as moving. **Max Wertheimer** (1880-1943) (1912), has demonstrated this. For example there are two lights A and B at a close distance of six inches. These lights are flashed one after the other. The time interval separating the two flashes is about I/ 10 of a minute (6 seconds). In this condition we do not see the first light and then the second light. But one light is moving from the field of vision A to B. This becomes the basis for electrician to produce electrical sign boards in which the arrow is moving. The distance between the two lights and the time interval between the two flashes are the crucial factors in this illusion. Within certain limits if the distance between the two positions is longer the time interval also should be optimum between the two flashes to produce the illusion of motion.

Perceptual Constancy

Perceptual constancy means the ability to perceive the world as stable and consistant. We always see the world as stable because, the world by and large is stable and permanent one. Perceptual constancy is interpreted in various ways. But a psycho-physiological interpretation is very difficult. Several factors are attributed as determinants of constancy. Early authors, like **Hermann von Helmhotz** (1821-1894) as early as 1886 (1924) and also **Ewald Hering** (1834-1918) around 1907 (1920) have attributed our *experience* with to our perceptual constancy. Since we have many opportunities to observe objects under different illuminations we develop a habit in our perceptual process, perception becomes a habit. Later on when we see the same object we are not conscious of our present sensations in perceptions, but make intelligent judgments of the objects from the habit experience.

Hering did not accept this view entirely. He states that it is the peripheral factors like contraction of the pupil and retinal adaptation which determine the constancy. According this view the brain is the receiving centre — the sensation. The sensorium is the area of the brain considered for receiving and integrating sensations from the outside world and also it is the entire sensory and intellectual apparatus of the body. This area (sensorium) gets modified when an object is seen by the eyes. When the same object is seen once again it is perceived by the modified sensorium and the sensation of the object is identified in the sensorium and we see the same object without any change. Since the brain centre, 'sensorium' is the centre of sensory and also intellectual apparatus the object is identified as it was seen earlier. This is how the object constancy is explained. According to Hering this is applicable to other characteristics of the object like, the size, colour and shape. Therefore the ability to identify all these is provided by the brain centre, the sensorium.

Perception is the process of organization of our sensory experiences by which we become aware of the worldly objects, their qualities and relationships and attach meaning. and stability to them and it is this perception that mediates human behaviour. One of the Gestalt psychologists, **Koffka** (1886-1941), (1950), states that the basic psychological problems in perception is concentrating on the world of sight. But the fundamental fact regarding the perceptual process is object achievement. In other words, it is the recognization of an object as enduring and necessary. If an object has been perceptually organized as stable and permanent it is stated as constant. This persists in spite of differences. Therefore the tendency to see an object with its permanent characteristics, under any varied altered conditions of light, distance and position is called *object constancy.* Similarly, we have the tendencies to see the objects with the same brightness, colour, shape and size.

But, how does the individual is able to interprete the constantly changing variables of *proximal stimuli* in such a way that these same stimuli accurately give back the constancies of the objects of the world? Let us discuss how this process is taking place with special reference to *brightness, size* and *shape.*

The organism makes use of certain characteristics in the process of complete perception. These characteristics are both internal and external which we have already discussed in the perception of space. *These characteristics* are generally called *cues.* In any visual perception the primary cue is the *visual field or frame of reference.* It is the complete view of the situation. In perceptual constancies the organism makes the comparison steadily between the specific object with the total frame of reference or situation.

Brightness Constancy or Lightness Constancy

In perceptual constancies our past experience plays greater role in identifying the quality of the object. Any new impression on our sensory apparatus is judged in terms of our past experience. For example a piece of coal is perceived as black. It is observed as black always. Whatever type of light is focussed on it, it is seen as black. The coal in a bright lighting situation may reflect good deal of more light but still it is seen as black. And also a piece of chalk is seen as white and it is always seen as white. Whatever type of light is flashed on it and even if it is taken to dark room, it is seen as white. It is because of our lifetime experience in dealing with these objects as black and white. Therefore we perceive them with their original characteristics. This we call as *brightness constancy.*

When some bright light falls on an object in a situation we can see that there are *two levels of light intensity* present in the situation. The first level is the intensity of light falling on the object — it is called (1) Illumination. This is also referred to as the incident of light. The object which received the illumination reflects certain amount of the incident of light. The brightness of object depends upon how much light is reflected from the surface of the object into our eyes. The percentage of the total light that an object reflects or the ratio of the reflected light to the incident of total light is called the (2) *Albedo of the object.* If the absolute intensity of illumination is increased the albedo remains constant. For example, if the albedo of a white sheet of Is 90 percent in sunlight, the albedo is still 90 percent when taken indoor where there is low illumination. Therefore even if the absolute amount of brightness is altered, the albedo of the object is not changed. The albedo is a physical property on the surface of the object and independent of the absolute brightness of light which falls on the object. Hence it is the *albedo that accounts for the brightness constancy.*

Hochberg (1964) states that it is not brightness constancy as interpreted by some of the authors, but *it is lightness constancy.* According to him brightness and lightness are not referring to the same thing. Lightness is a continuum running from white surface to black surface, whereas brightness refers to the apparent amount of light coming from the object. A piece of coal in the sunlight and a piece of coal in the shade are of equal lightness, but the coal in the sunlight will appear much brighter. The explanation given by Hochberg is that when light energy falls upon the surface of an object some of the light energy is *reflected* and some of the light energy is *absorbed* if the object is transparent. If the object reflects most of the light energy it is seen as white, instead if an object is not reflecting any light energy it appears black. The percentage of the light the surface of the object reflects is its albedo. Even though the amount of light from the object reaching the eye varies widely the physical characteristic (albedo) is observed accurately.

Colour Constancy

Our perception and judgment about colour of an object is based upon the fixed physical property of the object. In judging the colour of an object an organism takes into account the amount of light reflected by the object. Therefore illumination is a factor in judging the difference between two colours. Colour constancy is sometimes misunderstood with brightness constancy. The reason is that if there is any change in the illumination it may *alter the apparent colour* of the object. But the *colour is not altered.* For example, a green coat appears to be green and continues to look green wherever you carry it, whether it is to a bright sunlight area, or to soriiewhere indoors or to a dark room. Even if you place it in the midst of other colours, under an yellow light the coat looks green. *The quality of greenness is more fixed.* It may be light tint, or deep shade, you see it always green. Therefore, there is a tendency to see the familiar objects in the same colours, irrespective of altered illumination on the object. This is called *colour constancy.*

Since we are familiar with objects and their colours we see them in their colours under whatever variety of lighting conditions they are placed. Two different coloured objects under the same coloured light can be perceived with their original colours provided there are sufficient contrasting appearance and shadows. A black dog is seen by the owner as black whether he looks at it in bright sunlight, or in a dull illuminated room, or in the night under a yellow room light. He is depending on his memory about the colour of his dog.

The memory is one of *the contributing factors* of colour constancy. Unless the object is a colour absorbing one, our memory can be trusted in perceiving the same colour. If blue colour is projected on yellow wall it results in gray. This is because of mixing one colour with another colour. However we are able to identify the colours of the familiar objects, this effect will persist. But, whatever result takes place in perceiving the colour, the organism takes into account only the total amount of light in responding to the amount of light reflected by the object—the albedo.

That is, in judging the colour of an object, **Postman** and **Egan** (1949) state that perception neglects the factor of illumination atleast in part. Therefore the judged difference between the two colours of the objects is based upon the difference between the two colours of the objects is based upon the difference between the albedos of the two objects and not based upon the absolute light. Since it is only perceiving the colour, the illumination of the object is discounted and any changes in the illumination do not have any effect in our perception of colour.

Size Constancy

In our visual field we may observe tremendous amount of changes in the size of the objects. The changes of size in our normal visual perception are not noticed. Whatever changes in the size of the objects take place the observed size of the objects remain constant. Distant objects look smaller, while nearby objects look larger. Both the objects may be of the same size. Let us see a moving object, the motor car on a high way. When it is close to us it is larger with its normal size. As and when it moves more and more away from us it becomes smaller and smaller and ultimately it even looses its real shape and looks like an indistinguishable object. This means that distant object looks smaller and nearby object looks larger. Actually there is no change in the object. The change is taking place in our eyes. The perception of size is closely related to the perception of distance, The law of visual angle connects these two. The law is that an object must look smaller and smaller as it recedes from the eye and the decrease in size is proportional to the distance.

Boring and **Holway** (1942) have designed an experiment to demonstrate the size constancy in which the retinal image of the object remained constant in size irrespective of the object's distance from him. There are rows of trees on the road. A tree at 300 feet away looks smaller than a tree at 20 feet away. Why the trees of the same size look with different sizes? It is because of the *size of the retinal image in our eye varies inversely with the distance* of an object observed. As and when the object recedes from us the retinal image is going on *decreasing in size.* If the object is nearer, the image in the retina is larger and we see the object larger. When the object is *far away* the *retinal image is too smaller* and we see the object as smaller. But in our perceptual process there is a tendency to compensate the difference in the size of the retinal image. This tendency is to see the size of the object to remain constant by making changes in *the size of the proximal stimulus* is called *size constancy* (Fig. 10.4).

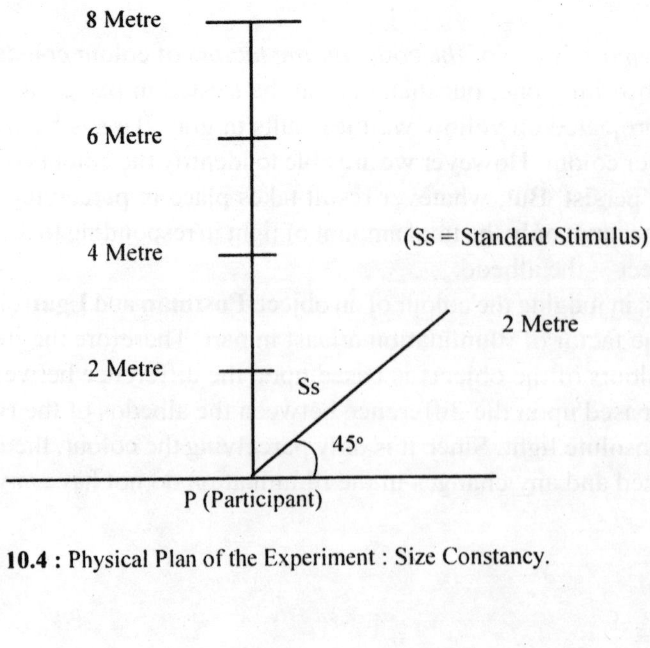

Fig. 10.4 : Physical Plan of the Experiment : Size Constancy.

The size constancy also includes certain related tendencies to see the larger objects as larger and smaller objects as smaller with the same proximal stimulus sizes or retinal images. It is because we knew the true physical sizes of the objects and also we are familiar with size cues. For example, for familiar objects such as people, animal, houses, automobile learning is important to judge the size. When we recognize an object we judge it with an apparent size irrespective of its distance. We have cues to judge distances. But if the cues for distance are eliminated then the size constancy breaks down. The shapes of the objects appear relatively constant irrespective of the position in space within certain limits. It has been estimated the real size of the object can be correctly judged upto a distance of half of a kilometre.

Shape Constancy

We observe objects often from different points or angles. The retinal image also vary accordingly. From whatever point or angle we look at an object we perceive it with the same shape. It is the point or angle from which have see an object causes change in the shape of an object. Actually the shape of the object is not changing. Changes take place in the retinal images. From whatever the angle the object is projected the object is perceived in the same shape. Hence there is a tendency in our perceptual process to observe objects with the same shape from whatever the angle it is presented to our view. There are changes in the retinal images as the shape of the object changes at every angle. The angles distort the retinal images. Generally the shape of an object is observed in relation to its adjustment within the total field.

We could notice several instances of shape constancies. The shape of an object may give us different shape while the object is moving either from left to right or right to left or receding farther away from us or approaching towards us from far away. At every position the object may appear with different shape. But we see it with its original shape and characteristics. For example, a door is seen with rectangular shape when it is seen in normal position. When it is opened, the door is perceived with the same rectangular structure or shape, eventhough the shape of the door as projected on the retina goes through series of twisting out of the shape. The *rectangular* shape of the door becomes a *trapezoid.*

The edge of the door towards us looks wider than the edge hinged with the frame. Slowly the trapezoid becomes thinner and thinner and all that is projected on the retina is a *vertical line* which is the thickness of the door. These changes we can observe and distinguish. But we do not accept it, our experience is that the door moves forward and backward on its hinges and is not changed. We see the door with its original shape. This *tendency to see* the objects with their *original shapes* irrespective of their angle positions, disregarding the apparent change of their shapes is called *shape constancy.*

A round rupee coin lying on the table may project its image on the retina as an *eclipse or an oval shape* and we see the coin on the table appearing like an eclipse. But the completeness of the round outline of the rupee coin is maintained and *we see it always as round.*

FACTORS OF PERCEPTION

We have stated in the preceding sections that perception is more than what the messages sent to the brain by our senses. Several factors effect perception and several factors are operating in perception. No two students of the same class perceive the subject matter of psychology in the same manner. The university tower is not perceived in the same way by a present student as by an old student.

In perceiving physical objects there may be greater degree of *unanimity*. But, the objects to be perceived are human beings we have troubles. In spite of it, if the background of experience and culture are not having too much difference, it is possible to obtain success in our perception. Children see the world differently from their parents. Women perceive world differently from men in several ways. Therefore, there are several factors or conditions that determine our perceptual process. These factors sometimes facilitate our perception and make it accurate, but sometimes make it inaccurate and lead to false perception. They are mainly *motivation, expectations, cognitive style and culture.*

Motivation

Perceptual process is highly influenced by the individual's desires and needs. When the individual is in need of something he will perceive the same which according to him, satisfies his needs. Experiments have been conducted to determine the influence of physiological need on perception. When a person is feeling hungry, an *ambiguous figure* is perceived by him as *related to food.* **McClelland** and **Atkinson** (1948) conducted an experiment to study the motivational process.

Persons who had not eaten food for several hours were made to see blurred pictures. They starved for varying hours ranging from one-hour to 16 hours. Persons who starved for 16 hours saw *images of food* in the picture *more often* than the persons who starved for less number of hours. The physiological drive for food made some persons to perceive images of food in the picture. In our every day life there may be several instances in which motivation may be playing important role in perception.

Expectations

Our perception is greatly affected by our advance knowledge — the, expectations. How the individuals respond to certain words have been tested by **Silipola.**

In this study the experimenter told a group of people that they would be presented with words related to animals. But instead of words related to animals some *combination of letters* that did not give any meaning were shown. The combination of letters, **Wharl** was perceived as *Whale, dack* as *duck* and sael as seal. The second group was told that they would be shown some words referring to *boats* and the same letter combinations were shown to them. They perceived the same as *Wharl, deck* and *sail.* Because they have been guided to expect something related to boats they perceived words related to boats even in nonsense syllables. In our daily life we may be expecting several things and we try to see them as we want to see them.

Cognitive Style

An individual as he grows, develops a cognitive style of his own in dealing with the environment. The person perceives the environment as a whole and does not see clearly the shape, colour, size and the qualities of individual items in the environment. He sees everything combining with the environment and not isolating from environment. These people are called *field-dependents.* But the *field-independents* on the other hand try to perceive elements separate from the environment and separate from each other,

Culture and Society

The culture in which one is born and adapts is having influence upon Ills perception. For example, the language one speaks may determine as to how he should perceive his environment. The individual may use various kinds of perceptual cues. All these are based upon his culture.

Our perceptual process may receive suggestions from outside in the form of section or words indirectly. We accept what others do in our society. The society in which we live determine many of our perceptions. **Sherif** (1936), an American Psychologist, has conducted an experiment to measure social effects on perception. It was a dark room experiment to study the *individual's autokinetic response.*

A point of light was fixed on the wall, and the individuals were asked to watch. Suggestion was given to them that the point of light is moving in small steps and everyone is asked to measure in inches the distance the point of light moved. Then all the subjects were put together to measure the distance the light moved. The judgment of the persons varied. This shows that the individuals modified their perceptions on the basis of the suggestion received from others or in confirmity with others opinions. This is how parents or society transmit to the children the cultural values and standards. Similarly the old members of the society transmit the cultural value to the new members of the society and these values determine their perceptual process.

ERRORS IN PERCEPTION

Our perception sometimes fails to agree with the actual stimulus. Error in perception may be defined as the disturbed or *mistaken perception* that does not agree with object. It is brought about by *conflicting sensory cues.*

Illusions

Human perception is not always accurate and perfect. Sometimes perception becomes easy and correct and sometimes difficult and erroneous. When the stimulus is ambiguous and sensory apparatus is not accurate we have false perception. The *sensory apparatus fails* to identify the object perfectly. This kind of *false perception* is called *illusion.*

Woodworth (1952), the American Psychologist observes that an illusion is an error of perception. If perception is for measurement, the observer judges the object with all his discriminatory power and if the judgment is not totally accurate something in the object is misleading him. This we call *visual illusion.* If perception is for identifying certain object, the observer observes, identifies and names it. Sometimes he does it correctly and sometimes falls to do it. When he fails to identify and name it correctly we call the process illusion. This kind of illusory perception may be due to several factors like the characteristic of *similarity* in the objects, the observer's certain *past experience, environmental condition and sensory apparatus.* For example, a white shell is mistaken for a silver coin, a rope is mistaken for a snake. The shell and silver coin have certain similarities. The rope and snake have certain similarities. The observer perceives the shell and rope with his past experience with silver coin and snake, Further, the place may not be a bright area as to see objects clearly and his *vision* may not be very acute. This kind of error in perception is general. Because many people may have the same error in perception.

According to **Kendler** (1962) an illusion is a representation of a discrepancy between *physical and psychological measurement.* It is also stated as a false representation of the qualities of the stimuli presented for perception and what is perceived, is not exactly related to the stimulus.

The most widely known illusion is the **Muller-Lyer illusion**. In the Muller-Lyer illusion, there are two horizontal lines, one consists of feather heads and the other arrow heads. They are of *equivalent length.* But human beings perceive the feather head line to be of greater length. The slanting position and the size of the arrows contribute to the illusion (Figs. 10.5A, 10.5-B).

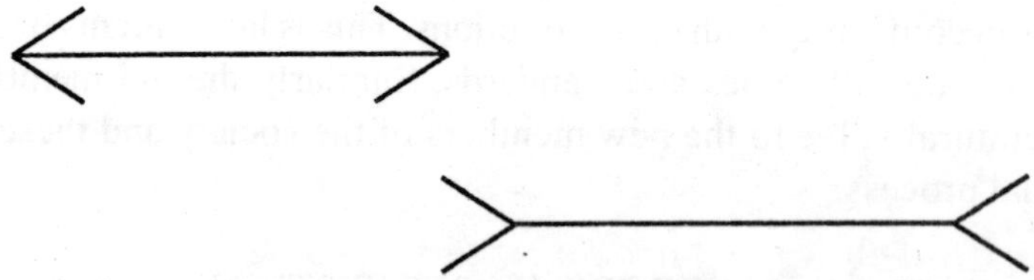

Fig. 10.5A : Muller-Lyer Illusion

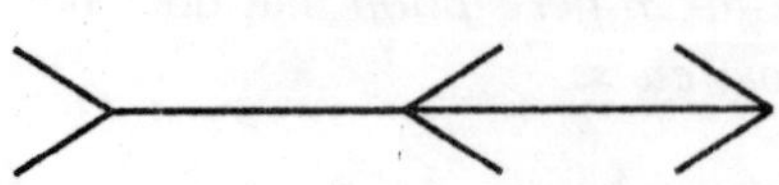

Fig. 10.5 B : Muller-Lyer Illusion
In Muller-Lyer Illusion the Feather Head always longer than the Arrow Head.

Another important illusion is the *moon illusion.* The moon appears much larger near horizon than it look at the sky. The moon illusion is due to the angle of the vision at which we perceive the moon (Fig. 10.6).

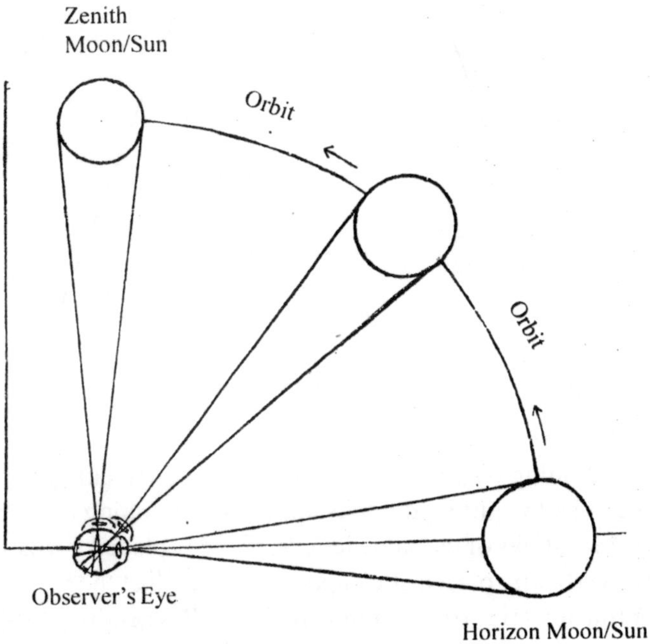

Fig. 10.6 : Moon-Sun Illusion

There is another illusion, called **Penzo illusion**. This is similar to Muller Lyer illusion. In Penzo illusion two horizontal lines are of the same length, but the upper line appears longer than the lower line (Fig. 10.7).

Studies on sensation and perception brought out many interesting and intriquing information. The illusion is a perceptual process. It is also stated as sensory experience, but not had correctly. It is always mistaking one thing for another. **Garrett** (1975) says that illusions are confused or misinterpreted perceptions. That is perceiving and interpreting one thing for another. Illusions may be also taking place due to some defect in the sensory apparatus of the person who receives the object. Some mentally unsound persons also may report things wrongly. Illusions may caused due to some confusions in our perception or due to blending of several stimuli. This quite often takes place particularly in perceiving sounds. We are often confused by the echo of a louder noise which comes from the direction opposite of the direction from where the sound is actually produced. Many persons have experienced the illusion of the direction and movement caused by the morrors in some amusement parks.

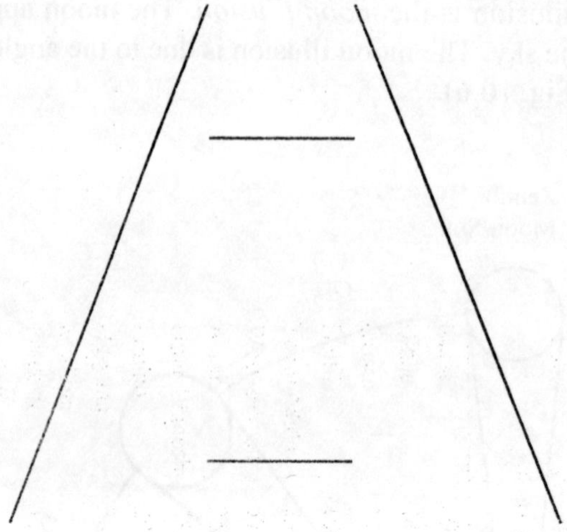

Fig. 10.7 : Ponzo Illusion.

Distance and combining objects of the same size in different pattern cause t see them differently. For example, the rail road tracks appear to be joining at a distance though they are not (Fig. 10.8). Two sticks of the same length, if one is placed horizontally and the other is put vertically, the vertically placed stick appears longer than the horizontal stick. This we call Height-Width Illusion (fig. 10.9). Similarly, in the Poggendrof Illusions we see a stick passing through two rods, one longer gives a distorted look and causes the illusion (Fig. 10.10-A). Another interesting illusion is a squire diagram inside the circle diagram. The squire diagram inside the circle looks as if it is having bends on all four sides in the middle (Fig. 10.10-B). These types of illusions we often come across in our everyday life.

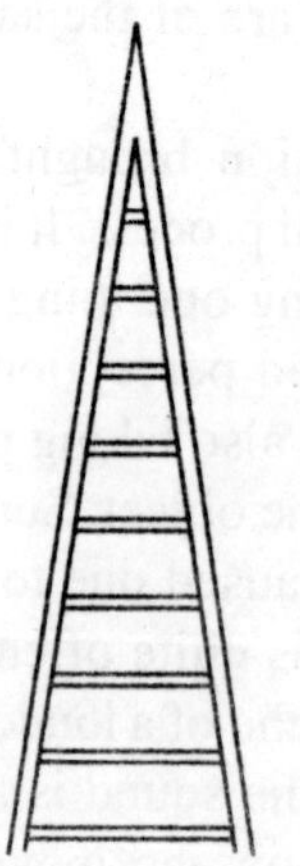

Fig. 10.8 : Rail Road Track : Ponzo Illusion.

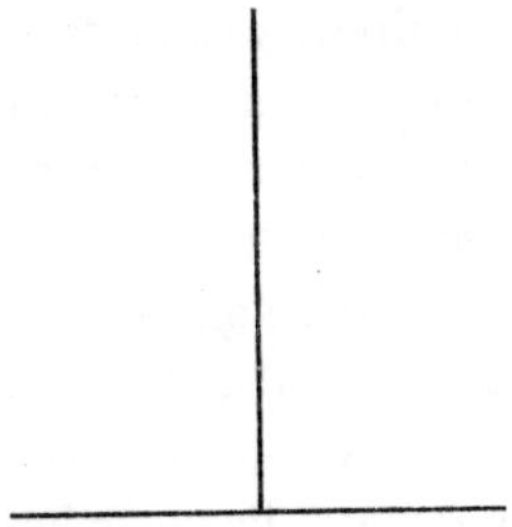

Fig. 10.9 : Height-Width Illusion.

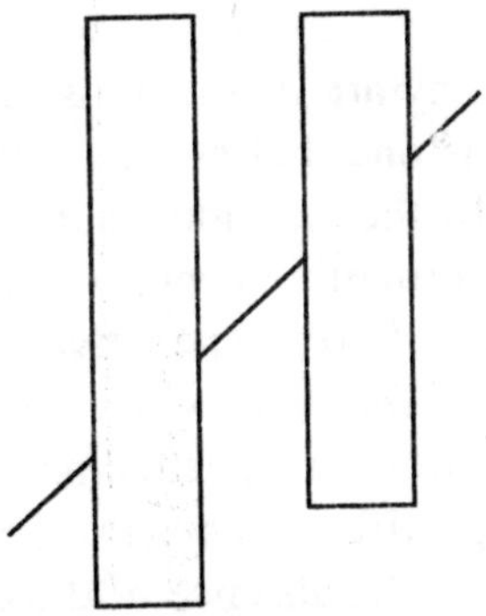

Fig. 10.10-A

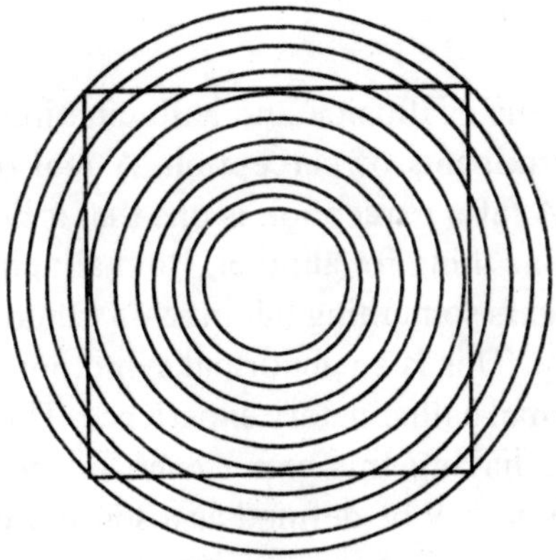

Fig. 10.10-B

Fig. 10.10 : 10-A and 10.10-B. Poggendrof Illusions

There are also illusions of *movement*, *shape* and *size*. We sometimes perceive movements in objects where actual physical movements do not occur. On the motion picture screen we see only still photographs. But we see movements of persons and objects in a continuous manner. A series of discrete still pictures are presented with so short intervals in such a way that we do not notice the tricks but we combine them in our perceptual process and see the figures as moving. This movement is only apparent and not real. Therefore, the movement is an illusion. This apparent movement is named as the *phi-phenomenon* by the German Psychologist, **Max Wertheimer** (1912), the founder of the Gestalt School of Psychology.

According to **Woodworth** (1952) there are two kinds of causes of illusion, one is due to *retinal physical cues* and the other is *within the organism*. The physical cues sometimes mislead us. For example, the light and sound may misguide our senses. The *mirror misguides* us and creates illusion. In the *mirror drawing apparatus* the subject drives the stylus along the pattern by seeing through the mirror and commits lots of errors. It takes sometimes for him to overcome the mislead given by the mirror. Even the mirror in which you see your face several times every day misleads you. By constant use of the mirror you learn to overcome the illusions.

Illusions are also caused due to familiarity of certain objections. For example, we have developed a habit to read every printed material rightly. If there is a printing error in a word we overlook the error because of the familiarity with the words. For example, *we read* the *words muoth* as *mouth* and *shrit* as *shirt,* ignoring the printing errors in these words.

The set you made already to perceive something would cause illusions in certain situations. You set yourself to perceive a particular object. Another object which may be having certain resemblance with the object you want to see is often mistaken. For example, *in a desert* a man sets up himself *to see water* because he is feeling thirsty. At a distance the *mirage appears like water* and he is disappointed when he goes nearby.

Hallucination

The problems of illusion and hallucination is in this way are interactions of the inner and outer determinants of perception. A real perception is one that agrees with the physical stimulus. A false perception is one that is determined by the individual's memory. It is also mistaken one thing for another. But hallucination is a form of perception in which a person sees and hears something but in reality there is nothing and no one perceives the same thing around him. This is an abnormal symptom. The person *hears voices and sees visions.* This is an abnormal hallucinatory experience. This differs from common hallucinatory experience. People are having this experience in their every day life. But it varies in degrees. A hallucination may be defined as a sensory experience in the absence of the corresponding external stimulus. It is a misinterpretation of memory or imaginary experience as true perception.

Sometimes we may feel that something is running across our skin but actually there is nothing. We may respond to some voice, but nobody calls us. These are all some kind of peculiar misperceptions and do not represent objective reality. Therefore, illusions and hallucinations are certain physiological and psychological states of the organism. We are responding to familiar signs in an inaccurate manner.

It is also viewed that hallucinations are self-induced perception without a known objective cause, seeing things and hearing voices. Hallucination is an extremely vivid images in any sensory area. Normally the term hallucination is used when a person states that he sees an object in the environment which is really not there. Therefore, hallucination in an alleged perception of an object when the object is not there.

We see sometimes that some persons in a most distressed and depressed state often feel that the world looks different and even unreal for them. Some may become hypersensitive to sights, sounds and smells and they may find torment to be touched. They find lights are blinding and sounds are agonies for them. Thus the most dramatic distortion of perception may be called as hallucination of sensory experiences of various types in the absence of any stimulation in the environment. They occur most often in the auditory sense modelity and less often in the visual.

ASPECTS FUNCTIONING IN THE DEVELOPMENT OF PERCEPTION

Having understood the various aspects of perception like the perception of form, perception of space, perception of depth, the various constancies and the errors in perception, now it is important for us to evolve the kernel aspects of perception that are contributing to the development of perception. Certain aspects are empirically evolved, certain aspects are theoretically evolved, still certain aspects are cognitively evolved. All these contribute to the process of perception for comprehensing the world at large.

Thus the study of perception is approached from different angles. Authors who deal with this subject matter choose any one of the approaches, because perception involves *two* important *aspects, one is the environment* where the object is to be perceived. The other aspect is the *individual organism* with its complete physical structure. Any physical object if it is to be perceived correctly needs the *vision,* whether it is human or higher or lower animal. This is the reason that the physicists, physiologists and the biologists place more importance on the *process of vision or visual perception.*

The biologists state this as *'retinal process'* to perception. The physiologists state this as visual perception. This means that it is the *function of the eye.* We see everything with our eyes. But simply seeing the objects with our eyes does not meant everything nor we understand the objects seen in the environment. There is *something more and complex.* This is the reason that some psychologists state this process as *'sensing and perceiving'.*

Therefore, it is important to know the variables that are functioning inside the individual organism when the organism coming to terms with the environment in the manner that we name it as perceiving. According to **Murphy** (1964) there are four variables which should be always taken into consideration. They are the sensitivity, quality, experience and the set.

Sensitivity refers to the organism's response to that stimulation which means the nerve cells carry the impulses from the sense organs to the cerebral cortex of the brain. The sensitivity factor of the sense organs may have differences in perception which may take place while seeing various colours. This decides whether one is colour-blind or colour weak or normal in seeing colours. In the case of hearing the differences may be seen in the nature of deafness, as stone-deaf, or partially deaf or normal hearing. The sensitivity variations may also occur in taste. Absence of taste experience occurs in certain people which is the inability to realize certain chemical components which is attributed to hereditary characteristics which run in certain families.

The aspect, *qualities* refer, to varied types of colours and blending of colours, designs, tones and fineness of materials. The quality aspects get registered in the brain cells and become a kind of experience. This may also mixed with other qualities and these become different shades of experience in colour, sound and taste.

The aspect of *experience* may be attributed to the level of familiarity one has developed with an objects, he has experienced with his sense of seeing, hearing and tasting. Perceptual experience varies from person to person on the basis of contact, training and familiarity with the objects in the environment. A musical instrument piano looks differently for different persons. Those who have familiarity with it see it differently from those who have no familiarity with it. The person who had contact and familiarity and training in the instrument perceives totally differently from another person who had no contact and familiarity with it. Any note from the Piano instrument is easily understood by a person who is familiar with it and enjoys and makes meaning out of it, while the other person gets confused and could not make out any meaning out of it. These two persons' experiences vary considerably. Thus in every perceptual field experience plays very important role.

The variable set generally refers to the tendency one has developed with regard to his personal problems. The set may function in various ways. The person is anxiously expecting something which may be pleasant or unpleasant. For example a starving man sees food materials in the passing clouds, the theif, after some nocturnal activities, while fleeing, sees every lamp post for a police man. These are some of the patterns of motives. These vary from person to person, depending upon the nature of the set the person has formed within him.

The way in which the sense perceptions are integrated in this way is a process by which all the four variables contribute to the *organization of perception.* The process of integration and organization depend upon the bodily organization of the sensory apparatuses which supply the raw sensory impressions partly by *natural means* into which the sense impressions fall and partly by the *role of set.* These processes are the variables of aspects of one basic process functioning in the organism which may be called as the *adjustment* of the individual to the environment in relation to the particular problem or task in which the individual is involved.

These four variables or aspects of perception seem to suggest that it takes place only in a fully grown adult person. In the childhood period all these four variable do not function with any purpose. The child does not respond to any normal stimulus. But if there is any louder noise or very bright flashing light the child shows only startled pattern of response which does not meant anything for the child. It is not only the sensitivity that is not having any meaning or any purpose. The other aspects, the quality or qualities of the object, experience and the set also do not function meaningfully for the child. Therefore, it is observed that the child at the age of three or four years will not have any clear view of the objects or persons. The child has only blurred perception in which there is only emotional and impulsive aspects of things involved. As and when the child grows and develops in age and experience the perceptual processes also became clear and meaningful.

It is true that even adult persons have the same type of blurred perception when they confront new and totally unfamiliar objects. Though age and gaining experience may facilitate the process of perception, perceiving the objects meaningfully in the environment involves also something else as these aspects are inadequate for meaningful perception.

Learning is one of the important aspects in the development of perception. It is not only learning to perceive the objects simply as objects, but it should involve also attributing meaning to the perceived objects. The learning a man acquires should help him to perceive things accurately and comprehensively. This type of characteristic depends upon the constitutional capabilities mostly. For example, the individual should have clear vision, sound hearing, verbal expression and sound mind. Whatever type of interesting and enchanting objects are infront of the person, unless he has good vision, interest and motivation the objects will not attract him and perception is totally nil. Similarly a man without a good sense of hearing and verbal expression and without any familiarity with music and musical talent will not be benefitted or will have any idea about the classical music. In such cases the objects are there, but no perception has taken place. From the foregoing discussions we can understand that apart from the afore said four aspects, learning also may be considered as an important aspect in the development of perception.

At every stage of our life we are confronting and experiencing quite lots of things and events. How much each one of us is benefitted by these things and events? We cannot say anything correctly. But something might have taken place. There are very big individual differences in what we do at every stage and it is very difficult to explain in terms of perception. **Murphy** (1964) states that we all pass through three states of perception, such as undifferentiated stage, differentiated stage, and the integrated stage. Individuals differ in each one of these stages. The individual differences that prevail in various aspects of human behaviour and in physical constitution prevail also in the processes of perception.

Man's perceptual process goes through first, the *undifferentiated stage.* It is quite similar to the child's blurred perception. It is a state of confusion and unclear sensation, on which the five aspects that we have discussed earlier should operate meaningfully. These aspects may operate or function at varied levels, one may be having better operation, the other may be having moderate functioning and one or two may be having very faint functioning or no functioning. It is the stage at which the individual could not make out any meaning out of the objects that the individual sees or hears. As he grows and develops in his constitutional makeup, capacities and age and experience he is able to *differentiate objects* with all qualities he observes in the objects.

This stage of perceptual process takes place fragmentally by understanding the qualities and the characteristics of the objects and events. Thus in the last stage he learns to *integrate* what he has partly observed and comprehended and makes out meaning for the objects and events he confronted in the environment. This process of integration that has taken place thirdly completes the processes of perception. At this stage the individual's perception is fully sensible, meaningful and acceptable. In every perceptual act these three processes, are immanent.

However, there may be some individual differences in all the three processes on the basis of the life experiences one has gained and the level of adaptation he has made to the development of Perception. The culture and tradition and the environmental factors may have some influence in this process. The process may be slow or fast depending upon the influencing factors.

PERCEPTION AS VIEWED BY NATIVIST AND EMPIRICIST

The psychological process of perception is mainly concentrated in perceiving world and the objects and understanding everything meaningfully. Many of the early studies indicated that the perception of the world varies in degrees. How does the world look like to the young infant or child? The recent research studies do not provide us with correct and definite, answer for this, **William James** (1890) said that the infant's perception of the world is "blooming, buzzing and confusion". But the présent day psychologists and the results of the recent studies do not agree with Jame's view. Most of the present day psychologists are very much surprised that the infant's provide evidence that they are able to make varied discriminations that are important for adaptive behaviour. But whether these infants could make out any meaning of the objects presented before them is not known.

There were difference of opinions and contraversies on this issue for more than two thousand years till the period of the emergence of the scientific psychology. **Hochberg** (1962) states that the controversies were between the philosophies of *nativists* and *empiricists.* The exponents of nativism strongly claimed that most of the perceptual functionings are innate and perception is the part of unlearned endowment of the organism. The organism is born with certain characteristics and the process of perception is one of such faculties. The important exponents of nativism are **Rene Descartes** (1596-1650) and **Emmanuel Kant** (1724-1804), Descartes was a French philosopher and mathematician, had formulated a mathematical basis for the philosophical theory of **Dualism** and he is the founder of modern philosophy. Kant, who came into the scene after a period of hundred years, was a German idealistic philosopher, sought to determine the limits of human knowledge.

On the contrary the empiricists Locke and Berkeley stressed that we learn our ways of perceiving the world through experience with objects of the world around us. According to them there is no innate perceptual capacities. (Though the controversies are there many of the philosophical orientation concerning the origin of knowledge is based upon the philosophical and logical analyses of the informal empirical observations). **John Locke** (1632-1704) an English philosopher discussed in detail about human understanding and influenced the political thoughts that prevailed at that time and contributed to the social *contract theory.* **George Berkeley** (1685-1753) an Irish Philosopher has expounded the system of *subjective idealism.* He expounded in his work concerning the principles of Human knowledge and he also formulated a new theory of vision, mainly about visual perception.

Bartrand Russell (1872-1970) the British Philosopher and mathematician of 20th century is of the view that perception is a species of a *wider genus.* This wider genus is referred to by him as *sensitivity.* According to him, sensitivity is not confined to living things. Sensitivity is also found in scientific instruments. A thermometer is sensitive to temperature. A phonographic plate is sensitive to light, Similarly there are many instruments we see in the everyday life reacting to on perceiving the stimulus to which it is sensitive. But we do not say anything that there is something more than what the instruments can do. But in our perception we say that there is something more than what the instruments do. Therefore, in perception what is that something more? That something more than perception may be stated as the *consciousness.*

But is consciousness subjected to experimentally /verifiable? We cannot give any correct answer for this, But we are concerned with the perception as one appears to an outside observer. An outside observer could see another person in flesh and blood arid nothing more than that. If you are a person, "do you really perceive a person?" You would certainly say, "Yes, I could perceive him consciously". Thus consciousness is an inference and the statement is mentally projected. Here we can compare the scientific instruments' reaction to the stimulus and the response of a human being. But here the answer is incomplete and incorrect to compare the dead instruments with living and functioning human being, Though it is possible to compare the scientific instrument or a sophisticated electronic machine with human. being regarding the response that takes place in both, man responds in the perception of an object or person in quite appropriately and individually, whereas the instrument or machine will not function in the same manner as the man does.

Human perceptual response has a variability and the response varies depending upon the type of relationship with a person the individual has developed. This response of the individual to a stimulus (probably the boy's uncle) is based upon the cognition and volition. The boy's reaction to the visit of his rich uncle is smiles. But when the uncle had lost all his properties and visits as a poor man, the boy's response is cold demeanour. Thus the reaction of the boy falls under two categories, one is purely receptive and sensory and the other is active and motor. Thus according to Russell perception is concerned with the end term, the *receptive and sensory* part the *reaction* while volition is the *active-motor* part of the reaction.

Perception may enable us to know something of the outer world, but it is not the direct revelation that is supposed to be. Perception is a form of reaction to the environment which is displayed in some bodily movement rather than in the form of knowledge. But if we want to know what knowledge is, we find that perception is a form of knowledge. Therefore perception can be obtained by an observer by reacting to something that displayed in the environment. Whether certain the objects stands in a spatial relation to a person's body, the person's body makes certain movement or a set of movements which we can observe as an external observer. We may describe this state that the person perceives the object.

In the case of an infant, the infant turns its eyes to bright light when flashed which is not the main field of vision for the infant. This incident makes us to think that the infant perceives the bright light. If the infant is blind, it *will* not turn its eyes to the light. The infant gets the visual sensation, but could not make out any meaning out of it. But an adult person could make out some meaning out of it because of his experience and knowledge. This means experience influences perception.

It is also observed that the human body is perceptually changing its reaction to the various given stimuli due to experience and also due to the *law of associations.* The human body is always doing something physically and mentally. We do not know to which stimulus the body is reacting or performing some task. Can we explain this phenomenon through the process of perception? We may also state that stimulus visually seen by a man sometimes back in some environmental situation makes him to reach him in this fashion? Or shall we state that some information processing is going on in the person's mind? We cannot say anything unless we verify it by some means.

On discussing the various philosophical view points of the nativists and the empiricists the present day psychologists were able to bring out a fruitful integration of these view points. The view that practice and experience affect perception has been accepted by modern. psychologists to some extent. But still a section of psychologists want to verify whether we are born with some abilities to perceive objects and space in our environment or whether these abilities are totally learned?

The modern experimental psychologists are of the view that the principles of nativists and the empiricists are too general and vaguely stated and have no predictive value. The scientific theory of perception is applied in this direction. In fact experimental psychology is providing data that are relevant to these issues and controversies. The general aim of the experimentalists is to identify the variables that can control the perception. The discovery they made is that the new born infant displays that the space perception is important and also something more they aim. They also identify the variable of optical stimulation or the central activity that govern the perception of space. They observed that subsequent experience affect the previous experience and experience affect the perception.

The experimental evidence as brought out by **Epistein** (1967) is that perception is modifiable and influenced by learning and past experience. The original appearance of the normal stimuli of perceptual discrimination does not depend upon the prior history of visual experience. Any such discriminations are innate, unlearned processing of visual input. The infant accumulates and history of his visual experience. This history starts exerting an important influence on the visual inputs and processed. The earlier experiences and learned matters are processed and interact with the visual inputs which determine the perception of the world of objects.

Though the experimental psychologists strongly criticised the views of the *nativism* and empiricism in the beginning, their experimental evidences obtained from their experiments on new born infants provide information quite similar to the view of the nativism and empiricism and thus we find that these two views are almost accepted and synthesized and integrated into the process of perception. What **Russell** has stated empirically in his discussion of perception that experience affects the perception has been experimentally accepted.

Extrasensory Perception

We have discussed in the preceding sections various aspects of perception. Perception. is a process of acquiring knowledge of the world. We all believe that knowledge of the objects of the world comes to us entirely through our known senses. Our senses so long as function normally provide information about the objects of the world and we understand what they and attribute meaning to them. This process takes place through sensory channels by seeing, hearing, testing, smelling and touching. But sometimes we often come across idea- that knowledge come to us also through some other channels. Personal experiences of some people seem to reveal communication through some other sources. This source they call as *extrasensory perception (ESP).*

It is true that some persons claim that they had some perception of objects spontaneously without any special effort. **Child** (1973) quotes that **Aldous Huxley** had his experiences through *psychedelic drugs.* Some writers' creative works are stated as highly fantastic and marvellous. From where these writers get such experiences and describe in their novels?

Some are of the view that it is through the *drugs*. Some state that it is through *extrasensory perceptual experience*. Since this phenonmenon has been occupying a part of the field of psychology for about hundred years right from the beginning of 20th century under the leadership J.B. Phine, we cannot reject or omit it from the field of our study. What is extrasensory perception?

Extrasensory perception is defined in various ways. Because of its controversial views a universally accepted definition yet to come. It may be defined as the supposed ability of certain individuals to obtain information about the environment without the use of normal sensory channels. It is also called *cryptaesthesia*. It may be also stated as the supposed faculty of perceiving things in the world by means of other than the known senses. Another possible definition of extrasensory perception is, the experience one acquires consisting of perception not mediated by sense organs stimulation. There are several concepts associated with extrasensory perception such as *clairvoyance, precognition, telepathy, psychokinesis and psrapsychology*.

The humanistic psychologists are found to be giving serious attention to account this. The possible explanation they are giving for this is that of humanistic emphasize on individualivity. According this view different persons may have different experiences. If one person is not having this experience another person may be having which should not be suspected. Some truly claimed that the incidents of the extrasensory communication are doubtless. But some are pretenses by some skilled persons who wish to entertain.

But a humanistic approach to extrasensory perception does not require one to be criticised and rejected. The exponents in this field provided evidences of extrasensory perception in their laboratory situations. The phenomena can be brought under two categories. *(1) Extrasensory perception (ESP)* and *(1) Psychokinesis (PK)*.

(1) ***Extrasensory Perception (ESP)*** falls under three parts such as telepathy, clairvoyance and precognition.

Telepathy : This is referring to though transference from one person to another person. It is a form of extrasensory perception. It is a form of communication between people of thoughts, desires, feelings etc., involving mechanism that cannot be understood in terms of known scientific laws. It is also called thought transference. This is the main subject matter of parapsychology.

Clairvoyance : Clairvoyance is the power or the ability of perceiving events, objects beyond the natural range of the senses. The person who is endowed with this ability is able to state the *number of cards sealed in the envelope* or pocket.

Precognition : This is the extrasensory perceptual process by which an event that is going to take place in future is perceived. It is the ability to foresee the future happenings.

(2) Psychokinesis (PK) : This is mainly dealing with some functions. It is claimed that a form of mental operation is said to affect a material body or an energy system without any evidence of usual contact. It is alteration of the state of an object with mental effect.

These are the main subject matters and problems of extrasensory perception.

Experiments on Extrasensory Perception

These problems stated above are investigated by some experiments in this field. The experimenters worked in accordance with the usual rules of science. For a common man the subject matter of' extrasensory perception give an impression of a magic play or some supernatural or occult doctrines. Modern scientists say, "simply, reject it stating that it is all superstitious and even do not accept the experimental evidences". But some psychologists accept the evidences as some of them are based upon scientific principles because the standard pattern of experiments in extrasensory perception are essentially in the laboratory tradition.

The participant in a laboratory is usually made to sit with much reduced stimulations to ensure quiet that might facilitate greater concentration. The person is fully isolated from the possible chances of perception at the time, or with some past knowledge. He is also free from the sequence of targets. If the targets in the experiment are *physical objects* or events and not perceived by the person now, or anybody else, the type of extrasensory perception is stated as *clairvoyance*.

If the targets in the experiment consist of another person's thoughts the type of extrasensory perception is stated as *telepathy.* Mostly these two possibilities are left open in the experiment. One is a set of *symbols in a list.* The other possible target is the *successive cards* in a deck. These may serve as targets.

But in the situation a second person is watching and concentrating on them with an idea or hope that his thoughts can be grasped through extrasensory channels. This person is called a *sender, or a transmitter.* The targets and the sender are in another room or in another building and he is carefully isolated from any sensory communication to the participant (subject). By some arrangement already made in timing, the subject is asked to arrange and record and guess about the successive symbols in the target list. Later on his guesses are compared one by one with the entries in the target list.

Generally the case for extrasensory perception is mainly based on experiments with cards under various conditions, in which the subject tries to guess the type of symbols in cards randomly arranged in packs which is not seen by him. The normal ESP pack consists of *25 cards* having each of *5 different symbols.* Thus a chance performance consists of 5 *hits* per pack. But successful performance is very rare. However successfully a subject performs the task he may take a minimum of 7 hits. But to reach the acceptable standard, subjects often serve above *5 hits*. There may be even about 200 trails are 40 (Fig. 10.11).

If the experiment or the sender *thinks* of the *symbol* at the time when the subject makes his response, the experiment becomes one on *telepathy.* If the experimenter does not perceive

Fig. 10.11 : Models of Cards used by J.B. Rhine in his Extrasensory Perception Experiments.

the card *(face down)* then the experiment becomes one on the *clairvoyance* of the subject. This is how the experimental results are explained.

Experiments in extrasensory perception reveal large individual difference. It is found that some individuals score consistantly above the chance level. But most of the subjects score close to the chance level and some score considerably below the chance level. But studies to find out the factors that determine the individual differences are very difficult.

What influences the individuals' scores? Investigations or experiments yet to decide the factors scientifically. Tentatively the investigators state some factors like intelligence, socio-economic status, vocational careers and knowing the details of the experiment prior to the experiment may determine the level of scores.

Forgetting the subject matter of extrasensory perception for sometimes, let us go into the community. We hear and see often that some persons perceive certain things without the functions of their senses. They give the correct time any time without the help of watch or clock, able to tell the correct date of birth, the age of another person without any information. There are some even go to the extent of stating what is going to happen next and what has happened at certain place to a person. How are we to explain these happenings? Shall we say that all these are extrasensory perception or genetic abnormalities? According to genetics, persons with one X and two Y (XYY) chromosomes can display something which others cannot do. This ability is not found in other normal persons. These problems need to be experimentally studied and verified whether they can be stated as extrasensory perception or something else. Genetic studies should also go into the details of the chromosome abnormalities of one X and two Y (XYY) and two 'X' and one 'Y' (XXY) chromosomes and their effects in behaviour.

Psychologists are skeptical to some extent to accept the present day experimental results of the extrasensory perception on the grounds that the experiments are not repeatable and verifiable and also a systematization has not been established which can be adapted by those who are interested in the study of ESP. There are opinions expressed by some that independent, investigators were unable to duplicate Rhine's findings in other laboratories and some were of the view that the ESP is real. It may be unconsciously influencing the subjects by giving some subtle cues.

SUPERNATURAL AND MYSTICAL PERCEPTION

A strong belief that is generally prevailing in our society is that some people have supernatural or occult powers which are having influencing effect on others. In the extrasensory perception this may be also having the possible effect. Magicians claim extrasensory psychic powers. They display many jokes like moving objects at a distance by giving some signals, or verba.1 orders, pulling a rabbit out of a hat and so on. The skilled magicians are said to be the *mind readers,* who often demonstrate their extrasensory powers in a crowd of people.

People all over the world in the lower socio-economic status group have strong faith that it is possible to predict the coming events by some mysterious techniques acquired by some secrete means and by controlling the spirits by reciting some mystical words and by sacrificing animals and birds, to the spirits and by using *human bones* in the mystical rituals. This has been going on since the time immemorial and still it is practised in many developing countries. In India this is, more common practice for various purposes. It is mainly practised for punishing enemies, for successful performance of their efforts and for getting cured of some serious diseases. The witch craft is practiced even now in Indian villages. The shaman practices witch crafts and makes the persons sick and suffer and makes them to submit to his commands. This is still going on in some places. People witness the effects of the shaman's activities and this is also counter acted through another shaman and the evil effects are eliminated. There is no means of perceiving as to how such things are taking place.

Magicians, occultists and sorcerers are claiming still that they are able to see the future and present occurrence of events and can also cause whatever events to occur. If a person in the family is missing these mystics claim that they can locate the person where he is and what he is doing and even can bring him back home.

All these phenomena stated here are beyond the scope of modern science and human perceptual capacities. But we cannot denay these. Things are happening here and there randomly. We are also witnessing incidents of death of sorcerers who could control the evil spirits and guiding them against some persons. The sorcerers control the evil spirits and get their wants through the spirits. But some times the sorcerers were not able to control the spirits as they are so wild and blood thirsty and kill the sorcerers themselves. People sometimes identify such cases and report.

Inspite of so much scientific advancement of the present time, whether one is educated or uneducated or with much scientific a outlook and rational the faith in *supernatural* powers still pervails to a large extent. Supernaturalisms generally means a type of special power that human beings can acquire by practising certain specific kind of rituals. The oldest and the best known faith about supernaturalism is that people with the special powers can communicate with the spirits and the spirit world. This is known as medium. Medium is the source of having contact with the spirits. A person who is possessed by a spirit is a medium, who can bring back messages from the dead persons. The dead can respond to enquiries through the medium. Any movement of the objects in the room, falling stones on the house premises, objects disappearing from the room all performed by the spirit under the control of the medium.

There are also other forces or powers which do not come under the purview of science and physical energies since they are highly superficial and not subjected to any perceptual demonstrations. There are also sonic accounts for *extraterrestrial beings* (EP) and *unidentified flying objects* (UFO) which are found to be beyond the boundaries of science and human perceptual abilities.

There are expert astrologers often publisher predictions of events by tracing the movement of the planets. By what means they perceive the functions of the planets and predict the events to occur. Are we to state that all these are the extrasensory preceptions or they perceive the world through Occult, powers or supernatural powers. We could also perceive that some persons perform *miracles* in the presence of their devotees. They claim themselves as god-man or god-inspired. Thus there are some medium through which some persons perceive the objects and events which are beyond the scope of human sensory channels. It is not possible to bring these perceptual processes under scientific experimental controls.

11

THE PSYCHOLOGY OF CONSCIOUSNESS

Introduction

Human beings are quite familiar with the terms *conscious, consciousness and conscience.* These terms are used invariably, both in the common sense world and scientific world. In the scientific world particularly in the field of philosophy and psychology the terms used more generally and also precisely for explaining certain complex experiences. But in the commonsense world these terms are used superficially referring to sensory experiences like, seeing, hearing, tasting, smelling and touching. As soon as a person experienced anyone of these sensations, he would at once state what he has experienced, In such cases it may be referring to *attention, perception* and *awareness*. A person may be aware of what is going on around him, if he is maintaining normal mental and physical health. We perceive the world only through our senses and we have understood the different conditions, of awareness in the normal waking state. But certain things we could not clearly perceive or understand in the normal situation and they become problems. The problems are *distorted, diminished* and even *clouded.* The normal level of our sense of awareness is found to be insufficient.

An individual most of his time deals with the environment in which he lives, it is important for him to know the environment and master it to the required level for his needs and demands. How he does this? He attempts to know it only by *using his senses* and he acts on them and masters the environment to some extent by using his physique and muscles. But how the individual knows and masters the environment? It is very difficult to provide a correct answer for this. Psychologists have made large amount of investigations on it. This involves all the human potentialities like his senses, intellect, learning, memory, thinking and all cognitive abilities. The individual observes the world through his senses. In understanding the full environment we will have to analyse our personality and which is found to be inadequate. This inadequacy is the trouble with us and it is due to *our poor observation.* Observation is the process of knowing the environment by means of the senses. There are *two steps* in observation. One is *attention* and the other is *perception.* Attending is selective observation. It is the process of getting a set to perceive a particular object. Attention can be focused on only one object at a time. Whatever is attended to most fully *conscious.* But we are not that much of conscious of other objects and activities at the same time. Yet still we are *somewhat conscious* or aware of' objects, acts-, thoughts to which we are not attending.

Suppose, for example, you are having some difficult problem which you find it not solvable. In that situation you are looking out through the window and you are most conscious of your problem. But you are only somewhat conscious of the scenes before your eyes outside and you are not *fully conscious* of the sensations in various parts of your body. Generally, Conscious means aware of responding to one's understanding. According to **Woodworth**(1952) some of the physiological activities of the *internal organs* are not within the limits of the *consciousness.* They are fully lying outside of it. The total activity of a person at any movement is divided into *three fields.* The first one is the *central field* of attention which is the full *consciousness.* The second field is the *intermediate field* which is having the *dim consciousness.* The third field is the *outer field* which is completely *unconscious process.* But the boundaries of these fields are not clearly defined or fixed. The central field of consciousness is narrow s6-metirnes. Because, for sometimes for some reasons we may be tempted to observe for a minute on object with great care. But at other times it becomes broad enough to include the whole area, since our interest is extended to wider area. In the case of the intermediate field which is the part of the dim consciousness, is generally quite wide, which includes everything visible except the object we are attending to. This may include the sounds we hear from outside while we are attending to an object seen. This also includes heavy bodily sensation involving, an undercurrent of feeling which may be more or less pleasant or unpleasant, more or less excitement and more or less tense. While all these were going on we are fully or partially aware of where we are and what we are doing. The third field is fully unconscious and no observation of any type through the senses is possible. But *conscience* is an internal recognition of standards of right and wrong by which the individual judges his own conduct. It is a person's moral sense of right and wrong chiefly as it affects his own behaviour

From the foregoing discussions we understand that the major factors that seem to be operating in consciousness are *attention, perception* and *observation.* Generally, consciousness is the state of awareness of an organism. It is the fully awakened state of the organism. *Conscious* is an adjective term, which is denoting or relating to a part of the human mind that makes aware of' a person's self, environment and mental activity and that to a certain extent *determines his choice of actions.* Conscious is only a small part of the mind. It is referring to alertness and awake, not relaxing. It is more of a particular fact or phenomenon. Conscious may also be stated as aware of and responding to one's surroundings.

What is Consciousness?

Generally, consciousness simply means the state of being conscious. It is one's awareness or perception of something. It is a noun term. It has a long History. This has been baffling the ancient philosophers for a long time. Even now consciousness like a riddle is prevailing in the minds of the philosophers and psychologists. Consciousness has been a common subject matter both for philosophers and psychologists for a long time, till the beginning of twentieth century. The psychologists wanted to study it objectively and wished that its functions should be observable as a response to a stimulus. But of late, however it has been accepted and used as one of the subject matters of psychology. Actually psychology was once thought to the study of consciousness and it was also accepted for immediate observation.

Behaviourists wanted to make psychology more like other sciences and because of this view they rejected the *introspective method* of reporting the conscious experience. But, however the facts of consciousness are pervasive and also too important to be neglected. The importance of consciousness was felt by both philosophers and psychologists at a time when there were great controversies about the functions of consciousness. **Bertrand Russell** (1872 -1970) (195 f) stated that **William James** startled the World with his article entitled "Does consciousness exist?" **William James** (1842 - 1910) in his *Essay in Radical Empiricism* has set out the view that there is only *one primal stuff* or material in the world" and that the word *consciousness* stands for a function and not an entity. According to him there are *thoughts* which perform the function of *knowing*, but that thoughts are not made of any different stuff from that of which material objects are made, James thus laid the foundations for what is called *neutral monism*, an idea advocated by most American philosophers. James states that consciousness does not appear to itself chapped up in bits. The words such as 'chain' or 'train' do not describe it aptly. It is nothing joined, it flows like a river or a stream by which it is most naturally described. In speaking of it we may call it as the stream of thought, of consciousness or subjective life. Having James' view in mind, Russell wanted to know, 'can we ask whether there is anything that we can call consciousness is any sense involving a particular kind of stuff or are we to accept what James has said that there is no *inner duplicity* in the *stuff of the* world as we know it and that separating, it into *knowing and known* does not represent a fundamental *dualism*.

According to **Russell** there are *two different* meanings attached to the word *consciousness* by those who use it. One is, we are said to be conscious of something. In this sense consciousness is a *relation.* The other one is that consciousness may be regarded as *a quality of mental occurrences* not having any relation to other things.

Thus consciousness and its significances have been the subject matter of endless arguments for thousands of years. Philosophers and psychologists continue to disagree on this matter even now. Some are of the view that consciousness means *self consciousness.* It is the awareness of one's behaviour, its causes and the effects. In this sense consciousness is something that most humans have in abundance. Some animals have it and plants do not have it. It is defined in another way.

Consciousness is one's personal experience, others cannot understand it. For example, we may become unconscious at different levels when we are asleep or when we receive a heavy below in our head in an accident and fallen down and after the treatment we become conscious and get up. When we wake up we become conscious again.

There are several arguments with regard to the meaning of consciousness, centering the question whether human consciousness represents the same kind of non-physical attribute that goes beyond the scientific investigation and understanding? Some of the scholars who believe that the brain and mind are two different entities, the first one is quantitative and the second one is qualitative and consider that the brain is very essential and not the mind, as the mind is not accessable for a scientific investigation. At the highest levels, according to this view, the human consciousness is not subjected to any research investigation. The mind is referring to the soul or spirit which is invisible, not perceived as sensory experiences.

Therefore many are of the view that we attribute the name consciousness to an *essential identity* that makes OLS what we are now which goes to show that naming something does not explain it. Many exponents who have devoted to the study of consciousness expressed their views independently. **Carl Jung** (1875-1961) states that man's task is to become conscious of the contents that emphasize upward from the consciousness. As for as the sole purpose of human existence is to kindle a light in the darkness of more being. According to him consciousness is a hidden source that needs to be kindled.

According to **Scott M. Peck** consciousness is the executive part of our *total being.* It is the conscious that makes decisions and translates them into action. If we were unconscious, we would be indeed like the new born baby, one with God, but incapable of any action that might make the presence of' God felt in the world.

The general view is that consciousness is the most remarkable product of' human brain. It is the *medium* for our thoughts and provides meaning to our experiences. Human beings could not experience sight, sound, taste, small and touch without consciousness. Our very sense of identity comes only from the consciousness. Consciousness is so multifaceted that there can be no single definition could satisfy us. The human brain has a built-in awareness of other people's thoughts. We could see colours in objects and in rainbow only through our conscious perception. Consciousness sets in action only when it is needed. A student who is sitting in the class room in the midst of one hundred students, when the teacher calls his name, he at once gets up and says "present sir". How this incidence takes place? The student is Conscious of the words uttered by the teacher and he finds that it is *referring to* him. The other students do not respond to the teacher's call. This student is in a conscious state and responds to the call, because of the function of his consciousness.

It is generally felt that consciousness is a word that has been always posing itself as a riddle, perplexing and paffling one. Human beings use it in their everyday conversations in an interesting manner. But when we begin to think about, what it really means, it seems impossible to make a clear view of it that is acceptable to every one of us. This is also the case with the exponents in the fields of philosophy, psychology and neuroscience. Their views and opinions vary drastically. Some are thinking that it is the fundamental stuff of the universe. Some are of the view that to be human is to be conscious of the world around him. But at the same time others see it as a fiction and nothing else.

At certain stage dramatic studies and investigations were made to answer the question, "What is consciousness"? These studies were made from varied perspectives. These studies did not bring our any clear cut answer to the question. The exciting works are still in a continuous process. Although it is not known exactly what consciousness is different theories of consciousness have been evolved. Some of the theories of consciousness are thousands of years old which go back to the ancient Greek philosophers, while some of them are recently evolved new ideas which are conceived by the explorations of philosophy, psychology and neurosciences. Generally, we use the terms conscious and consciousness, to mean many different things. Because conscious appears always as a puzzle and we are putting ourselves to investigate in a non-stopping manner.

These varieties of meanings and descriptions of consciousness and the difficulties of fixing it down, what we are referring to when we use the terms conscious and consciousness is the part of the *reason,* why it is so difficult to state exactly what consciousness is? It is not always possible to separate the states in which we are conscious from those in which we are not. A person is in a state of unconscious when he *is in deep sleep* and when he is wide awake he enjoys his experience of the world and he is conscious. This phenomenon is quite clear. But sometimes we may come across some borderline cases which seem to be inescapable. When a person is falling asleep, he is neither fully conscious nor fully unconscious. When we are sleeping we have dreams and we have same experience of something in which we were involved. But are we quite conscious in the same way as when we are in awakening state? This is the problem for which we yet to get the solution.

When we put the question, "What is consciousness?" there are varied opinions and views from the philosophers, psychologists and neroscientists who have been studying consciousness for a long time. One main reason is that consciousness is a puzzling matter which is depending upon the brain, It is a very different sort of thing, very difficult to assess what sort of things are going on in the brain cells. For example, when you are suddenly aware of a rose flower, it is present in your consciousness, may be also stated, is present in your memory. This rose flower is not actually in your brain in material *form.* But your mind connects your conscious thought or image formed by the earlier experience to the present situation and makes you conscious of it. The rose is not physically having any form in the brain cells. Therefore we may state that it is the memory image. But what sort of function is taking place in the brain cells, if brain is the basis for the consciousness? Further investigations may throw some light on this.

Theories of Consciousness : Mental and Physical Sources

Is consciousness a physical entity? **William James** stated that it is not physical entity, but a *quality of mental occcurrances.* Similarly some philosophers think that consciousness is a non-physical thing. It is a kind of *spiritual container* in which our thoughts, feelings and emotions and experiences take place. These philosophers' explanation is that man has unique access to his own consciousness container. He can work only inside of it and his knowledge of its contents is correct. There is no part of the physical world with which he can have any access. A *neuroscientist,* for example, can have much better knowledge of what is going on inside the human brain from physical point of view, but he has no access to a person's conscious experiences. This makes it clear that the human consciousness is not a part of the physical world. **Rene Descartes** (1596-1650) the French philosopher thought of consciousness in a different way. Descartes was a philosopher and a mathameticism. He provided a mechanistic basis for *philosophical theory of dualism* and he is regarded as the *founder of modern philosophy.* He also founded the analytic geometry and contributed greatly to the *science of optics.* The philosophical theory is named after him as *cartesian dualism,* which separates the world into *two distinct realms,* as the *physical realm* and the *mental realm.* The physical realm is inhabitated by human bodies, atoms, molecules, table and chairs. The mental realm is inhabited by the souls where conscious thoughts and experiences take place. Like Emmanual Kant (1724-1804) Descartes was strong scepticist.

The Cartesian dualism theory of Descartes has been viewed by some later and modern philosophers differently. The important problem they have raised was that the two realms postulated by Descartes cannot be regarded as entirely two separate entities. The main objection to it is that the body, as we know, affects the mind and the mind effects the body. "When we perceive objects our conscious experiences are affected by the events in our sense organs and when we determine the act, our conscious thoughts effect our muscles and limbs. As started by Descartes, if the *mental realm were fully separate* and different from the physical realm and did not exist in space how it could interact with the body and have the effect? What would take place if the interaction between the mind and body, the non-physical and the physical world is not taking place? Any *process* between the mind and body that could provide *a link* we call it as *experience.* This occurrance may be stated as the *material process* on one side and *non-material process* on the other side. This does not bring any ability between mind and body. Most of the modern philosophers seem to differ from the theory of Cartesian dualism. They are of the view that this theory does not provide any solution to the problem of relationship between the mental realm of consciousness and the physical realm. It is true that several philosophical theories advocate the *dualism* for explaining the problems concerning the scope and extents of knowledge. But linking the mind and body in the function of consciousness the contraversies prevail all the times.

Some of the Modern theories of Consciousness

Idealistic Theory **:** Idealism believes only in the existence of ideas. **Berkeley** (1685-1753) one of the exponents of idealism states that only ideas exist in our mind. It implies that whatever exists, exists only when it is perceived or thought of. Because as ***Locke*** (1612-1704) said that *ideas* can exist only in a mind. Historically, those views which emphasize mind or what relates to mind (or ideas) as the sole *universal reality* which has been called idealistic (or spiritualistic or mentalistic). Berkeley's idealism is one among the number of different forms. **Berkeley** states that common sense does not support the view that perceived objects exist independent of *consciousness.* But the common sense view is that they exist only when we are conscious of them, that is, when we have ideas and provides no evidence that they existed at any other time. According to **Berkeley's** idealism to be is to be perceived. The being of anything consists in its being perceived that is being in its idea in a mind.

The other idealistic view is that *mind* or *spirit* is the only thing that *really exists.* The material universe is an illusion produced by it. Therefore consciousness does not require to be explained in physical terms.

Functionalism : According to this theory consciousness is not a separate entity. But it is a way of ordering the physical world. The human brain is conscious and it is processing and coding information in certain manner and consciousness is the whole material. According to functionalism the form of a thing should be determined by its use. Any doctrine should stress utility or purpose.

Identity : According to this view the activity of the physical brain that is associated with consciousness is only the consciousness. Therefore, there is no need to see whether there is any link between *mind and brain.* Because there is no link and they are one and the same.

Eliminativism : This theory advocates that there is no consciousness and it does not exist in any form. Only we think that we are conscious. It is an illusion. On close observation of what we feel as experience is only from our senses. The senses reveal experience that it is in no way different from the knowledge we are holding unconsciously. Therefore, only the senses provide experience and knowledge which have nothing to do with consciousness.

Quantum Theory : Quantum refers to the smallest quantity of some physical property, such as energy that a system can possess. According to quantum theory the particle is with such a unit of energy, amount or quantity especially a specific amount. According to this theory consciousness is the manifestation of quantum effects in the brain and differs from any other observable part of the natural world. It is not subjected to any physical laws.

Otherviews about Consciousness : Some of the exponents who have deeply went through the study of consciousness are of the view that it is quite impossible to provide a valid answer to the question, 'What is consciousness?' **Colin McGin** and **David Chalmers** are of the view that there must be some *natural property of the brain* that is responsible for consciousness. But we are short of that part of the cognitive ability that could enable us to understand the relationship between the brain and consciousness. The adult person's mental development is to that of a three years old child. The only way that we can have the access to our conscious experience is by introspection. But introspection cannot provide any understanding about the brain. External perception and scientific investigation may provide some understanding about the brain. But there is *no faculty* that can facilitate introspection and external perception *together* to give us an understanding of the consciousness and the human brain together. Chalmers states that no theory of consciousness in terms of *neuroscience* is possible. But it is possible to construct a theory that could explain the relationship between the brain and the consciousness, if we use consciousness as a starting point and not the brain. This theory should include *experience as the basis.* But what required is a new type of science, if we want to explain the relationship between the brain and the consciousness.

In view of the above discussion we could understand that most of the present day thinkers seem to agree that there is no need to belief that consciousness is in fact something different from the activity of the physical brain. In other words, conscience can be regarded as an *emergency property* of the brain or function, a characteristic that arises out of the combined action of highly complex parts. It is not the feature of the brain that could be observed with our senses or with any minute instrument. The modern thinkers with regard to the consciousness appear to be totally pessimistic. What they state is that it does not exist as something over and above the brain and as separate qualities. With all the available information about the brain, the science of the brain has made it more difficult to accept **Rene Descarte's** *Cartesian dualism.* The existence of the soul also is not acceptable for the modern thinkers.

The Development of Human Consciousness

In the midst of the above discussion, is there any mentioning, whether there is any entity called human consciousness, or is it a substance or material component or only an unobservable functional aspect? We also wish to understand the evolutionary aspect of the consciousness. Consciossness is considered by many as the modern complexity, inaccessable for human understanding. Thinkers of both the ancient and modern World would agree to some extent that the development of human consciousness with all its riddle and baffling began to evolve information astonishingly. The period of time at which the process of consciousness *was* brought into the study is not known exactly. Experts with the help of some *quasi-evidence* dates it back to millions of years. In the evolutionary process of species on the earth at what stage the present type human beings were evolved? When the process of consciousness was traced and understand? A fit answer to these questions are impossible. If we want to ascertain anything several experimental investigations have to be done using the available fossils, natural sources and the tools and materials left out by all sort of species.

Experts state that the earliest human species had existed with large head and brain as bipedal apes for millions of years. The direct ancesters of human beings, *Home creatures* were able to produce fires and tools which were of very premitive types. It is assumed on the basis of the crude and indelible evidences that the human species, the *Homo sapieuns,* have evolved some one hundred fifty thousand years ago. This *Homo Sapiens* species gradually developed with some changes in all aspects of human characteristics. These *Homo Sapiens*, after one hundred thousand years, had transformed itself into *verbalized* communication contact within the species from *mass sounds of the voices.* The development of linguistic abilities, development of skills and talents, cultural and social developments have taken place after several thousand years. Since the human development in the evolutionary process was not sudden, but was very slow and gradual, at what period man had felt the needs for the *cooked food* and felt the necessity to conceal *his body* with some garments also not known. Thus a bigger and better and magnificient transformation has taken place in the evolutionary process. quite appropriately. The common sense world of consciousness expects that we must be awaken state to experience it. This experience is possible only when all our sensory appartuses are in sound condition. For example, a blind man cannot see anything, a deaf man cannot hear any sound, similarly all other sensations can be experienced only by sound sense organs. However, a man is mentally alert and conscious, his experience is not complete without sense perceptions. Therefore we may state that there is relationship between sensory experience and consciousness. But we do not have clear idea and sufficient information to ascertain the level of the brain activity during sleep. Experts state that no brain activity can be Just as busy during sleep as in waking state. In fact there are different states of consciousness.

But the dividing line between the states of consciousness and between consciousness and unconsciousness is by no means hard and fast. Further we can also alter the state of consciousness by means of various physical and mental procedures. It is also understood that behind all experience of consciousness there is some *wonderful power* of the mind functions *from within.* We are also experiencing this power in our dreams in some way. It is also possible to bring out the hidden realities by applying some technique. But even in normal waking, state this mental power or consciousness produces internally our thoughts and sensations to a large extent. This power is said to be the source of some other internal functions like *illusions and hallucinations.* Though these are false perceptions they are internally produced.

Type of Consciousness

Consciousness is stated to have varied states however an organism is in a quite awaken state. Thus the waking consciousness from which we can report accurately as to what is happening in the world around us or in a fluctuating state. In our perceptional consciousness the normal fluctuations include anxiety, excitement, stillness, fatigue, feelings and emotion and the experiences that may cause states of feelings of overpower and so on. Most people are of the view that they know what their normal walking state consists of, but find only the other states of puzzling or interesting for them. Generally, in fact, the ordinary waking state may be the most puzzling and exciting of all the states.

Therefore the waking consciousness consists of varied levels and it is not a single entity. It is not a simple, state, but a *state of complexity.* We may be attending and listening to or inattentive and not listening to something else. But at the same time we may be listening to someone's voice at a distance, hearing something else, mentally going back to some imcomplete task, the work assigned by the office manager and so on. Even while replying to someone's enquiry, we may also be planning to go to a place to meet someone else. Thus while we are doing something, we have too many things crowded in our mind, one interfering with one another.

Thus, there is no single normal state of consciousness. Every person experiences a huge range of mental conditions everyday as the brain adapts to the changing circumstances. The dividing line between consciousness and unconsciousness states is not clear and sharp. There are different degrees even within the conciousness state and the dividing line is not clear and sharp. In the state of dreamless sleep we are not fully conscious of anything at all. There may be some vegue thoughts and feelings. When we are day-dreaming we may not be fully conscious of the things around us or in the environment. For example, some one may be very close to us, talking something but we are unaware of it and we are living in our own world of imagination.

Experts who studied these aspects deeply speak of various states of consciousness and they evolved *three main types* of awareness. **David Rosenthal** (1964) a philosopher from New York describes these three main tvpes. *The first* and the most basic state, according Rosenthal is what we *perceive things through our sense,* but at the same time we do not have any thoughts about these perceptions. This takes place when we react or respond to a visual stimulus without being consciously aware of its perception.

The second type of consciousness is the *direct awareness* of whatever is attracting our attention. In this state we are consciously aware of the content of our mind. There may be inner thoughts to where to go next holiday. There may be also more sensory experiences such as to run and catch the train at 9 a.m. or we may be interested in seeing a motion picture, released shortly, acted by the favourite actors.

The third type of consciousness is called the *reflective consciousness,* in which we are aware of ourselves that we are having the experiences. In this type, we are consciously aware of ourselves as having the experience. It is not only we are perceiving something, but also perceive what it is our sense of self is brought to the front. Therefore we do not simply have thoughts, we also think about it as if we are looking it from outside. Here we apply *reflective consciousness* in different ways. For example in thinking we verify whether we have perceived something correctly or not. And also we think whether we should believe some information received. We also think whether our recall of the event in our mind is correct. This type of consciousness helps us to reflect on our perceptions, beliefs and memories as if they are the objects In our mind.

Reflective consciousness is also functioning in social thinking. It helps us to decide whether we should trust others and ourselves in social situations.

Apart from these three types there are also other types of consciousness. A person's readiness to a new stimuli which is called sometimes as the *state of vigilance,* which can be tested experimentally. An adult person's responses to the occasional appearance of a stimulus on a screen or in an instrument can be tested. **Mackworth** (1950) has designed an experiment to test the state of vigilance. He used a clock with a pointer that jumped regularly one step for every second, but at rare intervals, the pointer jumped double the length. The subject has to report these double jumps by pressing the key when the double jumps occurred. **Mackworth** observed that the subject did quite well for half an hour, but made many mistakes, when set the work for a longer period of time. Errors were reduced by reporting correct responses or by calling the subject up on the telephone and asking him to do better. This type of studies of vigilance revealed that subjects can be kept alert by several techniques.

Fatigue and Exhaustion : The waking consciousness can be subjected to certain modifications and degrees of atteration by excessive fatique and exhaustion or by making sleepless, or by disease and by drugs. Hard muscular task can produce several recognizable physiological changes, which may also include waste products in the muscles and in the blood streams followed by subjective experiences of pain, weariness and desire for rest and sleep. The person's ability to continue at work is reduced and also the quality of the work is made to suffer, more deep changes take place in exertion which leads to exhaustion. Our concern here is the results or actual-physical work in the case of fatigue and exhaustion. Many of the symptems in the subjects in this case is listlessness which are similar to psychosematic or neurotic cases.

The experiment was conducted by **Williams Lubin** and **Goodnow** (1959) who have obtained interesting results. The experiments were concerned with producing debilitating effects of sleeploss have shown that the sleep deprived subjects can perform satisfactory works of short duration. But a long continued work suffered with more interference. For example, in one of the tests little loss of vigilance was shown for the first seven minutes of testing after the loss of sleep for thirty one hours. Errors occurred in this state. After the loss of sleep for fifty four hours a good performance was observed only for *two minutes.* This is how *fatigue and exhaustion* could modify the consciousness of the subjects in work situation.

Fugue State : Generally fugue state is abnormal condition in which a person has *no memory* for his actions over a period of time. The fugue state may last for several minutes or even months. The fugue has something to do with the consciousness. The divisions of consciousness allow to do more than one thing at a time. This may be humming a tune while a person is reading something. This is not very much to be noted, because we may shift our attention from one task to another, if we wished to do like that. This type of divisions of consciousness is called *split consciousness.* In the state of split consciousness or in the *control systems* which directs what we should do and the experiences are more remarkable if one of the control systems is not aware of it. In *rare cases the split consciousness* is known as *fugue state.*

Some call this state as *multiple personality*, if a person continued his fugue state for more than the normal period of time he may engage in activities that are quite uncharacteristic of normal fugue. But when. fugue is over, the person may dency that he has done all these things. It is because of that he has totally forgotten everything. Fugue state may prevail for sometime which may be for a brief period. *The short period* is to sorric extent like *sleep walking.* In this state the person gets up from the bed in sleep and moves about and returns to his bed. On waking tip in the morning he is not aware cf the happening during night time.

Sleep and Consciousness

We have been discussing in the proceeding pages the various states of consciousness, which in other words may be stated as the states of awareness. A state that is most commonly contrasted with waking state is the sleeping. The sleeping state is a state of mostly reduced state of awareness and activity. Biologically sleeping state is a state of restoration like the waking state, Sleeping state is not simple function as we think. We do not go to sleep just because we have been engaged in physical activities and the processes got exhausted and they need sleep. But whether the bodily condition is an individual choose either, to sleep or to remain awake, every night we go to bed at a time we have fixed, lie down, close our eyes, leave the mind to go into free state and relax our bodily parts. While we are tempted, we treat sleep simply as a state unconsciousness. But some psychologists are of the view that sleep is really not far away from the waking state as it appears first. When one is in sound sleeping state it seems to give us a glimpse of what it would be like to be dead. Sleep no doubt appears to extinguish the light of experience. But however researches on sleep show that our response seems to be active affair in which consciousness is not really absent.

Experimental results on sleep brought the fact that the brain recordings reveal that we respond to each night more alternatively between two distinct types of sleep. One is the deep steady electrical rhythm of *slow wave sleep (SWS).* The other type is the frantic buzz of the *rapid eye movements (REM)* sleep, in which the eye balls seem chase phantom vision. Generally, we start with each night in SWS. At the first stage it is so deep that we are practically *inactive.* Then after sometime, mostly *after ninety minutes* we experience a short **interruption** of rapid eye movement which continues for about *ten minutes.* As the might time continues we switch over between the slow wave sleep (SWS) and the rapid eye movement (REM) continues to occur every ninety minutes. But the REM periods get much longer and the SWS periods progressively is becoming shorter. It has been observed in these type of experiments that on the average we spend about one quarter part of every night in rapid eye movement (REM) sleep.

It is observed that sleep does not make us altogether unconscious. For example, after waking up we are able to recall the dreams. When the mother is in deep sleep any type of crying sound makes hen, wake up at once. The mother set up an idea in her mind that she should see that the child is not disturbed in its sleep. In the same manner some people when they go to sleep set up an idea that they should get up at a particular time and get ready to catch the train.

There are some interesting information available to us while looking at the mental functioning in waking state and in REM sleep. While awake, our thoughts and intentions are generally clear and directed, but in REM sleep state they are confused and become illogical. Thought conceived earlier are strong in both states, though in REM sleep it is directed towards *internally generated* sensations rather than to the *external world.* It is to be known that memories on both the states are *long stored memories* rather than the recent memories. The different states of consciousness in sleep are characterized by marked changes in the electrical activity of the brain as seen in the *electro encephalogram (EEG).*

The slow wave sleep (SWS) appears to be very much needed for growth and maintenance. However, the main purpose of REM is very difficult to explain. It is viewed by some that the REM sleep is a natural way of keeping us in awaking state only to some extent, but out of some irregularity, kept us until the morning. It is possible that vivid dreams may occur which are merely the by-products of a sleeping consciousness.

Dreams

The Dreaming Brain **:** The rapid eye movement (REM) sleep was first traced by the American psychologist **David Foulkes** in 1950s. When it was described by him it was thought that it was only a sleep phase in which dreams occur. Since then it has been viewed that the mental activity of some type goes on all the night in the sleep. The subjects who were aroused from slow wave sleep (SWS) normally reported vague information. **Foulkes** described these *slow wave dreams (SWD) drowsy thoughts* rather than bright images. But these were having a hazy character. Their existence showed that the mind muses to itself all through the long time from the beginning to the dawn. Those who were working on sleep researches began to conclude that the brain never actually shuts down at night time. They strongly believed that the brain cells never switch off. But they fire a few time,, each second Just to stay alive and to maintain the health. Thus the brain goes into neutral when it enters into slow wave sleep (SWS). It puts a *black* on the *incoming sensations* to prevent external sights and sounds from troubling the mind. It also suspends the short term memory.

At this stage the state of consciousness is confused and illogical and thoughts or images disappear rapidly in the same speed as it is formed. In rapid eye movement sleep the brain functions in a quite different state. Activities in the brain stem are inhibited making the body paralysing temporarily. This is considered as *safety mechanism.* This ascertains that dreams are not occuring physically. But the block put on the sensation and memory remains. Because of this the conscious self in a confused way tries to make sense of some random images, but not successfully. But if any image occurs it is quickly forgotten on waking up. Thus the brain functions in the formation of the dreams. But studies in support of this finding have not taken place following this discovery.

Dreaming and Dream : Dream is defined in varied ways. Philosophers psychologists, neuroscientists, religious scholoars all define dream from their own point of view. But, whoever defines and describes dream, it is related to mind and body. Therefore without making reference to brain, body and mind we cannot speak anything about dream. Generally, dream is a mental activity. It is the form of an imagined series of events occurring during certain phases of sleep.

It is quite common that every person dreams of something everyday. The dreams are not alike in duration or length in content, in emotionality and in vividness. What people dream about? Are the dreams making any meaning? If we want to know answer for these questions we must ask some people to maintain diaries and make entries everyday when they woke up in the morning. **Hall** and **Van de Castle** (1966) have conducted an investigation among the college students. There were five hundred men students and five hundred women students a total of one thousand. The dreams of these students were very dramatic. The dreams were analysed from the diaries of these students making entries in a table. It was found that on the average, the dream of each person contained two or three persons and included one or two objects. These objects were quite familiar ones and the most often mentioned objects were given a descending order which consists of houses, automobiles, cities, rooms, doors, and stairs. It was observed that some of the events appeared in dreams were emotionally loaded, friendly, sexual or aggressive. These were the dream contents.

The experimental researches or dreams and dreaming throw some new light on the issue and revealed several new information. There were also several new psychophysiological methods which provided many interesting information by relating the dreamings to memory, consciousness and unconscious.

Theories of Dreams

Dreams have been puzzling and baffling both the common sense world and the scientific world for a long time. Dreams in the common sense world have been associated with spiritual and supernatural characteristics for a long time. Though it is a psychological issue there were no scientific or psychological or anatomical, genetic or bio-medical studies till **Sigmund Freud** (1856-1938) published his classical work, *"Interpretation of Dreams"*. The Indian Epics, religious scripts, myths and legends speak about dreams. But they have nothing to do with one's personal life experiences.

Freud's Theory of Dream : Dream has become a subject matter of study in psychology when Freud has published his facinating work, "The Interpretation of Dreams in 1900. Freud, an Austrain medical graduate, while practising medicine, observed many patients came with compliants of some neurotic troubles. This has made him to pay more attention on nervous and emotional disturbances and also on neurotic and psychotic troubles. He further observed in his patients that dreams were frequently reported by his patients and he suspected that this has something to do with the problems of the patients. While analysing the dreams, he found that the dreams were having some source in the minds of the patients which he called as unconscious. He believed that the unconscious impulses were mainly responsible for the dream and the dream was the gratification of some motives lying in the unconscious mind. Freud believed that the dream is the *wish fulfilment* of the person.

Freud, as a child was a voracious reader and he was *preconsciously intelligent.* His father, Jacob, once stated, "My Sigmund has, more intelligence in his little toe than I have in my whole head." Freud was driven to make a mark in the world, not only by his innate intelligence but also by his status as a Jew tried to forge a path into the European intellectual world. His early interest was in science which led him to medical school in *Vienna,* where he developed extreme interest in *neurology and physiology.* Then he went to France to work with some of the foremost *psychiatrists* of the time. Freud's early life had something to do with his theories of the mind. Freud's revolutionary ideas have the root in his interest in *family dynamics.* He applied his fertile intelligence to connect diverse information and to reveal the *hidden connections.* He was strongly influenced by the methods used in *Judaism* and made use of the scriptures. He used some of the terms and concepts found in the *religious scriptures* and *mythologies* for his purpose.

Freud was originally interested by the use of *hypnosis* for treating the hysteric and neurotic patients. On returning to Vienna he opened a private practice in *neuropsychiatry* with the help of **Breuer**. Freud's interest in dealing with psychiatric patients resulted in inventing several new techniques and through which he was able to explain many complexities. According to him *the real meaning of the dream is not directly expressed.* The *latent content* is the repressed desires lying in the unconsious mind. According to Freud these are mainly *sex desires* not fulfilled and they are trying to come out. They cannot come out in their original form as the censor is preventing them and they are pushed back to the unconscious. But, however, they come out during sleep when the censor is not alert. The *latent content* come out in a sublimated form, in disguised state. The original desires are converted into acceptable forms and *appear as dreams which* Freud called as *manifest content.* Since the original idea of the dream is not known it needs interpretation. Freud had associated certain symbols to find out the real idea or desire that has been repressed and sent to the unconscious mind. This procedure Freud had explained in his work "The Interpretation of Dreams".

According to Freud's theory the process of the dream *construction* occurs through the mechanisms of *condensation, displacement and symbolization.* Here *condensation* refers to the process of combining the ideas into more reduced form. This makes a single word to have multiple meanings of the dream *contents. Displacement* permits one thing to stand for another thing as one part of the body may represent for different parts. *Symbolization* is used in a more general term for representing ideas and events by making something else. It is also applied to persons. These symbols may be private, generally related to the individual experience of the dreamer. There may be also general or *universal symbols* which may be mostly sexual symbols. This is what Freud mainly had emphasized. Freud has expressed this view in 1880s in Vienna, when he started his practice. The Viennese bourgroisic from whom he drew most of his patients were *shocked* by his insistence that *sex lays at the root of neuroses.*

Much of the human life impulses, according to Freud, centre round the sexual wishes and sex has certain taboos associated with it. According to Freud's interpretation the male sex organ is symbolized by a snake or any long and pointed object and the female sex organ is represented by some container, such as box, vessel and basket, Freud often stressed that the dreams are the *guardians of the sleep* which he considered as the roles of the dreams. It is one way of getting rid of unfulfilled desires. Otherwise they may disturb the sleep. If the original materials of the dream were presented to the dreamer the desires or impulses would be often found to be terrifing and the dreamer would wake up suddenly. It was also observed that the dream often found to be failing in its functions. For example, the dreamer sometimes suddenly awakens severely frightened. This takes place because anxiety is aroused in the dream.

There were also many physiological and biological studies on dreams and these studies raise certain doubts on Freud's theory. These investigators studied the rapid eye movement (REM) sleep states in very young children and also in animals. These studies seem to correspond to the Freudian theory of adaptive nature of dreams. Around 1950s researchers in California put forward the idea that dreaming could have a role in animal memories. They discovered that rabbits, cats and other mammals emitted a previously unknown type of brain's rhythm called *thata-rhythms* from the hippoceampus of the brain region whenever they were exploring new territory in the look out for predators.

Generally, dream and dreaming is a mental process occurring only in human beings. We are not made aware of this phenomenon occurring in animal and in other species. There are several conceptions and theories of dreams evolved recently by psychologists and biologists after Sigmund Freud. They have evolved totally different meaning and description of dreams. Many psychologists agree that dreaming is intimately connected to the faculty of *memory.* But how it is intimately connected with the faculty of memory has not been very clearly explained. We have stated the discovery made by the American psychologist **David Foulkes**, in the preceding section. But the explanation needs some additional information with regards to the physiological aspects 'involved in the explanation. We hope that further experimental studies may throw some clear light over this specific issue.

One recent theory states that dreaming is a kind of remembering rendered confabulatory and confused by the special conditions of sleep. But many of the recent studies are of the view that most dreams occur during rapid eye movement (REM) sleeping state, the stage of the nightly sleep cycle characterised by increased electrical activity in the brain and rapid flickering movements of the eyes. It has been observed in the experimental study that the periods of REM sleep occur about five times a night and it is observed that most of us spend about two hours every *night in dreaming.* In working out the number of hours in a year, we spend about *seven hundred hours,* that is one full month in each year. This estimated figure appears to be very much interesting for us. But it is most surprising that we remember only five percent of these dreams. This normally occurs in the last round of REM sleep. But this theory does not explain why we remember only a fraction of the dreams and do not remember most of them. But every time we dream, our brain is functioning very actively.

***The Erasing Theory* :** As we have stated above the biologists also made some contributions to the study of dreams and dreaming with special reference to human beings. They are also testing whether animals like apes, chimpanzees are capable of dreaming. One theory formulated by the biologists is the *erasing theory.* **Francis Crick** and **Graemi Mitchison,** the American biologists postulated this theory in *1983.* Their hypothesis was 'why we forget ninety five percent of the dreams?' They state that our brains are not formed to remember all of the dreams. They have suggested that sleep is the time of *unlearning period* and it is not filling in memory everything we are experiencing during the day. The brain deletes or omits the vast majority of the information as useless.

According to these authors, as they postulated this idea, dreaming is a *side effect* of a nightly review of the vast array of stimuli that are poured in the brain through the senses every day. These information are joined into the dense of neurons network, of the cortex. This network in due course become overloaded with sensory imputs, and fail to function normally. Therefore they are required to clear the irrelevant information regularly. According **Crick** and **Mitchson's** conception dreams arise from the activity of the *brainstem* by firing the neural signals at the cortex (cerebral) for erasing the unwanted materials By this means it helps for storing the permanently useful information.

***The Filing Theory* :** The filing theory is one the recently evolved theories relating *memory to dreams.* It was **Christopher Evanes,** an American Psychologist, speculated and brought the theory. He suggested that sleep t4!fchas ac active role in the memory storage. He proposed that during the time of sleep the brain scans and arranges in an order all the innumerable number of new sensory experiences we have been receiving during the course of the day. The brain files them in the memory as efficiently as possible. Now and then during rapid eye movement (BEM) sleep the brain becomes briefly conscious of the scanning and cataloguing processes and tries to interprete the information in its own normal way in the same manner we were in the awaking state and responding to the sensory stimuli from the outside world. This type of function often detaches the streams of images. These *detached images* we call as dreams.

Memory Function and Dreaming : Experimental studies were made relating memory and dreams. Dreams occur during sleep and memory storage takes place in the brain in the awaking states. Recent psychological studies on dreams find that sensory experiences stored in the brain in the form of memory have close relationship with *dreams.* **Ari Karni** and **Dov Sagi** the Israbli research investigators have observed in their experimental studies the importance of REM sleep to memory performance. They have demonstrated this phenomenon with the help of their experimental subjects. They have trained volunteers to recognise the patterns that appeared for a short period on a computer screen. Some volunteers each might were woken up in REM sleep and the other volunteers were allowed to sleep without any disturbance. In the next morning when they were all tested on the previous night's task, it was observed that those who woke up performed less well than these who did not wake up, but much to be done as to how dreaming is related to the consolidation of memories. But the process that involves is the mixing of new and existing memories with the old memories. This is the reason why our dreams often consists of parts of recent experiences mixed up with the old memories or experiences.

The Conscious Brain and Its Organization

The emergence of the biological psychology has brought out several new findings about the functions of -the brain. The recent discovery in this field is that the organisms behaviour in the environmental conditions produces some *changes in the brain.* The brain cells are mainly responsible for the bodily functions and for the type of behaviour the organism exercises in the environment. They also relate the brain to many psychological functions like memory, learning, perception, attention and becoming aware of the events going on around us, Though consciousness was not accepted by the behaviourists, as the functions of it were not demonstratable, recent authors include it in their works and interpret it with psychological perspectives. The biologists and the neuroscientists associate it with the brain and state that the brain is mainly responsible for all the conscious activities of the organism, According to them the brain is conscious of everything. For them, the *highest activities of consciousness* have their origins in *physical* occurrances of the *brain.*

It is generally accepted to-day that the workings of *conscious mind,* that is, the thoughts perceptions, emotions and the awareness of the world around us are linked with the *physical events of the brain.* The neuroscientists have just started to discover how the activity of the brain cells translates into the rich and multifaceted experience of the consciousness. It is observed that some brain cells are responsible for producing awareness of some specific things. Experimental researchers use very sensitive electrodes to record the activity of single neurons in the *brains of the monkeys.* The researchers have noticed one group of cells that become active only when a hand was passed from the right side to the left side of the monkey's visual field. The cells did not respond to the hand alone, but also to other objects moved on the same course. Thus while *consciousness emerges* from the activity of the *whole brain,* some of its parts or contents are with the localised activity right down to the level of single cells.

It is also observed that whenever *we are conscious, our* brain cells, the neurons are *active* stimulating the near by cells with regular electrochemical pulses. But even when we are deeply conscious, most of the neural activities in our brain are linked to unconscious processing. The human brain works hard even when the person is not conscious. Then what is the significance about the neural activity of our conscious state? This depends upon how the *brain is organized.* The human brain, as we know, is a *massive network* of interconnected neurons. Each part of the networks processes a specific type of information. Some of them deal with particular kind of emotions. Other section of the neurons deal with sensory stimuli, memory, language and some cognitive aspects. Within each set of *sub-net-works,* there are individual neurons which are specialised to deal with more simple tasks directed. For example, in the case of the functions of visual system the particular cells, like the *rods and cones* respond to only to a particular *colour.* The *complexity of the human mind,* thought and emotion occur, when many of these *sub-works interact.*

Francis Crick, one of the great exponents in this area had evolved a theory regarding the organisation of the brain. He has explained what happens when information is fed into this network. This may be explained through an illustration. For example, there is an object on the *left side* of a person's *visual field.* The object is a blue vase. The neurons related to this information at once fires. These involve the cells that respond to the colour blue and the cells that respond to the *verticle shape* of the by object. The person Ram is reading a book and not paying any attention to the vase. The blue and verticle neurons in the person's brain fire relatively slowly and their activities ceased at once and the blue vase does not contribute to the person's conscious state. But, however the blue vase has drawn the attention of the person. The concerned neurons get excited and fire rapidly. When the neurons fire *forty times a second,* the blue neurons find out that the verticle neurons get agitated and the two groups of neurons pool their information. It results in perceiving the *blue vase and not just* the blue colour and the verticle vase. The blue vase as long as remains, messages will continue to flow back and forth between the two groups of neurons. This becomes the object of the person's attention. If this state of synchorny lasts fifth of a second the blue vase would become a conscious perception. But if the synchorny is dissolved in very short time the perception will not reach the person's consciousness even though the person's brain has registered the information unconsciously. This type of perception is called as *subliminal perception.*

Altered States of Consciousness

The human consciousness is very alert and active sometimes. But sometimes it lays dorment and not very active, does not provide any information. In between these two states there may be varied levels with different levels of feeding the mind with information. But most of Me times we take consciousness for granted that it is the visual window through which we perceive the world. This is not taking place always. Sometimes the view changes dramatically, if the perceiving window has become distorted or fall to open. These mysterious changes are known as *altered states of consciousness.* This is not totally a closed state, but it is like a *veiled transparent state.* The contents of our consciousness are undergoing changes moment to moment. We experience several experiences everyday. One sensation causes an emotion of anger or fear or love, then a thought, then a desire and then a painful feeling. Experiences vary in effects. Two experiences are riot felt exactly in the same manner. Whatever type of experience we feel we tend to remain stable for a specific type of experience. It is like watching a play in the television screen. The characters, the actions, the backgrounds, the events all are changing, but the screen is not changed. We notice only the action and the characters are changing like our experiences, but the screen is like our consciousness. The consciousness presents varied types of experiences.

Some of the scientists are of the view that all altered states of consciousness occur when different parts of the brain fall to interact in the normal way. Generally, the various information processing systems such a sensation, thoughts and memory and so on collect their information for creating *an integrated perception.* But this integration sometimes is not complete. The reason is that there is always a degree of distortion between the components that are being processed.

Induced Altered States : Altered states of consciousness may produce varied experiences. It may be a pleasant or a frighting experiences. The induced altered states of consciousness can be also intensely pleasurable. Some persons claim that they can produce things beyond the reach of everyday thoughts and emotions. On realising the experiences of the altered states of consciousness people have developed several techniques to explore these states. By using these techniques everyone purposely alters the states of consciousness many times a day. Some call this as *mind alter technique,* This mind or consciousness altering technique is used by people in various situations. When we want to make our body calm, we take deep breathing and wait for some times or when we watches something astonishing or wonderful we put our full attention on it and to keep ourselves up, we apply this mind altering technique. But however these types manipulation of the mind do not produce intensely different quality of consciousness.

But for achieving altered state of consciousness most people required extraordinary stimuli or must learn how to make use of the brain function in an unusual manner. Historically, it is well known fact that right from the ancient period to the present time, in every culture and society people have developed their own methods of using this. Altered states are produced by making changes in the functions of the brain. The altered state can be elecited in *three major ways* by adapting some techniques and by administering some substances.

Of the *three major* techniques, (1) the *first* one is by manipulating the normal activity directly *with drugs.* (2) *The second* is by changing the brain state using the *sensory stimui* such as *breathing exercises,* dance, physical rituals. (3) *The third* technique is by using the mind itself to change its own thought paterns or other psychological techniques like *hypnosis*. In some cases different methods are combined to cause the effect.

(1) Drug Induced States

There are people who resort to some kind of drugs to avoid some unpleasant experiences, so that they can be free from sorrowful or painful thoughts. Drugs have been used right from the ancient times. They were used for various purpose like the *poison and kill* some person, for cure of some diseases, for alievating some pains and also to produce sleep, for forgetting unpleasant experiences and to produce hallucinations. Drugs provide the shortest path to a visionary state. Many of the drugs are socially unacceptable and illegal, but many people in every society use them. These drugs include *synthetic chemicals.* They are alcohol, heroin and LSD (Lysergic Acid Derivatives). These are all called *psychoactive drugs.* Apart from these, people use caffeine, tobacco. Most of them are accepted in western culture, but the *opium derivatives* have been thought to be *socially undesirable* and lead to dangerous behaviour.

Many of the drugs are very popular among people time to time and also become objectionable from legal, moral and medical points of view. We are interested in these materials from psychological point of view. The psychological effects of these drugs are that they make *changes in consciousness,* in feelings and emotions of well-being and also cause depression. When many of these drugs are taken in, the subjective effect become dominent and therefore they are called as psychoactive drugs. The person who consumues too much of opium or cocaine becomes dependent upon them and he is called as the *narcotic addict.* Generally the drugs affect a persons mood or consciousness.

Alcohol : Alcohol is a colourless flammable liquid. It is used as a stimulant by some people for their purpose. Generally, it is taken in small quantities at home or in some social functions as a stimulant. Alcohol though primarily a depressant, the initial relaxation may turn the person to anger and aggression. Later on the depressing effects may appear again in drowsiness and sleep. One of the important reasons for the public disapproval is the increased accidents in the road transports which are driven by the intoxicated drivers. But many social drinkers do not drink excessively. They drink only for the social participation. Others drink for some expected effect of the alcohol.

Generally, the common alcohol consumption is that alcohol is taken as mild tranquillizer for *readucing the anxiety or tension* the person has developed. The motives of the persons who drink may be more complex. There is often a kind of pretended courage associated with drinking. In this state the intoxicated person becomes dominating the situation and behave aggressively. **McClalland** *et al.* (1972) have made chearful studies among the moderate alcoholics for ten years. These psychologists attempted to assess the motivation for drinking alcohol among those who were not extreme alcoholics. They used the familiar method of story telling pictures for studying motivation. In this experiment they have observed that the prominent motive was to get over the *sense of weakness,* to fell stronger and more effective. According to these investigators their motives were not simple, but were complex. They were also strong and drinking did not occur very frequently in their life.

Heroin **:** Heroin is a *white odounless bitter* tasting crystelline powder derived from *morphine* a highly addictive narcotic, opium is its derivative and related substances, all collectively known as opiates and classified as narcoties, Opium is derived from the juice of the opium poppy. In has both *morphine and codaine* as its derivative ingrediants. Codiane is very mild and morphine is about ten times stronger. Heroin derived from morphine, Heroin is prohibited in many countries. These drugs are taken in various forms. Opium is used in smoking form and also swallowed. Morphine is taken in various forms. Heroin is taken either by injection or by sniffing. The motives behind using heroin are complex. It is generally used for positive purpose of well-being, The users feel that there is a *special thrill* that occurs within few minutes after taking it through *injection.* The sensation one gets from it is intense pleasure. The effect produced by heroin is different from the effects of alcohol and other drugs. The users of heroin reveal any aggression or dominance very rarely unlike the alcoholics.

It was observed that young persons who use heroin for the first time through sniffing find pleasant feelings and describe the effects in various ways. **Chein** *et al.* (1964) have studied the effects of heroin among the young persons. Some of the statements produced by **Chein** *et al.* are : "It gave me a sense of peace of mind; Felt above everything else; I always wanted to feel the same and so on''. The youth felt only the pleasant feelings which made them to forget everything in this world for the moment.

But the striking effect the investigators have observed was that there were *no changes in consciousness.* It appeared that the *conscious state* of the young men, was not distorted. But there were changes in mood, self-confidence and anxiety reduction.

LSD (Lysergic Acid Derivatives) **:** Originally the LSD was considered as the most widely used *mind-altering drug.* Later on it was felt that it has no dangerous effects. One problem realized was its inclusion of the acid. The effects of this drug was that it was highly individual and unpredictable. On taking this LSD some persons had vivid hallucinatory experiences of colours and sounds. But other users had mystical or *semireligious* type of experiences. It is also possible that some advice reaction might be caused, in some users. Even those users who had pleasant experiences sometimes had experienced this adverse reactions. The adverse effects had severe disturbances in some of the users, they sought professional help. As result of this psychiatrists and other mental health personnel have developed great deal of experience with the adverse effects of LSD drugs.

There were some studies on the adverse reactions of the LSD. **Barr** *et al.* (1972) have studied those persons who have used LSD and found that the drug was very powerful enough to *shift awareness (consciousness)* in some respects in all of them. But the effects varied according to the drug dosages and mesurable personality characteristics of each person. It was also depended upon the context in which the drug was taken.

It was observed that the drug *induced pleasure* results from the changes to chemical path ways in the brain. Some drugs affect the neurotransmitters of the brain. The precise effects that a drug exerts on the consciousness depend upon which chemical, pathways are affected and these changes interact with the subjects mood and expectations and on the context in which the drug was taken. For example, a little alcohol drink in a party produces a feeling of mild *euphoria,* but the same amount of alcohol at the end of a stressful day produce *relaxation.*

(2) Brain State and Sensory Stimuli

The brain state may be changed by inducing *sensory stimuli.* The sensory stimuli may vary from culture to culture and also it depends on the situation in which the individuals are involved. Sensory stimuli which can induce changes in brain or consciousness may be anything like seeing, hearing, testing, smelling, touching and movements of bodily parts according to some techniques.

Altered states of the mind or consciousness can be reached by using the sensory or psychological stimuli. This may change the pattern of brain activity. Vision or seeing some highly exciting objects for sometimes naturally may influence the mind and consciousness. Man can produce or create wonderful things by virtue of his creative intelligence and also discover the objects or things or a new land which were not known previously. Any attention on concentration on such things generally cause change in the patterns of brain activity. For example, a spectacular natural landscape, a beautiful painting, arromantic or a sorrowful event in a play are all visual stimuli which can bring out *changes in the brain* activities and subsequently there might be alteration in the consciousness. This might also make changes in the behaviour of the individual.

The sensory stimuli, *sounds* might attack the *sensory apparatus,* the ear. The sound stimuli may be in various forms. *Music* is one of the important stimuli which might arrest the attention of the people very easily. A musical recital performed by a musician which his her seductive voice might *lull* the listeners or audience which may produce different effects in the minds of the people. The person might go with nodding his head and speaking to himself about the sweetness of the voice of the performer. This incident certainly have some change in his mental activities and also might alter his consciousness. (Fig. 11.1).

With regard to the sense of *taste,* we know that everybody is interested in eating tasty food. A birthday dinner of a person, prepared by a skilful cook with a tempting adour makes a person, to enjoy it by eating. The person may be even thinking about it whenever he goes near his home dinner table. The taste might have been highly fascinating and he may even go in search of such dishes in hotels for eating. A delicious dish always goes with *taste and smell* and there are people who give more importance to it which may cause changes in the person's patterns of brain activities (Figs. 11.2 and 11.3).

The *sense of touch* is a sensory stimulus which is highly realized by individuals. The child is highly sensitive for touch. In human body there are certain parts particularly the erotic zones of the body are highly sensitive for touches. Once a person begins to enjoy the pleasures of the touches in the erotic zones certainly there might be some changes in the patterns of brain activity and subsequently in his behaviour and consciousness. (Fig. 11.4).

Fig. 11.1 : Brain state and sensory stimuli : Sound Stimuli. Sound stimuli may be in various forms. Music is one of the very important sound stimulus. It may be Vocal and Instrumental and both the types have the power of arresting the attention of the people and may produce changes in the mental activities of the people and might alter the person's consciousness. Persons who are talented in some area of music take it up as a profession and spend their whole life on it and also make it as a service to human kind.

Fig. 11.2 : Brain state and sensory stimuli : Taste stimuli. The sense of taste is found in the food we eat. There are different types of taste. Man is very fond of eating tasty food. Festival, wedding and birthday dinners make human beings to be very conscious of eating tasty food. It may have some influence in the mental activities and also alter the consciousness.

The *movement of the body* to a persistent rhythm, when shut out all other thoughts, encourages the release of the *dopamine acid in the limbic system,* which gives similar effect of certain drugs. One common technique for this state is the *dance.* This depends upon the context or situation and the mental state of the dancer. The result of the movements of the body according to the rhythm may be *hightened awareness.*

In every culture dance is primarily a social event. The cheering effects are more to do with the ability to go beyond the social barriers. In some cultures the group dance is often performed as spiritual. In Hindu culture, God, Lord Siva is symbolized as the originator of dance. It is because of this the dance performances, *specifically 'Bharatha Natyam'* is always associated with Siva. Whether it is a dance performed by a single person or a group dance

Fig. 11.3 : Brain state and sensory stimuli : The smell. Smell is one of the important stimuli, which is mostly associated with the sense of taste. A delicious dish always goes with taste and smell. People give always some importance to these two sensations which may cause changes in the patterns of brain activities and also may alter the consciousness. Recent researches show that the sense of smell can lift the human spirits and can make a man sexy and also can reduce his weight.

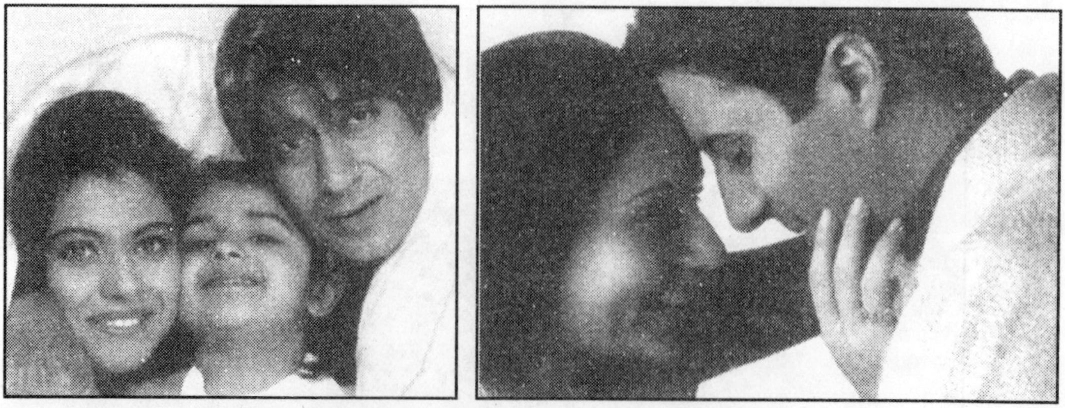

Fig. 11.4 : Brain state and sensory stimuli : The Touch. Touch is a sensory stimulus which often causes people to have a feeling of it. The child is highly sensitive for his mother's or father's touch. In human body there are certain parts which are highly sensitive for touches. If a person has experienced the pleasures of touch once, he wishes to have it often and enjoy it. This may have changes in the patterns of brain activities and also it may alter his consciousness.

the same procedure is followed. *Shamanic rituals* involve dancing a *repetitive rhythm* for hours and days, long enough for the body to know the music without having it to hold in mind. But movement becomes automatic, leaving the conscious mind free to become aware of the messages from the spiritual self. The dancer in this state often experiences *a separation* from his body and may feel as if he has moved into on alternative realitives, Then the movements of the body produce change patterns in the brain activity (Fig. 11.5).

Fig. 11.5 : Brain state and the sensation of movement of the body. The movement of the body to a persistent rhythm, release the dopomine acid in the limbic system which gives similar effect of some drugs. The rhythmic movement of the body increases the awareness of the dancer resulting in the changes in the patterns of brain activities and also alters his consciousness.He or she feels as if gone into the another world. Dance is the most accepted form of bodily movement which is practised in every culture in a more sophisticated manner.

(3) Using Mind and Psychological Technique

Rituals generally make the mind calm. Some cultures have developed rituals in such a way that the rituals, extend the effect and lead the participants into altered states. The *Tea Ceremony* for example, in *Japenese culture* is intended to induce relaxation and concentration. Many of the religious services are ritualistic and designed to free the mind to put in awareness of God. Some rituals dispense with physical actions.

Meditation **:** Meditation is another state in which the individual goes through a scheme of familiar mental behaviour to slow down the frantically rushing thoughts and the sensations which are similar to waking, the *waking consciousness.* This type of behaviour is known as *meditation* which is in practice in several forms. Meditation is an integral part of many religions. It is a practice that is used to employ the mind of *distracting thoughts.* Hinduism and Buddhism make it as an essential aspect of prayer. The priests of these two religions see it as the only way to achieve *spiritual enlightentment.* **William Blake** (1757-1837) an English poet states that if the doors of perception were cleansed, everything would appear to man infinite, represent the goal of many meditators. Blake was a poet, painter and mystic. His literary works include, "Songs of Inhocence and Sings of Experience."

Meditation first begins with focussing on the *breathing,* The simple way to do it is, to sit in an empty darken room with the eyes closed and for about twenty minutes silently pin down all attention to the act of breathing, It is important that the meditator should overcome the tendency of the mind to wander. This may help him to think of the words, *in and out, while 'in' stands for 'inhale' and 'out' stands for 'exhale'.*

Meditation is also used as a *therapeutic technique* to purify the mind in the *altered state* and also used for some *physical illness.* The *transcendental meditation* founded by the Indian Saint, **Maharishi Mahesh Yogi** became very popular at one time all over the world. *Trancedental meditation* is a form of meditation practiced by some who follow the Maharishi's Yoga. The meditative state is induced by repeating a particular sound or phrase called *Mantra,* over and over again. Each individual has his own *Mantra* selected as the most appropriate for him.

Herbert Benson, an American cardiologist, studied it and introduced it among his clients, the transcendental meditation (TM). They used the *mantra* a special word to concentrate their attention in meditation. The devotees meditate half of an hour, thrice a day. **Benson** observed that during meditation his subjects heart and respiratory rates were lowered, blood pressure lowered. During meditation his patients were *extremely relaxed,* awakened a state Benson called it as the *relaxation response.* His further studies indicated that meditation could *modify* a person's *reaction to stress.* Meditation also reduced headaches and chronic pains. Thus meditation is found to be a good therapeutic technique. Meditation in the name of *Yoga* had its origin and development only in India some thousand years ago. It is a technique of concentration of the mind on some specific theme. (Fig. 11.6).

Fig. 11.6 : Brain state and meditation. Meditation is a form of mental state in which the individual goes through a scheme of familiar mental behaviour. In this state the person who performs the act of meditation slows down the frantically threshing thoughts and the sensations which are similar to the waking state, called as *waking consciousness*. This type of behaviour is known as *Meditation.* Meditation is in practice in several forms. This is generally treated as part of religious ritual. Meditation is practiced to control the distracting ideas and thoughts and is an essential aspect of prayer. Exponents are of the view that it is the only way to achieve the spiritual enlightenment.

Meditation in some form is applied as the therapeutic technique. Maharishi Mahesh Yogi through his Transcedental meditation has demonstrated that it can be used to deal with some of the physical and mental illnesses. Meditation generally reduces stress and depression. Meditation, if applied according to the principle, it may fairly deal with many of the psychological illnesses.

It is possible that people can think and dream and in its highest level e they can *transcend* the everyday world. In the olden days in India people have isolated themselves by going to the mountain areas, lived on *roots, seeds are leaves* and indulged in some type of special exercises in breathing and in other physical exercises. Some lived in *herbs* and lived for long years by practising (Yoga) meditation. Sometimes these persons were called as Saints (Sanyasis) and seers. Their main task was to mediate all the times and discover the truth and practice and preach among their followers. In this wav the Indian culture has produced a community of persons, whose main task was to teach how to meditate and enjoy the benefits of it. Even no'", there are saints live in Himalayas in north and in the mountains of western ghats with the names of *saints and siddnars.* Many of the Indian medical treatments are based upon the Ayurvedhic, Siddha medicines, discovered by these meditating saints and seers. This caught the attention of the present day people and they too started practising the meditation and carried it to some of the foreign countries and taught them the benefits of the meditations. Thus it has received much better attention and became a modern technique. New and exciting interpretations were given to them. Another important characteristic of it is that it is linked with spiritual or religious values.

When it was spread out to foreign countries it was also subjected to experimental studies. Meditation can be carried not under various instructions and they fall under two main types. *The first* is an *opening-up meditation.* In this first stage the person clears up his mind for receiving new experiences. *The second* is the *concentrative meditation.* In this state the benefits are obtained through activity attaching to some objects, words and ideas. These two can be illustrated as given by **Navanju** and **Cornstein** (1971).

First of all the exercises are for relaxation and for *controlling breathing,* It is carried out by the person by sitting on kneeling on the floor. There may be also cushion for support. In this position the person must be in a position to relax himself without inducing sleep. It is generally called as *Yoga.* In this stage of yoga the *eyes* must be *closed.*

The Opening up Meditation : This approach starts with *a resolution* of not doing anything, not thinking anything and no any effort of his own for anything. He has to relax his body and mind completely and stopping all the ideas and feelings arising in his mind and body. According to **Chaudhuri** *(1968)* the meditator should not allow his mind to have any idea enter in his mind.

In one of the experiments, it was tested what takes place during the act of meditation. In **Deikman's** (1963) study the subject was made to sit in an armchair and at a distance of eight feet a vase of ten inches height was placed and the person was asked to see the blue vase as it appears without any connections to other things. After starting the session a number of sounds by tape recorder were produced for testing the persons ability to avoid distractions. The sessions varied first with *five minutes to fifteen minutes* and continued for three weeks. The common effects were, an *altered state of intense perception* of the vase, conflicting perceptions, decreased effectiveness of the external stimuli and a pleasurable state that the experience is valuable and rewarding. Thus meditation is found to be beneficial and rewarding if it is carried out with all sincerity. It takes away the person from this *complex world* to a state where there is *free mind* seeking nothing but *truth.*

Generally, meditation involves focussing the mind on a thought or image, or sensation like breathing or an imagined sound such a *'mantra'.* The aim of meditation is to free the mind from the content, so that only a *pure awareness remains.* It is very difficult to achieve this state of mind because the human mind is designed to think around constantly. Meditation is an act of thinking deeply on something. It is contemplating on some valuable matter with the view of registering it in the mind. In the meditating state the person sees the world with *many dimensions.* Meditative state often takes the meditator outside the everyday world. Brain scannings show like other mental states, there are distinct patterns of brain activities. In the normal state the attention awareness is very active. But in the state of meditation *brain has less activity* in the *sensory area* of the cerebral cortex. Specifically the *parietal lobes* which processes the information about the body is more active in the parts of the frontel lobes that are concerned with attention.

Hypnosis: An Altered State of Consciousness

Generally, hypnosis is a process of inducing artificial sleep. It is one of the techniques of inducing altered state of consciousness. Hypnosis is a subject that has brought out controversies among the experts. Some are of the view that it is a *trance state* produces genuine changes in the brain activities. Hyprotism was founded by **Franz Mesmer** (1734-1815), an Austrain physician in the year 1780. Mesmer, with his demonstrations of magnetic influences believed that he could cure diseases. Hypnotism has fascinated people ever since Mesmer has introduced it. The high society in Europe was very much excited with his displays of control over the minds of the people.

What is hypnosis? Hypnosis is considered as a special level of consciousness. It is observed that under hypnosis we loose ourselves in vivid imagery and behaviour to some one else. The present day hypnotists use a simple conventional technique evolved by an American *psychotherapist* **Milton Erickson** in 1940s. The term hypnosis is derived from the Greek *word Hypnos* which *means sleep*. The metaphor of sleep is commonly used in inducing the hypnotic trance. **Pavlov** (1849-1936), the Russian physiologist, was very much interested in his later years and wanted to apply this to animals to see whether animals could be hypnotised.

Pavlov considered hypnosis as a partial sleep. *Hypnosis* can be defined, as an artificially induced state of *semi consciousness,* characterised by an *increased suggestibility* to the words of the hypnotist used, clinically to several unconscious memories. Hypnosis is also stated as a **trancelike state** of behaviour resembling sleep characterized primarily by increased suggestibility and induced by suggestion.

Increased Suggestibility is one of the important aspects of hypnosis. The hypnotised subjects generally seem to be much more open to suggestion from the hypnotist than they would be in their waking state. Another aspect of hypnosis is the enhanced images and imagination. How this functions in a person? It is observed that the hypnotized subjects are agreeable to imagine virtually the sensory experiences that were suggested to them and also report that they are able to retrieve the images, even from childhood period accurately. These are some of the characteristics of the hyphotized subjects.

Some of the hypnotized subjects become annoyed when asked to do some planning on their own. But many subjects accept the suggestions that an animal is talking to them or some one is present in their room, when there was no body actually there, In the *trance state* the subject sees the world differently from what he is when he is in a waking state.

Hypnotic Technique : The psychotherapist who uses the hypnotic technique to deal with the patients begins his task with a short conversation with the client to establish a rapport. The hypnotist directs the client's attention inwards by asking simple questions like, whether his/her heads feel heavy. As the client is engaged in this thought, the hypnotist starts to assert his control on the client by suggesting that one hand is now very light that it will rise. Registering this idea in his mind the client may feel that it is correct and makes his hand to float upwards. The hypnotist at this stage can move on to more elaborate facts of suggestions, making the client to regress into his *past life*. To test the client's level of acceptance of his suggestion the therapist can tell the person that a beat-root a day is better than an apple. The client may imagine what the hypnotist has said, is responded by the client, thinking that it must be correct. When the hypnotist observed that the client is responding to the suggestion, he further goes on in his task of probing the client's inner life.

Who can be hypnotized? : An experienced skillful hypnotist can hypnotise successfully more people readily. But some persons who are highly critical of the hypnotizers action may not be susceptible for it. There is some doubt and uncertainity that all people can be hypnotized. There are also some reports that some persons underwent the hypnotic situation many times without any effect, but at last the hypnotizer has achieved his goal.

A scale has been designed to measure the hypnotic susceptibility of the persons by **Morgan** *et al.* (1974). The available scale for measuring susceptibility to hypnosis has more precise statements which could determine the persons who could be easily hypnotized. The person who responds to the statements most frequently like the hypnotised person scored high on the scale, is most susceptible to be hypnotized. The assumption is that those persons who score high often on the scale are having the same characteristics as the hypnotized persons. A follow up study was made on the same subjects who had taken the scale, after ten years and it was found that the correlation was +.60. between the earlier scores and the later scores. **Morgan** (1973) has observed that there was greater level of *similarity* of scores in the case of *identical twins* (monogygotic). This shows that the hereditary components may have the ability to be hypnotized. The early experiences of children have certain level of susceptibility for the induce of hypnosis. Children have the tendency to escape from the parent's harassment and go into the world of imagination and adapt a kind of dissociation which may cause to be susceptible to be hypnotised.

Theories of Hypnosis

Right from the time of **Mesmer** to be present time hypnosis has been used as a therapeutic technique in various forms. Mesmer who evolved the technique of hypnosis, used it first for treating hystrical patients, Mesmer helped many persons to get rid of their hysterical problems. Mesmer considered the *hysterical disorders* as strictly physical. He was generally considered as the earliest practioner of hypnosis. Since Mesmer was the first to apply it and also he was the founder of hypnosis, it is also called *after his name as Mesmarism.* The centemporaries of Mesmer once regarded Mesmer as a *quack* (an unqualified person) but later on when came to understand the values of hypnosis, started respecting his works and adopted his technique for many disorders. **Mortin Charcot** (1825-1893), a great French Neurologist, following Mesmer, studied the hystrical states, anesthesia and paralysis, blindness, deafness, convulsive attacks and the loss of memory caused by hysteria. Following these studies several physicians and psychiatrists, neurologists adapted the same technique. **Joseph Breuer** (1842-1925), an Austrian physician treated a young woman who had a number of hystrical problems. Breuer adapted a method called *catharasis.* Later on **Sigmund Freud** (1856-1939) has developed this method. Thus hypnosis underwent several changes and modifications while applying it on hysterical patients.

The earlier forms of hypnosis were to induce sleep in the patients and artificial sleep was produced by hypnosis. On practicing hypnosis for several years, gradually it under went certain modifications. It is now largely discredited by the physiological studies which bring out the difference between hypnosis and sleep and by the facts of *hyper alert hypnosis.*

A theory is evolved from psychoanalysis, which indicates that hypnosis is a state of partial regression in which the persons' impulsiveness and fantasy are revealed. This is a state in which there is control of ordinary consciousness of an adult person. **Gill** and **Brenman** (1959) have observed this state.

Another theory is also drived which is called *sociological* and social-psychogical theory of *role behaviour.* According to this theory the individual is seen playing different roles in ordinary behaviour, so that the person is like adapting the roles of some other persons like the mother, a teacher, a students. In this each one has his own needs and demands. This is how **Subin** and **Coe** (1972) explained the theory, but this theory does not seem to have hypnotic suggestion or character.

A theory has been very recently formulated which is a modified form of the earlier theories. It is called *Neo-dissociation theory.* **Hilgard** (1973) has described it in a different manner. It is observed that in each of those theories attention is found to be a significant feature of hypnotic behaviour. The method generally applied in studying the dissociation in hypnosis. The subject has to perform arithmetical operations and to write out the computations and at the same time to perform some other task, such as reading something loudly. In this study the subject is *not aware of one part of* his activity, the *arithmetical computations* and fully aware of other parts. How can an intellectual task, the arithmetic work, the voluntary behaviour, the writing, be carried, on out of consciousness, while some other conscious task is performed? According to **Hilgard** *(1969)* at a subconscious level there may be a process that is very much like consciousness but it is not available to observable consciousness. This is called *hidden observer,* an expression is instruduced as a metaphor to explain this hidden part that is *aware of things* that are not *open* to the *ordinary consciousness* of a person. The nature of this *hidden observer was* exposed in small experiments in which pain was hypnotically controlled.

There are many persons, regarded as characteristics of hypnotized persons. In all aspects of human behaviour we must know that not all people manifest all the characteristics in the same way. But it is also possible for the subjects who may appear to be deeply hypnotized not to show all the noticeable indications at any one time. There are long historical evidences right from the times of **Mesmer** in applying this therapy for hysterical disabilities. The subsequent works were performed by Characot, Janet, Breuer and Freud. During the same period there were also other works on hypnosis. The Surgeons were *mesmerizing* the patients *to get rid of pains.* **W.S. Ward,** a British physician, in 1842 amputated the leg of a patient by hypnotizing him. It appeared that the patient had not felt anything during this operation. But it is **James Braid** (1795-1860), a British physician, who was mainly responsible for coining the modern term hypnotism. He was also using the hypnotic technique for reducing pains in many persons. Thus the hypnotic theories were functioning in various forms. There is also *self-hypnotism* in use at present.

What is Self-hypnosis?

Self-hypnosis is a process of making autosuggestion. The person need not go in search of hypnotist to benefit from hypnosis. It is quite easy to put oneself into a relaxed state or hypnotic state. The person who wishes to have self-hypnotized should find a quite undisturbed place and ideally sit rather than lie down to prevent himself falling sleep. He can plan the kinds of suggestions he wants to practice. For example, a specific goal such as *giving up smoking* or *drinking* or *releasing tension* or anything the person should consider as important, may be included in the list of goals he wants to achieve. The pet-son can take three or four slow, *deep breathings*, holding his breath for a couple of minutes, then letting his breath go with the word *relax*. He has to focus his attention by concentrating on a print on the wall in front of him. He must speak to himself that he is going to take three special breaths. The first is to relax, the second to move towards hypnosis and the third to be in hypnosis.

After the breaths, he releases all the tensions from his body by systematically *releasing the muscles* from his toes to the top of his head. Now, the person can imagine a staircase with eighteen steps leading to a special place. Then he can count down from eighteen to one as he descends the stairs. Now he can visualize -his place and try to experience as to how it sounds, smells and feels. Now he can begin his passive suggestions relating to the goal of his self hypnosis. When he has finished the counting himself, back up from one to eighteen speaking to himself, that he has enjoyed a wonderful session or relaxation and that when he emerges, he may feel relaxed.

UNCONSCIOUS

Introduction

Any discussion of consciousness is not complete without making some reference to unconscious as it has been receiving greater attention in Psychological studies than consciousness

The terms conscious and unconscious are always associated with our thoughts, feelings, emotions and imaginations. We are conscious of many things around us in our everyday life, but we. are not aware of many things, that are taking place in mind. When we are not able to understand something we leave it aside and we do not care to pay any attention on them. But however we do not keep quite. We are trying our best to understand what was not understood earlier. Philosophers and Psychologists have been very curious of things and events occurring around us. In their search for the unknown they find some old stories which they called as *myths and legends*. Statesmen speak all about myths of a nation's invincibility. Scientists also pay their attention on primitive myths of sun, moon and gods with their astronomical knowledge. The *Greek* philosophers, **Plato** and **Aristotle** also spoke about myths, but did not give much importance for them. They placed *reason and logic* above the myths. Logos, the Greek word means logic which refers to rational thoughts and myths means *fictitious*, a narrative or stories involving supernatural beings like gods and spirits. Psychologists are of the view that myths (Mythology) and legends have played an important role in the life of people and thought process. According to them mythology is true and cannot be rejected as false. It is the collection of life events and thoughts of the people of several generations. Some psychologists are of the view that myths and legends are cultural heritages, *unconscious information.*

Carl Gustev Jung (1875-1961), a Swiss psychologist is of the view that all human beings carry the *potential ideas* or *archetypes* in their *unconscious* mind. Collective unconscious, according to Jung's theory, is a part of the unconscious mind incorporating patterns of memories, instincts and the experiences common to all humankind. Archetype refers to a perfect typical specimen, an original model or pattern of images. The collective unconscious is expressed in myths that have been repeated in various forms all through history. The basic themes are found in mythology for many of the *present day cultures and histories.* The cultural and social norms and ways of life of many vanished civilizations I 6re found in myths and legends. Thus **Jung** found the source for his archetype and collective unconscious in myths and legends and asserts his view by quoting the *Greek epics of Homer.*

What is Unconscious?

The term unconscious has become very popular and had drawn the attention of the psychologists only after **Sigmund Freud's** works were published. Though the term was in use among of people of various types, they did not go deep into the functions and manifestations. Generally, unconscious is a *state of unawareness* without any sensation or thought. In psychoanalytic theory it is the part of the personality, in particular the *id impulses,* or the id energy of which the *ego is unaware.* It is also defined as a state in which there is a lacking of normal sensory awareness of the environment, insensible, not aware of one's action, or behaviour. To be in precise, our unconscious minds are not known to us. But if we could think deeply over our unconscious mental processes they are no more unconscious. At times we all might have experienced the forces that our unconcious is exerting on our conscious thoughts and behaviour.

Sigmund Freud during the course of his practice often experienced some type of mental processes that have something with the patient's deep level of the consciousness. This experience let him to discover what is below the conscious level. It is the unconscious mind. At first he felt that it is a dark, seething mass of repressed sexual desires. But other thinkers built their theories *involving mystecism* and *surrealistic ideas.*

But present day psychologists view it in general that unconscious is involved in much the same things as the conscious mind. Some are of the view that the contents of unconscious are conscious of late psychologists seem to agree with Freud that our lives are ruled by thoughts and feelings of which we are unaware. We are not conscious when we are in deep sleep. When we are in this state, it is impossible for us to describe anything. It we want to know what is going on inside our mind we will have to look inside with our *inner ego.* The eighteeth century philosophers like **David Hume** (1711-1776) Scotish empiricist, and the English philosopher, **John Locke** (1632-1704) were of the view that all mental events like thinking, feeling and knowing are. taking place in our state of awareness. The German Philosopher, **Goethe** (1749-1832) is of the view that man cannot persist long in consciousness state; he must throw himself back into the unconscious, because his roots live there. Therefore, we are brought to a state to believe the -unconscious and to acknowledge that our *lives* are *ruled by unconscious motives.*

The person who has taken the first step to expose this phenomenon was the Austrian physician **Franz Anton Mesmer.** While experimentally applying *hypnotism* in his patients at the end of eighteenth century in Paris, Mesmer demonstrated that *unconscious thoughts* might be exerting some influence on the body which might be producing *Paralysis* or a state of *insensibility to pain.* His work was later on undertaken by the French neurologists, **Jean Mertin Charcot.** He used *hypnosis* to treat hysteria. It is a medical condition involving a combination of psychological and physical symptoms such as paralysis, blindness and amnersia. During this time Charcot was associated by a brilliant medical student Sigmund Freud. Later on Freud elaborated this theory with the cooperation of Charcot. Freud observed that the symptoms of hysteria were a hidden source of keeping emotionally charged memories under the mental control in the *unconscious mind.* This discovery let Freud to the development of the technique of psychoanalysis and to gain access to the unconscious and to find out curing methods.

According to Freud the unconscious thoughts are responsible for much of human behavioural problems. Apart from these, Freud strongly believed that the respectable members of society were all driven by *extreme sexual desires* as the result of the unlimited free will. These views of Freud were mostly accepted and became highly influential among those who were treating the hysterical and neurotic patients. The theory of unconscious as viewed by Freud brought an awakening among the psychiatrists and psychologists.

Psychoanalytic theory was not based upon experimental studies. But with all this it had profound influence on psychological thinking. The basic assumption of Freud's theory was that much of man's behaviour was determined by the innate instincts that are largely unconscious. Unconscious process, Freud meant thoughts, fears and wishes of what a person is unaware, but they influence his behaviour. Freud believed that many of the hidden or punished impulses of childhood period were driven out of awareness or from conscious level into the unconscious level, from where these impulses still affect the behaviour. Freud states that unconscious impulses find expression in dreams, slip of the tongue, mannerism and in the symptoms of the neurotic disorders. The unconscious impulses find expressions through socially approved behaviour patterns like fine arts, painting, music, dance writing and in scientific activities. Freud called these as sublimated forms of expressions.

Freud believed that all the actions of all human beings are based upon some causes. But some of the causes are often uncurious motives rather than the rational thoughts, the person states for his behaviour. Freud believed that man is driven by the same basic instincts as animals. This tendency can be seen in the sexual and aggressive behaviours. Man often struggles against the society that puts some stress and barriers and control over his impulses, which are constantly trying to fulfil the desires.

Modern Trends : Freud's unconscious theory is mostly accepted by the modern thinkers and psychologists. Though it was not accepted in the beginning on the ground that it is dealing with man's inner life and nobody wants to reveal his inner secrets. But this disagreement gradually disappeared and modern psychologists and others began to explain many of the human behavioural patterns in terms of Freud's unconscious. Generally, every matter we think and speak out is informed by our unconscious processing. The human emotional reactions, beliefs opinions, and even more carefully thought and evolved decisions are determined to a large extent by the mental processes out of our conscious awareness, which we may state as unconscious.

Solving Problem : One of the most fascinating aspects of the unconscious mind is the way in which it can solve problems, while the conscious mind is occupied with other matters in which the person is interested. In our everyday situation there are many problems solved by our unconscious mind. These are, some well documented cases where these persons have let to some important scientific break throughs with inventions and discoveries and aesthetic imaginations. found to be expressed.

Dream is the manifested content of the unconscious mind. Though most of the dream contents are suppressed sex desires that are not fulfilled in real life situation, they appear in some other forms which enhance in solving some life problems which are not ordinarily solved. There are several famous scientists, and inventors and experts who have solved problems through their dreams. The life incident of the German chemist, **Fiedries Kekule** tried to explain the chemical structure of *benzene* in 1860s, Kekule knew that *benzene* was an organic component which is made up of linked *carbon atoms.* He also had known A& formula. But its structure and the way the items were linked together were not clear for Kekule. He has stated the events related to this problem. One day when he was sitting and writing a book, he could not proceed further because his thoughts were elsewhere. Then he turned to the fire side and found that the atoms were moving before his eyes. He felt that his mental eye served more exactly and there was repeated visions of the kind that could distinguish larger structures of manifold confirmation, long rows, more closely fitted together, all were serving functioning, making like motion. He has stated that it was like a *flesh lighting* and he was awakened and spend the rest of the time in working out *to complete* the hypothesis he formulated earlier and he was successful.

Kekule was inspired by his dream and wondered at the idea flashed in his-mind was whether the benzene molecule was a *ring shaped, like a snake holding its own tall in its mouth.* When he tested his hypothesis, (the idea fleshed in his mind) he found that it was suited to all the known facts about the benzene and its chemical characteristics. In this way kekule's discovery has made an excellent contribution to the development of *Organic Chemistry.*

There were also other inventors and experts who have been benefited from the unconscious contents solved their problems. There were many artists and musicians have been inspired by dream contents of the unconscious. **Thomas Alva Edisim** (1847-1931) an American inventor did wonderful things by inventing more than one thousand scientific instruments. lie often used to sleep in his work bench holding instruments in his hands. When the instruments fell down they wake him up and he would recall his dreams and often directed him to some new inventions. The ideas stored in his unconscious mind which were not successful in waking state, found solution in the dreams and he successfully completed a new invention every time.

Just like the scientists, inventors and explorers, men in the literary fields also revealed their dreams of their unconscious contents in the forms of poetry, epics and novels. **Robert-Louis Stevenson** (1850-1924) was one of the most highly celebrated English novelists who brought out most of his writings, mainly novels, from. dream contents manifested from his unconscious mind. Thus modern psychologists take into consideration the unconscious mind as one of the sources for the solution of some obscure problems.

PART FOUR

LEARNING AND FORGETTING

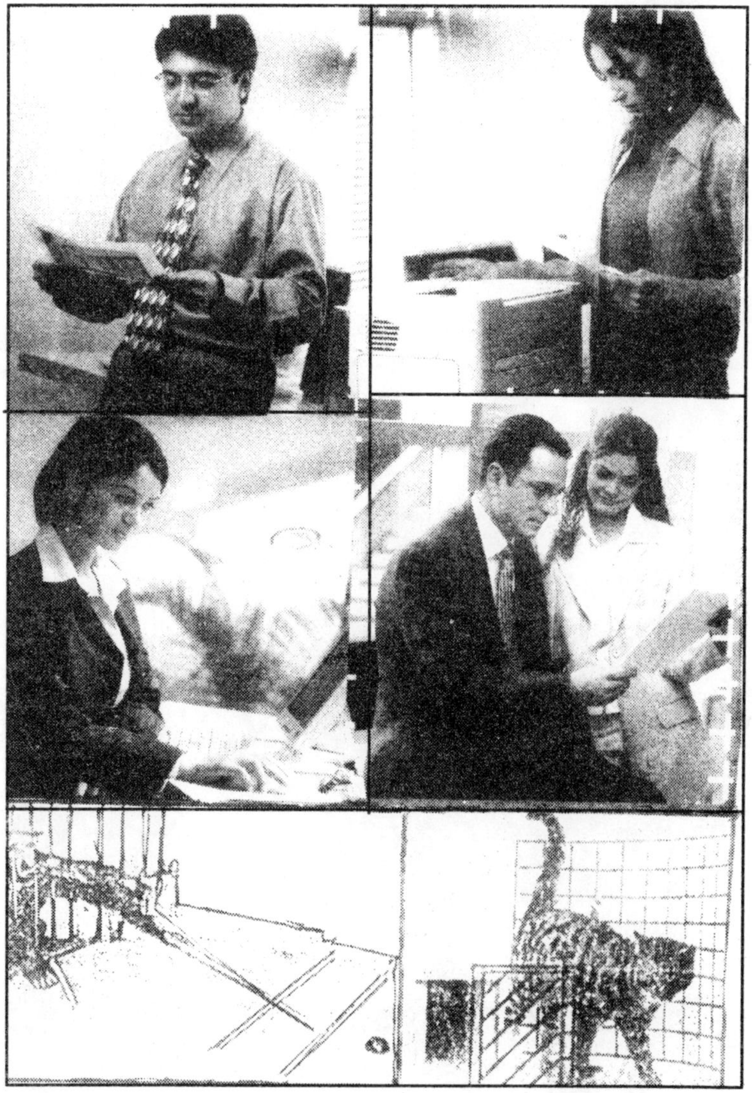

12

LEARNING

Introduction

Every scientific study now and then is undergoing vast changes on the basis of the outcomes of research findings, inventions, discoveries and creative thinking and the phenomenon of learning is not an exception. Learning does not mean the simple learning of the alphabets and symbols in schools. It is something more complex and intrigue. Learning and cognition are the fascinating focal points much of the present day research. A few of the recently developed sciences like biological psychology and physiological psychology are showing very much interest in learning and memory and their biological mechanisms and they are becoming most active and exciting areas of behavioural sciences.

We know that all the distinctively human aspects of our behaviour are learned. For example, the languages we speak, how we eat our food and drink water, how we dress, work and live, meet people and have conversations are learned. But much of a person's individuality in all these depends upon learning and memory. Researches are going on all the times on learning which provide us a better understanding of things around us and the matter we deal with. Every aspect of behaviour requires learning. How we perceive the world, and how we perform skilled activities, achieve our motivation and the ways of winning our goals, all are based upon our level of learning.

Animals also depend upon learning. They learn how to obtain their food, protect and escape themselves from dangers. Many animals learn how to mate and how to treat their young ones effectively. Very much useful researches on learning are being done at all levels of organization of the nervous system of various types of organisms. It has become important to find out the neural regions involved in learning and researchers are using a variety of methods and techniques, including studying the person with impaired memory and non-invasive brain scans of people of normal abilities.

It has become very essential in these days that if we want to understand behaviour we must study the neural and biochemical processes that are affected when an individual person learns. *The important* doubt that raised is whether there is a common mechanism of physiological events may characterize all learning, irrespective of the environmental and internal conditions that initiate the process of learning. Psychologists of the present generation are very curious whether there can be a common psychological and physiological process that could be assessed in different experimental situations. Search has been made to find out the specific physiological process that may be responsible for learning without paying much attention on different mechanisms. In this attempt it has been observed that there are diversity of experimental findings.

For example, the Russian investigators have followed the Pavlovian model of working with simple reflex responses and did not go further. The American investigators, psychologists and the physiologists followed the Watsonian tradition and also used much more complex learning situations such as approach -avoidance learning and operant conditioning. Some behavioural psychologists have used very simple learning tasks. The German psychologists adapted the Gestalts' phenomenon and the configuration techniques in their experiments on learning. But this technique is not acceptable to physiological psychologists. Because there is no experimental evidence. It is only an inferential result. In learning, it is also observed that there are further complications in the experimental studies on learning. The acquisition of new response in learning is much affected by the motivational, sensory and motor variables. These variables function frequently without any control in learning experiments.

The issue raised on this matter is whether this is having any effect on learning or it is an indirect effect on, the performance. But the investigators are of the view that unless some significant effect is observed in learning we need not take them into consideration. Since learning is a cognitive process it is very difficult to understand the changes that are taking place in the nervous system and in the other internal organs. Whether an organism, has learned something or not will be known only when it is manifested in a relevant situation. For example, when a student recites a piece of poem, indicates that he has learned the poem, a factory worker skillfully performed a task, a foreman completes a lathe-work accurately, all indicate that some type of learning has taken place.

What is Learning

Every living organism is showing some change in its performance or behaviour. Change is a general concept and basic to psychology. In determining the performance or behaviour of an organism the change is measured. There are two types of changes. The first is the change that takes place in the *sensory apparatus,* for example, in the eyes, ears, tongue, nose and skin. Change in these leaves permanent experience in the organism. The other change is independent of the sensory apparatus, and does not order any change in them but makes change in the *organism's* behaviour. *This change is called learning.*

It is true that all living organisms both animals and human beings are constantly put through a long process of interaction with the environment. All those interactions represent various things like fulfilling the basic needs for survival, finding great amount of pleasure and happiness and avoiding painful or dangerous situations. Whether the organism is to survive or to enjoy the life or to avoid the danger it has to learn to do it successfully. The organism's effort to do any one of those tasks successfully come under the heading of the process of learning. Most of us think that learning is a process of acquiring knowledge as something that takes place in schools and colleges and simply sitting and memorising certain passage or a piece oil' poem. But several kinds of learnings are taking place in us all the time.

What most of us are doing? What we axe not doing? What we are and what we are not? What we know and what we do not know? are all the results of learning. Many of our everyday 'activities like walking, talking, reading, writing, memorising a poem, typing a letter, operating a machine,— opening a door, playing a piano, driving a motor-car or an aeroplane, riding a bicycle are some of the clear illustrations of learning. We learn that fire would burn our fingers and live-electric wires would give us shock and therefore we learn to avoid touching them. In this way, we learn to avoid all possible dangers to our life, understand change of weather conditions, predict the election results, learn to adjust ourselves with others and also learn to plan our future prospects. Many of our interests, likes and dislikes, beliefs, preferences, attitudes and prejudices are all learned responses. On the whole life itself is a continuous process of learning. We learn to speak a language that enables us to communicate with others.

But the results of learning are not always doing good for us or useful for us. People who often learn more and more are also found to be unhappy. This is the reason why there are so many adjustment problems in our society, particularly among the educated people. We learn to hate people belonging to different culture, race or religion. Therefore, the psychology of learning plays a continuous role in the life of man whether his learning helps him or harms him in his interaction with the world in which he lives.

Learning is not subjected to direct observation. It is an inferred process. Learning may be defined as the process of modification of behaviour as a result of experience. Learning may also be stated as the modification of behaviour in the organism as a function of its interaction with the environment. Learning is the process by which the organism becomes able to respond more successfully or adequately to a given situation. Learning is also defined as the process by which the organism acquires experience and practice which modifies its behaviour permanently. The term learning is used in a very broader sense, covering every modification of behaviour and perception to deal with the requirements of the environment from learning simple habits like waving the hand to mean good bye to the acquisition of complex skills like playing a piano or driving an aeroplane and to formulating a philosophy of life. Man finds resources to all these things only through the process of learning.

It is very difficult to formulate a definition of learning involving all aspects of learning. But a working definition could be given in this way: learning is the process of reorganisation of an individual's perceptual, cognitive, motivational and emotional structure which functions as a guide to more adequate and satisfying adjustments both in the specific situation and in related problem situation. This definition by **Frandson** appears to include most of the aspects involved in learning.

Melton and **Munn** define learning very precisely. According to them, learning is a change in experience or behaviour. The change is produced by purposeful observations, overt activity and related motivational and emotional responses. Experience or behaviour implies combination of perceptions, cognitions, overt actions, attitudes, motivational and emotional feelings.

Moskowitz and **Orgal** are of the view that learning is the ability to profit from their own experience that makes some organisms, superior to others.

Fundamentally, learning is a behaviour change which is the direct result of experience rather than a consequence of inborn within the nervous system. For example, learning to play a violin and practising the new skill involves a permanent change in our behaviour. This new behaviour can be observed in an organism directly and can be measured. The change or modification of behaviour that takes place through learning can be observed in animals and in human beings. A dog can learn to identify his master's enemy and act quite appropriately to save his master. A rat can learn to press the lever and get its food. A cat learns to open the door in a puzzle box and comes out.

Kinds of Learning

Though man has been placed in the *highest place* of the *evolutionary process* he uses the same processes in learning as those of the lower forms of animals use. But what makes him different from other animals is the learned behaviour pattern. Man's learned behaviour pattern is flexible and varied. The nature of learning process has been discovered after more than *fifty years of research work* on the mechanisms of learning. After a lots of difference of opinions the present day psychologists have agreed that the learning process involves two basic forms. The two forms are different aspects of the same mechanism as all learning can be described with one set of rules. Here we shall deal with the two basic types of learned behaviour. These two basic types have some characteristics in common, but differ in other respects. The famous American psychologist **B.F. Skinner** (1938) had become the champion of this venture. He named the two basic types of behaviour as *respondent behaviour* and *operant behaviour.* All learning can, be brought under these two types. Those who do not agree with these two types add one or two more forms of learning.

Respondent Behaviour

In this behaviour, responses are elicited automatically. The responses are aroused by specific stimuli. It is a non-voluntary behaviour. The *activities of smooth muscles* and *glands* of the body are involved in this behaviour. In human beings the respondent behaviour refers to the *reflex activities* such as eye-winking, knee-jerk, sudden withdrawal from painful situation and so on. It also includes *emotional responses* of anger, fear, pleasure and so on. For example, a man is angry with his boss, or frightened by the sight of a tiger or aroused by a girl's look. At all these situations the man's emotional response is respondent behaviour unless he reacts to the situation. The parts of the body involved in these responses are *controlled by neural centres* that lie below the *cerebral cortex.* Therefore, it is assumed that respondent behaviour is related to the concept of *activation* or *emotion.*

Operant Behaviour

It is a voluntary behaviour. Operant responses are not controlled by external stimuli. It is not automatic or spontaneous type of behaviour. In operant behaviour the activities or movements of head, tongue, vocal cords, hands and legs are controlled by the *motor areas of the cerebral cortex.* The responses are provided by the organism itself unlike the respondent behaviour. For example, the sight of a tiger in a garden instantly produces a fear response in a man. It is a respondent behaviour. But what the man does about the situation will be determined by several factors, such as, whether he shouts, or runs away from the situation, or attacks it or not. In the case of a child the limitation of respondent behaviour is completed during infancy and after that his operant behaviour continues throughout his life time. These two behaviours are modified by experience.

On the basis of two types of behaviour, two types of learning have been formulated. The modification of behaviour that takes place through learning can be observed in animals and human beings. In the laboratory situation the behaviour of animals can be controlled and observed carefully. When we speak of learning under controlled situations the word conditioning is often mentioned. The word conditioning is a general term often used by the psychologists. It is referred to learning of a particular response. The word *conditioning is used synonymously with learning,* because conditioning produces modification of behaviour. The modified respondent behaviour is named as classical (Pavlovian) *conditioning* and the modified operant behaviour is named as *operant or instrumental conditioning.*

Classical Conditioning

Classical conditioning is a simple type of learning. It takes place in animals as well as in human beings. Classical conditioning provides lots of information for examining the learning processes in general. **Ivan Pavlov** (1848-1936) a famous Russian physiologist and psychologist was the first to study the respondent behaviour in animals. He was the first to perform the *quantitative and scientific investigation* on the basic nature of *learning process.* He called the process of his investigation as 'conditioning', the respondent behaviour itself is named as Pavlovian Conditioning. Since he was the forerunner of this kind of study he was called the *father of modern behavioural psychology* and his innovation is called Classical Conditioning.

Ivan Pavlov, in the year 1900 was concentrating on the activities of the salivary glands in the digestive processes of the animals as part of his physiological research. The results of his experiments on conditioning have earned for him a *Nobel Prize* and also changed the course of psychology all over the world. If we want to know more about conditioning and learning we must go through the details of the experimental procedures used in classical conditioning. While working on digestive processes Pavlov observed that *salivation is a reflex action* which can be produced by placing certain food substances on the tongue of the animal. He had also observed in several animals that salivary response can be brought under the control stimuli other than those used for salivation. His experimental animals were mainly dogs.

In Pavlov's experimental design a hungry dog was brought into a sound proof room and placed on the table-top comfortably. The dog was placed on the table in such a way that he was prevented from sitting or making any other gross movement. A small operation was made on the dog's cheek and a tube was attached to one of the salivary glands. Whenever the dog salivates, the saliva from the dog's mouth flows into the glass cylinder through the tube. The saliva is then measured.

Pavlov, before starting his actual procedure sounded a bell to the dog. The dog did not salivate, because there was no saliva flown in the cylinder. Then some food stuff with which the dog is familiar was placed in front of the dog. The dog then began to salivate. The bell sound did not evoke any salivation in the dog's mouth. But the food stuff did it. The food stimulus had the effect of arousing the response the saliva whereas the bell stimulus had no effect.

In the second stage of the experiment the bell was sounded and after a fraction of a second food was supplied to the dog. This kind of combined presentation of the bell sound and food was repeated several times. Pavlov had observed that saliva from the dog's mouth began to flow into the cylinder on hearing the bell sound before the food was supplied to the dog. That is, the bell sound which did not evoke saliva in the dog's mouth in the first stage began to evoke saliva in the second stage, because the bell sound was combined with the food. At *this stage the dog had learned* that the bell was the signal for the supply of food and therefore he salivated to the signals even when the food was not followed the signals. This was the stage at which the dog had been conditioned to respond to a new stimulus - the bell sound which did not cause salivation in the first stage. The essence of this experiment is that a meaningless stimulus (the bell sound) could evoke an internal response in an organism. After this study, physiologists and psychologists started studying the conditioned response of several types on animals and men. Some of the other responses studied were *protective reflexes, orienting reflexes, eye-blink* and so on.

The Conditioning Process

In Pavlov's experiment on conditioning we could observe four separate elements present. The reflex or the saliva in a response to the food. *The food* is the *first element.* Pavlov named this as the *unconditioned stimulus* (UCS). The *salivary response* to food is the *second element* which is called the *unconditioned response* (UCR) because it occurred before the process of conditioning took place. When the unconditioned stimulus (UCS), the food, was combined with a conditioned stimulus (CS), the bell sound (which did not evoke salivation in the beginning), and repeatedly presented for several times, the conditioned stimulus (the bell sound) began to gain control over the unconditioned response—the salivation. The conditioned stimulus (CS), the bell sound, was able to elicit the response (salivation) even when the unconditioned stimulus (UCS)—the food was not presented. When this process occurred the *neutral* (artificial) *stimulus, the bell sound* had become a *conditioned stimulus* (CS). This is the *third element.*

A conditioned stimulus may be defined as an event like the bell sound, a light, a pressure, or a shock that is a neutral stimulus and does not elicit any desired response originally. For example, ordinarily the bell sound would not elicit any saliva in a dog's mouth. But when the dog is conditioned in this way he salivates on hearing the bell sound. This kind of response is called *conditioned response* (CR). This is the *fourth element* in Pavlov's conditioning process. These four elements can be presented like this:

Before conditioning		During conditioning		After conditioning	
Conditioned Stimulus CS (Bell)	No response	CS (Bell)	UCR Salivation	CS (Bell)	CR Salivation
Unconditioned Stimulus UCS (Food)	UCR Salivation	and UCS (Food)			

Any response which is produced automatically by a specific stimulus can be conditioned. In other words, any unconditioned response to an unconditioned stimulus can be conditioned. Conditioning depends upon closely pairing of the conditioned stimulus (Bell sound) and the unconditioned stimulus (food). If these two were presented in an unrelated and haphazard manner conditioning will not take place (Fig. 12.1).

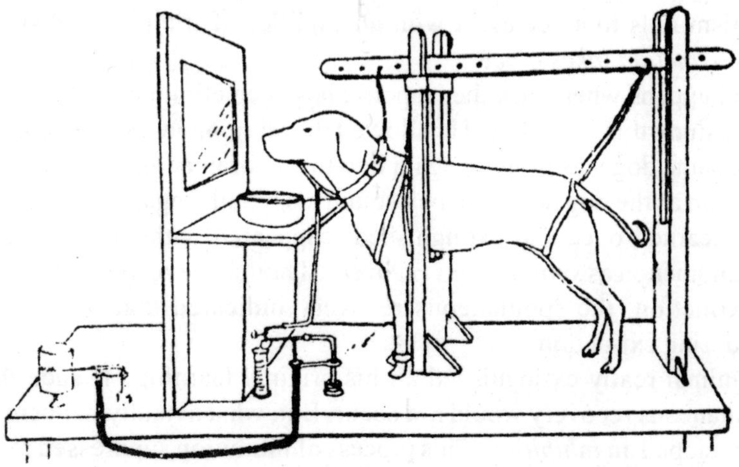

Fig. 12.1 : Pavlov's Experiment : Classical Conditioning

Classical Conditioning in Human Being

Human beings begin to learn from the time when they were born. Babies and small children learn to blink their eyes when they hear a sound. Adults also learn by classical conditioning. The eye-lid conditioning experiments have been conucted successfully with *dogs, monkeys,* and *college students* by Anrep. The results show that conditioning produces the same in all these three kinds of subjects. Another author, **Lipsitt** had observed that small babies only 5 to 10 days old learn to blink their eyes on hearing a tone. The *air puff apparatus* had been used by him. A puff of air-was blown into the babies eyes and babies naturally blink. The puff of air is an *unconditioned stimulus* and the babies *blinking* is the *unconditioned response.* It is a natural reaction.

The air *puff* was *paired with a tone* and it *was sounded* immediately before the air puff was blown. When this was repeated for several times the babies soon began to *blink their eyes when they hear the tone.* Lipsitt observed that the babies learned to *associate the tone with the air puff* and began to blink to the tone even when the air puff was not blown. In this way the babies produced a *conditioned response* to a conditioned stimulus (the tone). In our every day life we learn many things in this way and we are responding in certain situations even when there is no real stimulus that arouse the desired response.

Extinction and Spontaneous Recovery

We must know that classical conditioning is not producing a permanent change in the behaviour of the organism. What happens to conditioned response when the conditioned stimulus is presented alone for several times? The conditioned response (salivation) becomes weakened and gradually the *response is eliminated.* In the case of Pavlov's dog, he stops salivation when the bell alone is presented repeatedly.' This process of elimination of salivation is known as *extinction.* When the conditioned stimulus appears alone several times the learned organism fails to associate it with unconditioned stimulus and stops the salivary response and *extinction takes place.*

But, what happens when once the response has been eliminated? Has it gone for ever? Pavlov tried to study this in his dog. He allowed the learning in the dog to extinguish. After a few days the same dog was brought again into the laboratory and the same procedure was repeated. As soon as the dog heard the bell sound his mouth began to salivate. The response that had been learned once and extinguished reappeared without any retraining. Pavlov called this phenomenon as *spontaneous recovery.* Though the response was not as strong as before the extinction, the spontaneous recovery indicated that *learning was not lost permanently* during extinction.

Has the animal really extinguished all his original learning? If such thing had taken place the spontaneous recovery would not occur. But, what actually happening was that the animal had developed an *inhibition.* This process of inhibition suppressed response that had been learned by the animal earlier. At this stage the animal is in between a pull by learning in one direction and a pull by inhibition in another direction. If the *learning pull* is stronger than the *inhibition pull* then a response will be continued. If the strength of inhibition is equal to that of the learning the animal will not produce conditioned response and extinction takes place. But *as time goes* on the *inhibition becomes weaker* and weaker and *spontaneous recovery will take place.*

The inhibition is the process in classical conditioning that temporarily blocks the response that has been learned. A sudden change in the learning situation may also block a conditioned response. Once, in Pavlov's laboratory while some experiment was going on, Pavlov suddenly appeared on the scene. This event had disrupted the dog's routine performance. Then, he called this process as an *external inhibition.* That is, a strange person's appearance in the room caused the inhibition. But, the lost response can be rebuilt again by presenting the unconditioned stimulus (the food) along with the conditioned stimulus (the bell sound). This phenomenon is called *reinforcement.* It is a process of strengthening the conditioned response.

Other Properties of the Conditioned Response

(a) Generalization

Certain objects or events may resemble some other objects or events. The resemblance may be so close that the learner or the organism may react to one in the same manner as he reacted to another. In the case of conditioned response, a response which had been conditioned to a particular stimulus can also be *aroused by another stimulus* which had certain similarity. This spreading of conditioned response is called *stimulus generalization.* Pavlov's dog, which had been conditioned to ~alivate to the *bell sound* also salivated to the *sound of a buzzer* and to the *sound of a metronome.* The dog's conditioned response had been generalized to other sounds also. The man who got into some motor car accident and injured would develop a conditioned fear response to any transport like, train, bus, plane, lorry and so on. This kind of generalization is natural to all organisms.

(b) Discrimination

In the learning process, if we wish to *narrow down* the range of effectiveness of the stimulus more effort must be put. It is a way of preventing the organism form generalization and making it to *react to a specific stimulus.* This process, we call *discrimination training.* The organism learns to differentiate one among many stimuli and respond to it. Generally, discrimination training is a very important process in our every day life. If this is not taking place accurately lots of confusions would occur in the world. You are able to discriminate the college bell from the railway bell, the temple bell, or the church bell, the fire alarm bell and several other bells without any confusion or mistake. Discrimination is *mediated by past learning.*

(c) Higher-order Conditioning

This is another important aspect of classical conditioning. A conditioned response is built on the basis of a previously established conditioned response without any unconditioned stimulus. For example, the dog had learned to salivate on hearing the bell sound. But, Pavlov wanted to teach his dog to salivate on seeing a black square. Pavlov had shown the black square to the dog along with the supply of food. Next time had shown the black square to the dog following the bell sound and not following the supply of food. The dog had learned to salivate when he saw the black square. Here, the black square was used as a conditioned stimulus, and the bell was used as an unconditioned stimulus. This kind of learning is based on the *previous learning.* Pavlov called this process as the *higher order conditioning.*

Operant Conditioning

Classical conditioning involves learning to respond to something that does not ordinarily cause such a response. A *natural response* is caused by an *artificial stimulus.* The dog learns to salivate to the sound of a bell. It is passive and simple. The behaviour is elicited by some external events. There is another kind of conditioning called *operant conditioning* or instrumental conditioning. In classical conditioning events are anticipated in the environment. But, in operant conditioning events are caused or the *organism changes its environment.*

Most of our behaviour is emitted and voluntary. The behaviour is not triggered by outside events. For example, a child draws her mother's attention for care by crying. A school boy raises his hand in the class room as a gesture of informing the teacher that he knows the answer to the question posed by the teacher. A dog wags his tail and jumps in front of his master and gets patting or a piece of biscuit. A university student works hard systematically to get a pass with the first class. A passenger in the railway compartment pulls the chain and stops the train when he sees some robbers in his compartment. Actions of these types are called *operant behaviour.* These actions *are designed* to operate the situation or instrument in some way to gain something desired or pleasant or to avoid something unpleasant. Operant behaviour is learned behaviour pattern. It takes place through operant conditioning. Operant conditioning is another type or kind of learning.

Operant conditioning may be defined as a type of learning in which the desired voluntary (behaviour) response is rewarded and *incorrect response is ignored* or punished. According to **Kendler** an operant conditioning is an instrumental conditioning situation in which the organism is free to respond at any time. This type of learning occurs in animals and in human beings. Operant conditioning is a recent addition in the psychological studies. But animal breeders have been using operant conditioning for hundred of years. Dog and cat, breeders, monkey, horse, elephant and other animal trainers often speak of new tricks in animal behaviour. In circus companies there are expert animal trainers.

A wild animal like lion or tiger learns to sit on a stool in a circus on the appearance of the ring master into the ring with a whip and an iron-rod. All of you might have seen and enjoyed the dance performances of the animals like the bears, elephants, horses, n~onkeys in circus companies. Whenever the circus horse performs a desired response and pleases the audience the horse master puts a piece of sugar candy into the horse's mouth secretly as a reward. Horse is very fond of sugar candy. Similarly the dog trainers often reward the dogs when they perform the desired tasks. Whenever the horse fails to perform the desired task the master whips the horse with a loud noise. These are the *procedures involved in operant conditioning.* In operant conditioning, whether it is animal or human learning correct responses are reinforced by rewards and *incorrect responses* are ignored and punished. A correct response is reinforced and repeated and therefore it is learned.

In psychological laboratories several different kinds of instruments are used in operant conditioning experiments. The most widely used apparatus is the *skinner box.* **B.F. Skinner,** the famous American psychologist had designed the box and it is called after his name. He had developed several techniques of operant conditioning. The boxes were of different sizes and shapes according to the size of the animal. The rat was the animal mostly used in his experiments. When Skinner started his conditioning experiments in 1932 his experimental animal, the rat was allowed to be familiar with the box and it was not frightened in the new situation. Then the rat was made to starve by 24 hours without food and then it was put into the box.

It was a small box with bar walls and a protruding bar and a cup. There was a opening below the bar from which pellets of food can be drawn. The hungry animal began to explore the whole box. It wandered all around sniffing and standing on its hind legs and expressed several casual behaviours. After some time it put its weight on the bar by chance and pressed it. Immediately pellets of food were dropped into the cup by a mechanical device. The hungry rat ate the food and continued its exploratory activity. In its exploratory behaviour the rat happened to press the bar (lever) two or three times and food appeared all the times. The animal, then learned that it could get its food by pressing the bar every time. The supply of food by pressing the bar made the rat to repeat the bar pressing activity whenever it wanted food. In this way the *food getting response* was strengthened by *pressing the bar* repeatedly. This process, in operant conditioning is called *reinforcement* and it is the *primary concept.* (Fig. 12.2).

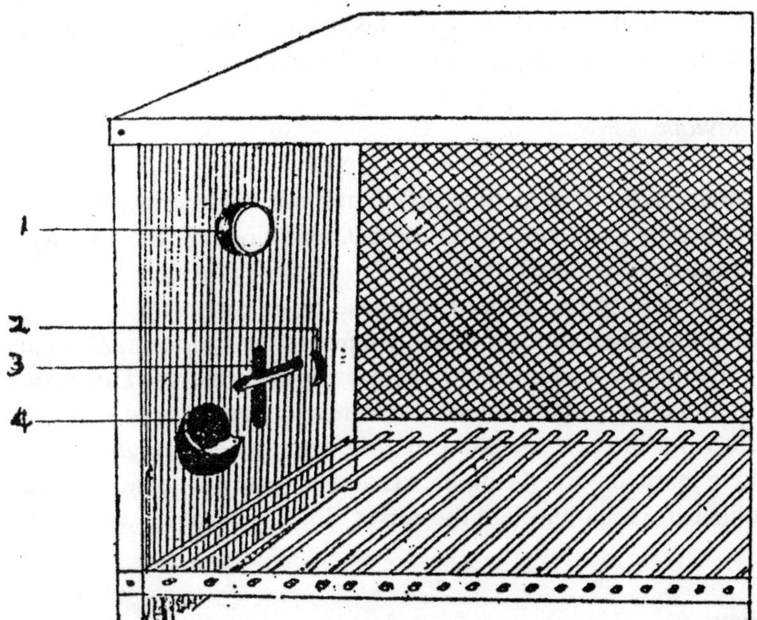

Fig. 12.2 : A Skinner Box (Learning)
1. Light 2. Water 3. Lever Bar 4. Food Cup

1. *Acquisition of Response*

In classical conditioning the response is elicited and it is easy to produce the desired response. What Pavlov did to make his *dog to salivate* was to *put meat powder* in the dog's mouth. Operant behaviour cannot be elicited in this way. A hungry rat will not press the bar naturally. On pressing the bar if the rat. did not get its food there will be no learning in the rat. Therefore, what is important in operant conditioning is, making the desired response to occur and then it can be reinforced and learned. All of us use several different methods in our every day life to evoke the desired response. It is a matter of common sense and sometimes we do it without any kind of thinking. Skinner include **Thorndike's** (1913) *trial and error method* of learning in his operant conditioning. We are dealing with trial and error method of learning separately.

(a) Motivation and Learning

Motivation is a dynamic process. By increasing motivation it is possible to speed up the process and increase the correct responses in an organism. A hungry rat will move around the Skinner box more swiftly than one that has been fed. This will make the rat to discover the bar in the Skinner box and. press it and release the food pellets. The *influence of motivation of learning* appears to be simple. But it is one of the *most complex problems in psychology.* It is generally believed that *increasing motivation* will always speed up learning. It is *not correct.* In certain situation learning is retarded by increasing motivation. Scientists are forced to give the idea that motivation can be taken as a simple causal agent in behaviour. Motivation has several facets and each demands different answer.

(b) Guiding Behaviour

In recent years attention has been paid to different kinds of experimental methods. One of the important and widely used methods is operant conditioning. The activity of the animal in the situation is more important in this.

Another way of getting the desired response in an organism is *forcing or guiding behaviour.* The experimenter himself actively participates in the situation and guides the organism or the learner through correct responses or movements. This procedure can be followed in animal learning and in human learning. If you want to teach your dog with new ways of behaviour the best method is to *guide him through right movements.* While making the child to learn to write the alphabets and to write his name, one way of teaching him is to direct his hand to go through the correct lines, so that he will do the same thing next time.

2. *Reinforcement*

We have been describing the various ways of producing~he desired response in an organism.

In operant conditioning it is also important that the desired response should be repeated correctly. In training the animal we saw that the *animal is rewarded for every correct response.* The reward strengthens the desired responses, and also increases the chance of repeating the same. Psychologists called this reward as reinforcement. Reinforcement is defined as an event like presentation of food that increases the tendency in an organism for a stimulus to evoke a response.

Reinforcement may be also defined as a process of strengthening a correct response produced by an organism and building up a tendency in it to repeat the same response often. Reinforcement is the primary concept in operant conditioning. Any response followed by a reinforcement is likely to be repeated and therefore it is learned. We have used this concept in Pavlov's classical conditioning. But, here we are using it with different meaning. In classical conditioning the concept is referred to unconditioned stimulus—the meat powder. It strengthened the conditioned response. The meat powder is not given as a reward.

(a) Primary and Secondary Reinforcements

Reinforcements are classified as primary and secondary reinforcements. A primary reinforcer is one whose reinforcement value is innate or unlearned. It is rewarding in itself. The primary reinforcers are food, water, sex and removal of pain. A secondary reinforcer is one whose value is learned and associated with the primary reinforcer of the past. It is secondary because it is learned. For example, a rat learns to press the bar and gets its food. One experiment conducted by **Gugalski** in 1938 is worth mentioning here to illustrate secondary reinforcer. The Skinner box was used in this experiment. In the *first stage,* the bar pressing was always associated with a click sound by the food dropping device. In the *second stage,* the subjects were divided into two groups. In the case of *one group,* the bar pressing response produced a click sound and also delivered food. In the case of the other group, the bar pressing response did not produce the click sound and no food was delivered. Here the phenomenon studied was that which of the two groups would *extinguish more rapidly.*

For the first group the click was closely and consistently associated with the primary reinforcer—the food. Therefore, it should acquire reinforcing properties. During the period of extinction food was no longer received on pressing the- bar. But the first group would be receiving the click sound—secondary reinforcer, whereas the second group received neither the click sound nor the food. In the case of the first group the click sound worked against the effect of extinction. In other words, even though food was not delivered on pressing the bar the animal pressed the bar because there was some response—the click sound. The click sound strengthened the animal's tendency was weakened by the absence of *the primary response—the food.* In the case of the second group, the bar pressing response was extinguished more rapidly during extinction because of the absence of the *secondary reinforcer—the click sound.* Money is used for different kinds of secondary reinforcers. Money itself is not food, but we can get all the primary reinforces like food, water, clothing, sex etc. through money. Therefore, it becomes a. powerful secondary reinforcer.

(b) Positive and Negative Reinforcements

Reinforcements are also classified as positive reinforcements and negative reinforcements. If the response action is pleasant or good to the learner it is called positive reinforcement. For example, food and money are positive reinforcements. Negative reinforcements involve the *termination* of painful responses such as shocks, uncomforts and so on. Both types of reinforcements bring about learning when they follow a response. A child might learn to solve his arithmetic problems for which he receives praises. The praise here becomes the *positive reinforcement.* By this means he also avoids getting scolding from his teacher if he fails to solve his problems. In this way it becomes the *negative* reinforcements, the child *could avoid unpleasant experience.*

(c) Punishment

Punishment may be physical or verbal from parents, teachers or leaders or from society if you violate codes of behaviour. Reinforcement is used to strengthen a response. If an organism is *punished* for producing a response it will *less likely repeat* the same response. A man is put in the prison for stealing another man's property as a punishment and we hope that he will not repeat the same action. But punishment often does not work successfully. Children often continue committing certain activities for which they have received punishment. Criminals repeat the crimes after coming out of prisons. In all these cases punishment had no effect. Therefore the effectiveness of punishment mainly depends upon when and how it is used. An *undesirable behaviour* should be immediately followed by a *punishment.* Courts take years to convict and imprison a criminal. Here punishment loses its effect, because punishment comes too late. The person himself might have forgot for which crime he receives this punishment.

Further, punishment should be adequate so as to correct the child from committing the same mistake once again. Strong punishment alone can reduce the response effectively and permanently. Punishment should also make the organism to be aware of the response (the mistake or crime). The mother tells her child not to touch the costly vase again. The words had no effect. She slaps him. The harder she slaps him quicker the effect it will have on him and he will learn not to touch it. Punishment will have effect if only it is continued. If it is not consistently applied the undesirable behaviour may appear.

There should not be any reward following the punishment. Parents often slap children and immediately hug them. Parents should avoid this kind of attitude. This would cause confusion to the child that reward follows punishment. Such kind of punishment and reward at the same time would create inconsistency in the child's behaviour. Punishment is actually a kind of reinforcement which strengthens the child's non-response to undesired behaviour. Punishment can suppress, a response so that other responses can be reinforced. It would make the child to be more aware of undesirable behaviour and instil a fear of certain situations. *Therefore, punishment* is an important *training tool* if it is properly used.

(d) Partial Reinforcement

Once a response is conditioned it can be maintained by a few reinforcements. A rat has been trained to press the bar for food and food need not be dropped for every response. But food may be dropped for every alternative response, every fifth response, every tenth, every twentieth until the bar pressing response is maintained with one reinforcement for every hundredth responses. That is, the process is continued until the rat presses the bar for one hundred times for getting food once. This is called *partial reinforcement.* Rewards are not given for every correct response, but for some responses. Skinner had conducted such experiments on pigeons and rats. The programme for choosing the responses for reinforcement is called the *schedule of reinforcement.* The reinforcement schedules are fixed *intervals* and variable intervals. These are based upon time. The other schedules are *fixed ratios* and *variable ratios.* These are based upon *the* number of correct responses.

Skinner observed that pigeons and rats are also higher organisms adapted to different reinforcement schedules and behaved differently on the basis of time (interval) (i.e. a response is reinforced once in 5 minutes) or on the basis of the number of responses (i.e. every twentieth response is reinforced).

3. Discrimination

This is the ability to discriminate between similar stimuli. Discrimination also refers to the ability of determining whether the right (rewarding) stimulus is present in the situation. It is important in operant conditioning that the learner should know when he should do the appropriate response. Discrimination training is given to the organism by rewarding one response and not rewarding the other. The discriminative stimulus is constantly associated with the reinforcement which develops incentive and reinforces the properties of its own. The discriminative stimulus becomes an incentive and activates the organism. Pigeons were trained *to peck at a disk.* Then they were presented with two disks one red and one green. The pigeon received food when pecked the red disk but not when pecked the green. Finally the pigeon had learned to discriminate between the two disks and pecked only the red disk. The baby also learns to discriminate between his mother and other ladies, his father from other men after committing certain errors in the beginning (Fig. 12.3).

4. Extinction and Spontaneous Recovery

The operant conditioned responses extinguish in the same manner as the effects of classical conditioning. A response which is not reinforced tends to lose its strength in extinction. Extinction is a graded process—the process by which a response is extinguished. This depends upon the degree to which the response had been reinforced previously. But a response that had been extinguished can be recovered spontaneously. Spontaneous recovery is the reappearance of the original learning after it had been extinguished. A learned effect that is extinguished long back can be brought back by a single reinforcement.

Fig. 12.3 : An Operant Conditioning Situation
1. Mechanism Compartment 2. Disk 3. Lever Bar 4. Food Delivery.

5. *Shaping*

Operant conditioning is used to teach some very complex behaviour patterns. The parts of the sequence have to be learned separately. In learning complex patterns we must integrate the bits of behaviour step by step into a complex response. The procedure of teaching complex behaviour pattern is to reinforce the partial responses. The bits of responses are made up into the whole. The complete response is shaped bit by bit. The method used to the behaviour of this fashion is called the *method of successive approximation.* In this process of shaping, every *bit of response that leads to the whole* is rewarded.

6. *Behaviour Chains*

Operant behaviour is a combination of several events. It is a chain and not separate events. One event functions as a cue to next event which in turn activates the next one and so on. In this way a *chain is formed.* Most of the operant behaviour functions are by connecting together the response, the stimulus, the response and so on in sequence. The chain ends with the reinforcement. Most of the complex motor behaviour is considered as the result of a process of chaining. When a discriminative stimulus is established it would help to reinforce other responses. In an experiment conducted by Skinner it was observed that the rat had learned only to press the bar when the light was on. The rat was trained to produce other responses also like pulling a string when the buzzer sounds. When the rat was placed in a dark cage it pulled the string to turn on the light and when the cage was illuminated with the light it pressed the bar to get its food. The *string-pulling* and *bar pressing form a chain behaviour.* This behaviour has been built up through reinforcement, discrimination and secondary reinforcement. Most of the human behaviour occurs continually in a chain like manner. For example, in teaching the child to learn the language, he is taught first letters or short words, then short sentences and so on.

TRIAL AND ERROR

Trial and error is one of the most common methods of eliciting the desired response. It is waiting for the subject's correct response. The child cries in a particular fashion. It is by an accident. But if he finds that his mother runs towards him and takes care of him, he learns to produce and repeat the same type of sound to draw his mother's attention. Whenever he wants her care he produces the same sound. Thus he has learned to produce the desired response in his mother.

Similarly, we can observe trial and error learning in rat's behaviour. *The Skinner box* may be used for this purpose. The box is constructed in such a way that the bar can be removed and inserted. This device would make the rat's behaviour to change in successive trials. A trial may be defined as the unit of practice in learning experiments during which an organism is exposed to a specific stimulus situation. A trial may also be the amount of time the rat takes for a correct response in a learning experiment.

In a series of successive trials in Skinner box the sequence of behaviour involves pressing the bar and eating the food and repeating the same. It is observed that certain important changes are taking place in this process. *In the beginning,* the rat takes *more amount of time to find* out the bar and to press it. In successive trials the amount of time for the task is gradually decreasing. Similarly the amount of time taken by the rat to eat the food is longer in the beginning, but the same is gradually decreased in successive trials. This shows that there is certain improvement in the behaviour of the rat. This is what the experimenters expect from the experimental animals. The animal gradually *discards the wrong and unsuccessful movements* and retains correct and successful movements.

But, the kind of improvement we see in the experimental animal is not always steady in every rat. On some trials the rat's response is slower than the previous trials. When large number of rats is used in such experiments the mean time of their responses decreases consistently on successive trials.

Trial and error learning involves the idea that an organism, as it progresses through its activity, continually meets with certain problems. The solutions to the problems require the combined application of generalization from the organism's *past experience* and the discovery of appropriate responses to the problem situation. Trial-and-error learning is also stated as *trial-and-success learning.* Trial and error learning is originally set forth by **E.L. Thorndike.** According to him the trial and error learning process is governed by *three laws.* They are, *the law of effect,* the *law of frequency* and the *law of recency.*

The law of effect refers to the activity that is successful and provides satisfaction to the organism. This activity is fixed and retained. The unsuccessful movements are discarded.

The law of frequency refers to the repetition of the activity for certain number of times. It is a tendency to establish an activity permanently. The repetition also increases the amount of strength of the activity.

The law of recency refers to the repetition of a recent and new activity. It is repeated for familiarity.

Thorndike used mostly *cats* as his experimental animals. Ee designeu-1 *puzzle boxes* with sides made of bars and doors. The door could be opened from inside by lifting a latch. Thorndike put a hungry cat inside the box and placed food outside and observed how quickly the cat learns to lift the latch and go out to eat the food. The number of trials for successful performance and the amount of time taken by the cat for each trial were taken into account. After conducting several experiments, Thorndike concluded that animals learn purely by trial and error method and this learning takes place according to the three laws explained above (Fig. 12.4).

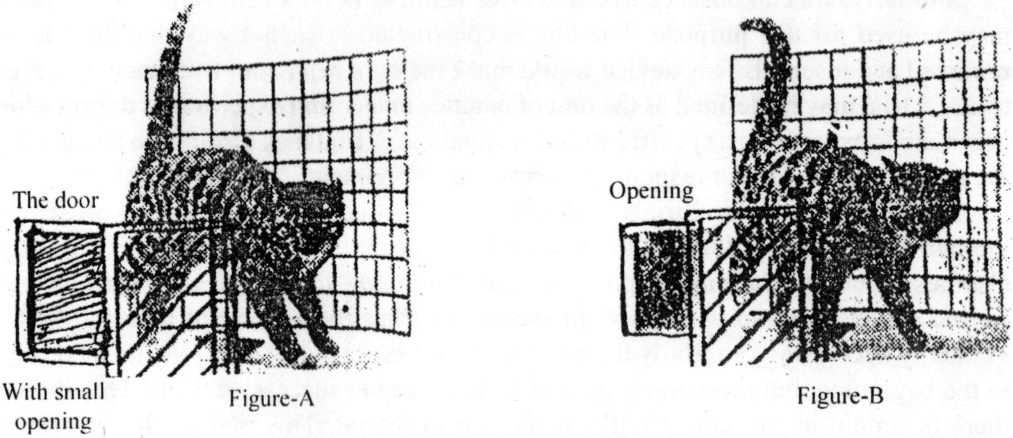

Fig. 12.4 : An Experiment on Thorndike's Cat.

The cat and the Puzzle-Box. The cat entered the box through the small door at the back. It also opens the front door by pushing. The same cat in two trials. It takes six seconds for the first trial and five seconds for the second trial. The cate once got out fixes its activity in the area of the door which was already achieved which may be stated as place learning. It operates the loop for its purpose, hence tool learning.

1. Mechanism Compartment 2. Disk 3. Lever Bar 4. Food Delivery.

Insightful Learning

Insightful learning is a concept used by the Gestalt psychologists to describe the process of learning. Gestalt psychologists are of the view that insight is present in all animals learning. Though insight is gradually dawning in learning situation its most characteristic form is its sudden appearance or flash. Insight means understanding. It is a way of finding out solution to a problem suddenly. Insight is comprehending the organisation and the interrelations of the significant parts of the problem in learning situation. It may be also defined as sudden graps of a situation or problem. It is the realization of the use of an object or situation. Psychologists for a long time have been using this concept 'insight' to describe one of the learning processes under the heading of 'learning'. But modern psychologists use this to describe the process of solving a problem under the heading of *thinking and problem solving.* It is better that we should understand this in either way.

In learning, students will have to pass through certain critical phases in the *acquisition of skills*. This involves the capacity to pass from *pieces to an integrated form*. Children pass from a letter habit to a word habit and from word habit and to phrase habit. The important process involved here is the *combination of different elements* into a meaningful composite whole. When an individual for the first time meets with different kinds of unrelated stimuli, he tries to relate them into a single whole pattern. He is in a state of helplessness. Then, suddenly he learns to have a control over the situation and gets mastery of the problem. This process we call insight. It is something going below the surface level of things. Monkeys, particularly, chimpanzees are capable of *insight learning.* **Kohler's** *experiment* on chimpanzees is a classical illustration to demonstrate insight learning in chimpanzees. They are capable of relating objects to the situation. Chimpanzees also like all other animals learn to some extent by fumbling and success.

In a number of experiments with apes Kohler had observed good evidence of insight in the animals. The animal was kept in a cage with banana lying outside beyond the reach of its hand. A stick was lying within the reach of the monkey's hand. The monkey was able to make use of the stick and pull the banana inside the cage. Then the animal was given a two stick problem. A short stick that is available at close range will not reach the banana. Only a longer stick could help the animal to pull the banana inside the cage. The animal learns the trick that no single short stick can be useful for his purpose. Two short sticks were placed close to the animal. Neither of them was long enough to reach the banana. One stick was smaller in diameter, so that it can be fitted into the open end of the other stick. (Fig. 12.5)

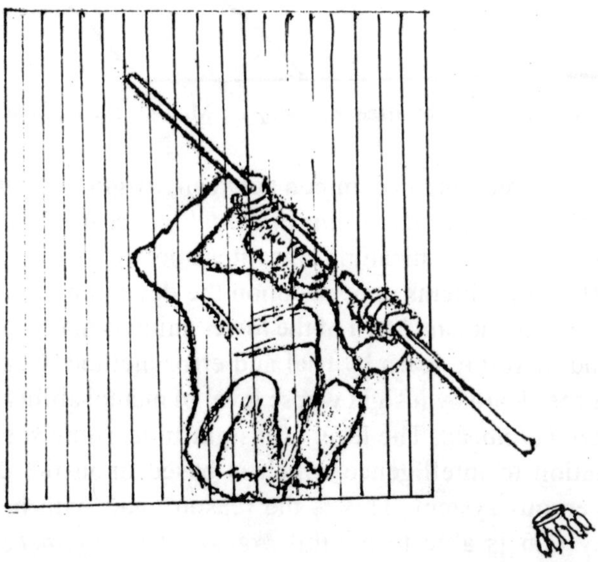

Fig. 12.5 : The Chimpanzee is Solving the Two-Stick Problem

After struggling for an hour with short sticks unsuccessfully the animal had learned that the sticks were too short for his purpose. He gave up the attempts and went behind the cage. While playing with the two short sticks end-to-end he pushed the one end of the smaller one into the open end of the longer one. On seeing the longer stick the animal jumped with joy and ran to the front side of the cage and tried to pull the banana with the joined stick. At one stage the joined pieces were separated as the joining was loose. He was able to put them properly once again. He then pulled the banana and ate it. The same experiment was repeated on him the next day. The animal was able to join the sticks within few seconds and used it in the same manner. The capacity to put the matters into new relationships which were never previously experienced is called insight. Kohler explains insight in terms of brain physiology. According to him animals learn by insight. This idea is contrary to Thorndike's idea that animals learn by trial and error method (Fig.12.6).

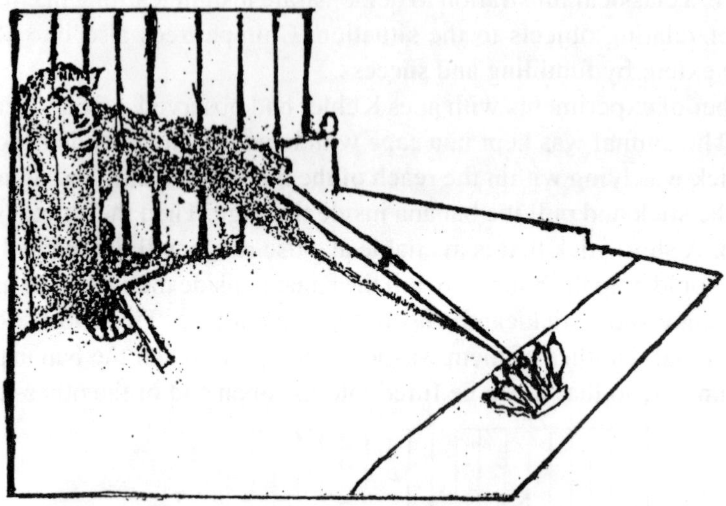

Fig. 12.6 : The Chimpanzee is Solving the Multiple Stick Problem.

There appears to be something in common between *insight and intelligence.* As far as learning is concerned both *Insightful learning and intelligence are involved to some extent.* But the term 'insight' has a wider meaning since it is applied to all types of learning. The capacity to learn difficult problems depends upon the development of the *complexity of the nervous system* in the organism. Most of the lower animals mainly behave *instinctively* and mechanically and therefore learn by trial and error method. Certain higher animals like, apes or chimpanzees learn by insight whose level of mental ability had been compared to that of *three year old children.* The level of insightful learning seems to be increasing in organisms in relation to intelligence which is based upon the development of the complexity of the nervous system. This is the reason, man with his highly developed complex nervous system is able to exhibit *higher order of insightful learning* and intelligence.

Animal Learning

In the preceding sections we have seen that most of the experimental works on learning were done with animals. Animals were used as subjects in physiological and psychological laboratories. Psychologists are very much interested in animal learning, because it is easy for them to conduct experiments on animals rather than on human beings. Further, they wanted to know whether animals are capable of learning. If they are satisfied with their experiments on animals, they hope, they can carry on the same type of experiments on human learning also. This is how lots of experiments on animal learning are available for us.

Animals learn by all the *four methods* we have described above like *classical conditioning, operant conditioning, trial and error* and *insightful learning*. Psychologists are of the view that the fundamental processes of learning can be seen more clearly in simple type of learning as animals do and not in complicated processes of learning as human beings do. After studying the simple type of learning processes it is easy to study the complicated processes of learning. Therefore, investigators on the nature of learning always begin with the simple learning processes of animals. Those who studied the learning processes of animals fall under *two groups* emphasising two different thoughts. The *first group* emphasises that *animals learn by habit*. It is the result of the process of trial and error and not insight. **E.L. Thorndike** (1874 - 1949) and **C.L. Hull** (1884 - 1952) were the exponents of this idea. The *other group* stresses that animals learn by insight. The level of learning may be low, but still it is insightful learning. They say that it also takes place in classical conditioning and operant conditioning. The animal in the beginning tries the problem only in fumbling manner with *wrong movements.* **W. Kohler** (1927) and **E.C. Tolman** (1932.) were the forerunners of this idea.

Whether it is animal learning or human learning the power to learn and reproduce is one of the most important facts in learning. This is present both in animal learning and human learning. If knowledge is of fundamental in nature it will be more useful in the life situation. Man solves his problems ill the same manner as animals do. For example, in the Skinner's experiments the rat learns to run the maze. Just like the rat in trial and error method o~ learning eliminates wrong movements and retains right movements man also solves his problems. **Lloyd Morgan**, one of the founders of animal psychology had laid down the principle of *Canon of parsimony.* According to his principle no action can be claimed as the outcome of higher mental faculty.

Lloyd Morgan's *dog* learned to open the gate by pushing the latch by his nose. The dog's behaviour was not based on any reason in finding out a solution. It was only by trial and error. But the dog had learned exactly the place where to work, though there were certain errors. This, we call place learning.

E.L. Thomdike, another pioneer in the field of animal learning, use(.-! mainly cats as his experimental animals. Thorndike's hungry cats learned to lift the latch in the puzzle boxes and learned to get out and eat the food. In this way the experimental rats and cats had learned the places and things. These animals *make responses to sensory stimuli.* They make *muscular movements or manipulate certain objects.* This is called *tool learning.*

In psychological experiments the favourite animals that are used for place learning are the rats and for tool learning the cats and monkeys or chimpanzees. In animals learning we get more information from monkeys. The monkeys have better observation of objects and manipulate the tools much better than the rats and cats. The monkeys have *much developed nervous system and larger brains.* They learn much quickly and handle the tools successfully quite similar to human beings using insight. Kohler's experiments with monkeys and chimpanzees were classical illustrations to demonstrate this process of learning.

Apart from place learning and tool learning an organism has to anticipate certain happenings in the environment and he must be ready to acquaint himself with them. It is applicable to both animals and men. An organism is capable of learning the sequence of events. He keeps himself ready to what is going to happen in the environment. There are several regular sequences occurring in the life of an organism and they become automatic in its behaviour, this kind of preparatory response becomes an involuntury part of its action. This principle was demonstrated by Pavlov in his dog's behaviour by making him to salivate to an artificial stimulus---the bell sound. Pavlov called this process as conditioned response or learning of sequences. The dog had learned to salivate (natural response) to a neutral stimulus (the bell sound). From these discussions we understand that a rich source of information had been derived from animal learning. The information collected from the experiments with Thorndike's cats, Lloyd Morgon's canon, Skinner's rats, Kohler's monkeys and Pavlov's dogs are also applicable to men as the same processes found also in human learning.

HUMAN LEARNING

Animals and birds have certain limited and simple needs of life. These needs are fulfilled by their instinctive behaviour pattern. The instinctive and sensory behaviour pattern of certain animals and birds is far superior to that of man. But man's intelligence and his unlimited capacity for learning can help him to conquer animals in all respects. He has to learn everything in this world. Man's capacity to form concepts and symbols and to develop language for communication is highly distinctive from animal learning.

Human beings learn in a number of ways. Learning occurs in man immediately after his birth and some learning occurs even before his birth. This learning in man is a continuous process and it goes on throughout his life. Experiments on human learning are so numerous that it would be impossible to enumerate all of them. Human learning is classified into several types. The four types we have described above are found in both animal learning and human learning. Some are of the view that these are associative process of learning. They can explain only simple form of learning. Perception and understanding play important role in complex form of learning. Human learning is of complex type. *Motor learning and verbal learning* are the most significant kinds of learning for human beings. Pavlov's conditioning experiments are possible with human beings. It is possible to condition a number of responses in man such as Knee jerk, winking reflex and so on. Some of them have been explained already. Here we will deal with the important human learning.

Multiple Response Learning

It is a kind of learning which involves more than one identifiable act, having the events fixed in an order as demanded by the situation. The *multiple response learning* is sometimes *called as motor and verbal learning.* Laboratory instruments have been designed by psychologists to study this kind of learning. the mirror-drawing apparatus, pursuit rotator, reaction time apparatus, memory drum are some of them designed to study motor learning.

The mirror drawing task is a form of sensorimotor skill. In sensorimotor skill muscular movement is important. This activity is taking place under sensory contact. For example, driving a motor car, riding a bicycle, playing a piano, typing a letter all are motor activities involving sensory control. While driving a car or riding a bicycle we are watching on all directions in the cross roads whether any one crosses the road and also we watch carefully whether there are any bumps on the road or red light on the traffic post. We are guided by all these factors. Further, the driver of a motor car must maintain the speed and time, the piano player must read the notes, the typist should copy the manuscript according to the specifications. In all these we could see more than one activity is involved in the task apart from muscular movements.

A good laboratory experiment to illustrate the *sensorimotor skin* is the *mirror tracing*. The subject is required to perform some activity requiring eye-hand co-ordination and his performance is measured. What occurs when the eye-hand co-ordination is accurate and not accurate? The subject has to adjust his task accordingly (Fig. 12.7).

Fig. 12.7 : Mirror-tracing Apparatus.
The participant traces the Star Pattern by following the pattern in the mirror. The apparatus is used in the Experiments on Transfer of Training.

In the mirror drawing experiment the subject has to trace the star pattern which is fixed on a wooden frame by seeing the pattern in the mirror which is fixed vertically at the end of the wooden frame. The star pattern which is perceived by the subject is reversed by the mirror. The subject has to run a stylus through the path without touching the sides and complete it. Every time the stylus goes away from the path it touches the sides and an error is committed. The amount of time to complete the tracing task of the pattern and the number of errors are taken into account. The task is very difficult. This experiment is suitable for studying the *transfer of training*. How many trials the subject takes to perform the task without any error and how much time he takes. This is the theme of this experiment.

Verbal Learning

In verbal learning the subjects' verbal behaviour is measured. It is by the *use of language.* It is one of the most significant kinds of human learning *involving words.* But whatever learning is taking place, the formal education is only verbal learning. Even informal learning that takes place in adults is only through verbal means—the use of language. Because of' this most of the experiments on human learning are conducted on verbal learning. Verbal learning has an important connection between simple non-verbal learning processes and language. Several experiments have been devised and different methods and materials have been used. The materials for the study of verbal learning range from *highly meaningful stories to meaningless non-sense syllables.*

Verbal behaviour gives accurate description of an individual's experiences. Verbal behaviour is not accessible to operant training. We have learned to respond to the verbal behaviour of others. All our thought processes are based on language. The effect of instruction is an important variable in verbal behaviour. The instruction must help a person to produce the desired response. The instruction may be a simple word when parents speak to small children. The same may be complicated with college students when the teachers speak. But wherever verbal instructions are given they must be clear and appropriate to arouse t he desired response.

TRANSFER OF TRAINING

Transfer of training is one of the' most important psychological problems of learning. The idea beyond this concept is that whatever you learn must be useful to you later on. Transfer of training may be defined as the application of the training received previously in one situation to other forms in another situation. The new learning may be applied sometimes more rapidly and sometimes slowly in another situation. Transfer of training received greater importance in the *field of education.* Educationalists in the early days believed that the study of mathematics led to the development of acquiring mastery over *Greek* and *Latin,* because Greek and Latin produced *attractive syntax* and good English vocabulary. They thought that it produces positive transfer in education. But researches conducted in the first quarter of this century have proved that this belief is wrong.

All learning, it is believed, is transferable to new situations. But the amount of transfer often may vary from person to person and situation to situation. Transfer of training may be positive or negative. *Positive transfer occurs* when the organism's performance is *benefited* by the previous learning in a new situation. *Negative transfer* occurs when an organism's performance in new situation is *hindered* by previous learning. Generally, the term 'transfer' refers to the effect of previous learning upon following similar performance (Fig. 11.7).

The important rule of transfer of training is the *degree of similarity* of stimuli and responses. The phenomenon of stimulus generalization may be applied in transfer of training. Therefore, if the similarity is greater the generalization also is greater.

Another important rule of transfer of training is the occurance of negative transfer when the desired response, is different from that of one learned previously in that situation.

The more dissimilar the new response is to the previous response the greater the negative transfer occurs.

In human behaviour the transfer of skill is important. The problem of transfer is not only changing simple responses or skills but also it is something more than this. What kind of skills can be transferred best to a new job in an industry? If an employee possesses a particular skill which is not needed to his new job, will this skill be transfered positively or negatively in a training programme? This kind of problems depend upon the attitudes and expectations of the learner. The emotional reactions conditioned in one situation form the foundation of prejudice may be transformed to another situation. Racial prejudices some times find their sources in the hatred developed by a person against a member of his own group. The psychological process of *scape-goat* functions in the same manner. Political leaders often make use of rage or fear responses, conditioned previously to activate people against the people in power.

We can design an experiment to study the transfer of training. Some kind of training is given to an experimental group and the performance of this group is compared with a control group which did not received any training or experience. If the performance of the experimental group is superior in similar task in comparison with the control group it is evident that the experimental group had the benefit of the training. There is positive transfer of training. But if the performance of the experimental group is inferior to that of the control group there is negative transfer. We cannot bring all cases of transfer of training under either stimulus similarity or response similarity. They both interact sometimes to produce transfer.

THEORIES OF LEARNING

In the proceeding sections we have explained how learning is taking place and various kinds of learning. Now let us know something about theories of learning. There are lots of information about theories of learning Each theory explains in its own way how learning is taking place. There may be some overlapping between one theory and another, which means that both the theories explain the same idea independently. **Bower and Hilgard** (1986) enumerated about fifteen theories of learning. All the theories they discuss under two major categories as (1) behavioural-associationist theories and (2) cognitive organizational theories. We shall see a few important theories of learning, the stimulus-responses theories and the field theories.

Behavioural-Associationist Theories

The *stimulus response theories* in other words are stated as *behaviour-associationist theories.* Though the founders of stimulus-response theory were Pavlov and a group of American psychologists, **Edwards** and **Scannell** (1975) state that stimulus response psychology still utilizes *Aristotle's ideas* with some valuable additions. According to this theory the response a person makes to a stimulus is the sign or indication that learning has taken place. The stimuli-is may be external or internal, but makes the organism to be active to do something. In other words the stimulus as the cause makes an individual to respond appropriately.

Thorndike's Connectionism

Edward L. Thorndike (1874-1949) a prominent American psychologist had his influence on learning theories for more than fifty years. Thorndike (1917) has used association as the basic concept in his theory. There is association between sense impressions and impulses to action which was later on stated as responses. This association is known as bond or connection. These bonds or connections between sense impressions and responses are making the habits strongly or weakly. This is learning according to Thorndike. Thorndike's system has been called for sometimes as *bond psychology,* or *connectionism* and this is the origin of the original stimulus-response (SR) psychology of learning. Thorndike has largely applied his theory to educational and social problems. He has published three volumes of books on Educational Psychology and he was called the *earliest educational psychologist.*

According to Thorndike's theory of intelligence, there are as many kinds of intelligence as there are behaviours. The theory states that the individual is constantly exposed to stimulation and he responds constantly. This response occurs in any given situation. In the occurrence of learning a *hierarchy of* responses is set up. Thorndike states this as the *'law of effect'.*

In the learning situation a response is made and it is satisfying to the individual, then the connection between the stimulus and response strengthened. But in the same situation or in any situation this response if necessary, in future may be drawn forth. But, instead, if the response is unsatisfying it is having an *annoying effect* which is weakening the response and recurrence is reduced by placing it at the lower level of hierarchy. There are several factors operating in this., The connection varies with *closeness, nearness in time, attentiveness to the situation, nature of response* and *satisfying event* in question.

Thorndike dealt with also other conditions in learning. He proposed the *law of exercise.* It is the *condition of practice.* But the law of exercise or practice is effective when the law of effect is at work. Thorndike *supplemented the law of effect* in his *law of habit formation* by the repetition of the task which he called the law of exercise. If the learner is benefitted the practice may be quite effective. But more practice in a situation has less effect on learning. If the practice is used when the effect is pleasurable then there is a reward to the learner. Therefore practice may be used when there is a positive result. There were *criticisms* against Thorndike's theory of the *law of effect* and the *law of exercise.* This made him *to revise* his law of effect. He has to *remove the portion,* "annoying consequences are weakening the effects of learning."

Pavlov's Classical Conditioning

This classical conditioning is always associated with **Ivan Petrovich Pavlov** (1845-1936), a distinguished Russian physiologist. Conditioned reflex were the concepts Pavlov (1927) has used after many years of research in laboratory. This has occupied an important place in 'Warning theory. Pavlov's influence upon learning theory has considerable effect in Russia and also in America. The behaviourists like Guthrie, Hull, Tolman and Skinner were also much influenced by Pavlov's conditioned reflex.

The important concepts of Pavlov's theory are unconditioned stimulus, unconditioned reflex, conditioned stimulus, conditioned reflex, reinforcement, extinction, spontaneous recovery and inhibition. Pavlov has conducted all his experiments on dogs. On placing meat powder in a dog's mouth, salivation is made to take place. The meat powder is an unconditioned stimulus and the saliva in dog's mouth is the unconditioned reflex. The dog's salivation in its mouth is a natural reflex for its food, the meat. After this some arbitrary stimulus, the ringing of a bell is combined with the presentation of the food (meat). This is repeated till the dog was able to associate the ringing of the bell with the supply of food. After certain repetition the dog learns to salivate on hearing the bell sound even if the food was not served. The sound of the bell becomes the conditioned stimulus and the salivation becomes conditioned reflex. But the American psychologists use the concepts *conditioned response* instead of *conditioned reflex,* used by Pavlov.

The food was always served to the dog only after the bell was sounded in its regular time. Pavlov referred this as *reinforcement.* Pavlov has discovered in his experiment that a conditioned response (the salivation in the dog's mouth) can be extinguished by repeatedly giving the conditioned stimulus (the sound of the bell) by not providing the unconditioned stimulus (the food). The bell sound usually associated with the supply of food. Now the bell rings, but the food not served. This results that the salivation in the dog's mouth is gradually diminishing and disappearing ultimately. Pavlov called this process as *experimental extinction.*

Pavlov has also discovered the *spontaneous recovery* of the extinguished reflex or conditioned response. As the, result of experimental extinction of the conditioned response (salivation) there was some recovery of the conditioned response (salivation) if the dog was taken away from the experimental situations and allowed to be free in its cage for few minutes. After few minutes if the dog was brought to the experimental situation and tested the conditioned response (salivation) was recovered spontaneously without any further conditioning. It was also found that this spontaneous recovery increased with the length of the rest time for the dog between sessions.

Pavlov's learning theory included *two more constructs* the *excitation* and *inhibition.* During the time of conditioning, the response is elicited by a number of stimuli around the specific conditioned stimulus. The conditioned response may occur also to a nearby stimulus, depending upon the *similarity of the test stimulus.* This Pavlov called as *stimulus generalization.* As result of training there is *generalization of excitation* and also *generalization of inhibition.* This is the *function of central nervous system* during conditions. There is interaction between these two which provide basis for the *operation of cerebral hemispheres.* The incoming impulses through afferent nerves and lower centres reach the special cell of the *sensory system of the cortex.* The excitatory process irradiates other cells over greater area from these special cells. This process of irradiation becomes basis for the generalization of conditioned reflexes among several stimuli. That is, the stimuli that are of similar type also activate the nearby cells I'll the cerebral cortex by overlapping. Therefore along with *excitations* there also irradiation. The irradiation must be counteracted by making the excitation to focus on the original special cells. Otherwise the learning process would be damaged badly. At this stage the inhibitory process comes in promptly as an aid. In this process the positive stimulus is reinforced and inhibits the negative stimulus. The irradiation effect in excitation is reduced. This means that responses are *not really extinguished.* They are only inhibited. But when appropriate aid is provided the responses reappear.

Skinner's Operant Conditioning

The *constructs 'operant conditioning'* are associated with **B.F. Skinner** (1938). It was he who proposed and formulated operant conditioning. For more than 50 years Skinner's learning theory has been occupying an important place in psychological and educational studies. He was born in, 1904 and earned many laurels for his outstanding contributions. Skinner *significantly differed* from tranditional stimulus-response psychology. He proposed *two classes of responses.* They are, a class of *elicited responses* and a class of *emitted responses.* The elicited responses are evoked by known stimuli. Skinner called this a *respondent behaviour.* For example, the pupillary changes to bright light, the knee jerk to a blow on the knee cap are classified as respondent behaviour. On the other hand the *emitted responses are stated as operant behaviour.* This is not evoked by known stimulus. The strength of stimulus is not measured according to normal laws of reflex. This is the function of eliciting stimuli. According to Skinner the rate of strength of response is measured which he called as the operant behaviour. The idea of using the word operant is that operant behaviour is operated in the environment to produce the desired effect. The organism is engaged in operating an instrument or something for eliciting some response in operant behaviour. The rat pressing the bar for pellets of food in Skinner's box, the cat by pulling the latch and opening the Thorndike's puzzle box, and a man driving a car, opening the door of his office, are all operant behaviour according to Skinner. Most of the human behaviour is operant conditioning. Human beings are conditioned to operate some instrument or handle something and the behaviour is evoked.

The operant behaviour is also stated as *instrumental behaviour or instrumental response* or learning (**Woodworth**, 1952, **Candland**, 1968). **Skinner** (1938) has described his theory of operant conditioning in his book "The behaviour of organisms". Skinner, based his theory upon the *two types of responses* and evolved two types of conditioning, the *type S conditioning and type R conditioning.* The conditioning of respondent behaviour is called Type S. In *type S conditioning reinforcement is correlated with stimuli.* The conditioned stimulus, the ringing of the bell is followed by the supply of food, the unconditioned stimulus and the salivation, (the response) is elicited. The reinforcement of the *unconditioned stimulus* is more *important* for Skinner and *not the response.*

For Skinner Type conditioning is more important. This lie called conditioning of operant behaviour. *In Type R response is correlated with reinforcement.* In the experiment Skinner's hungry rat's response is lever-pressing. This response is strengthened by food supply. The rat's bar pressing is the conditioned response and the following food is the reinforcing stimulus. It is the bar-pressing response (conditioned response) that causes the food (reinforce or reinforcing stimulus) to appear. If only there is conditioned response bar pressing there is reinforcement (supply of food to the rat). This is how Skinner differentiates his instrumental conditioning from the classical conditioning. The reinforcement (food supply) is dependent upon the response (bar-pressing). Pavlov's classical conditioning and Thorndike's trial error were considered to be totally two different forms of learning. **Woodworth** (1952) states that Skinner's operant conditioning experimental arrangement bridges the gap between Pavlov's classical conditioning and Thorndike's trial and error.

Cognitive-Organizational Theories

Several theories are combined under the heading of cognitive organizational theories. The Gestalt theory, Tolman's purposive behaviourism, information processing theories of behaviour, cognitive psychology, organization and memory, social learning theory and other recent developments in cognitive theories are brought under cognitive- organizational theories.

There is constant relationship between the environment and the organism. The field theory explains how learning is taking place in the organism in its relationship with the environment. The individual faces his environment with his feelings, emotions, abilities, motives and prejudices. The environment strikes on him and he in turn reacts to the strikes in terms of perceptions. The concept, field includes the environment and the individual. The environment and individual are having constant interaction upon each other and both are changing. The individual particularly is given more importance in the field theory. His reaction to the environment at any given moment is of great concern to the psychologist. Therefore one must understand the environment, the individual and his psychological perceptions of the environment.

Gestalt Theory

The Gestalt Psychologists were the first to give a systematic explanation for learning from the field standpoint. The concept, 'Gestalt' is a German word, which means configuration or pattern. The important process is the perception of the pattern by the individual. Gestalt theory is always associated with three German psychologists, Max Wertheimer, Koffta and Kohler. Gestalts opposed the stimulus-response psychology stating that it is atomistic approach. They state that it breaks down the whole pattern into small elements and then attempts to combine the elements into a whole. Many elements are simply combined and analysed. According to Gestalt the *whole pattern is greater than the combination of elements.* In the combined elements, the result is bigger or different from the combination of elements. Man is something more than the simple combination of elements in his behaviour.

Max Wertheimer (1923) and **Kohler** (1929) enlarged this theory of learning by emphasising it on the *law of proximity.* There are also other conditions constitued in proximity such as *closure, similarity* and *good continuation.*

(1) Proximity : Proximity means *nearness.* Elements very close to each other tend to make a pattern. For example putting series of *dots closely constitute a straight line.* We do not perceive them as dots, but as a straight line. In the same manner we perceive *objects as whole* and analyse them and not perceiving their *parts separately and create one* whole. (Fig. 12.8).

Fig. 12.8 : Proximity : Dots close together are seen as straight line

(2) *Closure* : Closure is anything that closes the gaps. We have a tendency to perceive objects as a whole and complete. Even if it is not complete or having a gap we complete it or fill up the gap and see them as full or complete object (Fig. 12.9).

Fig. 12.9 : Closure : The gaps in the figures are closed and the figure is seen as the Word END.

In many of our drawings there may be gaps in the circle or in the square. We perceive the figures of the circle and the square as complete or whole. There is a tendency to fill up the gap and wish to see a whole object.

(3) *Similarity* : Objects having similar nature are more quickly identified and learned. Words of synonym are learned easily and remembered for long time. The principle is that homogeneity of objects or words have greater access for learning than the objects of heterogeneous in nature. (Fig. 12.10).

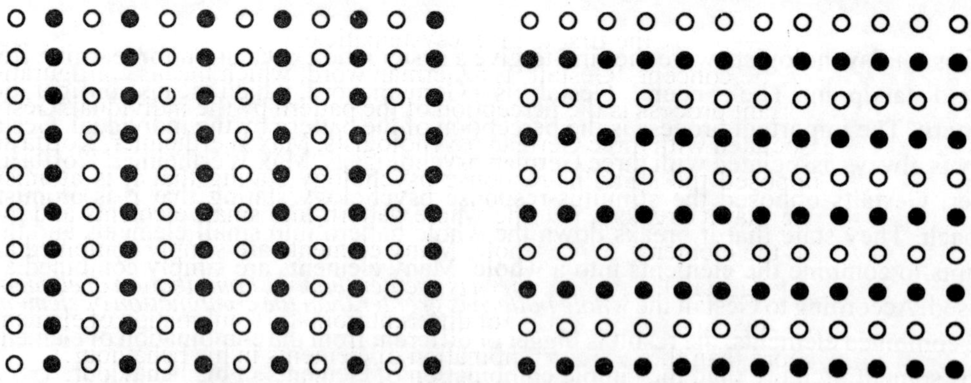

(A) (B)

Fig. 12.10 : Similarity : The Principle of similarity in Figure A the Direction of the surface is Verticle and in Figure B the Direction of the surface is Horizontal.

(4) *Good Continuation* : Objects which are having continuous process are easily learned than those which are discontinuous. For example, pair of words such as man Woman, boy : girl, brother : sister, uncle : aunt and objects such as tree forest, sea : water are easier to learn and remember than pair words like, fork : diamond, mouse : ambulance, shelf : road.

(5) *Insight* : One of the Gestalts, **Kohler** (1927) emphasised the importance of insight in learning. *Insight* may be defined as the *sudden awareness* of *finding out a solution to a problem.* It may be also stated as the ability to relate parts or elements into *meaningful whole* to solve problem. This ability is not taught. It is a *potential ability already present in the organism. Kohler's experiments* on apes brought out the *insightful behaviour in apes.* The ape in Kohler's experiment was able to *perceive the whole situation and learn to solve the problem.* (Fig. 12.11).

Fig. 12.11 : Insightful Learning : The Chimpanzee has learned by Insight to solve the Problem of getting the Banana by using the Sticks.

Tolman's Purposive Behaviourism

Tolman (1886-1959) was formerly a behaviourist. Behaviourism, a school of psychology, was founded in America by **Watson** (1878-1958). **Guthrie, Hull, Tolman** and **Skinner** were all behaviourists originally, but later on differed from Watson's theory of behaviourism and *each one founded a separate theory* of his *own.* In this way **Tolman** (1932) called his theory as *purposive behaviourism.* Later on he associated his theory with Gestalt theory and called it *sign-Gestalt Theory.* The emphasis was on *cognitive nature* which is distinguished itself from stimulus-response theories. According to Tolman mental process is a functional variable. Response in an organism may occur due to stimulation, or by initiating physiological state or past learning or heredity. It is *not possible to see the mental process,* but the *resulting behaviour can be observed.* Behaviour has certain aspects, which may be observed. (1) In the behaviour there is always some *attempt to achieve the goal* or escape from danger. (2) In this type of behaviour *a specific pattern of contact* is established. We always take *a specific route* to go to office from home. (3) Behaviour always *takes the short and easy methods* of achieving the goal or escaping from danger. Tolman states that all behaviour is *teachable, purposive* and cognitive. The *same is taking place in learning.*

COGNITIVE PSYCHOLOGY

The research methods adapted by the researchers to study *more complex phenomena* of the *human mind* resulted *in evolving a new field* of psychology called *cognitive psychology.*

Cognitive psychology is one of the fields recently developed in America. It finds its origin in Tolman's purposive behaviourism. It was founded in 1960 by a group of American psychologists, beginning with the writings of **Chomsky** (1957) and others. Cognitive psychology is becoming a major field of *contemporary experimental psychology.* It is concerned with how organisms acquire knowledge about their world and how they make use of the knowledge in their life situations.

The aim of cognitive psychology is to understand the *mind and its abilities* and achievements *through perception, learning, thinking* and in the *use of language.* Towards this aim cognitive psychologists have formulated theories about *mind's inner processes.* They follow *information processing approach* and consider that the human brain is functioning like a computer. Just like the computer, human mind has many clear levels of organization. Some of the organizations are *pattern analysers. sensory memories and conceptual memories and character analysers,* At some level it can be treated as a symbol manipulating system. According to cognitive theory a *mental symbol* can represent the environment, formulate patterns of symbols of events and scenes in the world. These *symbols* can be *stored in memory* and they can be transformed in activity and thinking.

The cognitive system is an *evalutionary process in human* history. It is occurring *in every human being's life.* The system acts like an instrument of adjustment and as an aid for guiding human actions which are purposeful. The cognitive processes by acquiring information about the world and serve these purposes. The cognitive system is formulating action plans on the basis of information acquired. These action plans are stored in memory like the computer program. These can be recalled into the working memory for guiding actions. These actions are in response to the contents of *short term memory* (STM). The short term memory is active. It has *limited capacity* of storing only a few items of information. In the information processing system the important components are *sensory receptors, effector units, memory, and central processing system.* The sensory receptors receive the stimuli. (input) from the environment. The effector units produce responses to the stimuli. The memory receptors store the data, the action programme. The major mental activities such as *thinking, judging and decision making* are taking place in the central processors.

The long *term memory* (LTM) is more *permanent memory.* The *verbal materials* on each rehearsal cycle is transformed into the long term memory and stored. There are *three kinds* of information stored in long term memory, such as, *sensory perceptual knowledge., procedural motoric knowledge and propositional knowledge.* Cognitive theory explains various aspects of human beings. The child passes through various cognitive stages in its development. *Cognitive development is a continuous process* that takes place over time.

DISSATISFACTION WITH THE LEARNING THEORIES

In the preceeding section we have discussed about operant conditioning as one of methods of learning. Operant conditioning in learning was formulated by **Skinner** which is considered to be one of the theories of learning. Operant conditioning or behaviour is characterized by the interaction between behaviour and environment. The organism operates on the environment and the results of the operation modify the behaviour of the organism. This is quite similar to the purposive behaviour.

One of the important cognitive processes for modern psychologists is learning. Psychologists have developed interest in learning from the research findings published from various sources and the theories formulated by eminent researchers. The Russian physiologist **Ivan Pavlov** (1949-1936) was the first to start experiments in learning on animals and brought out interesting results. What is popularly known as classical conditioning has been identified and described by him. He has also designed the principles for classical conditioning and demonstrated them in his experimental animal, the dog, normal or *unconditioned reflex* and the *conditioned reflex.*

When food is placed in dog's mouth, there is an increase of salivation which Pavlov called as *unconditioned reflex.* When there is an increase of salivation in the dog's mouth on hearing the bell sound it is called, *conditioned reflex.* The practice of performing the conditioning process is to present the conditioning stimulus, (the bell) first and after few minutes presenting the unconditioned stimulus, (the food). The essence of Pavlov's experiment on dog's learning is that the dog learns to respond with increased salivations to the sound of the bell on the idea that food will be supplied soon. Pavlov explained this process of learning from physiological point of view and evolved a theory.

Subsequently, the American psychologists, **Clark Hull** (1894-1952) has explained this conditioning process from psychological point of view and formulated a learning theory which may be called as *reinforcement.* According to Hull conditioning is the basic process. This results in the formation of habits. Hull is influenced by the doctrines of evolution. He emphasized his theory on the *biological drives* in behaviour of the organism. For example, hunger, thirst and sex are biological drives. According to Hull learning in an organism takes place only when a *response causes some reduction* in *the drive.* This means that some learning has taken place. The process of reduction of drive (learned response) may lead to *strengthening of the learned response.* This means that a definite learning has taken place in an organism which Hull called as *reinforcement.* Hull's theory is mainly characterised by *mathematic-hypothetic deduction* assumption and postulates.

Experiments on learning are going on enormoulsy and theories are formulated one after the other. Around 1950s there were lots of confusion prevalied among the psychologists who were interested in the study of learning. They were all involved in the *extensive theory building* and testing their theories in various ways. There were rival theories and the founders of the theories took the issue with them to their laboratories and to the forums for discussion, debates and arguments. During this time **Osgood** (1953) has brought out a book in which he has reviewed many contraversies. His book, "Method and theory in Experimental

Psychology", raised the validity of many of the theories in learning and in other areas. The great ferment an-long the psychologists further continued. Dissatisfied with the prevailing situation on this issue **Skinner** (1950) has published a paper in the Journal psychological Review under the title of "Are Theories of learning Necessary?" in 1950. Skinner's article was not based upon any empirical analysis. It was only a theoriestical assumption. Though this view was not acceptable to many of the researchers, this view is still prevailing. A section psychologists, like **Epstein** and **Shontz** (1971) did not include the topic, "Learning" in their works.

Though the importance of theories of learning is felt considerably, the dissatisfaction of the theories of learning is prevailing in several sections. It is felt by a section of psychologists that psychology is tl ie *science of behaviour* and therefore a science of behaviour must evidently deal with behaviour and relate itself to certain experimental variables. Skinner is not totally against the theories of learning. But he only wanted that theories, whether they are of neural, or mental or conceptual should speak about the intervening device in the.. experimental procedures., But what is taking place is that instead of permitting us to explore the relevant variables the experimental situations have the opposite effect. A theory should permit an investigator in whatever manner he wishes to conduct the experiment.

According to Skinner the principal function of learning theory of the present is not to suggest and appropriate research, but to create a false sense of security and undesired acceptance and the existing conditions should prevail. In such conditions research designs and experiments with respect to some learning theory becomes a waste. Unless the research result is valuable the theory does not prove its value. The value of a theory should be established by any experimenter with a suitable experimental design and desired or selected variables.

Skinner is of the view that a research designed for testing a theory for other reasons and the possibility of examining it may lead more directly to the type of result or information that a science normally accumulates. Thus Skinner's criticism on learning theory is not for complete rejection of theories. The same problem when tested with different theories or hypothesis should yeiled almost the same kind of information and a theory should be tested with different type of variables.

In the proceeding sections we have discured the major theories learning expounded by the intellectural figures who have made excellent contribution to the development of psychology. These eminent persons are associated with theories, experiment, and the experimental subjects. While discussing about learning and the scientists, it at once strikes in our memory Thorndik's cats, Pavlov's dogs, Skinner's rats, Kohler's chimpanzees and so on. This means that these scientists started their experiments first on animals. Animal learning was receiving more importance in the field of learning. They attributed their findings to human beings. What these scientists have done with their animals and what they have observed in animal behaviour were related to human. But to what extent the animal learning is agreeable to human learning? This has led the psychologists to a constant state of ferment with regards to the definition of psychology and learning. These were redefined now and then.

Thousands of psychologists are constantly engaged in verifying the existing theories of learning and proposed new perspectives, ideas, pheomena, experimental results and research methods at a confused headlong rush. Psychology is defind as the *science of observable behaviour.* Towards this experimental designs and methods of research, the selection of the subjects, formulation of hypotheses are planned. Similarly, with regards to learning, experiment are who deal with the animals or non-verbal humans need- sonic kind of *working definition* of learning *that does not require verbal questions and answers.* They have to observe some kind of *overt behaviour* of the subject that could be identified in a specific learning. Based upon the new ideas, learning is redefined by many psychologists. Therefore we may state that learning refers to the change in the behaviour of an organism to given stimulus or situation brought about by the organism's repeated experimences in that situation. This working definition does not involve the subjects native response tendencies, maturation or the physiological state of fatigue or drive.

Watson did not accept any activity that is not observable. For him even 1.1iinking and imagery are only verbal behaviours. Thoughts and images are communicated to others verbality. If it is not communicated it is sub-vocal and of low intensity that does not result in speech. Watson states that perception riot a mental experience. It is a behavioural discrimination of the stimuli in the environment. Emotions for the behaviourists are not feelings, but they are of various states of *functions of glands and body's* chemical *reactions.* Watson, Skinner and their followers reduce all psychology to the study of *observable stimulus-response mechanisms.* Learning for Watson is only *acculmulation of experimences* right from the childhood period and experiences the appear in the form of activities in response to observable stimuli in the environmental situation. Theories of learning dealing with *mental process*, only theoretically explained and interpreted. Each theory has its own research designing but unobservable and does not fit into the demonstratable phenomenon. It is because of this, according to Skinner that there was great confusion in defining learning and designing experiments. Thus a section psychologists around 1950s went against the formulation of rival theories of learning.

The controversal views that prevailed during this period not only caused confusion about the various theories of learning, but also with regards to the validity of various definitions, methods of experiments and the experimental results. This situation resulted, evolving some new ideas in dealing with the subject matter of learning. Authors who brought books on psychology dropped the topic learning and incorporated the subject matter under *"Remembering and Forgetting"*, and dealt with it in different manner. Now the topics are only "Remembering and Forgetting", New concepts are added in this topic and reinterpreted the processes related to learning and memory. The concepts *Memory and Remembering* are synonymously used.

13

MEMORY

Introduction

Psychological researches and the application of various methods of learning were gaining momentum after 1950s in the United States. Several new thinkings were emerged from the laboratory works. On learning *rival theories* brought out conflicating views and controversies. Human behaviour was explained through operant conditioning. *Operant conditioning* methods and the concepts which originated with the Skinner's laboratory rats in the box, began to dominate the research areas in learning and in the study of behaviour. Operant conditioning methods *originally applied on rats* in the laboratory situation are transmitted to schools, hospitals and industries and to other organizations and human beings served as subjects. Operant conditioning is characterized by *interaction between behaviour and environment.* The individual organism operates in the environment and the consequences of the operation *modify its behaviour.* This process of function is applied to behaviour in general and to learning specifically.

The enormous amount of researches in this area, concentrating on operant behaviour, resulted in ignoring the main subject matter, learning in their works by giving more importance to the topics remembering and forgetting. In place of learning, remembering or memory and forgetting jointly treated. Is this the outcome of the advanced research finding to deal with the subject matter *learning* differently, or is it a new approach to study the process of learning and memory to avoid controversies after **Skinner's** *criticism* on learning? Whatever the view is functioning behind this approach the student of psychology is very much interested to understand the fundamental principles of the subject matter of psychology and wanted to go further beyond this.

A correct understanding of the basic principles of a subject matter may serve as a tool for advanced thinking. Whether we use the concept memory or remembering, both are referring to the same subject matter. Memory is a noun form and remembering is a verbal form.

In our everyday conversations we may be often using the concepts *learning and remembering,* remembering and forgetting. Therefore, we may state that learning, remembering and forgetting are the continuous and interrelated mental processes. We have dealt with the first process, the learning. The immediate next process must be memory or remembering (or the stored learned material). The next process now, we are dealing with is memory or memories. The physiological psychologists like **Lenkel** (1968) dealt with memory under the heading of learning. They did not deal with memory under separate heading.

What a memory is really made of? Some are of the view that it is still remaining as a mystery. Precisely the researchers suggest that it is a faculty stored in the *hippocampus,* a structure close to the core of the brain, *maps and organizes information* and then it directs them to the other parts of the brain at once. There are several theories, try to answer this question. We shall discuss them here.

Memory is one of the *most indispensible abilities* that man is endowed with. There might have been occasions in the life of everyone to test his memory. You might have taken part in the quiz programmes conducted in your school or college and answered many questions rightly or wrongly. Your teacher is very much interested to know whether you remember anything from what has been taught in the previous class. He may put some questions on you.

Do you know who is the first president of the Republic of India? When and where the first atom bomb was exploded? Who was the first man landed in moon? Do you remember the name of your first teacher? Where did you spend your last holidays? What is the name of the story you read recently? Who were the guests you invited to your last birth day celebration?

These questions make you to go back to your past experiences and make you to remember something or to struggle with something. If you were able to remember them correctly you feel happy, otherwise you feel miserable. What is actually happening *when you remember* them and *when you do not remember them?* Memory is the concept we use to describe the *first process and forgetting* is the concept we use to describe the second process.

The process of *forgetting is always with us,* but the prospects of memory system in use *eliminates it.* Memory system is not a recent invention. It was indispensable to *Romans* and *Greeks* in the ancient times. The practice of rhetoric is facilitated by memory system. All human learning is verbal in nature. *Verbal learning involves language.* The importance of language in human behaviour has created interest in verbal learning and memory. **Hermann Ebbinghaus** (1850-1909), the German Psychologist was the first to investigate verbal learning and memory experimentally. Similarly, **Sir Francis Galton** (1822-1911) an eminent scientist of nineteenth century, was a pioneer in the measurement of individual mental capacities. He formulated a technique for testing *autobiographical memory,* in which a person is given a cue word and then he is asked to relate it to all those words that he could recollect, and could bring to his mind. These are rated as his characteristics such as *vividness of his memory* and for the *level of detailed information.* Galton found that older people typically recollected things from their early life. But at the same time younger persons recollected from much more recent events. Galton's findings are still held as true. From that day to the present time only *very few* substantial *experiments* were conducted. As result of it more valuable information are not available in memory on bar with other areas of psychological studies.

Researchers in the field of psychology, unlike the physical scientists, carry out their works according to their own plans and bring out the results with different conclusions.

This type of varied conclusions often create controversies among the theorists. This type of trends cause delay in further researchers.

But recent studies show that *memory changes with age,* which is in a complex way. According to the recent studies it is observed that some older persons have trouble with their memory tasks which may be due to their memory skills. The memory skills become *rusty* or they had *never reviewed* while the younger persons have stimulating situations and things, they often review their memories.

What is Memory?

The human *memory system* is a very vast, *every-changing web* reacts every new experience. Most of the experiences are too fast to provide noticeable traces on the webs. Memory is far more than the sum of the past events, experiences and knowledge. It is *memory that makes us what we are. The human self* can be explained only on the collective and private experiences and knowledge which *all rest in memory.* Like all other human abilities memory *does not always operate reliably.* Everyone of us may be having a collection of records of our *fallible memories* with the advancing years. We may be having the accumulated fantastic body of experience and knowledge, but we do not have access to the past experience and knowledge that we wish to get. The matter may be inactive or lay dormant. What we could recall and remember is only a fraction of the vast storehouse that our memory is comprised of.

The human experiences, mostly *sensory and motor,* are not lost totally and also not remembered fully. Every experience leaves some traces in the brain and some traces seem to become permanently framed into the brain tissues. Generally the events stored in the brain are highly selective. The events or information or sensory impressions are scrutinised and edited and highlighted instead of recording the complete events and experiences of our life. **Wilder Penfield** (1959) a Canadian neuroscientist has brought out a very remarkable result that seem to show that all our experiences are recorded somewhere in the brain. According to Penfield the brain tissues *do not have the pain receptors* and therefore the brain surgery can be carried out without the general anaesthesia. This information he has obtained while treating patients for *epilepsy. Every experience* realized through senses is *imprinted* in the brain tissues and these are *reproduced* when proper *stimulation is made* which we may *call as the memory.*

We may ask the question as: how much information the brain can hold? The possible answer could be that the brain can hold an unlimited amount of information. This means that the brain is with the *dynamic memory filing system.*

The basic unit of memory according to one researcher is the *remon.* This is the basic unit of memory. This guides the formation of other memories. It is also stated that *no single part of the brain* that is fully *dominating over memory.*

It is said that the *neocortex* part of the brain is maintaining the functions of the *long-term memories.* When it is stimulated electrically the neocortex may release a *push of sensations* such as *sights, sounds, smells, tastes and touches* from long-stored stocks. The neocortex is the *gray outer most layer of the brain,* which is mainly responsible to reproduce long stored materials. This means that the *learned materials get registered* or *form impressions* in the brain cells in the *neocortex region* and when there is any necessity *it is stimulated* and the learned material stored in the neocortex region is *pushed* in the form of sensation of the respective area.

In learning what actually is taking place is getting or *registering the sensory impressions in the brain areas.* But this is *not observable* function or activity as the process is going on in *neural areas.* But how to accept that learning of a particular matter has taken place? It is *known only from the behaviour of the organism.* Whatever the learning has taken place verbally or by any other sensory experience the organism reveals or exhibits on certain occasion when necessity arises. Where was this learned material till such time ? The material was in the memory, stored.

Memory may be considered as one of the *oldest* art in the world. Memory was the sole medium in the ancient times, by which all history and knowledge were recorded. Before inventing the language in written form memory was the only source of storing the information. The ancient Indians were trusting only the human memory for teaching, learning and recitation by the teachers and the students. But the invention of language in written form made the matter simple by putting everything in written form.

The development of good memory skills is very essential and also an endless fascination. The word *mnemonic* is an ancient Greek word which means *memory* or *memory aid* or remember. This concept is drawn from Mnemosyne, the name of the ancient *Greek Goddess of memory.* Mnemosyne was said to be knowing everything of the past, present and future. Mnemosyne gave birth to *nine Goddesses* each one presided over one artistic discipline, like literature, art and science. Each one is regarded as the spirit, inspiring a poet. Thus *Mnemosyne* was revered as *the source of creativity.*

According to the Greek mythology, when a dead person entered the underworld he should drink the water from *Lethe,* the river of forgetting. Then at once all his memories would be destroyed. These mythological descriptions reflect that *memory was a faculty* which the ancient Greeks were paying to it the highest possible regards.

The most celebrated classics in the Greek world were the *Iliad* and tile *Odyssey.* These two Greek epics are the earliest works celebrated in the European literature. The Greek philosphers of the Fourth Century BC have thought about *two types of memories* such as *natural memory* and *artificial memory.* It was **Simondes** (556-468 BC) a Greek Poet, who discovered the importance of *visualization of memory* and the *methods.* He used *superb imagery* in his writings, **Aristotle** (384-322 BC) the Great Greek philosopher of this period also had understood the values of the senses and wrote that it is *impossible to think* of anything *without the mental picture and memory* and these two are belonged to the same part as imagination. Aristotle was of the view that *natural memory depends upon association.*

One of the *techniques of memory,* the Greeks were using was the *rhetoric* which was considered as the *art of public speaking* for attracting the public. The ancient Greeks were also the *inventors of the theatre* in western tradition. Public speaking is an art for them, mainly depends upon the *memory of the performers.*

The Roman orator, **Cicero** (106-43 BC) had introduced the *Greek memory methods* in the Roman rhetoric world and the *technique mnemonic* became a crucial component of the rhetoric technique. Those who speak convincingly without recourse to written notes were highly regarded among the Roman orators. Cicero considered that *a good natural memory such as his own,* could orate *non-stop for three hours.* His writings are regarded as the model of *Latin Prose.* Cicero denounced Antony, the Roman ruler in his work **Phillippics** which caused his death by one of the Antony's Generals.

In the middle age period the *art of memory* and the techniques became the part of *Educational curricular.* During the period of *Renaissance* there were many progressive thinkers like **Bruno** (1548-1600 AD), an Italian philosopher, who developed pantheistic philosophy. He was burned at the stake for hereby.

But thinkers of modern world after Renaissance did not pay so much interest as to develop *Rhetoric* or *memory techniques.* This state was continued for about three hundred years. Around the fourth quarter of nineteenth century *memory* had become the *centre of attention* for the scientists and research and the new science of psychology has received greater importance. Theoretical and experimental studies on *learning and memory or remembering* have brought out enormous amount of information which attracted the educationalists and other scientists. Various types of instruments and methods were designed for assessing the *level of memory* in individuals.

Penfield's (1959) epileptic operational experiences brought out several new information *about memory* though later on some investigators raised some doubts on them. While performing one epileptic operation it has aroused his curiosity to know what would happen if an *exposed brain was stimulated* with electrodes directly. He obtained the consent of the patients for touching their brains more times. Penfield while performing his task, it appeared that he made quick sudden touches of the brain parts. It was so occurred that he *touched the memory areas.* Penfield found that when he stimulated certain *area of the cortex,* especially the *Temporal Lobe* of the brain, the patients reported the experiences in small parts. This looked like their *long forgotten experiences.* The patients started to speak about what had happened long back in their lives.

Penfield's realized that this must be the area of the brain where memories are imprinted. In such cases whatever experience a person had experienced may be lying down and they are waiting for some *suitable stimulation* to bring them back to the mind. This experimental results were verified by some investigators and found that Penfield's conclusions that *memories are imprinted in the brain permanently are not correct,* but the *memory impressions* are changing. Subsequently, the experimental results show that *memories are remembered from various parts of the brain.*

Memory is not a single event or matter. It involves several things. But it would be easy for us to think that memory is a kind of *mental storage chest.* But it is not. In the common sense *world, memory* is always *associated with brains.* Memory is a system of process by which learned material is retained. Memory is a faculty by which things are remembered. It is an act or instance of remembrance.

In our everyday life we use the word remember. For example, we remember certain names, dates and numbers. This is one way of remembering. The other way is to remember how to participate in the coming sports events, whether to go there by car, train or plane. The *first is verbal* type of remembering and the *second is muscular type.* Both are stating the same event, the revival of certain responses acquired earlier.

Memory may be defined as a *complex process* which is active and generative in nature and constantly *working with learned materials.* The more we learn and wider our experience, we acquire more and we feed our memory. All sorts of materials that come from our environment are the raw materials for our memory. These materials are often received and rearranged and kept ready for use in our memory. Therefore, **Smith and Smith** (1966) define memory as a *collection of responses* of a specialized and integrated kind which can be used when needed. Memory is often put into work. It may perform simple task as well as complex task. This task may be a simple task like recognizing a person whom you saw in a function or marriage or dinner or a *complex task* like *reciting a passage* from Othello backward or presenting the philosophy of Plato.

We see in our everyday life people often accusing of their own memory as bad when they *fail to remember events* that took place in their life some years ago. But if we go into the details we could see that we remember many things and we forget things rarely. Are you not recognizing your childhood friends? Don't you remember how you had learned to play cricket? You also remember the first motion picture you saw. Therefore, you remember many things. The function of memory is called retention. Retention is a *passive process* and remembering is an *active process.* Memory involves both the processes.

In our everyday life remembering is taking place in several ways. But we are not aware of it and we are not able to describe it properly. When you want to recall an event or the name of a person you first speak to yourself. Is it Ramasamy? Is it Kandasamy or Kumarasamy? You might have spoken even loudly sometimes. You may fail to recall the name sometimes. But, still you may remember it and not forgotten it. Sometimes you may recognize the event or name even without recalling it. We have learned something sometimes ago and now we want to remember it. **Woodworth** (1952) an American Psychologist has evolved a formula for this process.

L - I - R

Where L stands for learning, I for interval time between learning and remembering and R for remembering. The duration of interval time may play an important role in our remembering.

Where the Memories are stored?

The whole memory system is fully controlled by different parts of the brain. There are different types of memory. But where these are located and which parts of the brain are controlling them? This has been perplexing the psychologists for a long time. Several hypotheses and theories are formulated and each theory found a location of its own. One of the theories states that memory is stored as electrical charges in the brain cells. Another theory states that there are memory molcules and the whole brain cells hold the memories. A third theory states that memories are the remembered information which cause changes in the chemical substances of some cells. Whatever different views exist the brain is the basic source where memories are formed in various parts and monitored.

The human memory system is evolved in some fashion in which *new skills are built* from older skills, and facilitating the *growth in whichever brain area is most convenient.* This type of process results in various types of memories. But there is some kind of interaction going on constantly among the types of memories. The different types of memories formed are located in different parts of the brain. The natural process that is taking place is that the memories are *genetically encoded.* This takes place mainly Loin the part of the brain called the *caudate nucleus* which is one of the *basal ganglia* that has long tall or extension (Fig.13.1). All conscious memories involve the *hippocampus.* The hippocampus is a portion of the cerebral hemispheres found curled in the basal medial part of the *temporal lobe* that is considered as the important part for learning and memory (Fig.13.2). This is an *arch like* organ, is a part of the *limbic system.* This is an unconscious area of the brain, lying beneath the *outer layer* of the *cerebral cortex* of the brain.

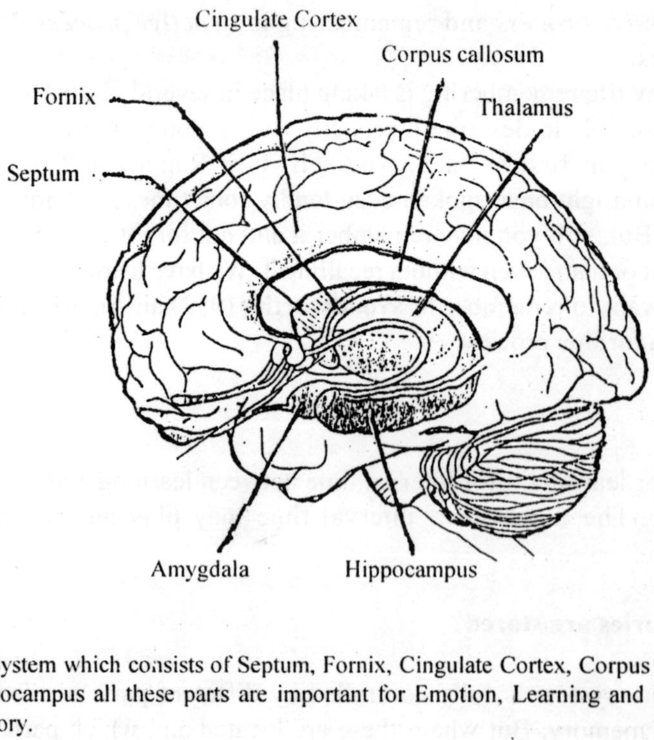

Fig. 13.1 : Limbic System which consists of Septum, Fornix, Cingulate Cortex, Corpus callosum, Thalamus, Amygdala and Hippocampus all these parts are important for Emotion, Learning and Memory. Mostly the Store-house of Memory.

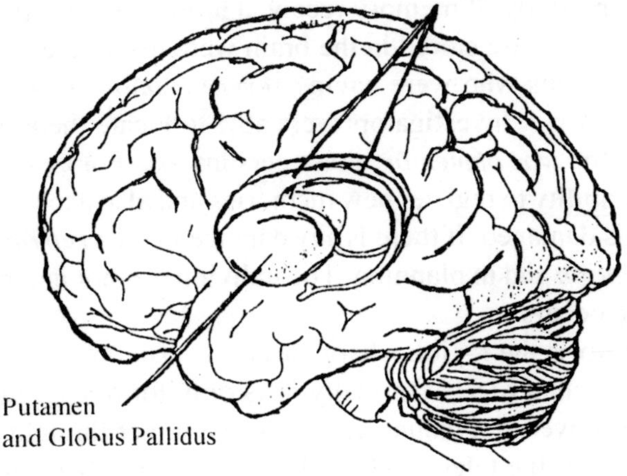

Fig. 13.2 : Basal Ganglia which consists of Caudate, Putamen and Globus Pallidus

When we are paying *attention* to the information conveyed through our senses, the information are *sent to the hippocampus* along the *nerve pathways* from the cortex. Some rapid activities are going on by replaying and making changes. At night times when we sleep the processes continue. During rapid eye movement (REM) some of the neural records in the hippocampus are replayed producing the curious snatches of the past activities that we experienced in dreams. The replaying takes the information to the deeper level. The replay of the hippocampus memories is transmitted to the *neurons* that are laying in the *cortex area.* Specifically in the temporal lobes on each side of the brain. Significant exciting events tend to be replayed regularly. After replaying for three years the *memories are lodged in the hippocampus and in the cortex areas.* These are the *long-term-episodic memories,* from where we can recall the events most as if we are having the memories all over again. If events *replayed less frequently* due to certain reasons, then the patterns that deal with them in the *hippocampus fade away* and are lost or become inaccessible.

But even if the memory of the events has disappeared, the events, *factual events stay* there encoded in the *cortex area, as semantic memories.* For example, you met one Mr. Suresh and this meeting and the person are stored in your hippocampus memory. This memory disappeared and you left with long-term knowledge of who is Mr. Suresh. But you do not remember when you were first introduced to him. You could recollect who is Suresh, but would not recollect when you first met him. This is hippocampus memory. The hippocampus memory plays a crucial role in *consolidating the long-term memories.* During the period of three years time it pushes the memories for a *permanent storage* in the cortex region. If they are stored there once they no longer dependent on hippocampus for a recollection. But one thing with the long-term-memory is, the cognitive maps of places and spaces may help us for finding out the way. These are held permanently in the hippocampus. Persons who suffer from *damage to the hippocampus* part of the brain *suffer from permanent loss of memory.*

Recent research findings in psychology and biology bring out new information and interpretations often. The study of memory received more attention than any other area of studies. Where the memories are stored in the brain were discovered gradually. This is by keenly observing the persons who were having *severe memory problems.* These persons had *head injuries and strokes.* Investigators were able to locate the area by comparing the *site of the damage* with the *type of memory* these persons *lost.* For example, the investigator was able to see that the ability to register new memories and also access to them is lost when the hippocampus area is damaged. If there is any damage in the *prefrontal cortex* area there is problem *in concentration* and in planning. The individual is not in a position to solve any problem if this damage occurs.

The *prefonatal cortex* is the region where the *short-term information* or memories are stored. The short-term memory (STM) usually lasts only for *few minutes* till the business for which it is intented is over. For example, an information or thought to be conveyed to a person in the next room, or the telephone number, the car number, the house number, all these last for only few minutes. These are short-term-memories. The *amygdala* is the *emotional centre* of the brain and it is the store house of *powerful emotional memories* such as rage, anxieties and phobias.

It is also the storage site of the *memory of facial recognition. The cortex is the location for storing the episodic memories.* The episodic memory is the memory of a particular incident or a particular time and place. These memories are the records of the past experiences often replayed for a period of upto three years in the *hippocampus region* and *end up scattered* throughout the cortex region. The *hippocampus* is the storage site of the *spatial memories.* This records our routes knowledge and places. It is the part of the *limbic system* and has a major role in consolidating the *episodic and the semantic memories* in the *cortex region* (Fig.13.1). *The limbic system* is like a ring of interconnecting pathways and centres, includes the *septum, cingulate gyrus,* hippocampus, amygdala and aenterior thalamus, important in memory functions.

We may be often receiving various types of information in our everyday life regarding some facts and figures and details, about some objects and places. Memories of this type are called *semantic memories* which are stored in the *temporal lobe of the left hemisphere of the brain.* We are also experiencing the *procedural memories.* This includes the *acquired skills* by training and practice. For example, learning to drive a car, or to ride a bicycle, learning to walk on a bar, are the memories stored in the *putamen part of the limbic system* of the brain (Fig. 13.2).

Now-a-days we find new methods of brain studies in operation. Investigators could see the precise brain areas that are activated during recall of different types of information. They are able to map the general areas that process the main types of memories. But it is strange that no two investigators encode the information precisely in the same manner. For example a musical note displayed by a musician is viewed by him as a *professional field,* while a sculpture views it as only a *sound.* Thus the highly designed *scientific techniques of scanning* the brain function in locating the memory storage sites remains still a mystery. However, information so far available with regards to the function of various types of memories and their locations fulfil the present day requirements in the study of memories.

Memories and Self-Identification

Our memories of the past and present provide us with the sense of *what we were* and *who we are now.* The preceptions of the events that we have experienced and the feelings and emotions we have felt are all stored in our memory. This process helps us to shape and to see how we are as individuals in the competitive world. Sometimes the individual is defined in terms of what *he remembers.* As we go through the life, year by year we are acquiring and storing layers of memories which tell us what we are and help us to decide what we will become in future. Every incident and emotion whether it is stored consciously or unconsiously it *can create new ideas,* ambitions or attitudes which will have the influencing effect in the behavioural patterns.

Let us imagine, that we have lost all our memories for sometimes. What remains? We are totally without any sense of understanding about the world and no experience and knowledge may help us to do anything. We will be without any thought about ourselves. Let us come back to our present memories in which state we have no any of our past experience and knowledge and also not able to make out anything new. What this state is actually? In this state there would be no chance for revealing any of the personality traits. We would not find in us any intellectual, emotional and social growth and development. Researchers believe that this would be the state of affair of a person quite similar to a person whose *brain was damaged,* specially the hypothalamus and neocortex areas. Thus the memories play wonders in the life of a man. A man's individuality is identified only through his memories.

What type of attitudes we develope to life? It is not easy or possible to state this. But one of the followers of **Sigmund Freud, Alfred Adler** (1870-1937) has some experience on this phenomenon. Adler, an Austrian psychiatrist, was of the view that the first memories of human beings show some identity to their fundamental attitudes to their life. Many persons who were ourstanding in various walks of life mostly the scientist, poets, creative authors, artists and Army Generals seem to have determined their fundamental attitudes to life by their first memories. It is the human tendency to remember certain things selectively. This memory goes back to the *early years* of life which may be the *first memories* and this selectively remembered things reinforce our ideas which naturally seem to *determine our life attitudes.*

Everyone of us might have had both the *pleasant and unpleasant memories,* and we may have the occasion to go into the past memories. It may be about the great achievements made in our life and we may feel very happy and be proud of them. Similarly, we may be having the memories of our unpleasant experiences and insults we had in our life.

George Gorden Byron (1788-1824), one of the great romantic English poets, noted also for his passionate and disastrous love affairs, expressed his *most unplesant and painful memories* of his life in his famous poem, **Childe Harold's pilgrimage.** Similarly **John Keats** (1795-1821) another great romantic English poet of the same period also had very unpleasant and *painful memorial* in his life. He prayed to God with all his serious devotional feelings to bliss him with the abilities to forget all his painful memories. Both the poets have identified themselves with their memories and revealed their personality characteristics.

Memory and Intelligence

Memory and intelligence were studied very widely right from the time of **Hermann Ebbinghaus** (1858-1909) the German psychologist and **Alfred Binet** (1857-1911) a French psychologist. Both Ebbinghaus and Binet started their research studies independently at different times. Ebbinghaus conducted experiments on verbal learning and memory. It was Fechner's work, 'Elements cf psychophysics' inspired Ebbinghaus to apply scientific methods for studying the mental processes. This was the beginning of the experimental study on memory. Following this study only some studies were carried out on memory and only certain amount of information were brought out.

Similarly, Alfred Binet, who has been contemplating on the problem of the backwardness of some children in schools, was entrusted with the task of finding out the cause of it and solve the problem. This resulted in the study of the level of intelligence of the school children and solve the problem. He designed tests for different age levels and measured intelligence. This was the beginning of the study of intelligence. After this immediately all over the world studies were conducted enormously.

But the study of the relationship between memory and intelligence is sill going on. However we cannot totally reject the idea that there is close relationship between memory and intelligence. In the common sense world the prevailing idea is that persons who have good memory are very intelligent. The other way also is said that more intelligent persons have very good memory. This is a hypothetical idea. It can be tested by conducting experiment and verified. A group of 30 or 40 college students may be invited and an intelligence test may be administered to them. The individual scores can be collected. After 3 days the same group of students may be invited and a memory test may be administered to them. The individual scores can be obtained. Now, we have the scores of intelligence test and the scores of the memory test. We can compare a student's two sets of test scores. Who are more intelligent by getting the highest scores and who are those getting the low scores. Correspondingly what their scores are in memory test? According to our hypothesis the students who have scored high on intelligence test also should score high on memory test and also those who scored low on intelligence test should also score low on memory test. If these two sets of scores correlate highly in the case of individuals we can confirm the hypothesis that those who have good memory are very intelligent and those who have poor memory are not very intelligent. A product moment correlation coefficient method can be applied to the two test scores to see at what level the association between those two variables exist. In this way we can assess the relationship between memory and intelligence.

The ability of memory vary widely between individual persons. Similarly, intellectual abilities vary widely between persons. It is also observed that a person with *good memory* need not necessarily be *more intelligent* than the person whose memory is not good. But however there is a connection between memory and mental abilities. Generally we used to classify people according to the perceived strengths and weaknesses of memories and intellectual achievements. In the community we find in our everyday life how memory is very much indispensible. It is mostly *individual* in its characteristics like the personality of an individual. Intelligence also is said to be the same.

In any intellectual performance memory certainly functions as an asset. In all human performances which need mental abilities, memory and intelligence function to the required levels *independently* in the activities and also *coordinatively.* But the coordination or link is *not straightforward.* In any mental activity when a person begins to solve a problem he cannot solve it instantly but waits and watches it for some time if it is a difficult problems. what he was doing during the time of waiting and watching? *He goes into his memory and traces the cues* which should be suitable to the problem. It is not thinking, *but searching the event in his memory.* It is not always the case that an individual of high intelligence has a good memory in all the areas he is dealing with.

A person who is very well performing in one area may be totally unreliable in other areas. He may not be remembering certain simple things which are very essential for his task. In classics we often come across that many creative authors, poets, novelists and painters describe certain characters behaving with unrelated activities. It may be for the purpose of amousing the readers and onlookers or it may be the real characteristics of the person who displays one thing for another, not of overlapping of the memories. The English play wright **Richard Sheridan** has introduced a character called Mrs. **Malaproup** in his play 'The Rivals' who often misuses unintentionally one word with another of similar sound. For example, *I am not under the affluence of alcohol.* Instead of "influence" the person uses the word "affluence". This is how memories function in some persons.

The study of intelligence now and then bring about several new information and new interpretations. Psychologists now speak about *multiple intelligence*. There are about *seven types of intelligence.* Each type speak about atleast *one aspect of memory into action.* Every aspect of memory deals with one of the areas of the tasks in which one is interested. In the types of intelligence, the skills of memory are associated. The mathematician, physicist, or astronomer may score *high in logical or mathematical intelligence.* Their memory can process rapidly in this type of information. The architects, painters possess more of spatial memory. These persons have intelligence of *visualizing paintings and drawings in three dimensional* manner. This may be called *spatial intelligence.*

Persons who have mastered and deal with several languages and speak them have high level of *linguistic intelligence.* These persons are guided by *superior memory for words and their sounds.* The professional musicians and singers have *exceptional musical skills* and talents guided by *musical memory.* They acquire this *musical intelligence* by *constant practice.* The ability to read other persons mind and behaviour is the *social intelligence. Recognition of skills* of others and *leadership skills* stem from *superior memory for names* and *personal details* of others and the recognition of behavioural areas revealed by others. But the common characteristic to all the different types of intelligence is the ability to fix one's attention on the task and to *bring out connections* between the *existing knowledge* and the *new information.* These are the ways of developing good memory. But however high intelligence and good memory one is endowed with, *interest, practice* and *dedication* in the performance serve as the important *motives* in organizational and industrial works.

Hormonal Effect on Gender and Memory

The whole human body is fully wet by the hormones secreted by the endocrine glands for efficient functioning of the body parts and for maintaining the metabolism. The brain also like all other parts of the body is shaped partly by the *sex hormones.* These hormones affect the functioning of the brain throughout the life. These hormones are partly responsible for the different ways in which the *memories function in men and women* and for the things they remember. This means that the memory functions differ in men and women which is mainly caused by the *genetic factors.* Men remember certain things quickly, similarly, women remember certain things quickly and accurately.

How men and women *differ in their memories* or in remembering things can he tested by a simple test. If we ask a person to describe the way to a particular place we can see the different ways a man and a woman describes the way. The man may provide a verbal description of the way to the place by giving the particulars similar to a *verbal route-map* which may be consisting of the name of the road, numbers, distances, turnings, left or right, the junctions and so on. Similarly, a woman also may describe the way to the place, by using the landmarks by stating the blue building, the tall tower building, the white gate at the right, proceed on to the tiled shed on the left, a small footpath and so on.

In guiding a person to a particular place some differences seem to appear between the man and the woman. It appears that men are more conscious of the land marks *where it is located. Man's approach to the place is, where?* They do not give importance to what. In contrast women pay more *attention to what* and therefore they remember the *specific features.* Thus there are two ways of perceiving and approaching the world. This difference occurs because the visual information is processed through *two different pathways in the brain.* It has been investigated and observed that the *'where pathway'* identified with the routemap like, the distance and position. This process is *stronger in the right hemisphere of the brain. This is in the case of men.* At the same time the *'what pathway' is stronger* in the *left hemisphere of the brain* which identified with the *landmarks* where the *long-term memories are stored.* This *'what' 'approach'* is in the case of women. This new information may be incorporated with the existing knowledge.

At the time of the general function of the memories, both the pathways are brought together *to the front part* of the brain and we are made conscious of the functions of both the pathways. The information fed in each pathway differ. According to the strength of the pathways, the activity of the brain is mostly functioning. The *left hemisphere is specializing in detailed information* while the right hemisphere takes into account the *whole memory stock.* Men use the right hemisphere more for recalling the learned things. This tends to be advantageous to *spatial memory* for men. Women depend upon the left hemisphere pathways which facilitate women to recall details much better than men. It is found that women are faster than men in finding out the small differences in things.

Studies show that women are more faster than men in indentifying small changes more accurately than men in *recognizing plants.* Further the face recognition area of the brain is situated on the *left hemisphere pathways* of the brain. It is because of this that women develop important social skills.

It is the hormone function that probably causes the subtle gender differences in the brain. The sex hormones also affect the general memory. Pregnant women sometimes do not remember many things in their everyday life, because of the severe fluctuations in their hormone levels. They also face memory problems due to the drop in the *oestrogen,* the hormone secreted by the overies during the period of menopause. The memory differences between sexes is only at a *very small level* compared to the individual differences in memory. A good memory depends upon the training we take up for improving it. The difference between the right and left hemispheres can be illustrated figmatively (Fig.13.3).

```
M M M M M M M M M M
M M M M M M M M M M
M M M M M M M M M M
M M M M M M M M M M
M M M M M M M M M M
M M M M
M M M M
M M M M
M M M M
M M M M M M M M M
M M M M M M M M M
M M M M M M M M M
M M M M M M M M M
M M M M M M M M M
M M M M
M M M M
M M M M
M M M M
M M M M
M M M M
M M M M
```

Fig. 13.3 : Views may vary among individuals. Look at the Large Figure. What you see? Look at the Small letters. What you see? The large F is composed of small M letters.

It we look at the large figure as a whole we see *a large F,* which is composed of small Ms. In this case our *right hemisphere* of the brain *is active.* But if we concentrate on the small Ms we *do not see the large F.* In this case our *left hemisphere* of the brain *is active.* This is a very good example to illustrate the *"Where"* and *"What" pathways* of the hemispheres of our brain.

The subtle difference in memory between men and women is caused by the action of the *hormones partly.* The action of the hormones takes place in the *male* individual *before birth.* The periodic doses of *testosterone* which is *secreted by the testes* is mainly responsible for the formation of male characteristics. This process mildly inhibits the growth in the left hemisphere of the brain and the activity is reduced. As result of this the males tend to have increased activity in the right hemisphere. Since the testoterone inhibits the activity in the left hemisphere the female hormone *oestrogen* begins to *process rapidly.* In the middle of the menstrual cycle the oestrogen hormone levels increases. Because of this women have more advantages over men in processing the memory tasks in detail. This is how the hormonal functions cause gender differences in memory functions.

Organization of Memory

Experimental psychologists have contributed very little to our knowledge about the memory system. There were too much of lethargic and aversive attitudes among them towards the development of scientific techniques for the study of memory. They were of the view that the mnemonic devices are unhealthy tricks used by only the wandering gypsies and the stage magcians. As result of this type of attitudes nothing valuable is known by the young psychologists about the wonderful faculty of memory. But in spite of this type of aversive attitudes, no doubt, the memory systems are becoming much more effective and the memory systems are well established. The memory systems are mainly applications of the methods of organization of memory. This phenomenon is a very recent development.

Skilful investigators can make amazing achievements in the study of memory. **Norman** (1969) an experimental psychologists has discussed the issue in detail. Most of the memory systems are the applications of one or two methods. They are the *method of loci* and the *method of analystic* substitutions. *The basic idea* of the method of loci is the old one. The basic idea of the method is to *impress on memory* a *mental image* of some architectural structure such as some magnificent building, Cathedral or something very impressive and then *store the information* in a *well defined manner.* Loci is a Latin word which means *places (or location).* Recall of any information is accompanied by *visualizing each place* or location and retrieving the contents of the place or location.

Though this *method of loci* is a recent application in the study of memory it has a long past history. The ancient Greek poet, **Simonides** was the first person who has given an explicit account of how to use the method of loci. Based on the method of loci elaborate memory systems were developed during the period of *Renaissance.* Detailed descriptions were worked out about the types of architectural structure best sucited to the work for *impressing the structure on memory.* The method of loci increases recall by fitting the material to be remembered into the previously learned organizational scheme. The method of analystic substitution is based on the same principle. But the organizational scheme of this method is different. This method is used to develop a system of memorization of unrelated list of items like *social security* and *relative service numbers.* This method is used only to provide support to the method of loci. Thus the first step is to prepare or invent a code that substitutes more easily the recalled symbols for numbers.

In this procedure of inventing the codes, numbers are converted into letters. In the case of some, the number is converted into the corresponding letter of the alphabet. For example,

5 is E. Some based on sound. For example, 6 for S, 9 for N and 10 is T. Now we have the numbers and the corresponding letters of the alphabets. 1 is A, 2 is B, 3 is C, D is 4, E is 5, 6 is S, 7 is G, 8 is H, 9 is N and 10 is T. Now the letters may be paired by adding one more letter are initials and presented here below.

1	2	3	4	5	6	7	8	9	10
AR	BR	CR	DR	ER	SR	GR	HR	NR	TR

1	2	3	4	5	6	7	8	9	10
AR	BR	CR	DR	ER	SR	GR	HR	NR	TR
AC	BC	CC	DC	EC	SC	GC	HC	NC	TC
AD	BD	CD	DD	ED	SD	GD	HD	ND	TD
AS	BS	CS	DS	ES	SS	GS	HS	NS	TS

Fig. 13.4 : Organization of Memory

Now the next step is to look into the *pairs of letters as initials* and think of a famous person or some one whom you know very well with the same initials. Now the *faces* of these persons are becoming the *visual images.*

For example

1. AR stands for Arcot Ramasamy
2. BR stands for Beena Rai
3. CR stands for Carl Rogers
4. DR stands for Dharma Rajan
5. ER stands for Edison Richard
6. SR stands for Samuel Ranganathan
7. GR stands for Ganapathy Raman
8. HR stands for Hari Ram
9. NR stands for Nathu Ram
10. TR stands for Tara Singh

The same principle may be applied to places also. This is how memory organization is made for retaining subject matter in our memory.

The *guiding aid* for the success of memory system is organization. The memory systems provide organizational structures *for storing* and *for processing* of the material that is learned by the organism, to be successful. Till the recent times psychologists did not show much interest on the *process of organization of memory* and did not include the phenomenon in their works. But the new approaches in the field of experimental psychology have brought out a *new stock of experimental* data which added new vigour to the organization of memory.

Investigators put more efforts in collecting data and to describe the phenomenon in various hedings like *clustering, subjective organization* and *memory.* **Bousfield** (1955) **Tulving** (1972), **Mandler** (1967) and **Underwood** (1964) have made notable contributions in this area. These contributions, no doubt may create new type of thinking and may provide interesting avenues for the young scientists who are aspiring for exciting performance in this area.

INFORMATION PROCESSING AND SENSORY MEMORY

Psychological investigations are continuous and non-stopping processes which from time to time reveal interesting information about human behaviour. One of the areas of the study is memory and psychologists often discover people whose memories found to be really astonishing. How a person is able to retrieve so much materials from his memory while another person is able to retrieve only few information? Why do we remember certain materials and forget many materials? By understanding certain basic information about the memories and forgetting we can answer these questions. At this context we must know that at present time the dominant approach the psychologists make to the study of memory and forgetting is the *information processing approach.* The information processing theory is a very recent invention in the study of memory and forgetting, though it is also mainly applied in the communication process.

What is information processing theory? It is a theory of memory modelled on the flow of information through a communication system. Though this theory is applied in several other areas of study we are concerned here with memory and forgetting. This theory is mainly concerned with *discrimination and detection* of correct *threshold* out of many. Psychophysicists call this as the problem of discrimination and scaling. It is determination of difference in thresholds (stimuli). It is the relationship between perceptual experience and the magnitudes of physical stimuli. The organism is invariably *sensitive* to many factors (sensory stimuli) apart from the stimulus energy, there is every reason to assume that the discrimination thresholds are also unstable.

A stimulus may have several characteristics, which may be weak or equal or stronger. Further, several sensory experiences might have been already stored in the memory and therefore *which of the particular information* out of the bulk should *be retrieved?* This is the basic principle upon which the information processing theory is formed. In the case of memory there are varied stages of memory.

One simple form of information processing theory is that there are distinct storage of memories through which information passes and certain control processes determine what information gets from stage to stage and which can be remembered later on. All the stimulations from outside reaching the *receptors* stay in bulk. But it reaches very fastly the *sensory memory.* The duration of the period varies from one sensory system to another, probably not more than four seconds for any of the sensory experience. Most of the information in the sensory memory during this period is lost but some of them pass on to *short-term memory*. The control process that determines the information is *attention*. The sensory memory is only *momentary* but is holding *vast amount of information*.

But the short-term memory in contrast has a *small storage capacity*. Its capacity is only seven items. The duration of this memory is approximately fifteen to twenty seconds. During this short period of existence in the short-term memory some of the information there are rehearsed and encoded. This means they get into the form of storing into the *long-term memory*. The information that goes into the long-term memory stay there *permanently* and stored meaningfully and can be retrieved when needed. One remarkable thing with this hypothesis is that the *acoustic sound* information plays an important role in the retrieval of *verbal information*. Thus the whole memory system is integrated in the information processing theory. The information processing functions *as stages of memory*. Information remains *briefly in sensory memory*. Whatever information attended to, enters the short-term memory. Rehearsal and encoding transfer the information to long-term memory. Recall retrieves the information from long-term memory and bring it back to short-term memory.

Some of the Types of Memory

The definitions of memory have given some descriptions of a process. This process is the expression of some behaviour. Therefore, memory is defined in terms of *expressions of behaviour*. This expression of behaviour can be experimentally studied. If memory is a store house of vast information how these information are fed? How do we get those information for use? It is believed that we use these stored information in different situations for different purposes. This assumption leads people to think of several sorts of memory. **William James** (1908), the famous American Psychologist, was the first to think in this line. He differentiated the *primary memory* from *secondary memory*. The primary memory or the immediate memory contains ideas that are *current* and *existing for the moment*. The secondary memory stores the past information.

From the time of William James to the present time many studies have been conducted and many information have been collected about memory processes and these lead to the idea of formulation of several types of memories.

We often find that the term *learning and memory* are used together and it also seems that the use of one term necessarily implies the use of the other term. We are not quite sure that learning has occurred unless *a memory* of the learned material is **elicited** later on. Anything the memory shows we understand that learning has occurred. Investigators often differ in their views on short-term and long-term memories. But these varied views are only simple matters.

Researches and experiments were conducted very recently on various aspects of memory and the researches have brought out several new interpretations about memory and the types of memories. According to the recent study there are several types of memories. These are called as sensory or Iconic memories, Intermediate memories, Short-term memories, Long-term memories and permanent memories. But the recent researches

Magaugh (1968) added two more types of memories. They are (1) Declarative memories and (2) Non-declarative memories. One of the recent additions to the types of memory is the *photographic memories.*

The Sensory or Iconic Memories

The iconic memories store the *sensory expressions of a sense.* The concept *icon* is derived from the Greek world *likon,* which means *image.* For example, a sense impression (i.e., seeing) is experienced only for a moment. We would be able to grasp only a part of the display. A fast running motor car on the road, a flight in the air, a running train on the rails, the runners in the olympic sports, an eagle in the sky are seen only for a second. The rest of the parts of these objects disappear in few seconds. This is with regard to *visual memory.*

A very short *auditory memory* is called *echoic memory.* The echoic memory may be ringing in our ears even in the absence of the stimulus. Similarly in the case of *smell, taste* and *touch* the iconic memories may be experienced only for a moment. These are also called *brief memories* which are considered to be reflecting *sensory neural activity* and therefore, they are called *sensory buffets.* This is supposed to be the first stage. The memories that are some what longer than the iconic memories are called *short-term memories (STM).*

The Short-Term Memory (STM)

The short-term memory is sometimes equated with sensory impressions, but it is much better than sensory impressions as sensory impressions are very short lived. The short term memory is somewhat longer when compared to sensory impressions.

The short-term memory is momentary, but active. It is quite similar to *span of attention.* The capacity of short-term memory is limited and it can hold only few items. Therefore, short-term memory can be defined as the components of the memory process that have limited capacity and will maintain information only for a short period of time in contrast to long-term memory.

The information or items are retained temporarily for a short duration. Therefore, the items are collected by the process of repetition. The repetition may be done loudly or silently. For example, when you write down an address of some person you use the *pincode number* of that place. You look the number from your diary and write it on the envelope and immediately forget it. You have remembered it only for a moment till you write it. Like this in your everyday life there are several instances in which you use this short-term memory.

If you want to speak to a person over the phone you find out the phone number from the directory and after dialing and speaking to the person you forget the number. Here too you remember the phone number only for a short period. So also you forget the name of a person who was introduced to you lastly or the number you have learned lastly. In all these instances we fall to reproduce the desired material presented to our memory.

An experiment was conducted by **Peterson, L. and M.J. Peterson** (1959) on short-term retention of individual verbal items. In this experiment a stimulus was exposed to the subject once for a very short time. A *three consonant nonsense syllable* such as *CHJ was read aloud.* Immediately, following the last letter J a three digit number such as 506 was presented. The subject's task in this experiment was to remember CHJ. But he was not allowed to do rehearsal of the syllable. He was asked to counter the numbers 506 backward until the signal was given. At the signal the subject tried to recall the syllable CHJ after the interval of 3, 6, 9, 12, 15 or 18 seconds. After 18 second, the recall was almost nil. This experiment shows that when there is no rehearsal the loss of memory is rapid. We may state that in this case either forgetting has taken place rapidly or learning has not taken place. This effect takes place after several trials and not during the first few trials. It shows that there has been *some interference* from the *earlier experience.* This process is called *proactive interference.* Proactive interference is the process in which previous learning interferes with memory of the *present learning.* Short-term memory involves remembering the learned material for a moment and then forget. But if the material is to be remembered for, it must be stored in the long-term memory for which it must be repeated often and rehearsed. If the material is *rehearsed sufficiently* it may be transferred to long-term memory *for permanent storage.* Information usually are dying rapidly in short-term memory.

Intermediate-Term Memory (ITM)

The memory that lasts some what beyond the Short-Term Memory (STM) is defined by the psychologists as the *Intermediate-Term Memory* (ITM). It is a form of or type of memory that lasts longer than the short-term memory and requires no rehearsal or recollection, but it does not last like the *Long-Term Memory* (LTM). According to **Rosenzweig** (1993) the intermediate-term memory is far from being a permanent memory. Beyond the ITM memories that last for weeks, months and years there are memories called the Long-Term Memories.

The *Intermediate-Term Memory* (ITM) is a very *recent development* in the study of learning and memory. Psychologists have observed that the process of fading away of the learned material is somewhat delayed but not lasts for days or weeks. The short-term memory fades away as soon as the business is over or as long as the rehearsal continued. But there is a type of memory that neither fades away as soon as the business is over, nor continues for long time just like the long-term memory. This memory we call as the *Intermediate-Term Memory.*

The Long-Term Memory (LTM)

The long-term memory is the record stage in the memory system. This involves remembering the material learned for a long period of time. It has tremendous capacity to store materials. Long-term memory may be defined as the permanent component of the memory process in contrast to short-term memory. The process of function required for long-term memory is *repeated rehearsal.*

Experiments have been conducted on long-term memory. But they all examined the process of *retention* of the learned material after the elapse of long intervals. If a particular information is not rehearsed it fades away after few seconds. But rehearsal is not unlimited. It is not possible to rehearse large number of items simultaneously. *Several rehearsals* should be made to transfer the information to long-term memory. In this process a task performed by STM is coding. *Coding is a process* of arranging and organizing the items of information into a compressed form in the LTM so that the information can be remembered if suitable *cue is provided.* But this process is taking place in such a way that we are not aware of it. By this process of coding, the information in LTM are so highly organized that they become permanent. The learned information attain this stage only after the processes of rehearsing, shifting and coding. But these information are often revised and recognized when new information are pushed into the LTM.

Further, the stored information in LTM may undergo a process of change in whatever manner we wanted. These information are permanent in LTM. Sometimes we may fail to recall as the cues are insufficient to do the task. Familiar information are easily assimilated and categorised then unfamiliar information.

The human memory is often compared to a library. In library, books are stored, but in memory vast range of information is stored. According to recent studies long-term memory is described as having two main divisions such as *episodic memories* and *semantic memories. Episodic memories* are concerned with events and experiences involving personal details. It includes our full life stories till today. But *semantic memories* involve our storage of *knowledge* and *facts* and related information.

In recent studies psychologists make distinction between *knowing that and knowing how to.* Our episodic and semantic memories are falling into the category of *knowing that.* For example, your thought for a moment may take you to the day before yesterday you witnessed a foot-ball match in Rome. This is *episodic memory.* At the same time you know the answer with some certainty to the question. "In which country and city the match was played?" You know it was played in Italy in the city of Rome. This is *semantic memory.* This phenomenon is also interpreted differently.

The long-term memory is considered as our ability to store things that are learned or acquired long time ago. But this term also is applied to the items or events learned very recently. It may be last week, or two days back or an hour back. These recent memories are getting fixed permanently into the long-term store house depend upon the complex interaction between the nerves and the brain. Storage in the long-term memory is a long way removing the memories from the encoded scene. The first part of the process of storing in long-term memory is *attention.* Once the attention is focused information is encoded at its arrival. Then the information is processed in such a way that it is stored in the long-term memory.

One of the most effective ways of processing information for storing with long-term memory is to attach meaning to the information. The more elaborate meaning we attach to the information the more strongly it becomes an established memory.

The last stage of in long-term memory is retrieval or recall. Retrieval is the ability to bring out the items or information from memory. When we remember something which we find not traced. Because it was never formed in the first place (attention) or it may be that it cannot be simply traced at that moment. It is because that other information stored in the memory is interferring with the retrieval. Most of the information or events that are stored in *the long-term memory are encoded without conscious efforts.* Some important events stored in the memory may suddenly hold our attention and effect significantly. These are the wealth of sensory and emotional information. But this occurs randomely. Certain incidents suddenly come to the surface that has occurred some fifteen or twenty years ago. Such incidents are very uncommon or mysterious types. It is very difficult to explain such things. But it shows that information stored in long-term memory are not died down after a long time nor it is an extinction.

Some investigators are of the view that long-term memories are not lost. Evidences of the stability of long-term memory are available from the studies of persons who suffered from stokes and mild brain damages. The strokes sometimes unlock and bring out lots of detailed memories of long past. The capacity of long-term memory is enormous. A fascinating study has been made by the Russian psychologist **Alexander Luria** (1987). The man whom Luria has studied had remembered almost everything he had experienced from his childhood. The investigator was very much wondered by the total recall presented by the person.

Rosenzweig *et al.* (1996) have observed that the American College students on the average can recognize about 50,000 different words. Many persons who have mastered several languages and the grammar of the languages can acquire more than hundred thousand items and they could retrieve the required item from their long-term memories. Many of us can recognize hundreds of faces and countless visual scenes and objects, hundreds of voices and many other familiar sounds and hundreds of different odours, all depending upon our interests. All such memories can be acquired rapidly and stored them well. This kind of memory mechanisms provide for the extensive storage and also for exhibiting them. Thus the long-term memory has enormous amount of capacity to perform the functions.

Long-lasting and Permanent Memories

There are certain memories functioning in our mind which are never lost right from our earliest years of life. These memories are protected from interference. Many *skills* we learn in our early years of life. Mostly the *motor skills* like riding a bicycle once learned never we forget it. There is no interference in this learned skills nor the passage of time would make us to relearn it. The irrational fears that we learned called phobias are also not forgotten for a long time. We retain such things as there are no interferences.

Some of the memories are permanently living with us, particularly *well learned.* These memories we call as *flashbulb memories.* These are often very *stressful and important events of our life.* These memories are *vivid and reappear* often. Some of the long-enduring memories are the passages from a lesson our teachers made us to memories in the class-rooms assignments. Some of the famous songs we heared in a music performance or in a picture performed by some famous songster are also long-lasting memories.

It is true that the long-lasting memories are stored permanently and they are considered as *exceptions.* But as time passes on, year after year the long-lasting memories and the permanent memories are gradually getting decayed. In brain surgery operations the last memories are *brought back* only in about *five per cent* of the patients (Loftus and Loftus, 1980). But all the early memories recalled raise the question of how far they are valid? This depends upon the importance and utility of the information.

Many memories that last for days and weeks or even months may become weaker, however they are LT memories. They may also fade away completely with the passing time. But some investigators are able to trace out a type of memory called *permanent memory.* **Rosenzweig** *et al.* (1996) have observed that the permanent memories appear to continue *without any decline for the rest of the life* of the individual. According to the recent researches information from long-term memory fade away as time goes on longer and longer. But the remaining information become permanent and never fade away. These memories remain intact as long as the organism maintains *good health.* Therefore, we may state that a permanent memory is a type of memory that lasts without decline through out the life of an organism. This may be also stated as the *residuals* of the long-term memory.

Declarative Memories

Several experiments and operations were carried in recent times on the impairment of memories. The experiments and the operations were mainly on patient who had brain damage. **Scoville and Milner** (1957) have studied thoroughly the effects of brain damage and the impairment of memory on a patient called H.M. Apart from medical anatomical, biological information, they also brought out the psychologically important information from memory point of view.

From psychological point of view **Rosenzweig** *et al.* (1996) state that important distinction was made between two types of memories. They were able to trace out (1) The *Declarative memory* and (2) *Non-declarative memory.* There is important distinction between the two kinds of memory and it is not between motor and verbal performances.

(1) The Declarative Memory

It is a type of memory that can be stated or described opposite to the non-declarative memory in characteristics. Declarative memory is directly accessible to conscious recollection. In other words declarative memory *can be reproduced.* It deals with *what we actually think of memory.* It is mainly concerned with *facts and figures* and acquired information through learning. It refers to what actually the memory is. It serves as store house. The declarative memory deals with *what.*

(2) Non-Declarative Memory

This memory is quite contrast to the declarative memory, does not deal with conscious collection of facts and information. The non-declarative memory is mainly concerned with *performance* or activities. It is also called *procedural memory.* The non-declarative memory *deals with how.*

The declarative and non-declarative memories can be measured by different type of tests. *The direct or explicit* tests of memory refer to a *specific episodic memory.* This is by giving a material consists of some items to a person and asking him to read it and then asking him to reproduce what he has learned after ten minutes. Or by asking him what did he eat in the morning day before yesterday. This involves the *conscious recollection* of the material.

In the case of *indirect* or *implicit* tests of memory it is not referring to any specific episodic memory and it does not require conscious recognition, the memory is inferred from the performance. The person might have performed various kinds of activities. The impressions of these activities are stimulated suitably and revealed in the form of memories. The memory for the recently presented words may be inferred by easing to recognize them. The presentation may be *unclear* or *fast as in word completion test.* Thus the main distinction between these two types memories is that declarative memory *deals with what* and the non-declarative memory *deals with how.*

Photographic Memory

While discussing about various aspects of memory or types of memory we often come across the concept photographic memory. In the common sense world the photographic memory is having a strong hold. Many persons claim to have known some person with a photographic memory. But is there such memory called photographic memory? In the psychological studies we did not come across this concept till the second quarter of the nineteenth century. This was developed only after this period and it is also becoming more popular. This is a *technique* that could strongly hold in time and capture events, incidents, experiences, faces, scenes, or information in perfect detail. We also sometimes tempted to have an idea that persons with superior memory abilities are able to catch any material photographically. It is not like brief after-images experienced in sensory memory. It is an *enduring* and *complete memory* of the materials or scenes in the pages of the books. But modern psychological researches do not provide any evidences with facts and figures of the photographic memory abilities in adult persons.

Psychologists while discussing about imagination, they mostly use the concept *eidetic images.* They are of the view that human beings have the capacity for developing *images* which may be quite *similar to photographs.* These images they call *eidetic images.* This eidetic imagery has very unusual clarity and vividness in character. A person with *eidetic imagery* is said to have a *photographic memory.*

However, a form of photographic memory is also called eidetic memory which is an ability to hold strong visual images in the mind. This is mostly found in young children. Two of the American psychologists **Lyn** and **Ralp Halber** have discovered that eight per cent of the elementary school age children were found to possess the eidetic memory. These children were able to describe a picture, that was presented to them, with all the details as if they are seeing the picture still in front of them. In one of the later studies it was observed that even children half of five years olds had some degree of *cidetic memory.* Eidetic or photographic memory ability fades away as and when they become older and it is extremely rare after adolescent period. But there are reports that there was amazing eidetic memory in adult persons.

There are instances that persons with photographic (eidetic) memories state that they can see a page in their 'mind's eye' and can easily read it. This type of claims have been demonstrated repeatedly. The eidetic persons were able to recite the whole (passage) pages from the books in the same way the events, situations and scenes were recollected with the same amount of accuracy. The classical painter, *Leonardo da vinci* was said to have the photographic memory. He was able to draw a detail portrait of some person after having seen him only once. It is also said that the French ruler, Napoleon Bonaparte had such photographic (eidetic) memory. Thus the eidetic memory is now called as *photographic memory.*

ASPECTS OF MEMORY

The concept of memory, according to experimental Psychologists, is referred to an act of reproduction or remembering. Whether it is referred to an act of reproduction or remembering we may be tempted to ask, 'what is to reproduce or to remember?' The answer is, what has been learned earlier. Therefore, if anything is to be remembered it implies that it has been learned, retained and reproduced when there is a need. You can reproduce the learned material if only you remember it. Therefore, learning, retention, recall and recognition are the aspects of a memory function. We shall discuss these aspects here in detail.

Learning or Memorizing

In the common sense world learning goes by several names such as memorizing, training, practice, experience, conditioning, fixation and registration. Here we are using the term learning in the meaning of memorizing, because in any memory experiment a person is asked to learn a given material and reproduce it. He at once puts himself actively in memorizing the passage. He will learn something if the material is unfamiliar and new. Since unfamiliar materials are not wholly found, meaningless words or non-sense syllables and numbers are used. In memory test meaningful and familiar materials are also, used if lengthy materials are to be learned or memorized. The basic idea in such test is that the subject *should make some errors* in his reproduction of the material. If the passage is difficult or lengthy this aim is achieved. However, the results obtained from different individuals with meaningful material would not be having equal influence in everybody. But the *non-sense* syllables such as BEW, ZUR, VOID, NAX, TEW *may have equal difficulty* to all the subjects as they are not familiar to them. **Ebbinghaus** believed that everyone will start with a blank mind in learning the non-sense syllables. He has designed 2300 non-sense syllables for his experimental purpose.

It is true that *meaningful materials* are *easily memorized* than nonsense syllables. The other factors in memorizing meaningful materials are *rhythm* and *rhyme*. This is the reason, poetry is more easily and quickly memorized than a prose passage. Meaningful prose passage is more quickly memorized than meaningless words. Materials which are *interesting* to us are more *quickly learned* than the materials which are not interesting to us. Psychologists tried to discover the principle that operates in this process. They call this learning as *verbal learning.* Any learning that is dominated by *language behaviour* is called *verbal learning.* There are two simple forms of verbal learning. They are *paired-associate* learning and *serial learning.* An apparatus called *memory drum* is generally used to study both the forms of learning.

The memory drum is an electrically operated apparatus. The drum is covered by a metal sheet. The drum inside is pasted with printed or typed material of non-sense syllables or words with meaning or numbers. The nonsense syllable is a three letter syllable that is not having any meaning. When switch is on the drum rotates and the syllables or words are exposed successively through a small window. The movement of the drum is so devised that *each item is exposed* only for a short period of *two seconds*. After two seconds automatically another item appears in the window. A subject is to sit comfortably in front of the apparatus and look at the small window. He is to memorize each item when appears in the window. Only *one item* is exposed *at a time* (Fig. 13.5).

Fig. 13.5 : Memory : Memory Drum : An Apparatus for measuring Span of Memory.

(The memory drum is an electrically operated and rotating apparatus with a small window for exposing test materials for few seconds and it automatically moves after few seconds. The subject is expected to learn material and fix in his memory. The simplest way of determining the efficiency of immediate memory or fixation is to test the memory span. How much materials correctly reproduced after one exposure of a fixed number of materials presented. Memory span materials consists of nonsense syllables, words, numbers, figures and objects. The digits are the easiest to fix in the memory. The experimenter first begins with the shortest items of 3 or 4 digits and goes on increasing the digits upto 9 to 10 digits. The normal memory span is even and it may be also with letters and numbers. For example, it may with three letters and four numbers.)

(a) Paired-associate Learning

In the method of paired-associate learning the subject learns to associate one syllable with another syllable or *one word with another* or *one number with another number*. Pairs of items are shown to the subject on the memory drum. *In the pair one serves as a stimulus* and *the other as a response.* The subject is first shown with the stimulus member or the first member of a pair alone. It is exposed for a short period of *two seconds* in the small window and within that time the subject should give the response *syllable (member)* or word or number.

Immediately after the elapse of two seconds the drum moves and the pair of the syllable or word or number —that is the stimulus and response members appear in the window. The subject is then to see whether he has made correct response or not. If he had failed to make correct response the appearance of the stimulus-response pair makes him to learn the association. Then the drum moves and the stimulus-response pair appears in the window. The subject has to give the response

Stimulus (S)		**Response (R)**
Soft Soft	— — — — — — — — — — — — — — — —	Hard
Happy Happy	— — — — — — — — — — — — — — — —	Fast
Minor Minor	— — — — — — — — — — — — — — — —	Few
Hot Hot	— — — — — — — — — — — — — — — —	Sweet
New New	— — — — — — — — — — — — — — — —	First
Safe Safe	— — — — — — — — — — — — — — — —	Green
Wicket Wicket	— — — — — — — — — — — — — — — —	Quite
New New	— — — — — — — — — — — — — — — —	Rare

Fig. 13.6 : Arrangement of meaningful words for paired associate learning. The stimulus member is presented first alone and secondly with response member. These words are adjectives, should be presented in different order.

Stimulus (S)	Response (R)
XEW	MUX
XEW	
CAJ	KEJ
CAJ	
VUQ	ZUY
VUQ	
DIJ	ZAJ
DIJ	
WOS	BOZ
WOS	
KEW	FUH
KEW	
BAW	PIW
BAW	
QUT	FOV
QUT	

Fig. 13.7 : Arrangement of non-sense syllable for paired associate learning

The stimulus member should be presented first alone for 2 seconds and then the stimulus-response pair should be presented for the stimulus. After two seconds the same stimulus member appears with the response member and the subject should see whether he has correctly responded or not. This process is repeated till the whole list of items is presented to the subject. There is a chance for the subject to memorize the order of the stimulus response pairs. In order to prevent it the order of the items in the list is to be changed in every trial. Therefore, the same list is to be changed in every trial. Therefore, the same list of stimulus-response pairs is arranged *in different orders.* In every trial, the pairs appeared in different order. The presentation of the list of items continued till the subject is able to respond correctly all the response members in one or *two successive trials.* Then only the subject is said to have learned the whole list.

(b) Serial Learning

The method of serial learning differs from paired-associate learning in certain respects. Whether we use non-sense syllables or meaningful words or numbers they are presented in a chain like manner. A series of non-sense syllables may be constructed for serial learning. Some symbol (X) is given at the beginning of the list. This *symbol* serves as *a cue for the first syllable.*

The same apparatus, the memory drum and the same procedure as explained in the paired-associate method is used here. In the method of serial learning syllables or words or numbers are presented one at a time for a standard time of two seconds in the memory drum. When the drum moves the symbol (X) is the first to appear in the window and the subject anticipates next the first syllable after two seconds. The first syllable is used as a cue to the second and the second to the third and so on. In this way the whole list is presented for the first time. When the list of items is presented one by one for the second time, the subject sees first the symbol in the window. Then he is expected to say the first syllable that is going to appear next. For example, a list of non-sense syllables is given here. The first is the symbol X. It is presented to the subject in the memory drum. He is expected to say the first syllable FEH within the standard time of two seconds before it appears in the window. When it appears he can check whether he is correct or not. Since he has already seen the list of items once, using the first syllable as a cue he is expected to say the second syllable that is going to appear in the window. He can check whether he is correct or not when the second syllable in seen in the window. The procedure in this manner is repeated till the subject could tell the whole list of syllables correctly in the order in one or two successive trials. He is said to have learned the list at this stage.

	X	
F	E	H
Q	A	M
C	I	J
D	U	Y
H	O	X
G	A	H
J	I	S
X	A	D
W	U	F
K	E	J
W	O	X
Q	U	V

Fig. 13.6 : List of non-sense syllables arranged for serial learning. Symbol X is the cue for the anticipation of the first syllable.

(c) Free Learning

Another method of learning has been suggested by **Underwood** (1964). It is called *free learning method*. Meaningful material is used in this method. The subject is presented with a series of words visually or orally. He is then asked to reproduce in whatever order he can. In this method the subject is not given any stimulus word. The same memory drum we have used in other methods can be used for visual presentation. The trial can be repeated till the subject reports the whole list of words correctly.

EFFICIENT METHODS OF MEMORIZING

Generally in schools students are asked to memorize certain passages and recite. Certain poetry lines are to be memorized compulsorily. In such cases students find the task very difficult. Because, memorizing is a *time consuming task.* It also takes *more efforts.* They even go to the extent of *swallowing memory pills* with a view of making the task more easy. Unless we conduct experiments with memory pills we cannot say anything about the effects of memory pills. But the idea they have in their mind is that they wanted to memorize the passage or poetry easily and quickly. They wanted to *save time and effort.* For this purpose they try to manipulate conditions to favour this task. One easy and possible method of making the task more efficient is to avoid all kinds of distractions while learning. Let us discuss some of the methods in use.

(a) Rote Learning

Sometimes we may use this method of learning, because certain materials are to be *learned by rote learning.* Rote learning is a general term referring to *repititive task* of memorizing of *verbal material.* It is learning the material verbatim or as it is. For example, *learning a poem by heart is rote memorization.* By hearting a poem or play or some famous quotations from eminent authors is rote learning. We do not make any change in the material. Particularly in learning non-sense syllables we use rote learning method. It is a mechanical process of learning by *repeating the lesson over and over again.* The learner makes slow progress and so he finds that this method is not very efficient.

(b) Learning by Observation

Since the learner finds that the rote learning method is not efficient, slow and consumes lot of time he tries to find out some other method by which he can learn the material *easily and quickly.* One such method is *learning by observation.* The learner tries to find out the characteristics of the lesson or material which can be remembered easily and puts emphasis on these characteristics. In remembering the faces of persons he puts emphasis on personal qualities of persons. In memorizing pairs of words he finds out the characteristics to make them as a unit. In case of numbers he observes the sequences of the numbers for memorizing. In by-hearting a poetry or a scene from a play or a famous speech the most important thing the learner should observe is the *meaning of the material.* If the meaning is not clear he must work out and find out the meaning for each difficult word or phrase and assimilate the whole material. If the idea of the passage or material is understood in the first reading further readings would facilitate the reader to put everything in the appropriate places.

(c) Spaced and Massed Learning

Spaced learning and massed learning are two contrasting methods, often discussed in memorization. *Spaced learning* is a kind of arrangement in learning that *allows for time intervals* between successive trials whereas *massed learning* refers to *continuous learning with no time interval between successive trials.* The merits and demerits of these two methods can be realized when one wants to memorize a lengthy assignment. If a car number or telephone number is to be memorized you can make a *single reading and recite* the same several times without any pause. But if you want to remember it for longer duration *a rest pause should be given* for few seconds *after every recitation.* This method of learning is more efficient than massed learning.

Hovland (1938) had conducted an experiment on 32 college students using *massed learning* in a laboratory setting. The subjects were given a list of *12 non-sense syllables.* They were exposed one at a time. Each syllable was exposed only for 2 seconds. When all the 12 syllables were posed one by one a pause of 6 seconds was given before starting the next trial. The trials were continued till the subjects learn the whole list correctly by the method of paired-associates. This is called massed learning, because there is *no sufficient interval between one trial and another.* The subjects have taken *15 trials* on the whole to master the list of syllables.

Similar experiment was carried out on spaced learning. The procedure was the same as in the previous experiment. But the interval between trials was increased from 6 seconds to two minutes. During the long interval period the subjects were not allowed to recite the syllables. They were given engagement in a test of naming colours. Later it has been found that the subjects have taken only *11 trials to master* the whole list of syllables. This is because of the long interval between trials. This experiment concludes that *spaced learning saved 4 trials,* but it has taken *more amount of total time* due to the longer rest pauses. Both the methods have certain advantages and disadvantages.

Another similar experiment with little variation was conducted by the same author. This was found to be more economical than the first two procedures. According to this procedure every syllable was exposed to the subject for four seconds with an interval of 6 seconds between trials. The subjects were able to recite the material with 7 trials. Both *trials and time were saved.* It is a more efficient method of memorizing. It seems that the subjects could make better performance with 4 seconds interval between syllables and 6 second interval between trials.

(d) Whole *versus* Part Learning

Sometimes we may have to learn a very long poem or lesson. The moment we think about memorizing such materials we feel worried as how to memorize it. Learning the whole material with several trials would be a waste of time. Therefore, we must find out an efficient method to master it. From our experience we could understand that dividing the whole material into certain convenient parts would be profitable.

Pyle and **Snyder** (1911) have conducted experiments on the economic unit for committing to memory. In this experiment a young man was made to learn a passage of 240 lines of a poem by whole method and by part method. The same number of lines from the same poem for each method was given. He spent each day about 35 minutes sitting and learning. In the whole method he made 3 readings of the whole material per day till he was able to recite the passage correctly. For this he has *taken 10 days* and the amount of time spent was *348 minutes*. In the *part method* he memorized 30 lines per day till he could recite the whole passage correctly. The number of days required was *12 days* but the amount of time spent *431 minutes*. It is understood from this experiment that the whole method was more economical than the part method. It has an advantage of 83 minutes over the part method and saved 20 per cent of the time. Therefore, the *whole method* seems to be *more efficient than the part method.*

(e) Recitation

Every material we learn requires certain amount of recitation. Recitation reinforces our learning. *Recitation*, generally, *means repeating the learned material silently*. It is a *kind of verification* of our own memory. After reading a passage two or three times the learner tries to recite it to see whether his learning is correct or not. If it is struck up somewhere he can read it once again and recite it. We can verify whether the method of recitation is advantageous or not. **Gates** (1917) had designed an experiment to test whether recitation is advantageous in memorizing on eighth grade children. Sixteen non-sense syllables and 5 short biographies with 170 words were used as reading materials. There were *five categories* in the distribution of *learning time*. In the case of first one, all the given time was devoted to reading.

In the case of second 1/5 of the study time had been devoted to recitation. In the case of third, fourth and fifth 2/5, 3/5 and 4/5 of the study times were devoted to recitation. It was found that *recitation* had *definite advantage* immediately and after 4 hours in the case of both non-sense syllables and short biographies. Recitation is much better than continued reading of the same passage.

(f) Mnemonic Devices

Mnemonic devices are the techniques by which *material is made easier to remember.* It is a system of improving memory. The technique involves a *set of symbols* that can *substitute the material* to be remembered. For example, a number can be used as a substitute for a word. The learning material is organized in such a way that it helps us to remember the material easily and quickly. The learner when uses the mnemonic device he purposely puts some order in the material to be learned. He may use any systematic order. He may use meanings or groupings or rhythm as means.

The simplest mnemonic devices are *rhythms* and *jingles* to remember the data and facts. The colours of the spectrum are remembered if we know the word **VIBGYOR.** The word stands for colours like Violet, Indigo, Blue, Green, Yellow, Orange and Red. Some of the psychological tests are remembered in the same manner. For example, tests like MMPI, FACT, EPPI, DAT, TAT all remembered by some mnemonic devices.

The *number-peg* technique is a mnemonic device that helps us in serial recall. Every *number* is imagined *as a picture on the basis of the shape of the number*. For example, 1 is as a pillar, 2 as a cock, 3 as a pitchfork and so on. It is found from the studies made by **Smith** and **Noble** (1969) that persons who had training in mnemonic devices were able to recall meaningless words better than those who are not trained in that line.

Retention

What is retention? Retention generally means the ability of retaining information or events in memory. Retention is the function of memory. It is holding on the information or events in memory. It may be also stated as the registration of information we have collected in the memory.

Retention is one of the important elements in human learning which is making us to function as the superior being. In the absence of memory man is almost a mute. According to **Gerard** (1953) man is not benefited by his experience or intelligence in the absence of memory.

Authors like Gerard define memory in terms of retention, because it is the way through which memory can be traced. Therefore, memory is defined as the *process of retention* that makes us to relate one thing with another and transfer things to new situations.

How do we know whether we have learned something? It is only through performance that we come to know of it. How much do we perform from the past learning? The procedure used in finding out how much we have learned is by measuring retention.

Measuring Retention : In testing the validity of our memory we measure only the retention. If a test is conducted to measure how much material is retained in our memory we could see that we don't retain everything we learn and we don't forget everything. But our retention is influenced by certain factors. All those factors that influence learning also influence retention. You might have prepared to deliver a speech in an organisation. After concluding the speech and leaving the platform you may remember an important matter you have failed to speak on the platform. How your performance is affected by certain factors? *Motivation* may be one of the factors, *influencing the performance*. A reward may have influence in the performance of a dog, or cat or a child. What kind of performance can reveal retention? Psychologists have devised certain methods by which the amount of retention can be revealed.

Suppose, for example, your teacher asked you to come with preparation to recite a particular poem next day. You have memorized the whole poem that night. Next day you recite the poem in the class room. You are able to do it word by word without any effort. That means you are able to recall the poem perfectly. But, suppose you are held up at certain place with some line not coming to you as the other lines. When your teacher helps you by telling what that line is with one or two words from that line you are able to know that line and proceed further. You are able to recognize the whole line when once you hear one or two words from it. This makes you to think that you are not perfectly remembering the poem and so you memorize it once again. *This is called relearning.* Now you are not taking many trials for memorizing.

You learn perfectly within few trials. This kind of relearning is easier than learning it for the first time. This relearning saves time and effort.

Recall, recognition and *relearning* are not three separate processes. They are three *methods of measuring retention.* These three methods differ in sensitivity to the material learned. *Recognition* is *more active than recall* because it identifies at once what is not recalled whereas recall is only recitation. So also *relearning is more active* than all of them. It reveals what we have retained.

Recall

Recall involves repeating what we have learned something already. It may be defined as the *reproduction* of some specific material learned previously. *It is reciting* the material as it is or *verbatim.* In recall there is also some kind of reconstruction. While you are answering an essay type of question in the examination you are recalling all that you know relating to the answer and reconstruct the material. You also try to arrange the material in an orderly fashion. For example, if two events are to be stated in your answer you try to put them in the order of occurrence of these events. This method of arrangement is taking place in our recall and reproduction.

Recall may be *serial recall* and *free recall.* In serial recall materials are often recalled in the same order in which they are learned. It has a specific order, the first will come in the first and last will come in the last. On the other hand, in free recall pieces of information are recalled. The information need not be in the order. Any matter may be put in any place or order. This is occurring when you want to quote some idea from a passage once learned.

While you are recalling the learned material you may succeed or fail. All unwanted material not relevant to the situation or purpose may come crowded when you try to recall. The material we wanted may not be appearing in our recalling. All of us might have had this kind of experience and we may be also using this term recall on several occasions. "Could you recall those days when we used to play under this tree?" "Do you recall the name of the old man who told us novel stories from *Ramayana* and *Mahabharata*? Such uses of the word 'recall' are quite common in our everyday life. But in experimental studies psychologists are interested in the amount of information recalled and the accuracy of recalling. The method of paired associates we have described earlier is used for this purpose. Studies of this type show that *recall is effective* if the set is to form a *particular association.* For example, if the set is to colours of the objects your recalling should be successful with regard to colours, but not to shapes of the objects. Your answer to the question in the examination is a method of testing your recalling of the learned material. You write everything you know about the matter. You use both *serial recall* and *free recall.* Here retention is tested by your ability to recall and reproduce the learned material.

Recognition

Measuring retention by using this method is becoming more popular in recent times. One of the uses of this method is the change in the pattern of question papers. One of the patterns is the introduction of multiple choice type of questions in the examinations. What is required here is whether the student is able to recognise the correct answer from the number of alternative answers. The student's ability to identify the correct response is tested in this way by the recognition method. Therefore, recognition may be defined as the *capacity of the person to identify the learned objects* or materials when presented along with other objects or materials. *Recognition is the measure of retention.* Certain materials we do not remember or recall when we want, but when we see them or hear them we at once recognize them.

For example, if you are asked to answer the question: "Who has set up the first Psychological Laboratory?" "In which year?" You may struggle to find out the correct answer even though you know it. But if you are asked to answer the same question by the *recognition method* you would readily give the correct answer. For example:

Who has set up the first Psychological Laboratory?

In which year?

Check the name and the year below:

I. 1. Watson
   2. Pavlov
   3. Wundt
   4. Galton

II. 1. 1879
    2. 1789
    3. 1899
    4. 1889

In questions like these you would immediately recognize correct the answers and check, **Wundt** and **1879** in your answer book.

Recall involves *more strain.* Recognition involves *less strain,* because the learned and stored material is before you and you are required only to decide the right one. The task is whether to accept or reject the object or information. In testing retention by recognition method the materials ordinarily used are pictures of human faces, words and things experienced before. The *ability to recognize* a material depends upon three factors such as the *structure* of the object and its relation to ground, *past experience* with the object and set or strength of motivation.

Experiments on recognition were conducted by **Edwards** (1941) by using *pictures of faces of Negroes and White students.* Seen Negro *faces were mixed with faces not seen* previously and *seen white faces were mixed with faces not seen* previously were presented to many white students. They were asked to pick out faces they had seen before. The scores in recognition test were mainly depending upon the *attitudes* the white students have developed *towards Negro students.* Long list of words also may be used in the same manner in recognition experiments.

Method of Relearning

The method of relearning refers to the process of learning again a material already learned when one has failed to recall and recognize it. It is also called savings, because relearning saves certain amount of time in learning. How time is saved by the method of relearning? Let us explain this. For example, you have memorized a poem or a passage from a book for which you have spent 100 minutes or 20 trials. After twenty days you wanted to recall and recognize it. But you are not able to do the whole material successfully. At certain places you had to stop and struggle because a particular line is not recalled by you. This shows that you have *not learned* the poem or passage *properly.* Therefore, you begin to learn or memorize it once again. This process is called relearning. This time you are able to learn and recall the whole material in 60 minutes or 12 trials. This shows that you are able to save 40 per cent of the time and effort by the method of relearning. Experimental studies show that the effect of *retention is greater* and the result of recall and *recognition is accurate* when the materials is *relearned.*

How to Improve Memory?

Everyone might have had occasions to speak about his own memory or somebody's memory. Either you might have boasted about your own memory or blamed it on such occasions. In the case of another person you might have admired his memory for the tremendous power or remembering many things or blamed him as a man of poor memory. Generally, people are worried over their poor memory. Particularly, the students at the time of examination worry about the retention of the learned materials. Many people complain about their memory power. There is nothing wrong in their memory. It is sound. But the method of learning and retaining the material may be wrong or defective. They must know the *right method* of attacking their problem. If the problem is with regard to retaining or remembering the learned material *what is the successful method* to do it? Psychologists have devised certain effective methods of learning to *improve retention.* Suitable methods suggested earlier can be used.

(1) First of all the material should be *liked by the learner.* He should *develop interest* in the learning task. Any amount of memorization will not have effect if material is learned without any interest or motivation. Continuous mechanical repetition will be less effective, whereas a desire to learn will have better effect with few repetitions. We

should not think that material learned for examination is *for a short time.* We must know that it is going to be useful for us in the long run. If we realize the use of it we will have better capacity to receive the material for longer duration.

(2) Secondly, the material should be *presented* in an *organized manner.* There should not be haphazard presentation of the material. The points should have logical sequence. Materials with *logical subunits* can be *easily learned* and retained for long time. For example, consider the following two sets of words:

I. Monday, April, West, Tuesday, Boy, January, East, Teacher, Wednesday, North, School, March, Girl, South, February, Friday, Book, Head-master, Thursday.
II. North, South, East, West, January, February, March, April, Monday, Tuesday, Wednesday, Thursday, Friday, School, Head-master, Teacher, Boy, Girl, Book.

The second set is easier to learn than the first. It is possible to recite the whole material by reading once because of its organization. The *whole material* is having *logical subunits.*

(3) It is possible to learn and remember the material if the meaning of the material is wholly understood and also if we could relate it to the material that we have learned previously. If the passage is with difficult words understand the meaning of the difficult words with the help of a dictionary.

(4) We can build up good memory by associating the learning material to some objects and events. While we teach a child with some new lesson the learning material must be presented with some objects. For example, the *alphabet* A may be associated with an object like *apple.* Thus **A for apple, B for book, C for cat, D for dog** and so on. Associations like these can be made even for difficult and complex materials.

(5) The learned material should be practised or utilized in some manner. For example, when a child has learned a word he should be made to apply it and practice it wherever possible. Because *immediate practice* builds up good memory. *Recitation is a kind of practice,* which is very much useful in retaining the learned material.

(6) The *mnemonic devices* are found to be good methods in helping us to remember materials quickly and easily. The mnemonic device is a method of organizing materials. A person, in mnemonic devices, consciously puts certain system or order in the material he learns. He can use *meanings, rhythms* and *groupings as means* of organising the material. It is said that the ancient *Greek and Roman orators* were using the mnemonic devices *to remember long speeches* as they have to deliver them *without written notes.* A simple mnemonic device is making sentences or words out of the material to be recalled. Another device is the *number-peg technique.* Every number is attached to *a picture* similar *to the*

shape of the number. For example, the number 1 is a pillar, 2 is a cock, 3 is a jug and so on.

(7) Improvement of memory mainly refers to the process of retention. It requires proper learning or reviewing of the material and also *avoiding interferences.* But for all these an organism must be kept in good condition. A good brain is very essential for a good memory.

It is believed in the early 19th century that *memory can be located in certain parts of the brain. Phrenologists* believe that memory centre is located in the brain *just behind the eyes.* They thought that a person with *protruding or bulging eyes* would have *bigger memory centre* and a greater capacity to remember things and events. But **Lashley** (1890-1978) a famous physiological psychologists in 1929 has stated that there is *no particular part of the brain* which is responsible *for our memory.*

Recent Theories of Memory

The information stored mostly in long-term memory is of *verbal or semantic in type.* Recent researches in the field of memory have brought out distinctions between various types of memories. These distinctions are evolved gradually out of experimental studies. The researchers also deal with memory differences in their subject matters from those that belong to other regular memories. The memory of tastes has been green for long time. This also brings to our memory the vivid image of the material. There are also contributions to memory with *visual factors.*

Dual-coding Theory

This theory has been formulated by **Allan Paivio** (1971). According to him the learned material is represented in memory by *two separate systems or codes.* But they are *interconnected.* These two are (1) system of *visual images* and *a verbal semantic system.* The system of images deals with *concrete, spatial* imaginable *objects and events.* These things are coded like *pictures in the head.* (2) The other, one is the *verbal or semantic* system which is using *words and associations.* This deals with *abstract linguistic* units and structures. In this way information is dealt with by both the systems. Therefore, the *learned information* according to Paivio is *represented by both the systems.*

Experiments have been conducted to verify dual-coding theory. Two groups of words were used in the experiment. The *first group of words are apple, arrow* and mountain and so on. The *second group* of words are *brevery, happiness* and *obedience.* The first group of words are *concrete* that can *evoke vivid images* of the objects. The *second group* of words have no such vivid images as they are *abstract semantic concepts.* It has been found that the words that can evoke vivid image of the materials are easily learned and retrieved than those words which are abstract of verbal system. Therefore, we have *better memory for concrete words.*

Episodic and Semantic Memory : A theory that has been evolved to illustrate the difference between *memory and knowledge* is called the *episodic and semantic memory.* These two differ in the nature of the stored information.

It is **Endal Tulving** (1972) who founded the theory, states the *personal experiences as* the *episodic memory* and the *knowledge* as the *semantic.*

The episodic memory deals with temporarily dated events, and autobiographical events in the individual's *personal life.* The *episodic memories* are referred to *time and place.* For example, "I remember now that I heard the lion's roaring sound in the zoo. I remember that I have seen the bus accident on the high way. On August 15th, I remember to have seen the march past of the Army." These are referring to particular sensory attributes.

The semantic memory on the other hand is referring to the organizational *knowledge about the world.* This includes verbal world of words and the way in which they are used. The semantic memory deals with *facts, principles,* relations and strategies. In this memory *there is a necessity for the use of language.* It is just like a mental treasure or *organized knowledge* an individual possesses about words and other verbal symbols, their meanings and references. It is also observed that semantic memory registers as *inputs only cognitive things* that are referred by words and *not perceptible properties* as inputs. Semantic memory permits the retrieval of information to leave from the system the *content unchanged.* This memory is not so easily susceptible for involuntary transformation of information like episodic memory. Further, information is *not easily lost as in episodic memory.* The semantic memory is independent of the episodic memory in recording and maintaining information. The semantic memory is typically the long term memory (LTM). It is a part of our permanent knowledge. There are *no autobiographical references* about their mode of learning.

Therefore, a distinction between these two memories can be brought out this way. But, however, the distinctions between the episodic-semantic memories are accepted and recognized, we must also accept that any memory theory must accept that the stimulus events that are recorded into the memory are also recorded along with some descriptions of their context of occurrence. The context may include the items perceptual characteristics and its physical mode of presentation. It also includes the incidental matters such as who said, when and where. All these involve several things.

When a person is asked to remember the event, all the contextual and incidental elements are associated to the event. In the study made by **Anderson** and **Bower** (1972) it was observed that the participants record automatically all sorts of contextual and incidental information. They were able to remember whether words presented visually and auditorily, spoken by a male or female voice, whether it was in the form of words or pictures, presented on the left or right of the display screen. They might have also *recorded the colour of the objects* presented on the screen. They could also remember the *passage of time,* how recently the event has occurred.

In these things it can be observed that though a distinction can be made between the episodic and semantic memories, they are in a sense interrelated in their functions.

MEMORY AND EMOTIONS

Psychological researches are carried out experimentally all over the world in memory and new concepts and interpretations are broughout everyday. Researchers consider certain psychological aspects are very important at one time and discord the same as not important at another time. Thus the research trend in certain areas is associating certain psychological aspects with the main subject matter and dissociating certain psychological aspects from the main subject matter. Memory is one of the subject matters recently attracted the attention of the researchers for dealing with the recent developments associating certain aspects which were not dealt with the memory in the past. Researchers of the present time associate the *function of emotion with memory*. They are of the view that *emotion is having influence in the function of memory* processes. Why the behavioural aspect, the emotion is associated with memory?

Emotions may be pleasant and unpleasant and it is experienced by all living creatures. The emotions that are familiar with us are anger, fear, love, grief and joy. Though emotions were one of the subject matters in psychological studies right from the time of William James, they were not studied so extensively as the present day researchers. Generally, it is considered that emotion is a disorganized state of the organism. Psychologists have not come to an agreement as to how to define emotion acceptable to all as it is found to be closely involved with other psychological processes like cognition, perception and memory. Since it is found that emotion is having interaction with all these three mental functions any explanation of emotion has to be considered from different angle.

Experiments associating memory and emotion have taken place only very recently and therefore, the available information on this phenomenon are very few. The human memory is a storehouse of information acquired in various ways. Therefore, *every experience is a stored memory*. Investigators are of the view that the functioning emotion has an important role in forming down memories. According to the recent information emotion is generated in the primitive part of the brain called the *limbic system*. But its effects are around the areas of the brain which *process sensation* and *generate thought processes*. One of the effects of the arousal of emotion is *to pay attention* to the events that stimulate it. *Attention is the first stage* of forming down the memory.

Any *exciting event* would certainly *draw our attention* and we may have more clear understanding than the routine occurrences. In this process one important matter is that emotion changes the *chemical balance* of the brain. Excitement is an emotion. This emotion of *excitement* pushes up the chemical levels of the brain, particularly the brain's chemicals *acetylcholine* and the *noradrenalin* in certain areas of the brain. At the same time the emotion of *pleasure* increases the production of the chemical substance, the *dopamine* and the emotion of *disgust* stimulates the increase of the chemical substance the *glutamete*. These changes create connections between the neurons involved in this function. For example, the noradrenalin promotes the *growth of axons* and the glutamete stimulates the release of *adhesion molecules* that connect the neurons together.

In this way emotion helps to maintain the pattern of neural activity connected with an event, thereby making it to replay later on. That is, what is registered or imprinted in the cells, or stored may be recalled or brought to the front when necessity arises. The experiences of the events stored in the memory may be recalled.

The memories *associated with emotions* can be *more easily recalled* than the routine information stored in the memory. This process may be stated as *emotional memories.* The memories associated with the *feelings of pleasure* tend to be recalled in the memory during the time when similar experience occurs. Our memory can easily associate the enjoyable experiences and also sorrowful or grieved experiences. In fact all the intense emotional experiences themselves are laid down or registered as memories. The more a particular experience is replayed whether pleasant or unpleasant, the more strongly it is imprinted in the memory.

Any event of our life emotionally charged becomes more advantageous to get into our memory. There might have been more significant events in our life which may be often causing anxieties or worries and we may be tempted to recaptulate the experiences of those events and feel happy since they are associated with pleasant experiences and feel desparate if they are associated with unpleasant experiences. From the practical point of view preserving or retaining in our memory such pleasant experiences and knowledge may help and guide us in our present and future life situations. The emotional memories which are unpleasant and painful may be also guiding us by means of warnings. We may make ourselves to get ready to avoid such events of life.

We might have met with our life situations several happy and unhappy events and experiences which are also engraved in our mind and memories. We may not be often recalling and remembering the pleasant experiences, but some of the unpleasant and painful experiences often may come to the surface of our memory and cause anxieties and worries. An intense unpleasant and painful emotional memory may cause more disastrous consequences then an intense pleasant memory. The type of most unpleasant and painful memories experimented by the famous Romantic English Poets, Byron and Keats are the classical examples for the consequences of the *emotional memories.*

Human beings generally do not wish to face difficult problems unless they are very indispensable for their everyday living. Everybody wanted only to avoid such trouble causing emotion involving activities. We naturally try to get rid of fearful and dangerous situations. We even go to the extent of advising others not to take risk in fulfilling the desires. Thus anything related to the emotions of fear, worry or anxiety and their consequences are already stored in our memory and we never fail to make use of them and save ourselves from the dangerous life situations.

Memory is not a single thing. It consists of many things, because it is our store house of facts and figures and our skills. Memory is also the store house of our personal recollections of life events, emotional experiences, unconscious knowledge and our conscious controlling devices of the immediate past experiences and memories. Every memory function is managed by different parts of the brain. The *prefrontal cortex* is the area where the *short-term information* are stored. For example, the phone number, car number or passing on messages and thoughts are held here. This is called the *working memory.* When the business is over the memory is lost.

Maintaining Good Memory and Health

We have discussed in detail about various aspects of memory and how memory is an *indispensable faculty* of the human mind. For the successful performance of any activity, mental or physical, the memory is a crucial functioning aspect. Therefore, the *memory should be clear* in an individual without any confusion. Memory is the central functioning aspect of our personal and professional lives. For the performance of any task we must maintain our physical and psychological well-being. Researches in the field of memory are going on enormously theoretically and experimentally. In the recent years researchers have adapted *an integrated approach for improving the memory.*

The mnemonic techniques, repetition and visualization are considered as the valuable *memory tools.* But at the same time mental functions are also influenced by many factors ranging from the physical and emotional states to our environmental and social interactions. We must be more aware of these influences that can cause something in our memory, either success or failure. Thus, apart from personal and psychological factors the social and environmental factors are also must be included in the study of memory.

It is imperative that persons who make use of memory in any personal or professional tasks should maintain their *good physical health.* Because *physical health* may affect a person's *memory strength.* It has been observed that people who are often suffering from some illness have significantly *more memory problems* than those who are healthy physically. Persons who suffer from some serious illnesses, generally, stokes of all types, the Alzheimer's *disease* may cause *loss of memory.* It is observed that even the common cold and headaches can have an effect in memory. The severe cold or flur may cause some difficulties in recalling learned materials.

The *physical and mental stresses* have significant effects on memory. Generally, human being when they are in positive or pleasant feelings, cheerful moods *recall more information* readily than in depressed state with more unpleasant feelings. Anxiety, frustration and stress usually reduce man's ability to recall the learned materials. But anxiety and stress reduce the ability for *direct attention* on complex materials. This provides some gap in the everyday activity and provides some relaxation. Therefore, a moderate amount of stress or strain is better for the memory than the continuous stiff words and to have severe stress which may have some bad effect on memory. If there is any amount of stress it may reduce a person's sense of steadiness to the extent of failure of paying attention to some important work.

Taking care of the physical health may produce positive memory gains. The brain needs continuous supply of *large amount of oxygen and glucose* for better performance. Therefore, it is important to eat healthy diets to keep the body well supplied with nutritious diets. A good night sleep is essential for all types of memory performances. Sensory experiences such as eyesight and hearing are also important for good memory. It is always better that we see and hear pleasant things and register them in our memory so that we may recollect the pleasant observations which may be the *healthy exercise* for the memory.

It has been observed that human beings often *experience more memory failures* when they are among other persons than they are alone. It is because of the *social pressures.* In a group of people some may be having the conversation in a rapid manner and some may speak softly and also the conversation or talk may be very violent and stiff and aggressive. The effect of such instances may distract us from the process of registering and remembering the substance. Further, we may be distracted from remembering by our emotions if the atmosphere is very much noisy. But at the same time *social situations* also may provide *better incentives* for improving good memory.

Persons who have often public contacts, particularly the orators of various categories, like the political, social, literature earn good reputation for presenting certain valuable information instantly in the public because of their memory. They learn and cultivate their memories for further efficient presentation. Persons who present information in the public with correct names, dates, events and the consequences are admired for their excellent memories. The young generation take those persons as model and they also try to cultivate such memories in keen competition with others. This is how the *social pressures* act on the memory of the young people.

The level of memory may vary widely among the people however efforts are taken to cultivate it. People differ in their competances. *Memory presentation* has become in these days a compatitive skill. The quiz programmes organized in schools, college are intended only to test the fertility of the memory of the students. Whoever is found proficient, is rewarded for the excellent memory he has demonstrated. But reward and punishment in *memory competency* may have both the encouragement and discouragement. The winner is happy and the loser is unhappy. But there is a chance for the loser to take some determination to cultivate his memory to win the admiration from others. This trend is often prevailing among the youths and they are becoming successful in their efforts.

The memory competency and the promotion is getting momentum in these days in the *Television Programmes* in our country. School and college students are very much excited by this programme and they take all possible steps to improve their memories in all possible ways for winning the laurels and expose themselves to the Television audience. The students who take part in the *Quiz programmes* collect information from various areas from which the Television programmes are concentrated and work hard and feed their memories and get ready to face the situation with *energized memory.*

We often hear from scholars that *a sound mind exists in a sound body.* The same principles can be applied to memory also. *A good and indelible memory* can be built in a *physically and mentally healthy organism.* Human beings can easily maintain this state and cultivate their memories. It is a general observation that human beings never stretch their brains beyond a fraction of their potentialities. By making best use of our brains to aim to reveal that potentiality through a wealth of fascinating new information as well as practical exercise that may help us to understand our brains and memories and extent their capabilities. Whatever the age is, once we realize what our brain is capable of achieving we will be able to use it to improve our memories, enhance our intelligence and increase our creativity.

In all these *memory is the foundation* for everything that we learn and act in our life situation. A good memory will help us to understand how our memory works and how to make the best of it.

We may discover the different types of memory and find out how they operate in all aspects of our lives. There are *practical exercises* that can improve the everyday memory power and help us to make *sure and correct memories* which will *not fail with age.* By using the techniques formulated by the memory experts, with practice, any one can improve is memory.

Researchers state that the *physical motor memories* are the longest lasting type. Psychologists generally accept that memories involving names, faces, dates, images and situations are *less indelible* than the physical motor memories. This type of perceptions first enter the immediate memory. There are *two functions* taking place. The *perception* or experience can be *last after few seconds.* But if perception caputres our *attention,* it is transferred to the short term memory where it can *remain from 30 seconds* to few hours. Then the memories are last.

Memory failures are quite common among the persons who *hurriedly carry out* their activities. This may be also due to some *behaviour disorders* or due to some *hasty decisions.* In the United States, it was observed during the periods of 1940s and 1950s there were severe *epileptic disorders* and the patients were treated by surgical operations in which the doctors removed certain parts of the brain. The doctors found that the electrical stimulations in the *hippocampus area* resulted in *provoking* lots of memories in the patients. Removal of or damage to the hippocampus made the patients *impossible to learn anything new.* But the older memories are left intact. This kind of loss of memory (amnesia) has proved to be permanent.

It is observed by investigators that progressive loss of memory is the disastrous consequence of *Alzheimer's disease.* It is the degeneration of the brain. This generally takes place in old age. This is also considered as an irreversible loss. A head injury also may cause amnesia, but the effect is temporary and it may be cured. *Severe emotional stress* is another cause of amnesia. Convicted crime victims, and battle scared soldiers may protest themselves from the horrors of to which they are exposed by blotting out their memories.

Studies were conducted very recently on *memory decline* and on the *hippocampus shrinkage* on healthy and cognitively normal people aged 55 to 87. Investigators also tested whether the *mild impairment in memory* is mainly related to the reduction in size of *hippocampal formation.* The hippocampal has been implicated in *memory formation.* The *loss of Hippocampal Formation* (HF) has been reported in the case of *mild Alzheimer's* disease and in elderly individuals with *cognitive impairment.* The HF can be accounted for *milder declines of memory* that many *neurologically healthy* older people had experienced in their life.

Researchers have observed that the mind like all other parts of our body *needs continuous activity for keeping it fit.* By keeping the *mind more active* in old age the *memory becomes sharper* and it will be active in *solving the problems.* It is not late to learn new things and keep the memory active. It is also observed that the older persons sometimes *suffer with overloaded information* in their memories. But the *loss of neurons* may affect the older person's *short-term memory.* However, once new materials are learned, older persons remember them as well as the younger persons. One of the best ways for the older persons to acquire new information is to activate their *long-term memory.* For this they may be allowed to choose new materials of their own. It is always better to see that there is no *outside interference* in their activities.

There are ample evidences that older persons have done wonders even after their retirement from service after the age of sixty five years, by keeping sound memories. It has been observed that many older persons have shown great intellectual growth, sound memories and achievements. Many of the great philosophers, scientists, poets and novelists made valuable contributions after sixty years of their age and they have *cultivated good memories* and were able to recollect and recite even their very early experiences. **George Bernard Shah,** the great English *play wright* of the twentieth century was writing something for printing even at the age of ninety and was *keeping a good memory.*

Psychologists are of the view that a good memory can be maintained if only a person is very much *attentive to the subject matter* he learns. The *enthusiasm* one has for the subject matter, apart from attention, makes one to develop good memory. In the common sense world it is often remarked in seeing a person with good memory that *memory is inborn.* It is like keen eyesight by which one can see objects clearly and can state all the particulars. Even at the old age state *without* wearing *spectacles* one can *see* and *read* everything. Similarly some persons have sound hearing irrespective of place, time and age. In the common sense world it is stated as *snake eye.* These persons' sensory perceptions are so *sharp and clear* that they can *perceive everything minutely.* These are said to be *inherited characteristics.* Many are of the view that *memory* also is *an inherited characteristic,* that is carried on genetically from generation to generation.

We witness sometimes that some persons display *monoact* by presenting the *long dialogues* of a play on a stage and create excitements among the audience. How do they get memorised so much materials as to speak out non-stop without seeing any written script? However practice or training may enhance the person's efforts, it is something, like the *inborn talent* that makes him to perform the monoact so efficiently. Some inherited talent may be functioning in his *memory faculty.* What required is to provide *suitable stimulation* to make the talent to function. In *Shakespeare's plays,* there are several instances in which the person who plays the role of a character, has to speak out long dialogues. Only certain persons who are endowed with good memory ability to present the material meaningfully in a realistic fashion can do it. This is a *special memory talent* that only very few can exhibit. The mental health of such person is so strong as to carry out the memory task successfully.

14

FORGETTING

Introduction

Everyday we are experiencing various types of sensations and these sensory experiences get into our brain and stored in the memory areas. All our experiences and gained ideas are not stored in our memory permanently. This phenomenon we experience everyday. Among the stored information some we need immediately, some after few days, weeks, or months and even only after few years. We try to remember only when we are in need of them. But our conscious attempts to remember are not consistently successful. Some type of information seem to be much easier than others. But it is also surprising for us sometimes that the buried information or thoughts come to the surface on their own accord without our attempts. When we urgently need some information it is not helping us.

The human memory is unsurpassed in its ability in storing the information and bringing them back and provide us when we need. Then why do we fail to get the desired information at times we needed. We say, we forget. Why do we forget? Psychologists state that *with time, rarely using* the learned materials, *adding* everyday *lots of new thoughts* and *ideas, difficult* and *unfamiliar materials, diseases, brain damage* all these may cause loss memory. Further, the psychological processes such as *motivation, interest* and *enthusiasm* may interfere in the process of learning and may cause failure in remembering the materials. All these may contribute to forgetting.

Forgetting the personal experience, the information that we collected and stored in our memory after learning something are becoming the everyday events in our life. What is learned, is stored in our memory and we remember the matter when we wanted. But what is *stored* in our memory, is *not remembered* which we call as *forgetting.* Thus the incident of learning is taking place first and then the next process is storing them in our memory. Then occurs the process of recalling and recognizing the materials. Thus the process of forgetting is the third stage in our knowledge getting activity. Therefore, we may state these processes as learning, remembering (memory) and forgetting. Forgetting occurs only in the case of learned materials. If nothing is learned then there is no memory or storing the information and there is no forgetting.

Generally forgetting is not a welcome mental process. Human beings only wish to remember everything that has occurred in their life. But forgetting occurs. Are we unconsciously wishing to forget anything? Yes, it is true. We wish to forget many things not only unconsciously, but also consciously. Just imagine for few minutes how painful and unhappy our life would be if nothing were ever forgotten. *Forgetting very unpleasant and disastrous consequences* of our life is *a wishful desire* of human beings. The unpleasant and painful events often may be appearing in our conscious mind and would make our life very *unhappy.* By forgetting such disastrous consequences our life would be going on normally and happily. We have already quoted elsewhere in this book how the famous romantic English poets **Byron** and **Keats** struggled with their painful feelings as they could not forget their painful memories. Therefore, we must forget certain things if we want to be happy in our life. **Epstein** and **Shontz** (1969) state that *like death, forgetting is always with us.* It is always associated with memory or remembering and discussed. Every author of a book on psychology discusses the subject matter as "Remembering and Forgetting". Therefore, forgetting cannot be discussed without any reference to learning and remembering or memory.

Human beings learn everyday many things. In fact man has been learning many things right from the time when he begins to learn in the childhood period. The learning may be casual or systematic. Whatever it is, he learns something which may be useful to him and sometimes it may not be immediately useful to him. But what are all a man learns are not remembered all the times. Sometimes it might have been very essential, but he might not be remembering the matter at the time when it is needed. For example, we do not remember the name of a person whom we met sometime back and had discussion with him on some important issue. We often do not remember the street name where we lived some years ago. This does not meant that we forget everything. We remember many things and we fail to remember many things. What are the things we remember and what are things we do not remember? Why do we remember and why we do not remember?

What is Forgetting?

Generally, we remember things in which we are very much interested and also the things and the persons with whom we had pleasant experiences. But we do not remember things and persons with whom we had unpleasant experiences. *What we do not remember is called forgetting.* It is also said as *loss of memory* or *failure to remember.* Forgetting is used for both good and bad purposes. It is the opposite of remembering and also *failure of retention.*

Psychologists made a thorough study of this process of forgetting as it is associated with *cognitive process.* We also *restore* what has been forgotten by frequent reviewing or relearning. One trouble we generally come across in life is that what we want to remember is not remembered and what we want to forget is not forgotten. According to one systematic study, one major source is, the *role of motivation* which is *blocking the recall.* **Sigmund Freud** in his work "The psychopathology of everyday life" emphasised that sometimes we frequently fail to recall things that we know perfectly well and we are put into embarrassment. The reason is the *expression of a motive* in us, but we do not accept it.

College students often forget the correct answer to the question in the examination hall. But they remember it after going home. **Robert Lind,** an American famous essayist, states that people often forget to post letters. Men often forget to bring back their walking sticks and umbrellas, when they go out for walking. **Murphy** (1954) says that husbands often forget wives' birthdays and Freud says that men forget often wedding anniversaries. We also forget promises, dates of appointments and obligations. But things can be remembered when they are really wanted, but was not recalled. You can find out why you do not recall it. One reason may be, what Freud called the motivation in the unconscious mind. If we do not accept it, the reason might be that when we try to recall, certain material come out easily while another with great difficulty. It is due to *sheer preference* of one over the other that things are recalled or remembered at once and others take longer duration. There might have been many painful events that might have occurred in everyone's life and everyone of us wanted to forget them. Keats, the famous English romantic poet had many unpleasant experiences in his life. He prayed to God to bliss him with strength to forget them. He was jealous of trees because they have the power to with stand all such things. Everyone of us on certain occasions might have had the same feelings as the poet, Keats had. We want to forget all the wretched things that we had done.

But we are happy when we remember and recall the information or material that we stored in our memory long ago. We also appreciate persons who remember many things.

A correct and accurate definition of forgetting that is agreeable to all is quite impossible. It is because, forgetting is considered to be an aspect of memory function. All those authors who deal with forgetting in their writings describe mostly only the aspects remembering such as short-term memory and long-term memory, passive forgetting, active forgetting Epstein and Shontz (1969) as aspects of remembering, while **Kimble** *et al.* (1984) discuss forgetting involving the memory aspects such as retention, interference, an autobiographical memory, long-term memory and permance of memory. Therefore, any definition or discussion of forgetting is impossible without involving remembering or memory.

On this basis we may state that forgetting is an aspect of memory or remembering of an organism's experiences after learning something. It is a negative aspect of remembering. Both remembering and forgetting refer to the effect which the passing of time has on something we have learned earlier. Forgetting may be tentatively defined as the *inability to retrieve* the specific material that has been learned earlier and stored in long-term memory. It is not the loss of memory. Loss of memory means the material is lost once for all. But this does not take place in the normal cases.

The study of forgetting is always associated with the study of memory right from the time of **Hermann Ebbinghaus.** Ebbinghaus was of the view that meaningful materials would have an effect in learning. The meaningful materials are easily learned and could be remembered very well without any distortion or loss. He tested this hypothesis with *three letter consonant vowel-consonant* sequences like LUN, ZIV, WEK and so on, which are not having any meaning. Ebbinghaus conducted test on himself by using the test materials he has designed with varied length of letters.

He made out quite large number of these syllables tack one in a separate slip of paper. Out of more than two thousand such slips he removed seven to thirty six of these slips randomly and made a lists of various lengths. Then he himself learned the syllables in the order from shortest to the longest syllables with beat of a matronome. Then he tested himself by recalling them after varied lengths of time. Ebbinghaus has made many contributions. But among them the course of *forgetting experiment* was the one very carefully traced. In this experiment the remarkable information about forgetting was *the speed* with which it has occurred. It was observed in this test that *after nineteen minutes* after learning *forty two per cent* of the material had been *forgotten and retention was fifty eight per cent.* But, though most of the amount of the materials were forgetten in a given period of time in the Ebbinghaus experiment it was also depended upon several other variables. It was also observed in many of the experiments that most of the forgettings took the course similar to Ebbinghaus' experimental result. Thus according to Ebbinghaus, forgetting is having the effects of meaningful materials, interval of time and other related variables.

Definition of forgetting can be also stated on the basis of these variables. In the common sense world it is stated as the failure to recall something or someone once known. It may also refer to the *inability to remember, making disuse of things.* Forgetting also may mean to neglect, not making use of the objects or things, as a result of an *unintentional error,* to leave behind by mistake or even disregarded intentionally. Thus the meaning of forgetting of the common sense world is not too much far away from the psychological world. The quintessence of the conception of forgetting of the common sense world is making disuse of things and it is an unintentional error of not making use of things. This definition seems to be also quite close to the psychological definition.

Psychologically, forgetting may be also defined as an instant of decay to retrieve the required material stored in the memory for the immediate use. As the learned material which is stored in the memory is not totally lost which may be also can be retrieved on a later occasion. The extensity of delayed retrieval or forgetting period may vary due to several factors in an individual.

In Ebbinghaus's study the test was really both the extensity of forgetting and also level of retention. In recalling test a person is required to retrieve something from memory. From this the extent of forgetting can be assessed. Actually forgetting something is what the *amount of material learned minus* the amount of *material retained,* retrieved after some time. The *span of memory* test is also in a way a test of forgetting in a short time. It is a test of short-term memory.

The *span of* (short-term) *memory* is a test to assess the number of items which can be remembered following a single, brief presentation of some items. *The span* tends to be about *seven* with a range from *five to nine.* This data may be treated as *span of memory,* level of retention and remembering and the extent of forgetting in a short period. Thus the phenomenon of forgetting is understood from different dimensions. In psychological studies we use the process of forgetting as a *part of learning and remembering.* If the learned material is not remembered with special reference to some object, person, place, time, pleasant and unpleasant experience, we normally call the state as forgetting.

Types of Forgetting

Forgetting as a process of experience can be classified into two broad categories. They are: (1) Normal forgetting and (2) Abnormal forgetting or pathological forgetting.

(1) Normal Forgetting

Most of our forgettings are normal forgettings. Normal forgettings generally occur due to three causes. They are (a) Decay through disuse or Atrophy, (b) Interference effects and (c) Motivated forgetting.

(a) Decay Through Disuse or Atrophy

This type of forgetting usually takes place with the *passage of time*. The idea that memories decay or that they fade away due to passage of time was first put forth by the German psychologist, **Hermann Ebbinghaus** (1850-1909) a pioneer, of the last quarter of nineteenth century. Ebbinghaus made *his own memory as the subject* of his 'experiment on memory and forgetting'. He used in his experiment lists of non-sense syllables consists of three-letter consonents and vowels. He learned them and tested himself at varying intervals of time. He observed that as soon as the learning is over and tested there was a period of *rapid forgetting* and more of the words he had learned were lost. Then as time passed away *more forgetting* took place at a slower rate.

From these results Ebbinghaus came to a conclusion that his memories were decaying and fading away over time. It is found from this experiment that the longer we hold something in our memory without the activity of retrieving the material in the brain by recalling it, the more likely we are to forget it. Having accepted this result we are also interested in knowing whether there were also *other factors* contributing for this forgetting apart from time. To examine this idea further we may also look into other activities.

A skill for example, is liable to be forgetten if it is not practised for a long time and kept in disuse. A geometrical theorem even if it was mastered thoroughly might be forgotten in due course, suppose it was not frequently put into use and practice in the event of solving a problem. It is evident that occasional reviewing of what has been learnt and also applying the general principles once understood, will immensely help the process of retention.

Here it is assumed that learning leaves a trace in the brain. This memory trace involves some sort of physical change, which was not present, prior to that learning. The normal metabolic processes of the brain, with the passage of time, cause a fading or decay of the memory. So, the memory traces of the material learned once upon a time gradually disintegrate and finally, disappear altogether.

We often forget pictures, faces and stories. This suggests that a process of fading with the passage of time is taking place. When a picture is perceived at first, it may show a wealth of detail. But as time passes away, the details are rapidly forgotten and only the general outline is remembered. Similarly, ordinarily learned material is rapidly fading. This kind of experience provides support to the point of view explained above.

It is true that some forgetting may occur because of organic changes in the nervous system with the passage of time. But, there is no direct evidence to support the view of decay through disuse. Critics remark that there are several instances which clearly convey that learning is retained over long intervals of time, even without intervening practice. For example, most of the motor skills such as cycling, typewriting, swimming, driving etc., are not easily forgotten. Generally, we do not forget how to swim or drive a car although for several years we might not have practised these skills.

It is surprising to note that some verbal material is retained for long periods whereas other material is forgotten. For example, you may be able to recall perfectly a poem which you have memorized long ago when you were in the 5th or 6th standard. But at the same time you are unable to recall a poem studied last year. Now the question that crops up in our mind is, why should the decay process affect only the later material but not the first material?

Another criticism levelled against the explanation of decay through disuse, is the recovery of memories which have been supposedly lost. It is mostly common among the people who are approaching senility, to often vividly recall events of their youthhood days when they can hardly remember the events of that day. Therefore, we are to conclude that unavailable memories have not necessarily 'decayed' or forgotten.

(b) Interference Effects

According to the theory of interference, what we do in the interval between learning and recall is responsible for forgetting. That is to say *new learning may interfere with* the recall of *old learning* (that which is learned previously) and *vice versa.*

The effect of interference is nicely illustrated by a story told about Stanford University's first president, David Starr Jordan who was an authority on fishes. As the president of a new University he began to call the students by name. But every time when he learned the name of a student, he forgot the name of a fish. So it is said that he gave up learning the names of students. Though this story has no foundation of fact, still it explains how new learning may interfere with the recall of old learning.

The term 'interference' here denotes that different memory traces are overlapping and disturbing each other. This gives a warning that we should not try to learn too many things at one and the same time.

There are two types of interference : (i) Retroactive Inhibition and (ii) Proactive Inhibition. In the former type, the *new learning interferes* with the *old learning* whereas in the later type *old learning interferes with new learning.*

(i) Retroactive Inhibition

The retroactive inhibition can be explained easily through an experiment. The experimental arrangement can be diagrammed as given below:

Arrangement for testing Retroactive Inhibition			
Experimental group	Phase 1	Phase 2	Phase 3
	Learn list A	Learn list B	Recall list A
Control group	Learn list A	Rest or unrelated activity	Recall list A

The experimental subjects first learn a list of nonsense syllables, from *list A,* and secondly learn the nonsense syllables from the *list B.* The control subjects learn the nonsense syllables from the *list-A* and they *kept at rest* when the experimental subjects learn the nonsense syllables from the *list-B*. This means that the control subjects *do not learn the list-B.*

After an interval both the experimental subjects and the control subjects were asked to recall the nonsense syllables from list-A. The amount of material recalled by both groups were recorded. In this experiment the control subjects should recall more materials than the experimental subjects. If the control group's (which has not learned list-B) recall of list-A is significantly better than the experimental group's (which has learned both the lists) recall of list-A, then we consider that the difference is due to retroactive inhibition. That is to say, we infer that the new learning (of list-B) has interfered with the recall of list-A. In other words, what we learn at present may make us forget, what has been learnt before.

(ii) Proactive Inhıbition

The Proactive inhibition is another kind of interference. What we have *learned previously* (old learning) *interferes* with the recall of something *newly learned.* The experimental arrangement can be diagrammed as follows:

Arrangement for testing Proactive Inhibition			
Experimental group	Phase 1	Phase 2	Phase 3
	Learn list A	Learn list B	Recall list B
Control group	Rest or unrelated activity	Learn list B	Recall list B

Experiments using the above design, have shown that the *control group had recalled better* than the experimental group. This is because of the proactive inhibition encountered by the experimental group. That is to say, the *prior learning* (previous learning of list-A) of the experimental group has *interfered* with their recall of *new learning* (of list-B). The control subjects in this experiment learn list-B and not list-A.

(c) Motivated Forgetting

The role of motivation in a person's remembering and forgetting is having a special significance. Pleasant engagements are seldom forgotten by the people whereas unpleasant ones are easily forgotten in day-to-day life. Therefore, motives of an individual play a vital role both in the process of remembering and forgetting.

(2) Abnormal Forgetting

Abnormal forgetting is also known as pathological forgetting which is rather quite peculiar and *unusual like amnesia.* Amnesia is a *special kind of forgetting* which should be properly differentiated from ordinary forgetting. The term 'amnesia' refers to a loss of memory resulting from damage to the brain. It is a total or partial loss of memory for past experiences. However, the memory is not completely destroyed in amnesia.

The amnesia patient does not forget everything. In fact, he carries out his current activities through his rich store of memories. But he forgets certain matters such as personal reference like his name, his family, his home address, his personal biography, etc. The *origin* of the amnesia can often be traced back to some *severe emotional shock* which the person suffered and back from which the amnesia provides him an escape. Amnesia cases, which come under psychotherapy, occasionally provide rather convincing evidence of repressed memories and later recovery from the repression.

Abnormal forgetting takes place due to certain unusual and striking circumstances. It does not depend upon the time factor. It has nothing to do with factors like meaningful learning, role of interest, effects of overlearning, reviewing etc. It invariably occurs soon after an experience.

There are two types of abnormal forgetting. They are (i) Exogenic forgetting and (ii) Endogenic forgetting.

(i) Exogenic Forgetting

This is occurring due to outside causes. For example, a severe injury to the brain, like a fall could damage the brain and disturb the memory process badly. Since this is more of a neurological type of case, it would require more medical attention.

(ii) Endogenic Forgetting

This kind of forgetting is arising from the causes *from within.* Many of our experiences are just completely forgotten in the flash of a second because of the working of a particular phenomenon called *repression.* Since the experience is simply pushed into the unconscious mind, the individual concerned will not be aware that it has been drawn from within. As a result, the individual will have complete recall amnesia. This occurs due to *emotional conflicts* leading to conditions of psychopathology. The forgotten materials under the situation can be brought out only by *psychotherapeutic methods like hypnosis, dream analysis* and free association.

Therefore, the phenomenon of abnormal forgetting is *brought out by repression* to individuals. ***Sigmund Freud,*** the founder of Psychoanalysis was the first to explain this. The abnormal forgetting is also called *intentional forgetting* and convenient forgetting but the intention of the individual is not a conscious one.

(3) Other Types of Forgetting

We have discussed in the preceding section more about short-term and long-term memories. These are based upon the objective characteristics of the experimental results such as length of retention of the learned material, retention level or duration of retention or remembering. The other distinction in our learning and remembering experience is forgetting which is commonly experienced. This experience of forgetting is classified into several types. Investigators make different broad classifications and use varied concepts to explain the same process of experience of forgetting. A second type of broad classification of the same matter, forgetting, falls into three categories such as *passive forgetting, active forgetting* and *not remembering.* **Epstein** and **Shontz** (1969) use this model to explain forgetting.

(a) Passive Forgetting

We could observe that most of the materials we learned are not fully remembered. We forget most of them and remember some. Thus most forgettings seem to occur without our effort or involvements. According Epstein and Shantz most of our forgetting generally is passive forgetting. How most of our forgetting is passive? It is because it occurs *without any conscious effort* on the part of the person. Today we learn many things and we keep quite without any retrieval for a period of time nor we made use of them. The end result is that we could not recollect anything and they are forgetten. Psychologists use the *disuse hypothesis* to explain this type of forgetting. According to this hypothesis if we do not make use of the learned material or make a rehearsal or retrieve, it fades away or decayed. This process of forgetting is stated as passive forgetting.

This passive forgetting involves the aspects of forgetting like the *interference* which consists of *retroactive inhibition* or interference and proactive inhibition. These we have discussed under normal forgetting. Thus we may infer and conclude that the normal forgetting is the same as passive forgetting.

(b) Active Forgetting

This type of forgetting seems to occur if the individual person is put himself in *vigorous activities.* People are engaged in exercising their physical and mental abilities, sometimes on an ill-planned manner which results in putting things out of mind or memory which results in active forgetting. Active forgetting may be of two types. One type of active forgetting occurs due to *delibrate suppression.* This involves putting the material out of memory wilfully. This is to acquire some new information the person makes himself ready.

At this stage if we wish to get the repressed or excluded material we find that the material is not available in our memory. It is lost or forgetten. The other type of active forgetting is *voluntary* or *intentional forgetting.* This forgetting is the central point in Freud's theory of repression. According to him forgetting is often motivated to repress the memories that may *produce anxieties.* This process of repression takes place unconsciously. *The second type* of active forgetting is motivated repression. According to **Freud** (1925) the essence of repression is the function of *rejecting and keeping* something *out of consciousness.* It is an unconscious rejection, motivated *to defend the ego,* the self, *against the anxiety causing* memories.

(c) Not Remembering

The third type of forgetting is not remembering. This statement *not remembering* needs certain explanation. In the common sense world we often hear that people in their conversations use the terms, "I forgot", "I do not remember". I forgot and do not remember logically do not refer to the same matter. These two indicate two different mental processes. When a person, says, ''I forgot", the matter ends there.

But when a person says, I do not remember, means it is something else. The phrase seems to suggest that he does not remember now, but if suitable aid is provided he can remember it. There are two processes involved in the statements. The phrase, I forget involves *problem* in the *storage system.* There is some disturbance in the memory organization. In the case of the second phrase there is *problem* in the *retrieval process.* The person finds some disturbance in recalling the material. People sometimes used to speak, 'it is in the tip of the tongue'. But not coming out. This means, the matter is not forgotten, but he could not bring it out as something interferes in it.

What is required here, is to have a suitable retrieval cue. The retrieval cues may vary in effectiveness, what are the characteristics of an effective retrieval cue. The retrieval cue is a part of the original learning task. The retrieval cue should have certain similarities with the learning material, other factors like the environmental situation and the type of material used in the task also may have some effect in the retrieval process.

MOTIVE IN THE UNCONSCIOUS MIND AND FORGETTING

Most of the discussions on forgetting put emphasize on the psychological process of *affecting the memory traces or of interference of old material into the new* by the processes of *proactive inhibition.* These theories do not pay much attention to the individual's personal motivation in remembering and forgetting processes. This subject matter is an omission in most of the studies and it should be seriously taken into account. It should be inferred what a person is wanting to do in remembering and forgetting. One of the processes of motivated forgetting is the process of *repression.* When this process occurs some memories become inaccessible to recall the stored learned material because of the way in which they are related to our *personal problems.*

After Ebbinghaus's quantitative studies on memory and forgetting Freudian interpretations of memory lapse had a hold upon the experimental testing. Psychologists first grasped only the *forgetting* aspect of *repression.* **Freud** distinguished between *primal repression* and the second phase called *repression proper* (1915a). According to this theory repression holds that the memories are not recalled because the retrieval of the material is in some way *inaccessible* to the person. It is because of the *anxiety* the material may cause or it may *activate the guilt feeling* the person has repressed. The nature of forgetting, if it occurs in amnesia it *provides aids* in understanding the repression. The amnesia patient does not forget everything. He has rich stores of memories to conduct current activities. The forgetten items are of personal references such as family, home, name, address and personal biographies. The starting of the *amnesia* may be traced to some *severe emotional shock* that the person suffered. But the amnesia provides an escape for him from such sufferings. This amnesia is a partial or total *loss of memory* for the past painful experiences. The person can be free from all such painful memories and sufferings by forgetting them. The person's *motive in the unconscious mind,* according Freud is *to get rid of the sufferings and painful feelings.* This motive is fulfilled by the occurrence of amnesia. Thus this becomes the motivated forgetting.

Transient Global Amnesia

This is a *neuropsychological disturbance* causes memory loss or forgetting. Memory has been studied experimentally in the middle of 19th century by Ebbinghaus. The 1880s have been stated as the revolutionary decade for memory research, because it marked the beginning of the description of memory disorder. *After one hundred years,* only in 1980 **Schacter** (1987) started a systematic study of memory and stated that memory comes in many different forms. This has been *a new approach* of investigating of the *neural basis of memory.* This period has been described as the *second revolutionary period* in memory study. All of us one time or other suffered of amnesia or memory loss. It has been discovered that the infentile amnesia is the most dramatic example of forgetting. The early years are generally regarded as the most critical importance in the child's development. In adulthood nothing is consciously remembered, but the information are not lost, still there, and no longer it can be retrived or remembered. It is because of the development of forebrains. This development of brain disrupts the memory and blocks its recovery. One seeming explanation is that one memory system is used by infants and another developed one is used by adults. Since the development of the new conscious adult system (brain) began to function, the early memories are lost. Not only the infants, adults also forget, who have no knowledge of the previous life.

The acute form of amnesia, discovered by **Fisher** and **Adams** (1958) is the 'transient global amnesia', which attacks suddenly, resulting in an inability to form new memories and also loss of old memories, which has no apparent cause. The symptom is linked with concussion (paralysis of the brain), migraine (a type of headache), hypoglycernia and epilepsy and also interruption of blood flow in the posterior cerebral artery from an embolism (air bubble or deposit of fat). It is said that the transient global amnesia can be a one time event. But **Markowitsch** (1982) says that there can be permanent memory loss or forgetting. **Mazzuchi** (1980) and his co-workers show that there is chronic memory loss in transient global amnesia, a form of Amnesia mostly caused by *head injuring.* The characteristics of amnesia depend upon the severity of the injury. There is also a typical transient loss of consciousness followed by confusion. But the period of forgetting usually reduces to one time. The amnesia existence is only for *few seconds* for the events. **Whitty** and **Zangwill** (1966) discovered in a series of patients suffered head injuries that 10 per cent of them had it for *less than a week,* 30 per cent had for *2 to 3 weeks,* and 60 per cent had for more than *three weeks.*

Amnesia can be produced by the use of minor tranquillizers or alcohol. One therapy for depression, electroconvulsive shock, can produce memory loss. Damage to restricted portion of the brain also cause amnesia, which often takes curious form. Clinical reports state that there are people who become amnesia for the *meaning of nouns* and some become amnesia for *animals* and some become for *human faces* and for other objects. There are also *everyday amnesia,* persons of this amnesia forget people's names, faces, keys, telephone numbers. This kind of amnesia becomes bothersome to people as they advance in age. For some people memory disorder occurs due to age and become incapacitating as in the Alzhemer's disease, which is characterised by the loss of many past memories and also followed by the *loss of neurons* and other pathologies *in the cortex.*

Attempt has been made to classify memory disorders. There are *two approaches.* According to *one approach* it is stated that there is *a memory centre* in the brain, and the *second approach* states that there are *many kinds* of memories and *each one* is having *a location in the brain.* The second approach is favoured by all.

When we forget new information, it is called a *interograde amnesia* and when we forget the old or previous information it is called *retrograde amnesia.* If we remember things only to very short time, such as telephone number, the street name, it is called short-term memory. When we remember things for a long period of time, it is called long-term memory.

The investigation conducted in 1980s indicated that there are many kinds of memory that exist in patients with anterograde amnesia. These memories are grouped together and named. One among them is the *implicit memory.* This is an unconscious, non-intentional form of memory. The other is contrast to this, called *explicit memory* which is conscious recollection of previous information.

A systematic neuropsychological investigation of memory and memory loss was started by **Karl Lashley** in 1910. His lifetime study brought out many interesting results about memory loss. Following **Lashley** (1950) several others, **William Scorille** (1957) made one of the most influential finding in neuropsychology by making operation in temporal lobe of a patient.

One of the regions of the brain is the *diencephalon,* which includes the hypothalamous, thalamous and epithalamus. The amnesia which finds its origin in this area is called *diencephalic amnesia.* Investigations reveal that the evidence for diencephalic amnesia is from two sources. One source is, patients having *focal injuries* in the *medial thalamus,* and the other is, patients having *Korsakoff's symptoms.*

The *focal injuries* (lesions) of the *medial thalamic area* mostly result from vascular accidents. This causes apparent memory problems. Examination of the anatomic localization yielded no clear information, but the alcohol-related diencephalic damage is clear, *Long-term alcoholism* accompanied by *malnutrition* produces memory defects. **Korsakoff,** a Russian physician observed the syndrome accompanied by *chronic alcholism.* The symptom indicated severe loss of memory. The loss of memory revealed itself in a severe amnesia of extraordinary type. This resulted in the main disturbance of the memory of recent events and those events just occurred. But the remote past is remembered well. This resulted chiefly in the person constantly asking the same questions and also repeating the same stories. It was difficult to note down the presence of psychic disorder, because the person makes an impression that he is in complete possession of his mind. He thinks about everything perfectly well, draws conclusions, makes humourous remarks, plays chess or games and *behaves* himself as a *sound person mentally.*

A long conversation with him alone may help a person to understand that *at times* he *utterly confuses events* and *remembers absolutely nothing.* He does not know whether he had his dinner and does not know where he is. The person forgets totally what has happened a few minutes ago. Persons of this type read often the same page in a book. The remarkable fact is that he has forgotten all events happened just. What he has forgotten shows that everything happened during his illness. There are persons exhibiting symptoms of this type: *Korsakoff's syndrome.* There are individuals with other forms of transient global amnesia. **Tallard** (1969) brought out the major's symptoms of the syndrome. They are:

(1) *Anterograde amnesia* : This causes inability to remember events as results of some disturbance in the brain such as *head injury,* electroconvulsive *shock* and certain *degenerative diseases.* Such persons are unable to form new memories.

(2) *Retrograde amnesia* : This causes inability to remember events that occurred *prior to* the attack of amnesia.

(3) *Confabulation* : It is a state of amnesia in which the person makes stories about past events and *refuses* to admit his *memory loss.* The stories are based upon past events.

(4) *Meagre content in conversation* : In this stage the person has little to say in spontaneous conversation, because of his amnesia, the loss of memory.

(5) *Lack of insight* : In this stage many persons are completely unaware of their memory loss.

(6) *Apathy* : In this state of amnesia the person shows indifference and unable to continue ongoing activities. The person looses interest in everything and appears indifferent to events.

The person is quite normal except these symptoms. The person has normal I.Q., is alert, attentive and appears motivated.

But the Korsakoffs syndromes may appear suddenly within the space of few days. The main cause is the *Vitamin-B (thiamine) deficiency* and also as a result of prolonged intake of *large quantity of alcohol*. The symptom can be arrested by massive dosage of vitamin-B. Current discovery is that there is *damage* in the *medial thalamus* and in the mammillary bodies of the hypothalamus. The damage to the *diencephalon* is a disconnection of the hippcampal (primitive cortical structure, the temporal lobe) circulatory and this disconnection produces the amnesia.

There are *two different reports* of memory loss, presented by the investigators, after a treatment for epilepsy in fourteen patients. The report was that amnesia did not occur in epilepsy patients. The other report was that damage to the fornix or mammilary bodies was not sufficient cause for memory loss. The diencephalon patients experienced the memory loss. Persons with Korsakoff syndrome have extensive loss of memories.

Apart from those cited above, damage to the cerebral cortex, caused memory loss in many cases, by destroying the memory stores and by preventing access to memory. Cortical damage may destroy the ability to recognize certain kinds of stimuli. **Endel Tulving** (1972) a cognitive psychologists states that *cortical damage* causes profound effects on memory. The person forgets his own history and all personal information.

Why do we Forget?

Forgetting is the concept that is used invariably by most of us on various occasions in our everyday life. Sometimes when we are able to remember something another unplesant occasion interfears and our memory puts us down. What we wanted to remember, something, not remembered, and what we do not want remember, appears. We often forget many things though we do not want to forget. People often discuss this matter in a funny manner knowing pretty well that they may forget an important thing to do, or to meet some person or to buy a birthday gift, people keep *some symbol* in front of them *as a 'reminder'*. But it is strange to think that the person forgets for what purpose the symbol is kept as a reminder. This sort of things are common occurrences in the life of many people. Why does this happen?

Forgetting occurs sometimes in our life, because we fail to register a part of information in the first place. For example, while you are reading the newspaper of that day, someone very close to you speaks to you and tells his programme, but you might not really taken notice of what is being said. This *did not form a memory* of it in your mind. Because you did not receive and register the matter in your memory with all seriousness. This becomes the most fundamental step for the occurrence of forgetting. *If a memory is not formed and stored, then it cannot be remembered* later on. Whatever efforts you take later on there is no effect.

But, however, the exact way in which we forget the stored information in our memory is a controversial issue. Several causes are attributed for this. Some researchers are of the view that memories simply *decay due to time*. Some others state that *other memories disrupt* them. This subject matter is discussed in a long way by several leading psychologists. Everyone discusses this matter under different heading using different concepts. There are several theories, formulated to explain the phenomenon of forgetting. Some explain the process of forgetting through decay theory, some through interference theory and some through motivation theory. Some discuss forgetting under two broad heading like *normal forgetting and abnormal or pathological forgetting.* All these theories explain the *basic causes* that function in the process forgetting.

SOME EXPERIMENTAL STUDIES AND CURVE OF FORGETTING

We shall now discuss a few experimental studies carried out by some eminent psychologists. One of the important and earliest studies was made by Hermann Ebbinghaus. He has published the result of his extensive studies on forgetting in the year 1885. One of the significant findings of the pioneer psychologist being that *forgetting is a function of time.* And this fact is illustrated through a curve of forgetting drawn to the results of the experiments on nonsense syllables after various time intervals.

Ebbinghaus found that the amount forgotten is very large, immediately after learning. But the amount forgotten per period of time decreases, when the duration of time continues to extend further. The relationship between duration of time and the amount of forgetting is represented by the curve shown in figure. According to this curve the amount of forgetting will be more in 10 days time interval than in 5 days time interval. In other words, the amount forgotten will be usually lesser in a shorter interval than it is in the longer interval of time. This statement follows from the fact, that the *forgetting curve* tends to *rise with the increase of time* (Fig. 14.1).

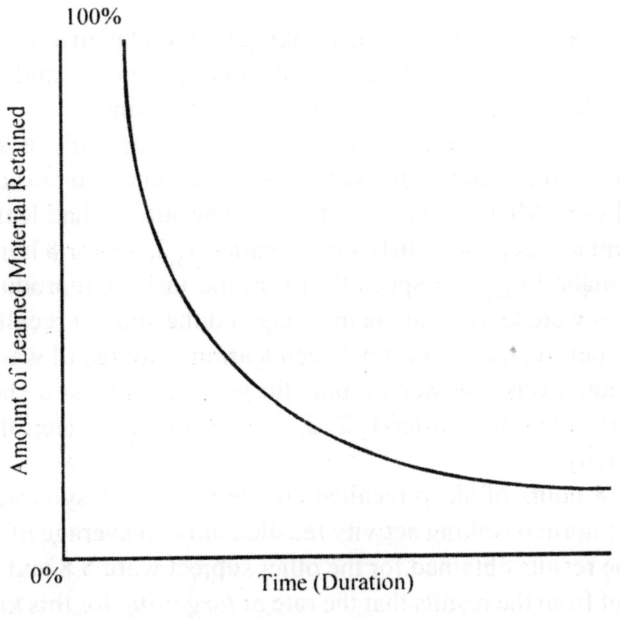

Fig. 14.1 : The Curve of Forgetting : General Nature of Forgetting

In another study, Ebbinghaus found that the amount of forgetting in a 24 hour time interval was *three times greater* than in the 15 hour interval. It is proper for one to expect more forgetting in the longer interval than in the shorter interval of time. But never one would expect forgetting in the 24 hour period to be three times greater than in the case of 15 hour period. This matter has been settled by Ebbinghaus himself later by suggesting that it might be due to the reason that *more is forgotten in a waking period than in a sleeping period,* when both the periods are of equal length (Fig. 14.2).

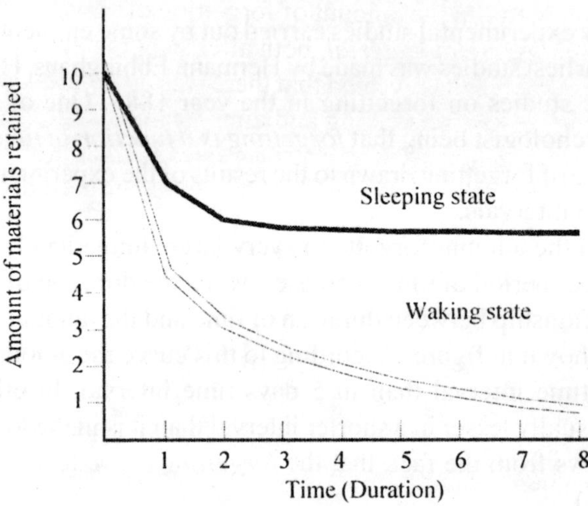

Fig. 14.2 : Curve of Forgetting during sleep and waking states

The hypothesis that 'more is forgotten in a waking period than in a sleeping period' was originally formulated by Ebbinghaus. This was taken up by Jenkins and Dallenbach forty years later, in 1924 by designing an experiment to verify the same.

Lists of nonsense syllables were given to two groups of college students and the experimenters asked them to memorize the same. Each list contained ten such syllables like 'BOQ' : ' KOG' : 'SEG' : 'MUC' : 'ZER' and so on. The subject had learned some of the lists just before he went to sleep. Such lists were recalled 1, 2, 3, 4 or 8 hours after learning and the subject was made to get up specially to do the task of reproducing memorised material. But other lists were learned in the morning and the subject recalled 1, 2, 3, 4 or 8 hours later and unlike before, the interval between learning and recall was without sleep.

This kind of procedure was followed for both the group of subjects. The results showed that both the groups recalled more after 1, 2, 3, 4, or 8 hours of sleep than after similar periods of waking activity.

One subject after 8 hours of sleep recalled an average of 5.5 syllables. But the same person after 3 hours of normal waking activity recalled only an average of 0.4 syllable. In a corresponding way the results obtained for the other subject were 5.8 and 1.4 syllables.

It is clearly evident from the results that the rate of *forgetting* for this kind of material is *lesser* when the *interval* after learning is *filled with sleep.* But at the same time the rate of forgetting is *greater* when the interval after learning is filled with normal *waking activity.*

Jenkins and **Dallenbach** have used lists of nonsense syllables in their experiments. Subsequently **Newman** tried to verify whether the results arrived at by Jenkins and Dallenbach will be valid even for meaningful material.

Newman (1939) in his experiment used three short stories. These stories were constructed in such a manner that each of them contained a number of items that were essential (meaningful) to the plot and a similar number of items that were not essential (less meaningful) to the story at all. Eleven college students were used as subjects. They read each story at a different time of the day and roughly 8 hours after reading they reproduced it.

Comparison was made between the amount of forgetting of *essential material,* and the amount of forgetting of *non-essential material,* both after sleeping and after waking states.

At the end of the study, the results revealed that there was *no difference* in the amount of *essential material* forgotten both after waking and after sleeping states. But in the case of *non-essential* material, the amount forgotten was *two times greater* when the interval was filled with waking than it was occupied by sleeping. In the 8-hour period, the percentage of forgetting was very much greater for the non-essential than for the essential material.

The Curve of Forgetting

Ebbinghaus in his experiments on memory observed that many lists of words of different length had retention at different intervals. He evolved an average curve of forgetting. Following Ebbinghaus, investigators tested the retention or the forgetting curves. **Luh** (1922) designed a study for investigating the nature of retention curve. He varied the methods of measuring the amount of return and also varied the degree of mastery in the original learning. He constructed ninety series of nonsense syllables with twelve syllables in each list. There were two groups and ten subjects in each learned the lists to one of the four criteria, such as 100 per cent, 150 per cent, 67 per cent and 35 per cent.

The retention measures Luh used were relearning, reproduction, recognition and reconstruction. Luh states that the *curve of retention varies* with the *method* of measurement and the *degree of original learning.* He further reported that individual variability increases with difficulty of the act of recall, but *decreases with frequency and recency of practice.* According to his study the *speed of learning* and *amount retention* were found to have *positive* correlation. He states that most of his results substantiated the conclusions of the study made by **Bean** ten years ago. It has been observed that for a number of years the general *shape* of the curve of forgetting *was not changed* (Fig. 14.2).

Do we completely forget everything? We do not completely forget everything. This is because we know from experience that we retain much, although we cannot recall them. More often students are able to recognize answers to the questions but at the same time they are unable to recall the same. This shows that something has been retained even though they are not recalled.

In some situations, events and experiences can never be recalled. These events may be recalled in exceptional circumstances such as in dreams; in somnambulistic state (sleep-walking): in the delirious condition of a fever; in automatic writing: in crystal gazing; in *hypnotic states* and in psychoanalysis.

How to Forget?

We can forget many things provided if we do not bother about them. But there are certain principles psychologists have suggested for forgetting. It is better not to follow any of the principles of remembering or retaining if we want to forget certain materials. Further, the learner should not indulge in reviewing or relearning the material after the incident of learning.

It is better to change the place at once from the place where the incident occurred.

It is better that the person should keep himself engaged vigorously in some physical work or some other work soon after the incident. The physical work should be so strenuous as to cause fatigue and it should produce sound sleep. By this means there will not be any chance for brooding over the matter.

It is better not to consider the matter as valuable on any account. By throwing away the matter mentally, forgetting becomes easy. Here the factor of preservation is lost.

Time is another factor that is contributing for forgetting, the longer the duration the greater the effect of forgetting. Further, anything that is associated with the material should be avoided.

It is true that time is one of the causes of forgetting. But mere passing of time alone cannot significally destroy the memory traces. It is not merely the time, but *the effects* the time has produced. If the time is simply blank without any incident in the life of the individual the time lapse has no effect on the memory. The time lapse without any significant incident is like sleeping. It is only like suspension for sometime. When life started again then everything would resume as if there was no blank interval, and no memory trace is lost. We must find out the causes of forgetting in the processes that go on in the time. The processes that facilitate forgetting are *behaviour and metabolism.*

Progressive loss of memory is the result of Alzhemer's disease. It is a degeneration of the brain that generally strikes in old age. This is an irreversible loss. A head injury can also cause amnesia. But the effect is temporary. For example, a man involved in a car accident, had head injury, may loose memories of events just perior to the accident. Biologists and neurologists call this *retrograde amnesia. Severe emotional stress* is another cause of amnesia.

Forgetting may be either retention amnesia or recall amnesia. Amnesia means total or partial loss of memory. In the case of retention amnesia, retention itself is very poor. So, to strengthen retention, further learning is needed. Whereas with recall amnesia, there is an inability to recall but at the same time retention is found intact. So, to overcome recall amnesia, a great amount of encouragement is required. Recall amnesia usually occurs because of *blocking* in the process of recalling. This blockage is often due to *emotional disturbances* such as *fear, anxiety or self-consciousness.*

We might be thinking that forgetting is a passive process, quietly taking place with the passage of time. But experimenters have found that forgetting is an active process. For example, whenever we recall a story, we do not accept generally that we have forgotten certain details. But more often additional details are supplemented in the places where we have forgotten, without any kind of conscious desire on our part. Further we have a definite feeling that we are recollecting exactly as in the original. Among several psychologists who had conducted experiments on this interesting phenomenon, it is worth to mention the name of **Professor Bartlett** of Cambridge University.

If there is any forgetting within us there is some unconscious effort to fill up the gap. There are *two views* to explain this process.

According to the *first view,* there is a tendency *to distort* the learned material. This tendency starts even at the time of learning. It is considered that this process of distortion is taking place in an unconscious manner. Some of the materials are distorted at the time of learning according to our likes and dislikes, attitudes and prejudices. We ignore even some of the important facts and substitute them with what we like very much. This process of fixing of substitute material in our memory is taking place unconsciously.

The *second view* is that it is generally believed that we do *not feel happy* with anything which is *incomplete* or incoherent. Therefore, there is a constant desire in us to make things complete. When we forget some points in a story, we become dissatisfied with the event and we take care to fill-up those gaps with some other points which may be more appealing and dramatic. Thus a completed story is produced ultimately even though it is deviated from the original. What is most important in this process is that we *drop out* all those *things we dislike* and *add* all those *things we like*. This process is taking place unconsciously without any effort.

Why do we forget? We forget learned material because of the lapse of time. We forget things as the *neural traces* of our previous experiences grow fainter and weaker. Time is an important factor in forgetting. Everyday we learn many things and come across several events in life. As the time passes on the past events become old and present day events become new. Therefore, events that took place recently make the neural traces of the past experiences weaker and thereby make us forget. But when materials learned in the past are drilled in thoroughly and repeatedly, they are not easily forgotten. It has been found that the rate of forgetting depends upon the thoroughness and adequacy of the learning.

(1) Forgetting is slow when the meaning of the material or passage is understood well.
(2) Forgetting is slow when there is active recitation or reviewing.
(3) Forgetting is slower in the case of the materials learned by spaced method (distributed practice) than to the unspaced method (massed practice).
(4) The process of forgetting is slower for the materials learned by whole method than by the part method.
(5) Forgetting is based upon the passage of time. The relationship between forgetting and time is such that when the time increases the amount of forgetting also increases, but at a decreasing rate.
(6) The rate of forgetting is dependent upon what one does after learning.

MODERN GENERAL PSYCHOLOGY

Thoroughly Revised and Expanded

VOLUME-II

About the Author

M. Rajamanickam is the Founder-Director of the Institute of Community Guidance and Research, which is one of the permanent Members of the Indian Council of Social Science Research. Formerly, he was the professor and Head of the Department of Psychology and the Dean of the Faculty of Education and Psychology of the Annamalai University. He was been teaching Psychology to the Post-Graduate students and guiding research for over three decades of the University and produced many Ph.D. and M.Phil. Degrees.

He has been elected to the Syndicate of the University three times. He was mainly responsible for introducing Psychology in the scheme of Distance Education, which has attracted more than 500 students every year all over India for the study of Psychology at the M.A. and B.A. Degree levels. He extended the Psychological services to the Rural people through the Institute of Community Guidance and Research which he founded in 1980.

He has been associating himself with a number of Academic and Professional Organizations and contributed quite large number research papers to the national and international journals. He founded the Journal of Community Guidance and Research in 1984 and has been the Editor of it for more than twenty years, which has benefited many young scientists in India and in other countries. He is a prolific writer, has published more than ten books on Psychology. Some of the important titles are, Contemporary Fields of Psychology and Experiments, Statistical Methods in Psychological and Educational Research, Modern Psychophysical and Scaling Methods and Experimentation, Modern General Psychology, Abnormal Psychology, Experimental Psychology with Advanced Experiments (in 2 Volumes) and other publications.

By the Same Author

1. Contemporary Fields of Psychology and Experiments
2. Statistical Methods in Psychological and Educational Research
3. Modern Psychophysical and Scaling Methods and Experimentation
4. Psychology of Perspective of HIV and AIDS
5. Experimental Psychology with Advanced Experiments (in Two Volumes)

MODERN GENERAL PSYCHOLOGY

Thoroughly Revised and Expanded

VOLUME - II

M. RAJAMANICKAM

M.A., M.Litt., Ph.D.

Formerly Dean, Faculty of Education and Professor
and Head of the Department of Psychology
Annamalai University

CONCEPT PUBLISHING COMPANY, NEW DELHI-110 059

Cataloging in Publication Data—DK
Courtesy: D.K. Agencies (P) Ltd. <docinfo@dkagencies.com>

Rajamanickam, M. (Muniappa), 1925-
Modern general psychology / M. Rajamanickam.— Rev. ed., thoroughly rev. and expanded.
2 v. cm.
Includes bibliograpahical references (p.)
Includes index.
ISBN 13: 9788180694219 (set)
ISBN 10: 8180694216 (set)
ISBN 13: 9788180694226 (v. 1)
ISBN 10: 8180694224 (v. 1)
ISBN 13: 9788180694233 (v. 2)
ISBN 10: 8180694232 (v. 2)

1. Psychology—Textbooks. I. Title.

DDC 150 22

Set	ISBN - 13: 978-81-8069-421-9 (HB)	ISBN - 10: 81-8069-421-6 (HB)
Vol. I	ISBN - 13: 978-81-8069-422-6 (HB)	ISBN - 10: 81-8069-422-4 (HB)
Vol. II	ISBN - 13: 978-81-8069-423-3 (HB)	ISBN - 10: 81-8069-423-2 (HB)

Revised Edition 2008

Published and Printed by
Ashok Kumar Mittal
Concept Publishing Company
A/15-16, Commercial Block, Mohan Garden
NEW DELHI-110059 (India)
Phones: 25351460, 25351794
Fax: 091-11-25357103
E-mail: publishing@conceptpub.com

PREFACE TO THE SECOND EDITION

In bringing out the Second Edition of the book "Modern General Psychology", I was very much profited and inspired by the suggestions made by many of the teachers who are teaching psychology in various Universities and Colleges in India. The increased acceptance of this book at a different levels resulted in bringing out the second edition, not in the original form, but by making great amount of changes in the light of recent development of the discipline. All the chapters are thoroughly revised, expanded and reinterpreted. The structure of each chapter is reshaped and enriched with many new information. The recent experimental results from various disciplines like biology, physiology, anatomy and neuropsychology are related to the subject matters of this book. While revising the chapters special attention was given to some of the chapters particularly to the brain and nervous system, intelligence and personality and motivation and behaviour.

The scope of this book has been increased by adding twelve more chapters in addition to the revision and expansion of the other chapters. These additional chapters include five in general psychology, namely Human Development, Consciousness, Frustration, conflict and adjustment, Nature of intelligence, and Personality assessment. The other chapters are Psychopathology, Psychotherapy, Cognitive and behaviour therapy and Health psychology under the Part of Psychopathology and Mental health. The chapters on Social psychology are Social behaviour and Attitudes and related traits. A chapter on psychological Statistics is also added.

Thus the subject matter of this book may suit to the syllabus of any University in India and I am quite sure that the students would be highly benefited by this book. The second edition is brought out in a quite different manner. The text is divided into eleven parts based on the functions of the psychological processes. Some Textbook authors do not give equal importance to all the psychological processes. They combine two or three themes under one heading and discuss the subject matter briefly. Some Universities include in their syllabus some aspects of Behaviour disorder and Social behaviour. This book would fulfil those requirements and may prepare the students for advanced work in the subject.

Psychologists may find that this book would survey the whole basic psychological processes of the behaviour of an individual and also an individual's place in a group. The student of psychology would have a comprehensive understanding about the role of the brain and nervous system in behaviour. This book is a sincere attempt to present the scientific biological basis of contemporary psychology and to explain the fundamental process of human behaviour. It is a modest attempt to present the origin and development of the human organism and how it functions through various stages of development. I have also attempted to present and maintain the biological emphasis which considered the human individual as a complex organism, existing in a complex environment and gradually adapting it. The student of psychology in this book is exposed, as far as possible, to the diversity of basic psychological research and also to provide him an understanding of the contemporary psychological approach with a scientific spirit and analyse the complex human behaviour.

I gratefully express my indebtedness to Mr. Ashok Kumar Mittal, the Publisher of the Concept Publishing Company for having brought out this book in an exciting manner.

Chennai
2007

M. RAJAMANICKAM

CONTENTS

Volume - I

PART ONE : Scientific Study of Behaviour, Theories and Systems

James, W. (1842-1910)

J.B. Watson (1878-1958)

I.P. Pavlov (1849-1936)

Chapter - 1

Chapter - 2

P.F. Skinner (1904-)

A. Maslow (1908-1970)

PART TWO : Biological Bases and Developmental Processes

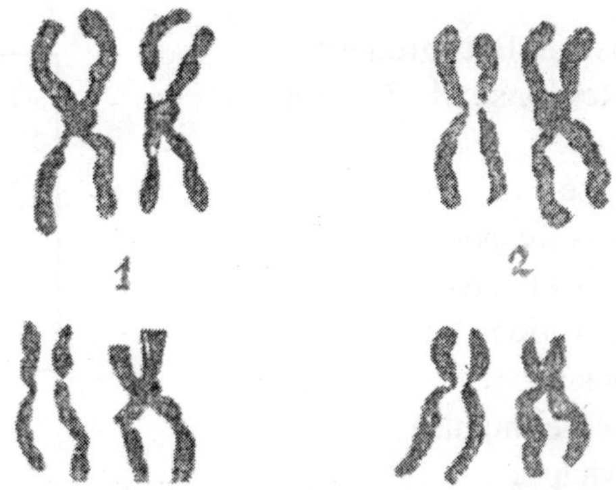

Chromosomes

Fertilized Ovum

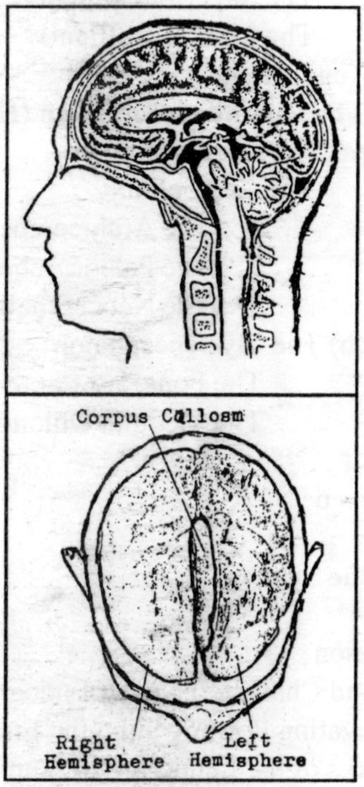
Corpus Callosm
Right Hemisphere
Left Hemisphere

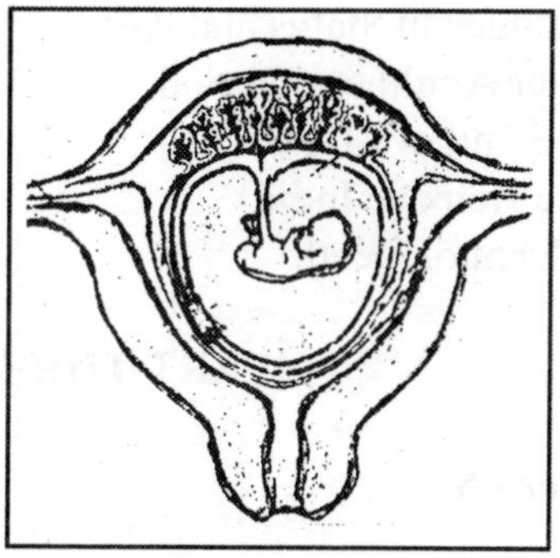

PART THREE : Perception and Awareness

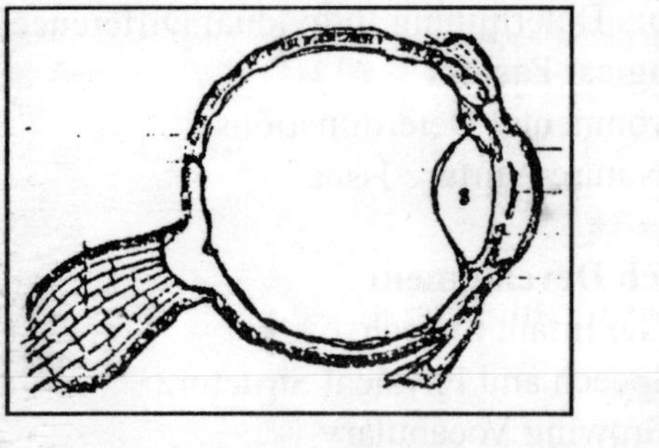

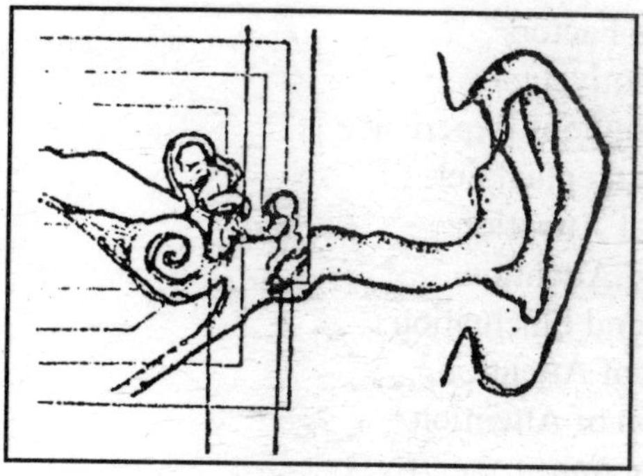

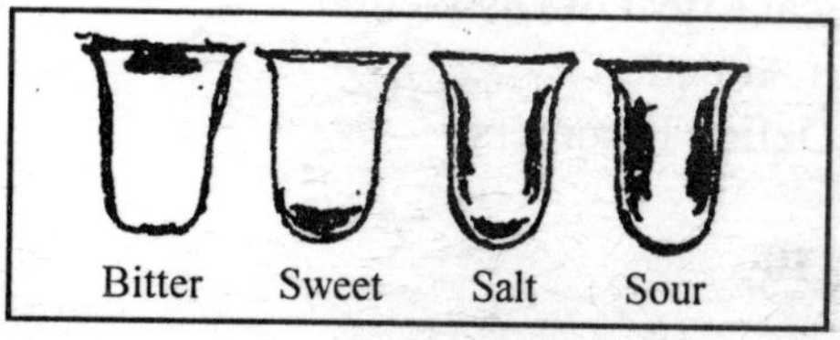

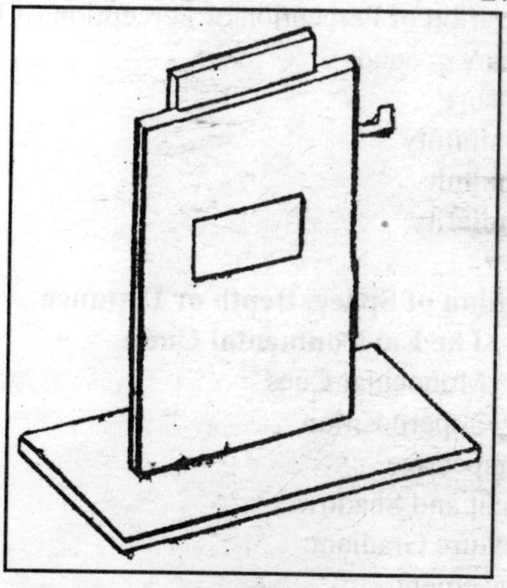

Tacistoscope

Figure - Ground

Meditation

PART FOUR : Learning and Forgetting

Volume - II

PART FIVE : Thinking and Imagination

PART SEVEN : Intelligence and Creativity

Binet, A. (1857-1911)

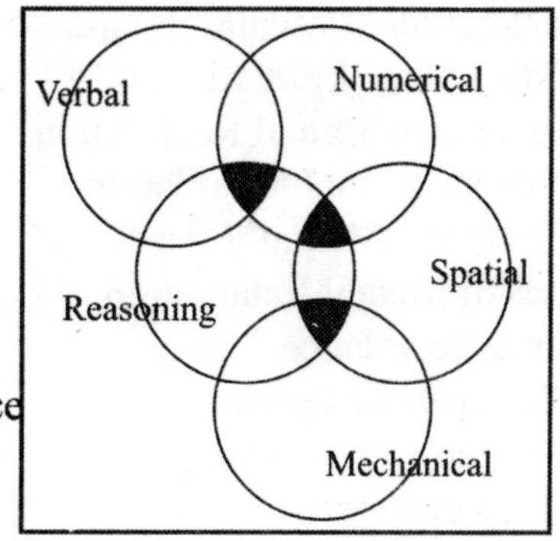

Sir Francis Galton
(1822-1915)

PART EIGHT : Personality and Assessment

PART NINE : Psychopathology and Mental Health

Chapter - 30 **911-947**

Attitudes and Related Traits

PART ELEVEN : Psychological Statistics

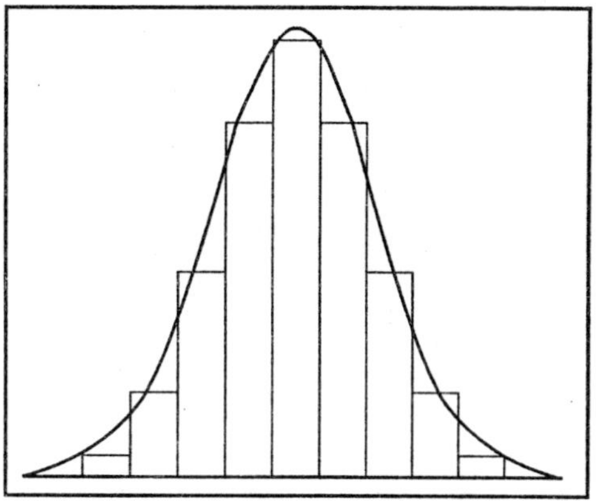

PART FIVE

THINKING AND IMAGINATION

15. Thinking and Problem Solving

16. Images, Imagery and Imaginations

Problem Solving

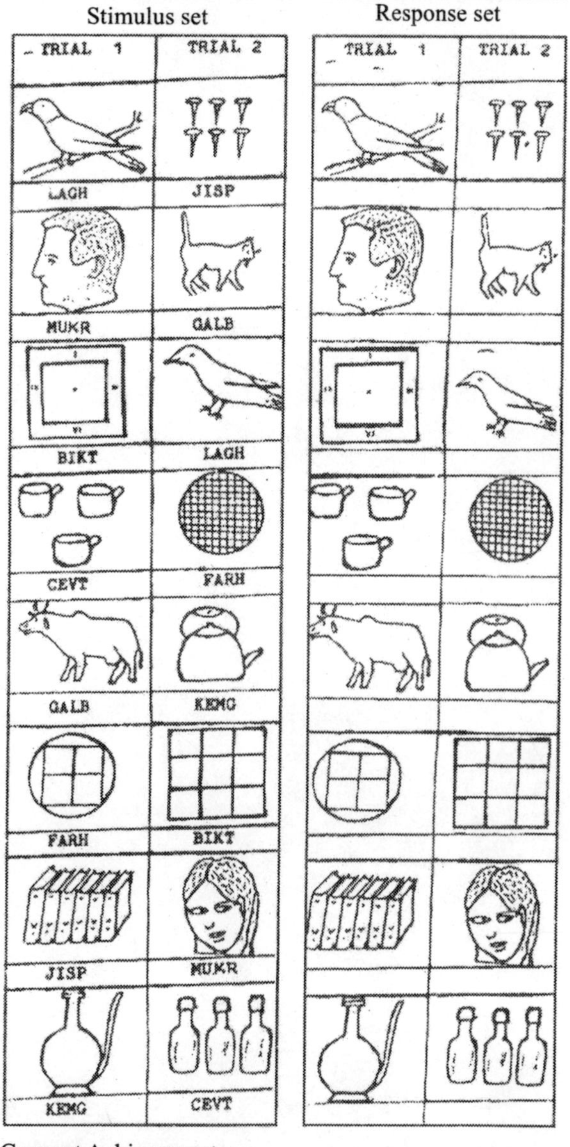

Concept Achievement

15

THINKING AND PROBLEM SOLVING

Introduction

Psychology as a science, like all other physical and biological sciences was undergoing changes and developments in its subject matter time to time. Particularly after 1960s psychological conception and treatment of the subject matters underwent radical changes. For example, a person who has studied psychology in 1940s using *Woodworth's psychology book* in his course of study finds himself totally in darkness on going through a general psychology book published after 1970s. Many of the topics he studied are not found in these books and two or three topics are combined and presented under one topic. The mental processes like learning, memory or remembering and forgetting are treated differently. Some of the subject maters like feelings and emotion, imagination, thinking and problem solving do not find place in some books. These subject matters are included under some headings without treating them separately. The basic psychological processes are not given due importance. The beginner is missing many things.

One of the reasons for such changes and developments is the *new thinking* based upon *behaviourism, operant conditioning,* the *gestalts' conception* of thinking and the *cognitive psychology*. The influence of these systems started questioning the validity of the knowledge of these areas and brought the matters as controversial issues. As result of this many of the investigators, researchers and authors started to give importance to *non-controversial* issues and a new line of thinking has emerged. Thinking and problem solving is one of the subject matters treated differently under cognitive processes *associating with language* (Kimble, Garmezy, and Zigber, 1985; Epstein and Shontz, 1969).

There need not be a rigid way of thinking that the subject matter should be dealt with only in a particular fashion. But there should not be any doubt or confusion on the part of the reader of the subject matter as to which area the particular subject matter should go into. For example, emotion is treated as an adaptive behaviour discussed along with stress and coping mechanism (Kimbal *et al.* 1985) while **Moskowitz** and **Orgel** (1969) and **Hilgard, Atkinson** and **Atkinson** (1975) treat motivation and emotion (activation) together. Some of the behavioural aspects of one area are having relationship with the behavioural aspects of other areas. It is true that there is some rational in the treatment of the subject matters of the present day's authors. Since human behaviour is an integrated outcome of several personal and environmental factors we can relate any one of the behavioural aspects with any other factors. It is because of this some authors deal with 'Learning and Thinking' together. (Hilgard *et al.* 1975), and seeing and perceiving together (Murphy, 1964).

The author has every right to discuss the subject matter in whatever manner he has planned. Psychology is a developing science and experiments are conducted and the hypotheses of different types are tested and new results often released with some excitement, as such results were not achieved earlier. It is true that when a new theory in some area is formulated several researchers instantly try to verify them with different approaches. The results of such experiments and the hypotheses sometimes end in serious discussion and bring out unresolved controversies. But in a developing science these types of phenomena are considered to be the healthy signs of progress. The Humanistic psychology founded by **Abraham Maslow** (1908-1970) is one such field of Psychology based upon the criticisms levelled against the existing fields of psychology (1970). Looking into the subject matter of a developing science with different approaches is a progressive thinking which deserves all appreciations. We shall now look at the nature and process of the mental function of thinking and problems solving.

What is Thinking?

The primary aim of education is training the people to think for solving their problems. Most of the children around the age of eleven or twelve years of age start to understand the connections that make it possible to put people, things or events into certain classifications. Children understanding these connections is the first step in learning to work with abstract concepts. Instead of thinking the school, in terms of some teachers and the classroom activities, a twelve year old boy is beginning to understand the broader abstract idea of education that connects them. **Jean Piaget** (1896-1980) (1926) calls this as *formal operational intelligence.* To him formal means being able to think about the forms of things and ideas.

Thinking is not taking place for nothing. It is taking place for some definite purpose. The purpose may be simple or very difficult. It has to be dealt with. For which one must try to findout ways and means to get it done. Unless one puts himself seriously in thinking as to what should be done, the purpose will not be gained or achieved. Every thinking will not be correct and will not serve the purpose. There may be erroneous thinkings in the beginning and the boy may correct it subsequently and may learn to improve his thinking. In every problem solving situation man first tries to understand the background and implications of the problem. Then he uses in his thinking the accepted and proved methods and avoids as far as possible unsuccessful methods.

Man establishes his superiority in the animal kingdom only by the process of thinking and reasoning. He excels all other species in a wide variety of cognitive skills. Physically and instinctively he is not superior to animals and birds, but still he could conquer them by his wide range of cognitive skills. Most of our thinking depends upon prior learning and most of our reasoning is a process of abstraction.

If you are not having any serious work to do, you keep yourself thinking on something or engage yourself in day dreaming. While you are engaged in some heavy physical work you have no much scope for thinking. You put yourself physically and mentally in the particular work. But when you have a problem to solve or have a plan to work out you keep yourself in thinking about those things.

Thinking is, generally, a kind of behaviour carried on in terms of ideas or symbolic processes in connection with solving some ideational problem. This process is distinguished from solving a problem through overt behaviour or manipulation.

Thinking may be also defined as the internal processes that are representing the previous experiences which could arouse images. Whenever we speak about thinking we never hesitate to use the terms conscious, mentalistic, private or subjective. This infers the idea that thinking may possess these characteristics. Since we could make very little scientific progress on thinking we say, it is abstract. According to **Garrett** (1975) thinking is a behaviour which is often implicit or inner, in the sense that it is not being readily observable. According to him, thinking is carried on by *slight muscular movements* and by *mental images* which represent different kind of symbols. Here *symbols stand for objects, experiences and activities.* Most of the intellectual activities are carried out only by means of symbols. While defining thinking, **Murphy** (1954), states that thinking is the process by which the *answer to a question* is found. He says that in thinking there is a definite answer to a question and we strive to find it.

Imagination also is a kind of thinking, but, there is *no one definite answer* in it. The mind may play in a large field and there may be several indefinite answers. The poets and novelists are fantastically dreaming and their thinking is not limited to one answer. Thinking generally means a process of thought, in which many things are associated, which are related in varying degrees. For example, when we think about school, we associate people like the teachers and students, objects like books, notes, blackboard, chalk piece, desk, bench, examination, library, classrooms, play materials and so on. People who are working in various occupations often think of things with which they are associated.

There is no thinking without a purpose or a need. A man ordinarily does not think, or does not want to think of anything unless he needs something. But man is forced to think to fulfil his needs. Since the needs are increasing often he has to think of getting his needs within his resources. Human needs resulted in inventing and discovering many things in this world. There is no invention or discovery of new things without putting one in serious or conscious thinking. **Piaget** (1961) states that thinking in an act. It is a dynamic act. The dynamics of thought or thinking deal with transformations which can be shown as an example in acts like disassembling a printing machine to understand how it functions. This type of transformations permit us to understand the nature of things. Therefore, the definition of thinking, according to Piaget, is understanding and the definition of understanding is the development of transformation, exemplifying in the reason for a state of affair. According to him thinking requires blending the operations of thought and integrating what is to be understood in dynamics of transformations.

Man is distinctively a thinking animal. **Aristotle,** the famous Greek philosopher said that man is a *rational animal,* because he is capable of rationalization. Man bends over his desk with his pen and paper to work with. While sitting in a large sofa sometimes he forgets himself and the world. What is that he is doing? We may say that he is in deep thinking. Everyone of us sometimes behave like this. We begin to live with our inner activity. We may be planning for some long range programme or some great achievement in our life by manipulating some change in the environment.

Man is also a *doing animal.* Whether the activity is simple or complex depends upon his previous thinking. He has planned something earlier and now he is executing the same. Therefore, planning is a form of thinking. Thirdly, man is a *talking animal.* He speaks out what he has thought little earlier or few minutes ago. Man's thinking resulted in inventing language through which he communicates his thoughts to others. Talking is also a kind of thinking. **Watson, J.B.** (1878-1958), (1930) says that talking is a louder thinking and thinking is silent talking. Ultimately man directs his own behaviour by his thoughts. Man plans his own behaviour and anticipates the consequences of his own actions.

The Nature of Thought

We are using the terms thinking and thought in an interchanging manner. Thought is a noun form and thinking is a verbal form. Whether we use thinking or thought, both are referring to the same process. Thinking is the process by which the answer to a question is obtained. Thinking *is a behaviour* carried on in terms of ideas or ideational *problem solving* activity as distinguished from solution *through overt manipulations.*

Thought is *an act* or the process of thinking. It is the power of reasoning. It may also mean idea. After the emergence of the behaviouristic school non-observable human behavioural processes took different stand. Thus the nature of thinking or thought process is viewed in different ways. The innate relationship between *thought* and *language* made the behaviourists to propose that these two skills are identical. Language for them is a *silently talking to oneself.* It is viewed that the essential mechanisms of thought is *not neural activity* in the brain, but it is a *muscular activity* in the *vocal apparatus.* There are several objections to this view of **Watson** (1920). One of them is based on an experimental result which is supposed to be a dramatic one. The experiment is considered to be a *heroic display* for establishing a scientific truth. **Scott Smith** an anesthesiologist has volunteered himself for this experiment and took a *chemical substance* called *cureare* to the level of *complete paralysis* and then introspected on his own thought processes. Cureare is a chemical that could stop all muscular and skeletal activities by closing the neural junctions by which nerve impulses pass on directions to the muscle.

This cureare is a chemical poisonous substance used by the South American Indian Tribal people who place this material on the tip of the arrow for hunting and for fighting among them. It would kill the animal or man immediately by causing breathing problems (Smith, 1947). Smith took this substance and he was placed in artificial respiration. At this stage he reported that his mind was clear and he could solve problems. He was able to recall the question put to him when muscular activity was absent. This suggested that *thoughts originate in the brain* and *language reveals them.* This experiment confirmed that thinking is a function of the brain. Thus the nature of the thought process is demonstrated in, an experiment. There may be also experiments on the other side of it, which yet to come out.

Tools of Thinking

In studying human behaviour psychologists use certain methods, tools and instruments. So long as behaviour is overt activity there is no problem. Thinking and imagination are also classified as human behaviour. But can we study these two processes by using the conventional methods and tools? We have dealt with imagination elsewhere in this book. But with regards to thinking we cannot study just like all other psychological processes. Of all the types of behaviour revealed by human beings, perhaps, the most difficult to define fully is thinking, because it is not an overt activity, although the behaviourists show some muscular movements while thinking is going on. An individual or an organism has to interact with the environment and should become aware of the interaction between them before it can develop the mechanism of thinking. Since thinking is supposed to be an implicit manipulation of symbols, where there is no concrete stimulus, it cannot be studied just like activities by the methods and instruments.

But thinking is a mental process, whether it is the function of the brains or of the peripheral area, which is not accessible to visible perception, just like all other overt activities. Ordinarily, we cannot understand whether a person is thinking or not unless the thinking or the thought process is expressed or revealed in some form. Thinking is expressed only through some medium. There should be an intervening object. Whether it is silent thinking or louder thinking (talking) a medium of expression is very essential. The medium may be some tools or instruments. The tools human beings use to reveal their thinking are *symbols, language and concepts.* These three are generally called the *tools of thinking.* But **Wordsworth** (1952) states that *memory* can also be stated as one of the tools of thinking. It is a better *tool chest* than the other three as we can draw our previous experience on required knowledge.

Your friend, sometimes might have asked you, "what you have been doing about?" Your answer, normally, might have been, "I was thinking about my teacher". What does this thinking involve? Let us discuss this. First of all while thinking about your teacher you might have had the image of your teacher, particularly, his face, eyes, voice. Sometimes your image might have been his clean and neat dress he often wears, his mode of approaching the students, his movements and his other mannerisms. You may compare him with other teachers you met in school or college and you place him into some category. "He is a good teacher. He gives all the best he has to the students." In this way you characterize and identify him with certain qualities in your thinking and reveal them in your language that "he is a good teacher". The teacher is not there. But you are bringing him to the context. Thereby you are making him present symbolically. The teacher stands here as a symbol. The term teacher is a word and the word is a symbol. In this process three tools are involved. They are the symbol, language and the concept. The teacher is a symbol, the word teacher is also a concept and you use language to indicate the object, the object, the teacher. Thus these three tools are combined in some form to express the process of thinking. The symbolic process is one in which we use symbols to represent some object or person or stimulus in solving some problems.

Symbols

Man is a rational and thinking animal. His enlarged brain is the seat of his mental activities. He has tremendous amount of capacity for learning and the ability to make use of what he has learned. More than that he has unlimited amount of capacity to *expand his learning ability* by the process of *symbolization.* Man's ability to learn and to remember is simplified by the symbolic processes.

A symbol is an internal representation of a stimulus or an aspect of stimulus or a response or a response aspect. This stimulus or response or the aspect is not present at the moment. A symbol stands for a stimulus or for a response. This might have already occurred in an event or situation or is about to occur in certain situation. For example, we might have made a symbolic statement that there was cyclone yesterday. In this case a response has occurred. You know the situation that has caused the cyclone yesterday. It was the depression of weather that has caused the cyclone. The depression of weather was the stimulus and the cyclone was the response. On the basis of this event already occurred, you can make a symbolic statement. Like this: "A cyclone is going to take place as there is depression of weather."

Further, a symbol also indicates the quality of a situation or event when that quality is not present or revealed in the situation. In this case you can make a symbolic statement like this: "The way of approach to solve the problem is not correct."

Therefore, a symbolic process may be defined as the utilization of symbols in making decisions, solving problems or adapting to the situations. We use symbols for places, objects, persons and even to ideas. Symbols help us to talk about New Delhi when we are in Chennai, to talk about the President of India when we are sitting in the class-room, to decide what we should buy before going for shopping, to predict how the student is going to solve the particular problem. Symbols help us to make our communication being easy and successful. Human activities in the everyday life would be much affected without symbols. It would not be possible for us to discriminate one thing from the other. Only with help of symbols communication through various channels make the world very busy. Without symbols it would not be possible for us to express our ideas, solve problems, and to make our future plans.

What we call symbols are only ideas and the symbolic process may be referred to as an ideational process. Ideas are often considered as synonymous with images. Images refer to sensory experience which cannot be experienced in the absence of an appropriate external stimulation. We have sometimes vivid visual, auditory, Kinesthetic experiences in our imagination. We utilize such images for our symbolic thinking occasionally.

It is often stated that the prefrontal area of the cerebral cortex (area of the frontal lobe) is mainly responsible for the formation and utilisation of symbolic processes. However, the efficient use of symbols increases as a function of the cerebral cortex in the development of the individual. In all human symbolic activities language is greatly involved. In certain situations you may use language but in certain situations you may use certain visual cues — a visual image of an area you wanted to visit. In either case you use symbols in solving your problem.

Man uses his sensation and perception as the basic elements of his thinking. Sometimes the perceived object may not be there. In such cases how can a man indicate what he has perceived. It must be also understood by others. Therefore, man uses conspicuously understandable ways and means to indicate what he actually perceived. We may call it as *Symbol.* Man uses symbols for indicating what he has perceived. It might have taken thousands of years for man to invent language and to indicate what he has perceived in place of symbols. Though the use of language has come into practice in our everyday life, still symbols are used and they are used in all parts of the world.

The ancient Egyptians and the Romans have drawn small pictures of the objects on the parts of some, plants and also designed them in clay. The Sumerians and the Semites have also used clay and shaped the objects, and the signs of the ideas were fixed. In ancient times people everywhere seem to have used only some materials as the indication of their ideas and thoughts. The archaeological discoveries in India show the type of media the ancient Indians were using for communicating their ideas. These media were only in the forms of signs and symbols. **Maslow** (1908-1970) states that the modern man has very little to think about the media or means of communication as most of his life aspects are the process of evolution and are functioning on habits and stereotypes. **Maslow** (1970) is of the view that the means of communication of the present times are results of the evolutionary process based, on the earlier man's requirements. Since the ancient man's life was very rough and uncultured, he might have struggled quit lot and spent much time of his life only in thinking about designing the way of his life.

Though the ancient man's requirements were few yet he had to toil and moil to collect his food either by hunting or by cultivation, build his shelter, for protecting himself from enemies he has to design and manufacture weapons. Thus all such necessities forced the ancient man to think seriously about many things and also he was forced to discover and invent many things. The ancient man passed through the stone age to metal age and he invented the use of iron and other metals for designing his fighting weapons. In the life of the ancient man, *necessity was the mother of invention.* All these made the modern man's efforts and thought processes simple and the modern man carried out all those things which the ancient man left out. Even in the case of modern man it was the necessity that made him to invent and discover all those things that we see today, and use them in our everyday life, and enjoy the comforts of life.

Man, generally, is driven to think, only when his experience, knowledge, learning and his stereotypical and habitual way of life fail to help him to solve the problems he faced. When man thinks, whether it is simple or complex way of thinking, something new emerges. It may result in rearranging or restructuring the existing old things and also adding something new to the old ones. Thinking is an internal mental process and it is individual. Unless the individual himself expresses, what he has thought, it will not be possible for others to understood. Thinking is an inexplicable mental activity. Therefore, the study of thinking is a very difficult task. Man's thinking is expressed through language and language is symbolically representing the objects, events and ideas.

Language

What is language? Language is a system for the expression of one's thoughts and feelings by the use of spoken sounds or conventional symbols. Language is specific to man, which has a distinguishing characteristic of man as compared with other animals. Every nation has a language of its own and every cultural group has a language of its own, whether the language is spoken or written or both. Sound as such has no meaning unless some specific meaning is attributed to it. Sound may be used as a systematic and unsystematic means of communication. For example, animals use sounds as gestures. In Human, every vocal sound produced, has some meaning. Therefore, language had its origin when man attributed meaning to his vocal sounds which indicated some object or event or person or place. These vocal sounds, which indicate some meaning referring to some objects, persons, places or events, are stated as symbols. As time went on, the symbols were given written forms.

Man had invented language upon symbols and symbolic processes. These symbols and symbolic processes in turn influence the development of language.

Language is one of the important tools of thought. Human thoughts are mostly *controlled by language.* Thoughts cannot be expressed without language nor communications are possible among human beings. Language is a *sound phenomena* in the *evolutionary process* of the species and human beings. It is having both direct and indirect influence upon the behaviour of others. You can command verbally a dog or a horse or an elephant. You can *make a person* to have *painful feeling by a verbal shaft.* This is how language is having indirect communication. In the beginning of human verbal learning the social aspects of language play an important role.

The function of language in *thought process* cannot be examined *without any reference to signs. Signs represent the objects* and events of the world. There are two kinds of signs, one the *linguistics* and the other *non-linguistics.* Waving the hands, nodding the head, raising the eye-brow-all these signs express some meaning. We could see sign posts, traffic posts in various places like road diversions, check posts, cross roads and in electric posts with a skull and two bones. Sign boards in all these places sometimes do not have written words, yet they express some meaning. We use *mathematical* and *statistical symbols.* Every symbol is used meaningfully in all our calculations. Similarly language is another *kind of sign.* In language a set of *letters* and *sounds* are *combined* in some particular fashion *to represent the objects, events* and our *experiences.* When we see an object we point it out, when we witness an event we explain it and when we have some kind of experience we express it. For example, the concept cow has no meaning to a child in the first instance, when the animal cow was *not pointed out* or shown to him. In the second time the child finds no difficulty in attaching meaning to the word cow. Therefore, *we use language to point out objects when they were not present.*

Language as a *tool of thought* is structured by certain formal rules. It is the subject matter of *linguists* to study the rules of language. The subject matters we have stated here are having certain *inter-connections* between *language and thought.* When the psychologists and the linguists wanted to understand these interconnections *a new field* of study called *psycholinguistics* was born. Psycholinguists analyse language *into its basic components* and the basic components can be put up into complex structures. The psycholinguists use linguistic principles to understand the ways in which people use, generate and comprehend language.

A language may have several components. The simplest component or unit of language is the *phenome.* This is *basic speech sound* and deals with rules of language sound. *Phonetics* is the *study of sound* made in speech. In English language there are *46 possible* sound or *phonemes.* Phonemes are combined and made into *morphemes. All words are morphemes,* but all morphemes are not words. The *morphemes* are the *smallest meaningful unit* of speech. A *meaningful word* may have *more than one morphemes.*

For example, the word *'fruit-sellers'* has *four morphemes.* The *noun* in these morphemes is *'fruit'* and the *verb is 'sell'.* These two are called *free morphemes.* They are *independent words,* could express the meaning without any combination of any other morphemes. But the *morphemes 'er'* and *'s' cannot express* the meaning *independently* and they are called *bound morphemes.* The *bound morphemes* always appear *connected with other morphemes.* In this way free morphemes in combination with other free morphemes and with bound morphemes make words in any language.

The *third component* is the *syntax.* It is the *study of grammar.* Syntax studies the rules *for combining words* and morphemes *into phrases and sentences. In the communication of any thought* or language the sounds of phonemes, the place of morphemes and the rule of syntax play an important part.

Another aspect of language is *semantics.* It is the study of the *meaning of a language. Semantics* mainly *deals with the formal structure* and the *emotion of language* employed in its use. In a particular context it depends on how an expression sounds and looks. For example, the phrase 'good morning' may be expressed in sounds in different contexts with different persons.

The meaning of a word and its various associations among them and various possible occurrences constitute the structure of language. How extensively a language is used depends upon the structure of language. There are *two kinds of meaning in a language.* The first is the *extensional meaning* and the second is *intensional meaning.* The extensional meaning *stands for the object.* A word points out the object. For example, we may ask a question, 'what is a cat?' The answer to this question can be by pointing out to a cat, 'It is a cat'. The animal 'cat' is an external reference of the word cat. Children in every culture learn new words in this way.

The animal 'cat' can be pointed out *in another way.* This is by describing the cat with words like this: The cat is a little animal, with four legs and a tail, found mostly at homes, walking very softly, producing sounds like "mia, mia". Anyone who has not seen a cat identify a cat *with such description.* This kind of meaning expressed by *verbal description* of words is called *intensional meaning.* The intensional meaning of a word may vary from person to person.

Symbolic Process and Language

At this stage every object had only symbolic form and was indicated by some sound and when language was developed with letters, objects and events were indicated by written words. This is the symbolic representation of the objects and events. Symbols are used in the absence of objects, persons or events either by vocal sounds or spoken words or in written forms. This procedure of using symbols for indicating objects or expressing one's idea had become very easy when the fulfledged language came into usage. The symbol has become the internal stimulus of the objects or ideas. In this way thinking takes place through symbols and expressed through the medium of language. At one stage we may even think that there is no thinking without language since every thinking is expressed only a language.

Words in any language are the symbolic representations of some objects, events or ideas. Symbols are of different types and are having different meaning. Generally, they are of two types, generalization and discrimination. This type of classifications are also made in learning which are called components of learning. In thinking there are *two important forms*, namely *directed thinking* and *associated thinking*. The first one is specific, aimed at solving a particular problem and the second is general, unlimited in scope, mainly referring to imaginations, fantasies, dreams, day dreams and are not directed to solve any specific problems. The process of thinking is taking place step by step from simple type, like symbols or signs through complex forms of thinking like concept formation and to the development of language.

There are *three important aspects* or components in the process of thinking, which are called *tools of thinking.* According to **Hilgard** *et al.* (1975), thinking is a cognitive process which is described by the symbols and symbols represent objects and events. In our everyday life we are using large number of symbols. We attach symbols to objects and events, and the ideas of attaching symbols to objects are formed in our mind. We project the ideas on the objects and events. In this way symbols become the internal representation of the objects and events.

When you are engaged in a conversation with your friend, you are making only symbolic processes. You might have described a football match, played between two nations, you have witnessed few days ago. During the conversation you might have described the interesting events, the scoring of goals, the saving of goals and the crucial and sensational situations at the match. You might have also mentioned the names of the players, the names of the referees and also their role in the game. But, now you do not see the match nor see the players or the referees, still you are referring them here with all details. This is symbolic reference which is stated as an idea or thought. Ideas or thoughts deal with objects, persons, places and events of the past. Symbolic representation also deals with recollecting and reproducing past events. In describing the football match today, which you witnessed two days back, it is only this symbolic process that is taking place. Everything about the football match you described to your friends today, is stated as the symbolic process or thought process or process of thinking.

Authors of modern psychology books use the phrases, symbolic process, symbolic behaviour, symbolic thinking, ideational process and so on. All these refer to the process of thinking. **Moskowitz** and **Orgel** (1969) state that symbolic process is a process of utilizing of symbols in making decisions, solving problems, in short, what we speak is through symbols. For example, what we did in the past, what we are doing today and what we are going to do tomorrow or in future all can be explained only through symbols. It is because, what we all speak about, whether objects, persons, places, or events are not there. They are all represented by words, and they may be the words of any language.

Human beings use often symbolic processes in their learning and talking activities. Moskowitz and Orgel state that man has the ability to learn and to utilize the learned materials. Man by using symbols in his activities, keeps himself in a process called symbolization and this symbolic process enlarges man's capacity for learning and also to utilize what has been learned. Therefore, according to them symbolic process becomes the basis for the development of language and in return symbolic process is influenced by the development of language. From all these what have been explained above, it is possible to arrive at an idea that there is certain amount of interdependence among the processes of learning, symbolic process, language and communication. **Piaget** (1896-1980) made extensive studies on various development process of children, their thinking and the development of language. **Piaget** (1926) states that we can clearly notice the relationship and the communication between the development of thinking and the development of language.

Therefore, while discussing the process we must involve the related processes like the learning and the development of language and the usage of the same. Hence we may state that a symbol is an idea or thought of an internal representation of an object or event. It is the representation of a stimulus or an event. Therefore, it may be also stated as a response, since it is representing for something. Strictly speaking symbols are the representations of the ideas or thoughts of some objects which are not there. Hence we may state that symbolic process or symbolic thinking may mean the process of identifying the idea or the response with an object or event. Generally our vivid sensory experiences such as seeing, hearing and tasting may also serve as symbols. For example, we could recollect and enjoy, a beautiful picture or figure, or hear a melodious voice, a delicious dish prepared by a skilful cook. Though these are symbolically present before our eyes when we think about them, we call them as images. There may be visual images, auditory images and gustatory images and so on. Sometimes we make images of these experiences in our imaginations. This is also a kind of symbolic process and therefore, we call it as symbolic imagination or symbolic thinking. Human symbolic processes or thinking activities are expressed only through language. In a sense, language is also constituted of symbols. For example, the words are symbolic representations of objects or events. Therefore, language is the medium through which human symbolic activities are expressed.

Concepts

Concept is considered to be the third tool of thinking. We use symbols or make symbolic representations of objects when the objects are not present there. A symbol is an indication of an object. We mention it by its name for which we use a word from a language. Thus the objects is made to be in the form of word. The word is a language symbol. Therefore, a symbol is an idea. The idea represents the object in the form a word from a language. When an *idea* is expressed *by a word,* we call it a *concept.* Therefore, a concept is a word which is having certain specific meaning.

The term concept is defined in various ways. Generally, concept is an idea or a general abstract notion. It is also stated as something formed in the mind. It is an abstract idea or thought. Concept is also stated as something conceived. It is a general idea of something formed, mentally combining all the specific parts and characteristic features. It may also mean a mental process of abstracting of an object and projecting a meaningful word on it. But the important feature of the term concept is that its meaning is universally accepted. It is basically a philosophical term referring to abstraction.

Therefore, we may state that a symbol stands for an object or a group of objects or events which have some common characteristics. Thus a table is a symbol and it is also a concept. Similarly, we may state teachers, students, men, women, boys, girls, are concepts. So also trees, leaves, fruits, carts, buildings are also concepts. In the same way circle, square, triangle, are symbols and concepts. Symbols like four, six, ten are also concepts. In the same manner we may state that large, small, fast, slow, are also symbols and concepts. We are also having symbols like green, blue, red, yellow as type of concepts. We use in our everyday life concepts like justice, faith, belief, sin and so on. All these are concepts, but vary in types. Thus we have various types of concepts like the person concepts, object concepts, spatial concepts, number concepts, colour concepts, concrete concepts, and abstract concepts.

All these concepts have certain common characteristics by which it is possible to classify them into certain categories. Among these certain concepts have simple characteristics which can be easily understood by children. But certain concepts are complex and are very difficult for children to comprehend correctly. Generally, concept is defined as a word having a definite meaning. It also has certain qualities, by which it can be identified.

Concepts may be referring to the concrete objects and abstract concepts. It is possible to identify the concrete objects by their common qualities and names. But it is a very difficult task to find out the common characteristics of certain objects which are having certain variations, and categories or classify them. It needs more amount of thinking to find out the common qualities in objects which are having certain variations.

We can easily classify certain objects which are simple type and also objects which are having few qualities and also objects having clear concrete differences. But, however, even in the case of concrete objects, if abstraction is required classification of the objects becomes very difficult. Human beings generally prefer objects with concrete properties, because it is easy to classify or categories such objects. Since there are various types of concepts and some concepts are required abstraction on the basis of common characteristics, they need deep thinking for classifying them on the basis of abstract qualities. We have stated above quite number of concepts. They are not of the same type. They are of varied types.

Types of Concepts

All the concepts mentioned above may be classified on the basis of their properties. Thus we have simple concepts, conjunctive concepts, disjunctive concepts and rational concepts.

(1) Simple Concepts

Simple concepts have few qualities. They may have one or two qualities. It is very easy to classify such concepts, since they have only few characteristics. For example, in classifying the objects having white colour, we find that the white colour objects have only one quality, the whiteness. White as a quality can be found in many things. For example, we see white paper, white chalk piece, white paint, white glass, white shirt and so on. There is no other quality combined with white colour. Therefore, it is easy to classify objects which have few qualities.

(2) Conjunctive Concepts

The conjunctive concepts have generally more than two qualities. These concepts can be described in terms of these qualities. For example, cat is a conjunctive concept. Everybody knows what is a cat. Since everybody knows it, the cat is symbolically present here when we mention it, because of its symbolic presentation in the form of the 'word' cat. Cat has more than two common characteristics. It has four legs and a tail. It has long mustaches on either side of the nose. Cat is an animal. Generally cats live in homes. Cat is a pet animal in many homes. Cat is very fond of catching rats and eat them. It belongs to the family of big cat, the tiger. The cat has all the characteristics of the tiger. Cats are in different colours. There are white cats, black cats, brown cats and also there are different breeds of cats. Any other animals which do not possess all these qualities are not called cats.

(3) Disjunctive Concepts

The disjunctive concepts are attributed with several qualities. There are also some common characteristics in disjunctive concepts. Out of several attributes to these concepts, only one must be the correct attribute. For example, we see many things in twos, or in pairs or as couples. But when we say couple, it implies several meanings. The concept, 'couple' refers to a pair of persons or anything two in number. It may refer to any two human persons, animals, or objects. We sometimes say, "I will return in a couple of minutes". This is the common characteristic of disjunctive concepts. This common characteristic, 'two' does not help a person to identify the object correctly.

For example, at a distance a couple of persons is coming. Since, it is at a distance, we could see that they are only human beings and not animals. As they come closer and closer we come to know that it is a man and a woman. A man and a woman as a couple of persons may have any type of relationship between them. They may be husband and wife, or engaged and not married, or partners in a game or in a dance group. They may be father and daughter, or mother and son, or brother and sister, or may be neighbours or two persons not known to each other. Thus there are many attributes. If it is verified any one of these relationships may be correct. This is occurring in the case of many disjunctive concepts.

(4) Rational Concepts

The rational concepts have similar qualities like the disjunctive concepts, but they differ. The characteristics of the rational concepts vary comparatively. The correct concept is decided only by comparing one with other objects. This is taking place in concrete objects. For example, there are three weights of the same shape and size. How is to decide the heaviest of them? It can be decided only by lifting one by one and judging them. One weight may be stated as heavy, the second may be heavier and the third may be the heaviest. If there are two objects, we decide the second in terms of the first. The second object is double the weight of the first. Similarly in the case of objects of different sizes, we decide the quality. These qualities are found only in concrete objects and children learn this type of size concepts and shape concepts easily. Since the differences in objects are perceived physically it is easy for them to learn and achieve them.

We could see that these four types of concepts are having different characteristics and they can be named differently, as person concepts, object concepts, size concepts, spatial concepts, shape concepts, number concepts, colour concepts. But all these can be brought under two broad types as concrete concepts and obstract concepts.

Concept Formation

Concept formation is a process of comprehending the essential qualities of a class of species. The concepts are the categories for classifying specific objects, persons, animals and events on the basis of common elements. A concept may be also stated as a response made to some aspects in a situation which can be later on generalised for a new situation or some other situation which contains the same aspects.

The world in which we live is full of objects of different kinds. These objects possess certain common qualities that make them different from other objects. When a person learns to respond to this common quality of an object and learns to ignore other aspects he is said to have learned a concept. For example, the child slowly learns to comprehend persons, objects and events. He sees a *pencil* and understands it as a pencil and it is used for the purpose of writing. He differentiates it from an ordinary piece of chalk or a pen. Secondly, when he sees the pencil, he sees it as *red in colour* and differentiates it from a green pencil or from other colours. In this context the child has learned the concept of redness and also had learned to differentiate the properties of the same object.

When a person develope a concept of a chair he deals with it as furniture. All those objects that have a seat, four legs, a back and two arms are called chairs. Whenever he sees such an object he calls it a chair. If he hears the voice of someone saying a 'chair' he forms the picture of the chair and is able to describe it. Other furnitures also may be having a seat and four legs. They are not called chairs. In this way, when correct generalizations are developed they are called concepts.

Concept formation is the process of discovering some characteristic that is *common to series of objects.* It refers to the description of this common element. It differentiates the quality of one object from another. Examples, we have used, the concept of a red pencil and a chair. Pencil may be red, green, blue or yellow. But all pencils are made of wood and lead. This common characteristic differentiates them from pens, ball point pens, sketch pens and other writing materials. Concept is formed like this. The concept is that all pencils are made of wood and lead. Children learn, colour concepts and number concepts easily. Concepts function to generalize, to differentiate and to abstract in our thinking process.

There are *two* important *psychological processes,* operating in the concept formation. They are *generalization and discrimination.* If we see a cow, we can assign it to our general concept, animal and mammalian. Just like other animals the cow has four legs, a tail and two horns. Secondly, the concept, cow helps us to differentiate this animal from other animals. From personal experience, by reading and from some other sources we understand that most cows are harmless animals. Therefore, we will not react to a cow with the alarm that we would show to an elephant or bull. Thirdly, we may use this concept cow for abstracting without making mention to any specific animals. In this process the cow has been discriminated from other animals that have common qualities. This is called *abstraction.* Abstraction is a process of isolating certain characteristic of an object and disregarding all other characteristics of the same object at the moment. When you say that your uncle is fat, you isolate this characteristic of your uncle from other characteristics. The adjective 'fat' indicates his characteristic and also other persons'.

Again in the case of cow, when it is assigned with a particular *word label* it is called abstracting and when the word label is consistently applied only to the cow the concept has been learned.

But, generalization in the beginning is very crude. A little boy on seeing a horse for the first time might call it a cow or a donkey if he has already learned little of the concepts of cow and donkey. The new animal is related to the old one in some manner. But when he comes to know that it is not the same animal that he saw earlier he is ready to learn the new name and differentiate it from the other.

I.S. Vigotsky had developed a test to study the ability to think conceptually. This test is very widely used in psychological studies. The test consists of twenty two wooden pieces of different sizes, heights and colours. The wooden pieces are spread out roughly *into a circle* (Fig. 15.1) and the subject is asked to arrange them into *four groups* on the basis of a principle that he finds suitable. The scores of the performance is based on the number of trials given to the subject and the time taken by the subject to arrive at a solution (Fig. 15.2 and Fig. 15.3).

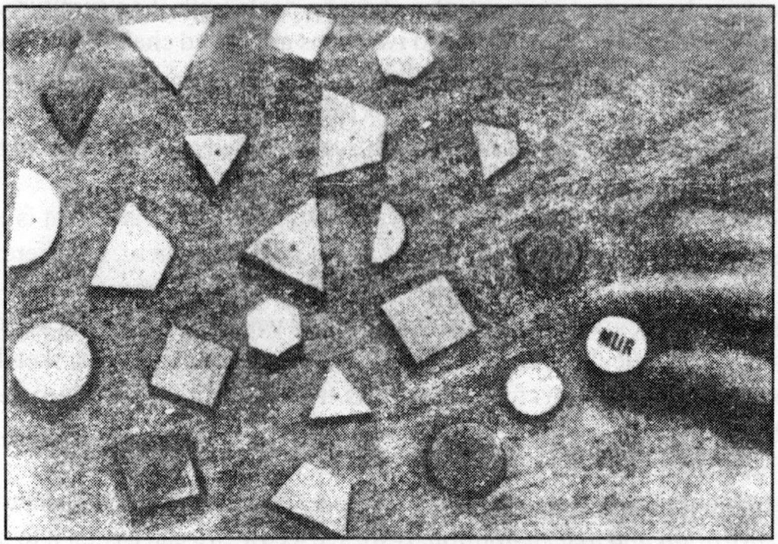

Fig. 15.1 : Thinking : Concept Formation Experiment.

(I.S. Vigotsky invented this test to study the ability of an individual to think conceptually. This is very widely used test in psychological experiments. The test consists of twenty two wooden pieces of different sizes, heights and colours. The wooden pieces spread out roughly into a circle for conducting the experiment.)

Fig. 15.2 : Thinking : Concept Formation Experiment

(According to the experimental procedure the experimenter wants the subject to arrange the twenty two wooden pieces into Four Groups by adapting whatever the procedure he/she finds suitable for the task.)

Fig. 15.3 : Thinking : Concept Formation Experiment
(Concept Formation Experiment in completed form. The subject has arranged the twenty two wooden pieces into Four Groups on the basis of heights and sizes.)

Concept Achievement

Concept achievement is the process of discovering some qualities or properties that are common to a series of objects. Concept achievement refers to the description of some common element in objects and differentiating the quality of one object from another. Concept achievement or formation takes place in children when they begin to understand objects and learn to differentiate and identify them. Concept achievement takes place also in adult persons. There are lots of instruments and tools, newly invented by scientific and technological advancements everyday and they bring out many new concepts. As such concept achievement is becoming a continuous process in the case of every adult human being. Children take longer time to learn and understand concepts, while adults achieve concepts quickly.

Generally, adults teach children to learn things and events. The child is slowly introduced with simple objects to learn and to understand their qualities. The child understands, what a pencil is, when it is shown to him. He also knows a pencil is red in colour. Thus the child understands two concepts, pencil, the object concepts, red, the colour concept. When he learns more and more things he also understands that pencils are red, green, blue and it is used for the purpose of writing. He learns also more about the pencil that it is made up of wood and lead. These are the common characteristics of a pencil. Any object similar to pencil when introduced, he is able to differentiate it from other materials. For example, other writing materials like the pen, ball point pen, sketch pen which are having different characteristics are differentiated from the pencil. At this stage the child has learned several concepts and can differentiate the pencil from other writing materials. The child has now achieved the concepts, pencil, which is used for writing, red in colour, made of wood and lead and different from a pen. Generally, children achieve colour concepts, number concepts and concrete concepts easily. The *spatial and abstract concepts* are more difficult for children to achieve and these concepts take longer time for children to achieve or learn. The functions of concept achievement are supposed to be the processes of thinking and also processes of learning. Generalization and differentiation are the processes generally involved in learning. But in thinking there are *three processes,* namely, *generalization, differentiation* and *abstraction.* All these three important psychological processes are involved in concept-achievement. Therefore, these processes need to be explained in this context and should be also explained how they are functioning in the process of concept achievement.

The Functions of Concept Achievement

The three psychological processes namely *generalization, differentiation* and *abstraction* are operating in children's concept achievement process. The complex process of abstraction is associated here. The child who has not seen the castrated bull, might put the question, 'what is this?' Then, he is told, "this is a bull". The child who has already known the domesticated animals like the cat, dog and horse, now understands a new animal called the bull. The bull is different from other animals having four legs and a tail. Now let us explain how the psychological processes generalization, discrimination and abstraction are applicable to the bull. The bull can be assigned to the general concept 'animal'. The bull belongs to the domesticated adult male of any bovine animal. The bull, just like all other animals has four legs and a tail. This is *generalization.* That is, the bull is an animal. But it differentiates from other four legged animals by having *two horns.* The bull is used to plough the fields and also for dragging cart. From personal experience and from the information obtained from reading materials we understand that bull is a harmless animal, since it is domesticated. Therefore, our reaction to a bull is not with any sense of fear that we normally show to an elephant or a tiger. In this way the bull is *discriminated* from other animals.

The concept, bull can be also abstracted without comparing it with any other animals. The bull is having certain specific characteristics which are not common to other animals. The bull is used for ploughing the fields and also for dragging loaded carts. In this process the bull is isolated from other animals in certain characteristics. Though the horse also is used for ploughing fields and for draught in certain places, horse is not used for dragging the luggage loaded heavy carts. Thus the bull is isoloated as it has certain characteristics not found in other animals. This process is called *abstraction.* Abstraction generally means, a process of isolating certain characteristics of a person, animal or object, and disregarding all other characteristics of the person, animal or object. For example, if you are guiding a person who is new to the place, to find out a particular house, you may mention, 'it is a two story pink building'. Apart from the door number, you are isolating the building by the concept pink. The adjective 'pink' indicated the specific characteristic of the house, and how it is different other houses. This is how abstraction is made.

The process of concept achievement, in the case of a child, may be causing confusion in the beginning. It is because of the process of generalization, which is not accurate and precise, but only a rough estimation. A little child on seeing a horse for the first time might call it a cow or a donkey, if he was not already told or shown the animal, horse or donkey having four legs and a tail. The child might have understood that all four legged and one tailed animal are cows. On knowing that it is not the same animal he saw earlier, he is ready to learn the name of the new animal and also differentiates, it from other animals. Like all other areas, experiments on thinking are not many as it is difficult to conduct experiments on thinking. Since studies made on concept achievement or attainment or formation are not many we shall discuss a few of them here.

Concept Achievement : Some Studies

Concept achievement or formation involves the process of thinking on the part of the person who is engaged in the task. Concept achievement at certain level can take place easily even without much thinking. It is because of the experience, knowledge or learned abilities some tasks can be performed easily or successfully. But it does not take place in all cases. Since concept formation or achievement involves abstraction, the task becomes difficult.

The earliest study on this topic was made by **Kent Goldstein** (1878-1965) in his clinical studies. He was a German psychiatrist went to United States in 1930s. He was the Director of a German Military Hospital which was specializing in brain injuries. He had contacted the brain injured German soldiers, that made him to formulate a theory called *organismic theory*. **Goldstein** (1935) had experience with patients who had different levels of brain defects. He observed that like all other patients the brain injured also tried to cope with their environment. He found that human beings have overall capacity, and the adjustment to the environment was the part of their overall capacity. Man's overall capacity was seen striving towards *self-actualization*. Only in this. context Goldstein had coined the term, *self-actualization* which he had found as the expression of a man's *basic motive* that centre rounded and organized all other motives. For achieving the self-actualization, man must cope with his environment. When a person had failed to cope with his environment his drives and motives were left with dissatisfaction and he was driven to anxiety level which also had become the cause of abnormality. Goldstein had observed that every organism was trying to get rid of tension causing situations.

At the end of the First World War, **Gelb and Goldstein** conducted experiments on brain injured persons with regard to visual perception. They have observed that injured persons have lost their capacity *to think in abstract terms*. Goldstein had also observed that his patients with brain injured may group things together, but they were incapable of genuine abstract thinking. Though Goldstein's studies were at the end of First World War, his works were published only in 1935.

One of the earliest studies in this area was made by **Kenneth Smoke** (1932). He used *figures* in his test. Smoke's test materials consisted of *figures of different shapes and sizes.* There were also varied number of dots. In smoke's study the figures were presented to the participants to develop concepts such as DAX. The test consisted of a circle in which one dot inside and one was outside. All the dots were located at different parts of the figures.

The participant should achieve the concept DAX, for which he/she was to indicate which of the characteristics were relevant and which were not. A circle with one dot inside and one dot out alone was the correct DAX. All others were not DAX. The participant should indicate it. This experiment, designed by Smoke, was testing five hypotheses one by one. It was presumed that the participant can perfectly identify the concept DAX only in the sixth trial. This experiment involved both *perceiving and abstraction.* A few years after the appearance of the studies of Goldstein and Smoke, another study was made on concept achievement by **Haufmann and Kasanin** (1937). These authors used in their study, the test material designed by **Vigotsky** (1962). Vigotsky had developed this test in connection with his earlier study on *thought and language in children in clinical practice.* **Haufmarm** and **Kasanin** (1937) have modified this test for the study of *concept formation.* The procedure used by these authors was very much useful for the study of concept formation experiments. Therefore, the test and the procedure attracted researchers all over the world and studies were conducted using the same procedure.

The Vigotsky test consisted of *twenty-two wooden blocks,* packed in a small rectangular suit case. These blocks are designed in five colours, namely green, blue, yellow, orange and white. There are only two sizes as larger and small and all the blocks are put into these two sizes. These wooden blocks are of two heights as tall and flat, but they have six shapes. At the bottom of every block a three letter non-sense-syllable is printed. These non-sense syllable letters served the purpose of classifying the blocks into categories and for testing them. There are only four types of three letter non-sense syllables such as MUR, BIK, LAG and CEV. The participant is expected to classify all the twenty two wooden blocks into four categories using any creteria. The important variables that can be used for correct classification are the sizes and heights. All other variables like, colour, shape, are distractors. The amount of time and the number of trials are taken into account in classifying the blocks. It is expected that the participant should take shortest time and fewest trials.

A different type of hypothesis has been formulated by **Heidbreder** (1947) and designed an experiment for studying the concept achievement. According to Heidbreder, concepts with concrete features are easier for the subjects for concept attainment than the concepts with abstract features and concepts with numbers. The test was called *concept attainment test.* This test consisted of *thirty pictures* and were put *in five sets.* In each set there were *six pictures.*

At the bottom of each picture one four letter non-sense syllable is printed. This non-sense syllable is used as the response word in testing. There are mainly six pictures. They are human face, building, tree, number two, number five and circle. The same six pictures were drawn in five different forms, but the same four letter six non-sense syllables were given at the bottom of all the pictures. The important characteristics of this test is that the pictures in every trial are drawn little differently. But the non-sense syllables, the response words are not changed. In all the five trials the same six response words are given under the respective pictures.

The test is conducted by presenting all six pictures one by one with the response, the non-sense syllable. Therefore, a trial is, presenting all the six pictures one by one, starting from top to bottom for five seconds. The subject is expected to associate the four letter non-sense syllable with the picture. When all the six pictures are presented, then the participant should call the response word for each picture. Then the experimenter reads the correct response word given at the bottom of each picture. This is the first trial. In the second trial the second set for pictures with same response words are presented in the same manner as the first trial. There are five sets of pictures and each set is a trial. Hence there are five trials. It is observed whether the participant has learned to associate the correct response words with the pictures which are having certain common characteristics and the concepts attainment has taken place. For example, RELK is the response word to human face and it is given under five human faces. The subject is expected to produce the same response word to all the five human face pictures in all five trials.

Rajamanickam's Concept Achievement Test

At present there are not many concept achievement or formation tests available, except the Vigotsky test. Since it was originally designed for children for the clinical purpose a test at the adult level is very much needed and therefore, a concept achievement test was designed and developed by Rajamanickam (2001) on the pattern of Heidbreder's test, but differently and with some improvements. Rajamanickam's concept achievement test could put the participant in deep thinking in the tests of concept achievements.

The design and development of this test is based upon the hypothesis that human thinking is generally more toward *object oriented* than towards abstraction. Therefore, in this test the concrete concepts are presented for inducing abstract thinking and for concept achievement. It was planned originally to have nine pictures and six sets. These nine pictures include three alive concrete concepts the human faces, animals and birds, two object concrete concepts, the jug and the pot, two spatial concepts, the square pattern and the circle pattern and two number concepts, three and six.

These four types of nine pictures are designed in six different forms. For example, the human faces are drawn with six different styles, the animal pictures are drawn in six different forms. Actually these six animal pictures refer to six different animals and the six bird pictures refer to six different birds. The concrete object picture, the jug is presented in six different styles. Thus the first form of the achievement test consisted of 54 (9 × 6) pictures, nine pictures forming a set and there are six sets (Table 15.1). Under each category of pictures a four letter-nonsense syllable is printed. But in the final form, after the validation of the test one category of pictures are dropped as it was poorly responded by the participants. Thus the test has eight category of pictures with six sets (8 × 6) 48 pictures on the whole. One extra set also is added for exceptional cases (Table 15.2).

The test is developed for *adult persons* which can be administered to the participants aged 18 years and above for testing the level of concept achievement. This test involves deep thinking for abstraction. Though concrete objects are included in the test they need to be abstracted. The same response word (4 letter nonsense syllable) should be applied to eight different types of objects presented. This is an individual test, should be conducted under the control of the experimenter.

Table 15.1 : Showing the sets and the number of pictures

S. No.	*Picture*	*Response word*	*Set-1*	*Set-2*	*Set-3*	*Set-4*	*Set-5*	*Set-6*
1.	Bird	LAGH	S – 1 / P – 1	S – 2 / P – 1	S – 3 / P – 1	S – 4 / P – 1	S – 5 / P – 1	S – 6 / P – 1
2.	Human face	MUKR	S – 1 / P – 2	S – 2 / P – 2	S – 3 / P – 2	S – 4 / P – 2	S – 5 / P – 2	S – 6 / P – 2
3.	Square pattern	BIKT	S – 1 / P – 3	S – 2 / P – 3	S – 3 / P – 3	S – 4 / P – 3	S – 5 / P – 3	S – 6 / P – 3
4.	Number Three	CEVT	S – 1 / P – 4	S – 2 / P – 4	S – 3 / P – 4	S – 4 / P – 4	S – 5 / P – 4	S – 6 / P – 4
5.	Animal	GALB	S – 1 / P – 5	S – 2 / P – 5	S – 3 / P – 5	S – 4 / P – 5	S – 5 / P – 5	S – 6 / P – 5
6.	Circle pattern	FARH	S – 1 / P – 6	S – 2 / P – 6	S – 3 / P – 6	S – 4 / P – 6	S – 5 / P – 6	S – 6 / P – 6
7.	Number six	JISP	S – 1 / P – 7	S – 2 / P – 7	S – 3 / P – 7	S – 4 / P – 7	S – 5 / P – 7	S – 6 / P – 7
8.	Jug	KEMG	S – 1 / P – 8	S – 2 / P – 8	S – 3 / P – 8	S – 4 / P – 8	S – 5 / P – 8	S – 6 / P – 8
9.	Pot	ZEVW	S – 1 / P – 9	S – 2 / P – 9	S – 3 / P – 9	S – 4 / P – 9	S – 5 / P – 9	S – 6 / P – 9

S for Set P for Picture

The participant should be comfortably seated in a well illuminated room where there should not be any visual and auditory distractions. The test has two sets, one *Response Set* (Figure 15.4) and the other *Stimulus Set* (Figure 15.5). The response set is having a four letter nonsense syllable at the bottom of every picture, whereas the stimulus set will not be having the nonsence syllables.

In the experiment, the response set pictures are presented first one by one for ten seconds with the four letter nonsense syllable and the participants is asked to carefully observe and associate the *Word*, (nonsense syllable) with the picture. Each picture with the word is exposed only for ten seconds. When all the eight pictures of the first set are presented the experimenter has to present the pictures of the stimulus set one by one for ten seconds and

Table 15.2 : Showing the final sets and the number of pictures

S. No.	*Picture*	*Response word*	*Set-1*	*Set-2*	*Set-3*	*Set-4*	*Set-5*	*Set-6*
1.	Bird	LAGH	$\frac{S-1}{P-1}$	$\frac{S-2}{P-1}$	$\frac{S-3}{P-1}$	$\frac{S-4}{P-1}$	$\frac{S-5}{P-1}$	$\frac{S-6}{P-1}$
2.	Human face	MUKR	$\frac{S-1}{P-2}$	$\frac{S-2}{P-2}$	$\frac{S-3}{P-2}$	$\frac{S-4}{P-2}$	$\frac{S-5}{P-2}$	$\frac{S-6}{P-2}$
3.	Square pattern	BIKT	$\frac{S-1}{P-3}$	$\frac{S-2}{P-3}$	$\frac{S-3}{P-3}$	$\frac{S-4}{P-3}$	$\frac{S-5}{P-3}$	$\frac{S-6}{P-3}$
4.	Number Three	CEVT	$\frac{S-1}{P-4}$	$\frac{S-2}{P-4}$	$\frac{S-3}{P-4}$	$\frac{S-4}{P-4}$	$\frac{S-5}{P-4}$	$\frac{S-6}{P-4}$
5.	Animal	GALB	$\frac{S-1}{P-5}$	$\frac{S-2}{P-5}$	$\frac{S-3}{P-5}$	$\frac{S-4}{P-5}$	$\frac{S-5}{P-5}$	$\frac{S-6}{P-5}$
6.	Circle pattern	FARH	$\frac{S-1}{P-6}$	$\frac{S-2}{P-6}$	$\frac{S-3}{P-6}$	$\frac{S-4}{P-6}$	$\frac{S-5}{P-6}$	$\frac{S-6}{P-6}$
7.	Number six	JISP	$\frac{S-1}{P-7}$	$\frac{S-2}{P-7}$	$\frac{S-3}{P-7}$	$\frac{S-4}{P-7}$	$\frac{S-5}{P-7}$	$\frac{S-6}{P-7}$
8.	Jug	KEMG	$\frac{S-1}{P-8}$	$\frac{S-2}{P-8}$	$\frac{S-3}{P-8}$	$\frac{S-4}{P-8}$	$\frac{S-5}{P-8}$	$\frac{S-6}{P-8}$

S for Set P for Picture

the participant is asked *to produce the response word* (nonsense syllable) *he has seen in the response set.* As and when the experimenter exposed a picture from the stimulus set the participant has to tell the four letter nonsense syllable. After presenting all the eight pictures of the first stimulus set the experimenter reads out the correct word once to the participant. Then he starts the second set after twenty seconds.

If the response word is right, the experimenter has to make a mark as 'R' in his Table already prepared (Table 15.3) and if the participant's response is wrong he has to make a mark as 'W'. Each right response receives one mark and the wrong response receives a zero (0) mark. After presenting all the eight Response sets one by one followed by each set of the Stimulus sets, the experimenter can sum up the total score for each set and for all the eight sets. This constitutes the individual score (Table. 15.3). This score determines the individual's level of concept achievement in this test. The experimenter can analyse the score and find out where the individual has failed to abstract the qualities of the objects. What are the objects for which he/she has given right responses? What are the objects for which he/she has been consistently giving the wrong responses? The reasons can be determined.

Fig. 15.4 : Response Set

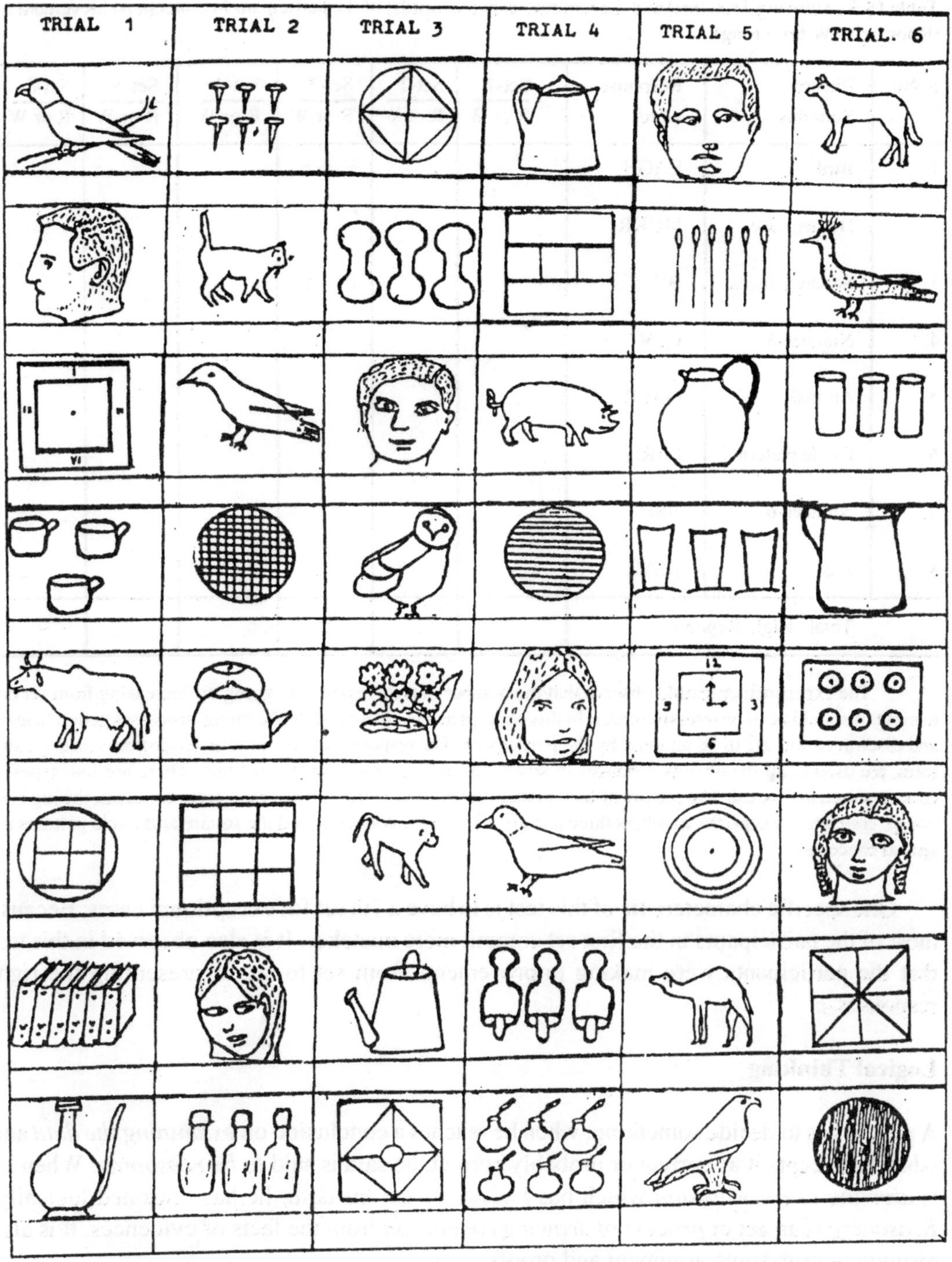

Fig. 15.5 : Stimulus Set

Table 15.3 : Showing Trial (set) and the reports (response words) of the participant in concept Achievement test (R for right, W for wrong)

S. No.	Picture Stimulus	Response word	Set-1 R or W	Set-2 R or W	Set-3 R or W	Set-4 R or W	Set-5 R or W	Set-6 R or W
1.	Bird	LAGH						
2.	Human face	MUKR						
3.	Square pattern	BIKT						
4.	Number-3	CEVT						
5.	Animal	GALB						
6.	Circle pattern	FARH						
7.	Number-6	JISP						
8.	Jug	KEMG						
	Total : Right Reports							

The experimenter should observe that the number of right reports are going on increasing from set one to eight. It should be in progressive order. In this concept achievement test, the learning processes generalization and discrimination are to be adapted by the participant. The response words, in four-letter-non-sense syllable form, are used to all the objects. Changes in small variation are made in the pictures. There are four types of concepts used in this test. The picture of human faces animals and birds are alive concrete concepts, the jug is an object, concrete concept, the numbers three and six are number concepts, and the square and circle patterns are spatial concepts.

One specific characteristic of this test is to have a 7th set for exceptional cases. Because most of the participants in the first set commit more mistakes. It is also observed in this test that the participants were making improvements from set to set in presenting the right responses.

Logical Thinking

A person has to decide something, when he reaches a conclusion on *examining the data* and when he accepts it as correct or probably correct, he said is said *to be reasoning.* When he is *arriving at a conclusion* which has *followed from his data,* he has reasoned logically. *Reasoning* is an act or *process of drawing conclusion* from the facts or evidences. It is also arriving at from some argument and proofs.

Reasoning is a faculty of *rational argument.* By which we are arriving at a conclusion by the *process of deduction.* Sometimes, when one reaches a conclusion which is not based upon any data, his thinking is not correct and the conclusion is illogical. It is because, his conclusion is not supported by any evidence.

Logic and Psychology

There is some relationship between logic and psychology. Logic is a branch of philosophy, mainly concerned with analysing the patterns of reasoning by which a *conclusion is drawn from a set of premises* without any reference to meaning or context. Logic mainly deals with *correct thinking*. It was well developed long before psychology began to examine the actual process of thinking, the growth of the reasoning ability in the child, the effect of training, the individual differences and other processes of psychological problems. Logic is concerned with the question whether the conclusion follows from the data ensuring logical procedures. The middle term in an argument is more belonging to logic than to psychology.

Therefore, when we speak about something, we mean that something should make some sense or meaning. When it is not, we say, it is not logical or does not make out any meaning. When we expect that something we speak to be logical, we are expected to follow certain principles or rules. Therefore, *logical thinking* is a way of thinking in which the *rules of reasoning* are to be *properly followed*. Logic has several branches such as *formal logic, deductive logic* and *inductive logic*. The formal logic is interested in whether statements correspond to the facts or not. It is interested only in connections — the logical connection of the statements with one another. Does statement A follow logically from statements B, C and D?

At the higher logical thinking these rules are highly complex and rigid. There are *several kinds of logical thinking* which are distinguished by logicians. In everyday life we are using several kinds of logical thinking even without learning the formal logic.

For example, you are waiting for a train in the railway station plateform with your friend. If you hear the sound of the railway bell then at once you tell your friend that "the train is going to come soon". He also nods his head as a sign of approval of your statements. You have applied a logic here. You need not go through all the steps, explaining to your friend the cause and effect of the sound of the bell and the arriving of the train that whenever the railway bell rings the train would come. When both of you are implicitly agreeing these two ideas there are logical steps between the two ideas, "the bell is ringing" and "the train is going to come soon". We use such kind of logical reasoning in our everyday life. But, when you move from everyday logic to formal logic to prove the truth you will have to follow the rules systematically. *In a formal logic we will have to use syllogism.*

The syllogism is the basis of formal logic. The syllogism is a *sequence of sentences* in which a major premise and a minor premise lead to a conclusion. It is a classic device for clarifying thinking. The semantic has been developed as a supplementary technique by modern researchers. *In a syllogism there are two sentences.* They are called *premises.* One is the *major premise* and other is the *minor premise.* These two are put together in such a way that a third statement called conclusion is inferred.

Let us illustrate the point with a well known correct syllogism.
All men are mortal (Major premise).
Socrates is a man (Minor premise).
Therefore, Socrates is mortal (Conclusion).

Another example :

All Anglo-Saxons are beautiful.
All English are Anglo-Saxons.
Therefore, the English are beautiful.

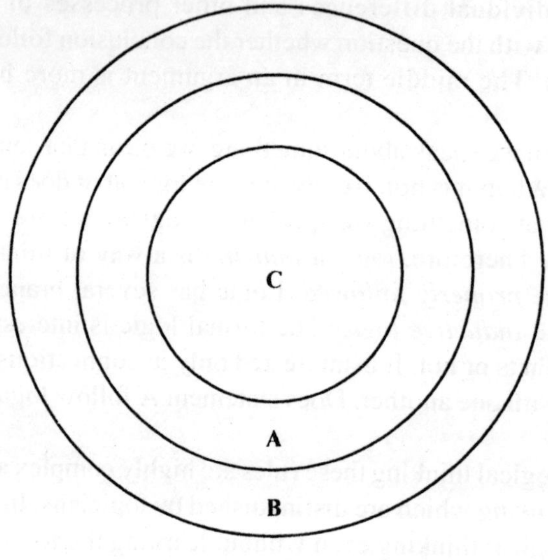

Fig. 15.4 : Thinking : Logical Thinking

But sometimes syllogism may be stated falsely if logical principle is not followed. The object of the minor premise should be the subject of the major premise and the subject of the conclusion should be the subject of the minor premise and the object of the conclusion should be the object of the major premise (Fig. 15.4).

Here is a faulty conclusion based on faulty premise :
All A is B
All C is B
Therefore all C is A.

This syllogism *appears to be correct. But it is not correct.* Because the *rule is not followed.* Let us apply the rule and set it right.

All A is B (Major)
All C is A (Minor)
Therefore, all C is B (Conclusion).

PROBLEM SOLVING

So far we have been discussing about logical thinking. It is one of the methods of solving problems. In our everyday situation we may have more complex problems to solve.

What is a Problem?

Problem means anything, a matter, or a person that is *difficult to deal with* or solve or to overcome. A child may be a problem, or a puzzle or a question set for solution may be difficult to deal with. A mathematical statement may be requiring a solution, usually by means of one or more operations, or it may be a literary work that deals with difficult moral question, all problems.

A problem occurs when an activity is blocked by an obstacle that cannot be removed by using the readily available habits. The *problem situation may be unfamiliar* and confusing to you. But if it is clear your habits or customs would be sufficient to deal with. The obstacle is something, must be changed or it must be absent. Sometimes all the necessary factors may be present, but you have to learn to look at them differently. For removing the obstacle you have to think. It is to see how the process works in solving the problem.

If there were any problems for which you already know the answers, they were not problems. Because there were no obstacles and they did not make you to think. Therefore, a problem may also be defined like this. A problem involves, a goal of yours that you see, is blocked by some obstacles and your habits were found to be *insufficient to solve it.* A problem is identified and accepted, only then the individual can work to solve it. There is individual differences in identifying and accepting problems. Because of the differences in perception, knowledge and skill and training, no two persons face the same problem in the same problematic situation. You can solve a problem if only the *obstacle between you and your goal* is within your capacity to remove.

Getting Clear Thinking

The student of psychology may come across this 'Problem Solving' in more than one topic. For example, in learning, in thinking, in reasoning and in creativity, problem solving is introduced. Therefore, problem solving is a behaviour concerned in all these areas. It is *a higher mental process* that is tested in monkeys and in human. *Gestalt psychologists demonstrated this problem solving behaviour in monkeys.* Problem solving as a category of learning is *more difficult than the maze learning* and *bar pressing* activities. *In human being* problem solving involves mainly *transfer of experience or knowledge.*

The mode of solution is a process of seeking and finding out some of the works that can be performed *by thinking* instead of motor and manipulating activities. In the case of human, in problem solving, the tools of thought are put into action. The symbolic process and the concepts learned earlier may be applied in the new situation to solve the problem. *Language helps the problem solver to talk to himself* about his problem and also *speak to others* and get the assistances. An insightful look at the problem situation may suggest the relevant concepts and known principles with which an appropriate solution can be easily achieved.

In a problem solving situation there is a beginning stage and there is a goal achieving stage. There is something given and something required. Certain data or information already given to the thinker. These are certain acts he knows and certain information he can use, but the available data are not sufficient and therefore more data must be obtained. The *thinker's task varies* with the *nature of the problem.*

Any problem solving situation has two characteristics. First there is an *initial stage.* In this stage there is low probability of getting the correct response or solution to the problem. The second is the *discovery stage* in which the correct process is expected to occur in problem solution.

One way of solving the problem is to set up a hypothesis and to work on it. The *hypothesis is a guiding idea* of most definite sort. But it should be based upon correct knowledge. If one wanted to win a competitive event in a sports, he must get trained himself by a good trainer. This is one way of facing the problem. It is a standard scientific procedure, setting up a hypothesis and the testing it. In most of the scientific studies and practical matters it is generally to have the guiding idea to suggest a definite hypothesis which can be tested by experiment or by objective observation. If the problem solver does not have a hypothesis, atleast he needs a good guiding idea which he can search in different directions. If he can correctly grasp the nature of the problem he is sure to get a good guide and solve the problem.

Problem solving is *a goal directed activity.* It may be defined as a goal directed, multiple trail-and-error approach to new situations and in learning task, in which the individual discovers new modes of response.

For example, let us consider this problem. Here is a four digit number 9999. Handle this number in whatever possible manner you like and make it 100. How do you solve this problem? The problem appears to be simple, but the method of solving it is difficult. A common reaction to this problem is, to make it into suitable components that should constitute into 100. The problem and the behaviour required for solving it would reveal some of the basic characteristics of problem solving behaviour.

The *first thing* in problem solving is that the correct response is not made immediately. If this takes place, then there is no problem. Therefore, there is a psychological gap between the problem and its solution. The gap can be filled up only by discovering the correct response to the problem.

The *second thing* is, that problem solving behaviour in human being mainly consists of covert responses. For example, in the case of the problem given here you may be trying to solve the problem in various ways talking to yourself. But no body knows what method you are using till you make some overt responses either by speaking it out or by writing the solution on a paper. This would reveal your problem solving method to others. In this way, most of the processes in problem solving method are not exposed.

The *third thing* in problem solving is that the correct solution to the problem sometimes *occurs suddenly* without much efforts. We sometimes solve the problems by sudden insight. For example, the author, when he encountered the problem given here, tried to solve it in several possible ways by manipulating the four nines. Suddenly the idea struck in his mind that what is required here is to add one to ninety nine to make the number 100. How to get one from the rest of the two nines? It is only by dividing one nine by another nine. By adding it to ninety nine we get 100.

That is : $99 + 9/9 = 99+1 = 100$

Steps in Problem Solving

Problem solving involves several basic psychological processes such as *learning, perception, motivation* and *sensation.* It is not a unitary process and no independent theory of problem solving can be evolved. Therefore, we must understand the behaviour involved in problem solving. Problem solving behaviour occurs in four stages (Woodworth, 1952). They are (1) preparation, (2) incubation, (3) illumination and (4) verification.

The *first stage* is preparation. During this period the person who solves the problem acquires the *necessary basic skills* for solving the problem and becomes aware of the particular problem and makes himself ready to face the problem. The *second stage* sets in when he finds that all his efforts to solve the problem were useless. He stops at this stage for the time being and engages himself in some other kind of work. He may even forget that he left a problem unsolved. This period is called *incubation.*

But when the person keeps himself engaged in some other kind of task he may have sudden insight into the solution which he left unsolved. Now the answer is quite clear to him. This period is called *illumination.* It is often thought that incubation leads to illumination and illumination need not be thought as an accidental phenomenon. It is rooted in the past experience of the problem solver.

It often occurs in the case of some problem solvers that they do not check their solution after the period of illumination. But it is important to verify whether the method of solution really solves the problem. Therefore, the problem must be attended once again with the new method. This stage is called *verification.* All the four stages in the process of problem solving behaviour are taking place in our every day life in such a way that we do not easily distinguish them. Modern psychological studies try to distinguish them in these ways.

Thinking takes place in an individual when the *customary methods* of solving his problem *was not successful.* A person thinks because he has encountered a problem and he has to solve it. Therefore, *thinking takes over when habit is insufficient to provide a solution.* Generally, thinking begins in the presence of a problem. Thinking does not mean day-dreaming or fantasy. It is a *delibrate application of our mind* to the solution of a problem.

Group Thinking

Individuals independently think to solve problem. There is no other person sharing in it, since it is exclusively the *individual's personal problem.* But sometimes, when an individual could not find out any solution to his problem, he may seek the help of others in getting his problem solved. In such case the *individual shares his thought* with another person or persons for the solution of his problem. Sometimes the problem he faces may be also a common problem, the effect of which may also affect others. Two or three persons jointly or co-operatively think in finding out a solution. For example, the husband and wife may jointly think and find out solution for the family problems. This type of thinking may be stated as group thinking which involves exchange of ideas, discussions and also getting some clue even from outside.

Similarly, a business organization, a limited company, a municipal corporation may face problem. In all these cases a group is engaged jointly and co-operatively and think to find out a solution to the problem. This sort of endeavour becomes very essential in certain critical situations. The problem may be simple or complex or difficult, a group consists of two or four or fifty may sit and discuss for evolving solutions. It appears that a **group** *thinking* can *evolve a better solution* than individual's thinking. School students often engage in group thinking in getting their arithmetic problems solved. **Garrett** says that *we think with our fingers and hands* and with *our brains,* and we learn with our *muscles* and with our *nervous systems.* According to him some activities involve more muscles or more physical energy than others. But *all behaviour* is depending upon *active brain and nervous system.* We cannot explain behaviour, mental or physical by terminating any one. Any reaction, physical or mental is a behaviour *depends upon the brain and nervous system.*

Theories of Thinking

There are two theories of thinking and they are trying to explain where the function of thinking is taking place. *Has the brain anything to do with thinking?* Or is it taking place in the *nervous system?* Or the whole human body is involved in thinking? We have stated in the preceding section that the brain is functioning in all kinds of cognitive activities. Therefore, we must see the role played by the brain in thinking and reasoning.

Let us discuss first the *central theory.* According to the central theory thinking and reasoning is *taking place only in the brain* and the brain is solely responsible for all kinds of thinking and problem solving activities. According to this theory, for example, a *sound stimulus* or an *order* (speech stimulus) is received or heard in the ear. The auditory receptors in the ear receive it. The *stimulus* is in the *form of nerve impulses* which are carried by the *sensory nerves from the receptors to the auditory area* which is located in the *temporal lobe of the brain,* which is just below the *fissure of sylvius.* In this area the *impulses are elaborated* through association areas and *transmitted to the motor areas.* Then these auditory impulses area transformed into *muscular responses,* in the form of words or sentences spoken. When we speak out in response to the auditory stimulus we are said that *we think with our brains.* It is often said that the *brain secretes thoughts.*

The other theory is the *peripheral theory.* The *peripheral nervous system* is mainly involved in this. *This functions outside the brain.* The peripheral theory is also called *motor theory.* According to this theory the brain is *one of the parts of thinking process.* Since all our learning are both mental and physical, we also think with ours fingers, hands, muscles and with brain and nervous systems — *with our whole body.* According to the peripheral theory the auditory receptors receive the stimulus, the sound waves. These are in the *form of impulses, transmitted to the brain* by the *sensory nerves.* The motor impulses go to the association areas, to the muscles of the body instead of moving fastly to the cortex area. Here the *motor impulses* initiate *new type of stimuli* which *go to the cerebral cortex.* Therefore, there is a continuous chains of *stimulus-response relations from brain to body and form body to the brain.* This is going on till the process of thinking takes place. The idea is that thinking may be carried on *by the muscles rapid movements.* Investigators are of the view that thinking can be experimentally demonstrated and it is through *implicit and tentative movements of the muscles.*

Evidences are produced for the peripheral or motor theory from observations and from experiments. Everybody can notice children's silent reading and people who are not well educated adult persons. Though they are reading silently they are *moving* their *tongues and lips* and also they are *mumbling.* Sometimes they even *move their fingers* and show *sign of gestures.* As they go on *reading silently* for one or two hours they show *signs of fatigue in throat* and wanted to *drink water.* Arrangement was made in one experiment to *attach an instrument* to the *tongue* and to record the movements. The person was asked to read silently or to think. The recording showed that there were clear tongue movements when the person read silently. There were also several experiments with various types of instruments to test the thinking process *through tongue movements.* All these results showed that there was often *sub-vocal speech* which is the *indication of thinking.* We may say in other words, there was thinking speaking silently. **Watson** (1914) the founder of behaviourism said that *thinking is silent speaking* as there was muscular movements in mouth and in throat and *speaking is louder thinking.*

According to peripheral theory, *humans think with their throat, the larynx* and *through the whole body* in general, atleast part of the time, as there are strong evidences that *muscular activities are present* during the *thought process.* Because of this we cannot reject the central theory totally. According to them, thinking is going on in the brain without the co-operation of the peripheral or motor systems. But, because of the end product of verbal and (bodily) muscular activities (evidence) the peripheral (motor) theory seems to be more agreeable than the central theory.

THOUGHT AND THE SELF (BEHAVIOUR)

Thought or thinking is a mental event. Thinking, imaging dreaming all are mental events. These *mental events* generally *precede human behaviour* or actions. They also *determine* the course of human actions. Since the time of **Aristotle,** the Greek Philosopher and the father of all sciences, it has been assumed that man is a rational animal and he *directs* his own behaviour *by his thoughts.* The western philosophers for a long time have been struggling with this phenomenon of explaining the thought process. What has puzzled them was *how this process is occurring.*

The problem they have confronted was by no means simple or insignificant. We believe in a person's capacity *to plan something* and the execution of his own behaviour. He also anticipates the consequences of his own actions. His *thoughts and actions* lie on the foundation of social and moral conceptions of responsibility and his free will. In this effort man may face certain inconveniences when he thinks that his behaviour was not correct. When he puts his thoughts in action some may be successful and some may not be successful. It is a common occurrence in a man's behaviour that a diametrically opposite thinking may take place on facing the consequences of his act. The first thinking was that the act he determined to do was correct. But after performing the act he finds that what he thought and acted was wrong. Thus for good and bad, for right and wrong our actions are *directed* and *determined by our thoughts.*

16

IMAGES, IMAGERY AND IMAGINATIONS

Introduction

Images, Imagery and Imaginations are the terms we often hear in our everyday life on different contexts. In the common sense world these concepts are used in the popular sense without any precision. These terms are generally referring to the mental processes involving thought or idea which identifies with some object, person or place. The term imagery is derived from the Latin word Image. Image refers to an artificial imitation, or statue, effigy, constellation or figure. Image may also mean a general impression a person creates about himself to the public. Image is a figure seen in a mirror. It is the presentation of the external form of a person or a thing in art.

An image may be also referring to a visual impression obtained by a camera. It is the general impression that a person or an object or organization or a product presents to the public. These conceptions are general in the common sense world.

Psychologically, an image is a picture of something in the mind. Imagery is, the plural form of image and we may also state it as imageries. Therefore, imagery means images in groups. When we wish to state the images collectively we state it by the concept imagery. A number of images collectively stated is called imagery.

Imagination is the act of imagining or forming a mental picture which is not actually present to the senses. It is a mental image or idea. Imagination is the result of the mental images. Etimologically we can explain the concepts in these ways. But it is only man who can have or form a mental picture of an object seen in the outside world and not animals.

Man differentiates himself from other living creatures in various ways by his intelligence and intellectual activities. All the intelligent and intellectual activities are the results of the development of his complex nervous system, the brain. The function of the human brain is a continuous process which, we may also state as mental function. This function is revealed in various cognitive activities like observing, perceiving, thinking, imagining, day dreaming and so on. Observation leads to perception and perception may lead to imagination **Garrett** (1975) says that there is not much difference between perception and imagination. According to him imagination is really perception because there is very little sensory control. Therefore, we may state that in imagination there is *greater* amount of *fantasy and less* amount of *stimulus* control. The observer experiences the specific stimuli like, seeing a moving vehicle, the sound of a running motor horn, the smell of jasmine flower, the taste of mango fruit and the surface of the floor and so on.

Imagination is *nourished by sensation and perception.* What one sees, feels and understands become rich source for one's imagination. Men with rich source of sensation and perception become highly imaginative and also become creators. A man could become an artist, painter, or scupture or a novelist with fertile imaginations. The creative artists spend most of their times in imagining something in relation to their experience and knowledge acquired from every day life events. Poets, novelist and painters often live in the world of imagination. They often produce mental images of what is not present or has not been experienced.

We have studied more elaborately about the various aspects of thinking in the previous section. All these subject matters can be brought under the heading of directed thinking — as they are aimed at solving a particular problem. This form of thinking is limited in its scope. There are also other forms of thinking which are unlimited in their scope. They are called associative thinking. In a word it can be stated as imagination. Fantasies, day dreams and dreams are of this type of thinking. All these are unlimited in their scope as they are not directed to solve any problem.

What is Imagination?

Imagination may be defined as an act of forming a mental concept of what is not actually present to the senses. It is also stated as the manipulation of images. An image is an artificial limitation or representation of the external form of any object.

An image is a collection or reconstruction of a sense experience, often visual. The important process that takes place in imagination is the rearrangement of the materials we have experienced in the past in relation to the present. The result of the arrangement is sometimes based on recency, frequency and vividness of our experiences. A recently perceived object can be combined with a similar object perceived in the past. This combination need not be real or in existence. In our imagination we combine several of our past and present experiences in a new or in a peculiar fashion. We wish to see something new out of the old objects. We don't care whether such combinations are possible in the world or not.

For example, Chimera is a fanciful animal with a lion's head, a goat's body and a serpant's tail. In Greek mythology the animal Chimera is often spoken of. It is not real, nor existed at any time in the world. It existed only in human imagination. Someone wished to see such a creature and he had created it in his imagination. Similarly, you can imagine a mermaid, combining a beautiful woman and a fish. A mermaid is an imagined creature in which a half of woman's body and a half of fish's body are combined. This idea may often arise in your mind, because your poets and novelists often present your reading materials, comparing women's eyes to fishes. You may also find certain amount of similarity between these two. Adolescent boys and girls have *fantastic imaginations.*

Children are highly imaginative. Every element of imagination is related to the child's experience. Children's imaginations may lack organization. They often have companions in their imaginations. Such imaginations express that children have needs and when their needs are not fulfilled they imagme companions and find comforts. Both in children and adults the main causes of imagination would be the *inner world of needs,* worries and pre-occupations with outer world of sensory stimulation.

IMAGINATION AND MEMORY

Imagination is a way, of presenting something to ourselves that has not existed previously and can never exist whereas memory is a way of presenting to ourselves something without any change or modification, that has been learned or experienced earlier. Therefore, imagination is different from memory. Imagination may vary from memory in degree to which changes or rearrangements are made in the original materials. The characteristics of imagination depend upon the materials experienced and retained by the person from the past. But, what new is the rearrangement of different materials in a particular fashion. The elements of imagination are old, but the combination of the materials presented is completely new. In memory you are bringing back the materials which you have learned earlier without any change. If you make any change or modification in the materials it becomes imagination.

Memory materials are true. They are experienced and presented to yourself as they are experienced by you. For example, in imagination a man with spectacles may be imagined with a horse's head and a human body, but in memory he is seen wearing spectacles. Imagination is *private* and *subjective* whereas memory is *objective.* For example, a material is presented visually to four persons. All the four persons may imagine four different things from the same material. But the same four persons, if their memory is accurate they will present the same subject matter.

It is very difficult to explain to blind men how an elephant would look like. You know the story of four blind men and an elephant. How fantastically they have imagined about an elephant? Therefore, it is possible to have fantastic imaginations from the verbal instructions one has received without any visual perception. **Gardner Murphy** observed that when the elements of imagination are combined in new ways by the application of the laws of association the same laws operate also in the case of memory. But with all that, the laws of association relating to memory are differently dealt with in imagination. Therefore, the study of imagination differs from the study of memory.

How to Study Imagination?

Psychologists tried to study imagination scientifically. The role of drives has been used in the study. Persons subjected to prolonged hunger often speak of rich feasts of which they dream. Persons put into more physical discomforts often dream of healthy and bodily pleasures.

Our imagination, sometimes is utterly free and uncontrolled. Our thoughts may run very wildly and fantastically. Nothing can prevent them. All our day dreams or fantasies are free imaginations. Though in our imagination our *ideas run riotously* there is a method in it. A set is operating, in it. The *role of drives* may be used to study the imagination.

It is not only visceral drives but also *activity drives* which play an important role in the day dreams. If you are interested in some kind of play and if you are deprived of it you indulge yourself in day dreaming of the play. Our sensory drives for seeing and hearing also guide our imagination. You may have a strong desire to *see and enjoy* a beautiful garden or to *hear and enjoy* a sweet music. When there is no chance for this you imagine about the one you have enjoyed in the past. So also the same *drive* would make you to *aspire for power* and prestige and you make great achievement in your imagination. The *sex urges* carry everyone into the world of imagination where he finds his desires are satisfied.

The process of imagination can be studied at close range by the method of *chain association test.* According to this method we present a single word to the subject and allow him to give us a chain of words associated with the word. By this means the subject uses each word he utters as a clue for the next word. For example, one may get a chain of words like this: Boy-girls-marriage-father-mother-family-prayer-temple-God-religion. The subject may stop for a moment at some word not knowing what to say and himself does not know that he is stopping because there may be *mental blockage.* This is called *interference* and psychologists are interested in such interferences. It reveals the preoccupations of the individual.

A child may be asked to tell a story of a bird, or an animal or a human being. A picture also may be shown to him to make a story of it. We can find relationships between his imagination and events of his life. Such stories reveal the subject's desires, worries and their preoccupations which they hesitate to reveal in ordinary face to face situations. The story events may reveal even the *materials* they have *stored* in their *unconscious mind.* **Morgan** and **H. Murray** (1939) have developed the *Thematic Apperception Test.* This test serves several purposes. It consists of a series of standardised pictures of one or more human figures in different poses. The subject is shown the pictures one by one and he is to make a story for each picture. This test would reveal the differences in predominant interests and worries of the persons on whom it was administered. The subject may make up a free story.

Day Dreams

In human beings there is going on a kind of activity, called *ideational activity* which has nothing to do with the present environment. According to **Woodworth** ideation is a mental activity or having ideas. In this activity there is neither muscular movement nor sensory stimuli present. You may be reading a new novel and may be absorbed in it as you find it very interesting. Every event or stimulus you come across in the pages of the book represents imaginary events or stimuli. They have nothing to do with your surroundings. At some point you may even stop reading the novel and relate those events, objects and persons to your own past and present experiences. You may recall both pleasant and painful experiences of your life and others and try to continue the life events further in the manner in which you are interested. Hence you begin to *live in an inner* world of *your own* which is free from the real world. This kind of mental activity is called *daydream*. Daydream is a kind of imagination in waking state.

Daydreams, reverie and fantasy all generally refer to same mental process. Daydreams and fantasies are used sometimes synonymously. *Fantasy* is the general term used to indicate free imagination. Fantasy in a highly developed form is called daydream. It is the result of forming representations of things not actually present. Daydream is a reverie. All the poetic materials are the results of daydreaming.

In daydreams you exclude external stimuli for the moment. At the time of sitting in the class room and listening to the teacher your mind may be wandering somewhere else on hearing some familiar song broadcasted from a distant place. Or while you are listening to a story told by your teacher you might have immediately gone into a daydream and associate it with the events of a story told by your grandmother or with a story of a movie that you have seen in a picture house a few days ago. For a moment you forget what the real world is demanding of you and do something else. You relax your mind and you allow your ideas to wander freely. In this process one idea drags another idea and one memory recalls another and a chain-like process is taking place. In this process of free association, if you review your thoughts and experiences in a sequential manner long forgotten life events are also recalled. Here your thoughts are not controlled and directed towards any specific purpose. You are manipulating your ideas in whatever manner you wanted. Walls, iron bars, seas and oceans cannot prevent them. You can fly in the air, live under the sea, build castle in the air and live there.

Daydreams occur frequently among children and more frequently among adolescents. When children are prevented from eating sweat-meat or playing with toys and weapons they find them in plenty in their daydreams. Sex starved youths gratify their desires in their daydreams. In daydreams a person achieves all those things that he cannot achieve in his real life. There is continuity of thought in daydreams and the thoughts are going on working out various themes in which the person plays the central role. In this kind of free imagination a man's long cherished hopes and desires are fulfilled. He sees himself always engaged in heroic activities and making great success. He defeats his enemies, wins the first prize of five lakhs of rupees in the lottery, wins the heart of the most beautiful girl, passes in the examination in the first class with a first rank, exhibits unchallenging ability of some kind. He wishes to play always the role of a *conquering hero.* But, sometimes a person's day dreams may lead to unresolved conflicts and he imagines all sorts of disappointments and worries and he plays role of a *suffering hero.* He becomes unpopular among his friends and not respected by them, loses good fortunes in the race or lottery meets with heavy loss in business. His girl friend disappoints him after giving hopes.

All these materials of the day dreams come from our every day experiences. Daydreams usually serve the purpose of *wish fulfilment.* There is always *a plot* in a daydream. It has continuity. Man can escape from difficulties and responsibilities of his real life for few minutes. It is also a kind of indoor game and he finds some kind of pleasure in it. But if the daydreaming goes to the extreme level it leads to several kinds of maladjustments. The person may be often playing the role of both the heroes and loose all contacts with the real world. Finally, he may develop *delusions and persecutions.*

Dreams

Dreams were considered as mysterious matters until the early twentieth century when **Sigmund Freud** has published his famous work. *"The Interpretation of Dreams"*. Dreams are regarded as a kind of personal experience man gets during his sleep. It is also a kind of imagination that goes on when one is asleep. Sleep is a state, generally, contrasted with the waking state. In the state of sleep there is a great amount of reduction of awareness and activity. Biologists are of the view that sleep is a *restroactive state.* A person may choose either to sleep or to be awaken. There are borderline stages between sleeping and waking states.

Dreams are considered as the private property of the individual. For a long time it is considered that dreams cannot be observed objectively and measured. But experimental psychologists have studied the dream experiences. According to them dream may be defined as the appearance of *rapid eye movements* on the *electrooculogram (EOG)* of a sleeping subject. *Electrooculogram* is a *graphic record* of eye movements obtained at locations around the eyes. It seems that the eye movements are reliable indicators of some type of thought process in subjects who are awake and who are asleep. This is an *operational definition* based upon experimental studies.

In the dream state the *activities* of the brain are *reduced* to a low level. There is no contact with the environment. But some stimuli may be having some kind of influence upon the organism which is mostly of unnatural manner. For example, a feeble tapping of some metal may be taken for a thunder storm, the sound of an alarm time piece for the play of a musical instrument and so on. Dreams do not have plots or sequences. The dreamer may find the events of the dreams are real at the moment and he may wish to continue it if it is a pleasant dream.

The dream is a kind of thought process often characterised by vivid sensory images mainly visual and auditory. Dream is associated with brain wave that is accompanied by rapid vertical and horizontal eye movements. There is correlation between eye movement and the dream content. The eyes are moving when there is a dream. In the waking state our eyes are moving in relation to the object. Similarly our eyes are moving when we watch an event or object or person in the dream. **Dement** and **Wolpert** (1958) studied various aspects of dreams experimentally. According to Dement there are *four periods of dreaming* in seven hours time. *The first period* lasts for ten minutes and begins within an hour after the onset of sleep. *The second period* may be lasting for an hour with an interval of about hundred minutes among them. Generally *the third and the fourth periods* seem to be longer than the first and the second periods. According to recent researches animals like monkeys, dogs and cats also have dreams of simpler types.

Every normal person must have dreams every night during sleep. A normal person in his twenties spends about twenty per cent of his night in dreaming. It is an important psychobiological function and *any suppression* of dreaming may cause serious *mental disturbances.* Persons who suffer from mental illness like acute paranoid, psychosis spend about fifty nine per cent of their total sleeping time *in dreaming.* Dream is a kind of hallucination. According to **Sigmund Freud** (1856-1939) dreams are the *guardians of our sleep.*

There are three important causes for our dreams. They are (1) wish fulfilment, (2) mastery impulse and (3) physiological discomforts.

(1) Most of our dreams are wish fulfilment. Children often dream of things that are desired by them. Starving people often dream of eating feasts. Sex craved people often dream of fulfilling their sex desires. Explorers of Arctic regions dream of warm weather and green fields. The thirsty man dreams of drinking gallons of water. The desires and physiological needs that are not fulfilled in the real life are fulfilled in the dream life.

(2) Several causes have been stated by psychologists for the dreams. Apart from the wish fulfilment, the mastery impulse also plays as one of the causes of dreams. Man's desire for dominence over others becomes active in dream. The servant who is teased by his master everyday dreams of teasing his master one day by saying 'shut up'.

(3) The man who sleeps by putting a heavy blanket over his feet dreams of getting into the hands of his enemies as his feet were tied together and was not able to run and escape. The physiological disturbance the person was experiencing during the sleep has become the cause of such dream.

Types of Imagination

There are two types of imagination. They are *reproductive imagination* or memory and *productive or constructive imagination.* When our imagination simply reproduces our past experience without making any change in it, it is called reproductive imagination. The process of reproductive imagination is the same as the process of memory. In the productive imagination our past experience is completely rearranged in a different fashion by adding something new. All our scientific inventions and discoveries, buildings and dams, plays and novels, painting and sculptures are the results of productive or constructive imaginations.

Images

What is an image? We acquire knowledge of the world through the processes of *sensation and perception.* We acquire knowledge also through the process of imagination and the formation of images. We have studied more elaborately about imagination. We must know that imagination is not possible *without the formation of images* and the formation of images is not possible *without keen sensitivity* to objects, places, persons and events. We see often objects, places and persons and witness events and hear voices. After few minutes or hours or days we would like to know them. What we had was only sensory experiences and now we like to recollect those sensory experiences by forming a picture of those objects in the absence of those objects. This is called mental image. Mental image is a process of visualization of our past experience *in the form of an object.* An image may be also defined as an artificial representation of the external form of any object. It is a mental picture or impression of something.

Imagery

Imagery means images in groups. Generally imagery refers to mental images collectively. A series of mental images may be called imagery. It is generally stated in literary works as imageries in plural form referring to symbolic representations of objects, persons and places. When we wanted to state images collectively we use this concept imagery. Therefore, a group of collective images is called imagery. Imageries are the materials involved in the general processes of imaginations.

Imagery is the distinctive type of mental images formed in the mind of a particular individual. *Mental imagery* psychologically may be referred to sensory experience of something that is not immediately present to the senses. We acquire various kinds of mental images through our sensory experiences such as visual images, auditory images, olfactory images, gustatory images and tactual images.

Generally the auditory and tactual images are dominent in blind persons. The painters and the artists have more dominent visual images. Musicians have more dominent auditory images. Therefore, there is no chance for a man to form any type of mental imagery without sensory experiences.

Mental Images

The human sensory experiences are the primary sources of the formation of mental images. We move and act through the functions of our senses and every sense is registering some experience whether we wish to have it or not, it is occurring in our everyday life. Thus every sensory experience form a mental picture and when we are in need of it through some cue or stimulus it becomes a mental image drawn from our memory.

We are forming images of persons, objects, places, events and experiences. Therefore, we may state that mental image is a personal experience of a person about a person, object or an event that he has formed. These mental images may be referring to visual, auditory, olfactory sensations and the sensation of touch. Apart from these we are also experiencing many new things of different types everyday.

Visual Images

Among the sensory experiences the visual sensation and experience has distinctive quality. We see with our eyes every object in front of us. But we cannot give detail description of all those objects we have seen. Because every object does not possess the stricking characteristics in such a way as to attract our attention and form the image of it. The sensation we get through our eyes is called the vision. Vision means the act of perceiving with our eyes. It may also mean the power of seeing. Though the concept vision is mainly referred here for the sensation of seeing or sight, it has also other meanings. We may say that, he is a man of vision, which means he can foresee things. Vision generally refers to the vivid mental image, vivid picture of what has been seen. It is also used to state a person or object of extraordinary beauty. The concept vision has also philosophical and religious meanings. Thus visual image we may have about many things, a person, place, an event, or an object.

Generally human beings are fascinated by the objects or persons of extraordinarily beautiful. It is not only referring to beautiful human beings, but also to man's creations like, the paintings of very sensuous and erotic types, the sophisticated sculptures, beautifully designed buildings, architectually designed and built dams and bridges, machines, cars, planes, ships, arches, printed pictures, books — all these once seen get strong impression in our memory and when we wanted to speak, about any of these things we get the *mental picture or image of it.* All these may appear in our image as we have seen them many days ago.

It is also happening in our life experience to combine the vision to visual sensation and auditory sensation and we may get the image of both visual and auditory. This is what happening in these days in the case of *motion pictures* and *television transmissions.* The physical appearance of the actors with the combination of sweet songs in some scenes may cause both the visual images and the auditory images. Even after many years when we hear the song at once the physical images of the artist appear in our mental picture.

Auditory Image

The ear is the sensory apparatus of our auditory images. You might have attended a good musical performance played by a famous songster or might have seen a beautiful picture drawn by a renowned painter, or a statue carved out by a reputed sculptor, or a scene from a famous play or enjoyed a delicious dinner prepared by a skilful cook. All these experiences sometimes make you to think of them or there might have been an occasion for you to narrate it, for which you are making a mental picture of the event and you enjoy it. Actually this is the past experience about the event. Though the object or the event is not present in front of you, you make it to appear before your mental eyes and describe it in detail. Similarly in the case of the musical performance vocal or instrumental performed by a noted musician or a violinist, witnessed by you often make you to have the mental image of the sweet notes ringing in your ears and enjoy it while describing to others.

Gustatory Images

The tongue is the specialized apparatus of realising the taste sensation. The mental image of gustatory sensations may make us to go back to our past experience to the days when we ate a dinner with several delicious dishes. The taste of the food often makes us to think of it and have the mental image of the incidence. We may have also the images of the taste we disliked. Sometimes you may have discussed with your friends the specific dishes you enjoyed your hostel day dinner, in a marriage dinner, in a birthday dinner, or some marriage anniversary dinner and so on. In all these you were able to bring the image of the food from your memory and discuss it. Therefore our sensation of taste once we relished and enjoyed is not lost and we were able to make the image of it mentally.

Olfactory Images

In our experience of enjoying the pleasant sensations by making mental images of them, our olfactory sensation is not falling behind all other sensory images. The nose, particularly the part, *olfactory epithelium*, the smell sensory area is often associated with taste. All the food dishes we eat are *tasted with smell.* Everything smelt and then tasted. *Smell and taste* are called *chemical senses.* The fragrant smells of the flowers, the ethereal smells of the fruits, the smells of the delicious dishes prepared by a good cook, the smell of chicken roast or the fish roast, or the smell of the *turkey biryani* and the smells of fresh food can make you to salivate in your mouth. The smells of these dishes and foods may cause salivation in your mouth and they are good appetizers. Generally a tasty food with smells stimulates the appetite. All these smells make good mental images of the scenes.

The garden of the east is noted for its smells. The smells of scents and perfumes of the Arabs were spoken of in the long past. **Shakespeare** (1566-1616) the Great English dramatic poet, made reference to the *perfumes of the Arabia* in his play Macbeth. Lady Macbeth had developed the abnormal symptom of *fixation* and Shakespeare made Lady Macbeth to speak of her abnormalities. She states in one place : "even all the perfumes of Arabs could not change the smell of my hand". Thus Shakespeare had the images of the smells of the perfumes manufactured by the Arabs.

We have the images of not only of the pleasant smells, also of the *unpleasant smells,* the stinkings of the garbage in the rods of some cities often make us to have the images of them.

Tactual Images

The touch sensory images are as effective as any other images in our life experience. We touch every object and examine before we purchase at the shopping centres. Why do you want to touch any object before you buy? It is because, only by touching we can tell whether the materials is soft or hard, smooth or rough. We also test the texture of the shirting or suiting cloth only by touching. The touch sensation is felt by our skin. Our palm is very sensitive for touches and touching by our palm any object we can say what sort of surface it has, rough or smooth and so on. The human skin is capable of realizing four sensations such as cold, warmth, pleasure and pain. The sense organs associated with our sensation of touch are called touch spots. These spots are laying below the outer surface of the skin. Some very smooth objects give us pleasant touch sensation and we may often have the image of this touch. A deep pin prick causes painful feeling and at once we jerk back. Sometimes when we go for a medical checkup, one important test is the blood test and blood is taken by fastly pricking our one of the finger tips with a sharp needle with causes painful feeling. Therefore, whenever we hear about the blood test this incidence forms a mental image. Young men are very much inspired by the touch of young women and they often have images of such touches. Hence we have the images of the pleasant and painful touches of our personal life experiences.

Organic Images

Every sensory experience is capable of arousing the mental image if it had strong impression. The organic sensation is felt in our internal organs. The sensations of hunger and thirst are called organic sensations. The sensation of hunger is felt by us when the stomach muscles are contracted. When there is water deficiency in our body we feel dryness in throat and mouth and we feel thirsty. Persons who were without food for several days often have the images of food and food materials. The Caravans pass through desert areas for many days become short of water and they often get the image of water springs. It is their feeling of thirst for long time makes them to have the images of water. The European empire builders when they were on long expeditions on the high seas often have the images of seeing lands out of anxieties. **Columbus** (1451-1506), an Italian by birth later on became a navigator and an explorer of Spain, who discovered the new world (America) in 1492, had many times the images of the new lands out of fear and anxiety.

The idea we can evolve from the above discussions is that the formation of image is possible only with keen sensitivity of our sensory instruments to objects, persons, places and events. All of us do not have the same type of vivid visual auditory and other sensory experiences all the times. But everyone will have the vivid sensory experience in some areas at some occasions which might occur even without our will or intention and we might utilize such vivid experiences for the formation of images and the same may be used in our symbolic thinking. Authors like **Moskowitz** and **Orgel** (1969) have stated that the symbols are the ideas and symbolic processes are referred to ideational processes and ideas are equal to images.

Images may also be classified into three categories as memory images, after images, eidetic images and associative images.

Memory Images

Text book authors in psychology of the earlier period had been introducing two topics definitely in their books, namely learning and memory. But modern authors write the same topics as learning and remembering. The reason might have been that there is no theory evolved which can be universally acceptable to account for all the phenomena of memory. Therefore, memory is dealt with as a process of remembering what have been learned previously. Memory involves learning, retention and remembering. According to

Garrett (1975) memory is classified under four headings such as fixation, retention, recall and recognition. **Woodworth** (1952) states that memory consists of learning, retaining and remembering what has previously been learned. What Garrett referred as fixation is the efforts taken by the individual for memorizing or getting the impression, which may also stated as learning or retention. Retention is the maintenance of the impression or the fixed matter for a period of time. Recall and recognition involve the revival or bringing back to use or re-establishing the learned material on stimulation. All the four components function together and none of them is independent of the other. But according to Woodworth retention is not an active process, whereas remembering is an active process and remembering involves recall and recognition. But both the processes, retention and remembering are included under the general heading of memory.

In these mental processes involving a memory, where does the memory image stand? Woodworth states that *memory image* is the result of *recalling the experience* or the fact of the past which has *sensory character*. Therefore, according to him *memory image* is the *recalled fact* or experience which is having the realistic sensory character as if we are actually seeing or hearing or smelling an object. The memory image essentially should have the actual *Sensory experience*. Individuals may differ in the vividness or realistic character of their memory images.

We see everyday many things in the outside world. All those objects, persons, events that we saw, were not completely disappeared from our remembering or memory. It is true that we do not remember all in detail, but we remember many of them. All these what we have seen or learned are stored in our memory. But we are not successful in making use of them when we are in need of them. Because it depends upon the person's ability to recall and identify the object or person, or event. Students who read the text books fully fail to recall and reproduce the learned materials at the time of the examinations. But when they come out of the examination hall they are able to recall the correct answer. Therefore, it depends upon the individuals ability to recall.

We restate, sometimes, our past experience after few days in the form of images. We may even narrate what has taken place with all the details. This we may state as the image, it is the mental picture of the event. Sometimes the image may be clear to us and sometimes not clear. But there are persons who can describe everything very well. These persons have very vivid, sharp and clear picture of the objects and events. Some persons may have only some vague impressions, and the memory image may be very ambiguous. They may be going on verifying either is it this, or that. Hence memory images of the people may vary very much depending upon their keen sensitivity in seeing or hearing. It is true that no two persons who have witnessed the same event can give the same statement of what they saw. Certainly, there may be different versions with regard to seeing and hearing. This depends upon their memory image of what they saw and heard. Psychologically, the formation of memory image depends upon several factors like motivation, interest, learning, age and other social contact and the person's personal activities. The verbal statements may vary in describing an event. Both are describing the same event. They remember the event and so there is no change in their memory. But the vividness may vary. If is because of the formation of the mental image and memory image. Therefore, we may state that all images are mental images and also memory images.

After Images

After image is a kind of image, formed in a person *after the disappearance* of the object. We know that when our eyes are exposed to a bright light for a few seconds and immediately experience a short dark interval the resulting sensation is the persistence of that stimulus with less brightness. This is called *positive after image.* This image is *short lived.* The positive after image is a reproduction of the original sensation after the disappearance of the stimulus. Therefore, an after image may be defined as the sense experience that occurs after the stimulus had ceased to exist. Several psychologists like **Swindle, Galton, Binet** and **Terman** have conducted experiments to study after images. Persons who have recurrent after images report of ghosts and apparitions.

Sir Francis Galton (1822-1911) was the first to conduct experiments on this. He had collected data on frequency and *vividness of imagery.* His studies dealt with *memory images* and not after images. Memory images depend upon voluntary recall. The after image does not depend upon any effort to recall it. But, however, we differentiate these two, the memory image has the same basis as the after image.

In Galton's study the subjects were asked to recall the appearance of 'this morning break-fast table' to collect data about the 'pictures in the mind's eye'. He has designed a scale to measure the degree of vividness of the image. He has given zero mark for no image and 5 for an image which had equal clarity to the original object. The results had a very wide range. It was a great surprise to Galton that some had no imagery at all while the others had as vivid images as hallucinations. In Galton's study and in other studies it was observed that visual imagery was very vivid and most common. Other imagery like auditory, tactile, kinesthetic, gustatory and olfactory were in this order and approximate. It was also found that those who were high in one type of imagery also found high in other types. People who were in jobs of *abstract thinking* had their images *below the average.*

In **Binet's** (1857-1911) study it was found that musicians who dealt with rhythms had more *vivid auditory images* than the psychologists had. Variations in imagery have been observed in people. Children have more vivid imagery than adults. *Concrete objects* produce *better vivid imagery* than abstract ideas. But the vividness of the imagery considerably deteriorates when we deal with abstract ideas and meaning. People engaged in *routine activities* have *poor imagery* about their performances. But the same persons when engaged in new kind of activities have better imagery.

It has been observed in the studies of thinking and problem solving that *images* are likely to *occur more* if there is any *blockage* in the *flow of thought.* The problem-solver has more imagery when he finds himself struck up and unable to proceed further. One may find *more imagery* when he *goes to sleep.*

These observations seem to suggest that *training, interest* and *motivation* may have some influence upon imagery. Training appears to improve the quality of imagery and motivation seems to influence the content and intensity of imagery. **Te.man** cites a case of a boy. The boy was very much interested in music and he was strongly motivated to become a musician. He was practicing music for a long time to improve his auditory imagery and he was very much successful. But it was also found that the *training had no effect* if there was no interest and strong inner motivation.

In one of the studies made in the University of Minnesotta on starvation the subjects often reported thoughts of food and their imagery for foodstuff was more vivid than normal human beings. Studies in the clinical field show that men who have strong sex drives have more visions of women and images of sexual embraces. Imagery may be simple and exact copy of one's past experience. They are modified in various ways according to the conditions of the whole organism.

Eidetic Images

In the previous section we have read that there are variations in the imagery. Some persons, particularly children have better image than the adults. But human beings on the whole have the capacity for developing *images* which may be quite *similar to photographs.* These images are called *eidetic images.* Persons who are having such experiences are called *eidetics.*

An eidetic imagery has *unusual clarity and vividness* in character. A person with *eidetic imagery* is said to have a *'photographic memory'.* Therefore, an eidetic imagery may be defined as an usually sharp and *detailed visual representation of something* that has been seen.

This something may be a picture, or a sculpture, or a scene or a page in a text. It enables a person to describe the features of an image in *minute details.* It has been observed that children between ten to fifteen years of age seem to show eidetic imagery frequently.

Psychologists have been showing much interest in the study of eidetic imagery. It is closely related to both after images and to the *actual percepts.* The word *percept* refers to the object to be perceived. **Meenes and Morton** (1936) jointly and independently studied the eidetics and non-eidetics. Meenes and Morton have compared the after-images of the two groups one the eidetics and the other non-eidetics. It was observed that the after images of the eidetics were superior to the non-eidetics in several respects. The images were aroused at once, remained for *longer duration, more positive* and with *more details* in the case of eidetics whereas in the case of non-eidetics they were not. After a week's time the eidetics were found to be again superior to the non-eidetics in their ability in reproducing and reporting their images with details.

A recent study in this line has been made by **Haber** (1969). He has discovered 20 school children out of 500 with eidetic imagery. Every child was asked to look at a picture part by part for 30 seconds. Then the picture was removed and the child was asked to look at the blank frame of the picture and report what he saw in the picture. The children required 3 to 5 seconds to look at the picture closely to produce an image of the picture even though it was familiar to them. The images fade away after describing the picture. They could not prolong the images and report them. They could erase the images by looking away from the blank frame of the picture. There was individual differences in the quality of eidetic imagery. Haber has observed this in his study.

One of the girls served as a subject in Haber's study was able to move and reverse images could reproduce them after several weeks of the study. Further, three children were able to produce the eidetic images of three dimensional objects. Some children were able to form a new combination of the picture by super-imposing the eidetic image of one picture into another. But one interesting result in this study is that the *eidetic children* were found *no way better* than the *non-eidetic children in memory tests.* Eidetic images can be modified by voluntary effort. The two studies made by the authors seem to differ in their results. This shows that there may be several factors that determine the eidetic images of the individuals. Further studies on this phenomenon may provide us with much better knowledge.

Associative Images

We have been mainly looking into visual type of imagery. Imagery may be visual or auditory type. It is also possible to have imagery of every sensory experience. But it is also likely that every image may be associated with every other image with the same modality and also with the image of other sense organs. It is easy to associate the visual and the auditory images of an object. For example, if you are asked to form the image of a musical instrument, the violin, first you may have the visual image of the instrument — the physical appearance and secondly, the auditory image of the instrument. You might have witnessed a performance in which a famous violinist might have made you to form the image of this instrument the moment you hear the name of it. Those two images may be aroused more promptly.

A mental picture of a delicious dish like a chicken roast or a sweet meat like jankery is followed by gustatory experiences. Here the visual image is associated with taste image. There are incidents in which three images like visual, gustatory and olfactory may be associated. A skilful cook may prepare a delicious dish for a dinner colourfully combining it with palatable taste and pleasant smell. A person who ate such dinner might have formed all the three images when he had the occasion to remember the dinner.

Therefore, we could see that it is only the *sensory experiences* that are *arousing images* of various kinds. Experimental evidences show that one sensation may arouse an image of another sensation. This may take place as a result of conditioning. If you have learned to associate a good taste of a dinner with a pleasant smell, the very smell of it even without seeing the food would arouse the image of the food and taste. In this case you have been conditioned to the smell and the smell arouses the image of the food and taste. In learning experiments of various kinds shocks are introduced. Students who serve as subjects in such experiments often hallucinate to shocks even when shocks are not administered in certain learning experiments. With regard to vision and hearing this kind of association between sensory modalities is quite common. There are people who could translate the sounds heard by into visual pictures. A *sweet song* heard in a radio programme could *arouse the image of the person* who had sung the song. Some people have the tendency to *associate colour* with musical notes. This kind of tendency is present in every culture for a long time. But, what colour is to be associated with a particular musical note requires elaborate studies. These kinds of associations are called *cross-modality associations.*

Psychologists have shown special interest in this kind of phenomenon. They want to build up special terminology in this area. *Synesthesia* and *chromesthesia* are the two important terms they use to indicate the cross-modality associations. Synesthesia is a general term used to indicate the form imagery in which a person experiences one sense in terms of another. Synesthesia refers to the vivid image aroused by a stimulus of one sense in a different sense modality. Chromesthesia refers to the occurrence of the image of colour in association with non-visual stimulus. In this state colour is seen when sounds are heard. Chromesthesia is also called *colour hearing.* Among these two forms of imagery synesthesia is widely found among children and they are also called eidetics. People with colour hearing often associate high tones with bright colours and low tones with darker shades. Some psychologists are of the view that creative artists have synesthesia to a greater extent than the average. It is assumed that the artistic mind is having greater familiarly in arranging these sensory experiences. Science and technology facilitate this state of mind.

Imagery and Brain

Is there any relationship between images and the neural structure? Where the images are located in brain? There are three possibilities through which we can answer this question. According to experimental results of **Roland** (1993), the very same *neural structures* that produce the *movements* could also *produce the images.* The other possibility is that only *a portion* of the structure that is responsible for *movements* could produce images. *The third* possibility is that there are fully *independent areas* of the brain, that **produce movements,** produce also images. Once it was thought that the third possibility was most likely to produce images. But recently, it is felt that among the three, *only the first two* are of serious concern in producing the images, Roland performed the first experiment in which the *imagined movements* and the *real movements* were compared when the region of cerebral blood flow was regulated. In this experiment subjects were made to imagine a series of finger movements from the *premotor cortex* was active. When they performed the movements, both the *premotor cortex* and *motor cortex* were found to be *active.* This experiment revealed that imagined and the real movements are represented in the premotor cortex area.

It is observed in this experiment that the second possibility that only *a portion* of an area is involved in forming the images of that movements. Similar results were obtained in one of the case studies of patients with visual system damaged. **Moscovitch and his coworkers** (1994) conducted experiments in patients who had severe trouble in identifying real objects, but they were able to maintain visual images. These studies suggest that the *cortical visual levels* are entangled in perception and the *later visual levels* are entangle in *forming images.* **Roland** and **Gulyas** (1994) provided some more evidence that it is the *higher level of cortexes* that are responsible in forming both memory and images. A different view has been expressed by **Farah and his coworkers** (1990). They found three kinds of evidences from their studies of visual system. According to them the *same structures* that are involved *in perception* are also involved in *forming images* in the absence of visual stimulation.

Are there kinds of images? In testing this in the case of **Ski Jumpers,** it is found that there are *two kinds of images.* These players can imagine in two different ways. They can use a process that has been stated as *internal imagery.* This is a first person process in which the players imagine that the movements were made by themselves. They can also use the process alternatively which is called *external imagery.* In this they see *some person is making the jump.* In the first case, the imagery they had was of the movements they themselves have made, In the second case they were imagining themselves of some other person. They see that person, they see the jump. **Jeannerod** (1994) refers to the first kind of imagery as *motor imagery.* In this it was the self in action. The second kind of imagery was really the *imagery of object,* the jumper and the surrending area. It is not known the part of the brain involved in these two kinds of images. But it cannot be the same region producing the two kinds of imagery. The internal imagery of movements may be the use object — *location system* in which the *parietal frontal cortex* system is involved. The external imagery of movements which is in relation to objects may use *object-recognition.* System in which the inferior temporal frontal cortex system is involved in it. Roland says that each brain system is involved in *a group of three functions* such as *action, memory and imagining.*

Active Imagination

Imagination is generally taking place silently and passively in the mind of an organism. But it may be also sometimes vigorous and active. **Jung** (1875-1961), one of the associates of Sigmund Freud used the term *active imagination* for the process we need to cultivate to understand the unconscious. It involves letting *ideas unfold without interference.* This seems to be easy. But it is one of the hardest things for us to achieve. It is because consciousness usuality cannot help but only interfere.

One way of engaging in active imagination is to take some clay or plasticine and try to mould it by our hands. But the shapes take place *spontaneously,* instead of trying deliberately to make a particular objects, when we feel that a satisfactory shape has emerged, better to leave it at this stage. After a day or two let us go and see what it represents. Just think of a creature or some scenes you might have encountered in some stories, myths or films. What does it tell you about your work. The shape evolved has taken place spontaneously. This means our idea has taken this shape which is the result of your imagination. Since we see *something substantially* we may state that it is an *active imagination derived from the unconscious mind.*

Strength of Imagination

The human imagination flows sponteneously which has no boundaries. It is as fast as anything else. The details and vividness of our imagined or remembered imagery depends upon mainly on our *brain's activity* to the imaginative process. Because it is observed that many areas of the brain are involved in creating an imagined scenario. Any competing cognitive activity may divert resources away from the task. In order to have a daydream successfully we must *keep our mind clear* and attend to it. Because internally generated experiences or ideas are rarely as rich as those created by the outside stimuli. The brain mainly functions to give *primary attention* to the *outside world.* If daydream were as tangible as the real experience then we may fail to observe the things that are real in the outside world. The mind's creation may take fantastic shapes, sometimes may not be in line with the real world.

Daydreaming may be considered as passive imagination. But daydreaming is *difficult in busy exciting environment* where there are many distractions. It is easy where there are no or very few distractions. When attention is distracted no imagination takes place. A creative imagination needs *concentration of attention.* A passive imagination may turn into active imagination, which results in creating something which did not exist previously. The *creative imagination* gets more *power and strength* in mind of the person which may lead to some constructive work.

Eye Movements and Imagination

The recent experimental studies conducted on sleep and dreams bring out information relating to images and imaginations. In the experiments it was observed that the *eye movements* reflected the *mental activities.* But the experimental subject's were conscious of it. The eye movements reflect the subject's thoughts while they were in awaken state. When the person looked at the objects, the eyes were moving. A group of psychological researchers like **Grunin and Mostofsky** (1968) and others were very much interested in the study of eye movements in *visual perception* since every *thought process is influenced by visual perception.* Another group of psychologists, **Antrobus, Antrobus** and **Singer** (1964) also studied the "Eye movements accompanying daydreaming, *visual imagery,* and thought suppression." The subjects were 24 women college students. Each student was laid on a bed and was told that it is only a research on thinking and nothing else. They were led to believe that the electrodes attached were to measure the brain waves.

These students were first asked to choose an important *secret wish* and to freely *realize* the wish and they need not reveal it. Secondly, they were asked to suppress it or get rid of it from the mind. Each wish and each suppression has taken one minute. This activity was repeated *three times* for every subject.

When the subjects were engaged in the realization of their secret wish there were on the average of *37.8 eye movements* per subject for *each minute* (60 seconds). Thus for three minutes ($3 \times 60 = 180$ seconds) the *average eye movements* for 24 subjects while they were realizing the secret wish were 37.8. When the subjects were asked to suppress the wish the difference in the eye movements was not so great. Greater activity of the eye was associated with the effort to suppress the secret wish.

Then in the next stage each subject was asked to engage in three one minute periods of *relaxed thinking* and three one minute periods of active thinking. During relaxed thinking the subjects were asked to allow their thoughts to *go slowly,* but during *active thinking* period to make their thoughts *as fast as possible* that is to solve many problems. The eye movements occurred on the average during active thinking were 51.6 and during relaxed thinking the average eye movements were 41.8. But the *eyeblinks* data during *active thinking* were on the average of *72.7 blinks*. But during the *relaxed thinking* the average *eyeblinks were 38.7.*

In the next stage of this experiment the subjects were asked *to imagine some specific scenes.* Some of the scenes they imagined were involved *motion* scenes such as tennis match or busy traffic intersections and so on. Some others imagined *static scenes* such as *orange on a table* in a dark empty room or a mountain on a distant horizon. Each scene was imagined two times, one with *eyes open* and one with *eyes closed.* In the case of *motion* the eye movements occurred on the average of 63.9 times, while the *eyes were open* and for the *static* scene on the average was 33.6. times. When the *eyes were closed,* the eye movements for *motion* on the average were 92.4 and for the *static* scene the average eye movements were 67.2.

In both cases imagining the *moving scene* produced higher scores for movement than for imagining a *static scene*. This means that imagining *movement* scene required *more eye movements,* while the eye movements for imagining *static* scene were *less*. In the case of eyes open the score is less than the score for eyes in open state. While imagining *scene of motion* the score for the *closed eyes* is *higher* than the score for the open eyes.

In imagining *motion* scene the *eye blinks score* (44.4) is higher when the *eyes* the are *open* than the score when the *eyes* are *closed* (27.6). While imagining *static* the subjects' the *eye blinks* score is *higher for open eyes* (27.6) than for the *closed eyes* (16.2). This experiment highlights certain important results. The human beings when involved in the mental process of imagination two important physical changes are taking place. When there is *active imagination* the *eye movements are higher* and the *eyeblinks* are also *rapid.* This is experimentally verified by using electrodes and recording the changes that took place in the eyes while imagining.

PART SIX

EMOTION AND MOTIVATION

17. Feelings and Emotions

18. Motivation and Behaviour

19. Frustration, Conflict and Adjustment

Emotional Expressions

Physiological Motives

Psychological Motives

Human Motives

17

FEELINGS AND EMOTIONS

Introduction

All living creatures are behaving in some manner according to the nature of their physiological components and capacities. Man is a behaving organism, bestowed with much better physiological structures and capacities than any other organism. Man's behaviour is expressed in various kinds of cognitive, conetive and emotional activities. All these activities are based upon a psychological process which may be stated as set or motive. The organism is set to behave or act. It is for achieving its goal. An organism has a 'desire' or Wanting'. It is a desire to achieve a goal. The goal may be physiological like getting its food, water and shelter or psychological like escaping from some danger or attacking the enemy. **Woodworth** (1952) states that there are two problems inherent in the desire, one develops.

The first problem is the problem of motive. This involve what the individual wants to achieve? The second problem is the feeling and emotion. This involves the internal state of the organism. The first problem is rooted in the bodily condition of the organism which is physiological. The second problem is rooted in the feeling aspect of the organism which is more of psychological. The motive or set is directed to achieve the goal seen in the activity of the organism. Since the second is the feeling aspect it is short lived in the organism. Therefore, we may state that the desire one develops includes both the aspects and they are interrelated.

The feeling aspect of the behaviour of the organism is common to all living vertebrate type of organism. The term is used often in the common sense world and human beings are using this invariably on all occasions. How do you feel about the incident? He was sick for sometimes, how does he feel now? My feeling is that Ram is wrong. These are the ways the term feeling is used in the common sense world. The early psychologists were invariably discussing this phenomenon in their studies and books. For sometimes, feeling and emotion were jointly discussed. But modern psychologists, particularly after 1960s, they dropped this term, 'feeling' in their discussion and deal with only emotion. But generally all human beings are invariably using various feelings and shades of feelings in their everyday life. There are hundred of words used in the everyday life by human beings synonymously to the term feeling. Modern psychologists interprate all the terms referring to feeling in terms of emotion. Therefore, these two terms are jointly stated. All the names of feelings and emotions can be brought under two broad categories as pleasant feelings and unpleasant feelings or feelings of pleasure and feelings of displeasure.

We may state here some of the names of feelings and emotions. There are several pleasant feelings with different shades of feelings. Similarly, there are several unpleasant feelings with different shades. Let us enumerate a few of them.

Pleasant feelings : Feelings of happiness, pleasure, joy, love, affection, delight, elation, satisfaction, desire, wish, longing, hope, eagerness, merry, lively, agreeableness, gratifying, mirth, amusement and so on.

Unpleasant feelings : Feeling of unhappiness, displeasure, pain, hate, dejection, dissatisfaction, discontent, sadness, sorrow, grief, agitation, anxiety, worry, fear, hatred, anger, fright, terror, shyness, doubt, horror embarrassment, disappointment, rage, resentment, disgust, aversion and so on.

In order to explain the different shades of feelings and emotions we have presented only some words referring to the pleasant and unpleasant feelings.

Wilhelm Wundt (1832-1920), the German experimental psychologist, who set up the first Psychological Laboratory in 1879, studied the phenomenon of feeling deeply and formulated a three dimensional feeling system. According to him a man's feeling at any moment may be discussed in anyone of the three levels. Wundt's three-dimensional feelings involved both pleasant and unpleasant feelings at each dimensional level. The three dimensions are pleasantness-unpleasantness, excitement-inertness and expectancy- release.

Pleasantness-Unpleasantness : These feelings are generally felt by all and they are universally accepted levels. The pleasantness-unpleasantness feelings can be associated with any of the pleasant and unpleasant feelings stated above. There cannot be any difference of opinions with regard to the feelings involved in the dimension of pleasantness-unpleasantness feelings. In the everyday life every human being one way or other might have experienced one or two pleasant feelings and one or two unpleasant feelings. This dimension is the expression of natural characteristics.

Excitement-Inertness : These feelings are the expressions of different levels of feelings. Excitement generally refers to the process of exciting or state of being excited. A person or thing is by some stimulus. It mainly refers to stimulated state or thrill. The other feeling, inertness refers to the absence of inherent abilivity to move or act. It is mainly referring to inaction, laziness or aluggnish. It is a limited ability to react enthusiastically. Thus we could see the difference between the feelings of first dimension and the second dimension.

Expectancy-Release : This is the third dimension, mainly relating to something hypothetical in nature. Woodworth states that this dimension refers to selected tenseness-release. The third dimension is different from the first and second dimension. The organism is activated for some action vigorously. The organism is anxiously waiting for some action. It is just like the second or third runner in a relay race getting ready to receive the stick from the first runner and to start. It is expecting for something anxiously, loaded with some urgency. It is a state of stress, tightly. A rigid or taut state the organism is facing. The state of feeling of release is the opposite of the tense or rigid state in which the organism releases itself from the grip of the tenseness. It is a state of relaxation free from tension or stress (Fig. 17.1).

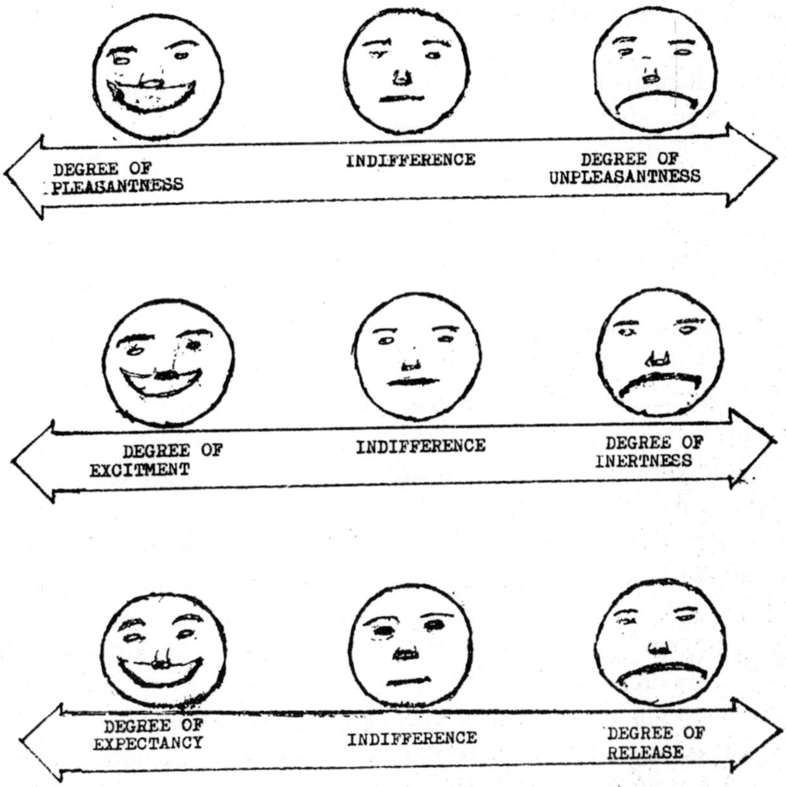

Fig. 17.1 : Judging the Feelings of Pleasantness and Unpleasantness

This is how Wundt's three-dimensional feelings can be explained. These three dimensions of feelings differ from each other and they are carefully analysed and classified. These three dimensions can be experimentally illustrated and distinguished (Fig.17.1).

Garrett (1975) states that feelings accompany complex of sensations. Technically, feelings refer to *affects, affectivities.* It is assessing the affection in an organism or individual. It may be aroused in any of the areas of tactile, organic, visual or even auditory. This aroused sensation is always judged to be more or less as pleasant or unpleasant. We might have experienced various types of feelings in our everyday life.

Generally, in the common sense world and in the psychological world the term 'feeling' is referred to various shades of emotional expressions in everyday activities. But all our experiences possess some feeling-tones which is mostly pleasant or unpleasant. In any of our life activities whatever we experience and express has an emotional value in the situation in which a person is involved. Many of our emotional experiences when expressed are called feelings and the feelings expressed in some form, are attributed to emotions. However, we try to bring out a distinction between feelings and emotions, a clear cut difference is very difficult. It generally rests on the way in which a response or reaction is described.

The Origin and Development of Feelings

There is some difference of opinion with regard to the origin and development of the feelings. Human beings react or respond to some event or incident in the life situation in some manner either with pleasant feeling or with unpleasant feeling. How does it arise? The question often raised is whether the expression of the pleasant or unpleasant feeling is innate or inherent in the organism or learned or as the result of training?

One of the everyday experiences that human beings face is the taste in eating and drinking. Whether one is a child or an adult the taste is realized and expressed either it is agreeable or disagreeable or pleasant or unpleasant. Some of the stimuli we come across in the everyday life are the sweet tastes, attractive colours, harmonious tones, soft touches. These are liked by us because of the characteristics of the stimuli. At the same time some of the stimuli we come across in our everyday life are bitter tastes, dull colours, loud noises, irritating touches. These are not liked by us because of the nature of the stimuli. These likes and dislikes may be applicable to the child and also to the adult person. Psychologists who have conducted experiments on these phenomena state that it is due to the inherited characteristics that we like certain stimuli and dislike certain stimuli and it is inborn. We may also state that it is due to the physiological make up. The physiological make up of the organism is such that whether it is human or animal liking or disliking a stimulus occurs automatically.

The liking or disliking of the feeling tone of colours and other stimuli is determined to some extent by the age, experience and culture of the individual person. Experiments have been conducted on colour preference among the people of different age groups. It was found that the feeling tones of colours seemed to change to some extent with age of the people. Red colour was preferred by most of the people of both sexes to other colours. Red colour was liked by them because of its bright and stimulating qualities. Purple colour was disliked by most of them because of its dull and depressing qualities. They disliked purple colour because it is associated with unpleasant event of mourning and old age. Blue and green colours had pleasant effect in both sexes. Because of this, bedrooms and hospital rooms are painted with light green or pale blue-green colours. These colours normally cause pleasant feelings and they are visual, pleasant to see and to feel.

It has been observed that young children and primitive people generally like warm colours like red, yellow and orange. Adult persons prefer green, blue and grey colours. Just like colours produce pleasant feelings the taste of different food materials produce pleasant and unpleasant feelings. Generally, children prefer mostly sweet taste to sour, bitter and salt tastes. It is not only children, but also adults of both sexes prefer sweet to bitter and salt tastes. Smell is another sensory stimulus which can cause pleasant and unpleasant feelings. The odours of fragrant and spicy are felt more pleasant then the odours of putried and burned. In all the cosmetic materials all sorts of pleasant smells are added. People feel to be smelled pleasantly by others and they also wish to have harmonious sounds and produce these in various ways.

Human body is always exposed to some kind of events, incidents or activities in some situations. In all these man somewhere may get into some painful event or action. Pain is normally rated as the intense unpleasant experience and people usually try to avoid it. But some type of mild painful experience is felt as pleasant. It is always viewed by some psychologists that the more familiar experiences are natively pleasant to young children as they are associated with them right from the time of birth. Similarly, the unpleasant experiences are natively felt as unpleasant as these experiences were not felt right from the time of birth. From the foregoing discussions we may understand that in all these situations when the pleasant and unpleasant feelings are felt, learning seems to play an important role, though certain feelings appear to be innate or inborn. The origin of feelings, therefore, seem to rest on both innate and learned.

With regards to the origin of feeling, studies were conducted to find out the origin or the source of the feelings of pleasure and displeasure. Certain things are pleasant even when they are not desired for them in advance. When we are thirsty and hungry a cool drink and food may give us pleasant feeling after drinking and eating. Here there is a motive, when a motive is set up and if it is fulfilled there is a pleasant feeling. When the motive is not fulfilled there is an unpleasant feeling. In such cases, there is personal pleasure or displeasure.

There are certain things common to many people and the same incident may cause pleasure to some and displeasure to others. For example, in a cricket match two teams are involved and thousands of people are witnessing the match. If a player hits a boundary or sixer half of the spectalors yell with joy and when the same player was bowler and made out the same of spectators are groan in agony with feeling of disappointment. First there is an arousal of motive which may give pleasure on gratifying it and there is a displeasure when the motive is thwarted. Any achievement brings pleasure. After setting up a motive one must strive hard to have the joy of achievement. First, one must strive for the achievement and not for the joy. It is because one must concentrate on the task he has planned. Then the person must be fully observed on the task and stride through success and then joy of achievement comes as a surprise. Studies have not been made in getting the complete list of common pleasures of human life. But a list of names has been collected from the studies made by several persons. **Cason** (1980) had undertook the task of gathering the list from many persons the feelings of common annoyances. The list is so lengthy that it cannot be presented fully in any contest. **Woodworth** (1952) presents a short list of the names of the feelings of the common annoyances.

The expression of the feelings of common annoyances are mostly individual's personal reaction in a situation where the person had the occasion to contact another person or a group of people collected in a situation. Every statement given by an individual is different from every other individual in varying degrees of annoyance. We may quote a few of the common annoyances expressed by some individuals; continually criticizing others, dominating the conversation, always pulling the child, always talking about his illness, assumption of superiority, always giving advices to others, paying no attention to what others say, "always interrupting when I am talking," always nagging at me, trying to tell me something I don't want, untidy room, cloths etc., a window rattling, starting the motor car with land noise.

But everyone like this might had expressed some feeling of annoyance and we are no doubt making everyday one or two such reactions. In our everyday life, our expression of the feeling of annoyance may go into several categories. A correct classification of these feelings, whether it is annoyance or pleasant is not easy. These feelings may have different shades and also may reveal the personality traits of the individuals.

Types of Feelings

There are occasions in our everyday life to react with the same type of feelings what the other person has expressed. It is a natural tendency to have the same feeling what our companions have felt. We react with pleasant feeling to the happy companions and red or gloomy feeling to the companions with gloomy feeling. When others look very tensely and silent, you have the feeling of curiosity to know why they are looking like this. At this situation you too look like them and reveal the same feeling. "It is something very sad and painful that has occurred and they are in a very depressed state". On knowing the consequences you also get into the feelings of those people. This we may call feeling for a thing or feeling of those people. You also feel what others are feeling in that situation. In other words, we may call it as *sympathy.*

Feeling of sympathy : The feeling of sympathy for a thing or for a person is taking place in a person without any direction from others. It is a spontaneous reaction that any human feels. Even animals have this sympathetic feeling on seeing another of the same species in trouble or in danger. Studies have not proved whether the feeling of sympathy is innate, or instinctive or imitative response or learned behaviour pattern.

But we cannot deny the positive response that spontaneously arising in us on knowing that our friend is in a very difficult situation. The motive involved in this is the desire to have the same feeling as our friend is feeling, whether it is for liking something or disliking. In sympathizing fully with the feeling of another person and putting yourself in his place and to feel what he feels, needs certain amount of imagination and you must know how to deal with the situation which is affecting your friend. Sympathy does not end with the feeling for the person, but also involves helping the person to get rid of him from the troublesome situation.

Empathy : The other feeling similar to the feeling of sympathy is empathy. Human beings everywhere are having the experience of the feeling of sympathy and empathy. Empathy is a feeling into a thing or person with whom one is associated. Empathy is manifested in the behaviour of a person who is involved in the situation. Empathy may be also defined as the feeling of a person into the feeling or into a work of art or into the activity of another person. In empathy we identify with the person in his feelings and works. This expression of feeling is visible. We can observe the behaviour of the person visually who identifies his feeling with the other person.

It is common among the spectators of a foot-ball match, played between two important teams or nations. In certain cases the spectators become highly excited and enthusiastic that their favourite team should win the match. Every spectator puts himself in the activities of his favourite player and when the player kicks the ball into the goal the spectator becomes so much excited that he kicks the person who is standing before him by putting himself in the place of his favourite player. This is how the feeling of empathy is expressed physically. This action may be stated as what the spectator has imagined and expected in the player. It is not the player in the ground who is playing, but the spectator himself is playing in the ground in his imagination. We can generally observe in all the exciting games and athletic contexts that many spectators express their feelings of empathy in some form.

While seeing a motion picture or a drama or play and also while reading some classical novel it is quite but nature that we identify ourselves with the feelings of the hero or heroine or with a favourite character and become worried when the hero or heroine is in difficulties or in danger and get elated when events end happily. The type of feelings one has developed can be called as empathy.

Boys are very much interested in soaring the kites in the sky and there may be keen competition in taking kite to a very high place in the sky. The boy gets so excited that his kite is above all other kites and he feels that he himself is in the sky and not the kite. The boy puts himself in the kite and feels very proud. This type of manifestation of feeling is empathy.

Aesthetic Feeling : The art and beauty are supposed to be the objects of appreciation and they arouse man's deep sense of creative artistic skills. When a person sees the ancient monuments which are preserved for centuries arouse his feelings of artistic taste and also arouse his imagination and make him to go into the details by putting himself in the minds of the artists and enjoy them. Beauty is always associated with the feeling of pleasure. Beauty is the source of pleasure. The ability to understand the qualities of beauty and to appreciate and enjoy may be said as the aesthetic sense. Everybody is endowed with this ability, but it varies in degrees. Some may have this to a greater degree and some may have it at a smaller level. The primitive man was not lacking in his aesthetic sense and activity, caves were curved out in mountain rocks and decorated them with colours and they were used as living abodes. The vanished civilizations left out behind them unimaginable architectural monuments.

Any design artistically drawn may reveal beauty, Generally art is defined as the revelation of beauty. Art is always associated with sculpture and painting. A skilful painter could paint a human figure in a fantastic manner that it could arouse the feeling of empathy in any human observer. Similarly the sculpture could chisel out a human figure, the figure of a horse or an elephant, a cart in a most fascinating style which could arrest the hearts of the outlookers and make them to get into the work of the art, revealing their feelings of empathy. The aesthatic interest of a person or the feeling of empathy may spring out from the internal needs of the organism. But it is not always taking place on all occasions. It is the objective world that provides stimulations and arouses the organism to reveal the feeling of identification with the event or situation.

Some psychologists discuss the phenomenon of empathy as the process of perception. What factors govern our choices in perception? This is the basis under which the aesthetic values are discussed in perception. It is observed from the experimental results that symmetry in forms and objects people prefer. Symmetry combined with decorations are regarded as most pleasing. The most pleasing rectangular figure should have the ratio of 3:5. The same ratio is also applicable for ellipse figure (Fig. 17.2A, 17.2B).

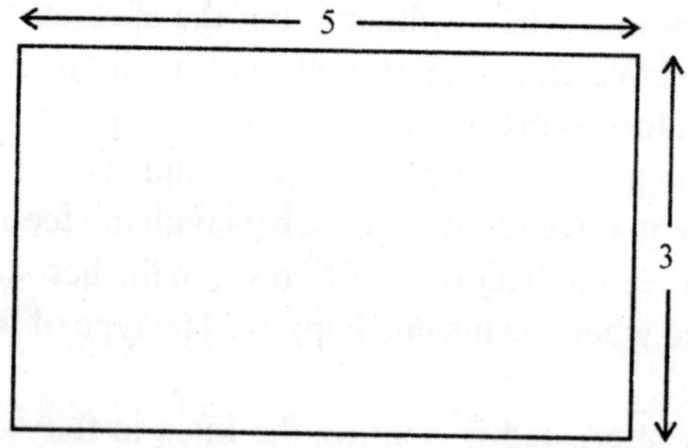

Fig. 17.2A : Rectangle Design of Golden Section — Symmetrical and Balanced based on the Ratio of 3:5.

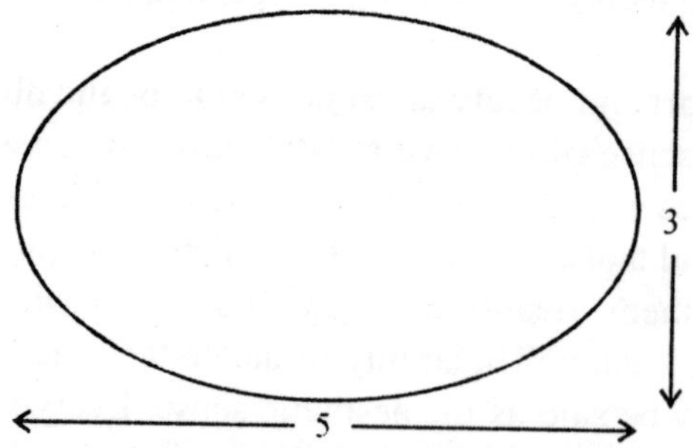

Fig. 17.2B : Ellipse Design of Golden Section — Symmetrical and Balanced based on the Ratio of 3:5.

This means the length should be 5″ and the width 3″. This gives a pleasant symmetrical figure. This is aesthetically appreciated. This relationship of length and width is known for a long time and this is generally called as *golden section.* The ratio of 3:5 in a rectangle and in an ellipse appear more symmetrical than in any other ratio. It also gives an attractive and pleasing look. Artists for generations have been using this ratio in their artistic works.

The preference for the artistic design of the golden section has been accepted and well established for very long period. But modern artists consider that this golden section of artistic design is based upon the deep rooted Greek civilization and the period of Renaissance culture and not based upon the human needs and the theory of balance and symmetry. Modern artists are strongly reacting aganist the golden section and the *Greek 's' curve.* But the principle of the golden section is not totally rejected from the aesthetic world. Activities in the aesthetic world show that the architectural designs which follow the principles of symmetry and proportion are still very much preferred and applied.

The principles of symmetry and balance when applied in artistic designs were found to be arousing the pleasant feeling of empathy in the hearts of the onlookers. The observer in the state of empathy feels himself into the objects perceived and experiences the pleasant feeling. Unbalanced unsymmetrical objects arouse the unpleasant feelings of strain and tension. The works of art and natural objects with more stability provide the observers the pleasant feelings of joys and hopes. Therefore, empathy, according to Garrett, is the emotional identification of a person's self with the perceived objects. The attraction of the natural objects and the symmetrical and balanced artistic works arouse in us the pleasant feeling, which is the same as the spectators have while witnessing an excited football match. Thus the feeling of empathy is observed in them also visually.

Social Preference : In any culture or society people express certain likes and dislikes on the basis of their preferences to certain groups of people. In the preference of other people or people belonging to other culture they identity their native characteristics. The more similarities they find in the characteristics of the people of other culture or nation they show more liking or prefer them and also develop friendship with them. But when they find more variations in the lifestyles of the people of other country they prefer them less or dislike them. Social preferences or likes and dislikes were studied by some psychologist and sociologist. Begardus (1925-1959) social distance scale is well known and this scale has been used by many psychologists for assessing the racial and colour prejudices.

Though there were several works on prejudice and published by the psychologists and sociologists, the voluminous and the monumental works of **Gunnar Myrdal** under the heading of *''An American Dilemma''* published in the year 1944 is considered to be the outstanding one in this area and revealed the nature of likes and dislikes based upon the pleasant and unpleasant feelings of the American people towards the racial groups. Subsequently several works on different titles were published out on likes and dislikes and on prejudices. But some of the important works referring to social preferences are **Allport's** (1954), *The Nature of Prejudices.* **Ehrlich's** (1971) *The Social Psychology of Prejudice* and **Arnold Rose's** (1950), *The Negro in America.*

The Kernel of the subject matter of these works are referring to only social preferences. The term mainly used in these works has been *prejudice.* Generally prejudice means an opinion formed already against a person or a group of people or object or race without verification of the facts and expressing the feelings of likes and dislikes or pleasant or unpleasant feelings. This shows the type of social preference a person or a group of people expressing against another person or another cultural or social groups. This may be pleasant feeling or unpleasant feeling. It is the ill feeling or displeasure an individual has developed against a person belonging to a different racial or social group. The individual's feeling of displeasure or hatred, if continued for long time, it may also influence other persons of his group and they also may project the same ill-feeling or hatred against more number of people belonging to different social or cultural group. In the long run the individual's feeling of displeasure or hatred becomes a social or national problems. This was how the racial or colour prejudices developed in America.

The Social Distance Scale designed, developed and used by Bagardus brought out many interesting results in the issue of social preferences. The scale has seven steps ranging from strong liking to strong disliking. The liking indicated acceptability and the disliking indicated rejection. The most favourable state is : "*Would admit to close Kinship by marriage*". The most unfavourable state is : "*Would exclude from my country.*"

This procedure can be applied to study the type of feelings of any racial, national, religious and communal groups and the nature of the pleasant and unpleasant feelings can be estimated. The different levels of feeling tone in the society have caused diasaster consequences to individuals and to the groups. Persons with strong determining characteristics are stated as *Authoritarian Personalities.* They mostly reflect with strong unpleasant feeling or hatred towards other people who are not of the same type. During **Hitler's regime** in Germany, the **Nazis** were trained to develop strong unpleasant feeling or hatred against the Jews, which resulted in the mass killing of the Jews with the view of destroying the whole racial group of the Jews. The same type of feelings was practised in America but in different style against the Negroes, expressing the feelings of hatred by **Iynching.** The consequences of the expression of unpleasant feeling in wild forms are revealed in action by individuals against other persons. Thus the social preferences of feeling have played havoc in the life of human beings.

Feeling and Motor Activity

Our feelings as they are not having any motor function. Feelings are innate in the organism. But muscular movement or act can produce some external results. The act or behaviour is dealing with the environment. Feeling is always internal and itself cannot produce any external result. But the feeling an individual has felt in a situation makes the individual to move and act in a specific way. But there is always a tendency in man to associate feelings with motor action.

For example, when we are in the midst of an unpleasant feeling situation we try to avoid it by some means. We try to avoid some unpleasant smell, some ugly scene, undesirable persons and so on. If we cannot physically remove them, we avoid them by moving ourselves from the unpleasant situation. When we want to find out a house for rent, we usually look the surrounding area, if it is highly congested area we avoid it and go in search of some pleasant area, where we see everything pleasant. Thus the feeling tone activate us to avoid anything unpleasant and we wish to have everything pleasant.

Feeling and Sensory Activity

Feelings and sensory activities are associated in some contexts in human life. Some psychologists distinguish feelings from sensory activities, but some relate these two for some purpose. Feelings are mainly referring to *affectivities,* followed by complexes of sensations mainly organic, tactual and also visual. These are judged more or less as pleasant or unpleasant. Sensory activity is an indicator, which indicates what is taking place in the environment, but feelings will not provide any information about the environment.

Studies in this area state that some sensations may indicate something about our own body. For example, the feeling of pain indicates something to us. Pain is an unpleasant experience. The question is, whether pain is a feeling or sensation? It is a tactual sensation and also an unpleasant feeling. The stomach pain is an organic sensation and also an undesirable and unpleasant feeling. But sensory activities dealing with the environment are indicators of what is going on and happening in the environment and anything internal in this function is a feeling.

EMOTIONS

What are Emotions?

Human beings are acquiring in their everyday life various types of experiences through their sensory perceptions, reacting to various kinds of stimuli and by associating themselves with others. A man's personal life experiences are expressed in some kind of feeling tones and with some emotional shades. Behind every aroused emotion there is a feeling. Feelings are described in most of the times as emotions. Poets, novelists and artists are using many adjectives to describe the shades of emotional experiences.

They describe their characters with emotional experiences like joy, mirth, anger, grief, love, fear, contempt, pity, repulsive, jealousy, suspicions, wonder, curiosity, distress and so on. A poet, or novelist or an artist can never describe his characters without attributing any of these emotions to them in some shades and make them to express in some situation. Therefore., emotions play important roles in our daily life. Life must be filled with some emotions. Emotions serve as spices to our life. They brighten our lifestyle and also darken our lives.

Emotions have both pleasant and unpleasant end. **Kimble** (1985) states that emotions are the richest aspect of our mental life. The evidence for this is that all languages have enormous number of words referring to emotions and to their different shades. **Emmannuel Kant** (1724-1804), the famous German philosopher, said, *"Emotion is the disease of the mind."*

Scientists are of the view that emotions are different from rational thought. This they state from the information known through our sense perception and from the information stored in the memory. But from our experiences we come to know that emotions interact with all our mental functions. While discussing about emotions we must take into consideration three factors such as, what arouses an emotion, on arousal how the body and mind respond and express their reactions and how our physical and mental responses colour the emotion in the way we experience. Some psychologists treat *emotion* synonymously with *affect*. But it is viewed that affect simply refers to a part of an emotion. Garrett says that *feelings* refer to *affectivities*.

Any attempt to identify and define emotions is a difficult task as emotions are subjectively felt feeling of an individual. Emotions in the form of feelings occur as reactions to some external stimulus events. An operational definition can be evolved if only the same stimulus could arouse the same type of response in more number of persons. In such case we could disregard the subjective experimental aspects of emotion and obtain an acceptable classification of emotional responses relating the stimulus categories. But even such definition evolved on the basis of stimulus categories will have to depend upon the subjective experience.

To have a reasonable basis, experimental studies were made. **Corssman** (1967) states that there were wide individual differences in the experimental results of some specific stimuli and also in the behavioural reactions in the case of a particular feeling. Because of the perversive influence of learning, many behaviourists are of the view that emotion like consciousness and other mentalistic processes are not within the purview of scientific enquiry and therefore, should be dropped from the psychological enquires (Duffy, 1934, M.F. Meyer, 1933). But, **Dunlep** (1932) states that in the case of emotion there may be some question with regard to some specific emotions, but we cannot doubt and reject the realities of the phenomenon of emotion from the field of study.

Every civilization right from the ancient period to the present time has made clear distinction between the emotional behaviour patterns and the intellectual functions in the society. There were no controversies between these two phenomenon at any time. Human emotions cannot be studied in the same manner as we study the material objects. Emotions are expressed in the behavioural patterns with pleasant and unpleasant characteristics. An object or stimulus can arouse either a favourable emotion or unfavourable emotion in an individual depending upon his learned experience he had with the object. For example, a snake can arouse the emotion of fear in an adolescent or an adult but not in a child. The child may even like it for its bright look and may even wish to play with it not knowing that it is a poisonous creature. As the child grown into an adolescent boy he learns the characteristics of the snake and develops the emotion of fear whenever and wherever he sees it. Thus fear is aroused in a person only on experiencing the characteristics of an object.

But the physiological development of the individual organism may arouse the feelings of love and affection towards the opposite sex and on perceiving a member of the opposite sex. The arousal of sex desire and the action for fulfilling the desire may fall under two categories, like love and lust. Love is a general term, used to indicate various types of likes or different degrees of loves and affections. Love as an emotional expression projected not only towards a person of opposite sex whom the person likes very much from sexual point of view, but also towards an object place, occupation, a pleasant house, garden, dress, furniture, food or some specific quality of drink, other human beings with whom a person is associated for long time.

Lust is also a type of emotion, aroused in a person due to physiological development and maturation of bodily factors in which the function of the sex glands become very vigorous and violent. This important characteristic of the emotion of lust is to have the physiological and psychological gratification of extreme sex desire. The emotion of lust may be defined as the extreme form of sex desire or love which stimulates the person for immediate gratification of his sex desire by whatever means which may vary or range from unwilling acceptance to forceful rape. Thus the definition of emotion may involve various factors like the perception of the stimulus object and its effect on the individual organism, physiological development and maturation, the functions of the internal systems like the nervous system, the glandulor muscular and the circulatory systems and the psychological motive of fulfilling the desire of overt action of the organism expressing the particular type of behaviour.

The characteristics of emotions and their various forms have been studied right from the ancient times and they were expressed and demonstrated in literature, art and sculpture and also in various forms of lifestyles expressing all the culture and civilization of that period.

The ancient philosophers have ignored one of the most important problems of modern psychology, because of their extreme love for matters of intellect. The problem is of the *psychodynamics.* Psychodynamics are the basic sources of power behind human behaviour. What impels a man to behave as he does? Why does one person work harder than the other? You may find answers to these types of questions in the psychodynamics of human behaviour. You may find explanations to these questions in terms of hunger, thirst, love, anger, fear, anxiety, ambition, joy, jealousy and similar powerful *human motives.* The simplest dynamic factors arise from the nature of man's biological structure. There are certain constant biological states which must be kept uniform if the organism is to survive. The organism must be constantly supplied with food, water, oxygen and other things. The *organism mobilises energy* and puts forth vigorous efforts to maintain these constant *states for survival.*

Much of human activity is directed to the achievement of goals. But what are the *other dynamic factors* that are *mobilizing energy? They are emotions* such as fear, anger, love, hate, joy, surprise, jealousy and the like, personal motives such as ambition, prestige, power, self-expression and social values such as religion, search for knowledge. Modern psychologists treat *emotions as a kind of motives* and so they deal with these two concepts in the same topic. Recent authors reinterpret emotion in terms of activation and arousal.

Activation refers to general body tension or arousal. It enables *body to move and act.* It is very difficult to study the cause and effect. Because it deals with complex set of internal physiological changes. The patterns and *degrees of activation* are primarily responsible for the feelings and emotions that play a very great role in our daily life.

According to activation theory, emotion is a *state of general excitation.* Activation occurs whenever physiological factors activate a motivated response. The stronger the internal need or incentive the greater the degree of activation. The more desire one has for a job the greater the degree of tension a person develops. Therefore, different forms of motivation produce different patterns and degrees of activation. For example, a *painful stimulus* may produce one pattern of activation called *fear.* A frustrating situation may produce another pattern of activation called *anger*. A sex-related motive may produce a pattern of activation called *love*. Therefore, the primary basic biological drives make the organism to move and act to satisfy the needs and maintain the constant biological states for survival. The second type of dynamic factors are energy mobilizing processes. They are emotions such as fear, anger, love, hate, joy and jealousy. Activation is essential to most motivated behaviour in complex organisms. Activation not only releases physical energy for activity but also it produces the feelings which we call emotions.

Nature and Characteristics of Emotions

We can understand from the introduction given above that modern psychologists treat emotions as motives since emotions show all the characteristics of motivated behaviour. Recent researchers interpret emotions in terms of activation. They are of the view that activation in varying degrees produce different patterns of emotions and therefore variety of feelings accompany activation.

Emotions such as anger, fear, love, hate, joy, jealousy, worry and all their variations are becoming a part of our daily life. Our conversations are not complete and interesting if they are not loaded with some of these emotions. Poets, novelists and artists could not present their subject matters in their works without love and hate, anger and fear, and joy and jealousy. What do we mean by love and hate or anger and fear? It is very difficult to define and describe an emotion, because the characteristics of an emotion is not found in the same manner in everybody. It varies from person to person and from situation to situation.

For example, a smile is the indication of friendship. It expresses a pleasant feeling. This is the general opinion held by people. But, it is also an expression of enemity, jealousy and hatred. Some people express their unpleasant emotions through their smiles. But psychologists try to study the nature and characteristics of emotion from real events of life and natural situations. Several definitions and characteristics of emotions have been brought out from such studies.

Emotion may be defined as a complex and conscious affective experience involving physiological changes and overt behaviour.

Emotions are the stirred up states of the organism. The stirred up states are expressed in some parts of the body like dilated eyes, twisted fist, pounding of heart and sweating.

It is also defined that emotion is a *stirred up state of consciousness,* masses of feeling and impulses. Emotion is not a local response, but a general response of the whole person. Therefore, the whole person, the whole body and the whole consciousness are involved together in the response. Emotion also involves imagination, memory, thinking and perception. Therefore, emotion may also be defined as the stirred up states or responses of the bodily parts of the organism involving physiological changes that express in overt behaviour patterns. But on certain occasions most of us conceal our emotions to some extent. We do this to protect our self-respect and to confirm our social manners or conventions. In such situations certain clues help us to determine another man's emotions.

Emotions possess all the characteristics of motivated behaviour. Every emotion is a goal directed one. In the state of intense fear men are capable of feats of strength and endurance for escaping from dangers. They would be capable of the same even under normal circumstances. Anger motivates great amount of energy in an organism for combating against the enemy. In the case of hate, energy is motivated towards the destruction of the hated person.

Emotions vary in intensity. It may range from mild antagonism to extermination or destruction of the enemy. In the extreme emotional state the person may become panic and refuse to accept any reason or logic.

There is certain amount of persistence and variability in emotional behaviour. In specific emotions such as fear and love and in generalized emotions such as anxiety and worry we realize the persistence of tension and this forces the individual towards action and thus reduces the tension. In certain respects the strong emotions are having reaction in our physiological make up. For example, pounding of heart, stomach disorder, flushed face and the like are the physiological reactions. Usually in the case of emotions tension is aroused by outside stimulus or external situation. The outside pressure is applied on the individual and great amount of tension is collected in the striped muscles and stomach muscles. Some horrible noise, physical injury or some unpleasant thing may throw us completely out of balance. We are upset and enraged. We at once rush to avoid them. All these we call emotions. But emotions differ from person to person in certain situations and time to time in the same person.

The duration of emotion is very short. It exists only for a short moment and then disappears. But during this short period it disturbs the equilibrium of the organism. If it exists for longer duration it may affect the mental health of the organism.

Emotions are not inherited. They are aroused by external stimuli and circumstances. Emotions are developed right from the childhood period.

Physiology of Emotions

When a man is in a highly emotional state he is not in a position to observe it. Because he is out of control of his actions, he is up in the air and he has lost his head. Things may even go to pieces. There is utter confusion. That is why emotion has been defined as the 'stirred up' state of consciousness, masses of feeling of impulses. **Woodworth** (1952) states that emotion is a moved or stirred state of the individual. According to him it is a disturbed muscular and glandular activity, that is how it appears to an observer. Therefore, it seems that the body as a whole is involved in the emotional state. There is tension in the body of the individual, he is clenching his fist, shouts, sets tears and so on. In these there are activities of the muscular, glandular and nervous systems. The body as a whole is upset and agitated. Therefore, let us see what happens in the body in the intense emotional states where there are internal and external changes.

The bodily changes form an important part of emotional reaction. These bodily changes have been widely studied. Emotions are not related to a single bodily part. They involve complex bodily response patterns. Several bodily changes both inside and outside are taking place. All the visceral or internal responses become a part of emotional response. It is experienced by us that emotions are associated with flushing, pounding of the heart, paling, tumbling, shaking, clinching the fist, grinding the teeth, erection of hairs (Goose flush) and so on. There is also change in blood pressure, breathing rate, heart rate, intestinal contraction and glandular secretions. All these changes are aroused by fear, anger, love, hate and similar states of emotions. Several researches have been conducted on the physiology of emotional responses and found that the intensity of such bodily changes goes well with intensity of emotion felt by the individual.

In studying the bodily changes psychologists use the methods of expression. By this means they were able to reveal physiological changes and also measure. The aim of the measurement is to relate physiological (bodily) changes to emotion. These are the muscular activities in the person who is highly excited and these call for increased breathing rate and heart beat. The increased breathing rate can be recorded through the chest movement by an apparatus. From breathing record the individual's emotional state can be detected.

In the intense emotional state there will be increased blood pressure level. This is measured by an instrument called *sphygmomanometer*. This is ordinarily used by doctors in their clinics to measure the patient's blood pressure for diagnostic purpose. But it will not be possible to state what specific emotional state determines the increase of blood pressure. Blood pressure also indicates changes in blood distribution and pulse rate. Blood pressure means pressure in the arteries. It is a complex physiological function. The most important one is the *out put of blood* from the heart. Blood should pass through the small arteries and there is resistance in these. Since these arteries are small the blood pumped forcibly from the heart is not easily passing through. Hence there is resistance. This shows hightened level in the instrument which is the indication of blood pressure (Fig. 17.3).

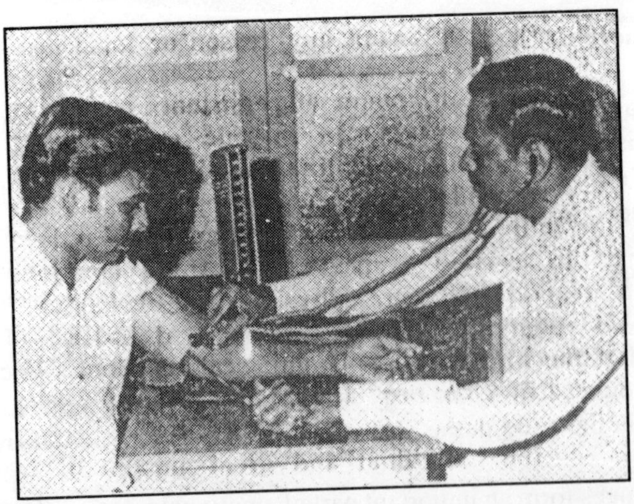

Fig. 17.3 : Emotions : Sphygmomanometer, an instrument for measuring Heart Rate and Blood Pressure. (Emotions can arouse various types of physiological and bodily changes like flushing, pounding of the heart, trembling, shaking, clinching of the fist and also influence the blood pressure, heart rate and glandular secretions.)

Blood pressure increases when the individual is highly excited. Man get excited in his life experiences such as in witnessing and in participating in competitive races, athletic meets, sexual events, sudden electric shock, loud noises, when he is without food or engaged in some heated arguments in controversial issues. **Inban** (1942) states that in speaking lies on some important matter blood pressure rises in 80 per cent of the cases. In the strong and unpleasant emotional state like anger or fear pulse rate and breathing rate increase, blood pressure increases, stomach contractions decrease or stop fully and digestion is affected. The mouth gets dry and also there is muscular trembling.

In one of the studies conducted by **Kubis** (1962) it was observed that there were various physiological changes following intense emotions. The *polygraph* which is a combination of several recording instruments was employed in the study made by Kubis. The polygraph is also called the *lie detector.* This instrument now-a-days is used for criminal detections by the police. This instrument records simultaneously the person's breathing rate, blood pressure, heart rate and electrical resistance of the skin (Galvanic skin response). The important feature of the polygraph is the measurement of *electrical resistance of the skin.* The electrodes are attached to the fingers of the person and changes in the electrical conductance are recorded by the Galvanometer (Fig. 17.4).

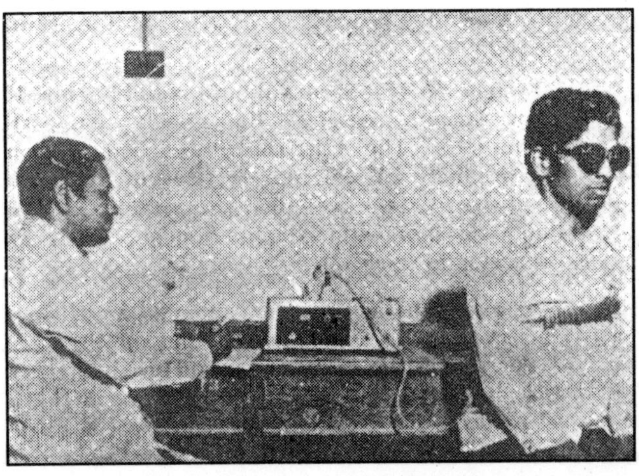

Fig. 17.4 : Emotions : Polygraph, an instrument that measures several emotional reactions.
(All the bodily changes are aroused by the emotions like, fear, anger, love, hate and similar states of emotions which are controlled by the autonomic nervous system. Experimental results on physiological responses of emotion are found that the intensity of such bodily changes goes well with the intensity of emotions felt by an individual.)

The procedure adapted in using the polygraph is to record first the person's normal reactions in the relaxed state. Then the person is asked to answer to carefully worded series of questions. He has to give *the first answer* that comes *to his mind* with 'yes' or 'no' words. Among the neutral questions, some critical questions are put here and there. In between questions usually one minute time is allowed. The recordings of breathing rate, blood pressure and galvanic skin responses are made continuously. It is supposed that the person's guilt is revealed in the recording by *heightened lines* in the graph when answering to the questions relating to the crime especially in the record of Galvanic skin response. Thus the polygraph as a lie detector uncovers the lies through the heightened emotional feelings.

But this method of detecting the lie or crime by using the instrument, polygraph is not foolproof. The recording is not a sufficient evidence to distinguish the culprit from the innocent. An innocent man may react in the instrument more emotionally than a criminal to a critical question. Further the experimental situation may create in the subject a feeling of fatigue and anxiety and this may also have similar reaction in breathing and heart rate. Therefore, the polygraph may be used as a supplementary device or in the preliminary investigations for detecting the lie or the guilt feelings. But the polygraph is a very good instrument for research purposes for measuring emotional reactions to stresses.

Autonomic Nervous System and Emotions

Large number of researches have been conducted on the physiology of emotions and it has been ascertained that the physiological responses of the body are controlled by a branch of central nervous system called autonomic nervous system. The autonomic nervous system is lying outside of the central nervous system. It is not inside the bony skull. Only the central nervous system (brain and spinal cord) is inside the skull. But the autonomic nervous system is connected with the spinal cord and brain through an intermediate nerve centre called ganglia. The autonomic nervous system is not completely independent of the central nervous system. But the autonomic nervous system carries out independently many of the glandular and smooth muscle responses without receiving any stimulus from the central nervous system. But in a sense it is dominant over the central nervous system as the activity of ANS may coerce the activities of the central nervous system.

For example, all the reflex actions, digestion in the stomach, breathing, sweating, circulation of the blood, heart beat are taking place independently without any conscious control of the brain. Primarily the autonomic nervous system consists of bunches of nerve cells and nerve fibres connected to the brain. There are functionally three main parts in the autonomic nervous system. The first one is the cranial division, the second is the sacral division and the third is the sympathic division. The cranial and sacral divisions are called parasympathetic nervous systems. The middle part is called sympathetic nervous system. These two systems function opposite to each other.

The cranial division is located near the top and at the bottom of the body (Fig. 17.5). The cranial is connected with the brain stem, the upper part of the spinal cord and mid-brain. This part plays an important role in controlling the glands of the stomach for pouring out gastric juice in the muscular walls of the stomach and also is active in digestion. It also controls the duct glands and the unstriped muscles of the vital organs. The cranial and sacral divisions are anatomically distinct but their functions are the same (Fig. 17.5).

The parasympathetic nervous system is a system of energy conservation. It functions to build up the reservation of energy of the body and regulates the energy expenditure of the body. It slows down the action of the heart beat, reduces the breathing rate and speeds up stomach and intestinal functions in digestive process and stores the glycogen. The glycogen is a carbohydrate that is converted into energy by the liver. In this way the parasympathetic nervous system is mainly concerned with urgent needs of the organism such as nutrition, oxidation, elimination of wastages, growth of self-repair, hunger and sex. In other words it deals with vegetative activities and accelerates digestion.

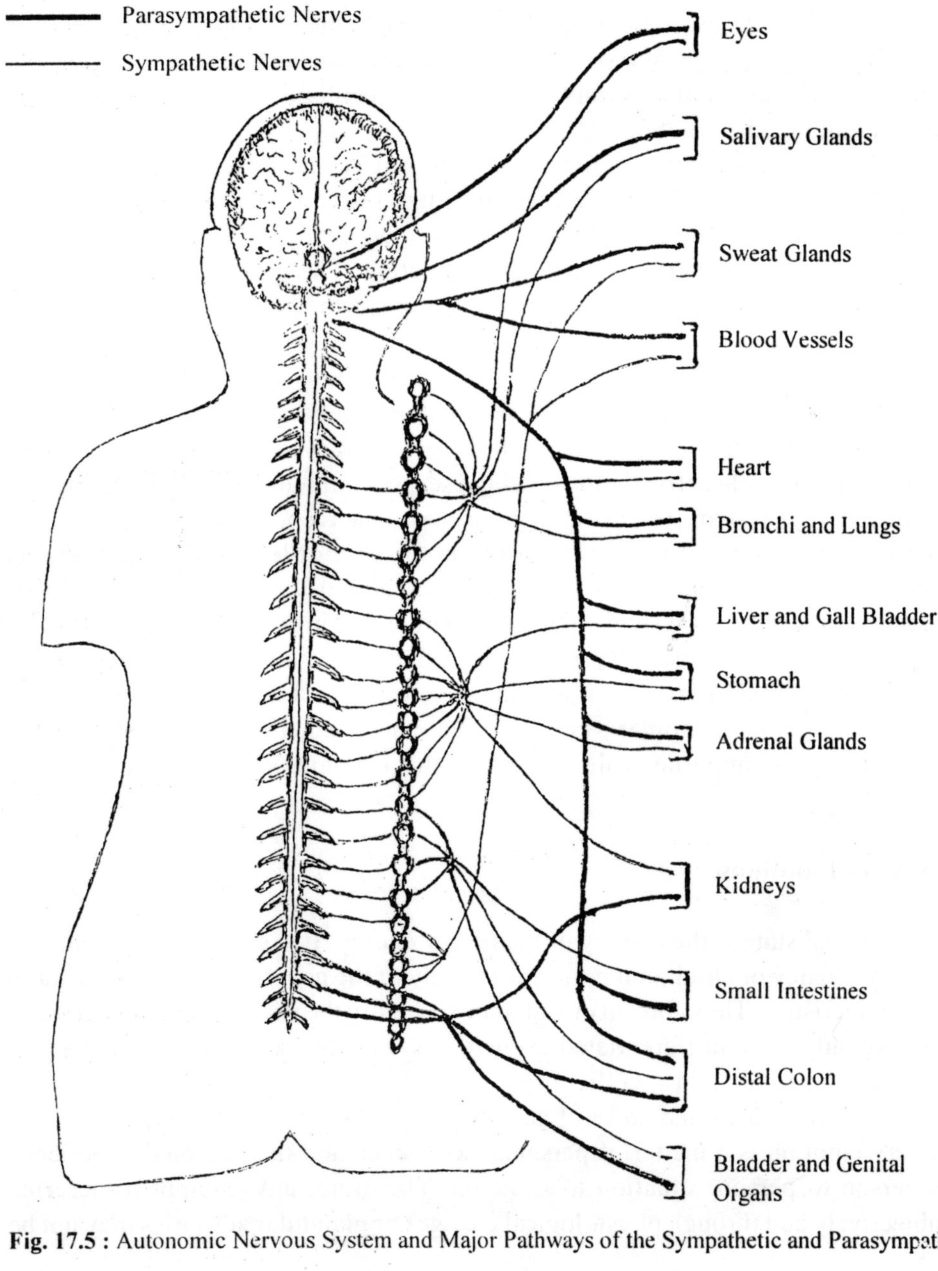

Fig. 17.5 : Autonomic Nervous System and Major Pathways of the Sympathetic and Parasympathetic Nerves

In between the cranial and sacral divisions of the ANS lies the sympathetic division. This is connected with spinal cord near the chest. The sympathetic nervous system is a system of activation. It releases energy for immediate use of the body. It is specifically active in emergency situations. It increases the heart rate, breathing, blood pressure, releases glycogen from the liver into the blood streams and converts glycogen into energy and forces the energy rich blood into muscles of arms and legs. The sympathetic nervous system is concerned with emotional responses as fear and anger. It interferes with digestion and some of the endocrine glands. On emergency situations, it accelerates the heartbeat, increases blood pressure and stimulates the adrenal glands to secreate the hormones called *adrenin.* Secreation of more adrenin in blood means more sugar in the blood and more strength. The bodily energy is mobilized for action. The adrenin is a very powerful hormone which is having much influence in the emotional life of the organism. A very little amount of adrenin in blood may produce several effects in the organism like faster heart rate increased blood pressure, dilation of the eye and so on. The organism gets ready to face the emergency situation. It may be fearful situation or aggressive situation. The person gets emotionally excited or get tension. If it is fear he takes a flight to escape from danger and if it is anger he attacks the object of his anger.

Since the sympathetic and parasympathetic systems function relatively independent of each other, different forms of activation may be dominent in each of these systems. For example, activation following hunger is likely to involve parasympathetic (conservation) system to a great extent while activation following escape from hunger is likely to involve the sympathetic (energy) system. The differences in emotional responses are also due to the differences in the action of these two systems. But in the case of love and sexual activity there seems to be some kind of nervous activity involving both the systems simultaneously.

Thus the autonomic nervous system plays an important part in controlling the endocrine glands and the duct glands. Since the glandular system is mainly responsible for the maintenance of the internal environment any interference by autonomic nervous system affects the functions of the glandular system. In the emotional life of the organism the endocrine system plays an important role. Therefore, this becomes a special interest to the psychologists.

Characteristics of Emotions

The general emotional state is the display of the bodily energy. It is sheer excitement. The emotions of early infancy or childhood period are mostly of the general excitement without any specific characteristics. Therefore, in our attempt to classify the various types of emotions we could observe only the undifferentiated excitements. But the excitements may vary in intensity. This may reveal certain specific responses and we may name them as fear, rage, thrill, surprise, irritation, love and so on. Each of these emotions is a feeling and also a motor set. Every emotion is a matter of personal experience and the personal experience varies from person to person, situation to situation. Therefore, any attempt to describe emotions subjectively and through physiological changes or glandular activities may not be accurate and complete. Let us discuss the characteristics and the expressions of the standard or the primary emotions such as fear, rage or anger, love or affection, joy and grief with available information.

Fear

Fear has been classified as one of the emergency emotions in psychological studies. According to **J.B. Watson** (1878-1958) it is one of the three primary emotions (1930). The other two are rage and love. The instinct theorists like **William McDougall** (1871-1938) and others are of the view that the emotion of fear finds its roots in the instinct of flight which is an inborn tendency in all living organisms (1908). Watson says that fear is a learned response and it can be aroused in a person. Fear is a response to a painful stimulus and the organism by taking a flight tries to avoid it. It varies from person to person and from culture to culture.

Several studies have been made on children and found that they were showing wide variations in their fear responses. Some children were extremely fearful of everything while some were free from fears. If a child met with a few painful events he showed signs of fear, otherwise he was not. Fear can be aroused in a child easily. A loud noise at close range would frighten the child terribly. Watson had demonstrated that children did not have inborn fears of snake, fire and other stimuli believed to arouse fears. Therefore, emotions depend upon learning.

In the developmental stages every child must learn to walk and he is likely to have some painful experiences. He might have fallen down several times and had painful feelings as he had not gained sufficient strength to stand up and walk without help. Therefore, he might have developed some fears, the fear of loss of balance, lack of bodily support, absence of other people or loneliness and so on. While the child has learned to walk he might have developed fear of high and low places as he might have had hurting experiences. Similarly, when one is alone or far away from mother he has the fears of loss of comforts and hunger.

Several names like terror, dread, horror, fright, alarm are also associated with the emotion of fear. Many kinds of responses have been observed in studies made on fears. Some of the responses of fears are startle pattern, turning head, twisting the trunk, shrieking, shrinking, shivering, shaking, trembling, sweating, vomiting, loss of control of bowels and urinating, increased pulse rate and rapid breathing.

As the child grows into an adolescent and adult he learns to deal with fearful situations and threats. Then the fears disappear. But if they reappear at maturity they symbolize as adult problems. Psychologists call these as symbolic fears. It is not possible to remove symbolic fears easily. If we want to remove it we must find out first the underlying basic fears.

The level of fear arousal is not unlimited. There is certain level of limit for our emotion of fear. If the fear causing stimulus is carried out beyond certain point the organism gets disorganized. Both the central and autonomic nervous systems are fully put into action very vigorously. The organism may fail to respond to the situation. R.S. Woodworth says that life would not be interesting without the emotion of fear. According to him fear is a kind of thrill. Amusements, adventurous activities and sports will have no fascination if there is no thrill. There is a joy when you escape from a dangerous situation which had caused you fear.

Anger or Rage

Rage or anger is an emotional response which varies widely. It is a response mostly to interference or frustration. No organism is free from this emotional response of rage like fear. You get angry with your friend when he fails to understand your argument. The child gets angry if you interfere with his play or toys. The mild animal, cat gets angry if any one goes near her kitten. The hen gets angry even with her mistress if she touches the chicken. Even Saint Vishwamitra was not spared from this emotion.

If you snatch a toy from a child's hand he at once shrieks, shouts, jumps, kicks, bites, rolls on the ground and so on. Thus the emotion of anger is expressed in various ways when a person's self-esteem is injured.

Rage is one of the many mobilizing emotions, drives man to strong action. The instinct theorists consider that anger is the outcome of the instinct of pugnacity. It is an innate, hereditary impulse to fight and destroy the enemy. **Sigmund Freud** (1856-1939) says that anger is a form of death instinct (1920). The instinct theory fails to convince us with its explanation that anger is an instinct like hunger and thirst. But the frustration-aggression hypothesis explains that the emotion of anger is a consequence of frustration to some goal seeking activity. If you are feeling hungry and food is not served you get angry. Your anger mobilizes more energy in your body to attack and remove the obstacles on the way of your goal. This anger is the consequence of frustration. In your everyday life you might have experienced several such angers.

In some cases anger is a potential aid in dealing with the problem. In winning a game, in solving some difficult problems you may become angry with them and aggressively deal with them. Sometimes you even throw a challenge that you are going to win the game.

But it is very difficult to determine the manner in which anger is expressed. There is no inherited facial pattern of anger. It is very difficult to distinguish one emotional response from another. Emotional responses vary in intensity and human beings conceal emotions. Therefore, the activation theory holds the view that only the emotional intensity would be relevant to distinguish one emotion from another. It is true that we express our anger in some way in our everyday life. But that is not the universal way of response. Like all other unpleasant emotions the emotion of anger involves threats to bodily equilibrium. The sympathetic nervous system comes into action and there is higher pulse rate, blood pressure, respiration and also interferences with digestion. It is observed in several studies that chronic high blood pressure is associated with persistant anger.

Anger often functions as a motive and is called as the psychological motive. It mobilizes energy and supports vigorous activity. It intensifies threats and actions in goal seeking activity.

Love or Affection

The emotion of love or affection is considered to be one of the primary emotions. Human beings and animals express the affectionate behaviour to their young ones. People express their love or affection towards their children by hugging, embrassing, kissing, patting and so on. Love may be defined as the great attachment and affection for a person, place or an object. It is to have a passionate desire, longing and feelings. It is also meant liking or a desire to do something very much. It may mean also to make love. Love is an intense emotion, affection, warmth, fondness and regard towards a person or thing. It is whole hearted liking or pleasure in something. It is a deep feeling of sexual attraction and desire. It may also meant a man's attitude of reverent devotion towards God. Thus the emotion of love or affection may be stated in various ways. Normally young children long for love and affection of their parents.

Children as they grow and develop, the overt responses tend to be replaced by verbal expressions of endearment. Among adolescents and adults sexual excitements become the major function is inducing affectionate behaviour. In general most frequent expressions of affection are found in the lower socio-economic states groups where there is lack of strict social pressures. Love is always associated with sexual excitement which leads to marriage in most of the cases.

Joy or Pleasure

Joy is a pleasant emotion expressed in various forms of merriment or gaity. Joy means deep feeling of pleasure. It is synonymously used with pleasure, delight, happiness and merry. It is experienced in many situations. Joy is an undifferentiated or generalized excitement response occurring in many of our life situations. Babies cry when born, expressing to the uncomfortable feelings. They smile when they become two months old, laugh at six months expressing their joy.

The situations where the young children express laughter or joy are home situations in the midst of the family members. While playing with parents and brothers and sisters the young child is running, jumping, leaping, playing with toys, singing and also imitating. Laughter is an expression of joy and humour for the child. The child in a joyous mood or emotion becomes very active. The child may express his joy in some practical jokes.

Adult persons find or mirth when their efforts are successful and when they achieve laurels and appreciations. Adults normally express their joy overtly sometimes and sometimes silently feel within themselves. The joyous responses may be also passive and relaxed which is the indication of the general well-being.

Grief or Distress

Grief is a deep or intense sorrow or distress at the death of some person who might have been the individual's close relative or friend or very well known to him. It is expressed in overt response by weeping in the case of close relative and passively felt in the case of others.

Human beings generally express their sorrow in various ways, like weeping or crying and lamenting. In the case of death of the close relative the deep sorrow or grief may continue for many days. There are instances reported in newspapers that the death of the husband was so deeply felt by the wife that she also died on the same day by shock or by committing suicide. Thus deep grief or sorrow may cause such painful results.

Young children express their grief or distress by weeping and crying. This arises in children due to mainly deprivation of the parents for long time, restrictions, punishment and loneliness. Young children's expression of grief or distress is mainly related to anger and remorse. But it is observed that deep sense of grief in young children rarely occurs. But response similar to grief appears in young children during the period of first six months. As he completes the first year the child expresses the feeling of helplessness. However, it takes a minimum period of two years for the children for expressing their real emotion of grief or distress.

Emotion and Motivation

Modern authors associate motives with emotions in their treatments. That means, there is some relationship between emotion and motive. What type relationship there two psychological processes have? The physiological state of the organism may exhibit some type of emotional expression. But unless an organism is associated with an initiating stimulus and other influencing factors in the environmental condition, one cannot judge from the physiological condition alone whether the organism is in the emotional state of anger or fear. Anger and fear are two different emotions and are aroused on different motives. The motives determine the emotional responses.

Suppose, for example a dog chases a eat. The eat in a high state of fear with the motive of escaping from the danger runs fast. When the cat was cornered by the dog, the cat has no other means of escaping from the danger. In such a situation the cat turns swiftly and starts pounching on the clog ferociously in strongly aroused emotion of anger. The dog which has chased the cat became the object of chasing. Here we can find that the cat's emotion has changed suddenly from fear to anger. There was a sudden change in the physiological state of the animal in a very short time. The difference between these two emotions depends upon the motivations of escape by flight to avoid the danger. Thus there is a motive to escape and a motive to fight. The emotions of fear and anger are aroused by these motives. The motive determines the character of the emotion.

Therefore, we may state that there is a motive behind every emotional behaviour. Similarly the emotion of hate and love have some underlying motives. The situation in which an individual is involved and the event related to him determine whether the emotion is hatred or joy. People reveal their emotion of hatred by denial or refusal or non-cooperation or withdrawing from the situation. The motive behind hatred is to carryout one's intention, which was denayed. Behind joy or happiness the motive is fulfilling the desire, or achievement of the goal. When the desire is fulfilled or the goal is achieved the individual's emotion is joy or happiness.

THEORIES OF EMOTION

We have discussed the functions of the endocrine system and the action of the autonomic nervous system, under the heading of physical aspects of emotion.

The autonomic nervous system plays an important role in the emotional life of the organism. But we are also interested in knowing what role the brain is playing during the emotional responses the organism is making. Only on the basis of this information a theory of emotion could be formulated. Modern psychologists were able to formulate several theories of emotion from the information available to them from physiological studies. We are interested in three important theories of emotion. They are (1) James Lange theory (1927), (2) Cannon-Bard Theory (1929) and (3) Cognitive Theory.

James-Lange Theory

This is one of the most influential theories of emotion. This was first announced by **William James** (1842-1910), (1891) a famous American Psychologist in the 1880s. Almost at the same time the Danish physiologist **Carl Lange** (1834-1900) arrived at the same conclusions. Therefore, this theory is known as James-Lange theory. The general statement of James-Lange Theory is that emotion results from physiological changes rather than that emotion makes physiological changes. Therefore James-Lange Theory runs reverse to what was believed by scholars previously about how emotions are aroused. The thought held previously was: You see a tiger, you feel afraid and you run. According to James-Lange Theory the thought was held like this: You see a tiger, you run, and you feel afraid.

The James-Lange Theory holds the fact that when you see an emotion arousing stimulus, the tiger, a number of physiological responses occur immediately in you. Then only you feel the emotion, the fear. This emotion of fear is caused not by the direct perception of the tiger, but by the physiological changes. The changes such as pounding of the heart, dryness in throat, shivering or shaking of the body and so on are caused oil seeing the tiger. The James-Lange Theory does not relate the emotional responses to any special centres of the brain. According to this theory a receptor is stimulated by an object and the nerve impulses are sent to the cortex (brain) where the stimulus (the tiger) is perceived. The nerve impulses are transmitted to the muscles, skin and visceral parts and changes are caused in them, When these changes are communicated to the cortex (brain) the resulting emotion is fear. This would mean that physiological changes precede the experience which we call an emotion. Therefore, an emotion consists of these sensory impressions and these impressions are initiated by the physiological changes. The physiological changes which arise upon perceiving the stimulus is the basis of the emotion. The James-Lange Theory is called a peripheral theory because it emphasises body as the basis of an emotion and not the brain.

Cannon-Bard Theory

The peripheral theory which we have described above raised certain controversial issues. The theory was verified by certain physiologists, concerning the centres of emotion. In certain experiments conducted on animals the nerve fibres which bring impulses from the vital organs to the brain were cut to see whether the animal expresses emotional responses even after the removal of the impulses.

In the year 1927, an American Psychologist **Walter B. Cannon** (1871-1945), (1929), questioned the validity of the James-Lange Theory. He was of the view that when some physiological changes like heart beat and sweating were experienced for emotions like fear, anger and excitement how can we differentiate one emotion from another? The same type of physiological responses occur in different emotions. Cannon's strong point from his experiment to reject the James-Lange Theory is that if the viscera of an animal were separated from its central nervous system — that is if the nerve impulses were cut off, the emotional behaviours of the animal remain unchanged. Similar view has been expressed by **C.S. Serrington** in 1906. Sherrington had conducted an experiment on a dog. He cut the nerve fibres going to the brain from the dog's vital organs. Even then the dog continued to show the emotion of rage. Therefore, emotion is not depending solely upon viscera. Cannon strongly believed that emotion depends upon hypothalamic activities and not on peripheral activities.

During the same time **Bard** was working on the thalamus and reinforced Cannon's views. In this way additional knowledge was accumulated concerning the centres of emotion and the Cannon-Bard Theory was developed.

The Cannon-Bard Theory has proposed that there are certain brain centres which are involved in the emotional responses. These brain centres are thalamic and hypothalamic centres. This theory also states that emotions and bodily responses occur simultaneously and not one after the other. According to this theory, when an organism perceives an emotion arousing stimulus nerve impulses pass through the thalamus part. At this point the impulses are splitting and some of the impulses go to the cortex and some of them go to muscles and viscera. In the cortex the stimulus is perceived and the emotional response is felt. At the same time in the muscles and viscera physiological reactions are taking place. This is how an emotion is experienced.

Cognitive Theory

The cognitive theory of emotion is a very recent development. The founder of this theory is **Stanley Schachter** (1971). He accounts visceral changes of James-Lange Theory in this theory. He proposes that the interaction of cognitive processes and physiological changes produce emotions. According to this theory, we feel an emotion when there is a physiological state of arousal and the awareness of this state in us together. The interpretation of emotions depends upon our past experience and the present environments. The experiments conducted by Schchter and Singer to test the effect of cognition on emotion would explain the theory very well.

Another cognitive theorist **Stuart Valins** (1970) holds that a bodily change like increased heartbeat itself leads to emotional reaction. When there is increased heartbeat suddenly in you, immediately you wonder and ask yourself why it is happening and you become frightened if you do not know the reason. **Magda Arnold** (1960) has made a thorough discussion of contemporary views of emotion. Arnold stresses the importance of cognition of emotions. She had formulated a theory called Arnold's Theory. According to this theory, the emotions we feel depend mainly on our cognitive appraisal of the situations. Arnold says that there are five stages that occur from perceiving a stimulus to putting an action on it. The stages are perception of the stimulus, appraising the situation, arousal of emotion, expression of emotion and the action. Arnold's theory appears to be more realistic than the other two. The theory is not interested in what is going on inside the body, because the inter-relation between physiology and emotion is a very complicated process and no part of the body is solely responsible for the emotional responses.

The Control of Emotion

Human beings aspire for emotions and also afraid of emotions. When it is pleasant we welcome it and when it is unpleasant we reject it. People carry heavy load of emotions due to unpleasant events in their every clay life. Carrying heavy load of emotion may affect health of the people. Prolonged intense emotional state of anger or fear creates chronic tension which may cause psychosomatic problems. Many people are riot aware of their heavy load of emotions and the consequences of it. It may affect their adjustment profoundly. Therefore, emphasis is placed on reducing chronic emotional states.

There are several methods :

(1) One method is to make the person aware of his chronic tensions in his body and to teach him how to relax it.

(2) The chronic tension of emotional state can be reduced by making the person to build up his self-esteem gradually and thereby he may be free from the state of fear and build up confidence in life.

(3) The tension of emotional state can be also reduced by having free discussion with the person and with his parents and friends.

(4) Since emotion is unruly and unpleasant it is better to avoid situations that are likely to arouse unpleasant emotion. Generally poor health, family quarrels, impoverished state all susceptible to irritation. In the case of youngsters interference in their activities, sleep, feeling hungry may provoke anger and tension in them. By preventing such occurrences tensions may be avoided. People who are highly socialized use various methods like religious, social, educational and political issues as distractors of emotional tensions.

(5) It is not always successful to counteract emotional tension with another emotion. Particularly it is not correct approach in human relations. The teacher should deal with the angry student separately instead of punishing him in the presence of other students. Verbal abuses and scolding may not bring expected result. When emotion is subsided a good talk may have better effect. Individual variations are there in dealing with emotions. It depends upon the person and the problems. The person may be very rigid always self-defensive and may not accept anything himself. In such cases suitable method may be applied.

(6) Emotional tensions may be drained down by providing suitable outlets. Social gatherings, functions, recreational facilities, providing roles in some functions,

pleasant conversations, music, dance, sports and visiting new and exciting places may serve as outlets for unhealthy emotional states and may drain down the stresses and tensions.

Emotion and Mental Health

Emotions, just like feelings have both pleasant and unpleasant characteristics. But some scholars are of the view that emotions are unpleasant and harmful. We often find people reacting with strong emotions of anger, disgust, contempt, rude and harsh. Such unpleasant emotions arise due to certain reasons which they could not avoid. Then what happened for a moment is that the conscious mind becomes out of control. The autonomic nervous system begins to function. Several visceral reactions occur which result bodily changes. If it is intense and continued in the state or prolonged it becomes similar to stress.

At this state the heart beats faster, the liver is stimulated by the secretion of hormones to produce more energy giving sugar into the blood streams. There is change in breathing which results in pumping more oxygen to the body. The stomach and the intestines get tighten. The eyes become wide open or dilation of the eyes, the saliva gets dried up. There is more sweating all over the body. The skin surface contracts and there is goose flesh, the hairs get eracted. The bodily changes registered very quickly in the conscious mind.

How the mind reacts in the next stage depends upon wholly on the memory processes and thinking of the individual person. There may be individual differences in this process. A situation in one person may cause powerful feelings in his mind. The same incident may cause simple feelings in the mind of another person. The expression of the aroused emotion may result in some action and experience. How people express their emotions or how they behave when arosued by some emotion? This has aroused the curiosity of the psychologists whether the intense emotions or mild emotions have any significance in the mental health of the persons?

However, It is observed that there is very wide range of behavioural variations in the emotional states. This varies from normal level to the level of mental illness of some type. If the strong emotional state is prolonged for longer duration or if the individual very often gets emotionally imbalanced then mental health of the person is likely to be affected and the individual may express or reveal abnormal symptoms of some type. If the person expresses a response in an environmental situation in a way that falls within the normal range of behaviour, psychologists state that the person shows *appropriate* emotion or *affect.* But if the person expresses less range of emotion he is said to have *constructed affect* or emotion.

But we may see in the society that an apparent reverse of this affect in people who seem to ride an emotional abnormality. The person laughs at one time and cries at another time and next moment feels friendly, at the next moment quarrels using abusing words. This type of condition is stated as *instability affect.* It is a symptom of *unstable personality* or *intense stressful state.* But if the emotional conditions are revere and long standing they are said to be *affective disorders.* When people constantly face unadjustable situations in the society they are likely to develop the affective disorders. Intense emotional state causes maladjustment which in due course develops into some type of mental ill health.

18

MOTIVATION AND BEHAVIOUR

Introduction

People always find it a pleasure to associate themselves with some occupation or work. It is not only for earning their livelihood, but also to keep themselves engaged in some activities which they liked. They might have got into the occupation or activities by choice or by accident. Whatever the reason lays behind this their self-esteem is closely related to the work. When we meet some known persons after a long time the first enquiry we are tempted to make is, "What job you are doing?" Many people generally wish to take up a job what their relatives have been doing. Once a person has taken up a job which will be continued for about more than half of his adult life. He may retire from that job only when he finds himself old and weak.

Psychologically, it has been observed that the work a man does, has a significant influence on his outlook and behaviour. Studies have been made and found that people use their decision and judgements to perform some complicated tasks at work, tend to develop their intellectual ability that they can carry over into the other parts of their lives. They tend to become independent and wish to have some new experience than to perform the same routine jobs. They also wish to select jobs which require intellectually engaged. Some may select and choose jobs which should suit to their leisure pursuits such as learning, writing and achieving fame and also something useful for their life.

Generally, we find in the work situation that a happy worker is found to be more productive than a person who is not satisfied with his work. It has become customary to associate job satisfaction and good performance. This is only hypothetical. But it is the sense of accomplishments from doing well on the job that makes the worker to feel that he is competent and happy. It is true that many dissatisfied workers remain absent and also likely to change their jobs often than the contented employees. The primary incentive for most people is the need to get more income. It is by having a job. Apart from money and income there are also other factors that play important roles. Some of them are the needs for the development of the skills to perform the task efficiently, the needs to be a creative worker and to get the admiration of the organization where they work. There is also the need for a job which can build up the feelings of competence and self-esteem in a competitive society that expects that individuals should attain a social status by performing valuable service. But all these factors depend upon the individual and his backgrounds, expectations, aspirations and the goals he set.

There are several incentives in an organization or in an industry to motivate employees for efficient performance of the work which is supposed to increase the production and profit. But what are other factors besides money that would motivate people to work? They are generally achieving the goals, fulfilling the ambitions and aspirations. There may be physiological needs, psychological needs and also social needs. In the common sense world there are several concepts used synonymously with motivation, like the drives, desires, needs, aspirations, ambitions, incentives, inducements, interests, inclinations, wishes and so on. Therefore, motivation or motive is an important psychological process, playing important roles in all aspects of human behaviour.

What is Motivation?

Human behaviour in certain situation is directed towards getting or achieving an object or a goal. The object or the goal, a man tries to achieve in a situation may be intended to fulfil the man's needs. The man is seriously wanting it. He may achieve it or fail to achieve it. If he achieved it he is achieves satisfied, if failed he tries again and again till he achieves it. If he has failed to achieve it with all his possible attempts, he is dissatisfied which may lead to even frustration. At this stage he may even drop the idea of getting it. What is that directs the man to behave in this fashion to achieve something? What is that initiates the man for such behaviour? The behaviour may be moving, or walking or running or jumping or manipulating some tools or instruments. He may sometimes seek the help of others to achieve it. In all these processes, we can observe that the man has a desire, feeling, need and wanting. What is that impels him and determines his behaviour? It is the motive that impels and determines his behaviour. Motive is an inner drive, impulse that causes one to act. It is an incentive causing motion.

The motive is a concept, it is not an object. In other words it is a psychological process functioning in an organism to move and act in a particular fashion. From the nature of behaviour of the organism, man or animal, we can infer that the organism has a motive. What is a motive or motivation? Motivation is generally defined as that which initiates and directs an organism for an act or a type of behaviour. Motivation may be also defined as an act or an instance of motivating an organism for certain behaviour pattern. It is a desire to do, an interest or a drive. Therefore, motivation is the process that arouses, sustains and regulates the behaviour of human and animal. A motive becomes the reason for certain course of action whether conscious or unconscious. It may also be stated as the source of energy used to produce motion or behaviour in an organism. Motivation generally, is the cause of behaviour.

Why does a man or an animal respond in a particular fashion? The answer to this question can be found in the continued interaction between the external stimuli or the immediate environmental situations and the internal conditions or the physiological mechanisms of the organism. For example, a student works hard night and day with his book and other materials.

This means, he has a strong desire to achieve a better grade in his examination. Another student goes to the college field both morning and evening and practices in running and participates in various competitive sports activities. This student has strong determination to win certain sport events. A hare got into a hunter's net, tries in all possible ways to escape from it. A boy in Ruggs is standing in front of a bakery shop and looking with wide open of his eyes at the stacked bread loafs. This means that the boy is homeless, expressing his hunger. Diverse view are attributed to these instances, but according to the meaning of the typical concept, motivation, these instances are due to the strong desire to get better grades, strong determination to win the sports events, fear of death and hunger. Every action arises as the result of certain antecedent cause and it is revealed in certain expression of behaviour which is physiological in origin. Therefore, every motivated behaviour has an antecedent cause.

Every behaviour, if overt, makes some change in the environment. We can see the change. Therefore, we see change in every observable behaviour. The behaviour is a response to certain stimulus, it may be also stated that a response if it makes any observable change in the situation it may be stated as behaviour. A behaviour is based upon a purpose. The purpose may be moving, walking or running, or lifting something, opening the door, or opening the pages in a book, smiling or speaking. In all these a response is involved and it is also a pattern of behaviour. Any response that makes a change in the immediate environment and fulfils a demand or reaction to a cause may be stated as motivation. Since, psychology mainly uses the stimulus-response mechanisms to explain the behaviour of an organism and every motivation has an antecedent cause, motivation is the keystone or the central point in psychology. In explaining the process of motivation there are certain concepts used synonymously. For example, a set, an incentive and initiative, impulse, drive are sometimes used in the same meaning.

Motivated behaviour is based upon our needs and it is directed for achieving a goal. Generally motives are expressed in three forms. The most clear expression of any motive is complete fulfilment. The behaviour is directed for the fulfilment of the immediate need of the organism. For example, eating, drinking sex are for hunger, thirst and for copulation. The second expression is instrumental which involves action upon the environment. The act which should secure what will satisfy the motive. The stronger motive makes the organism to act vigorously to have it satisfied. The politician may work very aggressively to get the power. The power motive is so strong that he becomes aggressive in attacking the opponent to get the power. The third expression of the motive is substitute in nature which may be also indirect. The motive is not going to be fulfilled immediately. For example, in the case of some people who were deprived of good food for a long time, may express their motive for food in some other forms. They may be engaged in some conversations or talks with some persons on some occasion and mention the delicious food they had eaten. They do not bother about other motives. They often may dream of good food, some times even engaged in daydreams and see good dinners. This is how their extreme hunger for food substitute their hunger drive and satisfy themselves. The feeling of deprivation makes the people to find out any alternative through which they express their hunger motive.

Behaviour may be explained in terms of motives and the agreeable explanations are relating behaviour to the environmental conditions and to the physiological conditions. The organism is directed towards the goal and the strong desire provides more energy for the behaviour. The motives are expressed in various forms of behaviour. The physiological motives are so complex that the organism is put into various stages of the processes like perceiving strong motivation which is reflecting in feeling, emotion and action.

Motivation and Evolutionary Process

Modern definition of behaviour is thought in terms of response of the organism to two types of stimuli, the internal stimuli generally known as need, drive, motivation and the external stimuli are the environmental factors which are acting upon the organism. The behaviour of a single-celled amoeba also acts on motivation. The single celled amoeba also responds to the environmental stimuli by moving in the direction. The food for the amoeba may be certain chemical substances, towards which it moves and responds. Anything other than its food, is not responded. The receptor mechanism of the amoeba is a natural structure in the organism which initiates it automatically. But if the organism is of complex type the motivational process also becomes complex. In such case there are variations in the response or behaviour (Detheir and Steller, 1964).

The behaviour of the organism is not fixed and not easily influenced. The influence of the external stimuli is not very much explicit and also less self moving. The influence of physiological conditions are becoming more complex. Therefore, the organism should have sole experience and greater role in deciding the behaviour, but the evolutionary development is going on slowly which provides little scope with which it is very difficult to state in detail.

Tropism : A Form of Motivational Behaviour

The simplest form of *Motivational behaviour* that we may observe in all plants, insects, and in some lower animals is called *tropism.* This form of behaviour is the *movement of the organism* towards the specific *stimulus.* Or it may be also *moving away* from the stimulus. But these types of responses are possible in some higher organisms. Generally *tropism means the response of an organism,* especially, the plant to an external stimulus by growing in a direction determined by the stimulus. It may be also stated as the *indicating tendency* to turn or develop *in response to certain stimulus.* There are various types of tropisms. *The geotropism* is the response of a plant part, the roots, to the *stimulus of gravity.* The *plant stems,* which grow upwards irrespective of the position in which it is planted, show *negative geotropism.* The *hydrotropism* is the *tendency to move* in the directional growth of the plant in response to the *stimulus of water.* The *heliotropism* is the indication of the growth of a plant in response to the *stimulus of sunlight. Phototropism* is the growth response of plant parts to the *stimulus of light. The Galuanotropism* is the directional growth of an organism, especially, a plant in response to *electrical stimulus.* The responses of these types are identified by their *respective evoking stimuli.* Though the external stimuli are dominent in the behaviour, the internal factors also play important role. Even the *lowest insect* the moth is also directed by tropism in its behaviour.

Animals of various types like *daphnia* which generally move *directly forwards light* were found to move in circle *around the source of light,* and *receive stimulation* regularly. This type of simple oriented movement is an increasing influence in a more complex form may be treated on bar with *reflexes.*

Reflex

Reflex is a type of response to a particular stimulus. It is an *automatic reaction in an organism.* Reflex action is involuntary and also innate in human beings. When we touch a very hot vessel unexpectedly we suddenly give a jerk and remove the finger from the vessel. This we may call reflex action. A reflex is a *rapid and unleaned response to a sensory stimulus.* The automatic reaction is taking place in the effectors. The effector may be a *single muscle or gland.* This muscle or gland is capable of producing a response. One of the essential characteristics of reflex action is that it is not an act of will in the performance. The awareness of the stimulus or the response occurs only after the act of response. The natural pathways are responsible for the simple reflex action. The receptors (cells) which receive the stimulation carry directly *to the effectors* through the *spinal cord.* It is not necessary that the stimulation should go to the brain or cortex region. *A message from the spinal cord* is sent through *motor nerves* and the response takes place. The reflexes are found *in all complex organisms* including *human beings.*

The variables in organism influence reflex behaviour much more than in the behaviour of the *tropistic responses.* The nervous system is flexible and fresh synaptic connection can take place and reflex activity is modified by the experiences of the organism. Unlike tropistic response, reflex behaviour can be conditioned.

The concept of conditioned reflex was first used by the Russian physiologist, **Pavlov** (1849-1536), and it is always associated with his name. While working on digestive system, Pavlov observed the process of conditioned reflex in his experimental animal, the dog. According to Pavlov, all human activities or behaviour may be conditioned. Human beings plan each and every activity or work to perform in a specific time. Thereby human beings are conditioned to move and act according to some specific time. College students and teachers are conditioned to respond to the College bell. When the bell rung at 10.00 a.m. all the students and teachers rush to the classrooms and when the next bell rung they move quickly to a different place for different work. Industrial and factory workers respond to the industrial siren. Railway passengers respond to the railway bell. Thus human behaviour is conditioned to certain stimuli. Similarly, the temple bell, church bell, the police man's whistle on the road, the refree's whistle in the playground or in the sports activities make the respective persons to respond to the stimuli. Thus human behaviour is conditioned for everything including eating and drinking.

Instinctive Behaviour

It is said that animals and birds live on instincts and man lives on intelligence. It is believed that certain behaviour patterns of human beings are also instinctive. If reflex action is unlearned, innate and automatic in man, instintive behaviour also is innate and unlearned and biologically based form of behaviour. **William McDougall,** the daughty *champion of the instinct theory,* has elaborately dealt with the instincts.

Though reflex action and instinct behaviour are taking place automatically without being guided by consciousness, instinctive behaviour can be differentiated from reflex action. In. the *evolutionary process tropism is the first step,* the *reflex second,* improved step over *tropism* and the *instinctive behaviour* is the *third step.* Though instinctive behaviour is unlearned and innate pattern of behaviour, *it is modifiable* to some extent to cope with varied environmental conditions. In the course of evolution, certain species are able to adapt to the changing environmental conditions. In the case of certain organism, the chances for survival were improved by the chain of complex behaviour. Certain organisms' behaviour vary according to the change of seasons. For example, the behaviour such as *migration, nest building, mating* and taking *care of their young ones,* the organisms *behaviour is instinctively* for survival. But these species *do not retain their experiences* as the *nervous system* in them *did not develop.* The chains of behaviour were initiated by the *structure of the reflex arc* or the *receptor-effector connections* of the nervous system and *passed on genetically from one generation to another.* On any account there would not be any change in the behaviour of these organisms. There are succession of responses towards the goal. This type of successive responses are same for all organisms. This is a *fixed behaviour pattern* or stereotyped (Fig. 18.1)

Fig. 18.1 : Instincts as Motives : Unlearned Patterns of Behaviour

There were several investigations on the instinctive behaviour of animals. Many*ethologists analysed animal behaviour.* Ethology is a science that studies the behaviour of the animals in their natural environments. **Tinbergen** (1957) studied the instinctive behaviour in*stickloback,* a type of*fresh water fish* elaborately by conducting experiments on a wide variety of animals.

Imprinted Behaviour

Imprinted behaviour is a type of experience found in *young chicken, ducks, young gooses.* These species have a strong tendency to*follow their mothers.* In the yards where they grow follow behind their mothers in a single line. Though this type of behaviour is a *universal tendency* among these species, studies on them show that this*following stimulus* is *acquired* or learned by experience. Experiments show that the newly hatched chick follows something. It may be a toy chick, or a white small ball, or any moving object in the front, chick follows. If a man in the presence of a chick moves forward and backward slowly, the chick also begins to follow him. Normally, the *first moving object* these chicks see is the mother duck and the chicks move behind her. In the absence of the mother duck, the chicks follow any mother object. This type of response is *adaptive behaviour.* This imprinting behaviour is different from other complex type of behaviour. There may be several responses in a situation in higher organisms. But in the case imprinting response or behaviour there is only one type of response that is following. Even in the case of young chicken and ducks the imprinting behaviour occurs only to certain period and not always. It takes some time for the young chicks to identify the objects that are different. After sometimes the young chicks do not follow the stimulus object except the mother duck. This period is called *'critical period' in the imprinting behaviour.* This *imprinting behaviour is also classified as one of the motivating behaviour* according to the evolutionary process (Fig. 18.2).

Fig. 18.2 : Imprinted Behaviour : Young Chicken have the Uni versal Tendency to follow the First Moving Object whether the Mother or any Moving Object.

Spalding quotes from **William James** (1890) that it was observed that many young animals become more attached to their parents. The new borns of many species tend to follow the *first larger moving object.* Imprinting takes place in *young animals of many species* like, *goats, deer, zebra, buffalo and seals.* **Eckhard Hass** (1958) used mallard duckling as subjects and found that it was easy for them to imprint on a circling wooden decay of the male mallard. The decay contained a heating element discharges some warmth.

PHYSIOLOGICAL MOTIVES

In the evolutionary process, the nervous system in higher species is developed to certain level which becomes possible for the species to learn certain things easily and develop certain unique experiences. This process of development mainly responsible for the organisms *to preserve the self.* This level of the development of the *preservation of the bodily needs* is characterised by the existence of *physiological drives.* It may be also called physiological motives. A physiological motive may be defined as the bodily or *organic need* such as the need for food, water and shelter. Modern psychologists use the terms physiological drives. Hence according to **Moskowitz** and **Orgel** (1969) a drive may be stated as an *internal state of tension* or arousal triggered by some physiological need or imbalance and this acts joining together with an environmental incentive which is a stimulus that is acquired by learning the characteristics of a goal.

Organic Needs

The needs which are very essential for maintaining the internal environment of the body very smoothly may be stayed as organic needs. In the physical environment the body has to maintain a balance or equilibrium among its internal physiological states. Physiologists use the term *homeostasis* to explain the body's tendency. The basic organic needs are hunger, thirst, sex, the respiratory needs, the elimination needs the needs for activity and rest, and so on. All these are required to maintain the internal environment smoothly.

The body should maintain its temperature at the normal level. This is *self-regulation* of the body. The body's temperature should be always at 98.6° degrees Farenheit. If an individual's body gets too warm, he sweats and the evaporation of the sweating makes the body cool. If the physical environment is too cold the body shivers. Then the body generates extra heat and maintain the normal heat. The level of blood sugar and the oxygen supply are regulated automatically. The chemical composition of the blood is also regulated by the bodily mechanisms. The *hormones of the pituitary gland* regulate the *potassium and calcium* balance of the body fluids. The regulatory behaviour is a part of the homeostasis process. Behaviour is usually aroused by the internal changes called needs. This arousal is a drive.

The needs for rest after some activity and the needs for sleep at the interval are only organic needs imposed by the bodily conditions. Continuous *muscular activity* causes *collection of lactic acid in muscles and joints* which causes a strong *desire to stop the work which is an organic need.* The circulatory system is very important, which functions as an interacting system between the internal and external environment. The homeostasis that maintains the internal environment is carried out only by the *circulatory system.*

Physiological Basis of Hunger

Hunger is a basic physiological motive and it is often experienced by all living creatures. Physiologically hunger is a function of *energy balance.* Psychologically hunger is a *bodily sensation,* which we all may experience in our stomach if we delay in taking our food in time. It is a physiological need and every living creature has to fulfil it. Hunger is a physiological need, but the reaction of it is psychological by *causing painful sensation* and making the organism *stressful.* If the hunger sensation is severe it results in painful feeling in the stomach. This is an indication that something is lost or deficit in the body, specifically the energy balance, and therefore there is a demand by causing the painful sensation and putting the organism as a whole in a disturbed state. As the time passes on in this state the organism becomes weak. In an extreme state the organism may sink down below the normal level. This is what generally is happening outwardly. But what is happening inside the organism?

The physiological observation is that the physiological and biochemical mechanisms which are constantly functioning, contribute to the energy balance of the organism. The ingestion of the nutritious substance is an essential condition for the survival of the living organisms. The organism maintains a constant relationship between the energy output and the energy input. The regulation of food intake is the universal property of all living organisms. When the energy input is lacking, the mechanism of the organism fails to maintain the energy balance, and the organism is driven to stressful state, the hunger.

In the state of hunger, there is *stirred up* and *vigorous movement* of the *stomach muscles,* peripherally. Therefore, hunger appears to be directly related to the contraction of the stomach walls. Gradually the stomach muscles start contracting vigorously and a feeling of pain (pang) is produced. The relation between hunger and the contraction of stomach walls was stated only hypothetically. This has been tested experimentally by **Cannon** (1912) and his collegue, **Washburn** at the University of Chicago. The subject was asked to swallow a small balloon. The balloon was inflated to press against the stomach walls. Whenever the stomach contracted, it pressed the air up in the balloon tube. This air pressure was measured and *recorded on a smoke paper* placed on a *revolving drum* of the *kymograph.* The report of the subject of the hunger pangs (sharp pain) and the stomach contractions corresponded closely on the upper line of the instrument. The *contractions* were the stimuli informing us that *the man is hungry.* But some are of the view that the *stimuli are not the absolute indicator* of the hunger. People who have had their stomach removed reported normal eager for food. It is believed that the best evidence obtained in recent times indicates that hunger is caused by the *deficit of certain, substances in the blood streams.*

It has been observed that the deficit of *blood sugar* (Glucose) is the *specific substance responsible for hunger.* By injecting insulin which can reduce the glucose level in the blood causes the individual to feel hunger suddenly. Actually there are more than one type of hunger, studies show that there are *specific hungers.* The drives for food depend, upon certain specific food material in the normal diet. There may be sugar deficiency, or salt deficiency, protein and vitamin deficiencies. When there is any of these deficiencies the normal diet is not preferred. It was observed that animals deprived of salt show increased demand for food and they do not prefer even if there were surplus saltless food.

Wald and **Jackson** (1944) conducted studies of cofeteria feeding system, in which a group of the young children were fed for several years. At the same time another group of children served as control group, fed on *regular diets* planned by a dietician. In this system, the children were permitted to eat whatever food they preferred from a varieties of food presented to them. But in this, the overall diet is concerned, each meal was not a balanced food. The child preferred only one kind of food out of several meals and it was observed that each child has choosen only a balanced diet. At the end of this experimental study it was observed that the children fed by *cafeteria system* were as *healthy as the control group.* It was because that the food supplied were all good and nutritious.

When the blood sugar was low, the motility of the stomach increased. The contractions tell us that we are hungry and also in extreme cases it becomes hunger pangs. In the case of specific hunger this mechanism does not operate. The insulin is having crucial role in mobilising and distributing the food energy and the brain monitors circulating the insulin levels to decide when it is time to eat and when it is time to stop eating. Suppose, for example, if the insulin level is high there is food in the stomach and fat stores will be increased and hence the brain produces the *sensation of satiety* and we stop eating. Experimental results show that forcing an animal to low insulin level causes the animal to become hungry and it eats more food. If the insulin level is moderately injected the animal eats much less. Therefore, it is clear that insulin is a satiety signal.

Discoveries show that there is no *single molecule* in the blood stream triggering hunger of satisfy and also no single brain region has complete control of these sensations. But according to a *dual-centre-theory,* the hypothalamus has the primary control centres for hunger and satiety. A hunger centre in the *Lateral hypothalamus* (LH) facilitates eating and the satiety centre in the *ventromedial hypothalamus* (VMH) *stops eating.* **Anand** and **Broback** (1951) have conducted experiments on rats and cats and found that bilateral destruction of the lateral hypothalamus (LH) caused the animals to stop eating. The animals refused to eat and in some cases the animals died of starvation even when their normal food was present. Therefore, the authors suggested that *the LH contains a feeding centre* and that the VMH acts as brake on feeding by preventing the LH.

By damaging the lateral hypothalamus (L.H.) two important features of the effects soon emerged (Teitelbaum and Steller, 1954). The first one is that the rats refused not only to eat but also refused to drink. When the experimenters placed food or water on the lips or in the mouth of the rat, the rat spat out the food or water as if it is not tastee. The second is that these effects *aphagia* (refuse to eat) and *adipsia* (refuse to drink) are not permanent.

It was observed that the VMH destruction did not totally abolish all inhibitions of eating and LH destruction did not permanently prevent eating and drinking in these animals. A few rats began to eat and drink spontaneously after a week. Therefore, there must be some other *brain region monitoring the feeding* and signal satiety in these animals.

But these results are challenged by some investigators. Whether the damages are in the lateral hypothalamus (LH) or in the ventromedial hypothalamus (VH) they are effective, because of the integrative centres, (Marshall, Rishardson and Teitelbaum, 1974). Many brain regions outside the L.H. are involved in the regulation of the feeding behaviour and they do not contribute to L.H. These suggest that hunger and satiety are not centred in any single point of the brain. *A network of cell* distribution across the brain seems to be responsible for these operations.

Hunger traditionally is considered as *prototype* (first) of *primary motivation.* Physiologically it is considered as one of the most important mechanisms, which makes the organisms to maintain a constant internal environment. But it is not yet made clear that the contributions made by many peripheral and central physiognic mechanisms to correctly explain the *earlier occurrences that cause the eating behaviour.* **Grossman** (1967) states that psychological theories or observations have generally disregarded the important relationship between hunger and the energy balance of the organism. Therefore, generally hunger may be defined as a *"Complex of sensations"* (**Grossman,** 1955). If this definition is accepted, what about the lower animals which lack the development of nervous system for experiencing the complex sensation. If this is so, how about the insects? Do they have complex nervous system to experience complex sensations? Hunger is a *feeling of pain, emptiness,* or *weakness induced by lack of food,* an appetite desire, need or craving. According to physiological theories (Carlson, 1916, Cauzon, 1934), hunger is an assumption that the peripheral, gastric mechanisms regulate *food intake* on basis of the information from the *gastrointestinal tract* and they are independent of chemical or nerves changes elsewhere in the body. According to **Carlson** the gastric-hunger mechanism is primarily autonomic, independent of blood changes and the central nervous system influences. The relationship between food intake and environmental temperature was suggested by some **(Brobeck,** 1947-48). The food intake and hunger may be regulated by the organism's temperature as a part of function.

The Peripheral Theory

The hunger and eating behaviour has been a subject matter to the natural scientists, the physiologist, of eighteenth and nineteenth centuries. The two important and contrasting theories are the *peripheral* and the *central theories.* Some aspects of both the theories are discussed above. The earliest theory is the peripheral theory, proposed by **Hall** (1778), suggests that hunger appears to be the sensation of peripheral in origin. Hunger is caused as a direct stimulation of sensory nervous from the stomach. The hunger nerves in the mucose of the gstrointestinal tract may be stimulated by the grinding or rubbing (tritus) of the delicate folder of gastric mucose against each other through the contraction inherent in the stomach. **Darwin** (1801) supported this view, but did not agree the mechanism proposed by **Muller.** Muller, a German physiologist and anatomist, supported this view that hunger is purely *peripheral* and said that hunger sensations show merely the absence of some positive sensation which is generally caused in the *gastrointestinal tract* when food supply is not taking place. **Stiller** (1915) another investigator also accepted the *gastric origin* of the hunger sensations.

The Central Theory

The central theory of hunger is quite opposite to these peripheral interpreations. **Magendie** (1826) formulated a theory of hunger. According to him the *neural centre* in the brain causes the periodic sensation of hunger, regardless of other sensations from the stomach. This theory, having the neural centre of the brain as the cause of hunger, has been accepted by many famous psysiologists of the nineteenth century. **Wundt** (1802) and others were influenced by the central theory and it prompted them to conduct experiments and the result of which provided empirical evidences to the central theory of hunger. **Bardier** (1911) was able to indicate a *hunger centre in the brain.* According to this theory some changes take place in the blood of the food deprived organ which stimulate directly or indirectly a hunger centre in the brain. This view has been accepted by many physiologists.

The clinical studies that took place during the early twentieth century supported the idea that the *hunger centre* is at the *base of the brain.* There were extensive studies of the role of the hypothalamus which is having control over the hunger and food intake.

Obesity

Obesity is stated with several concepts. Thus, obesity means *excessively fat* or *fleshy,* physically bulky, corpulent. Obesity is a psychological problem caused by *over eating.* It is mostly concerned with eating. The hypothalamus centre responds to some receptor system that regulates the body weight and causes the individual to eat enough to maintain the weight. The fat cells of the body which store the body's fat are called *adipocytes,* is enlarged by over eating. Sometimes the body increases to five times of the normal size. These *cells never disappear* and in adulthood *new cells* are developing. The number of *fat cells and their sizes* are the basis of the individual's *'set point'* for weight. Set point means, a fairly constant level at which the hypothalamic centres maintain an individual's weight for a long period of time. *Once it is developed the set point will not change.* From early childhood the obese persons will have high set-points. Whenever the obese persons weight fall below setpoint the hypothalamus centre of the brain receives signal from the fat cells for increased food intake and *restore the weight* of the obese persons.

Obesity is defined as the body weight that exceeds the average for a given height and body build by 20 per cent. An *over weight* person is more *susceptable to heart disease, high* blood pressure and diabetes than a normal person. Attempts for reducing weight are only partially successful and they regain after some time. One view is that obesity is the result of *emotional conflict* in early childhood period. Obese persons may not be responding to the same hunger cues as the non-obese persons. Recent studies show that obese persons eat as much as in full stomach as they do on an empty stomach. The fear of obese does not decrease their food consumption (**Selachter** *et al.*, 1962). These studies show that the eating behaviour of obese persons is not regulated by internal cues. The external cue for eating is the passage of time. After 4 to 6 hours of time the obese feels that it is time for the next meal. The obese persons could not indicate their hunger state. This is because of their needs which were handled by their mothers when they were infants. Further, the fat persons is *sensitive to crystal food-related stimuli* than the non-obeses. One of the reasons for the obesity according to **Thomas** and **Mayer** (1973) is the disturbance of the long-time *weight control mechanism in the hypothalamus area.* The obese person does not trust the internal signs of hunger, but mainly depends upon sensory information.

Some young people become much observed with their body weight and become extremely weak. They eat very little, take laxatives or drink large amount of water to suppress appetite. This condition is more common among adolescent girls and women than among the males. It is called *'anorexia nervora'*. 'Anorexia nervosa' is a syndrome in which individuals were severely deprived of themselves of food This results in the *stopage of menstrual cycle* of the female altogether. It is because that the person has no appetite and that the origin of the disorder is in the nervous system.

The hypothesis that obese persons will not respond to the same hunger cues as non-obese persons has been verified in an experiment. The frequency with which both the groups associated with the feelings of hunger when the stomach contractions, were taking place. In **Stunkard's** (1959) experiment both the obese and non-obese persons came to the laboratory in the morning without eating the breakfast on that day. Each one swallowed a gastric balloon. This was attached to the *recording drum of an apparatus.* The balloon in the subject's stomach was inflated with air to the extend of the balloon touching the stomach walls. The recording of the gastric movements were made. The recording of the gastric movements went on for about 4 hours continuously. After every 15 minutes the subject was asked whether he was feeling hungry and the report was recorded.

On assessing the responses of the two groups it was found that the two groups did not differ significantly in the number of times they reported hunger when the stomach was not contracting. But there was significant difference in the reports during stomach contractions. The hunger report of the non-obese persons agreed with stomach contractions for 71 percentage of the times. But the obese persons agreed only for 47 percentage of the times. In the experimental situation the hunger cues were different for the non-obese persons and to the obese persons.

PHYSIOLOGICAL BASIS OF THIRST

All living organisms invariably need water and food for their survival. Organisms can go without food for some days, but not without water even for a day. Water in the body of an organism is spent in various ways such as perspiration, evaporation and evacution. Because of these, water deficiency is taking place in the body and water must enter into the system periodically in sufficient quantity. An organism which has water deficit is motivated to drink water. Even if there is small quantity of discomfort of water in the body due to giving off, it is reflected immediately and water intake must occur.

The common sensation of thirst for human beings is the *dryness in throat and mouth. Why this should occur?* It is because there is imbalance of water within the body. This imbalance occurred because of some reasons such as *more a salty food* eaten, or the body might have been put into *severe physical exercise* or because of the *continuous lecture for 2 or 3 hours* in the classroom. While making *long speeches by the political orator intermittently drinks soda water* to quench the thirst.

Sportsmen who take parts in running races, cricket players who stand in hot sun for longer duration, tennis match players, football, hockey and basket ball match players often go in search of more water, for drinking and also for pouring on them to reduce the thirst and fatigue. Sometimes the cricket match players go into the state of *dehydration* and fall down as the *body water fully exhausted* in them. In such cases the body's water deficiency should be immediately restored, otherwise it may result in death.

After a night's sleep one may observe that his urine is more concentrated and looks dark and has a strong odour than at other times during the day. While sleeping we are not drinking water, but the body needs water for normal functioning of the mechanism and the body has recovered water from the kidneys where the water is stored and compensate the deficiency. This kind of homeostasis may occur in the body only on certain occasions and not at all the times. If the water deficit is very severe thirst becomes very strong and the organism is forced to go in search of water.

The human body requires sufficient amount of water for maintaining its balance. When there is dryness in the throat and mouth we feel like drinking water or something in liquid form. A thirsty animal is strongly motivated to seek water and its behaviour is directed towards this goal, because the tissues of the throat give the signal that the body is feeling thirsty. Deprivation of water for longer duration, may result in death of the organism.

Thirst can be eliminated when the dryness in mouth and in the throat tissues become wet by intake of water or any liquid. Though a mouthful of water is sufficient to wet the mouth and the throat we often go on drinking much more than that. Sometimes we drink even five or six glasses of water. This means that there is some physiological mechanisms operating in us. How much water exactly the body needs? In case of hunger, the assumption is that the *lateral hypothalamus* (LH) (though it is controversial) is the *centre for hunger.* But in the case of thirst we must know the centre for thirst in brain. The human body is continuously loosing water in perspiration, inspiration, and in the formation of urine. The body has no control over the water loss in breathing and sweating.

Many of our body functions constantly require water as we use up much water, for example, when we pass urine to get rid ourselves of *waste molecules.* By giving up some water molecules, we shed down the excess body heat. We are obliged to loose some water molecules and therefore, we are obliged to get some more water. The water in a body moves back and forth between two major compartments. The whole water in our body is contained within the *billions of cells* that make up our body. This is the *intra-cellular compartment.* But some fluid is outside of these cells, in the *extra-cellular* fluid compartment. The extra-cellular fluid compartment is a *buffer, a reservoir* which provides and accepts water molecules. By this means the cells can maintain proper interval conditions. The extra-cellular fluid (plasma) is inside the body that surrounds all cells. Further the nervous system is responsible for making sure that the extra-cellular fluid compartment has sufficient amount of water and allow the cells to absorb or shed water molecules according to the conditions.

The extra-cellular fluid compartment, apart from being a reservoir, is also an indicator of conditions in the intra-cellular fluid compartment. Actually the nervous system carefully regulates the extra-cellular fluid compartment to determine whether we need water. There are two cues as signals to indicate whether we need more water, one is the *lower extra-cellular volume (hypo-volemic thirst)* and the other is *higher extra-cellular solute concentration (Osmotic thirst). Solute* is the solid component that is dissolved in a liquid.

The Hypovolemic Thirst

This is activated by a *loss of water* volume. There is also *loss of blood.* This includes both solutes and water. But the blood loss is not experienced. Any organism that looses a significant amount of blood has a total low blood volume. The *blood vessels* which should normally contain, full, but now it is not containing to the full capacity. The blood pressure drops and the organism becomes thirsty. *Water is lost from the blood loss,* which does not change the concentration of the extra-cellular fluid. It is because *salt and other ions* are lost with the fluid. In these processes only the volume of extra-cellular fluid is affected. But the continued loss of fluid in the extra-cellular compartment makes the fluid passing out of the intra-cellular compartment. Therefore, the brain at this stage is informed through the sympathetic portion of the autonomic nervous system, of the initial drop in the extra-cellular volume, and the brain initiates thirst to replace the water and salt hunger to replace salts.

Osmotic Thirst

Osmotic thirst is activated when the total volume of water is constant. The *osmotic thirst* is aroused by a change in the concentration of extra-cellular fluid. The more common way of thirst than by the loss of blood is by the loss of water through *respiration, perspiration* and *urination.* In this process though some salt is lost with water, more water is lost. Already there's some reduction in the volume of extra-cellular fluid and so the *solute concentration* of the extra-cellular fluid increases. Since the body has to maintain the balance of the intra-cellular and the extra-cellular concentrations, water is pulled out of the cells of the body. The solute concentration can be increased without increasing the volume of the extra-cellular fluid by taking more salty food. Thus the increase in the solute concentration triggers a feeling of thirst which is stated as the *Osmotic thirst according to Epstein* (1971). This thirst is independent of extra-cellular volume. This osmatic thrist makes us to seek water for protecting the intra-cellular fluid compartment from becoming deficit of water which may damage its cells. Therefore, the organisms will have to take sufficient water.

SEX DRIVE

The discussion of physiological motives or drives in psychology is not complete without the inclusion of the sex drive which is considered to be one of the important physiological drives. The status of sex as the physiological drive has been recognized 40 years ago and it is considered as one of the important subject matters that the present generation should know something about it. The very production and survival of the species in the world depend upon sex. Sex is a very, complex motive. It is one of the organic needs. The primary sex organs, *overy* in female and *testies* in male produce the *reproductive cells ovum and spermatzoon.* They also secrete hormones that are important for the growth and development of behaviour.

The human body secretes several hormones. Some of them are present both in male and female. The male hormones guide the development of masculinity and the female hormones guide the development of femininity. At the time of puberty these sex hormones facilitate the development of the genital organs. In women the sex characteristics such as development mammary glands and in men growth of beard and deep voice. The sex hormones increase during adolescent period and they are essential for the development of sex interest.

The sex drive is regulated by the hormones in animals. The male hormone, androgen secreted by the male testes and the estrogen secreted by overies of the female. The antrogen stimulates the male to become active sexually, and the estrogen stimulates the female to become heat and accept the male. But in human beings sex is something more than the hormones. The initial drive for sexual activity, after puberty comes from the *nervous tensions* within the body set up by the sex glands. The *urge for sexual activity is physiological and organic.* The relative strengths of the drives of hunger, thirst and sex have been studied experimentally in animal's behaviour using an experimental box (Garrett, 1975). An electrically charged grid in the box is used which the experimental animal should cross. The box is kept at one end and the animal at the other end. The animal is made to cross the electrically charged grid to reach the food at the other end of the apparatus. The rats are deprived of food, water and sexual partner for some period of time. Then the animals were allowed to cross the grid and the number of times counted as the strength of the drive. It was observed that the animal had *thirst,* the strongest drive. Then the *hunger* drive for food and thirdly, *the sex drive for mating.* It was also observed that deprivation for longer period of time made the animal's (rat's) *physical condition get weakened* and they become very *aggressive.* In the case of human being the *strongest was the hunger drive* and secondly, *the sex.* Since eating is essential for servival of the individual, hunger becomes the strongest.

Human sexual drive behaviour is determined by the environmental factors such as social, culture and economic status. Society has certain conventions, rules and regulations in varying degrees in the fulfilment of many drives, not only in sex, but also in other drives like eating and drinking, for example, in eating the type of food, mode of eating, and time of eating, the items used for eating the breakfast are not used in the noon or night for lunch and dinner. So also, there are restrictions on sex, the kind of man or woman, the nature of relationship between the man and the woman in extending love and affection. The man should love and marry a woman not within their blood relations, i.e., brother and sister and cousins. Fulfilling the sexual drive has several taboos in Indian community, like moral, cultural, religious, communal and sects. One should not break away or *violate these taboos* and have the sex drive fulfilled, with all these, some privacy is needed for sex drive.

Candland (1968) states that experimental psychology has not spoken openly giving evidence to the view that sexual relation as a basic drive along with hunger and thirst. **Sigmund Freud** (1856-1959) says the *sexual energy* is one of the drives. Sexual energy is an *innate drive,* but it is capable of changing the characteristics through *learning and experience.* Thus sex drive is given important place along with the basic drives, hunger and thirst. But sex strongly differs from hunger and thirst in one respect. A man may die without food and water but man may die without sex, has riot occurred. Therefore, there are more compelling reasons to restrict sexual drive from the list of primary drives than the reasons for including it among the primary drives. **Beach** (1956) states that sex drive is *not a homostatic mechanism* as hunger and thirst. Experimental information are available. There are also a number of statistical information available from the studies made by **Kinsey, Pomeroy, Mertin** and **Gebhard** (1953) on the frequency of various forms of sexual behaviour of human beings, on the basis of age, education and economic states.

The human urge to fulfil all other drives is not covered by the society with taboos and controls and they are given social significance. But in the case of sexual urge man must observe certain conventions. Sexual behaviour is more important for *young women than for young men.* Actually immorable behaviour in women refers to conventionally *improper sexual conduct.* This double standard of morality has a long history. But now-a-days, it has become much less, as women are taking more important roles in government, business, politics, education, art and religious matters. Society recognized it as socially acceptable ways of reducing the tension arising from the primary sex drive.

Sex Drive : Hormonal Factors

The *sex drive* seems to be *under the control of nervous system and the hormones.* In the matured male person, the *gonads* and the associated glandular tissues produce semen and it is stored in the *seminal vasicle.* The accumulation of the semen should be drawn by intercourse. Otherwise emissions may occur periodically. The *occumulated semen* if it is not *spent* by regular *copulation* and ejaculation it may *act as a sensitizing* stimulus and arouse sex drive in male members. The function of the *endocrine glands* show that the sex drive in male and female and the production of germ, plasm are under the control of the endocrine glands. The sex drive in man is dominated by learning. But the sex drive in women is mostly independent. Because of this, sexual behaviour continues in both sexes with aging even after the ceasation of the germ plasm and production of sex hormones. But with the past sexual experience the level of sex hormones to arouse and sustain sexual behaviour is not necessary in human adults.

The human sexual behaviour appears to be the *joint functions of the hormones and nervous system* but the important area of interaction of these two factors is the *hypothalamus* and its connection with the *master gland, the pituitary gland.* At the adolescent stage the secondary characteristics appear because of the increased output of the gonadal hormones. The sex hormone level regulates the sex drive in lower animals also. The level of androgen output and the sex drive is constantly maintained by the testes in adult males. In females the *fluctuating estrogen level* of the monthly cycle causes them to come into heat when the female egg or ovum is produced which is to be fertilized by copulation with the male.

In lower species sexual behaviour can be abolished by *removing the gonads* either before or after puberty. But normal sexual behaviour can be restored by injections of sex hormones. In human or in lower animals *early puberty* can be *induced by injecting sex hormones* in either sex. The development of nervous system in higher species is as important as the hormones for the sex drive in primates and in human. The development of the central nervous system is having increased influence on the sex drive and is having more effect in learning sexual behaviour. One of the important aspects in sex drive is that the *gonadal hormones are regulated by the pituitary hormones* which in term is controlled by the hypothalamus.

Sex Drive : Neural Factors

The effects of neural factors in sexual behaviour play important role. The eraction and ejaculation are taking place at the lower level of the nervous organisation. They are mainly *spinal reflexes.* They can be elicited by genital stimulation in the human spinal cord. The *erection* is mainly *regulated* by the *parasympathetic nervous system and the ejaculation by the sympathetic nervous system.* Therefore, these reflexes come under the control of *hypothalamus part of the brain.* The hypothalamus controls the autonomic nervous system. It is also likely to occur that the hypothalamus regulates the output of *luteinizing hormone* (LH) which is a hormone of the anterior pituitary gland that stimulates the gonads to produce sex hormones (androgens or estrogens) and the hypothalamus also regulates the *follicle-stimulating-hormone* (FSH) which is also a hormone of the *anterior pituitary gland* which *stimulates the gonads to produce the reproductive cells, sperm* or *ova.* The hypothalamus is sensitive like the sex hormones in the internal environment and controls the sex drive. The hypothalamus stimulates gonads which release sex hormone amount via pituitary luteinizing hormone and the sex hormone level evokes the sex drive in turn by stimulating the hypothalamus. In this process more comprehensive neural structures are involved. In all these the hypothalamus plays a very important role.

Harmones' Role in Sex Drive and Sexual Response

Sex hormones play a permissive role in human sexual behaviour. In the case of animals, particularly rats, little amount of *testosterone* (hormone produced by male gonads) is in circulation for the male to initiate mating. Any additional testosterone has no effect, on the mating. Therefore, there is no relation between the amount of androgen (male sex hormone produced by testes) produced by an individual and his more to copulate. What is taking place in an animal is also taking place in a human male. In the study of **Davidson** *et al.* (1979) it was observed that some boys fail to reach the adult stage of physical development and unable to release *gonadotroins* (hormones that stimulate the gonads to produce hormone and reproduction cells), prevent the testes to produce testosterone. These persons should be treated with androgen but it is not clearly known whether the androgen effect the human copulatory behaviour. But however, these studies show that *androgen really stimulates man's sexual activity.* It is observed in these studies that low androgen level sometimes causes infertility in man. The level of *testosterone gradually declines* in men over 60 years, which indicates the decline of function of gonads.

Some women after manopause report of reduced interest in sex. This may be due to several *hormonal changes.* But estrogen treatment produce increased lubricant (reduce friction). Some investigators report that there is some increase in sexual behaviour during the time of ovulation in women.

Sexual Responses in Men and Women

The sexual responses in males and females vary considerably. Sexual excitement increases as the result of stimulation in males. The stimulation may be physical or mental or both. This is followed by blood glutting in the penis which makes it erect. The level of increase of excitement varies due to sexual functions. It may also vary from person to person. What causes excitement in some persons, may reduce the excitement in some persons. The level of excitement may reach a maximum level and some persons may wish to retain the level as long as possible. Finally, *orgasm* is reached following the *ejaculation.* After this the excitement is vanishing slowly and the erection of the penis is lost. Generally the excitement stage and the process of action stage are the longer duration. The maximum erection stage lasts for only few minutes and the climax of the sexual experience (orgasm) exists for less then a minute. In the *male sexual experience* there are *two sex complaints* commonly reported. The man often fails to achieve the expected *erection* (erertile dysfunction) and the other is premature ejaculation, that is the male or female partner ejaculating before achieving the desired excitement in the sexual behaviour. **Masters** and **Johnson** (1966) state that in most of the cases the *anxiety was the cause* of the failure of achieving erection and the premature ejaculation. Masters and Johnson provided men with some techniques to reduce their anxieties and tension for achieving the full erection during their sexual activities (Fig. 18.3).

Masters and Johnson observed similar pattern of behaviour was most common among women. In the excitement stage the *clitoris* (a small heep of tissue just anterior to the vaginal opening) expands due to accumulation of too much of blood in that area and the lubricating fluids increase. The muscles around the vagina are relaxing. The rhythmic muscular contraction in women is the indication of orgasm. The stimulation of the clitoris makes the orgasm easy in most women. *Women are some what slower than men to reach orgasm* during intercourse. Just like men, women have two most common complaints. One is the discomfort experiencing during intromission (dysparevnia), that is, insertion of the erect penis into the vagina during copulatory behaviour and the second is the difficulty to achieving orgasm. The muscles around the opening of the vagina may experience sudden involutary contraction *causing pain*. But all women do not feel the need to achieve orgasm to be satisfied with intercourse. Some who seek more frequent orgasm need the treatment of relaxation that may help easy orgasm.

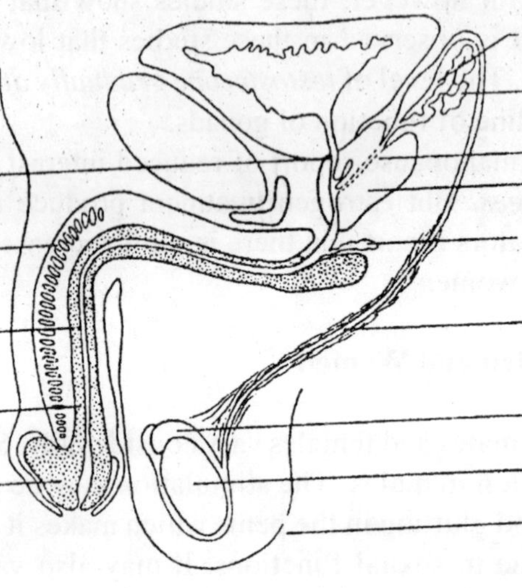

Fig. 18.3 : Sex Drive : The Male Reproductive System and the parts

Generally in sexual responses there are similarities and differences. In some women the difference may be due to *genetic make up* and in some it may be due to *hormonal levels* (Fig. 18.4).

REST AND SLEEP

The need to rest and sleep are also considered as *physiological drives*. They are primary needs. The body, when get tired of continuous work, needs rest to get rid of the fatigue. There are always some fatigue products in our blood. The *collection of waste products* in the muscles and *in the blood stream* reach a stage that they interfere in our activities and we feel *drowsiness.* **Murphy** (1951) says that sleeping is a matter of degree, because there is *no absolute sleep* and also absolute awakening.

In sleep the central nervous system is involved. The high level of central nervous system activity is accompained by much restless behaviour. Therefore, sleep may be defined as the *periodic state of physiological rest* during which unconciousness is suspended and the *metabolic rate is decreased.* Certain amount of sleep is required for the function of normal central nervous system. Sleep deprivation results in bodily discomfort and the alertness for motivated activity suffers. Experimental studies on animals show that *forced and long continued wakefulness ends in death.* Therefore, sleep is a definite need for existence, like

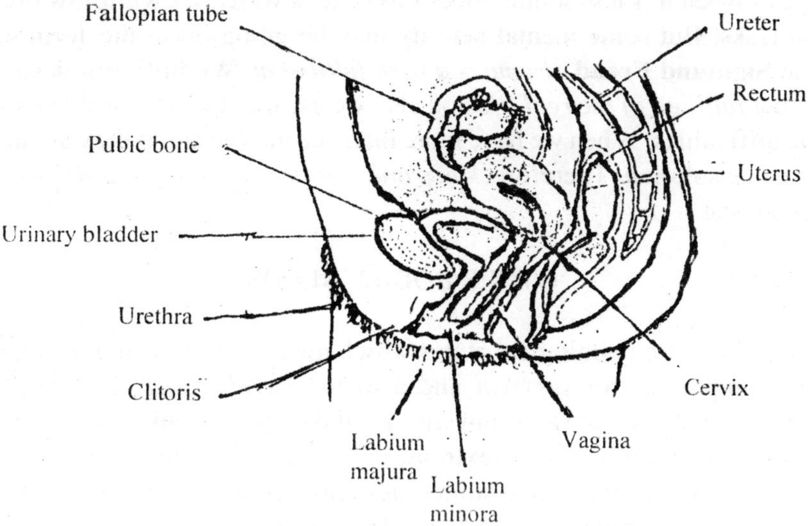

Fig. 18.4 : Sex Drive : The Female Reproductive System and the parts.

food and water. According to Kleitman's (1963) evolutionary theory of sleep as wakefulness, *sleep is more natural* or homeostatic than that of waking activity. It is natural for animals and children. However animals and infants sleep unless they are aroused by bodily needs such as hunger, thirst and discomfort. A human child sleeps most of the day and night, except during the time of fulfilling the needs of hunger and thirst. Kleitman says that the wakefulness is, 'Wakefulness of necessity' since it is absolutely necessary for the infants to awake to satisfy his hunger or to remove the source of discomfort. With growth and development the child sleeps for longer periods. The *wakefulness of necessity is 'Polyphasic'*. This means that the wakefulness occurs *in several phases* during the course of 24 hours day. The continuous waking period, Kleitman calls, *wakefulness of choice*.

The cerebral cortex develops in the evolutionary process with an increase of learning ability. This varies from species to species. In this need satisfaction is associated with daylight. There are more stimuli during the daylight for awake the animals. The brain cortex controls the centre which regulates sleep. The animal learns to remain awake during daylight hours. During this period the cortex is stimulated by the environmental factors. The cortex ceased to stimulate the sleep centre during night time.

Whatever factors are associated with sleepiness, the depth of sleep varies with individuals and from time to time. We fall asleep rapidly for an hour and then slowly wake up during the rest of the night. Sleep is a *complete rest for both mind and body. After a good sleep* and rest we start our work fresh and with *pleasant feeling*. Sleep is always associated with darkness, as we sleep during night time, in a comfortable bed. When we go to sleep we give farewell temporarily to other needs and conflicts. *A good evidence of our sleep is the dream*. There may be *dreamless sleeps* if it is *very sound sleep*. Dreams occur mostly when you reach the waking stage or when it is less sound. Sleep is a state in which we withdraw ourselves from the physical tasks. But some mental activity may be going on in the form of dreaming. According to **Sigmund Freud,** *dream is a wish fulfilment.* We fulfil our desires in dreams which were *not fulfilled in the real life* and also we escape from the problems to be solved and from the difficulties. When we have more than one motive and if they are in conflict our dreams find some solution. Therefore, *sleep and rest are physiological drives* as important as hunger and thirst.

OTHER BODILY NEEDS

There are a number of physiological states which make us to respond adequately. These are extremely important for *survival* and also *to provide safety and comforts* to the organism. We often try to get rid of physiological discomforts like loosing balance at an uneven flooring. We try to get rid of extreme cold and hot conditions to our body as they are causing very uncomfortable and sometimes make us disorganized. There is always a physiological need to eliminate the waste products from the body. We try to avoid painful instance. Anything that we try to avoid is called *aversion drive*. We move away from the pain causing objects, unlike hunger, thirst and sex where the movement is towards the objects, food, water and mate. Pain is an *uncomfortable sensation* we experience. Painful feelings may occur in our parts of the body, like headache, stomach pain. We are taught to endure certain pains. For example, in playground or in a match played, the player get injured and experience painful feeling. Similarly when we take up an injection from a doctor, we are enduring pains.

Activity Drive

It is also experienced that some parts of our body often crave for activity which were called the *activity drive.* The physical basis of these activity drives lie in the central nervous system and in the large muscles of legs, arms, neck and turnk. These muscles are voluntarily controlled. They are often getting restless, if they are kept without any action.

They are always contacting and relaxing in their own rhythmic ways. To keep the body perfectly still without any movement is a punishment. Children are given this type of punishment. Soldiers though normally trained to 'stand-still' in attention position, as a punishment for committing some mistakes, made to stand in attention position for longer time to experience the hardship. **Murphy** (1951) says that life indeed may be regarded as an activity stream in which the vital organs play important part. **Wenger** (1947) studied this aspect and found that the amount of general *muscle tension* is related to the total personality pattern, while one person is able to relax another person is not able to do it.

Another type of activity drive is the tendency to keep on doing something which we have started to complete. We find it hard to leave some work when it is about to finish. Every part of our body has its *own tension for its activity.* Every activity that is performed by the part, contributes to the bodily response for fulfilling the needs or drive.

At times our sense organs are stimulated and *the corresponding brain parts* are aroused and the response may be *excitement or thrill*, delight or some kind of pleasure *through the sensory apparatus*. We find pleasure when we are stimulated by some beautiful scene, colours, musical tone, taste, smells and so on, when the specific brain areas were aroused. Children have wide *varieties of eagar* and show *vivid responses* to things with excitements.

Learned Drives

The physiological drives are *unlearned drives*. These are the body needs for its *survival* and *existence*. But the learned drives or acquired drives are needed for the individual for the *assertion of the self*. For example, an individual desires to become a musician, a dancer or a painter. These are all *acquired drives*. In these there are large individual differences. Thus, the type of drives are *both unlearned and learned*.

A person in the company of others may desire to have the same drink or snack as others are having for a light refreshment or witness a game of tennis or a person may develop interest and desire to start playing the game. He may develop strong desire to master it, because it brings him appreciation, fame and if it is a professional line he gets also money. Thus it is a *learned drive* and also he *gets pleasure* in it. Further, when one gets into the company of some people, he desires to keep himself likable by others in all respects. Hence he loves himself when he becomes proficient in some area where he can exhibit his skills and talents and strive to be successful and register himself as a man of distinguished characteristics. In this way one slowly *builds up a status and enjoys his status in the society*.

Thus the motives for status, prestige, fame and power become very important in *social life. The motive for status, prestige and power are acquired slowly in the society*. These learned or acquired motives *are complex* as they involve several factors outside of the individual. *These drives are fulfilled* in the *social structure* and *in the company of others*.

PSYCHOLOGICAL MOTIVES

Emotions as Motives

The psychological motives are not as strong as the physiological or organic motives. Psychological motives are mainly based upon feelings and emotions. The expression of any feeling or emotion is a psychological expression. For example, if there is a sudden pin prick on some part of our body, we at once give a jurk in our body and also raise a voice 'ah'. This means that there is some incident regarding our feeling aspects. For example, *emotions are treated as motives*. Emotions at times act as powerful motives and often *direct human behaviour*.

Generally, we classify emotions as *pleasant emotions and unpleasant emotions*. Pleasant emotions motivate people to perform acceptable and expected activities. For example, a person is *strongly motivated* to become a *songster or musician, another motivated to become a professional dancer*. These persons' behaviour are accepted and the audience appreciate by raising voices, and clapping hands. Here the musician and dancer are performing the pleasant activities and the encouragement of the audiance *motivate them for further performance.* Both the players and the audiance express the feeling of joy, pleasure and happiness, *joy, pleasure and happiness are emotions* and they are strong and tend to express in these levels in the *release of energy.* The persons and the audience are highly *excited.* In these days the music is associated with dance. A group of people in pleasant mood sing and dance and express their excitement in high tones. The rock and roll or rock 'n' roll is a form of popular music originated in 1950s as a blend of rhythm and blues. It is a kind of dancing, performed to such music with exaggerated body movements stressing the beat. This type performance makes both the players involved in it and audience to express their joy, sometimes with very louder voices. The *more louder their tones, the faster their movements* are and *greater their excitements.* They are *emotionally stirred* up and *thrilled* very much. The milder states of their happiness lead to a sense of well-being and relaxation. A physically healthy child who is very comfortably laid down and nurished may try to express his joys by his *threshing activities.*

The Emotion of Love

Similarly the *emotion of love,* which is tender produce *affectionate responses* in concerned persons. *Love* is extended by a person to his close relatives, kith and kins, friends, liked human beings, and also to animals. We express our love to human infants and animals in various ways. To smiling, petting and fondling gestures the infants and animals respond with very fondling feelings. The emotions of love and affection have generally *reciprocating characteristics.* There would not be any offending response to love and affection. The child which receives the love and affection from his mother and others, slowly learns to extend the same to others by his smile and other gestures and makes them happy. Human adults respond with the feelings of joy and pleasure when they are honoured with praises, rewards from friends, organisations and community. Persons who are rewarded for gallentry services rendered to the society and nation are very much gratified, which may motivate such persons to further heroic performance. Therefore, we may state that the exercise of the psychological motives of love and affection provide us or receiprocate with *a sense of affectionate behaviour.*

Grief

In the everyday life humans are likely to experience the emotion of grief or sorrow when they *face with some unhappy events.* We often speek about joys and sorrows. The emotion of grief or sorrow may be caused by some disappointments, discomforts, deprivations, fear of punishment and also death of our close relatives. The, grief or sorrow is generally *expressed by our feelings* of mourning and weeping and sometimes by keeping utter silence. In adult persons the motive for mourning or weeping behaviour may be caused by some *irrepairable loss, severe frustration,* feeling of *helplessness and conflicts of motives.* Persons, suffering from *incurable diseases, physically disabled* are unable earn their daily bread, and *those who are unable to take care of their dependents* and the destitutes often mourn and weep. Shedding tears, weeping and mourning are the expressions of the *motives of grief and sorrow.* But some shed tears out of extreme pleasure or happiness.

The Emotion of Anger

Anger or rage may cause increased physiological activity. Since this produces some type of behaviour, it is also called as a psychological motive. The *manifesting behaviour for anger is struggle.* The motive, anger may also *mobilize* the body's *physical energy for action.* The organism gets more strength or energy when it is highly motivated by anger. In young animals and in human infants the *struggling behaviour* can be witnessed. When we restrict the movement of a young animal or even the grown up animal like a dog, or cat or cow, brought up at home they try to get rid of the restriction by behaving in various ways like pushing, pulling, tracking and jumping. In these ways they struggle to express the anger. Sometimes some pet animals may even bit in the way of struggling. Similarly, the human infant when he is restricted from his movements by holding the arms and legs, the child would begin to jump, kick, bite and scream. In all these ways the child behaves and expresses his anger. The *response of struggle against the restriction* is also *an act of stimulus to arouse anger* which is a reinforcement of the struggle. In adult human beings restriction of any type may result in various forms like *protesting it ideologically* and also *behaving violently, even damaging prosperties.* This may result even in social, economic and other physical disturbances. Modern citizens show their anger in various forms such as *strikes, walkouts, fasting, refusing to do the duties* and so on. In a very angry state a person is *physically disorganized.* He is in a *stirred up state* and he mobilizes his bodily strength for *fighting physically.* He gets more energy or strength for *violent attack.* Thus the emotion of anger as a phychological motive, manifests in the *struggling behaviour* of various ways. **Woodworth** (1952) says that pure excitement is neither anger nor fear. An *anger is an excited feeling,* it has its own impulsive quality which is different from the impulse of fear. These impulses generally are directed towards *overt activity,* it may be towards attacking the object or person of the cause of anger or towards escape in fear.

The Emotion of Fear

Fear is another *unpleasant emotion* serves as a *psychological motive,* manifesting in various types of behaviour patterns. The emotion of fear may be also stated as *startled pattern, surprise, frightening, and alarming.* Excitement also is an accompanying behaviour. But the most *common manifesting behaviour* in animals and in human beings is the *act of escape from the situation.* The psychological motive of fear in a startled pattern facilitates the organism to run and escape from the danger. Fear in young child may be aroused by some *painful stimulus.* Generally the *psychological drive of fear arouses safety measures.* In a dangerous situation, like, a tiger suddently appearing in a garden, an individual's immediate action is *to run way* from the scene and escape from the tiger's sight. On seeing the tiger some physiological changes are taking place, mobilizing the energy or strength for the escaping behaviour. **Woodworth** says that it is *a mass sensation from all over the body,* mostly from internal organs. In the case of a child a sudden and intense painful stimulus may cause him to cry, whimper, avoid, withdraw or escape. These may be the immediate responses or behaviour of the child because of the fear. Fear may also cause anger. In case, if escape is not possible, then the organism may use all the forces and attack the object which is the cause of fear. The people of present communities, if face any fear causing stimuli, if unavoidable, make themselves ready to *face the situation by violence.* In such instances the emotion of *fear becomes a motive to arouse the emotion of anger.*

Pain

Pain is a feeling physiological in nature, but it is a *psychological motive.* The *pain drive* is called *an aversion or avoidance drive.* Pain may be defined as the *sensation of acute physical hurt* or discomfort caused by some injury or illness. It is *an emotional suffering* or even mental distress. A man who committed some crime is subjected to the penalty of pain. He is punished with some painful act. Since the consequence of pain is emotional suffering it has to be avoided.

The person who is subjected to pain should find out some means to remove the source of the physical discomfort or annoyance. For example while walking in the garden with bare foot a thorn might have pricked your foot. It may cause some *irritative pain* and the *immediate reaction* to it is *to remove it.* Always there is a movement away from the pain causing incidences.

Generally pain is an unpleasant emotional experience that may *cause suffering.* Pain guides *adoptive behaviour* by indicating potential harm. The experience of pain makes us to remove the body swiftly from the source of injury. It makes us to avoid and protect us from getting into injury.

Pain is the most *potent stimulus* known to arouse and to *maintain behaviour.* Therefore, it is important in the study of drives. **Leukel** (1968) states that pain stimuli set off *somatic and visceral reflexes.* The human body makes *first line of defence* against injury is the *somatic withdrawal reflexes.* The *second line of defence* is from the sympethetic nervous systems which may be called *emergency responses.* These are through the viseral, glandular and autonomic systems. *The third* stage is the *realization of sensation* which directs to drives and it results in the perception of the sensation of pain.

Melzacs (1984) has studied extensively about the pain drives. The mechanism of human pain can be explained through pain receptors. The *nociceptors* are the *pain receptors.* Wherever we experience pain, there the nociceptors are found. Therefore, in human body there are *cutaneous pain receptors, organic pain receptors* and *kinesthetic pain receptors.* The cutaneous pain receptors also *exteroceptors,* the organic receptors are *interoceptors.* Pain always gives the signal of injury. So that we can try to avoid it. In this way pain serves as *an aversion* or *avoidance drive.* Therefore, *pain is a sensation as well as a drive or motive.* Leboratory studies show that pain can be measured by employing suitable methods.

Social Motives

Man is a social animal. He lives only in the society, and makes his presence in society in various ways. *Social motives* indicate a person's involvement with other people in the society. He shares the activities with others co-operatively. The *social motives* may be also *directed towards the welfare of the members* and for the *progress* of the *members of the society.* **Campbell** (1963) uses the term *"acquired behavioural disposition"* in place of social motive. Psychologists describe the social motive and the related motivational states in a number terms which are somewhat synonym to motive. Campbell has given a long list of seventy six terms which run alphabetically.

Some of the important terms according to Campbell are, acquired drive, attitude, frame of reference, belief, idea, need, set and so on. *Motives produce behaviour.* The social motives are considered *as needs,* specifically *psychological needs.* A need is basically lack of something vital, important, for the survival of the organism. It indicates a deficit or need. The need also implies *a given goal.* Man must act or behave in certain specific ways to achieve the goal. Behaviour vary considerably from person to person in fulfilling the needs.

There is a tendency in man to be friendly with his family members, relatives and friends. The same tendency one may extend to be friendly with the members of the community. Hence he loves his parents, brothers, sisters, wife, children, and also outside members. This tendency to love people is stimulated by the *motive of affection.* The individual's motive of love or affection may be extended to a much bigger circle of people which may be his society or the nation. We have discussed love and affection under psychological motives in the preceding section. Though love and affection are psychological motives they have social significance, as the *feelings* may be also directed to the *outside larger group of people* or society as a sign of the *individual's social motive.*

People are interested to live in groups, because they will have to do so many things together. For example, conversation, play, games, work, participation in common activities are the major goals of individual's existence. Therefore, there is a tendency in every human being to make something in his everyday life, towards which man strives hard and puts all his skills, talents, knowledge and experience. Man's social participation and team work contribute to achieve his goal. The individual is stimulated by the social motive to achieve his goal. The man's goal may be anything. In some tasks, people with high skills and talents do well much better than others and achieve the goal. Such persons assert their selves as victorious. In a society some one motivated strongly and some are moderately, because some may be highly skilled and talented. Therefore, it is possible for those *who are strongly stimulated to achieve the goal easily* and others may struggle hard to achieve the goal.

Further, those who are strongly motivated may learn the task easily, faster than others and those who are more skilled may be motivated strongly and hence they achieve the goal easily. These people assert their self and also identify themselves in the society. They often *receive appreciation,* from others and *also rewards.* This further stimulates their social motives.

There is also relationship between *goal achievement and intelligence.* More intelligent people may be strongly stimulated to make achievement in the society and assert themselves. It is also possible for some to develop the fear of failure of achievement. A person who is fearful of failure to achieve may try some easy task and may not be interested in attempting hard task. There may be reasons for this type of goal achieving attempt. Some might have been given training right from the early years, while some might have been discouraged or restricted in attempting hard tasks. The individual might have tried independently and got trained well. **Morgan** (1938) states that the *achievement motive* is strong in such persons. Further, it has been observed that there is some correlation between training and achievement.

Children often play in groups and they may develop more competitive tendency than those of less socialized in the controlled situations. From such experience in this group, children develop the *social patterns of behaviour* like gregariousness, co-operativeness and competitiveness. In these situations the self in relation to others realizes and become more aware of the relationship with others (Goodenough, 1938). There is *interaction between social motive and social learning. Attitudes, values and beliefs are also social motives,* learned through contact with others and they determine the type of behaviour people should display.

The social motive is highly influenced by the position, status and role. The status one enjoys in a community is always dependent on how others perceive and value him. What amount of prestige others are willing to grant him. The person is able to influence others by the amount of status and dignity possessed by him. The status of a position may be described only in terms of respect one commonds from others and also by the prestige, values, occupations held by the individual. The position one holds is valued by the social roles attached to it. Once an individual is attracted by these *position, status* and the *social roles* he is *highly motivated to achieve them* and *he enjoys them.*

In any society there is *an order in dominance.* The individual tries to assert his self in various ways. Even among the lower organisms this tendency is there. For example among the hens, the dominent hen pecks all others hens beneath her. There may be also next, most dominent hen and down to the most submissive hen. Similarly the order of dominance is found among the *pigeons* (doves), among the *wild horses* and among the *wild asses. The same tendency* also prevails *among human beings.*

The *mastery motive* in human beings reveals the desires of the person to *dominate over others.* The urge to dominate over others determine his behaviour. This urge of a person makes him to perform a dominent role through his style of life, the role he displays and so on. *It is the motive* that produces *these behaviour patterns* in the society *to be dominent.* The *dominance-submission* is an *instinctive tendency* found in almost all creatures, insects, bees, birds, animals and also in human beings. Among the human beings there is a desire to have dominance over others and also to exhibit the dominance. There is a *pride to show* that *he is the chief of all. This is stated as the mastery motive.* The person demands respect from everybody and expects that he should be worshipped by all.

The mastery motive is very complex, because it involves many things. It is not only in making one as the dominant person in a group or society, but also it involves *restricting other's* physical, economic, social, political, activities and also determining and *limiting others' liberty* of movement. In the attempt of asserting one's dominance over others, it often results in the *expression of aggression by punishing others,* on finding *negligence in the submissive behaviour.*

When an individual tries to assert himself in the society there may be *blockings* for the assertion of the *social motive.* This leads to aggression. The person projects his aggressive action against the person or situation blocking his behaviour. The aggressive action may vary in several ways. The child exhibits all sorts of temper tantrums. The *adolescent makes violent revolt, the adult argues strongly.*

Aggression as response to frustration is expressed directly, sometimes even by violent attack on the person or object which stands on the way, blocking the attempts. **Coleman** (1976) says that the *typical reaction to aggression is counter aggression.* Further the reaction to one's fear aggression is getting ready for **counter aggression.** Aggression is not the only motivation which leads to group conflicts. There are also other personal reasons for aggression which may be competitive in nature in the struggle for power.

The political leaders who become rulers, become very aggressive and power crazy and always try to *eliminate their political opponents* by whatever means possible for them. This kind of mastery motive was prevalent right from the ancient times onwards and the present generation of rulers are no exception from such techniques, in fact, *the leaders of the present times follow more sinister techniques than the ancient leaders.*

Among the social motive, the *motive for power* which is intended *to establish one's supremacy is more dangerous than any other motive.* The ambition for power, greed, rivalery and ideological fervour are key factors for political warfare with opponent leaders. *The power motive* often *hatches great conspiracies* for eliminating the political opponents. Once a man has enjoyed the taste of power, he becomes more and *more greedy* and nerver wanted to give it up at any cost. The *power motive* has both *psychological* and *sociological significances.* From the psychological point of view people differ in their feelings of being powerful. **Winter** and **Stewart** (1978) state that the western civilization greatly enlarged the scope and the range of social power and to control the human behaviour. *Power motive* means the quest for power, the desire for power or seeking of power. The power motive is not conscious. It does not always leading to a single or a set of behaviours, because the behaviours that lead to power depend upon the situation.

The *power motive* may be in *conflict with other motives,* such as affiliation, play and autonomy. The essence of power motive is the ability to make material world out the social world and to conform to the *individual's own image or plan.* One must be visible above all to attain power. The power motivation in political life often directs the individual towards certain positions, the one with high power potential and activities where the power greed is likely to be enhanced.

THE HIERARCHY OF MOTIVES

The term motivation is applied to various psychological processes. Everyone who is specialized in certain area of study tries to incorporate motivation and explain the subject matter. For example, in the area of learning, motivation is used invariably, like motivation and learning. Every theory of learning involves motivation to some context. In classical conditioning it is presumed that some sort of drive reduction is involved. It is also stated that **Freud's theory** is chiefly a psychology of motivation and he elicited motivational control in behaviour.

In the school of *functionalism,* **Woodworth** was the first to give *motivational treatments.* **The Gestalt school** accepted the role of *reward and punishment, motivation* was viewed as putting the organism in the problem situation, reward and punishment acted to confirm the attempts to solve the problem. In *Guthire's learning theory,* it is stated that the *motivational state* is important because it determines the presence of movements which may go into associative connections. The motive is important for the *stimulus-response-sequences* like eating, drinking and mating.

Motives are important for providing and maintaining stimuli which *keep the organism active until the goal is reached.* In neurological theories, it is stated that the neurological mechanisms are involved in the common *biological motives.* In **Skinner's system,** reward or reinforcement is necessary to increase operant strength, and the motivational terms, interest, apathy, refer to the typical consequences. In **Thorndike's theory,** it is stated that the *reward as a motive* is used to strength the act. **Tolman** states, learning is used when drives are activated and when the drives are aroused a state of tension results which is a demand for goal objects. Even in personality development motivation is invariably used.

Thus *motivation cuts across all topics in psychological studies,* like attitudes in social psychology. Recent studies also put more emphasis on motivation. **Abraham Maslow,** (1908-74) the founder of Humanistic psychology, deals with motivation in a more elaborate manner involving all aspects of human behaviour. He has given a psychological interpretation to science and states that *science is a human creation* rather than autonomous, because it has its *own intrinsic rules.* According to him *science has its origin in human motives* and the goals of science/scientists are *human goals.* It is created and often renewed with new discoveries and maintained by human beings. Science discovers the *nature of reality* and also *discovers the nature of human beings.* Scientists are motivated like all other human species. The species have several needs, the needs for food, needs for safety, needs for protection and the needs for care, the needs for gregariousness and the needs for affection and love, the needs for respect, status, and self-respect and the needs for *self-actualization.*

Thus the species have several needs and all the needs are *organized into a hierarchical structure* and the basic physiological needs are at the base and other needs are at higher level. According to Maslow *human motivation is a search for self-actualization.* The individual is an integrated organized whole. Therefore, the whole individual is motivated and not as parts in him. There is need only for the individual and not such as the need of mouth, need of stomach or the need of genital (Fig. 18.5).

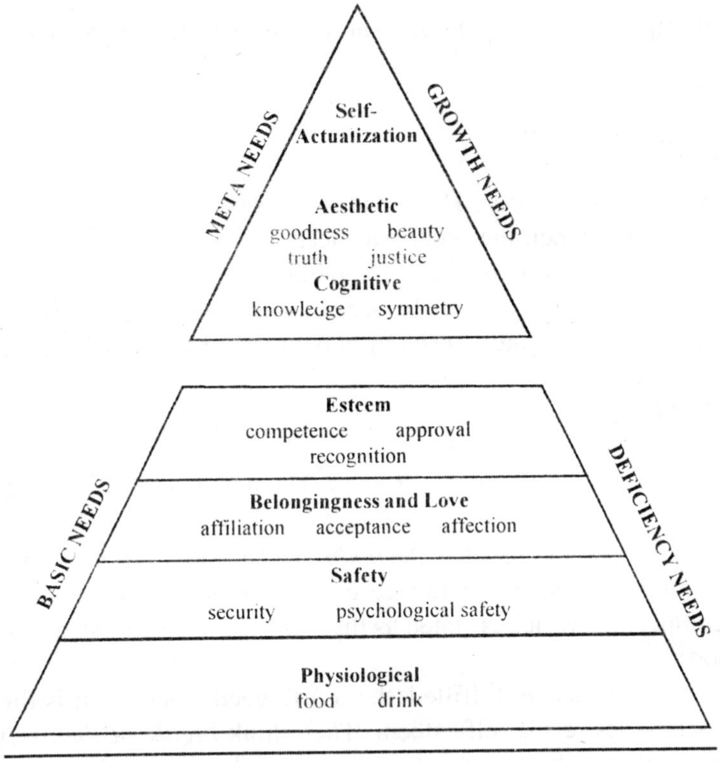

Fig. 18.5 : Maslow's Hierarchy of Motives

(1) Physiological Needs

Maslow in formulating a motivation theory, states that the physiological drives are the basic needs, the first of which is the *homeostasis* and the second is the *apptite* (food). Homeostasis means the *body's automatic efforts to maintain normal state of blood streams.* This includes food water, comfort, blood, salt, sugar, protein, fat, calcium, oxygen and also fulfilling the *basic needs.* According to Maslow these motives are not very important, because the human species are satisfied with these, all the times. But the *motives* that are at the higher level are expressed as important needs.

(2) Safety and Security Needs

This is the second in the heierarchy of motives. When the physiological needs are fulfilled, there emerges a new set of drives or needs which is stated as *safety and security needs.* Human beings need *safety for lives* and *security for their belongingness.* These needs include, security, stability, dependency, protection, freedom from fear and anxiety. Human beings must be safe from wild animals, natural disastrous and assalts from enemies. The safety needs are of great importance for children. The failure of safety needs causes *fear and a sense of insecurity.*

(3) Love and Belongingness Needs

The humans need *love and affection.* Here love is not sex, because *sex is a physiological need.* These needs emerge when the other two needs, *physiological* and *safety needs* are fulfilled. The secured individual will be able to go out for love and affection, with relatives, friends and other human beings. This may be even extended to a larger group of people. The person feels hungry for affectionate returns from people and a loving place in a group.

(4) Self-esteem Needs

The *self-esteem* is an essential aspect of *self-hood.* The human infant lacks self-conscious. *Self-consciousness is a gradual attainment of self-hood* which takes place during the first six years of life. First, the self of oneself is the *bodily-self* and the *self- identify.* This depends upon one's developing capacity for language usage. The third aspects of the self is the development of competence which is stated as the *self- esteem.* Self-esteem is the essential aspects of selfhood.

When the first three needs are fulfilled, the fourth need emerges. It is the *self-esteem.* All people in a society desire for self-esteem. They think *highly of themselves* and have *others think highly of them.* They need *self-respect* and also admiration from others. This includes high evaluation of themselves, self-respect and self-esteem. Humans strive to achieve status and recognition for the accomplishments. The satisfaction of the self-esteem leeds to the development of the feeling of self-confidence, worth, capability and adequacy. He is useful and necessary in this world.

(5) Self-Actualization Needs

According Maslow this is the *highest human motive.* It is the *need of self-fulfilment,* which one is capable of achieving. The need for self actualization results in *discontent or restlessness.* This state will continue till the individual is fitted with what he is. He wants to show how great he is. He must make himself to his own nature either by becoming a dancer, or a musician or a poet, or a novelist or a great scientist and so on. The *individual drives* to actualize himself in something which may fetch him fame. Self-actualizatión, according to Maslow, means *the highest order of the needs in hierarchy,* the desire for *self-fulfilment and realization of one's potentialities.*

Apart from all these, human species have *cognitive needs* for knowledge and for understanding the *philosophical and theoretical values.*

19

FRUSTRATION, CONFLICT AND ADJUSTMENT

Introduction

Human life is with full of resources for meeting with the needs and demands, desires and motives, achievements and goals. But however resourceful the human life is, it is often subjected to some unresolved stress and conflicts. The circumstances of life cause storm and stress. If human beings are to develop fulfil their needs and demands and to survive, they must continuously adapt to the changes that are taking place in their physical and social environments and in themselves. The human organism's adaptation to the situation is the type of behaviour that the organism's should follow. This adaptation to the situation and to the changing world is termed as *adjustment*. It is the process that describes the adaptive behaviour of the organisms. Adjustment refers to means by which the individual person copes up with his own constantly changing needs, demands, motives and habits on one end, and with often changing life aspects, opportunities, obstacles and restrictions of the environment on the other end.

The human motives are not easily satisfied. There are obstacles on the way which should be overcome. The desire are not fulfilled or delayed which need to be tolerated or with stood. Every human being develops certain characteristic ways of responding to the obstacles when their attempts to achieve the desired goal are restricted. These results cause *frustration* and the responses to frustrating situations determine the level of their adjustment of life to a larger extent. In the preceding chapters we have dealt with some aspects of adjustive behaviour. Intelligent way of behaviour reveals only man's adjustment to the situation in the everyday life. Similarly, learning to perform certain tests successfully and perceiving the self and the outer world correctly are also indicating the successful human adjustment behaviour. Now, we are dealing with much broader area of adjustment behaviour. It is the process that involves the whole individual person in relation to his environment.

How does an individual person satisfy his needs, demands and motives in a world that is often yeilding less chances for achievements. Man develops more than one motives which sometimes may be incompatable. How are the individual persons' different motives resolved and how the barriers to goal achievement reconciled? How the individual's intelligence and other potentialities applied in solving the everpresent problems which are in crowded form in the environmental situation? These problems we shall deal with in this section.

Frustration

Unlike other living creatures man is pushed in all directions for fulfilling his needs, desires and motives. He is a goal-oriented and highly motivated organism. Generally, human behaviour is constantly directed towards achievement and satisfaction of a large categories of physiologically oriented learned motives. But the goal-oriented motivated behaviour is not always achieving the goal. It is experienced as very difficult and also not possible to achieve as desired by the individual. Something on the way of the efforts is blocking the person from achieving the goal. The person's efforts become useless. At this stage what type of feeling the individual person has developed? The person may again attempt to achieve the goal, but in vein. The type of feeling, the individual person has developed, is called *frustration*. Frustration is a condition which indicates the sustained motivation without achievement or satisfaction.

What is frustration? Frustration is defined and described in various forms. In the common sense world frustration refers to a feeling state of a person when his efforts are found to be useless in achieving or fulfilling something he most desired. This is the view that is generally pervailing among the people about frustration. But what most people view about frustration is that it is a special or undesirable circumstance. The psychologists view frustration as a condition of man's life as a constant companion that provides the stimulus for some of a man's most distinctive human behaviour (**Moskowitz** and **Orgel,** 1969). The result of frustration is the inevitable. It is because of man's complexity and because of the nature of the environment in which man lives and because of his structure and psychological limitations. According to **Henry E. Garrett** (1975) frustration is a sense of defect (annoyance confusion, anger) when a motive is blocked in conflict. **Hilgard, Atkinson** and **Atkinson** (1975) state that frustration is an event that thwarting circumstances that block or interfere with the goal directed activity. It is also stated by them as a state of the annoyance, confusion or anger engendered by being thwarted, disappointed and defeated.

According to Hilgard *et al.*, frustration has two different cannotations in everyday speech. In one sense it refers to the blocking of the motive and preventing the satisfaction. When there is progress towards the desired goal it is interfered or delayed. This stage we call as frustration as the person encountered it. The other cannotation is that frustration may be also stated as an unpleasant emotional state, that results from the blocked goal. For example, a person wants to catch the train at 8 O'clock in the morning and he has also made the reservation. But he is ten miles away and drives his car very fastly as the time is very short. On the way the back wheel of his car gets a flatten and held up on the road and there is no other means of going to catch the train. He became very anxious and the feeling he developed is frustration. Who is responsible for this? The frustration here is treated as *internal state*. The thwarting circumstances caused his frustration. The *external events* which are unexpected that occurred, caused the frustration.

In our everyday life we are encountering minor difficulties and obstacles of some types. We may be able to surmount most of them easily. Anything that stands on our progress of the desired goal, anything that interfers with the satisfaction of a motive is an obstacle. The most natural response to the obstacle is to put in greater effort towards the goal. If this greater effort does not remove the obstacle the immediate natural response is to vary our attack to explore the situation in the hope of finding out a way to surmount the obstacle. This type of problem solving behaviour may be consisting of moving around and looking around and thinking about the best way of handling the difficult situation. But if all these things result in failure what takes place next? What is the reaction to the failure in spite of all these efforts? It is the arousal of the feeling of frustration. The state is described through the process of frustration.

The term frustration is used in two ways. According to **Roberts, Woodworth** (1952), the states that the term frustration refer to, are the stimulus and sometimes to response. Sometimes it refers to unsurmounted obstacle or failure and sometimes it refers to the concerned person's reaction to failure, especially if the reaction is highly of emotional type. Therefore, we may state that if a person is frustrated, he may be one who is simply blocked in all his attempts to solve a problem or who is one responded his failure by anger, worry and anxiety. But some individuals continue their problem solving behaviour longer than others before becoming emotionally frustrated. They are said to be *more frustration tolerants.* The emotional response to failure is revealed in many forms.

Environmental Obstacles : In attempting to achieve the goal, the goal is not easily attained and the person gets satisfaction. Because there are many obstacles which make the efforts useless. The physical environment presents obstacles through some restrictions and norms imposed by other people by means of customs and traditions of social life. Children are pervented by parents' denials and delayed approvals. Children normally make many claims which may be beyond the capacity of the parents. Because of this some children get frustrated. Every society, however simple or cultured, places restrictions on its members, if the needs and demands, motives and intentions cause some strain on it.

Apart from the *physical and the social environments,* the obstacles to the goal satisfaction may lie within the individual's *own personalities,* deficiencies and limitations. Some persons are handicapped by deafness, blindness and paralysis. Every person can never think of becoming an outstanding sportsman, or an eye specialist or an ENT specialist. Persons sometimes fix their goals beyond their mental and physical capacities and when not able to achieve, get frustrated.

Conflict : The primary *source* of frustration is *conflict* between two or more opposing *motives.* It is the most *complex source.* The essence of conflict is that the individual is unable to obtain the desired satisfaction for one of his motives without frustrating one or more of other motives. Therefore, conflict is a kind of frustration within the individual, because of his own irrational response tendencies. *Conflict* may be defined as the mental state that is caused by the presence of several opposing or mutually exclusive inpulses, desires or tendencies. Conflict sometimes arises because there are limits to the number of goals that can be achieved at the same time and sometimes more motives arise at the same time, each with a different type of goal seeking behaviour. To simplify and to illustrate this complex source of conflict. **Moskowitz** and **Orgel** (1969) quote an example. We cannot make two telephone calls at the same time nor we can take two equally attractive girl friends out on the same evening. In these cases the achievement of one goal must be postponed, while the other is sought. According to them postponement constitute frustration, but it is only for a short period. Thus we can make one telephone call before dinner and one after the dinner and we can date one girl friend one day and another after two days. In this way the frustration will not have its full effect.

In the case of even one motive there may be different ways of approaching the goal and conflict arises when the specific approach to achieve the goal is employed. Out of several paths, which path is the most suitable and successful? For example, a student has decided a particular course of study in a professional college. But there are several colleges offering the same course. Which college is better? This choice of college presents a conflicting situation. After joining one college, he cannot change it. Though the goal is finally achieved the progress towards it is disturbed. Generally almost all the conflicts involve goals that are simultaneously desired and undesired. They have both positive and negative aspects. For example, mango fruit is tasty for eating, but it adds too much of sugar in the blood, spending few days in a hill station on hot days is pleasant, but the study is much affected. In both cases the choice is anxiety producing. The attitude the person has developed towards a goal is wanted at once and also not wanted, liked at once and at the same time disliked. This is called *ambivalent attitude.* Ambivalent attitudes among young men and women are very common.

A young person is confronted by an attractive goal and at once he likes it, but feels that it is dangerous and so he is vacillating what to do. The danger is less real when the goal is at a distance. This is inviting approach which leads to *approach reaction.* But the sense of danger increased as the goal is approached. The nearer the incentive a person has a tendency to withdraw. This type of simultaneous tendency "to and from" leads to vacillation at some point near to the goal for one to be aware of dangers, but is distant enough to be safe from them.

Thus conflict becomes an increasingly dominant source of frustration in human life. It starts with the acquisition of primitive social motives and goes through increasingly complex motives. Psychologists have put forth a great deal of efforts to analyse and experimentally investigate the conflict of behaviour. Since the conflict of behaviour falls under different levels with varied emotional aspects a classification is felt and **Kurt Lewin** (1935) has proposed and described *three basic types of confict* such as approach-approach, avoidance -avoidance and approach-avoidance.

Approach-Approach Conflict : When an individual is simultaneously motivated towards two equally attractive goals approach-approach conflict occurs. For example, a student who wishes to continue his studies in a college finds that both the engineering and medicine are equally important and he is attracted by both the courses. But he can take up only one course. Now he faces the conflict. It can be illustrated with a figure (Figure 19.1).

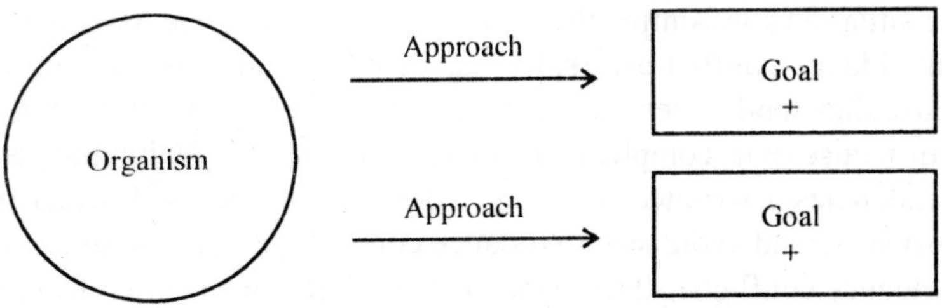

Fig. 19.1 : Approach-Approach Conflict

In the figure, the plus signs indicate the alternative goals. The lines and the arrows indicate the motives and directions of their response tendencies. The circle at the centre signifies the individual in conflict. If the response tendencies in the approach-approach behaviour were fully equal in strength and if the goals were also fully having equal in strength the person would certainly experience frustration for a long time. But actually this is rarely taking place. Generally, the person who experiences the conflict has unequal response tendencies towards the two goals. It is because in the past he has been concentrating more often and more intensely on one of them than on the other. However, even the response tendencies were exactly the same, the approach-approach conflicts are easy to solve. Because the individual person does not stand at equal distance from the two goals physically and psychologically. When the equal distance is disturbed the relative strengths of the two response tendencies automatically change. This was observed by **Miller** (1944). The approach-approach conflicts occur often in individuals and also they are resolved.

Avoidance-Avoidance Conflict : The individual person is confronted with *two aversive situations* in avoidance-avoidance conflict. The person is motivated to avoid both the situations. But he cannot avoid one without avoiding the other one. The college student has to choose one between *two equally unattractive courses* like earth Science or Marine Engineering. The avoidance-avoidance conflict can be represented in a schematic manner (Figure 19.2).

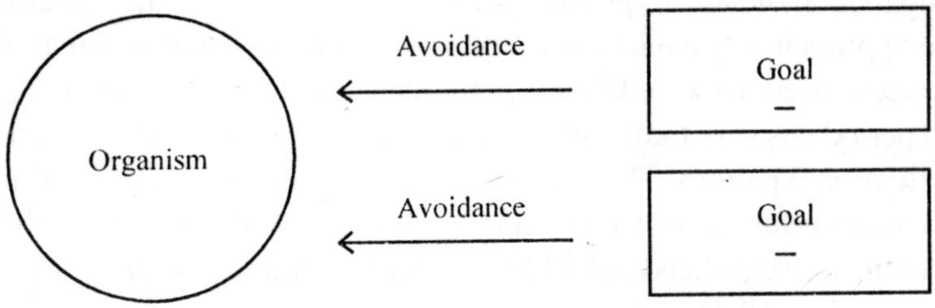

Fig. 19.2 : Avoidance-Avoidance Conflict

The substance of avoidance-avoidance conflict is that it always involves an imposition from outside rather than a conflict between the individual's own motives. The conflict is generated by the situation confronting the individual and with in the person himself. When the avoidance-avoidance conflict are really inacceptable and unresolvable they became important, because they tend to generate *other motives* that have to do with *fear or anger*. These drives in turn raise some complicated motives that originate within the person himself. The response tendencies associated with anger develop conflict with other motives. But however the most sustained avoidance-avoidance conflicts resolve themselves to the form of approach-avoidance conflicts. At this stage the conflict is *withdrawn* from the situation for a *positive effect.*

Approach-Avoidance Conflict : In the approach-avoidance conflict the individual faces with a situation which consists of both attractive and aversive type. The approach-avoidances response tendencies are evoked simultaneously. For example, a college student is strongly motivated to play cricket along with his classmates. This is an approach response tendency. But at the same time he is also motivated to stop it because of the punishment that awaits for him as he could not come home in time for the dinner. This is an avoidance response tendency. This approach-avoidance may be illustrated in figure (Fig. 19.3).

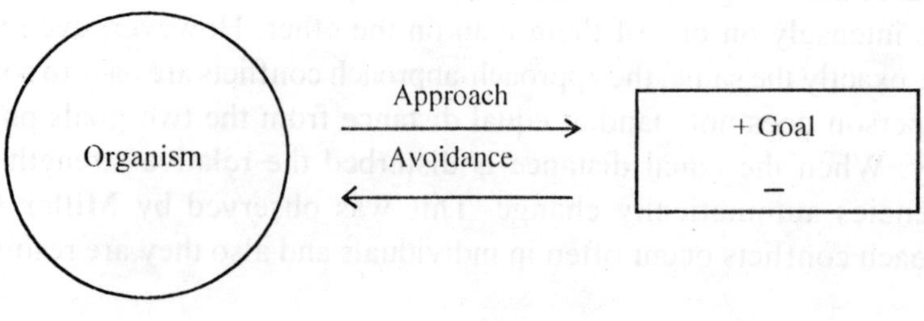

Fig. 19.3 : Approach-Avoidance Conflict

This approach-avoidance conflict unlike the approach-approach and the avoidance-avoidance conflicts tend to increase the frustration. In this type of conflicts both the response tendencies are directed to the same situation. The person can neither postpone one motive as in the case of approach-approach type of conflict, nor can escape from the conflict as in the case of avoidance-avoidance type of conflict. Any response to the situation will cause frustration. The appraoch-avoidance conflicts are important source of frustration and they are more often experienced in the childhood, adolescence and adulthood periods. Particularly, in the adulthood period it is very frequent and severe. This is because of the greater complexity of the adult motivational pattern. This involves a greater number of personal motives. This causes greater potential for inconsistency and contradiction. But some psychologists refer the complexity of the true conflicts as *double approach-avoidance conflict.* This is illustrated in the form a figure (Fig. 19.4).

Reaction to Frustration

The reaction to frustration refers to the consequences of frustration. This type of behaviour results when the achievement of the goal is blocked. This may be the result of environmental obstacles, personal limitations or social restrictions or conflict. Thus there are a number of possible consequences. The frustration produced behaviour seems to be impossible to analyse fully. Because there are differences in the ways people respond to frustration. When a student has failed in his examinations, several causes are attributed to by the students. But the real cause may be something else. It is true that some students work hard and some do not. Some of them blame their teachers, and some blame themselves. Some students may respond to frustration in different ways at different times. Thus the responses to frustration are varied. But the underlying processes are the same.

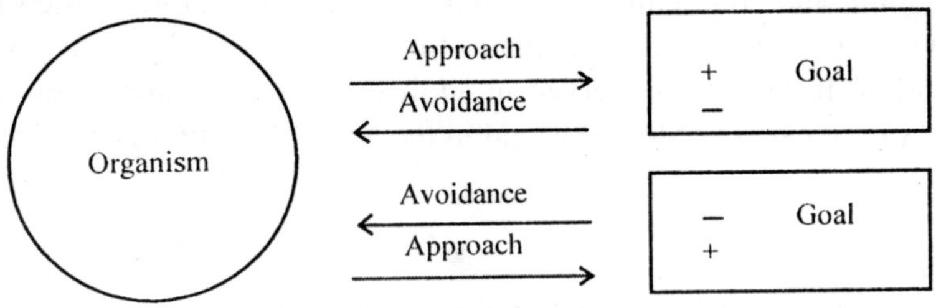

Fig. 19.4 : Double Approach-Avoidance Conflict

In all the figures the PLUS signs indicate Approach and all the minus signs indicate Avoidance.

Some of the psychologists (Mosokowitz and Orgel, 1969) are of the view that the reactions to frustration have three *different sequences* such as *source, drive and coping behaviour*. The source may be physical and social environment and also personal limitations that accompany frustration which may serve to activate and direct the behaviour of the organism. The behaviour refers to the instrumental responses that are generated by drive and formed into habits by reinforcements. In this way the frustration produced behaviour is thought of as a particular kind of motivated behaviour which is different from other motivated behaviour.

Experiments have been conducted an reactions to frustration, to see whether it is the result of physical or social, environmental obstacles or personal limitations or conflict. The classical experiment was said to be conducted by **Barker, Dembo** and **Lewin** (1941) on young children. There were two sets of young children. They were put in a room where there were several toys and the parts were missing in the toys. They were asked to fill up the parts imaginatively and continue their play. They did accordingly and played.

On the second day, another set of children were introduced in the same room where they found the toys without complete part. In this half-toy situation, these children behaved differently. They were unable to play constructively unable to fit the toys into meaningful and satisfying activities. They also behaved in an unruly manner. The children of the second group were actually the same as those of the first group. They were showing symptoms of frustration. This stage was purposely created.

Now, the problem is, why was the half-toy situation satisfying the first group and frustrating the second group? The answer is that the goal seeking behaviour was satisfying them in the first time and the children played happily with the available toys, whereas in the second day the children were not satisfied with half-toy situation. This experiment is an illustration which reveals a number of immediate responses to frustration.

Restlessness and Tension

One of the reactions to frustration is *restlessness* or *tension*. In the experimental situation the first reaction to frustration exhibited by the children was an excessive movement and restless behaviour. This restlessness was associated with many actions of showing unhappiness, sighing and complaining. The unhappy actions were recorded for less than twenty per cent of the children in the free play situation, but over eighty five per cent in the frustrating situation. It was also observed in adult persons when they were blocked and thwarted, there was an excess of tensions and in the level of excitement.

Aggression

There are several other unpleasant emotional states associated with *tension and restless* movements, feeling of anger that may lead to destructiveness and hostile attack on others. In the half toy play situation experiment, the children displayed knocking, kicking, breaking and destroying. These activities greatly increased as the consequence of frustration. There are several types of aggression, like *direct aggression, displaced aggression.* Aggression is a behaviour intented to harm other persons.

Direct Aggression

Aggression is one of the reactions to frustration which is expressed sometimes directly against the object person who is responsible for this. The source of aggressions is frustration. Aggression may be *verbal* and also *physical,* expressed by human beings in their everyday life. Animals, birds and other living creatures also express aggression in some form when they are disturbed by other creatures. The anger engendered person when blocked tends to find expression for the aggression in some form direct attack. When an obstacle is put by another person in the activities of a person, it results in direct attack on the person, treating that the person was the barrier on his way of achievement of the goal. But aggression in response to frustration always will not be only in this form of direct attacks. Aggression expressed also in other forms.

Displaced Aggression

Aggression is expressed in a displaced form in most of the situations. The frustrated person cannot express his aggression against the source of the frustration. The source of his frustration sometimes mostly may be vague and intangible. The frustrated person does not know where to project his aggression and does not know who was the real cause of his frustration. But he is very angry and seeks some one or something to attack. He may find sometimes that the person who is the source of frustration may be a very powerful and he cannot attack him. Any such attempt may be dangerous for him. When circumstances are the causes of his frustration direct, attack on the source is impossible. Since the aggression is functioning in the person vigorouśly, it has to be displaced.

Displaced aggression is an aggression against a person or object other than that which was the source of frustration. Displaced aggression is an aggressive action against an innocent person or object other than against the actual cause of the frustration. A person who is highly reprimented at his office by his manager may be very much frustrated and he cannot show his aggression against him. But when he returns home he expressess his aggression against his family members.

Displaced aggression is sometimes projected against innocent persons, *Scapegoating* is one such displaced aggression. An innocent person is victimized and blamed for one's troubles and becomes the object of aggressive attack. Racial and communal prejudices against minority groups that are practised in some countries are based upon displaced aggression to a large extent. Sometimes the economic frustration, if it is very severe, it tries to find out an outlet by displaced aggression or scapegoating against the persons who have nothing to do with the real cause of frustration.

In United States the Negroes were scapegoated sometimes for any unfortunate events in the life of the white race. In some places for some reasons the blacks were the object of displaced aggression and it also expressed in the form of *lynchings.* In Germany, during *Hitler's regime the Jews* were subjected to displaced aggression and scapegoating. Thus *displaced aggression* or *scapegoating* activities are mostly irrational.

Apathy

Apathy is one of the reactions to frustration expression in the behaviour of human beings. Apathy may be defined as listlessness, indifference which is one of the consequences of frustration revealed in human behaviour. It is the tendency found in different individuals to respond to simillar situations in a variety of ways. The common response to frustration is active aggression. But the *other reaponse is pathy.* This is functioning *opposite to aggression.* It is not active but an act of *indifference, withdrawal and silent.*

It is very much surprising to see why one person in the frustrated state behaves aggressively and violently and why another person is apathy to the same situation? It appears that *learning* is an important factor operating in the behaviour of a frustrated person. For example, when children find that expression of aggression for their frustration results in punishment by the parents or come to know that nothing would take place, the consequence is pathy. Therefore, reaction to frustration can be learned in the same manner as other behaviours.

It is also observed that under certain circumstances frustration serves to reduce activities. This generally takes place when frustration is very prolonged and when the drives involved in it are very intense. The condition may lead the organism to fatigue and exhausion and the subsequent result is decline in activity level. Frustration also involves extreme deprivation, pain fear and people sometimes respond with a type of emotional withdrawal and inertia. This state is called *apathy.* This type of reasponse occurs when there are several natural calamities like storm, tempest, earthquack, tsunami and during wars. **Selya** (1956) states that it has also been observed in animals that are driven to severe *prolonged stress.*

The state of apathy was studied by **Strassman, Thalar** and **Schein** (1956), among the prisoners of war and in concentration campus. They found that prisoners developed the attitudes of *detachement* and *extreme indifference* in the case of continued deprivation, torture and the threats of death. Generally, apathy is a normal reaction to frustrating conditions. These investigators have observed in their interview with the American soldiers released from prison after the *Korean war* that all of them experienced at some time during their *imprisonment, listlesiness* and *total lack of emotion.* But these prisoners speech and behaviour did not have any symptom of neuroses or psychoses. But the reaction was only apathy. The most severe apathy often resulted in death of the person.

Fantasy

Human beings often experience difficult life events and problems which one could not easily solvable. In such situations they often go into the world of imagination and try to be free from such problem situations for sometimes. Fantasy is a type of day dreaming, fanciful and uncontrolled imagination. It is a consequence of frustration. When problems become too much for a person, he seeks the solution of escape into a dream world. A solution is found in fantasy and not in reality. Children who do not get their play materials, often go into fantasy and find them. Adolescent boys and girls often go into fantasy and meet the most liked ones in their fantasy. Since fantasy is an uncontrolled daydream, there is no any type of restrictions for this. Fantasy is *wish-fulfilling* daydreaming by the use of imaginary thoughts. It is a compensation for the personal limitations and also it serves as a substitute for the person's failure of an effort in solving the problems. It is a tool to resolve the person's real frustrations. Thus it serves as one of the reactions to frustration. Fantasy is also used as a *defensive mechanism.*

Stereotype

Stereotype is also considered as one of the consequences of frustration. It is revealed in the behaviour of an individual. It is a tendency that the frustrated individual is repeatedly exhibiting the fixed behaviour. *Stereotypy* may be defined as the continued repetition of behaviour that appears to serve to the realistic purpose and may be punished *inflexible* behaviour which may be the consequence of frustration. Generally, any ordinary problem solving behaviour requires flexibility. Because when the original path to the goal is blocked new directions should be sought for. When the goal seeking person meets with repeated frustration he is very much baffled. In this situation some type of flexibility seems to be lost and the person goes on repeating the same action again and again foolishly eventhough the experience finds no progress.

Experiments have been conducted on animals to test the stereotypic behavioural reaction to frustration. **Maier** (1949) had demonstrated this in white rats. He designed an experiment in such a way that the white rat gets frustrated in getting its reward, the food and started the stereotypic behaviour. The essential aspect of this experiment is that the rat gets confused in its action to get its food in the experimental situation. A pair of stimulus cards attacked to the windows. It is so arranged that the rat finds its *food* behind the *positive card,* which *is a black circle on white background.* But when the rat jumps to the *negative card* which is white in black background gets punished. The experimental tools are so arranged that when the cat knocks the *positive* in (black circle) card it gets its reward and when kncks the negative card (in white circle) gets punishment. When the rat learned the correct stimulus, the experimenter varied the position of the cards to make the rat to learn to knock consistently the positive card regardless of its side. This experiment is arranged in the frustration situation by making the problem unsolvable easily.

It is so arranged that each of the two cards provides reward half of the times and half of the times punishment regardless of the position on the left or on the right. This means that whatever choice the rat makes is right only half of its attempts. The result revolved in this experiment is that the rat is forced to jump and knock to form a *stereotyped behaviour* by jumping and knocking repeatedly one side, either to the right or to the left, and does not pay attention to any specific card. Till the end, the rat was rewarded half the times and punished half the times after adapting the stereotyped behaviour. The important charactertic of stereotyped habit is that once it is formed and adapted it is not *easily changed.* This type of behaviour is called *abnormal fixation.* Similar experiments can also be conducted on human beings for testing the nature of stereotyped habits as the reaction to frustration. **Woodworth** (1952) observes that in maze learning the moments of the rates are nothing but stereotyped movements. Stereotypic behaviour pattern is quite common among individuals and in some groups unconsciously. Generally stereotype is a fixed set of beliefs and action held by members of a group. Shakespeare, in his play *Macbeth,* has presented a character, Lady Macbeth with the abnormal symptoms of *fixation* by washing her hands several times everyday, *stereotypic behavioural habit.*

Anxiety

Psychological interpretation of anxiety is always associated with the *emotion of fear.* Like anger and other emotional responses fear can easily generalize more number of emotional situations. Frustration when often repeated in the life of a person may result in a spread of the fear response to even irrelevant stimuli or to other frustrating situations. This is how fear is generalized. This type of generalized fear response, especially when it is evoked by irrelevant or objectively non-fearful situation is called anxiety. *Anxiety* is essentially a *conditioned response.* Anxiety is generally not followed by any awareness.

The anxiety ridden person knows that he is fearful of something, but does not know why he is fearful of something. But even if he can identify the objects, persons or situations of his fears he is not aware of the reason for his fear. The general tendency for the anxiety situation is as far as possible to avoid the object of the anxiety. According to **Farber** (1954) anxiety, like fear functions as a drive and as a stimulus in the life of a person for escape and avoidance behaviour. The main function of anxiety is to activate an organisms to behave and to direct it towards *flight.* When anxiety is found to be very intense in response to objectively benign stimuli, it can evoke responses that are not effective and self-defeating. It is similar to *phobia*, an irrational fear for something objectively not harmful.

At times some people become afraid of closed doors and windows, which we call as *clasustrophobia.* Some are afraid of *dark places* whom call as having *nytophobia.* And also some people are fearful of seeing crowds who are said to be having *ochlophobia.* Thus fear response takes different forms in the behaviour of the people once when they are subjected to severe frustrations. Generally, anxiety involves an intense state of *tension and discomfort* and the person is motivated to avoid or to reduce it. In a person's lifetime, he develops various methods of coping with an anxiety producing situations and for the feelings of anxiety.

Theories of Anxiety

The psychological process of anxiety is an emotional aspect which has been given importance in all the studies right from the time of **Sigmund Freud** to the present time. Everyone is interested in evolving a theory of his own in the study of anxiety. We shall discuss the two important theories of anxiety briefly.

The two important theories of anxiety are, the *psychoanalytic theory* and the *behaviouristic theory.* Theorists have demonstrated in their studies how anxiety functions in the individual's behaviour. Anxiety is generally considered as the *emotion of fear.*

The Psychoanalytic Theory

It considers that anxiety is an *unconscious conflict.* According to Freud the neurotic anxiety is the result of an unconscious conflict between the id impulses and the constraints imposed by the *ego and the superego.* The id impulses function as threats to the individual as they are contradictory to the personal values or in opposition to what the society sanctions.

The Behaviouristic Theory

The behaviourist is of the view that anxiety is a learned response. This theory was evolved by the behaviouristic psychologists in contrast to the psychoanalytic theory. Anxiety is a learned response to the fear arousing stimulus. This has been demonstrated in a young child. The young child has developed fear on seeing the rabbit, because of its *association with loud noise.* Having developed such fear it is very difficult to estinguish the fear in the child. The child has associated this learned response with all other animals even though they are not harmful.

Defense Mechanisms

We have discussed in preceding section about the reactions to frustration. The reactions to frustration are the consequences the concerned individuals experience in their life situations. But there is a psychological process that is helping individuals who are subjected to the harmful effects of frustration. These processes help the individuals by directing them to cope with the frustrating situations. This process **Freud** (1949) called as the defense mechanism. There are varieties of coping behaviour in defense mechanism.

What is defense mechanism : A defense mechanism is an adjustment behaviour made often unconsciously, either through action or the avoidance of an action in order to escape recognition by the self of personal qualities or motives that might damage the self-esteem or cause heighten anxiety. Defense mechanism may be also defined in a simple manner. Thus defense mechanism is a type of behaviour of an individual often elaborately disguised, designed to safe one's face or to avoid the loss self-esteem. The defense mechanisms are generally not purposive responses to the source of frustration itself. These are not direct respsonses to frustration like aggression and avoidance. But defense mechanisms are directed and reinforced by escape from the aversive emotional responses like anxiety, hostility, guilt and shame which are accompanying frustration. Therefore, their relationship to the frustrating situation is always obscure.

Generally, defense mechanisms occur without any clear-cut awareness on the part of the individual who applies them. Further, the defense mechanisms are originating unconsciously. Because the person who behave defensively is not having any insight into the reasons for his behaviour. The defensive behaviour of the organism is both unconscious and irrational. It is the self-protecting function. Freud was the first to describe this phenomenon. Thus the defense mechanisms are the coping behavioural habits that are evoked by the classically conditioned emotional responses to frustration. The major defense mechanisms are many. We shall discuss some of them here briefly. These are *repression, rationalization, projection, reaction formation, regression, repression.* According to Freud repression is the most basic among all the defense mechanisms. Freud states that it is the motivated forgetting. It is the failure of remembering or recognizing the fearful, painful, or guilt provoking incidence of an individual's experience. Repression may be also defined as the conjectured process by which unpleasant memories or guilt feelings are exculted from consciousness. Repression is a defense, mechanism in which an impulse or memory that might provoke the feeling of guilt is denied by means of ıts disappearance from the conscious level. Generally, each defense mechanism is a method of protecting the individuals from the awareness of impulses or memories that they would deny. Actually repression is not forgetting, but it is a directed inhibition of memory. In all possible ways it involves a continued attention to irrelevant stimuli.

The childhood sexual experiences are often repressed. A person by not remembering these experiences, avoids the intense guilt feelings. This person is not aware of the reason for his memory blacks since the repression itself protects him from the guilt feelings. If repression is successfully effected it results in complete absence of the awareness of the unacceptable motives and the total absense of behaviour evoked by such motives. But actually, repression is not completely successful and impulses find often indirect expression in some way. Repression can be distinguished from *suppression.* Suppression is self control which keeps the impulses or wishes in check and holding them secretly and denying them openly. In these cases the individual person is aware of the suppressed materials.

Rationalization

Rationalization is a type of defense mechanism which also protects the concerned person 'from damaging his self-esteem. Therefore, we may state that rationalization is an interpretation of one's behaviour in which the real motive is concealed in favour of an acceptable motive. Rationalization is a defense mechanism in which the self-esteem is maintained by attributing acceptable reasons for conduct entered upon or impulsively or for less acceptable reasons. Rationalization means here *not to act rationally.* It is attributing logically and socially desirable motives to what one has done that he seemed to have acted rationally. It is explaining one's behaviour in socially acceptable and reasonable terms for avoiding the awareness of guilt or anxiety provoking motives. Rationalization is a response for coping with frustration. It employ logical explanation and ideas as tools.

Generally, rationalization like all other defense mechanisms involves distortion of facts. Because the individual in order to justify his faults, has to omit certain real facts and add points which would support his actions. According to **Hilgard** *et al.* (1975) rationalization serves two purposes. One is that it makes one to relax his feeling of disappointments when he failed to achieve a goal. The person rationalizes in such a way that he is ready to disown what he attempted. He may state, "I am not very much interested in it". The other purpose is that it provides one with acceptable motives for his behaviour. If one acts impulsively or for motives that he does not wish to acknowledge even to himself, he may interpret what he has done it so as to place his behaviour in a more favourable place. Rationalization is used when individuals are totally unaware of the reasons for their conduct. This means that rationalization is unconsciously motivated.

These types of results were obtained in some experiments, **Hilgard** (1965) states that a subject under hypnosis was asked that when he wakes up from the trance he would watch the pocket of the hypnotist when he removed a handkerchief from his pocket, the subject should raise the window. The subject was told that he will not remember what the hypnotist told him. On waking up from the trance the subject felt drowsiness, but went round the people and had conversations. While the subject watching the hypontist's pocket, the hypnotist in a casual way removed the handkerchief from his pocket. At once, the subject had an impulse to open the window. He took a step in that direction, but resisted to do so. The subject wished to be a reasonable person unconsciously. He sought a reason for his impulses or wishes to open the window and said, "Is the room stuffy?" He was able to find a suitable reason, the subject opened the window and felt more comfortable.

Projection

Projection as a defense mechanism is the process of avoiding undesirable motives or deficiencies in the individual, attributing them and then condemning them in other people. The person by projecting characteristics on others protects himself from self-awareness by finding faults with others. It is true that all of us have some undesirable traits that we do not wish to have them. Projection is one of the unconscious mechanisms that protects individuals from accepting the traits. In projection we protect ourselves from accepting our own undesirable qualities or traits by attributting them with some exaggerated quantities to other people.

Projection is a form of rationalization. But the tendency to projection is very pervasive in every culture and its merits deserve for consideration. Experiments have been conducted on projection among the university students which revealed the pervasiveness of projection observed by **Sears** (1936). In this experiment the members of each fraternity were asked to rate the other members on some undesirable traits like, *stringeness, obstinacy* and *disorderliness.* Each one was also asked to rate himself on the same traits. It was found that those students who possessed an undesirable trait to a high degree as rated by others, but they were unaware of possessing it as also rated by themselves. These students assigned undesirable traits to others to a greater extent then what others did. It was observed that the correlation was low, but it was consistant in the direction that was expected. Thus the result of this experiment indicates that instead of accepting an impulse as his own, it is attributed to some one else. For example, the unpleasant emotions, hatred in a person is projected on to some person.

Projection was originally conceived by Sigmund Freud as the prime defense mechanism of paranoid persons. The paranoids believe that others are seeking to harm them. But actually the paranoid person has harmful thought towards the suspected enemies. Freud was of the view that projection as well as displacement was involved in parania. The impulse threatening the paranoid person is thought to be homosexuality. While analysing one of the important cases, **Freud** (1933) observed that the homosexual impulse was first transferred into that of aggression, from sexual assult to physical attack and then projected into the person towards whom the paranoid patient had the homosexual impulse. Thus the paranoid person's own homosexual impulse is transformed in his mind into an idea that the object of the sex impulse in reality is seeking to attack him.

It is observed that projection like repression is involved in selecting the perceptual set that could divert the attention from oneself by reverting the same on the behaviour of others. Further, projection like rationalization finds out suitable excuses and finds justification for the expression of socially disapproved motives. And also projection like fantasy is involved in essential denial of realities. Those who project unpleasant things on others usually generalize their unpleasant impulses more and more on others. Projection as defense mechanism is of highly malignant type because it has harmful consequences and others are scapegoated at the times of social unrest and communal clashes projection freely applied on innocent people and they are scapegoated and penalized. Projection finds an important place in racial prejudices in which we can find how it could cause social destructiveness and possesses socially contagious characteristics of the defense mechanism. In Germany during the period of Nazi regime there were loss of life for Millions of people due to scapegoating which is result of the qualities of the defense mechanism of projection.

Reaction Formation

Reaction, formation is one of the defense mechanisms in which a person denies a disapproved motive through giving strong expression to its opposite. The threatening impulse is expressed verbally and by action by its opposite. The impulse here is reversed from its normal threatening character to an opposite one. The person, instead of expressing hostility to other person, expresses love. The stronger the hostility the greater the intense the positive expression of love. Reaction formation is generally the avoidance of anxiety provoking self-awareness by behaving in an opposite way to one's real motivation. Generally, reaction formation is a tendency to conceal a motive from oneself and giving strong approval to its opposite. The present of reaction formation in some persons is not meant that motives cannot be taken at their face value. It is not correct to state that all social workers are engaged in their work with hidden impulses. But those who are defending against unapproved impulses can be identified from the social workers by the excessive interest with which they persue their programme.

The important quality of reaction formation is exaggeration of the opposite theme. Reaction formation usually infered on the basis of excessiveness of the approval or affirmentation. It is because of this that we mistrust the person intutively when he prises us extremely. Some of the examples we often come across in our everyday life as excessive are the religious affiliation, excessive expression of goodness, irrational aggressiveness, excessively posing charitable views and extreme exposition of passivity. All these are the *characteristics of hypocracy.* The reaction formation is a tendency of *concealed hypocracy* practiced by some people in society and exploit innocent people.

Regression

Regression as a defense mechanism is the description of the reaction to frustration of an organism. But literally the term means going back to the time when the person was enjoying his life happily. Regression is an attempt to deal with severe frustration met by a person. It is by reverting to some earlier days. Once the person had effective form of coping behaviour. The main characteristic of regression is its *infentile qualities.* It involves distraction of the realities like all other defense mechanisms. The individual's own social role, and responsibilities are not responded. Regression is reinforced by avoidance of the aversive feelings. When the young child throws *temper tentrum* the mother responds at once with her love and affection. It is because the child's regression causes her to feel good and sad. But if the mother reacts with anger it leads to reinforce the child's temper tentrum. In this process of expressing her anger she is attending fully on the child and by this means she is showering some of the child's dependency motives. Generally regression is equated with immature behaviour. It causes irritation and also results in destructive behaviour.

Regression, since taking back a person to the earlier days of developments, when the person was coping well with everything and enjoying a free life without any strict control over his behaviour. Generally persons in maladjusted state when they are frustrated often think about their past life experiences. They often speak of those days when the elders were at their beg and call and attending on them with deep love and affection, when had painful feelings attended them with deep sympathy and pity.

Apart from these defense mechanisms described above there are also many other defense mechanisms observed and described by the clinicians and personality theorists. **Freud** (1946) stated that these defense mechanisms sometimes vary from person to person. Every defense mechanism has certain specific characteristics depending/upon the nature and level of frustration. With all that, all the defense mechanisms have certain characteristics in common. All of them involve some level of distortion of realities and *self-awereness.* The basic characteristics of defense mechanism is avoidance-response. There are some commentaries on effective coping behaviour of the individual. Generally, these defense mechanisms function at certain level both in normal and abnormal persons. The important functions of them are *protective in nature.*

The school student who has failed in his examination projects his blame on others by the process of rationalization. He blames his teacher for his failure. He represses his own fact of inadequancy. There is also certain amount of reaction formation. Thus, we find the defense mechanisms such as repression of guilt feeling, projection of putting the blame on the teacher and reaction formation of denying his own fault and strongly expressing the opposite.

Defense Mechanisms and Adjustment

All the defense mechanisms described above are the reactions to frustration. Frustration arises when man meets with conflict in his goal seeking activities, in conflict of motives and in various problems solving situations. Generally people are capable of solving their problems rationally. People may face problems squarely and they judge and weigh the alternatives according to their probable or expected consequences. Then they take action as guided by the results of deliberation. Our understanding about the defense mechanisms guide us that some behaviour that seem to be guided for action by our conscious reasoning.

In problern solving process sometimes people find it easy when the way is clear. But sometimes they find it very difficult to face and solve the problem, because of some limitations in the problem situation and in the problem solver. It is possible to perceive the problem clearly and exercise an attack and solve certain personal problems and we solve also other simple problems. Students generally have to solve some problems in mathematics, science and also in conducting experiments. They find out the correct answer among several alternatives and so on. But sometimes we may face certain difficult problems which may be possible by applying ordinary procedures and methods. Sometimes we, may even give it up.

Hilgard *et al* (1975) state that there are two chief reasons, for not solving our personal problems in a rational process.

The first reason is mainly dealing with the nature of motives. The motive may be so strong that they distort the problem and that the person is incapable of solving. Secondly, the problem is the equation which consists of many new and unknown aspects and the world in which we live is not in an orderly fashion and permitting us freely for rational problem solving. People must take risks on the basis of the best estimates about the future. Thus the limitations of the individual persons sometimes are on the way of solving problems.

Satisfactory Adjustment

The main function of defense mechanism is to reduce and to avoid anxiety and to protect the self-esteem without any damage. If the defense mechanism is not successful at any level it means that it is not persisting in the frustrating situation. Generally, human beings learn more successful ways of solving problems. The importance of defense mechanism fades away when we do not require the functions of defense mechanism. By providing alternative, behavioural pattern to frustration the individual's coping behaviour is assured. Thus by this means the defense mechanism is helping the individual person for a satisfactory adjustment.

The defense mechanism helps the individual to rationalize his failure and thus provides him satisfactory adjustment in the society. If this is not taking place, the individual meets with despair, which may even lead to maladjustment. Secondly, the defense mechanism provides new modes of adjustment. The person himself in a frustrating situation searches for a way of life which may provide him ways and means of adjusting himself to the situation.

The process of rationalization generally searches for a rational conduct by which the individual can find a solution to the future way of life. It is to be known that all the processes of the defense mechanism unconsciously function for the successful adjustment of the frustrated person in the society.

PART SEVEN

INTELLIGENCE AND CREATIVITY

20. Intelligence and Its Measurement

21. Nature of Intelligence

22. Creativity and Its Measurement

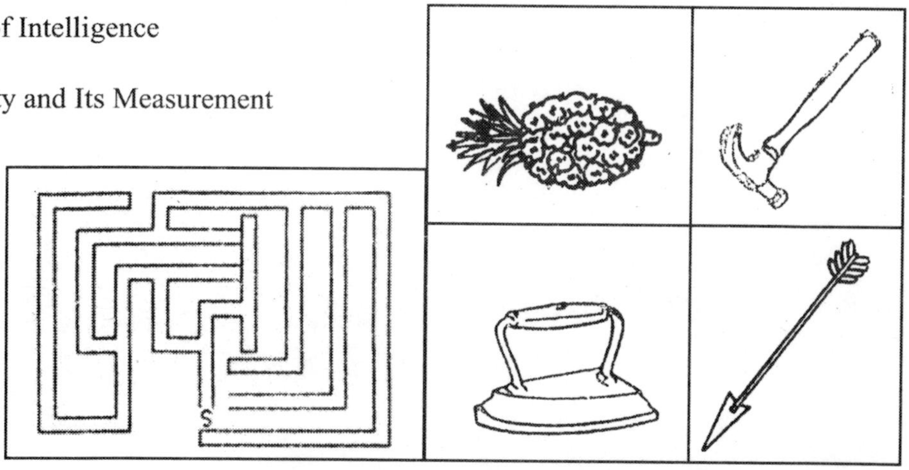

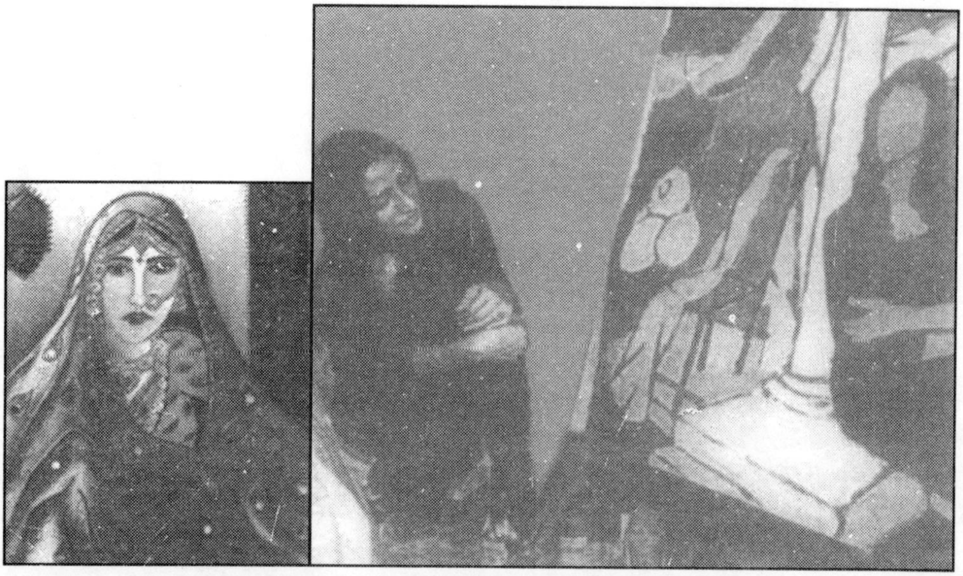

20

INTELLIGENCE AND ITS MEASUREMENT

Introduction

Man distinguishes himself from other vertebrates only by the functions of his mental abilities. Though he is physically not strong compared to many animals, he is able to control the mighty and ferocious animals only by his mental abilities, in other words by his intelligence. Man's superiority over other animal kingdom is well established by his varieties of achievements in the world. But what makes him superior over other animals? Scientists are of the view that it is the evolutionary process that went on for millions of years, put man in a very advantageous position. One such physical advantage is the development of large brain and its functions. Man alone among the living creatures has the largest brain. It is the function of this brain that made him superior in the world.

What goes on in the brain, which is packed inside the skull, is not visible or observable. But the function of the brain whatever it is, is observed in the behaviour of human beings. Human behaviour may vary from simple motor task of picking up a tool from the ground, folding a paper, posting a letter to very complex activities like solving a problem, operating and working on a computer and predicting the consequences of the nuclear weapon. The human potentialities are not limited. There are tremendous variations which are revealed in various types of behaviour.

As and when we go up in the evolutionary ladder it is clear that organisms get improvement in their structural, specifically in neural complexity which facilitates the degree to which their behaviour can be modified by the increased capacity for learning. The increased capacity for learning makes the organism for increased flexibility for adjustment. Generally, organisms undergo changes on gaining more and more experiences and also get better equipped to adapt to the changes in the environmental conditions. But in every species there will be individual differences.

Generally human beings learn and adapt to the situation much more than other species. Even in human species some individuals learn certain kind of materials more rapidly than others, while some others learns slowly. Some persons learn enormous amount of materials and perform wonders in life situations while other persons perform nothing. Among the human beings the extensity of individual differences in physical and mental activities, in learning and performing complex skills are found to be striking and remarkable. This has become highly significant and it has become a subject matter of philosophical and scientific analysis and modern psychologists are very curious on this issue.

It is important for a man to recognize and make use of the natural resources and human potentialities for learning and performing the complex tasks. Modern psychologists' major concern is understanding the individual differences in learning ability and performances. It is not for their interest in education, industry and social planning, but for advancing their understanding the vital aspects of human behaviour.

What makes the behaviour of the organisms to have extraordinary potentialities for variations? Scientists are of the view that in the evolutionary process the organisms increase in their complexity of the neural systems which increase the ability to learn. The structural complexity varies from organism to organism and also in the same species from individual to individual. The interaction of the complex nervous system with the environment and the unique combination of the hereditary factors cause infinite diversity in human experiences. The interactions of these sources of variations cause the human species differ enormously and the range of individual differences exist among the members of every community. The range of variations can be seen in human physical appearances like the size of the physique, colour of the skin, strength, performance, subjecting to ill health, and the psychological variations like feelings and emotions, way of thinking, interest, attitudes, habits, skills and talents, values and in endless ways. These variations or the individual differences stand as the challenge to the physical scientists and the social scientists. Psychologists with the available information try to explain these phenomena.

What is Intelligence? Definitions and Descriptions

Children who were born and brought up in the same homes or environmental conditions, do not behave in the same manner. Each one behaves differently from the other in most of the activities. One child perceives an object and narrates all the particulars of the object, while the other child has a hazy look at it and does not narrate anything. In school, some children learn the lessons quickly while other children take more time to learn and also fail to repeat the same next day. In performing certain task each one takes his own time and also some perform accurately, while others perform the task erroneously. There may be aggressiveness and hostility in the personal relationship of some persons. We can see everywhere keen competition or rivalry for status, failure in school or college, variations in the grades. Thus people *differ from* one another in *aptitude, ability, honesty, sensitivity, friendliness,* seriousness and performing duties and holding responsibilities. There is some thing behind all these differences. Psychologists attribute to all these differences as the individual's psychological make up, 'the trait' that each one possesses in varied degrees and the way in which the trait is functioning in human beings. In a way these existing differences among the human beings are for good and they serve some useful purposes and also solve several problems. For a moment, imagine that all those people who are living in a big or small town were looking exactly alike as everyone else in all respects. What would happen? There would be nothing, but too much of confusions, disappointment and unhappy ends. But though these differences solve this problem, they are also sources of much unhappiness. The unpleasant trait, *aggression* or *hostile* may cause *provoking remarks* and *physical attack.*

The trait distinguishes one from another in every respect. It is not a concrete reality that is present in a person, but it is *a functioning ability,* or distinguishing characteristic that is differing from person to person. The trait is revealed only in behaviour or in performing some task. The concept *habit* is also used in terms of personality characteristics like *extrovert, introvert,* hostile, aggressiveness, warm, generous, reliable, timid, shy, dominant, submissive, cheerful, sorrowful, friendly, suspicious and so on. The trait is also thought in terms of abilities or aptitudes.

Generally, the concept, *ability* is used relatively with *aptitude, capacity* and *achievement.* **Woodworth** (1952) says that ability has two meanings. It may mean actual ability or potential ability. According to him the actual ability refers achievement as it is applied on some task, whereas potential ability refers to capacity or aptitude. It may be what the individual is capable of. It is his natural ability. In capacity or aptitude a number of abilities are grouped together. A person may have more of mechanical aptitudes, less of artistic aptitude, moderate clerical aptitude and more of scholastic aptitude, which is also stated as intelligence.

Intelligence is generally known as the ability to solve some problem or handling a situation successfully. It is the ability to learn, to adapt to an unusual situation, or to get along with society or environment. It is more often used in the common sense world and also in the scientific world, and the value of it is realized and appreciated in every successful performance of some task. It is true that there is no other concept in the field of psychology that is better known to the public or common man than the concept intelligence. A person, whether he knows the meaning of the concept or not, never fails to appreciate another person or himself by using the word intelligence on findings some excellence in another person or on himself. We admire intelligence in another person and we may describe the whole man with this.

Man manifests himself in a variety of ways. There is no person in this world exactly like yourself nor he would be born. Therefore, what distinguishes one person from others is the way in which he behaves in different situations and also the way in which he organizes his motivational, perceptual and learning processes. The more an individual learns the more unique pattern of capacities he develops. The increased capacities to learn is the result of increased tendency to adjustment. Generally intelligence depends upon the knowledge one has gained and the way in which he is using the knowledge in performing some task or dealing with a situation. *Learning is one of the tasks* in which *intelligence is revealed.* But some persons learn things very quickly while others learn the same thing very slowly and few others nothing at all. Some people can apply their knowledge more successfully while others do not. This kind of *individual differences* in learning, application of knowledge and reacting to situations are great concerned of modern psychology. Intelligence is the concept used to describe all these phenomena.

If intelligence is defined as the way of using the intellectual abilities in handling a situation, everyone handles a situation in his own way, differently from others. Therefore, intelligence is defined in several ways. It is referred to varieties of skills. Some say it is *inherited* and others say it is *learned capacity* through education. Still others say it is a *combination of both.* In whatever manner it is defined intelligence is a characteristic of behaviour. This characteristic can be separated and measured. **Wickens** and **Meyer** (1961) define intelligence as the degree of proficiency with which the individual solves problems, deals with abstract materials, verbalizes and learns new material. It is also stated that intelligence is an overall ability which includes skill as perceiving relationships dealing with abstract materials, verbal facility and learning and retaining concepts. **Moskoweitz** (1969) states that intelligence refers to an individual's ability to perform certain kinds of *complex tasks*.

The earliest study on intelligence was made by **Binet**. According to Binet intelligence is a general capacity for comprehension and reasoning that manifests itself in various ways. **Stagner** and **Moskoweitz** (1952) are of the view that intelligence is not physical object. It must be understood as a complex characteristic of human performance. It has *two dimensions, speed* and *power*. That is, how fast the person solves the problems and how difficult the problems are when he fails to solve them. Psychologists offer various definitions of intelligence. These definitions generally are focused in different kinds of performance such as learning ability, ability to manipulate abstract symbols, ability to use learning in new situation and ability to solve problems, etc. Learning is a function that we find both in animals and human beings. But there is enormous amount of difference in speed and complexity of learning. At lower extreme we have feeble minded cases who are not developed to learn the acts of normal human beings. At the other extreme we have persons who deal with extremely complex abstract functions. **Gardner Murphy** (1951) says that intelligence means the ability to learn and the ability to increase steadily with the complexity of the nervous system. He states that the ability to profit from experience and environment is called intelligence. His definition of intelligence is from biological point of view.

The evolution of the human brain was so complex that it was revealed in the behaviour which was far superior to those exhibited by other forms of life on the earth. This behaviour was commonly referred to the reflection of intelligence and therefore what revealed was an intelligent behaviour.

Spiken and **McCandless** (1954) say that intelligence is what the intelligence tests measure. It is known as operational definition. Because the psychologist measures what he calls intelligence with his tools. As there are many kinds of tests of intelligence there may be different words of intelligence. Each test measures one intelligence. Therefore, we must specify the particular intelligence we speak about at the time of measuring. The operational definition is not satisfactory one for some. From the sensory field, intelligence is referred to some fixed structural or genetic potential for learning. The structural potentials are not measurable. The only thing that we can measure is, an organism's current behaviour. In this sense an intelligence can be interpreted as an index of potential learning ability.

Most of the definitions and descriptions of intelligence have academic basis. In every culture there is a tendency to evaluate people's intelligence from the academic activities. The academic activities and achievements provide us with substantial data to show what an individual is capable of. We can see that only those people who can withstand physical and mental strain could make progress not only in academic life but also in every walk of life. How are we to describe this behaviour phenomena? Naturally, they are intelligent people. They plan, work and achieve. They are also responsible and dutiful. Otherwise they could not make any achievements in life. The higher form of intelligence, particularly creativity is determined only from the achievements one had made.

In the preceding section we have stated that intelligence is sometimes referred to *learning ability*. But some are of the view that learning ability is not fixed nor a unitary organismic variable. Therefore, learning may be referred to the ability the person possesses at the time of performing a particular kind of task. After sometime, learning ability may change and there may be a time in a man's life that he may not be having equal abilities in all walks of life. A man's successful performance may also depend upon his motivation. A person who is very brilliant in language abilities and games may be lacking in mechanical abilities and his acquired abilities in various areas may change when he advances in age. This may also change his motivation and interest.

Therefore, learning ability is subjected to change. If intelligence is referred to learning abilities, intelligence also is changing with age. Therefore, intelligence cannot be called as a unitary and wholesome characteristics of an individual and he would solve all the problems at all times.

Intelligence is conceived in varied ways by various authors. Every author or every investigator defines intelligence in his own way basing the importance on some specific variable or on some functioning factor. Whatever variable is functioning in intelligence the outcome of it is acceptable to all of us. Because the outcome of intelligence is providing us some benefit whether it is a simple task of getting the food or constructing an abode for living, or a complex task like learning to speak a language or a complex task like inventing a technical instrument or a printing press or a computer and solving an arithmetic problem. In all these simple and complex activities the use of intelligence is mostly involved. Only man is capable of doing all these and not animals. Anatomists, biologists and physiologists state that it is the consequence of the evolutionary process which determine the nature and the quality of the performance in man and the *brain is the centre* of all these outcomes. Man is an organism with fully developed neurological components which facilitate the human beings to achieve whatever they require in various environmental conditions. The function of intelligence is the result of the well developed neurological components.

The various physical and mental activities that the human beings perform for fulfilling their requirements are the *expressions* of the *advanced development of the neurological structure,* the brain, which we call as the intellectual activities. In a *single word* we may state this as *intelligence*. Time to time human thought process may vary and also improve by putting oneself in constant action. Thus intelligence has been receiving varied interpretations and demonstrations. *The size* and the *normal development* of the *brain* is having corresponding *effect on the function of intelligence and behaviour.*

Some progressive thinkers in the field of psychology even go to the extent of omitting some of the important psychological processes like *emotion, learning* and *intelligence*. They are of the view that certain psychological concepts raise controversies. Whatever controversies prevail, the function of intelligence in the individuals and in the societies can never be considered as controversial and inaccessible accurately. Modem authors involve *cognitive abilities* like the capacity of thinking, reasoning, remembering and understanding in the definition of intelligence. Generally intelligence involves socio-cultural bounding in its function and outcome, though it is rooted in the *structural components*. Modern authors associate intelligence with society's needs. One of the most important characteristics of intelligence is the individual differences in the performances.

Intelligence has *two extreme ends* and has wider range from end to end. The two extremes are the *mental retardation* or the feeblemindedness and the *giftedness*. These two terms are used in the society invariably by all those who speak about intelligence, persons who score *low IQs* in intelligence tests are stated as *mentally retarded* and those who score *very high* I.Qs are stated as *gifted* or very bright or very intelligent persons. *In both groups* there are *individual differences*. However, two persons have the same intelligence quotient either in the low IQ group or in the very high IQ group or in the average IQ group *they vary* significantly in their *everyday behaviour and achievements.* In this context, it is important for us to know how and why these individuals differ in their intellectual performances. Therefore, experimental investigations have constantly tried to understand the nature of human intelligence and the factors that determine the individual differences in the performances of the people in various situations. It is understood that several factors are functioning in intelligence and we are to know the specific trait that dominates in the intellectual activities of human beings.

The phenomenon of individual differences in intelligence has been much debated in recent years, whether the differences are due to heredity or environmental factors or due to the interactions of both the factors. Another similar issue was also raised regarding the causes of differences in the personality characteristics, which is also equally important. These are fascinating scientific problems and also carry practical implications.

The controversy over the issue of the individual differences is a continuous one, as both the hereditarians and the environmentalists produce evidences to justify their stands. Whether a person is superior in intelligence or mentally retarded is mostly due to *biological factors*, the *structure of the nervous system*, *the brain* and *the genes* and therefore the differences in the intellectual performances are due to heredity. The *chromosomal abnormalities* can never be ignored. While the environmentalists putting stress on the environmental factors they state that hereditary factors have *very little Influence* in all human affairs, including individual differences in intelligence and personality. Any healthy child can be made superior in intelligence and in personality by providing *suitable training* and *rich environmental* background, *food and nurishments*. Therefore, the individual differences often aroused as an important aspect of intelligence.

VARIOUS CONCEPTIONS OF INTELLIGENCE

Most of the expert scholars in the field of psychology accept that intelligence has two *aspects*. **Woodworth** (1952) states this in terms abilities as *potential abilities*, in other words *capacity* and *actual abilities*, in other words stated as achievements. **Kimble** *et al.* (1985) state that one aspect of intelligence is the individual's intellectual potential which is functioning in the *chemistry of the brain* which is not directly accessible. The second aspect is the *intellectual behaviour*, which is *observable and assessable* to some extent. These two aspects of intelligence are identified in two different ways. **Donold Hebb** (1966) discusses intelligence as intelligence-A, and intelligence-B. Intelligence-A is the innate potential for the development of *intellectual capacities*, while intelligence-B is the level of a person's *intellectual functioning*.

Another expert, **Reymond Cattell** and his associates (1963) have developed a theory of intelligence. According to them intelligence has *two aspects* such as *Fluid intelligence* and *Crystallized intelligence*. According to Raymond fluid intelligence is the *cognitive analytic ability*. This ability is relatively uninfluenced by previous learning which mainly deals with the tasks requiring for adaptation to *new situations*, whereas the crystallised intelligence is involved in *skills* and *talents* that is more depending upon *learning* and *experience*. Some other experts, like **Moskoweitz** and **Orgel** (1969) though did not formulated a full systematic theory of intelligence, have dealt with intelligence in a more practical sense. Further the common sense world have their own conception about intelligence and identify people with different types of intelligence. **Stemberg** and his coworkers (1981) also studied intelligence and explained the nature of intelligence through their theory.

According to Sternberg and others, the common sense world's conception of intelligence cannot be totally rejected. They compared the common sense world's conception of intelligence with that of the experts and discovered that these two groups of persons describe only similar basic components of intelligence. These are *three components*, which include, first the *verbal intelligence* which refers to *general learning and comprehension abilities*. This also includes good vocabulary, verbal influency and good reading comprehension. The second component of intelligence is *problem solving ability*. This includes *abstract thinking and reasoning*, mainly referring to the application of knowledge to problems to be solved, raising problems in an optimal manner and identifying the connections between ideas. The third component according to them is the *practical intelligence*. This involves *dealing with the situations successfully*, deciding how to achieve goals, showing interest in the community at large, becoming socially competent person. These three types of intelligence should be interrelated. In an intelligent person these three types are supposed to function. All these three types spring from the general intellectual source.

General and Specific Factors

The study of intelligence always raises the question whether intelligence is a single ability which includes all cognitive functions or it is a combination of several independent abilities. In the beginning stage it was viewed that if a person performs well in one intellectual task he must perform well in another task also. But later on there was a change in this view. A person who is proficient in one intellectual task may not be equally proficient in another intellectual task. Several studies revealed contradictory results.

If intelligence is general it also means that whatever level of intelligence you are endowed with also must be found in another person. For example, a man who is very proficient in running a business and is making enormous amount of profits must be with high level of the general intelligence. At the same time another person carrying out the same business meets with heavy loss and he starts a different type of work and he is found to be very much successful. This means that general intelligence is not functioning in the same manner in two persons. This means there is some variation in the effectiveness of thinking of the two persons. It is partly a matter of *more interest and motivation.*

A student who is very proficient in mathematics is considered to be a bright student. In spite of interest and hardwork he is not good in language studies. Therefore, we cannot say that there is a general intelligence which may cover up all the areas of intellectual activities in the same manner. But we cannot also accept the idea that everything depends upon a particular area of intelligence. It may be inferred that there is a *positive relationship* between the brightness in one field and the brightness in another field. This may be stated as the *general intelligence* running through all the different types of intellectual tasks. This may be also stated as *common factor*. In addition to this general factor there are also *special abilities*, the abilities to deal with the particular kind of problems.

This has aroused the curiosity of many scientists in various areas. **Charles Spearman** (1862-1904), a British psychologist was the first to explain in detail about the *general ability* and the *specific abilities*. Spearman was an Army Officer and he was very much surprised to see the *behavioural differences* among the soldiers. He was dealing hundreds of soldiers everyday and was observing the individual differences among them eventhough they were provided with the *same type of food and shelter, training and treatment* and exercises. They were also demonstrating *different levels of mental abilities* in spite of the same type of a education. He felt that the *mental development* of the soldiers must be the main cause of the differences in the mental abilities of the soldiers and he wished to deal with the problems.

Spearman left the Army and started studying psychology. *After twenty years* he came out with his famous work, "*Psychology Down to the Ages*" in two huge volumes in which he dealt with human intelligence more elaborately. Spearman had suggested that intelligence is *more than an accumulation of specific skills*. In 1904, he described his findings about intelligence in terms of variables which he called as *factors*. He stated that there is *some kind of mental energy* that flow into everything what an individual does. The individual who is endowed with this *mental energy* or *intelligence* behaves intelligently in varieties of situations. He also found that some people are *quick and successful* with this intelligence in many activities, while some do not act well and not successful in certain areas. He felt seriously that there must be *two aspects in intelligence* which may be functioning *in combination* in certain areas and *independently* in certain areas.

Therefore, he tried to clarify the issue and solve the problems. He formulated a *theory of intelligence*, which is generally called as *two factor theory of intelligence*. According to him there is a *general ability* with which people used to perform successfully a variety of intellectual tasks. Spearman named this as *general ability* and indicated it with the capital letter G, which is *General factor*. The general factor is considered as the *basic mental energy*. Spearman thought that if an individual fails to perform equally well in other types of *cognitive tasks* like mathematical, verbal, logical and so on, then such type of tasks require *some special intellectual abilities*. Spearman called such special abilities as *specific factors* and indicated this with letter S. There may be several specific factors. Spearman, thus, conceived that intelligence is composed of the G factor and several S factors (Fig. 20.1).

Spearman had presented his conclusions by fomulating a theory called two factor theory and also evolved a formula:

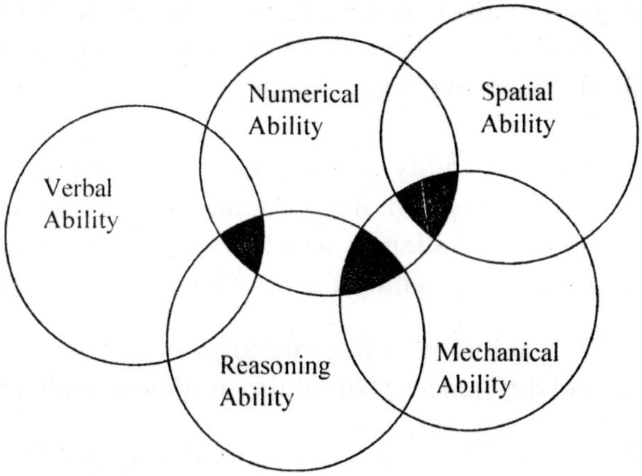

Fig. 20.1 : Study of Intelligence : Spearman's Model of General and Specific Abilities.

$G + S_1 + S_2 + S_3 = A$, where G is for general factor, $S_1 + S_2 + S_3$ are specific factors and A is for the *total ability* the individual is endowed with. Further researches in this area added something to what Spearman had formulated.

According to the recent development in the study of intelligence we are also made to know that there is *something in between* the general factor or intelligence and the specific factors. It is believed that the something in between these two may facilitate to deal with *some particular situation* or problem. These abilities are found to be *less than the general factor* and *less specific* than the *specific factors*. These factors are named as the *group factors* which mainly deal with *words and numbers* and the ability to imagine the *spatial relations*.

Spearman's two factor theory was callenged by some American psychologists like **Thorndike** (1874-1948), **Thurstone** (1889-1955) and **J.P. Guilford** (1959) and each one advanced a theory of his own. Some of the psychologists were of the view that there is a general ability which functions in all intellectual activities.

Thorndike (1926) states that intelligence is a difficult term to define with any precision. This term is used in *general usage*. Most of the intelligence tests are limited to the *abstract intelligence*, that is mainly referring to the abilities to solve problems involving *ideas and symbols*. According to Thorndike, intelligence consists of a large number of *independent abilities*.

Thurstone's view of intelligence falls in between Spearman's view and Thorndike's view. Thurstone (1938) studied intelligence and published it under the heading of *Primary Mental Abilities*. According to him, every ability is having some relationship with every other abilities. He worked out the *inter-correlations* between the scores of various abilities and found that there were *high and low positive correlations* between general and specific factors and also there were some *negative correlations* between some factors.

Thurstone after an extensive study on intelligence was able to identify the abilities that *deal with words, numbers* and the *ability to imagine spatial relations*. These abilities are traced, identified and measured by Thurstone. The method he adapted in his study was the *Factor Analysis*. His factor analysis revealed seven important factors (abilities). These are (1) Spatial ability, (2) Perceptual Speed, (3) Numerical ability, (4) Verbal meaning, (5) Memory, (6) Word Fluency and (7) Reasoning.

(1) ***Spatial Ability*** : The spatial ability, according to Thurstone, represents two different factors such as covering *perception* of *fixed spatial* or geometric relations. The other factor is *manipulating* visualization in which *changed positions* must be visualized. The spatial ability also involves a third factor of *kinesthetic imagery*.

(2) ***Perceptual Speed*** : This ability is with regard to *quick* and *accurate grasping* of visual details, similarities and differences. This ability is almost similar to speed factor.

(3) ***Numerical Ability*** : This ability is closely identified with speed and accuracy of simple arithmetic computations.

(4) ***Verbal Meaning*** : This may be also stated as verbal comprehension. The important factors involved are the *reading* comprehension, *verbal analogies*, irregularly arranged sentences, *verbal reasoning* and proverbs matching.

(5) ***Memory*** : This factor involves the test of role memory. For this the *paired associate* test material is used. It is a broader factor for all memory aspects.

(6) ***Word Fluency*** : This ability involves *anagrams*, *rhyming* or naming words in a given category. For example, words referring males and females or providing words beginning the same letter.

(7) ***Reasoning*** : This ability refers to general reasoning. This is to test one's ability in syllogistic reasoning.

Thurstone observed that these abilities are *independent of each other*. This means, a person who is *high in perceptual speed* may be *low in word fluency*. It may be also possible to find that one or more abilities may be functioning in an intellectual activity.

Thurstone is of the view that *intellect* is composed of *specific abilities*. But few of the specific abilities have broad general characteristics which he called as the *primary mental abilities*. Thorndike and other researchers are of the view that Thurstone had sub-divided the *general factor* the G, traced by Spearman, into *seven abilities*. These he called as primary mental abilities of intelligence. There were also criticisms against Thurstone's primary mental abilities. But with all that Thurstone received all appreciations for his work on intelligence. It was laborious task that Thurstone had understood and set up a model in the psychological measurements. He used 59 tests and collected the data, worked out inter-corrections for all the 59 test scores and applied *factor analysis which was the first* of its kind in psychological measurements.

Thurstone extracted *seven significant* factors which had *high loadings* out of 59 factors and he named them as mentioned above. These seven factors are the most important abilities of intelligence. Everything can be included in these seven primary mental abilities and interpreted.

Following Thustone's study of *factor analysis* of intelligence there were volumes of researches on the organization and structure of human abilities for the past seventy years from 1938. Most of the researchers employed the technique of factor analysis and tried to trace out the underlying mental factors. The factor analysts attempt to identify a small number of underlying factors that can be accounted for the complete set of relationships among the test variables. Normally each test may have a loading on each factor and the analysts tried to interprete the pattern of factor loadings whatever they are *psychologically meaningful.*

Researches in this area brought out the facts that it is possible to distinguish clearly quite a number of special ability factors like the verbal comprehension, word fluency, numberical fluency, perceptual speed, mechanical knowledge, spatial visualizing and inductive and deductive reasoning. As it has been stated above, most of the abilities to some extent are related to each other. All these are essentially cognitive or intellectual abilities. In short all these are aspects of intelligence in general. In factor analysis some minor factors also can be extracted which a like small branches in a tree. **P.E. Vernon** (1960), an English psychologist emphasized this aspect.

Vernon put forth an interpretation more vigorously is the *individual differences* in intellectual abilities. According to him the human intellectual abilities are quite similar to a branching tree. The main trunk is the *general intellectual ability*. There are *two main branches,* which represent the *school-related ability* and the *practical ability*. The smaller branches of the tree represent more *specialized abilities* like the verbal, numerical, spatial, mechanical and other abilities (Fig. 20.2).

Many of the psychologists tried to identify the main intellectual abilities and the specialized abilities. Some wished to establish them by *test methods* and some *experimentally*. One of them who worked very extensively on the nature intelligence was **J.P. Guilford** (1967) an American Psychologist and also a psycho-metrician. Guilford felt that both **Spearman** and **Thurstone** have not described intelligence correctly. Hence, he followed a different approach for the study of intelligence. He adapted a model to explain intelligence. The model he formulated was an interesting one which represented the human intellectual functions.

Guilford's model of intellect might serve to illuminate the processes involved in both *intelligence and creativity*. The model of intelligence is organized aroned *three main aspects* of human functioning. He stressed that there are three kinds of mental abilities. They are:

(1) *Operations*, (2) *Contents* and (3) *Products*. A specific ability involves a *combination* of all the three aspects.

The functions of these three aspects specifically are : (1) What kind of material is acted on, (2) What action is carried out with it or (3) What product is produced. Guilford has proposed a 4 ×5 × 6 structure that constituted in 120 cells or abilities. He also carried out a monumental research programme of *Test Development and Analysis* to see whether tests

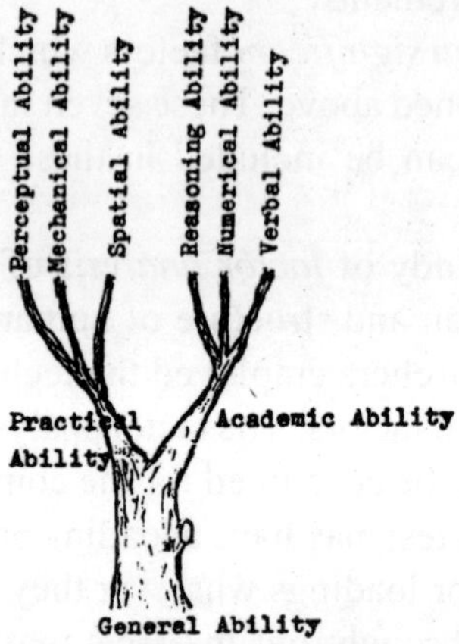

Fig. 20.2 : Study of Intelligence

Hierarchical Structure : P.E. Vernon's Model of the Structure of Human Abilities.

can be generated which may correspond to each of cells or abilities which must be logically identified.

(1) *Operations* : For Guilford operation is the major intellectual process. This process includes such aspects as knowing, or discovering or being aware of things and events. It is *cognitive in nature*. It is also referring to retrieval process from storage or memory, the generation of multiple responses or *divergent thinking*, arriving at a unique and accepted solution or covergest thinking, and judging the appropriateness of information or decision (evolution).

(2) *Contents* : An operation is performed upon certain types of *information*. This information is called *content* which may be *words and symbols*. The content information also may be semantic or behavioural. *Figural content is concrete information*, which may be *images*. There may be symbolic information with denotative signs which are *numbers* or *codes*. The semantic content of information is in the form of *word meanings*. But the behavioural content of information is non-verbal, mainly involved in human interaction. The *emotional expression* is of this type.

(3) *Products* : This is the *outcome* of *applying operation to the content*. The product is the form of the information when it is processed. Product also includes *single separated items* of information. It is also formed of sets of items grouped by virtue of their *common* *qualities*, due to connections between items of information, by the organization of the information, into systems, by changes of information like transformation and predictions of information.

Thus, Guilford stated his theory of intelligence in a *three dimensional model*. Within each category there are several factors or abilities included. (Fig. 20.3). On the basis of these descriptions we may state that there is some kind of *general intelligence* and also *special abilities* depending upon the structure, development and knowledge of the individual.

Having spent quite number of years in the study of human behaviour, intelligence and personality and after gone through the works of expert scholars, it is quite but natural for a psychologist to express his own view. This view may be endorsing a previously expressed

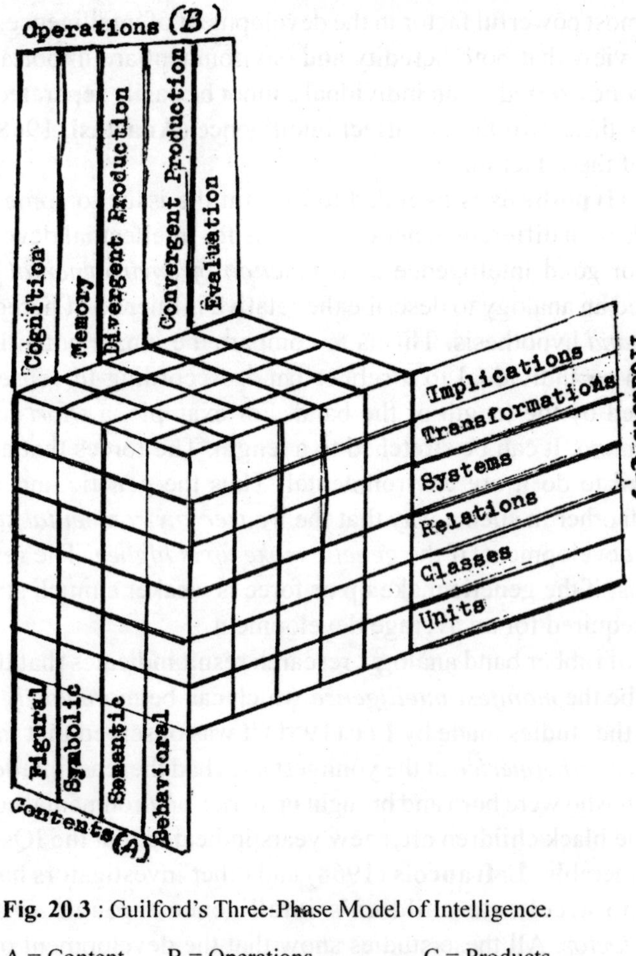

Fig. 20.3 : Guilford's Three-Phase Model of Intelligence.

A = Content, B = Operations, C = Products

view or totally different view. In the stream of psychological thoughts **Jensen** (1968) had his own share. According to him, Intelligence is *like a disease*. It does not just happened. It is caused. It is clear that the causes of intelligence are also the *causes of stupidity*, since one is present in the absence of the other.

The assumption is that human abilities or characteristics result from the *innerplay of heredity and environment*. If this assumption is accepted then the causes of intelligence are heredity and environment. But a question that has been causing confusion and puzzling the psychologists for more than hundred years is this: How much each one contributes to the development of intelligence? At this stage **Jensen** was able to collect a great deal evidences for both sides. According to him the question is how much difference each variable makes?

Another relevant question also was raised by some psychologists. That is, can intelligence be increased by manipulating the experience of an individual? A relevant answer to this question and to the general *nature-nurture* question is presented by **Lefrancois** (1979) through a debate between **Galton** and **Watson**. In the debate Galton strongly stressed that heredity is the most powerful factor in the development of intelligence. Most of the scientists now are of the view that both heredity and environment are important, but the amount of influence each one exerted on an individual cannot be easily separated. Therefore, now that question is how these two factors affect intelligence (**Anastasi**, 1968) and not what might be the effects of these factors.

The **Stern Hypothesis** is intended to explain the issue to some extent. According to this, individuals with different genetic potentials for intellectual development can manifest poor, average or good intelligence *as a function of environmental pressures* or factors. Jensen advanced an analogy to describe the relative influence of heredity and environment. It is a *rubber band* hypothesis. This is to compare the innate potential for the intellectual development in an individual to a rubber band. According to Jensen intelligence at any point is reflected by the length of the band. For example, a *short piece* indicates *poorer genetic background*. It can be stretched to a length. The forces that are used to pull on the band or that fail to do it are environmental. Thus the genetic and environmental forces interact on each other in such a way that the *weaker environmental stimulation* is required for an average development, if the *genetic make up is higher*. The reverse function also is possible. That is, if the genetic make up or force is weaker a much stronger environmental stimulation is required for an average development.

In the case of rubber band analogy, research result indicates that the rubber band which is supposed to be the *manifest intelligence* which can be more *easily stretched at younger ages*. In one of the studies made by **Lee** (1951) it was observed that the black children who had *migrated to North America* at the youngest ages had the least intellectual level compared to white children who were born and brought up in rich environmental conditions. Intelligence test scores of the black children after few years indicated that the IQs of the black children increased considerably. **Lefrancois** (1968) and other investigators have demonstrated that it is possible to accelerate children's intelligence through *direct manipulation of environmental factor*. All these studies show that the development of intelligence at very early years can be accelerated by providing rich and advantageous environment without causing any damage to genetic factors.

MEASUREMENT OF INTELLIGENCE

It is extremely difficult to define intelligence accurately though Intelligence is defined as what an intelligence test measures. This is one of the definitions we have discussed earlier. Here measurement appears to be the *primary matter*. The interest to measure intelligence was aroused by the *evolutionary biology of nineteenth century*. It tried to explain the causes of individual differences. **Darwin's** theory, *the survival of the fittest* stimulated the psychologists to measure the *ability of men to learn*. During the third quarter of nineteenth century Experimental physiological laboratories were started in Germany and in other European countries.

The experimental physiologists became very much interested in the operation of the senses and studied intensively *seeing, hearing* and other senses. Following this in 1879 the first laboratory for experimental psychology was established by **Wilhalm Wundt** (1898-1920) *in Leipzig* and studied the *sensory processes*. At the same time **Sir Francis Galton** (1822-1895) an English Scientist has shown much interest in the *individual differences in intelligence*.

The earliest intelligence tests were developed in the last quarter of nineteenth century. They were based on the idea that *intelligence has a physiological basis*. It was also felt that the *speed of neural impulses* determine intelligence levels. Based on this idea Galton as early as 1880 started an *anthropometric laboratory* in England. At the same time **J.M.C. Cattell** (1860-1944) an American (1890) Psychologist also had shown greater interest in the measurement of *simple abilities* like *sensory activity, reaction time, memory, motor skills and head size* on the idea that these are the *components of intelligence*. He has combined physical strength in the measurement of intelligence. But the method and the result of his study did not attract many men in this field. Only after 1904 systematic intelligence tests were designed and developed.

Nature of Intelligence Tests

Intelligence is a human faculty, which is measurable like all other faculties. The intelligence tests that are used in these days for measuring are mostly abstract intelligence. They are the general intelligence tests. Apart from the general intelligence tests there are also other specific intelligence tests like the mechanical intelligence which includes motor skills, perceptual speed, verbal comprehension, numerical ability and so on are also developed. The abstract or the general intelligence tests were first developed for using in the schools. These are very simple tests used to determine the abilities of the children in school situation mainly dealing with reading and writing abilities. How easily and quickly school children read the verbal materials and understand and use the numbers and other symbols. This type of abilities are considered to be important at school level.

The tests that were designed and developed for measuring the abstract abilities are said to be as the *general intelligence tests*. These general intelligence tests are of three types such as *individual tests, performance tests* and *group tests*. The individual intelligence test is administered to one person at a time whether he is a child or adolescent or an adult. The individuals tests normally take forty-five minutes to sixty minutes for answering. These tests should be administered by an experienced psychologist and the scores should be interpreted.

The *performance tests* are mostly *non-verbal tests*, words are not used in these tests. These tests are designed to test the problem solving ability and only material objects are used, mainly wooden blocks, pictures, designs and some type of physical tasks. The performance tests are designed and administered to small children, persons who do not know the language, illiterates, persons with speech defects and mentally retarded persons. These subjects need only to handle the test materials according to the directions.

The *group intelligence tests* are having certain advantages over the other two tests. These tests are administered to a group of people or children at a time. We can save time and efforts in these tests. The group tests are of two types, one being the verbal test and the other being non-verbal test. The verbal group tests are mainly testing the problem solving abilities in reading and writing. But the non-verbal tests are also dealing with problems solving ability by means of mazes, puzzles, pictures, diagrams and the like. In these type of tests the examinee needs only to perform the task, or just to make a mark with a pencil or pen and nothing to read and write. We shall now see in detail the kinds of intelligence tests.

TYPES OF INTELLIGENCE TESTS

Binet's Test

The earliest intelligence tests were developed by Sir Francis Galton in 1880s and almost at the same time an American psychologist, J.M.C. Cattell developed some tests for the measurement of simple mental ability tests for assessing sensory and motor skills. But a systematically developed tests for measuring intelligence came into existence only after 1911. It was **Alfred Binet** (1857-1911) who had constructed the first individual intelligence scale for the measurement of intelligence. Binet was the professor of psychology in the city of Paris. The school students in the city of Paris were very much backward in their classroom works which had drawn the attention of those who were interested in education.

The school authorities were worried over the poor performance of the school works of the children and wanted to find out a remedy for the problem. They entrusted the task to Alfred Binet, who was already interested in the matter of intellectual development of the school students and wanted to find out some method to measure intelligence. Binet's assumption was that the dull child is like a normal child, in all respects, but backward in the mental development. Because of this he would perform on the test like a normal child but like a child younger than him. He also wanted to separate the mentally retarded children and given them separate training.

Binet had the idea that children do better in their performance, as they grow older. Therefore, on the basis of this idea he devised intelligence tests for *different age levels from third year to seventh year* with increasing difficulty in the test items. Binet designed and developed the tests with the assistance of the French Physician, **Theodore Simon** and published the first edition of the famous *Binet-Simon Intelligence Scale* in 1905. The first scale appeared with 30 items and it was in a crude form.

Binet and Simon revised the scale and published it in a better form in 1908 and 1911. Binet's scale had sharp differences from most of the mental tests that were developed at that time. While other tests were measuring simple mental abilities, which were functioning narrowly, Binet's scales attempted to discover the child's abilities to solve problems given in the form of words and numbers. It also included items to follow simple type of reasoning. The child was also expected to show some comprehension and a fair judgment of the question found in the test. Binet thought that instead of simple and relatively easy problems, fairly complex were found to be better indicators of intelligence of the children. The tests were systematically arranged in groups suitable to different age levels from *three years to fifteen years* old and extending the tests up to the adult persons.

Binet has devised a scale of scoring system and used a concept, which he called as *mental age*. The intelligence test items were arranged for different *age levels*. A child's intelligence was determined by the age level test, which he could pass. When the 1908 and 1911 Binet-Simon scale was found to be measuring intelligence systematically, it has become a model for most of the tests. The scale consisted of heavy weighting of verbal items. The items were arranged according to the age of the children. Thus the scales were developed for each age level. The scale has a mental age concept. These aspects of the Binet-Simon scale at once attracted the American psychologists, and they incorporated these aspects into their subsequent tests that were published in the United States and in other countries.

The Stanford Revision of Binet Intelligence Scale

The original Binet-Simon intelligence scale developed in 1908 and 1911 has underwent a number of revisions, of which the best known revision is the *Stanford Revision*. In the University of Stanford, professor **Lewiss Terman** revised the Binet scale and produced it in 1916, which was called the *Stanford-Binet Intelligence Scale*. This scale was again revised and restandardized in 1937. This revised version of the scale was very widely used and had become the most wanted intelligence test ali over the world.

The 1937 revision of the Binet scale was almost like developing a new intelligence test. **Terman** and **Merrill** have collected thousands of items that were used in the previous intelligence scales and each item was administered to not less than twenty subjects whose ability levels were known. These items were used for the construction of the scales. The items of these scale are supposed to differentiate the children of different ages and their ability levels. These items were believed to be the easiest and interesting to the subjects. It was also found to be taking short time. The rough scales were administered to about three thousand subjects by some trained investigators. With these collected information Terman and Merrill had constructed two alternative forms of intelligence tests as *Form L* and *Form M*. There were 129 items in each form. They were carefully arranged on the basis of age levels and difficulty levels. On treating the test scores by the method of correlation, the two Forms had produced almost the *same results*.

The Stanford-Binet Form L had the same type of items for the age levels, four to seven years. The 1937 revised scale included some of the items of the Binet-Simon Scales. Another revision of the Stanford-Binet scale was attempted by Merrill in 1960. He updated the scale with some new items and also combined the two Forms (Form L and M) into a single test and called it as *Stanford-Binet Scale* which is considered as the most respected intelligence scale for children in spite of some criticisms.

Mental Age

While working on intelligence test several new concepts were evolved, one is the concept of mental age. The test consists of series of sub-tests arranged according to age level. Upto 5 years there were two sub-tests for each year and from 6th year to 14th year one sub-test for each year. The experimenter while administering test, starts with each child with easiest item at the top of the test and works about till the child answers no more questions. After collecting the data from large number of children he could see a pattern. That is, in each age group large number of children were able to answer all most at the same level (the same number of questions) while a small number of children stopped at the bottom of that level and some children went beyond that level. The average scores of the normal is determined.

If a 4 year old child answered all the items of a 4 year test his mental age would be 4. If the same child answered all the items of the 5th year test his mental age would be 5. If he passed all the items at 8 and half of the items at 9 and none of the items at 10 his mental age would be 8½. His over all performance is equalled the performance of an average 8½ year old child. The mental age is obtained from the scores and the chronological age is the individual's actual age. A child, whose chronological age is 8, has a mental age of 8 if he can do the test of an average 8 years old child can do. But if he would do the task of an average 7 year old child his mental age would be only 7. He is supposed to be a dull child. If a child's performances were better than the average of his age he is superior in intelligence.

THE CONCEPT OF AN INTELLIGENCE QUOTIENT (I.Q.)

The intelligence quotient speaks of relative intelligence. The I.Q. is a comparison of a person's performance on a test to the performance of other persons of the same age. It may change due to time. Mental age increases at the same rate as the chronological age in normal children. But the M.A. of a bright child, who is far above the average increases faster than his own chronological age and the chronological age of a normal child. But the M.A. of a dull child, who is below average, increases slower than his chronological age. For example, two children have the same mental age of 9, one is 6 year old and the other 9 year old. This means that the 6 year old child is brighter than the 9 year old child. The brightness of the children can be expressed quantitatively. At this stage only we came across the Intelligence Quotient (I.Q.).

The I.Q. is expressed in the form of a ratio between mental age and chronological age. In order to avoid the decimals, the ratio is multiplied by 100. Therefore the formula is :

$$\text{I.Q.} = \frac{\text{Mental Age (M.A)}}{\text{Chronological Age (C.A.)}} \times 100$$

We have stated above that two children of different ages 6 and 9 have the same mental age of 9. Let us apply the formula on their M.A. and C.A. to determine their I.Qs.

The I.Q. of the first child is $= \frac{9}{6} \times 100 = 150$

The I.Q. of the second child is $= \frac{9}{9} \times 100 = 100$

Let us assume that a 10 year old child also has the same mental age of 9.

The I.Q. of 10 year child is $= \frac{9}{6} \times 100 = 90$

The child of 6 year old with a mental age of 9 has an I.Q. of 150 and the child of 9 year old with a M.A. of 9 has an I.Q. of 100 and the child of 10 year old with a M.A. of 9 has an I.Q. of 90. The first child is extremely bright, the second child is average and the third child is below average or he is mentally retarded. The I.Q. indicates the rate of intellectual growth of the children. If a child performed the tests and obtained a M.A. to the level his C.A. he is an average child. Woodworth points out that the mental age is the measurement of the level of intellectual achievements whereas the I.Q. is an index of brightness.

Table 20.1 : Examples of some of the items in Stanford-Binet Test

Age	*Type of item*	*Description*
4	Naming objects from memory	Toy, automobile, dog, shoe etc. These are presented and named. Then one is covered and he is asked to name it from memory.
	Picture identification	Cards with pictures of objects were presented. Child is asked to 'show me what we look on', 'what do we carry when it is raining' etc.
7	Similarities	In what way wood and coal alike?
	Copying a diamond	Ship and automobile etc. Draw three diamonds (Printed in the record booklet).

Construction and Standardization of Tests

Intelligence tests must provide scores that are convertible into mental age and I.Q. In constructing such a test care must be taken in the selection of test items. First of all it must be assumed that intelligence increases with chronological age from birth to maturity in a continuous process. The first step in test construction is the selection of items that will reveal the intellectual abilities of the individual. The items of the test must cover various areas like perception, discrimination, learning, memory, manipulation of numbers, vocabulary, reasoning and other elements of human abilities.

In the second step the population is selected. It must be from very wider range covering different geographical area, age, sex, different socio-economic status and rural-urban areas. This population is used as a standard. The purpose of selecting this kind of wider range of population is to get a normal or average score for each age group with which the individual score is compared. The normal or *average score* is the *middle range* of the distribution of the scores of the whole population. The items devised for each age group (for example, 4th year, 5th year to 14th year) are administered on a large number of children. The percentage of the children who pass each item of each age level is determined.

At the final step, the *earliest age* at which *each item is passed by 50 to 75 per cent of the children* is determined. Therefore, every test is given to a large sample of children for whom it is developed. The average score, the range of distribution of scores also are determined for various age groups. Unless the test is standardized in this fashion an intelligence test is of no use. Test developed in this manner lead to the concept of mental age.

Verbal and Performance Tests

Intelligence tests are designed to measure the general intelligence or certain special abilities of human beings. These tests are administered to individuals and to a group of people. They are respectively called individual tests and group tests. Individual tests are administered to one person at a time. The Standford-Binet test is a good example for this (Fig. 20.2). The group tests are administered to a large group of people simultaneously. The Army General Classification Test (AGCT) is a group test. Intelligence tests are further classified as verbal tests and performance (non-verbal) tests.

Verbal Tests

The verbal tests require the skills in using language. The verbal tests are designed to measure the general intelligence or some special abilities. These are contrasted with performance tests. The verbal tests are also called paper-pencil tests, because most of the tests are written language in giving instructions to the examinee and also in the questions to be answered by him. The verbal tests are used to test the ability to deal with ideas by the use of word. One of the important verbal tests is the *Thorndike CAVD test*. It has four sub-tests such as Comprehension (C), Arithmetic (A), Vocabulary (V) and following the Direction (D).

The items are carefully graded in all four areas. The items run from very easy to very difficult. The difficult items are so hard that everyone will have different level in scoring. Therefore, nobody would answer all the items. Because of these characteristics the test mainly measures the individual differences in intelligence. It can differentiate those who are superior in an activity. The test puts at the top those who are really very bright. In judging the intelligence of the individual we must put: all the tests together.

The Army General Classification Test (AGCT) is another important verbal test widely used during the Second World War. This test brings out three primary functions of the individual. They are (1) Vocabulary, (2) Arithmetic reasoning, (3) Block counting. In block counting the examinee is shown with piled up blocks. He has to find out and tell how many blocks were in sight and know how many are behind out of sight. This test is not suitable to College situation. Therefore, the American Council on Education Test (ACE) is used in College situation, which is the most representative test. It emphasises the necessary skills dealing with words and mathematics.

It is felt in recent years that the adults who are to be recruited to various jobs must be tested with intelligence tests. The Standford-Binet test was found to be inadequate for adults. The need has been fulfilled when **David Wechsler** (1958), a psychologist designed and developed specifically for use with the *adults*. Originally the test was published in 1939, but it was revised and published in 1955 as the *Wechsler Adult Intelligence Scale* (WAIS). The items of the test are arranged as sub-tests and there is no age levels in it.

The WAIS consists of *eleven sub-tests, six of them are verbal* tests and *five of them are performance tests*. The verbal tests are : (1) Information, (2) Comprehension, (3) Arithmetic, (4) Similarities, (5) Digit span, (6) Vocabulary. The five performance tests are (1) Digit Symbol, (2) Picture completion, (3) Block Design, (4) Picture Arrangement, (5) Object Assembly. Wechsler has developed two more tests one for children (WISC) and one for Pre-school and Primary school (WPPSI).

Wechsler scales yield *three kinds of I.Qs*. He believes that the verbal scores should be supplemented by performance scores. The three I.Qs are *verbal, performance* and *verbal and performance*.

Performance and Non-verbal Tests

The performance tests were developed for those who could not take up verbal tests. These tests do not involve words. Culturally deprived group, children of foreign speaking parents, unschooled persons and those who do not fit for verbal test would take up performance tests. For these reasons a number of intelligence tests have been developed to reduce the dependency on verbal skills. These tests are called performance or non-verbal tests.

The most widely used performance test of intelligence is the *Draw-A-Man Test.* The test was developed by **Goodenough** (1926) and revised by **Harris** in 1963. The idea behind this kind of test is that the ability to draw human figure in correct proportion with all the details goes with *age and intelligence.* We can administer this test upon children upto the age of 15. It is easy to administer and also interesting. It correiates well with other tests.

A second important performance test is the *Arther Adaptation* of the *Leiter International Performance Scale.* It is a very interesting test. It can be administered to children aged from 3 to 8 years. There are blocks and they are randomly presented to the children. They are asked to match the blocks with stimuli, which are visually presented in cards. The child is expected to exhibit the ability to grasp the sequential prinicple in the stimulus cards. Since the test direction is through pantomine or signal, words are not used at any stage. Words are not used even to express the correct responses. Therefore, it is almost a non-verbal test correlates highly with Binet test. This test is highly valuable in testing deaf children and those who have language difficulty.

A widely used performance test is the *Pintner-Patterson Performance Scale.* This test reveals the general performance ability of the child. The *Healy Picture Completion Test* is a performance test, which can measure the child's complex intellectual functions without any verbal questions. The child in this test is required to put the blocks in the appropriate holes in the picture. In the task he has to grasp the idea to fit the block into two holes and something more than that — that is understanding the physical and social situations for the child. The Healy Picture Completion Test is very much useful in clinical situations and in dealing with children problems.

The *Cornell-Coxe Performance Ability Scale* was developed by two authors in 1934. It is individually administered. The test includes diverse tasks such as the arrangement of form boards, completion of missing parts, assembling of discrete parts into wholes. The amount of time taken by the subject to complete the test and the number of errors made by him were taken as the scores.

There were also other very effective individual performance tests, developed for the measurement of intelligence of the child. Among the most widely used individual test is the *Kohs Block Test* and the *Porteus Maze Test.* The Kohs Block Test is included in the more extensive performance test batteries. In the Kohs Block Test, the subject is required to match a series of designs with blocks of different colours. The procedure is starting with the simplest designs and going through the most complex designs. The simplest designs require a minimum number of four blocks, and the most complex designs require upto *sixteen blocks*.

The Porteus Maze Test is a Maze test. It consists of a series of pictures of mazes and the subject is required to trace each maze. The mazes become progressively more difficult. Each is to be solved correctly. The test may be administered on a population ranging from three years of age to the adult level. There is no time limit for the test. The subject's mental age is determined by the relative difficulty of the last maze the subject is able to trace correctly (Fig. 20.4 - A, B, C).

Performance Tests : Maze Tests. The subject has to run the Maze starting at the entrance (S) with a pencil should trace the pathway to complete the Maze by avoiding blind alleys.

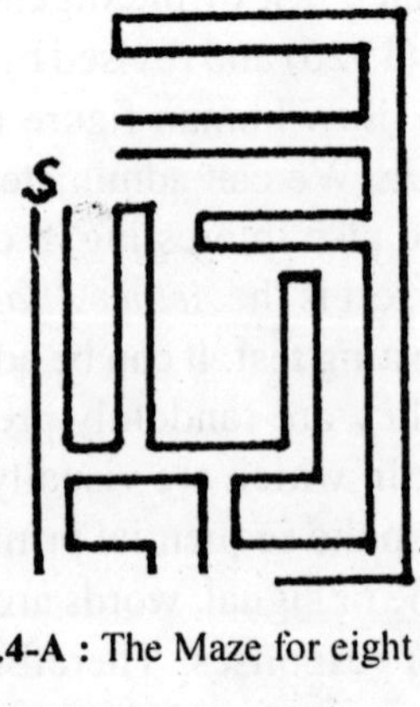

Fig. 20.4-A : The Maze for eight year old.

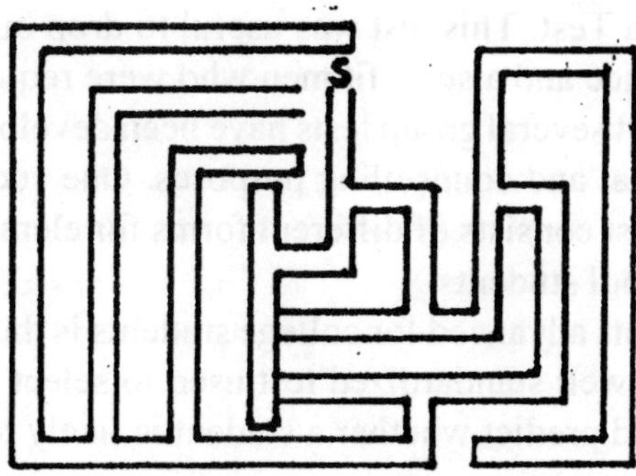

Fig. 20.4-B : The Maze for ten year old.

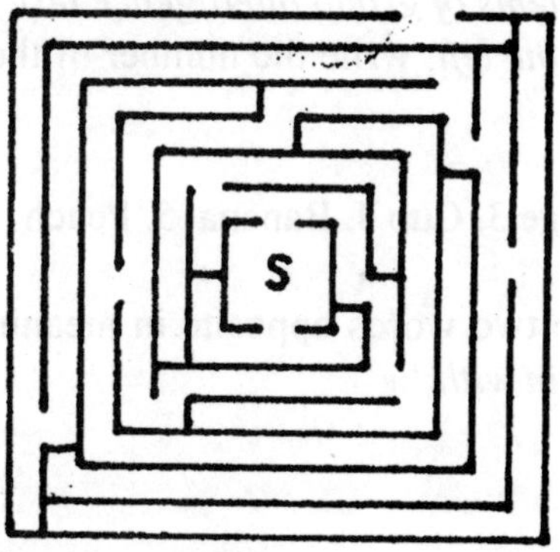

Fig. 20.4-C : The Maze for fourteen year old.

There are also several other performance tests such as the *Alexander Pass-along Test, Cube Construction Test, Block Design Test, Form Board Test, Matrices Test* etc.

Group Tests

We have been so far discussing the individual tests. Most of the tests can be administered to a single person. The administration, scoring and interpretation should be made only by trained examiners. Therefore, these tests are very expensive and time consuming. But the group intelligence tests can be administered to a large group of individuals at one time. During First World War thousands of men had to be tested for their fitness for military service. Therefore, there was an urgent necessity to develop group test of intelligence. A committee of Psychologists were appointed to devise the group tests. The committee has developed two group intelligence tests called the *Army Alpha Test* and the *Army Beta Test.* The Army Alpha Test was a *verbal test* devised for those who could read and write English language. The Army Beta Test was designed as a *performance and non-verbal group test* for those who had language difficulty and for those who were unschooled.

During the Second World War, the American Psychologists have developed the Army General Classification Test for testing large number of individuals. This was found to be better than the Army Alpha Test. This test was useful to drop out men who were severely retarded from military service and also to fit men who were required to modem warfare.

Following the Army test several group tests have been developed to meet various needs of educational and industrial and counselling purposes. One such test is the *Otis-Lennon Mental Ability Test*. This test consists of different forms for elementary school, junior high school and senior high school students.

A group intelligence tests advanced for college students is the *Scholastic Aptitude Test*. This is a very reliable and well standardized test used to select students for admission to Universities. The test would predict whether a student is likely to succeed in college level work or not.

Examples of some of the items of group intelligence test

In the *space provided on the left,* write the number of the words that is different from the other words

(______) 1. Apple 2. Orange 3. Cup 4. Banana 5. Peach

★ ★ ★

In the blanks below write two words opposite in meaning to cold. One should *begin with 'h'*; the other should *begin with 'w'*

h ________________

w ________________

The numbers in this row follow one another according to some rule. You are to findout the rule and fill in the blank to fit the rule :

2 ———— 2 8 ———— 32 64 ————

If the following argument is good reasoning, mark it plus (+). If the argument is faculty reasoning, mark it minus (—) on the line left

All people who vote are twenty one year old.

John Smith did not vote.

———— Therefore, John Smith is not twenty-one year old.

Indicate by making a cross (X) mark against the picture that is different from the rest of the pictures.

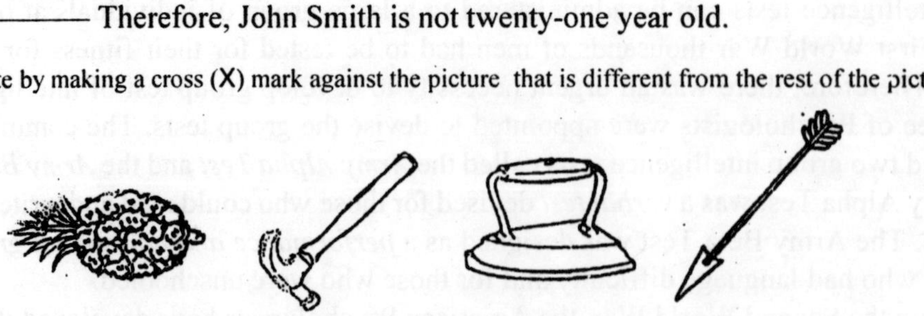

Fig. 20.5 : Performance Test

Group tests have certain advantages of saving time and money. But there are also certain disadvantages. They are not as meaningfully measuring the intelligence as the individual tests. In the group test situation the examiner may not be having control over the examinee. A person may fail to understand the correct meaning of the instruction given to him or may not be motivated to take up the test in right sense or he may be in a depressed state or fatigued. These conditions may have effect upon his performance. Group tests often yield low scores and so persons are classified as being of low intelligence.

Write the correct numbers in the square on the right in such a way that the picture is complete.

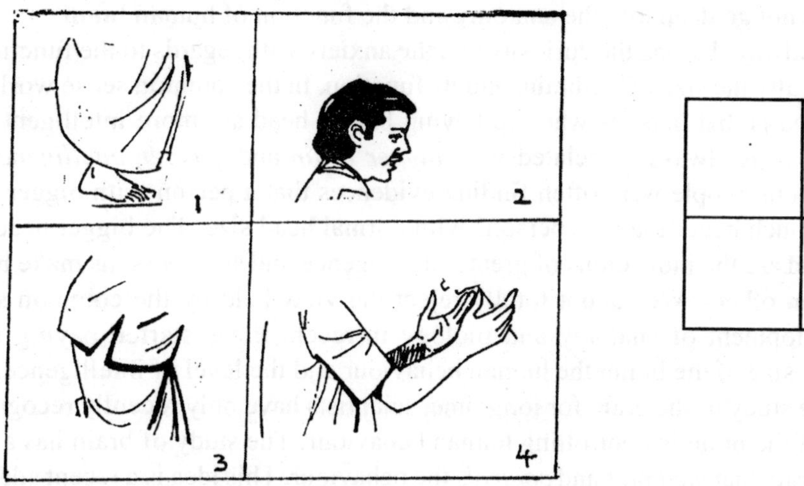

Fig. 20.6 : Intelligence Test : Completing the picture.

Write the correct numbers in the square on the right in such a way that the picture is complete.

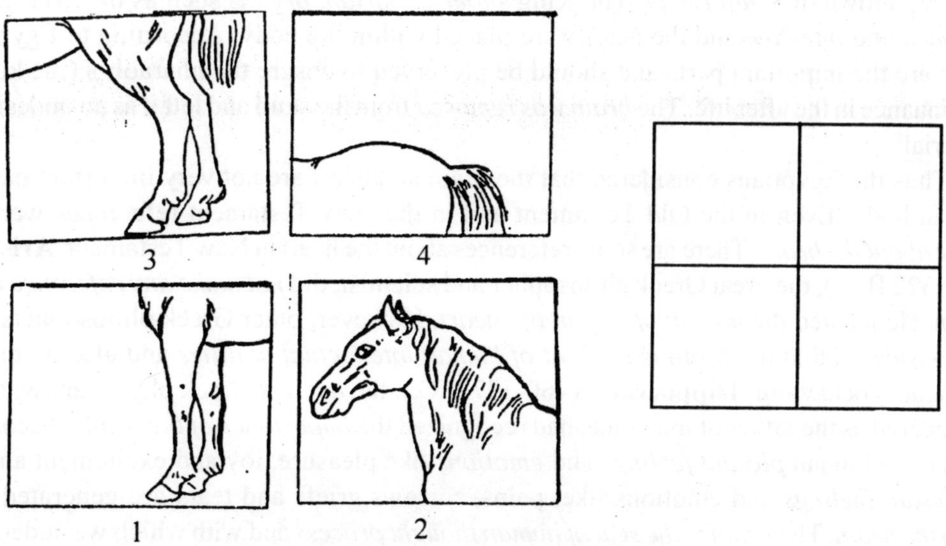

Fig. 20.7 : Intelligence Test : Completing the picture.

Measuring the Brain Size and the Level of Intelligence

Is there any *relationship* between the *brain size and the level of intelligence*? This is a perplexing question not answered well till very recently. Since the study of psychology has been concentrating on human behaviour, the *observable activities*, psychologists did not show much interest either in the anatomy or in the biological aspects of the brain. Psychologists were mainly concerned with the successful performances of everyday activities, the achievements and solving life problems, which man was often confronting in his life situations and they strongly believed that all these are the function of intelligence. Hence psychologists did not go deep into the anatomy and the function of human brain.

But man was always having the curiosity and the anxiety with regards to the function of the brain, specifically the size of the brain, and its function. In the common sense world the most accepted idea is that persons who are having bigger head are more intelligent than others. *Bigger head* is always associated with *bigger brain* and *greater intelligence*. In certain life situations people were often finding evidences that a person with bigger head does everything much better than the persons with normal head size. The bigger head and the broad forehead are the indications of greater intelligence and such persons make better achievements than others. We cannot totally reject the view held by the common sense world. The development of anatomy and biology in recent times started paying more importance to the size of the brain, the human behaviour and the level of intelligence.

In spite of the study of the brain for long time, scientists have only recently recognized the central role of the brain in controlling human behaviour. The study of brain has a long history. It is the brain that *mediates* and *controls* the behaviour. This idea is a recent addition to the knowledge about the brain and was not considered as an important part of human body for very long time. When the *Egyptian King, Tu-tan-Kha-men* (1361-1352, BC) of the 18th dynasty died, his body was mummified and preserved in alabaster jars in his tomb in *Luxore*, a town in *South Egypt*. The King's *four important organs* such as his *liver, lungs, stomach* and *intestine* and the heart were placed within the body. According to Egyptians these are the important parts and should be preserved to ensure the pharaoh's (the king's) continuance in the after life. The *brain was removed* from the skull and left it as an undesirable material.

Thus the Egyptians considered that the brain and heart are not very important parts of human body. Even in the Old Testament and in the New Testament references were *not made about the brain*. There are some references about the heart in New Testament. **Aristotle,** (384-322 B.C.), the great Greek philosopher and scientist, did *not make any reference to the brain*. He located the *mental ability in the heart*. However, other Greek philosophers later on considered that the *brain is the seat of human intellectual abilities* and also it controls the human behaviour. **Hippocrates** (460-377 B.C.), the ancient Greek physician, who was considered as the father of medicine, had recognized the *importance of the brain*. According to him the human *plesant feelings* and *emotions* like pleasure, joy and excitement and the *unplesant feelings* and emotions like, pains, sorrows griefs and tears are generated only *from the brain*. The *brain is the seat of human thought process* and with which we understand everything in the world. We *see with our brain,* hear and differentiate between right and wrong, good and evil, pleasant and unpleasant. According to Hippocrates all our *cognitive activities emerge only from the brain.* Thereby it seems that he *located intelligence only the brain.*

The controversy between those who located intelligence in the heart and those who located it in the brain was going on even after two thousand years after Hippocrates. Then came **Herophilus** (280 B.C.) a Greek physician who was called as the father of Anatomy, had advanced a theory on the functions of the *nervous systems.* According to him each part of the body was connected to some separate nerve centre. *A third phase of thinking* about human bodily systems emerged with **Galen**. This was the beginning of contemporary thought. After the fall of Graeco-Roman civilization the writings of **Galen** (130-200 A.D), a disciple of Hipprocrates, were the sources of information about both *physical and mental systems.* The death of Galen, the last major physician of the classical era marked by the beginning of the *Dark Age* for all medicines and treatments. Galen advanced a theory that a mysterious type of animal spirits pass through the nerves to all parts of the body. According to him the nervous system is the controller of human behaviour.

Group Test : Tracing the Maze

Trace the Maze starting from left and ending at the right.

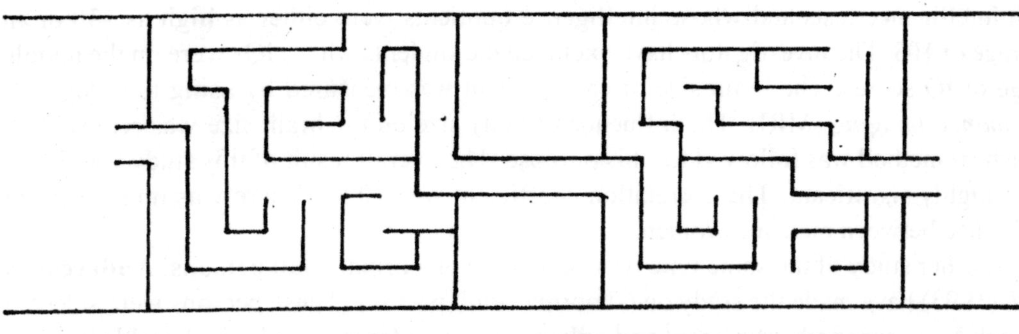

Fig. 20.8 : Group Test : Tracing the Maze.

The foregoing discussions arrive at a conclusion that the *brain is the seat of intellectual abilities* and the *controller of behaviour*. But has the *size of the brain* anything to do with the *level of intelligence*? This phenomenon has been investigated by several researchers and scientists for the past two hundred years. **Sir Francis Galton** (1822-1911), who made break through several areas of study and an accomplished Genius, attempted to find out the significant relationship between the *size of the head* and the level of intelligence. His hypothesis was based upon *bigger brain indicates greater intelligence*. In his attempt he was not successful as there was no proper tools available to test this hypothesis. He wanted to measure the brain size, but he could measure only the *head size*. Further, when he undertook the task there were no *good tests to measure intelligence*. He has to depend upon the teacher's estimation of their students' intelligence. It was the middle of nineteenth century. The investigators of this period measured the size of the skulls of the different groups and made the *estimation of intelligence,* based upon some criterion with which the subjects were associated.

Standard intelligence tests were developed only in the first quarter of twentieth century which provided an excellent opportunity to measure intelligence of various groups. But on the other side, till very recently only *head size* was measured as the estimate of *brain size* of the *living persons* or animals. This type of measurements generally should provide *positive relationship*. But there were only small *correlations* between the brain and intelligence. In one of the studies made by **Van Valen** (1974) it was found that the correlations between the brain size and intelligence ranged from +.08 to +0.22. According to his study after giving allowance for the *errors in the estimation of brain size and for the measurement of intelligence,* the correct correlation between the brain size and intelligence was +0.30.

A technique was invented for the measurement of the brain size which is called *non-evasive technique*. It is used on living subjects to measure the brain size. It was possible with this technique to have direct approach to establish the relation *between the brain size and intelligence*. A team of psychologists recently in collaboration with other investigators **Willerman** *et al.* (1991) started direct study in this line. They have selected a group of college students who were *free from neurological problems*. These students were tested with intelligence tests and whose intelligence quotients were either as high as 130 or an average of 103. The investigators have excluded the students whose IQs were on the middle range of IQ scores. The brain size of each student was measured by using the *Magnetic Resonance Imaging* (MRI). The influence of body size on the brain size was removed. A statistical method was followed for this purpose. The overall result of this study was found to be highly significant. The correlation coefficient was +0.51. There was no significant difference between men and women.

Another study of the same type was conducted on normal adult persons. **Andereasen** *et al.* (1993) have made the study on 67 normal adult persons. These persons were selected through Newspaper advertisement and only persons free from neurological problems were selected for this experiment.

All these persons first took the *intelligence test* and then the *MRI scanning*. After correlating the body size the correlation between the brain size and intelligence quotients (IQs) was made. The correlation between the brain size and the intelligence quotients (IQs) was found to be +0.38. Thus the modern technique of measuring the size of the brain has helped the psychologists to solve the long standing problems. The controversy on this issue at last settled that *there is significant correlation* between the *human brain size and intelligence*. Therefore, the conclusion may be drawn from these experimental results is that "the bigger the brain size greater the intelligence". Though the correlation +0.38 is modest size, it is highly significant. It is also important in determining intelligence, some additional factors which are having influence must be added in the measurement of intelligence. It has been observed in some cases that some scientists have *misused information* of the size of the brain. One of the exponents involved in these types of investigations was **Gould** (1981). He has stated that some scientists out of *extreme liking* or *prejudice over estimate* or *under estimate* the brain size which results in bringing out incorrect information. The racial or ethnic groups show overlapping and widely varied size of the brain. To get rid of such irregularities more *accurate instruments* must be invented or foolproof techniques must be designed for measuring the brain size.

The studies made by **Andereasen** *et al.* (1993) throw some new light on *brain size-intelligence correlation.* Whatever might have been the plan in mind used to *section the brain, two separate shades* of colour are seen. In Fig. 20.9 the *whitish area* contains tracts

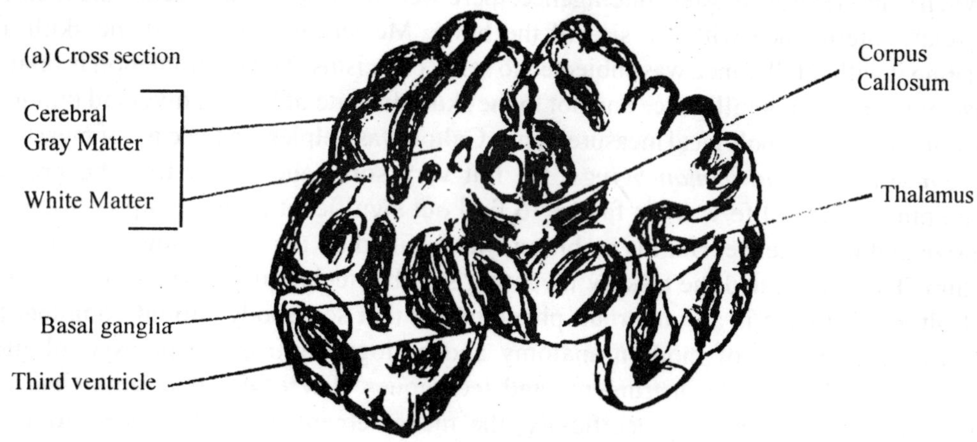

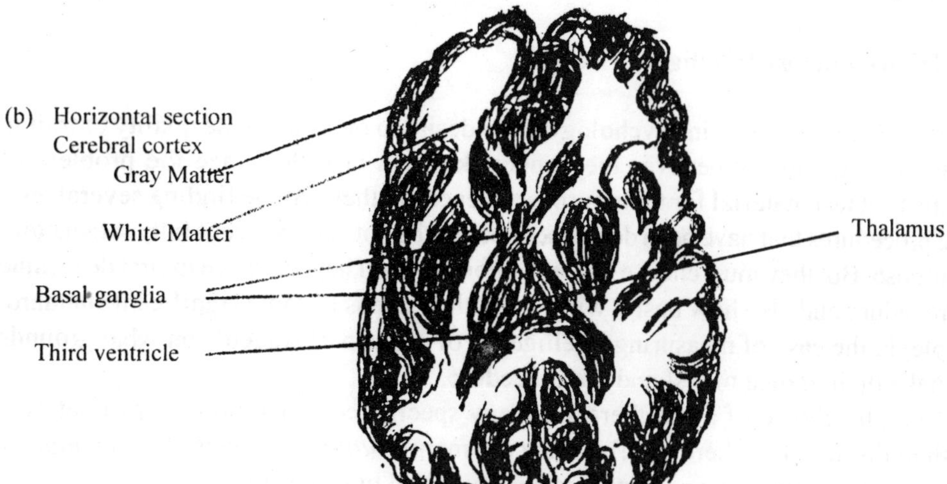

Fig. 20.9 : The Human Brain : Inside (a) Cross-section of the brain. (b) Horizontal section of brain. The light shade indicates the White Matter which is formed by fatty (white) myelin surrounding, the axons in the major fibre tracts. The gray shade indicates the Gray Matter which consists of cell bodies which the outer layers of the brain, called the cerebral cortex, mainly dealing with *higher intelligence levels.*

of white because of the content (fat) of *myelin* covering the axons. The *light gray* area mostly filled by *cell bodies* which are *free from myelin*. The white matter consists of bundle of nerve fibres and the *gray matter* consists of areas filled with *rich nerve cell bodies*. Andereasen made it clear the importance of the distinction between the *white* and *gray matters*. According to him the correlation between intelligence and the *volume of gray matter* in the cerebral hemispheres of the brain is much higher that that between intelligence and the volume of white matter of the brain. Therefore, the correlation between the size of the brain and intelligence is established. The *gray matter* of the cerebral cortex area is highly correlating with intelligence than any other area.

When Galton started to study intelligence, there were no proper instruments and methods to correlate intelligence with the size of the brain. Measuring the size of the skull and correlating it with intelligence was subjected to serve criticisms. The phenomenon is relating the size of the brain to intelligence and not to the skull. In spite of having invented hundreds of instruments for psychological measurements, Galton was helpless in the case of measuring the *brain size in living human beings*. He felt very sorry and stated that the greatest disappointment in his life was his failure to find out significant relationship between the brain size and intelligence. But his ambition was fulfilled by the psychologists who came after him. They continued the task by inventing suitable techniques to measure the brain size. Galton felt that it is an important phenomenon that somebody should complete the task. The development of modern anatomy and biology contributed to psychological measurement with suitable instruments and techniques. Therefore, it is high time that psychological measurements, specifically, the measurement of intelligence should be correlated with the size of the brain for substantiating the valuable studies.

Qualities Desired in any Intelligence Test

Researchers and investigators in psychology often desire to measure some quality or trait in individuals or in groups. Whenever they undertake such task they face the problem of choosing the best test material for their purpose. Generally they will be finding several tests and testing procedures that have been developed seem to be not cent-per-cent fully satisfactory for their purpose. But they must choose the best among them. They will have to first determine the best procedure and also how best it will satisfy their needs to some significant standard. For example, in the case of measuring intelligence of a group of students on what grounds they can make their choice of test and the procedure.

In making the choice of a test there are many specific considerations in evaluating a test. We shall discuss them here briefly. They are mainly *stability, consistency, reliability* and *validity*. Some researchers add another quality called practicality.

In any measurement of intelligence of some individuals or groups the investigator should see the suitability of the test. After choosing the test he should carefully revalidate the test before administering it to a population. It is because the test has been designed and standardized some years ego on some population. Therefore, there is every chance that there might have been sons changes in the population. An intelligence test standardized in one population a few years ago may not yield the same IQ scores in the present situation. Therefore, a revalidation is an essential thing to test the consistency of the intelligence test scores. It is to test whether the same test yields the same IQ scores of the population to *avoid greater error variance*. The test user also must see whether the test maintains the stability in measurement. It is to *provide confidence* for the investigator that the test is very stable and he can make up his mind to *go ahead in his effort to measure the intelligence of the population.*

Since the test user selects a test for his purpose of measuring intelligence of a population, the test should have been already standardized according to the principles. Therefore, the test user need not go into the laborious task of standardizing any test. But an important aspect of the process in standardization is testing the extent to which the test meets *two statistical creteria,* the *reliability* and the *validity.*

Reliability

Reliability of a test is the extent to which a measuring instrument yields consistent result at each time when the same individuals are tested. Thus a test is *reliable* if it measures anything *consistently.* An individual's traits remain constantly and therefore, psychologists attempt to test the consistency and the stability of their tests.

An intelligence test would not be considered as reliable and useful if it measures an individual's intelligence quotient as 120 at one time and 90 on testing the same individual for a second time. Therefore, the reliability may be defined as the consistency with which a test measures whatever it measures. A perfectly reliable test should yield the same scores on different occasions for the same individuals provided what it measures has not changed. The most often applied three methods of testing the reliability are, the test-retest method, the split-half method and the parallel form method.

1. The Test-retest Method

The procedure adapted in this method is to administer the test to the same group of individuals of a population two times with a short interval. It is to see whether the scores of the group are quite similar on the two testings with same test. The statistical method of correlation is computed between the two sets of scores and if the correlation coefficient is highly significant then the test is quite reliable. It is expected that the correlation coefficient should be above +0.80. Now this test is reliable and can be used for measuring intelligence of the population.

2. The Split-half Method

According to the split-half method, the scores of one half are correlated with the scores of the other half. The same test items are divided into two parts, one part with *odd number item* scores and the other part with *even number item* scores and these two are correlated to test the reliability of the test. If the correlation coefficient is above +0.80 and significant then the reliability is sound and accepted.

3. *Parallel Form Method*

According to this method two tests of the same type are administered to a group and the scores of these two tests are correlated. The coefficient of correlation between the two test scores should be more than +0.80 and should be also highly significant.

Validity

The Validity of the test is also a very essential characteristic of a test which should be tested on a population before administering the test. The validity of a test is defined as the extent to which the test scores should measure what they are intended to measure. An intelligence test would not be considered as a valid one if the performance of the subjects on the test has failed to agree with another similar independent test scores on the same trait. Psychologists mainly apply four *different types* of validity.

(i) Predictive Validity

It is one of the statistical technique the psychologists use in validating the tests. It is to see how far a test can predict an individual's performance *in some task later on.* Intelligence tests were originally designed to predict the performance of the students in schools. A large number of studies have been made on intelligence and the correlations have been worked out between the Intelligence Quotient scores and the achievements in school performances among the students. If the correlation between these two variables is found to be high and significant then we may assume that the test is having the predictive value.

(ii) The Concurrent Validity

This is established by comparing the scores of one test with the scores of another test which is having well established validity. If the scores of the new intelligence test correlate highly with another highly validated test like, Stanford-Binet or Wechsler Intelligence quotients then the new test is having the concurrent validity.

(iii) The Content Validity

The content validity of a test is established when a test has adequately and appropriately include the chosen area. Generally, an achievement test in mathematics or in a language may serve the purpose of determining the content validity. The main aim is whether the individual is capable of demonstrating his knowledge in the specific area.

(iv) Construct Validity

Construct is an important concept in this type. It is concerned with the validation of the test that measures a *general construct* or *trait*, for example, intelligence or neurotic tendencies. The construct validity is a complicated one. There is no simple technique for assessing the construct validity. Information may come from various sources for this validity, for example, from correlations with other tests. In general, construct validity may be expected to be as accurate as possible if only the construct is well defined. It depends upon how well the construct is defined.

PROBLEMS IN INTELLIGENCE MEASUREMENT

Psychological studies are mainly based upon testing human adjustment and productivity for classifying people into various categories in which the intellectual functioning is used as a criterion. The major factor used in the testing is the *learning ability*. If the intellectual functioning is very low the person cannot learn to avoid even simple dangers of life or may fail to obtain his primary needs of his life. He needs regular guidance at every level. Therefore, the problems of identifying and guiding people who are at the extreme of intelligence become the challenging task for the psychologists who are engaged in the measurement of intelligence. Thus, this becomes sometimes serious problem.

The scientific inventions and discoveries of the present times made human life filled with more *tensions* and *stresses* and life has become highly *competitive*. The present human life is driven to the state of "survival of the fittest". Vocational choices are based upon the intellectual levels and persons who complete for a vocation are subjected to categorization on the basis of intellectual development. In this context the role of psychologists and the *measurement of intelligence* become very crucial as they have to identify persons by fair means by using correct and accurate tools. If the test material is not fairly applied there is every chance that a normal person who is endowed with average intelligence may be pushed into the category of low scores. Therefore, in determining the extreme cases as feeble minded or mentally retarded a retest on the same persons must be made. Any error in categorization may cause severe damage to the life of the persons who become the victims of the testing procedure.

In the cases of the other extreme, even if any minor error is occurred, it may not cause any serious consequence as they are always considered as above the normal level of intelligence. These men make their achievements easily in areas in which they are interested and they need no regular guidance or close supervision. Though these persons can adjust to the situations and get along they also need suitable guidance on the basis of their IQ scores obtained in the intelligence test. They must know the specific area where they can easily and successfully perform their task whether it is of intellectual type or of physical type. Apart from intelligence tests, some aptitude or vocation interest tests also need to be administered and on the basis of the scores they may be guided and the area in which they can easily prosper can be indicated.

Most of the intelligence tests that are existing at present have been standardized and are reliable and valid. But there are unsolved problems observed in using them to measure intelligence. Another problem observed is that the composition and function of intelligence undergo changes with age. The tests that tap the intelligence of the infants are quite different from the tests that tap the intelligence of the older children and the adults.

Moreover many of the measurements of intelligence is not fully flawless or free from error. Pure and absolute measure of intelligence test is not possible as it has to deal with mental functions.

One of the problems psychologists face is at what early age the difference in intelligence measurement can be noticed in infants. Most of the items in the infant tests are like those of the adult tests. But the important characteristic of the infant intelligence test is the emphasize on *sensory motor abilities* which may yield much better result than the intelligence tests conventionally used. Therefore, the importance of specific tests to measure the infants' intelligence was strongly felt. To fulfil the requirements **Arnold Gesell** (1880-1961) has developed the first test of infant intelligence. It was a *developmental schedules* contain items with age progression similar to *Binet tests*. In Gasell's developmental schedules the age levels are from *one month to two years*. The items in each level are divided into *four catergories* such as *motor, adaptive, language, personal* and *social*.

The motor category consists of items to assess the behaviour like the head balance, standing, walking jumping and the child's ability to reach for grasping and manipulating objects. *The adaptive tests* assess the infant's reactions to objects such as toys, dangling ring and the infants' ability to fit differently shaped forms on a board. *The language items* deal with prelinguistic vocalization of infants, their comprehension of talks others made in front of it. It also includes item to test their ability to point to objects and pictures named by the investigator. The *personal* and *social* items are used for conducting interview with the mother of the infant. The investigator makes entries in the schedules. These items cover the behavioural aspects like smiling, self-feeding, toileting, play and the way in which the infant makes the demands for things.

The test designed by Gaesell is viewed by him not as an intelligence test, but he called it as schedule for measuring *Developmental Quotient* (D.Qs). He calculated the D.Qs with the same formula used for I.Qs. The age is in terms of months and the performance is expressed as the *Development Age* (D.A.). This is divided by the chronological age in months (C.A.) and multiplied by 100. For example, if a fifteen-month old boy passes all the items of eighteenth month old his Developmental Quotient (D.Q.) is 18/15 x 100 = 120.

An important doubt has been raised on the basis of the concept of D.Q. It is whether the same infant when he becomes an older boy or an adolescent, would it be possible for him to obtain the same level of I.Q. when an intelligence test is administered? Or would the infant continue the same level of brightness when he becomes an adolescent? Children vary widely in their rate of development. Studies show that the D.Qs. and the I.Qs. did not correlate significantly. The failure to find relationship between the infant D.Qs. and the later I.Qs. shows that these two scores reflect different abilities.

Therefore, we may state that intelligence is not a component or capacity, but a *collection of abilities*. Some of them may be found in young children and others found only in older children. **Hofstaetter** (1954) had observed in his study that early test scores mainly depend upon primarily *sensorimotor alertness*, but the test scores of two to four years olds depend upon *persistence*. The findings of these studies indicate that when intelligence develops, the mental functioning changes both qualitatively and quantitatively. Several studies Gaesell brought to light the wide changes that take place on the basis of age levels.

INTELLIGENCE AND ACHIEVEMENTS

All those psychologists who were engaged in measuring intelligence right from the time of Alfred Binet have agreed that it is important that intelligence tests should measure intelligence specifically and should *leave achievement and knowledge separately*. Achievement means something that has been accomplished especially by hardwork or successful competitions of some task. Achieve means to bring to a successful conclusion. But intelligence generally means the quality of behaviour of an individual of a given time or the ability to skilfully deal with a complex problems or situation. Thus intelligence is different from achievement. According to **Woodworth** (1952) achievement is *actual ability* that is put into action in a task and completing it successfully. Thus intelligence and achievement are two separate human potentials. Each one is tested with separate tests as intelligence test and achievement test.

An intelligence test developed in one culture, using the items of that culture does not assess intelligence fairly well in another culture. An item that is very familiar in a culture might have been included in an intelligence test. The test developer might have thought that the item known is to all the people in that culture. But a person who is born and brought up in another culture may find the item quite alien to his culture and tradition. If there were many items, like that individuals belonging to different culture score poorly in the test. Therefore, intelligence tests should be culture free and also should be based upon age level. Further, the language used to design the items are also found to be ambiguous. There should be clear language in writing the items.

Generally culture biased tests and test items with twisted language cause serious consequences in the measurement of intelligence. Such tests may lead to wrong or incorrect diagnosis and categorization of a normal person as a mentally retarded. **Mercer** (1975) strongly asserted that this is more likely happening in the case of children of minority groups, lower socio-economic status levels and in racial groups. Therefore, attempts have been made to reduce the influence of culture on test results and a quite number of culture fair intelligence tests have been designed and developed. In these tests the *verbal materials* have been removed and in place of them *spatial reasoning items* were introduced. Children are asked to complete the incomplete drawings in the test.

One of the important test constructed on this basis is the *IPAT Culture Fair Intelligence Test by Raymond Cattell* (1963) at the Institute of Personality and Ability Testing (IPAT). Investigators state that this test is only partially successful in bringing out the innate ability. Though the test results show some comparison between the subjects in United States and with the subjects in France and Taiwan, the scores did not show relationship with the scores of other nationalities. There were marked differences. Therefore, it was felt that the test is needed some revision and revalidation if it is to be fully a culture fair test.

MOTIVATIONAL AND EMOTIONAL ASPECTS IN INTELLIGENCE TESTING

Researchers in the area of intelligence testing have observed that many intelligence tests are *sensitive* to other psychological processes. The test scores are influenced by mainly personality, motivational and emotional factors. It is observed that some test items in intelligence tests require close attention. Some of the researchers are of the view that *emotional factors have drastic effect* on intelligence test performance. A team of researchers like **Zigler, Abelson** and **Seitz** (1973) attempted to demonstrate the effect of emotion on intelligence test performance among economically disadvantaged children. They have administrated an intelligence test and retested the same children after a few weeks.

It was found that there was *an increase of ten points* in the average score of these children in the *retcst*. A second group of children was also tested by the same test. The IQs of the second group's first test were compared with the scores of the first group. There was no much difference between the first group's second score and the second groups' first score. The children of the second group have *played some games* with the *test user* in a warm and friendly situation. Then the children of the second group have taken the intelligence test. The scores of the children were found to be significantly higher. These results suggest that in an unfamiliar situation the children's *emotion* or *anxiety* had impaired their performance of the intelligence test. When the children have become quite *familiar with the testing situation and friendly* with the test user the performance became much better and the scores increased significantly.

These investigators have tested this hypothesis in some more experiments. In all their experiments they have observed that emotions can affect the performance of the children in intelligence testing. Children in familiar and comfortable testing situations do better than in unfamiliar and anxiety situations. Since anxiety interferes with test performance, motive and anxiety influence intelligence test performance and also school performance of the children.

Uses of Intelligence Tests

One of the most useful tools available for a teacher of a school or college, counseller, educationist, business and industrial organizations is the Intelligence Test. People who are engaged in grading the pupil, counselling the clients, recruiting persons to various jobs should have some knowledge in the administration of intelligence tests and in the interpretation of the scores obtained from the tests. There is no trade secret in this. The key of the test is there for their use.

Individual Diagnosis and Guidance

Intelligence test serves as an aid in diagnosing pupil's difficulties and problems. A good teacher is interested in finding out whether the particular pupil is working to his potentialities. If it is not, what is the real cause behind his low achievement. He requires some help. To determine his actual scholastic aptitude an intelligence test must be administrated to him. The scores would reveal where he stands in comparison with other pupil. Then he may find out the fact whether the low achievement is due to lack of capacity or something else.

One of the important purposes for which intelligence test may be used is the educational and vocational guidance. The pupil may require guidance in various areas. The knowledge of the scholastic aptitude test is very much useful in helping the pupil to select the proper course of study at high school and college levels. Guidance can be also provided to the pupil whether they can proceed further on their studies at college or take up a vocation after the completion of the high school studies. If the decision is to take up a vocation what type of vocation he should choose also may be determined.

Further if the school teacher knows the intelligence scores of a group of children at the beginning of the school year he may use the data along with other information in grading, guiding the pupil, assigning the level of texts and other learning materials and he can set the goals and expectations of the students. The curriculum planning, the grouping also can be done on the basis of the I.Qs. of the students. Intelligence tests may measure apparently what the teacher wants. The test scores may be useful for him in predicting certain traits. For example, verbal competence is related to achievements in academic subjects. If the teacher knows it in advance he can provide certain opportunities for the pupil to develop the verbal skill if they are lacking it.

Placement and Programming

Ever since the development of the intelligence tests one of the important uses of them has been the determination of placement of the students in schools and colleges. Which student should be admitted to regular school study and which student should be placed in special course? In the United States many schools and colleges have their own standard of I.Qs. and mental age levels as a limit for admission to their institutions. Intelligence test in this respect is of great help to the educational institutions in admitting students in various courses.

Colleges and Universities may find difficult time with their programmes to execute if some of the students, in every course of study fall below the average level. These institutions may go ahead with their programmes successfully if admissions are done on the basis of the scores of certain intelligence tests.

Intelligence tests may be used for diagnostic purposes. We can find out the individual's strengths and weaknesses not only in the academic activities but also in the over all performances. Further, intelligence test may be used to measure individual's present abilities. People may have different opportunities in learning and developing their skills. The measurement of the skill may be of immense help to certain organizations who are selecting men to various trades and occupations. Selecting men to various jobs and placing them at different levels may be done successfully on the basis of the scores of intelligence tests. The

disadvantaged persons may be traced out and proper care may be given to them to develop their skills. The student unrest in the Indian Universities may be reduced to some extent if the College and University authorities would encourage Psychological testing programmes in their institutions.

21

NATURE OF INTELLIGENCE

LEVELS OF MENTAL ABILITIES

One of the major aims of the study of psychology is to understand the behavioural differences in individuals. Individuals differ in several aspects which are determined by heredity, environment, culture and society. Individuals differ mainly in *learning ability* which is considered to be the crucial factor in human adjustment and productivity. Human beings differ from one another in mental abilities and in personality characteristics. These individual variations take place in learning abilities, intelligence and personality characteristics. If the mental functioning in a person is very low he cannot learn anything. He cannot learn his livelihood, cannot avoid any dangers to his life and could not find out a shelter for his living. Such persons cannot do anything without constant supervision and proper guidance.

On the other end a person with high level of mental abilities makes himself fit to perform any productive activity by proper training, education and successful adjustment and behavioural patterns. Such persons in the society make great achievements and also useful for the society. Thus there are two extremities. The problem the human kind faces is to identify and discover suitable techniques of treatment or placement for the persons who are at the extreme ends of intelligence. These *extreme ends* in intelligence are caused by the interaction of heredity and environment.

The problem is whether individual differences and also the extreme levels in intelligence are due to heredity or environment. This has been much debated and discussed in recent times (Woodworth, 1941). If superiority and inferiority in intelligence are mostly due to heredity, the prospects of mankind lies in making *improvement in human stock and not allowing it to degenerate.* But if the two extremities in intelligence are due to mostly environment, society should take preausious measures and provide better nutrition and hygienic and better opportunities of all types. Whatever might be the cause behind it, the major concern at the present time is the extremities in intelligence.

Feeble Mindedness

Long before any attempt was made to measure intelligence it was observed that there were some persons in large communities *mentally deficit.* They were too stupid to manage their own lives. They were generally called *feeble minded people.* They were classified as *idiots, imbeciles and morons.* The idiots were showing greatest amount of deficiency. Those somewhat better than the idiots are imbeciles. The least mentally deficit persons are called morons.

The *idiots need* close supervision and care, otherwise they get into some dangers very easily. They cannot avoid even common dangers of life. They may walk into the deep water, go against the moving motor cars, stretch the hands into the fire. They are not capable of learning anything, do not know how to dress themselves. They need to be fed by others and do not take care of their bodies. They cannot learn any language, except uttering few monosyllables. There must be somebody always by their side.

The imbeciles are little better than the idiots in the mental deficiency. The imbeciles do not get into the common dangers of life, like the idiots. These persons can speak few words, but cannot learn to read or write. It will not be possible to teach these person to do any useful work.

The *imbeciales* fall into different levels. The lowest level imbeciles do learn to perform any work. Those who are little higher than the last category on the scale may learn few useful works under supervision. The imbeciles on the upper category can learn to eat and dress and also learn to wash. But the imbeciles of all categories cannot be left alone to perform any short and simple tasks without constant supervision. Some elders must be there by their side.

On the other hand among the feeble minded persons, the *morons* are much better than the other two categories. They learn to eat themselves, dress on their own and learn to wash. Morons can be taught to perform to do some simple tasks without close supervision. They may be taught to make beds in some institutions and also work as office boy to carry files from table to table. The better grade morons can also perform some skilful works, like taking care of small animals looking after small children.

Some *high grade morons* can be taught to perform simple skilled tasks like some filing works in lathes, some type of wood works and sawing works. There are institutions taking care of feeble minded persons and training them on the basis of their physical and mental capacities and they were found to be successful in training and sending the high grade morons for some occupations in the communities. But they require that somebody should be watching and guiding them understanding their mental limitations. If there is no supervision the morons may mishandle their money and spend their leisure time foolishly. The male morons may get into the thievery and other delinquent activities. The female morons may be easily tempted into prostitution. Generally the nature of morons is not carefully handling the situations and they are incapable of dealing with the problems of everyday life.

A *feeble minded child* may be defined as one who is not able to learn and master the legitimate school subjects within the normal course of time and with the required thoroughness. Similarly, a *feeble minded adult* may be defined as one who due to his mental deficiency is unable to support and protect himself and keep himself out of danger in the community. It has been estimated that in any population there are feeble minded persons. In America the number of feeble minded has been estimated as one to two per cent. But it will not be possible to estimate accurately because there is *no sharp difference* between the little low normal and the feeble minded. **Tredgold** (1937) states that the observable difference between the low-normal and the feeble minded depends upon partly on the environmental factors with which they will have to cope with. **Abel** and **Kinder** (1942) who studied this aspect also endorse Tredgold's view. Because apparently it will not be possible for us to judge whether a person is subnormal or normal unless we observe his behaviour in certain situations.

Authors who dealt with intelligence in a more extensive manner are of the view that intelligence is the *achievement of different kinds of abilities.* But it is also better to know whether the ability is a specialized one or generalized one and also we should know whether intelligence tests measure the individual's all-round abilities or certain specific area of intelligence. Generally when we describe a person's characteristics we may state that the person is very clever or intelligent, able to carry out everything in an excellent manner, while stating about another person we may state that he has not developed his ability to the expected level and often commits mistakes in his works. While speaking about a third person we may state that he is good in certain things but poor in other things. The idea behind these statements is that individuals differ in their abilities.

In our everyday life we observe that some persons exhibit enormous amount of abilities in certain areas like sports, games, acting, writing stories and poems, public speaking music, art, painting, sculpture and so on. But those who are very efficient in these areas are not proficient in other areas. This means that these persons have special abilities in certain areas and therefore they are very brilliant. But the question here is whether there is any *general ability* which could make a person to be an all-round expert. A systematic investigation may be made to test whether different abilities may be functioning in relation to each other and make a person efficient in different areas. This process we may state as generalised function of abilities and a person may be proficients in more than one faculty. *Individual differences are revealed* in various ways in an act whether the activity is performed by an average person or by a gifted person or by an eminent person.

Arithmetic Prodigies

There are also some other type of persons little far from feeble mindedness, but *moderate intelligence make extraordinary achievements* in manipulating numbers and in mental education. This type of persons consider this ability as a *big gift* and demonstrate it in public places and receive appreciations of the people. They accept challenges from people, even from the computer scientists in working out big calculation and arithmetic problems. Psychologists carefully studied these persons and found that they had *excellent memory* for numbers. In any population, however persons are highly educated, they can repeat maximum only seven to ten numbers once heard, but these *arithmetic prodigies* are able to repeat *twenty to twenty five* numbers after hearing once. It is very much surprising that they have very *unusual memory power for numbers.* They could reproduce the same numbers even after a week without any error.

These persons have wonderful ability for multiplying large numbers in their minds. They are able to tell the squares for any big numbers, square roots for any numbers and also tell the cube roots and square roots. This type of mental ability is very unusual and make the audience to think that these persons have some *supernatural power* in performing these types of tasks. Some are of the view that it is the work of the *unconscious mind.* But it cannot be the work of unconscious mind. This may be due to some *short-cut* method *they have invented.*

Disability in Reading

Some persons are endowed with one special ability which is very much superior to the general abilities. But some persons may have lack of ability in a particular line. Some students who do well in all other subjects feel bad that they could not do better in a subject and they think that they may fail in mathematics. **MaCullough** *et al.* (1946) wanted to know the level of reading ability of school children. Some children in the school had great difficulty in learning to read. But they are normal in their abilities in other lines. It was observed that an *eighth grade child with an intelligence quotient of 115 not able to read* anything *better than* an average *third grade child.* This child is having the *reading disability.* He was able to get along with his study by carefully attending to the teacher's *oral presentations* and by satisfactory work in arithmetics. He had the specific *disability in reading.* Even if any special training is given sufficiently earlier for improving his reading ability, he would be always very slow. There are many adult persons in every population having the reading disability and feel unpleasant and never accept any assignment which require reading. This is a specific disability with which some suffer.

Learning Disabilities

The intellectual level of certain children fail to develop to the degree expected in a specific academic, language or motor skill areas. The American psychiatric Association (1987) in its Diagnostic and Statistical Manual (DSM-III) made several of specific developmental disorders, one better known is the *learning disabilities.* **Bryan** and **Bryan** (1979) studied this learning disabilities. Learning disabilities are normally identified with school system. These disabilities are more *common in male children* than in female children. Similar to *mentally retarded children* these children with learning disabilities require special educational attention. These children, unlike the mentally retarded are on the average or even above average in intelligence. But the problems of these children are in the specific areas of learning such as reading or mathematics.

The learning disabilities specifically refer to the children who show a major *discrepancy* between achievement and the ability, that is not caused by other circumstances. Children who have the learning disabilities are those who have no difficulties caused by mental retardation, social deprivation or visual, and auditory or motor handicaps. The well known learning disabilities are the *dysiexia* and *dyscalculia.* These are disturbances in the children's ability to learn and read. Children who suffer from dyslexia have unusual difficulties in learning to repeat the words and sentences into their components and then to integrate *the components* to form into the wholes. They also have the problems in writing the words with correct spellings. They have problems with *letters* and *numbers* and these children often get confused with directions. For example, dysiexia children substitute letters and numbers which are having certain similarities. In place of 'b' they write 'd' in place of 'g' they write 'q' and 'c' for 'e', 'b' for 'h'. In case of numbers they write 14 for 41, 17 for 71 and so on.

The other learning disability children suffer is the *dyscalculia.* This is a type of difficulty in dealing with mathematical concepts and computations. This also refers to the learning disability of *space and size relationships.* **Bryan** (1978) states that children sometimes have both types of disabilities. But it does not necessarily occur at the same time.

IDIOT SAVANTS

While discussing about the feeble minded persons we made mention about three categories of persons such as idiots, imbeciles and morons. These persons show lots of differences in their behaviour. But these persons in the society are in between normal persons and feeble minded persons. We cannot take such people into confidence as normal and cannot entrust any responsibilities to them. These people are *called Idiot savants*. Idiots are extremely feeble minded or retarded and savant means *man of great learning or sage.* But here the term idiot savants means *talented idiots.* This term is used to indicate certain feeble minded individuals. These individuals are *not very low in intelligence* as to classify them as idiots.

The idiot savants sometimes reval *remarkable high ability* in certain specific area. Every idiot savant may show different type of ability in certain line. They may differ in the performance of certain activity and in the *level of performance.* Some idiot savants show their *musical abilities* at a very great level with good memory for musical notes. Some perform some type of musical instruments. Similarly savants reproduce exactly what they heard or witnessed. A musical concert which is very difficult can be reproduced by a savant after a single hearing. Some idiot savants are able to reveal their abilities in *drawing* and *painting* and also in the *construction of machines and tools.* One idiot savant was able to construct a model ship. The capacity of an idiot savant is *specific.*

An interesting phenomenon generally regarded as a *special form of mental retardation is the idiot savant,* a retarded person with superior functioning in one small or a *narrow area* of intellectual activity. There are several case histories to support this view. An idiot savant, therefore, may be defined as an individual with a rare form of mental retardation or subnormality, being *extraordinary talent* in one or few limited areas of intellectual achievement. The performance of these persons are extremely rare. The type of activities vary in nature, generally very peculiar and remarkable.

The idiot savants are treated as feeble minded inspite of the expression of certain special talents. A feeble minded person of this type could correctly state the *day for any date* of the preceding *months and years* up to *three to five years.* Another one could tell the time exactly with minutes sitting in any place without a watch or clock. **Tredgold** (1937) who deeply studied many savants stated that one feeble minded idiot savants can tell the day promptly for any date even within the past thousand years. He has stated that such capacity is not pure facts of memory, but effect of some genetic factors. These persons are very conscious of their special abilities and feel very proud of them. They spend most of their time in improving their capacity and techniques.

The idiot savant is a *special form of mental retardation.* This is one of the most interesting phenomena in the study of mental abilities. An idiot savant is a person with a superior functioning in one narrow area of intellectual activity. Generally, the idiot savant appears normal and may be carrying out all his activities as a normal person. But his intelligence quotient is below 70. In the infinite variety of human minds there is none more fascinating then that of the idiot savant. These persons show *astonishing skill* in a specific area, yet they are *mentally retarded* as their I.Qs. are low. There are both *male and female* idiot savants. But the male idiot savants out number the female idiot savants in the ratio of three to one.

Every idiot savant is endowed with a wonderful complex skill, most of them are dealing *numerical values and time values.* Among the mental defects known are the hereditary the **Down's** *syndrome* or mongolism and there are few other rare forms of mental retardations. Many of the scientists in this area suspect that the *genetic factors* may underlie these mental abilities. It is also felt by some scientists that the environmental influences such as the *family* in which one is born and grow up from the childhood period is also an important cause.

But some *biological and genetic scientists* find the cause behind the *chromosome abnormalities.* The type of chromosomes that function in the intellectual activities of the idiot savants are not accurately explain by them. One chromosome abnormality according to genetic studies is the combination of one X and two YY (XYY) chromosomes in the case of males and two XX and one Y(XXY) chromosomes in the case of females. These combinations of chromosomes are the causes of abnormal or defective physical bodily structures. But the nature of their behaviour and their intellectual development with the levels of intelligence quotient were not studied in detail. Therefore, it is also possible to assume that one type of chromosome abnormality is revealed in the *physical growth* and another type of abnormality is revealed in the *intellectual functions* which we call as *gifted idiots* or *idiot savants.*

Ordinarily it is quite impossible for a normal person of any amount of brilliancy to deal with the arithmetic calculations as the idiots savant can work out, like the calculations such as multiplication of any big numbers, squares, cubs, square roots, cube roots and so on and similarly in the case of correctly telling the day for any date for any number of past years. In 1980s an Indians lady has thrown challenges to the American computer scientists in the calculation of numbers and she won. For all these, a correct scientific explanation has not been brought out. Therefore, there is nothing wrong in accepting the type of superior intelligence demonstrated by the idiot savants and consider that it is a gift for them due to rare *chromosome combinations* rather than chromosome abnormality.

INDIVIDUAL DIFFERENCES IN TEST PERFORMANCE

In our attempt to measure the level of intelligence we must see that there should be normal distribution of scores. All the intelligence tests are designed and developed with an assumption to reflect that mental abilities are normally distributed. Any intelligence test should have this characteristic. The scores should spread out in such a way that a *curve* drawn to the scores should be *normal* (Fig. 21.1). The distribution of intelligence quotients (I.Qs.) in this example is not having any sharp breaks between various levels of intellectual function. The I.Q. scores are spread out continuously on a wide range from the lowest to the highest which is the indication of individual differences in the test performance. If there is no such wide spread of the I.Qs. then the curve would be flat and will not have the *bell shape.* This asserts that there are definite individual differences in the I.Qs. of the subjects. The same result should occur for any *intelligence test* which was carefully designed and administered to any population.

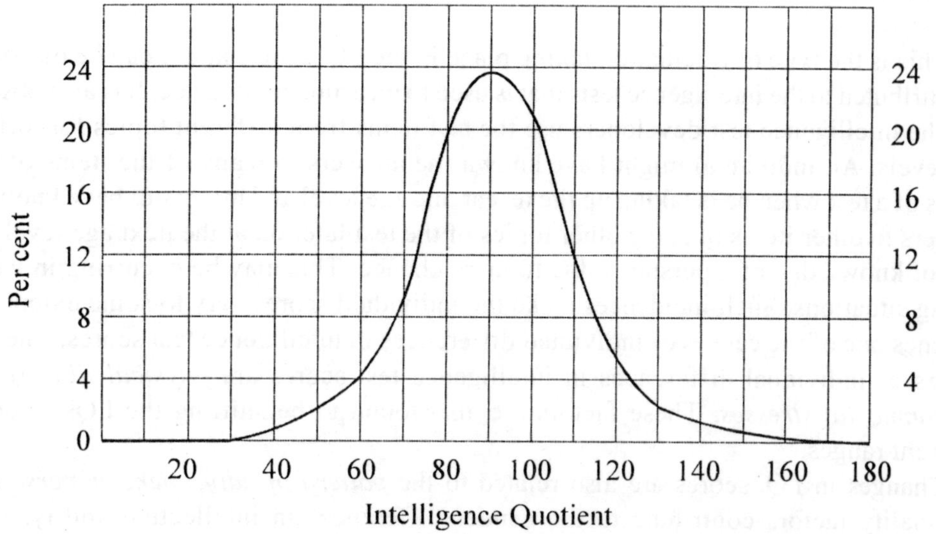

Fig. 21.1 : Distribution of Intelligence Quotient and the Individual Differences

Such intelligence tests must be used normally for identifying persons for putting them in certain categories. For example, if a school or college is very keen in admitting only those students who can fulfil certain requirements by achieving the I.Qs. above 110, then the students who achieve the I.Qs. 110 and above are eligible for admission to that institution. Suppose, for example, a student's I.Q. falls at 108, he is below the category level which the institution has specified. There may be several students in this category. In such cases we must also consider the other factors like, whether he/she is *highly motivated* and also *free from emotional problems.* If he is of such category he may be considered even better than a student with a higher I.Q., who may be poorly motivated and may have emotional problems. Therefore, several other factors must be taken into consideration in categorizing individuals, instead of deciding the selection on the basis of a *single factor,* the test score. The I.Q. should be the main factor and other personal desirable factors should be the supplimental factors. Individuals who score *very high* I.Qs. are referred to as gifted persons and whose I.Q. scores may range from *140 to 180.* This group constitutes *two to four per cent* of the general population. The gifted person's intelligence differs from that of the general population.

Psychologists' efforts to study intelligence in recent times have brought out several new findings. One of the important findings is the *growth of intelligence as a steady process.* This process is *rapid* in the early years and goes down slowly as and when the person grows older. The patterns of the growth of intelligence are considered as much more a type of behaviour of an individual. In the growth of intelligence there may be *spurts, plateaus* and even *drops* in an individual's intelligence test scores as a function of age. A person's I.Q. score may generally remain with in the same range. But it may be higher or lower from time to time. This means that the I.Q. may be going on changing as the individual advances in his age.

This is the type of fluctuation taking place in the I.Q. scores and it can be discovered and attributed to the intelligence tests that is used to measure intelligence. It is also observed that the intelligence test developers use the test items from different topics for different age levels. An individual might have known the answers to many of the items of some topics of a test when he is taking up the test at one age level and might not have known the answers to other items of some other topics of the test later on at the next age level. This type of knowledge of a person is due to mere chance. This may be occurring in various testing situations. Such incidences make the individual scores vary to some extent. This becomes one of the causes of individual differences in intelligence test scores. The other factors of individual differences in intelligence test scores are *physical illnesses* and *environmental stresses.* These factors are also found to be shifting the I.Qs. scores to different ranges.

Changes in I.Q. scores are also related to the *pattern of adjustment or personality.* Personality factors contribute to the gender differences in intellectual ability. *Social deprivations* play important role in the changes and the differences in the I.Q. scores of the individuals. The *socio-economic status* also was found to be having influence in the I.Q. scores of the individuals.

In the final level of the distribution of the intelligence quotient scores we *find two extremes,* one at the left end with few lowest scores and another at the right end with few highest scores, having most of the scores in the middle ranges. This type of I.Qs. distribution provides information to the psychologists to classify people into three categories as dull or mentally retarded, normal and the brightest or gifted.

DISTRIBUTION OF INTELLIGENCE

Testing the level of intelligence in a group of people generally brings out wide variations from lowest scores or I.Qs. to the highest scores or I.Qs. at the ends and most of the scores falling in the middle. This would indicate wide individual differences from extremely poor to extremely rich levels of intelligence. In other words most dull to most bright individuals may be found in a population. Any adult intelligence test, well tested in a population, would reveal the different levels of mental development. If any statistical treatment is given to the scores or I.Qs. of the population and if a curve is drawn, it would be a bell shaped curve (Fig. 21.2).

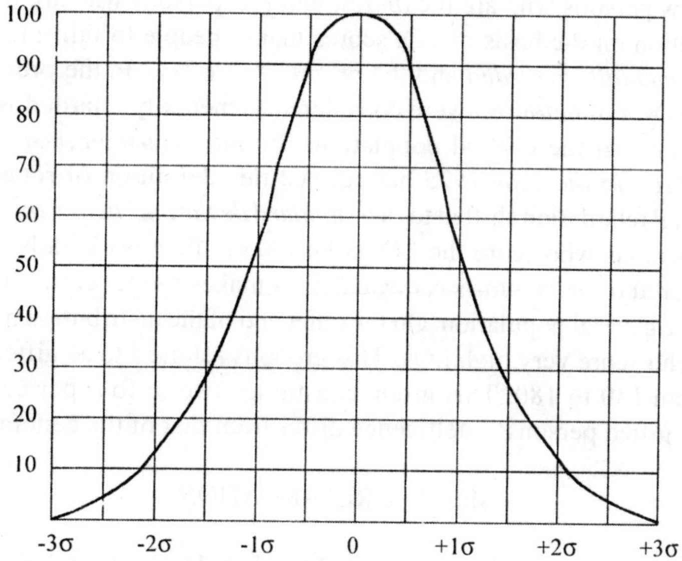

Fig. 21.2 : Intelligence : Normal Probability Curve

This type of distribution normally is obtained mostly when complex human mental abilities are tested. The tests may be referred to physical, social or intellectual areas. Even when we measure people's height, weight, speed of response, strength of a particular group's muscle or even the personal qualities, we may find typically that there is crowding or filling up of the cases near the middle of the distribution and a tampering off at the left as fewer and fewer cases, showing fatter and fatter cases at the right. In any population the test scores (I.Qs.) in intelligence may be having only this type of distributions.

It is better to study more closely to know how the frequency distribution of this type arises. From the studies we come to know that intelligence is normally distributed. This means that the number of cases within a range (may be 100 to 110) is with 10 points, will be greater near the middle of the distribution than when we move out to the ends of the distribution. Normal curve forms when great many different things enter in the determination of intelligence. Human development is determined by several factors. Whether it is heredity or environment or bringing up or social factors all have influence in the development intelligence.

Any accurate measure of the *centre of the distribution* is said to be *the average.* In the distribution we can see how for each individual *deviates up and down* from the average. This system of measuring intelligence is very convenient for indicating an individual's relative position in the group. In the distribution at the left end the individuals who had obtained the lowest scores fall which is the indication of greater amount of *dullness,* while at the right end we see individuals who had obtained and highest scores fall which is the indication of greater amount of *brightness.* In this type of distribution of the scores of intelligence tests, always there will be only few cases at the extreme ends, which mean that there are only few persons who are the *dullest* and few persons who are the *brightest.*

The distribution on the basis of I.Q. scores makes people to fall into three categories such as *normal, mentally retarded* and the *brightest or gifted.* In the previous distribution persons whose I.Qs. fall *below 85* were considered as mentally retarded which constituted about *three per cent* of the general population. But the *American Association of Mental Deficiency (AAMD) in the year 1973* has revised this definition of retardation. The I.Q. creterion of mental retardation shifted to *two standard deviations below the mean.* According to this change persons who score the I.Q. *below 70* on the most widely used intelligence tests are considered to be *mentally retarded.* This makes the level of retarded persons at *2.3 per cent* of the general population. On the other end of the distribution a small percentage of individuals who score very high I.Qs. This group is referred to as gifted and whose I.Q. scores range from 140 to 180. This group constitutes two to four per cent of the general population. The gifted persons' intelligence differ from that of the general population.

MENTAL RETARDATION

Psychologists have been applying the terms feeble mindedness, idiots, imbeciles and morons in the study of mental abilities of human beings and classified people into these categories on the basis of the low scores obtained by them in the standard intelligence tests. Those who scored high were classified as superior and very superior and genius in intelligence. In between these two categories fall the persons who made average scores. This type of classification was underwent radical changes and different type of classifications are made. The terms feeble mindedness, idiots, imbeciles and morons seem to be very unpleasant and derogatory concepts. Therefore, it was strongly felt by the American Association of Mental Deficiency and (AAMD) replaced concept feeble mindedness with mental retardation and dropped the derogatory terms idiots, imbeciles, morons and introduced the terms average, high average, superior and very superior, low average, border line and mentally retarded. The mentally retarded are classified into mild mental retardation, moderate mental retardation, severe mental retardation and profound mental retardation.

The concept *mental retardation* is defined in various ways. Generally mental retardation is the subnormal intellectual functioning associated with impairment of social adjustment which is identified with early years of the person. But the definition of the American

Association of mental deficiency is that mental retardation is a mental state which is significantly sub-average in the general intellectual functioning, existing concurrently with the deficits in adoptive behaviour which is manifested during the developmental period. This state may also continue further in the later periods of an individual if suitable treatment is not given.

Usually mental retardation is arbitrarily defined by drawing a line through the distribution of the intelligence quotient scores by that individuals with scores above the line are considered as intellectually normal and those I.Q. scores fall below that line are considered as mentally retarded. As the individuals advance in years the position of this line undergoes changes considerably. Always the classification is made on the basis of *intelligence test scores.* Generally psychologists depend upon the performance of the standard intelligence tests for determining the intellectual functioning. The level of performance on such tests is used for determining the *mental age.* The test items are arranged in an order of difficulty. A norm is established to determine the age levels. An individual's mental age is the age level that is the norm for the items they are able to pass. For example, if a boy aged twelve is able to pass the items of the test designed to 11 year old and did not pass the items of 12 years old, his mental age is 11. The I.Q. scores are designed in such a way that the mean score is 100 and the standard deviation is 15. This means that about two-thirds of the population scores fall between 85 and 115 (100 – 15 = 85 : 100 + 15 = 115). A score of 70 is two standard deviations below the mean of the population. Thus a *score of 70* is considered to be significantly *sub-average of general intelligence* functioning. It is estimated that about 2.5 per cent of the general population of any country fall into this category with different levels of mental retardation.

Adaptive Behaviour and Mental Retardation

The salient means of identifying mental retardation is the deficient of intellectual functioning. It is the criterion currently used by the *American Association of Mental Deficiency.* In considering mental retardation an individual must show deficits in *adaptive behaviour* in addition to the I.Q. level 70 and below. Adaptive behaviour refers to the effective functioning in areas such as *social skills, communication* and *daily living skills* and *how far well the person meets* the standards of personal independence and *social responsibilities* expected from the individuals *at varied age levels.* This is what the American Psychiatric Association (1987) has devised.

The adaptive behaviour the child is expected to follow are *taking care of the self,* learning the *concepts of time and money,* knowing to *use tools,* knowing *shopping,* learning to *travel* in public transports, becoming *socially responsive* and should know how to direct the self. The *adolescent* is expected to make use of academic skills, should know reasoning and judgment in his daily living and to participate in group activities. The *adult person* is expected to become *self-supporting* and also assume *social responsibilities.*

The adaptive behaviour of the mentally retarded persons are as assessed by administering several *adaptive behaviour tests.* The best known is the American Association of *Mental Deficiency Adaptive Behaviour Scale* (ABS) designed by **Nihira** *et al.* (1974) and the

Vineland adaptive behaviour scale designed and developed by **Sparrow** *et al.* (1984). This scale assessed the adaptive behaviour at five age levels such as 2, 5, 8, 11 and 16 years. These scales have recently established norms of mental retardation.

The scales included several areas of variables, but failed to include the environment to which the person should adapt. The problem with the many assessments of adaptive behaviour of the mentally retarded persons is that a person who lives in *small rural area* where all of them are accustomed with *not many needs,* whereas in the *urban* area people have *many needs* and they live in a complex environmental conditions. The live style in rural and urban environmental conditions would certainly have significant effect on the behaviour of the mentally retarded persons.

Investigators are of the view that the assessment of adaptive behaviour is not having any clear cut distinction from intelligence. It means that there is no so much difference between these two. **Charles Worth** (1976), an ethological psychologist believes that the ability of the individual to adapt to the environment is the indication the level of intelligence. According to him the adaptive behaviour scores are more raliable measure of intelligence than the scores of the I.Qs. tests. But some investigators do not agree with this view.

Period of Onset of Mental Retardation

Some investigators are of the view that mental retardation may manifest before the attainment of adulthood, that is, before the individual attains the age of eighteen. This view is rejected. It is observed that majority of children who are identified as mentally retarded are not diagnosed until they go to school. These children are not having physiological, neurological and physical impairments. These children face problems only when they are not able to keep up themselves with their peer groups in school. But the mildly mental retarded children when leave the school and become observed in the community and also in the work situation, their mental retardation may become less evident as they were in the beginning.

Classification of Mental Retardation

Mental retardation is identified by administering the standard intelligence tests. Those who score below 70 intelligence quotient are designated as mentally retarded. The intelligence test scores also identify the various levels in mental retardation. The current American Psychiatric Association in its *Diagnostic and Statistical Manual DSM-III-R* recognized *four levels* of mental deficiency on the basis of the scores of the intelligence quotient. These four levels of mental retardation are, mild mental retardation, moderate mental retardation, severe mental retardation and profound mental retardation (Fig. 21.2).

Mild Mental the Retardation: The mildly retarded individuals comprised about *85 per cent* of the subjects whose *I.Q. scores* are *less than 70.* The I.Qs. of the mildly mental retarded people are 50-55 to 70. These are very young persons, not always distinguishable from the normal young persons. They are identified *only in school situation* when they are not able to perform the school assignments *of their age.* When they reach the *late teen age* period they learn the school academic skills of the *sixth grade level.*

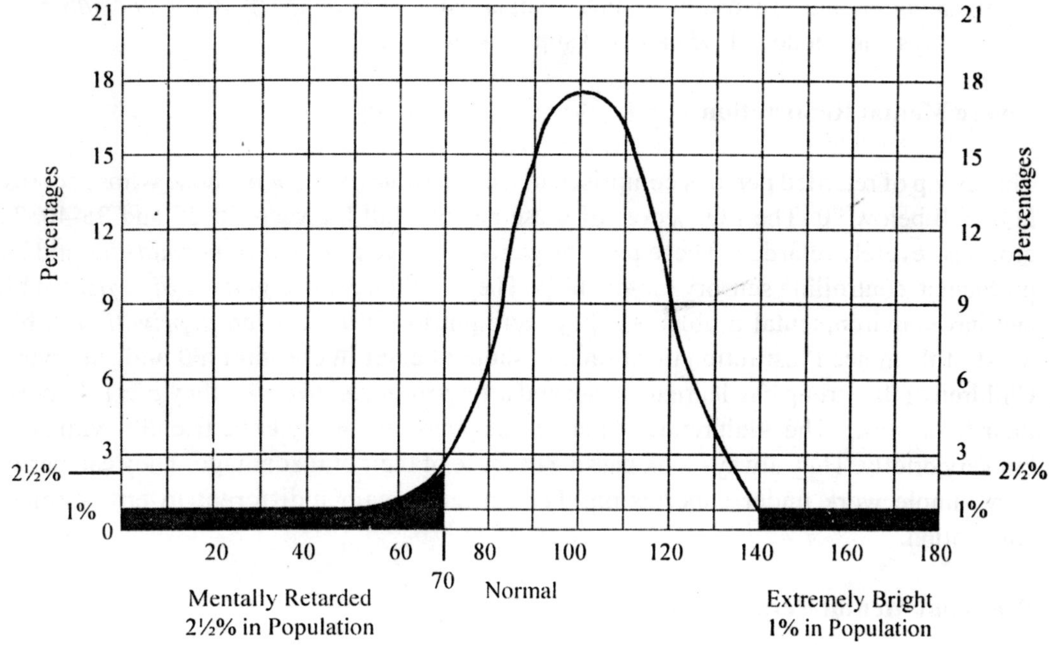

Fig. 21.2 : Two Extremes of Intelligence, and the levels of Intelligence Quotient

Mild Mental Retardation	:	I.Q. Level	:	55-70
Moderate Mental Retardation	:	I.Q. Level	:	40-55
Severe Mental Retardation	:	I.Q. Level	:	25-40
Profound Mental Retardation	:	I.Q. Level	:	20-25
Extremely Bright Intelligence Level			:	140-180
Normal People : Average I.Q. Level			:	70-140

At the adult level they could maintain themselves only in the unskilled jobs. They *always need help* in financial and social problems. They may also get married and have children. About *one per cent* of the adolescents need *institutionalized care.* Most of them do not show any sign of brain pathology. Members of these families have low level of intelligence and socio-economic status.

Moderate Mental Retardation

This group of persons show retardation little higher than the first group. This group comprised of *ten per cent* of those whose I.Q. scores are below 70. The I.Q. scores of these persons fall *between 35-40 and 50-55.* These persons suffer from *brain damage* and pathologies often. The moderately retarded have physical and neurological defects which *affect their motor skills.* These individuals during childhood period *need special classes* which should be consisted of development of *self care skills* rather than academic achievements. The moderately retarded persons do not progress beyond the *second grade* that too *at the adult level.* Most of them need institutionalization. Most of them can carry out useful work in the family situation and in work place under supervision. The *sensory motor coordination* is poor always and needed close care and supervision.

Severe Mental Retardation

This group of retarded persons comprised of *three to four per cent* of those whose scores of I.Qs. fall below 70. The I.Q. scores of these persons fall between 20-25 and 35-40. This group is severely retarded. These persons generally have *physical abnormalities* and have problem in controlling sensory motor skills. They might have had *genetic disorders* at birth and have environmental problems. They were put in *severe oxygen deprivation* at birth. Most of them need institutionalization and should be put in constant aid and supervision. Children of this group can learn to speak and after *prolonged training* they *can take care* of their *basic needs.* The adult retarded persons may have tendency to be friendly with others, but very briefly. They always look passive because of their brain damage. They can perform very simple work under supervision. They often remain indifferent to environmental stimulation.

Profound Mental Retardation

This level is supposed to be very severe mental retardation and it is the lowest level. It is estimated that *one to two per cent* of those whose I.Qs. fall below the level of 70 come under this category. *The I.Qs.* of the profoundly retarded persons fall between 20 and 25. These persons require close supervision all the times and they need to be nursed all the times. There may be some improvement if severe training is given *in motor skills and self-care.* Most of them have severe *physical defects and brain damage.* The death rate of this retarded group is very high in the childhood period.

Causes of Mental Retardation

The mental retardations range from mild form to the profound form of retardations which are not without any cause. What are the causes behind these mental retardations? It is observed that these individuals have mainly low level of intelligence and also *inadequate adoptive behaviour.* In the preceding pages we have discussed various aspects of intelligence and the kinds of mental retardations. But what must be the causes of mental retardations. Etiologically it is said that there are at least two hundred different etiologies or causes of mental retardation. These many etiologies are classified into *two basic groups*. The first one is retardation associated with recognized *organic disorders.* The second cause is, retardation for which *no organic causes can be detected.*

Organic Retardation

It is observed that organic retardation covers twenty five per cent of the total retarded population. Organic retardation is caused by some *abnormal physiological process.* The known major organic causes are five as diagnosed by the American Psychiatric Association. They are: (1) the genetic conditions, (2) Infectious diseases, (3) prematurity, (4) chemicals and (5) environmental insults.

1. Genetic Conditions

Mental retardation is caused by genetic or chromosomal abnormalities. It is estimated that *four percent of* retardations are due to-this. Many of the pregnancies end in abortions. According **Smith** *et.al* (1978) only *half percent* of the *born babies have chromosomal abnormalities.* Considerable number babies *die within few days after the birth.* Most of the babies born have *Down's syndrome.* The physical appearance of the infant is sufficient to diagnose the Down's syndrome. The infant is having *broad and short skull,* round face and low bridged nose. The eyes are slanting upward and outward.

Down's Syndrome

This syndrome was first described by *Lagdon Down,* a British physician in 1866. The genetic basis of this syndrome was noticed by Jerome Lejcune and his associate, the French geneticists. There are *forty six chromosomes* in human beings, receiving *twenty three from each* parent. But the individuals suffering from Down's syndrome have *forty seven chromosomes.* Down's syndrome is found in *one infant out of 700 to 1200 babies.* There is one extra chromosome. In the maturation process something wrong takes place. The most common form of *cognitive disorder* due to chromosomal abnormality is Down's syndrome. Most of these cases are due to an extra chromosome *number 21* for a total of three (2+1). Technically it is stated as Trisomy 21. This is most related to the *age of the mother* at the time of *conception.* The behavioural disorders are much varied. Most of the individuals with Down's syndrome have very low I.Qs. but some *rare individuals attain the I.Q. about 80.* The *brain abnormalities* in Down's syndrome also vary.

The cerebral cortex part, the *biopsies* of Down's syndrome shows abnormal formation of *denductie spines.* **Epstein** (1986) has observed that a *mouse model* that involves an extra chromosome results in structural and behavioural changes in humans. Investigators of this model provide valuable insights into the chromosome which causes *structural and behavioural abnormalities.* The most frequent cause of *inherited mental retardation* is the condition of *Fragile X syndrome* which is produced by *fragile site on the X chromosome.* Thus the fragile X chromosome is the main *inherited mental retardation.*

Generally women *above thirty five years old* have the chance of more than 13 per cent pragnancies with the Down's syndrome. The Down's syndrome is after called as *Trisomay 21.* The egg cells are as old as the women and with age the egg cells may suffer from damage, because of the exposure to the environmental influences like viruses, irradiation and chemical are known to cause damage. These effects may cause the chromosomal abnormalities resulting in Down's syndrome in the infants.

Critinism

It is a form of mental retardation stunted physical development usually caused by an inborn failure of the individual 16 synthesize thyroxin. Critinism is a condition beginning in prenatal or in early life characterized by mental retardation and physical deformities caused by deficiency in the output of the thyroid gland. All persons who suffer from critinism resemble one another. They have round and yellow faces with wide, flat noses, thick lips and protudig tongues. Critinism also causes sluggish disposition and severe mental retardation. This condition is caused due to *thyroid deficiency.* At one time critinism was widely prevalent in many countries. Now the cases are much reduced. Now it is mostly caused in regions where the soil, air, water have low concentration of iodine. Now a days iodine is added to table salt to prevent such conditions. Low thyroid function may result in *diminishing intelligence.*

2. *Infectious Diseases*

There is increased risk of *mental retardation* from maternal infectious diseases for the fetus. The effects of these diseases are more serious during the first three months of pregnancy. In the state of infectious diseases the *first three month* period is a *very crucial* one for the development of the brain. The infectious diseases like the rubballa, syphillas toxoplasmosin are all maternal diseases that may cause both the physical deformities and *mental retardation* of the fetus. The mother may not experience any symptoms from the infection, but it will have serious effect on the developing fetus. Woman who is contemplating on having a child can take blood test to determine whether she is immune to the disease rubbaella. If a fetus contracts rubbaella from its mother it is likely to be born with *brain lesions that cause mental retardation.*

It is also observed that even after the birth the infectious disease may affect the child's development of the *brain.* Generally *encephalitica* may cause irreversible brain damage and *even death.* There are also several forms of childhood diseases in which the brain membranes are actually inflamed and severe fever is immenent and may even *become fatal.* But if the child is survived and not severaly retarded it may result in *mild mental retardation.* The other after effects are epilepsy, paralysis and deafness.

3. *Prematurity*

Sometimes it occurs that a mother may given birth to a child prior to the expected time. This we call as *prematurity.* Thus a child is considered as premature if delivered two or three weeks before the time. The premature infants who survive are having increased risk of developing *mental retardation.* There is high relationship between prematurity and mental retardation. But it is very difficult to specifically identify prematurity as a factor. There are many factors like *poverty, teenage motherhood,* nutrition deficiency, lack of parental care. These are all attributed for mental retardation. It is also observed that *premature infants* may develop other problems like medical which may put them in the *high risk of developing mental retardation.*

4. *Chemical Substances*

There are certain chemical substances cause damage to the fetus if the *mother is drug addict.* The born child may develop mental retardation. Pregnant Women who are experiencing *sickness* in the *morning time* were advised to take a harmless drug called *thalidomide.* Thus pregnant women in some of the West European countries and Canada took the drug during the critical embryonic period. It was observed later on that many of the babies born to these mothers hand *some defect* in the development of *arms and legs.* These parts looked like buds, a condition called *phocomelia.* On observing this type of development in children the medical authorities immediately stopped the drug. The limp development preceded in the embryo. By the time they stopped the drug *thousands of infants* already *met with the damage.* The drug thalidomide caused extreme havoc, but it has cautioned the pregnant women that *any chemical substance* may diverge the developing fetus in a serious manner from its normal course of development. It is also observed that alcohol is one such chemical substance.

Pregnant women taking one or two glasses of *wine* or any other alcohol during the *first three months* may give birth to infants who will be born with *fatal alcohol syndrome.* This syndrome consist of physical symptoms like *small head, widely spread eyes, flat nose, deep upper lip* and a *small stature* and *mental retardation.* These disorder symptoms increase with the amount of alcohol consumption.

5. *Environmental Hazards*

In the environmental situation there are several factors which may cause *mental retardation.* The environmental *pollutions* cause poisoning and *mental retardation. Lead poisoning* can cause *kidney and brain damage,* anemia mental retardation and death. The increasing chemical and other technologies like the *industrial smokes,* the wastage, garbages on the streets, all cause environmental pollutions. The idea of medical treatment may do good and also improve the health of the people, but the unforeseen consequence is, causing *mental retardation.* The *diphtheric pestoussis-tetanus vaccine* is intended to protect children from the diseases, stooping cough. But it causes at high rate of *adverse effects* in the children who had the vaccination. Some children get their *brain damaged* and also the cause of severe *mental retardation.*

Familial Retardation

The familial retardation is not organic in nature. Person with familial retardation sometimes called *cultural familial retardation.* These persons are mildly retarded, whose intelligence quotient levels fall above fifty. This category of the mentally retarded people constitute *seventy five per cent* of the total retarded people in a population. The cause of retardation of these people are not yet clearly traced, though it is felt that the causes are found in the combination of environmental and hereditary factors. But it is very difficult to determine the relative contributions of each factor. In the lower socio-economic status groups familial retardation is the most prevalent one. Some are of the view that the familial retardation is environmentally caused. Many of people live in homes of poorly designed not affluent, have to fulfil their children's developmental needs often. The parents who care their children seem to be that they themselves *do not score high in intelligence tests.*

Children born and brought up in unstimulated poor environmental situations cared by parents who score low in intelligence tests naturally have developmental problems and they are likely to face both physical and mental problems. Therefore, these children may develop *cultural-familial retardation.* **Kugal** and **Parsons** (1967) studied the level of the effects of children deprivation. According to them mothers in impoverished circumstances are often found in poor physical condition. They are frequently malnourished and often get only poor medical care. As result of this, their children who were identified as *familial-retarded* may actually have organic damages or *organic mental retardation.*

Personality Traits and Retardation

Investigators have observed in certain individuals that certain distinctive personality traits are found in many retarded persons. The view held by them is that mentally retarded individual's behaviour is an unpleasant product of their *low intelligence.* Many studies show that the *social adjustment* of the retarded persons depend upon as much as on their personalities as on their *levels intelligence.* Some of the personality traits are found to be associated with *poor social adjustment.* These are anxiety, over dependency, hostility, hyperactivity, resistance and failure to follow orders. *These traits are within the area of intelligence* of the individuals. It is observed that retarded individuals are not so much proficient in cognitive activities. They do not act better than what the intelligence test scores predicted. They often fail to perform any tasks on bar with other normal children.

The retarded children are almost certain that they fail more than the non-retarded children in most of the tasks. These many failures create in the minds of the retarded children *a feeling of expectency of failure.* Therefore, they often attempt harden to avoid failures rather than making achievements. In their anxiety to avoid failures they even do not attempt the tasks that are well within their intellectual abilities. They go to the extent of *distrusting their capacities* and try to get guidance from outside. The retarded children, compared to that of the non-retarded of the same mental age, are *more sensitive to prise and appreciation* and to other cues from adult persons. They follow the adult persons in behaviour and receive appreciations from their peer groups. There is great amount of accepting suggestibility in retarded children. This type of behaviour is called *outer directedness.* Many mentally retarded children are suspicious, mistrustful, fearful and try to avoid to see new persons. Thus retarded children differ in their personality traits from non-retarded children.

GENIUSES AND EXCEPTIONALLY INTELLIGENT PERSONS

There are varied opinions, prevailing among the general public about the *genius* or the *gifted person.* Most of the opinions are only stereotyped continued for generations. Though certain things are correct, most of them are harmful. Sometimes the popular opinions are used for joking purposes in many of public media. One common opinion held by the public is that persons of great intelligence are *physically weak* and *sickly* and their brains and bodily health are inversely related. Another belief held by the people is that gifted intelligent persons are socially *maladjusted, socially misfitor, delinquents.* Another belief prevailing among the public is that persons with gifted intelligence are extremely *narrow minded.* They are brilliant in a specific area, but totally lacking in all other areas. They also *lack common sense.* An intense and very wider study was made on gifted persons. All the studies have shown that all these beliefs and opinions held by the public are *totally false.* It is out of some prejudice that some people were continuing some false stereotyped views on the gifted persons.

It was observed in the studies that one of the extremely intelligent groups found to be *physically superior, matured emotionally, faster, taller, stronger and healthy* and also less sickly than the general population. They were emotionally less disturbed than the other average persons. Further, it was also observed that the gifted persons had considerably *lower rates of crimes, insecurity, divorce,* and maladjustments *than the general population.* The gifted persons were found to be *highly versatiles* and also revealed high competence in more different areas of skill than the average intelligent people. Their talents and interests are always placed at a considerable higher levels than the average intelligent people. On The whole the studies on the gifted persons have indicated that extremely high intelligence and pleasant traits are positively correlated, and the beliefs and opinions prevailed against the gifted are found to be *false and highly prejudiced.*

Out of several studies of the giftedness the most meaningful has been the longititudinal study made by **Lewis Terman** and **Oden** (1963) to be worth mentioning. In this study an attempt was made to identify giftedness, with the standard intelligence tests, and then to study its development over a period of years. Terman was the person who designed the *Stanford-Binet Scales.* During the period of 1920 and 1921. Terman and his associate Oden had selected thousand five hundred California School children. These children were tested with the Stanford-Binet Scale and their *intelligence quotients* were found to be *above 140.* The age of these children was below the High School age. This means that all the subjects had scored in the *upper one per cent* of the *general population.* Further the subjects were studied intensively with case histories, achievement tests, personality tests, medical and Anthropometric examinations and also a variety of rating scales.

In the next stage *regular follow studies* were made for evaluating the subjects in the areas of educational, social adjustment and vocational. Each follow up study consistently indicated that this gifted group of the persons have *maintained a superiority over the general population.* This was found not only in the intelligence test performance but also in educational achievements and in personal social and in mental adjustments. This study commenced in 1920-21 and the latest was performed in 1955. At this stage the *average age of the subjects was forty four*. The group maintained superiority in intellectual, scholastic and vocational areas. At the end, the accomplished subjects were eight hundred male subjects.

Terman's longitudinal study is remarkable one which has established the fact that really gifted intelligent persons maintain their *superior status all the times* in whatever occupational status they are put up. In the case of Terman's subjects almost after a period of thirty five years they were solely in the ability level to score on the top one per cent of the school population in the intelligence test. These subjects, 86 per cent of them were in the professions and in higher business occupations. The giften men were not in the lowest levels of the occupational hierarchy. Some of the men of this study have made notable contributions to the *physical, biological* and *social sciences.* Three of these men have been elected to the National Academy of Science. This is supposed to be the highest honour accorded to the outstanding scientists. Apart from this, many of their notable achievements are found in nearly all fields of endeavour. Some of these men's names are found in the biographical histories. Many men's names are found in the list of American men of science. in the directory of American scholars, in the volumes of biographies of persons with outstanding accomplishments in humanities.

In the said documents the names of the men who were subjects in the Terman's study and their outstanding contributions and works were highly attractive. It is astonishing to see the type contributions made by the men who were subjects in Terman's study. The productivity and the creativity of these men are found in their publications. There are about two thousand scientific and technical research papers and meritorious articles and about sixty books and monographs in *science, literature, art* and *humanities* have been published by the Terman's subjects when the follow study was made in 1995 after 35 years when the men reached the age of forty four. Other productivity includes thirty three novels about 375 short stories and plays, more than 60 essays, and 265 miscellaneous articles on variety of subjects. Terman's study is a clear cut indication that persons who score high to the level of top one per cent prove to be the outstanding persons and they establish their all round superiority over the general population. Those whose intelligence quotients are above 140 are certainly the gifted persons, extremely intelligent, highly productive and creative persons and they maintain their accomplishments throughout their lives. Therefore, we can assert that intelligence tests really identify the intellectually gifted persons who are extremely intelligent and also may be called as geniuses.

There were also women subjects in Terman's study. They were tested with the same intelligence tests. In the original sample the attainments of women were not very much impressive for Terman. They were treated as separate group and assessed. But the gifted women maintained, as it identified by the tests, considerable amount of superiority over the general female population. In every culture the women's roles are different from that of men. Therefore in Terman's study it was observed that women could not equally employ their potentials in academic and vocational achievements with men.

It was also observed that those who were identified as gifted children by the intelligence tests by obtaining the intelligence quotient above 140 did not show continued excellence in their performances. Terman had observed that some of his subjects ultimately turned out to be failures. This means that some of the subjects of his study in the beginning were found to be school dropônts. They were from low economic status, shown poor social and emotional adjustments than the average of the general population. Therefore, giftedness did not serve as criterion for the later development. Terman wanted to find out the reason for this problem. Why did some children fail when they revealed so much confidence in the beginnings? Terman and his associate conducted a sub-study for tracing the causes.

This sub-study revealed some facts on this important issue. This study commenced in 1945 and it was made to compare the most successful of the gifted subjects with the least successful for determining the variable that functions in the two groups. Three psychologists independently were assigned the task of examining the records of 730 gifted men of 25 years old. Out of this 150 most successful men and 150 least successful men were selected. They were mostly top 20 per cent and bottom 20 per cent of the subjects. The primary criterion used in this was, the extent to which a person had made use of his superior intelligence. The first group has been designated as A, and the second as C. These two groups were compared on 200 items of information which revealed many interesting information. They compared on educational and occupational records. It was found that 90 per cent of the subjects in group A were graduated from a college and from group C it was only 37 per cent.

In the case of *occupational pursuits* there were 70 per cent of men from group A and in group C it was only 9 per cent. In case of *salary drawn* by these groups, men in group A were drawing more than the men in group C.

Going into the childhood records and test scores, groups A and C during the period of elementary school years group A had shown higher than group C. There were a number of significant differences between the two groups in the childhood data in the *emotional stability, social adjustment* and in various personality traits. In physical health there was no significant difference between the two groups. Group A maintained reliably better mental health than the group C. There were more evidences of social maladjustment in group C. The marriage rate was higher in group A then in group C. The divorce rate was lower in group A than group C. The wives in group A were more college educated and it was three times greater then the wives of group C.

Men in group C had more adjustments problems than the men in group A. Men in group C had frequent unemployment problems and also change of jobs without any benefit in their positions. The A-C group differences were brought by the method of self-rating by the selves, their wives, their parents and by their field workers. It was observed that the ratings of the subjects, their wives their parents showed remarkable agreement in the ratings of group A, which was far higher than the group C's ratings. The ratings were made on the groups performance, self-confidence, integration towards goals and absence of inferiority feelings.

It was observed that there was greater contrast between the groups A and C. The contrast was found in their drive to achieve and in all round social adjustment. In the case of the gifted group of this study, superior success was associated with the stability rather than instability. Thus the sub-study highlighted many interesting factors that were functioning in the groups and specifically group C was found to be not successful in many areas of life. Such longitudinal study with many follow up studies for a period of thirty five years and also a follow up substudy to trace out the causes of the failures of the gifted group are very rarely conducted in psychological field. Gifted persons are really extremely intelligent but due to circumstance some of them fail to maintain it continuously through out their life.

Persons who obtain the highest I.Q. scores are identified as the gifted persons or geniuses. They establish superiority over the general population by exercising their intelligence in all their rare achievements. These persons are supposed to be having I.Qs. from 140 to 180. All the creative works, scientific inventions and discoveries and other highly valuable productivities are the outcomes of the works of these gifted persons or geniuses. Among the gifted persons a small number of individuals are considered to be the geniuses (Fig. 18.2).

Who is a genius? **Kimble** *et. al* (1985) define that a genius is a person of unparalleled achievements. Formerly, a person with an I.Q. exceeding 150 to 180 was considered to be a genius, but the more recent discussions impose an additional criterion of supreme accomplishment.

But at one time the convention was to consider a person as a genius who scored the I.Q. above 150. But later on it was considered that a genius is something more than that. Academic world is filled with faculties and students whose I.Qs. are so high enough as to quality them to this criterion. But we could observe that these academic faculties rarely qualify themselves to become geniuses. A person to become a genius must be an exceptional pattern of basic talent and must achieve transcendent accomplishments. These persons' accomplishments should not fade away with them, but must conquer the time and should be eternal. The coming generations must be benefited by their achievements. Persons who receive wide spread appreciation and recognition for their accomplishments in their own life time may be more appropriately called as eminent persons or recognized as *Men of Eminence* instead of geniuses. The term genius is generally applied to the persons whose achievements of extraordinary types in specialized areas such as science, art, literature, music, painting and so on.

There are something extraordinary talents in the persons who are geniuses. Their special talents and qualities often observed in their early years of their lives. There is a view held by the people in general that children with exceptional talents may fulfil and express their talents if only their abilities are recognized early and have all the good facilities to spend their childhoods in well advantaged homes where parents diligently cultivate their children's gifted talents. But it is quite obvious that many geniuses were not blessed with such environments.

The geniuses are those who are on the other extreme of feeble mindedness or subnormal, making outstanding intellectual achievements. They are the persons highly accomplished in the society and are becoming models for intellectual performances. They are famous for their achievements in some special fields such as sciences, literature, art, sculpute, music, expeditions or government sciences. These persons apart from the achievements in some specific field, they might have made very high scores in the general intelligence tests and also in some specific tests. These persons always seem to have been very bright.

Many examples we can quote to justify this view from the past histories. For example, **John Stuart Mill,** the English philosopher was so brilliant that at the age of thirteen he was able to get very good command over the realms of literatures, mathematics, history and law and at the age of sixteen he mastered philosophy and was ready to discuss everything with his father, John Mill, a great British Philosopher and also made distinctive contributions in the *fields of economics, logic and psychology.* On the basis of his early years achievement his intelligence was tested and it was found that his intelligence quotient was 195. Mill's achievements and his performances in these areas were highly remarkable for his age. The same type of exceptional brightness are also found in the childhood histories of all great men and women.

Terman and others (1825-1930) and **Hollingworth** made extensive studies about the, *genetic backgrounds* of the geniuses and were able to get astonishing information about the great men of the world. These individuals' other qualities, besides intelligence, stand out in the records are *persistence of efforts, confidence in their decisions, force of character, ambition* and the *desire to excel the previous records* and often a *passionate love for their cherished plans.* From these descriptions we can understand that the persons who are called geniuses as may be stated as mentally or intellectually *gifted persons.* They are the people superior in learning abilities.

The term gifted person is coined very recently in place of genius. People are classified on the basis of the intelligence test scores and the classification of genius was based arbitrarily on intelligence quotient level. Some investigators considered that persons who score the I.Q. level at 160 to 180 and above are genius. **Moskowitz** and **Orgel** (1969) state that actually the word genius involves *value judgment of the creativity and originality* of the achievements of an individual. It refers to the nature of contribution made by a person to some area of human benefits. Generally, originality and creativity factors are of high intelligence and not really based upon the abilities measured by intelligence tests.

Apart from the characteristics stated above a person who is stated as a genius is also endowed with the variables such as *strong motivation,* experience, opportunity, personality traits, all play important roles in the person who is deemed as genius. **Mackinnon** (1962) states that a genius is truly identified by his accomplishments and after the facts by evaluating his achievements. Some are identifying the gifted persons in terms of *superiority* in intellectual functioning. Those whose I.Qs. are above 140 are said to be the gifted persons. They are supposed to be superior in intelligence. Such persons are found to be one per cent in a population. But the terms genius and gifted in the common sense world cause some confusion and misconceptions.

There are several cases quoted by **Burt** (1955) that a quite number of geniuses were not nurtured by superior environments. For example **Issac Newton** (1643-1723), the great physicist, was born to a humble farmer. Newton's father died before Newton was born. **Koyl Gauss** (1777-1855) was the *greatest mathematician* of the time. His father was a brick layer. **James Watt** and **Abraham Lincoln** were sons of carpenters, **Martin Luther** and **John Knox** were sons of peasunts and **Johansness Kapler** was son of a druken innkeep. **Goertzel** and **Goertzel** (1962) quote another genius who emerged from a most unpleasant situation. He is **Lovis Armstrong** as a mark of true genius, who made lasting contributions. Armstrong was a neglected child whose father deserted the family and whose mother was always out of town. As a child he was caught firing blank cartridges from his step father's revolver and placed in the New Orleans coloured wolfs Home for Boys. There, he was taught to play an instrument and he specialized in *folk music* and made many spectacular and outstanding performances and was considered as *the genius* of the *special field.* In these geniuses accomplishments neither genetic factors nor the superior environments had any influence in their achievements.

The life histories of some eminent men bring out very fascinating information about them. The life history of **Sir Francis Galton** (1822-1895) is a very good example for this. Galton, a British Psychologist who made many valuable contributions to psychology. He dealt with the *genetic potentiality* of the human species. He coined the concept *eugenics* in 1883. Galton believed that the fine tuning of the nervous systems of intelligent people benefited their bodies and made them more physically vigorous persons.

Galton as a small boy had demonstrated his extreme brightness in various ways. He started writing letters at the age of four. Galton's extreme prococity in a number of intellectual spheres has been colourfully described by **Terman** (1917). Galton read everything at an incredibly early age. His understanding was not mechanical, but with real comprehension. Galton has attained a scholarly understanding of the *Greek epics, Iliod and Odyssey* at the age six.

When Galton was six years old, a visitor at his home had an occasion to speak about *Homer's works* and wished to test his knowledge. Instantly, Galton recited passages from 12th book of the *Odyssey* and surprised the visitor. Galton had grown up in an advantaged home circumstances when he was young. His intellectual interest in the early years and his family's support were good evidences for all his intellectual achievements. At the age of ten he wrote a letter in such an out standing manner that it was very unremarkable. What Galton was able to read and write at the age of five suggests that he had then a mental age of ten which worked out an intelligence quotient of 200. He was considered as the child prodigy.

Galton's intellectual feats have earned for him an excellent reputation right from his childhood years to the end of his life. Galton made several breakthoughs in the fields of Meteorology, Anthropology, Genetics and Psychology. It was he who formulated the Eugenics. *Euggeneics* is the science that studies the methods of improving the quality of human race. In the field of statistics, Galton designed the *coefficient of correlation method.* He invented a scientific instrument and method for sending messages to distant places. He also invented the technique of *identifing individuals through their ringer prints.* The Indian Police Departments and many countries all over the world are using this Finger-Print Technique for tracing the criminals. He had invented *hundreds of instruments and test materials* in the fields of psychology and anthropology. The famous *Galton's Bar* that is found in the psychological laboratories all over the world is his contribution. Galton had established his *intellectual superiority* over the general population throughout his life, right from his childhood period. But his accomplishments were recognized only in his later life to be referred to as the genius. However, he was a great genius in the real sense (Fig. 21.3).

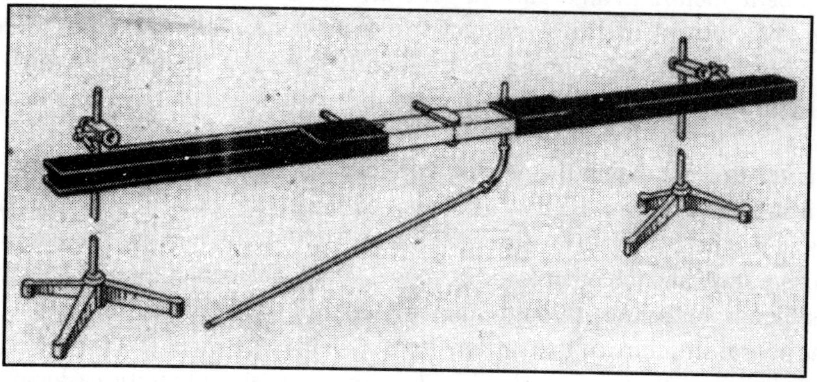

Fig. 21.3 : Galton's Bar : It is an instrument designed and invented by Sir Francis Galton. This is one of the hundreds of instruments invented by him for psychological experiments. Galton's Bar is generally used in the study of the perception of small distances.

But what made them-geniuses? Their live histories show that they were all *dedicated individuals.* They had a mind to overcome all obstacles for fulfilling of the inspired burning destiny. The view held by some is that a genius is *simply highly intelligent person.* His accomplishments are only the final step in the work. This view is also correct in a sense. The individual's personal characteristics such as intelligence, devotion, interest, enthusiasm, motivation the ability to withstand all sorts of inconveniences, dealing with problems sincerely and honestly and also the strong mind to face all sorts of criticisms, all these characteristics function in an integrated manner in an individual to make him a genius. There is also an ambition always in such individuals to do something outstanding. Therefore, it is true that geniuses are born rarely, but in most of the cases the geniuses are made by inspiration, perspiration and perseverance.

THE INFLUENCE OF HEREDITY AND ENVIRONMENT ON INTELLIGENCE

Psychologists are very much interested in the relative functions of heredity and environment in the intelligent behaviour of human beings. Many varied opinions have been expressed about the issue of nature and nurture. But all those opinions were not scientifically tested. One of the assumptions is that genetic and environmental variables are operating in increasing fashion in producing behaviour and therefore the influence of these variables can be identified in behavioural analysis. **Moskowitz** and **Orgel** (1969) state that though this view appears to be correct it is not a well conceived view. Because this view may easily lead a person to conclude out of extreme liking for one, that either one or the other is attributed with greater importance in producing intelligence. For example, the behaviourits are of the view that it is the environment that is the main cause of producing intelligence. But actually, except on extreme cases, the *interdependence of genetic and environmental factors* are essentially indispensable and also inseparable for the successful function of intellectual activities.

The *genetic factors* in an organism generally set up broad base of potentialities for the behaviour of the organism. But it is a matter for consideration that to what extent the genetic resource is reaching and functioning independently in the organisms behaviour. This process depends upon the extent to which the hereditary potentialities have satiated through the *interaction with the environment.*

Let us see to what extent the genetic variable is providing information by its action on the organism's intelligence. First of all let us see the neural or structure interaction and the outcomes of it. There are two levels on which the interaction may be considered. One is the *contemporary level* and the other is the *historical level.* The contemporary level of interaction describes human behaviour through immediate stimuli that are acting upon the organism and on the neural structure of that organism. But on the historical level it is to be known that the neural structures that are involved in the contemporary interactions have already been *modified* and *affected* by the past experiences. Because of this the structural functioning become the end products of the histocal interactions. Therefore, it is *on the historical* level the *genetic factors* are playing their important role by determining the functions of the neural structural developments.

The human neural structure is having constant interaction with the environment by means of instantly (responding) reacting to the stimuli in the environment. The reaction of the organism to the environmental stimuli is revealed by the organism through its *physical and mental activities* or behaviour. This type of behaviour may be referring to anything, any aspect of the organism's personality characteristics. But, what **Moskowitz** and **Orgel** (1968) question is this. While discussing about the hereditary and environmental factors in intelligence, is it possible to separate the contributions made by these two variables independently? Is it possible to state from the test performance the amount of contributions made by heredity and environment? According to them it is impossible. It is because the intelligence test performances or scores would always indicate a type of achievements which are educationally and culturally defined. The achievement is a learned skill which is acquired by the organism as there are favourable environmental conditions for learning and there are neural structures for getting benefits from these conditions.

From the test performance (scores) we only make assumptions and inferences about the learning abilities, in other words, intelligence. This is the ability that underlie learned behaviour which may be interpreted in different fashions. Moskowitz and Orgel strongly assert the view that the influences of genetic and environmental factors on intelligence are inseparable which is always not accepted and recognized. This resulted in great deal of research works to determine which of the two variables is more important in determining the Intelligence Quotient (I.Q.) scores in the individuals. These researches contrary to the expected result only impeded the progress of knowledge and nothing genuine, but only total misceptions are brought out.

Intense research in associating hereditary and environmental factors on intelligence is a complicated area which has been felt as extremely useful. It has been felt at last that heredity and environment are inseparable factors and indisputably important in determining and functioning of intelligence and intelligent behaviour. The researches carried out in this area fall into three broad categories. The first one refers to the family histories which should be examined for the incidence of high or low intelligence. Secondly, the experimental studies which should manipulate heredity or environment as independent variables. Thirdly, the naturalistic and correlational studies which should attempt to determine relationship between the measures of intelligence and the level of genetic make up.

FAMILY HISTORIES

It has been felt by many researchers that studies associating hereditary and environmental factors in intelligence are very complicated. But the importance of the issue is felt very much for the study of human behaviour. However, efforts were made to collect evidences for the genetic influence in behaviour. The researches were with regards to examining the family histories of few generations and behaviour. Evidences for examining this issue are very rare. One classical evidence available for this is the study made by **Goddard** (1913). For this study he named a family as *Kallikaks.*

Goddard strongly believed that this family is more suitable for his study of the inheritance of intelligence and behaviour. There were two groups in the Kallikaks family. One group of the family was evolved from an illegitimate contact of the revolutionary war soldier with a *mentally retarded prostitute.* The other group of the Kallikaks family raised from the offspring of the same soldier who later on married *a woman of a decent family.* The immediate descendents of the mentally retarded wife included many more cases of mental retardation (feeble mindedness), alcoholism, prostitution and so on than the descendents of the decent family wife. Goddard was of the view that this disorder was mainly due to inferior genetic factors that had been passed on by the prostitute women.

Goddards study was not conducted methodologically. The way in which he assessed the intelligence levels and emotional adjustment were subjected to the question of validity. Because some of the subjects were dead long back. Goddard did not include in his study the environmental variables that operated differently in the two family generations. A mentally retarded prostitute, who had an illegitimate son by her lover, was abandoned. She was not in a position to provide her son an intellectually stimulating environment. This condition cannot be equated with the case of a woman from a decent family, married, financially sound and enjoying a social status. Here there were both the genetic and environmental factors operating in intelligence and in other behavioural traits. The genetic differences existed in the two family groups. In spite of the drawbacks and the severe criticisms this study had its own value as it had revealed the naturally, occurred results and it has some psychological importance. But we should bear in mind that even any rigidly and methodologically planned studies also have some shortcomings in this area.

EXPERIMENTAL STUDIES

The influence of heredity and environment on intelligence can be experimentally investigated methodologically. Either one of the variable may be manipulated as an independent variable and other variables may be treated as dependent variables. But any rigid experimental method on human is difficult and complex and accurate result may be lacking. Therefore, psychologists conduct experiments on animals and try to carry out the same correspondingly on humans. **Tryon's** effort on rats is a classical example to illustrate the consequences of the experiments. **Tryon** (1940) and conducted the experiments to find out the effects of the performance of rats on maze running. He tested his hypotheses by systematic breading of rats in maze learning. Tryon had selected a large number of rats for his experiments. There were sets of bright and dull rats.

These rats were made to run a complex maze, the rats were given 19 trails. It was found that the lerightast rates made only 14 errors in running the maze, whereas the dullest rats made 174 errors. Then the bright female rats mated with bright males and the dull females mated with dull males. Then the offsprings of these rats were tested on the same maze. Again the bright female rats mated with bright male rats and the dull female rats with dull male rats. The offsprings of these rats were made to run the same maze and the errors were noted down. This breeding was made on the basis of their ability to run the maze. This kind of selected breeding of experiment was carried out for 18 generations. But the performances of these rats were tested in the 3rd, 7th and 9th generations. After the 9th generation it was observed that the dullest of the bright set of rats were brighter than the brightest of the dull set of rats.

At this stage changes were made in the environmental conditions. All the rats were given the same care. The pups of the bright mothers were cared by the dull mothers and the pups of the dull mothers were cared by bright mothers. After two years **Searle** (1949) in Tryon's laboratory, tested the learning abilities of the rats. He found that the bright rats were not bright in everything and the dull rats were dull in everything. There were different patterns of abilities in the rats. It was concluded that the changed environmental conditions have altered the learning abilities of the rats. The environmental conditions had definite influence upon the learning abilities of the rats apart from the changes due to genetic factors by the mixed breadings. In studying the learning abilities in rats Tryon found that the capacity to learn the particular complex maze problem was not linked with the capacity to solve other problems. Further the characteristics of brightness and dullness of the rats were applied to the particular task involved. Therefore, it was observed in this experiment that the dull rats were not lacking in general intelligence.

The influence of heredity and environment on intelligence was further tested on rats by **Cooper and Zubeck** (1958) continuing Tryon's experiments on rats. These authors raised one group of bright rats and one group of dull rats were mixed. They were brought in plain surroundings and another group of the same type in *stimulating environmental* conditions. When these sets of two groups of rats were grown they were made to run the maze. The investigators have observed that actually there was no difference between the rats that were bright and dull formerly. This shows that the inherited abilities of the bright rats had not shown any difference. It is observed in this experiment that hereditary and environmental factors have definite influence upon the learning abilities (intelligence) of the rats.

In Tryon's experiments it was observed that heredity and environment could alter the learning abilities of the rats. His experimental results also indicated that his selective breeding affected the sensory and muscular functioning. Having these results in mind we are to know whether human intelligence also could be altered by this type of programme of selective breeding. This type of carefully designed experimental results guide us to the conclusion that the ability to learn depends upon both hereditary and environmental factors. Hence, a sharp alteration in either heredity or environmental factors may certainly affect the performance of an organism. But at the same time we should not assume that either one of them is more important in determining the behaviour of the organism.

Influence of Heredity and Environment on the Mental Ability of the Twins

The development of human physical and mental faculties are determined by various factors. Intelligence is not an exception for this. **Watson,** the founder of behaviourism states that we can bring up a child in an environment in a manner what type of person he should be when he becomes an adult. Therefore, it is quite appropriate to discuss the effect of the nature and nurture in relation to intellectual abilities. Let us look into the recent developmental factors. We may consider the influence of nature by following the development of the *identical twins* and look into the types of variations in the performances. According to this experimental study these two twins are separated and put up in *two different environmental conditions.*

At the same time *another set of identical twins* were put up together in the *same environmental condition.* These children are subjected to the same type of pressures from parents and from the siblings (brothers and sisters). It has been observed from several studies that such identical twins treated alike to a degree more than the treatment given to ordinary brothers and sisters and also better than the treatment of non-identical twins of the same sex. The *identical twins are reared together* side by side with identical twins who are separated from each other in the first few weeks of life. **Newman** (1937) and his associates have been showing interest in this problem for a very long time. They were able to collect data on twenty pairs of identical twins separated during early infancy period. They were given medical, educational and psychological tests and were carefully observed. It was found that their physical resemblances were about the same as that of twins who had grown up in the same home. Even little different type food and other things did not make any change in their physical developments.

Considering the intellectual traits the evidence was found to be much puzzling and complex. In a number of cases the twins who were separated in the early infancy period and grown up in different homes continued to reveal the same level of intellectual traits as if they had been reared in their own homes. But, however, there were some cases in which the pairs had become very dissimilar. There were differences of 25 to 30 points or more in intelligence quotient appeared sometimes. The average difference in intelligence quotient of the separated identical twins was significantly greater than the identical twin reared together. These evidences show that the despairty in the environment seems to be getting into this effect. The difference varied from pair to pair. But the dissimilar was not as much as the non-twin members of the family.

The investigators also tested the social and personal qualities apart from physical and mental resemblances of the twins. The investigators found that the resemblances were not as close as the intellectual performances. The Twins brought up in very dissimilar environments, one was growing up in a form and the other was growing up in a busy urban environment. It resulted in significant differences. The observers used personality tests and found that the environmental conditions appeared to have produced marked differences in the twins. These results suggest that when heredity held constant and the environment is changed, a wide variability is caused by the environment in social than in intellectual attributes.

Influence of Extreme Environment on Intelligence

Human beings are naturally living in varied types of environmental conditions, ranging from extremely impoverished conditions to extremely rich conditions. People who live in such conditions are very much influenced by them and there may be remarkable differences among them in every walk of life. People who live in such impoverished environmental conditions do not find any stimulation for their physical and mental developments and they accept and live with whatever facilities available for them. But on the other end people live in extremely stimulated situations they find all possible resources to develop their physical and mental facilities. Our hypothetical view is that people who live in the impoverished conditions have low mental abilities or they are mentally retarded while on the other end people are highly intelligent. This view can be tested. These two extreme environmental conditions seem to influence the development of intelligence. If the conclusions stated in the proceeding section are correct, then we can expect that where the environmental conditions are demonstratably impoverished, the intelligence test performance also may be poor.

It is possible to conduct experiments to verify the hypothesis stated above in many of the developing countries. India is a live laboratory for testing such hypothesis. But such studies were not conducted by the Indian psychologists so far. Therefore, information regarding this issue are not available in India. **Gordon** (1923) has made an investigation in England among the group of *canalboat people.* People of this group *lived as nomadics the canal boats.* They were isolated mostly, travelling in boats in the river Thames. They were in isolation from the main stream of English culture. Their children did not go school to learn anything and there was hundred per cent illiteracy prevailed. The average intelligence quotient (I.Q.) of the canalboat people was much lower than the intelligence quotient of the general English population. Thus these people were considered be as mentally retarded.

The canalboat people, just like most culturally isolated people, had inbred for many generations among themselves. It appeared that there was some kind of genetic inferiority feeling prevailed among them because of their low intellectual achievements. Gordon had continued his studies and a longitudinal study also was made among the canalboat children. This has aroused some doubts on the genetic level of the children. It was found that the average intelligence quotient of these children was declining progressively as they became adults. It was, because, at the school age level the canalboat children scored close to the norm of the English population. But when grow into the age level of five the average intelligence quotient was going on decreasing until they reach the adulthood.

The same type study was conducted in America on a culturally isolated group of people. **Sherman** and **Key** (1932) have observed in their studies the same phenomenon. The younger people in these groups have shown considerable amount of intellectual deficincies than the older people. These investigators used intelligence tests among the isolated mountain children who were culturally separated from the general population.

The intelligence quotient of children born to normal and feeble-minded mothers was tested in one of the studies conducted by **Speer** (1940). They were all from intellectually impoverished environmental conditions this study was from little different standard point. The children of mentally retarded mothers have shown progressively *decreased scores* in their intelligence quotient. The Longer they stayed with their feeble minded mothers the more the scores decreased. The results were very much surprised. At the age of one year the average I.Q's of these children was 100.5, at the age of four the average. I.Q. was 83.7, at the age of seven the average I.Q. was 74.6 at the age ten it was 71.5 and at the age of 13.5 the average I.Q. was 53.1. Thus the average I.Q's of these children were decreasing progressively.

It appears that proper development of intellectual abilities in the children of feeble minded mothers was found to be *blocked by the environmental* deficiencies. It was also observed in this study that the decreasing effect of the environment has no limit. The level of intellectual deficiency induced by the environment may depend upon the *degree and duration* of the impoverishment of the environmental conditions. This means that we can anticipate that if the environmental conditions change there would be remarkable improvement in the intelligence test scores.

In a number of studies on this phenomenon, it has been demonstrated that when the impoverished environmental conditions change, the I.Q's tend to increase on an average of ten to twenty points. The institutionalized children have shown increased I.Q's in their tests when they were put up in good faster homes. Children of feeble minded mothers have shown increased average I.Q's in their tests when they were adapted by normal parents in good homes.

But what is good environment? It is not clearly defined in these studies. From the discussion we made in the preceding section we may state that a good environment is which can improve the level of mental abilities and should be free from any type of impoverishment. We can take into consideration the moderate to high socio-economic status areas, reduced unhygienic situations, the area where all essential materials are available, absence of unemployment, congenial parental care of the children and so on. According to **Baylay** (1966) the environmental condition should promote the highest level of intellectual interest, motivation and achievement. There should be the fullest exploitation of genetic potentialities.

Groups Differ in Intelligence Test Performance

In psychological studies measurement of intelligence has become one of the major theoretical and experimental studies. Right from the beginning of the intelligence testing movement psychologists have been showing very much interest on various aspects of testing and interpreting intelligence test scores. In the administering of intelligence tests people are tested individually and the groups as a whole. Individuals in a group may vary in their scores of intelligence quotient. The group as a whole is taken and average is worked out. Thus we have the average I.Q. of the group. But the problem we are facing in this is that a group consists of not of the same category of people. In a public school the students might have been drawn from various categories of families. The students may be from varied socio-economic status families, different social, religious, cultural and ethnic groups. Individuals from these groups may show varied intellectual performances.

Enormous amount of studies have been made right from the time of **Alfred Binet** to the present time to determine whether all these groups show significant differences in their intelligence quotients. To answer this phenomenon a number of explanations have been presented when the differences were occurred. The explanations offered by the interpreters of the I.Q's of the groups are referring to biologically-oriented theories, some are genetic interpretations, some are from environmental point of view, some are sociological interpretations and some interpretations are partly genetic and partly environmental. All these explanations do not give a clear cut idea about the issue. The matter is some what confusing. **Moskowitz** and **Orgel** (1969) solve this problem by touching the basic fact of testing intelligence. According to them intelligence tests do not measure genetic potentialities, but measure only achievement of the individuals and the groups. Therefore, while measuring intelligence the group difference we find in I.Q's cannot be attributed to genetic factors unless the group had identical environmental advantages.

It is obvious that within the same society there are different groups and they develop different interests and roles and achievements which they cannot neglect. They may also differ in their I.Q's. In India we know that there are different religious, racial, communal and cultural groups. Within the same religion, for example, in Hinduism there are number castes, communities, each one having different customs and traditions, living habits and marital habits and so on. Intelligence test scores certainly may reveal different I.Q's for each group. In any school group, occupational or work group we find all these religious, racial and communal or caste group individuals. In every group we can find rich and poor, young and old, male and female, apart from religious and communal variations. On certain intellectual works one group of people may do better than the other groups. Particularly in the Indian Communities it is very difficult to explain the group differences on the basis of I.Q's.

EMOTIONAL INTELLIGENCE

The emotional intelligence is a concept which is introduced very recently in the study of intelligence. It is a general idea that every human behavioural aspect is interrelated to some extent with every other behavioural aspects. Some psychological processes may be more interrelated with others and some may be moderately and some may be rarely interrelated with others. Intelligence is a mental energy which is having influence in every behavioural aspect to some extent. It is viewed that emotions are the species of human life and also a wild force in human intelligence. But there are not many studies on this phenomenon yielding more information.

Emotions can be aroused by thinking or thoughts as well as by the sensory perceptions. Emotions can also create or produce thinking on experiencing some sensory perceptions. For example, when we hear some unusually loud sound, at one we begin to imagine what has caused this. No doubt it may cause the emotion of *fear* and also *anger*. We try to use the emotion to identify the cause or causes with the environment and the brain or the memory makes use of all the available information to the context. This process becomes highly critical and important part of an individual's thought process with which he decides and concludes the cause of the incident. In this incident several processes are involved. Apart from sensory perception and experience the individual's past experience, knowledge and intelligence are involved. Emotional activity or behaviour is intelligently performed most of the times.

According to **Thurstone** (1939) there are seven mental abilities and every ability to some extent is related to memory. Similarly intelligence is involved in every aspect of human behaviour whether it is over behaviour or silent inherent behaviour. Therefore, the emotional functions are generally related to intelligence at varied levels. Emotionally related intellectual activity may be quick and fast and sometimes slow and delayed, depending upon the nature of the emotion. Any emotionally loaded intellectual activity brings out success quickly. But sometimes it may cause adverse consequence if the emotionally loaded activity functions rudely and violently.

22

CREATIVITY AND ITS MEASUREMENT

Evolutionary Stages in Creativity

The purpose of education is to maximize human ability to actual practice in life situations by enhancing the adaptability of the individuals. Humans generally assume, that of all the animals on earth, and immitate their most intelligent acts and the most inventive spirits. The evidence humans produce for this is the increasing mastery over nature and the more successful survival than those of other creatures. The successful human adaptability to the environment and protecting himself from the natural disasters indicate clearly that man can make and unmake many things on earth. Man, right from the very beginning, wished to see things in new ways. Since all the existing things become quite familiar to him in the long run, he is fed up with the existing old things. Hence he wishes to see everything in a new way. In this attempt to see things in a new way, man puts up his observations, experiences and knowledge in action in different manner. This action is thinking in relation to the environment and objects around him. In his thought process many things which are not seen anywhere or anytime may be projecting out and he wanted to give new shape to all of them by utilizing the materials available within his reach.

There is a saying that an indolent mind is the devil's workshop. When your mind is free, free from any purposive thinking, it wanders and muses. You gently push it out and it begins to travel beyond the accepted and familiar world. Nothing can prevent it. You may imagine anything. You win the competition, you conquer your enemies. You meet the lady love of your own choice. This kind of day dreaming is the rich source of much of creativity in human life. There are many difficult things in this world, which you cannot deal with them. But when you come out with a better and an easier way of doing your job, you reveal your creative ability and you are becoming a creative.

All these days you might have heard that, 'it is the way we have always been doing things'. But now, you no longer tolerate such explanations, since you are thinking of creativity. What our ancestors had been doing was, converting the stones and bones as instruments and tools for cultivating their crops and protecting themselves from the wild animals. By using the stones and metals as weapons were all based upon the creative works of that period. The continued day dreaming and creative processes from generation to generation brought perfection and improvements in human creative thinking and works.

The present progress the humans made in various areas cannot be considered as the sudden emergence of human effort. The continued creative daydreams that humans were having time to time resulted in the development of enormous amount of artistic, scientific and technological achievements. The development that took place in every area is the result of creative thinking that is passed on from one generation to another generation. Similarly progress in the industrial and technological fields is a continuous process of creative imagination. It is just like the relay race in the sports ground. Every generation continues and carries out the progress made by the previous generation and adds something new to the existing progress by the creative artists. The history of progress shows that all the developments that took place in this world today are not in the piece-meal fashion, but it is a continuous process.

During the period of renaissance there was a great revival and awakening in art, science, culture and thought. The characteristics of renaissance are considered to include intensified classical scholarship, creative imaginations, the painting, scientific and geographical discovery, a sense of individual human potentialities, creative talents, the assention of the active and secular over the religion and contemptative life. This Italian renaissance had much influence all over Europe and also the world at large. There were many scientific inventions and discoveries of new lands. The drawings and paintings of Leonardo da Vinci (1452-1519), 'The vigin of the Rock', 'The Mona Lisa' and, 'The Last Supper' revealed his scientific, creative precision in observation with intense power of imagination. Then the eighteenth century famous paintings of William Hogarth, 'A Rake's Progress' revealed the great artistic and creative talents of the period. This trend continued till the end of nineteenth century. The nineteenth century was the beginning of industrial and technological development and also enormous amount of researches in biological, neurological and anatomical sciences were carried out by the scientists. The twentieth century is noted for its scientific inventions and discoveries and creative products of individuals, and the inventions of scientific war weapons which were used in First World War. The scientific creative inventions were continued in keen competition among the developed nations and they were utilized for both constructive and destructive purposes. The Second World War which was faught between 1939 and 1945 was the test for supremacy.

After the Second World War there has been an increasing tendency among the developed nations and they continued the competitive spirit with each other in the scientific, artistic and technological inventions and discoveries. The real competition has been the utility of intellectual resources in various areas successfully. Therefore, individual's intellectual abilities and talents were encouraged to the maximum possible extent, so that they may be useful to themselves and to the nation. It is an accepted fact that we need more brains to survive in the competitive societies or nations. Therefore, it is important to identify many kinds of talents by means of psychological tests.

What is Creativity?

Many artistic, scientific and technological achievements have been made in modem times.

Whether it is a development of a theory of relativity, creation of a new music, tone, invention of a new type of railway engine, or an artistic painting or sculpture all have the same psychological process. Creations are made by individuals who have put themselves in the field for work. They make achievements only after very long time of vigorous work. They might have made lots of unsuccessful attempts and received ineffective responses. Several times they might have failed to achieve a correct response to solve a problem. Therefore, it implies that correct response is not available at once. Several trials, several errors might have been the incidents in an invention or creation.

In this task of creation the individual himself requires to judge his own skill whether he can make any creation or invention. The artist must judge whether a particular colour could impart something new in the whole product. Similarly, a scientist may be making some judgment about his exploring activity with an expectation of getting some promising result in a particular direction.

Creation and invention are most complex activities which require much greater degree of schematic and sustained activities and critical abilities. Therefore, creativity may be defined as the capacity to discover something new, a mixture of realistic and imaginative thinking often resulting in normal solutions. **Gardner Murphy** (1964) says that it is a capacity to produce through thought and imagination; capacity for original work. According to him there are two kinds of mental activities, that which reconstructs the past and that which initiates the new.

How do we judge things? What prompts men to make something which appears beautiful to him and to others? The answer is the everyday *impulse and not intuition.* There is an impulse in man to search for something new in art and science. The impulse to invent or to create something new was present in the cave man as well in the modern man. The cave man was the first inventor or creator. Otherwise he would not have carved out the caves nor painted the walls with beautiful colours and written his thoughts. The impulse to invent or to create new is also based upon necessity. This is also an impulse or an urge to discover something new.

We have used here three concepts such as *discovery, invention* and *creation*. These three concepts appear to be similar, but they are different. The concept, discovery refers to finding out something from the existing things. The object or product is already there. The discoverer only finds out as to what it is and where it is. For example, Cook discovered Australia and Columbus discovered America. Australia and America were already there. Similarly in the case of invention, the inventor only finds out the product and explains what it is. The inventor is not finding out something out of nothing, but out of something. He puts the existing materials in different combinations and finds out the product. For example, James Watt invented steam engine and Graham Bell invented telephone. In all these things the basic materials for invention are already there. Therefore, in discovery and invention there is nothing personal and effective. If Columbus did not discover America and James Watt did not invent steam engine someone else would have discovered and invented. In biological sciences there is only discovery and in physical sciences there is both discovery and invention.

But in creation, the creator makes out something out of nothing. It is from his own thought, experience and imagination that he makes out something. For example, a play like Shakespeare's *Hamlet* or an epic like Milton's *Paradise Lost,* or the Paintings of Leonardo da Vinci is a creation. The act of creation is the same in science as in art. It is a natural, human and living one.

How does a poem or a play differ from a theorem? Both are creations, brought out from nothing. The units in both differ because they match human experience in different ways. Every child can rediscover pythagorean theorem in the same form. His experience is intellectual and can be matched exactly. In art this does not happen. Many people are painting women but not like Leonardo da Vinci. Many people write play but not like Shakespeare. It is impossible to match the experience of one person with that of another. Creativity is a multifaceted phenomenon. It involves developing something unique. Everyone may become a creator under the stress of strong feeling. Therefore, creativity is not limited to artists and inventors.

A child may devise a creative play. The clay is easily available for the child and he may create a new type of toy out of it. A housewife creates a new kind of taste in cooking to please her husband. Therefore, creativity may be defined as the sense of ability to perceive something new in the external world through past experience and imagination and creation is the expression of such ability in some form so as to arouse the attention, curiosity, feeling and emotion of others in different degrees on the basis of their experience and knowledge. The form of expression of creativity may be a poem, play, painting, sculpture or even a musical tone which was never in existence in the same form.

Therefore, creativity may be defined as seeing and expressing new relationships in objects through thoughts. Creativity is possible in thinking in the manners of social interaction and also in studying, working or playing. In every creative work what you are doing is, you shift the old relationships into some new way and also add some new touches to the old forms and bring it out with new shape. A man can read creativity and also remember it creatively. A person, when every time solves a problem, becomes a creative. Any new problem cannot be solved without being a creative. It is true that many of our problems are solved in our everyday life by customs and habits. But customs and habits will not work all the times. In such cases we will have to add something new.

But some investigators maintain that creativity is nothing but problem solving, carried out under difficult and novel conditions. What they consider as novel condition is learning task in computer programing and problem solving (Gruber, 1962). Every idea or expression that is original for the person who works on it is an example of creativity. Others also might have had similar idea or expression, but for the concerned person it is new and a creative work. Creative behaviour, like painting or sculpture, just like any other behaviour is having complex or multiple determinants. Creative activity may be compensating or economic. **Maslow** (1970) while discussing the degree of fixity of the heirarchy of basic needs, states that, there are innately creative people to whom the drive to creativeness is more important than any other basic needs.

According to **Mednick** (1962) creativity is forming of associative elements into the combinations for meeting the specified requirements which are in some ways useful. If the new combination is more mutually remote, the more creative solution it is. This type of creative works are often found in our everyday life. Any work involving physical skills with tools and instruments may be brought under this category. In this way creativity may also be defined as the ability or power to create something new, characterised by originality of thought or inventiveness having or showing the fertile, imagination. It may be also stated as the designed to or tendency to stimulate the imagination or invention. Generally creativity leaves current knowledge behind and produces something that is new and not in existence. It may be also stated that creativity puts the old ideas or familiar things together in a new way. Creativity is a free wheeling imaginative things that make fresh insights and revolutionary ideals and ends with useful products. In creativity the creative thought or imagination goes beyond the normal route to see something new or strange. A great amount of work has been put up to find out the origin of the drives towards creativity and also to identify the conditions that appear to be inducing or motivating and also the conditions that are blocking it. Psychologists were able to formulate two theories for the origin and development of creativity.

Creativity is a special talent an individual is endowed with. It is an insight to see something new in everything. Whatever it is an existing things or, whether it is old or current, the creator wishes to see them in different manner which could attract the attention of others. It is putting some revolutionary ideas or imagination into some useful products. Imagination and dreaming are quite common among all human beings. But all human beings do not indulge in creating something new, only some are very curious in doing something in a strange and wonderful manner. Creativity is one way of manipulating human mind to meet the needs and demands of the time. The ancient man, no doubt had the sense of creating something new of the time. The stones and bones, the woods and metals were shaped and turned into tools and weapons. The cultivation of crops and using certain plants for eating and surviving, all constitute creativity. Right from the time of stone age some type of creative activity was going on continuously without any break or discontinuity. There would not have been any progress if everyone had done the same thing or produced things in the same manner they were produced before.

Therefore, there has been a continuous urge to go to a better stage than what was in existence. Imagination or creative daydreams must have been responsible for all the continuous and coherent activities and achievements that humans brought into this world. Millions of years of such activities only resulted in the present day industrial revolution, the scientific and technological development, the invention of the computer and the exploration and conquering the space. Time to time some kind of new tools and machines make life very easy and comfortable. Man finds pleasure and happiness in his everyday life by making use of the new inventions and discoveries. All inventions and discoveries are the results of imaginations and creative thinking.

People often relate intelligence and creativity. If intelligence is the ability to solve problems, creativity is the ability to see things in different ways. In creativity there is more of receptivity to nevel ideas and a kneck for making connections which had not occurred to someone else in the past. But creative individuals have trouble with problem solving activities, since problem solving needs strict adherence to rules and principles.

Characteristics of Creative People

Creative people are always hardworking whether there is any reward for it or not. They are, most of the time, engaged in daydreaming and also manipulating their thoughts and experiences and try to a evolve something fascinating. Creative persons are strong non-confirments. They do not stick to the rules and restrictions. They always violate them and formulate their own way of life principles. They are also guard their private moments and jealously maintain their programmes. The creative persons need the stimulation of their co-workers and also interested in group discussions. A creative person does not bring out an appreciable piece of artistic work over a night. Most of the creative works take time to bring out something valuable. Some of them spend even years on some artistic work. First of all creative people in any field take years of time for learning the intricacies of their field and apply their skill appropriately in actual work. Innovations in any field need accuracy, perfection and aesthetic values.

The creative persons may be interested in many areas and they may collect rich knowledge in various aspects of innovations. They are always skeptical about the correct answers and solutions. They try to look upon the problems from different angles. The creative persons always trust their instinctive stimulations. They are very much persistent, persuasive in solving their problems. They also invite criticisms on their works. But normally they do not accept and carry out them.

Creative persons strongly believe that the child has natural creative ability. It is based more upon fantasy. It is to think beyond the correct answer. The characteristic of wonder in the child is, 'why not this, and why not that way'. This type of curiosity should be cultivated in the child. It is a trait, should be developed in all possible ways. Experts often stress that it is the parents who should expose the child to new situations, ideas and activities. The child should be left alone in the situation to experiment with the new ideas in the situation. The child should be provided with opportunities to put his ideas into action by the way of experimentation.

The child can acquire more information by reading various types of books and magazines and can store them for creative works. But contrary to what is expected, many of the adult persons who are creative do not know what contribution their parents had made in their works. Neither their parents were creative persons nor they put their children in any of the creative ideas and activities. The creative persons were carrying out their creative works under several calamities such as financial crises, broken homes and other family misfortunes and distresses. These were found to be common in the life situations of the creative persons. Children from happy homes may turn out to be creative persons, but children from miserable homes also come out with as creative as the children of happy homes. The type of interest one develops right from the childhood period and the environmental situation that provides rich sources and the parents' encouragement are functioning in a dynamic manner to make one a creative person.

Creative Process

The creative person or the creator always tries to link the elements that are not in anyway could link naturally. For example, a painter could create a species which may be a combination of both human and fish and he could call it a mermaid. It is a new combination not existing in reality. Creative process involves greater amount of imagination and thinking based upon learning and experience and continuous pressure, the pressure that J.S. Mill (1806-1873) had during his adolescent period. Murphy says that the creative process involves sensitiveness to the environment and love of its colours, tones, and relationships of space and time. All these enter into the trend of the mind from which creativeness springs.

The more sensitive a person is to the environments, forms, colours, tones, shapes, and contours, the more he makes them as the roots of creativeness in his later life. It is the early joy in spatial and temporal relationships and mechanical laws that made **Sir Isaac Newton** (1643-1727) to become great mathematician in his later life. Murphy observes that to become creative, the individual should be sensitive and selective for certain kind of materials in two ways, first by the way of directing his attention to them and the second by the way of enjoying and making new combination with them.

Mackinnon (1962) emphasises five important stages in the creative process. They are: (1) a period of preparation posing a problem, (2) a period of concentrated effort to solve the problem, (3) a period of withdrawal from the problem, (4) a moment of insight, (5) a period of verification, evaluation, elaboration and application of insight. The tendency to see a problem and constantly making enquiries about it are the processes with which creation starts.

The basis of creativeness is one's awareness to the situation. There is something lacking in it and therefore, there is a problem. This is one of the specific characteristics of a creator. Therefore, the creative person is too sensitive to the situation. According to **Roslansky** (1970) the creative process is not a unitary process. It is a combination of several processes like cognitive, motivational and emotional processes. This combination of processes involve a group of psychological processes like perception, memory, imagination, appreciation, thinking, planning and deciding. These processes may be present in everybody but varies in degree, probably in greater degree, in a creator.

Guilford (1959) considers that creativity is a form of intelligence. He has associated a number of special abilities with creative process. They are flexibility, originality and fluency with words associations and ideas. These are associated with divergent thinking. The divergent thinking is one of the five operations of his thinking aspect of his theory of intellects. The convergent thinking emphasises on a single solution whereas divergent thinking on a number of different solutions. Guilford says, there are also other abilities and they are lying outside of divergent thinking.

A creator or an inventor may be always engaged in some silent activity which is stated as one of the creative processes. Psychologists generally bring this issue under 'problem solving'. Facing a problem in a life situation is a common occurrence and every human being faces a problem when the way to achieve a desired goal is not plain. A problem exists when the means for the cherished result is not fully clear. A problem is solved when the ways and means are discovered.

But what is a problem? A problem is an unwelcome or harmful matter needing to be dealt with or a thing that is difficult to achieve. A problem also may be stated as anything, matter, person etc. that is difficult to deal with, solve or overcome. The nature and type of the problem depends upon the situation or matter with which the aspired person deals with. A problem is solved when the ways and means are found or discovered or invented. Therefore, the process of solution is a process of seeking, searching and finding. It is observed that most often some of the works can be done by the way of thinking rather than by motor exploration and manipulation. The tools of thought are put into action. The earlier experiences are applied to the new problem.

In the problem situation there is a starting point and also there is a goal. In this process something is given and something is acquired. Problem solving behaviour takes place in novel or difficult situations. In these situations solutions are not obtainable by the habitual methods. **Woodworth** (1952) states that there are five steps in the problems solving process. According to him, certain problems contain obstacles. These should be cleared by clear thinking. Secondly, there are some guiding ideas which determine what kind of attempts should be made at the solution. The third step is a hypothesis. It is stated in the form of a question and it is tested. The fourth step is an incubation stage. It is a period of interval during which period the problem is left aside after making the preliminary attempts at the solution. The fifth step is the rational thinking. It is something to do with trial and error and finally, to put up a final attack on the problem.

Incubation is an essential aspect of creative process which is the result of several attempts on the problem. Incubation is not simply a silent cessation of attempts. It is a continuous process and there are four steps in this, such as preparation, incubation, illumination and verification. The Roman philosopher, Lucretius said that nothing can be created from nothing. Scientists, artists and explorers spend several years in exploring their fields. This is the period of preparation which they try to find out an answer for the question, which stands on the way of solving the problem. The preliminary achievement may be a small beginning. It would be a big challenging task for an inventor or for a creator to complete the other parts of the work.

Any creative work, however, excellent in its quality, however, the problem is recognized, the correct solution may be eluding from the creator in someway. At this stage a short interval may take place in the work. The creator indulges himself in internalizing the problem and some kind of work unconsciously is going on. The actual work may take place after the internalization of the thought process is over.

The third step in the incubation is the illumination. This is sudden appearance of an idea which the creator has been searching. The writer or a creator or an inventor may be searching and waiting for a final conclusion of his work for finishing it in a most enchanting manner. The stage of illumination is the result of the stages of preparation and incubation.

The fourth and final stage of incubation is verification. At this stage the creator or inventor wishes to test whether, what he has done is really a new price of work and, has he correctly solved the problem or achieved the cherished goal. Generally creators put their creative protects into some kind of test, like conducting exhibitions and calling people to witness and to certify and getting their opinions. This method of testing may help the creator to get some new ideas which may help him to reach the solution.

Creativity and Heredity

People believe that creativity is appearing in generation after generation in families. Examples can be produced to show that creativity is inheritable. But such examples of creative talent spanning generations are quite rare. The evidences for creative gene are not strong. There may be a general kind of creative temperament running in families. The members may be showing creativity in diverse field such as arts, business, politics. In such cases heredity would not be the only function that predisposing an individual to creative thinking. The influence of family encouragement might have been equally important in it.

Bringing up : Childhood period is the most creative period in the life of a person; pre-school children make dolls, play, talk, build blocks, construct houses and towers. Some games also become creative for children. At school the child must learn the correct answer or solutions, by the approved methods, for the child every question has a right and wrong answer. A part of the child's natural creativity is the ability to fantasize or think beyond the right answer, to wonder, why not? One way to cultivate this trait for the parents' is to allow the child to new situations and exposing him to new ideas, and activities and allow him alone to experiment with toys and games. Reading various materials should be encouraged. The more information that is stored up, the more the child can be drawn on for creative efforts (Fig. 22.1).

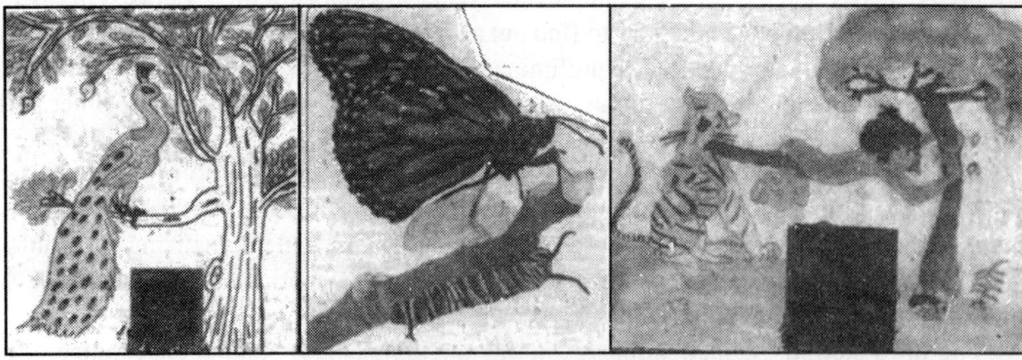

Fig. 22.1 : Children's Paintings

In every culture children are exposed to creative activities and they are stimulated and encouraged by giving them rewards. Children are made to participate in painting competitions which are often conducted in schools and in some organizations. These paintings are made by children below the age of 15 years. These are the manifestations of the children's creative imaginations. Many children who were proficient in paintings in schools take painting as their profession in their later life.

Many people who are creative as adults have had no loving parents would have chosen to give them the chance. Financial problem, separation of parents and other family difficulties are common in the background of creative people. The reason may be that on the intellectual development the child may have to cope up with disappointments and hardships which make the child to think differently than his less troubled peers. Happy children may turn out to be just as creative as miserable ones.

Brain : Studies have advanced in physiological, neurological and anatomical areas to test whether the human brain has anything to do with creativity. What part of the brain is having influence in a person's creativity? It has been found that there is some evidence that the right hemisphere of the brain has more to do with creativity in most of the people. The left hemisphere handles day-to-day business of life, like talking, writing and reasoning.

The right hemisphere has the task of organizing experience, finding pattern in events and responding emotionally. Imaginations and creative ideas are fired upon the right hemisphere. The corpus callosum, a thick bundle of nerve fibres connect the right and left hemispheres. The right and left hemispheres make communication only through the corpus callosum. Studies have been conducted on accident victim in whom the corpus callosm have been cut. It has been found that there were indications of divergent functions in the hemispheres. Studies in this line of functions of the brain have to bring out more information for correct and full understanding of this phenomenon.

This question is often raised, whether creativity is inheritable? It is not genetically proved that creativity is inherited. But we could see in families that creativity is found to be running for generation after generation. This has happened in the family of a music composer. **Johann Sebatian Bach** (1685-1750) was a German composer. He was also musical director for **Prince Leopold of Kothen** (1717-28) and Musical director for the city of **Leipzig** (1728-50). His output was enormous and displayed great vigour and inventions within the northern European culture. His works included nearly 200 cautaties (compositions) and oratories (dramatic musical works) settings of the passion. He was born in the family that had produced musicians for the previous hundred years. Three of his children also became composers. Though we may quote more instances as evidences to the hypothesis that creative talents exist and continue in families for generations, it cannot be fully tested and accepted. In these days whether the sons and daughters of the artists and composers are endowed with such talents or not, they take it up as a profession by means of practice. But the real creative talent is not found in them.

Creativity and Intelligence

Psychologists often relate creativity and intelligence in their studies. But there are two kinds of view. Some are of the view that creativity is a form of intelligence and some are of the view that creativity is a special kind of ability distinct from general intelligence. The two concepts have attracted several psychologists and several studies have been made. But all these studies have failed to establish that creativity is independent of intelligence.

It would be quite relevant in this context to discuss one of the important studies made by **Getzels** and **Jackson** (1962). The test was conducted by them in one of the private schools in Chicago. They have administered intelligence and creativity tests on a group of students. These students were from 5th grade to 12th grade. On the basis of this testing the students were divided into two groups, as those who scored high in the creativity test and those who scored high in intelligence test. These two groups performed regularly in academic achievements.

The authors observed that these student groups varied in personality traits. Those who had high I.Qs. found to be conscientious, careful and self-controlled. Their goals appeared to match with those of their teachers. Those who scored high in creativity test were playful, expressive and independent. But they worked harder than the children with high I.Qs. and they did not adjust well with their teachers.

Another interesting study in this field has been made by Wallach and Kogan in 1965 relating intelligence and creativity. The subjects were 5th grade students in one of the Sub-urban Schools. These authors felt that the conventional intelligence tests would not serve their purpose and so they developed tests consisted of game like activities to evaluate creativity. There was no time limit for this test. The students were divided into four groups on the basis of their scores in intelligence and creativity tests. The students who scored high on both the tests performed well in both free and controlled situations and well with child like and adult like behaviours. The high creative and the low I.Qs. were frustrated in the school environment, but performed well in a free context. The low creative and high I.Qs. valued school achievement highly and worked very hard to excel other students. The fourth group, the low creative and low I.Qs. wanted to avoid the school work. They are involved in social activities and expressed regression. The study suggests that there are different modes of thinking in creativity and in general intelligence. But in many diverse activities both can be found.

Development of Creativity

How this creative product is evolved? It is the product of the creative thinking through perception, experience and practice that man has acquired and applying his creative intelligence on the objects around him. Creators and inventors have developed a philosophy of taking the failures in their attempts as stepping stones for success. This philosophy they have evolved only from their experience. This idea seems to be correct as many of us also might have met with many failures in our life situations and at last we might have been successful.

A person who desires to become a creative artist, he should have forms of flexible thinking. During the course of serious practice a creative person might have got this idea and might have realized it. It is true that many persons may not be aware of this when they first started to perform the task. But still some persons, eventhough may not be knowing this phenomenon, are in possession of it to some level. The popular mythology is that creatives are born. This idea is widely prevalent and accepted in every culture and society on seeing some persons revealing and exhibiting some extraordinary talents in some artistic areas.

For example, Vincent Van Gogh, a Dutch artist, was a very brilliant painter. He was an eccentric and also a victim of a mysterious neurological disorder. But this did not stand on his unique vision. compassion and on the dazzling creativity that were reflected in his powerful and enchanting paintings. As an artist Van Gogh worked intensely to develop his use of colour and texture into a unique and profound language of emotion. His neurological disorder had nothing to do with his fiery intelligence, compassion and creativity that revealed in his paintings. The life history of Gogh shows that his extraordinary creativity was neither hereditary nor systematically learned and developed. Shall we take it for granted that he is born genius?

With regard to the development of creativity different views are expressed. The brain is often spoken of as having influence in the creative skills. To the question, 'is there any part of the brain that influences a person's creativity', we may state, 'yes, there is a part in the brain that is having influence in the person's creativity.' Psychologists and biologists state that the right, hemisphere of the brain has something more to do with creativity than the left hemisphere of the brain. The left hemisphere is very much involved in the day-to-day business of the people, like speaking, writing, reasoning and problem solving. But the right brain has the task of organizing the experiences, taking into account the patterns of objects and events and responding to them emotionally. Only in the right brain our imaginations and the creative ideas are formed.

The two brains, the right and the left hemispheres communicate to each other through the corpus callosum. The corpus callosum is a thick bundle of nerve fibres which connects both the right and left brains. When corpus callosum is injured or damaged divergent functions occur. This is observed in some cases in which the corpus callosum was cut or damaged. Some are raising the issue whether there is any obstacle to creative thinking. Investigators in this area are of the view that the education one receives is aimed at the development of the ability for learning various skills. In this, various methods are applied to acquire the skills. Some of the methods are rote learning or repeating the same task several times which may result in acquiring the skills. It also involves following the formulas, principles and the rules in acquiring creative thinking and the ability to perform the task.

It is observed that many from very early age begin to value reasoning, practicality, accepting the rules and principles over the creative faculties, imagination, intuition, insight and fantasy or daydreaming and dreaming. In these efforts to acquire the creative thinking there seems to be no obstacle on the way.

Right from the ancient time to the present day man is always involved in designing, creating and manufacturing something for his life requirements through experience and practice. He was able to create and invent his tools and machines. All the materials that we use today in our life situations were not suddenly brought out creation of the present time. All these were gradually and slowly designed and created and manufactured through thousands of years. The evolutionary processes in nature have very much influenced mental development of human beings. All those creations and inventions that we see today are the manifestations of man's perception, experience and thinking. All the scientific and the technological developments of the present time are based upon some basic principle of the evolutionary variations and the complexities of nature.

What has been taking place in the evolutionary processes is also taking place in the modern technological processes at present time. The simple evolutionary process in the beginning resulted in the complex process after millions of years. In the same manner what is roughly brought out in simple manner has become a very complex one. For example, the television, computer systems, the multicolour printing technology, the present development of the telephone system and many others are the results of man's creative intelligence and experience. All the human creative achievements are based upon man's innate creative thinking and experience. Apart from learning and thinking, we must also bear in mind that practice brings perfection in every work. The tasks performed thousands of miles away are seen and heard in the form of music, dance and play by sitting in a room anywhere in the world.

But there are some special cases like Vincent Van Gogh, which we cannot explain easily. Analysing the performances of the artists of the present time we find that they all became the creative geniuses by perspiration and by inspiration. The skill for a creative work can be acquired by intensivily cultivating and practicing a particular art, whether it is sculpture, painting, music or dance. But however exponents in the field of creative works have evolved some rules and procedures and they wanted that these should be followed with all sincerity and devotion. The rules, principles and the procedures are more strictly followed, particularly in music and dance than in sculpture and painting.

First of all one must decide what he wants to do, then secondly, he should develop a positive attitude towards it. This means that the person should have a strong desire for the art. Thirdly, he should learn all he can do and perform his task successfully. Then fourthly, he should think hard on what he has decided. If he finds it very difficult to carry out the task, he has to wait for sometimes. This is called incubation. The person has to incubate for sometimes with the same thought process. Then in the fifth stage he has to review the ideas that he has already formed in terms of pros and cons and has to make changes by adding something and deleting something, if necessary. He can introduce variations and whatever modifications are needed until he arrive at the best one. Then in the sixth stage he has to put his ideas in the work and test them whether they are functioning well to the satisfaction. The development of human creative thoughts and processes involve all these characteristics. All the creators, inventors and also explorers face and experience these chatacteristics.

Persons who have some interest and skills to create something may start the creative work. But after sometimes they put it off. The reason is that they develop some sort of fear. They begin to think that what they produce creatively would not receive the appreciation from others and would be of no use in anyway. But some persons overcome their fears by starting the work on some small aspect of the main work first and if they are successful in it by any means they continue the work and produce more and more continuously and find at last they are successful in their creative works. Many of the creative writers, whether one writes poems or short stories or novels or one who makes paintings or making sculptures become successful in their creative works only by adapting these procedures.

Throughout history, we find that creative artists have made *sculptures and paintings* to record the *historical events* of the world to tell the posterity the stories, to explain the religious beliefs and faiths and to illustrate the objects of great significances and rich and rare beauties of the time. Art is also an expression of the artist's own personality and the spirit of his age. Even an abstract arrangement of colours and shapes can communicate a message that may be passionate and mysterious. The ancient Egyptians, Greeks and Romans were very keen in presenting their historical events of their countries and their achievements in sculptures and in paintings (Figs.22.2 and 22.3).

Fig. 22.2 : Paintings : Expression of creative thinking and imagination. A new mode of enquiry came up in Italy which spread across Europe in 14th century. Artists and painters began to reappraise their world. This change is known as *renaissance* or the rebirth of art and culture. A new style of painting came into existence.

Ideas about art and culture, imagination and thinking, style life and religious faiths and worship all transformed into renaissance style.

New methods were developed in all aspects of life. Religious paintings were given new expositions. The painting presented above is the classical architecture reflecting the Italian renaissance. It is said that it also has the basis of the Greek and Roman creative thinking.

Fig. 22.3 : Modern Painting
The painter's imagination finds no bounds and barriers. The painter expresses his imagination, sometimes in such a way that it would be very difficult to understand the meanings. It needs deep thinking and imagination to understand the theme of it.

Creativity and Awareness of the Mind

In creativity ideas and emotions are illustrated in some form. This is one of the many practical ways in which we can encourage the mindfulness or awareness to develop. These ways do not demand any extra time. There are different ways or carrying out and thinking about everyday task. These activities will help us to recognise that everything around us is part of a continuous process of change. Every time when we look at a particular object we see it in somewhat different way. That is, everything we see, we see it as if we see it for the first time. Thus mindfulness is helping us to observe everything in a subtle manner and the experience we gain reminds us that we are fully aware of everything and that we feel that we are fully alive and alert.

Development of Creativity in Children

Since we are aware of the events and objects around us, the mind always desires to see something new out of something or out of nothing. Children from childhood are stimulated and encouraged tc cultivate the creative habits. They are put in such a situation as to be creative, to use their imaginations and discover some innovative solutions.

Can children learn and develop creativity? Can they create something new? Or do they have innate ability or nature's gift bestowed on few? Or, is creativity hereditary, the ability passed on from parents, grandparents or from great grand parents? Psychologists who have investigated this phenomena of creative process may provide answers for these problems. Psychologists are of the view that we can study the process of creativity and assess the level of creative ability of a person in the same way as we assess the intelligence quotient (I.Q.). For this they may use certain standard tests. Persons with high creative ability can pass through test and score high and those who are lacking the creative ability may fail to score high and fall behind others.

For example, a test of creative ability may be consists of different shapes and sizes and colours of wooden pieces. They are placed before the child aged six years and he/she is asked to handle them in whatever manner he/she wanted and to make out something. The child with some creative ability can make out some designs, structures on the basis of his/her creative imagination. The same materials may be given to some more children. These children can be both boys and girls of the same age. The test may have some scoring system and accordingly the experimenter can get the scores of each child and can assess the high scorers and the low scorers and analyse the level of the creative ability of each child and also of the group. He can also find out whether the boys score high or the girls score high and the family and social backgrounds. By this means the experimenter can find out whether the high scoring boys and girls have any hereditary background. The way in which the children handled the materials and the scores would indicate the nature and the level of the creative intelligence of the children. As and when the child grows and becomes a matured he/she may be tempted and develop interest to create still more complex type of creative works.

Albert Einstein once humourously remarked that shaving is a risky task because every time he takes a shave a good idea is seized. He also said that he has no special gift, but he is only passionately curious of events and things. Psychological investigators are of the view that there is a contrast between the hard grafts of intellectual labour and the completeness of their moments of sudden inspiration that takes place at times. When a person is involved in a problem solving task, is it by inspiration that he solves it? Or, is it different from inspiration? Or, is it possible to stimulate that faculty or the power whenever we are in need of it or wanted to solve the problem? Psychologists are of the view that creativity is accelerated when we relax our attention on the set up way of thinking and get ourselves free from the event and also from our mood and also by allowing our mind to be free and turn to an alternative task. This type of arrangement is freeing the mind from the clutches in which the mind is involved. This leads to the involvement of some simple interplay between the right and left hemispheres of the brain through corpus callosum.

We have discussed the functions of the right and left hemispheres elsewhere in this book. But here we are relating creativity to the functions of the hemispheres. The idea that we have two brains functioning independently in our head seems to be misleading one. But it has been clarified by the scientists that each hemisphere specializes in certain functions. The left brain specializes in spoken and written language, logic, number skills and scientific concepts. Its functions mainly involve writing, book-keeping, selling and laboratory works.

But the right hemisphere functions mainly in recognizing patterns and shapes and how they relate to each other. The functions of this hemisphere are mainly instrumental in remembering faces. It is the right hemisphere that is contributing to insight imagination and for understanding sense of humours. It is this part that appreciates the art and culture. The works of the architects such as the paintings, the sculptures encraved in stones, music, dance and other works of excellence draw mainly on the right hemisphere.

The right side of the brain mainly functions in the creative works and involved in perceiving the nature of emotional expressions in speech. Thus the right hemisphere is mainly the source of creativity. According to the results of the recent researches the right hemisphere plays an important role in perceiving the spatial relationships and in many kinds of creativity. It is mainly involved in thinking and imagination and in the creative processes.

We may face a problem when we want to start a fresh work. In such problem situation the right brain forms a general mental picture of it which is the image of the task. In such situation the left brain sets to work logically by considering an alternative. If no correct solution or answer is reached then we need to give it up the existing plan of the picture and find the ground in another situation and search for the correct solution. At this stage the *right hemisphere gets relaxed* and starts again to solve a problem. The mental picture is set up and the intelligence is exploring the situation. At this stage creativity is changing the situation and permitting the brain to reach a problem from a different angle. One of the investigators in this area is David Gelenter, a cognitive scientist, is of the view that the secret of making the mind to have creative relations is to relax the mind in a significant way. He speaks of two styles of thinking in creativity. One is high focus state, in which we concentrate on logical substance of ideas. The other one is a low focus state, in which we think of concrete images. These are based on an unclear emotional connections. Creativity falls in between these two states, in which the mind is relatively focused. It is not on a specific and fixed thought process. This state takes away the mind from the fixed task that we deal with. This takes the mind to an unresolved active thought. This we may call as the relaxing state. During this period we may carry out some other work like walking or some home work. By this means we allow our mind to cool down without any disturbance. Then we come back to the problem with fresh creative thoughts and successfully perform the task. In this process the mental activities are reframed with creativity (Fig 22.4).

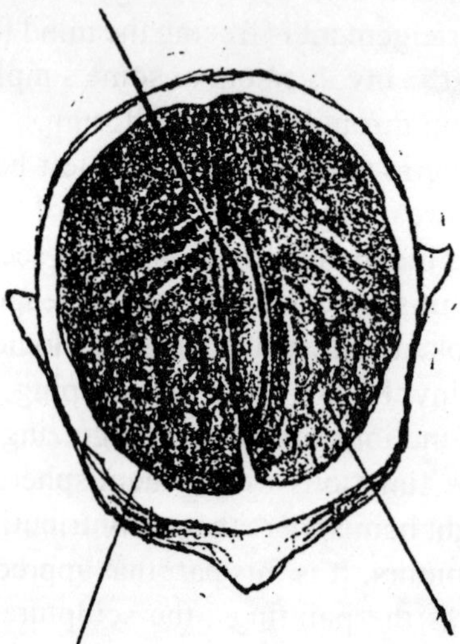

Fig. 22.4 : Brain and Creativity

Areas of Specialization
Organizing Experiences
Recognizing Patters and Shapes of Objects and Events
Remembering Faces
Contributing to Insight
Sense of Humour
Appreciation of Art and Culture, Works of Architecture,
Paintings Sculpture, Music, Dance
Main Source of Creativity, Appreciating works
Excellecnce, Perceiving
Spatial Relations, Thinking and Imagination
Perceiving the nature of Emotional Expressions
Forming the Picture of Images of the task of
new creations

Areas of Specialization
Day-to-day Business,
Spoken and Written Language
Logical, Number skills,
Scientific Concepts, Book-keeping,
Selling
Laboratory works
Logically approving the task designed
by the Right hemisphere

Development of Creative Mind : Some psychologists who have spent much of their times in studying the development of creative skills have observed that to master the creative skills one has to practice continuously at least for six hundred days in a specific field. The person should have chosen the field with positive attitudes. He should like it very much and have strong motivation that he must become an expert in that field.

The popular opinion about creativity is that it is a gift. But psychologists do not agree with this idea. There may be some innate tendency in a person for a specific task or field of study. It is observed that most of the great achievers in some specific area started to perform the task from very early years of their life. Some have the family background which might have stimulated them to take up task. They might have received rigorous training and also practiced for years before they becoine experts in the field.

Persons who are carrying out the creative works have lived with their works the whole life time to become a creative genius. The life histories of great exponents in various areas tell us that they have taken the training and practice for years. For example, take the cases of the painters, sculptures, musicians, the dancers and other artists have taken up the training for many years before they enter their profession. These artists first practice the conventional skills and then go for developing innovative skills of their own and then they become creative geniuses.

Psychologists have observed that there are certain stages in the creative processes. The English Psychologist, Graham Wallas states that there are *four stages* in the creative process. *The first stage* is preparation. It is the intense frontal attack on the problem that familiarizes the thinker with all its characteristics. *The second* is the incubation stage. It is period of relaxation stage. It allows the way of looking at the problem which may lead to an alternative ideas. *The third* is the illumination stage, in which there is some mental flash of sudden stroke of new idea to solve the problem. *The fourth* is the verification stage which indicates that the sudden flash of idea is the *insight* which produces the right answer to the problem. This is from the right hemisphere of the brain. This has to be ascertained logically by the left side of the brain. In any creative activity or problem solving situation these four stages may occur.

Theories of Creativity

These two theories are opposed to each other. Sigmund Freud, the founder of psychoanalysis proposed that creativity is the consequence of conflict and frustrations. According to Freud creativity is an emergence of inner confusion. The life histories of great creative persons are full of familiar conflict and frustration, personal struggle and often have shown evidence of maladjustment. These creative persons find solution for their maladjustments by themselves by taking up a creative work like writing poems, stories, taking up painting, sculpture and similar work. They give expression, in their creative works, to all their personal unresolved problems and conflict. They need not be branded as mad people or psychopathic persons. If we try to find out meaning for their creative works from psychoanalytic point of view their real motive behind this can be traced. There are sufficient evidences in the literature for this view.

For example, the famous English romantic poet, **William Wordsworth's** (1770-1850) poem, 'Tintern Abbey' was the outcome of expression of his emotional frustration.

Wordsworth's personal life was full of frustrations. In his lifetime he had experienced social, economic and emotional frustrations. Similarly, the personal life of the poets, Keats and Byron had severe emotional frustrations and maladjustments which they tried to resolve by expressing their sorrowful life incidents in their poems. We can collect more evidences for this theory in these days from the persons who are in the field of motion picture. The personal lives of the motion picture actors and actresses, the film makers and others in this field are full of personal maladjustments and frustrations mostly rooted in sex. There are instances that these people commit suicide and die as they could not resolve their sex rooted emotional frustration in their life situations. They meet with sad ends.

The second theory is that creativity is the product of a resolved conflict-free mature self. **Maslow** (1950) states that when all the basic needs are fulfilled, man goes for higher needs. It is the tendency towards self-actualization. It is the unfolding and the fulfilment of the self. Self-actualization is a process of growth of becoming. It is a process of promoting healthy experience of human being. Every person in one way or other shows a special kind of creativeness that he has certain peculiar characteristics. The person is especially endowed with a drive and a capacity that may have little relationship with the rest of the personality. The creativeness of the self actualized person appears to be natural, simple and universal creativeness of innocent children. People once were more spontaneous, more natural and more human, but these people have powerful set of inhibitions and the spontaneity is checked and it does not appear often. If this is not occurring, every human being would show some special kind of creativeness. Studies show that fear and threat produce rigidity, but self-confidence and freedom from threat produce flexibility. Creativity is not a simple phenomenon. It is a product of many diverse forces and conditions and no single factor operating in it. Evidences are there that many conditions may foster creativity and also many other conditions are inhibiting it. Therefore, an individual becoming a creative depends upon the balance of these two conditions in the case of individuals.

Psychological Analysis of Creativity

Psychologists for many years have been engaged in this task of identifying talents of school children by using intelligence tests. But the conventional intelligence tests were not sufficient to exhibit certain special talents in human beings. At present there is a great need for mathematicians, scientists and linguists all over the world. Even though these abilities are included in the intelligence tests probably these professions require little different abilities. If the psychologists search for these kinds of talents, they should not search them with their existing intelligence tests. Therefore, it is a challenging task for the psychologists to develop tests that would identify the special talents of creative type. Who are the potential creators? Who can make something in original, in artistic and scientific worlds? Persons who are making extraordinary achievements are called genius. Therefore, the concept 'creativity' is always associated with genius or creator or inventor.

The primary requirement for one to be creative is to have keen sensitivity. Sensitivity is a quality or characteristic that an individual should possess. It is the quality of being sensitive.

It is a state or condition, a quality of reacting or being sensitive to an external world or stimulus. There is extraordinary individual differences in the sensitiveness of men. Children while playing show certain sensitivity to the objects around them. All children do not prefer the same object or toys. Stimuli in the children's play group may vary. The stimuli may be of sights or sounds, smells or even tastes. One may be sensitive to sights, another may be sensitive to sounds a third my sensitive to smells or tastes. It is because, one is sensitive to the particular object, through which he can make out his own world. Probably in the later life, when he becomes an adult, he may become a painter, another may become a musician, the third may become a perfumer. It is such sensitiveness to the environment that makes one a creative person. Psychologists' study of the individual differences in various aspects provide them interesting information. These information arouse the curiosity in them to go deep into the individual differences and probe the behaviour of the individuals and they try to bring creativeness into their field of study.

Psychologists see every activity of an individual from the behavioural point of view. The behaviour of an individual is expressed in some form. Just like intelligence is expressed in solving some problem, so also creativeness is expressed in some form of producing something new which did not exist previously. It may be a simple thing, but it is new. It is evolved out of one's creative skill.

Psychologists have paid more attention on creativity for the last four decades after Guilford's work on creativity was published. If a person's work is original it may have remarkable result. There may be manifestations like fluency, flexibility and originality. This is what **Guilford** (1962) states. It is sometimes viewed whether a creative person is an eccentric, non-confirmist or radical or like all other persons, normal. It is a surprise that they are of both types. Creativity is not a magical quality. Creative persons do not make out something out of nothing. It is an accepted fact that creativity like intelligence is a quality that every human being is having to certain degree. It is a human behaviour and every human is having it. There is variation in intelligence like, the genious, normal intelligence, low intelligence like the feeble minded with idiotic, imbacile and moran qualities, in creativity also there may be persons of very high or extraordinarily creative and also stupid creative. A person is judged whether he is a creative or not from the products that are thought to be creative. But how to judge whether a person is a creative or not. Several methods have been followed to test the creative skills of a person. **Gellagher** (1960), **Taylor** and **Holland** (1964), **Guilford** (1950), **Yamamota** (1964) and **Torrance** (1966) have used various methods and tested the level of creativeness in students and other persons.

Psychological studies often associate development or maturity of individuals in assessing their performances. One such quality is creativity which is believed to be based upon the maturity of the individuals. Maturity requires adjusting and participating creatively in an environment. Persons in these days to participate creatively must be literate in politics, social, economic, scientific, artistic and cultural spheres. A mature adult in the present society must do more and become adjusted to like environment. The mature mind is a creative mind with courage and foresight and must have willingness to work. In these days much has been said, written about the values of creativity and also the importance and necessity to foster it in schools. Guilford, who has done very extensively in this area has found that a person may be much more fluent with one type of material than with other types.

According to **Maslow** (1970) creativeness is a universal quality present in all people. Every person in one way or other shows a special kind of creativeness or original or inventions that has certain particular characteristics. Therefore, it is obvious that creativity is not confined only to genius. Few have originality like Leonardo da Vinci. But we all have new ideas and insight. It may vary from time to time and also vary in the ways of doing things. It seems that it is fundamental human characteristic and it is a common human nature. It is the potentiality with which that every human being is born. Maslow states that most of the human beings on becoming encultured loose this. But some individuals by working at life directly retain this quality fresh and naive. If it was lost as some people do, they were able to recover it in the later life.

Creativeness may appear in any form. A person expresses his creativeness in writing books. Another person expresses it in the form of painting, or composing musical tone, or try the work of sculpture. But a person's special type of creativeness is the expression of his healthy personality. He may project this out in the world of activity in which he engaged. He may not be claiming that the output of his work is a creative products. Any person carrying out some work is a means for him in which he can show his skill in a fine manner and in this way his creative skills may be expressed.

The ancient Egyptians and the Greeks presented their culture and history in the form of sculpture and monuments for the posterity to know their creative skills, art and imagination, curved in stones (Fig. 22.5).

Fig. 22.5 : Sculpture : Sphinx

The Sphinx sculpture (Statue) curved out of stone which stands as the Egyptian civilization. Some of the great monuments are the pyramids and the Sphinx. The Sphinx is having the lion's body and human head, sculptured on huge rocks. The great Sphinx is the portrait of a Pharaoh Khafra and stands in front of the pyramid of Giza. The Sphinx is the symbol of ancient Egyptians' creativity.

Any creative work either in the form sculpture or painting is for others to see and find out what type of creative skill is expressed in it. For example, a carpenter, in his designing and manufacturing furniture may use all his imagination and creative skill and manufacture a new type of furniture which may be appreciated by others, the fine artistic work in wood. Similarly a pot maker, who prepares different kind of pots out of clay may attract the attention of the passers by and they may be tempted to buy it for its artistic work. But the pot maker does not claim that he is a creator or creative genius. We see on road side footpaths toys made out of different materials, painted very attractively and kept for show. All these are the expressions of the healthy personality. The Egyptians illustrated their historical events like the victories over others in the form of sculpture, curved on the walls of rocks in various places (Fig. 22.6).

Fig. 22.6 : Sculpture : Egyptians' Art and Culture

This picture is a symbolic representation of the Ancient Egyptians' art and culture and also the history. The Egyptians were celebrating their victories over the non-semantics. The non-semantics were taken as prisoners during 12th century B.C. The sculptures were curved on rocks.

Predicting and Measuring Creativity

Investigators, who have been studying creativity for a considerable period of time, state that the measurement of creativity is not the same thing as intelligence. Therefore, creativity tests are developed as open-end type and it involves not asking for a correct solution of the problem. The question usually is how many solutions the subject can do? Scores are obtained on the basis of the number of solutions and on the original answers. Though there is a positive correlation between these two, the correlation co-efficient is very low. According to **Torrence** (1960) it is from 0.22 to 0.32. But a highly creative person shows above average in intelligence. But it is not all creative persons show above average in intelligence. Many persons with high intelligence quotient have low on the measure of creativity. Therefore, a high score on intelligence test should not be taken for granted high level of creativity. Real original thinking involves non-intellectual factors such as receptivity to novel ideas and a knack for making connections that had not occurred to any one else.

Tests were developed to identify creative talents. The tests show that the top scores on creativity test had on the average 23 I.Q. points lower than the top scores on intelligence test. It was observed that the most creative individuals had the trouble often in following the rigid rules to find out the answer. Authors who made several studies in this field observed that the nature of creative process is, diversified. Some of them found that the projective tests were not much useful in testing creativity of the individuals. Therefore, group tests, in spite of their limitations were developed by some authors like **Guilford** (1950), **Torrence** (1962), **Getzels** and **Jackson** (1962). **Guilford** (1962) included factors like word fluency, expressional fluency, association fluency, idiational fluency, semantic flexibility and originality in his test.

The Remote Association Test was devised by **Mednick** (1962). It was devised to meet certain specific needs. The test consist of 30 items. They are in the form of sets and each set contains three words. The subject is asked to bring out a link among the three words by giving a fourth word. The time to be taken by the subject is 40 minutes. The number of right links is taken as the score. It is very widely used test. Wallach and Kogan brought out a battery of creativity tests in 1965. This test was used very widely in India by Prof. Paramesh and others. **Thorndike** (1963) states that Guilford's tests did not measure the real creativity.

Creativity is measured in several ways. But one simple way of identifying creative talent is to have the teachers rating the students. Another simple method is to have the individuals to rate themselves. **Taylor** and **Holland** (1964) found that this method was very effective. A more scientific method was designed and developed by **Guilford** (1950). **Torrence** (1966) developed the test of creative thinking. This test was developed in the assumption that creative ability is comprised several separate factors, such as fluency, flexibility and originality. Many tasks have been designed that make the subject to produce a variety of responses. These responses were scored in terms of responses and other factors. The example most of the cited of the test is the unusual uses of the test. The subject is asked to think and state as many uses as they can of the ordinary objects, such as a brick or nylon stocking. The responses for the item are counted and an index of fluency is arrived at. The second factor, flexibility is measured by counting the number of shifts between classes of responses. Originality is scored on the basis of the number of responses that are rare which are judged as unusual by the experimenter.

There was also a related test to the unusual uses of the test. This is called the 'Product Improvement Test.' This test requires the subject to think in as many ways as possible to improve a toy or an object shown to him. The Minnesota Tests consisted of several non-verbal measures of creativity. These are picture construction task and a Figure Completion Task. The Battery Tests include the test of creative writing. All these tests are speed tests and timed.

Extensive studies were made by **Bavron** (1958) on creative people and observed various characteristics like (1) ability to keep many ideas in focus at a time, (2) independence of judgement, (3) Preference for complex phenomena, (4) self-assertiveness, (5) ability to integrate diverse elements, (6) a high energy level and (7) having more unconscious materials readily available. This last idea refers to a lack of inhibition and freedom of fantasy with rich associations. A person who is highly creative in one area is quite uncreative in another area.

A factor analytical study has been made by **Guilford** (1959) for identifying the factors in creativity. He has identified five intellectual operations. They are (1) cognition which deals with discovery, rediscovery or recognition, (2) memory which refers to retaining what one has known, (3) divergent thinking, which refers to going beyond the given information to imagine new elements and having more possible answers, (4) convergent thinking, referring to find the one right answer, or drawing conclusion, from the data, and (5) evaluation, which refers to arriving at decisions regarding goodness, correctness. All these or any one of them may be used in the creation of new ideas or forms. Out of these, the one that is very much central to creativity is the divergent thinking. This is going beyond the given element to the conception of new ideas. According to Guilford, a great talent for divergent thinking, in the absence of corresponding ability for evaluation, may result in instability. This has been observed in some creative geniuses. Guilford's work in the field of creativity has become an excellent model for the researchers and also as a much valuable tool. He devised the tests to see whether convergent thinking is taking place or not. All the studies in creativity are now processed for exploring the divergent thinking. Guilford's work has brought the process of creativity and divergent thinking close together and made these two concepts as global concepts. This has created an impression in the minds of the researchers that there is no creativity without divergent thinking. These two must go together.

It is presumed that fluency and originality are sub-factors of general creativity. **Thorndike** (1963) has pointed out that these two do not seem to be related to each other significantly and either of them does not seem to be overlapping on one another. Therefore, Throndike states that we must be cautious in constituting the sub-factors as global concepts.

A creative person may have some doubts sometimes as to how much creativity he is endowed with. He may have also some desire to test this or make others to judge his creative skills. He has to find out his own creativity by some means. Generally all human beings may have some experience of excitement or occasional bursts of inspiration that may give them some new insight into the old problems and also on the experience. This mental state we may call as the creative thinking. Such creative thinking may emerge in some people naturally. But it is also something that everyone can nurture or evolve. Such occasional inspirations may not transform everyone into a great artistic genius. But they are not without any effect. They can in someway guide us to develop our natural creativity and may take us towards some successful problem solving situations at home or at work place. At the time of performing the task we ourselves could see how far we are successful in performing the creative task. We can also compare our creative work with the works of other artistically talented persons and we may feel happy on our creative achievement.

PART EIGHT

PERSONALITY AND ASSESSMENT

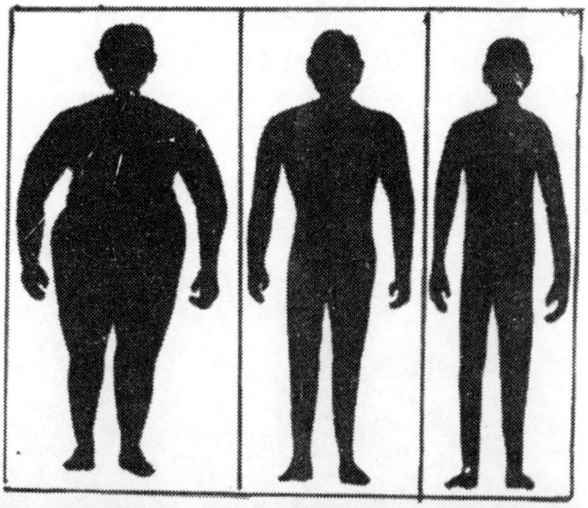

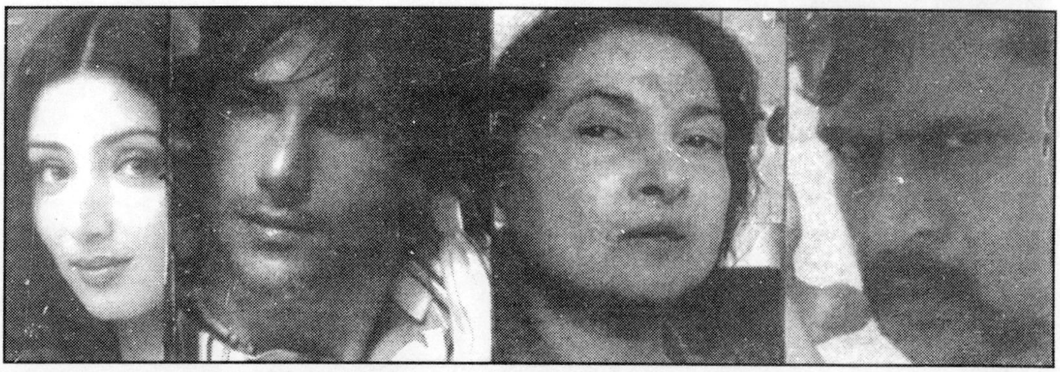

23

PERSONALITY

Introduction

The term personality is used by the psychologists to describe the integrated behaviour of a person. Personality, generally is considered as the *unique aspect* of a man. There are certain aspects or qualities we may find to be *appearing* in more than one individuals. In psychological studies the topic, personality occupies an important place as it is considered to be the unique characteristic of human beings. Psychological studies of personality traits identify persons with similiar and dissimiliar traits for its purpose. We can observe that in some situations all human beings share some common interest and act similarly. Everyone cooperates with everyone else and finds some pleasure in it which is the psychological nature found in all human species. Every human being naturally becomes a member of some community or society and makes his contribution to it and also shares the benefit of it. But this type of behaviour we may find only in certain life situations not in all life situations and make their characteristics to manifest with all their personality make ups.

If we analyse the human behaviour critically in the history of human kind of the past and the present there has never been a person quite similar to you and never anyone would be born. The way in which you feel, perceive, think and act or behave has a *specific pattern* which in its perfect characteristic manner can never be matched with the behaviour pattern of another person. In short there cannot be another person most like you in all the behavioural characteristics, nor can any one duplicate your pattern of behaviour in all respects. Therefore, we may state that human personality is a complex phenomenon and psychologists tried to understand the *complexity of human personality.*

Personality in a broader sense refers to the *behaviour of individual person* whether it is right or wrong good or bad, pleasing or offensive, to other persons, or agreeable or disagreeable to his fellow beings. In recruiting candidates to various jobs, it is customary to have a personality test of the candidates, apart from other particulars already collected. Organizations may have their own methods of judging a particular person's ability and experience. They may be wanting to know whether the candidate has a pleasing or forceful personality, whether he is energetic and persistent, whether cheerful or even-tempered, whether he is self-reliant or selfish, cooperates well with others or quarrelsome and so on.

There are also hundreds of words that are used to describe the characteristics of an individual personality. These personality characteristics or qualities have immense importance in various situations like, playground, work place and in all situations where social life persists.

All these words stated above would indicate how an individual behaves in a situation. One person behaves in a pleasing manner, another behaves in humiliating manner, one person performs the task energetically, while another behave slugguishly in the same situation, one behaves cheerfully, while another behaves irritatively. All these words are not referring to the names of activities, but they are referring to the *qualities of behaviour.* Any activity performed by an individual whether it is a smaller one or bigger one may reveal the personality characteristics of a person.

The study of personality in a wider scope involves *all the topics* that we did *in psychology,* because personality components exhibit all the psychological processes in various behavioural situations. It is because of this that personality is considered by some psychologists who have specialized in it claim that it has become a *separate field of science* and call it as *personalogy.* Personalogy means the science of personality and personalist claim that every person reveals a distinctive organized integrated patterns of behaviour and thus every person has a personality of his own.

The personality patterns as exhibited by the individuals provide ample information for the personalist to describe every aspect of behaviour fragmentally and organize them in an integrated whole and also is able to predict the future behaviour. The complexity of human personality cannot be assessed easily with one or two methods. The psychologists have to develop accurate methods and techniques and formulate theories as any one single method or technique and theory is inadequate for assessing personality. We shall discuss in the ensuring pages the various aspects of personality and the methods and techniques that can be applied for assessing personality.

Personality Defined and Described

Psychology is defined as the science of behaviour. Behaviour generally refers to the way in which an individual or an organism moves and acts for fulfilling the needs, since there occurred some *deficiency in the organism* due to utilization of energy which was stored in the body. For the very existence to be alive some energy in the body must be spend. What amount of energy must be spent for a day or for an hour? This is not known. The nutuition experts were able to tell this. But some spend more bodily energy and some spend less amount of energy. This is known from the way in which an individual keeps his body functioning. It may be very vigorously or it may be very slowly or smoothly. For conserving energy in the body, the individual or the organism must feed itself or eat food according to the everyday prescriptions or according to the habit developed by the organism. The human body, specifically, not only needs the food and water to keep alive, but also needs all comforts normally required, such as sleep and rest and also the required activities or exercises. In all these functions the individual or organism has to contact different type of people and perform various kinds of activities or behaviour such as feeling and emotion, sensation, perception, motivation, learning, adjustment, intellectual activities and so on.

The way in which our study of behaviour is to be dealt with has scientific approach. We have analysed the subject matter into elements or small components which may be stated as molecular. The small components of behaviour though appear to be in isolation they are having functional relationship. Behaviour is emitted by an individual according to the needs and demands in a discrete manner, but with *dynamic function* in relation to the environment. By studying the components of behaviour we can obtain more information. Thus behaviour finally is an out right expression of an individual person or an organism. When all the components of behaviour put together and function in a dynamic way, it becomes a constituent of an individual person with all complexity and richness. Psychologists, thus not only interested in the study of molecular or separate points or components of behaviour, but also the organism's integrated, or *organised molar aspects* of the individual. The parts are organized into an individual person and in any behavioural aspect the individual is revealed as a whole. Therefore, the individual's organized behaviour totally refers to his personality.

The term personality is used invariably by all sections of people on certain occasions, during the course of their conversations about a person. The concerned person may be young or old, male or female, a person of ordinary type, or person of great eminence. Whether they have known the person very well or known simply from what other say. They attribute some characteristics to the person and admire with some pleasant words that he is an attractive, pleasant, agreeable, uncontroversial, dynamic, friendly, person of outstanding ability and so on. It may be also stated that he is a very good personality, attractive or the person is not good looking, very forceful, adament, not friendly, quarrelsome and so on. Thus the person is described interms of some traits or characteristics. This is what we observe in the common sense world in utilizing the concept personality. But it does not meant that what we hear from others about the description of personality is wrong. We cannot say that the common sense observations are totally wrong or incorrect. This is what the psychologists state as social attraction. Whatever it is, this is not the all with which the psychologists are satisfied.

The psychologists accept these and also go beyond all these and include many things and describe the individual psychologically. The definitions of personality include the characteristic ways an individual conducts himself with his personal factors such as *physical appearance, sensitivity, intelligence, aptitudes* and other traits in his everyday life situations, which may be *pleasant* or *stressful.* The personal characteristics of a person may be of varying degrees. But all these contribute to the individual's total disposition with which the individual creates an impression on others. The individual's total quality is represented in all his characteristic way of behaviour. A person's behaviour in one situation may be different from another situation. One may be problem solving situation, where he is expected to behave intelligently. Another situation may be a helping situation where he is expected to express his feeling of sympathy and help a person who is an danger.

In a third situation the person might have been asked by his organization to carry out a task successfully in a given time and he carried out the task and completed it successfully even before the specified time. Here the person was found to be very responsible and dutiful. In all these three situations the persons have revealed their personality characteristics of *intelligence, sympathetic attitude* and the *sense responsibility and dutifulness.* A person's other side of personality traits may be *mediocre, unkind* and *irresponsible and undutiful.*

The definition and description of personality cannot be limited to a few traits. Personality in a sense, is having very broad scope, as broad as the psychology itself. It is because the subject matter of personality cuts across all the topics that a psychology book deals with. Therefore, the study of personality cannot be limited to a few information. From the layman's point of view personality may refer to a person's popularity in the community or in the public, or may refer to an eminent scholar who attracts people by his erudite scholarship, eloquence, and powerful writings. The layman's assessment of this person is that he is a man of many sided personality. A political orator, with all his eloquence, alliterations and analogies may attract and convince the general layman public very much and create an impression that he is with lots of personality. In the field of games and sports, person's are celebrated by the admirers as the man of many-sided personality. Generally, in these days the motion picture actors, the cricket players, tennis players, and the football players by their impressive or spectacular performances, are able to create enthusiastic fans and admirers all over the world and they are attributed in a characteristic way with multiple personalities.

But at the same time if a person is sky, or withdrawing, or retiring or refuse to make his appearance in the public is said to be having no personality. Whatever characteristics are attributed to persons of highly talented and to persons of no talent, psychologists are of the view that a person has a personality of his own, because he displays a distinctive characteristic patterns of organized behaviour in various life situations. Therefore for a psychologist there is no negation of personality.

There is no one single way of defining personality. Whatever an investigator observes in his study of personality and what type of information he gets in his study becomes the subject of matter for defining personality. Therefore, personality is defined in various ways. According **Murphy** (1952) personality is, all the qualities, modes of reaction, which set off an individual as distinct from all others. Personality is the integration of these qualities in a unified system.

Garrett (1975) defined personality as the organization of characteristic behavioural traits which serve to describe an individual. According to **Moskowitz** and **Orgel (1969)** personality refers to the *organized behaviour* of the individual. Personality may be defined as the *sum total of all the behavioural and mental characteristics* by means of which an individual is recognized as being unique **(Callins, 1984).** Personality is the *distinctive character* of a person that makes him *socially attractive.* It is also defined that personality is a unique organisation of fairly permanent characteristics that sets the individual apart from other individuals and at the same time determines how others respond to him **(Kimble** *et al.* 1985).

In these definitions, formulated by various authors, we can observe that there is some general idea running through all. That is *unique organization, distinct characteristic.* As we have discussed in the beginning that small components of behaviour are organized into unique pattern which is the sum total of the individual's quality. In the unique organization of small components of behaviour, there may be certain amount of *similarity contiquity,* and *differentiation.* But we cannot say how much in each of these. In view of these we may state that every individual in certain respects is like all other people or look like some people, or look no other people. It is quite, but natural that all human beings have something in common or share with everyone physiologically or genetically. Every individual shares certain general characteristics of the people in his society, and not having complete uniformity of the whole make-up with everyone.

In the long past history of human race there has never been a person quite like you not only in physique, but also in feeling, thinking, perceiving and behaviour. An individual can never be matched with another individual in all respects. In other words, a living human being can *never duplicate* exactly like the another person in all characteristics and behavioural patterns.

In psychological studies the general theme is the individual differences. In the individual differences one of the important characteristics of the individual may be *character and personality.* Character is a combination of traits and qualities which distinguishes the individual's nature or a thing. One such distinguishing quality may be *integrity* or *morality.* We may also say that he is a man of character, man of good reputation. Character generally refers to *conduct* that can be said *right or wrong, good or bad* that may receive or fail to concord with accepted social standards. Thus personality is a *consistent of many desirable* or *undesirable components.*

Personality is potentially the most interesting part of psychology. But it is very difficult to give an adequate description or complete characterization of an individual's personality. The human personality is *dynamic in nature.* An individual's *genetic factors* in combination *with the environmental* and *social processes,* his intelligence, learning and experience, his ambitions and achievement, the types of social interaction, his success and failure, all determine what he should do or how he should behave in a particular situation.

One way of defining personality is that it is the sum of all the feelings, thoughts and traits that make one unique as a person. Personality also may be defined as the *quality* or *trait* of a being, a person, that quality makes a being personal. In the same quality or trait there may be variation in degrees. The trait is emphasized in the most of the definitions and descriptions of personality. Therefore, any definition or description of personality without making any reference to traits is quite incomplete.

Personality Traits

What is a trait? A trait is a characteristic feature or quality distinguishing a particular person or a thing. It is also a means of distinguishing one person from another. A trait may also be stated as an enduring attribute of a person that is manifested in variety of circumstances. Traits of various types constitute the whole personality.

Traits are often spoken of in the study of personality and in the categorization of personality. Different types of personality are evolved from the characteristics of human behaviour. In every expression of behaviour the traits are revealed. Therefore the traits become the important components of personality. The concept of the trait of personality is having a long history right from the time of **Hippocrates** (450 - 370 B.C.), the ancient Greek physician, who was considered as the father of medical sciences and also modern medicine.

According to Hipprocrates sickness is an *imbalance* of the *body fluids* which he called as *humours*. There are four fluids such as *blood, phlegm, yellow bile* and *black bile*. Each fluid or humour is believed to be interacted with a particular temperamental characteristic or personality trait. A man's *temperament* is determined by the *proportions of his bodily fluids*. Any imbalance of the fluids may produce an imbalance in the four basic traits of *sanguine, phlegmatic, choleric* and *melancholic* with anyone of the fluids dominating on the behavioural tendency.

This was the view held by **Hippocrates** and on the basis of this idea he categorized *four personality types as sanguine, phlegmatic, choleric* and *melancholic*. Each type of personality is composed of several traits. All such personality traits are put into a wheel. The wheel describes personality traits. What makes *good trait* and what makes *bad trait* is a matter of *balance maintained by the fluids* or humours (Fig. 23.1). Probably Hippocrates used the term fluid or humour in place of *hormones* secreted by the endocrine system.

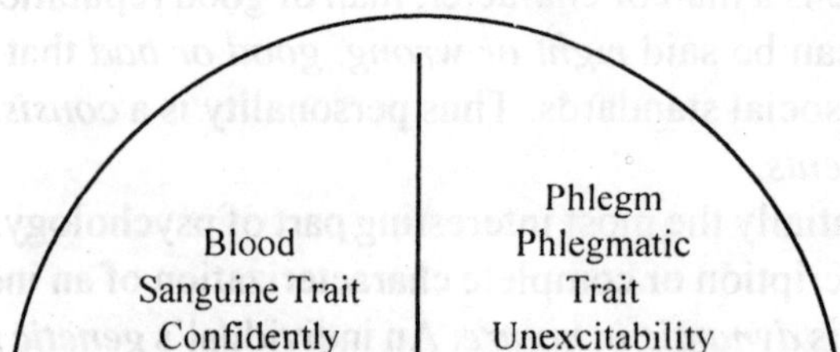

Fig. 23.1 : Hippocrates' Model of Personality Description, based upon the four Liquids (Humours) Blood, Phlegm, Yellow Bile and Black Bile

(1) ***The fluid of blood*** is the cause of *sanguine personality trait*, the person with this trait behaves *confidently*.

(2) ***The fluid phlegm*** is for forming the *phlegmatic trait* and the person who is with this trait behaves *unexcitably*.

(3) ***The yellow bile*** causes the *chloeric type* of personality trait and the person with this trait behaves *irritatively*.

(4) ***The fluid of black bile*** is responsible for the *melancholic type of personality trait* and the person mainly behaves *sorrowfully*.

According to the classification made by Hippocrates the sanguine and the phlegmatic type of persons behave confidently and unexcitably. While the choleric and melancholic persons behave irritatively and sorrowfully.

Modern psychologists'study of personality and personality traits are not far away from these types of personality and the personality traits. They adapt some of these types in their studies. The description of the traits of the earlier studies are quite similar to the modern description of the traits. Any imbalance in any of the bodily fluids makes the person unstable and sick or *disorganized.* Modern psychologists speak of the *imbalance of hormones* of the endocrine system which causes disorders in the bodily parts and in behaviour.

Gordon G.W. Allport (1897 - 1967), (1936), one of the foremost and highly influential personality theorist states that a trait is an enduring and stable attribute of a person which is exhibited consistantly in various situations. Allport believed that there are *two types of traits* functioning in human beings, one type is the *common traits* and the other types is the *individual traits.*

Allport states that a *common trait* is one that manifests in an individual to some extent in a friendly manner or in an antegonistic manner. These traits are generally, generosity, friendliness, kindness, dominance, aggression and responsibility. These traits are based upon *social adjustment.* The extent to which people express these traits can be assessed by suitable personality inventories.

The individual trait is specific, possessed and revealed by few people in social situations. The individual traits, generally provide some unique characteristics to the personality. Individual traits are studied only by careful observation and by going through the *personal records, dairies, personal letters* and *reading materials.* The individual person also may react with quirky wits, a cutting sarcasm, a certain infectious ebullience and so on.

These common and individual traits were studied by Allport by the *rating method.* He wished to see how far these traits are pervasive in personality. By the way of rating the traits of the people, Allport recognized another type of traits which is different from the other two traits. This type of trait is rare among the people. He called this type of traits as *cordinal traits.* This is the trait through which the individual organises his whole *lifestyle.* The cordinal traits are trying to achieve power in the society, ambition to go high in lifestyle, makes greedy, self-sacrifice, humanitarianism, social attraction and so on. It is true that it is not possible to exercise the cordinal virtues by all people but only a few persons can strive. Persons endowed with cordinal traits receive inspiration and motivating force generally from mythology, history, literature, poems and novels. These persons' motives are determined by these sources.

Allport had a belief also in another type traits, which he called them as *central traits.* He was of the view that there are about five to ten such traits in some persons. These traits motivate persons for determining important values and these are interest patterns which raise lifestyles and describe their personalities very well. Persons endowed with the central traits often interested in beautifying themselves, the surrounding areas where they live, making the living area more greenish by gardening, often preoccupied with sexual matters, dancing, quarrelsomeness, an urge to simplicity being artistic, sentimentality, wishing for independence and so on.

In addition to all these traits. Allport observed that people also have some *secondary traits.* These traits are found in people only in some particular situations. This type of traits in individuals reveal being grouchy at one time specially in the mornings, wishing for a different type of food, expressing strong dislike for crowds and so on. In his extensive study of personality, Allport was aroused by some interest for probing into the *unusual traits* that are demonstrated. by some people. He felt strongly that an individual's whole personality could be described only by some *intensive long range investigation* and analysis, because *each trait is unique* in the way it functions within the total personality make up of the individual. Allport considered that the study of personality is the study of single individual persons.

By way of making his study an extensive, Allport tried to enumerate the words or the adjectives that are referring to the personality traits. He had indicated about 17,953 human personality traits. From these he was able to classify the types of the traits that are generally attributed to human behaviour.

Therefore, we may state that there are *common personality traits* and *individual personality traits.* In psychological studies psychologists try to identify the most significant traits. These traits are manifested in the behaviour of the individuals.

Personality trait according to **Woodworth** (1952) is a particular quality of behaviour such as cheerfulness or self-reliance which characterizes the individual in a widerange of his activities which is also consistent in an individual over a period of time. The sum of all these traits constitute the total personality. Every person has his own style of life, which is often associated with his name that is, the person is very pleasant and cheerful. Generally traits are attributed to the individuals in the interpersonal or in social situations.

A trait also may mean a more or less a permanent characteristic which the individuals possess in greater or lesser degrees like the generous, reliable or irritability, hostile, shy and so on. Generally, traits distinguish one person from another. A trait is an enduring attribute of a person which may be manifesting in a variety of situations. Therefore, trait refers to a specific way of behaviour that occurs again and again which tends to characterize a person and is useful in predicting the future behaviour of an individual. For example, a person donates money to some charitable organizations liberally or to some poor persons who are badly needed help and also makes use of his leisure time for some worthy cause. This person could be described as the person who possesses the *trait of generosity.*

If we are quite familiar with the trait names, and the characteristic of the traits we could easily identify the behaviour of a person with these traits. Following Allport's study of personality, the traits and human behaviour and the long list of the trait names, **Warren Norman** (1963) had developed interest in the study of traits. He was able to bring out another long list of trait names with which we could describe any human behaviour. Norman's list of *trait names* consists of about 40,000. Using this long list of trait names Norman applied experimental and statistical methods to identify the most often used traits in the everyday life situations and reduced the number of trait names drastically.

Norman had designed his test by forming sets of paired polar opposite trait words which he picked up from **Allport-Odbert** (1936) list of trait names. For every trait dimension the polar opposite trait words of a minimum of three sets were arranged in an order. For example, against Aggresiveness the three polar opposite trait words are : *mild, gentile-Headstrong; good natured - Irritable; Not Jealous - Jealous.* Against the Trait Dimension *Emotional Stability* the three polar adjective words are: *Poised-Nervous, Tense; Composed-Excitable, Calm-Anxious.*

The model followed by Norman in his test is:

Trait Dimension	Polar trait adjective words
Aggressiveness	1. Mild, gentile - Heardstrong
	2. Good Natured - Irritable
	3. Not Jealous - Jealous

The persons who participated in this test were asked to rate their friends or peers whom they know very well on these polar traits words. Norman applied factor analysis and extracted five factors which had high loadings. These five trait Dimensions were, extroversion, aggressiveness, conscientousness, emotional stability and culture. These were according to him, the *basic traits* of personality.

Subsequently **Passini and Norman** (1966) have conducted similar study of traits among the college students. These students were brought together and they had acquaintance for fifteen minutes and each one was asked to rate the other. But everyone had known that the other person was also a student. These students described each other and these were *identical basic traits.* These traits are being used by some persons for judging other persons whom they did not know. The investigators found that there were too much of *similarities* in the assessed traits. It was a sterotypical way by which people speak about others.

There was also another important study to classify people on the basis of *traits of personality.* It was **Raymond Cattell** (1905 - 1973), (1965) who designed and developed the famous Sixteen Personality Factor Questionnaire (16. PF). Cattell developed his test after making an intensive analyses of the judgements that people made about their close associates. He was able to make out a list of thirty five trait clusters. He named these clusters as *surface traits.*

These surface traits were named like this, because they were *overt expressions* of personality traits. By applying factor analysis Cattell extracted sixteen more of the basic traits. He called these traits as *source traits.* He developed a self-administered personality questionnaire for measuring the traits. He administered this questionnaire to many different type of groups of people and brought out interesting results. He found the *source traits* of personality are the underlying *basic traits* which are expressed in a number of less fundamental surface traits, Cattell's personality questionnaire underwent several revisions. The final form is called the *sixteen personality Factor Questionnaire* (Fig. 23.2).

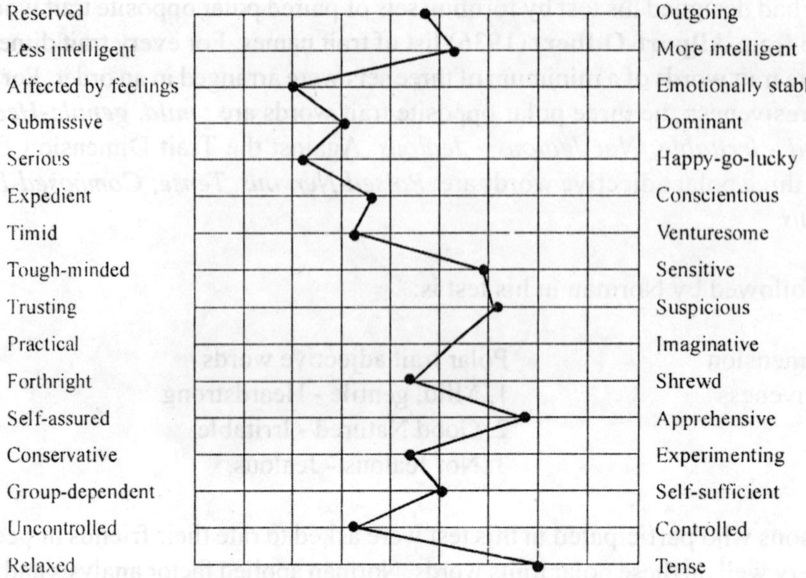

Fig. 23.2 : Assessment of Personality : Cattell's Sixteen Factors. Profile showing the General Neurotic Tendencies.

Charismatic Personality

We sometimes hear charismatic and this often attributed to some person, that he is a charismatic leader. Charisma, a Greek word which means 'gift' or 'favour'. It is believed that the divine grace is bestowed on prophets or saints. Charisma is a special personal quality or power of an individual making him capable of influencing and inspiring large number of people. Theologically charisma is a *divinely bestowed* power or talent. It is also designating of a religious group seemingly inspired by God, as in healing powers. Generally charisma is a special inspiring quality of leadership. Charisma is also attributed to some outstanding inspiring leaders and they are admired as *charismatic personalities.* Personality suggests a charismatic quality that tends to attract the attention and admiration of others. We often regard charismatic personalities with deep respect or love time to time. **Max Weber** (1864 - 1920) a German economist and sociologist, best known for the protestant ethics and the spirit of capitalism called, charismatic, a quality by virtue which one is set apart from ordinary men and treated as *endowed with supernatural or superhuman* or exceptional powers and qualities. Some are of the view that charismatic personality is *inborn* and some view that it is a developed personality quality.

Brain and Personality

Is personality located in the brain? But where in the brain these qualities are organized and orchestrated? The maps of the brain of the 19th century showed the specific areas labelled with such personality traits as secretiveness and loyalty. Today some functions like touch and language are assigned to places in the brain. But the whereabouts of personality in the brain is not known.

The attempt for locating personality or personality traits in the brain is not yet completed. The search is still going on. But do people with different personalities have different brains? Modern physiologist and geneticist are in long way from connecting every human trait to a *tall tube* (revealing what is meant to be kept secret), Wrinkle in brain tissues. Researches have discovered enchanting links between brain structures and the ways in which different people behave.

One of the leading contemporary British psychologist, **Eysenck** (1916 -1988), (1967) has suggested that the answer to the difference between the extroverted (outgoing) and the introverted (reserved) personality lies in the *cerebral cortex,* a part of the human brain much involved in learning, reasoning and planning. Activities can be viewed in brain scans. In the extrovert the cortex area is quiet and seems to receive noisy, exciting, situation that may arouse it. The introvert's level of cortical arousal, is already high (which is in contrast). Therefore, the *introveted* personality does not require much outside stimulation and seems to prefer peace and quiet. According to Eyscnck extroverts can fall asleep quickly and introverts cannot. The extrovert's cortex slips easily into a quiet state. Extroverts tend to be less sensitive to pain which indicates that their cerebral cortexes are hard to arouse.

According to Eysenck (1965) individuals differ in their excitation/inhibition balance. This difference is mediated by some parts of the central nervous system. Eysenck on revising his theory in 1967 was able to postulate the specific relationship between *personality dimensions* (intervension-exterversion) and the *Ascending Reticular Activating System* (ARAS) of the brain. He further stated that the personality dimensions of *normality-neuroticism* (dysthymia and hysteria) are rooted in the central nervous system structures. A group of subcortical structures such as *hypothalamus,* the *gyrus cinguli* (raised surface area between the *sulci* the cerebral cortex), the hippocampus a portion of the cerebral hemisphere, (a primitive curved cortical structure lying in the anterior medial region of the temporal lobe, important for language memory) and their interconnections, are the mediators of bodily arousal associated with emotion. These structures are responsible for changes in heart rate, respiration, blood pressure and other emotional activation. The whole group of structures are named as *visceral brain and limbic system.* The relationship between the brain and dimensions of personality may be summarised in this way : Introversion/Extroversion-mediating differences in ARAS arousal, normality/neuroticism differences in visceral brain activation (Fig. 23.3).

Thus Eysenck's theory of biological basis of personality shows how personality dimensions are located in the brain structures. Eysenck did not produce experimental evidences. It is hypothatecally, Eysenck located the personality dimensions in the brain structures. Many accepted Eysenck's conclusions. But the search to locate personality in the brain is still going.

Development : Is Personality Inherited or Acquired?

This question, whether personality is inherited or genetic make up or acquired through family and social environments has a simple answer, that personality is the interaction between these two variables. **Woodworth** (1952) states that physiological psychology (genetic) and social psychology meet in the study of personality. The meeting of these two sciences sometimes have the appearance of a battle on one side physiology is claiming that personality is the product of genetic make up and on the other side social psychology claims that it is the development of social factors. A child brought up under strict discipline may grow up with passive characteristic, while the child brought up under liberal circumstances may grow up with ambitions and dominent characteristics. Allport (1937) states that the same fire that melts the butter also hardens the egg.

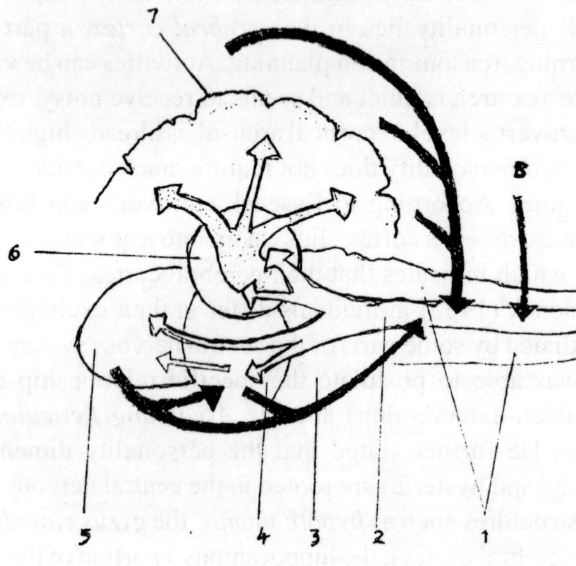

Fig. 23.3 : Visceral Brain and Limbic System

1. Brain stem. 2. Ascending Reticular Activating System (ARAS). 3. Medial Forebrain Bundle. 4. Lateral Hypothalamus. 5. Rhinencephalon. 6. Thalamus. 7. Neocortex. 8. All Sensory Systems.

There are views that personality development depends upon glandular system. The type of hormones secreted by the endocrine glands greatly influence the personality development. Another prevailing view is that the individual's physique is a potent factor in the development of personality dimensions.

Some of the recent researches suggest that the *genes play* important role in determining certain traits or temperaments such as aggressiveness, shyness, impulsiveness and social dominance. Individual's have inherited physiological differences that are partly responsible for the development of different temperaments among the children.

Human personality traits are generally *shaped by biological and social factors.* It has long been recognized, the two factor origin. Psychologists sometimes describe personality traits primarily as the product of *biophysical* and *biosocial.* The biophysical traits are observed in a variety of behaviours as inherited and basically organic. For example, a person's aggressive behaviour, cruelty, passivity and quick temper are *biophysical traits.* On the other hand, the biosocial traits are revealed particularly in the interpersonal relationships and mostly determined by the social and environmental factors. For example, rudeness, persistence, tact and sense of responsibilities are primarily biosocial traits. These are built up in a social situation.

THE BIOPHYSICAL FACTORS IN THE DEVELOPMENT OF PERSONALITY

The individual's behaviour is much influenced by the chemical substances. A few chemical substances are introduced from outside in food and drink. Chemical substances are also produced in the body which are having significant effect in growth and behaviour. The chemical substances are depending upon the circulation which takes place through the nervous system. The circulation is very rapid. The influence of physiological factors in shaping the personality has been experimentally studied. The effects of drugs, blood substances and the diet have been investigated. For example, the sedative drugs cause reduction in the general activity of the body, damaging concentration, muscular in coordination, drowsiness and sleep. These effects are caused due to depression in certain parts of the brain. Alcohol has the same effects. Alcohol sometimes has stimulating effect. Instead of depression the person becomes highly talketive and more sociable (Emersin, 1982). The other drugs with which the individuals are familiar which may have certain specific effects depending upon how far the brain is influenced.

For the normal functioning of the brain the *blood-sugar* must be maintained above certain level. If the blood-sugar falls below the critical level mental function is damaged and there is alterations in the personality of the individual. There are changes in the individual's mood, increased irritation and lack of understanding things. Further, *oxygen deficiency* also may cause *personality alterations* (Shock, 1939).

Experimentally it was established that persons suffering from *prolonged deprivation of malnutrition* by war or famine are psychologically affected. There were significant alterations in the personality characteristics of the persons who were subjects in this experiment. The subjects were preoccupied with dreams of food. Their normal interest in sex, athletics and motion picture were reduced. They were gradually depressed, irritable and unsocial and not having any sense of humour. These are the marked alterations in the personality traits of these persons and when they were taken to normal life, the changes gradually disappeared.

Severe illness also may alter human personality. Some of the changes left permanent alterations in persons suffered from severe illness for longer period. Therefore, alterations in personality traits occur due to drugs and other bodily chemistry.

The Endocrine Glands

The endocrine glands constitute a system, consists of several glands. These glands secrete certain chemical substances called hormones directly into the blood stream. All glands are organs made up of a variety of tissues. All secrete substances needed by the body. These harmones are having the *power to raise* or *lower the activity* level of the body and thereby having influence upon the personality dimensions. The hormones of the endocrine glands *directly affect the growth and development of the organism* and also cause emotional disturbance. These hormones strongly influence many of the personality traits. The endocrine glands constitute an interlocking system. If there is any *malfunctioning in any of the glands, it affects all other glands* and *certain glands directly affect the particular personality trait.* Both, the over secretions and the under secretions of the glands affect the normal function of the body and the development of the personality.

The most important hormone is *thyroxin,* secreted by thyroid glands. Any *over secretion* in this affects the body activity and causes nervous tension, irritability, restlessness and emotional instability. The *under secretion* of the glands causes chronic fatigue, laziness and listlessness and any malfunction of thyroxin often interfers in personality development. If thyroxin is severely under secretion, the growth and development of the organism is badly affected. There will be low intelligence level and the person, is lead to a disease called cretinism. The person would become abnormally small person (Fig. 23.4).

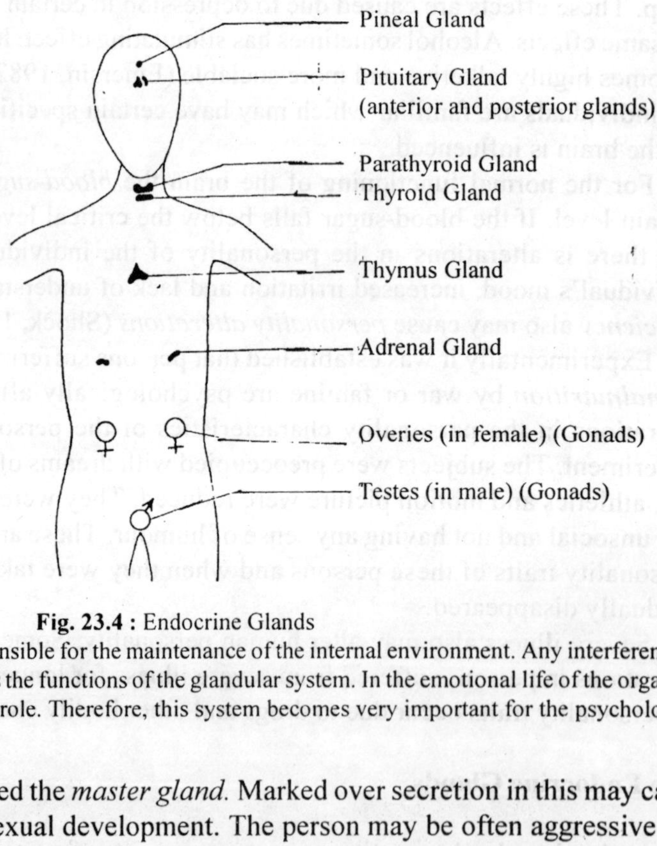

Fig. 23.4 : Endocrine Glands

The glandular system is mainly responsible for the maintenance of the internal environment. Any interference of the autonomic nervous system affects the functions of the glandular system. In the emotional life of the organism the endocrine glands play important role. Therefore, this system becomes very important for the psychologists for the study of human behaviour.

The *pituitary gland* is called the *master gland.* Marked over secretion in this may cause gigantic physique and early sexual development. The person may be often aggressive and dominating over others. Under secretion may affect the body and the muscular structure. The person lacks stemina and develops a tendency to get fatigue easily. Thus the *pituitary gland* is having a definite influence in the development of personality.

The *adrenal glands* are located close to the *Kidneys.* The outer part is called *Cortex* and inner part is called *medulla.* These two vary in structure and function. The cortex secretes *cortin* and the medulla secretes *adrinin.* The *adrinin* is a very powerful hormone. Even a small amount of adrenin in the blood may cause rapid heart beat, high blood pressure, stopping of activities in stomach and intestine, releases the stored sugar from the liver, widening the air passage in the lungs, delay in muscular fatigue. The other gland, cortex secrete several hormones apart from cortin, contisoni which are very essential for life.

The *gonads* are the sex glands. Gonads secrete hormones in conjunction with the hormones of adrenals and pituitary which are essential for normal development of the sex organs and the secondary sex characteristics. These glands secrete several hormones and some of them are found both in males and females. A balance of male hormone regulate the development of masculine characteristic and a balance of female hormone regulates the development of femininity. The sex hormones at puberty promote the development of genital organs, mummary glands in women and the growth of beard and deep voice in men, any lack of gonads in either sex may develop a neutral human without normal sex characteristics. Regarding the development of personality traits some individuals seem to be over sexed and some under sexed. These differences are due to hormone levels.

PHYSIQUE, BODY TYPE AND TEMPERAMENT

An individual's physical size, the structure and form and the general appearance with regard to shape, and the muscular development constituted called *physique.* The characteristics determine the way in which an individual reacts to other people and the way in which others react to him. There is a strong view that the physique is related to temperament.

Generally *temperament* refers to a person's *emotional state.* It is a characteristics of emotional level, the mood. It may be also stated as the natural tendency of a person. Recent research findings suggest that the *genes* play significant role in determining certain personality traits or temperaments such as aggressiveness. Shyness, impulsiveness and social dominance. A person who is habitually cheerful is said to have an *optimistic temperament.* Further, the often described temperaments, are happiness, gloomy, fratful, lethargic and indifferent. If a person often exhibits rapid alternative moods from happiness to sadness he is said to be a person of *unpredictable temperament.*

There were studies correlating temperament and physical factors. This resulted in categorising people into *three groups* as, the *tall and thin person* at one end, the *short and fat* at the other end and the *more harmonious* persons in between. **Kretschmer** (1925), a German psychiatrist, published a book in 1925 under the title of *Physique and Character* (Fig. 23.5). He found four general types of body build, the pyknik, who is soft, round and over weight. *The second* is the *asthnenic,* who is thin and long limbed with weak and under developed muscles. *The third* is the *athletic* who is strong and well muscled and *the fourth,* is *dysplastic* person of thin type often suffering from some kind of endocrine gland disturbance, resulting in disproportionate physique (Fig. 23.5).

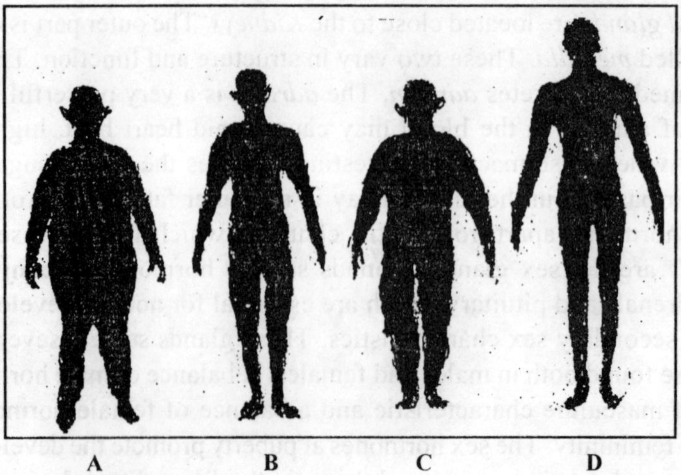

A B C D

Fig. 23.5 : Kretschmer's Model of Body Type Personality.

Type-A : Pyknix type characterised by soft, round and overweight.
Type-B : Asthenic Type characterised by thin, longlimbed, weak and under-developed muscles.
Type-C : Athletic Type characterised by strong well-muscled.
Type-D : Dyspladtic characterised by suffering from some kind of endocrine gland, disproportionate physical body build.

Following **Kretschmer, Sheldon** and **Stevens** (1942) have published a book *The Varieties of Human Physique.* This is a systematic research work, relating *physique and temperament.* Sheldon proposed *three basic somato types* of personality. The *endomorph* is a person of soft, round, large body and short arms and legs, the *mesomorph* is a person hormoniously proportioned in bones and muscles and the third is the person *ectomorph* with long arms and legs, skinny and small body build. The physique and temperament relationship existed even in early life. The adult personality reflects temperament (Fig. 23.6)

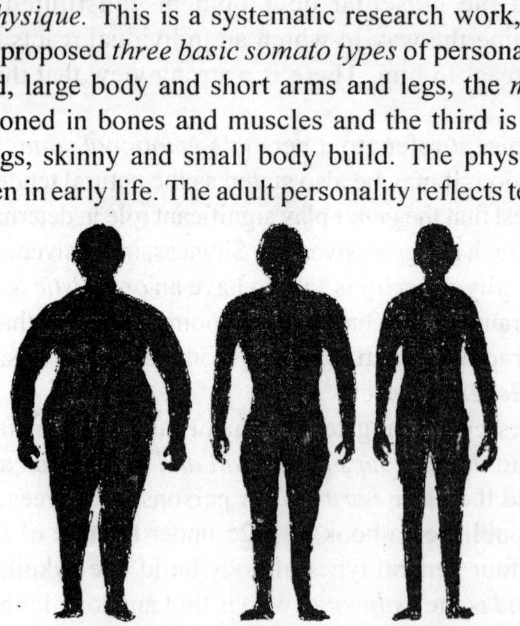

A B C

Fig. 23.6 : Personality Types : Sheldon's Body Types
Type-A : Endomorphic, Type-B : Mesomorphic, Type-C : Ectomorphic

The results of these studies seems to confirm that the glandular, biological constitutions of individuals are having influence in the physique and temperament and hence we may also expect the same corresponding *relationship* between *physical types* and *personality* characteristics. People differ greatly in physique, body type and temperament, which propose that these differences are associated with the *differences in personality characteristics.* Therefore, the influence of biological factors in the development of personality is strongly stressed.

SOCIAL FACTORS IN PERSONALITY DEVELOPMENT

A human child is born in a community and this community is the smallest community, the family, and only in such community a human child is possible. The family may be consisted of the parents, the child and other close relatives who are also around him. Therefore, he is born only in a social set up, lives and grows only in a community which takes care of him throughout his life. Therefore, we may state that a person is born, lives and grows in a society and at every stage receives the *social influences.* Though biological influences are contributing to the development of personality of an individual these biological factors can flourish only in a social environment. Therefore, we may state that the personality of a child is shaped and moulded by the community in which he is born and lives.

In the beginning the child's community is consisted of only the family members and later on his community is extended to school, street children, temple or church, co-students, teachers, playmates and other companions — all exert their influence upon him. The home is mainly responsible for inculating all the habits, eating, sleeping, dressing, the habits of personal cleanliness and toilet habits.

The child at home receives the training in useful social behaviours which determine the biosocial traits. The child is encouraged for exhibiting *socially acceptable traits.* The child may develop emotional instability, shyness and hostility if the father is domineering and if the mother is of negging type. These reactions may leave permanent mark on the personality of the child. The child may develop a feeling of insecurity if there were constant quarrel between the father and mother. Any incident of death or illness may cause fears and anxieties and a *feeling destitute,* all may have some influence on the development of the child's personality. An adult trait like, aggression, hostility, operativeness are not the sudden appearance in him. These are having their roots in the person's *childhood experiences at home.*

The birth order of the child is also having some influence on the child's personality development. The eldest child may be dominating over other children and he may be developing a feeling of superiority which he may try to *exercise in a social group.* The youngest one in the family may be always depending upon their parents for everything as he is not capable of doing anything independently. The *feeling of dependency* later on becomes a *permanent personality trait.* The only child and the over protected child may develop selfishness and also becomes egoist and also may be having *indifferent attitude* towards others.

The individual has to develop the personality which should be quite appropriate to the roles he adapts in his society. There are *various social roles* the child has to play. The roles in the family are determined by the parents. The child's *role in the gang* may cause conflict with his role in the family. Sometimes the child has to develop a negative personality when he finds his roles in the gang are in conflict. Thus the development of personality is greatly influenced by biological and social factors. The individual's personality is a product of the interaction of the biological and social factors.

Theories of Personality

Psychologists who were interested in the study of personality were searching to find out a framework with the view of comprehensing the intricacies of human behaviour as a whole and specifically certain areas fragmentally, resulted in developing certain rational for their purpose. Everyone tried to discuss and analyse the human behaviour and one of the most important areas of study specifically was personality which resulted in evolving and formulating several theories of personality. Every theory tried to describe the structure and function of personality and also assessed in whatever manner that could be assessed. Every theorist had observed several dimensions of personality and enumerated number of qualities or traits and characteristics in his studies. On the basis of these dimensions, traits and behavioural characteristics classifications were made. People are classified into different categories.

Modern psychologists like **Kimble** *et al.* (1985) classified all theories of personality under two broad categories like *descriptive theories* and *developmental theories.* There are also other categories of theories named as dynamic theories, psychoanalytic theories humanistic theories and so on. Every theory tries to define, describe the nature, characteristics and development of personality and suggests the methods and the procedures of assessment of personality. The descriptive theories include the trait theories and the type theories. The traits attributed to human beings are in words of adjectives are so enormous that it will not be possible for us to enumerate them and identity each trait specifically in human beings. Therefore in the categorization several traits are synthesized into several dimensions and whoever possesses a group of certain traits brought under certain types of personality. It appears that in the categorization of types of personality some traits are involved. Therefore, some are of the view that traits are the basic factors used in the process of classifying the people into different types.

The human personality is highly complex. It is easy to state and predict the behaviours of lower animals. But humans cannot be brought under the category of lower animals and predict. Though all human beings seem to have the same physiological structure they do not feel, think and behave in the same manner. The *personality make up* of the individuals would never match with other persons'. Psychologists for long time have been searching for a framework through which the complexity of human personality can be explained. Though they *could not arrive at* a unanimously *acceptable* way of dealing with the complexity of human personality, they were able to evolve the important approaches to describe human personality. There are three approaches mainly used to study personality. They are (1) traits approach, (2) types approach and (3) structure or dynamic approach. Modern psychologists add few more approaches (theories) to study personality.

(1) Traits Approaches to Personality

Individual differences among the people appear in several ways, for example in aptitudes, perceptions, intelligence and physical characteristics. But a more conscious difference can be observed in personality characteristics or traits. Personality traits are considered as the fundamental units of human personality. We have seen clearly in the preceding section that a trait is a stable and permanent characteristic of a person which is exhibited by him consistently in a variety of circumstances. Personality traits indicate the distinctive behaviour of an individual which may be more or less permanent for a particular individual. Personality traits are neat and clear ways of tracing the manifold aspects of behaviour. The traits are the most useful approach to the study of personality.

The study of personality by the trait approach is an attempt to analyse personality of an individual. The results are not used for categorization of individual's behaviour. The trait approach to the study of personality would help to identify and measure the components of personality. The components are the traits. The traits psychologist tries to determine the smallest number which are taken together and characterize as the uniqueness of personality. Personality traits are sometimes closely connected together into a distinctive patterns. Sometimes one trait may be dominant and sometimes several traits may be dominating. **Norman** (1963) went deep into the number and developed a pool of 40,000 traits. Norman reduced these traits dramatically. From the Allport list of traits Norman arranged sets of paired polar-opposite adjectives and conducted an experiment on a group of peers and they were asked to rate the peers whom they knew well on these adjuctive word pairs. He applied factor analysis to the data and extracted five personality traits dimensions such as extroversion, agreeableness, consciousness, emotional stability and culture. Each trait dimension defined through three sets of paired polar-opposite adjectives. For example, for the extroversion trait dimension, the paired polar-opposite adjective are :

Trait Dimension	**Polar Adjectives**
Extroversion	Talkative - Silent
	Frank - Secretive
	Sociable - Reclusive

After sometimes **Passini** and **Norman** (1966) repeated the study on a group of college students and obtain similar result. In the analytic approach of the study of personality the central concept is the *trait* which is the important attribute of the individual behaviour. Personality traits also refer to the important qualities of behaviour that are consistant with the individual's behaviour which are revealed in a variety of situations. Another important study was made by **Cattell** (1965) for classifying the personality traits. It was on the judgments made by people of their associates. He used a list of 35 trait clusters. Since the subjects have expressed overtly these traits, Cattell called these as *surface traits*. On using factor analysis Cattell was able to isolate sixteen basic traits and he named them as source traits.

Having these results as the source information Cattell developed a self-administered personality questionnaire. This was the Cattell's famous sixteen personality factor questionnaire. The personality traits, like the intelligence, are evolved from the individual's performance on a personality test.

Personality traits at the outset should be differentiated. Which are the basic traits and which are real? Similarly which are apparant and which are spacious? A timid man may be rude, because he is self-conscious. Another person may be aggressive because he is fearful. The difference between surface behaviour and the basic behaviour should be revealed. The common saying that appearance is deceptive may be also applicable in the case of some traits. One of the important tasks of psychologists is to define and describe and measure the basic personality traits. All the 18,000 trait adjectives are used sometimes to describe some aspects behaviour. But all of them do not define and differentiate the personality traits. In practice we can observe many of them are referring to the same behaviour pattern. But a basic personality trait may reveal relatively the permanent behaviour which may vary along the trait dimensions from one extreme to another extreme. For example, friendly-hostile, aggressive-complacent, dependent-independent, kind-cruel are trait dimensions. Many of us may be revealing these in some situations. Some of the psychologists made systematic attempts to reduce the unwieldly larger number personality descriptions to a smaller number of fundamental traits. One with which we are very familiar is the "Cattell's sixteen personality factor questionnaire".

The *clinical approach* has been used by some investigators to find out the traits. **Murray** (1938) was one who attempted by using interviews, questionnaires and projective tests and made very extensive study on a small sample of normal persons. Murray and his associates analysed their data in terms of personal needs that were the characteristic of the behaviour of the subjects. Needs were observed as the internal processes which yielded meaning, organization and direction to behaviour that was interacted with the persons' environment. **Maddi** (1963) stated that the needs were the ending characteristic of a person. Murray and his collegue have identified twenty different persons' needs that were determinants of the individual behaviour.

According to Murray, the story in the test is in constant interaction and expressive of two things, going on within the individual. One is *his need* and the other is *the pressures* or the environmental stresses acting upon him. The interaction of a *need* and *a press* is called a *'theme'*. The way in which the theme directs the process of perception is called '*thematic apperception.'* **Morgan and Murray** developed *"The Thematic Apperception Test"* which consisted of several standardised *pictures.* This test is widely used in clinical practices. We shall discuss this test in the following pages.

(2) Types Approach to Personality

The world in which we live consists of many different things and while reacting to them we cannot deal with them accurately. We cannot specify each one of them individually and describe the object and all those with which we are having contact may have certain amount of similarities and differences. Because of certain amount of similarities and little differences we may not be in a position to point out an object correctly. In order to solve this problem we classify objects. Generally we group most of the objects in our everyday experience into broad categories, so that the whole category of objects or things can be handled without any delay and we can easily identify them. The same procedure we are following in dealing with human beings. Psychologists who study human behaviour, classify people on basis of certain characteristics, so that they can easily describe the individuals.

While mentioning about certain persons' behaviour pattern, we say that they are of friendly-hostile types, affection-cold types, frank-secret types, aggressive-complacent types and so on. In making such classifications we ignore certain minor differences and put them into large categories. Classifying people on the basis of types is the oldest and most popular approach in describing personality. Classifying people according to types and describing and personality is having a long tradition. The ancient Greek and Roman physicians based their classification of men into types, upon four body fluids called humours such as blood, phlegm, yellow bile, and black bile. Hippocrates who was considered to be the father of modern medicine, proposed that human personality was the function of these basic body fluids.

Hippocrates believed that *each fluid* was found to be *correlated* with the *particular temperamental characteristics.* Later on **Galan** (130-200 AD), Greek physician, anatomist, and physiologist, codified the existing medical knowledge. His authority continued till the *period of renaissance.* He devised the *temperamental theory* of personality and human conduct. He systematized the vague notions of biological conditions and psychological character types into acceptable fourfold typology of temperament. An excess of anyone of the four humours determine the emotional temperament of the individual.

When *blood* predominates the temperament is sanguine which means that the person is volatile, cheerful and optimistic. The predominance of *phlegm* causes the person to be phlegmtic, and the person is calm, slow and univolved with world. The *yellow bile* causes a person to be choleric temperament which means the person becomes quick tempered and irritable and the *black bile* causes a person to be melancholic temperament and the person becomes depressed, pessimistic, gloomy and dejected (Fig. 23.7).

Later on **Immanuel Kant** (1724-1804), a German philosopher, developed more fully the *classical temperament theory* and he also proposed separate groups of trait descriptions for each of the four temperaments and described the sanguine personality, as one who is careful and full of hope, good natured and sociable.

Therefore, it is believed that a man's temperament is determined by the proportions of his body fluids (humours). If there was any imbalance in these four basic fluids there would be imbalance in basic four characteristics. Anyone of these fluids or two may dominate at a

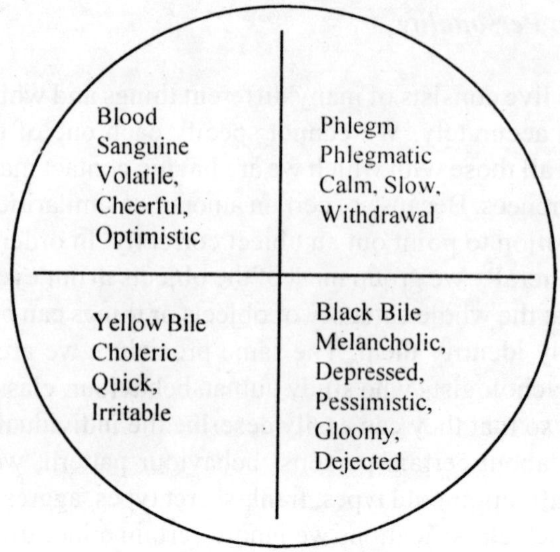

Fig. 23.7 : Galan's Personality Model.
Personality and Human Conduct.

time and the behaviour may be dominated by the humour. **Murphy** (1952) says that modern man still believes that *these four fluids* are causing some of the *basic law of life*. Modern scientists do not fully accept the typology formulated by Hipprocates as the temperament and personality involve many more types than the four types. Human personality characteristics also reflect the environmental and the physiological influences. Present typologies are much more scientific than they were in the past. **Sheldon** and **Stevens** (1942) ascertain the relationship between behavioural characteristics and the quantitative assessments of body structure. **Williams** (1952) believed that there is good relationship between hormones and human behaviour.

Some are of the view that every typology has certain deficiency. One great deficiency is that there is no typology which can provide a rich vocabulary for characterizing individuality. Huge number of characteristics are not found in every typology. For example, Hippocrates categorized person as angry and nothing more about him, whether he is intelligent or dull, sociable, withdrawn, dependents or independent. Similarly in the case other characteristics like, depression, optimism and calmness (Moskowitz and Orgel, 1969).

Psychologists are quite familiar with the famous types of introversion and extroversion. **Carl Jung** (1985-1961) described the personality characteristics with these concepts. Jung states that the *extrovert* is an outgoing exuberant, lively and often inclined towards direct action. The *interovert* is the *opposite of these* behavioural patterns. He is more prone to thoughtful reactions. In this typology the individuals are put into extreme categories which may apply only to a few persons. Most of the people fall in between these two types. Further, this typology did not take into account the most important factors about personality, the multi-dimensional, which consisted of several characteristics.

These defects were rectified in the works of **Eysenck** (1916-1998) a famous German born, English Psychologist. Eysenck, as typologist, proposed a smaller number of personality types. He defined *three major personality dimensions* such as (1) introversion - extroversion, (2) neuroticism-stability and (3) psychotism. During Second World War, Eysenck analyzed more than ten thousand individuals and also neurotic soldiers.

He was able to identify the end points of the introversion-extroversion dimensions. He found that the *introvert* exhibited anxiety, depression, apathy, ruminative thinking and having psychologically physical symptoms. The *extrovert* had poor occupational histories, imaginary physical complaints, sexual difficulties and poor performance of the intelligence test. *Neurticism-stability* is unstable, when emotions are reliable, easily aroused and strong. These persons are moody, touchy, anxious and restless. At the other end of the dimension the individuals are with emotional control. They are also reliable even-tempered and calm (Fig. 23.8) .

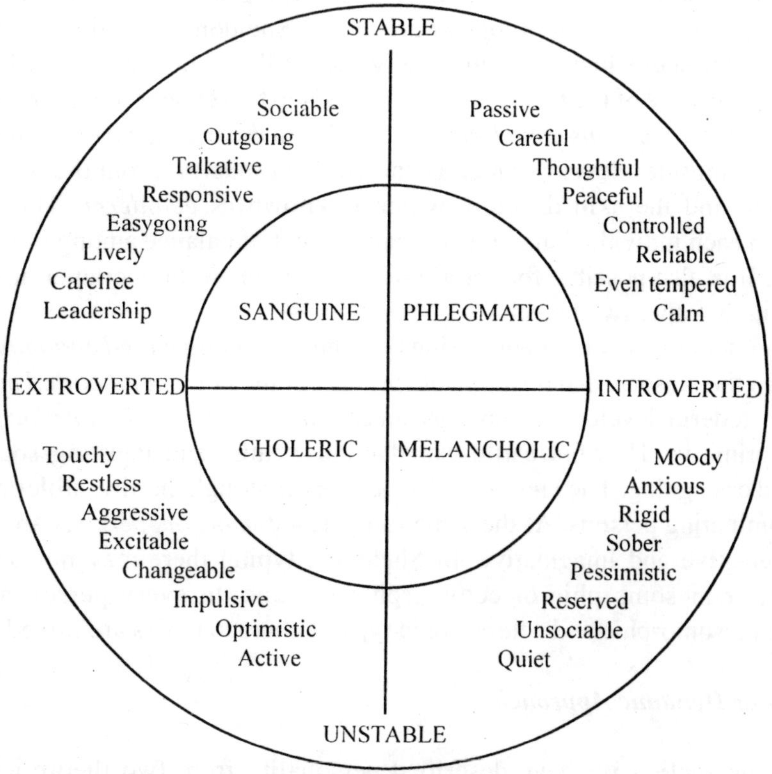

Fig. 23.8 : Personality wheel describing the Traits evolved from the Factor Analysis made by Eysenck based upon Hippocrates' and Galan's Dimensions of Personality.

Personality types are also classified on the basis of attaching *importance to values*. **Allport-Vernon** study of values determine what a person values most, what he is interested in and what he considers to be the maximum important for him. In the study made by

Allport-Vernan there are *six fundamental life areas* such as theoretical, economic, aesthetic, social, political and religious. The assumption is that the *theoretical* man's dominant interest is in determining truth. The economic man's dominant interest is in what is useful mostly, the *aesthetic* man's dominant interest is in *beauty,* the social man's dominant interest is in love of people, the political man's dominant interest is in power over others and the religious man's dominant interest is in understanding of human life in the light of religious and *ethical values.* The Allport Vernan's questionnaire will identify the subject's dominant value-system and accordingly persons are classified into types as economic man, political man and so on. It is also possible that a person may be interested in more than one value area. But it is a question of degree. The degree of theoretical value may vary from large or small through gradations and also the theoretical value may coexist with other strong values.

There are some interesting studies to establish the basic physical types of human beings, their relationship with personality types. An important study was made by Sheldon in collaboration with **Stevens** (1942). The individuals typing were made on the basis of the individual's *anatomical and physiological attributes.* Sheldon stressed the fact that there are three layers of tissues which shape the bodily parts in the embryo and in the later growth. It may occur that anyone of the layers is likely to develop to a greater or lesser degree. From the *inner layer the vital organs endoderm develop.* From the *middle layer the bones tendon* or sinew and the muscles develop which is known as *mesoderm.* From the *outer layer* the nervous system and the skin develop, which is known as *ectoderm.* The anatomical development in each individual maintain certain amount of balance among the three layers of the tissues. It is also possible for certain over development in one layer and an under-development in the other two layers.

From the endoderm development, springs out the *body build called endomorphy.* In the same way from the mesoderm development springs out the *body build called mesomorphy* and from the ectoderm development springs out *ectomorphy.* The *endomorphic persons* are dependent on primitive. They find earthly satisfaction of eating and digesting, socially warm, impulsive, and easy going. The *mesomorphic persons* are tough, hardly, athletic, noisy and find joy in dominating persons. At the same time the *ectomorphic persons* are thoughtful, withdrawn, sensitive and imaginative. In Sheldon's typing there may not be *extremely* endomorphic, or mesomorphic or ectomorphic persons. In every person a degree of endomorphic, mesomorphic and ectomorphic type of characteristics are mixed together.

(3) Structure or Dynamic Approach

In the preceding section we have described personality from two theoretical point of view, the trait theory and the type theory. Each theory views personality as a finished product. Personality was described in terms of quality. Psychologists and psychiatrists view that personality as *unique finished product.* It is a unity and it is more than the sum of its parts. The psychologists are of the view that traits can be understood only in the light of the whole structure. Persons are described cross-sectionally in terms of qualities of behaviour.

The two approaches, trait and type are not opposed to each other, but they are relatively complementary to each other. An individual's personality can be described only in terms of his *dominant traits.* The dynamic theories view personality as a process. These theories observe on the interactions between the individual and the environment and on the antecedent causes of the individuals behaviour. The functional expressions of the particular personality characteristics are the interesting subject matters for the dynamic psychologists. *The primary interest* of the dynamic psychologists is *"how it is happening and why it is like that".* For them behaviour is the product of *interaction between the individual* or the organism and *the environment.* The interaction is a continuing process.

One of the dynamic approaches puts more emphasis on the role of learning in the development of behaviour. Other approaches put more importance on the processes like perception, motivation and heredity. In applying dynamic theory to study personality the differences can be readily seen. This is because the behaviour of an individual is not repeatable since there are many causes operating at the same time. One investigator may see an individual more aggressive and another may see an individual very sensuous still another sees an individual as very intelligent. Hence there is a great deal of diversity as in the case of trait and type theories.

Psychoanalysis is considered as one of the *dynamic theories.* **Sigmund Freud** (1856-1939) is the founder of the school of psycho-analysis. Psychoanalysis was evolved from clinical practices and not from laboratory studies. The trait theory tries to find out stable dimensions of personality by studying groups of people. The information collected were from self-report, whereas psycho-analysis is an in depth study of the individual personality. According psychoanalysis *motivation* is totally *unconscious.* Freud's personality consisted of three major components or systems such as the *id, the ego* and *super-ego.* All these three systems function independently, but at the some time they are interacting to express behaviour of the individual.

The Id : According to Freud, the id is the original source of personality, from which later on the *ego* and *super ego* are developed. The id consisted of all the *primitive* and all *biologically derived drives* and impulses with which the *human life starts.* The primitive drives are the *instincts* and they demand *immediate gratification.* It is a pleasure seeking instinct. The id like a new born child operates on *pleasure principle.* It *avoids pain* and gets pleasure, irrespective of external conditions. Freud believed that the *id forms the core of personality.* It is the raw material and it is the *source of an mental energy.*

The id is the *oldest component of personality.* It is born with child. The rest of the aspects of personality values, conscious thought, fantasies etc., develop gradually. Freud included in id all that we call biological drives and gave important position to sexuality. The guiding principle of id is the pleasure principle, specially the gratification of *erogenous, sexual instinct* which is named as *libide.* Freud placed aggression only next to sexual instinct. Formerly Freud believed that aggressive instinct is derived from *frustrated libido.* But later on he asserted that *aggression* also like, sexual instinct is an *in born* characteristic of human nature.

Freud affirmed that man is basically motivated by two opposing tendencies such as the sexualized *pleasure oriented* and *life related* instinct called *'eros'* and aggressive-destruction oriented **death-seeking** instincts called **'thanatos'.** The id knows no judgement of value, no good and bad, right and wrong and no morality. The quantitative factor which is linked to *pleasure principle dominates in all its activities.* The id in its striving for satisfaction does not show any regard for goodness, evilness, realness and appropriateness of the objects. Whatever brings satisfaction is good. The id is an unchanging powerful force.

The Ego : The id, is facing continued frustration when the pleasure is not gratified and therefore the id learns more realistic and rational thinking and develops perception. This is from the *primary process of thinking,* hence a secondary process becomes the greater content of the conscious and preconscious mind. This provides basis for the development of the ego.

Thus *the ego develops only out the id.* The ego differentiates itself from the id and becomes specialized in response to the external world. The ego serves to mediate between the primitive instincts of the id and the harsh realities of the world. The ego though derived from the id, not like the id, operates more on the basis of the reality principles. The ego functions to get control over the expressions of the id instincts. The ego decides whether the instincts of the id to be satisfied immediately as it demands or the satisfaction should be postponed for the time being. The development of the ego as organized part of the id, the organism increases in the degree of sophistication with which it meets the life tasks. The *pleasure principle is removed by the reality principles* and the *ego provides safety* and suitable *pleasure for the id.* Thus the *ego stands for reason* and *good sense* while and *id stands for irrational* and *untamed passion* (Freud, 1953).

The Super Ego : In the secondary process, the ego is mediated between the id and reality. But gradually a *third source of conflict develops.* The child develops *a system of values and taboos,* which is called the *consciousness.* The conscience is composed of all the *learned motives.* Freud called this as *superego.* The components of superego are acquired before the ego is fully developed and therefore it has *irrational and non-verbal qualities like the id.* Many of the qualities are unconscious like the id instincts. The superego emerges between the ages of three and six years. The superego judges whether the action is right or wrong according to society norm. The id seeks only pleasure, the ego tests the reality by having contacts with outside world and the superego makes efforts for perfection.

The *superego* suppresses *all sexual and aggressive impulses.* A second part of the superego is the *formation of the ego-ideal,* which is *an image of what the individual ideally can be.* The ego-ideal pushes the ego in the direction of moral action and self-sacrifice and *urges the ego to strive for perfection.* In this way the *ego part of the personality serves three masters,* the pleasure-seeking id, the reality world the ego ideal and the moralistic superego. In this way the main functions of the superego are to inhibit the social prohibits such as the sex and aggression, to urging the ego to substitute moralistic actions for realistic ones and to make efforts for perfection.

Development of Personality : Psycho-Sexual Stages

Personality, according to Freud, is a course of development that takes place right from the period of early infancy through *oral stage, anal stage, genital stage,* and *dedipus complex.* Freud believed that personality develops as a result of what recurs at certain fixed stages during the first five years of life. The individual, according to Freud, has a certain amount of *psychic energy,* which is called *libido,* which is devoted for a number of different interests and enterprises. The individual differences in the adult personality are the reflections of the manner in which the individual coped with the conflicts that might have happened during the stages of psychosexual development.

Psychosexual development forms the history of various activities and objects to which the libido is attached itself. According to psychoanalytic theory the *psychosexual development takes* place through the stages of oral, anal, phallic, latency period and genital. Each stage has a focal bodily point for stimulation, that is, in the *oral stage the mouth plays a central role* and specific external objects for sexual attachment. According to Freud we are born with two basic instincts (motives). The first one is **Eros** (life) and Freud described it as *motive force* behind all our positive strivings and behaviour. The *sex drive* (libido) is the *primary part of Eros,* but lesser then the drives such as hunger and thirst. The second basic instinct is the *thanatos* (death) represents all other destructive motives such as aggression and hostility. *In a normal healthy person Eros dominates* and *thanatos is simply present.*

The child's basic personality is formed by the end of fifth or sixth year of life. The sex drive is an important part of Eros and therefore an understanding about the stages of libidinal development in the child is necessary for a knowledge of personality. But this view has caused displeasure in many people during 1900.

Oral Stage

As soon as the child is born, the libido is diffused and spread out and unlocalized in the infant. The child now enters the *first stage* of development which is called the **oral stage.** In the oral stage the *libido is localized* in the *region of mouth.* The child enjoys the pleasurable satisfaction through sucking the milk, the mouth, lips and tongue are manipulated. In the *later stage* of the oral-period, the child gets teeth and *satisfaction was obtained through biting.* Thus the mouth is the first sexual or *erogenous zone* and the child keeps its sensitivity. The child *finds the pleasure of holding something in the mouth.* The child drives this pleasure from the act of kissing. Next, the child finds these pleasures by biting something as an element of gratification and also chewing food, gumming. The normal infant goes through this stage with an amount of sexual sensitivity in the oral stage.

Anal Stage

This is *the second stage* of psycho-sexual development. The infant *derives pleasure* from the *process of elimination.* The libidinal energy is *feces in the organs of excretion.* For the *first ten months* the elimination process is automatic. The infant is unaware of it. Parents at this stage initiate the toilet training. In the *later anal stage* the child resists *toilet training* and finds a different pleasure in withholding his feces and even get constipated.

Genital Stage and Oedipus Complex

The third stage of psycho-sexual development is the *genital stage.* It is at about *four years of age.* The child's libidinal energy is directed towards genital activities. This stage is called *phallic period.* It was named for the phallus or *male sex organ.* When the libido is localized in the sexual organs the child tries to seek a member of the opposite sex in an object of affection. The most easily available sexual partner for the boy of this age is his mother, who was gratifying all his needs. *At this stage the father* is looked upon as a rival and he is *fearing on seeing him* that he is a potential castractor who will remove his penis and make him a girl. During this period the *male child develops the Oedipus complex* and develops a desire to have sexual relation with his mother. Freud derived this idea from *Greek mythology.* Oedipus, the king of Thebes slew his father and married his mother. Freud thought that the early sexual desires of the body are towards his mother, but rejects it out of fear for his father.

At the *age six* the boy reaches the *period of latency.* The boy's *sexual interest is suppressed* until the period of *puberty.* At the puberty period there is a sexual reawakening in the boy and the real genital period of *libido is traced.* This period continues and ends in the satisfaction of marriage. Freud believed that the normal development of Oedipus complex problem is solved through the *mechanisms of sublimation and identification.* The child uses the mechanisms of sublimation and represses the lust for the mother and develops affection for her.

The boy unconsciously puts his libido energy into other activities like play, sports or other recreational competitive activities. At the same time the *boy finds some excellence in his father* and *tries to identify himself with his father,* which Freud called *identification* and the boy finds vicarious pleasure.

Freud was able to find out an equally comparable complex for the girl. The little girl has a comparable *Electra complex* and has to resolve. In *Greek mythology, Electra* is said to be the daughter of Agamemen and Clytemnestre. Electra persuaded her brother to slay their mother and mother's lover for killing their father, because Electra desired to have sexual relations with her father, Agamemen, who was killed by her mother's lover.

The little girl enters this stage and becomes highly devoted to her father on finding out that he has a penis. She blames her mother for not having a penis and *becomes rebellious* and *develops intense feelings of fear and guilt.* She wishes to settle for a baby from her father in place of a penis and focused the sexual instinct upon her father and *death wishes towards her mother.* The genital stage is the final stage of development, because of this stage the child is able to transfer his/her original feeling towards the opposite sex parent for a suitable hero-sexual object. The genital stage is achieved only when the oral, anal and phallic stages have normally progressed. According Freud the *genital stage is quite similar to emotional maturity.*

Evaluation of Freudian Theory

Psychoanalysis as a system of psychology has been accepted in the first quarter of twentieth century. It was not only a *system of psychology,* but also a *therapic technique* to deal with emotional disturbances, hysteria, neurosis, psychoneuroses, conduct disorder, behaviour disorder and various abnormal behaviour patterns. Though the valid of the services rendered by psycho-analysis is questioned at some stage, Freud's psycho-analystic theory and his works were considered as the revolutionary approach to deal with human problems. Freud's approach is dynamic in its nature. Because of this, his theory has spread out for beyond the fields of psychology, psychiatry, sociology and literature and also touches all aspects of human life. Freud's insight into the human mind, the unconscious has never been compared to any other work. Freud strongly stressed the idea that childhood experiences can influence adult person's life. He recognized that the sexual problems are often involved in personality development, maladjustment and mental disorders.

Contemporary Freudian, new-Freudians and psychologists find the psychoanalytic vocabulary is rich and meaningful. The concept *unconscious, the ego, repression, anxiety, sublimation, symbolization, regression, projection, ego strength, Oedipus complex, Electra complex, genital stages, phallic stage* all signify important dimensions of personality. Freudians find that the unconscious is a very useful explanation of behaviour. Freudians state that many aspects of the adult personality are shaped in early childhood period. But the critics say that the Freudian concepts are *fuzzy and unwieldy.* The words are difficult to define and they are limited in their predictive capacity.

After Freud, a group of followers of Freud, accepting some aspects of the core of Freud's psycho-analysis and developed psycho-analytic theory independently. Each one developed a system of his own. They were called as *neo-Freudians.* Neo-Freudian is a term applied to the generation of psychoanalysts who came after Freud's theory was well established and who modified the theory to include social influences. The no-Freudians are of the view that experiences of adolescence and early adulthood are crucial periods and we cannot consistently determine the behaviour of the individual, without experiments. But contemporary psychologists accept that modern man has become aware of his unconscious and also aware of the value of rational and cognitive factors in shaping his own behaviour. They say that psychology should *now reconstruct the conscious mind in the light of Freudian thought.*

With all the objections raised by modern psychologists, psycho-analytic theory had enormous impact upon the science of personality. Effort are being made to explain the theory as a testable concept that can be verified in a laboratory situation. It was Freud's emphasis on sexual factors which brought awareness of their role in adjustment problems and created the atmosphere to scientific study of sexuality. Many are of the view that Freud overemphasized the role of sex in human motivation. It was the Victorian period, during which period sexual matters were restricted. Hence, many of Freud's patients, out of guilty sexual feelings, had conflicts, and problems centred round sexual desires. But now guilty sexual feelings become less, but the cases of mental illness are the same. This shows that sexual conflicts are the causes of emotional and personality disturbances. Many of the lasting contributions are attributed to Freud's outstanding works. According Freud human behaviour is largely irrational and man can never achieve control over himself by observing his own conscious experience and much of human behaviour is illogical and non-purposive. Freud emphasized *the historical antecedents of behaviour* and wanted to put these in the *naturalistic context.* Freud wanted to put an end for dualism and for the supernaturalism and wished to place human behaviour in the world of physical sciences. Freud has strong view that *sexuality* is an important determinant of *human behaviour* which he also demonstrated it. Freud was of the view that there is no qualitative difference between normal and abnormal behaviour patterns.

Freud in his later writings has replied to the criticisms levelled against his theory. He strongly stressed that his psycho-analysis is not completely *devoid of social aspects.* Further, Freud's view was that the childhood experiences can influence the adult person's life very much. This may be observed that *sexual difficulties are often involved in the personality maladjustment* and *disorders in adult life.* Modern psychiatry, psychology, psychotherapy and clinical psychology place the behaviour disorder on the naturalistic line, because Freud was of the view that behaviour disorder of this type can be modified and the maladjustment of individuals can be improved and the normal way of life can be revoked in such persons.

(4) Other Psycho-analytic Theories of Personality

We have discussed above, Freud's theory of personality in terms of *structure and development.* Though psychologists named it as psychosexual theory of personality, the individual in the normal course of development goes through a number of stages which involve the history of various activities and objects in which the libido (psychic energy) is invested. Psychologists classified Freud's theory of personality under *dynamic theories,* since it tends to view *personality as a process.* The dynamic personalogists are primarily interested in the 'How' and 'Why'. Freud had many followers and all of them made contributions for the development of psychoanalytic theory. Some of his early important followers were Carl Jung Alfred Adler, and Karen Horney.

Carl Jung

Freud's important work, *'Interpretation of Dreams'* was appeared in 1900. On reading this work, Jung (1875-1961) became the most outspoken supporter of Freud. In his strong support to Freud, Jung was even ready to relinquish his academic career. After sometime Jung had difference of opinion with Freud in certain points. Jung disagreed Freud's sexual basis of neuroses and the concept of unconscious. Jung believed that repression may lead to the forgetting of any frightening event and such forgotten experiences and expressed thoughts form the complexes in the individual. Jung called this as *personal unconscious.* Jung was also interested in myth and mysticism. According to Jung there is also an *impersonal collective unconscious,* which is the *image of humanity.* This is an *inherited archetype* which is the mythical ideas of the past and repeated experiences of human kind. *Archetypes are recurrent impressions* made by subjective reactions.

Jung's collective unconscious consists of all the memories and patterns of behaviour *inherited from man's ancestors.* All human beings have the collective unconscious which is *repeated life events and experiences of many generations.* It is different from the personal experiences of the present life. Jung's writings are influenced by philosophy, religion, more than psychology and psychiatry. *He was the first to distinguish between introverts and extroverts type of personality.* He stressed the importance of man's goals, aspirations, achievement, maturity and *self-actualization.* **Abraham Maslow** (1908-1970) one of the founders of the humanistic psychology, elaborated the concept 'Self-actualization', given by Jung. A number of modifications have been proposed by later psycho-analysts in Freud's theory. **Jung** (1968) started his own school of psycho-analysis which was known as *Analytical psychology.*

Alfred Adler

The neo-Freudian's criticisms of Freudian theory was that Freud neglected the *social influences* and placed two much of emphasis upon the *instinctive and biological aspects* of personality. He failed to recognize that people are the *products of society* in which they live. These neo-Frendians, Alfred Adler, Karen Horney, Erich Fromm, Harry Stack Sullivan, and Erik Erikson say that personality is shaped much more by the people, and society and culture than by instincts. **Alfred Adler** (1870-1937) an early follower of Freud and one of the important contributors to psycho-analysis did not agree with Freud's notion of the *infentile sexuality* and the *unconscious* and departed from Freud and formed his own school of psycho-analysis.

Adler was of the view that it was the *sense of inferiority that underlie maladjusment* and not repressed sexuality. It was the influence of Adler's early life history that made him to disagree with Freud over the issue of the *central motive force* in human personality. **Adler** (1927) wished to have a new model of personality which should emphasis the fundamental human strivings. All children are born with a deep sense of inferiority because of their small physical size, weakness and dependence upon the adults. He used the term *inferiority complex* to identify this feeling of the child. An individual's sense of helplessness, or powerlessness in the face of death, physical inferiorities or damaged or diseased organs that prevent successful competitions with peers, all make him that he is inferior. Always the *goal of children is to develop superiority to accomplish self-perfection.* One person tries to become superior by developing his intellect, another by improving his athletic skills, a third tries become a painter. In all these cases daily routines, relationships and social activities are involved to achieve the ends.

Adler is of the view that *people have creative self,* with which they strive hard and with full awareness to develop potential skills to overcome obstacles. In this way people *create their own personalities.* People have social interest and social responsibilities towards others. They do not feel happy unless they are actively engaged in some activities and make improvements in life.

Adler uses the phrase *'individual psychology'* to identify his system. He emphasized the individual's nature of goal strives and the *innate creativity* of human psychological adaptations and the wholeness of the individual's unified personality. The individual's *style of life* is the product of a *creative personality trend to overcome* his unique sense of inferiority. Adler developed a *typology of personality* organized on the principles of activity level and degree of a social interest. He described the personality types such as *ruling dominant* (choleric), *getting leading* (phlegmatic) *avoiding* (melancholic) and the *socially useful* (sanguine).

Karen Horney

The neo-Freudians, having the theoretical assumptions of the psycho-analytical theory as the basic and their personal clinical observations as the personal contribution to psychoanalysis, developed their own school of psychoanalysis and formulated theories of personality. **Karen Horney** (1885-1952) developed psychoanalytically oriented *interpersonal theory of personality* and the neurotic conflict or the neurotic personality. Horney based new personality development and its distortion by neurotic strivings on the basic emotions, the *basic anxiety* and the *basic hostility.* If the parental attitude towards the child is cold, *abusive and inconsistent,* it lays foundation for the child to develop a sense of basic hostility towards the parents and toward the people in general.

The child cannot openly express his anger or hostility for the fear of the powerful parents and therefore, *he represses his feelings, the hostility* that is boiling below the surface, the child develops an imposing and pleasing loveableness. But unconsciously he fears of his potential chaos of his own helplessness, the basic anxiety shapes his dealings with significant people in his life. The child's personality development is blocked from its natural direction of growth. The neurotically disposed child begins to develop compulsive, indiscriminate and unrealistic needs for controlling himself and others. Horney, therefore, grouped *ten needs into three character patterns* moving towards, away from or against others.

Horney's account of personality is to understand the neurotic strivings for safety and controls over his world. **Horney** (1942) postulated the *neurotic personality* which is governed by one or more of the ten needs or trends. Each of these needs or trends is directed towards interpersonal control and coping. The ten neurotic needs are, the need for affection and approval, the need for a partner who will takeover oneself, the need for one's life, the need for power, the need for getting better of others, the need for social recognition and prestige, the need for personal admiration, the need for personal achievement, the need for self-sufficiency and independence and the need for perfection. There seems to be overlapping of certain needs, and also similarities.

The individual develops an idealized self-image which is composed of many things. Horney speaks of three selves, the *real self*, the *ideal self* and the *actual self*. Horney though did not specify her system of psychology, it constitutes a *truely holistic approach* to psychoanalysis, that is the *whole person in a social environment*. Horney did not accept Freud's concepts like penis envy, overemphasis of sex, oedipus and electra complexes (1925) and also rejected Freud's scheme of personality built, around the id, ego and super ego.

(5) Ego Psychology

One of the most important developments in psychoanalytic theory after Freud's death has been a new theory called *Psychoanalytic Ego Psychology*. The founder of this system of psychology is **Erik Erikson** (1902). Erikson was born in Germany to Denish parents and became an American citizen in 1939. He mainly emphasized the relationship between the individual and the society in which he lives. Erikson also emphasized the importance of personality development of the child, and its interaction with the social environment. In 1968 Erikson brought out his famous essay, "Womenhood and the Inner Space". Erikson tried to blend biological, cultural and psychological interpretations of his findings. He asserted in his purely biological interpretation that preadolescents are overtly concerned with their bodies and their developing sexual parts. Therefore, it is a natural consequence that play activities reflect their structure of the sexual equipments.

Erikson restricted the social interpretations to the impact of learned attitudes and performances. Thus according Erikson boys learn to enjoy the outdoors, girls the indoors. Boys learn to seek adventure and excitements, girls lean to seek the family love and home making skills. In psychological (Psychoanalystic) view, he emphasized the unconsciously symbolic implications of the child's construction of play space. Erikson accepted Freud's conception of psychosexual development, but made small modification in it.

Some of the neo-Freudians did not accept that the ego emerged from the id. The ego has its own origin and course of development. They provided the ego a *conflict-free sphere* in which it uses its cognitive process of perception, learning, memory and attention. Its *satisfactions* are exploration, manipulation and skilled tasks. This idea of ego shows that the *ego is autonomous*. Freud has given primary importances to id, where the ego psychologists gave importance to ego. Erikson declared eight stages theory of psychosocial development (1963, 1964). He was best known for this.

Erikson believed that in each stage the individual passes through some major crisis. This must be resolved if a healthy development is taken place. At each of these points of conflict, society provides the developing individual a specific task and particular component of personality faces the crisis and finds a solution to the crisis. Erikson knows that the conflict resolved in the early stage may reappear at a later stage of development and this has to be again resolved. But a *conflict unresolved* in the early stage can be *resolved successfully* at a later stage. Erikson does not want to give a restricted definition of ego identity. It was his personal and clinical experience that have shaped the formation of the concept, ego identity and ego psychology.

The eight stages of personality development according to Erikson range from first year to old age. The first stage taken place in the first year of life. In this stage the child develops basic trust *versus* basic mistrust, children develop a sense of being loved. The children relay in and trust the parents. If the child's world is chaotic the parent's affection cannot be accounted for and the child develops a sense of mistrust. The second stage passes through two to three years of life, the attitude is autonomy *versus* shame. The child wants freedom of movement and action and when it is hintered it results in shame. The third stage is from four to five years. The attitude of the child is initiative *versus* guilt. The child wants to explore the situation and if the child is prevented, the resulting attitude is guilt. The forth stage is from six to eleven years of life. The attitude, the child develops is industry *versus* inferiority. In this stages children widen their skills and wanted to be accomplished through industry and expect, praise and reward and when there is no chance for this, the child develop inferiority complex.

The fifth stage is from the age of twelve to eighteen years. The attitudes are identity *versus* confusion. It is the period of adolescence and is expected to increase the play activities to create impression on others and also, identify with excellence. But if there is any danger, the result is confusion. The sixth stage is young adulthood. The corresponding attitudes are intimacy *versus* isolation. The person forms close association with persons of both sexes. If a danger is posed, then the resulting attitude is isolation. The person develops disappointment or hostility. The seventh stage is the middle adulthood age. The corresponding attitudes are generativity *versus* stagnation. The person expresses concern with his own children and also with future generations. When this is not possible the person shows no care to people and life is without movement and it is dull. The eight stage is old age period. The corresponding attitudes are integrity *versus* despair. Everyone will take stock of what they have achieved and the type of person he has been. The person looks back in his life with a sense of integrity and satisfaction or with a sense of despairs. Thus Erikson classified the span of life with eight stages and attributes. The characteristics a person possibly could develop both in favourable and unfavourable life situations. Erikson, while promulgating these eight stages theory of psychosocial development also assigned particular components of personality. *Time and appropriate experiences* have more positive changes for the change of personality components.

Erich Fromm : Healthy Personality

The rise of *modern capitalism* resulted in an *expansion of man's personal freedom.* Erich Fromm (1900-1980) found psychoanalysis, a rich fertile field of concepts by which the human behaviour can be understood. Fromm sought the broad psychanalystic instinct theory with a conception of man as a social being in a world of interacting organisms. Erich Fromm, another neo-Freudian, added social emphasis to Freud's psychoanalysis. He had the same background as Horney. A German born, on completing his analytical training at the Berlin Psychanalystic Institute, where Horney was on the staff, went to America in 1933 and started practicing and teaching. He obtained his doctorate at the age twenty two and at the age thirty three he became the authority in Chicago Psychoanalytic Institutes.

Fromm focused his psychanalystic theory to the conflict between the individual and his culture. His theory is mainly on the analysis of *human freedom.* For Fromm, human being is both good and evil in proportion to how well a society permits him to satisfy his biological and human needs. Sex, he considered as an important drive, but is the centre of everything. For Formm man is both an animal and a human being. The body of man possesses all the animal motives and drives.

Fromm (1941) believed that to attain the desired status of strength or power man may seek to fuse himself with another, somebody outside of the self. In this effort, in the extreme form man becomes submissive, passive and masochistic. Feeling powerless, inferior, the individual struggles to escape from his own freedom. In his struggle he submits to the power of others with belittling. An opposite authoritarian mechanism is seen in those who attempt to become the authority, rather than submitting to authority. A strong sadistic desire, in place of masochism to obtain absolute power over others, becomes the way of escape. This type of authorization desire to have others dependent on him makes him to exploit others ruthlessly and makes others to suffer physical pain. The masochistic and sadistic forms and authoritarian solves the problem of freedom by destroying the individual self.

Fromm attempted to make *character structure,* that is the nature of human nature. For Formm, the *most potent danger* is that *future will* become *feelingless, loveless* and *unthinking robot* (1955). Acquisition, consumption and material accumulation become *man's loyalties.* Therefore, the pressing question is a humanized psychology, which may ask, what is man's true nature? Fromm (1968, 1947, 1964) suggested that the most basic trait of human nature is a *tendency towards self-actualization* (1940). It is important to understand his unpleasant situation if we want to know man's true nature. Formm defined a man's true nature in terms of six needs. The first need is the need for relatedness and unity. For him love is a union with somebody. The second need is the need for transcendence and a sense of effectiveness. Man needs to surpass his own fear. By actively striving to master his world he accomplishes this. The third need is the need for Rootedness. Man through out his life has a need for roots, sense of stability and permanency. The fourth need is the need for a sense of identity. Man must control his own fate. He must say, 'It is I' who make decisions. The fifth need is the need for a frame of orientation and an object of devotion. It is necessary for maintaining the sanity. Men need an objects of devotion, a goal, or God. The sixth need is the need for excitation and stimulation. Human beings are continuously in action even if there are reduced stimulation. Man needs interest provoking environment. He needs stimulation that can induce in him to actively participate in life. In this way Fromm described a healthy personality without a mask.

(6) The Humanistic Approach

The humanistic approach to the study of personality involves several theories that share a common view on *human potentialities* although vary in several aspects. Humanistic psychology has emerged as a *vigorous third force* in psychology. passionately expressing its protest against the limited scope of two other major schools of psychology the psychoanalysis and the behaviourism.

The humanistic tradition is of the view that there is no person in this world who has achieved complete psychological health. It is also possible that, indeed, no person can claim that he approximate the total personality integration. Most of the people assumed to be only relatively healthy. Personality theorists put more emphasis on abnormal personality. The philosophical position of humanistic tradition places man, the human person at the centre of its study. In American psychology it is identified as an alternative to psychoanalysis and learning theory (behaviourism). The most prominent persons associated with the humanistic movement are **Carl Rogers** (1902) and **Abraham Maslow** (1890-1970).

The humanistic theorists stress man's positive nature, man's stretch toward growth and self-actualization. Rogers stresses that humans react to the world as they perceive it. Human perceptions of the world, the environment depends on things potentiality available to human's awareness at any moment. The most important concept in Roger's theory of personality is the self. According to him the self consists of all the ideas, perceptions and values. Roger's theory of personality is called *self-theory and person centred theory.* The well adjusted person has a self-concept which is consistent with his thought and behaviour. The self is *not rigid but flexible,* so that it can change. There is also an ideal self in Roger's theory. This is similar to Freud's ego-ideal. The more closer of the two solves the more needs fulfilled and the individual is happy. Any discrepancy between the two results in unhappiness.

The organism seeks to fulfil its potential with its limit as it grows. There is a tendency towards the fulfilment and towards actualization. The basic force which is motivating the organism, is the *self-activilization.* The innate tendency towards self-actualization often meets with conflict by two needs, the need for *positive regard* and the need for *self-regard.* According to Rogers the positive need is universal. This is with regard to love others and to be loved by others. The self-regard is internal, develops later on. This is like the *superego.*

Abraham Maslow's theory of *Healthy Personality* strongly connected with his theory of man's intrinsic nature and potential for growth. Man persues life in the context of an *intinction need hierarchy* which covers full spectrum of human desire and potential. The hunger driven man seeks food to human joy in the experience of aesthetic ecstasy. The basic needs for physiological satisfaction, safety, love, self-esteem and self-actualization emerge in an ordered sequence as each in turn is gratified. The human potential is viewed from a concept of *progressive need gratification.* The acceptance of need gratification theory rests upon the inner core of personality.

The hierarchy of needs progress from the basic biological needs through the psychological needs and result in the need for self-actualizaton. Self-actualization, for Maslow, means the development of the full individuality involving all aspects of the personality that function hormoniously.

Maslow formulated his theory of self-actualization by carefully observing the personality characteristics of eminent persons who made-achievements in several fields. He took samples from the historical personalities for self-actualization, both men and women. Some of the personalities whom Maslow set examples for his self-actualizing are, Abraham Lincoln, Thomas Jefferson, Jane Adams, William James and so on. The important characteristics of self-actualization are perceiving the reality efficiently and tolerating uncertainty, accepting themselves and others for what they are, spontaneous in thought and *behaviour problem centred* and not self-centred, good sense of humour, creative, appreciating basic experiences of life, setting up happy interpersonal relationship with few viewing life from objective point of view.

According to Maslow, self-actualization is the peak of experience. It is one of happiness and fulfilment. In his study Maslow interviewed more than three thousand persons. Out of which he discovered to have potential for self-actualization only in a handful of persons. At the end, he found only one person who was self-actualized in the fullest sense. Maslow developed a schematic picture of self-actualizing personality. Maslow's self-actualizing pattern consisted of *fifteen characteristic of positive* and *five negative traits.* Self-actualizing persons are characterized by *independence and self trust.* They are accepting others. They are the persons in whom observed the basic needs for food, shelter and intimacy with an opposite sex partner satisfactorily. Self-actualizers function in response to higher needs. Beauty, Truth, Justice and other virtues are the core of the self-actualizer's existence. Man strives not merely for survival, but also for a meaningful existence out of an innate need to the whole.

24

PERSONALITY ASSESSMENT

What is Personality Assessment?

In the preceding chepters we have dealt with personality by defining and describing personality, personality traits, theories and types of personality. The problem we are now facing is how these aspects of personality are assessed in an acceptable manner. Assessment in other words also may mean measurement. The general meaning of assess means evaluated value and assessment means evaluation. Hence personality assessment may be construed as the process of evaluating the value of personality and also personality traits. In the evaluating process since some assessing instruments or tests are used and *numerical* values are assigned to the responses of test results. Some psychologists name this process as the measurement of personality. Just like the intelligence measurement in the case of personality assessment also the responses or the ratings are converted into numerical values.

Therefore in most of the personality assessments, qualities or the traits of personality are judged in terms of values. In personality assessment the important purpose is *describing the individual's characteristics* or the behavioural patterns. Personality assessment is made for various purposes. Any study of some subject matter should be based upon some purpose. All the personality assessments after the study, classify people into various categories on the basis of the traits traced in the study or on the basis of scores.

We have discussed in the earlier sections about various theories of personality which are based upon the characteristics and the behaviour patterns of the people. The categorizations of people as dominant-submissive, aggressive, introvert or extrovert are made only on the basis of the traits or characteristics that are assessed by using some instruments or tests or techniques. For assessing the personality characteristics psychologists use various kinds of techniques and procedures.

Assessing the characteristics of the individuals by using various tools and techniques and describing them have certain theoretical and practical reasons. There are certain assessments of the individuals for some general purposes and also for some specific purposes. From practical points of view we need certain type of persons for some specific purposes.

Governments, public and private organizations need men of various types for performing intellectual and physical works of various types. They need individuals with special skills and talents. Therefore, they advertise in newspapers specifying the required general and specific qualifications and other requirements. They receive applications from individuals in response to the advertisements. They are all supposed to be qualified according to the specifications. But still governments and organizations conduct examinations and judge the standard of knowledge of the individuals. All those who are found to be qualified in the written test are called for an *interview* which is in other words called as the *personality test.*

The Government of India through the Union Public Service Commission recruits candidates for various services like the IAS, IFS, IPS and other services. They conduct the competitive examinations and the successful candidates are called for personality test which is conducted in an interview. Similarly, the state governments conduct the personality test. What they do in the personality test? They only test the candidate's suitability for the specific post. They test the candidate's level of intellectual ability in general and the knowledge and the skills and talents specifically.

The same type of procedures are adapted by the business and industrial organizations in selecting individuals for their purposes. In defense service the recruitment procedures are followed very systematically. In all these areas after selecting the candidates they are put in the training programmes which may vary from six months to two years and the candidates are expected to exercise their full mental and physical abilities to get mastery over the performance of the job for which they are selected.

In all these processes right from the beginning till the individual gets into the job independently the personality traits and other characteristics are carefully assessed. In all these cases the services of the psychologists are made best use of in assessing the personality characteristics and the behavioural patterns.

Apart from these areas, the social psychologists assess the effects of culture, economic and the social factors upon the development of personality characteristics of the individuals and their type of reactions to other groups. So also the clinical psychologist is very much concerned with the personality functioning of the individuals for diagnosing and solving their problems. It is to be known that some personality tests systematically get at the personality as a whole and not at any specific traits or behaviour. The idea behind this view is that in an activity the individual puts himself in action and reveals what he is capable of. The individual's potentiality is made to manifest in various type of personality characteristics which may be assessed by using suitable test materials.

Methods of Personality Test Construction

Personality assessment was first attempted in England by Sir Francis Galton at a smaller level by some method. The study of personality in the beginning was merely describing the observable characteristics and the same were attributed to others, when similar traits were observed. A systematic psychologically oriented personality test was designed by the American psychologist, Woodworth. Subsequently several tests for assessing personality dimensions were designed, developed and validated. At present roughtly there are about five hundred personality tests used in various parts of the world for assessing personality.

Personality tests are developed by various methods. The construction of personality test is a laborious task. It not only causes physical and mental strain, but also expensive and time consuming. Even after constructing a test with so much efforts, it is doubtful whether the test measures reliably for what it was intended to use. In constructing and developing personality tests psychologists use various methods. The most often used methods are identified in recent time. There are three techniques psychologists follow in developing personality tests. These are *rational construct method, empirical criterion method* and the *factor analystic method.* Generally test designers use all the three methods.

The Rational Construct Method is generally used by the test developer for measuring personality traits. It is implied that the test developer should have comprehensive understanding about the human behavioural traits. Since the personality traits are in thousands he should have a clear understanding about the *Dimensions* and able to classify them into categories. In developing the test, he mostly makes assumption and inturtively decides certain traits. Some one who develops a personality trait like *Dominance* by this method might begin with an assumption that dominant persons are strong, dominant, highly influential, initiate leadership and so on and write down quite number of items and design the test to trace out these traits. Test designed by using the rational construct method could not claim that it is highly reliable. The validity has to be determined. The reliability and the validity of the tests are very essential aspects of personality tests. Therefore, the validity has to be determined by using some other method.

The Empirical Criterion Method is totally different from the rational construct method. The test development has to be done by obtaining a criterion of validity. For developing a test by this method for Dominance the test designer should start by selecting two groups of persons who must differ in this trait. He can choose 200 students from arts and science college and 200 students from an engineering college who are *supposed to differ in this trait.* These two groups of students should take up a large number of items from which he has already prepared as the *potential test items.* He might have written these items by the rational construct method.

The test developer in the final stage should retain only those *items which were responded* by these two groups *differently on the trait of dominance.* Thus the test developer is having the items responded differently by two different groups of students. The chosen items out of several hundred items are those that should distinguish the two different groups of people. This is an empirical criterion method because the sample consisted of two different groups who provided data which *differ in the trait of dominance,* which is to be assessed in some population. The kernel point in this method is that an item accepted by one person is rejected by another person.

The Factor Analytic Method is considered to be one of the techniques of test development. Test construction according to this method is made with large number of items on the personality dimensions planned to study. Therefore, a large number of items must be selected from the statements already written by the test developer. He has to administer the items to several different type of groups of people. By applying the method of factor analysis it is possible to find out the items that are having relationship with one another. By this means the items cluster together in the sense that items come to closer on the basis of similarities. The items which are having high loadings in the factor analysis are *identical in nature.* Such items are selected and the final scale is formed. The items that have the high loading in factor analysis indicate the interrelations among the items. Whichever item is having high loading is selected for the final scale. In the case of dominance, whichever item is with dominance oriented trait will have high loading and whichever item is having *low loading* or *minus loading* is supposed to be submissive oriented item. The items which are having high loadings in a personality dimension get grouped or clustered indicating that they are belonging to a single dimension. Similarly the items with low loadings get clustered indicating the opposite trait, the *submissiveness.*

We have given here only a *brief summary* of the methods used in the personality test construction and it is not a detail description of the procedures involved in the test construction and development in general.

Thus in every area the need is felt to understand the individual's behaviour pattern and if a correct assessment is made of the person the task may be easy and successful. Towards which quite number of techniques have been devised and hundred of test materials have been designed and produced for assessing the personality characteristics. We shall describe in the following pages some of the important techniques and personality tests.

The Interview Method

The interview method is one of the techniques very widely used for personality assessment. It is also a very old method of study. The interview may be an informal way of having conversation with the concerned person and getting information and also observing his/her way of reaction in the situation and other mannerisms. Or it may be a formal way of meeting the concerned persons with standardised questions and the interview tries to get certain definite information from the subjects. Normally in these days, an interviewer is always followed by an application filled by the subject. The application from may be fully relating to the specific purpose for which the candidate or the interviewer has applied. It may be for a college admission or for a professional training or for certain job determined from the academic records, his character, his work history or experiences and personal references and various other pertinent information are collected in an interview.

But, what is an interview? Generally, an interview may be defined as a conversation between the interviewee and the interviewer in a face to face situation, with some definite purpose of getting information from the interviewee and providing information to him and assessing the personality characteristics of the interviewee. The usefulness of the information depends upon the skills and experience of the interviewer and upon the co-operation of the person who is interviewed.

An interviewer is highly skilled and he is creating a warm atmosphere, a persons of acceptance and understanding. This type of atmosphere may encourage the person who is interviewed. Then he may speak freely about himself and also may reveal willingly about himself, his attitudes, experience and feelings and other characteristics of his personality. The skills in interview involve the ability to ask right and answerable questions and also the ability to listen carefully the interviewee's words and mannerisms and also the ability to interprate all the answers as part of his personality traits. The interviewer should also assess how intelligently the person deals with some difficult questions and also his emotional reactions.

An experienced interviewer may gain many valuable insight with the interviwee's personality traits from his expressed interests and ambitions, his way of dressing, his manner of expression, hesitations, emotional blockings, corrections and confusions. An interview is also used as a means of aiding people who have personality problems, for example, those who are anxious or worried about themselves, their school or college work, their job, their social or personal relationships. In personal interviews the subject is encouraged to talkout his problems and he is guided to discover himself where his difficulties are in his life.

There are some good, rich and potential interviewing techniques. One such technique is called *open-end-questioning*. It is by way of recording the information in a style that encourages the person, the interviewee has to respond in an elaborated manner. In the intensive, open-end-type of interviewing, the person is permitted to have *free conversational responses,* rather than general questions. The open-end question is so designed as to elicit a free response not restricted to predetermined categories. Essentially, the interviewer asks general rather than specific questions. The formulation of the questions are in such a way that they cannot be answered with certain specific bits of information. The interviewer does not ask the questions that can be answered with 'Yes' or 'No, but ask questions that require an expression of feelings or attitudes.

The unique feature of the open-end question is that the interviewee is given not anything or predetermined for his reply and hence is permitted to answer naturally and spontaneously in whatever term or reference he chooses. The interviewer makes simply verbatim record the subject's answers. The open-end question may be said have developed from the psychological clinic. The clinical interview demands a free informal questioning atmosphere in which the patient is encouraged to talk in his own terms and thus reveals most easily the underlying the dynamics of his attitudes, desires and emotions in the unconscious mind. The role of the clinician is not using cross questions but that of a friend who is interested in his problems, a listener, guiding in the desired direction. The interviewer's responses to the subject's answers should create an impression in the mind of the subject that his problems are really listened to and he may find solution to them.

One of the important advantages of the interview technique is the healthy background, because the interviewer is free to touch a very wider area and ask questions on a number of topics and elicit useful materials from the person by way of expression. It is also possible to find out the area of his interest and the interviewer may go deep into it through which the person's interests and his aims can be explored. The subject may be a different type of person, knowing that it is an interview, he may be well guarded, suspicious and also reluctant to give correct information, or may be also not co-operative and may be a person of uncompromising type. Therefore, the willingness of the subject in the interview to express himself is the most important and the interviewer should create the situation.

The interview's important task is to interprete the material provided by the subject objectively. Since the interview information are not based upon standardised test material, they are likely to receive a biased interpretation. The interviewer, however, skilled and experienced he may be inclined to interprete the responses differently from what they really meant, if the subject reacted in the situation in an irritating manner. Then, there may be subjective or personal biases creep into the interpretation. Therefore, there is lack of dependence upon the interviewer's personal skills. This is one of the great disadvantages of the interview.

But these drawbacks have been recognized and steps have been taken to reduce them by introducing objectively recording procedure of the interviewer's response in tape records. Several methods have been introduced in recent years. **Roger** (1942), **White, Dollard,** (1966), **Stone, Babes** and **Namenwirth** (1962) have used computer analysis for the interview materials. Because of its subjectivity the interview method is often supplemented with more objectively assessing methods like rating scales, personality inventories and so on. But however, there is a tendency or desire to *see the person and talk to him.* How the person looks like, how he speaks and reacts to humiliating questions. There is also a desire to know whether the person is honest, loyal and sincere. A swifty look cannot be mistaken for a dishonest behaviour and *the contours of face* of a person cannot tell us anything *about his conduct.* Therefore, interviews are often supplemented with application blank and personal history of a person in assessing personality.

Rating Scales

Rating scale is one of the techniques used for judging a person's certain traits. Rating scale in other words is a marking device on which a rater can record judgements of himself or others. It is a device for obtaining systematic judgements of the extent to which an individual possesses certain traits of behaviour, which may be judged in terms of degrees.

One of the most widely used devices is the *graphic rating scales.* Any rating scale may be either relative or absolute. Rating scales are also mainly used in comparing people with one another in any psychological trait. The *rating may be spread* out along a scale *from one extreme to the other extreme.* For example, it may be from very high through high, average, low and very low in judging the painting quality. We may also use any terms which may indicate levels. We may use a three point scale, or a five point scale or even a seven point scale or a nine point scale. If the rating is to be made in absolute terms, i.e., without making any reference to group norms, it requires to judge and place a check make at the appropriate point on the graphic line which runs from one end to the other end. The simple graphic scale provides lables for the end points and may also have a neutral point. A check mark is placed at any point on the line.

In graphic rating, the findings indicate the degree to which the person possesses the trait by putting a check mark (✓) at some point along line drawn to represent the personality trait or dimension. Rating may be expressed in number by dividing the rating line into three or five equal parts, indicating the sections 1, 2, 3 or 1, 2, 3, 4 and 5, for example:

High (3) Average (2) Low (1)

Rating Punctuality of the Person

Very Good (5) Good (4) Average (3) Bad (2) Very Bad (1)

The judge may place a check mark on any one of these categories corresponding to his judgement about the trait punctuality. In the graphic rating scale for personality assessment, we are using certain personality traits and judgements made by persons on the trait or personality dimensions on anyone of the categories which may be converted into numerical values. For example, a person is judged on a *trait of punctuality* in the graphic line at the point 'good' which carries 4 marks. Another person on the same trait may be comparatively judged as bad which carries 2 marks. Any number of traits may be used for judging several persons comparatively.

One advantage of using rating scales is that the judgments made by persons in categories may be converted into numerical values and the data may be treated statistically. Further the data may be presented quantitatively. The disadvantage is the validity of the rating of the traits. This depends upon the skill and experience of the rater. Rating scale also may be subjected to personal bias like the interview if it is not done objectively (Sarason, 1966).

Example : Graphic Rating Scales

The rater can place a *check mark* on any one of the *three categories below :*

(i) **Performance on a Technical Job**

High Average Low

The rater can place a check mark on anyone of the *five categories* below.

(ii) **Steadiness and Self-control**

Excellent Good Satisfactory Fair Poor

(iii) **Sense of Humour**

The rater can place a check mark on the line below anywhere from extremely high to Extremely low.

Extremely High Extremely Low

Personality Inventories

Personality inventories are used for assessing personality traits or dimensions. Personality inventories are designed in various forms. Modern psychologists wish to use the term *inventories* rather than tests. Personality inventories are in the forms of questionnaire or behaviour inventory or behaviour test or performance tests and situational tests.

But what is a personality inventory? Personality inventory may mean a list of questions or items each one referring to some trait or characteristic of an individual's personality, the total responses to all the items reveal the whole personality of an individual which may be consisted of several personality traits. Therefore, the concept inventory generally refers to a detailed list of items which are supposed to describe something which may be some articles, goods or properties. Applying this meaning to personality, we may state that personality is the sum of or collection of various types of traits and each question or item is a description of some personality trait. The personality of the individual is the total product and each item or question is the description of a trait. *Several questions or items* may be referring *to a single trait* each one expressing *a part of the trait.* Thus the term inventory has a *wider and deeper meaning* than the terms test and questionnaire in the case of personality assessments. The whole personality inventory may also be consists of *several scales.* The scale indicates a personality trait or dimension.

Personality inventories are designed to yield *quantitative information* about personality. These inventories are quite similar to intelligence tests. The objective *verbal behavioural responses* are converted into *numerical scores* and statistical treatments are given to the data and interpreted.

Types of Personality Inventory

Personality inventories are of two types, one *direct approach,* and the other *indirect approach.* In the *direct approach* the specific information is obtained and no attempt is made to conceal the meaning of the question. A subject may be asked whether he/she is shy in the midst of the opposite sex people, whether he/she is a member of gambaling/group, whether he/she wish to be scientist. The test user must get the cooperation of the subject in answering these questions. A clever subject may understand the true nature of the questioning and answer in a correct way.

If the real purpose is *to be concealed*, the test user may take up the *indirect* or *disguished approach.* One indirect technique that has proved successful is that of giving the subject a choice between the alternatives i.e., which of the two occupations you prefer? (Fruit seller or Toy seller), which of the two activities you prefer? (Plumber or Mason), which of the two recreations you prefer? (Football or Basket Ball). The person is asked to choose one of the two alternatives. If the subject finds none of the two for his choice, he is to choose one which is *less undesirable.* A person's consistent traits and attitudes are often revealed in these choices. The subject may find it very difficult *to deceive* in this type of questionnaires. He cannot guess the choice that is relevant to personality traits.

The *first personality questionnaires* were developed to study the inner world of perception and feelings of the individuals, by **Sir Francis Galton** in the 1800s. The use of questionnaires to assess personality again was designed later on by **Stanley Hall** in 19th century. He used the information given by the sample grouop of an adult persons. The questionnaires served for different functions. Galton and Hall used *self report* method. One of the earliest personality inventories is the *Woodworth personal data sheet* (1919). It consists of 116 items derived from various symptoms of the psychopathology and *emotional instability.* Each item is answered with 'Yes' and 'No' and the total score is the total responses.

The questionnaire presents a list of statements or questions and the subject has to answer by checking one of several possible answers. The questionnaire, psychologists, generally use for *three objectives* having in mind. The first is to obtain *systematic information* concerning the individuals' anxieties, feelings of inadequancy, worries and the like. This type of questionnaires are called *personal data sheet.*

The *second objective* of the questionnaire is to obtain information concerning *attitudes,* and beliefs about social, economic and religious matters of the persons. The *third objective* of the questionnaire is to show that it is a valuable means of assessing a *person's interest* in a variety of things, like, vocations, sports, books, people and so on. For example, the Bernreuter Personality Inventory, Kuder Performance Record are some of the inventories enable us to judge individual's personal problems, individual's interests in various areas.

The *Strong Vocational Interest Blank* is one very widely used inventory. We can judge the individual's strength of interest in fifty occupations, professions, including, law, medicine, engineering, teaching, salesmanship, accountancy and so on.

The *Minnesote Multiphasic Personality Inventory* is considered to be one of the important inventories and holds an important place in the personality assessment ventures. This inventory was originally designed and developed by **Hathaway and Mckinley** (1943) for clinical purpose. The items for this inventory were selected by these authors for empirical and logical reasons. The authors of this inventory have collected first a large number of items and administered to normal people and also to groups of people of different psychiatric diagnoses. The items that were found to differentiate the various groups were selected and included in the final scale. This scale consisted of 550 items in the form of questions which are answered with *true, false,* and *cannot say.* Thus the inventory has three response categories.

There are *fourteen basic scales* which are used for the analysis of personality. The first four scales are the validity scales which measure the test taker's attitude and provide evidence for the subject's attempt for taking. The remaining ten scales are called clinical scales, eight of which are for clinical diagnoses. One scale is *masculinity-famininity* and one measures *social introversion.*

The four validity scales are : (1) cannot say scale, (2) lie scale, (3) validity scale and (4) correction scale.

The individual's responses to the inventory are reported in terms of separate standard scores which indicate the person's *relative standing on ten personality scales.* Each scale describes several characteristics. The ten clinical scales are : 1. Hyprochandriasis (Hs), 2. Depression (D) 3. Hystecia (Hy), 4. Psychopathic Deviation (Pd), 5. Parania (Pa), 6. Psychasthania (Pt), 7. Schizophrenia (Sc), 8. Hypomania (Ha), 9. Masculinity-Femininity (MF), 10. Social interoversion (Si).

The four validity scales are intended to assess the attitude of the person who is taking the test. If the person who takes the test and answers to the questions of the scales in a different manner like defensively or just answering without any real feeling or falsely replying simply by making marks on the response categories, then the procedure of the MMP1 is to be revised. There should not be any faking answers. A few sample items are:

(1) I believe I am being plotted against : T-F-?
(2) I wish I could be as happy as others seem to be : T-F-?

An individual may be given the whole inventory or sometimes a part of it. The scoring keys are prepared to identify the patients with respect to psychiatric states like *hysteria, parania* and so on. Patients of each type are compared item by item. The MMP1 is having several control or correction keys to identify or to make allowance for exceptional responses. Because the response *"cannot say" with? mark* may be many. This indicates *excessive evasion* of the questions. This makes meaningless to compare the individual's responses with the standard group of persons. Individual's MMP1 responses are reported in the form of a profile with his score on each sub scale.

Profile showing *high (cannot say)?* scores are *invalid.* An individual's MMP1 score is reported in the profile and the interpretation of the scores is a complex procedure. It is based on detailed analysis of the total profile. The MMP1 is a more sophisticated instrument than the inventories developed earlier. It has several advantages. It is an attractive instrument for assessing personality by describing it in detail and also very much useful for research. Many studies were made for describing the personality traits of the normal people.

The MMP1 has also certain disadvantages with all its rigid way of testing procedures. It also depends upon the individual's *self-perception.* It provides a *particular kinds of measures* of empirically related to psychiatric categories. Therefore, investigators often make it a point of using also other techniques like interview or projective tests.

The development of the various types of tests to assess the general disposition and the psychological adjustment of people has been going on activity for generations. It has been estimated that there are at present more than five hundred personality tests in use. They all differ in their nature and characteristics one from the other in some way. The MMP1 which we have stated above has become a model for many personality inventories. But they all differ very much from the model inventory. The one which is similar to MMP1 is the *California Psychological Inventory* (CP1).

The California Psychological Inventory is one of the important personality inventories very widely used for assessing personality. It consists of *four hundred and eighty items* of which *178 items* were *drawn from the MMP1.* The CPI has different purpose from the MMP1 and it has a specific purpose and serves a *better need.* The MMP1 was designed and developed for a specific purpose of diagnosing psychopathology. But one disadvantage is that the MMP1 has too many items which describe people with pathalogical characteristics. In contrasts the CPI was developed for use with the general populations specifically from the age of thirteen and above. It is also mostly applicable to *high school and college students.*

The California Psychological Inventory has items of *true-false* responses and it consists of *eighteen scales,* of which *three are the validity scales* like the MMPI. These *three validity* scales measure the test taker's attitude. These validity scales assess how far the subject *honestly has taken up the test.* The rest of the *fifteen scales* measure for what purpose the inventory was developed. The CPI has symptom related content. The language used in the inventory and the interpretative scales are quite suitable for ordinary people.

The development and the standarizations processes of the inventory were very much taxing. The inventory designers have collected *responses from 6000 men* and *from 7000 women of different ages, occupations, educational levels* and *varied socio-economic status and also from different geographical regions.* The scores of the inventory are reported in the form a profile. **Gough** (1960), the inventory author had divided the scales into *four classes* for making the interpretation easily.

The psychological significance of the *class one scales* are measures of poise, ascendency, self-assurance and interpersonal adequacy. The *class two scales* are the measures of responsibility, socialization, maturity, and intrapersonal structuring of values. The *class three scales* are the measures of intellectual efficiency, and achievement potential. The *class four scales* are measures of intellectual and interest modes.

The California Psychological Inventory has been very widely used and quite number of researches have been conducted by using the inventory. Many studies have been found very significant. The inventory mostly administered in elementary schools, high schools and colleges and also in various organizations like the defense forces, in police training centres, medicine, dentistry, nursing and in teaching centres. In all these studies there were significant associations between this inventory (CPI) and various measures of *achievements* among the subjects.

Further the CPI is able to identify people who are likely to take part in extracurricular acivities and also could identify the students who cheat in examinations. The inventory also relates persons to leadership, managerial positions, occupational interest and adjustment in various situations. The CPI is capable of predicting the juvenile delinquency and can reliably discriminate alcoholics from marijuana and also smokers. It can also relate confirmity, creativity and those who are physiologically responsive to stress. The other aspects the inventory can identify are politically motivated, marital adjustment and the persons who prefer family planning methods.

Gough had designed the CPI as the measure of folk traits or those traits which people generally use in describing others. **Kish** (1971) in one of his studies found that the 18 scales of the CPI correlated highly with the SS, SSII general scale in a sample of 100 hospitalized male alcoholics. The pattern of correlation of the CPI suggests that *sensation seeking* is related to *impulsive type of extroversion personality* and the *high sensation seeker* is characterized by *non-confirmity, uncoventionality* and also *little concerned with social mores,* responsibility and self-control. The *less sensation seeker is overtly confirming, conventional and controlled.*

ASSESSMENT OF AUTHORITARIAN PERSONALITY

Development of the Trait of Authoritarianism

The concept authoritarianism refers to the personality syndrome which has received the greatest importance in recent research works. The publication of *Authoritarian Personality* by *Adorno and others* in 1950 aroused an exciting spirit among the psychologists who were interested in the study of personality dimensions. Authoritarianism finds its root in the *political philosophy of Fascism and Nazism* that were started functioning in 1920s and 1930s in Italy and Germany. The development of authoritarianism provides a link between *personality and ideology,* and between *character and prejudice.* A wide variety of psychological, social, political and economic phenomena have been incorporated in the study of this construct. The formulation of authoritarianism stands as a *classic in Psychology.*

What is authoritarian personality? Psychologically, authoritarian personality may be defined as a type of personality identified by *rigid adherence to middle class values,* extreme awareness of authority relationships, denial of personal sexuality, a tendency to exploit others but also to feel exploited, categorical thinking and a tendency toward projection. The initial formulation of the concept authoritarianism and its expression in action owes much to the *German National Socialisitc Party,* headed by **Adolph Hitler** (1889-1945). He was appointed as the Chancellor of Republic of Germany in 1933 by **Hidenberg,** then Chancellor of Germany. Hitler did not win the election in 1932 in the presidential election. But by gaining powerful public support he was able to become the Chancellor by appointment.

No sooner Hitler became the Chancellor, than began a long eventful journey that started affecting the world at large for many years to come as a result of the extreme cruel activities of Nazi Germany under his leadership. Most of these cruel activities required the complicity, cooperation of large number of German people. The actions deliberatily against the Jews in Germany, Poland and in France were very conspicuous. These actions and the responses of the German people against the Jews prompted psychologists, sociologists and other behaviour scientists to study the behavioural characteristics of the German people. The people of Germany at that time were seen relatively passive, but many accepted the cruel and violent actions of the Government against the Jews. It was a strange situation in Germany that was prevailing there. A well developed democratic country, which gave birth to great intellectual giants, going for propaganda targeting the Jews as the source of social, economic and political struggle, identifying the Jews as the evil which is to be removed from the path towards the struggle against the difficulties the country is facing. The aim of the Nazi Governments policy was to exterminate the Jews from Germany. How this *antihumanism could occur?*

This was what really happened in Fascist Germany. **Enrich Fromm** in his famous work, *"Escape from Freedom"* tried to answer this question. This work was considered as the *historical and sensational work* providing information about the German characteristic structure of that period. The character structure that prevailed in Germany at this time has been viewed as having long history.

It was the result of the consequences of the century old changes in the *religious experiences* of the individuals *since Reformation* and the alteration that took place in *economic behaviour* caused by the *rise of capitalism* and the changes in the religious experiences and the *growth of protestentism* all diminished the influence of traditional moral principles and gave importance to the individual freedom. All the changes that took place after the reformation, according to Fromm, resulted in rise *strong capitalism* which provided potential for strength and independence. By setting men free, it produced a sense of isolation or loneliness and fear. This new character structure which resulted from the social and economic changes was further intensified by the religious doctrines.

The consequence of all these changes, centuries later resulted in the *change of personalities* of the people of the western Europe and particularly in the lower middle class people in Germany. The main change was the development of *Authoritarian character structure* in personality. This character structure is a type of personality dimension which is the *idealization of authority* along with *fear and submission to the authority*. The authoritanian owes unqualified emotional allegiance to the authority and in turn he too desires the role of authority and adapts, the same kind of exploitation of those who are under him. The concept authoritanian character refers to the personality structure which is the *human character* in the philosophy of Fascism.

Study of Authoritarianism

A research work was carried out at Berkeley in California University by **Adorno, Frenkel-Brunwikm Levinson and Sanford** on authoritarianism and it was published in 1950 under the title of *Authoritarian Personality* in a 900-page volume. When the authors started the studies it was focused as *Anti-Semitism.* During the period of Nazi regime in Germany there was rapid extermination of the Jews in Germany. There were also other racial groups subjected to this mass murder. The world was watching these atrocities. This prejudiced attitude and the actions against the Jews were not only found in Germany. The hatred against Jews existed for centuries in many countries around the world. The United States was not free from the practice of anti-semitism. There were ample evidences that anti-semitism prevailed in the form of prejudice. The American Jewish Committee established a Department of Scientific Research to support the investigation of prejudice, with the Authoritarian personality. The Berkeley study is one of the first product of the problem.

The Berkeley investigators had the assumption or the hypothesis right from the beginning that the dispositions of anti-semitism lie deep in the framework of personality. This assumption set the course of their study. The plan, the investigators used was to identify the persons who had *high anti-semitism* and individuals who had low is such prejudice and to compare these two groups in terms of their *personality dynamics.* Towards this the investigators need to measure prejudice, conceptualize and integrate personality. Prejudice against the Jews was conceptualized by the investigators in terms of ideology about the Jews and they viewed prejudice from psychoanalytic perspective for which they needed a scale.

Therefore, a scale to assess anti-semitism was developed. The scale consisted of items about the Jews and it was called as the *A-S scale.* The idea was to know the extent of negative sentiment which people would express towards the Jews by approving the statements that were anti-semitic in nature. The A-S scale stands for *Anti-Semitism Scale.* The A-S scale was comprised fifty two items and each item expressed an idea which was a characteristic feature of prejudice against the Jews. With the A-S scale the researchers could assess, the ideology about the Jews. They could identify with this those who are most prejudiced in their beliefs and those least prejudiced in further studies.

A few sample items of A-S scale

(1) Gentile does not have a fair chance of competition.

(2) It is wrong for Jews and Gentile to inter-marry.

Ethnocentrism and the E-Scale

This research has been extended from anti-temitism to deal with *Ethnocentrism.* The researchers therefore broadened their investigation for examining ethnocentric ideology by using some of the items relating to Negroes, Okies, women, criminals and other ethnic and other identifiable sub-groups in American society. This scale is called E-Scale.

A few of the item used in the E-Scale are :

(1) Negroes have their rights, but it is best to keep them in their own districts and schools to prevent too much contact with whites.
(2) There may be few exceptions, but in general, Jews are pretty much like.

The items of the E-Scale were once again reworded in such a way that a person's agreement with the statement should indicate ethnocentrism. It was also made in some cases that the scores on the E-Scale were used as a substitute for some of the responses to the A-S (Anti-Semitic) scale for identifying the prejudiced and unprejudiced individuals. This was considered as an intensive follow up study. It was observed that among the samples included in the Berkeley research conducted by the investigators the scores of the A-S scale and the E-Scale were correlated. There was high correlation between the scores of these two scales ranging from +0.63 to +0.75 in the earlier forms of the scales.

One of the studies in the series of researches conducted by Adorno and his co-workers was the *political and economic ideology.* The reason incorporating this study was that the investigators had the conviction that the political and economic forces play an important role in the development of ethnocentrism in both institutional and individual *psychological forms.* For assessing and confirming this conviction the investigators developed the *Politico-Economic-Conservatism Scale* (PECS). They used the same procedure as they used for the development of the Anti-semitic (A-S) and the Ethnocentrism (E) Scales.

The investigators had observed that the scores of Politico-Economic Conservatism Scales were found to be not highly correlating with scores of the A-S and the E scale in general. But they demonstrated relationship between the PEC scores and the A-S and E scores and it was high enough to support the hypothesis that prejudiced persons are of *conservative type of personality.*

Authoritarian Personality and the F-Scale

The investigators of the Berkeley study, Adorno and his co-workers felt that a scale to measure prejudice, without mentioning the name of any minority groups, the aim and the meaning of the study was very much needed. Therefore, they developed a scale called the F-Scale. The F stands for Fascism. *Fascism* means any *right-wing nationalist ideology* or movement with an authoritarian and hierarchical structure that is fundamentally *opposed to democracy and liberalism.* Thus according to Fascistic ideology *authority is the supreme entity* that governs the country in hierarchical structure. The Italian Fascism under **Benito Mussolini** (1883-1945) and the National Socialistic ideology, under Herr Hitler adapted the some political ideology. Thus the F-scale was designed and developed in such a way that the real purpose was not revealed in the items they used for the investigation. The scale was intended only to identify the authoritarian dimension of personality.

The scale could be used without any chance for *resistance to honest self-disclosure and hostilities* towards the measuring tool that is *assessing the prejudice* that the individual has developed. Therefore, in view of this the evolution of the F-Scale and its idealization as authoritarianism has become an experimental research tool. The F-scale is a tool to measure the personality syndrome underlying prejudice. It is also considered that it represents a vital motive force behind the sentiment of prejudice. Adorno and his co-workers believed that the degree of *Prejudice* in a person was the result of *implicit* or *inherent personality dispositions.*

At the final stage of the F-scale they retained *thirty items.* Several revisions and modifications were made in the scale to arrive at this stage. The assumption is that an individual who takes up the scale has to express his agreement and disagreement with the idea of the statement. The individual's greater agreement with the item the higher the authoritarianism. The reliability of the scale was tested by administering scale to a number of groups and it ranged from +0.81 to +0.97 with an average of +0.90. This scale is called the California F-Scale.

A few sample items are:

(1) Obedience and respect for authority are the most important virtues children should learn.

(2) An insult to our honour should always be punished.

Adorno and his coworkers by analysing their studies have observed the syndrome of authoritarianism and evolved nine variables functioning in it. These are :

The Nine Variables of Authoritarianism

(1) **Conventionalism** which are characterized as rigid and adherence to conventional middle class values.

(2) **Authoritarian Submission,** which stresses submissive, uncritical attitude towards idealized normal authorities of the in-group.

(3) **Authoritarian Aggression :** This is characterized by the tendency to be on the lookout for and to condemn, reject and punish people who violate the conventional values.

(4) **Anti-interaception:** This is characterized by opposition to subjective, the imaginative and tender-mindedness.

(5) **Superstition and Stereotype,** which is characterized by the belief in systematical determinants of the individual's fate, the disposition to think in rigid categories.

(6) **Power and Toughness** refer to *preoccupation* with dominance-submission, strong-weak, leader-follower dimension, identification with power figures, overemphasize upon the conventionalized attributes of the age, exaggerated assertion of strength and toughness.

(7) **Destructiveness and cynicism** which are characterized by generalized hostility, villification of the human.

(8) **Projectivity,** which refers to the disposition to believe that wild and dangerous things go on in the world, the projection outwards of unconscious emotional impulses.

(9) **Sex :** This is characterized by exaggerated concern with sexual behaviour.

According to the investigators these nine variables represent a *single personality syndrome of authoritarianism.* If a person exhibits any one of these trait variables or several of them it is taken for granted that he would display all the other trait variables.

These investigators verified the existence of certain other human faculties. They verified them with repeated studies and correlated them with the authoritarian dimension. One of the faculties repeatedly verified was *intelligence.* They observed that *higher intelligence* was associated with *lower authoritarianism.* The Berkeley investigators have observed in some of their samples that the *F-Scale scores* correlated between −0. 13 and −0.48 with various *intelligence* test scores. They used the AGCT and Otis Higher Form-A. This indicated that those who score high in intelligence tests are low in authoritarian personality dimension.

There were several criticisms made by the contemporary scholars against the theory of authoritarianism formulated by Adorno and his co-workers and also for the methods used to assess the authoritarian personality. One major criticism made by **Shils** (1954) is that the conceptualization of the authoritarian as a rightist, conservative profascist personality type is too limited. The items of the scale were framed on the theoretical basis to measure one or more of the variables which characterize the syndrome and they were developed out of conceptualization of the personality syndrome and not on the basis of statistical analysis.

PROJECTIVE TESTS OF PERSONALITY

What is Projective Test?

The projective test is a personality test in which the subject reveals or projects himself through his imaginative productive ideas. The projective test gives the subject too much of freedom of expressions than the fixed alternative personality inventories. Examples of projective tests are the *Rorschach inkblot test* and the *Thematic Apperception test* which are with pictures that elicit stories. Projective personality test allows the respondent to freely responding to the suggestive materials on the assumption that interpretations of them will reveal the important aspects of personality. Projective test of personality is in contrast to the objective test of personality. The important characteristic of the projective test of personality is that the individual who takes up the test projects throws out himself into the situation of the specific picture of the test. Projection generally may mean the process of attributing one's own qualities to others. It may be also defined as the process of attributing one's own better qualities to others. In the case of projective test the individual who takes up the test is made to reveal what are all he could say on seeing picture.

The projective test kindles the fertile imagination of the test taker. It may be stated that some of the potentially most powerful means of measuring personality come from the judgement of the test materials aroused in the imagination of the person who takes up the test. The technique that is adapted to tap the test-taker's imagination is called *Projective technique.* The individual's freedom of expressing his own ideas is not limited in the projective test. In other type of personality tests or inventories the test-taker has to answer certain specific questions or he is asked to follow some directions and therefore, he has to answer the questions on some problem or issue. This type of problem does not arise in the case of projective tests.

The main aim in projective test is to see the whole individual at a time and to see the traits and other characteristics and their interrelations. It is also to see the important features of the specific object figure-ground interactions. It is believed that the project test, whatever test it is, may expose both the external and the deep internal characteristics of the person.

Characteristic of Projective Tests

Projective tests are designed for assessing the personality traits or dimensions. Projective tests are often unstructured rather than providing fixed alternatives. The subject in his answers may reveal his own wishes and needs. Generally projective tests are with ambiguous or indefinite stimuli. It is true that many people experience anxieties and a sense of fear when they were asked to discuss intimate personal matters. They try to conceal the type of behaviour like blame worthy. These people are of the characteristics of insecure and emotionally maladjusted individuals. In these cases the only test that can bring out their characteristics is the projective tests. The projective tests present ambiguous situations, into which the individual is given chance to see and read his own wishes, ambitions, fears and fantasies. The subject projects himself into all the ambiguous situations and also identify himself with the situations. In all these the subject projects his personality trends.

The problems in the projective tests are often intentionally ambiguous. The subject's responses are analyzed for the purpose of revealing his interests, motivation, perceptions, problems, values and modes of adjustment. The projective tests are used largely in the study of personality. **Murphy** (1952) says that the aim of projective test is to see the individual all at once, to see the traits and their interrelations, to see the reliance and the figure-ground organizations and also to get both at surface, and at deeper dynamics. **Frank** (1939) has phrased, that projective tests reveal the way in which the individual understands, takes hold of a task. They reveal his inner or 'Private World'. The private world of one person is unique and different from that of any other individual.

Psychological tests are of various types and the way in which each type of test is administered varies from other type of tests. Every type of test varies in the skills, procedure and experience in the administration. That is the reason sometimes we use the term technique. Generally when we speak about the projective tests we say projective techniques. Technique literally means the method or procedure in artistic work, or scientific activity, or the degree of expertness in technical or artistic work.

It is the practical method, skill or art applied to a particular task. It may be also stated as the proficiency in a practical or mechanical skill, a special knack. For example, we may say that, he has the technique of turning everything to his advantage. The term, projective technique also carries all these characteristics, because the administration of this test is a skilful job. It is scientifically dealt with. It is also an art. The dealings with every aspects of this test is an art, i.e., the direction, administration, the scoring of each response, and the art of getting into the mind of the examined and interpretation of the responses in a meaningful way and understanding the real personality is an art of scientific skill.

The projective techniques refer to several techniques as each test is having a technique. We shall discuss here a few of the projective tests and the corresponding techniques. There are projective tests for children and also for the adults. The important children projective test are the drawings of a man test, the Brush-paints tests and the personality projection in the Drawing of the Human Figure.

Children Projective Tests

Children projective tests are the materials with which children play. Children play with toys, mud pieces, clay, stones, pieces of wood, paper toys and other available materials. Children often play with finger paints, simply on the floor, water, colouring materials and so on. Children working with finger paints may reveal some interesting information about children.

Based upon children play and the materials with which children play, tests for measuring intelligence were developed by *Goodenough* as early as 1926. The idea is what can be done with paintings, can be also done with drawings. *Goodenough* used drawings of a man as a measure of intelligence. She viewed that such drawings may be also useful as an index of a child's conception of human beings. It is also possible to know the child's understanding of male, female, adult, and the attitude to himself. The drawings reveal what actually the child thinks about people around him. Generally the child's deep feelings may be revealed. Hence it is understood that the drawing of man by a child may reveal the child's various characteristics, like his intelligence, his conception about men, women, others around him, his personality characteristics like love, affection kindness, aggression dominance, submission and also what he is capable of.

A few years later **Alschuler and Hattwick** (1947) have shown a great variety of production of *Brush-paints* by *three* and *four year old children* which broughtout the ways in which the personality dynamic in children are expressed by using the colour paints and brush. It also revealed the relation of personality to painting. The gay and happy children may use the *vivid colours* and the *insecure children* tend to use *black and brown colours.* From the drawings it is understood that an impulsive outgoing child with no sense of order has drawn *repulsively,* while a child who has got everything in order, and followed rules, adjusted with others, drawn something *smoothly* (Fig. 24.1).

Machover (1949) has attempted a deeper dimension in his study. It was symbolic interpretation of the world as a child sees it. His study was personality projection in the Drawing of the Human Figure. The drawings in his study were vivid, creative imagination, and also deep at home in social relationships. There were also orderly, well-planned structure of juvenile architect and also there were examples of rigid ways of avoiding mother's disapproval. The child simply reveals firmly established habit which was built up at home in the play group. Personality is also understood by the way in which the individual twists the materials presented to him. From the drawings, the story telling procedure also was developed as a projective test. In this method the child is made to listen to a story and then, he/she is asked to tell it back and it is recorded. While *telling back* the child may make some *commissions and omissions* in the original story. *What type of materials were omitted and what type materials were added* were studied carefully. All these based upon the child's *likes and dislikes* and the feelings he/she developed toward them.

Fig. 24.1 : Model of Children Projective Test

Adults Projective Test : Rorschach Inkblot Test

Projective tests are designed to assess personality of an individual as a whole. These tests are like the intelligence tests used both with children and with adult persons. Though the techniques are the same, separate tests are designed and developed for children and for the adults. We have discussed in the preceding section about projective tests with children. Here we shall deal with the adults projective tests, and assessment of personality. One of the important characteristics of the projective test is to make the subject to imagine in whatever possible manner he could deal with the test material. For this purpose the projective tests are designed in a more ambiguous manner.

Thus projective tests use ambigous and indefinite stimuli for personality assessment. The test users require the subjects to respond to the test materials and interprete them. This is supposed to elicit the maximum amount of response information on the material. In some projective tests perceptual tasks are utilized. This refers to the *inkblot tests.* The subject perceives some picture like material and interprets. In some other tests some designs are used which evoke fantasy expressions like story telling. Still in other types the individual's creative productions are elicited, mostly painting and drawing responses. Tests of these types are expected to evoke the complex responses in individuals which reflect organized functioning of the whole personality of the individual instead of some specific aspect of personality. Thus the responses to any projective test are extremely varied and also individual and unique.

The projective test that is widely used and well known is the Rorschach Inkblot test. This is the earliest adult projective test developed by a Swiss psychiatrist, Hermann Rorschach (1884-1922) (1942) Rorschach tested the reactions of the patients to thousands of inkblots prepared by him beginning with 1911. Finally, in 1921 he selected ten inkblots because they provoked the most vivid and emotional responses of the patients he tested.

Having done so much work in designing and finalizing the Inkblot tests for more than ten years, Rorschach died within a year after finalizing the tests.

All the ten inkblot tests consists of *pictures and all of them are symmetrical.* Half of the pictures are *in black, white* and *shades of gray. Two contains red splotches.* Three of them are in full *soft pale shade colour.* Thus the ten pictures are in varied shades. These are called inkblot tests. The ten inkblots are given to an individual one at a time. These are presented in a *specific order.* The subject is given the instruction to tell *what the inkblot looks like.* Or what it might be, you see in this? The responses given by the subject are recorded. Then the examiner or the test user and the subject go through each card again. Now the subject explains which part of the inkblot made him to give the *specific response and why* he/she has given such responses?

This type of administration of the Rorschach test is called the *free association.* The examiner carefully records all of the subject's responses such as his *reaction time, his questions, his movements,* and all the explanations of the inkblots. An average person is generally expected to give thirty to forty responses to the blots (Fig. 24.2). The figure shown here is not the real Rorschach Inkblot test as there were objections that if the real picture of the Rorschach are shown by reproduction the purpose in the clinical field would be spoiled. Therefore, what is shown here is only a model of Rorschach inkblot test.

The responses of each test are scored in three columns, such as *location, determinants* and the *content. In the first column* the responses regarding to the whole blot or to its large details or small details and the white spaces within the blots are given. If the subject has given the response that he/she saw the card as a whole a 'W' is written. For large details a capital 'D' written and for small details a small 'd' written. A response to white space a large 'S' is written. In this way in the first column the respective responses will have W.D., D and S (some of the minor distinctions are ignored).

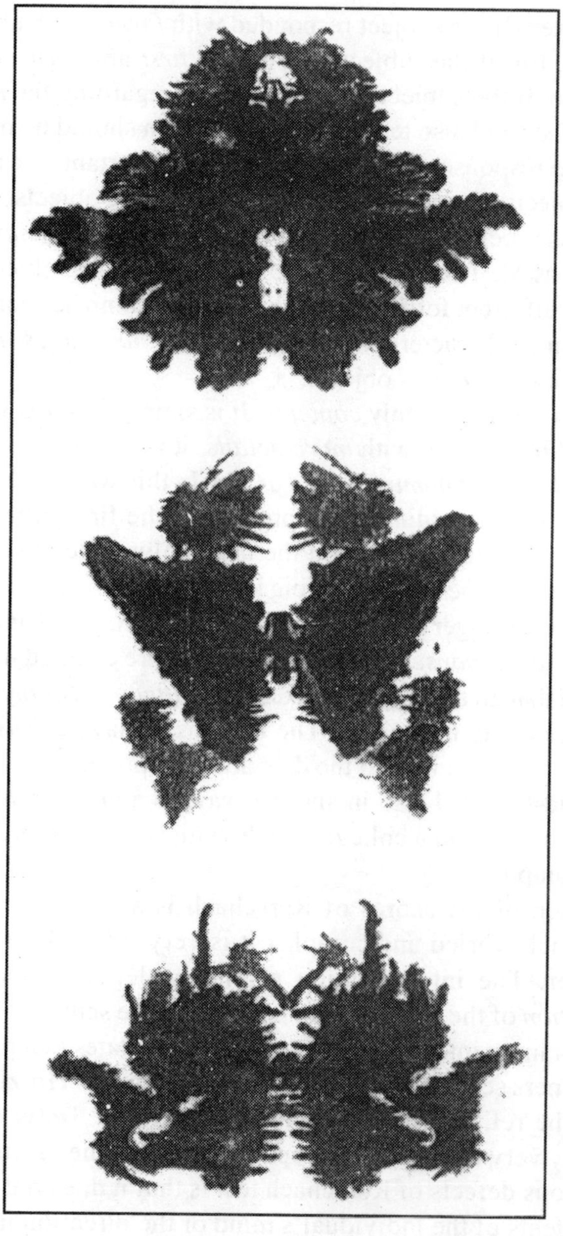

Fig. 24.2 : Projective Technique : Rorschach Inkblot Model Test.
The ambiguity of these Inkblots make different people to see and think different things. The Researcher can understand a great deal of personality of the perceiver.

In the second column the score of all the responses are with regard to *determinants or attributes,* whether the subject seen the form of the thing or column or the combination of the two? What determined the response? If it is the *form a large F* is written in column two. If the response put stress on *colour* of the blot, (When the subject said that it is blood or fire) then *a large C* is written. If the subject responded with *form first* and *column second,* then, *large FC* are written. But if the subject said *colour first* and *form secondly,* than *CF are written in column* two. If the subject give any answer regarding the *shading,* for example, *rough or smooth surface* and also texture some provision should be made in column two to make entries of these responses. There may be also important out responses like *human beings, animals* or objects in cards having movements of the objects. There is every chance for giving the response seen like *running, walking, dancing* or *jumping* and *flying.* These responses are scored as M. There may be also moving animals, object like water or fire or wind, these scored in different forms. Thus the second column has dealt with the perception of objects with dynamic characteristics like *forms, colour shades, movements* of various types of *animated* and *in animated* objects.

The third column deals with only *contents.* It is simply, *what is seen.* If animal is seen, it is *scored as A,* and if it is seen with *more details,* it is *scored as Ad. The human form is scored as H* and the *details of human scored as Hd.* In this way all the three columns of the Rorschach test are scored according to the procedure, the first column dealt with location the second column with determinants and the third column dealt with contents. The next step is to count the *total number of wholes* (big letter), details, *small details* in every column. This is not like the tests of finger paintings or drawing a man. The Rorschach is a quantitative test. *The total number of responses* given by the subject are counted in all the three columns. It has been estimated that an adult person takes from *fifteen to twenty* and also up to *forty to fifty minutes* for responding to the test. The responses may be *between ten and sixty.* A quantitative treatment can be given to the date and interpreted.

The Rorschach test is used also in another way, as a *group test.* It may be taken how well a group of new students in a college will do with college work and how well they can adjust in a college campus.

The interpretation of the scores of Rorschach is a summary procedure. Since the materials are extremely varied and complex it is very difficult to define in rigorus and quantitative manner. The interpretation primarily depends upon the clinical skills experience and *intution* of the persons who interpreted the scores. The same interpretation is not made by two or more interpreters. **Masling** (1960) states that personal characteristics of Rorschach examiners seem to influence the test responses. **Holzberg** (1960) states that as the result of it, the reliability of Rorschach test is low. **Detel** and **Generalli** (1955) state that the validity very much depends upon the skill of the examiners. But, it has been viewed that the serious defects of Rorschach test is that it does not provide us concretely much about the contents of the individual's mind or the direction in which the individual is inclined to move about.

Picture Tests

In view of these, and since *we are in need* of knowing *more about* the *individual's characteristics, thoughts* and *feelings* and *imagination,* it has been felt that the use of drawings, picture, photographs and paintings may provide us with more information in a variety of situations. A picture is given to a person and he is asked to make out a story and through which to explain what he/she saw and also predict what is likely to happen in that situation.

One best known procedure is the *cards* prepared by **Morgan** and **Murray** (1935) for assessing the personality characteristics of persons. The test is called *"The Thematic Apperception Test"*. This test consists of *standardized pictures.* The examinee is shown a set of pictures and is asked to report what is happening in each picture. People often *project* their own *feelings* and *experiences* into the themes of the picture. The pictures are usually ambiguous but *potentially emotional scenes* which the examinee can interpret in the light of personal experience and fantasy. The person being tested is asked to make up a story on each. The instruction is that the story should describe the events that led up to the scene of the picture. What the participants in the story are thinking and feeling? What is presently happening? How the story will end? (Fig. 24.3)

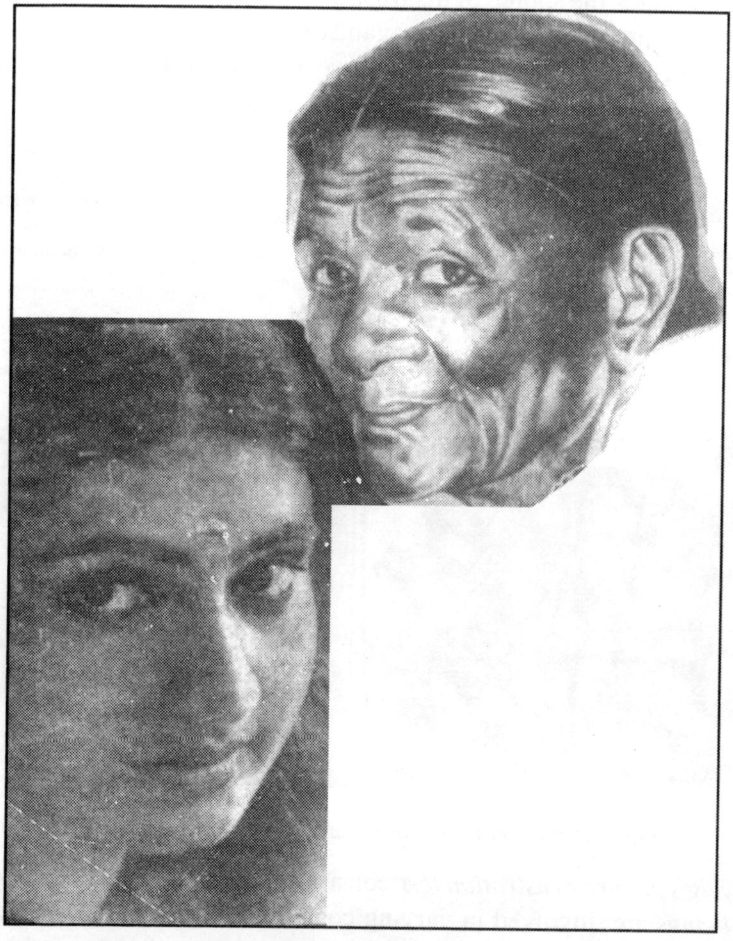

Fig. 24.3 : Morgan-Murray Thematic Apperception Test.
A Model Picture

The test is used to *both sexes* and at various age levels and of intelligence levels. The test may bring out differences in predominant interests, anxieties and worries.

It is possible that the examinee may make up a story in such a way that he would be more free than that in which he/she would carry on a direct conversation about himself. The TAT, may be useful in getting life histories and attitudes towards jobs. It may even *create an awareness of needs* in which he had never been known. Therefore, the test may go into *deeper self-understandings.* Sometimes imagination may directly lead to action. The TAT may give an objective picture of the often occurring pre-occupations of the examinee. The fantasy materials revealed in the TAT are agree with other fantasy materials to some extend, like the daydreaming and dreaming at night. But the validity of the TAT depends upon its relation with the clinical materials.

The other important picture test is *Rosenzweig's picture Frustration Study* (1948). The purpose of this test is to assess the tendencies to respond to frustrations. When a person is frustrated he may attack the source of his frustration, or he may turn his aggression towards himself or he may direct his aggression against none. These three types of responses of frustration are called *extrapunitive, intropunitive* and *impunitive.* (Fig. 24.4) .

Fig. 24.4 : Model Picture of Rosenzweig Picture Frustration Test

Rosenzweig's picture Frustration test consists of *twenty four pictures* and in each picture *two human beings* are involved in varying contexts. There are two boxes in each picture and in one box what one person speaks about is given. The other box is blank. The subject who takes up the test is instructed to write in the blank box above each picture the very first reply that comes to mind. When all the twenty four blanks are completed by the subject the responses are scored according to the procedure. The scoring procedure is very laborious task and one must have some experience in scoring this test. On the bases of the scores for the twenty four picture, interpretations are made and the person is identified. There are important functioning relationships between the ways of handling the response to frustration and to various other dimensions of personality or personality traits. It is an interesting test, but difficult to score and interpret.

Hellerberg (1945) has developed an interesting *picture test.* The test consists of *incomplete pictures* and the subject is asked to complete it in whatever manner he wants. The test needs motor responses. The results are interpretated in terms of needs and imaginations.

The *handwriting* analysis as a study of personality is not developed. Since there is difference of opinion among the graphologists with regard to the basic principles, the *method of graphology* as a projective method is yet to come out.

Behavioural Assessment

Psychologists have been applying various techniques and methods for assessing personality. One of the techniques widely used was the projective technique. Researchers who have been using this technique in a number of research works are not able to successfully identity the personality dimensions in their subjects. They were generally facing two problems mainly. One is the subjects were *not able to get a clear idea* about the picture or the design and they get confused. Secondly the researcher finds it *difficult to interpret the responses* provided by the subjects and to identify the personality traits. Therefore, there was some increasing dis-satisfaction with the projective techniques. The most important problem they faced was the *ambiguity of the test* designs. The Rorschach Inkblot Test and the Thematic Apperception Test are the two important tests subjected to the criticisms of the researchers. Therefore these psychologists searched for an alternative and substantial method for assessing personality. They have developed a technique called behavioural Assessment.

The method of assessing personality by observing the behaviour came into use as early as 1920s. But this new attempt is based upon the dis-satisfaction with the projective techniques. According to **Ciminero** (1977) behavioural assessment technique concentrates on what an individual does in a specific situation, will also do the same thing in similar situations. Behavioural assessment of personality is made by *two methods.* One method is the *self reports* and the other is the *direct observation* of the behaviour and recording.

The self report procedures have become more popular and many researchers resort to this method. According to this method the individuals are asked to record the various aspect of their own behaviour. This is a *self monitoring technique.* For example a college student takes up three examinations. The answer papers are generally valued by the teacher who teaches the subjects *without entering* the markers *in the answer papers.* Then the same papers are given to the student who wrote the examination and he is asked to value his own answer papers and to record the marks in a separate sheet for each answer separately. Thus the student assessed his performance by the self report method. The teacher's valuation and the students' valuation are compared. By this means we can judge how the student *honestly* made the *self-assessment.* How for the behaviour of the student is honest can be determined in this way.

The second method is the *direct observation* of the individuals in the performance of some task. In this method an independent observer records the individual's behaviour. Ciminero called this method as *"The hallmark of behavioural assessment"*, and he states that this method has a long history in psychology.

Assessing the personality of an individual person as a whole or certain traits specifically can be made by systematically observing the behaviour of the individual. In what type of work the behaviour is exercised? *Behaviour test* is designed on the basis of the type of work in which one is engaged. Tests of this type are available as the other type of tests. In applying behaviour test individuals must be put into frame of an experimental situation which may bring out individual differences in behaviour. In this experimental situation individuals are exposed to direct observation and each one's performance may be compared with everyone else. What sort of skills and talents or traits or characteristics each one exhibits can be observed and recorded. Comparatively it is one of the earliest testing procedure adapted in personality assessments.

The first of this series of behaviour test was designed and used by **Hartshorne** and **May** (1928) for their study of *Character Education Inquiry.* The testing procedures were very interesting and valuable results were also brought out. The behaviour test was conducted by Hartshorne and May among several hundred school students of both boys and girls. They all were from the *fifth grade to eight grade.* The investigators first had oral conversations with the boys and girls. They were asked quite number of questions about their ideal views and attributes regarding the matters of *honesty, generosity, self-control* and *persistence.*

In the second stage of the test, the boys and girls were put into an experimental situations where they had free chance to exercise the same *ideal behaviour patterns,* honesty, generosity, self-control and persistence. Several honest tests of this type were conducted among the school children. It was observed that they had an opportunity to *improve their marks* on arithmetic tests *by cheating.*

A procedure was followed by which the *chating behaviour* was easily detected. In the test of generosity they were measured by the degree to which they were willing to *give up* their *simple satisfaction* and were ready to go to school earlier and *make up things for sick children in hospital.* In all these tests there were changes between the fifth gride and the eighth grade children. These changes were due to *growing up factor* and also environmental situations in which the children were growing up.

It was observed that there was a rapid increase of honesty among the group of children who come from the background where *honesty was stressed by the parents* and the *teachers* but there was a decline in the honesty of the group of children who come from *slum areas* where the importance of *honesty in behaviour was not stressed.* It should be also kept in mind that such behaviour tests have certain advantage and also certain disadvantages. Here the behaviour tests is self-rating and rating by others. It cuts only *a small segment* of the total personality. We cannot attribute the result of the segment to the total personality of the individual. Therefore, we may state that *it is the test of a trait of* an individual. The *predictive power* in this type of testing *is much less.* Therefore different type of behaviour tests should be administered in different situations for assessing the total personality.

In the case of administering behaviour test for assessing personality it is often better to give *both the verbal and behaviour tests* and also to combine both the results in the same investigations. During the *Second World War* an effective combination of the these two methods was made by the British Defense psychologists for the British soldiers. Later on the same procedure was adapted by the United States defense service for rating men for difficult and dangerous overseas military assignments. By this means it is possible to assess the total personality of the individuals.

In behaviour test mostly single items are used to assess personality of the individuals. It is also true that industrial and business organizations make it a point of testing *specific task oriented behaviour* of the individuals. But generally in the industrial and business organizations and in educational situations they want to know simply what will be done in narrow specific situation which yield themselves to behaviour testing and it is done *in a short time available.* Since it is referring to a specific task or to a specified area elaborate behaviour testing with many items not feasible for them. They expect proficiency in the area for which they wanted men and look out for men with some pleasant personality traits. In such situations behaviour tests may exhibit only the specific task or area related personality traits and *not the total personality* of the individuals.

But what is needed generally in the behaviour testing is the broader view of personality as a whole. In the wider view we may get the type of overall composition of the person from which there is every possibility of specific predictions. The specific aspects of the personality can be evolved as and when occasions require. However, it is always better to have the two aspects of the testing, *the verbal and the behavioural methods with several items, taking more amount of time* to avoid mistakes in the identification of correct personality traits in the selection of men to various occupations.

There is always the stereotypic way of selecting men to various occupations by the organizations. They want persons whose broad dispositions something valid which is decided on the basis of a test conducted in a short period of time and they also supplement the person's dispositions with other personal information. But the behaviour tests that are administered to assess the whole personality of an individual serve the real purpose in various situations provided the tests are conducted with all seriousness. Instead of using single items, it is better to use more number of items and the verbal and the behaviour tests should be combined if we want to assess the whole personality.

PART NINE

PSYCHOPATHOLOGY AND MENTAL HEALTH

25

PSYCHOPATHOLOGY

Introduction

Human life is essentially a process of function which is intended for coping up with the existing environment. If the human organism is to servive it has to adapt itself successfully and continuously to the changing situation and environmental conditions. Thus an adaptive behaviour of the organism becomes an essential process for survival. This adaptive behaviour of the organism or the individual is generally described as adjustment of the individual. The individual in his adjustment or adaptive behaviour mainly copes with his ever changing motives, needs, interests, habits, beliefs and goals on one hand and also has to cope with changing demands of the society, circumstances and environment. The human organism has to grow and develop by achieving all its needs and demands and resolving all the social, psychological and environmental problems.

Generally most of the individuals in a society or culture are able to withstand the demands and successfully achieve their goals and everything goes normally with them. But this is not the case with all individuals. Individuals' motives and goals vary enormously. What is good for one is not good for another. That which is very essential for one is not essential for another. What is achieved by one very easily is not achieved by another with all his efforts. The type of relationship one maintains with others or with the society also determines the degree of adjustment one has developed towards others.

There are wide variations and differences with the type organisations with which the individuals are associated. Individuals' mode of behaviour and temperament also are needed to meet the demands of the organizations.

In a society some people most of the times perform everything very efficiently and achieve their goals easily, their adaptive behaviour is very realistic, socially acceptable, productive and their personal satisfaction is always pleasant. But we also find some people function and behave erroneously and succeed partially and do not find their achievements to the expected level and some in spite of several efforts fail to achieve their goals. These people spend their energies in most of the times in defending their behaviour which were not successful rather than solving their problems. Their acquirements are very limited and distorted. In all their efforts they find personal dis-satisfaction and experience unpleasant feelings.

The third category of people function with great difficulty and their contribution is totally nil. They are not able to make any type of adjustment or cope with others and with their own problems. They depend upon others and on society for everything. Their behaviour pattern is also found to be harmful to society. The personal life of these people is totally miserable, painful and destructive in spite of their efforts to adjust themselves with the society.

All these three types of people behave in certain way. It is by making efforts to adjust. Psychologists view that all behaviour patterns are efforts for an adjustment. But all behaviour patterns do not result in successful adjustment and produce satisfactory results. Psychologists try to differentiate between satisfactory and unsatisfactory adjustments, between real and unreal coping efforts, between normal or sane and abnormal or insane and attempt to deal with the problems of life. It is true that every behaviour is caused, every action has a reaction and every stimulus causes a response. But all behaviour is not unavoidable, or unchargeable or inexpliable. Therefore psychologists are trying to understand the causes of maladjustment and adjustment in the human behaviour so that they believe that behavioural dis-satisfaction or abnormal behaviour pattern can be prevented as far as possible. It is strongly believed that only by better understanding of human behaviour we can achieve better control over the behaviour disorders of human beings. Much efforts were taken in the past few decades for studying various aspects of human behaviour. The branch of psychology that is specially interested in the study of maladjusted and disorganized human behaviour pattern is called *Abnormal psychology* or *psychopathology*.

NORMAL AND ABNORMAL BEHAVIOUR

What is normal and what is abnormal? It is very difficult to bring out a strong clear and distinction between these two states of human behaviour. There is only a thin line that divides these two states. Abnormal psychology is the study of deviant behaviour of human beings. This branch of psychology is also referred to behaviour pathology, behaviour disorder, mental disorder, and abnormal behaviour. All these concepts or terms indicate the conditions of a departure from normal state. Generally it is believed that deviant behaviour is considered as the extreme state of abnormality. We see persons in the society border line cases like impulsiveness, extreme irritability, suspiciousness associated with extreme sensibility which all cause a sense of fear that these may damage the cordial relationship with others. These states may also endanger the happiness of life of the persons who suffer from these types of emotional disorders and also the happiness of others who are related to them. Strictly speaking, it is very difficult to draw a thick line of demarcation between the states of normality and abnormality.

The behavioural characteristics that we observe in persons are in an unsuspectable manner and are going on continuously without any scope of separating one extreme state from another extreme state. While critically observing the whole situation we find that things are going on normally as they should be. But the differences that we see in the behavioural characteristics they seem to be a matter of different levels or a matter of degree and not a type or kind. What we generally attribute as abnormal behaviour is applicable to this. What we speak of a person as mentally disturbed, is an exaggerated version of certain behavioural patterns revealed by the person in different degrees are only the patterns revealed by a normal person. What we consider a person who is cent-per-cent accepted as a normal, may sometimes behave very impulsively on an incident of unpleasant result.

It is often stated that mentally deviant persons apply defensive mechanism for their inconsistent behaviour. But this process of defensive mechanism is very common among the most of the normal persons, either to justify their mistakes or to reject them.

If that is case, how can we attribute the defensive mechanism only to mentally or emotionally deviant persons? In the normal life situation no person in the society fully makes use of his intelligence, physical skills, social innovations in all his goal seeking efforts. His non-achievements of the goals are not attributed to any degree of abnormality. But society labels them as normal and mentally healthy. **Moskowitz** and **Orgel** (1969) state that there is no universally accepted definition of abnormality at present. They state that experts in this study have different views based upon theoretical description of mental abnormality, emphasizing effective and healthy adjustment of the individuals. But the psychoanalysts description of normality puts importance on conscious control of behaviour. Persuasion of behavioural differences theoretically always create controversies among the experts.

But, however differences exist, many of the basic behavioural qualities of the normal persons cannot be totally neglected or omitted from the studies. **Moskowitz** and **Orgel** (1969) define the ideal normality and an ideal normal person, the normal individual behaves in a realistic manner in his problem solving and in his coping efforts, he is effective in meeting his needs and satisfying his motives, is flexible to the changing demands of reality, is highly socialized, acts according to the requirements of society, enjoys his life, derives satisfaction from the effective and productive use of his intellectual, emotional and social resources, is a matured person, lives in commensurate with his age, intelligence, social maturity and physical skills.

Abnormality or abnormal behaviour is a matter of degree. Abnormal generally means not normal, it is deviating from the usual way of life. Abnormality may mean extraordinary or odd in behaviour or in appearance. Abnormality may mean abnormal feature, event or act or behaviour. It may mean mental malfunction, deformity, deviation from usual. Abnormality also may mean physical malformation of a person who is an abnormal or deviant, exhibits unusually or highly disturbed behaviour whose personality characteristics deviate to a greater or lesser degree from that what we regard as normal. Mildly deviant people do not make up a species different from the normal.

The behaviour of an abnormal person is not realistic, not effective in his efforts to achieve his goals, he is rigid, not flexible, he is socially isolated, not capable of learning new responses. The abnormal person is not capable of deriving productive use of his intellectual, emotional, physical and social resources. He fails to act in accordance with his age, experience, intelligence and physical capacities. But all these descriptions of abnormalities often subjected to criticisms and different explanations are offered time to time.

Then how to distinguish abnormal behaviour from normal behaviour. According to Sutherland (1976) there are three criteria by which the distination can be made. The first criterion is the statistical. Abnormal behaviour is statically assessed. It is the extreme departure from ordinary behaviour, seen with curve. It is also unusual and uncommon. The second criterion is the culture in which the individual lives. The society expects the behaviour of an individual at certain level. But when the behaviour is different from what the society expected, it is stated as abnormal. The third criterion is the individual's personal adjustment. When this adjustment is impaired the individual's behaviour is different from the normal.

Since the theoretical description of abnormality and the characteristic of abnormal person is not universally acceptable, the clinical approach is used to describe abnormality. In clinical approach the symptoms found in the person are taken into consideration. Mostly the psychiatrists who worked on it had collected the symptoms and described the various types of abnormalities. The descriptive and clinical approaches are also called by a group of psychologists and psychiatrists as descriptive psychopathology and developmental psychopathology.

The problem based upon the study of abnormality is, what is the best way of describing and categorising disordered behaviour? Secondly, what are the causes and origins of disordered behaviour? These questions were answered by evolving descriptive approach and developmental approach by two eminent psychiatrists, **Emil Kraepelin** (1856-1926), a German psychiatrist, and **Sigmund Freud** (1856-1939), an Austrain psychiatrist. Both of these have made painful and enduring contributions to the study of behaviour disorders. Kraepelin was a descriptive psychiatrist, made excellent contributions for the development of classification system for mental disorders then any other medical scientists. He evolved the descriptive approach to the study of mental disorders. His contribution is generally stated as descriptive psychopathology. Kraepelin strongly believed that personal investigation and continued observation of the largest possible number of patients alone can provide scientific information for founding psychopathology.

Information of this type may help to identify the patterns of behavioural or syndromes of characteristics of varied types of patients. These syndromes, according to Kraepelin, would help us to evolve a system of classification of the mental disorders.

But, later on it is felt that description of the symptoms and classification alone cannot answer the question, as how different type of mental disorders develop or originate. At this stage another eminent psychiatrist Sigmund Freud came for rescue and to answer the question. Here, these two psychiatristis' contributions vary in two different lines. Freud's contributions are based upon developmental approach, thus it is called developmental psychopathology. In other words, Freud's approach was from clinical source. Freud started his career as a neurologist, dealing with neurotic patients. He was a practicing clinician directly dealing with the neurotic and psychotic patients. He founded the school of psychoanalysis and his psychoanalytic works constitute twenty-four volumes. He first published the book under the title of *'Studies in Hysteria'* co-authored with Joseph Breuer based upon the information collected from five patients in the year 1895. In the beginning this was went through severe criticism though Freud did not differ very much from Kraepelin. Freud was able to set forth the developmental approach in which he emphasized that the symptoms he observed in patients had sexual origin. Though Kraepelin did not fully accept Freud's stress of sexual origins, Freud continued his clinical observation and verified his hypotheses that the symptoms of the discese had sexual origins. He observed this in many of his patients while treating them. Thus the developmental or clinical approach was evolved.

The clinical approach to abnormality may be also stated as medical approach or practical criterion. In the clinical approach abnormality is observed by the appearance of symptoms. The impairment of personal adjustment is noticed in the personal distress and in the behavioural disturbances. Medical scientists identify the disease on seeing the signs appear on the body. In the same manner clinical psychologists and psychiatrists depend upon the practical signs they observe in the patient. The signs may be mainly referring to the psychological disturbances like depression, anxiety, aggression, frustration, phobias and so on. These symptoms of abnormal behaviour are used for classification. They are of two types, as subjective symptoms and objective symptoms.

The subjective symptoms are the reflections of the patient. The patient's sufferings, discomforts and distresses are reported by the patient whenever the patient felt any discomfort and distress or depression, he reported the same and they were recorded. The patient may sometimes report the feelings of insecurity, helplessness also feelings of destitute. All these are subjective symptoms.

The objective symptoms are those that can be observed by the clinical psychologist or psychiatrist. These symptoms are observed overtly. These are the signs of maladjustment, eccentric behaviour, impulsive behaviour, blaming others without any evidence, expressing the emotion of fear and so on. It is also observed that abnormal behaviour is often followed by any one type of symptoms and sometimes both the subjective and objective symptoms may be present in the patient. But a person's expression of depression or anxiety cannot be taken for granted that he expresses an abnormal symptoms of anxiety and depression. Any normal person on receiving a very sorrowful information may express depression. This cannot be interpreted as the symptom of abnormal behaviour.

Though the clinical approach like the medical, is a practical criterion, it has its own weakness in evaluating the significances of the symptoms. The clinician is making only subjective judgement of the symptoms. It is strongly believed that the clinician is the correct person to judge the behaviour of the abnormal persons. But, however a clinician is well qualified, experienced, sensitive to the sufferings and distresses of the people, being human, is likely to make errors in his judgement of the symptoms, since he makes a subjective judgement. Therefore, the clinician in actual practice may make use of different techniques in evaluating the symptoms of abnormality. Sometimes a group of clinicians jointly judge the symptoms for arriving at the final decision.

Diagnosis

What is diagnosis? Diagnosis is a process of identification of diseases from the examination of the symptoms. It is an opinion or conclusion so reached. It is a thorough analysis of the facts or problems in order to gain understanding and aid future planning. Diagnosis is a detailed description of an organism especially a human body for the purpose of classification. This is mainly practiced in medical and psychological fields.

The psychiatrists and the psychologists in dealing with mental disorders or in dealing with mental patients the main objective of their job is to diagnose the nature of the disease on the basis of the symptoms they observe in the patients to whom treatment is administered.

In descriptive psychopathology, or approach diagnosis is the principle task of the psychiatrists and psychologists. They observe and collect and describe the symptoms of mental disorder and classify the forms of deviant behaviour on the basis of the description. Diagnoses are valid only if they are reliable just like the psychological tests. Reliability means consistency in diagnosis. It is agreement among two or more clinicians.

Classification of Abnormal Behaviour

Classification is an important characteristic of any science. It is the systematic arrangement of the subject matter of a science in an order for the purpose of classification. Any specific scheme that is to be adapted should be based on purpose for the classification. In the field of psychopathology or abnormal psychology the process of classification is very important, for identifying the individual disturbances. Only then proper treatment can be arranged. Classification is also necessary for preparing mental health programmes, for teaching and practicing various courses of study in the academic and in the public service field. All these goals can be achieved only through the system of classification which is routed in the causative factors.

Several classifications of the mental disorders have been published. The most widely used classification list of mental disorders is the one published by the American Psychiatric Association in 1992. But this scheme of classification was revised later on. As and when the scientific advancement take place, the understanding of behaviour disorders also undergo a process of change and revision time to time. Mostly they were dealing with the area of functional disorders. These orders listed in the classification are primarily originating from psychological factors and not on organic factors. It is also possible to discover the physical basis of these disorders. Thus there is some recognition with regards to the types of mental disorders such as functional disorders and organic disorders.

But medical sciences were progressed very rapidly by the end of nineteenth century and methods of testing, diagnozing and treating the patients were throughly revised and it was realized that different treatments should be given to different diseases. There were improvements in the diagnostic procedure and on the basis of these, diseases were classified and treatments were provided. Investigators, in the field of mental health were very much impressed by the advancement that took place in the field of medicine, had also attempted to make use of those techniques in dealing with mental disorders. However, there were much inconsistency and confusion in classifying the behaviour disorders. They grouped the mental disorders on the basis of the symptoms.

Around the same time, in Britain in 1882, the Royal Medical Psychological Association had produced a classification system. This underwent several revisions. But this was not adapted by the Americans. The American Institution for the insane had published a classification system in 1866. This was supposed to be the forerunner for the American Psychiatric Association for the later classifications. In the year 1889 the Congress of Mental Services in Paris had adapted a single classification system. This classification system did not attract many and was not adopted. But the World Health Organisation (WHO) in 1913 accepted a new classification system based on Kraepelin's ideas. Then came the role of the American Psychiatric Association which had published its own first classification system. It was published in the year 1952 under the title of Diagnostic and Statistical Manual — DSM -1.

A second, revised Diagnostic and Statistical Manual was also published in 1968. It is the DSM-II. The same American Psychiatric Association has published a third revised classification system in 1980 and it is known as DSM-III. An extensive revision of the classification of mental disorders was published by the APA in 1987 which is called DSM-III-R. The DSM-III of the American Psychiatric Association's classification system is widely used at present. All the later revisions of the classification systems are mainly based upon the first classification system of mental disorders of the 1952 classifications.

Types of Mental Disorders

The 1952 classification of the mental disorders consisted of psychoses, neuroses and character disorders. These three major categories consists of sub-categories of mental disturbances. This classification system differentiates psychoses, nuroses and character disorders mainly based upon the symptoms. Generally the symptoms are the manifest signs of the abnormalities, whether they are subjective or objective types of behaviour, the clinicians apply to identify the behavioural disorders or maladjustments. The 1952 classification is the basic for all other subsequent divisions made later on. By understanding the type of disorders in each of these three major categories, it will be easy for us to know the enlarged classifications of DSM-III.

Psychosis : According to this classification psychosis is a disorganization of personality in which the intellectual, vocational and social functions are impaired.

It consists of four major divisions or categories such as (1) involuational psychosis, (2) affective psychosis, (3) schizophranic reactions and (4) paranoid reactions.

Neurosis : In neurotic state there is also impairment of vocational and intellectual functions, but has no personality disorganization and loss of contact with world of reality. There are six sub-categories of neurosis such as anxiety, dissociation, conversion, phobia, observsive — compulsive reaction and depressive reaction.

Character disorders : This is mainly concerned with the deficiencies in the development of personality characteristics in which mainly the coping mechanism is distorted. There is trait disturbance and hardly sociopathic personality disturbance and also emotional disturbances.

Types of Mental Disroders as Presented in DSM-III Classification

The creteria designed for diagnosis of the mental disorders are provided in the third revised edition of the Diagnostic and Statistical Manual of Mental Disorders. It is DSM-III (1980). This is most widely accepted classification. This manual was designed to explain clearly in great detail specific behavioural definitions of many of the more severe mental disorders. In defining this much attention was paid with regard to the reliability of the diagnosis. This was very much stressed by many outstanding scholars like **Eysenck, Wakefield** and **Friedman** (1983) and **Stobar** (1981). Eysenck and his co-authors have criticised the DSM-III Manual as many of the behaviours do not have medical relevance. But with all the criticisms this manual had its own merits. The description of schizophrenia, affective disorders and psychotic descriptions are given more importance.

The criticism made by Eysenck and others is that the DSM-III classification includes non-medical problems like gambling, antisocial behaviour, academic and occupational problems, parent-child problems, marital problems and substance use disorders. The DSM-III classification system is presented in five axis classifications system. Axis is an imaginary line or area into which the relevant subject matters are included.

Axis-I : This axis consists of disorders right from the period of childhood stage. Disorders are usually first evident in infancy, childhood or adolescence, organic mental disorders, substance use disorders, schizophrenic disorders, paranoid disorders, affective disorders, anxiety disorders, somatoform disorders, dissociative disorders, psychol-sexual disorders and psychological factors affecting physical condition.

Axis-II : This axis includes personality disorders such as paranoid, schizophoid, schizophal, histrionic, narcissistic, antisocial, borderline, avoident, dependent, compulsive, passive-aggressive. Further, it includes specific developmental disorders such as reading, arithmetic, language and articulation.

Axis-III : This axis includes physical disorders and conditions.

Axis-IV : This includes seven psychosocial stressors such as, none, minimal (like small loan), mild (like argument with neighbour) moderate (like pregnancy), severe (like serious illness), extreme (like death of close relative), and catastrophic (like natural disaster).

Axis-V : Axis V includes highest level of adaptive functional past year. This involves seven levels such as,

Superior, usually effective in social relations, occupation, and use of leisure time (e.g. single parent in reduced circumstance taken excellent care of children, has warm friendships and a Lobby).

Very good : This is better than average in job, leisure time, and social functioning (like retired persons doing voluntary work, sees old friends).

Good : It is no more than slight impairment in either occupational social functioning (like individual with many friends, does a difficult job, extremely well but finds it a strain).

Fair : This refers to moderate impairment either in social or occupational functioning or some impairment in both (like a lawyer has trouble in carrying through assignments, has almost no close friends).

Poor : This indicates marked impairment in either social or occupational functioning or moderate impairment in both (e.g. man with one or two friends has trouble holding a job for more than a few weeks).

Very poor : This level is marked impairment in social and occupational functioning (e.g., Women is unable to do any house work and has violent outbursts),

Grossly Impaired : It is the level of gross impairment in virtually all areas of functioning (e.g. elderly man needs supervision in maintaining personal hygiene, is usua!ly incoherant).

All these types of mental disorders presented under five Axis as given in the Diagnostic and Statistical Manual of Mental Disorders can be brought under three broad categories such as neuroses, psychoses, psychophysiological disorders and personality or character disorders. But each one may be divided into sub-categories as shown in the diagnostic manual. We are mainly concerned here only with some of the major categories and in the course discussion we may discess some of the sub-categories which are most relevant or which are of essential types.

NEUROSES

Neurosis is one of the major disorders and it has to be distinguished from other major disorders like the psychoses. It is one of the less severe form of psychological disorders. Neuroses (plural form of neurosis) sometimes become severe and cause more trouble and also may require hospitalization. But neuroses generally do not cause personality disorder and also do not result in the loss of contact with the world of reality. Neurotic persons do not isolate themselves from society. They can get along with others easily. But since they are in the extreme state of anxiety they may fail to make use of their potentialities.

The term, neurosis refers to a state of behaviour in which the individual suffers from a form of maladjustment. The individual at this stage is unable to cope with his anxieties and conflicts and develops certain abnormal symptoms. The disturbance the individual suffers is not so much severe as to produce profound personality disorganizations similar to psychotic reactions. Neuroses are generally referred to a group of disorders. Neuroses include anxiety reaction, conversion reaction, phobic reaction obesessive-compulsive reaction, and neurotic depression.

A Neurosis may be also defined, according to **Garrett** (1975), as a relatively mild personality disturbance, marked by excessive fatigue, anxiety, tension, lack of self-confidence and feelings of inadequacy and inferiority. Since neurotic persons are physically diseased people do not die of neuroses. They are rarely hospitalized. Neurotic persons have deep insight. They are capable of realizing their behaviour that they behave unusually and unsatisfactorily and because of that they have disturbances and also disturbing others. The individual in his life fails to achieve a realistic solution, he is driven to the state of severe anxiety or resorts to a defense mechanism, a state in which he finds some partial satisfaction by reducing the severe anxiety state. The defense mechanism in a neurotic person may either reduce the anxiety state or may create further problems if the defense mechanism does not alleviate the anxiety. In the neurotic disorders, anxiety is the central part of the functioning.

Persons in the server anxiety state are found to be very much strained and also in the state of tension. The other symptoms are insomnia, indigestion, diarrhoea, inability to concentrate on anything, feeling sexual importance. Anxiety is not sometimes apparant in a person, but it is possible to observe from the person's behaviour of maladjustment. The individual may find out some means of defending himself from the maladaptive behaviour. Neurotic behaviour sometimes is found to be having the reverberating effect. That is, a solution found for a problem becomes the cause of another problem. By these means the individual is not able to cope with many of everyday problems. Therefore as far as possible the neurotic person avoids then by defense mechanism. By this process, he feels guilty, and unhappy as he fails to meet the situation successfully while others are successful.

Moskowitz and **Orgel** (1969) make the psychoanalytic interpretation for neurosis. According to this, it is an excessive development of super ego. Super ego is considered to be the keystone of socialized behaviour. In normal people the super ego development is normal and their behaviour is self-controlled and socially acceptable. Persons who have character disorder have deficit of super ego. The behaviour of such people is poorly controlled and socially not acceptable. But when the super ego development is in excess it generates intense conflict and the person is unable to satisfy his personal motives and without the feelings of guilt and anxiety. This state is stated as over socialized which is the characteristic of the neurosis.

Classification of the Neuroses

Sometimes it is said that neurotics become the psychotics. **Garrett** (1975) states that it is very rare. The continuity from neurosis to the psychosis is based upon the increasing severity of the neurosis. It is also observed that some of the symptoms of neurosis and psychosis appear in the same manner. For example, the dissociative type of hysterical neurosis has such reactions as amnesia and multiple personality, the depressive neurosis has abnormally prolonged dejection, associated with internal conflict, interpersonal loss and environmental set back. These are similar to the symptoms of functional psychosis. In the general behaviour, there is maladpative behaviour in neurosis and in functional psychosis. There is mild impairment of personal and social functioning in neurosis and in functional psychosis. There is marked impairment of contact with the world of reality and in personal and social functioning. Therefore, the classification of neuroses in a strict sense is very difficult at the functional level. However the important neurotic reactions can be easily identified. Neurosis may be also defined as a relatively mild mental disorder characterized by the symptoms of anxiety, hysteria obsessive behaviour, is also called psychoneurosis.

Anxiety Neuroses

Anxiety neurosis or reaction is one of the most common form of neuroses. It is characterized subjectively by intense apprehension and dread. This is the common neurotic symptom. There are physical symptoms accompanied by anxiety. These are rapid and irregular breathing, intense sweating, heart palpitation and feelings of pressure in the head and sinking in the stomach. A neurotic person lives everyday with a level of tension much greater than a normal person. The person is subjected to acute anxiety attacks which may occur several times everyday or once in a week or month. Anxiety also may be defined as a state of uneasiness or tension caused by apprehension of possible future misfortune, danger, worry etc. It is also an intense desire or eagerness. Psychologically an intense state of apprehension or worry is occurring in a variety of nervous and mental disorders.

When there is acute attack, the neurotic person develops an overwhelming feeling that something dreadful is going to take place in him. This feeling is followed by some physiological symptoms that we have stated above and also the symptoms like muscle tension, faintness and nausea. It is all because the sympathetic division of the autonomic nervous system becomes excited. The same type of symptoms may occur when a person is extremely frigatened. The over activity of the endocrine glands like the thyroid gland may also produce the same type of symptoms like the anxiety attack.

The neurotic person may not be having any clear idea as to why he is so much frightened. The frightening is not associated with any specific stimulus or object. Therefore, **Wolpe** (1969) calls it as 'free floating.' This means that eliciting stimuli, are obscure, complex and pervasive, not external, but must be the feelings of internal conflict within the person who suffers from neurotic anxiety. According to **Levitt** (1967) most of the anxiety neuroses are lying between diffused state, free floating anxiety and highly specific dreads. For example, the mother who is very much worried over her child's bodily safety and security imagines all possible dangers to her child and very much anxious and develops severe anxiety neurosis (Fig. 25.1).

Apart from these free-floating anxiety indications, there are also other observed symptoms of anxiety neurosis. The anxiety neurotic person suffers from the inability to concentrate on anything, unable to make decisions. He is extremely sensitive to certain sensations, he is very much discouraged, unable to sleep peacefully. These symptoms are the characteristics of the anxiety neurotic pattern.

It is quite but common even among the normal persons to become very anxious, tense while feeling some difficult and dangerous situations. The normal person is put into severe stress. While witness an exciting Foot-ball match between two nations the people often become highly excited and stressful and develop too much of anxiety or worry as to the end of the match. This anxiety and stress exists only for a short period in the people who witness the match. But in the case of a neurotic person such situation becomes a thread to his normal way of life and gets break down and is subjected to the attacks of anxiety neurosis and fails to get relaxed from the tense situation unlike the normal person. Several studies have been conducted on anxiety neurotics and found that they had high standards of performance. But they come from the families where their parents had high expectations in their children and wanted that they should make high achievements. **Jenkins** (1968) states that however successful the anxiety neurotic he cannot relax from his state and free from the feelings of apprehensive of his ability to deal with his demands of his life and society.

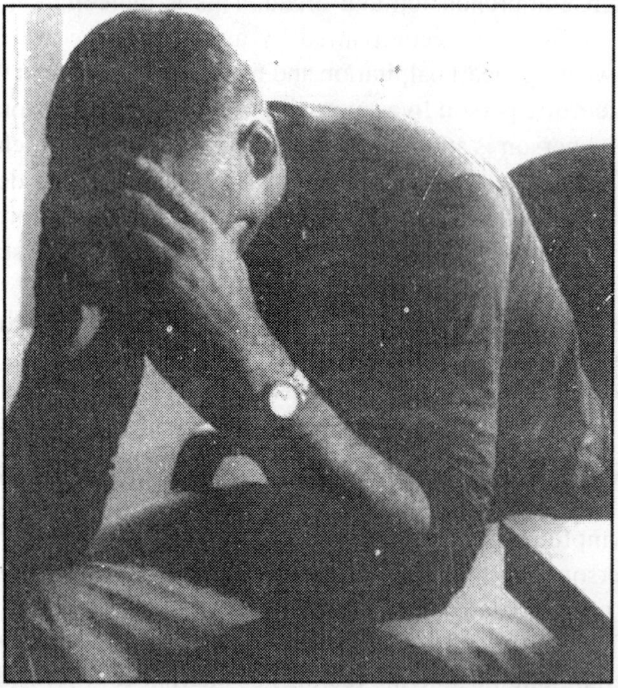

Fig. 25.1 : Anxiety Neuroses : The classification of neuroses made today includes six main forms such as anxiety neuroses, dissociative neuroses, conversion neuroses, phobic neuroses, obsessive-compulsive neuroses, and neurotic depressions. The anxiety neurosis (reaction) describes the most common neurotic symptoms. The patient often cannot identify objects, event and persons in severe state. This type of anxiety neurosis is called free-floating neurosis. It is because that it often goes on shifting, not attaching to any single situation. It can be chronic or acute. the anxiety neurotic patient frequently may be experiencing muscular tension, emotional strain and insomania. In acute state the person may face several attacks everyday or even once in a week or month. It is a state of sevrere uneasiness or tension caused by apprehension of possible misfortune, danger or failure in his efforts to achieve some worthy thing. Psychologically an intense state of fear or worry occurring in a variety of nervous and mental disorders.

In the life situation the person may face some sudden stress which may cause an acute anxiety attack. The situations may be the failure to get a promotion in his job which may require greater responsibility, the death of an important person in the family or failure to get the job which he has been trying for a long time. These are the types of threatening situations which may cause acute anxiety attacks. In his attempts to achieve the desired goal he feels that he is incompetent and gets frustrated and provoked and becomes hostile and also violent.

Obsessive-Compulsive Neuroses

The obsessive-compulsive neurosis is a type of neurotic pattern in which the neurotic person is indirectly forced to perform certain actions and also forced to think certain things which he did not want to think about. It is a disorder characterized by persistent intrusion of unwanted desires, thoughts and actions. It may be also defined as a neurotic behaviour pattern characterized by one of the three forms of behaviour such as recurrent thoughts, often disturbing and unwanted (obsessions), irresistable urges to repeat stereotyped and ritualistic acts (compulsions) and also both of these two functions combined.

The obsessive-compulsive neurosis is an anxiety disorder in which the mind floated with persistent and uncontrollable thought and the person is compelled to repeat certain acts again and again which may cause actue distress and interference with the everyday functioning of the person. In the obsessive-compulsive neurotic state there are persistent urges to repeat the acts though there is no, specific purpose in it. The neurotic is given to irrational and uncontrollable thoughts and actions.

Psychoanalytically the obsessive-compulsive neurotic builds up a faulty and incomplete super-ego or conscience in the childhood period. But some are of the view that this neurotic builds up a very strong conscience. This neurotic person becomes also hostile and very much worried by sexual feelings. The neurotic may have obsessive thoughts in a wide areas. They are most often dealing with immoral aggressive and sexual acts. For example, a man may be having recurrent thoughts of exposing his genitals in public places. A man may be obsessed with the idea of poisoning his wife to death on a charge of unfaithfulness. Actually the person is not carrying out these thoughts in action. It is because the person is not having any control over these thoughts. He is also very much horrified by these uncontrolled or obsessive thoughts. He is also worried that why such thoughts persist in him. Persistent such thoughts often in the person may even make him to think that he has become an insane.

It is observed that compulsive neurotic acts vary from mild type of superstitious behaviour to ritualistic behaviour. People of normal type find certain amount of comfort in the ritualistic behaviour in times of stress. But the obsessive-compulsive neurotic person exaggrates the ritualistic. behaviour while facing difficult situations. The obsessive - compulsive acts can be seen in some persons as and when they walk through the streets, by touching every lamp post and also counting the number of houses in the street. Their compulsive act is such that if they miss anyone lamp post they go back and touch it and walk. English poet, William Shakespeare in his play *Mecbeth* has explained this neurotic behaviour through

Lady Macbeth. She was presented with the compulsive act that she washed her hands several times everyday to wash her blood stained hands.

Phobic Neuroses

Phobia means a morbid irrational fear held by a neurotic person in defense against anxiety. Phobia is a compelling fear or dread of a particular object or situation. Thus phobic neurosis is an anxiety disorder in which the intense fear and avoidance of specific objects and situations, based upon irrational fear the individual has developed. Phobic neurosis may also be defined as the disorder characterized by intense fear or an object or situation which the individual has developed irrationally which in reality causes no danger. Janet while classifying psychoneurosis put every neurosis under the heading of psychasthenia leaving hysteria. Psychasthernia means weak mind or weak will, which also means affected will power of an individual. Thus psychasthenia includes all phobias, obsessions and compulsions. Generally phobia is concerned with the feelings of fear of some object, place or situation of harmless nature. Since the harmless objects are the objects of fear for the neurotic, it is called irrational fear.

Formerly each phobia was labelled with a Greek word. In this way we have each phobia is associated with Greek word and we call them as *claustrophobia* which means fear of closed places. The fear of open places is called *agoraphobia* and the fear of high places is called *acrosphobia.* According Strange (1965) in the formation phobia some threatening impulse is repressed. In this process some harmless external object is symbolically present in the original impulse. The represented impulse might have been the strong emotion of desire to have a large genital organ and the horse is the cause of the phobia which symbolically represent the large genital organ. In every phobia there is a tendency to avoid the feared object and to keep away from the overwhelming anxiety. If the repression fails to function the phobia begins to dominate and the person becomes more and more afraid of the objects.

There are a number of situations and objects around which phobias may be developed. **Coleman** (1974) has enumerated a long list of fourteen kinds of phobias. The most feared objects for most of the people, not only for the neurotics but also for normal people are darkness which is called *nyctophobia, phyrophobia* causing fear for fire. Many people are afraid of crowds, which is called *ocholphobia.* **Eiduson** (1968) in his observations, found that among the neuortic patients who sought clinical assistance, twelve per cent of them were suffering from phobic neuroses and this occurred more commonly among the young adults and more women had phobic neuroses.

Phobic neurostic persons show also other symptoms in addition to phobic symptoms such as dizziness, stomach upset, back pain, headache and also have the fear of breaking down. One of the main reasons for the development of phobias in individuals according to **Terhune** (1961) and the Kerry (1960) is the current culture in which the individuals live.

Phobias are mostly the childhood neurotic symptoms. They decrease as and when they grow into adolescence and adulthood. But in some cases phobic disorders increase and become more complete state of psychoneurosis. Phobic neurotic patients may be treated with psychotherapy and psychoanalysis.

Hysterical Neuroses

The term hysteria is derived from the Greek word 'voripa' which means womb and also from the Latin word, 'hysteric', 'hystericus'. Thus hysteria, may be defined as a functional disturbance of the nervous system characterized by anaesthesia, hyperesthesia, convulsions etc., and usually attended with emotional disturbances or perversion of the moral and intellectual faculties. Women being more liable than men to this disorder. It was originally thought to be due to a disturbance of the uterus, and due to unhealthy emotion and excitement. Hysteria is a type of psychoneurosis characterized by dissociation or conversion reactions. Hysteria was considered as the women's disease and it was caused due to sexual difficulties. Because of the sexual desires and for wanting children, the uterus (woman) is frustrated and wanders. This may be revealed in several symptoms like chocking, sensations, spleen and temper tantrums. Hysterical symptoms may vary from person to person. Hippocrates thought that immediate marriage is the best remedy, since most of the hysterical patients were young women. **Freud** (1985) who was treating mostly the hysterical patients, who were mostly women, modified these ideas and stated that the hysterical symptoms are due to repressed and deviated sexual desires and energy. According to Freud the psychosexual conflict in the person was translated into bodily disorder. Freud had stated that paralysis in the bodily parts might be the expression of repressed deviated intense sexual energy. Freudian view that the sexual conflict is converted into bodily disorders is not fully accepted by the modern investigators. But, however, they could not fully disagree with Freud's view and they retained the psychological process, conversion. Therefore, the hysterical neurosis is stated as conversion reactions and dissociative reactions (Fig. 25.2).

Fig. 25.2 : Hysterical Neuroses : Hysterical neuroses find the root in the emotional disturbances. Any severe emotional shock due to some disappointment, specially of sexual type causes hysterical neurosis. It results in the impairment of moral and intellectual faculties. Hysterical neurosis is mainly an emotional disorder and women are more liable than men to this disorder. Originally, it was thought that it is the disturbance of the uterus and due to unhealthy emotion and excitment. Hysteria is a type of psychoneurosis characterized by dissociation or conversion reaction. When women's sexual desires are not fulfilled and when there is no chance for giving birth to children, they get frustrated and succumb to hysterical attacks. This is revealed in the symptoms of choking, sensations, spleen and temper tantrums. These symptoms vary person to person. Hippocrates stated that immediate marriage is the best remedy, since most of the hysterical patients are women. Freud is of the view that hysterical symptoms are due to repressed deviated sexual desires and energy. The psyosexual conflict in the person is converted into bodily disorders. This hysterical neurosis is stated as conversion reaction and dissociative reaction.

Conversion Hysteria or Conversion Disorders

The conversion reaction according to Freud was the real conversion of the sexual energy of the neurotic person and the repressed sexual impulses were symbolically expressed in the form of symptoms. The conversion symptoms have the effect of providing some advantage to the patient by reducing the anxiety when the person's defense mechanism fails to act. The conversion hysteric symptoms are purely psychogenic and has no organic basis or cause, except physical weakness. The blind person cannot see, but there is nothing physical defects in his eyes, nor in the nerves or in the brain centres that serve the eyes. The paralysed person has no physical defects either in his muscles or in his nerves. This type of hysteria may impair the speech ability. This type of histerical paralyses do not have physiological sense.

Conversion hysteria may cause blindness, deafness, impair the speech and also may involve in some sensory areas of taste and smell. The food may be tasteless and the pleasant smell is not realized. Conversion reactions occur in persons who have developed extreme form of stress. It is found mostly among the soldiers who are put in the battle fields where there is intense fighting. The soldiers confront with extreme stress of combat. The consequences may be paralysis of the legs, hands, blindness, deafness which prevent the soldier from returning to the battle field — the combating area. This is not out of cowardice on the part of the soldier.

But recent studies on hysteria brought out several changes and modifications. What was formerly called hysteria, now termed as conversion hysteria or reactions, and dissociative reactions. Now they call it as somatoform disorders. Somatoform disorder may be defined as the disorder in which physical symptoms suggest a physiological cause. Therefore, they are believed to be linked to psychological conflict and needs. Thus we may also state that somatoform disorders are with physical symptoms for which there is no evidence accompanying condition for strengthening the assumption that psychological conflict is underlying them. This includes somatization disorder and conversion disorder.

In somatization disorders there are complaints of bodily symptoms which mean, there is a physical disorder. Conversion disorder is formerly called hysterical neurosis. The serious ailments (symptoms) described above suggest neurological diseases but have no organic or phychological basis to be demonstrated. Therefore, it is assumed to be the expressions of psychological conflict.

Hysterical Neurosis or Dissociative Disorders

The dissociative dissorder involves some change in the consciousness temporarily. Generally there is loss of memory of the past incidents, which is technically called amnesia. The dissociative disorders are invested with strong emotional expression in voice. There is some primary gain like other hysterical case, the anxiety reduction. This patient's amnesia is associated with fugue, a flight from the living place, in other words, the patient runs away to an unknown place and starts a different type of life. He wanted to conceal his identity totally by taking up a disguised way of life.

The patient in the dissociative state performs the function of sumnambulism or sleep walking. It is a translike state. In this state some repressed event is revealed. The patient's strong repressed sex desire is symbolically expressed in his activity of going up and down the stair case.

The disruptive state of the hysterical neurosis also turns into a very complex and intriguing state, in which the person goes into dual or multiple personality. It is a rare type of dissociative disorder. In this state each personality is independent of the other end each personality has its own stable thoughts and feelings. But one of the personalities may be dominant over other personalities.

Origin of the Neuroses

Neuroses are caused by a number of factors. Many of the neuroses are prevalent for a long period of time. The real causes of neuroses can be traced only by intensive psychiatric and psychological treatments. The general observation is that neuroses originate from two sources. One possible source of neurotic states is that it is inherited. The person inherits the bodily systems and the nervous system which are susceptible to emotional stimuli. The bodily parts instantly react to the emotional situation.

The second possible source is the person's environmental situation. For the potential neurotic persons the environmental pressures are so intense and stressful and continued for a long period as the cause severe emotional breakdown by destroying the power of resistance of the individual and cause neurosis. Therefore, it is probable that the hereditary factors and the environmental factors in this have sufficient reasons for causing the neurotic disorders. In some cases both may be causing the problems. Generally these conditions are found among men as they are put in fearful situations like the wars. **Garrett** (1975) states that the form of a neurosis and its characteristics and direction seem to be an accident on the part of the individual. The continuous harassment, scolding and abusing a person, however, he is emotionally strong he may breakdown emotionally. Once it takes place similar harassing situation even in the mild form may cause the same emotional breakdown in the person. Such recurrent incidents no doubt may lead to neurosis. The person may resort to alcoholism or drug addiction and also may become an anti-social personality.

Treatment : Neurotic states may vary in degrees from mild form to severe neurotic state. Neuroses of mild forms may diminish and disappear gradually even without treatment. Persons with good physical health may get rid of it easily, poor financial position if improved may relief the person from mild form of neurosis. Marital status, occupational status, achieving the cherished goal, a relief of conflict and frustration, emotional stress, change of place, avoiding the tension causing situations, persons, incidents, all may gradually reduce the tension and relief the person from the anxieties and even cure the person from the problems.

In the case of severe type of neuroses systematic treatment is needed. It may be a long standing neurotic case and causing problems to himself and to others. Such cases must be attended by psychiatrists or clinical psychologists. The treatment in such cases may be prolonged to the person for bringing him back to the normal emotional state.

There are some Destitute Homes in Tamil Nadu, where the inmates are mostly destitutes, no body to take care of them and no place to go into. Most of them, except few, are psychiatric cases. The cases range from mild neurosis to severe form of neurotic cases. They were not neurotic by heredity, but caused by environmental conditions. They were working in houses as servant maids (most of them women), were treated crually by the housewives, by scolding and abusing them with filthy words right from morning till night and also beating sometimes. All these resulted in neurotic problems in the servant maids. An organization called 'Udavum Karangel' in Chennai restored them to their Institute Homes and treated them with the help of psychiatrists. Within few days they have regained their mental equilbrium. The organization feels very proud of their services which is not done even in psychiatric clinics or mental health centres. The most important reason for the quick restoration of normal live of these destitute was kind words and kind treatment.

PSYCHOSES

Psychoses are the serious mental disorders or illnesses. Psychoses are extreme form of behaviour disorders. It is estimated that about one per cent of people in any culture suffer from psychoses of some type, specifically organic or functional type. Psychoses may be defined as the severe mental disorders adversly affecting mental functioning and social competence. In most of the psychoses there is a loss of contact with the world of reality, extended withdrawal from social relationships, marked deviance in emotional expression, severely impaired thought process and pervasive regressive behaviour.

Psychosis may be also defined as the mental illness in which the patient shows severe change or disorganization of personality, often accompanied by agitation or depression, delusions, hallucinations, commonly required hospitalization. Psychoses are of two types as functional psychoses and organic psychoses. The psychotic patient is more severely disturbed in his behavioural impairment than the neurotic patient. All his normal social functions are mostly cut off and often goes into depressive state. There is an unresolved controversy over the functions of neuroses and psychoses as to whether psychosis is an extreme form of neurosis or these two types of disorders are distinctly different. Some experts are of the view that there is a continuous process of function of disorders from normal level to neurosis and to psychosis. The differences can be traced from the severity of the symptoms. But it is generally observed in the society that the neurotic person turns to normal state when the atmosphere or the environmental conditions become very congenial for him and also by taking some simple treatment. In the case of a psychotic person there may be severity of symptoms and the person always found in deep depression, hallucinations and delusions and requires hospitalization and serious treatment.

The psychotic person does not reveal the same abnormal personality traits or symptoms in a more distorted form than the neurotic person. There may be more physical symptoms in a neurotic then in a psychotic, for example anxiety neurosis, phobias and hysterical attacks, are specific to neuroses. Psychoses are really serious mental illness.

Origin of Psychoses

There are two general broad categories of psychoses, such as organic psychoses and functional psychoses. Organic psychoses are caused by the diseases and other degenerative processes in the body, like the damage to the central nervous system, head injuries, brain tumours or infections, hardening of arteries, poisoning by toxic substances like lead, carbon monoxide and drugs. The functional psychoses have no known organic cause or component. The term functional implies that there is no organism involved or damaged and disorders are caused. It mainly refers to the impairment of personality traits or characterstics. The functional psychoses are the mental disorders that are the results of psychological causes. But there may be also some genetic and biological factors playing important role in causing the functional psychoses.

Classification of Psychoses

The two major types of psychoses are : the organic psychoses and the functional psychoses. The organic psychoses include scme sub-divisions of disorders. They are parasis, senile dementia and alcoholic psychoses. The main functional psychoses are of two types such as *affective psychoses* and *paranoid psychoses.* The affective psychoses are manic depressive psychosis and involutional psychosis. These are characterized by abnormal symptoms of mood and emotion. The other type paranoid psychoses consist of systematized delusions, mostly false ideas. The third one is schizophrenia. But in the recent treatments the functional psychoses are manic depressive psychosis, personoid psychosis and schizophrenia.

Organic Psychoses

The organic psychoses some times back are also called as structural psychoses. Some experts like Strange (1965) used the terms, somatogenic disorders. However organic psychoses include parasis, senile dementia and alcoholic psychosis. It is believed that if there is any type of personality disturbances there may be chances for the diseases of the central nervous system or some degenerative, processes which may result in physical and physiological impairment. These problems are mainly medical type and also must be taken care of.

Parasis is one of the organic psychoses caused due to syphiletic infection of the brain and nervous system. Parasis may be defined as the consequence of syphiletic attack in which organic damage to the brain the invasion of micro-organisms producing profound mental and neurological symptoms. Physically it may cause muscular weakness, slurred speech. The personality changes which may take place in parasis are boastfulness, euphoria, and imposing ideas and unwanted exaggerations. The advancement of therapeutic techniques and treatments have reduced the syphiletic attacks and the disorders of parasis are also becoming rare in the society.

The *senile dementia* causes damage to the brain and to the nervous tissues because of old age. It is a form of psychosis caused in part by the deterioration of brain changes due to age. The level of brain deterioration may vary in degree. Women suffer from this disorder more than men. The on set of this disorder start occurring from the age sixty years.

The clinical assessment of the senile psychosis varies in severity, ranging from mild organic brain syndrome to severe mental impairment and to a passive withdrawal state and death. The senile psychosis occurs gradually resulting in physical and mental disorganisation. It has been clinically assessed that senile psychosis falls into five types. The *first type simple impairment*, in which the patient gradually stops contact with the world. The *second type* is the paranoid senile psychosis, in which there is gradual development of delusions of persecution. The *third type* of senile psychosis is preslyophrenia type, which is clinically assessed as the state of jovial and amicable mood and loss of memory. The *fourth type* is depressed and agitated type. The senile psychotic patient is very much depressed and agitated and suffer from delusions. The *fifth type* is delusions and confused state in which the patient suffers from severe mental impairment and becomes restless.

Alcoholic Psychosis

Alcoholic psychosis occurs due to extreme addiction to alcohol. In alcoholic psychosis there is impairment of brain tissues due to the effects of over indulgence in alcohol. The patients of this psychosis are found in a confused state, suffer from forgetfulness, often boastful and very much disoriented. They do not have any clear idea about what is going on around them. Alcohol psychotic persons often found in depressed state and go into delusions and hallucinations.

FUNCTIONAL PSYCHOSES

Functional psychoses are the severe forms of mental disorders attributed mainly to psychological stress, though some genetic and biological factors are found to be often playing a significant role in the etiology of these disorders. Generally functional mental disorder or psychosis is a form of psychopathological reaction without a known psychological basis. This is in contrast to organic psychoses or mental disorders. It is also stated as a form of mental disorder that results from poor social learning or other psychosocial causes, similar to psychogenic and psychosociogenic disorders. These disorders are divided into three major types. They are manic depressive psychoses, paranoid and schizophrenia.

Manic Depressive Psychosis

Manic depressive psychosis is a mental illness in which the behaviour of the person goes through cyclic manner from elation (manic phase) to depression and melancholia (depressive phase). The manic depressive psychosis is characterized by extreme fluctuations of emotional mood. At the manic state the patient is exhilarated and excited, swears, shouts and talks continuously and at another state, that is, in depressed state cries, refuses to eat, feels extreme fatigue, dependent, feeling extreme sad and accuses himself that he has committed various sins.

In the mild form of manic state the patient exhibits great energy and enthusiasm. He has great confidence in his ability, moves fastly from one activity to another. In a more severe form of manic state which is called hypermania, the patient behaves like a more popular person, singing and shouting with obscene words. Mentally the patient is in a confused state to have experiences of delusions and hallucinations. Some hypermania patients violate all

Fig. 25.3 : Manic Depressive Psychosis : The manic depressive psychosis is marked by elation and depression. It was first described by Kraepelin. It is a mental illness in which the patient goes through in a cyclic manner from elation (manic state) to a state of depression and melancholia. It is characterised by extreme fluctuation of emotional moods. At the manic state the patient is exhilarated and excited, swears, shouts, and talks continuously. The other is the depressed state in which the patient cries, refuses to eat, feels extreme fatigue, becomes dependent on others, feels extreme sad and accuses himself that he has committed serious sinus. He is highly dejected and feels that it is not worth living in this world. He refuses to speak, eat and drink. Clinical observation is that the patient gets periodic recovery from this illness and turns to normal life.

moral principles and get involved in unrestricted sexual behaviour and violent activities. The hypermania person's extreme excitement state can be reduced by the use of sedative therapies (Fig. 25.3).

The depressed patient's behaviour is the opposite of the manic patient's behaviour. The person's physical and mental activities are much slow and incorrect compared to that of a normal person. The patient is very much defected in life and feels that it is not worth living in this world and feels guilty of being born and living. He is very much indifferent to all those who are going on around him and neglects everything which are very essential for living. He refuses to speak, eat and drink. He is fully at the care of others. Clinically it was observed that most of the manic and depressive patients get periodic recovery form the illness and turn to normal life. But some patients go in between the manic and depression states in a cyclic way.

Paranoid or Paranoid State

Paranoid is a mental disorder in which the patient's symptoms are persistant delusions of persecution or delusional jealousy and unfounded conviction that his/her mate is unfaithful. The person is often found to be seclusive, eccentric, contentions and given to litigation, but he is free from thought disorder. The paranoid patient has no justification of anything. He is hightly suspicious and not trustworthy. He is hypersensitives, quickly taking offense, and makes everything big out of nothing.

The terms paranoia, paranoid, paranoid disorder or psychosis, paranoid personality and paranoid schizophrenia are used invariably by scholars in their books. Are these refer to the same subject matter or to dipparent subject matters? As these terms cause some confusion it is better to clarify these terms here. Paranoia is a noun, refers to a mental condition characterized by delusions of persecution, unwarranted jealousy, or exaggerated self-importances, unjustified suspicion and mistrust of others. It is also stated that it is a general term for delusions of persecution, grandously or both, found in several pathological conditions like paranoid disorders, paranoid schizophrenia and paranoid personality disorder.

Paranoid here is an adjustive term characterized by or suffering from parania. Paranoid refers to a person suffering from paranoia. The person is a paranoid patient. Therefore, the correct term to indicate the disorder of paranoid psychosis is *paranoia.* **Coleman** (1974) used the correct term. According to him paranoia is a psychosis characterized by a systematized delusional system.

Clinical assessments state that in paranoia state the individual feels that he is isolated and not trusted by others. He complains that people are planning to kill him, his properties are stolen and he is suspected by others. The paranoid psychotic patient has one major delusional system which is centring round his life.

It is mainly referring to his financial matters, occupation, discovery, unfaithful wife and other matters related to his life. Even in clinical observations it is found to be very difficult to differentiate the paranoid schizophrenia and the manic depressive psychosis from paranoia or paranoid state as these disorders have a common symptoms of delusions. But in the classification only delusions are taken as the main symptoms of paranoia or paranoid state.

Origin of Paranoia

In the etiology of paranoia, investigators are not able to find out the root cause of paranoia in the genetic and in biochemical areas. The available evidences are only from the psychological field. Only on the basis of the functions of psychological factors, paranoia has been classified as one of the functional psychoses. Among the psychoses the paranoia and involutional psychosis are stated as purely functional in origin. The home and the early childhood period experiences are found to be the common factors of the origin of paranoia. There are several other factors enumented by Coleman (1974) as the causes of paranoia. The faculty learning and the personally development in the early years, the individual's failures in the life situations and his inferiority complex, because of the failures in life.

Lemert (1962) found in his studies that the paranoia often misconstured by the people of the world. He thinks that others are against him. He is alert, very much to real and imagined injustices. He creates a pseudo-community real and imagined for his purpose, (Cameron, 1959). Another cause attributed to paranoia is sexual maladjustment. In cases, paranoid reactions are caused due to guilt feeling of real and contemplated immoral sexual behaviour.

Treatment : Treatment for paranoid patients in the earlier days was found to be successful with psychotherapy. This was found to be effective if the patient sought the professional assitence. Salzman (1960), Schwartz (1968) were of the view that individual and group theraphy were found to be effective. These investigators state that paranoid delusions can be altered by combining averse conditioning method. Hospitalization also can alter condition and the patient may realize that it is time to stop his actions and ideas, otherwise he has to be in the hospital for a long time.

SCHIZOPHRENIA

Schizophrenia is a type of psychosis characterized by breakdown of integrated personality functioning, withdrawal from the world of reality, emotional blunting and distortion and disturbance of thought and behaviour. Schizophrenia may be defined as a functional psychotic disorder in which there is a lack of harmony or splits between the aspects of personality functioning, especially between emotion and behaviour. This is characterized by the symptoms delusions, hallucinations and autism. According to Strange (1960) schizophrenia is one of the functional psychoses characterized by apathy and social withdrawal.

The functional psychosis, schizophrenia is spread into four forms, all of them are marked by dissociation which is spliting of the personality. The schizophrenic person is highly introverted and shut himself within him. He shows his real emotions very little. He often experiences auditory hallucinations and lives in a dream or fantasy world of his own imagination. The schizophenic makes jerky movements. The other characteristics are grimacing and giggling and expresses sometimes stupor and catalepsy. Schizophrenia is a widely prevalent form of mental disorder. Patients suffering from schizophrenia fill up more than fifty per cent of the hospital beds in any population.

Schizophrenia mostly occurs among the adolecents and among the young adults. It is the most serious and widely prevalent type of psychosis. The frequency of occurrence of this disorder is so great that one in every four patients admitted in mental hospitals is dignosed as having some type of schizophrenia (Fig. 25.4).

According to the U.S. Department of Health (1961) the majority of these patients are under the age of thirty five years. This mental illness was first noticed by French psychiatrist and called it as 'demence precoce'. Later on Kraepelin named it as dementia praceox which meant in those days as *Youthful insanity.* The term Schizophrenia is derived from the Greek word Schizein means 'to split'. The term phren means 'mind'. Therefore, *schizophrenia* means the 'split of the mind' or split personality. Thus schizophrenia may be stated as a form of mental disease in which the personality is disintegrated and detached from its environment. It is a long term mental disorder involving faulty perception, inappropriate

Fig. 25.4 : Schizophrenia : The Split Personality. Schizophrenia is a type of psychosis characterized by breakdown of integrated personality. It is a functional psychosis, a disorder in which there is a lack of harmony or there is split between the aspects of personality, specifically between emotion and behaviour. This state is said to be having the symptoms of delutions, hallucinations and autism. Schizophrenia is one of the functional psychosis marked by apathy and social withdrawal. It is highly prevalent among the adolescents and young men and women. Among the psychotic disorders schizophrenia is highly prevalent in any culture and society. The person is highly introverted and shut himself within him. He shows his reals emotions very little. He often experiences auditory hallucinations and lives in a dream world of his own imaginations. Kraepelin said that it is a youthful insanity as mostly the young men and women suffer from this disorder.

actions and feelings and withdrawing from reactivity into fantasy and delusions. Originally the term 'skhizein' is a Latin word, which means to split and 'phren' means mind. These two terms are used in Greek as schizein and phren — which constitute the term *schizophrenia.* Thus the concept has a long history and received greater importance in the medical field than any other concepts.

Schizophrenia refers to a group of psychotic disorders. This reveals several symptoms and on the basis of the varied symptoms schizophrenia is classified into several types. The American Psychiatry Association had classified into twelve types on the basis of the symptoms. But originally it was classified into only four main types. They are simple schizophrenia, catatonic schizophrenia, paranoid schizophrenia and hebephrenic schizophrenia.

(1) Simple Schizophrenia

The simple type of schizophrenia is exhibiting many difficult or complicated symptoms like delusions and hallucinations. The individual in this state is not interested in what is going on aroused him, he is not bothered whether he has passed in the examination or not. But he is fully department on his family. The simple type of schizophrenia is slowly progressing in the childhood years. In the adolescent period there is loss of interest in the external world, contact with others and conversations are reduced to a considerable level. Concentration is much affected and there is loss of ability to answer any question in the situation. Kant (1948) has stated that nearly two-third of the cases had family history of functional psychosis in the family. Almost the same number of cases are from very poor homes, broken homes and homes with hatred and unpleasant situation. The onset of this disorder in the individuals was gradual and was going progressively for a long period of time. It begins in puberty period and early middle age, that is, from the age of twelve or fourteen to thirty-five and/or forty years. The symptoms in most of the patients are shyness and hypertension in dealing with others. Antisocial activities, when observed in them, they were admitted in hospitals.

Studies of the patients in the hospitals revealed that there were instancess of bizarre behaviour, delusions and hallucinations. It was found in them that attempted rape was very common. There were various patterns of behaviour among these patients, but the symptoms varied from patient to patient depending upon the environmental backgrounds from which they were drawn. In the treatment by introducing suitable circumstances the behaviour of the patients was altered. Effective treatment programmes can be evolved for the schizophrenic patients just like for other disorders, on the basis of the strength and the needs of the particular problem.

(2) Catatonic Schizophrenia

The catatonic schizophrenia develops much more suddenly compared to other forms. The catatonic patient behaves in a very strange manner. His appearance also looks differently. His appearance and behaviour creates an impression that he is a patient of uncurable disease. But catatonic schizophrenia in some respects is not that much of serious disorder as that of the other three types. These patients are easily available for treatment compared to paranoid and habephranic disorders.

In catatonic schizophrenia there is extreme withdrawal behaviour and also extreme excitement. Sometimes one may be prodominant over the other. Sometimes catatonic patients remain motionless in a stereotyped posture for hours. The catatonic patient is extremely stubborn and he resist strongly any effort for a change. This patient may suddenly pass from the state of withdrawal to extreme excitement and openly indulge in sexual activities and also attempt to suicide and becomes impulsive and even attempt to kill persons within his reach.

Catatonic literally means abnormality of movement and behaviour arising from a disturbed mental state, an abnormal state of immobility and stupor. It is also stated as a form of schizophrenia characterized by muscular rigidity and stupor with at breaks of excitement. It is the function of reduced muscle activity and complete lack of motor activity. In some catatonic patients ordinary adjustment is interferred with hallucinations, but there will be no memory impairment. It is also found in some patients that they may shift from the state of stupor to a great excitement and back to the stupor state.

(3) Paranoid Schizophrenia

Paranoid is characterized by delusions, a form of schizophrenia. A paranoid is a systematized false belief. In this state the person has thought disturbance which is characterized as schizophrenia, and not able to make consistent logical thinking. It has been observed that nearly half of the schizophrenic patients admitted in hospitals are paranoid schizophrenics.

Clinical studies show that the symptoms are illogical and changeable delusions. The most common delusions are persecutory type and these delusions are most often taking place. The paranoid schizophrenic patient is highly suspecious of his relatives and friends and often complaints that they are watching him and also they are trying to poison and kill him. In the prevailing situation he begins to think that he must be honoured as a great personality and he thinks that he has remarkable capacity to do anything. Because of these, he thinks that his enemies are going to persecute him. The patient claims that he can talk to God. The patient's behaviour mostly centre round these types of delusions and hallucinations. His behaviour becomes errotic and inconsistent.

The paranoid schizophrenic patient sometimes becomes violent and hostile and never hesitates to attack a person whom he thinks that he is persecuting him. Investigations and studies were made on the paranoid schizophrenic patients. In the etiology of the disorder, **Klaf** and **Davis** (1960), **Moore** and **Selzer** (1963) have observed greater incidences of homosexual conflicts. This conflict among these patients prevailed both among men and women. The delusions of the paranoid schizophrenic patient provide an insight into his motives and feelings which are not possible for the patient to control. These patients often develop ideas of reference. They incorporate events that are not important into their delusional framework, such as very simple activities of others.

(4) Hebephrenic Schizophrenia

This type of schizophrenia is the severest one among the types. This type of schizophrenia sets in the early years of a person and progressively becomes more severe resulting in disintegration of personality factors than the other three classical types of schizophrenia. The patient reveals odd and unscrupulous behaviour patterns. He becomes very serious on very small matters. He shows some signs of interest in religious and philosophical matters. He is often found to be involved in delusions and fantasies while his schoolmates are engaged in social activities and playing games in normal ways. The habephrenic schizophrenic patient isolates himself and found to be in seclusion.

As and when his disorder state progressing to severe state the patient develops emotional indifference and reveals infantile reactions in his behaviour. He speaks inconsistently and incoherently and speaks like a baby.

The patient experiences delusions and hallucinations. The delusions commonly are of sexual, religious and persecutory in nature. The hallucinations are of usually of auditory nature. The patient is always highly suspecious of others and accuses them of murders. He even goes to the extent of the state that he has been killed by his enemies several times. The other characteristics of behaviour of this patient are utter indifference to modesty, by behaving in obsence manners. He is very much indifferent to real life situation in the state of outburst of anger. However in some cases of habephrenic schizophrenic patients remission of symptoms occur showing in their cases that disintegration is only in symptoms and not in their real life. In short, the behaviour of the habephrenic schizophrenic patient is highly fantasy oriented, childish and emotional distortion.

In the recent classification of psychoses there are about twelve types of schizophrenia. But all of them are based upon the four classical types that we have discussed above.

Etiological Considerations of Schizophrenia

While discussing about the causes of mental disorders we have discussed some of the causative factors of neuroses and psychoses. The disturbance and ego functioning had significant role. Strange (1965) states that severe ego disturbance is forced to be one of the primary characteristics of schizophrenia. But what are the factors that cause the defective ego development to cause schizophrenia? To what extent the psychobiosocial settings function in the process of defective ego development must be the question we will have to answer in this section.

Several causes have been attributed as the cause of defective ego development which underlie in the onset of mental disorders. Schizophunia, a major type of psychosis is found in every culture all over the world which is impairing or destroying the normal happy life of human beings. Medicines and methods of treatment provide relief and cure. But it has to be prevented from disturbing the happiness of life.

Biological Factors

Experts have extensively investigated the etiological factors that are functioning in causing schizophrenia. There are several biological factors such as genetic, neuro-physiological and biochemical factors that are interferring in the normal functioning of the brain and nervous system and cause the disorder, schizophrenia. Recent studies have brought to understanding that certain biological or physical indications occur more frequently among the schizophrenic patients than among the patients suffering from other wide range of diseases and among the normal persons. This is as observed by **Roessler** and **Greenfield** (1962). Investigators strongly believe that schizophrenic incidents are found in the families disproportionately and they stressed the importance of genetic factors in psychotic disorders. In one of the earlier studies it was observed by **Kallmann** (1953, 1958) that more than eighty per cent of identical twins had the incidence of schizophrenic disorders. But in other studies the concordance rate was not so high.

Schizophrenia has genetic influence, though there is some difference of opinion on this issue. Evidences are produced that genetic factors function in the prevalence of schizophrenia. We cannot conclude that all schizophrenic cases are having genetic or hereditary component and also cannot say that there is no genetic component. But it is a matter of degree to which schizophrenic cases are reached and suffering. It may be stated that persons who are exposed to unfavourable and uncongenial environmental conditions, if they are fully normal, may not be influenced for any mental illness, but persons from the family backgrounds of schizophrenia in such environmental situations may develop mild degree of schizophrenia. Persons from the family of high degree of prevalence, no doubt may develop high degree of prevalence of (unfavourable) schizophrenia in the unfavourable and uncongenial atmosphere conditions. This hypothesis has been tested and verified by **Gottesman** and **Shiedldsz Shields** (1966). They have observed that the concordance rate for the twins was higher with severe schizophrenic disorders than for the twins of mild schizophrenic disorders.

Biochemical Factors

Researches have thrown some light on the schizophrenic patients but the biochemical factors have produced profound mental changes. The presence of some chemical substances in the blood stream can cause some disorganisation or impairment of thought process and variety of symptoms which are identified as the indication of psychosis.

There are investigators who strongly argue that all schizophrenics have an inherited characteristic that eventually result in the disorder irrespective of the nature of the early family background. Generally a stressful childhood life may lead to a severe illness corresponding to the process of schizophrenia. Children brought up in a placement family situation will not be having such disorders, until they face, severe stressful situation. Thus there may be a number of causative factors of schizophrenia.

Psychological Factors

Serveral factors are traced as the psychological factors that influence the person with schizophrenic disorders. All these factors are in contract to biological and biochemical factors. Studies to find out the origin of schizophrenia are concentrated on early childhood stage, the rearing practices of the parents and others and on interpersonal relationships in homes. The parent-child relationship may vary from home to home. The rearing practices of the parents determine the development of pathological conditions of their children. Mothers were found to be overprotective and also rejecting and neglecting their children. Some mothers exercise two much of restrictions and some too much of liberal. Both the approaches may put the children in some disadvantageous state which may in the long run cause some mental aberrations and finally resulting in schizophrenia. Fathers of the children always expect their children to come up in life to occupy some lucrative job or to occupy some high prestigious position. Towards which they are very anxiously training their sons rigidly and keep directly under their control. But this may put up too much of strain and stress on the children and they feel that what their fathers are doing are too much and therefore, they try to escape from it by unhealthy means and may loose their mental health.

Schizophrenia might have been existing in the family, either the father or mother might have been characterized as weak mentally and kept themselves aloof and not very much attached with their children. All may contribute to the poor mental health of the children. **Klebanoff** (1959) conducted an investigation among three groups, mothers of schizophrenics, brain injured and mothers of mentally retarded children and about their child rearing practices. The mothers of all the three groups were found to be more possessive, overcontrolling than the mothers of normal children. The results indicated that these mothers began to reveal the same attitudes when faced with abnormal behaviour in their children irrespective of the nature. Therefore, it is concluded that maternal attitudes towards the schizophrenic children were found to be the effect of disorder and not the cause. Therefore, we may state that disturbed home life and early years of life are found to be in the background of schizophrenics than the normal homes. The home environment with intense conflicts, with both the parents may leave the children with emotional outbreaks.

The early years of life, the interpersonal relationship, parents attitude towards their children in bringing up may have some effect upon the mental health of the children which may contribute for the slow development of schizophrenia. The pathogenic family pattern is not without any effect upon the children. Pathogenic state in a family may prevail due to several reasons. Recent studies have shifted for finding out etiological factors in schizophrenia from parents to the family interaction. The factors that cause schizophrenic disorders are found to be the excessive stress that prevails in the family situation.

Learning

One of the psychological factors in the etiology of schizophrenia is the family learning and self-structure. The explanation offered for this is that behaviour disorders are based on learning faulty responses for the needs and demands of life. The individual must learn the correct response that will lead him from the dependent state, that is form the autistic world of the children to that of the adults, which may be called the world of reality, where the social and interpersonal demands are recognized and met with.

Treatment and Prognosis

The advancements brought out in the field of medical sciences have enhanced the methods and techniques of treatment to mental disorders. However, the prediction of success of the schizophrenic cases in the mental hospitals cannot be accurately determined. The schizophrenic patients need not loose their hopes of getting cured of their disorders. The days of hopes are becoming more and more bright. Chemotherapy and other treatment methods made the out patients to go with the satisfaction of being comfort. The patient who is admitted in the hospital for treatment gets discharged within few days. **Yolles** (1967) collected the information about the results of the treatment. According to his study more than ninety per cent of the patients get discharged from the hospital within few days. But there was about forty-five per cent of the patients get readmitted after one year. The overall view has been that about one-third of the schizophrenic patients recovered from the iliness, another one-third of the patients have shown partial recovery and the third one-third of the patients were totally disabled.

Chemotherapy

Chemotherapy is a treatment given to psychotric patients in which the use of drugs in the treatment is an essential aspect. In this type of treatment chemicals are used for influencing behaviour for the past many years. It is found that narcotics are for to reduce pain and alcohol and sedactives are used to reduce anxiety and induce sleep. Stimulant drugs are administered for relieving depression. But in the treatment of mental disorders two major chemical tranquillizers such as reserpine and cholorpomazine have been used extensively in the treatment of psychoses. These two types of tranquillizers have excellent effect in making the patients calm and relaxing patients without inducing sleep. But there may be some amount of drowsiness. These drugs have been found to have effective particularly in the case of schizophrenic patients. These drugs calm down intensely agitated schizophrenic patients and also gradually stop the patient having hallucinations and reduce delusions. Another important effect caused by these drugs is the decrease of emotionally withdrawing tendency. This effect facilitates the psychotherapist to attend on him.

The drug that has been recently discovered is lithium, which is used in treating mannic disorders. Chemotherapy has been found successfully in reducing the serious mental disorders like the psychoses. It has been observed that in the case of psychotic patients they need not seek hospitalization for treatment as these drugs alleviate the seriousness of the pathological disorders. Some of the investigators like **Goldstein, Rodnick** and **Ludd** (1971) have stated the limitations of these drugs in the acute cases. But however modern chemotherapy has improved the conditions of most schizophrenic cases.

Somatotherapy

Somatotherapy is one of the therapies used to alleviate mental disorders. It is specifically used for the treatment of personality maladjustment or for mental illness by the use of drugs, electric shock or other methods directly affecting bodily processes. Somatotherapy is developed for the treatment of the functional psychoses. But it is not as successful as it is expected. The medical technique assumes that psychopathology, specifically the psychoses is caused by neurophysiological malfunctioning and can be best cured by treating the soma, the body rather than the psyche. Vitamin treatment is found to be reducing the prevalence of mental disturbances associated with pellagra, antibiotics for treating syphilis, reduces the organic psychosis, known as general paralysis. The epileptic symptoms are alleviated by barbiturates.

Psychosurgery is used to reduce uncontrolled emotional behaviour by cutting the nerve fibres, connecting the hypothalamus with the frontal lobes. But this is not very much successful treatment. But in chemotheraphy the somatotherpeutic treatment is found to be most successful method.

26

PSYCHOTHERAPY AND OTHER METHODS OF TREATMENT

Introduction

We have discussed in the previous section about various types of behaviour disorders. Behaviour disorder is the term used by the psychologists in recent times. Generally it is the mental disorder, or mental illness that is referred to as behaviour disorder. It has been felt in some circle the terms mental disorders, mental illness, madness and abnormality all cause some kind of disgust or illfeeling. Therefore, it is felt that some terms which do not humiliate or insult the person who suffers from some mental illness may be used. As result of this the most prevalent terms to indicate the mental disorder or mental illness are behaviour disorders, psychopathology, emotional maladjustments and so on. But in using different terms the correct meaning of the subject matter should not be sacrificed.

The most prevalent term that is generally used in treatment at present time is psychotherapy. People suffer from both by physical illness and mental illness. *Treatment Procedures* and methods for both were designed right from the time of the Greek physician, **Hippocrates** (460-357 B.C.). Medical treatment is given to physical illness and psychotherapeutic treatment is given to mental illness. Generally the treatment to illness is stated as *therapy.* The medical practitioner's treatment to the illness of a person is called therapy. There are several types of therapies in treating the physical patients. The medical therapies include the use of drugs maintly which may be stated as *chemical therapy.* The other types of therapies are *surgery, exercise* and *massage* which is generally called as physical and physiotherapy. In medicine there is radiological treatment which we call as *x-ray therapy.*

PSYCHOTHERAPY

The treatment for those who suffer from *mental illness* or *imotional maladjustment* is called *psychotherapy.* This is the treatment mostly required to cure the psychoneurotic patients. But psychoneurotic patients in addition to psychotherapy also used to get other therapies.

These patients usually first go to the medical treatment. Then they are given the *drug therapy* and *electroshock therapies.*

But in the case of behaviour disorder there are generally *two categories of treatment.* The *first category* of treatment is called *symptom alleviation.* This is the first step of the treatment which mainly is involved in removing or reducing the symptoms that very often manifest in the patient. The symptoms of abnormality may be anything like totally away from the normal way of life. The symptoms may be observed in the *impairment* of intellectual, social, and vocational functioning. The patient could not concentrate and perform in any of these areas. The treatment involves in dealing with these symptoms and these are either reduced to some extent or removing it.

The second category of treatment is called *psychotherapy.* This involves to deal with the patient's mental illness directly by focusing the attention upon the origins and the causes of the symptoms. Generally the attempt to treat the behaviour disorder is stated as psychotherapy. But authors have developed different views with regard to the treatment methods, mostly about psychotherapy.

What is Psychotherapy?

Psychotherapy is a technique of counselling and other interviewing techniques used by the *clinical psychologists* and *psychiatrists* in the treatment of behaviour disorders. It is defined in several ways. Psychotherapy may be also defined as a *verbal means* of helping mentally and emotionally troubled individuals to change their thoughts, feelings and behaviour to reduce their distress and to'achieve better life satisfaction.

A simple and basic definition of psychotherapy is offered by **Strange** (1965) is that it is a conversation between two persons, a patient who suffers from some degree of emotional disturbance, and a professional therapist, trained to help the patient to change such condition. These persons who treat the patients who suffer from behaviour disorders are those who have better training in psychological theories and techniques and qualified themselves as psychotherapists are *the clinical psychologists, psychiatrists* and *psychiatric social workers.* There are quite number of varied techniques applied by the therapists to deal with the persons who suffer from emotional or behavioural disturbances. But at present the therapists fall under two main groups. One group of psychotherapists follow the system developed by **Carl Rogers** (1920). The other group of therapists follow the psychoanalytic techniques originally developed by **Sigmund Freud** (1856 - 1939).

The Aim and Scope of Psychotherapy

The main aim of psychotherapy is to reduce or cure the emotional disorders and mental illnesses. People who suffer from these disorders get cured or relieved of them. This cannot be done suddenly all of the times. There are various steps during the course of treatment. The progressive steps in the treatment are gradually helping the patient to attain the normal emotional state.

Since the patient's emotions and motives are mainly involved in his problems, he must first understand and accept the cause behind his trouble. Once he understands and accepts the role of his true feelings and emotions there is a chance for the patient to change or modify his emotional state and he slowly learns to readjust himself to the situation. By this means the patient slowly learns to get rid of his state of maladjusment.

A process of relearning also may be brought in, if necessary, in the case of some patients. In the peck of the state of emotional disorder the patient might have been highly destructive and violent. In the learning process such behavioural patterns are removed and a constructive behavioural pattern is induced. By this means a healthy emotional state in the patient can be produced. This process of handing the patient may be called as the process of *emotional reeducation.*

The treatment should not be concluded at this stage. The patient is in a vulnerable state and he may be easily succum to further emotional disturbance. Therefore, the patient has to strengthen his recovered state. The patient must be put in a state of meeting further stresses successfully. The individual must develop a state of self which should provide him a feeling of self-confidence to meet the world of reality.

The progress of the treatment that is functioning in the individual should be revealed in the everyday activities successfully. There should be some development in the patient's ability in performing the tasks. He should face and resolve his inner conflicts and develop his skills to meet the external stresses. By this means he is slowly building up within himself a strong confidence to solve the problems he is facing in the outside world. Then he is slowly building up of the normal happy way of life, and is able to adjust successfully with normal way of behaviour.

In applying the therapeutic techniques there are several factors functioning generally in all the methods. In some of the techniques those factors or elements are strongly stressed. In some methods they are not given more importance. In psychotherapeutic methods the clinical psychologist or the psychiatrist has to expereince these elements in the situation where therapy is exercised in the therapist's clinic.

Since psychotherapy is basically a conversation with a purpose between two persons, the problem-person and since the therapist and since the therapist deals with the problems of the persons, *certain human factors* may be naturally emerging in the situation. These are establishing rapport with the problem person, the resistance of the problem person to cooperate with the therapist, the process of transference, catharisis and insight and the coping with life. These five elements are quite common in any psychotherapy.

Rapport : The first step in psychotheraputic technique is to establish good rapport. This is generally taking place in the first meeting. Before commencing the therapy, the patient and the therapist come together and make a formal meeting. Then the therapist begins his business with some informal conversations by making enquires about the patient's welfare. This conversation should create a sense of confidence in mind of the patient. This meeting might have taken place on the part of the patient voluntarily. Or some of the relatives of the patient might have brought the patient to the therapist's clinic or some agency involved in social work might have brought him.

The client goes to the counsellor with some problems. It may be mostly emotional problems of some motivational conflicts or even mental aberrations. The therapist is a professionally trained and experienced in this line and is supposed to be well equipped to deal with such problems. If the patient and the therapist come to a state of good agreement that they can work together and solve the problem, then it may be taken for granted that a *good rapport is established* to the expected level. A rapport is a feeling of mutual trust and confidence between two persons, the therapist and the patient. Thus a rapport is an expression of similar feeling by both the counsellor and the client.

Resistance : Resistance is one of the elements that may make its appearence during course of therapy. Resistance is a Freudian concept which is a defensive mechanism, often the therapist meets with during the course of therapy from the client. The patient makes efforts in psychotherapy to avoid the therapists interference into his unconscious motives. It is an indirect refusal of the patient to reveal secret matters. The resistance hinder for the time being the therapeutic progress. As the therapist goes into the repressed materials of the patient which are likely to come out, the patient reveals his unwillingness by means of resistance. This may impair the process of therapy but the therapist's task is to relinquish the resistance with his expertise knowledge and experience and continue the therapy and help the patient to resolve his problems (Fig. 26.1).

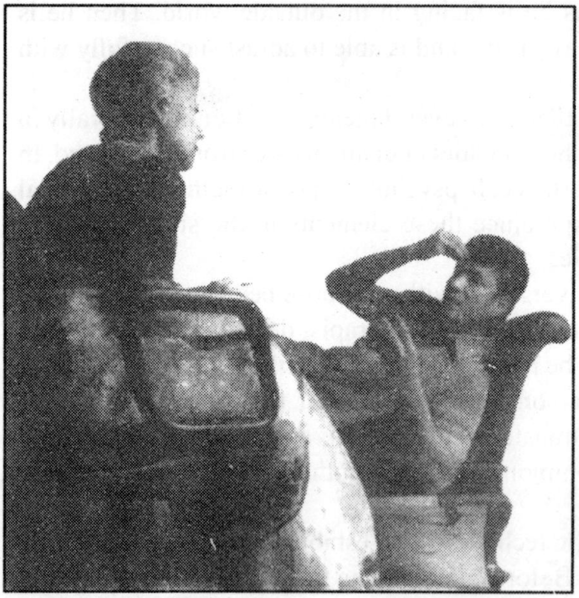

Fig. 26.1 : Psychotherapy : Emotional maladjustment and behaviour disorders are reported everywhere in big cities in India from 1990 onwards. These pathological incidents are found in mass media of communication everyday. Out of self-esteem and family prestige many of the cases are not reported. Young men and women between the ages of 20 and 35 are put into serious consequences. The most common causes are found to be extreme sex deviations caused by the present day motion pictures, difference of opinion, conflict, incompatability among the husband and wife, frustration and apathy. If care is not paid on this problem, it often results in divorce and separation. If the person is of highly emotional type, it results in committing suicide and death. This type of cases are taking place everyday in India among men and women below 35 years of age. If timely effort is sought, psychotherapy comes to rescue them from disaster consequences and the problem can be resolved.

Transference : One of the most important elements of psychoanalytic therapy is the transference. During the course of therapy the patient reveals his/her emotions. These emotions may be positive or negative or love or hate. The patient considers the therapist as the agent of some terrorist. The patient becomes angry with the therapist to release his emotions, actually the patient felt toward his/her father. The emotions transferred stem from the patient's unresolved childhood conflicts. In the counselling situation both the therapist and the client may get into some emotional involvements. It is the human tendency that no body wishes to get his/her self-esteem hurt. In case the therapist could not resolve the emotional state the therapy may fully be disrupted.

It is observed that in psychotherapy most of the times the patient either loves the therapist or aggressively hates. Freud calls this situation as *transference.* The patient's childhood experiences might have been pleasant or unpleasant and he/she might have developed the emotion of anger or love. In the counselling situation when the therapist touched those experiences the patient begins to project the same emotion against the therapist. This process in psychoanalysis is called *transference.*

Catharsis : The element catharsis is one of the important components of psychoanalysis. It is a process of bringing out what are all stored in the unconscious mind. Actually catharsis means purging out. It is by way of talking out the emotional conflicts are brougtout. Catharsis is a therapeutic procedure introduced first by **Breuer** and it was further developed by Freud in the last quarter of nineteenth century. According to this procedure, patient recalls and relieves all the earlier emotional conflicts and reexperiences, the emotional tension and unhappiness.

An experienced and a well trained therapist is able to receive all these negative emotional expressions and can deal with them in such a way that the patient gains insight and understands his trouble with some emotions. Gradually the patient's negative emotions from negative state are turned to *positive stage.* Once this stage takes place his problems also slowly resolved.

Roger's Client Centred Therapy

Roger's client-centred therapy is one of the therapeutic techniques very widely used in dealing with persons who suffer from emotional disorders and mental illnesses. It is also called as *non-directive therapy.* **Carl-Rogers** is one of the founders of the *Humanistic Psychology*. He formulated his therapeutic technique on the basis of humanistic psychology. **Carl Roger's** (1902), an American Psychologist, has developed his therapeutic method out of his long experience of practice in the field of teaching (1942).

Rogers followed a technique which is less directive and also less intensive form of treatment than psychoanalysis. Rogers made its aim to help the patient for *achieving more of self-conscious* and more of understanding. Rogers wanted in his therapeutic technique to be involved in it dynamically through strong understanding.

Rogers strongly believed that each person within himself has a potentiality for growth and self- actualization. It is because that every individual human being is intrinisically worth and significant. Rogers is of the view that a skilful therapist should proceed in his task of treating a client with hypothesis that the client has the capacity to deal with whatever he can bring to awareness is constructive. The therapist is helping the client to clarify his feelings.

In this task the therapist is very active in creating an environment in which the client becomes an alert person to bring out the long repressed material to the conscious level. This type of task on the part of the counsellor is stated as the *strongly reflecting of the emotional* materials of the client's responses. It is also believed that the counsellor with his efforts provides the environment and it is the client's own effort rather than by the direct intervention on the part of the counsellor (Fig. 26.2).

Fig. 26.2 : Psychotherapy - Client-centred

The client-centred psychotherapy is little different from the classical psychotherapy in the sense that it provides more freedom to the client and it is a passive approach. The rapid development of science and various technological advancements made human life fast and restless. Men are put into severe strain and stress. The common occurrence due to these states is impairment of intellecutal, social and vocational functionings. The person could not concentrate and perform anything in these areas. More than ever before young men and women in India specifically after the period of 1995 started seeking the help and treatment of Mental Health Exponents, the Psychotherapists. The reasons behind this are many. It is observed that these men and women often face severe stress, conflict, frustration and depression due to maladjustment in the family, in the interpersonal contact, and in social situations and suffer from behaviour disorders. The only alternative at last they find is to meet the psychotherapist. On meeting the therapist in three or four sessions they find themselves that they are relieved from the mental aberrations. The works of psychotherapists in India are found to be increasing astonishingly.

The client-centred therapist like the psychoanalyst mainly depends upon the establishment of the significant relationships between the patient and the therapist. It is passive understanding and acceptance. Rogers did not accept the traditional transfered relationship as he felt that it may cause sometimes unnecessary complications in therapy and also it is like disregarding the alient's *self-esteem.* In client-centred therapy the client's individuality is respected. For the client-centred therapist the *essence of the therapeutic relationship is the respect* for the potential *self-esteem* or individuality.

Reflecting Feeling

Rogers put emphasis on the *reflecting feeling*. It is a process by which the therapist conveys his understanding and sensitivity to the patient. It is a technique which the client-centred therapist applies. The therapist mainly responds to the emotional content of the patient's statements rather than to his intellectual aspects. The therapist's response is not an interpretation or judgement or approval or disapproval of what the client has expressed. The therapist simply makes a restatement or makes a clarification of the feelings of the patient.

The therapist makes his responses more emphatic in encouraging the patient. By this means the patient is made to elaborate his feelings. He becomes less defensive, avoids irrelevant statements and also avoids making escaping statements. In this way the patient now is in a *very comfortable atmosphere,* as a result of this the patient is in a position to explore his own motives and conflicts with clear feelings honestly and openly. Then the patient develops a more realistic and agreeable attitudes towards himself.

Rogers believes that an atmosphere of this type in a client-centred therapeutic session, therapy may be said to be successful. Then the patient achieves correct understanding about his state and develops self-confidence. At this stage the client may start behaving in a more matured manner begins to be more constructive and normal.

Client-Centred Therapy

Rogers client-centred therapy takes considerably more shorter period of time than the standard psychoanalytic therapy. But however though it is non-directive therapy, it also takes several months to complete the therapy. The main reason for this state is that the therapist has to face several stages during the course of therapy. In the client-centred therapy most of the counsellors find that the rapport is established with the patient in the first session. It is because the client begins to realize quickly, as and when the session takes place, that he is having conversation with a person who correctly understands, recognizes and accepts his deep feelings.

The patient in the client-centred therapy when comes to know the therapist's approach he feels free to talk everything that he has experienced. Then the therapist does not worry about the patient's resistance or abuse the transference process. But if any such thing occures the therapist adapts the *reflecting feeling* technique and clarifies the issue.

Eliminating Negative Feelings : Promoting Positive Feelings

The expression of negative feelings by the patients in the non-directive therapy may sometimes occur inspite of allowing the patients to release and re-evaluate the expressed feelings and emotions. This may take place in the therapeutic session in the early stages. As and when the therapy progresses the negative feelings gradually are reduced and the positive attitudes gradually increase. Rogers has indicated that in the early period of the session it was observed that about seventy per cent of the statements were in present tense which are of negative feelings. But such negative statements of feelings gradually reduced to thiry per cent of present tense at the close of the therapy.

It is also felt that apart from tranining and practice on the part of the therapist, the *personal skill* of the therapist in the counselling situation is very much needed for creating an atmosphere in which the patient should avoid to express his negative feelings and emotions. What the patient felt and stated in many instances had not been faced and admitted by the patient himself. But in the presence of the trained therapist those feelings are revealed by the client. If the counsellor does not understand and accept these emotions from the facial expressions or voice, the client feels shy or guilty about what he has revealed. In such instances the skilled therapist correctly reflects the patient's feelings and re-evaluates the patient's negative feelings.

When the negative feelings are re-evaluated and desensitized the positive impluses emerge. First such impulses are not very much expressive. The expressed negative feeling accusing some person is made by the client. Immediately he/she realizes the mistake and makes another statement which is neutral, not blaming the person and subsequently states that he was also helpful in several situations. In this way the therapeutic situation creates an atmosphere for expressing positive feelings. Rogers is of the view that producing positive feelings by client is an important stage in client centred therapy. The positive feelings are one of the aspects of every person's attempt to get ready to maintain and actualize himself. Towards such improvement the therapeutic atmosphere provides congenial situation for positive expression of feelings and emotions.

Stage of Correct Understanding or Insight : The development of positive feelings gradually get strengthen and begin to dominate in the person. He further develops the desire to go deep into his problem and understands. The client understands his weakness and also the values of his assets and resources. The client's understanding is mainly based on rationalization. This emotional and intellectual factors are involved in his insight. Though the client hated the person first, gradually he learns to develop more positive emotional feelings and begins to express them. Since the client's problems were developed from his childhood painful experiences, the person who might have been his mother generally. He may even go to the extent of justifying her stand and desensitize his violent emotional experience of hatred. The client's justification may be, after all she too is a human being and she also may be having her own problems and difficulties. The client also begins to realize that any more violent emotional expressions may damage his life and therefore he begins to feel to give it up for a better way of life.

Dealing with Successful Life : One of the important aspects of counselling is that it must have successful result. It must be seen only from the effectiveness of the counselling. It can be assessed from the client's ability to cope with life situation. The patient should show increased efficiency in every walk of life outside the therapeutic situation. The effectiveness of counselling by showing increased ability by the client is not enough within the counselling situation in a session with the counsellor.

The client-centred therapist generally expects that his client should recognize and accept his true feelings and also the client should take initiative of his own to change his behaviour towards others. It is observed that once the client understands his problems and started to behave in accordance with the feelings of others we can take it for granted that he is turning to normal way of life. At this stage the counsellor feels that the person does not require any more counselling as he is acting with emotionally well adjusted manner.

There were different views expressed by those who were engaged in the treatment of behaviour and emotional disorders. **Moskowitz** and **Orgel** (1969) state that the client-centred therapy is most useful like the psychoanalysis only with certain types of people. It is because the client is put up with greater responsibility in the therapeutic situation for dialogue about his own feelings and problems. The client-centred therapy is mostly used with mildly and moderately neurotic patients and not appropriate for treating psychotic and character disordered patients or for treating patients who are unable to and unwilling to express or communicate their emotional problems.

Rogers and his followers in addition to client-centred therapeutic technique have also made many extraordinary contributions to the whole field of psychotherapy. They have also stressed the need for empirical study of the process of psychotherapy and worked out to discover the basic principles of effective psychotherapy. Their empirical and quantitative orientation made the psychotherapy as a systematic subject that can be experimentally studied.

Freudian Phychoanalysis

Psychoanalysis as a therapeutic technique was developed by **Sigmund Freud** (1856-1939) after several years of treatment for persons who were suffering from various types of neurotic and emotional disorders. Freud's psychoanalysis is a system of psychotherapy and also a theory of personality. In spite of many critical attacks levelled against the theory of psychoanalysis and also its limited effectiveness, Freud's psychoanalysis and its many by-products stand as important forces operating in psychiatry and in clinical psychology fields in many countries. We shall discuss in this section the important components of psychoanalysis.

What is Psychoanalysis?

Basically psychoanalysis is a technique for treating the neurotic patients. Psychoanalysis is a method developed by Freud for treating neuroses. It is a system of psychological theory grown out of the experiences with psychoanalystic method. Psychoanalysis is a term applied primarily to the therapy procedures pioneered by Freud entailing free association, dream of interpretation and working through the transference neurosis. But as and when the system was applied for its business more recently the term has come to encompass quite number of variations on the basic Freudian theory. Who is a psychoanalyst? Psychoanalyst is a therapist who has taken a specialized course of study at post-doctoral training in psychoanalysis after earning either a Ph.D., or M.D. degree. Several psychoanalysts who have been practising psychoanalysis started and founded therapeutic techniques of their own and have called them by separate names.

In this way there were several offshoots, evolved out of Freudian psychoanalysis. All these offshoots came into force after the death of Freud, mostly after 1940s. The important psychoanalysts who evolved varied therapeutic techniques are **Alfred Adler** (1870-1957), **Karen Horney** (1885-1952), **Erich Fromm** (1900), **Harry Stack Sullivan** (1892-1949) and **Erik Erikson** (1902). These psychoanalysts are called as *Neo-Freudians.* They differed from Freud stating the reason that Freud did not give importance to the influences of environment and culture in his theory. According to them Freud placed too much emphasis upon the instinative and biological aspects of personality.

In United States important modifications were made in Freud's psychoanalytic theory. This was done by a group of psychoanalysts called themselves as the *ego analysts.* The views of these analysts are often stated as *Contemporary Freudian Psychoanalysis.* Some of the Neo-Freudians are also encluded in this category. These *ego analysts* are Karen Horney, Anna Freud, Erik Erikson, David Rapport and Heinz Hartmann. Actually Freud did not ignore the interaction of the organism with the environment and culture. But he treated them in different fashion.

Techniques and Processes of Psychoanalysis

Freudian psychoanalytic techniques set up a convention as how to begin with the procedure. Freud's psychoanalysis includes his theory of personality and the method of psychotherapy. His method of psychotherapy mainly includes free association and dream interpretation. But Freud's theory of psychoanalysis includes several processes. The main purpose of psychoanalysis is derived from the assumption that *psychopathology develops when people are not aware of their underlying motivations and fears in their minds.* One of the important aspects of Freud's method is the physical setting up for conducting the therapy. The patient must be made ready for the session. Physically the patient should be free from any inconveniences. There should be free movements. The patient should be fully in a relaxed state without any tight fittings of dress, shoes, belt and trousers. He is relaxing on a cot and the room is with dim lighting. The therapist is *not in the visible sight* of the patient. He is at the back side of the client. In this situation the patient is not seeing the therapist. The therapist during the course of the session makes it a point not to interfere with the patient's performances. He only speaks to the patient softly and quietly and makes note of everything that takes place in the situation and what the patient spoke about himself. This type session situation does not cause any embarrassment and humiliation to the patient. He is with all freedom to speak anything. Freud felt that this type of situation is very essential in dealing with patients who are suffering from neurotic and emotional problems.

Free Association

Psychoanalysts apply a number of techniques to help the patient to get relief from conflicts and emotional maladjustments. One of the important techniques is the free association. The problem causing materials are repressed by the patient sometime back or long ago. *What is free association?* Free association is a key to the psychoanalytic procedure in which the analyst encourages the patient to give free expression to his thoughts and feelings by speaking out whatever comes into his mind without my censoring its content. The assumption is that over time by this means repressed material will come out for examination by the therapist and the patient. The therapist is expected to follow the important principles and record the thoughts and feelings as accurately as possible. There should not be any screening for the material on the idea that it is not important for the purpose or think they are shameful or hurting.

Freud felt that the thoughts and feelings may occur in an associative manner. A chain like process may take place. The recent incident reported first may have a link with the past or some earlier crucial events. The therapist in order to get the earlier events should carefully deal with the matter and guide the patient into the past experiences.

Resistance

Freud in his sessions had observed that suddenly the patient stops his expression of thoughts and feelings. Patients may suddenly change the subject matter or stop it. The patient may not be remembering all those things that had happened long ago and how the matter ended. This may interrupt the session. The patient may remain silent and even get up from the couch and may start speaking something totally different from the therapeutic conversation. He may even go out on his own and come to the next session late or even forget it. This is an obstacle to the free association techniques. Freud called this stage as *resistance.* Resistance takes place during the session of psychoanalysis, which is the function of the *defensive mechanism* of the *unconscious part of the ego* to get rid of the probing of the therapist. It is the threatening of the suppressed material. Freud developed the *concept repression* from this process. He felt that interference with the free association can be traced or attributed to the control of the unconscious mind in case of sensitive matters. These resistances sometimes in free association provide the therapists some important serious information about the patient. The patient's self-respect (the ego defence) begins to manifest in the form of resistances.

Dream Analysis

Freud brought out his famous work, *The Interpretation of Dreams* in 1900 after many years of rigorous research work. This work has made him the foremost theorist in the field of psychotherapy and brought to him international reputation. In this work he explains how the *dreams provide a key* to understand the *unconscious.* Freud has stated that during sleep the ego defenses are not active. As result of it, the repressed materials come to the surface in disguised forms which Freud called as dreams.

What we find in the form of dreams are called as the *manifest content* which is the fulfilment of repressed wishes, the *latent content.* All the unfulfilled desires try to come out which is not possible when we are in an awakened state. Therefore, they come out in disguised form. In free association the disguise can be removed. The inaccessible material is brought into consciousness that has been a factor in the motivation of behaviour.

It is possible for a trained therapist who has been familiar with the dream symbols that were often appeared in dreams. Freud had stated that a dream in which a person is found be a king or a queen is to dream of a father or mother. Dreaming of small animals are symbols of little children. Any object like snakes, daggers, poles symbolize penis of the male sex organ. Dreaming of object like covers, jars, baskets symbolize vagina or womb of the female. Climbing the trees or stair, running are symbolic representations of sexual behaviours. A man who fell in love with his neighbour's wife found in his dream driving the neighbours' motor car. Freud put more emphasis on the correct interpretation of the symbols that are found in dreams.

Steps Involved in Psychoanalysis

The psychoanalyst first arranges the therapy, situation quite congenial for his purpose and helps the patient for establishing the initial rapport. Further, he briefs the techniques of free association and dream analysis and makes the patient to know what is expected from him. The therapist, during the course of his session, resistance may occur from the patient. When the first resistance occurred, therapist tells the patient about the nature of it and helps him to overcome it. Once the therapy begins, he proceeds systematically and the patient starts to *talk out* his problems freely. Freud calls this process as *catharsis.* Catharsis literally means *purging or cleansing.* Freud used this term for the process of talking out by the patient about his emotional conflicts, tensions and unhappy experiences he had in the past. Cathersis occurs through free association and dream analysis.

In psychoanalysis when the analyst deals with the patient, the transference neurosis is evident when the patient could bring out all the repressed childhood emotions that were directed toward the parents and transfered them to the analyst. Freud observed in therapy situation that the patient at first does not realize that all these are his old emotions that he had experienced, but feels that they are his loves and hates (with his positive and negative transferences) for the analyst. Having this type of experiences which he did not expect, Freud developed his psychoanalysis in such away that working through the transference neurosis becomes the essential aspect of his method. Once the patient removes this hollow effect of these emotions he could use his energy to solve the real problems that are present in his adult state. The patient slowly gains insight and emotional relearning is taking place in him. The therapeutic experience and the type of relationship he had with the therapist also terminated.

At this stage the patient ceases to be a patient and gets ready to face life problems without the feeling of guilt that have been troubling him. There is no difficulty with the super ego. The ego part of his personality is strengthened and now the ego can mediate between the id and the super ego and also with world of reality. Since Freud himself had several unpleasant experiencess in counter transference in counselling situation with patients, he advised the analysts to be careful to avoid such situation. In the counter transference the analyst himself feels strong love or hate for the patient which is coming up from the analyst's own unresolved childhood emotional problems. On the basis of such instances Freud with all seriousness advised the therapists to undergo a thorough training in therapeutic techniques before starting any counselling with the patients.

Merits and Demerits of Freud's System

Freuds psychoanalytic techniques in treating the persons who are suffering from neuroses and emotional problems have been viewed from several angles. It has been strongly felt by some therapists that Freudian system has some limitations. Freudian psychoanalysis has more problems than the other systems. It takes much more longer time than any other system. In some cases it takes even three to four years with several hours for each session in a weak. Patients themselves get tired of their visits to the therapeutic activities. Secondly, it has become more expensive because of its lengthy period of time. Many could not meet the expenses and give it up in a year. Further, there were not many psychoanalysts available. There are only very few trained therapists who could meet the emotionai problems of many patients who seek treatment.

Many of the Rogerians raise the values attached to the dream interpretation. They are of the view that the dream interpretations do not reflect the true problems of the patients. The patient's true problems are not elicited by the interpretations given to the dreams. Further, in psychoanalytic therapy the analysts go too deep into the problems of the patients. But the Freudians state that the Rogerians do not go deep into the problems of the patients. They stop everything at the surface level and never deal with the real problems of the patients.

Varient Systems

These criticisms resulted in finding out some other systems of which varied approaches in psychoanalysis have emerged. Several of the orthodox Freudians broke away from the Freudian systems and started their own techniques and they were called as Neo Freudians. One of the Freudians, Adler has replaced Freud's emphasise on sex and changed the technique involved in counselling by putting more emphasise on *striving for superiority.*

Adler : Adler's change of the technique involves treating the patients differently. First of all he preferred to greet the patients in the face to face situation and has shown more interest in the patients and actively attended on them than what Freud has done. Adler made his role as a social role. The contact between the therapist and the patient he wanted to be like a teacher and a student who needed a re-education. Adler had directly indicated to his patient where he had failed to make adjustment properly. The patient had used the deceptive defence mechanism which was the cause of his problem, Adler's system of therapy is similar to cognitive behaviour therapy.

Adler considered people as inextricably tied to their society, because he believed that fulfilment of desire is found in doing things for the social good. He stressed the importance of working towards the goals and much of his theories anticipated to later development of humanistic psychology. It was observed in Adler's work one central point that it was focused on the individual's phenomenology. It is the key to understand that person, an emphasis that is anticipated for the contemporary development in congnitive behaviour therapy. Another important point is that Adler worked to help the patients to change their illogical mistaken beliefs and expectations to make them feel and behave better for which one has to think more rationally.

Jung : The different views that prevailed on Freud's system of psychoanalysis resulted in defection of **Carl Gustav Jung** (1887 - 1961) and **Alfred Adler** (1870 - 1937) from Freud. Jung and Adler were Frued's important followers. Following Jung and Adler, a large number of orthodox Freudians also defected from Freudian systems and each one started building up his own system of psychoanalysis and methods. **Harry Stack Sullivan** (1892-1949) though was originally referred to as Neo-Freudian, he was influenced by **Adolph Meyer** (1866-1950) and adapted a system based upon psychosocial approach. He strongly felt that the meeting of the counsellor and the patient is a human interpersonal relationship.

Sullivan : Sullivan did not adapt all the formalities of the orthodox Freudian system of changing the therapeutic session situation. He called his sessions with his patients as *psychiatric interviews.* **Sullivan** (1954) made the therapeutic session as creative situation. He made himself ready to meet even the most difficult cases which were considered by even Freud as impossible to deal with by psychotherapy. Sullivan performed excellent works with hospitalized schizophrenic patients and persons who suffered from obsessive-compulsive neuroses. He attended and set right some of the cases which were given away by other psychotherapists. He found that the Freudian free association technique did not bring out the expected level of cure. He found in his therapeutic sessions that the face-to-face direct approach to the patient was very much needed.

The psychiatric interview was divided by Sullivan into *four steps.* The first step involves the *formal inception.* The second step is the *process of obtaining* information from the patient. The third step is obtaining detailed enquiry and the fourth step is the *termination.* In Sullivan's system the therapy begins with formal interview in which the therapist observes closely the patient and notes down everything what he observed in the patient which includes the *voice, gestures, ways of expressing his feelings* and his *movings.* During the course of conversation the therapist builds up the interpersonals relationship upon which Sullivan believed that a successful therapy can be established. In the process of getting information the therapist is able to collect the personal case history of the patient by questioning him deeply. While getting the detailed information in the third step, there is a two-way conversation between the therapist and the patients.

At this stage the therapist is able to explore from the conversation the main causes of the patient's problems. The patient's words provide sufficient information with which the therapist is able to make out the correct meaning of them and takes decisions. In the last step therapist is going to close the session and the relationship between them. The therapist is confident that the patient has understood the nature of his difficulties and also comes to know what steps should be taken outside the counselling clinic. Sullivan's system of Psychiatric interview has become very popular in United States after his death.

Existential Therapy

This is also called *humanistic existential therapy.* This is a general term for an insight psychotherapy that emphasizes the individual's subjective experiences, free will and ever present ability to decide on a new life course. *Existentialism* is a system of philosphy, a modern philosophical movement stressing the importance of *personal experience and responsibility* and the demands that they make on the individual who is seen as a free agent in a deterministic and seemingly meaningless universe. Those most commonly associated with existentialism are the *Danish* philosopher **Soren Kierkegaard** (1813-1855) and the German philosophers **Edmund Husserl** (1859-1939) and **Martin Heidagger** (1889-1976) and with existential therapy of Swiss psychiatrist **Ludwig Binswanger** and **Medeed Boss.** In United States the main proponents of existential therapy or existentail analysis are **Rulle May** and **Abraham Maslow** (1908-1970).

As viewed by Ford and Urban (1963) existentialism advocates that man has the capacity for being aware of himself. He knows what he is doing and what is happening to him. As a result of it he is capable of making decisions about these matters and is capable taking responsibility for himself. The individual is also aware of the chance of becoming isolated and left alone and he might be driven to nothingness and death. *This innate fear* is with the individual. He knows that he is constantly changing. His important role lies not what he has been in the past, but what he is now, and his development which goes in the direction of fulfilling his innate a potentiality. Like *humanism* existentialism emphasizes the personal growth of the individual.

But the humanism as advocated by **Rogers** (1951) puts more emphasis on goodness of human nature. If the individual is left free from fetters and social restriction he will develop into an exceptionally normal person. Existentialism, according to Rogers, is gloomier in having much strain within it. Though existentialism advocates *free will and responsibility* it stresses anxiety which cannot be avoided if important choices are made, similar to be or not to be.

What required is that the individual should be protected from anxiety and choices should not be made. Because the choices and anxiety deprive him of having a life with meaning and also becomes the important cause of Psychopathology. At the same time the humanistic approach is found to be quite good and agreeable, while the existentialism is found to be coloured with sadness and anxiety.

The Goals of Existentialism : Existentialism makes concentration upon the individual's progress. According to this system an individual person is the sum of the choices he makes. There may be difficulties in making choices. This can be realized by understanding the experience involved in it. The existential therapist provides help by getting into the feeling of the patient and helps him, explores his feelings and emotions and what his life means to him. The therapist encourages the patient to face the past and the present choices and to clarity them. Since people generally identify themselves and define their existence in terms df their personal relationship the existential therapist helps people to relate authentically to others. The existential approach is highly subjective in nature, it puts strong emphasis on the relationship with others. It is because, the individual person is *threatened* with alienation if he is left alone. Therefore, the therarpeutic relationship should be a strong encounter between two human beings which may encourage the patient to carry out the same relationship with other individual persons with more plesant feelings.

Existentialism stresses that the therapist should exercise his honesty and self-disclosing character in dealing with the patient and help him to learn everything realiably. This also includes the therapist's disapproval of what the patient does openly and correcting him. This in any way should not damage the client's self-respect.

One of the important goals of existential therapy is to create a situation in which the patient should be more aware of his own potentialities for making choice and development of the self. Existentialism strongly view that people create at each moment new existence of their own. It is obvious that there is also potentialities for disorder and for growth. Another important goal of existentialism is that the individual persons should be encouraged to accept the responsibility for their own existence. The client should set up his existence with in certain limits. If necessary the individuals should redefine themselves at any time and feel and behave differently from the previous period within their own social atmosphere.

But it must be also realized that the *freedom to choose and responsibility* that follows with it are not easy for the individual persons to accept and work with them. Generally people are afraid of this freedom, because their wantings are not livily and happy. Many may develop greater awareness of their freedom to choose. In such cases the goal of the existential therapy is to strengthen and encourage their resolution and ability to cope with the unavoidable anxiety and help them to continue the development. **Prochaska** (1984) states that the interpreters of the existential therapeutic techniques do not clearly show the technique that help the client. It appears that the therapist acts upon the patient as if he is an object o be manipulated.

While evaluating the merits and demerits of the existential therapeutic technique **Davison** and **Neale** (1989) state that any research on existential therapy is quite impossible as it is not an accurately definable therapeutic technique. According to them though large number of case reports of success with a variety of clinical problems have been broughtout no research actually has been done. Scientific researches cannot be conducted by applying the existential therapy as it lacks concrete operations. The thought processes that explain the existential therapy seem to be attractive as they are based upon the existential philosophy. But these thought processes should be suitably applied scientifically.

Gestalt Therapy

Gestalt therapy is a *humanintic therapy,* developed by **Fredarich Perls** (1893-1970) that encourages the clients to satisfy the emerging needs and desires so that their innate goodness can be expressed through them to increase their awarness of unacknowledged feelings and to reclaim the parts of their personality that have been denied or disowned. The Gestalt therapy is one of the humanistic therapies that has been developed during 1940s by a German born psychiatrist. On receiving the medical degree in Germany, **Perls** became a psychoanalyst. Due to unavoidable political and ethnic circumstances he has to migrate from country to country till he finds a place in United States, where he practiced the Gestalt therapy.

Basic Principles of Gestalt Therapy

Perls was of the view that people have innate goodness and this is the basic nature of humankind which should be allowed to express itself in social situations. When this innate goodness of human nature is denied the individual human being is get frustrated which results in originating psychological problems. Like all other humanistic approaches Gestalt therapy puts emphasize on the creative expressions of humankind and not on negative and distorted features.

One of the basic principles of gestalt therapy is that all human beings bring their needs and wants to any social situation. Man does not see the situations as they are. Human beings engage their social environment by projecting their desires and needs or fears. For example, according Gestalt therapeutic principles, if a person came and saw us, we do not see or perceive him as he is as he exists. We react to him in terms of our needs. In this context our needs might have been filled up. But if the needs are not fulfilled, the task is unfinished with an important person. This may be also stated as disappointment. In such situations how a person reacts? This may affect a person's feelings. In what manner the person deals with such situation. Gestalt therapists work with groups of people and observe how each one reacts to such situation.

Generally, gestalt therapy is apparently related to Gestalt psychology, Gestalt psychology is a school of psychology mainly concerned with perception of *objects as whole.* The close relationship between the school and therapy is in their concern *on the wholes.* Perls wanted to make the individuals as wholes by making them to become more aware of their unaccepted and unrecognized feelings and making them to claim and accept the disowned parts of the personality factors. The Gestalt therapists are more concerned about what the client is doing in the therapist's room here and now. They do not go into the past life experiences. According to them what is most important in the life of a client is, what is happening at the present moment.

For the Gestalts, what is happening at present and what exists now in the case of the client. if the past is giving trouble it is brought to the present.The gestalts do not encourage questioning the clients, for rearching the causes of the past. Because it is one way of escaping responsibility for making the choice for the present. Attempt is made in all possible ways to make the client aware of what is happening at present. For the gestalts creating a keen awareness in the client is an immediate and direct thinking and reasoning. The clients must know and keenly sense what is going on about them. They must also think and sense, feel and can also expose whatever they sense in the situation. These are the ways by which a sense of awareness is created in the clients. Perls felt that awareness is creative process which make people to think and feel what they are at the present moment and move away from the past.

The Technique Adapted

Generally Gestalt therapy puts emphasize on the unity of mind and body by placing strong emphasis upon the integration of thoughts, feelings and actions. Gestalt therapy is commonly used on a group setting, but the emphasis is on the individual. The therapist works very intensively with one patient at a time, he is helping the patient to perceive those aspects of himself and of his world which are closed for him in the past. The therapist is helping the patient to increase his awareness of the self and also the areas of avoidance. The client is asked to act out his imaginations fantasies about his feelings and conflicts in one chair and then change over to another chair to take part of the adversary. This is the technique adapted by the Gestalts which is called taking care of unfinished business which is worked out through unresolved conflicts.

According to *Perls* we are all going through unfinished or unresolved or shock producing harmful effects and conflicts. We carry with us excess package of these unfinished situations into new relationship and begin to act out our tensions in our relationship with others. If we are able to complete our past unfinished business, then we are having less psychological tensions and get more awareness to cope with the world. The client or the individual is totally on his own in the therapeutic process. There may be also other in the clinic.

Sometimes, the clients may deny the therapist's technique of creating self-awareness in the situation or refuse to accept the procedure. In such instances, the client is brought to a state in which he must face his unpleasant feelings and conflicts. **Perls** (1967) state that in the emergency therapeutic situation the neurotic person discovers that the world does not break into pieces if he gets angry, sexy, joyous, mournful. In this way he can get beyond the safe emergency therapeutic situation on his own.

Perls (1967, 1970) put more emphasis on dreams. According to him the dream is not a rich source of symbolism relating to unconscious processes. The image that appears in the dream whether it is human, or animal or an object is taken to represent an *unfriendly portion of the self.* By re-experiencing and retelling the dream over and over again in the present tense from the point of each image, the patient can begin to reclaim these *unfriendly parts* and accepts them, lives with them and expresses them more correctly. For **Perls** (1967) all the different parts of the dream are fragments of the dreemer's personalties. But the aim of the therapist is to make every one of the clients a wholesome person, a unified person without conflicts or the therapists puts the different fragments of dream together. The client has to reown these projected fragmented parts of the personality and reown the hidden potential that appears in the dream.

Perls, in the Gestalt therapy session makes focus on the elements of the client's behaviour. These sessions are often called as *gestalt awareness training.* It is because, the therapeutic results of the experience stem from the process of making him more aware of the client's total self and his environment. Gestalt therapy is very widely used in United States because of its humanistic-existential orientation which promotes the importance of *self awareness and personal growth.*

Interpersonal Therapy

Psychotherapy includes also the interpersonal therapies. The interpersonal therapy is mainly concerned with marital and family problems. The maladaptive behaviour emerges in the *role of faculty communications,* interactions and relationships. These are having an important impact on other approaches to therapy mainly on humanistic and behaviourstic therapies. In many of the interpersonal relationships the individual's problems become the centre of attention. We shall discuss in this section the marital therapy and the family therapy.

Marital Therapy : This has become a growing field of therapy in these days with the problems prevailing around people's marriage situations. In many cases the partners are seen together, but therapy is highly needed. The therapy is focused on *clarifying and improving* the interactions and the relationships between them. A very wide range of attempts and procedures should be included in achieving the goal.

This marital therapy mainly emphasizes the mutual gratification, specific social roles, type of communication, performing the specific duties and responsibilities. Generally marital therapy is directed towards improving the communication and interaction between marital partners. The marital problems vary from culture to culture. Marital problerr arises when the married couple, the husband and wife develop differeiice of opinion in personal and family matters. It is said that "marriages are made in Heaven". This suggests the idea that the two married man and woman are expected to live a happy life and they are supposed to lead a happy life. The marriage might have taken place by the arrangement made by the parents of the man and woman or the marriage might have taken place by love. In Indian culture it mostly takes place by the arrangement of the parents of the man and woman. But there are also love marriages which may take place privately when both the man and woman agreed to live as husband and wife (Fig. 26.3).

But in western cultures it is mostly taking place when a man and woman love each other and decide to live as husband and wife. Whether it is a love marriage or arranged marriage both are expected to live happily. But in a human life, however happy it is, instances may occur which may lead to difference of opinions which may manifest in the expression of feelings and emotions, disagreement, strongly adhering to one's decision and so on. This feeling of disagreement or non-acceptances of the life-partners start internally work for sometimes and may manifest at some circumstances which may result in some wordly quarrel and may end in maladjustment. This maladjustment will not stop at this stage, it may develop into even big quarrel leading to seck some therepist to deal with this problem. This type of consequences are quite common in many cultures. Some may even go to court and seek divorce (Fig. 26.4 and 26.5).

Fig. 26.3 : Marital Maladjustment : Marital maladjustment is becoming a serious problem in these days in India. The married young husband and wife live happily for one or two years and in some cases only for few moths. Then difference of opinion begins to take place often. What the husband says, the wife does not agree and instead she makes a counter suggestion and wanted that it should be accepted. Mainly finding fault with each other has become a regular feature. There is a conflict on a very minor matter which led both to a state of frustration. Each one felt within that he/she is disappointed and deceived. Difference of opinion and conflict day by day increased and also became serious which caused them severe frustration. This made them each one set aside.

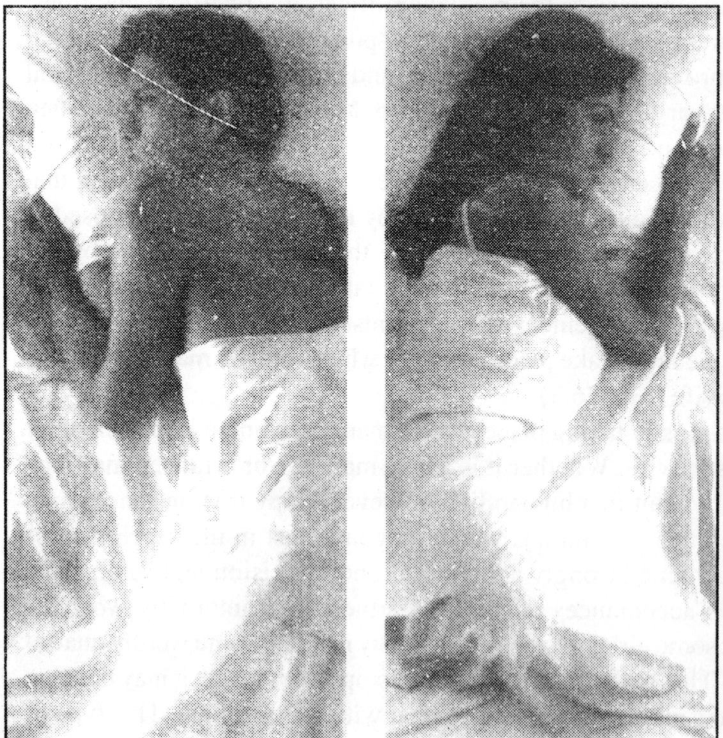

Fig. 26.4 : The difference of opinion, conflict and the frustration which made them each to set aside, resulted in severe depression and made them to think of the situation very serviously. Each one found separate bed in separate rooms. None of them tried for any effort to alleviate the difference of opinion. Instead, their feelings of frustration became intensified. The wife began to speak openly that she lost her happiness in life and the only alternative is to go away or to get separation.

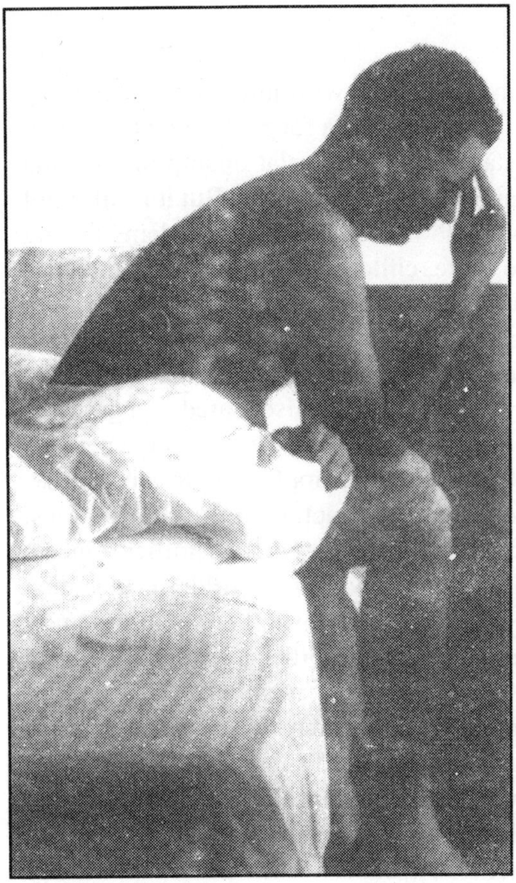

Fig. 26.5 : Severe Tension and Depression : On knowing the decision made by his wife the man became very much depressed and left himself in an isolation. He found his wife was very much rebelious and strong in her view. He strongly felt that separation or divorce should not take place in his family. He felt very much ashamed and became nervous and began to deeply to solve this problem. He thought about various methods of dealing with this problem amicably. At last he decided to meet the psychotherapist on the hope that he alone could solve this problem.

Marital problems, thus, may take this shape sometimes. Therapy in such cases if successful may promote compromise and make the marital couple happy. **Navran** (1967) in one of his studies had observed that his clients have shown more sensitivity to each other's feelings and needs. **Paul** (1971) has observed that faulty role expectations often play havoc with marital adjustments. Paul had cited an interesting case of a couple who came to him for marital therapy which has been resolved. He explained how a thirty-nine years old husband, who wanted to divorce his wife to marry a much younger woman, started setting tears and weeping of greif while narratings his problems in the therapeutic session. At the end of the session that was a reconciliation with his wife and she has become more understanding of her husband's needs.

According to **Coleman** (1974) the intense emotional involvement of the marital partners is one of the difficulties in marital therapy. This makes difficult for the marital partners to perceive and accept the realities of their relationship. Coleman is of the view that wives often see what is wrong with their husbands, but the wives do not see what is wrong with themselves. Generally husbands have remarkable insight into their wives' faults. Several marital therapies have been developed. One of the therapies found to be very much successful is the **Rogerian** non-directive technique. Olson(1970) had reported the effects of this technique in marital therapy situation.

Family Therapy

Family therapy is a form of therapy which mainly concerned with interpersonal therapy focusing on the relationship within the family. It is also stated as a form of group therapy in which the members of a family are helped to establish a better relationship among one another. The family therapy appears somewhat similar to marital therapy. But it is different from it. Marital therapy mainly deals with couple who had adjustment problems among them. But a family may be consisting of husband, wife, children if joint family, includes parents and in laws. Therefore, there are more chances for problems of adjustment.

What is a family? A family is a primary social group consisting of parents and their children, the principal function of which is provision for its members. Another definition can be, one's wife or husband and one's children. A family may be also stated as a group of persons related by blood, a group of descendants from a common ancestory, all the persons living together in one household. These general definitions are applicable to any family. This may be applicable to all the cultures. But time to time the structure of the family varies. In the present generation in some culture we find joint families, separate families. Once a man get married he sets up a separate family life with his wife and children. In Indian culture though this type of family life prevails in most of the families there is joint family life consisting of husband, wife, children, husband's parents, brothers, sisters, aunts and other in-laws. Thus the family is consisting of large number of members. This type of family life is often subjected to difference of opinions, which may arise on the basis of individual needs and demands, likes and dislikes, dominance and submission, exercise of authority or seniority, choice of food, time of eating, sharing of household works and so on. All these matters may create varied feelings which may result in emotional maladjustments and quarrels. Once an intensive emotionally involved instance takes place, the family happiness would be jeopartized which may also result in disorganization of the family structure. In a joint family, in India, disorganization of family structure has become a common occurrence. Due to difference of opinions, varied individual needs and, demands, ambitions and aspirations, some type of guidance from outside the family, suspicision that one is enjoying better status than others and so on may contribute to the disorganization of the joint happy family structure. This is what has taken place in Ambani's family. Ambani's Business and Industrial Organization is one of the richest in India and also at the International level. The death of Ambani, the Founder Managing Director, resulted in difference of opinions among his two sons with regard to the management of their property, resulted in the disintegration of their family structure. This type of situation is quite often occurring in Indian families. The only remedy is to have an intense family therapy (Fig. 26.6).

There are a number of specialized therapeutic techniques for use with children which may be used either in connection with family or in individual treatment programmes. **Haley** (1962) is a pioneering therapist in the field of family therapy, dealt with several problem families. **Haley** carefully studied the family of disturbed child and understood that the *child is reflecting* the pathology of the family unit. When the child is at the therapeutic centre he is able to workout some of his problems. But when goes back home he has to face the same pathological circumstances.

Fig. 26.6 : The husband met the therapist with a strong hope of setling the issue. He had three sessions with the Psychotherapist and the developed some confidence that the problem would be solved. His wife in the beginning has strongly protested this and behaveed in a very rebellious manner and when met the therapist she has become somewhat sober and reduced her agressive stand. On having three sessions with the psychotherapist the young wife turned to normal state and her rebellious attitudes gradually diminished. She accepted hat are all the therapist had suggested and agrred to live with her husband happily without causing any difference of opinion.
[This is based upon a case study of a young couple. This is quite similar to the Marital problem case treated by G.L. Paul (1971). The illustrative figures are step by step development of the problem the couple experienced.]

In family therapy, the members of the whole family should be seen together in the centre, but there are variations in the family therapy. After making an extensive survery of the field of family therapy it was concluded that family therapy at present is not a therapeutic technique in the normal sense. There is no one accepted procedure of family therapy.

Group Psychotherapy : One of the important aims of psychotherapy is to treat the patient and restore him for normal social functioning and to make him to behave adjusting with others. In this area an inevitable development has taken place which is called *group therapy.* Thus a group therapy may be defined as a system of therapy in which one or more therapists work together with several patients at the same time. Group therapy is also defined in a little different way. It is the method of treating psychological disorders whereby several persons are seen simultaneously by a single therapist. Group therapy was found to be more useful during the *Second World War* in dealing with the soldiers who were suffering from *war neuroses.* More therapists were required during the war period. Since there were only few therapists available for treating them the group psychotherapy technique was applied and it was found to be successful. Thus the group psychotherapy came into force.

In group therapy *each patient* has a constant check and feedback on his *social needs.* He is encouraged by the group to view his relationship with the therapist with more objectivly.

This means that he is less likely to be over dependent on the therapist. It can be stated that the group therapy reduces the time of the therapist and also reduces the patient's expenses. At present it is found that wide diversity of groups are getting group therapy to some extent. But they do not call it as group therapy. For them it is *club*. One of such group club is called as *Alcoholics Anonymous* (AA). It is also stated as *half-way house club*. These clubs serve as the *rehabilitation centres* for the persons who have been taking treatment in the mental hospitals. The alcoholics anonymous group members also not prefer the professional therapists. They seek self-help from the psychiatiric social workers (Fig. 26.7).

The half-way house club is used as a temporary relieving centre for the former patients which may serve for their half-way back to normal way of life in the society. But these

Fig. 26.7 : Group Psychotherapy : Alcoholic Anonymous : Club.

The Group psychotherapy is a further development from classical psychotherapy. It is system of therapy in which one or more therapists work together with several patients at a time are dealt with. Another development in group therapy is treating the psychological disorders in which several patients are seen by a single therapist simultaneously. Group therapy is found to be very useful when sufficient therapists are not available. During Second World War the soldiers who were suffering from War Neuroses were treated by few therapists. Wide diversity of groups are seeking group therapy. They call it as Club (Therapy). One such Club is called *Alcoholic Anonymous* (AA). Another name for this is *Half-way House Club*. These clubs serve as rehabilitation centres for those who are taking treatment in Mental Hospitals. These alcoholic anonymous club members do not prefer professional therapists. The most important step in group therapy is to make the alcoholics to accept that he/she has the alcoholic problem and there are sources to help them and to deal with the problems.

alcoholic anonymous club expects the members to join the club as lifetime members. It is because the alcoholics may be often tempted to drink again. In order to prevent or to resist the temptations they must be members of this club for lifetime. But those who adapt psychotherapy as the only means of their problems take up either Freudian or the Rogerian.

The most important step in group therapy is to make the alcoholic to accept that he is having the drinking problem and there are resources available to help him and to deal with his problems. In the group therapy by giving and taking technique the alcoholic is made to face his problems and to recognize its possible disastrous results. The patient's this type of recognition always opens the doors towards the positive results to cope with his drinking problem. In some cases the patient's wife and children also are invited to join the group therapy sessions.

General Principles of Group Therapy

Frank and Powdermaker (1959) are of the view that irrespective of the theoretical considerations of the therapist, there are certain general principles that are applied to all group therapy sessions. A problem person may need social stimulation and support which he may be expecting to have it in a group therapy. With the care and help of the therapist the patient can speak out his problems freely, can also make comments about the problems of others. In the therapy situation he is able to interact with others and also he shares with the emotinal experiences of others. Once the patient accepts his problems and the consequences he is highly amenable to the therapy.

Choice of the Patient : The therapist is expected to be very careful in choosing patients. The exponents in this area have laid down broad guidelines. The antisocial person will not fit into the group situation. Generally neurotic persons suffering from psychosis find very difficult to adjust to the group therapy situation. The more seriously ill patients may make very wild remarks and comments and the neurotic patients in the group may get threatened. These patients in their emotional detachment may comment brutally on the motivation of the group members. The therapist should see that in the group therapy situation individuals should not be subjected to embracing situation because of some personal characteristics.

The social environment outside the group therapy situation may be consisting of different categories of people. They may vary on the basis of sex, religion, race and age. A therapeutic session may be composed of both men and women equally. An ideal therapy group may be consisting of the admixture of both sexes, different religious affiliations and wide distribution of revels. The individual person in the group therapy session will not be left alone or isolated on the ground that he/she is the only male or female member in the group or the person is the only from different religion, or the individual is the only person very young or very old.

But in the case of any specific problem or purpose the groups may be constituted with husband and wife, in the case of marital problem, parents and children, in the case of family problems. An ideal group therapy may be consisting of two or three and more members depending upon the problem. According to the group therapy principles close relations and friends are not included in the same group for therapy. But, however, all left to the approval or disapproval of the experienced therapist to decide the formation of the groups with which he/she is going to deal with. On the commencement of the group therapy after constituting the group in the first few days, the therapists may be experiencing some pros and cons with the group. In such case the therapist can add some more patients to the group or drop some patients as he/she has to maintain some balance for achieving his/her goals. The persons who are included in the group therapy session should cooperate and maintain a congenial atmosphere till the session comes to a successful end.

The Group Therapy Session : Having chosen the patients for the group therapy session, another problem the therapist faces is the number of patients in the session. Experiences in this counts a lot. Generally the number of patients for a session may vary from five to eight. Many therapists prefer five to eight patients for a session for reasons that less than five for a session would be intimate and more than eight would be devoid of probing the problems fully. Therefore five members constitute an ideal group for a session. In a five member group session their is more chance for every member to see each other in a face to face manner. The five to eight member group is able to sit comfortably around a table. The members of the group are allowed to sit according to their choice.

In group therapy one of the principles is that the members of the group therapy should not meet outside the session without the therapist. This practice may leed to several problems. But a member of a group therapy can meet the therapist outside the session and can have some discussion about his personal problems. In a *group session a patient* who is a dependent always choose to sit by the side of the therapist which is the indication of seeking protection. But at the same time a patient who is of paranoid tendencies may also prefer to sit by the side of the therapist for the purpose of watching the therapist closely. But the therapist observes the reactions of every member of the group and makes note of them for his purpose.

In the therapy session one problem arises is whether an outside person can be permitted to observe the session. An observer can be allowed to get into the session and can observe and may collect information for some research purpose provided if he is not intruding into the session. The observer can give a better view to the therapist of what is taking place. Generally a group of neurotic patients may accept the presence of an observer and they can go on their own ignoring his presence. But however the psychotic patients may question the presence of an observer in the therapy session. But between group sessions the therapist and the patients can meet and discuss the problems if anything serious consequences take place. This type of individual sessions outside the therapy clinic should be a avoided. It is better to have everything in the group session. The group sesston normally should meet once in a week.

The therapist on any account should not disclose anything in the group session about the individuals session that took place outside. The individual's sessions outside the group therapy session has been viewed by most of the therapists that sessions should not go more than one hour as it may have some influence in the group session.

The Functions of the Group Therapist : Every psychotherapist is a trained person under some experts and they carry out the therapeutic techniques in the therapy sessions. Most of the therapists are trained only in the individual therapy. But when come across more individuals with problems they develop some techniques for helping the problem individuals applying their knowledge and practice in psychotherapy. Group therapy sometimes may be very difficult for them as they become emotionally involved. The main reason is that he is shifting from individual session to a group session.

In the early sessions the therapist should indicate the patients their purpose and objectives and clarify. By these means every member should develop a sense of security and try to work out on his problem.

The therapist must make himself to fit to the therapy situtation in which he is dealing with more number of patients. The therapist should see that the patients should act on the expressed feelings. The expression of feelings sometimes may be verbal and sometimes it may be facial and gestures by the therapist. When the therapist interacts with one of the patients is watched very keenly by all other patients of the group. This means, the group expects something more from the therapist. If the therapist is able to exercise more skills in the group situation successfully his achievements are greater. In the next session the therapist sits and watches each member whether he is trying to understand his own problems and tries to help the members to resolve their problems.

We have discussed in the preceding section about Roger's client centred or new directive therapy. There were many therapists adapted Rogerian client centred therapy. Hobbs (1951) one of the Rogers followers of client-centred therapy and experienced the necessity of developing a group-centred therapy. The exponents who have contributed to the development of the group-centred therapy were all from Rogerian's school of thought. They had all rigorous training in the client-centred therapy. Therefore, they applied the same principles in group centred therapy also. They put more emphasise on group situation,the complete acceptance of all those what were said and done in the client-centred therapy. They followed the same content, clarification and the reflection of emotions in the group centred therapy situation.

The therapists who followed the client-centred therapy in treating the patients have increased the non-directive aspects in the sessions. The non-directive therapist has to learn some new skills for the group-centred therapy, other than the skills required for individual sessions. The therapist is expected to accept and reflect each member of the group in such a way that he should encourage the participation of the group members. They must interact with the therapist, freely in the session. A lengthy dialoque between the therapist and a member of the group may produce a harmful effects on the group. The therapist's exercise on the members of the group and his intervention in their expressions may disrupt the session. Therefore the non-directive group therapist must be very careful in following everything that goes in the group inter-actions. If the members of the group are brought along the line step by step without any intervention the therapist can handle even the most difficult situation successfully.

The therapist can get the help for his purpose from the members of the group as they become a more accepting group and may express better in clarifying each member's emotional states. In the case of group centred therapy there is no much different from other type of group therapy with regard to the length of the time of a session, the number of sessions, the number of members in a group and the number of sessions required to terminate the therapy. It is to be stressed that all the therapy groups here are highly group-centred. It should be known that highly leaderships exercised or leader-centred group psychotherapy may not bring out the expected success. Though the *group therapy* and the *group-oriented therapy* seem to have certain similarities they are different in certain respects. It is developed on the pattern of Rogers client-centred therapeutic techniques.

Psychodrama : A Group Therapy : Psychodrama is a psychotherapeutic technique developed by J.L. Moreno in which the acting of various roles is the cordinal virtue of the therapy. Psychodrama is a form of group therapy in which individuals act out before an audience situations from their past life instances. It is an interesing form of group therapy based upon role playing techniques. The patient is encouraged to act out problem situations in a theatre-like setting and he is assisted by the staff members and by other patients. This *role-play technique* makes patient to get free from the problems. The individual expresses his anxieties and hostilities and get relieved from traumatic experiences which he had met with in a situation which stimulates his real life. The purpose is to help the patient to achieve the emotional stability, increased understanding and to improve the interpersonal activities. **Sundberg** and **Tylor** (1962) state that this form of therapy is proved to be very much beneficial for the psychiatric patients who constitute the audience and for those who participate in the play on the stage. Psychodrama as a form of therapeutic technique makes use of several category of people as participants.

There is an interesting incidence in the origin and development of the technique, psychodrama. **Jack Roy Strange** (1965) cites how *Moreno* has been working it out for many years to originate the psychodrama. He has been making it perfect and distinctive therapeutic technique which he finally called it as *psychodrama.* The idea of psychodrama came to *Moreno* in 1924, when he founded the *Theatre of Spontaniety* in Vienna. In the theatre of spontaniety the participants as mentioned by Moreno were the Protagonists who were the *patients,* the director who is the *chief therapist,* the auxiliary egos, who were the *other therapists* and the group, which consisted of the audience made up of the *other patients* (Fig. 26.8).

In psychodrama the protagonist (patient) is encouraged to take part and act a role that should be from *real life problem* to him in which he is expected to reveal the *source of anxiety and anguish.* The auxiliary egos (other therapists and patients) take roles that are required to reveal the situation as given by the protogonist. A protogonist, for example, who is having trouble in his role as a father made, needs someone to play the role of a wife and mother and someone take the role of a child. Moreno stressed the point that the acting should be spontoneous and creative and the actors must throw themselves in the part and the dialogues should be natural. Moreno observed that those who were poor actors were quite good in speaking out the dialogue spontaneously. There was *free acting without any rehearsal* similar to free association, resulting in releasing the repressed emotion loaded material.

Fig. 26.8 : Psycho-Drama : A Group Therapy

Psychodrama is one of the psychotherapeutic techniques in which the acting of various types of roles is the cordinal virtue. J.L. Moreno has developed this technique after several years of study. Normally, psychodrama is a group therapy in which individuals act out in the situations on the basis of their past life experiences. It is based upon the role play technique. The patients are encouraged to act out the problem-situations in a theatre-like setting and they are assisted by the staff members and other patients. This role-play technique makes the patients to get free from the problems. In the situation the patients express their anxieties and the hostilities and get relieved from them. The purpose of this therapy is to help the patients to achieve the emotional stability, increased understanding and to improve the interpersonal activities. In this technique the patients are encouraged to act the roles that should be from their life problems to them if they are expected to reveal the sources of their anxieties and the anguish. Moreno had stressed that the acting should be spontaneous and creative and the actors must throw themselves in the part they played. Psychodrama is very much useful for those who refuse to attend the psychotherapy centres.

Moreno states that psychodrama is very much useful for institutions, where generally the patients refuse to attend therapy sessions of more orthodox in nature. A patient who had with'drawn from a therapy session when he sees the drama for sometimes, is encouraged for playing a small role. There were several modifications in the Moreno's psychodrama technique outside the institutions, according to the needs and type of the patients. In schools and industries where frustration, conflict and emotional malajustments prevail, psychodrama may be of great use in dealing with problem individuals.

Encounter-Group-Therapy : Encounter group therapy is a very recent one, developed by **Aronson** (1972). It is also called by some as *sensitivity training* and encounter group therapy. Sensitivity encounter group therapy is a therapeutic techinique in which a small group of people who spend a period of time together both for therapy and for educational purposes. In this, the participants are encouraged or forced to examine their interpersonal functioning and their often ourlooked feelings about themselves and others.

Coleman (1974) states that the entire field of group therapy has been changed by the development of the encounter group therapy and it appears that it is better to close our psychotherapy on account of this important form of group therapy. It is said that some encounter groups arise from the sensitivity training. Some encounter groups are guided towards helping normal individuals and guide them to learn more about how their feelings and behaviour affect themselves and others. But the other groups are directed mainly towards the treatment of personal problems and thus it becomes a part of *transitional group therapy.*

The founder of encounter group therapy *Aronsol* had formulated some format and goals. The formulation of encounter group is not very rigid. There is always great deal of flexibility and change in the structure of the group. Different type of formats also often emerged. In the *drug addicts groups* there are no leaders. The other important type is the *marathon groups*. The participants of marathon groups are clothed according to convention. But they are also in nude. Thus there are conflicting groups in the community. The encounter group normally consists of six to twelve participants which may have one or two leaders. The *physical setting* is considerably *unclothed.* The group members are permitted maximum freedom to move and act among themselves. The specific goals to be achieved in the encounter group therapy very considerably; depending upon the nature and function of the groups. But generally the main purpose is to provide intensive group experience which provides change for the members to work through emotional problems and to gain more adaptive experience. The encounter group situation is mainly based upon intensive *give and take* interactions, which is considered to be much more closer of social reality than that is practised in the traditional group threapy or individual therapy.

The encounter group therapy in fact is breaking all the conventional psychotherapeutic techniques and goes beyond the social and cultural norms to provide a healthy atmosphere to the participants of the groups. The interpersonal interactions, among the participant of the groups are in terms of *honesty and frankness.* There is *free expression of emotional conflicts.*

The leader of the encounter group is considered to be the *leader-participant.* He does not exercise any authority over the participants. His responsibility rests on protecting the participants' secrets and arranging the meeting. One of the main responsibilities of the leader-participant is to create a climate of *psychological security,* by which each member of the group is feeling safe. The participant could express his feelings freely and can find out new ways of interacting technique with others outside. The leader participant also helps the participants to be a model figure by way of plainly and honestly expressing his own feelings honestly and also accepting the hostile feelings or other negative feelings expressed by others *without developing any offensive feelings.* The leader participant also helps participants of encounter group by giving them the feedback of the experience they have gained in the therapy situation. The participants in any situation should not make any unpleasant statements while having conversation with others. The leader helps the members to resolve any anger expresson or feelings among the participants.

With regard to the *group process the non-verbal* technique is used as *warming-up exercises* to create awareness and group interaction. Varied techniques are used for this.

The verbal techniques are used to help the group's interaction processes. Generally it is directed on the personal problems. There are variations in this technique. One technique is the *positive and negative bombarment.*

Participants of the encounter therapy groups also have the, liberly to leave the group and after sometimes can get *re-entry* into the group for feedbacking. When they go out without fully reso!ving their conflicts and emotional disturbances their unpleasant feeling reappear and they get hurt in the interaction with others. To get rid of such situations they seek *recently* to the group situation. The encounter group therapy also has certain advantages and disadvantages like all other therapeutic techniques. But however the potential adverse effects can be reduced by carefully screening the participants in the specific type of encounter-group therapy session.

There are varied types of encounter-group therapy sessions. **Aronson** (1972) differentiates the sensitivity training or *T group* from the encounter-group therapy stating that certains encounter group therapy sessions are more radical. He states that it is very popular only in the West Coast areas. According to him the T groups or sensitivity training mainly depends upon verbal procedures and it avoids the physical contact among the participants. Many encounter group therapy sessions are based upon physical contact touching exercises, unrestricted expressions of emotions, and nude appearence of the participants in sessions. Many did not accept *Aronson's* statement and distinction between the two types of techniques. But, however, Rogers (1970) traced the development of the encounter therapy in his counsellors training programme in 1946. He stressed the personal growth and improved interpersonal communication. Many therapeutic centres accepted the judgment made by Rogers and they all resolved that the T groups and the encounter groups can never be distinguised and differentiated from each other.

The sensitivy training (T-group) was found to be mainly for educational purposes and it was not developed for those who are emotionally disturbed. But however the T-group successfully promotes the personal growth and brings out better understanding in persons who are functioning better. Individual persons are drawn often to T-group even though they are not having many problems in their lives. But the problem with them was that they feel as if they are missing something. The days they lived are without intensity and intimacy.

According to Aronson, the general goals of T-group are promotion of general welfare of the persons who are dissatisfied with their lives. The T-group is interested in developing a willingness in an individual to re-examine his behaviour and to test with new ways of behaviour. It is to learn more about people in general. It is to make more authentic and honest in interpersonal relations. It is also to make the individuals to work cooperatively with other people without assuming, dominating or submissive state. The group therapy aims to develop the individual's ability to dissolve the conflicts by means of logical and rational thinking and not byu manipulating and forcing others. Generally, the T-group and the encounter group therapies provide a healthy background in which people are encouraged to behave in a free manner. Sensitivity group therapy and the encounter group therapy vary in a number ways like all other therapies.

Psychotherapy with Children

Children are often emotionally disturbed in families. They sometimes become big problem for the parents. In middle socio-economic families and in lower socio-economic families in Indian community the emotionally disturbed children problems are going on increasing. Generally children in Indian communities at the age of eight to fourteen become very sensitive and any quarrel with the parents causes them to run away from home. Particularly in the case of school performances when parents question their low marks in the examination, children become emotionally disturbed and run away from homes. In the case of girls they commit suicide and die. Instances like these are taking place everyday and the mass media and the Television announce these unhappy instances. Therefore, a systematic psychotherapy is a must to counsel the emotionally disturbed children (Fig. 26.9).

Fig. 26.9 : Psychotherapy with Children : Klein's Play Analysis : Verbal Technique : It is observed that children are becoming more and more problem for the parents. Children from middle and low socio-economic communities becoming more sensitive and easily get disturbed emotionally. When there is some minor quarrel in the family children are very much perturbed and they even run away from home.The mass media of communication is everyday reporting such incidents. The children are between six and fourteen years of age. Specially when the parents question their children's school performances regarding the low marks in the examinations the children become very furious and sometimes commit suicide and die. Parents become helpless with such children. The only alternative for the parents to do is to take the children to the psychothrapist when problems arise. The therapists who are specialized in child therapy can bring the problem child to normal state in three or four meeting sessions. The therapist follow some special techniques to manipulate the problem child quite appropriately.

Psychotherapy for emotionally disturbed children is in many ways quite similar to the adult's therapy. But in many important aspects it is quite different from it. The child is totally, a dependent person. He is not a free person. The child is dependent upon their parents, or guardians, his problems are inextricably related with other persons. Therapists mostly advice the parents and guardians if the child is having serious emotional problems along with their parents. The child has much difficulty in explaining his problems to the therapist. Parents' statement may vary, because it may be from their point of view. But the child's problem may be different.

Therapists in treating the psychologically disturbed children' meet with some problems of getting the correct cause of their problems. The therapist has to follow some special technique to manipulate the child in the same manner he deals with an adult patient. Varied techniques of psychotherapies have been developed to deal with children. Of which the play therapy is one of the important therapies often used on children. **Malanie Klein** (1937) has developed therapeutic technique which she named it as *play Analysis.* In dealing with the child's problem she used both the *verbal and play responses* for her interpretation. Her main aim was to relieve the pressures of the Id and *super ego* and to strengthen the ego. She followed the *Orthodox Freudian therapy.* Later on several child psychotherapeutic techniques were developed (Fig. 26.10).

Fig.26.10 : Play Therapy and Family Therapy

Play and family therapy are the two methods that are used to treat the childhood problems. Why these methods are needed? It is because that children are generally low in verbal skills and are reluctant to speak about their complaints directly and openly. Therefore, the psychoanalytic and client-centred therapies are needed. It was through the pioneering works of **Malanie Klein** (1932) and **Anna Freud** (1946) play became an analytic tool for delving into the children's consiousness, taking place of the free association and recounting of dreams. It is assumed that in play a child will express his/her feelings, about the problem parents, teachers and the peers. Klein named it as *Play Analysis* (therapy). A play therapy room is the most important aspect of this therapy. It should be equipped with puppets, blocks, game buzzles, drawing materials, paints, water, sand, clay, toy guns, inflated rubber baloons and other playing materials. The therapist can arrange the play therapy room according to the situation. These toys may help to draw out the inner tensions and the concerns of the problem child. The child may be allowed to demonstrate his skills through the materials placed there.

27

COGNITIVE AND BEHAVIOUR THERAPIES

Introduction

Psychopathology has been treated by several approaches with modern techniques right from the period Sigmund Freud and Carl Regers. Researchers in this field have classified the therapeutic approaches with several categories. The psychoanalysis and the client-centred therapy and other related methods were named as insight therapies. Therapists who applied these techniques to treat their patients had their goals to help the patients to achieve a better understanding of their problems, feelings, conflicts and motives. These therapists believed that once the patients reach this stage they can cope with their problems better and may lead a normal life.

Psychoanalysis is mainly concerned with the past conflicts and frustrations of the patient as the present behaviour of the individual is very much influenced by them. By understanding the cause of the present day problems the individual can get rid of his problems. But the client-centred therapy mainly is concerned with the present day problems and the types of interactions that are taking place in the patient. The therapist goes into the insight in dealing with the problems and the treatment.

Behaviour Therapy

These two insightful techniques were developed into several therapeutic approaches in dealing the psychotic patients. But there were also severe criticisms against the insight therapies that these therapies are very expensive and take longer time in the treatment. This resulted in the development of some new approaches which focused on the problem behaviour and not on the insight and past conflicts and feelings. These therapeutic techniques are called *behaviour therapies.*

Behaviour therapy may be defined as a method of psychotherapy based on learning responses which refers the techniques of counter conditioning, reinforcement and shaping to modify behaviour. The main aim is behaviour modification. The behaviour therapists use a number of different techniques based on the principles of classical and operant conditioning, *to modify behaviours.* It has taken four decades from 1920s to 1960s for treating psychopathology in a new way. The behaviour therapy in the beginning was limited to *classical and operant conditioning.* The present approach of behaviour therapy is to change abnormal behaviour, thoughts and feelings and emotions. Behaviour therapy, if scientifically applied on patients it can proof to be an effective method of therapy, by alleviating the problems of the patients.

The therapists who have been using the operant conditioning in treating the patients have been using for sometimes the term *behaviour modification* in place of behaviour therapy. But some therapists were of the view that behaviour therapy can be differentiated from other therapies. The view of these therapists was that every therapy, whether it is psychotherapy or psychoanalysis seek only behaviour modification, which they considered as the ultimate goal. There is no correct date when behaviour therapy came into force. In some techniques several changes and modifications were going on for a long time and finally, we find the behaviour therapy. According to **Davison** and **Neale** (1989) there are several areas of therapy overlapping, but there are *four theoretical approaches* on which the behaviour therapy was developed. These four distinguishing approaches applied in behaviour therapy are *counter conditioning, operant conditioning, modellings* and *cognitive restructuring.* All these four approaches were functioning at some level and treating the psychological patients. For several years people who were in the clinical field started formulating a new type of hypothesis and test about the best means of dealing with the problems that they have met with. They tried this in spite of several approaches were in practice at that time.

Counter Conditioning

In the recent past a number of therapeutic methods and techniques have been developed by the clinical and experimental psychologists which are the outgrowths of learning forms. This has given expression to behaviour therapy and behaviour modification. These two terms were applied to the outcome. It is because in the beginning it was established that they are based upon experimentally and clinically tested laws of learning which were developed by the behaviourist.

The counter conditioning is a technique of changing the behaviour form. *Anxiety* is classically conditioned in a person. According to *mediation theory* there are effective ways of *eliminating* this anxiety or fear this involves associating the conditioned stimulus with a *non-anxiety* or *fear* causing stimulus. This process is called as *counter-conditioning.* It is based on mediation theory. *What is mediation theory?* The general view in psychology is that certain stimuli do not directly initiate or arouse an overt response or behaviour in an organism, but activate an intervening process which in turn arouses the response. It explains thinking, drives, emotions and beliefs in terms of stimulus-response mechanism.

Then, what is counter conditioning? It is the outcome of the mediation theory. Counter conditioning is a form of relearning, achieved by eliciting a *new response* in the presence of the same stimulus (particular stimulus). This can be illustrated with an example. A child is fearful of seeing an animal, the cat. The cat(s) arouses fear (R1) in the child is response. The therapist creates a playful situation and a different type of response (R2) is made to arouse in the child by a playful reaction and the child now is not afraid of the animal, cat. The central point is that the animal which caused fearful reaction (response) is made as non-fearful reaction (R2) response. This is called *counter conditioning.* The clinical and experimental evidences suggest that this counter-conditioning or changing the response can *remove the fearful response.*

This counter conditioning has been systematized by **Joseph Wolpe** (1958) and it has played an important role in thinkings, drives, beliefs and emotions. He devised a set of behaviour therapy techniques. In applying the counter conditioning to a patient who is suffering from anxiety disorders, the therapist first collects the list of feared situations. He starts with the situation which arouses the least or the minimum anxiety and proceeds on the fear situations gradually increasing to those that are the most frightening situations. The patient is taught to relax deeply. During the course of the treatment he relaxes step by step and also he imagines the graded series of anxiety-provoking situations. The relaxation tends to inhibit any fear causing situation that might be elicited by the imagination. The fearful person is able to tolerate increasingly more difficult imagined situations as and when he goes through a number of therapy sessions. Clinical and experimental evidences show that this technique is able to reduce a very wide varieties of anxieties.

Systematic Desensitization

Behaviour therapists are strongly of the view that insight, that is used by the psychotherapists is not necessary in treating the persons who suffer from emotional disorders. Further they felt that it is not sufficient in the treatment of behaviour change. The behaviourists use a number of different techniques based on the principles of classical and operant conditioning for modifying the behaviour. One technique used by behaviour therapists is the principle of counter conditioning which is discribed above. The other technique applied by the behaviour therapists is the *systematic desensitization.* It is a behaviour therapy technique for eliminating anxiety disorders. Generally systematic desensitization is a major behaviour therapeutic procedure. A fearful person while deeply relaxed imagines a series of increasing more fearful situations. The relaxation and fearful responses are incompatible and fear is dispelled. This technique is very much useful for treating psychological problems in which anxiety is the important and principle difficulty.

What is systematic desensitization? Systematic desensitization is a behavioural therapeutic technique that is applied to a specific approach developed by **Wolpe** (1958). The assumption on which **Wolpe** has developed this technique is that most neurotic patterns are fundamentally conditioned anxiety responses. Wolpe attempts to train the patient to remain calm and relaxed in situations which had caused anxiety previously. The effects of deep relaxation are automatic and accompany the deep relaxation. These effects are diametrically opposed to the characteristics of anxiety.

Wolpe's systematic desensitization therapeutic technique falls under *three processes such as training for relaxations, construction order and desensitization procedure. The first* step in the therapy is training the patient to relax. This normally takes six sessions. During these sessions, the training consists of training the person to contract his muscles gradually to a state of complete relaxation. The principle adapted for this is the *progressive relaxation.* Other techniques followed for relaxation are the *meditation and drugs.*

The construction order is the arrangement of the patient's anxieties in an order of the nature of severity. The stimuli that cause anxieties which are arranged in a desending order. The stimuli that caused *very intense anxiety* placed at the tap and all others are placed gradually step by step to the stimuli that caused very low anxiety. This construction of hierarchies is in the reverse order arranged in the technique of counter-conditioning.

Thirdly, the procedure of desensitization begins only after mastering the relaxation technique and the establishment of the appropriate anxiety hierarchy. At this stage the clinet is made to relax completely in a comfortable chair and he closes his eyes. The therapist narrates a series of scenes of the client's anxieties and directs him to imagine the experiences he himself had and the situations. The therapist presents to the client, first, the natural scene. If the client is calm, the lowest scene in the hierarchy is a presented. Then the therapist goes progressively in presenting the scenes and when the client states that he experiences anxiety, the therapist stops the presentation of the scenes. The treatment continued till he remains calm and relaxed. In this state vividly imagined scenes are those which have caused the *greatest anxieties.*

Each session normally may take fifteen to thirty minutes and the number of sessions may be two or three per week. The completion of the programme may take a number of weeks and even months. A number of variations in this therapeutic techniques were evolved by the contemporary behaviour therapists. **Paul** and **Shannon** (1966) applied group sensitization techniques among college students who were manifesting severe anxiety and disablement in public speaking situations.

Implosive Therapy

This is one of the methods of behaviour therapies used very widely in recent times. This is also quite similar to *desensitization* therapeutic technique based its central theme on *anxiety arousal.* In this method the therapist attempts to elicit the *implosion of anxiety.* The client is put in a safe and comfortable situation and the stimuli that formerly caused anxiety are exposed repeatedly for several times and the stimuli that were causing anxieties previously *loose their powers* in causing anxiety and neurotic behaviour is estinguished. **Stampfi** and **Levis** (1967) were the behaviour therapists who have developed this approach. Generally, *implode* means cause collaspe inwardly in a violent manner. The term explosive is derived from implode. Thus according to the behaviour therapists the anxiety state takes place in the individual in such a violent manner that the person get collasped. Based upon this anxiety state a therapeutic technique is developed which is called *implosive therapy.* It is a type of behaviour therapy in which desensitization eliciting is achieved by the massive flood or implosion of the anxiety.

During the course of the procedure of the implosive therapy the therapist takes every effort to encourage the client to loose himself in the part that he was playing and lived in the incidence of the scenes with real emotion. The scenes are with assumed cues which are described at first by the therapist. The more vigorously the therapist is involved in describing the scenes the more realistic the therapist's presentation becomes, the easier for the patient to participate in it.

At each stage of the therapeutic process, attempt is made by the therapist to obtain a maximum level of anxiety the patient clear out. When a high level of anxiety is achieved, the patient is held at this level and the therapist expects from the patients some sign of spontaneous reduction of the anxiety inducing value in the cues. The procedure is repeated again at the first sign of the spontaneous reduction of fear. Varied new procedure is introduced for eliciting an intense anxiety response. This procedure is carried out until a more significant dimension in anxiety level is brought out. In between the sessions the patient is instructed to re-enact his imagination of the scenes which were described in the sessions of the treatment.

What is more important in behaviour therapies is, making the patient to accept and carry out all those suggestions made by the therapist. It is enhancing suggestibility that the therapist does. For this the more valuable procedure is the *hypnosis* and the drug administrations. We could observe that all the psychotherapeutic and behaviouristic techniques mainly depend upon hypnosis, what Sigmund Freud first used in his treatment to neurotic patients. In this implosive therapy as in the case of systematic desensitization hypnostic procedure is adopted and applied on patients who did not imagine the anxiety scenes realistically. For example, a patient who had developed a severe anxiety, when he had an experience of a road transport accident not imagining the anxiety causing scene. The therapist had demonstrated to the patient by exposing the client to the anxiety causing stimulus that the consequences he had experienced earlier did not occur.

This implosive therapy has become so popular that there are large number of reports with regard to the effectiveness of the therapy and many therapists found that this therapeutic technique is surperior to other behaviour therapies in dealing with neurotic and phobic cases.

Aversion Therapy

Aversion therapy is one of the behaviouristic techniques which involves the modification of undesirable behaviour in an individual. In the treatment punishment is involved. **Coleman** (1974) says that it is old fashioned method of punishment. In this therapy attempts are made to attach negative feelings to the stimuli. **Sandler** (1986) considers that aversion therapy is a form of punishment and it should be placed under the general rubic operant conditioning. *What is aversion therapy?* Aversion therapy is one of the behaviouristic procedures that pairs *noxious stimulus* such as an electric shock with situations that are undesirably attractive, the intent being to make the later less appealing. Punishment in aversion therapy involves either in removing the positive reinforces or the use of aversive stimuli. But the basic aim of the aversive therapy is to reduce the temptation value of the stimuli that cause undersirable behaviour. The most commonly applied aversion stimuli are the *electric shocks*. The other generally applied stimuli are the drugs. But with all these, we must bear in mind that punishment is employed as an important method of treatment. This method is somewhat similar to desensitization, but has the opposite goal. Here the new response of anxiety or an aversion reaction is to be substituted which should cause the behaviour modification.

Aversion therapy as a formal method of treatment was used by **Kautuwich** (1930). He used the electric shocks to alcoholics associating the sight, smell and taste of the alcohol with it. After this time the aversion therapy has been used in the treatment of several types of maladaptive behaviours. Apart from drinking, smoking, overeating, drug addiction, gambling and also various types of sexual deviations. Aversion therapy has been used among the extreme drinkers and causing nausea in the clients. The nausea producing technique has become very popular among the therapists when **Voegtlin** (1940) has introduced it for the first time. In one of the approaches alcohol was administrated in combination with an emetic, a drug that causes vomiting, when the clients were presented with undersirable stimulus. It may be also stated as chemical aversion.

In recent times in the application of aversion therapy the use of chemical aversion with alcoholism and nicotine dependence provide support to the proportion that emetic combined with the taste of alcohol or with smoking a cigarette can produce a standard conditioned aversion. This results subsequently the absence of taking alcohol and smoking cigarette or any other tobacco products. There were several cases treated with these formulas and studies also were made on these treatments. But, however, there were severe criticisms made by **Wilson** (1987) and others and debates also were going on this. Aversion therapy has become a controversial matter from ethical point of view.

There are also very louder voices against this as it is considered as a technique inflicting painful feelings and discomfort on the clients. It is observed that the severest objection and anger were exhibited by the joyful liberation organizations in the United States. Those therapists who use the aversion therapy do not use it alone. They also institute positive technique to teach the clients new behaviour to replace the undesired one and to eliminate it. The drug related aversion therapy is believed to make the client continued avoidance of the alcohol for a period of years. This is found to be occurring in more than fifty per cent of the cases. Some of the investigators found that in the aversion therapy the *intensity dillemma* becomes a problem. According to them the aversion effect heavily depends upon the intensity of the *noxious stimuli* administered in the aversion therapy. But **Lazarus** (1968) is of the view that it is not the intensity of the aversive stimulus, but the *incompatibility* of the response produced by the aversive stimulus and the maladaptive behaviour.

The use of punishment in the aversion therapy is accepted willingly by the client or strongly hated depends upon how the client perceives the idea of the punishing agent. Is it for the convenience of the therapist or for seriously helping the maladaptive client? It is important that the client should be fully satisfied with the aversion therapy with punishment. Aversion therapy is basically intended to eliminate the maladaptive responses for a period of time and then during this period there is a chance for changing the lifestyle of the client and turn to normal life.

Operant Conditioning

In the history of psychotherapy we may come across quite large number of formulation of new therapeutic techniques with some major and minor modification in the original therapies. The evoluation of behaviour therapy in 1950s brought out several changes and modifications in the therapeutic techniques practiced by many therapists. **B.F. Skinner** (1953) the founder of operant conditioning of the learning theory stressed that the therapists should attempt to shape overt behaviour through rewards and punishments. The therapists should have the faith that they could exercise some control over the complex, puzzling, very much disgusted behaviour of hospitalized patients through operant conditioning method. Experimentally oriented psychologists may try to bring the severely distrubed persons into practical order. The use of praise and reward are quite familiar. But they can also use other means in the treatment.

The main needs of the patients like the tasty food can be used as a positive reinforces. The verbal and physical punishments may be used as negative reinforces. The operant conditioning therapists have evolved other reinforces also. **Premach** (1959) has evolved a reinforcer technique called the *Premach Principle.* According to this principle, in a given situation for a problem behaviour, a more probable behaviour can serve as a reinforcer. A more probable behaviour that is decided by the therapist can be introduced in the situation which may produce some effect in the problem person. For example, if a person is very much interested in seeing Television rather than the gardening work, seeing Television can be made as a part of the gardening work. It is commonly observed that many persons applied this principle to their own behaviour many times. The main task is associated with another task in which the person is interested. This is how the Premach Principle is applied in operant conditioning.

Operant conditioning is the acquisition or elimination of an undesirable response as a function of the environmental contingencies of reward and punishment. Several investigators have worked on this problem and broughtout valuable results. **Agrin** *et al.* (1973) state that if a destructive child tears the bed sheets of his bed instead of making the bed the therapist has to make him not only his own bed, but also others' bed. They have observed that operant conditioning technique works more successfully with the clients whose intellectual potentialities are limited and in situations where there are considerable amount of control can be exercised by the therapists in treating them.

Cognitive Therapy

Cognitive therapy finds its origin in *cognitive psychology* which is a recent development in United States. Cognition is a *process of knowing, thinking, judging, reasoning* and *planning activities* of the human mind. Behaviour is often defined and explained in terms of these processes. All these mental processes when *experienced overtly* in some manner, and we call them as activities or behaviours. Cognitive therapy is a therapeutic technique which aims at the *modification* or *restructuring* behaviour. The term, cognition or cognitive does not directly deal with overt behaviour. Basically it deals with mental functioning mainly, the *mental activity of thinking.* Generally any overt action or behaviour is the result of some thinking process. What the person thinks, is revealed in his actions. Therefore, there cannot be any type of behaviour without thinking.

Cognitive therapy finds its basis only on this hypothesis. Therefore, *cognitive therapy* may be defined as a behaviour therapeutic procedure that attempts to alter the manner in which clients think about life, so that they can *change their overt behaviour and emotions.* Among the behaviour therapies the procedures now *collectively called as cognitive behaviour therapy.* These procedures are evolved to help the patients to understand how the thoughts and beliefs that they stem about themselves and others contribute to their problems. Therefore, the goal of the cognitive behaviour therapy is to help the clients to develop more realistic judgements as a means of coping with their problems.

The cognitive behaviour therapy takes its own way of perceving the problems of the clients. Every exponent in this field tried to evolve something new and wanted to apply it on the problem persons. One of the important therapeutic technique was evolved by **Albert Ellis** (1962). Ellis believed that people are governed by *unreasonable hidden expectations* such as, "I must make every one like me". Ellis helps patients to find out their own problems.

Rational-Emotive Therapy (RET)

The rational emotive therapy was evolved by **Robert Ellis,** who is one of the leading exponents in the field of cognitive behaviour therapy. According to Ellis cognition and emotion are closely related. He is of the view that if one's thoughts are to be controlled one must control his emotions. Generally the neurotics think irrationally and engage in self-destructive behaviour and they will not behave more visibly.

The rational-emotive therapy as stated by **Ellis** (1977) is more a cognitive therapy as has been called by him as ABCs of the method. According to him the letter 'A' stands for *activating experience* or *event.* For example, a person is removed from his job. The letter 'C' stands for the emotional or behavioural consequences that follow. He now can go home in a *depressed state* and sit in his home and his self esteem is reduced to a zero point. Is it that the event 'A' caused the consequences 'C? According to Ellis it is not. It is because that the effect of 'B' interpolated between the event 'A' and the consequences 'C'. It is 'B' that caused 'C'. Since the person loved the job he is depressed when lost. If the person would have disliked the job the consequences would be non-committal or relieved of the consequences. Therefore, beliefs about something are the essential elements that cause the irrational views that normally people develop about themselves and the situations they meet.

Beck's Cognitive Therapy

Ellis applied his cognitive therapy on depressed patients. On the basis of Ellis work several forms of cognitive behaviour therapy have been brought out. One of the outcomes is the method developed by **Aaron Beck** (1976) who is a psychiatrist noted for his contributions to the cognitive behaviour therapy with depressive patients. **Aaron Beck** was trained in psychoanalysis. But the experiences he had gained in clinical practices made him to think that the main task in treatment is changing the ways in which people think about themselves. The patients who were depressive had broad negative views about themselves. These negative views reflected in their painful emotional states. Self evaluations indicated by self-blame and self-dislike. The future life was invariably perceived by bleak and punishing. *Beck* holds the view that there are numerous disorders like depression and caused by the negative patterns of thinking in which individuals think about themselves and about the world and their future. These negative patterns of thoughts (Schemata) are maintained by more errors in logic. The goal of Beck's therapy is to provide the patient with experiences within the consulting room. This will disown the wrong conclusions drawn through errors in logic and finally, alter the negative patterns of thinking in a favourable way.

Clinical Involvement

The main clinical involvement is to change the negative thinking. It is done both at the behavioural and cognitive levels. One behavioural technique that was useful for the patients has revealed that the patients were depressive all the times and they become more deeply repressed because of the beliefs. Their moods were recorded at regular intervals during the day time. It was observed that these records had some variability. This information ascertain their general belief that life is *always miserable.* The change in thinking ultimately bringout change in behaviour. This may include getting up of the bed in the morning, praying with some hymes and even going to work place. Generally, depressed persons often do very little because they think that the task is very difficult and they believe that they can do very little or nothing.

Therapists wanted to test this hypothesis that the task is unsurmountable. With the aim of verifying the hypothesis, therapists break down the particular task into small components and encouraged the clients to focus on only one step at a time. If this tash is skilfully and successfully handled than here is a good relationship between the therapeutic technique and the clients' performance. It is a good technique to create an idea in the minds of the clients that they can accomplish something in the task. Once it was thought that the task was beyond the clients' efforts. Then the clients began to change the thought. The client began to think that the task of greater difficulty appears less negligible and success can be achieved. Thus a change was brought out in the thought processes of the clients about themselves and the world.

Another technique evolved by **Beck** is *collaborative empricism.* According to this therapeutic technique, the therapist and the client work together as co-investigators to uncover any maladaptive understanding about the world that was aggravating the client's depression and the life conditions in general. The therapists try to uncover the automatic thoughts and also the dysfunctional assumptions. Automatic thoughts refer to the things that people tell or present the picture to themselves as they go about their daily business, the running conversations or dialogue they have with themselves in situations. Clients generally need practice in taking note of such thoughts and images, especially that are associated with their depressed mood. For example, a school boy tells his father that he has failed in his examinations. This at once makes the father to think, "What an unfortunate parent I am?" After this he feels bad and depressed. The therapist helps the client by dealing with such thought and jointly they examine the validity of the incidence. The therapist asks the father of the boy, "why should you feel depressed for your son's problem at school? What else affects your son's failure in the examinations? Let us find out whether your son did well in all school works." In this way the therapist checks the man's thoughts against the information received by him. The reason is other than the man being the parent of the boy.

This process of identifying and modifying the automatic thoughts is accompanied by another phase of thinking that underlie the dysfunctions. Dysfunction is an impairment or disturbance in the functioning of an organ, organ system, behaviour or cognition. The father of the boy on coming to know his son's position may begin to realize that he is mainly responsible for the welfare and happiness of the family which includes better performance of his son. The therapist may counsel the man the implications of the most unpleasant event and the evidence indicates that he is a bad father. The man is more concerned about his son's case and he desires to do something about his son's problems, are understandable. The reaction the man has developed energises him for a new action and not making him to go into despair. What the man had was the *dysfunctional assumption.* Now the therapist has to help him. How he can help the man?

The therapist may adapt some verbal persuasion. The therapist puts pressure on the person to behave in an inconsistant manner with the problems. This means the man should not give any importance to the problems of his son. Once this view the man takes the problems may slowly get resolved that he has nothing to do with his son's performance in the school works.

Research Study

There were several research works on Beck's cognitive therapy. In one of the intensive studies **Rush, Beck** and their co-workers (1977) have reported very effective results. In their studies there were two groups of severally depressed men and women with an average of age of thirty six. They were treated as out-patients. One group was treated with the psycho-active drugs, *Tofanil* daily at bedtime dose. This drug is widely used for depression and was found to be effective. The other group was treated with *twenty sessions* of Beck's therapy. The out-patients had their depression for about twelve months. There were two therapists attended on them unsuccessfully. *Seventy-five per cent* of the patients have *thought of committing sucide.* After twelve weeks of Beck's cognitive therapy treatment, *seventy-nine per cent* of the cognitive therapy patient group were very much improved and some completely recovered. But the patients who were administered with *Tofanil or Imipremine,* the psychoactive drug, have shown improvement of only *twenty per cent.* There were also few dropout cases from the cognitive therapy group. The patients who were treated with cognitive therapy were found to be less depressed after six months. This was observed in the follow up study.

In many of the cases it was demonstrated that Beck's cognitive therapy was found to be much better then the drug therapies. The cognitive therapy works so well, because it helps the patients to change their cognitions and it takes the patients away from the intense depressive state to the world of reality. **Hallon and Beck** (1986) in their studies have observed that predictable changes in cognitions are occurring in cognitive therapy. But **Rush** *et al.* (1982) state that there are successful results in treating the depression by administering the drugs. According to these investigators cognitive change may be the result of the change caused by other means also. In the case of the disorder, depression, cognitive therapy has brought out the change in cognition which may be the therapeutic improvement but these investigators state that this cognitive change can be brought out by any therapy. The change may be through *Beck's cognitive therapy,* or psychoanalysis or drug therapy.

Cognitive Therapy and Social Problems

Therapists, some of them have observed that people often face some problems in the society which are driving them to some distressed states. Some of the psychological stresses seem to be the reaction to some problems. People often feel that they have no solution or remedy. for them. The psychological stresses that persons face may be anything. It may be a school student's problem with regards to his inability to complete his school assignments, or a problem faced by a girl student who has scored poor marks in her examination and feels very much distressed to go home with the mark list and show it to her mother who is very particular about her daughter's school achievements. Or it may be a problem with an office assistant who had committed a grave mistake in his work which resulted in great loss of money to the company and very much afraid of going to the office and see the manager. Such problems may occur in families, in institutions and in work place or in some social situation.

The person concerned blames himself for what has taken place and get very much depressed. Therapists on observing such occurrences, have developed a technique of dealing with such occurrence of the problems called as *Social Problems Solving* (SPS). **D. Zurille** and **Goldfried** (1971), **Kander** and **Busenncyer** (1982) and others have shown much interest in this and devised the technique. One of the important aspects of their efforts is training the clients in Social Problems Solving (SPS) which consisted of a number of steps. The first step is to regard their distress as a reaction to unsolved problems or regarded as the challenges. It is also treated as opportunities rather than as threats.

The clients are then taught to identify what the problems might have been to the extent of becoming a *brainstorm.* It may be allowed to generate as many alternative solutions as possible without assessing their possible effectiveness. With the view of assessing the possible consequences of each solution and to implement a decision and to evaluate the effectiveness for achieving their specific goals, the clients are made to go back to the earlier stages. Because their initial solution may not work out successfully. Clients with even serious problems often have another chance to solve their problems. Some therapists in clinical researches find the *social problem solving training* very much useful. They have observed that *depressed older adult persons* in nursing homes throw away more of their depressions after the social problem solving training. This was found to be much better than the patients who were treated behaviourly based techniques.

Some of the therapists like **Nezu** (1986) have made good achievements by this technique. He found greater reduction in depression among the clients who have received the full social problem solving training to treatment compared to a control condition, where the group discussions about the problems did not have the systematic procedures for solving the problems and assessing the effectiveness of the solutions. In one of the studies made by **Weissberg** *et al.* (1981), school aged children acquired problem solving skills which generalized to different situations from those dealt with the SPS training. The best of all achievement is that the clients learn a general attitude and set of skills which they can apply to a wide range of future situations they may face. By this means they can improve their general wellbeing and a healthy way of life.

Metacognition

It is another aspect of the social problem solving technique evolved by **Mcichenbaum** and **Asernow** (1970). This refers to what people know about knowing. Metacognition is the knowledge people have about *the way they know their world* such as recognizing the usefulness of a map if one has to find out his way in a new place. In applying metacognition to a problem, how do we react to a problem, which is very difficult to solve it. Since it is like a challenge to our efforts, our consciousness tells us, 'go slow, think of as many possible solutions as possible without deciding any particular solution, the next step is to carefully test each one of the solutions till we find any of them quite suitable.'

Some of the investigators like **Goldfriend** (1980) criticised the social problems solving training programme for its overall approach to life. The SPS training suggests the idea that people should strive hard to take effective action against the problems for gaining or regaining control over their problem situation. It suggests the idea that life is full of challenges and we must strive hard to overcome them.

Behavioural Medicine

Methods of treatment to various types of psychopathologies are emerging very rapidly in the therapeutic areas. The newly emerging methods and the techniques of treatment find their basis in psychotherapy and cognitive behavioural therapies. The reason, the therapists attribute, for their new techniques is that the existing therapies do not meet the needs and demands of the disorders. Since individual person's physiological make ups are varying considerbly from person to person, what is applicable to one person is not applicable to another person. A universally suitable and acceptable to all the disorders to all the persons is impossible in these days. Therefore, varying methods and techniques on the basis of the nature of the distress are designed and developed and applied. In this way we all have the *behavioural medicine* which is a very recent development in the area of treatment of disorders.

What is behavioural Medicine? Behavioural medicine is a new area of behavioural therapy procedures that are applied to psycho-physiological disorders. Behavioural medicine is viewed as a speciality that attempts to alter or change bad or irregular living habits of the individuals in the society, disturbed psychological states and abnormal physiological processes by psychological methods for preventing and to treat medical illnesses. Behavioural medicine is based upon *two roots,* one was very earlier and the other is very recent. The early root is called *psychosomatic medicine* which was adapted mainly by psychologists and the physicians. The other root is very recently developed in behaviour therapy, primarily by psychologists. This is what **Brownell** (1982) observes.

The psychosomatic medicine contributed by psychiatrists and physicians, inspired the mental health workers to the role of psychological factors in the treatment of mental disorders. But the contributions of the psychologists are primarily in the techniques of change, the instruments by which the psychological factors can be analysed and manipulated to help us to understand the illness and to prevent it. Thus behavioural medicine is a combination of two separate fields, one field appears to be functioning very well. Workers in this field apply a variety of procedures. It is not limited to a particular technique or principle of changing behaviour. All of them are directed to achieve the goals of altering the bad living habits, distressed psychological states and abnormal physiological processes to achieve a healthy impact in the physical condition of a person.

Behavioural medicine mainly deals with psychophysiological disorders, the problems of children seriously affected by sickness, certain addictive disorders that damage the health of the people due to alcohol and smoking. The way in which the behavioural medicine is developing indicates that it is becoming a *multi-disciplinary field* of treatment and research.

Problems and Issues in Behaviour Therapy

Behaviour therapy is found to be functioning very well and also expanding its scope every year and thus becoming a therapeutic area in which many cases not dealt with successfully by other methods of treatment were attended and rescued the patients. Every effort is being taken by the behaviour therapists to deal with various categories of ailments both medical and psychological. Behaviour therapists are generally sought the source for their therapy in behaviourism founded by **Watson** and **Skinner.** But some are of the view that radical views formulated by the behaviourists seem to be highly objective and depend upon the overt responses of the organism. But the radical position of behaviourism does not help the therapists to make inferences of the internal processes of the individual persons.

Any therapeutic technique, whether it is psychotherapy or behavioural and cognitive therapy has to depend upon the organism's internal processes like feelings, emotions, thoughts, ideas and imaginations. Behaviour therapy sometimes take into cognizens the symptoms observed in the client's behaviour. But psychoanalysis and related therapies are mainly concerned with the roots or the underling causes. Many therapists believed that the cause that determines the behaviour lies in the unconscious or in the past live experiences and not in the current environment. This view is against the conception that science always searches for the significant causes of the behaviour in the individual.

The cognitive therapies had their important source for the development, the underlying causes of the problems. Therapists like **Ellis and Beck** and their followers also have shown more concern with the form and nature of beliefs, ideas, assumptions and scheme that can be accounted for the client's disordered behaviours and emotions, but these were not openly expressed always. The clinicians generally inferred all these from the clients' actions and thoughts that could be detected. Some of therapists speak of "protective belt" behind which one has to search for the core beliefs which are generally related to one's idea of one self, which may be the *negative self-image.* **Mahoney** (1982) states that the *core, central cognitions* are extremely difficult to change even if it is uncovered. It is because they stem from the patient's earlier developmental history. **Kelly** (1955) distinguished the core central cognitions into two types as the *core constructs* and the *peripheral constructs.* The core construct is related to the client's basic sense of the self or identity. The other one is related to the client's physical aspects. These are important core cognitive processes required for the therapist for assessing and for the attempt to bring out change.

It is experienced by the therapists that bringing change in the core beliefs is a very difficult task for the patient and for the therapist. This is very essential for the cognitive therapy provided if its effects are positive and of enduring in nature.

Broad Spectrum Behaviour Therapy

Behaviour therapists apply several techniques in dealing with the patients. They apply all of them at a time or one by one in their attempts to deal with all the important controlling variables. This type of application of the techniques generally are stated as *broad spectrum behaviour therapy.* This therapeutic technique was suggested by **Lezarus** (1971). This therapy primarily deals with expression of emotional aspects of the patients who have developed the sense of fear in the early years of life.

Psychoanalytic and Behaviour Therapies

This technique of broad spectrum behaviour therapy as suggested by **Lazarus** is not easy to apply on any patient who is afraid of going out for unknown reason. There were not many studies applying this method as it is not only difficult for the therapist, but also the effects and the results are not sure to bring any change in the client, unless the patient cooperated with the therapist.

From the time the behaviour therapy has emerged, there has been severe incompatibility with the psychoanalysis and therapists from each area have been challenging each others achievements in dealing with the patients of various types. In both the areas there were new techniques and methods evolved to deal with the patients. Though **Rogers** client-centred therapy in United States was very rapidly growing and many followers tried to adapt it, the behaviour therapy in some areas was gaining more importance over psychotherapy. The controversy between these two areas was going on for long time till some therapists tried to bring a approachement. **Paue Wechtel** (1977) has offered a scheme for establishing some agreement between psychoanalytically oriented therapists and the behaviour therapists. The ego therapists claimed that their approach is functioning very well compared to Freudian technique. **Sullivan** claimed that the patients feel better about themselves if they focus on the current problems in the interpersonal relations, **Wachtel** was of the view that therapists should deal with the patients in such a way that they should change their behaviour.

The primary aims of the therapies are to make changes in the current behavioural patterns by bringing out the underlying problems by pointing out that a direct alteration of behaviour may help the patients to attain a more better understanding of their past repressed conflicts. This is one of the primary goals of the psychoanalysis.

The behaviour therapists state that the problems of young men appear in the beginning fearful of hetrosexual relationship. Taking this as a cause the therapist works to help him to reduce the fears. They combine the techniques such as desensitization and the rational emotive therapy and social skills training. But according to psychoanalytic theory the young men are basically angry at the women. This is not taken into consideration by the behaviour therapists. The infantile sexually is basic for the psychoanalysts which is ignored by the behaviour therapists. By this means they miss any important phenomenon. But **Wechtel** states that the problem of young man is *not just fearful* of women, but *angry with* them as he associates with women with a mother who has been the object of *hate and love* right from the early childhood period. The behaviour therapists add the sense of fear as an additional information to the classical psychoanalysis. This additional is a behavioural intervention. **Wachtel** wished that the behaviour therapists should be at tune with the client's unconscious motivation. Because a person may be motivated by some events which he/she not conscious of the reason stored in unconscious.

The psychoanalytic thoughts of the present generation also help the behaviour therapists in understanding the meaning of this intervention. Psychoanalysts consider more about the unusual concerns, wishes, and fears of their patients then the behaviour therapists. Having this view in mind that emotional problems are derived from the repression of their conflicts, they consider that the psychological difficulties reflect the infantile wishes and fears, sad miseries of primary though if process. Some of the psychotherapists are of the view that behaviour therapists commit the mistake of accepting the patient's complaints at the face value and do not go deep into the patient's problems. They infer the unconsious processes by the outward appearances and symptoms they observe in the patients. Thus the rapproachment is going on between the psychoanalysitc and behaviour therapies.

28

HEALTH PSYCHOLOGY

Origin and Development of Health Psychology

Psychotherapeutic methods and techniques are widely prevalent in dealing with neurotic and psychotic patients. It has been observed that there were more persons come with the complaints of emotional disturbances than with neurotic and psychotic problems. But in treating the emotionally disturbed people there were more contraversies over the effectiveness of the therapies. Some of the therapists were thinking of evolving some other methods. At this time a new and a wider approach to intervention programme has begun to function in varied names. This approach has been called sometimes as medical psychology, health care psychology and behavioural medicine. **Millon** (1982) called this new approach as *health psychology.* This is supposed to be a deviation from clinical psychology. Clinical psychology was designed to deal with behaviour disorders. An interface between physical health and behaviour disorder, a new field of psychology has emerged with the name, *Health Psychology.*

Health psychology, is a mass of distinct things gathered into a total whole. According to **Matarazzo** (1980, 1983) it is an aggregate of the specific educational, scientific and professional contribution of the discipline of psychology. It is intended for the promotion and maintenance of health prevention and treatment of illness and the identification of etiologic, diagnostic correlates of health, illness and related dysfunction. **Materazzo** is one of the pioneers of health psychology. He states that a new discipline has directed its attention to a wider range of physical disorders. Health psychologists now are contributing their might extensively to the entire scientific fields, and to the profession of psychology.

Scope of Health Psychology

Health psychology at present attracts the interest of physiological, cxperimental, developmental and social psychologists. Some of the psychologists at present are working in the clinical field with laboratory experimental studies. Common problems now have brought together. Two decades ago (from 1960-1980) those who were going away, have now sharing the interests in a common field of study. The areas of research and treatment which are included in the list of health psychology have covered many subjects that come together under the tree of health psychology.

It is very much surprising for the present day psychologists to see the areas which were included under health psychology. Some of the areas are the stress control and therapy for people with Type-A patterns of behaviour and therapy for various learning disabilities. Application of the methods of classical conditioning to enuresis, behavioural methods of treating obesity, alcoholism, drug abuse and smoking programmes are designed to apply psychology's knowledge to one of the most important present day medical problems of getting people to fill the prescriptions, take their modifications and the diagnosis of sensory impairment as an aid in therapy. It also includes the memory training for old people and those with brain damage that causes loss of memory and Genetic counselling which helps people to understand the contributions of inheritance to diseases.

Thus the scope of health psychology has become so wide that it involves every aspect of human physical well-being and illness and psychological well-being and disorders. It is functioning like a connecting bridge between physical and mental health problems.

What is Health Psychology : Health psychology is a branch of psychology that is dealing with the psychological factors in health and illnesses. It tries to relate mental health with physical health and *vice-versa* and and evolves suitable techniques for treating.

Thus psychological studies concentrate mainly on all the aspects of hurnan health and behavioural aspects. Its concentration rests not only on healthy individual's behaviour, but also it takes care of human beings both mental and physical health aspects. If only man is physically healthy, his behaviour would be normal. An impairment in his physical health may have some influence in the behaviour of an individual and if it becomes serious it may affect even his mental aspects which may lead to behaviour disorder and may develop into psychopathology. Therefore, psychologists give equal importance to various aspects of health. This approach resulted in evolving a separate field of psychology called *Health Psychology*.

Modern physical and health scientists accept that health and illness involve more than simple physical factors. It is true that same personality factors predetermine that the person is having certain illnesses and because of these he seeks the help of a physican whose efforts recover him from the illness. **Gondole** (1985) states that psychological factors not only affect a person's physical well-being but also increase the physical and mental fitness which results in an improvement in psychological characteristics like creativity and inventions. Thus health psychology is the field that studies the psychological processes, physical and mental health aspects. By providing physically and mentally *healthy atmosphere* health psychology tries to avoid the treatment of physical illness.

There are quite large number of researches conducted in recent times which have provided a great deal of information in health care. A research investigation was conducted by **Amerbeck** (1953) on psychological stress. In his study it was observed that there were negative physiological changes in the subjects due to severe stress. In one of the studies

Krantz and Glass (1984) verified the hypothesis that personality traits predisposes individuals to develop specific illnesses. They observed positive results in their study. In another health psychological study **Levy** *et al.* (1985) investigated patients who are well adjusted to their illness have relatively *ineffective cancer fighting* physiological responses. Reactions such as anger are associated with rnorc effective control of cancer spreading. These experimental results show that psychological processes are involved in physical illnesses.

Prevention of Illness by Behaviour : We are quite familiar with the proverb that "prevention is better than cure". Therefore, it is better to get rid of illness without any systematic treatment. Several approaches may be suggested to achieve this goal. But one best way is to motivate people to take interest of their own. One important reason we have stated above is the personality variables.

Human Personality Traits and Preventive Behaviour : Individual persons vary very much in their personality traits. Any medical treatment that is agreeable to one person may not be agreeable to another person. While analysing the health problem of the individuals, **Kirscht** (1983) has observed that there was a development of health beliefs among the individuals. Because of this health belief model individuals develop *beliefs* about health and the *threat of illness.* Thus an individual's health related behaviour is predicted by this belief. One such model was suggested by **Weinstein** (1984). According to this, college students had unrealistically optimistic beliefs about their health risks.

The college students felt that they are not likely to have heart attacks as they do not act and drink anything that cause heat attacks. But some students perceived as having high risks for such outcomes as they assumed that such factors are based on family history and they have given effect for the given illness. They ignore their own health practices or exercises that could affect the illness. They saw the relationship between their risk for heart attack and what amount of exercise they practiced and the quantity of consumption of meat, smoking and the high cholesterol foods like eggs and other foods. These results suggest that though the students have some knowledge they have not correctly understand everything about the risk of health.

In one of the studies made by **Tunk** *et al.* (1984), college students, high school teachers and qualified nurses were asked to give rank-order to several activities as to how much health protective they are. The result was very much surprising as all three groups had the same type of knowledge. They all have the basic knowledge about the prevention of health and preventive health practices would bring out certainly valuable result in an individual. But one must take proper step for achieving this goal.

Preventing Smoking : Smoking cigarette and other related activities cause lung cancer. This has been well established medically and most people are aware of the dangers of smoking. But there are people, argue in favour of smoking, knowing pretty well the potential harms, stress and tensions. But investigators have observed that even when persons strongly decide to give it up the use of addictive substances find it very difficult if it has become a behavioural habit. Those who are smoking in the adult years were influenced strongly by their peer groups. Several research studies were made on smoking. These research results suggest that the most effective aspects of prevention programme is trying to learn how to resist the pressure for smoking and getting the information about its effects. In this, the effect of the health risk should be emphasized (Fig. 28.1 a, b and c)

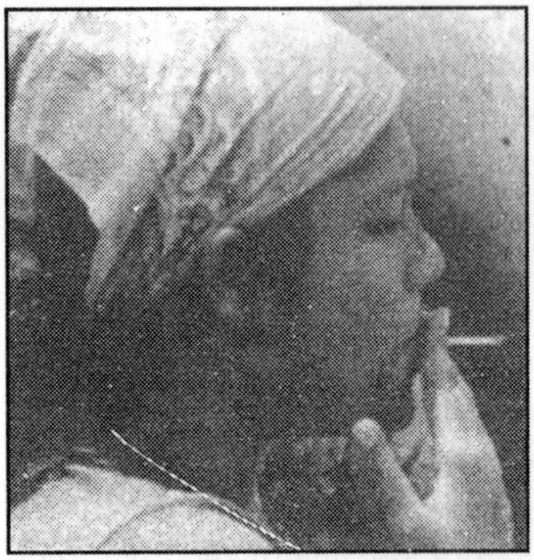

Fig. 28.1(a) : Prevention of Smoking and Physical Illnesses.

Fig. 28.1(b) : A section of psychologists in United States were very much astonished at the death of many people while taking up the medical treatment around 1990s and they were very much surprised at this and investigated the causes. They observed that many were not suffered from some type of physical illnesses. But to their surprise there were psychological causes, had influence in their physical illnesses. Smoking in various forms and the use of tobacco were found to be having influence in many of the cases of the deaths.

Fig. 28.1(c) : Students, young men and women made smoking as a regular habit and fashion. The students have developed several inffectious diseases. Men and women developed stomach pain, headache, appendix and lung cancer. According to recent discovery smoking also causes early manopause resulting in divorce and separating young couples. Health psychology, a field of psychology, recently developed area evolving suitable techniques to deal with such problems.

The Illness Episode and Behaviour : All human beings are subjected to the attack of some illness from time to time and all of them seek some method of getting it relieved. The main cause of the illness is not recognizing the psychological factors that are having much influence in it. But if it is not recognized and accepted, it causes illness in some way. Once an individual succums to sorne illness, he seeks several choices to get rid of it. First, the individual attempts with self-diagnosis of the physical symptoms which precedes whatever treatment or action is taken later on. It may range from doing nothing to seeking some medical advice and treatment.

While seeking medical advice and treatment from some person, whether that he decides to treat him or not, the individual should cope with the symptoms of the illness. In addition, any formal or informal cure, the individual is required to cope with the treatment process.

Students generally believe that they could develop the infectious diseases like cold than a disorder that causes on their own behaviour. Those who suffer from invulnerable illness try to avoid by keeping themselves in appropriate preventive behaviour. It is also possible that, one can imagine some possible diseases in himself. It appears that preventive health programme is expected to do well to describe all relevant symptoms in a substantial way that could be easily recalled. Until the symptoms indicate some illness he has nothing to do about seeking treatment. But people out of anxiety try to do something when they feel some uneasiness or stress. They try to make self-diagnosis. Some people, for example, believe that some type of uneasiness is the symptom of fever or headache and they try to make self-diagnosis by some common sense method from the past experience.

This type of self-diagnosis from the past experience may be correct at one time. But it will not be correct in all cases. Based upon this type of diagnosis some persons may avoid or neglect medical consultations. Sometimes this may become a serious health problem. For example, an individual is having an experience of *stomach pain* and vomitting. He may think that it is a simple intestinal problem, a virus which will go away in few minutes. But it is actually a more serious problem like *appendix,* if he would follow the same method as he did in the case of headache or else, the internal organ may bust before he goes for a medical treatment. In the same way a person's wrong belief about some serious illness like high blood pressure or hypertension may lead to the avoidance of medical treatment may result in heart attack and death. In the absence of clear symptom of some diseases, people resort to self-diagnosis and avoid systematic treatment. But behind any such illness, the psychological factors are found to be influencing in the development of the illness.

Behaviour and Physical Illness

One of the important themes of health psychology is that behaviour is a contributing factor to physical illness and the physical illness is a response to behaviour. This hypothesis is tested by several investigators in the field of psychology. Hurnan beings generally experience the response of stress whether they are feeling ill at a particular time or not. It may be true that they might have done some work. They might have been engaged in some kinds of physical works. These are the external events. These external events produce some effects on the internal physiological states. This causes some changes in the internal organs. Researchers of the present day direct their attention on this phenomenon which is named as *psychoneuroimmunology*. This is a new area of study which tries to formulate some systematic concepts. What is psychoneuroimmunology? It is generally defined as the study of the immune system and its interaction with the *nervous system and behaviour*. It is also defined from health psychological point of view that it is the study of *the way the human responses* to the *external events affect the internal physiological states* which influence the immunity to illness. In this way there is relationship between psychological responses and the illness.

One of the effects of the relationship between the psychological responses and the illness is the *release of the stress hormones*. **Johnson** *et. al* (1984) in their studies have checked it in an oral examination of the students in the process of examing their doctoral dissertations. These individuals have shown an *increase in the hormone of adrenlin* production during the weeks *before and after* the examination. Adrenlin secretion generally increased during the times of flight for safety and fighting. Here in the case of examination, it has increased. This is a *reaction stress*. The behaviour of facing the oral examination product changes in the internal organs which result in the increase of hormone. Research findings show that immunity to disease is reduced in certain situations when exposed. But in this case there are individual differences in perceiving the situation as the source of stress.

It has been observed in several studies that one of the incidences in which the immune system is very much affected is the stress caused by the loss of loved ones or separation from loved ones. There are several family situations in which people are facing severe reduction of immunity leading to diseases. Getting seriously ill, feelings of helplessness, disappointments in getting a suitable job, failure in the examinations, constant quarrels in the family, children problems, all damage the immunity system of the body and cause severe stress and serious diseases.

If an individual comes to know the incidences or events that cause negative effect in the body like the stress, he can learn to avoid it and the reduction of immunity can be avoided. What is required, is that the individual can learn to respond physically in a less self-defeating manner or even he can go to the extent of taking it as not a defect. The adult persons can take up some *relaxation training*. This has been found to be increasing the immune system. The human feelings and emotions, thoughts and attitudes need not have any effect on their health or negative effects. They can be taught to get rid of the effects of behaviour on the internal organs of the body. This may result in increasing the immunity system and reduce the stress and disease.

Control of Distress

The basic human needs, apart from food and shelter, are to control events and situations that cause distress in one's life. Man's psychological well-being and physical health get benefited when a person has *perceived control* over what is happening to him. Health psychologists use this new term perceived control. What is perceived control? Perceived control is the extent to which an individual believes that he is able to influence the course of event that cause distress. The persons who believe that they have control over the evasive stimuli like loud noises which cause shock or sudden flash of heat radiating bright light feel less stress and greater tolerance for the unpleasant feelings than those who state that they are helpless and very much distressed in the situation. This was investigated and valuable results were obtained by **Getchal** and **Proctor** (1976). Persons who have *low sense of control* over the aversive stimuli *develop chronic depression.* It was also observed that changes in moods may also result in changes in perceived control and the stress can be overcome.

The human life is full of perils and individuals have to face them and try to conquer them. A failure in an attempt to achieve something valuable may lead to miserable consequences. In our everyday life lots of uncontrollable distressing events such loosing one's occupation by some means who has a big family to maintain may be followed by severe consequences of mental aberrations and getting into a psychiatric hospital, running away from home leaving the whole family members in a most distressed state, committing suicide and dying and so on occur due to the failures of perceived control. Sometimes even minor events in our everyday life such as failing to pass a departmental test for getting promotion, failing to complete the assignment in time and getting harsh or remarks of the manager. All these are uncontrolled distressing events and they have negative effects on our physical health. Every person's life situation may be having some psychological processes having adverse effect on the physical health which are beyond the perceived control.

Perceived control is an important factor in promoting physical health. Generally people try to explain the causes of their illness in terms of internal or external factors. Internal factors can be controlled by the individuals by following the technique, perceived control. But the external causes are perceived by the individual as beyond his control. If some one on the road suddenly hit you with his scooter and injured you, you are helpless and it is beyond your control. This is an external cause that has hurt you and put you in distress. But if any distress caused by some internal causes, it is because of your failure to take proper care of yourself and you are responsible for the distress.

A person who perceives internal causes has to take personal responsibility *about preventive behaviour, self-diagnosis and restoration of health.* **Taylor** *et al.* (1984) believed that a person even with a serious illness like the heart attack can have the hope in his ability to control the course of the illness, can act more activily during the treatment.

Accidents on the roads have become common occurrences in these days. It is true that no one wants to get into an accident and injured. It is happening unexpectedly. But persons who got into accidents in work situation or elsewhere, return to their works as soon as got cured from the injuries. They may develop a *positive feeling,* if they perceive that they have caused the accident and therefore they can be careful and prevent any further accidents (**Brewin,** 1984).

An individual may decide the nature and the cause of a symptom or a set of symptoms and as result of this his next behaviour may help him for recovering the illness, but sometimes it may prevent it. It all depends upon the individual's personality traits. **Kelley** *et al.* (1955) are of the view that persons of *chronic self-destructiveness* wish to respond with avoiding potentially helpful activities like medical help and also may be engaged in potentially harmful activity, something injurious to health. Kelley states that *self-destructive* individuals may put the blames on external causes for their difficulties and their *self-control* is poor.

Interaction with Medical Personnel and Treatment : Persons who seek treatment for their illness have to meet different situations where they may have potentially powerful experiences, such as meeting the physicians and having painful feelings and diagnosis in some threatening manner. First of all the *Doctor-patient interactions* take place. The patient meets the doctor for treatment. The patient generally develops anxiety and fear and also certain amount of hesitation to speak out freely, fails to speak about certain essential symptoms he has experienced. The patient also fails to ask certain questions related to his illness and get his doubts cleared. Because of this there is certain amount of *dissatisfaction in his medical consultancy* and returns with some unpleasant experience. But it is possible to get rid of this problem partially.

There were several experimental investigations on this problem. **Roter** (1984) in this study asked the patients to spend ten minutes with a health educator before they go and see the physician. The aim of this procedure is to find out the areas of interest of the patients to formulate a list of questions to be put to the physician. Some of the patients were instructed to prepare a list of questions before meeting a physician to ask more number of questions during their visit than those patients who were not instructed to ask questions. The patients asked the physician both medical and non-medical questions in greater numbers. The patients, for example, asked the physician questions such as, "Are there any side effects with this prescription?" and non-medical questions such as, "How are you coming to the clinic, Doctor?"

This procedure made the patients to increase the number of questions put forth to the physician. In this experiment the investigator has taken into account the behaviour of the physicians. Some of the physicians perceived as more comforting and has a better bed-side manner than others. The physicians had more of' non-verbal behaviour. The physicians were instructed in one experiment to *lean forward* in interacting *with some patients* and not with others. There was a positive evaluation from the patients when the doctors leaned forward and did not fold their arms and simply nodded their heads. **Harrigan** and **Rosenthal** (1983) have observed these results in their studies. It is easy to teach medical students to behave in a way to promote positive and comfortable feelings towards the patients.

Dissatisfaction with Medical Advice : The patients on meeting the physician in the clinic after some formal and informal conversations, formal medical dignosis was made. The treatment quite appropriate to the illness includes the physician's advice to change some of the usual behaviour of the patients. Then the physician makes prescriptions for administering the drugs and if necessary to undergo some surgical treatments. One important aspect is whether the patients agrees to all these and carry out or not, it depends upon the patient's beliefs about them. If any physical exercise is needed he has to carry out the same. Producing the expected result is an important element in the adpative behaviour. There is every chance for the patient to take the medicine prescribed by the physician and written in the case history of the patient or the patient may take it irregularly or incorrectly. Taking the prescribed drugs is a function of information and the patient needs to know more about them, more than the physician or the pharmacists.

A serious patient is more concerned about the side effects of the medicine than the professional is and may decide to discontinue the treatment if any unexpected symptoms are appeared. If the patient is provided with more information, he may be satisfied and does not make any negative complaints.

Taking up the Treatment : After all the formal procedures of diagnosis and therapeutic techniques the patient is to undergo the interventions that are painful and sometimes even dangerous. The *process of coping* with these events is said to be the behaviour experiencing the feelings and thoughts that make him to tolerate and master the threats. This process in the cause of' treatment is referred to as coping with treatment. Coping with treatment means, those things a person *does, feels and thinks* in order to master, tolerate or decrease the negative effects of a threatening event or situation. Several studies were made in this area. **Suls** and **Fletcher** (1985) are of the view that in the early stage of the illness one of the best short term strategy is to avoid thinking or doing anything about the illness. But if the illness progressively increasing, it is better to go for a treatment at the earlist.

But there are some variables functioning in the individual's everyday life which are having influence in the individual's ability to cope with the illness. These factors also interfere in the treatment procedures. **Baum** and **Singer** (1984) have observed these results in their studies. Some investigators are of the view that they are benefited by their friends who encouraged them for restoring their good health **Ruth** and **BI Booley,** (1985). There is also the fear as to "how long time the treatment will take place and what will happen and how much painful feelings with which we will have to suffer". Even the feelings of children are hurt if they know about what is going on their parents **Jay** *et al.,* (1983). Thus we may come to know from these few experimental investigations that psychological factors are very much involved in various aspects of human health from preventive behaviour to the levels of dealing with the illnesses and diseases and to their treatments **Baron** and **Byrne** (1987). Thus, health psychology, though was evolved very recently it provides much valuable information about human physical and mental health aspects from the experimental investigators.

Ailment of chronic pain : People everywhere suffer from some kind of ailments and pains which are common occurrences in every society and in every culture. If proper care is not taken to alleviate it, it becomes a chronic ailment. *Chronic* means some illness or disease has a *lengthy duration* or recurring frequently, often with progressing seriousness. Any physical ailment may have a chance of developing into a chronic state. Pain is one such physiologicial aliment becomes a chronic. Pain may occur in any part of the human body and it is considered to be one of the physiological in nature. The consequences of chronic pain in human are beyond the scope of estimation, but generally it is not estimated in any culture as people do not take it into a serious phenomenon, **Bonica** (1981) defines chronic pain as the pain that persists beyond the healing time of the affected parts of the body. It accounts for millions of rupees of lost work time and disrupt family life, as well as keeping the sufferers in continuous and extreme discomfort. Behavioural medicine believes that it can be treated by psychological and by medical means.

Any severe physical pain in any part of the body puts the person in discomfort and in impairment of activities. There are two disadvantages in this ailment, one is, the person suffers and the activity or work performed by the person is lost and the other is the gain by the activity is also lost. In the ease of attack of severe back pain or hip pain, for example, a person is reluctant to keep himself engaged in any physical activity more vigorously than to getting into bed. In an acute state this is the sensible result. Several investigations have been conducted on this issue. **Fordyce** *et al.* (1986) in a comparative study, have demonstrated the superiority of a *behavioural programme* over a *traditional medical* one for management of acute back pain. In the traditional programme patients have exercised and moved about only until they felt pain. But the *behavioural management programme* encouraged them to exercise at a predetermined intensity for a predetermined period of a time even if they experienced pain. There were low back pain patients who were given relaxation training and they were encouraged to state their pain as tickling. A cognitive restructuring procedure was adapted according to **Rubstein-Blinechik** (1979). The patients were not pressed to go beyond the level of their bodies could exercise the task. The information obtained in this study involved is that the traditional medical practice has under estimated what the chronic pain patients could perform [Fig. 28.2(a), Fig. 28.2(b)].

Fig. 28.2(a) : Prevention of smoking and chest pain.
There are several psychological causes behin the physical illnesses. Recent observations brought out the fact that people do not speak about the real causes of their illnesses like influenza, tuberculosis, pneumonia, diseases of the heart and cerebrovascular diseases.

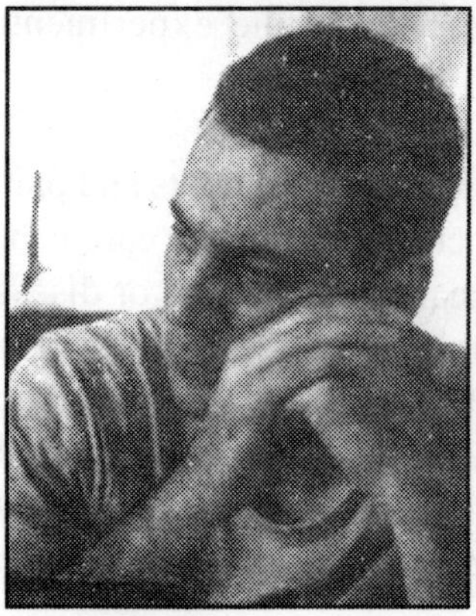

Fig. 28.2(b) : The behaviour of the people, their lifestyle and their life-time are impaired due to lack of nutrition, lack of care in time, proper movements and exercises, isolation and social detachment. All are attributed to some illnesses and early death. Failure in lifestyle has become one of the causes of mental depression which may impair the happiness of the family life. Psychological interventions are designed for such problems and they are found to be having favourable effects on these problems.

Studies were also carried out on other pains and the psyhological intervention had favourable effects. **Wells** *et al.* (1986), have chosen twenty-four patients for their study. The patients were prepared for variety of electric-surgeries and one hour of stress innoculation training. The training session consisted of deep breathing, relaxations, pleasant imagery and the encouragement to use positive coping self- statement. The patients carried out those exercises for a week prior to the surgery. These patients were compared with the control group. These patients were *less anxious* before and after the surgery, had less feeling of post-surgical pain and they were discharged on an average of three and a half days. This, technique saved large savings of'medical costs.

Chronic Illnesses and Lifestyle

Psychologists while dealing with the psychological problems often observe that many of them were found to be suffering from some type of physical illness. After several decades of observation and research studies they came to a conclusion that the psychological causes were found to be having some influence in their illnesses. The evolution of *behavioural medicine* finds its root in this. Behavioural medicine is primarily concerned with the alleviation of physical illnesses and pain but also interested in the prevention of such illnesses. When the purpose of the behavioural medicine is assigned, it is named as *health psychology.*

Health psychology emerged as a separate branch of psychology in the United States in 1980s. The reason for this new field of psychology is the *death incidents* that caused due to some serious diseases such as *influenza and tuberculosis.* With these illnesses largely under medical treatment many people succumed to the deseases of heart, cancer, cerebrovascular diseases like strokes and accidents. The behaviour of people, their lifetime and lifestyle were impaired in each of these diseases. Several causes like *high cholesterol, lack of nutrition,* lack of exercise and lack of proper care are attributed to the severe illnesses and deaths.

Medical scientists for several years have struggling in dealing with some of the chronic diseases and treating patients and advised people to do what is best for their health in the interest of them. There were several studies on this issue and many interesting and also serious results were brought out. The study made by **Epetain** *et al.* (1980) among the university women students needs to be worth mentioned here. They were asked to run one or two miles a day for five weeks and they agreed for this programme and also deposited five dollars. This was returned to them one dollar on completing the programme for a day. The college women students ran more consistently and systematically than the control group. They were told that *"your heart needs exercise, and here is a chance for you to execute".* Such programmes were carried out in several ways with the view of promoting their health.

On knowing the value of such programmes in the interest of *promoting health* and for *avoiding illnesses* many organizations have started in maintaining their own *health and fitness programmes.* When people come to know the *operation of psychological factors* in their health and illness, they readily accepted and carried out the programme to prevent their chronic illness. By this means they can avoid medical treatment. *Health psychology* is in a very infant stage. It requires more research studies, procedures and techniques for its development.

PART TEN

SOCIAL BEHAVIOUR AND RELATED TRAITS

29

SOCIAL BEHAVIOUR
(Social Psychology)

Introduction

Psychology is mainly interested in the study of the individual behaviour. The individual person is born and brought in a home situation and moves out in the outside world as and when he becames a functioning organism independently. The human individual first is having *close contact with the members of his family.* He gets his needs and demands through his family members. His simple personal motives are fulfilled by the help and co-operation of others. When he becomes a competent person he, himself tries to achieve whatever is required. Thus the individual gradually is going out into world and extends his *contacts, interactions* and *mutual give and take* adjustment and behavioural activities.

When the individual goes out and meets persons of various categories he finds much difference between one type of persons and other type of persons. He goes out with his own thoughts, feelings and emotions and motives and projects them on others and also accepts some from others. Thus there is constant interaction between the individual's behaviour and other's behaviours and he is mostly influenced by them. Social psychology is interested in these types of *behavioural interactions* and the individual's influence on others and other's influence on him.

The soical psychologists generally look into the social influence on the individual from a different perspective and not in the way in which the developmental psychologists and the personalitists are ideal with. The social psychologists focus their attention first of all more on the *current situational influences* on human behaviour than on the factor that influence the development and personality of the individual. Secondly, social psychologists differ from others in dealing with the subject matter. They emphasize the individual's own point of view, that is from the *subjective point of view* and they do not make any objective measure of the situations. For example, an individual behaves in certain manners because the behaviours of *others•are hostile* and aggressive to him, though it is not correct. That is the way a person looks at the society if proper social development has not taken place in him.

Similarly others see him as a *social outfit* or an unsocial as he is not adjusting himself with others.

Emergence of Social Psychology

Having this background as the basis for the development of social behaviour we may define social psychology. The social behaviours of human beings are going on changing in its structure and content because of the changes often taking place in human social atmosphere. The rapid development of science and technology and the impact of them on human life, the social conflicts at the international level which cause great wars between nation, the conflicts arising from racial and religious prejudices bring out changes in the social structure and the behaviour of the people all over the world.

A science that takes cognizance of these human behavioural patterns and studies in detail is *social psychology.* Social psychology is mainly interested in the study of social interaction between persons and between groups and nations. Social psychology makes its subject matter the behaviour of human groups. The groups may be of any type and may be found in any situation. Whatever type of groups, situations, there are human beings and they come into contact and interact either for good or for bad. Everyone projects his thoughts and feelings, emotions and motives, and attitudes and prejudices. These are the subject matters of social psychology in a broader sense. Social psychologists go deep into the behavioural interaction of the human beings in the group situation. Unlike other scientists social psychologists can comprehend their subject matter by manipulating the independent variables or condent experiments in the laboratory situations. It is a very completed task for the social psychologists to study the social behaviour of the individual and the human groups.

Group Behaviour : What has aroused the curiosity of the social psychologists is the individual's behaviour in a group or in a social situation. An individual when he is alone at home or at work situation how behaves? How is that, that his behaviour becomes totally different when joins a group of people. Whether he projects his own thoughts and feelings on others and forces them to accept or he silently or agreeably accepts the thoughts and feelings of other persons? What type of role the individual plays in a social group? How does he perceive other individuals or the group as a whole? These are the hypotheses the social psychologists wished to verify or test. Social psychology has emerged from their phenomenal situation.

A work that was more directly dealt with the social behaviour was brought out by a French sociologist, **Gusteve Le Bon** (1895) under the title of "The Crowd". This book mainly dealt with *crowd and their behaviour,* **Le Bon** and **Durkheim** , another sociologist, maintained that human behaviour is dominated by a *group mind.* Le Bon put forth the idea that *limitation is the central principle of group life.* The individual in the group has a *sense of solidarity.* The individual has desire to act for himself, but to *follow the activities of others.* The individual is also in a *highly suggestible state.* The individual accepts the ideas emanating from the leader of the group or fellow-member of the group *uncritically.*

In a group, the responsibilites and *duties* are *divided.* When an individual is alone he should bear the consequences of the deeds. But in a crowd it is possible *to escape* from the responsibility and the blame can be easily passed on to some one else.

According to Le Bon, group behaviour is the operation of a *collective mind.* A group of people acts, feels and thinks as a distinct beings. The group as a whole tends to behave in the same manner as directed by the group leader.

But the book that has put foundation for the subject, *social psychology* in its modern form was under the title of *Introduction to Social Psychology* written by **William McDougall,** in 1908, a British psychologist. Are the earlier books published before 1908 were not dealing with the subject matter systematically? In this book **McDougall** tried to provide psychological basis for the social sciences. He stressed the importance of the *group mind* for understanding the social behaviour. It is the *herd instinct* that makes the group mind that is mainly responsible for the *human social life.* It is the tendency in man to be in the company of his fellow beings. Man is miserable when he is alone. It is not only human beings but also animals and birds are impelled to seek the company of its kind. The functions of herd instinct in human beings and in other creatures can never be explained adequately by using any other factor. What is that compels a living creature including humans to seek the company of the own fellow beings? It is the *innate tendency,* in other words, *instinct* that is *inherently forcing* the living creatures to seek the company and associate with it. This view can never be refuted logically or scientifically by the social scientists. The social behaviour of the individual explained by McDougall through his instinct theory was accepted first by most of the psychologists at that time. His stress was on groups *as organic wholes.*

But the American psychologist **Floyed H. Allport** who brought out a book under the title of *Social psychology* in 1924 refuted McDougall's view. Allport accepted the individual's approach (instinctive force) to social behaviour proposed by McDougall, but did not accept the *group mind idea.* Allport also refuted later on McDougall's *instinct theory* and in its place he proposed *prepotent reflexes theory,* which is based on behaviouristic psychology. Some of the researchers of the time viewed that it is *only replacement of terms,* because human social behaviour, strictly speaking, cannot be explained by reflexes and conditioning. Human social behaviour is something *more than reflex actions.*

The group mind theory formulated by **Le Bon** and **McDougall** seems to be in agreement with *Machiavellian theory.* Scientists are still having the doubt about the brain. Why the human brain is so big? Until 1970s most of them have assumed that the expansion of the human brian was driven by adaptation to changes to varying ecological conditions and changes in diets. However, a different view was expressed by the psychologist, **Nicholas Humpiray** (1976). He argued that *human brain grown bigger* because of changes of *social environment,* the most important being that we have started to live in *larger and larger groups.* This theory is known by the *Machiavellian Intelligence Hypothesis.* It is after the name of **Machiavillian** (1469-1527), an Italian statesman and a political philosopher. He was also secretary to the war council of the Florentine Republic during the period of 1498-1512. His most famous work is *Principle* (The Prince) brought out in 1593, relating to the alleged Political Principle of Machiavellian. This work became very famous notoriously and was the handbook for those *who wished to manipulate and exploit others.*

Machiavellianism : It is a *human behaviour style* which is characterized by the willingness of the individuals who *wish to handle others* and who do not withdraw from the use of power in interpersonal relations. Machiavellian intelligence hypothesis puts a *key role on memory.* As the human group size increased our ancesters needed to keep track of increasing members of people. They needed to remember everything what they look like and what their characters were like. Above all they needed to keep in the memory as record of how everyone else in the group had treated them in the past. What the theory suggested is the *reciprocal altruism.*

Persons who come into contact with more number of people often are driven to the necessity of having the names and of the images of the faces. The more and more *crowds of people one meets* the more names he has to store he his memory. Thus the *enlargement of the social groups* make the person's brains to enlarge as he has to store all those names. According to *Machiavellian philosophy* persons, *particularly the statesmen,* who meet often larger social groups develop larger brain and they become *more intelligent* than others who do not come into contact with larger social groups. This view expressed by Machiavelli seems to be correct to some extent on seeing the extraordinary memory power of the *great statesmen* which may be also called as *social intelligence.* There are some political parties in India, knowingly or unknowingly the Machiavellianism, have emerged and they are functioning without any basic principle of political ideology, merely on seeing the number persons assembling to their meetings. Some political parties in India, first stand on the name of their community or caste and on seeming crowds, rename the party with a political touch and try to capture the power of the state without any ethical principles, exploit people and amass enormous amount wealth and power, and live like a prince and dictator. (Figs. 29.1 and 29.2).

Fig. 29.1 : A Group

The above picture is an example for a group. A group consisted of a number of persons gathered or conditioned together. A group is a collective unit or team. It is bound together by a common social standards, interests, etc. The members of a group work together in an ineracting manner under the guidance of a leader, who determines the policies of the group. A group may function temporarily or may continue for long.

Fig. 29.2 : A Crowd
A crowd is an aggregate of large number of individuals on their own in a place. A crowd is formed of separate individuals in large number in a place on their own and they do not have any common interest and they do not interact as group. For example, a large number of persons are seen in a beach and they do not have any common interest. They are collected for some time and disperse after some time on their own. If an accident occurs on the road, suddenly large number of people are aggregated and in few minutes they go away. They had no common interest.

The Machiavellian principle has been experimentally tested by **Richard Christia** and **Florance Geis** (1968, 1971). They constructed a number of scales for taping various dimensions of Machiavellian principles. It has been observed in their studies that persons who scored higher in the Machiavellian scales are those who are involved in working with large number people, that is those who are having contact with big social groups. College students who come from *prestige colleges scored higher* than others. College students who scored higher on the Mach Scale out manipulate other students and also they consistently win laboratory games. In all the tests the students who scored higher in the Mach. Scales were found to be *superior in all the dimension tested.* This experimental results show that Machiavellian hypothesis is found to be correct. Those who have contact with larger social groups have *bigger brain,* than those who do not have such characteristics, and also they have higher social intelligence.

Following the publication of the social psychology books on scientific lines, written by **McDougall, Ross, and Allport** quite number of books on social psychology were brought out. Every author in his book referred the views expressed by the earlier authors. Controversies, discussions and criticisms were common occurrences till third quarter of twentieth century. Certain disputes and controversies were resolved and others are still continuing.

Definition and Meaning of Social Psychology

"The most perplexing problems faced by man today are social problems", thus observed by **Miller** and others (1963). The term social as used by psychologists and other behavioural scientists refers to the phenomena resulting from the interaction among individuals and/or groups as well as to the influence of individuals and groups on one another. A correct definition of social psychology which is acceptable to all is a very complex task. It is because, according to **Baron** and **Byrne** (1987), the field is having a very wide diversity and its rapid rate of change and development.

With all the broad topics, psychologists prefer to study the behaviour of individuals in social context. Most of the psychologists are primarily concerned with comprehending *how and why* individuals *behave, think, and feel* as they do in situations involving the presence of others. This may be taken as the important factor and account for the definition of social psychology which may be acceptable by most of the investigators, who are interested in the study of social psychology. According to **Baron** and **Byrne** social psychology is the *scientific field* that seeks to understand the nature and causes of *individual behaviour* in social situation. The definition is too general and not grisp. It needs elaborate explanation.

The term scientific refers to some specific fields of study such as physics, chemistry, biology and so on. Social psychology also study its subject matter in the same manner as these sciences do by experimentally investigating and verifying its hypotheses. Any systematic study of a subject matter done according to the principles becames a science. Social psychologists in their efforts to understand social behaviour depend upon mainly on the same basic principles and methods as the physical and biological scientists do. The subject matters of social psychology cannot be put into the test-tubes experimentation and studied. Its subjects is very wide and no easy solution can be evolved to the perplexing human problems.

The gathering of several human beings constitute the group or crowd. Though the characteristics of a group and a crowd vary considerably they may be called as social groups in which the individuals have their own roles. Social psychology focuses its attention on the behaviour of the individuals in the group. There may be different social or cultural groups in a country. These groups differ in their interests motives, goals and also at various levels of behaviour. The groups have something in common. But the individual persons have the specific group character, *experience, positive feeling* to the *ingroups* and *negative feelings* towards the *outgroups.* Thus it is the specific individuals who express the feelings likes and dislikes, love and hate, friendship and antegonism within the group and with the members of outgroups. Thus ultimately it is the individual persons perform the social behaviour.

It is with these basic facts in mind social psychologists focus their attention upon the feelings, thoughts, attitudes and beliefs and social interactions in which the individuals are involved. This type of behaviour always takes place in a socially and culturally shared situation. In addition to these the social psychologists exercise their major interest an understanding the factors that shape and guide the actions of the individual human beings in a social situation.

What Causes Social Behaviour in an Individual?

Any behaviour must have a cause. So also the social behaviour of an individual is caused by several factors. Social psychologists are of the view that the individual is physically, emotionally, cognitively and socially impelled to act and behave in a particular fashion. Social psychologists mainly concerned with understanding the wide range of factors that shape the social behaviour of the individuals. The factors that influence the individuals behaviour in a society are in quite large numbers. An individual's reaction to his family members may vary from his reaction to his neighbours and to the outside members. The simple reason is that every individual is having different thoughts, feelings, traits and attitudes. Therefore, there are many factors influencing social reactions of an individual in a group. These factors which are condisered to be the main causes of social behaviour of an individual. According to **Baron and Byrne** (1987) these factors fall under five broad categories.

These factors are : (1) the behaviour and characteristics of other persons, (2) the social cognition which involves own thoughts and feelings, memories and attitudes about the people around us, (3) the ecological factors which involve direct and indirect influences of the physical environment, (4) the socio-cultural situation in which the social behaviour occurs and (5) the factors that involve our biological nature which is relevant to social behaviour. This means that social behaviour of human beings is stongly influenced by the actions and characteristics of other persons in the social situations.

It can be observed that no two individuals think alike or feel alike individually. But in a group or social situation these two fellows accept and follow certain group or social norms. The social situation is one in which many persons participate in some common issue. Each one has different view, but as a member of a social group each one gives up certain personal likes and dislikes and accept what is good for the whole members of the group. What causes them to behave like this? As we stated above there are many factors that work as the causative factors in the individuals social behaviour. But all of them brought under five categories by Baron and Byrne.

Werner and Sugarman (1986) are of the view that the observable characteristics of other persons also exert strong influence in our thoughts, feelings and subsequent behaviour. According to them, generally we have a tendency to react differently to highly attractive persons than to unattractive ones. We react to well dressed and well groomed persons differently from those who are not so well turn-out. We have also tendencies to react to and respond to others style of speech and manners of conducting themselves with others. While we accept that the behaviour and characteristics of others play important role in shaping the behaviours of the individuals we also must see that it is only a part of the total development and not the whole. Further the individual is also influenced by *social cognition* which involves our own thoughts, beliefs, attitudes and memories about other persons.

The Physical Environments

The influence of *physical environment* upon the social behaviour of the individuals can be observed in various environmental situations. The influence of hot and cold atmospheres, noise, pollutions of various types, over-crowding in public places all exert influence upon the individual's social behaviour. Since people always live in a society and every society maintains some norms the individuals who live in such societies are expected to adapt all the social norms formulated by the society. The individlaal can never go out of these norms. The individual is further expected to follow the cultural way of life such as in eating, dressing, important functions like marriage, death and so on. All these influence the behaviour of the individual and make him culturally and socially oriented.

Just like the socio-cultural variables, the biological factors like the *heredity* and/or the *genetic backgrounds* put certain limitations on the bodily development of the individuals. For example, the individuals' physical appearence is an inherited aspect which has a built in limited capacity for processing social information. This may have certain influence in social situation. All these factions play important role in the social behaviour of the individual and serve as the causative factors of social behaviour.

The individual while moving and acting in a group he learns many things by associating himself with others in his social situation. One important aspect of his social behaviour is identifying himself with others. According to **Lindgren** (1974) there are two basic ways through which interaction with others can *provide meaning.* One way is through *identity* and the other through *definition.* Identity indicates who we are which is having a positive meaning, *whom we resemble* in particular in a group. But definition indicates who we are, but in more *negative meaning* by showing differnece between ourselves and others and between those whom we resemble and *those whom we do not resemble.* It is the identity that divides the *ingroup members* from the *outgroups members.*

Based upon these points of view we can evolve a definition of social psychology. Social psychology may be defined as the systematic study of the social behaviour of an individual which involves also comprehenting the effects of the causative factors such as the behaviour and characteristics of other persons, the social cognition, environmental factors, socio-cultural factors and the biological factors and how the individual exercises his feelings and emotions, thoughts and imaginations, likes and dislikes, beliefs and attitudes, the needs and motives in social situation and also how the individual *actualises his self.*

Subject Matter

The subject matter of social psychology has been going on changing and developing during the course of seventy years, that is from 1930 to 2000. Exponents in the field of social psychology have been also changing the titles and substances of the subject matter. Therefore, there are lots of differences in the subject matters in the books published between the years of 1930s and 1990s and 2000s. However, we are here dealing with the important subjects matters precisely that are very much useful to the present gestation of students.

The main subject matters include the social perception, social cognition, social process which include public opinion, persuasion and propaganda, structure and functive of social groups, group morale and leadership. Attitudes and related traits, ethinic relationship, prejudices and discriminations.

Methods of Social Psychology

Social psychology studies various aspects of social life of the individual in relation to social group situations. Since psychologists are always interested in developing the field, by adding more and more information obtained by conducting experiments and by applying various techniques and methods, social psychology has grown very widely with enormous amount of information and adding them to the existing knowledge. The methods they mostly apply are the experimental methods and correlational methods.

Experimental Method

The experimental method has been used by the social psychologists in the form of research. Though this method in social psychology has been criticised by some scientists as vague and mysterious method, still social psychologists have demonstrated that how it is an experimental research method. In any scientific study the experimentalists formulate the hypothesis and design the experiment, choose the population or subjects and apply the instrument, to collect the data. The same basic principle is adapted also in social psychology by the researchers. The social psychologist determines whether the given variable can influence some aspects of social behaviour of the individuals. In order to get the answer for this question the experimenter follows two basic steps in his experiment.

First, the experimenter varies the strength of the variables in a systematic manner, and secondly, the experimenter tries to determine whether these variables have any mpact upon the aspect of social behaviour of the individuals under investigation. The main idea behind these procedures can be stated like this. If the variables exert any effects on the individuals who are exposed to different qualities or levels of the variables, they should show different patterns of behaviour. That is, exposure to certain smaller quantity of the variable should result in one level of behaviour; exposure to larger level of variables should result in larger level of behaviour.

The experimenter varies the variables systematically in his experiment to produce the expected result. These variables are called *independent variables* as they determine the behaviour, while the behavioural pattern or the effects of the independent variable on the subject is called as the *dependent variable.*

One of the problems that exists in a population is that people fall under different socio-economic status and their style of life varies on the basis of their financial positions. Children from these families go to schools and colleges. These children's performances in schools and colleges vary considerably. Why they vary in the educational performance? What is the factor that determines the better and poor performances? These questions need reliable answers. If we want to know the correct answer for the problems we must conduct some experiments. For conducting experiments, we must formulate hypotheses and define the variable, select the sample and conduct experiments and verify the hypothesis. In this specific problem we state the hypothesis like this. Children who come from higher socio-economic status families perform better in school assignment than the children who come from lower socio-economic status. In this hypothesis the independent variable is the socio-economic status and the dependent variable is the school performance which may be stated in terms of marks they obtain in the examintaions. The hypothesis may be accepted if we find difference between these two groups of students. If the difference is not found, the hypothesis is rejected. This type of *experimental methods* are used in social psychology.

The Correlational Method

In this method, attempts are made to determine whether two or more experimental variables are related to each other in a significant manner. In such experiments the human traits can be used as variables or some characteristics. The problem may be stated any hypothesis may be formulated and the variables may be assigned. The problem is that individuals who are highly religious are more *conservative.* This means that religious affiliation and conservatism go together and they are related. In the same hypothesis another two characteristics can also be included and the hypothesis may be stated that people who are less religious or those who are *agnostics* or *atheists are radicals.* In a population we may find people with different attitudes, mainly *favourable, unfavourable* and *neutral*. In the study of social behaviour of the individuals we can assess what type of attitules and beliefs they have developed and exercise them. By conducting experiments we can classify persons into categories such as *radicals* and *conservatives, liberals* or *reactionries.* The statistical methods may be applied to treat the data obtained by the experimental method and correlation methods can applied to the data obtained for different variables. The levels of association as revealed by the correlation method between variables would tell us whether we have to accept the hypothesis/hypotheses or reject.

Social Perception

What is social perception? Social perception is the process through which we seek to know and understand the persons who are moving around us. It is knowing others and ourselves in a social situation. How do we conduct ourselves in the presence of others and how do we perceive others and understand in the social situation. Social psychologists try to explain this process in a wider perspective which constitute the social perception. Social perception is an important part of our daily life. We find that other people sometimes behave mysteriously. They speak and do things that we do not expect. We could not understand their motives readily. They are all very different from ourselves. But since they play important roles in the society and interacts with us we cannot leave them aside unsolved.

We are also very much interested in knowing what is going on around us. Otherwise we would be missing many things in our life. What the other people are really like and why behave in this fashion. Social psychologist generally explain this process of our knowledge about others as *social perception.* But we are also to know that *our efforts to understand* the people around us and about ourselves are directed on many different issues. Many of them mainly fall under two important areas.

Firstly, we are interested to grasp the current or immediate presently existing causes behind others' behaviour which may include their present moods, feelings and emotions.We may know all about these issues through the *signs and signals and by non-verbal means* related to others' facial expressions, contact through their eyes, bodily postures and movements and actions.

Secondly, we attempt to understand the more permanent causes behind the actions of other people, their firm characteristics or traits, intentions, motives, beliefs and faiths. Information pertaining to these issues, can not be obtained by simple means as these are very complex issues and we must apply only some complex process which is generally stated as *attribution* which means, the assigned characteristcs of the nature of behaviour by observing other's behaviour and then infer the causes behind it and state them.

The general processes of understanding others' behaviour as we stated above are the non-verbal means and the attribution means. Generally we see the people in some social situations that people express their feelings and emotions not by speaking. The non-verbal expressions are made by facial cues. Any non-verbal expression is a behavioual aspect. For example, eyewinking, mouth and lip twisting, signals and signs by fingers and hands, gazes and gestures, pending bodily parts, moving legs, all are non-verbal expressions of people's feelings and emotions, approval and disapproval, accepting and rejecting. These gazes and gestures, signs and signals are communicating some information of the people in the social situation. We also understand as human beings the meanings of these non-verbal communication and react to them if we are related to them in some way.

These are the ways in which some information is conveyed to and we understand the nature of behaviour of those persons who conveyed them and also about ourselves as we are reacting to the meanings of those non-verbal means of communication. This type of understanding the behaviour of others may be stated as primitive type of communication. It is because, this type of non-verbal communication prevails not only among human beings, but also among animals and birds. Social psychologists like **Baron and Byrne** (1980) are of the view that human behaviour is strongly affected by these temporary current causes. These causes shift moods, fleeting emotions, fatigue and various drugs can all exert strong effects on the behaviour of the individuals in the society. **Isen** (1984) states that for example, many people are more willing to do some favour for others when they are in good moods then when they are in bad moods. **Geen and Donnerstein** (1985) have observed in their studies that many people are likely to loose their tempers and lashout others who seek some favour when feeling irritable than when they are feeling mellow.

But many forms of basic behaviour that we think should be normal in some person are strongly affected or turn to be hostile and violent by the effect of alcohol and other widely used drugs. These temporary influences in such persons produce serious effects upon social behaviour. It is better to know something about them. But how can we know whether others are in good tempers or in bad tempers? How do we know whether they are in angry mood or joyous mood or whether they are in friendly mood or in tense mood? **De Paul** *et al.* (1985) suggest a strategy for this. According to them, the only, way is to ask them directly. But it is very much doubtful whether this strategy will work out successfully. Sometimes some people may be willing to reveal their moods and feelings and sometimes they may not be willing. Some may be misguiding us and some may even deceive us. But, take your own case, if you have tried to conceal your own feelings or anger or hatred, likes and dislikes, you are aware of the fact and you need not try in that fashion. But we need not put out efforts in understanding the behaviour of other people. But the behaviour is revealed by another source which may indicate in what mood one is by. These temporary causes of behaviour can be assessed by non verbal cues. It is possible to know much about the current moods and feelings through this silent language. This non-verbal expression of feelings and emotions always accompany them and they are with them. This means of communication is not connected with language.

The Means of Non-verbal Communication

These temporary behaviour patterns that take place through *non-verbal* communication, is a complex factor, very difficult to assess certain behavioural patterns accurately. But, however, we can assess and understand the meanings of the non-verbal communication by adapting certain techniques that are quite familiar with human beings in social situations of various types. One way of understanding the meanings of *non-verbal communication* is by knowing the basic channels through which the communication takes place in social situation. Secondly, the nature of role, that is functioning in social situation.

The basic channels of non-verbal communication are mainly the human gestures through the bodily parts. It is possible to communicate many ideas, feelings and emotions through bodily postures and gestures. They are both *instinctive and learned.* Right from childhood period and through all the other developmental stages man has been using these instinctive and learned gestures and postures to reveal himself and to communicate to others.

It is said that the "face is the index of the mind". The idea behind this statement is that the face reveals what the man thinks and feels. In human beings the faces express everything what they are thinking and feeling. These thoughts and feelings may be revealed by some expressions and sometimes they are masked or concealed. Modern researchers suggest that it is possible to understand more about the current moods and feelings of other persons from their facial expressions. It is observed that *six basic emotions* are revealed clearly on the human faces, such as *fear, anger, happiness, sadness, surprises and disguest.* This is what Buck (1984) had noticed in his studies. It does not indicate that we could express only six different facial expressions. Emotions occur in many combinations. Every emotion with every other emotion in different combinations may be demostrated. The human facial expressions infact reveal the inner feelings. But it is also possible to assume that they are not true expressions always. We cannot attribute universal characteristics for them. The reflections may vary from culture to culture. But, however, when the basic emotions are expressed people all over the world show similar facial expressions and the meaning for such *expression is universal.* Therefore, we may state that the facial expression is a language which does not require any interpretations (Fig. 29.3).

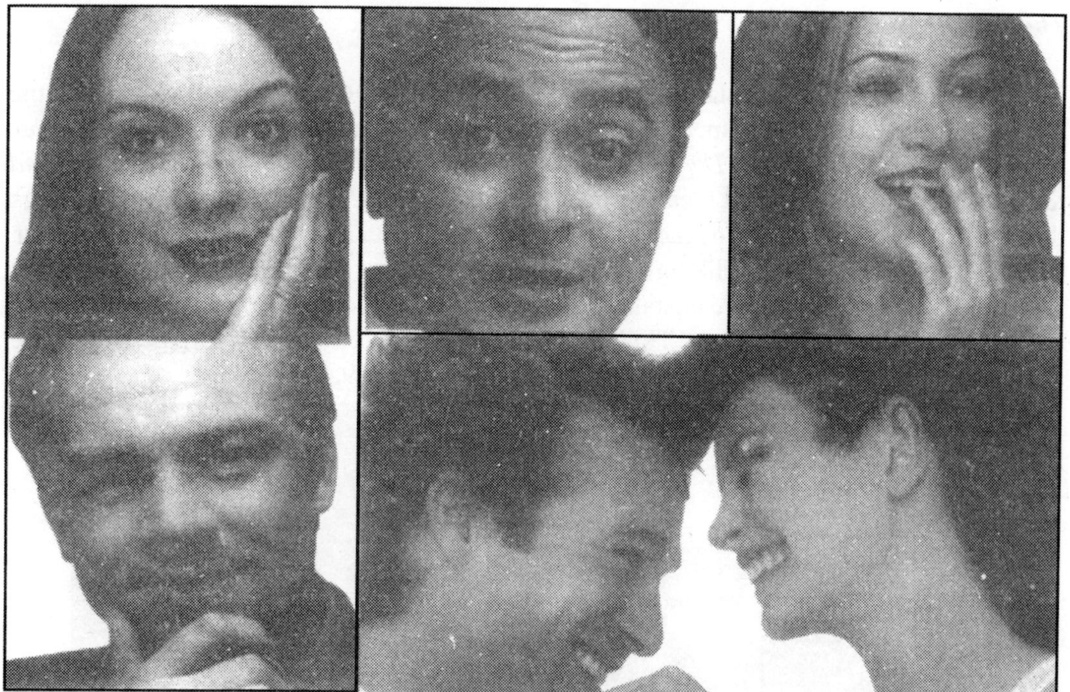

Fig. 29.3 : Social Perception

Non-Verbal Communication : Facial Expression : The basic channels of non-verbal communication are mainly the human bodily gestures. It is possible to communicate many of our ideas, feelings and emotions through the bodily parts. These characteristics are both instinctive and learned. It is said that the *face is the index of the mind*. It means that the face reveals what the man thinks and feels. The human face expresses everything what he feels and thinks and wishes to tell others. The thoughts and feelings one expresses through his face sometimes are masked and concealed. Psychologists are of the view that it is possible to understand more about the current moods and feelings of other persons from their facial expressions. The face can express six basic emotions such as fear, anger, happiness, sadness, surprise and disgust. There can be even more than six emotions in combination of one emotion with other emotions. The face reveals the human inner feelings, but it may vary from culture to culture. The facial expression is a non-verbal language which does not require any interpretation.

The Language of the Eyes

The language of the eyes is staring and gazing. It is a common expression of most of the human beings sometimes to speak through their eyes. We can use our eyes to reflect our inner feelings and emotions. *The eyes may be stated as the windows of our emotions and feelings.* We often understand great deal about other persons " inner thoughts and feelings through their eyes" and we also react to it. From the eye contact we can understand whether the eye expression is a sign of liking or disliking. The eye may reveal pleasant or unpleasant feelings. If you commit a grave mistake in your office work, your superior officer may stare at you but may not speak to you. This means, he did not like what you did. Staring is the expression of unpleasant feeling and also disapproval (Fig. 29.4).

Fig. 29.4 : Social Perception

Non-Verbal Communication : The Language of Eyes : In our social perception our eyes are playing crucial roles in revealing the feelings and emotions, thoughts and imaginations. The language of the eyes is tearing, gazing, winking, widening and closing. It is also used for an all-eye look. In some social situations people speak through eyes. We can use our eyes to reflect out inner feelings, emotions and ideas. The eyes are stated as the *windows* of our thoughts and feelings. We often understand more about the persons' inner feelings and thoughts through their eyes and we too react to them. From the eye expression we can understand whether it is a sign of liking or disliking love or hatered. Staring of the eyes is a sign of disapproval of our ideas and actions. Parents star at their children when they commit mistakes and teachers stare at the students when they fail to complete their home works. Generally staring is a sign of treatment of hostility. Everyone of us might have experienced the staring of the eyes from persons with whom we had close contact. Thus our eyes are the media of non-verbal communication.

Staring at us may come from any source. It may be from the father or mother on the boy or girl who commits the mistake of not getting good marks in the examination. The teacher may stare at a student who had failed to do his home work. The senior officer may stare at his subordinate for having not completed files. In all these instances staring produces negative effect. The staring at all these instances is a sign of treatment of hostility. The reaction to it generally is the development of displeasure and avoiding it and fleeing from the scene. Staring also may produce positive effect. Everyone of us in some context in some situation might have experienced the staring of the eyes of some persons with whom we might have had some contact.

Generally staring is a non-verbal expression of one's inner feeling, can be revealed without hurting the feelings of others.

The Language of the Body

Human beings generally make use of their bodies to express their thoughts and feelings by means of some *gestures and postures* which are supposed to be thought as the language of the body. Generally human beings use their hands, fingers, legs, head and other bodily parts in their conversation with others. We could clearly observe that some people move their hands and head very vigorously while explaining something. Their heads and hands are moved much more than what they speak with their mouth. Similarly, they move their bodies and demonstrate the incident more than their words spoken. Similarly they move their legs in different directions.

The non-verbal expressions from these sources generally are stated as *body language* which are communicating several important information about many things and also about the behaviour of others. Other persons' emotional states are revealed by the body language. These bodily expressions may be scratching, stroking and other forms of signs. Certain bodily gestures and postures convey some meaning of the person's feelings and ideas. The movements of some bodily parts express greetings, requesting for some help, expressing the inability to earn and live, expressing the disabilities such as deafness and blindness, shaking hands vigorously and nodding head are also signs of expressing some meaning in a social situation. Nodding head with speaking may express two meanings like acceptance and rejection, but these types of nodding vary in characteristics. All these types of bodily language are taking place in social situations, expressing some kind of behaviour of the people (Fig. 29.5).

Understanding the Causes of Behaviour

Social perception is the process of understanding the behaviour of individuals in a social situation. Our interaction with other people in a social situation mainly depends upon our understanding the current moods and feelings of other persons. Further, we must also know the stable traits or the characteristics of others and how they interact in a social situation. Why do they interact and behave like this? We wish to understand the causes behind this type of behaviour. Social psychologists are very much interested in knowing the causes behind their behaviours as to why they behave in that fashion in certain social situation.

Based on these inferences we may formulate the hypotheses about the dispositions of other people, their motives, feelings and emotions and their intentions.

Fig. 29.5 : Social Perception

Non-Verbal Communication : Language of the Body : Human beings make use of their body and bodily parts to express their feelings, emotions and thoughts to others in social situations by means of some signs, signals and gestures and postures. Generally human beings use their hands, fingers legs, and other bodily parts while having conversation with others. Some people move their heads and hand vigorously while expressing feelings and ideas. Some make bodily demonstrations without speaking any word. People express love and affection by bodily contacts without speaking. They also set tears silently to express their extreme sorrowful feelings. These non-verbal expressions and communications are stated as *bodily language*. These bodily expressions are communicating important information to others in social situations. *Bharatha Natyam* is a type of dance is fully based on non-verbal communication. Every movement made by the dancer is made with fingers, hands, legs, face and other bodily parts and the feelings, emotions and thoughts are expressed to others only by movements and action. There are also other signs and signals to express greetings, sadness, helplessness and the inability to earn and live. Thus the language of the body serves several purposes.

We can observe that the behaviour of others as caused by these dispositions. **Fritz Heider** (1958) states that our perception of the causes of the behaviour of the people is just as much a part of perception as our perception of their physical characteristics. **Heider** has proposed that the perception of the cause and effect in human behaviour is primitive and basic. On the basis of this, he evolved a theory known as *attribution theory.* It is concerned with the ways in which we attribute behaviour to causes.

What is attribution? Attribution may be defined as the process by which we seek to determine the causes of the behaviour of other people and collect the information and gain knowledge of their stable traits and dispositions. Attribution is a *complex social-process* and several theories tried to explain the operation. One of the theories formulated by **Jones and Devis** (1965) is based upon acts and dispositions. These authors used others' behaviour as a guide to their stable traits and characteristics. These investigators are of the view that how do we decide and infer about the stable characteristics or permanent traits of the people from their behaviour. This type of inference is necessary because people do not carry the signs of their traits outwardly. The trait is manifesting only in the behaviour of the person in certain specific situation. Therefore, if we want to understand the persons around us we must observe their behaviour and use the information that observation has provided, which is the basis for arriving at the conclusions about them. This procedure appears to be simple. But generally the behaviour of the people does not lend us with valuable source of input. People behave in certain ways not because of their own traits and the dispositions, but because of the influence of the factors in the situation outside of their control.

If that is the case, how do we decide whether others' actions reflect their true characteristics or alternatively that these actions stem from other factors? The answer is provided by **Jones and Devis** through their theory of *correspondent inference.* We complete this difficult task by turning our attention on certain types of actions, most likely informative in this regard. First of all, we consider only those types of behaviour that seem to have been choosen freely and those that seem to be some what forced to be ignored. Secondly, we pay attention carefully to the behaviours that produce *unique or non-common effects.* These effects would not be produced by any other actions. Thirdly, we also pay greater attention in our efforts about other persons and the actions they perform that are with lower in social desirability than the actions that' are in high on this dimension.

In fact we learn more from the actions of others that differ from the ordinary and from that are not encouraged by society. To conclude this theory we may state that other's behaviour reveal their permanent traits. We arrive at the *correspondent inference* about those people. When these actions occur they are by choice which yield clear non-common effects and are of lower social desirability than the actions of higher on this dimension.

Why other persons behave in that way? This is one common question many people raise in certain situations. Social psychologists try to find out answer for this question. One of the attribution theorists, **Kelley** (1972) has visualized this problem and provided the answer in a convincing manner. In most of the life situations the causes of human behaviour in social situation is the *central attributional task* we are facing. We want to know why other persons had behaved in such a way as to draw our attention? A correct information for this becomes a crucial important for the understanding of the causes behind their actions may make us to adjust our own behaviour accordingly and create a sense of understanding about the social world around us. But it is not easy, as the number of causes behind the behaviour of other people are so many that it is very difficult to deal with them specifically.

The task of dealing with this problem can be made easy by raising the simple question, does the behaviour of others stem from the internal causes like their own characteristics, feelings and emotions, motives, intentions and conflicts or from the external causes like the aspect of the social, physical or environmental world or stemming from the combination of both the factors? For example, you are working in an organization for a very long time and you did not get any promotion. One day your director calls you and informs you that you have been promoted as an assistant manager and take the orders. Here we can apply both the causes to the event. How to explain the director's behaviour? It is because of the internal causes that the director is kind, benevolent, respects the services of his subordinates or some external causes like the pressure from the organization to give a lift, or a *combination* of both the causes? **Kelley's** *attributional theory* tries to explain the causes behind this type of behaviour. According to him there are *three causes* behind this type of behaviour. The behaviour stem from internal and external causes.

While trying to answer the questions "Why" about the behaviour of others, **Kelley** focuses his attention on *three major dimensions.* These are, the *consensus, consistency and distinctiveness. Consensus* is the extent by which others react to some stimulus or event in the same manner. In other words, the extent to which actions shown by one person are also shown by other persons. Generally consensus means an opinion held or *accepted by most persons or by all.*

Consistency is the extent to which an individual responds to a given stimulus or situation *in the same way on different occasions.* Generally consistency refers to the extent to which an individual reacts in the same manner to the same stimulus or event *at different times.*

The third dimension is the *distinctiveness* which refers to the extent to which an individual responds in a similar manner to *different stimuli* or *different situations.* If an individual reacts in the same manner to a very wider range of stimuli the distinctiveness is low. Wider range of stimuli may differ in their qualities or characteristics.

According to **Kelley's** attribution theory we are attributing another person's behaviour to the internal causes under conditions of *low consensus, high consistency and low distinctiveness.* But at the same time we are most likely attributing another person's behaviour to external causes under conditions of high consensus, high consistency and high distinctiveness. **Kelley** is of the view that we generally attribute other person's behaviour to a combination of these three factors under conditions of low consensus, high consistency and high distinctiveness.

Fundamentals of Attribution Error

The attribution theory that we have discussed above is mainly based upon some *logical sequences.* According to the views proposed by **Jones and Davis** and also the view offered by **Kelley,** the basic idea is much the same. That is finding out the *causes behind other's behaviour.* It has been viewed by some investigators that we are made to omit serious errors about the causes of other's behaviour.

First of all we must know that the fundamental error is that it is *over estimating* the role of dispositional causes. While expressing the view about the attribution theory, **Lee Ross**

(1977) states that it is the fundamental attribution error and it has been demonstrated in collaboration with the coworkers. The human tendency is to describe others actions in terms of *dispositional* rather than *situational causes.* We perceive others behaviour often as they are behaving in a particular situation. For example, a person may play a practical joke on another person by pouring some colour water in a social situation. This type of behaviour cannot be taken as joke as it puts the other person in an embracing position. This, we may state that "he is that type of person". How to explain this behaviour? It is internal cause? or external cause? It is only situational. Many situational factors which may also have affected the self-esteem of the person who becomes the object of this type insults. It has to be ignored and no substantial cause can be found for this and attributed.

There is a tendency to *overestimate* the dispositional causes and *underestimating* the situational causes. This tendency is getting stronger. There are also observations that other's actions are not under their own control, but elsewhere. One explanation for the fundamental attribution error is that when we observe another person's actions we generally look at his actions, to the context or to the situations in which they occur. But the situation fades away. As results of this the potential influence of the situational cause is not taken into consideration.

Another possible interpretation is that individuals notice the situational factors, but fail to attach proper importance to them. Authors like **Gilbert and Janes** (1986) and others are of the view that individuals do not perceive them as important as they really are. The attribution theory inspite of the fundamental error, it has important implications. According to the theory even if individuals are made to be on the situational forces that affect adversely, individuals and other minority and disadvantaged groups like poor dieting, broken families, they still perceive these groups and individuals are largely responsible for their own plight. They do not see that the situation as the cause of the important social miseries.

The Egocentric Biases or the Self-serving Biases

There is always a tendency in human mind to claim that unlike others we do no wrong. This is tested by **Miller and Ross** (1975) as the self-serving bias. The person who commits something wrong does not accept his mistake, but attributes the mistake to same external causes. This view has been confirmed from the results of many different experiments. The functions of attribution bias were observed in the experiments conducted by **Baumgardrev, Hepner and Arkin** (1986). These investigators presented *six hypothetical* problems to a group of individuals and they were asked to solve them. These problems were referring to their personal problems such as wanting to be more popular, feeling nervous, being lazy and so on. On obtaining their solutions, the subjects were provided with a false feedback, stating that they had done the problems either *very well* or *very poorly* on the task. Then they were asked to indicate the extent to which their performances were stemmed from various causes such as their ability, their efforts, their difficulty on the task, the luck and so on. The two groups, those who have done very well and very poorly, revealed a self-serving bias. They have *attributed the success* to internal factors so as *to their own ability,* but attributed their *failure to external factors* like task was difficult and to luck.

There was also other evidence which suggested that the self-serving bias arise from two different and related sources. There is a tendency to protect the self-esteem. We claim the positive outcomes and not ready to accept the blame for the negative outcomes. Secondly, according to **Greenberg** *et al.* (1982) self-serving bias permits us to improve our *public image,* that is, we must look good to others. This type of attribution error also becomes the cause of *inter-personal frictions.* For example, if three or four persons are engaged in some task and when the task is very successfully completed, each one claims that because of his own personal efforts it was successful and not due to the work of others. Instead, if it is not successful, each one claims that because of others inefficiency the task has failed. Each one claims that he is right and others are wrong. This tendency undergoes further co-operation.

Self-Perception

We have been discussing the manner in which we come to know and understand other persons. But how do we know and come to understand about ourselves. By turning our attention towards us we can understand our own feelings and emotions, motives and intentions. By this means it is easy for us to obtain information about us. Several investigators have attempted to study this phenomenon. **Wisbett and Ross** (1980) stated that though it is easy, there are two complications in this process of understanding about ourselves.

First of all, we are not aware of some of the factors that affect our own behaviour. We have been behaving in a given manner. This may be uncertain and also wrong. We may feel later on and ask, "Why I did like that?" Secondly, it is difficult and also not possible to evaluate our own traits, abilities and attitudes without making any references to the persons who are around us. Are we intelligent or stupid, or successful in life or not successful? We cannot speak out this by looking into us inwardly. But to get information about us we must ask others or depend upon social information, the remarks and comments made by others. Social psychologists have been trying to get a correct answer for this for long time. **Deryl Bem** (1972) was able to provide a satisfactory information for this. He formulated a theory called *self-perception.*

A Theory of self-perception : According to this theory the individual's behaviour is the source of *self-knowledge.* The idea put forth by **Bem** is that we do not know our own feelings and emotions or attitudes directly. Since we want to find it out and it is necessary, we *infer them from observations* of our own behaviour. For example, if we acted on some occasion in some manner we try to reason out and then hold an attitude or feeling consistent with that behaviour. In what manner our action serves as a useful guide to *what is happening inside.* We also *draw inferences* about our own actions in the same manner as we infer about the behaviour of other persons. Therefore, it is clear that we come to know about ourselves in the same manner the process we apply to understand about others. **Bem** states that we use observations of our own behaviour and *infer our feelings and attitudes.*

Bem's *theory of self-perception* received support from several investigators. In their experiments they have observed that individuals perform some action in relation to their attitudes and they come to hold it even more strongly than they were holding it earlier. **Kiesler** *et al.* (1969) having Bem's theory as the basis, evolved a new idea called *intrinsic motivation* and added to the theory of self-perception, which is named as self-perception and intrinsic motivation.

Self-perception and Intrinsic Motivation : The added idea to self-perception puts more emphasis on motivation. Intrinsic motivation is the motivation to perform various activities, simply because they are enjoyable in themselves. According to the investigators, this includes many things right from simple passtime activities or hobbies through many enjoyable activities like competitive sports activity. These activities are performed not for rewards but for pleasure. Any pleasure giving activity is intrinsically motivated and the individual performs them spontaneously. When individuals perform some activity, whatever type they are, they enjoy the activities in the absence of external rewards. Their intrinsic motivation to keep them in this behaviour is maintained. But when they receive extrinsic rewards for performing this activitiy *their percetions* about it were why they do so many changes? As result of this the *individual's intrinsic motivation was reduced.* This means the individuals do not expect any rewards for their intrinsically motivated activities.

There is also another interpretation for the theory of *self-perception* offered by **Pretty** and **Seligman** (1984). According to this view, offering external rewards for the individuals who perform activities for enjoyment causes them to experience *negative effects like unpleasant feelings* or *moods*. The individuals feel that they are being coerced in this way and also they felt that their activities are rated in terms of rewards. This type of negative effect interferes with the individuals' enjoyment of activities and therefore, tend to reduce the intrinsic motivation. The negative feelings experienced by the subjects eliminate the effects of over justification.

However, the theory of self-perception and the added interpretations seem to be not very clear. All these are trying to explain how self-perception occurs. In other words, how the individual perceives himself or his self? It is the process through which we seek to understand our own feelings, traits, attitudes and motives. It is by self-observation, going deep into our feelings and ideas and the activities performed pleasantly or unpleasantly. If this is the process, for self-perception what is the difference between *self-perception and introspection?* A short and simple definition is given by **Henry E. Garrett** (1975) for introspection. Introspection is a psychological method in which a person reports his own ideas, sensations experienced and all the images about him.

Introperaction as explained by the early psychologists is that it is a procedure of looking into one's own experiences, feelings and emotions and reporting them verbally. Generally introspection is looking with in one's own experiences whatever they are and reporting. That is what the individual is. The individual is perceiving himself, which may include everything about him. It was **Wilheim Wundt** who founded this method. According to him introspection is self-observation and human beings can introspect into their various sensations, feelings, elementary motives and cognitive processes. **Wundt** did not use the term conscious or consciousness in his definition of introspection. According to him it appears that introspection is a process of self-perception. We can understand the self by the process cf introspection much better than the process of self-perception without any confusion. It is very unfortunate that modern psychologists have totally neglected the introspective method to understand and perceive the individual and apply a more vague process to understand the individual. If introspection is subjective and does not yield reliable information about the self, *self-perception* also has the same characteristics.

The causes of our own behaviour if we could understand by the information provided by the process of self-perception we could also obtain the same information from the introspective process. Self-perception seems to be some what controversial in the study of social perception. It is brought under the study of personality and social phenomenon involving, self, self-concept, self-esteem. Generally individuals self-evaluations vary widely from the opinions that others have about them. For example, a person has strong potentialities and yet he is derogating himself and not taking any efforts to develop them. Everyone tells him, he has talent, yet he strongly refuses to believe that he has it. Another person claims that he is an extraordinarily talented, but no chance to develop or exhibit it. Thus there are such *distortions in self-perception.* Probably, there are factors that are unique in each individual's experience that may lead him to *misperceive himself.* Others in prasing or criticising us, they are trying to influence us by telling us that we are wrong in our self-perception. This attitude of others show that they know us better than we do which may lead to develop a sense of defensiveness in us.

SOCIAL COGNITION : SOCIAL WORLD

What is Social Cognition?

The world in which we live is full of living creatures like reptiles, birds, animals and human beings. Apart from these animated living beings there are also inanimate objects like the environmental, and the natural resources all around us. The moment we come out and open our eyes, we see the social world which is filled with human beings moving about and interacting among themselves in a pleasant or unpleasant manner. Social psychologists who are watching the behaviour of human beings in social situations find their subject matter in the social world. They are very much interested in understanding the social world and observe as what is going on there. This phenomenon we may call as social perception. The subject matter for social perception is the behaviour of human beings in the social atmosphere. Therefore, we may state that soical cognition is the study of how people perceive, interprete, analyse, remember and use information about the social world.

Social cognition may be also stated as the individual's *understanding of the social world* of other people and of the *personal relations* with other persons. There are several variables involved in social cognition and we shall discuss a few of them briefly.

Heuristics : Biases and Fallacies

In our attempt to understand the social world the information related to the functions are so much overloaded that we must be constrained to same extend in presenting the information without sacrificing the main theme. For this, some short cut formule may be applied. This is generally called *cognitive strategies* which fall under *two features* for the people to use. By the *first* feature, people may provide a quick and simple way of dealing with social information. According to *the second* feature, they must provide information reasonably and accurately all the times. If a cognitive strategy often leads to wrong decisions, better to change over to the other strategy. Whichever feature among the two is functioning accurately and quickly that should be adapted.

Social psychologists often use a term called *heuristics* in their social cognition researches, which means *biases and fallacies.* Heuristics according to them means *decision making principles.* This is applied for making *inferences or conclusions* easily and quickly. Biases and fallacies refer to the errors and distortions that may come up in the way people use social information and think about them. The terms related to heuristics involve often biases and fallacies.

Some major cognitive strategies for understanding the social world : There are many different *short-cut* strategies applied in understanding the social world. One of the important strategies is the representativeness *heurisitic strategy.* This mainly involves judging by resemblances.

It is making judgements or inferences by resemblances to some specific cases. For example, we are meeting a new person in a cosmopolitan club and we find him as regular visitor and is also a member. We wish to know about him and whose identity is revealed by him. We judge or try to inter what his occupation should be? We judge him in terms of common occupation of the place. What type of occupational persons mostly join that club. In this case we mainly rely upon *base-rate information.* This means how common pattern is in the general population. Thus we infer or decide the person, using the *base-rate-information* and decide that the person may be a *lawyer.* But the person is a *Physical Director.* Here our decision or inference is wrong. We commit here the *base-rate fallacy.*

Several experiments have been conducted by the social psychologists to verify varied types of hypotheses on base-rate information. It was found that people prefer to use *representativeness instead of base-rate-information* procedure. This has been revealed in the experiment conducted by **Tversky and Kahneman** (1973).

Some exceptions to biases, and fallacies : In the use of cognitive strategies in getting information through any of the strategies, biases and fallacies are occurring in our decisions and inferences. These biases and fallacies can be overcome. It has been observed in the experimental results that people generally overrate or go to the extreme level in some kinds of information and underuse or reduce the value of some information. In this way the fallacies take place. Some of the investigators suggest that these can be overcome or avoided if people learn the occurrence of base-rate fallacies.

Feelings, thoughts and cognition : Psychologists strongly believe that the emotional states affect our thoughts and judgements and also our memory process. The human feelings play important role in our social cognition. This view has been tested by an experiment conducted by **Baron** (1987). In a job interview, two sets of interviewers were engaged, one set of them with *good mood* and the other set with *bad mood.* It was observed that the interviewer's mood affected their evaluation of the candidates. With good moods there was more favourable evaluations than with bad moods. The bad mood had effect on their evaluation negatively. The difference in the judgements was due to the *effects of mood on memory thoughts.* The same type of information was used in the interview. Interviewers with good mood did not remember the candidates' *weaknesses and drawbacks,* but the inverviewers with bad mood did not remember the candidates *good points.* Thus emotional states had effects on thoughts and feelings. Moods and emotions generally influence mental processes. *Pleasant moods and emotions produce positive and optimistic judgements* and they accelerate memory process for good actions.

Schematic Guide

Schematic guide is one of the main phenomena in social cognition which is an organized collection of a person's feelings, beliefs and thoughts, about a person, a group of people and events and objects. Schemata include *stereotypes,* preconceptions and generalizations about everything. The basic idea of schemata is that the human mind organizes very clearly and carefully all its contents. The schemata consist of the wealth of information about the world. The types of schemata include *self-schemata, persons schemata, role schemata* and *event schemata.*

The Self Schemata : What is the self? The *self* is the process of socialization and it is an important psychological organization. Like other psychological structures the self-emerges only in the interaction with other members of a group. It is a *social protect.* It is the most important structure in the psychological field and under normal condition, it is the strongest strucutre. The infant does not have any conception of *himself* as a person. **Jean Piaget** states that the infant lives in an *undifferentiated absolute.*

In social psychological study the self is a schemata helps us for processing information. That self is something that helps us to interprete certain events and to understand their implications. For example, you are appearing for a competitive examination and you did not find your name in the list. What it indicates? It indicates your level of general knowledge, intelligence, your future prospectus, your family life and so on. Does the event make any change, how you feel about your self? If it has some change what the implication is?

This person uses his self-schemata. The self-schemata is a powerful factor through which people understand the social world. The events make the person or bear on the self, attract more attention, more thoughts evolved, and are remembered better than the other events of life. Investigators observed the way in which the self-schemata of a depressed person differ from those of normal persons.

Person Schemata : Every person we know has a person schemata and that organizes what we know and feel about this person. A person sometimes may behave in a surprising manner.

it may strike in you that he is perfectly normal. It is a surprise because the behaviour is a violation of the act what you expected of him. This makes you not to include this person in your schemata you take into consideration, the perons' role which is not desirable.

Role Schemata : This referes to the organised sets of expectations but how individuals supposed to act in certain roles. A person is expected to play his role in his working centre. For example, you go to an eye specialist for the first time. Instead of testing your eye he begins to dance with his lady assistant. Your role schemata indicates you that something wrong with the person's roles which are not as we expected.

The Event Schemata : The event schemata refers to the organized beliefs or ideas about the normal or some peculiar course of event in different familiar situations. These events schemata help us to know in advance what will take place when we go to see a football match without purchasing a ticket or to a picture house to see a motion picture without a ticket. All these suggest that event schemata should be specific and in detail.

Processing of Schemata Guide Information : Schemata generally mean organised collection of feelings, beliefs about something. The functional effects of schema can be observed on three cognitive processes mainly on *attention, encoding and retrival. Attention* is a general cognitive process which refers to what we see in the outer world. *Encoding* generally means to put out something from stored memory. *Retrival* means remembering something stored in the memory.

The main principle is that some information organized earlier get lost after sometime. In the case of attention we notice only a part of the world that we put out and only a part of what we noticed earlier and we remember only a small part of the information collected and stored in own memory. Thus, of all the three processes there is a loss of information. The loss of information is also thought in another way in each of these three processes.

It is by the process of *selection.* The mind is always choosing what to attend, put out and remember. The schemata begin to function at this stage and guide the selection processing of *social information.* The schemata function as a main instrument for coping with information that once overloaded. But the schemata are not filtering the information. They function actively by changing or distorting incoming information and fill up the gaps in the input and provide complete form to our understanding of the social world. Thus schemata work in general as the guides as to what should be stored in memory. Sometimes information contradicts the schemata get encoded and the schemata guide what should be stored in memory.

Researchers like **Conway** and **Ross** (1984) have dealt with this phenomenon and demonstrated the schemata guide processing information in their experiment. According to them people have an event schema for taking part in a study skills improvement programme. That part of the schema states that a person's skills do improve. The schema, therefore, produce selective and distorted recall. The participants of the programme in **Conway** and **Ross** experiment have *rated their skills before* and after the programme. There were also *control subjects* drawn from waiting lists for the programme and rated themselves. The experimental and control subjects were found to be average on some skills. After the programme with the groups were rated on the same study skills. This was consistant with the general pattern and the study skill programme did not work as expected. But when the subjects were asked to recall their earlier ratings there were significant differences between the two groups.

It was surprising that the control group subjects recalled their earlier skill levels accurately, while the experimental groups recalling levels were very poor before the programme. The study skills programme did not increase the participants' ratings of their current skills. According to the schema they should improve their skills but they did not. But where they were persuaded their skills improved.The other findings of these researchers is that the schemata stated that their grade would improve after their study skills programme and it has happened. The problem, one researcher observed, was that in reality it did not occur. This is how the schemata guide information processing through attention, encoding and retrival.

Awareness of the self : Awareness of the self is the process of looking inward. We have been focusing our attention on the external world. But we are also interested in focusing attention into the self. Psychologists are also interested in knowing the causes and the consequences of self-focused attention, that is, the process of self-awareness. **Duval** and **Wicklund** (1972) studied this aspect of social cognition. According to them, the schemata process of attention can be focused outward or inward and sometimes human beings are struggling for focusing their attention outward and inward.

But what is self-awareness? Self-awareness is one's mental state in which a person's attention is focused on himself. That is the person directs his whole attention on his self. In the preceding section we have discussed about the self and self-perception. Here we are interested in self-awareness. What is happening when one turns his attention within himself This is the subject matter of self-awareness. Self awareness involves comparing oneself to one's ideals and goals.

People feel pleasant on seeing pictures in television and big photographs and also on seeing their *whole image* in the mirror. All these increase one's self-awareness. Self-awareness involves checking our current self against our goals and then altering our behaviour to the level of achieving these goals. According to **Scheier** and **Carver** (1991) self-awareness can be distinguished between two distinct types self-awareness or self-consciousness. *The first* is public self-consciousness and *second* is the private self-consciousness. The first is ourselves as social object. That is how we are appears to other people. The private self-consciousness involves awareness of our feelings, attitudes and values. We have discussed that self-awareness, involving attention to the self.

Attention : **Hull** and **Lavy** (1979) state that attention is not the only cognitive process involved in self-awareness. According to them self awareness is *often involved with encoding* and not attention. Self-awareness is a matter of processing information about the self and not noticing the self. Self-awareness can be approached also through personality trait.

Applications of Self awareness : We could observe that some people are often becoming self-conscious, thinking always about themselves, feeling happy by looking their images in the mirrors, speaking only about themselves and everything related them, and often consider how others perceive them. But other people are not like that. While studying about self-awareness we also must know something about the applications of self-awareness. There are several ways, but we shall go into the details of three applications of self-awareness. These are *alcoholic use, paranoid thinking* and *performance under pressure*.

Alcholic use **:** Alcoholic use is a common behavioural aspect among some people. Why do they drink alcohol? According to alcohol users alcohol *reduces a person's self-awareness*. But alcohol users give several reasons for their alcohol use. Everyone may find some reason for his alcohol addict. If a person meets with some unhappy events in his life and could not withstand the loss or missery he resort to some way of escaping from it which results in taking up alcoholic drink. At this stage he does not bother about any thing in the world life. From the time of this incidence he makes it as a regular habit of taking various kinds of alcoholic substances and does not think about the self and its status and respect. The person ceases to think about how the troubles reflect on his self. It was **Hull** (1981) who studied the effects of alcohol in self-awareness in detail. People who are highly self-conscious drink more alcohol after meeting with failure in some efforts than after success.

Paranoid thinking and the self **:** It is a tendency to *exaggerate the self.* **Fenigstein** (1984) states that this tendency is the self-target phenomenon, a tendency to exaggerate the belief that external events are directed at the self. The self as the target phenomenon although normal, resembles the thought processes of the paranoids.

The paranoids think to an extreme degree that many external events are directed especially towards them. A person's self-awareness creates situation in which the individual self is made as the target of attack. Self-consciousness is described by some authors as the personality trait. If a person focuses his attention on himself or becomes aware of himself, he begins to think that all the external events are directed towards him.

Acting under Pressure **:** Here pressure refers to some internal force acting upon the individual persons. It may be also external pressure in some work situation. Whatever it is when the pressure is on, people try hard to perform well. But what must be the result? Do they come out successfully after completing the task or get hurt and do the work badly? **Baumeister** (1984, 1985) conducted research on this phenomenon. His results show that self-awareness helps to determine the outcome of the performance *under pressure.* According to **Baumeister** pressure is referring to the importance of performing the task well and hurting or choking, under pressure means performing the task badly when it is very important to perform the task well. He stated that *pressure increases self-awareness* which interfers with skills. **Baumeister** states that skills perform the task automatically without conscious attention and attending to them may damage the quality automatically.

A person for example, is highly skilled in typewriting. While performing the task, the typist normally does not pay attention to his finger movements while typing. If he pays such close attention he would be making more mistakes than what he usually does or the speed of his typing would be reduced. This is what would occur in any task under pressure.

The social self : The child differentiates himself from others by his self-image. It also differentiates the attitudes others have towards him. The attitudes of others toward him mould his own attitudes towards himself. This constitutes an important aspect of his self-evaluation. Self-esteem is mainly a person's attitude toward other people's opinions about him or the way one perceives himself in relation to other people's opinions. When his ideas about the relationship to other people change he also changes. The individual does not take into account the past experiences fully to determine his behaviour. The individual takes into consideration the contemporary events, restructure his perceptual processes of the individual in relation to other people which becomes an imporant factor in his self-perception. **Carl Rogers** has observed in his client centred theory that the individuals behaviour changes as the perception of self is altered. The same procedure can be adapted to the individual's social perception and the change of the social self.

Thus we may state that the *social-self* of the individual is *shaped by the group.* The individual's self-image, self-esteem, attitudes and his personality all undergo changes in relation to the social groups. Experimental studies on the self have demonstrated that an individual's self is an inextricable aspect of the selves of other people. A person's self is always identified with the self of other people. Thus one individual's self is determined in terms of the selves of the other people with whom he is interacting in his everyday life. If there is any basic root to the understanding of the psychological man that root lies in the group centred character of his activities.

The Role of Language

Language is an essential factor that promotes social behaviour among human beings. Man distinguishes himself from the lower animals by his linguistic behaviour. In applying language, the individual person arouses a response to some stimulus and also raises some stimuli. The linguistic stimuli affect the speaking individual as they affect those with whom he is speaking. The words an individual uses have a degree of commonness or universality. This means that they are essentially the same thing for other persons also. A person cannot speak anything absolutely different thing.

Therefore, an individual person apart from the group acquired symbolic self and other responses the *self cannot arise.* This is one of the clear evidences that the self is an integral component of the social groups and acts only in accordance with the social groups with which he is associated. Language is a system of socially accepted significant symbol.

Propaganda

What is Propaganda?

Human beings are generally very much interested in seeing something new and fascinating objects, ideas, performances and in all those things that are useful for their living in the ever changing world. People come across in their everyday life several scientific inventions and discoveries which are converted into very useful material is to make our living more easy and comfortable. Thus any thing that is new and more comfortable for living are made known to the people at once. Anything that is new, exciting, useful and easily reachable attract people and they come to know all the details. How they are made known? They are made known by means of the information propagated in an organized manner. Propaganda is a publising psychological technique that modern politicians, industrial organizations, businessmen, researchers and advertisers are using for their purposes.

Propaganda, in a sense, is a skill of influencing the behaviour of individuals and groups of people, and makes extensive use of the skills for controlling the thoughts and ideas of the people. Propaganda may also be stated as a contrived attempt to influence the attitudes, beliefs and the behaviour of people in the direction desired by the propagandist. According to **Lasswell** (1935) “Propaganda in the broadest sense is the technique of influencing human actions by the manipulation of representations”. Generally propaganda means an organized dissemination of information or allegations to promote or to drainage the cause of a movement or government. Propaganda is an organized scheme for the propagation of a principle or doctrine. Propagation means an extension of the field of operation of a thing or theme.

The modern business and industrial society faces keen competition in selling their goods in the market. Each industrial organization follows some technique of attracting people towards their products. Industries bring out the same commodity in different patterns to the market. This puts the customers in the market situation in a confused state. Every customer is made to think, “whether to buy this or that?” Goods must be sold in the market. Every organization brings its goods to the market with more attention getting attractive manner. Different technique are adapted to attract the customers. There is much publicity for a new product even before it is brought to the market. Advertisements in familiar mass media of communication often appear. Advertisements are the forms of propaganda. The propaganda forms of advertisement are getting neck breaking competition. For example, the Television programmes are followed by crowded advertisements in varied forms. Advertisements in mass media and Television are heavily associated with model girls and sex. There is no exception for any material product in propagating their goods associating attractive ladies.

Political organizations, religious propagandists, educational organizations, medical centres all are now resorted to propagating not only their ideas, but also are very much strongly involved in attracting people and building up their organizations strongly. Thus propaganda has become a more powerful weapon in the modern world and man can never escape from the grip of propaganda and he is influenced very much by the ideas and principles propagated by the various organizations. In order to convince people and control their behaviour propagandists who are the agents and representative of various organizations tempt people by promising things. Thus propaganda is a technique of influencing people and controlling their behaviour and directing them in their desired direction and making them to accept the ideals, principles, objectives and goals, mainly by changing the attitudes and beliefs of the people.

Propaganda, according to **Kimball Young** (1966) is of two kinds such as *general propaganda* and *special propaganda.* The *general propaganda* may mean any form of proselytising, publicity, advertisement, education designed to change attitudes and opinions through the technique of *suggestion.* In this type of propaganda suggestion is an important factor.

The *special propaganda* is an effort to create deliberately popular opinions and attitudes and thereby to control the popular conduct and behaviour. The implication involved in this type of propaganda is that the aims of the propagandist are concealed. In reality, it is the propagation of ideas, opinions, attitudes and beliefs, which are not exposed plainly to the people who listen to the verbal expression or read the matter in printed form.

Medium of Propaganda

Propaganda is made only through some medium as it needs to be made known to the people. *Language* is the most important medium through which propaganda operates. Language is used to propagate many things in this modern world. It is used both in verbal form by speaking and in written forms. The human beings are made known the specific information when it is presented through language they use for speaking and for writing. Language is the most powerful weapon that human beings are using to express their feelings and thoughts. Propaganda has been psychologically interpreted by many psychologists. According to **Biddle** (1931), language is man's the greatest achievement, his crowning glory and the sign that first and foremost one which distinguishes him from the great ape. He further states that philosophers, psychologists, anthropologists, philologists and the educators have regarded language that which makes possible the formation of societies among men, the creation and transmission of cultures and the sharing of experiences. Man's supreme, intellectual achievement is revealed in the invention of language and made it a form of communication of his feelings and thoughts through speaking and writing.

Man has learned to manipulate language for his purpose in several ways. One of the important areas in which man applied his language is *propaganda.* Propaganda has become a *unique force* through the use of language. Propaganda as a unique force with which man can tap the powers of force that control the beliefs and attitudes of the people much better than other psychological processes. Some authors equate language with propaganda in controlling beliefs and attitudes. But some do not agree with this view. 'Language is operating as a social factor in individual thinking', thus observes **Krout** (1942). According to him language is the primordial stuff of mental behaviour. He states that man can no more get away from language, once trained in its use than he can dismiss his memories. Thus language plays its greatest role in social control and in persuation and it is the greatest interest for the social psychologists, so observed by **Krech** and **Crutchfield** (1948).

In using language in propaganda the patterns of speech of the speaker are considered to be important. The words, the phrases the sentences and the examples he quotes and the illustrations are having the influencing effect on people who hear the speech. The style of the speech should have first the influencing effect upon the audience. The people who hear the speech should reconsider the idea that they have already fixed in their memories and must develop some favourable view on the matter the speaker has said. Any propaganda speech in an oratorial style may attract the attention of the people who are listening to the speech. The personal appearance and his other fectures of the speaker may have some effect upon the audience favourably or unfavourably. Language is the chief medium of propaganda. Therefore, the pattern of speech and the nature of language should be given important place in propaganda.

Experiments have been conducted on these aspects of propaganda. The results of these studies indicate that the listeners judge the personality traits of the speakers in terms of their individual speech patterns. These patterns were found to be having influence in persuasion and the generalization is suggested that the efficiency of speech in influencing the beliefs and attitudes is to some extent dependent on the type speech patterns employed by the speakers and that is generally accepted by the audience or listeners.

Techniques of Propaganda

Propagandists and advertisers use several techniques in propagating their views or ideas and attitudes. They are designing techniques of their own according to their subject matter to be propagated. We can discuss here a few of the techniques used by the propagandists. These techniques vary according to the purpose for which propaganda is made. For example, mainly the leaders of political parties design technique to catch the attention of the people to get them in their parties. Religious leaders design techniques welcoming the people of different sects into their fold. There are also efforts made to convert people into different religions. Business and industrial organizations design techniques to suit to their products. These are the main areas in which propaganda operations are going on everywhere.

Application of quality : Any propaganda technique must make it a point of emphasizing the *qualities, characteristics* and *services* of their subject matter. They must give wide publicity about which propaganda is being made. It must always take into account only desirable contents and as for as possible undesirable elements in the propaganda material should be omitted. The attractive aspects of the matter, the genuine characteristics and the type services rendered by the organizations must be emphasized in the propaganda.

Attractive Display : The propaganda should be associated with attractive organizational functions like meetings, processions, exhibitions, cultural programmes, music, dance, competitions and so on. In all the techniques applied, the public should be attracted by some means towards the propaganda objectives.

Making concrete ideologies : The valuable ideas associated with the object of propaganda must be constantly presented to the public by some attention getting channels, like banners, posters and display of some achievements made earlier (Fig. 29.6).

Fig. 29.6 : Propaganda

A Technique of Publicity : Political : Propaganda generally means widespread propagation of a particular item which may be some political idealogy or religious principle or some commercial products. The ancient Greeks and Romans made wide publicity for their political ideas and spoke about their leaders valour and achievements through specially engaged orators. Propaganda techniques carry ideas and information to the people. Modern political leaders make use of them to propagate their ideas in various forms like public meetings, demonstrations, conferences, agitations, posters and banners to attract the people.

Repetition : Any propaganda about an ideology or principle or material may not be very much successful at the first time. The propagandists must be going on repeatedly carry out their efforts in presenting their propaganda materials to the public. Repetition of the same thing several times would certainly catch the attention of the public.

Prophylaxis of Propaganda

People are persuaded through propaganda towards some ideas, beliefs and attitudes. Various techniques are adapted to attract people to the proposed scheme. Propaganda and advertisements always do not provide us with correct information. Sometimes they mislead the people by providing exaggrated information and put us in a confused state. Thus there are good and bad propaganda and we have to protect ourselves from bad propaganda and get benefited by good propaganda. People must carefully analyse the nature and characteristics of propaganda and accept those that are having educative values and reject those that are of cheating type. Therefore, it becomes necessary that distorted, unjust, harmful and misguiding propaganda must be eliminated and controlled from our attention. A healthy and true public opinion is one of the ways of getting this result. The law of the state also may come for rescue from the harmful propaganda.

As propaganda techniques are of varied types, harmful propaganda may take disguished shapes like whispers and rumours. The real purposes of such propaganda are covered by sweet words and they easily attract the people's attention. Therefore, people must be made aware of such sugar coated bitter information. False ideas in the form of pity and sentiments, sometimes may prevent people from seaking protection through law.

Every propaganda made by various organizations, whether it is for propagating some ideas and principles for political, religious and commercial, goes through some advertisement agencies. The advertising agencies for all these organizations associate ideas and principles with attractively printed posters. The posters are designed in such a way that they produce demoralizing effect among the people. Particularly, the posters of motion picture advertisements on the walls in every street and in the daily, weekly and monthly mass media of communication and in the advertisements in the Television screen are good examples of corrupting the minds of young men in the country. The model girls in nude are used in all types of propaganda materials advertised. People must be protected from the effects of these types of displayed propaganda materials. These types of propaganda are slowly destroying the morale of the society (Fig. 29.7).

Fig. 29.7 : Propaganda

Commercial : Techniques of Advertisements : Commercial and Industrial Organizations also make use of propaganda to introduce their products to the people by means of advertisements with the view of promoting their marketing. These organizations present their products in attention getting attractive manner by associating attractive modelling girls with their products. Mostly the products like textile goods, garments, cosmatics, two wheelers, motor cars, computers and so on are all associated with attractive modelling ladies and men. Present day newspapers, television and magazines display large number of advertisements everyday to attract the people.

There are certain propaganda at times of crisis create panic among the people of the country. During the Second World War at certain stage the whole world was put into constant fear and anxiety. *Goebbels* the German propaganda minister in Hitler's government had invented several techniques of propaganda in favour of Germany and against Britain, America and Russia. People in countries like India have believed all such information. In contrast the Allied nations (Britain, America and Russia) have made counter propaganda against Germany and to protect the people from the false propaganda made by the Germans. This has become very essential at that time to save people of the affected nations from fear and anxiety. Some of the pieces of German propaganda literature indicated how each piece of literature had used certain propaganda devices. The material used for propaganda was issued by the German library of information and consisted of beautifully printed picture magazines, which suggested that the *Nazi leaders are friendly, benevolent* and often expressed common human sympathies. Pictures are shown of Nazi leaders kissing babies and growing flowers in their gardens at home. Nazi leaders are strong and wise and Nazis have more humanitarian plans for the European communities. This type of propaganda is intended to promote friendly attitudes towards the Nazi leaders and Germany. This is one way of protecting themselves from others aggressive attitudes.

Leaders and Leadership

People, we could see, all over the world live only in groups. A group may be smaller or it may be a bigger having large number persons in the group. Whether the group is smaller or bigger it is guided by some from the group who claims himself as the chief or head man or leader. People live in groups and they always striving hard to satisfy their needs and demands. This depends upon a number of factors. The efficient functions of the group depends upon largely their ability to find out a workable of structure, a structure that makes it clear who shall be the guide and direct the people and who shall be the subordinates? The group's structural arrangements function better if the key positions are occupied by persons who are suited for them by the virtue of certain personal characteristics.

Several investigators studied about leaders and leadership qualities. **Marvin E. Shaw** (1971) has made more extensive studies and defined many aspects of leadership and group structure. Shaw has defined the leader as one who is a member of the group and exerts more positive influences over others than they exert over him. **Katz** and **Kahn** (1966) state that the term leadership has *three major meanings,* such as an attribute of position, a characteristic of a person and a category of behaviour. These three meanings are employed at the same time, that is, they are used to refer to a person who possesses certain *qualities,* who occupies a certain *position* and who *behaves in certain ways.*

Social psychology is dynamic in nature. As result of it time to time the concepts, definitions, functions of some social psychological processes also are going on changing, modifying and redefining. Certain social psychological phenomena described and interpreted in terms of culture and group. Therefore, some are culture based and some are group based. It may be primitive culture and primitive group. Some may exist in developed culture and some in developing culture and some may be in underdeveloped. In all these groups the development of leadership may not be the same. The evaluation of leadership and functions depend upon the nature of the group. The qualities and characteristics of leadership in one culture cannot be attributed to another culture or group. The system of leadership function vary. But however social psychologists try to evolve the leadership qualities as far as possible applicable to all cultures through some common characteristics required for leaderships. We shall discuss certain general characteristics required for leadership.

Everyone of us right from childhood period to the old age period pass through various stages. At one time of our life we become a member of some group. At each stage we call one member who is more influential than others. In all likelihood every group contains one person who holds some power and better quality than the other members. Such individuals are generally labelled as *leaders.* The process through which they exert influence over others and guide the group activities is known as *leadership.* Leaders often exert powerful effects upon other members of the group who are the followers. Leaders often exert their influence upon their followers whenever there are some urgent needs. In emergency cases they may induce subordinates to accept the views that make some sense to others. Generally, a person either by his extraordinary qualities or by the choice of other members of a group might become a leader. A group as a whole may face some crisis, because of this the existing leader has to relinquish his position temporarily for few days. At this stage the leader can nominate some one in the group who is supposed to possess the same ability to maintain the group. By this means another member of the group by virtue of his characteristics becomes a subordinate leader. This practice is widely prevalent in most of the groups. While speaking about organized groups we must refer to leaders. In fact group morale depends to a large extent on leaders. When an organized group faced some crisis the members of the group should look up to a leader for proper guidance and solve the problem. What is happening in group may also happen in a nation. When the Indian freedom struggle movement started, the idea that, India should get independence was not raised in an organized manner. Agitations and struggles were going on without proper organized efforts. There was no proper communication channel. The freedom struggle movement began to function in an organized manner when Mahatma Gandhi took the leadership of the movement. This is a good example of leadership function and the effect of organized group struggle had led to the achievement of the goal.

There are different views about the emergence of leadership. Some are of the view that leaders are born. The common sense world sticks to this view. The other view is that leaders are made by the people. But the life histories of great leaders tell us that many of the great leaders become great leaders by hardwork. The other view also is possible that leaders emerge due to situational circumstances. Whatever reasons are attributed to the emergence of the leaders, the persons who are called as great leaders possess certain qualities and characteristics which are not found generally in all human beings. A psychological observation on the performances of leaders suggest that all the leaders possess a general capacity called *social intelligence.* This social intelligence is consisting of social perception, social cognition and social motives with wider perspectives. Great political leaders and statesmen are endowed with social intelligence but not all politicians.

Some of the investigators who were interested in the study of leaderships formulated a theory called the *great person theory.* These investigators are of the view that great leaders possess *key traits* that distinguish them from most of other persons. These traits according to them remain stable/across time and across groups, so that all great leaders share the same characteristics, irrespective of *where and when* they lived. There are some crucial ideas revealed by some, seem to agree with the present situation, **Geier** (1969) states that they have not been agreed upon the list of the *key traits* shared by all leaders. Decades of active researches. Geier states have failed to yield the expected result. A few findings emerged in them, that is, leaders tend to be slightly *taller and brighter* than their followers. But these traits cannot be generalized.

Research results suggest that persons possessing certain patterns of motives like *high needs for power* plus *high degree of self-control* are more successful as great leaders in various types of organizations like political parties, large business and industrial organizations, international institutions and so on, than persons not possessing these patterns, **McClelland** and **Boyatzis** (1962) have observed these results in their studies. **Costantini** and **Craik** (1980) found in their studies that political leaders do appear different from non-leaders in several respects. They are found to be higher in *self-cofidence and dominence.* Further, individuals possessing certain traits, especially the ability to adapt to changing situation also raise to the level of leadership in a wider range of different social situations. In several recent studies made by **Fleisaher** and **Chartkoff** (1984), it was observed that individuals of high dominance were chosen to be the leaders of small groups by other members much more frequently than persons of low dominance. These findings suggest that certain traits play important role in determining the leadership.

The other view about leadership is that it is not mainly a function of the traits that the leaders possess. But what involves is the *situational approach.* What is more important in this theory is that it replaces the age old theory of the *great person theory.* The *situational theory* recognizes the fact that different situations often call for different types of leaders. This theory is applicable to several small social group situations mostly apart from large organizational and political situations. There may be conflicting and uncompromising large national problem situations. In such cases a tactfully negotiating persuasive, more insightful leader who can delicately deal with the situation successfully at national and international levels emerges as a leader. Such leaders could attempt to conclude a delicate as well as complicated international agreement without sacrificing their own national interests. These situations need a leader who is calm, persuasive and charming. Thus according to the situational approach theory **Bass** is of the view that a person who becomes a leader should be with full understanding of the whole situation and also must possess all the traits of a potential leader involving careful attention to the situational constraints. He must also make use of the available resources in the situations.

One of the currently accepted views by most of the social psychologists is the *transectional approach.* This approach was evolved because the situational approach was found to be not giving importance to attention *to the role* and the *impact of the followers.* Because of this weakness it has been replaced by a more sophisticated theory called transectional approach. According to this theory while the leaders exert influence upon their followers, the followers in turn frequently exert some *reciprocal influence* upon their leaders. **Hoilander** (1978) while stating this, also observes that leaders are often strongly affected by their followers attitudes, preferences, perceptions and values. This is because, the leaders pay little attention to the wishes and needs of the persons they lead. Leaders must pay careful attention to the views and preference of their followers. Thus, they often seek feedback from these persons.

We have discussed the role of traits, situations and followers and how leaders emerge from the point of view of recent development of theories of leaders and leadership. The earlier authors and investigators have analysed the characteristics of leaders. Some of those characteristics we can cite here for information concerning leaders. The *personal prestige* is one of the characteristics of a leader. It is the respect that a leader commands from his followers. The achievements a leader makes in a functioning group successfully and encouraging his followers under him for the cooperation they offered in all his attempts increase his prestige. The good qualities the leader exposes in situations may add his personal prestiges.

Some of the personal characteristics are *patience, endurance* and *scholarship* and becoming himself an authority in some area of knowledge.

Understanding the followers fully and quick response to their needs and demands, and carefully listening to the problems of the followers are some of the characteristics of the leader.

Domination is an essential characteristics of a leader. He must establish his dominance over others in such a way that they must accept his dominance.

Drive and ***Courage*** are the characteristics of a leader in the sense that what the leader promises or speaks out must be executed without any delay and face the group problems couragiously.

Types of Leadership

Generally a leader is one who leads a group of people for a common cause. The common cause of a group may be anything. It may be a political one, social, group oriented, needs and demands for the well being of the group, health, education, shelter, guidance and so on. Therefore, leaders may emerge in all these areas and as result of it leaders fall into certain types. We shall discuss here a few of the types which are essentially considered.

When we speak about leaders it at once strikes out in our mind the political leaders. Generally leaders and leadership always go with political leaders, though there are also leaders in other areas. The political leaders fall under the types of *democratic leadership,* and *authoritarian leadership.* There are also informal and temporary groups formed to meet certain immediate needs and demands which also have the leaders. Sometimes suddenly a number of people are thrown into close dynamic relationship over a period of time under a leader and go away as soon as the purpose is over. The leadership exists only for a short time in such groups. In a more formal and stable groups such as office, business and industrial organizations leadership functions may be operating. There is also imposed various freely choosen leadership which is intimately related to the contest between authoritarian and democratic leadership. We shall deal with some of them here.

Authoritarian Leadership

The authoritarian leader holds more absolute power (Fig. 29.8). The leader alone determines the policies of the group and he alone takes the major plans for the group or for a nation. He

Adolf Hitler (1889-1945) Beneto Mussolini (1883-1945) Joseph Stalin (1879-1953)

Fig. 29.8 : Contemporary Authoritarian Leaders (Dictators)

Adolf Hitler was born in 1889 in Austria, but he believed in the German's excellence and settled in Germany. He was a soldier in the First World War which was faught between 1914 and 1918 and Germany was defeated. He was outraged by it and became a staunch member of the National Socialist German Workers' Party and became the president of it. He attempted to overthow the Government of Bolivia and was imprisoned. In prison he wrote the *Mein Kampf* expressing his philosophy. Then he was the Chancellor of Germany in 1933. He was a very powerful public speaker, was able to get the support of millions of Germans. He transformed Germany from Democratic Republic into a Totalitarian State and became the Dictator in 1934. He annexed Netherlands and Austria in 1936 and Czechoslovakia 1938 in and invaded Poland in 1939 which precipitated the Second World War. He eliminated his rivals and political opponents and put to death millions of Jews and Jipsies. In 1945 Russian soilders captured Berlin City and Hitler and his wife committed suicide and died.

Beneto Mussolini was Italian, Socialist Dictator, dismantled the Italian Democracy and used the *Fascist Black Shirt militia* to intimidate other politicians. Mussolini was a Journalist before he founded the National Socialist Party. He was appointed Prime Minister by King Victor Emmanuel III in 1922 and assumed dictatorial power. He annexed *Abyssinia* and allied Italy with Germany in 1938 and entered in First World War in 1940 and captured *Sicilly* in 1942. In the War, Italy suffered disaster defeat and Mussolini was deposed in 1943 and executed in 1945.

Joseph Stalin was the Soviet Leader, General Secretary of Communist Party of Soviet Union from 1922 to 1953. He succeeded Lenin as head of the party and created a Socialist State by crushing out all opponents. He was born in Georgia, a man of steel, became the Secretary-General of the Communist Party's Central Committee in USSR in 1922. He became the head of the State in 1927 presiding over a region of terror in the 1930s, that intended trails, purges, prison camps and execution of some ten million people. From 1941 he let the Soviet Union's titanic struggle against Nazi Germany and dominated the Allied settlement and imposing Soviet rule over Eastern Europe. He played crucial role in Second World War in defeating Germany in 1945. Stalin built the Soviet Empire in his time and made Russia as the most powerful nation in the World. But those who came after him dismantled the Empire.

has the power to decide the policies of the nations at the international levels. For the success and the failure of the functions of the group or nation, he alone is fully responsible. He alone dictates the activities of the members and patterns of interrelations among the members. He alone is responsible in making judgements of rewards and punishments for the individual members of the group or nation and therefore the fate of each individual member is within the group structure. The authoritarian leader deliberately introduces all the functions of the group by himself and strongly resists any modifications in them.

The authoritarian leader protects his leadership and status by following various techniques. The authoritarian leader claims that his direction and guidance will be indispensable for the stable motivation of the group as a whole or for the nation. In this type of control the motivation of the individual members is segmental and dependent. The authoritarian leader promotes a segmented group structure in which the intercommunication among the divisions is held to a minimum level and wherever possible the avenues of communication are only through him or under his immediate control. The socio-economic structure of the authoritarian group is more likely to be of *star shaped* with the leader at the centre of the radial strands. There is always tension in the group because of the structure.

Winston Churchil (1874-1965) Charles de Gaulle (1882-1945) Frankiin Delano Roosevelt (1882-1945)

Fig. 29.9 : Contemporary Democratic Leaders who played crucial roles during the Second World War, faught between 1939 and 1945.

Winston Churchill was a British Conservative Statesman, an orator and writer, noted for his leadership during Second World War. He was the Prime Minister of Great Britain in the coalition Government in 1940 at the age of 64. He was the opponent of the British Parliament for a long time. He inspired determined resistance to Hitler through stirring speeches, political resolve, a solid group of strategy and skilled diplomacy. He guided Britain to victory in the Second World War. But he was defeated in 1945 election. He returned to power again in 1951 election at the age of 73.

Charles de Gaulle was an Army General served in First World War and retired with very brilliant records. He refused to accept *Petain's the Armistice* with Germany and fled to England and founded a French Government in exile in 1940. Petain was French Marshal, noted for the victory in First World War and he was running the French Government for the unoccupied part of France and he was a pro-Nazi, from 1940 to 1944. He was running a provisional Government and he became the first President of the 5th Republic of France. After the War de Gaulle was hailed as the national hero and in 1958 he was made the President of France. During his years in office de Gaulle turned France into a neculear power and rebuilt the relations with Germany. Petain was imprisoned in 1945 on a charge of treason. President de Gaulle resigned his presidentship in 1969.

Franklin Delano Roosevelt (1882-1945) was the President of United States continuously for three periods. He was the President for the third time when the World War was broke out and the USA entered the War. He was a forceful administrator and brought out major reforms in United States. He was the 32nd president and was in office from 1933 to 1945. He introduced the New Deal to encounter the economic crisis of 1930s. He prepared the country for an anti-Japanese war from 1939 to 1945 and led the national war to victory and finished it prior to the German war by dropping atom bomb in Nagasaki and Hiroshima two Japanese cities.

Democratic Leadership

The democratic leader is having the same amount of power as the authoritarian leader (Fig. 29.9). But he enjoys the power with the consent of the group members. At certain level the democratic leader exercises more severe rule than the authoritarian leader. For example, the American President is a democratic leader. The Soviet Union was an authoritarian nation and it had a President. At international level the American President exercises more powers than the Russian President. Realizing this power status the Russian President Gorbachev dissolved the communist regime and introduced democratic pattern of Government in Russia. At times of crisis the democratic leader can exercise more severe power with the consent of the parliment and can control any situation in the country.

One democrative leader seeks to evoke the maximum involvements and the participation of every member of the group in activities and in the determination of the objectives and goals. The democratic leader seeks to distribute or spread out responsibilities and duties to all the members of the group and tries to get the hierarchical group structure and also promotes interrelationship among the members of the groups. All other types of leaderships can be brought under these two types.

30

ATTITUDES AND RELATED TRAITS

Development of the Concept, Attitude

In behavioural sciences, particularly in social psychology we often come across the concept 'attitude' which often covers a very wide area of human experiences and has become a major field of scientific study for the social scientists.

This concept has been employed to explain the manner of thought of a man towards some place, person or thing. A person's stand on any of the political or social issues is often referred to as his attitude. Writers invariably employ this word to explain an individual's or a group's approach to or avoidance to certain problems.

The origin of this concept has been attributed to Muller and Schumann. The nineteenth century philosophers like Alexander Bain, Herbert Spencer and Wilhelm Wundt have used this concept to refer to the mental processes. Bain (1968) has pointed out that "forces of the mind may have got into a set track or attitude".

When reviewing the psychological researches of the past two decades in U.S.A. no one can ignore the important position the study of attitude has been occupying. Its measurement has become a major psychological industry of the United States.

The rise of the Behaviouristic movement under the leadership of J.B. Watson gave an impetus to explore psychological concepts with sceintific accuracy. Every interpretation on the basis of mental process and consciousness has been examined and its validity tested. New techniques and methods have been introduced and the quantitative aspect has been strongly emphasised. Very few concepts remained unaffected by these revolutionary trends.

Thomas and Znaniecki (1918) were the first two sociologists who made use of this term very comprehensively. Their study of Polish Peasant in Europe and America after the First World War attracted widespread attention. Around 1920s and 1930s it became fashionable for the social scientists to make attitude studies. Bogardus (1928) attempted to study social distance. Following his study, Thurstone and Chave (1929) published monographs on attitude measurement. Allport, Newcomb, Likert and others have made extensive studies in this direction. A large number of articles, reports, pamphlets and books have been published on this subject. Information on attitudes is freely available in United States.

Allports remark : "The concept of attitude is probably the most distinctive and indispensable concept in contemporary American Social Psychology. No other term appears more frequently in experimental and theoretical literature. Its popularity is not difficult to explain. It has come into favour first of all because it is not the property of any one psychological school of thought and therefore, serves admirably the purposes of clectic writers ... concept has been so widely adopted that it has virtually established itself as the Keystone in the edifice of American Social Psychology."

Definition

Man's social behaviour can be explained in terms of motivational, emotional, perceptual and learning processes. It is impossible to explain man's social behaviour if only one of these four processes is taken and examined. Man's social behaviour is evolved from the dynamic nature of these four processes functioning together. Human behaviour is neither momentary nor immediate. It is a long term arrangement. Therefore, it is always better to see man's social behaviour from higher level. This higher level of arrangements of motivational, emotional, perceptual factors is called beliefs and attitudes. An attitude can be defined "as an enduring organisation of motivational, emotional, perceptual and cognitive process with respect to some aspect of the individual's world". An attitude does not always demand emotional excitement on perceiving the object. It mainly depends upon circumstances and situations.

Attitudes themselves are not actions though often their possession leads to action. They are not directly perceived. We perceive only actions or overt behaviour. The overt behaviour of an individual is the manifestation of his attitude. Therefore, Hartely and Hartely (1958) define attitudes, "as the inferred bases for observed consistencies in the behaviour of individual". It may be expressed either in overt or verbal behaviour. A man may reveal his attitude verbally on his own accord or in response to a question or on the arousal in him of a feeling of irritation or excitement.

Allport (1935) has defined attitude as "a mental and neural state of readiness, organized through experience, exerting a direct or dynamic influence upon the individual's response to all objects and situations with which it is related".

Ruch (1958) says, "an attitude is an emotionalized system of ideas which predisposes us to act in a certain way under certain conditions". According to F.H. Allport (1924), "an attitude or preparation in advance of the actual response, constitutes an important determinant of the ensuing social behaviour". It is also considered as a tendency to act towards or against something in the environment which becomes thereby a positive or negative value.

Doob (1940) opines that "an attitude refers to an implicit response that is both anticipatory and mediating in reference to pattern of overt responses, that is evoked by a variety of stimulus patterns and that is considered as socially significant in the individual's society. Guilford (1948) puts that "an attitude is a personal disposition common to individuals, but possessed to different degrees, which impels them to react to objects, situations or propositions in ways that can be called favourable or unfavourable". Macrone (1937) states that it is the subjective aspect of the manners and morals, the customs and traditions, the mental and material culture of the group to which the individual belongs. Droba (1934) declares that an attitude is a mental disposition of the human individual to act for or against a definite object. Broadly speaking, an attitude denotes an adjustment of an individual towards a selected aspect of his environment or of his own conduct.

Newcomb (1952) makes, exhaustive study of motive and attitude. To him an attitude is the state of readiness for motive arousal. To Jung (1924) an attitude is "a readiness of the psyche or to act or to react in a certain direction".

A careful examination of these definitions suggests one central point, that is a sense of readiness of a person towards something outside of him. Therefore, an attitude can be defined as a positive or negative responsive tendency towards a person, object or situation. The important factor implied in an attitude towards on object or person is response or reaction to it or him. It need not necessarily be a favourable response. However, it involves some action either overtly or inherently.

Personal and Social Attitudes

The individual's adjustment to the problems of day-to-day life involves many issues. He develops a large number of attitudes towards object, person, or situation. Certain attitudes are found to be common to a group of people. Every individual may find an interest in it. These attitudes are called social or public attitudes. There are also attitudes in a group of people may not find a common interest. They centre round an individual's personal and private affairs. These attitudes are called personal or private attitudes.

According to Allport an individual possesses two sets of attitudes. One set is kept confidentially as private and personal and the other which is socially determined is publicly acknowledged. There may be some conflict between these two sets. A magistrate finds his two attitudes in conflict when sentencing a prisoner "as a man, I pity you, but as an official I must show you no mercy" (James, 1890).

Every individual has attitudes towards his parents, children profession, dog, etc. These are private and are not shared by anybody. These personal or private attitudes are very important with regard to the individual's adjustments. Public attitudes concern "objects, persons, and issues known to the members of a group and to the society in which they live" (Morgon, 1961).

Schank has studied the public and private attitudes of 51 Methodists and 46 Baptists, He came to three important conclusions. (i) People's public attitudes are quite opposite to their private feelings. (ii) The public attitudes are found to be the same for all members whereas the private attitudes are not. (iii) Lastly the private attitudes of the members of one group are found often similar to other people who are not members of their group.

Attitudes and Other States of Readiness

(a) Attitudes and Interests : Interests are distinctive states of readiness of the individual towards an object. Attitudes, on the other hand, are states of readiness projected towards some objects, persons, or situations. Attitudes are subjective whereas interests are objective. Terms such as love, hate, sympathy and pride can be grouped under attitudes and terms such as friend, enemy, family, discovery and God can be grouped under interest.

All social adjustments made by an individual involve both attitudes and interests. A burglar, a citizen, a Policeman and a judge are all interested in law. But their attitudes to it are quite different. The burglar wants to violate the law and cheat the citizen, police and the judge. A citizen wants to safeguard his property by the help of law and the policeman and judge try to maintain the law. Similarly, a person may not be interested in witnessing a cricket match but he may have a strong attitude towards it since it maintains his national prestige. A conservative man does not agree with principles of communism, but at the same time he is interested to know how far the communist party functions and progresses. Interest is associated with one's likes and desires and attitudes with one's approval and evaluations.

(b) ***Attitudes and Traits :*** Allport (1934) differentiates a trait from an attitude thus— Firstly, a trait has no definite reference to an object whereas an attitude has a definite object of reference which is either material or conceptual. A person's views towards liquor, marriage, birth control, divorce. Fascism or Communism are attitudes, but a person's conservatism, radicalism, introvertism, extrovertism and asceticism are traits. Secondly, a trait may be narrowly limited to a state of readiness for response whereas an attitude may be specific as well as general to the world at large. Thirdly, an attitude indicates the acceptance or rejection of an object or the concept with which it is related. But a trait has no such definite indications. A man may love his dog; he has a loving attitude towards the dog, but if the person possesses a sort of sympathetic outlook towards animals and human beings he has the trait of kindliness or love.

(c) ***Attitudes and Motives :*** Attitude is a state of readiness for motivated behaviour of performance, perception, thought and effect. This definition by Newcomb (1952) suggests that there is some relationship between attitudes and motives. Motive is a term implying a need and direction of behaviour towards a goal. It is often used synonmous with need or drive. Any condition of the organism that affects its readiness to start upon or continue in a sequence of behaviour can be a motive (1957). Motivated behaviour is directed to one goal only. But an attitude may have many goals to be fulfilled.

(d) ***Attitudes and Beliefs :*** An attitude is acquired through the process of learning, but it does not involve any higher mental processes. A belief is a statement or proposition accepted on some evidence. It requires some amount of thinking. The qualities of beliefs and attitudes are different. Krech and Crutchfield (1948) distinguish these two concepts clearly. "A belief", according to them, "is an enduring organization of perceptions and cognitions about some aspect of the individual's world. A belief is a pattern of meaning of a thing; it is the totality of the individual's cognition about a thing". But an attitude is defined as "an enduring organization of motivational, emotional, perceptual, and cognitive processes with respect to some aspect of the individual's world.

(e) ***Attitudes and Opinions :*** In between attitude and belief falls a reaction called opinion. An opinion like a belief is an acceptance of some proposition or statement. Unlike a belief, a person who holds an opinion does not have a strong desire to prove it. It has neither a verified characteristic like knowledge nor unverifiable characteristic like a faith. Opinion is a go between knowledge and faith. On verification it becomes knowledge. Otherwise it is a faith. So also one cannot say that attitudes are always verified. When facts are sure and evidences are strong men hold knowledge and when facts and evidences are weak they hold opinions. But this distinction is not always correct and true.

Thurstone and Chave (1929) clarify the terms attitude and opinion. According to them an attitude is a complex affair and it is quite impossible to describe it wholly by a single numerical index. The concept attitude denotes the sum total of a man's inclinations, feelings, prejudices, preconceived notions, ideas, fears, threats and convictions about any specific topic. It is purely a subjective and personal affair. The same authors say that opinion is a verbal expression of attitude. If a person said that his people committed a mistake by waging war against Germany that statement is called his opinion. So the term 'opinion' is limited to verbal expression, which expresses a person's attitude. An attitude is symbolized in an opinion. Hartley and Hartley (1958) seem to endorse the view of Thurstone and Chave.

Formation and Development of Attitudes

To explore the factors in the formation and development of attitude is a major concern in the study of experimental psychology. Any attempt on the part of the psychologists to foretell or to regulate human behaviour involves the necessity of understanding the formation and development of attitudes. Political leaders, manufacturers, businessmen and advertisers wanted to implant a new attitude in the minds of people concerning their goods.

There is no human being who is born with attitudes, nor are they formed of their own accord. They are formed in relation to a person, place or thing. Attitudes are formed in an individual in his interrelation with human beings with some social institutions. The most important factor that operates in the function of attitudes is selectivity. If the individual is interested in certain issue he willingly tries to acquire some attitude which may be either favourable or unfavourable.

As Sherif and Cantril (1949) put it "the first stage in the actual formation of an attitude is a perceptual stage, with the internal factors of the organism and external (objective) factors of the stimulus situation coming into play". They say that there is always a subject-object relationship in attitudes and attitudes are precisely related to stimulus situations. The stimulus situations may be home, automobile, souvenir, eating place, persons such as one's own body, mother, father, brother, some friend, rival, teacher, sweet-heart, wife or groups of people such as classmates, playmates, the community, institutions such as school, college, church, club or socially accepted concepts, values or norms such as flag, constitution and democracy. This kind of subject-object relationships do not have any biological significance. This subject-object relationships are developed towards the items or objects from the stimulus side. Any kind of relationship between the individual and the object is possible if only there is contact with the outside stimulus situations. This description implies that attitudes are formed. They are formed by the contact the individual had with the objects, persons and values. Any palatable food may satisfy an individual's hunger. But every individual develops some special liking to some special kind of food and to a particular restaurant. When this kind of affective nature becomes permanent it forms an attitude towards the object (Fig. 30.1).

Attitudes are formed in relation to some objects, persons, places, or values and are not innate states of readiness. In this process of their formation the individual must first of all

Fig. 30.1 : Attitudes : Formation and Development.

An attitude is a person's stand in any one of the political, social, religious and behavioural issues often referred to as the person's attitude. This view is also applicable to a group. Attitude is a mental process functions towards other persons or groups positively and also negatively. The positive function of attitude in social situation is attraction and the negative function of it is prejudice. Generally attitude is a sense of readiness of a person towards something outside of him. The subject matter in this context is referring to the members of the social groups. In any culture there may be in-groups and also some out-groups. These groups may be of social, communal, political, religious, ethnic or racial groups. Among the members of these groups there may be both positive and negative attitudes may prevail. Human beings are not born with the mental set of attitudes. Attitudes are acquired and learned through experience and from others. Attitudes develop in a person in the process of socialization for liking a group and disliking another group. The figure indicates a group of people's attitudes about the problem that prevails there, which includes the feelings, ideas, thoughts and actions.

come into contact with the objects. Contact is possible only with perceptual situation. This is the primary stage for the formation of attitudes. According to Sherif and Cantril (1949), perceptual stage has psychological importance. It has no motivational basis in formation of attitudes.

Doob (1946) considers that an attitude cannot be inherited property of an individual. It is learned through experience. Children do not have any prejudice towards any race to which their parents are antagonistic.

Allport stipulates four conditions for the formation of attitudes. (1) Attitudes are formed by integration of numerous specific responses of a similar type. A single experience has not sufficient strength to create an attitude. (2) The integration of numerous specific responses in formation of attitudes is influenced by the mechanisms of individuation, differentiation or segregation. (3) A single strong emotional experience may be the root-cause of some permanent attitude in an individual. For example, a hostile attitude may be cultivated towards Negroes on finding a Negro committing a crime against a member of the Whites. But, a favourable attitude may be cultivated towards Negroes on finding a Negro saving a member of the White race. (4) An attitude may be formed by adopting ready-made views of parents, teachers or playmates. A child may have many strong attitudes towards race, religion, marriage, profession and sin which he may have mainly taken over from priest, friends or parents without any personal experience.

(a) Culture

The formation and development of an attitude is a continual process of experience. It starts at childhood. One of the main sources of the development of attitude is the home. Parents and teachers are the cultural forces that play important part continuously on the child's life. They are mediators between the child and his culture. All social contracts of the parents, and teachers exert direct influence upon him.

Every culture exerts a great amount of force upon the individual in various ways. But the individual's selective nature does not accept all the cultural patterns as they are. It limits the cultural effects to certain extent according to the needs of the individual. The individual limits the formation of attitude in number, kind and content. But it is possible to predict whether the cultural influence is actually taking place in the formation of attitudes only if the individual perceives, interprets and uses them.

In some cultures the formation of attitude is not left to his selective nature. The cultural situation has absolute control. The cultural situation as a powerful political force may set limits to the individual's attitude. For example, the limitation put upon free thinking and expression in a totalitarian regime may be compared with that of a free democratic regime. Sargent and Williamson (1958) quote what was actually seen when one of them made a trip to certain iron curtain countries in 1949. In Czechoslovakia, they observed some freedom of. speech and movement. Certain influencing factors such as higher standard of living, history of civil rights and closer access to Western norms were in operation. It is possible to measure the amount of influence a culture has exerted upon the individual by correlating the attitudes of the individuals from different cultural groups. When we compare two cultural groups if one is found to have more frequency than the other it is assumed that the difference in attitudes is due to the cultural factor. This has been illustrated by the Social Science Research Council in a research conducted on an attitude study towards armed imperialism.

The investigation was conducted among one thousand persons. They were asked the question, "some people say we should use our Army and Navy to make other countries do what we think they should. How do you feel about that?" Table 30.1 illustrates that there is higher percentage of general approval from lower educational and low income groups. People with college education and those of higher economic status disapproved of the view. There is significant difference in the views expressed by groups belonging to different cultural backgrounds. This study concludes that education and socio-economic status have determining effect on the people's attitudes towards Armed Forces.

Table 30.1 : Relationship between various cultural determinants and attitudes towards Armed Imperialism

Respondents	*General Approval per cent*	*General Disapproval per cent*	*No Response per cent*	*Number of cases*
Education				
Subject with :				
Grade-School Education	19	57	24	500
High-School Education	13	77	10	455
College Education	8	83	9	213
Income				
Subjects earning annually :				
Under S 2,00019	58	73	440	
S 2000 to S 3.99916	73	11	478	
S 4000 and more	8	86	6	216
Religion				
Protestant	14	70	16	855
Catholic	18	68	14	245

Research studies have made to find out the influence of cultural factors in the formation of attitudes. It has been found that educational status, income and religion, seem to furnish very interesting results. The Catholics were found to be more conservative than the Protestants on the issues like birth control, war and political issues.

LaPiere and Farnsworth (1949) say that most of the verbal attitudes of an individual are cultural in origin. They have been derived directly from the group or groups by which the individual has been socialized. Macrone (1937) seems to endorse this view. He states that an attitude is the subjective aspect of the manners and morals, the customs and traditions, the mental and material cultures of the group to which the individual belongs.

(b) Family

Another important factor that influences the individual in the formation of attitude is the family. The child in the family has to grow under the care of his parents and other elders. He has to respond and react to the wishes and directions of his elders. He cannot feel or act against the likes and dislikes of his elders. Parents and teachers play the mediating role between the culture and children. Morgan (1961) considers that the parents and teachers are more subjected to the cultural influences that the children. Parents and teachers are having more social contact than the children. Their profession, entertainments, their study, social visiting and other cultural contacts are more frequent and more wider than the children. The parents are having more direct contact with various cultural factors. Therefore, they are more susceptible to the cultural influences and they in turn influence their children with their acquired cultural influences.

Hischberg and Gilliland (1942) attempted to discover the parent-child relationships in attitude formation at North Western University. The subjects were 200 college students. Their attitudes towards the New Deal administration, economic depression and towards God were correlated with that of their parents. The correlation figures ranged from .29 to .58.

Newcomb and Svehla (1937) have studied the influence of home in the formation and development of attitudes. The attitudes of parents and children on Religion, War and Communism show a remarkable correlation in intra-family relationship.

Table 30.2 : Parent-child relationship

	Church	*War*	*Communism*
Mothers - sons	.57	.45	.58
Mothers - daughters	.69	.43	.49
Fathers - sons	.64	.45	.54
Father - daughters	.64	.43	.62
All siblings	.59	.39	.47
Mothers - fathers	.75	.43	.57

Parents influence in the, formation of attitudes on the part of children is more powerful with regard to racial prejudices. Horwitz (1936) has studied the development of attitudes towards Negro. The subjects were White Grammar School children from first to third grades and they were interviewed and asked to reveal their opinions towards Negroes. The following statements of the children interviewed clearly bring out the fact that the family is one of the important factors in the formation of many prejudicial attitudes.

First-grade girl : "Mamma tells me not to play with black children. Keep away from them. Mamma tells me, she told me not to play with them ...."

Second-grade girl : "Coloured children. Mother does not want me to play with coloured children.................. I play with coloured children sometimes and Mamma whips me".

Second-grade boy : "Coloured children, mother and daddy tell me. They tell me not to play with coloured people or coloured person's things".

Third-grade girl : "Mother told me not to play with them because sometimes they have diseases and germs and you get it from them".

According to Lasswell (1938) the interactions of the family are the most powerful influencing force in the formation of attitudes. The children who are politically radical express hostile attitudes towards family authority—particularly that of the father.

(c) Personality

An individual's personality traits are said to have considerable influence in the formation of attitudes. Persons of different personality traits may be correlated with social attitudes. Dexter has attempted to correlate the personality trait with conservatism and radicalism. He discovered that the women students who are radical in their beliefs and attitudes were more introvert, more self-sufficient and more dominant than women students who were conservative. But this kind of correlation between personality traits and attitudes cannot be generalized always to all cultures. What is considered to be radical in one culture may be looked upon as conservative in another culture. Therefore, a positive correlation between personality trait and radicalism in one culture may turn to be negative correlation or may correlate with conservatism.

(d) Learning

Every human being learns many things in his everyday life. He learns both in school and outside the school. This learning does not leave us without any effect in our behaviour. It determines our likes, and dislikes, favourable and unfavourable responses to persons, situations, institutions and values. According to Ruch (1955) attitudes are also learned behaviour of the individual. We learn our attitudes towards persons, institutions, situations and social issues. The existence of attitude towards something is quite impossible without having some knowledge about that thing.

Imitation and identification are the two processes through which learning takes place in an indivictual. The child grows up in the immediate presence of his father and mother and slowly learns to imitate and identify himself with them. In high schools and colleges, students try to imitate or identify their teachers and professors in some of their ways of life. It is because of the regard and respect they have for them. Experimental evidences show that learning has some influence in the formation of attitudes. Vetter and Green (1932) have studied the formation and development of anti-religious attitudes of 350 members of the American Association. Many of them stated that they derived their attitudes from a study of history and science, some of them from a study of materialistic philosophy. Others traced their attitudes to the feeling and experience of grief and disgust. A few of them traced them to the influence of friends who are atheists.

(e) Suggestion

In the opinion of F.H. Allport (1924) an attitude is a kind of physiological effect produced by verbal suggestion, which operates in a way not unlike hypnotic suggestion. For example, to a hypnotised subject it is suggested that he should call upon a certain person at six O'clock the next day. In this way a motor setting is prepared to perform a certain act at a certain hour. The subject later on performs the act automatically at the appointed hour even when he is no longer under hypnotic suggestion. The motor set built up in this way is called an attitude. Allport declares that in everyday life attitudes are built up in the same fashion. There are, already, some deeplying reaction tendencies in human beings such as interests, emotions, sentiments and potential reactions on which every normal suggestion builds up an attitude.

Change in Attitudes

An attitude is not only developing in an individual but it also undergoes a process of change according to circumstances and experience. In Morgan's (1934) words, "it is characteristic of attitudes to be modified by every experience which relates to them". Allport considers that the most favourable occasion for the change of attitudes is the period of emotional disorganization. The existing attitudes are found to be unsuitable and new suitable attitudes come up. This view has been supported by Young (1931) that the emergence of new attitudes is "a resolution of a crisis, the solution of a difficulty — the end of a period of chaos, the termination of a moment of disorganization". An attitude is very strong when it suits the conditions and fulfils the desires of the individual. It becomes weak and yields to change when it lacks these characteristics, G.W. Allport (1935) says that conversion, the shock of grief, economic disaster and falling in love are the times favourable for new attitudes to come into being. Therefore, our attitude change, for him, is generally an orientation to contexts and current events. It attempts to systematic old experience with the emergence of a new attitude.

"Attitude change means," says Sherif "change of the individuals stand on a given issue. How can we know that a person's attitude on some issue has changed? We can measure a change of attitude by the same means used to assess its existence. An attitude is revealed and measured through a characteristic mode of behaviour, verbal or non-verbal. A change of this attitude is assessed through significant changes in this characteristic mode of behaviour"(Fig. 30.2).

Attitude change does not take place spontaneously. It takes place in relation to a person, place, thing or some values. Several factors, both internal and external work actively when an individual is exposed to change. Attitude change is found to take place in various areas of human values. Economic depression in the country, unemployment problem, insufficient income, social inequality can be said as some of the factors exerting influence in the individual's reaction to certain issues. Psychologists are very much interested to assess the amount of change and the factors that influence such change. The individual's experience and knowledge, social contacts, personality, the interacting situation in which he actively participates, prestige, communicating forces like newspapers, magazines, radio, television and motion picture are some of the factors constantly contributing towards attitude change. As Krech and Crutchfield (1948) put it, "despite the significant influences of cultural factors on the development of beliefs and attitudes, despite the deep rooted functional nature of many belief, and attitudes, and despite the inherent tendency of beliefs and attitudes to preserve themselves, beliefs and attitudes can be changed. The experimental literature in social psychology presents a number of successful "instances of such changes".

Fig. 30.2 : Change of Attitudes : Protest Against the Existing System

Attitudes are developed through experience, learning and by the process of socialization. These attitudes are also subjected to change due to the prevailing conditions in the situation, mainly political, changing in the life aspects and due to the experience of some hardships in the life of the people. When people are not satisfied with the existing lifestyle, generally they wish to change the attitudes of the present administration or government by means of showing their dissatisfaction in staging agitations, protest and demonstrations. Generally political views work very vigorously in bringing about changes in the attitudes. The most favourable occasion for the change of attitudes is the period of emotional disorganization. The emergence of new attitudes is the indication of the resolution of the crisis and solution of some difficult problems.

INTERPERSONAL ATTRACTION : FRIENDSHIP AND LOVE

What is Interpersonal Attraction?

The outcomes of social perception and social cognition have positive and negative processes. We have discussed above the development of social and personal attitudes which are having both the positive and negative characteristics. Here we shall discuss the positive characteristics of attitudes and feelings and then we go far the negative attitudes and feelings. Interpersonal attraction is a type of social behavioural pattern which promotes positive and negative attitudes among the members of social groups, acquaintances and friendship among men and women and love between males and females.

In our everyday life we spend more time in our interaction with other persons known and unknown, strangers and friends and with those persons with whom we are most intimate. There are some specific determinants in our choice of friendship, in falling in love and in establishing and trying to maintain close relationships. The main psychological process involved in all these functions is *interpersonal attraction.* The question involved in this is, why an individual likes one person, dislikes another person and remains indifferent with a third person. On the basis of these attitudes only we begin to behave with others. There is something inherently tempting us to exercise on the outside object. It may be a person, an object or a place. What is functioning in our behaviour, is the feeling of attraction. When it is projected on a person, it is called *interpersonal attraction.*

Generally, attraction is the degree to which we, like others. Attraction according to **Baron and Byrne** (1987) is based on the direction and the strength of *interpersonal evaluations.* Thus interpersonal attraction indicates *the attitude one holds* about another person. This type of attitude is based on the characteristics of positive-negative dimension that may vary from love and hate. These attitudes reveal how we are responding to other persons emotionally and also the level of the strength of influence on our interpersonal behaviour.

But this type of interpersonal attraction or relationship may develop and also change in an individual. Several studies have been made to find out as to how friendships develop out of acquaintanceships and but of several casual meetings and how people fall in love and get married. **Duck** (1985) and **Hatfield and Walster** (1981) have clarified this phenomenon. It is also important to know why some close relationships are maintained over a long time, while some of the persons do not maintain them. (1) It is observed that the variables such as *proximity or closeness, emotions* and the *needs to affiliate* determine acquaintances with strangers. (2) It is also occurring that some acquaintances become close friends in response to the factors such as the physical attractiveness, similarity and reciprocal judgements of each other. (3) We are also to know what the complexities behind fall in love and affection.

How Strangers Become Friends?

We sometime come across strange incidents in our life and they become in due course important matters in our life. We may acquaint with a person accidently, but it may also cause a series of contacts. It may be in any situation. A stranger may find a seat next to a student in a class-room and this may continue till the course of study is over. A stranger may rent a room next to ours and we have to see him and greet him everyday several times. In a working place a stranger, who is appointed recently, may find a seat close to us. While going to the work place, when we catch a bus, a stranger comes at the same time and gets into the same bus sit close to us everyday. This type of proximity in these instances leads to a gradual increase of familiarity. In such cases we feel more comfortable in the familiarity and often go to the extent of exchanging greetings. **Levitt** (1980) states that even children after several exposures to the same children in a situation respond positively to the strangers.

Proximity : Proximity is referring to physically close to each other. **Festinger** *et al.* (1950) have observed in their studies that any aspect of the environment that increase the physical proximity of two persons regularly, there is even probability that they will gradually get to know each other and begin to like one another. This type of incidents occur in class-room, work place, recreation places and in evening walking situations and so on.

Repeated Exposures : Repeated exposures of a person to others occur due to one's placement or where he is working. A classroom, a work place, an apartment where one is put up all serve as locations where persons make exposures repeatedly and have the chances of seeing or meeting others frequently. Here, in the beginning everyone is a stranger to everyone. But the environmental arrangements make the strangers to experience the repeated exposures. The spatial arrangements are such that these individuals have more chances to see one another time after time in their daily routines. This leads to strengthen the familiarity between or among persons which leads to increased attraction in the long run. Thus the effect of the repeated exposures in the physical environment results in acquaintanceship, development of friendship and then results in mutual attraction.

Feelings and Emotional Responses : The feelings and emotional responses play important role in the development of mutual attraction. We may come across positive and negative incidents in our daily lives which may have strong and immediate effect in our feelings and emotions. In the studies made by **Stone and Neale** (1984) it has been observed that our mood in turn affects many aspects of our behaviour, mainly the interpersonal attraction. **Littlel** (1981) states that *sad songs* in music performance lead to an increased alcohol state, while *fast and loud songs* lead to very little drinking. According to him music is an emotion arousing stimulus and it has an effect in our interpersonal behaviour.

We may experience in our everyday events both positive and negative effects. The *positive effects increase* the interpersonal attraction and the *negative effects decrease* the attraction. For example, in hot summer we experience irritation and depressed persons make us to feel uncomfortable in the situation. Sad news in television cause negative effect in our interpersonal attraction.

Conditioned Emotional Responses : The human emotional responses are conditioned on the basis of the effects caused by the consequences. **Clore and Byrne** (1974) have observed in their studies while explaining the interpersonal attraction have stated that our likes and dislikes are based upon the nature of emotional responses. They have also developed a theory, called *reinforcement affect model of attraction.* According to this theory all our likes and dislikes are based on emotional responses. For example, if any person makes us feel good, we respond with liking him, we dislike whoever makes us feel bad. If these incidents are repeatedly experienced the response is conditioned. We respond frequently to the person whose interaction is pleasant.

Affiliation Need and Desire for Friendship : Generally human beings spend much of their times in the interaction with other persons. We make new friends, spend time with friends, share in their personal feelings. Man has intrinstically positive feeling for having friends. According to **Wright** (1984) human beings differ in their need for affiliation. Some people prefer to be in the midst of many people, but some prefer to be alone. It is because of the personality differences. It is observed that the need for affiliation is a trait. **Murray** (1938) has studied this need for affiliation extensively. Male members who are having high affiliation need more often speak of attractive females than the male members who have low affiliation need. Then affiliation need may be defined as the motive to seek interpersonal relationships and to form friendship. Individual differences are observed in the motives involved in affiliation and friendship. It is in the social skills. It is rooted in the birth order. The last born is having high motive for affiliation than the first born. Acquaintanceship and friendship are the important determinants in the development of the affiliative needs of those involved in it (Fig. 30.3).

Fig. 30.3 : Interpersonal Attraction : Need for Affiliation

Interpersonal attraction is a social behaviour pattern. Human beings strongly desire to be in the company of other persons and enjoy. There is an *intrinsic need* in human beings to be with other persons who have similar mood and temper. Affiliation is the attachment and association with like minded persons. The feeling of affiliation is expressed by the people who have been away for a long time without having any contact. When such persons meet they express their extreme feelings of affiliation by holding each other strongly and embrace each other.

There are situations that can increase the need for affiliation. In a fearful situation any individual person may desire to affiliate with other persons who are nearby whether they all friends or strangers. The effect of fear, seeking affiliation seems to be that frightened persons wish to communicate with others about the situation. In fearful situation people normally seek other persons for help, even strangers are sought. In such situations we do not see whether the other person is known or unknow or totally stranger. In the frightened situation when some strangers come for rescue normally it happens and they become acquaintances later on.

Factors that Determine Acquaintances and Friendship

There are several factors that determine acquaintances and friendship. We have discussed some of the factors. We shall look into the other factors pertaining to the individual's physical characteristics and the reaction of others. How friendship between two persons develop? Whether the friendship development depends on physical attractiveness of each member of the future pair, or the similarity in various characteristic and the level to which each expresses the reciprocity with respect to the positive evaluation of each other?

Physical Attraction : The physical attraction of the other person is judged by his/her appearance. How the person appears? The answer in one word is *attractive.* The opinions are based to a larger extend on appearances. When we come across other persons for the first time, we generally look at his face first, then the other parts of the body such as the height, body build, skin colour, weight, dress, hair style, approximate age and so on. We have various ideas and opinions on these factors for liking and disliking the physical appearances. Apart from our culture which is having stereotyped views about appearance and attraction, we have our own ideas and prejudices developed on the basis of our experience and knowledge.

But whatever factors that determine our likes our dislikes of a person to the promotion of acquaintance and friendship, the most powerful factor is the *physical attractiveness.* This aspect of human beings becomes the strongest primary determinant in the development of relationship between persons.

The Outcome of the Physical Attractiveness : The physical attractiveness brings out several favourable effects. Several experimental studies were made on the effects of attractiveness. **Folkes** (1982) and **Hatfield and Sprecher** (1986) have observed in their studies that when some male, and female members interact in a situation, the attractiveness of the members affect the behaviour of both the male or female members. Some of the business and industrial organizations indirectly make it a point to select and employ attractive males and females in their organizations for promoting the market of their products. Even customers in attractive dress desire to visit such commercial shopping centres for displaying their physical attractiveness. Once they contact the attractive sales man or woman they are convinced by their words and develop a desire to buy the goods whatever the cost of the materials. Thus physical *attractiveness promote* positive reaction in social situation. But the low attractiveness causes negative responses. For example, a sales man or woman, if he/she looks very unpleasant with an unfair physical look may not attract customers in business centres. Customers try to avoid going there.

Attractive male or female members have more social interactions than others. This may also lead to successful love and marriage. Attractive young men and women get into any highly paid vocation easily than the persons who are low in attractiveness. For example, the Indian Airlines make it a point of selecting only attractive fair skinned young ladies for the posts of Air Hostess. This, they execute it in a concealed manner and not openly. Attractive young men and women expose their attractiveness in various forms and earn huge salaries and lead princely life. The modelling has become an area of profession and attractive young men and women play their roles in whatever manner they could expose their attractiveness. Mostly young modelling ladies expose their physical attractiveness to the needs of commercial and the organizational requirements. The present day mass media of communication, the television and the motion picture are all interested very much in making use of the physical attractiveness of the young ladies for their benefits. These are some of the effects of attractiveness.

Attributions of Physical Attractiveness : Having discussed the effects of physical attractiveness and the positive responses it could promote, we are also to know what do we call physical attractiveness. Experts in this field state that it is difficult to pin point any single variable that is used to make the judgment. There are a number of physical and behavioural factors taken into consideration. **Beck** *et al.* (1976) have conducted an investigation to ascertain the variables involved in this phnomenon. According to the study of Beck, females tempted to respond positively to males who are with thin legs, thin waist, broad shoulders and small buttocks. The general opinion is a that tall male appears to be an ideal romanic partner for attractive females. But in the study of **Graziano** *et al.* (1978) the female college students prefer the males of medium height. In one of the studies it was observed that males are attracted by females with medium-sized breast and medium - sized legs and medium-sized buttocks **(Wiggins** *et al.* 1968**)**. But both the males and the females consider that obesity is unattractive. In all these studies it was observed that the behavioural aspects play important role in addition to physical attractiveness. Generally smiling is accepted as an attractive external behavioural aspect and sad look is an unattractive aspect of behavioural expression. For some persons the name of the person seems to be the attractive aspect. Thus physical attractiveness is associated with these characteristics (Fig. 30.4).

Similarity and Attractiveness : In every culture we could see that a person may be having the same physical appearance as another person. In the physical make up two persons look similar. Persons of this type may like each other very easily. We learned to like persons who agree with us and we tend to reject those who disagree with us. Acceptance of our ideas and opinions make us to like others. Here we find certain amount of similarity of ideas and opinions and attitudes lead to attraction.

We generally like persons who are similar to us in physical appearance, in expressing the ideas and attitudes and judgments similar to ours. Since others' ideas and attitudes are identifying with ours we like them and take them into confidence.

Fig. 30.4 : Interpersonal Attraction : Physical Attractiveness

The interpersonal attraction is process of social behaviour. Interpersonal attraction is the most powerful and essential factor in the social behaviour. In making friendship the physical attractiveness of other persons becomes the primary determinant.

The effects of physical attractiveness can be observed when the males and the females interact in the situation and the attractiveness of each person affects the behaviour of both persons. Physical attractiveness has strong effect on males and females. The effect may be friendship, love and affection and it may also end in marriage and family life. Commercial organizations make use of the physical attractiveness of the males and females and associate their products with these attractive persons and advertise to turn the attention of the people towards their goods.

Mutual Attraction or Reciprocal Evaluations : In the process of interpersonal attraction the most influential factor is the *reciprocal judgment.* A favourable judgment of other person promotes attraction and an unfavourable judgment causes rejection. We normally like a person who likes us. What the high opinion and regards we have about a person must be reflected with the same feeling from the other person. A male and a female may have different attitudes in some issues. But when meet in a place accidently, have eye contact first and the female member leans toward the male member and also starts some conversation pleasantly. He is attracted by her manners. The physical signs and the verbal conversation promote mutual respect, interest on one another, wanting to meet again, wish to have communication, wanting to be together, and wish to express affection. Generally everyone likes to have only positive response from others. Once this type of contact and relationships are established one can take it for granted that the person likes him as she has evaluated him positively. But if there is any disliking attitudes from her he may develop the negative attitude. The positive mutual evaluation promotes friendship between the two persons and in the long run it may strengthen the relationship and also mutual attraction which may end in love and marriage.

Love and Companionship

Human feeling aspects involve several expressions. These feeling expressions are love, liking, disliking, hatred, fear, anger, disgust and so on. Human beings express these feelings in certain situation. Expression of the feeling of love and companionship involves two persons of different sex. In the function of love one male and one female are strongly involved. Generally, when two persons of opposite sex perceive each other intensely they may become a potential sexual partners, which may promote a relationship that may go beyond the simple friendship and the acquaintanceship resulting in love and marriage.

The concept of love is always combined with the feeling of affection. These two feeling aspects generally associated with sexual partners. The interpretation of love and affection vary from culture to culture. Though the expression of love between the partners is the same in all cultures the occurrence of love and the development may vary in cultures. In some cultures love takes place first and then the marriage takes place. But is some cultures love takes place only ofter the marriage. Therefore, a common explanation of love may not be possible. We can go into only the general occurrence of love between a male and a female persons (Fig. 30.5).

Kinds of Love: There are different kinds of love prevailing in human society. Roughly we may state a few kinds of love as passionate love, companion love, conjugal love and so on. Love is an emotion expressed with strong affection on an opposite sex member. Love involves great attachment and affection. It is having passionate desire, longing and feeling for something very much. Love is an intense emotion of affection, warmth, fondness and regard towards a person. Love is a deep feeling of sexual attraction and desire. It is a whole hearted liking for enjoying the physical pleasure. Love is an expression or outcome of interpersonal attraction. In the case of love two members of opposite sex are involved.

The most important characteristic that is involved in love is the physical attraction of the opposite sex member. Since love is a feeling aspect, it relative in characteristic. It means, what is most attractive for one person is not so attractive to another person. The physical characteristics we have described above are only relative. The college girls' judgments cannot be accepted as the universal judgments of the physical attraction. Further the definition of physical attraction in the case of love may vary from culture to culture and individual to individual. It is quite impossible to provide a correct and stable definition of sexual love as it is built on physical attractions and the momentary pleasure one desires.

Passionate Love : Passion means ordent love or affection. It is an intense sexual love between a man and a woman. It is a strong affection, expressing enthusiasm. Thus passionate love is a process of manifesting or exhibiting intense sexual feelings or desires. It is an expression of overwhelming emotional state in which a person thinks about the person of opposite sex

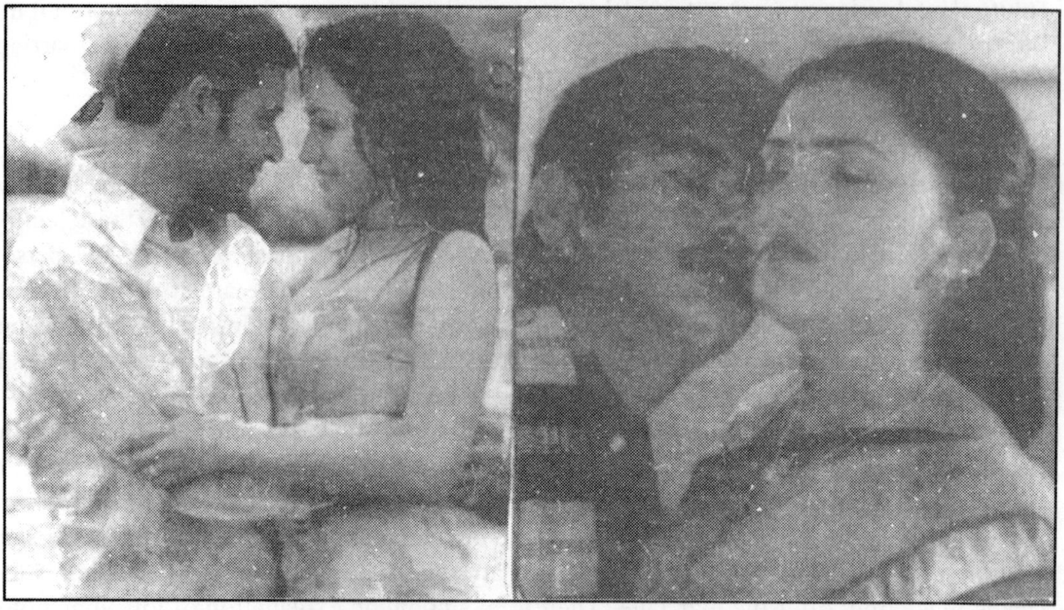

Fig. 30.5 : Interpersonal Attraction : Love and Companionship Expression
Human feeling aspects involve several expressions. These feeling expressions are love, liking, disliking, hatred, fear, anger, disgust, sorrow and so on. These feelings are expressed in certain situations. The most intensively expressed feeling is the feeling of love and companionship which involves two persons of opposite sex. In the function love and companionship one male and one female are strongly involved. Generally when two young persons of the opposite sex see face to face each other intensely they may become a potential sexual partners, which may promote a relationship that may go beyond the simple friendship and acquaintanceship resulting in love and marriage. The expression of love between the partners is the same in all cultures, but the occurrence of love and its development may vary in cultures. The interpretations vary on the basis of cultural and social norms. We read about love and marriage in Epics and Mythologies and wish to have the same type of life. But the modern concept of love and marriage and the type of life are taking different directions.

constantly whom he/she loves and intensely wants to spend as much time as possible with him/her. The person who is involved in passionate love affair is not realistic in judging about the person whom he/she loves. Since the passionate love is based upon physical attraction for fulfilling intense sexual desires and for the company to spend the time it may not end in marriage to live as husband and wife. The passionate love at one stage may be transferred to another person in both sexes.

Companionate Love : The companionate love unlike the passionate love is mainly based upon close friendship between a male and a female. This love affair slowly begins with friendship. The friendship in the course of time develops into companionship. These two persons often enjoy the friendship. Both are interested in one another's welfare and also develop mutual give and take in every issue. Thus companionate love develops in a strong feelings of friendship, having common interests, mutual respect, intense care on other person's welfare, happiness and progress.

Conjugal Love: Conjugal love is an intense love relating to marriage. It is also the love expressed by the husband and wife in a family. Unlike the other two love affairs conjugal love is firm, puts more responsibility on both persons and each one shares in the other person's welfare, likes and dislikes, happiness and misfortunes. It is firm and study in all life aspects.

PREJUDICE AND DISCRIMINATION

What is Prejudice?

The term prejudice is invariably used in the common sense world and in the scientific world of social sciences. It is used to indicate the type of social behaviour one has developed out of his experience. *Prejudice is a feeling aspect* and it is revealed in various aspects of feelings such as likes and dislikes, loves and hates, pleasing and unpleasing, friendliness and unfriendliness, approaching and avoiding and so on. It is an unfavourable attitude often practiced at various cultural levels. Prejudice, as an attitude, is also practiced in every culture in some form. The study of prejudice has received greater importance in social sciences when the racial admixture began to take place in United States and later on in some of the European countries and it has become a major social and psychological problem.

Generally prejudice means pre-judgement of persons, objects, events and situations without any verification. It is the negative attitude towards the members of a specific social group. It is the outcome of our false perception of others. It is the human tendency to behave towards others in such a way as to make them valid in some manner either positively or negatively. We tend to act towards another person or group of persons in accordance with the information we are concerning about him and the type of idea we are developing about him. It is logically inferred that favourable idea may evoke favourable reaction toward the person and unfavourable idea may evoke unfavourable reaction towards him. Our perception always tries to justify the idea that we have already formed in our mind eventhough the information we have received earlier has no basis in fact.

This means that prejudice seems to be having something to do with *self-fulfilling prophecy* and sterotype. These psychological processes determine a person's ideas about other persons and group of people. According to the principle of self-fulfilling prophecy a belief or an expectation that a person had already fixed in his mind is expecting every time that will happen and works towards the fulfilment of what he had expected. Similarly, the psychological process of stereotype refers to the expectations and opinions a person had already formed about some specfic group of people is generally negative. Thus stereotype can be viewed as a negative schematic relating to such groups. Thus self-fulfilling prophecy and stereotype seem to be functioning in the same manner as prejudice. The type of behaviour that is involved in the self-fulfilling prophecy is characterized by bias or prejudice. In the same manner the behaviour involved in stereotype is also referring to prejudice. In view of these we may state that the negative attitudes or prejudices towards any group are often justified and reinforced by stereotype and it is generalized and values are assigned the people of one group or culture. Investigators in the field of social psychology use this term in this sense in interpreting the nature and origin of ethnic prejudices.

"Prejudice etymologically refers to pre-judgement and as such it is an intellectually labour saving device," observes **Adinarayan** (1964). Prejudice is an attitude functioning mainly in a social situation and it is a way of coming to terms with certain perplexing and unpleasant problems. Prejudice as it functions for the time being seem to be comfortable, but in the long run it is definitely very much injurious. Prejudice by itself is anti in nature or against a person or group of people. But it may also function as anti-prejudice which means not having the prejudice against a person or a group. But prejudice is accepting the prejudice and acting against a person or a group of people. Both positive and negative aspects of the prejudice may function in a group of people or in a culture or race. We may also observe in a group or culture that every member of a group by virtue of his membership, the person acquires all the qualities of the group to which he belongs to. All other factors however good or valuable are ignored.

The prejudiced person of a group or culture is blind to all other virtues of a member of the prejudiced group. It is because of this that prejudice is sometimes described as a form of artistic thinking, a type of thinking that is not subjected to verification. Prejudice involves certain intellectual in capacity to think about a problem from another man's point of view. A prejudiced person only reacts in a social situation what his group has assigned to him. If we have formed an idea already we try to stick to it and react whatever it is, right or wrong in a social situation. We do not take any trouble to re-examine our view, or find an evidence for it. This type attitude is strongly functioning in the prejudiced person.

By developing an idea about others due to certain reasons, whether it is genuine or false we try to cling to it. We do not bother to verify it. A certain type of prejudice is natural and inevitable in our thoughts and feelings. This may find its root in our everyday life experience. It is quite impossible to envisage a world where there is no prejudice of some type. It is a human feeling develops internally experiencing some discomfort or painful experience mentally or physically. Many reasons may be attributed or discovered for the development of prejudice against a person or against a group of people or some ethnic groups. The psychological process of dislike or hatred or disgust is born with man and man never hesitates to project it when faces or experiences something against his wish or life comforts.

If man is subjected to the development of the feelings of prejudice against other persons it is individual in nature. In such case we cannot take him for granted that he is innocent. It is personal, within him. But the ofter review is that man is a social animal. He learns everything by becoming a member of a social group. He finds pleasure by being in the midst of other human beings irrespective of race, religion and culture. In this case prejudice is said to be learned. Man requires some type of knowledge about others and projects the idea he has learned in the society which may liking or disliking attitude. Therefore, prejudice may be said to be originating in an individual both internally and externally. Varied reasons have been attributed to the origin and development of prejudice person (Fig. 30.6).

Fig. 30.6 : The Development and Prevalence of Ethnic Prejudice

The term prejudice is used to indicate the type of social behaviour one has developed out of his experience. Prejudice is a feeling aspect and it is revealed in various feeling aspects such as likes and dislikes, loves and hates, pleasing and unpleasing, friendliness and enmity and approaches and avoidances. Thus it is a negative attitude towards the members of some specific social group. Generally, in every culture or society there may be in-groups and out-groups. The in-group members always blame the out-group members for every misfortune they expeirenced. Thus some type of negative attitudes may prevail among the members of the social, communal, ethnic, racial and religious groups. The outcome of prejudice is discrimination which is practiced in every culture in some form. Prejudice and discrimination in the extreme form may cause aggression, violence and riots. Prejudice has received greater importance in social sciences when the racial admixture began to take place in United States and in some European countries and it has become a major social and psychological problem. The practice of prejudice and discrimination is a never ending social problem in India.

Stereotypes may be both positive and negative. But a negative stereotype may have greater impact in the actions and self-concepts of the members of the targeted groups. Allport (1954) states that prejudice often creates traits of discrimination. Discrimination is one of the traits created by prejudice. But however discrimination need not be treated synonymously with prejudice. Prejudice may be functioning within the groups, between groups. It is more serious at racial, religious and communal levels. According to **David Krach and Richard S. Crutefield** (1948) prejudice means prejudgement, a man who has prejudices, has beliefs that are established prior to the revealation of the pertinant objective facts and that by their strength tend to predetermine the way in which new perceptions will emerge. They are also of the view that in the very nature of the case all beliefs partake to some degree of prejudice. This is the *universal characteristics of cognitive process.*

Prejudice is mainly functioning in social situations. A socially oriented definition may fulfil the needs. Therefore, we may also state that prejudice is an attitude towards the members of some group based mainly on their membership in that group. When we state that a particular person is prejudiced against the members of some social group it is generally means that he wishes to judge its members in some characteristic manner because they belong to that particular group or culture or race.

Discrimination : The Outcome of Prejudice

There is a tendency to treat prejudice and discrimination with the same manner though the functions of these two processes are varying significantly. Attitudes and prejudice are overtly functioning in the behaviour of the individuals. But persons who are having prejudice against the members of various groups cannot state their views directly against them in some social situation. There are certain factors that prevent them from expressing their hatred openly. The laws of the state, the social norms, feelings fear and hesitation, others hostile reaction all may make them not to reveal the negative attitudes openly. In certain situations the overt functions of prejudice are restricted by the prejudiced group members. But in certain situations the restricting forces are not found. In these situations the negative beliefs, feelings and attitudes and the behaviour tendencies may be put into action overtly. This type of discrimination in behaviour may be observed in different forms.

The action of prejudice is discrimination. Some of the forms of discrimination are in mild forms and sorne are moderate forms and some very severe. Some of the mild forms are simple avoidance which may take place by not seeing, speaking and not doing any work together cooperatively and so on. At moderate levels excluding the persons of the prejudiced group from occupations of the same category, same educational institutions, not allowing them to live in the same residential areas and so on. At the extreme level the discrimination go to the extent of expressing aggression overtly against the targets of prejudice. This may result in harmful incidences of some form which may result even in beating, lynching and killing the targets of prejudice. These types of discriminations are quite common in *racial prejudices* that prevailed in United States. In places where the white race and balck race put together these types of discriminations are quite common. But strong protest, agitation and political policies putting pressures on such discriminations have reduced the incidences.

The subtle forms of discrimination are many in practice. The user of the subtle form of discrimination is permitted to conceal his underlying negative views or prejudice. Though there are many forms, three of them seem to be more common. *The first* of them is withholding aid from the members of disliked groups which is often done in a delicate manner. Generally prejudiced persons do not engage to assist the objects of their bigotry. The prejudiced persons are more likely to refrain from aiding members of the groups they disliked. *The second* type of subtle form is *tokenism* which means small benefits. The individuals of the disliked group perform trivial positive actions and they are avoided for more beneficial works. Instead of denying the groups for more important works they are engaged in minor works. *The third* type of subtle form of discrimination is *reverse discrimination.* This is like giving with one hand and taking away with another hand. This is little severe form of discrimination. This is the tendency to evaluate or treat the members of outgroups, the racial and ethnic groups more favourable than the members one's own group. Though this type of discrimination appears to be not fitting into the definition of prejudice, in the long run it is harmful for the individuals of the prejudiced groups. These individuals who are the objects of prejudice are checked in many situations such as promotion, recreation and enjoying other benefits and promote negative actions toward their target groups, when the situations cause emotions the prejudice many run high.

The Occurrence of Prejudice : The attitudes of prejudice exist in all social groups. This is not a new psychological phenomenon. It has been constantly occurring all over the world in every social group right from the time when society came into existence. The consequences also very conspicuous and every social group had experienced this. History of the world tells us that no two human groups lived in a friendly manner for a long time. Difference of opinions, conflicts, aggression, violence and wars were going on endlessly. There are varied views expressed by the philosophers of *social contract theory* about human nature. Thomes Hobbes (1588-1679), an English materialistic political philosopher was one of the founders of the social contract theory. His greatest work is the *Leviathan* which contains the expansion of his theory *absolute sovereignty.* He was of the view that man by nature is short, brutish, nasty, crual and fearful. In the state of nature he is at war with each other and he becomes civilized only on becoming a member of a social group. He is always in a state of fear and this fear makes him to seek security and joins with others for getting protection.

This fear in everyman results in the constitution of a state or society. To become a member of the society or state he surrenders all his liberties. This society is the state of law in which man enjoys all kinds of protection. Thus Hobbes' description of the nature of man is not friendship, but enmity. This nature often projects in man whenever he meets with discomfort or obstatcle on the way of his rights and motives and he develops jealousy hatred, enemity and stress, revenge to exterminate the object the stands on the way of one's happy life. He does not think about other's inconveniences, but his pleasure and happiness. This is the origin and development of prejudices. Exponents in the field of social psychology may differ from this view and attribute varied reasons. But they also may aware of the basic human nature. **Hobbes'** theory is a good evidence to explain the origin and development of prejudice.

The organized society or the state provides security for the life of the individual, for this property, wife and children. When he is alone, he is isolated and always in a state of fear. Thus it is the sense of fear that forces an individual to become a member of an organized society or state and the state with all its laws provides him security. Man develops prejudice against some individuals or groups only out of fear that he will one day loose the pleasure of his life and therefore, he must be extreminated or prevented by some means. Prejudice is the powerful weapon that modern man uses against the individuals or groups of people whom he suspects as his rivals. Because of the state law the prejudiced person cannot projects his enmity or jealousy directly and do harms. Therefore, the modern man who is supposed to be powerful, projects his weapon *indirectly* and in a *disguised manner* without any chance for suspicion.

Attributing Social Character : Prejudice, psychologically, is *individual in origin.* But it is propagated and spread out in a group and it is made to prevail among the *members of the group* against the object of the prejudice, generally it is always another group of people. Then it becomes, *we* versus *they* or *us* versus *them.* It may sound as *my men* versus *your men, our group* versus *your group, our culture* versus *your culture, our religion* versus *your religion, our race* versus *your race, our nation* versus *your nation.* Thus starting with individuals it is ending with race and nation (Fig. 30.7).

Fig. 30.7 : Prejudice and Discrimination : An ever present

Danger : Causing Social Problem : Prejudice in Action.

Prejudice is a negative attitude towards the members of some specific group. It is the expression of disliking and hatred projected towards the members of some specific out-goup. Prejudice is an emotional aspect of human behaviour. Because of this people are easily aroused against the members of the out-group. The development of prejudice is not based on real genuine causes. It is an irrational behaviour pattern based jealousy and suspicion and not on verified facts. Though the development of prejudice is attributed to the period of racial intermixture it is seen that it had been a part of human society right from the time when society came into existence. History tells us that at no time two types of human groups lived in friendly manner for a long period. There were endless conflicts, aggression, violence and wars all the times. Attempt to combat prejudice was not successful at any time. Thus prejudice is an ever present discomforting problem.

Every man belongs to some social group. There may be different social groups. Difference of opinion, conflicts, aggression and violence may develop between groups which may not come to an end in spite of several attempts to resolve it. Ultimately it develops into some big issue and continue for years and also may cause violence between the groups. The cause of this group conflict may be very subtle, but it may develop with a very big issue. If one group is affiliated to some communal or religious faction, people from other places also may come for supporting the groups. In most of the developing countries this type of occurrences are quite common. Particularly in Indian communities this type of prejudicial incidences are occurring very often which lead to communal and religious clashes.

Several causes are attributed for the emergence and manifestation of prejudices. The causes may have different basis in every country. One of the important causes attributed to the development of prejudice is the direct intergroup conflicts based upon competition in all life aspects. People generally everywhere give importance to economic and social status. This is based on good job with high salary and occupational status with power and prestige. But situations do not provide chances for everyone to achieve what one desires. Members of one group may have more chances then the other groups. This becomes the basic fact of the interpretation for the development of prejudice. This is stated as the *realitic conflict theory*. Prejudice according to this theory stems from the competition between various social groups.

The prejudice, which is originated in the individual's feelings of fear and jealousy, is made to have social or group characteristics and it becomes a social attitude with likes and dislikes against the target group. If there is any illfeeling, fear, jealousy and hatred it finds place in the minds of a set of persons due to the failure or disappointment in their attempts to achieve some personal gains, they internally develop some hatred in the beginning and they may be waiting for a chance to project the illfeelings and hatred against groups which made these persons to be devoid of their prospects. In the long run this type of attitudes are revealed in some form of prejudice which also get aggravated and may function in various walks of life. This may get into the feelings of the like minded people and may develop into a social attitude. Groups of this type may constitute and a type of social categorization is formed. They may have even some names for the groups. Motives are formulated, contrasting beliefs and feelings are attached to the members of the groups. When the groups became larger and larger they may attach some political motives.

In Indian communities these characteristics are quite common. In India there are many political parties, have their origins only in the social and occupational categorization. If the categorization is stopped in the beginning by some means, then the chances for the development of prejudice can be eliminated. Several investigators studied these processes. **Tajfal and Turnea** (1979), **Lucksley** *et al.* (1980) are of the view that there were strong tendencies to devide the social world into contrasting groups. In their studies they have observed that the subjects generally expressed more negative attitudes towards the members of the outgroups and treated them in more unfavourable ways then the members of their own groups.

Prejudice and Cognitive Factors : There is a tendency for human beings to think about others in terms of the idea that is already formed in them. This is the important process of *social cognition.* We develop this attitude on the basis of something which we like or dislike. This is the basic aspect for the development of prejudice. There are growing evidences to the conclusion that several basic aspects are contributing to the development and maintenance of prejudice. We never verify any aspect of prejudice and reject it stating that it is not the true cause. Instead, we retain every weak evidence with us to justify prejudice. Some of the most important factors are the stereotypes and illusions of the outgroup characteristics.

Stereotypes : Stereotypes generally refer to the beliefs and attitudes and expectations we have formed about the members of some specific social group these are all generally in negative terms. Thus stereotypes can be viewed as negative schemata relating to such groups. Stereotypes may be also stated as the opinions and ideas about a large groups of people, allowing the persons holding them to disregard individually. A stereotype is overtly concrete, specific and resistant to disconfirming evidence. Stereotype is a kind of cognitive framework for interpreting and processing social information.

An individual who is strongly loaded with stereotypes behaves negatively towards a specific social group about whom he has already acquired some information negatively. Stereotype thus to a large degree is self-confirming. Several studies have been conducted to verify these views, **Dovidio** *et al.* (1986) in their studies presented to the subjects the *words with white and black* and asked them to decide whether various traits could be ever true of each of these racial groups. They presented the traits to the subjects one by one. All the *subjects were whites.* They *fastly decided the positive traits* to whites and in the case blacks they decided the *negative traits fastly.* This shows the cognitive expectations of the subjects of the whites and blacks found to be very sharp among the white subjects.

The Illusions of Outgroup : This is supposed to be another important cognitive factor involved in the development of prejudice. This is the tendency people often form in their minds about the members of the groups. We perceive the behaviour of the outgroup member with all what we have learned about them more differently from the ingroup members. Generally illusion of outgroup characteristics is a tendency to assume that the members of various outgroups are more similar to another than are the members of the ingroup or one's own group. We perceive members of the outgroup differently because we have more contact with them and develop more richer and differentiated cognitive processes about them. *The homogenity* that we perceive in outgroup members that we perceive is an illusory perception. We do not verify it.

Apart from the homogeneity there are also other factors that contribute to our prejudice against the outgroup members. There is some gender variation in the homogeneity. For example, the *males perceive women as more homogeneous* in their attitudes and behaviour than the men. But the females perceive *men as more homogeneous than women.* But all these reflect our basic aspects or social cognition and also the intergroup conflicts, early social learning and our wish to increase one's *self-esteem.* All these processes seem to suggest the ways in which we should fight prejudice and reduce the negative aspects of our behaviour.

Combating Prejudice

The occurrence of prejudice wherever and in whatever form causes serious consequences. The people who are the target groups become viatims of prejudice always. The prejudice, whether it is in a mild form, existing only in the minds of the prejudiced people or in a violent form projected outside aggressively, it results in serious problems. Cnce it takes place in one area it would immediately spread out like the wild fire all over the places and the socially categorized people who are branded as outgroups are put into troubles which may take place in violent attacks and killings, and destroying the properties of the target groups. Though prejudice of some type was prevailing in different forms, the prejudice that practiced in United States, and in South Africa, and in Germany caused serious consequences against the black people and against the Jews. Strong protesting voices were raised everywhere as it is against the humanism. This created a sense of feeling that prejudice should be challenged and faught. Ultimately prejudice has become a national and international problem.

Approaches

The intensity of the effects of prejudice may be experienced in various forms. In its least harmful forms it is like a thorne in the toe and also like constant pin pricks producing undesirable annoyances, frictions, irritations and humiliations. At its worst form it is like deep cuts, endangering the cultural values and the prestige of a nation, apart from violent attacks and mass killings and exterminations. Instances of such consequences have taken place against the highly hated people in United States, Germany and in Iraq very recently. Therefore, fighting and reducing prejudice and combating its negative effects are the important activities, prejudice can be faught through several avenues. Bu no single avenue can bring out the expected result of eliminating prejudice. It is better to combine all the strategies and directed towards the prejudice if we want to achieve the goals of eliminating prejudice and discrimination.

The avenues or the strategies should be planned in such a way that they should make the people to realize the essential values and create a sense of respect for the strategies. This should be one of the ways of life that people should accept and respect them. Such avenues are mainly *Religious, Ethical, Legal* and *Psychological.* These four avenues for fighting prejudice can be through the appeals of the eminent individuals and through organized institutions.

With the view of combating prejudice in a systematic manner, the United States Government wanted to know objectively about the problems caused by prejudice, for which they appointed **Gunnar Myrdal** (1944) to study the prevailing serious problems of prejudice. After several years of study Myrdal had published his work under the title of

American Dilemma. Myrdal is a sociologist from Sweden. To make the study more objective without any extreme likes and dislikes the United States had appointed an outsider to study the problem. The book, *American Dilemma* is an exciting work which helped the Government and the people to understand the prejudice that was causing serious problems in America. Several changes at various levels took place on the effects of prejudice and discriminations. This is one way of educating people for acquiring right knowledge about prejudice and discrimination. Following this, there were quite number of research studies made and publications were brought out. This is one way of combating prejudice and discrimination.

Religious Approach : The religious approach to combat prejudice and discrimination is mainly based upon the *inner conviction* of the individuals. It is a way of arousing the *sacred feelings* of the people and make them to realize that all men are created equal and no distinction should be made against any individual human being and groups of people whatever characteristics they are endowed with. Once this feeling is aroused which may function as the force for the annihilation of hatred and ill-feelings against other social groups. If this conviction that prejudice is basically wrong and irreligious takes place in the minds of the people then it may lead to change of heart which in due course way produce valuable results. The idea that hating and harming other people are sinful acts against the wishes of God may be aroused in the minds of the people by a systematic religious approach.

Ethical Approach : The ethical principles may be made known to the people who are prejudiced against other groups of people. It is the moral duty of every person, not doing any harm to an innocent person. Human beings must aim for the greatest happiness for the greatest number of people. Generally ethical principles emphasize one's duties and responsibilities and when men realize their rights, duties and responsibilities there is no chance for harmful thinking and discrimination. The ethical ideas and principles may be propagated through eminent individual persons and through some organizations interested in the moral behaviour of the people. One important moral principle that a person should not be put into hardship and sufferings. Generally, ethics teach us what is *right and wrong* and what is *good and bad* and what is *truth and falsehood.* Man is expected to accept the positive aspects and reject the negative aspects of these principles. Ethics also asserts that any deviations from rights, goodness and truth may result in punishment and there is reward for some type for positive actions.

Ethically prejudice is conducive to the *greatest happiness of the greatest number of people.* In fact it tends to cause the *greatest sorrowfulness* to the *greatest number of people.* This idea should be made known to the people who are religious and ethical. The ideals of both, the religious and ethical may be made known to the people and should be shown that prejudice is an evil force stands on the way of human life and prevents him from achieving the ideals. By appealing to the man's better nature through religious and ethical ideals change of heart can be sought. By this means people can be made to think that prejudice and discrimination are leading the man to commit sin against the ethical principles which is against the wishes of God. According to ethical principles prejudice is a *social injustice.*

Legal Approach : But the religious and ethical approaches are internal in nature. But the legal approach is external, something from out of us is acting upon us for performing something. In legal approach the law of the state is involved. Law is intended to correct the behaviour of the people in the social situation and tries to punish the persons who fail to carryout the law or violate the law pusposely. Legal approach in the case of prejudice and discrimination is to watch the law transgressors and take suitable action on them. The *legal approach* is making a *general appeal* to the people and it is not making any personal appeal. In the case of prejudice and discrimination legal approach also is not fully successful. Once prejudice outbreaks in a social group, the members of the group do not hesitate to violate the law. They deliberately behave against the law and cause damage to life and property. When the law breakers are got, law punishes them and they are ready to undergo imprisonment.

The law is intended to correct the society and to regulate orderliness in the state. It is *man made for man to correct himself.* In all the democratic countries it is the people's representatives who make the law and the people in the state are willing to obey the law. But sometimes people may feel that a particular law is injurious. In such cases people raise strong protest in the beginning, but ultimately they accept it. States may pass laws against the practice of some kind of prejudice and discrimination. In the eyes of law there is no discrimination. But it is man who executes the laws. The law executor may use his own personal likes and dislikes. Thus prejudice develops. It may be against the members of a specific group. Persons belonging to this group could not get into any employment. Thus conflict may arise between the persons in power and their group and the groups that are neglected from getting jobs. In United States in some states there was a law to keep away the Negroes from employments. It is the result of the effect of prejudice and discrimination. Legal action seem to be having only indirect bearing upon the reduction of personal prejudice. Law can be used to control only the outward expressions of prejudice. Therefore, legal approach can be successfully implemented if only there is a *change of heart* among the prejudiced people. It may be functioning indirectly wherever and whenever chances are occurring.

The Psychological Approach : The psychological approach is mainly based upon making changes in the attitudes of the prejudiced people against the target groups. Since prejudice is an attitude all those approaches that are used to change the attitudes can be applied to prejudice also. The pscyhological approach includes *learning not to hate any person.* This must take place right *from the childhood period.* We know that children are not born with hatred for some persons belonging to various social groups. In fact children acquire prejudice through their mothers and fathers. There should be changes in the child rearing practices and the way in which the members of various social groups are presented in the mass media of communication.

There should be increased contact between persons belonging to different social group. These group members must be met under favourable social situations. Individuals must be taught to be friendly and not to behave antagonistically irrespective of situations. All these may yield beneficial results. The psychological approach puts more emphasis on realistic manner. It makes appeals to the essential goodness of human nature and the wisdom. They are of the view that prejudice makes very costly consequences, by wasting valuable human resources and talents.

Prejudice is an attitude and psychologists feel strongly that it could be changed and the energy and knowledge wasted on it could be cannalized into some beneficial fields. Psychological approach mainly concentrates in exposing the evil consequences of prejudice and wanted to *purge the minds* of all kinds of *bias functioning* in terms of *pro* and *anti* or *we* and *they* or *ingroup* and *outgroup.* Psychology is fighting against prejudice from several angles.

PUBLIC OPINION

Introduction

Having discussed about the various aspects of attitudes and prejudices in detail, it becomes customary to know also about the nature of public opinion and its relation to attitudes. It is because, public opinion is a force that functions in society as a regulative social force that controls the attitudes of the people. The general public have some idea of what public opinion is, but they may not be having clear idea of its social, psychological character. Public opinion is as old as the organized society, right from the time when man had learned to become a member of some society for security of his life and property. But as a subject of scientific study it is relatively new phenomenon.

In human society, whether it is a larger group or smaller group, every group is having a leader. Whether the society is an ancient one or modern one there is a leader of some category who controls the group only through public opinion. Public opinion is an instrument in the hands of the leaders to regulate, educate and correct the people. Thereby the leader of a group creates a healthy and well informed followers. An opinon, whether it is good or bad, in a small group may spread out like a wild fire and may create a panic atmosphere if it is wild, and pleasant feelings if it is a good one.

What is Public Opinion?

Public opinion is the opinion held by the people in a society about some person, object or incident. Thus an opinion can be defined as the opinion held by the people of a group with a consenus of judgments arrived at through some means of communication. Public opinion is a group phenomenon. It is the crystalized attitudes of an issue or problem of a group of people, a class, a political body, an economic society or some other group of people. In this way it refers to a position or stand of an issue taken or shared by a number of people. In the behaviour of a society, public opinion functions as a *social trait.* Public opinion generally is a momentary one and it is not an enduring one for a longer period of time.

Generally public opinions are influenced by social psychological forces and interested groups, mostly by political parties. The members of the political parties are often excited by some opinion spread out by the party workers. More than any other information the political information spread out very rapidly and the opinion sometimes does not carry any truth. It may also be a strong rumour. Some of the social psychological factors function as powerful forces. For example, stereotypes, deep-seated human experiences, the cultural background of opinion holders, the personality traits of the interacting individuals and the dominant social and religious myths and preconceptions of the groups. Interested persons manipulate the group members with some exciting information and influence the public.

There are different types of public opinion. These are the public opinion of the community, public opinion of the nation and the public opinion of the world. The public opinion of the community functions within a limited circle which mainly concerned with the people of the community. The second one functions at national level. At the times of crisis public opinions are quite common, based upon some incidence or calamity. The third one may function at international level. For example, war incidents are quite common information for spreading out opinions for and against a nation which is involved in war. For example, during the Second World War there were alarming public opinions about the nations involved in the war. **Ginsbergi** defines public opinion as the mass ideas and judgments operative in a community which are more or less definitely formulated and have certain stability and use the results of many minds acting in common. In this sense it differs from the personal opinion. But it is not totally different from the personal opinion. Personal opinion provides the necessary background for the formation of the public opinion. The interaction of several people in a social situation provides scope for the development of public opinion. Public opinion in some way controls the behaviour of the individuals.

Attitudes, Opinions and Faith

Attitudes and opinions are related in several ways. But yet they are different. An attitude is a tendency toward something, a person, place or an object or situation. When an attitude is shared with other members of a group and accepted by most of them it becomes an opinion. An opinion is an acceptance of some proposition or statement. A person who holds an opinion does not have a strong desire to prove it. An opinion has neither a verified fact like knowledge nor unverifiable characteristic like a *faith.* Therefore, *opinion is a go between knowledge and faith.* When an opinion is verified and accepted it become a knowledge and when it is not verified it is a faith. When facts are sure and evidences are strong men hold knowledge and when facts and evidences are weak men hold opinions.

The phenomenon of attitudes and opinions were widely discussed by social psychologists. **Thurstone** and **Chave** (1929) have attempted to clarity attitudes and opinions. According to them an attitude is a complex affair and it is impossible to describe it by a single numerical index. The term attitude refers to the sum total of man's inclinations, feelings, prejudices, preconceived nations, ideas, fears and convictions about any specific object. It is purely subjective and personal affair. But for them opinion is a *verbal expression of an attitude.* For example, if a person said that America made the mistake of waging a war against Iraq, that statement is called *his opinion.* Thus the term opinion is limited to the verbal expression which is the person's attitude. An attitude is symbolized in an opinion.

Public : *A Public* constitute a *number of people,* who are spatically, widely separated, but at the sametime *interact* with one another through various media of communication. The *public is a unitary body* of people established through the interaction of feelings and emotions, ideas and thoughts. Thus public opinion is the opinion about something held by a group of people who are interacting on some common issue or interest.

Development of Public Opinion : Public opinion is an interesting subject matter in social psychology, and many exponents in this area have dealt with it in detail. According to **Kimball Young** (1946) there are *three stages* in the development of public opinion, such as the rise of the issue, discussion about the issue and proposed solutions as *pros* and *cons* and the arrival at consensus.

The Rise of the Issue : When there is a problem for the people at large in an area some type of opinion related to the problem develop. As long as things go in a normal way everything will be going on in a routine manner. People carry out their activities in their own way attending to their personal matters and nobody bothers to think about other matters.

But if an issue or problem that night affect the normal routine way of life there is every possiblity for the emergence of some opinion. In the beginning the process of public opinion formation may be underway. When the issue or problem begins to affect the life aspect of the people more and more people begin to think about it loudly. The issue must be general and should arouse the serious attention of the community. A community of people in their everyday life might be facing several issues. But all the issues may not be causing serious concern to the people.

Generally people often face in our country the water problem mostly. As long as there is regular water supply to the people of a city by the corporation and if there is water in their own wells people do not pay much attention to the water problem. But when the corporation water supply is stopped and when the wells are dried up the problem emerges. This has become an issue that affects most of the people living there.

Discussion About the Issue and Proposed Solutions : The said issue has become the subject matter for the public to make it as a common problem and a solution has to be sought, when people began to think about the issue loudly. Some one from every area becomes a self-styled leader and such leaders meet the representative of the constituency and make representations to find out solution for the issue. As the discussion goes on there may be different or varied types of talks, consultations and possible repercussions on other aspects of the community life are also involved in the discussion. Opinions far and against the issue and solutions may arise. There may be agreement and disagreement on the solutions expressed by the various sections of the people and the opinions are crystalized. Different opinions voiced by rival groups are synthesised and a compromise is evolved to deal with the issue (Fig. 30.8).

Fig. 30.8 : Public Opinion : A Body Taking Decision after Discussing the Issues and Problems

People form opinions in various aspects of their life. In a democratic society public opinion is taken as one of the aspects of community life. People freely express their views, attitudes and prejudices and also react to them. Public opinion is a social force that controls and regulates the functions of the society. Public opinion was found to be valued right from the time when society came into force. The public is a unitary body of people established through the interaction of feelings emotions, ideas and thoughts. The public are interacting on some common issues and when there is any difference of opinion they meet and discuss at a meeting. A community of people in their everyday life might be facing some difficult problems and issues. Persons representing various areas assemble, sit and discuss about them. This stage takes place when the problems and issues affect the life aspects of the people and they begin to think loudly. They may come to a decision and present it at a larger body of the community.

The Arrival at Consensus : In setting right the issue every one comes to an agreement and approve the resolution arrived at. However apparently there might be some consensus and difference of opinions prevailing among some smaller section of the people. But a majority opinion is evolved. Generally social forces do not always operate according to the formula evolved. It is easy to get the consensus in some cases, but in some cases it not possible. In certain cases culture and social norms might determine the consensus in small communities. But in larger communities a feeling of dissatisfaction always may prevail and public opinions become a continuous process in a society whenever issues and problems emerge (Fig. 30.9).

Rumours and Public Opinions

Rumour means a general talk not based upon any definite knowledge. It is an unconfirmed report or story in general circulation. While discussing about public opinion it becomes inevitable for us to state something about rumours. It is because rumours in some situations become public opinions. Psychologists like **G.W. Allport** (1935) and **Postman** (1949) have dealt with the nature of rumours very extensively. Rumours about a subject matter might have been circulated in a group of people. It may be to the importance of an individual member of a group. It may a good information which may put the person on the limelight or it may be a bad information which may damage the reputation of the person. On verification it is found to be not correct, a false information. This is the function of the subject matter of

Fig. 30.9 : Public Opinion : Decision to Resolve Problems : Arriving at Consensus
The representatives of the various sections of the people take a decision after discussing the pros and cons of the problems and issues prevailed and present it to the general body for approval. Sometimes the problems may be crucial and difference of opinion may be expressed by a section of the representatives. In such cases majority of opinion is sought. Generally some forces do not always operate in agreement with the formula evolved by the committee. Then one way of solving the problem is to go in for *consensus*. In some cases the cultural and the social norms determine decisions in small communities. But in larger communities a feeling of dissatisfaction may prevail. At this stage the public opinion is sought for the approval. The functions of public opinion are to get the decisions approved.

the rumour. The subject matter will be more ambiguous and full of imaginative words which do not fit into reality. Generally rumours are built on imagination. Rumours may be emerging about an individual person of a group or about the whole group of community. It may be also involving a nation when there are some international crisis like revolution in a country or war between nations.

The phenomenon of rumour was very extensively applied during the period of *Second World War* by the *Allied* and the *Axis Nations,* the two warring groups of Nations. Each group of Nations imagined new techniques in spreading out rumours against each other. At that time there were very interesting and exciting information in the rumours. *Goeb-bals,* the German propaganda minister (1933-1945) in Hitler's Government had become the champion of this venture of spreading out rumours against the Allied Nations during the Second World War. The technique of spreading out rumours was successful for sometimes. But at last it was defeated.

One of the important reasons for circulating rumours is jealousy, the other reason is enmity against the persons who are prospering in some area and live with name and fame. Thus it is a feeling aspect of human beings. When the rumours are circulated very effectively and if it causes some harm to the person who is the object of the rumours there is some relief and relaxation of the emotional tension for the person who circulated the rumours.

Generally we can observe in our everyday life that the public opinion that is circulated turns to be a rumour. Sometimes rumours also called as public opinions. It is the people who spread out rumours. Rumours influence the thought process and control the activities of the public. An opinion takes longer time to reach the public. But a rumour spreads out fastly like the wild fire. Because of this characteristic rumours are called as *emergency public opinion* as the acceptance takes place quickly, while public opinions go through laborious task. Both the rumour and the public opinion are releasing some information about something for good or bad. These two are functioning only in some social situation, controlling the activities of people. Public opinions and rumours are subjected to change.

Efforts may be taken to change the public opinions in a systematic manner. Because, public opinion is a time serving factor. When the purpose is over it changes automatically. The life of a public opinion is very short. The *instrument* human beings use *to change* the public opinion is *propaganda.* Since the effects of public opinion are short lived and there is no any feedback for it, public opinion meets with natural death. There is no organized institution to maintain the public opinion nor no individual human being comes forward to organize human groups to maintain the particular public opinion.

Psychologists apply the same technique, that are used for changing attitudes and prejudices, for the changing the public opinion also. **Katz** (1938) states that an individual adapts certain opinions as he has strong desire for conformity. Because, the members of the group should keep him in limelight and he wants that others should always think of him and. keep him as a good member. An individual who has earned a good name wishes to maintain it by some means. The frame of reference in which he was liked by many, gradually looses its effect and therefore he creates a *new frame of reference* and takes people with him *to a new idea* and maintains his status. This is the type propaganda the present day political leaders in India are making and maintaining their status and power by *exploiting people* with some *sugar coated words* and *by exaggerating* certain specific *weaknesses* that are prevailing among the helpless people.

PART ELEVEN

PSYCHOLOGICAL STATISTICS

31. Statistics for Psychological Measurement

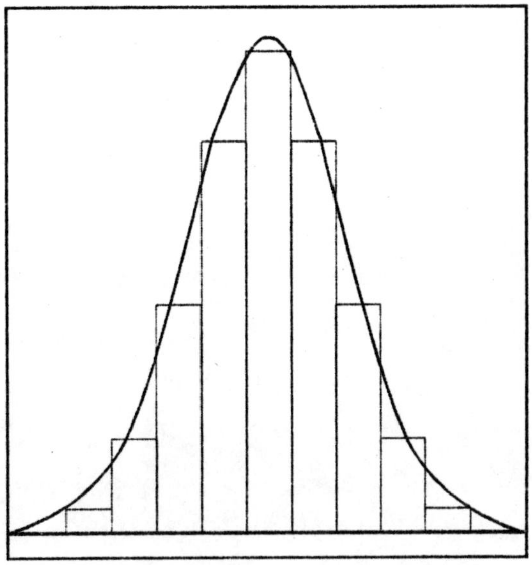

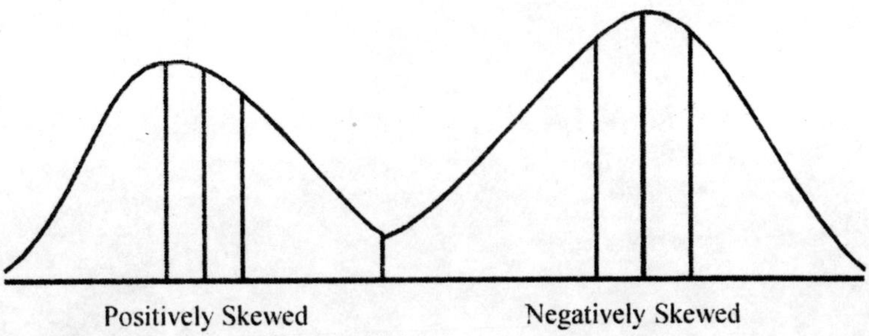

PART ELEVEN

PSYCHOLOGICAL STATISTICS

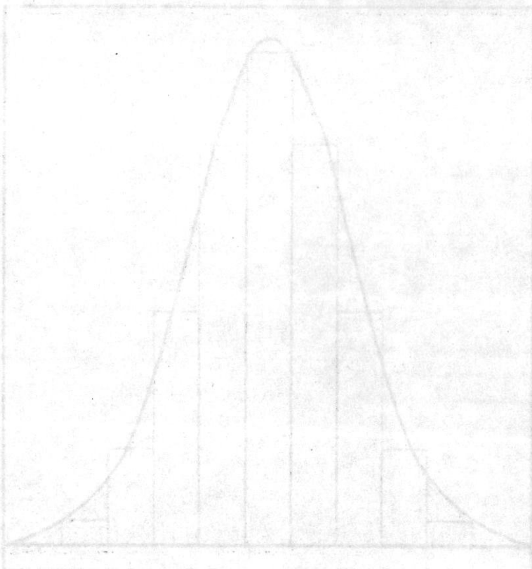

31

STATISTICS FOR PSYCHOLOGICAL MEASUREMENT

Individual Differences

Human beings are born and brought up in a most complex structural and environmental situation and constantly involved in the process of interaction with several types of people and social situations. Though they are born and brought up in the same manner each one differs from the other in some respects. Psychologists are very curious in finding out why this difference occurs. They have also conducted experiments and found out sufficient reasons. It was observed that the *genetic and the complex environmental* factors interact on the individual persons in such a way that they produce human behaviour to have an extraordinary potentialities for variations. We have discussed in the earlier chapter that human species are drawn upon some enormous amount of *genetic factors* with each individual person receiving in a unique combination of hereditary factors. Further, the human experience has infinite diversity. An individual's life experience is mainly based upon the quality of life one leads, the economic and social backgrounds, the nature of the occupation, the type of social interaction, all are having some influencing effect in the life of an individual person.

Generally, human beings differ from one another in appearance, in physical strength, size of the body built, colour, susceptibility to illness, nature of activities in which one is interested and in quite number of other characteristics. The human individual's behavioural differences are the other important striking characteristics such as the emotional aspects, perceptual processes, thoughts and imaginations, values, beliefs and attitudes, habits, skills and talents, desires and motives, mental abilities, all occur in an endless frames of references in unique combinations and function in diversified ways in every person. We can see that no two persons are totally alike. This is the fact within the *natural law* which throws the greatest challenge to the psychologists who are trying to assess and describe the human behaviour precisely and as accurately as possible.

Psychology as a Science

Psychology is the science of human behaviour. If psychology is a science it should possess all the characteristics of a science. But what a science is? A science is the systematic study of the nature and behaviour of the material and the physical universe, based on *observation, experiment and measurement,* and the formulation of laws to describe these facts in general terms. Psychologists ascertain that psychology possesses all these characteristics and in all psychological studies of some materials these characteristics are found. Information is obtained by observation experimentation and by measurement in psychology.

Psychology is interested in individual differences. Psychology measures individual differences by using psychological tests and quantifies the results in a meaningful way. This quantification enables the psychologist to describe and to classify his observations (results) to bring out relationships between important variables and to make predictions about his subject matter with maximum *accuracy and precision.* The psychologist in designing his experiments has to depend upon a knowledge of quantitative methods to have control over the experimental situation and the independent variables and also to obstain the maximum information, Thus the psychological studies are carried out in the same manner as any other physical or biological sciences. By these means psychology qualifies itself as a science and the measurements are accurate and reliable.

Psychological Testing

In any psychological measurement quantification of the observation or the result is obtained by administering some instrument or test among the subjects of the study. Experiments may be conducted in a laboratory situation or in a field situation. Whatever type of study is made some psychological scale or test is administered. If suitable scale or test already designed and standardized for measuring the particular characteristic it can be used. Otherwise, a scale or test to measure the specific characteristic should be designed and developed. The construction and the use of the scale for the measurement of the individual differences in some specific trait is called *psychological testing.* Fundamentally, **Anastasi** (1961) states that a psychological test is an instrument for obtaining an objective and standardized sample of behaviour. The psychologist is not simply interested in the sample of behaviour, but also interested in the class and behaviours which the test material is designed to represent. But his basic aim is to obtain the measures of categories of behaviour. This is used to generalize the characteristics or modes of response of the individuals. **Thorndike** and **Hagen** (1961) state that in psychological testing measurement may be direct or indirect process which depends upon the nature of the characteristics that are to be evaluated.

In direct measurements the psychologist is concerned with classes of behaviour that are observable which involves the individual's *responses.* For example, a person is driving a car to a distance in which we are also sitting and watching how he is performing the task of driving. How he is handing the gears, breaks, how he follows the traffic signals, how he drives at the turnings and what is the speed he maintains, all we can observe and determine the efficiency with which he is performing the task. We directly observe and judge the performances. This is really a test of achievement. In this type of task the psychologist attempts to measure what an individual can actually do at the time of testing. There are many different types of tests of achievements have been developed for use in the fields of vocational guidance and selection, education and industrial organizations where persons are tested for recruitments. The individual's various skills are tested objectively.

The other type of measurement is made by *inference or indirect method.* Whatever test is used in a measurement the result must be meaningful. If the test results are meaningful then the test results must elicit the responses that should represent a class of behaviour that is measured. This must be a sample of the behaviour with much scope and precise. It should be the typical behaviour of the individual who is being examined or tested. In this way the psychological testing is almost like an experiment. This involves what actually takes place in an experiment, like control of the variables, manipulation of the independent variable and so on. In such experiment the *items* used are the *stimuli* which could be manipulated in the planned manner. The individual's test responses are the dependent variables. This is supposed to be the behaviour of the individual observed and recorded. In the experimental situation there may be also other variables which might influence the subject's responses. These variables must be controlled as far as possible.

Psychological Measures : Statisics

Psychologists like all their scientists measure their subject matter either by experiments in the laboratory situation or by using tools and instruments or in the field situation by using some tests or scales designed for the specific purpose. Psychologists whether they use instruments and tools or tests and scales, they record the responses of the subjects whom they are testing. These response provided by the subjects are given *numerical values.* By this means either responses are in the form of verbal or written are converted into numbers. Once the psycholgical test responses are converted into numerical values they are subjected to some kind of treatment. The most commonly, psychologists treat the data (results, or responses) statistically. This type of treatment makes the data in precise form, so that the behaviour can be easily described, compared and interpreted. The numerical scores drawn by the psychological tests do not provide us any significant meaning unless they are collected, organized and treated in a systematic and meaning manner. For this, psychologists choose the statistics. Statistical techniques are very important to many areas of psychology.

In a laboratory situation the psychologist who is interested in the learning habits of rats, faces with a problems similar to those who are testing the children's school performance for placing them into different levels of performance. These two kinds of researchers have the problem of organizing and interpreting the data of their experiments. In both cases the proper designing of the experiments depends upon adequate statistical knowledge on the part of the experiments for the quantitative treatment of the data. Even in the case of selecting the sample for the experiment, the choice of variables and the procedure of manipulation, experimental control and the data recording are all to be dealt with a knowledge of statistics. In this way like all other sciences, statistic has become a major tool of psychological studies. Therefore, a fair knowledge of statistical techniques has become very essential to understand the present day psychological subject matters.

Then, what is statistics? A simple definition of statistic is that it is a number that describes a sample. The value of a statistic is known when we have taken a sample, but it can change from sample to sample. We may use a statistic to estimate an unknown parameter. A parameter is a number that describes the population. A parameter is a fixed number, but in practice we do not know its value. *Statistic* is a singular form and its plural form is *statistics*.

Statistics also may be defined as a science concerned with the collection, classification and interpretation of quantitative data and with the application of probability theory to the analysis and estimation of population parameters (Collins English Dictionary, 1979). In general, psychologists distinguish between two types of statistics. According to Gullford (1956) these two are Descriptive statistics and Inferential statistics.

Descriptive Statistics : Descriptive statistics are employed in the collection, organization and description of the data. Descriptive statistics enable the psychologists to characterize the large amount of information in a simple precise form. Psychologists also use them to work out the relationships that are existing between the various kinds of data that they are collecting.

Once the psychologist has organized and described his data, he is facing with the task of generalizing from them. In his final analysis his data is representing only a sample of his observations and his final goal is to infer from this small sample some characteristics of the general population from which he has drawn the sample. In achieving his goal the psychologist can apply a varieties of statistical procedures that are called as *inferential* statistial procedure.

Inferential Statistics : Inferential statistics enable the psychologist to generalize from his data and also it helps him to determine the probability that his processes of generalizations are correct. It provides a quantitative estimate of the confidence level which the psychologist can add in his table of results. It is also called sampling statistics or technique, as it is mainly based upon the sample drawn from the population.

Descriptive Statistics

The descriptive statistics are mainly concerned with the organization and description of the data obtained from the sample. The scores of the subjects of a population are collected and they are arranged in a systematic order. The normal technique for organizing and arranging the data is called *frequency distribution.* Large amount of data are arranged in few categories. Frequency distribution is one of the components of descriptive statistics. Descriptive statistic is a procedure of simplifying or summarising statements about measurement made in a population. Generally stating, descriptive statistic should be applied specifically to a population rather than to samples and the term is used for summarising statements or scores of the samples. To be precise, in descriptive statistics we make some type of summary of the data by arranging all the scores or the findings from highest to the lowest or from lowest to highest. These summaries make everything easy to remember and think about the data. This type of simplified statements are called descriptive statistics.

In the process of frequency distribution, the scores are placed in a vertical column from highes to lowest and the frequency of its occurrence is recorded. These data, if larger in number it is impossible to arrange them in an hierarchical order. Therefore, they must be grouped. For grouping, the simplest way is to find out the highest we get difference between the highest score and the lowest score. This difference should be divided by the total number of cases. The product of this may be used for the class interval. The advantage of the class interval in grouping is that it makes easy to tally the scores. For example, the following scores represent 20 students attitudes towards birth control on summated rating scale. These scores as they are have no meaning.

In this measurement the highest score is 87 and the lowest score is 42, the range of the scores is very wide (87 – 42 = 45), 45. Therefore, a range of 45 score unit is required to include all the scores.

Table 31.1 : Showing the Scores of the College Students on the Birth Control Attitude Scale (N-20)

62	72	83	66
87	68	67	58
60	76	50	46
70	78	61	42
66	63	52	50

Table 31.2 : Showing the Frequency Distribution of 20 Scores Data from Table : 31.1

Class Interval (1)	*Scores / Tallis* (2)	*Frequency* (3)	*Deviations* (4)
40-49	//	2	+2
50-59	///	4	+1
60-69	~~////~~ ///	8	0
70-79	////	4	–1
80-89	//	2	–2
		N 20	

It is desirable to condense the frequency distribution by grouping the individual scores into intervals. Grouping the scores into classes simplifies the frequency distribution without sacrificing the accuracy. In this case since the range of scores is 45 it is better to have five class intervals with ten scores in each. It is better to start with the lowest scores and multiples of these to the size of the class intervals. In this sample to have the class intervals we should divide the range 45 by a suitable class size. Dividing the range 45 by ten we get five class intervals. On this basis, class intervals are set up and presented in Table 31.2.

There are five class intervals and in each class interval ten scores are included. Against each class interval the scores that fall within the class are recorded since the frequencies are not too many in each class interval. Instead of entering the scores directly tally marks are used. But if the scores are not many, the actual scores can be entered (Table 31.2). In this case it is not done as it would be crowded. Therefore, the total number of scores fall in each class interval are entered under the third column, *frequency.* The frequencies can be added and verified whether there are 20 or not. In the frequency distribution always there is a tendency for the scores to concentrate more at the middle of the range and gradually going on smaller and smaller at the two ends (Table : 31.2).

Histogram : A visual presentation of the frequency distribution is obtained, by drawing a histogram to the scores given in Table 31.2. It is also called a *column chart.* This can be drawn by placing the scores along the horizontal axis (X-axis) and placing the frequency numbers on the vertical axis (y-axis). By this means a histogram becomes a serious of vertical bars drawn to the frequencies with a common frequency line between two adjacent frequencies. The heights of each bar represents one frequency. Each bar occupies certain area and the sum of the areas of all the bars is equivalent to the total number of cases or the whole area of the histogram. If the histogram is divided by the N then there will be 20 small rectangles, one for each score. (Fig. 31.1).

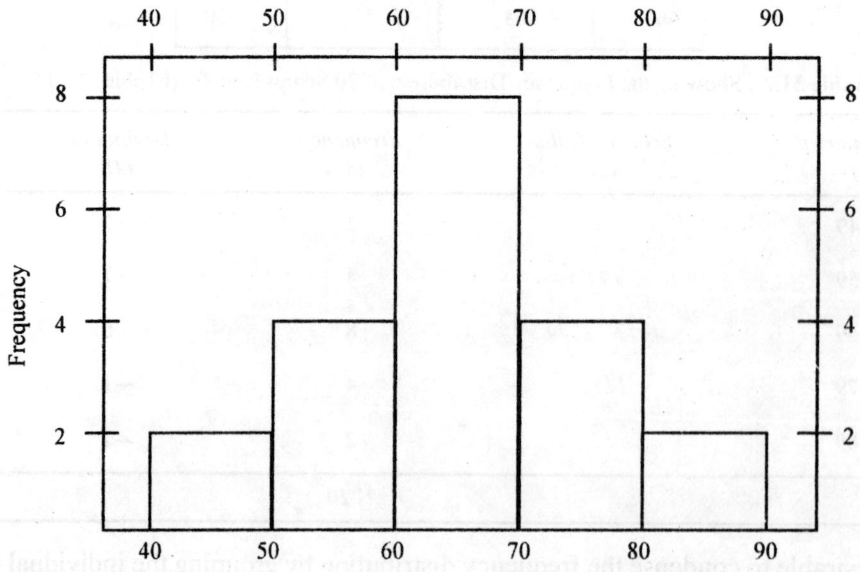

Fig. 31.1 : Histogram for the distribution of scores given in Table 31.1

Frequency Polygon : The same frequency distribution scores (Table 31.1), with the horizontal and vertical scales can be used for drawing the *frequency polygon.* The frequencies are marked by dots directly above the midpoint. These dots are connected by straight lines. The area of the frequency polygon is the same as that of the histogram if they are drawn to the same distribution (Fig. 31.2). The frequency polygon is a many sided figure, not having smooth shape (Fig. 31.2).

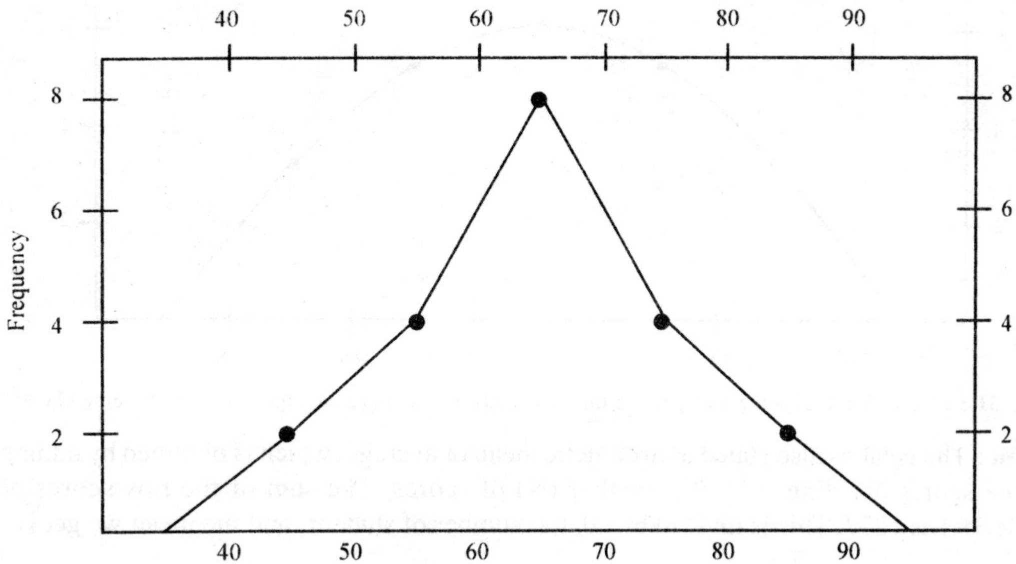

Fig. 31.2 : Frequency Polygon for the distribution of scores given in Table 31.2

Smoothed Frequency Curve (Polygon) : We could see that the histogram and polygon are having irregular shapes. This might be due to chance errors. The irregular shape of the polygon can be made a regular shape, by the process of smoothening the curve. It is by getting the average score of three intervals. The frequencies of three interval values, one above and one below are taken up and the average is worked out. For example, in Table 31.2, the smoothed value for the interval 50-59 is worked out like this. The frequency value for the interval 50-59 in 4 and for the interval below 60-69 is 8 and for the interval above 40-49 is 2. By summing up these values we get 4+8+2=14, which is divided by 3 and we get 14**3=4,66. The other values can, be obtained in the some manner and these can be applied on the frequency polygon and the, smoothed frequency polygon can be obtained (Fig. 31.3).

Measures of Central Tendency

A measure of central tendency may be stated as the centre of the frequency distribution. It is the representation of some point on the scale, a central point on the distribution. It is a central point with scores scattering on either side of the distribution. There are three most commonly used measures of central tendencies. These are *mean, median and the mode.*

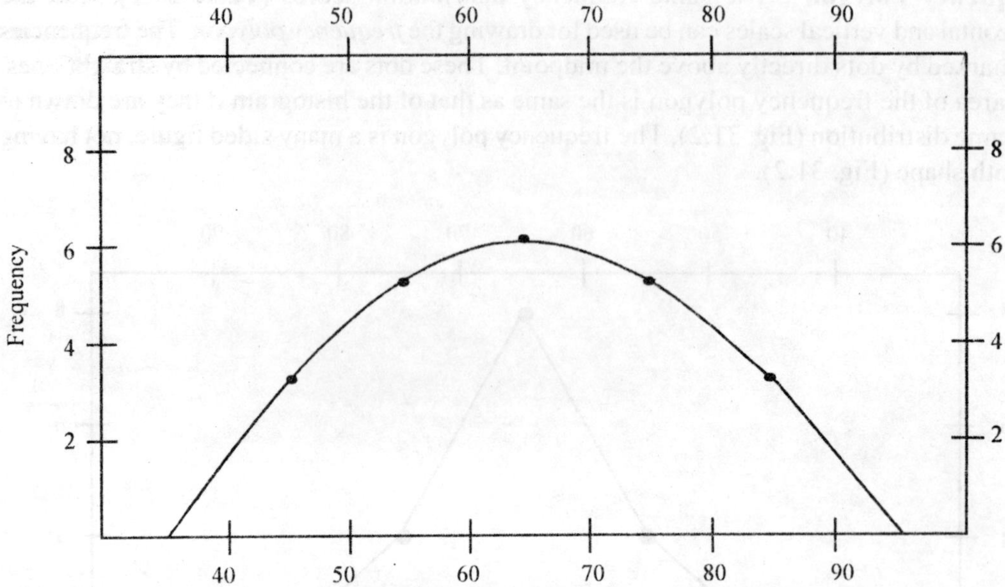

Fig. 31.3 : Smoothed Frequency Polygon for the Scores given in Table 31.1 Figure 31.2 is smoothed here.

Mean : The mean is also stated as arithmetic mean or average, which is obtained by adding all the scores dividing it by the number (N) of scores. The sum of the raw scores of Table 31.1 is 1277. This is divided by 20, the number of students and the mean we get is 63.85. The formula is $\frac{\Sigma x}{N} = M$. The equation, Σx stands for the sum of scores. N for number of cases and M for the mean. In any psychological measurement the mean is an important measure.

Median : The median is the score of the *middle case.* It is a better measure for characterizing the central tendency of the skewed distribution. It is the most middle point in the distribution. Normally there will be 50 per cent of the scores above and 50 per cent of the scores below of it. It is obtained by ordering the scores from the lowest to the highest and then it is determined the score which falls at the *middle point* (Fig. 31.3). The median is computed in a group date. In the model given here (Table. 31.1) the number of cases are 20 and the median falls between 10 and 11 as the arithmetic average of the tenth and the eleventh cases.

Mode : The mode is a measure of the central tendency which can be easily computed. It is the most frequently occurred score. The mode can be obtained by arranging the scores in a frequency distribution and noting down the score with *highest frequency.* For the data of Table 31.1 the Mode is 63.85. The scores must frequently occurred between 60 and 69. The mode generally may fall somewhere in the centre of the distribution if the distribution is unimodel. Sometimes it may coincide with mean and median.

The mode in some cases may be estimated from ungrouped date. There may be some difference between the modes obtained from the grouped data and ungrouped data. The true mode is the greatest concentration in a distribution, the point at which more measures fall than at any other points. The formula for estimating the mode roughly is :

$$\text{Mode} = 3 \text{ Medians} - 2 \text{ Means.}$$

The mode is a crude measure and not a stable measure of the central tendency. But in the perfect normal distribution the mean, median and mode may coincide. In a *positive skew* the median occurs on the *left of the mean.* In a *negative skew* the median occurs on the *right of the mean.* In the former case the value of the median is lesser then that of the mean and in the later case the mean is lesser than the median (Fig. 31.7).

Measures of Variability

The central tendeneies have the characteristics of clustering around the central values and scattering away from the centre. There are several measures to obtain from the clustering and scattering of the scores in a distribution. For example, we are in need of a measure to tell us whether scores clustered closely around their average or the scattered widely. A measure of the spread of scores around the average is called *measure of variation.* The measures of variation are useful for two purposes. First of all it shows how representative the averge is. If the variation is small we come to know that the individual cases are close to it. If the variation is larger we can make use of the mean as respresentative value with less assurance.

Generally, variability is the manner in which the scores vary within the distribution which is called variability and this may differ considerably for distributions that have identical means medians and modes. For example, there are two distributions as *X* and *Y*. Distribution *X* has the scores as 9, 11 and 13 and *Y* has the scores as, 3, 11 and 19. These two distributions have the same means and medians. But these two means are different in some respects. In the case of first distribution the scores are very close to each others, while in the second case the distance between the scores are very wider. To state it statistically, distribution *X* has *less variability* or dispersion, while *Y* has wider variability. The variability is also stated as dispersion and variation. The possible measures of *variabilities are range, quartile deviation* (Q), the *average deviation* (AD) and the *standard deviation* (SD).

Range

Range is the simplest measure of variablity. It is the difference between the highest and the lowest scores. In other words, it is the intervel between the highest and the lowest scores. It provides an estimation of the degree of spread of scores within the distribution.

However, the range depends upon the distance between the two extreme scores of the total group. It does not tell us anything about the individual scores spread out within this distance.

Quartile Deviation

The quartile deviation is also called as *semi-inter quartile range* and indicated by the letter *Q*. The semi-interquartile range is half of the scale distance between 25th percentile and 75th percentile. It is generally the middle cases of 50 per cent in a group, fall between 25th and 75th percentile. Just as the median divides the frequency distribution into two equal parts, the quartile divides the frequency distribution with four equal parts, The quartile represents the point on the variability scale. The first points is Q_1, the second Q_2 and the third Q_3. The second point Q_2 is the median of the central point. Therefore, the formula for the quartile deviation is

$$Q_2 \text{ or } Q = \frac{Q_3 - Q_1}{2}$$

Therefore, the distance between the quartiles Q_3 and Q_1 is called the inter-quartile range and the statistic that is stated as the measure of variability in the *semi-interquartile range* Q.

Average Deviation

Generally deviation means dispersion, going away from the central point. In every distribution, scores deviate from the mean on either side to certain distance. The deviation is *zero* when the individual score is *exactly the same with the mean.* If the deviation is above the mean, then it is taken as the *positive distance* and if the deviation is below the mean then it is taken as *negative distance.* To what extent every score deviates from the mean? This is estimated and the *average* of all the deviations is taken as the *average deviation.* In other words average deviation (AD) is the mean of all the deviated scores. The formula to find out the average deviation is

$$\text{AD} = \frac{\Sigma \text{x}}{N}$$

Where x is the deviation of a score from the mean, Σ is the total of all the deviations and N is the number of cases.

Standard Deviation

The standard deviation is the most reliable measure of variability (SD or σ). In experimental research studies it is most frequently used variability. This serves as an indicator of the degree of disperson. The important information of any study in psychology often rusts only upon the stardard deviation. It cannot be explained in any simple terms with pictural or geometrical illustrations like other measures. The SD is nothing but the representation of deviation of scores. It increases proportionately with the increase of range. If the scores are spread out widely the SD also increases proportionately. There are two methods used to compute the standard deviations. One employed to *ungrouped data* and the other to grouped data. The formula for the *ungrouped data* is :

$$SD = \frac{\sqrt{(X - \bar{X})^2}}{N}$$

Where X is the individual score, $\bar{X}$ is the mean, Σ is the sum of, N is the number of cases.

The formula to compute the standard deviation (SD) *for grouped data* is :

$$SD = \text{or } \sigma = h\sqrt{\frac{\Sigma f(x)^2}{N} - \left(\frac{\Sigma fx}{N}\right)^2}$$

Where N is the total number of cases, f frequency, x is the deviation score. h number of cases in each class interveal, Σ sum of. We expect that the distribution should be normal having the proportion of area as given in Fig. 31.4. But if, it did not occur then the distribution is not normal and we go for Skewness and Kurtosis.

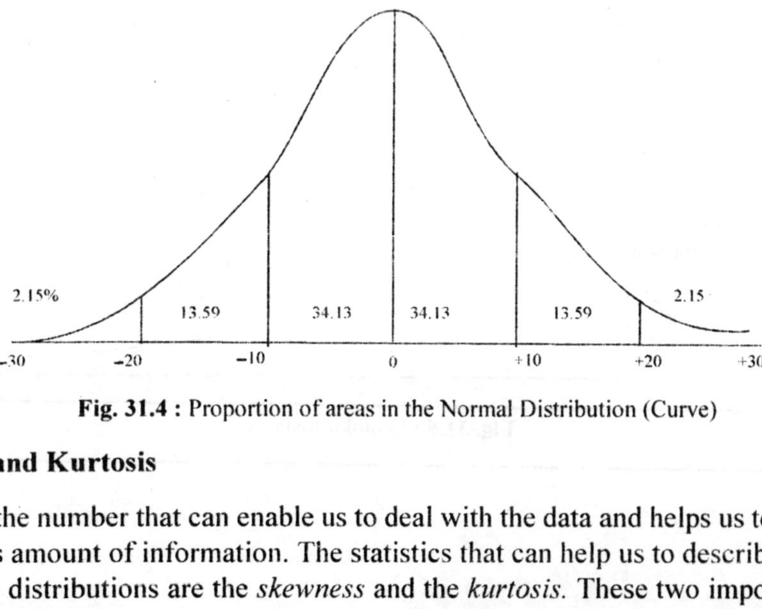

Fig. 31.4 : Proportion of areas in the Normal Distribution (Curve)

Skewness and Kurtosis

Statistic is the number that can enable us to deal with the data and helps us to extract from it enormous amount of information. The statistics that can help us to describe the types of non-normal distributions are the *skewness* and the *kurtosis.* These two important types of non-normal distributions that are possible from poor sampling precedures, are skewed distributions and multi-model distributions. If the curve is *asymmetrical* or *non-normal* we can discribe the positions of the means and medians through skewness and kurtosis.

Generally skewness is a distribution. Therefore, we may state that a skewness is a frequency distribution that is *not symmetrical.* It is named for the *direction in which the tail lies.* For example, there may be many small incomes and few large incomes and the distribution is skewed in the direction of the *large ones.* In psychological researches if the distribution is not normal then the data cannot be used for descriptive purposes. It is because the characteristic of the trait that is being measured is not distributed normally in the sample. It can be observed that sometimes if the sampling distribution is small, the samples may depart from normal distribution in various ways and the statistic may reveal them. The statistic that reveals the departures of the normal distribution of scores and the lack of symmetry is called skewness.

The skewed distribution can be of a *pecked type* in a distribution or of a *flat-topped* type of distribution. This charactertistic of the skewed distribution is called Kurtosis. Therefore, Kurtosis means the sharpness of a peck on a curve of a density function in comparison with the normal curve. In other words, according to **Guilford** (1987), it is the degree of steepness of the middle part of the distribution. This generally occurs in skewed distribution and it is of three types such as **Leptokurtic** distribution, **Mesokurtic** distribution and **Platykurtic** distribution. If the distribution is having higher pack than the normal distribution is called Leptokurtosis. If the distribution is flatter than the normal distribution is turning towards rectangular shape and it is called platykurtosis. If the distribution is normal it is called mesokurtosis (Figures 31.5 and 31.6).

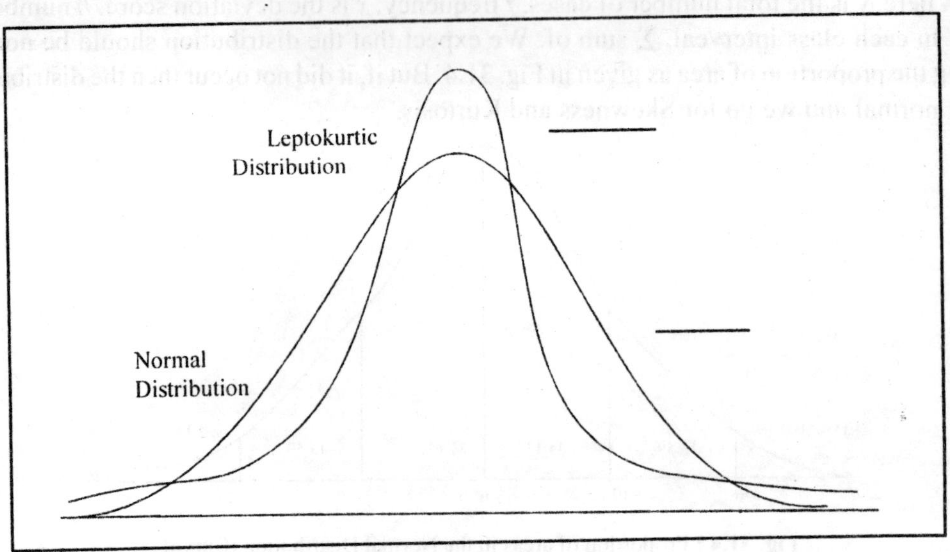

Fig. 31.5 : Leptokurtosis

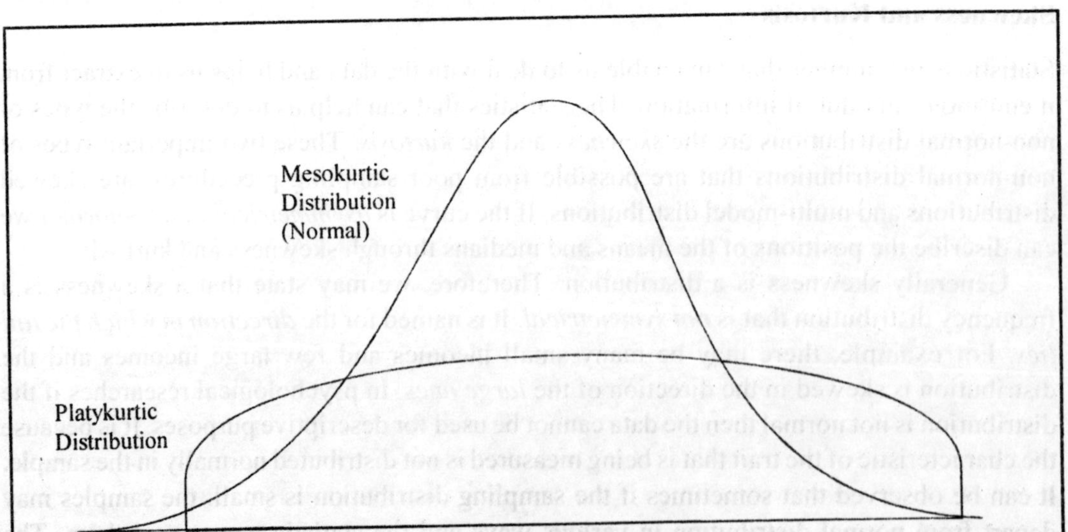

Fig. 31.6 : Platykurtosis and Mesokurtosis

Skewed Distribution

Generally, a skewed distribution will have *long tail* or a *low end* on one side and *a short and high end* on the other side. These two types of ends may occur on either side and therefore in a distribution there is a long tail, the distribution extends towards the low end or left is stated as *negatively skewed* (Fig. 31.7). In a distribution if the longer tail of the distribution extends forward the *high end or right side* is stated as *positively skewed* (Fig.31.8). Therefore, a skewed distribution may be *defined* as non-symmetrical distribution in which most of the scores are closer to one end of the distribution which will have only one mode and the tall extremes are unequal in length.

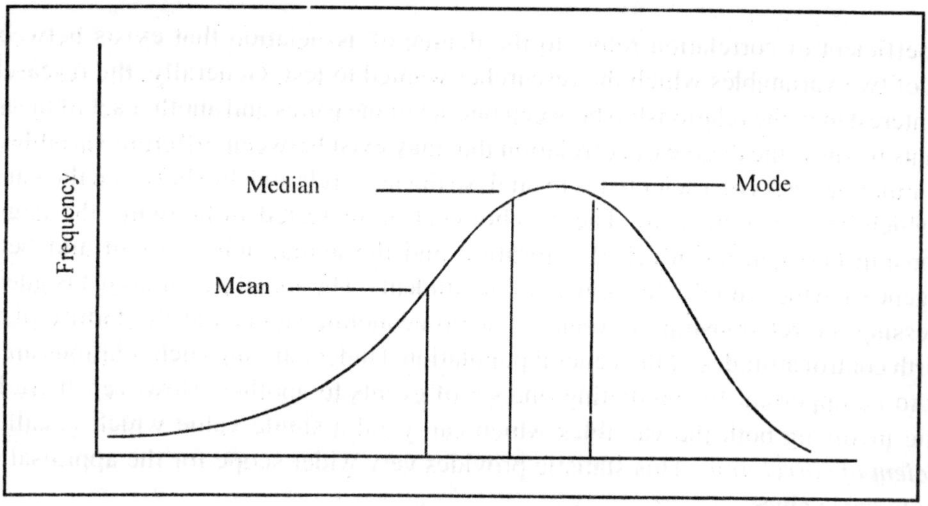

Fig. 31.7 : Negatively Skewed Distribution

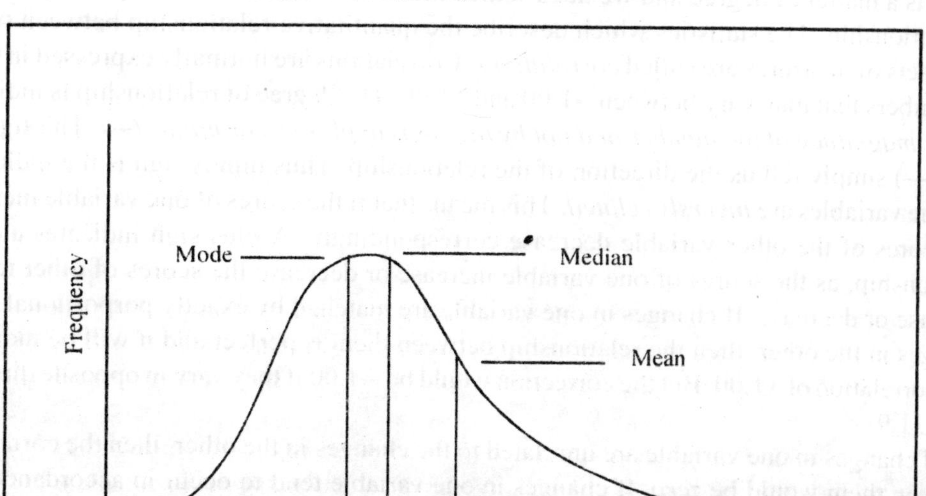

Fig. 31.8 : Positively Skewed Distribution

It is also important for a researcher to test whether his investigation is reliable. This skewed curve can tell him to what extent he is correct. The *skewed curve* for the scores will indicate that the sample he has collected is of non-representative or the sample was not collected according to the random sampling technique or the instrument has been defective.

The skewed distribution may occur if the tool used for the collection of data has been defective, eventhough the sampling technique might be correct. This is known by the skewed distribution which is a departure from the normal distribution.

MEASURES OF CORRELATION

The coefficient of correlation refers to the degree of association that exists between the scores of two variatables which the researcher wanted to test. Generally, the researcher is often interested in the relationship between one set of measures and another set of measures. He wants to know the degree of correlation that may exist between different variables. One of the important tasks of a scientist is to analyse the inter-relationship between the variables with which he is dealing with. The psychologist is interested in knowing the degree of relationship between the level of aspiration and the actual achievement and between intelligence and the school achievement of the students. The social psychologist is interested in assessing the relationship between the socio-economic status and the family planning and birth control attitudes of the general population. Understanding such relations are often found to be important for predicting one set of events to another. However, there is one statistic involving both the variables which can yield a single value which is called the *coefficient of correlation.* This statistic provides very wider scope for the appraisal of the psychological values.

It is a matter of degree and we need some kind of staistical index to express the degree of relationship. The statistics which describe the quantitative relationship between two or more sets of measures are called *correlations.* Correlations are normally expressed in terms of numbers that may vary between –1.00 and +1.00. The degree of relationship is indicated by the *magnitude of the number and not by the signs of plus (+) or minus (—).* The signs of, (+ or —) simply tell us the direction of the relationship. Thus minus sign is the indication that the variables are *inversly related.* This means that if the scores of one variable increase, the scores of the other variable decrease correspondingly. A plus sign indicates a direct relationship, as the scores of one variable increase or decrease the scores of other tend to increase or decrease. If changes in one variable are matched by exactly porportional to the changes in the other, then the relationship between them is perfect and it will be indicated by a correlation of +1.00. But the correction would be –1.00 if they vary in opposite direction (Fig. 31.9).

If changes in one variable are unrelated to the changes in the other, then the correlation between them would be *zero.* If changes in one variable tend to occur in accordance with changes in the other, then the correlation between them would be somewhere between zero and +1.00, and the exact value is depending on the extent to which they covary. One of the most often used methods of correlation is the *person product moment correlation coefficient.*

The often correlation methods are Spearman's rank-order correlation, Kendall's coefficient of rank correlation, contingency coefficient of corrrelation, the Phi-coefficient of four fold paint correlation, Point biserial correlations, Biserial correlation and the Tetrachoric correlations. Among these the Pearson Product Moment Correlation and the Spearman's Rank-order Correlation methods are very often applied in psychological and educational researches.

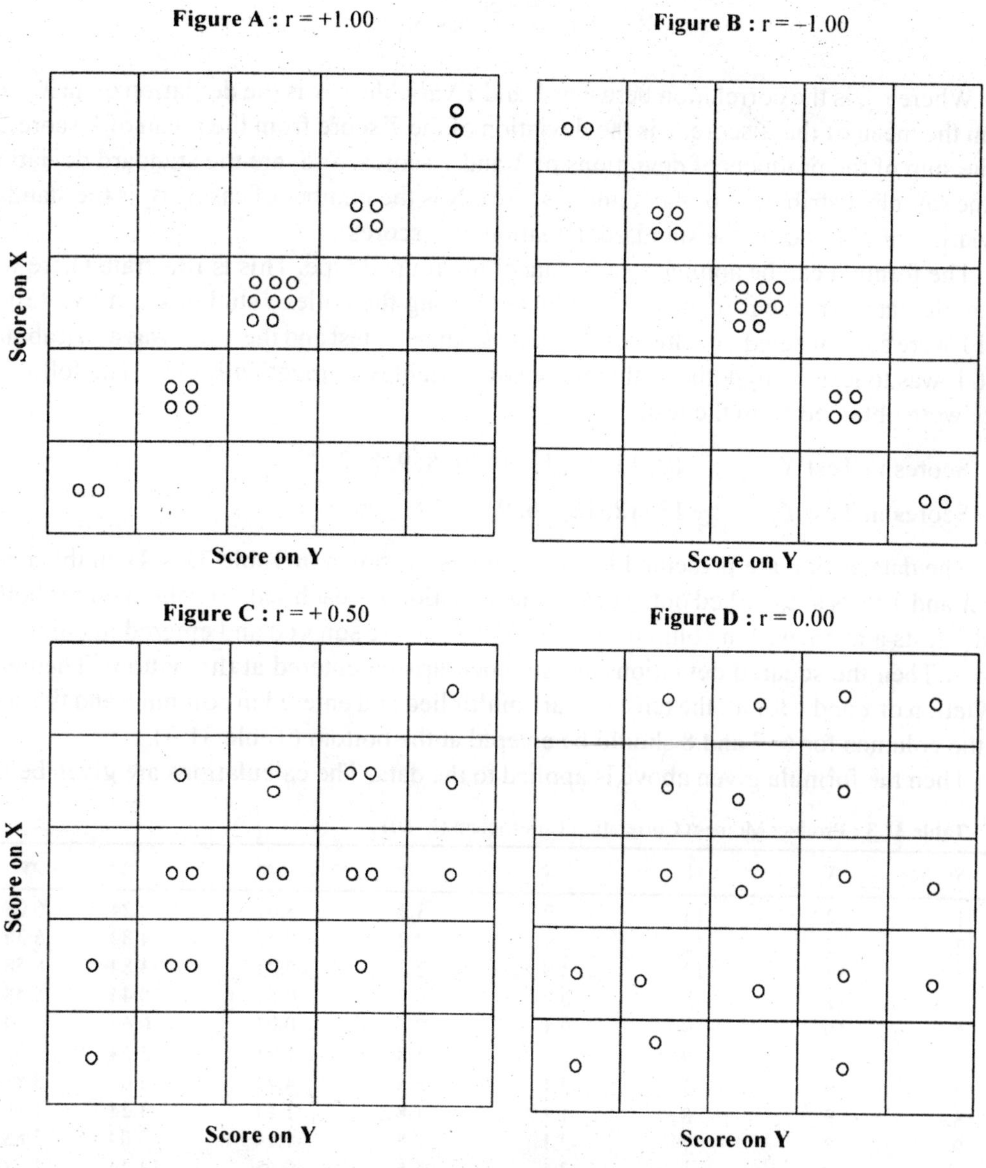

Fig. 31.9 : An Illustrative Model for Scatter Diagram Correlation

The Pearson Product Moment correlation coefficient is a very widely used correlation method. In this method the scores are used as they were obtained from the tests. But in the Rank order method only the ranks of the scores are taken into consideration and the scores are not directly applied. The product moment method is used when the measures are in quantitative terms, the basic formula applied to obtain the correlation coefficient value is :

$$r_{xy} = \frac{\Sigma xy}{N\sigma_x \sigma_y} \text{ or } \frac{\Sigma xy}{NS_z S_y}$$

Where r_{xy}, is the correlation between *X* and *Y* variabless, x is the deviation of an *X* score from the mean of the *X* score, *y* is the deviation of the *Y* score from the mean of *Y* score, *XY* is the sum of the products of deviations of *X* and *Y* scores. $S_x S_y$ are the standard deviations of the sample distribution of the *X* and *Y* scores, N is the number of cases, σ_x is the standard deviation of *X* and σ_y is the standard deviation of *Y* scores.

The formula can be applied to any data of ungrouped type. This is illustrated here with a simple data. An experiment was conducted among the college students and two tests *X* and *Y* were administered, of which one was an Arithmetic test and the other was a Vocabulary test. It was to test the hypothesis that these two variables go together highly. The following data were obtained from the test.

Scores in Test *X* : 12, 13, 14, 11, 10, 9, 8, 9, 8, 7

Scores in Test *Y* : 13, 12, 12, 16, 9, 8, 7, 8, 7, 6

The data at first are presented in two columns as shown in Table 31.3 Then the means for *X* and *Y* tests are worked out. Thirdly the deviation for each score is obtained for both *X* and *Y* tests and entered in columns 4 and 5 and these are squared and entered in columns 6 and 7. Then the squared deviations are summed up and entered at the bottom. Then each deviation of *x* and *y* for all the ten cases are multiplied and entered in column 8 and the sums of the columns for 6, 7 and 8 should be entered at the bottom (Table 31.3).

Then the formula given above is applied to the data. The calculations are given below.

Table 31.3 : Product Moment Correlation Co-efficient (N-80)

Sr. No.	*X*	*Y*	*x*	*y*	x^2	y^2	*xy*
1.	12	13	1.9	3.2	3.61	10.24	6.08
2.	13	12	2.9	2.2	8.41	4.84	6.38
3.	14	12	3.9	2.2	15.21	4.84	8.58
4.	11	16	0.9	6.2	0.81	38.44	5.58
5.	10	9	-0.1	-0.8	0.01	0.64	0.64
6.	9	8	-1.1	-1.8	1.21	3.24	1.98
7.	8	7	-2.1	-2.8	4.41	5.04	5.88
8.	9	8	-1.1	-1.8	1.21	3.24	1.98
9.	8	7	-2.1	-2.8	1.14	5.04	5.88
10.	7	6	-3.1	-3.8	9.61	14.44	11.78
	101	98	0.0	0.0	45.90	94.00	54.72
MEAN	10.1	9.8	0.0	0.0	Σx^2	Σy^2	Σxy

$$S_x = \sqrt{\frac{45.9}{10}} = \sqrt{4.59} = 2.14$$

$$S_y = \sqrt{\frac{94.0}{10}} = \sqrt{9.4} = 3.06$$

$$r_{xy} = \frac{54.72}{10(2.14)(3.06)} = \frac{54.72}{65.48} = +\,0.8257 = 0.836$$

The method of correlation explained above is applied only when the number of cases are very few. The formula is already stated above. Where, r_{xy} is the correlation coefficient, xy is the sum of the squares of the deviations of X and Y multiplied. Σx^2 is the sum of the squared deviations of x, and Σy^2 is the sum of squared deviations of y, and N is the number cases.

The method of correlation described above is applied only when the number of cases are very few. There is also another alternative correlation method, a shortcut method, for computing the correlation. This method does not require the computation of standard deviation. What is required in this methodis to work out the square root and one division for obtaining the correlation coefficient. But the number of cases should be very few. The formula is:

$$r_{xy} = \frac{\Sigma xy}{\sqrt{(\Sigma x^2)(\Sigma y^2)}} = \frac{54.72}{\sqrt{(45.9)\,(94.0)}}$$

where r_{xy}, is the correlation coefficient, $\Sigma\, xy$ is the sum of the squares of the deviations of X and Y variables multiplied.

Σx^2 is the sum of the squared deviations of x

Σy^2 is the sum of the squared deviations of y

Applying the formula to the data we have the correlation value:

$$r_{xy} = \frac{54.72}{\sqrt{(45.9)\,(94.0)}} = \frac{54.72}{\sqrt{4314.6}} = \frac{54.72}{65.68} = 0.836$$

This formula yields almost the same result. The calculation work is somewhat reduced and therefore it is more economical from the computational point of view than the other methods.

There is another method of correlation for the raw scores of ungrouped data. It may deal with any number of cases unlike the other two methods. The statistical procedure is also very simple and many researchers prefer this method, if the number of cases are many. This method has several advantages over other methods of correlation. The data need not be grouped, the raw scores are used as obtained from an investigation from any population and the method can straight away yield the correlation coefficient.

The formula of this method is :

$$r_{xy} = \frac{N(\Sigma XY) - (\Sigma X)\ (\Sigma Y)}{\sqrt{\left[N\Sigma X^2\right]\left[N\Sigma Y^2 - \left(\Sigma Y^2\right)\right]}}$$

Where N is the number of cases, X is the original or raw scores of X variable, Y is the original or raw scores of Y variable and Σ is the sum of the measures.

Table 31.4 : Correlation coefficient obtained from ungrouped raw scores

S.No.	*X*	*Y*	*X*²	*Y*²	*XY*
1.	12	13	144	169	156
2.	13	12	169	144	156
3.	14	12	196	144	168
4.	11	16	121	256	176
5.	10	9	100	81	90
6.	9	8	81	64	72
7.	8	7	64	49	56
8.	9	8	81	64	72
9.	8	7	64	49	56
10.	7	6	49	36	42
Sum	101	98	1069	1056	1044
	ΣX	ΣY	ΣX^2	ΣY^2	ΣXY

$$\gamma_{xy} = \frac{10(1044) - (101)\ (98)}{\sqrt{\left[(10\times 1069) - (102)^2\right]\left[(10\times 560) - (98)^2\right]}}$$

$$= \frac{10444 - 9898}{\sqrt{\left[(10\times 1069) - 10201\right]\left[10560 - 9604\right]}} = \frac{542}{\sqrt{(489)\ (956)}}$$

$$= \frac{542}{\sqrt{467484}} = \frac{542}{683.73} = 0.80$$

According to this method the correlation is 0.80. This is little lower than the other methods.

This formula can be applied to any ungrouped raw scores. We need not work out the deviation scores in this fomula. It is straight method and does not require much statistical calculations.

There is also another method of correlation coefficient which is called the *Scatter Diagram Method* of correlation. This method involves much laborious task in working out the coefficient of correlation (Fig.31.9).

Spearman's Rank Order, Correlation Method

The product moment correlation method described above is mostly preferred by the researchers. When their data are very wide and.the number of cases are many. But it will not be always. possible to apply Pearson Product moment correlation method for all kinds of data. There are certain data for which we apply the Spearman's Rank order correlation method, specifically when the sample size is small. It is generally stated by the Greek latter *rho* (P). The formula for the Spearman's rank *w* order correlation method is:

$$P = 1 - \frac{6(\Sigma D)^2}{N(N^2 - 1)}$$

Where *D* is the difference in ranks for the scores for any one subject, N the member of cases. We can apply this fornula to the data given in Table 31.5. Σ is the sum of the difference squared, *p* is the Spearman's correlation coefficient.

Table 31.5 : Showing the Rank Order Correlation Coefficient between the scores of Aptitude and Achievement tests for 10 persons

S.M.	Aptitude	Achievement	(Ap) R_1	(Ach) R_2	D	D^2
1.	45	73	8.5	6	2.5	6.25
2.	69	77	3	5	2	4.00
3.	50	83	6	4	2	4.00
4.	46	48	7	9	2	4.00
5.	33	47	8	10	5	25.00
6.	39	57	10	8	2	4.00
7.	45	68	8.5	7	1.5	2.25
8.	74	89	1	2	1	1.00
9.	70	91	2	1	1	1.00
10.	63	95	4	3	1	1.00
					ΣD^2	52.50

$$P = 1 - \frac{6(\Sigma D)^2}{N(N^2 - 1)} = 1 - \frac{6 \times 52.50}{10(10^2 - 1)} = \frac{315.00}{10 \times 99} = \frac{315.00}{990} = 0.318 = 0.682$$

The correlation coefficient between these two variable is 0.652 which is highly significant.

Inferential Statistics

Inferential statistics are the statistical processes by which conclusions are drawn in some way from certain facts. By applying the inferential statistics we can obtain conclusions for our observations. We have dealt with the descriptive statistics in the proceeding section. But there are certain facts that do not come under descriptive statistics. Therefore, we resort to the inferential statistics. The researcher is interested this much more than the descriptive data. He wishes to interpret his data in terms of general principles which may enable him *to make predictions.* Moreover he wants to know how for his interpretations are reliable. The process of generalization from the sample of the data and to measure the certainty of the researcher's generalizations involve the use of inferential statistics. The most basic aspect of the inferential statistic is the *representative sampling.*

Sampling : The scientific measurements are limited to a finite number of events called *samples.* From samples he believes that he can generalize much larger class of any events, called *universe or populations.* For example, if we want to know the family planning attitudes of the people of India, we cannot test all the hundred million Indians. Therefore, we must limit ourselves to some sample of the people that we believe is the representative of the total population. But if only we are sure that our sample of the Indains is representative of the total population in all respects. Then we have no need to have inferential statistics, because the characteristics of the sample would be identical with the characteristics of the population.

But, however, carefully the sample was drawn, their means and standard deviations would be rarely identical to the means and standard deviations of the total population from what the sample was drawn.

There is always *the sampling error* and the nature of it depends upon the sampling technique applied. Basically there are *two sets* of sampling errors in any sampling, These are the chance error and biased error. The *chance error* refers to accidental differences between samples and the population that they are supposed to represent. The other one is the *biased error.* This biased error refers to the result of some *non-representative method* of sampling the population.

All the elements of the population that are being studied do not find proper representation in the biased sampling. For example, if a researcher wants to assess certain personality traits of a population, like *dominance submission*, might have conducted the investigation in an urban area with a view that he can collect the data easily and believed that the traits are highly prevalent in that area. In this case it is apparent that any conclusion drawn from this sample might not apply to the general population. Since the sample is taken from a specific urban area, mostly consisted of high socio-economic status people, the sample might not be true representative of the traits, dominance submissiveness. Therefore, this sample is a *biased one.* This is a poor selection procedure in sampling, which may be called as the *experimental error,* which occurs generally due to the failure in experimental control.

To avoid such errors statistians have evolved two general techniques to avoid the biased selection. The first one is the *random sampling* and the other is called *stratified sampling.*

Random Sampling : It is the procedure that can assure that all the elements of the population will have *equal proportional chances* of being included in the sample. This is the most preferred method of sampling when the characteristics of the population is not known. This method consists of mostly *chance factors* such as *tables of random numbers,* or based on *tossing of coins.* The chances of any paticular element entering the sample depends upon its proportional representation in the population and no chance of bias.

Stratified Random Sampling : This is also called *controlled sampling.* This method of sampling is applied when the characteristics of the population is known and when it is not possible to select a sample randomly from the whole population. In such cases the researcher selects elements in the populations according to the known proportion in the population. Out of several characteristics the researcher chooses any one or more than one and assigns for his sampling. His collection of sample pertains to one single characteristic or more and the sample is supposed to possess the specific characteristic in which he is interested. It may also be called as sub-group in the population identifying the same characteristic as the parent population possesses.

Suppose for example, the researcher wants to collect a sample of 20 per cent of the students from a college of various classes to study their attitudes towards higher education. It is a degree college having only three year courses. The college authorities admit in the first year 1000 students and due to drop out, in the second year it was reduced to 600 and in the third year the strength reduced to 400. On the whole he has to collect a sample of 400. This is 20 per cent of the total strength and 20 per cent of each class strength. Since the same percentage from each class is maintained the sample is proportionately distributed. The composition of the college population is 1000 + 600 + 400 = 2000. The proportion of students in each class is worked out.

Table 31.6 : Showing the composition of population and the distribution of population in every year.

Year	*Population*	*Proportion*	*Sample size*
1st Year	1000	0.50	200
2nd Year	600	0.30	120
3rd Year	400	0.20	80
Total	2000	1.00	400

Sampling Error : The sampling, either it is random sampling or statified is not without error. There is always some error involved in generalising the result from the sample. The chance distribution of the sampling error can be worked for every sample.

Standard Error of the Mean : There is a statistic with which we can proceed further with the standard deviations of the distribution of the sample mean. The statistic is the standard error of the mean. This is obtained by working out the *square root* of the numbers of measures of the sample minus one (N–1). The member of measures in the sample has the importance in determining the size of the standard error since it is the denominator of the formula. The smaller the number of cases the greater the standard error and more likely *it is the sample mean* deviates from the true mean. Therefore, we may state that the samples must be large and also must be representative for drawing valid conclusions. The formula for the standard error of the mean is :

$$SEM = \frac{SD}{\sqrt{N-1}}$$

Where SD is the standard deviation, *N* is the number of cases.

Critical Ratio : There is also another important use of the normal distribution of the sampling error. We have from the means of several different samples and also we have the differences between sample means. In these measures there is a chance of probability error. These facts can be used to determine the probability that two different samples have been drawn from the same population and suggest the idea to what extent these two means are different from each other. Statisticians have provided us a statistics called *critical ratio or the 't' test* which allows us to test the possible difference between the sample means whether it is true or not. The number of cases in each sample has crucial significance, The larger the number of cases in each sample, it is more probable that the difference between the two samples is true.

The standard error of the mean difference suggests to the experimenter to test the probability that the mean differences from various experimental situations are not due to chance. The importance of the use of the standard error is felt very much at this stage. In the inferential statistics the important subject matters are mainly the sampling methods and their related statistics, the standard deviations of the samples, the standard error and the critical ratio.

The formula to work out the critical ratio is:

$$CR = \frac{M_1 - M_2}{\sqrt{SEM_1 + SEM_2}}$$

Where :

M_1 and M_2 are the mean.scores of the two samples.

SEM_1 is the standard error of the first mean.

SEM_2 is the standard error of the second mean.

In the inferential statistics the critical ratio plays a crucial role in deciding the level of difference between two measures or two variables. A high critical ratio of 2.58 and above indicates that the difference between the two means is significant at one per cent level and if it is 1.96 it is significant at five per cent level. Any value less than 1.96 indicates no difference. But for a much smaller sample a t-test is worked which is also similar to the critical ratio.

DICTIONARY OF PSYCHOLOGICAL TERMS

(Glossary)

(A)

Ability: The skill or potentiality an organism possesses for performing some task.

Absolute threshold: It is the minimum amount or strength of a stimulus to arouse response in an organism.

Achievement: It is the successful performance of a task by an organism.

Adaptation: The act or the process of adapting or the state of being adapted, something that is produced or created to suit to the situation.

Adjustment: It is the way of making oneself to meet the situation without any humiliation. It is a way behaviour of an organism amicably with the individuals or to the environment.

Action identification: It is the process of labelling and interpreting one's own behaviour.

Adolescence: It is the period in human life after the childhood period, mostly the teenage period from the age of 13 to 19 years.

Adrenal Gland: It is a part of endocrine system situated below the kidney. It secretes the hormone called adrenin. A very powerful hormone for meeting compact situations.

Adrenin: It is the hormone secreted by the adrenal glands.

Adrenal Cortex: It is the outer cells of the adrenal glands that produce cortiods.

Adrenal Medulla: It is the central portion of the adrenal glands that produce the hormones, epinephrine and norepinephrine.

Affective experience: It is the feeling aspect. It may be pleasant or unpleasant feeling tone.

Afferent: Those nerves are carring impulses towards a cortex as when sensory nerves carry nerve impulses towards the brain or spinal cord.

Affiliative need: It is a motive to seek interpersonal relationship and to form friendships.

After image: It refers to the process of end of sensation. The sensation of the stimulus may continue for sometimes even after the disappearance of the stimulus.

Age norm: Average score of a test conducted on a population.

Aggression: It is an emotional behaviour pattern expressed by a person in a hostile manner. It is a hostile action towards some person or object.

Albedo: The percentage of light falling on a surface that is reflected by the surface.

Alcohol: It is non-addicting depressing drug contained in wine, beer, brandy and whisky.

Allocortex: The oldest area of the cerebral cortex, orginally devoted to smell.

All-or-None Law : This is with reference to the functions of the cell of the neuron. The nerve cell or gland cell either it may respond to its fullest extent to change the electrical and chemical conditions or does not respond at all.

Alpha Cells: Endocrine cells of the pancreas secrete the hormone gulcogan, a hormone that stimulates the storage of blood sugar in the liver and muscles.

Altered States of Consciousness: The human consciousness is very alert and active. But sometimes it is not. It is the visual window through which we perceive the world. But due to some dramatic changes the visual window is distorted or fail to open. These mysterious changes are known as altered states of consciousness. The altered states may occur in three ways such on inducing drugs, influence of sensory stimulation and psychological method of hyphosis.

Ambiguous Figure: An outline figure which may be perceived in more than one way.

Amnesia: Partial or total loss of the memory in a person. The memory is often regained.

Amygdala: It is a nuclear complex buried in the temporal lobe of the brain, important in the organization of emotion by the limbic system.

Anal Stage: According to Freud it is the stage of psychosexual development in which one or two year old child's most intensive pleasures are derived from the activities associated, with elimination.

Anxiety: Intense uneasiness, about the future.

Anatomy: The branch of science that deals with the structural relationship of parts of the body.

Androgene: It is male sex hormone, produced by the testes of the male.

Anemia: Lack of red blood corpuscles.

Anoxia: A condition in which the tissues are not receiving enough oxygen from the lungs via the circulating system to sustain their metalotic activity.

Antagonist Muscles: The muscles that would oppose a given movement, if not relaxed, for example, the extensor muscles of the arm are the antogonist muscles for arm flexion.

Anterior Chamber: The cavity of the eye ball that lies between the cornea and the iris.

Aphasia: It is the impairment in language skills, generally caused by brain damage. It is the inability to recognize words by sight or by sound.

Apparant Motion: Perception of motion under condition in which no physical movement occurs.

Apptitude: It is the potential ability in an organism, which can be exercised by an organism where there is a need. It is the capacity inherent to profit by training on exercise in some particular area of skill.

Aqueous Humour : A fluid that is transparent to light, filling to space between the lens and the cornea of the eye.

Arbitration: It refers to the intervention of a third party in bargaining in which recommendations of the person intervening are binding upon the parties involved.

Attachment: It is the emotional tie of an infant to mother or other care takers.

Attention: Active selection of some aspect of complex experience by the perceptual process. Selecting one and disregarding others.

Attitude: It is the tendency to respond to positively or negatively to other individuals, objects, places, occupations.

Attraction: The degree to which an object or a person or persons make or to develop strong liking or love with the beautiful characteristics.

Attribution: It is a process through which one seeks to determine the causes of others' behaviour and acquired knowledge of their stable traits and dispositions.

Authoritarian Personality : It is a type of personality characterized by rigid adherence to middle class values, extreme awareness of authority relationships, denials of personal sexuality, a tendency to exploit others but to feel exploited and a tendency towards projection.

Autonomic Nervous System (ANS): The part of the nervous system that consists of the sympathetic and parasympathetic divisions and is involved in the control of the body's internal environment.

Axon: An extension of the cell body of the neuron that carries nerve impulses away from the cell body and to other neurons.

(B)

Balance Theory: It is a cognitive theory of interpersonal attraction. Attraction is assumed to be based on the relationships among cognitions about another person and about various objects.

Basal Ganglion: One of the four deeply placed masses of gray matter within each cerebral hemisphere the caudate neucleus, the amygaloid nuclous, the claustrum.

Base-rate Fallacy: In social judgments people tend to ignore or underuse base-rate information, which is about how prevalent or frequent something is in general.

Basilar Membrane: The membrane in the cochlea of the inner ear that supports the organ of corti.

Behaviour Modification: There are varieties of therapies based on learning principles, classical and operant conditioning and modelling procedures used to shape and adjust the behaviour of the emotionally troubled people. This may be also called behaviour therapy.

Behaviour Assessment: It is a procedure of personality assessment. Evaluation of personality is made by observing behaviour in representative natural situations.

Behaviourism: One of the schools of psychology founded by J.B. Watson. According to this school, psychology is a behavioural science and studies the observable overt behaviour.

Behaviour : Any observable act of an organism.

Bilateral Transfer: The effect of practive in a task with one hand in learning is performed with the other hand.

Bimodel Distribution: A frequency distribution with two modes.

Bipoler Cell: In the ratina a cell connecting the rods and cones with ganglion cells.

Blind Spot: An area in the retina where a person is effectively blind. Fibres leave the retina at this point to become the optive nerve.

Body Language: It is the cue provided by the position, posture and the movement of others' bodies or bodily parts.

Brain: It is the whole of central nervous system except the spinal cord.

Brain Waves: Rhythmic fluctuations of voltage between paired locations of the electrically active brain, resulting in the flow of an electric current.

Brainstem : All parts of the brain except the cerebral and cerebellea hemispheres.

Brainstem Reticular Formation (BSRF) : Mass of gray matter of the brainstem made up of mainly short branching colgi type-II cells. The BSRF arouses the brain to activity and stimulates or inhibits extensor motorneurons.

Brightness : It is the dimension of visual experience that depends for the most part on the energy of the dimension from black to white. It is the intensity of the visual experience.

Brightness Constancy : It is the tendency for the objects to be seen with some relative brightness under various levels of illumination.

British Empiricism : It is a school of philosophy which advocates that knowledge is gained through experiences of the senses and there are no innate ideas.

Broca's Area: It is the area in the frontal cortex area, hypothesized by Broca which is responsible for speech. Broca's area mainly deals with motor functions.

(C)

Caffeine: A mild and nonaddiciting cortical stimulant, present in coffee, tea and in some soft drinks.

Calories: A unit of heat, commonly used in diets. The unit is 1,000 times the amount of heat necessary to raise the temperature of one gram of water by 1°C.

California Psychological Inventory (CPI): It is a personality inventory for the general population of young people to assessing the personal adequacy, social maturity, interest, and thought efficiency.

Cancer : It is an abnormal tissue growth (tumour) aroused by uncontrolled cell multiplication.

Candle: A standard measure of brightness of light source.

Capillaries: Thin walled tiny blood vessels, through whose walls oxygen, corbondioxide and waste are exchanged with the extra cellular fluid.

Cannon-Bard Theory : It is a theory of emotion. According to this theory, one emotional stimulus activates the cerebral cortex and bodily processes at the same time. Bodily changes and the emotion occur together.

Cordiac Muscle: It is the muscle that for the heart, intermediate in structural and functional characteristics between striate and smooth muscles.

Case History Method : One of the methods applied in psychological studies. It is a way of conducting investigation of psychological problems through individual's biographical information of various types.

Castration Anxiety : According to Freudian theory the little boy developes a fear that his father will castrate him, because of his sexual attraction towards his mother.

Catharsis : According to psychoanalytic theory it is the elimination of a complex by bringing to consciousness and allowing it to expression. Generally it is reduction of an impulse or emotion through direct or indirect expression, mostly by verbal and fantasy expression.

Cathexis : It is the psychic energy invested in an idea, object person or action.

Central Fissure: It is a fissure of each cerebral hemisphere that separates the frontal and the parietal lobes.

Cell : A protoplasmic body which is the unit of life. A cell can be an independent living organism or a specialized unit of a complex many celled organism.

Cell Body : It is the part of the neuron that contains the nucleus and nutritional mechanism.

Cell Membrane: The membrane that separates the cytoplasm of a cell from the environment of the cell.

Cell Metabolism: Chemical reactions of a cell that are required for life.

Central Nervous System (CNS): The brain and the spinal cord constitute the central nervous system.

Central Sulcus : The sulcus divides each cerebral hemisphere into an enterior one third and the posterior two thirds. It separates the somesthatic sensory and the motor projection areas.

Cephalon: Cephalon means brain. Cephalon has three main divisions such as prosencephalon (forebrain), the mesencephalon (midbrain) and the rhombencephalon (hind brain).

Cerebellum: A structure with two hemispheres of the central nervous system just below and behind the cerebrum and above the medulla. It is covered with transversely fissured gray matter. The cerebellum is involved in the maintenance of muscle tonus and in the coordination of movements.

Cerebral Cortex: Surface layers of gray matter of the two hemispheres of the cerebraum.

Cerebral Hemisphere: One of the two nearly symmetrical halves of the cerebrum. The left hemisphere controls the rightside of the body. The right hemisphere controls the left bodily functions. The hemispheres are deeply separated by the longitudinal fissure strongly joined by the corpus collosum.

Cerebrum: The main division of the brain in vertibrates consisting of a right and left hemispheres. It is active in thinking and learning.

Character Disorder : It is a defect of personality structure, characterized by immaturity, antisocial activities and lack of social responsibility.

Cerebrospinal Fluid : The tissue fluid surrounding the brain and the spinal cord and filling the ventricles.

Choroid coat : The vascular pigmented middle layer of eye tissue that lies beneath the sclerotic coat.

Chromosomes : The microscopic rod shaped bodies in the cell nucleus. Chromoscmes contain the DNA molecules that govern hereditary characteristics, cell specialisation and cell function. They are paired bodies in the cell of the body, contain the genes.

Chromosomes X and Y : Sex Chromosomes : An XX pair produces a female child; an XY combination produces a male child.

Chronic Disorder : A long lasting persistent disorder.

Chronological Age : Age in years and months it is used for computing the I.Q.

Ciliary Muscle : Muscle which controls and focuses the lens of the eye.

Clairvoyance : It is a form of extrasensory perception in which the subject or perceiver is expected to recognize a stimulus that is influencing neither his own sense organs nor those of another persons.

Classical Conditioning : A learning theory formulated by the Russian physiologist, Pavlov. In this process of learning an organism is presented with a natural stimulus (CS) and a biologically significant stimulus (US) in fixed order without repeating to the organism's behaviour. This is in contrast with operant conditioning.

Cocaine : An additive stimulant for the central nervous system.

Cochlea : The coiled fluid filled structure of the inner ear. The cochlea contains the structures that transduce sound vibrations into nerve impulses in hearing.

Clinical Psychologist : A psychologist trained to administer tests, conduct interviews, make diagnoses and administer psychotherapy, works in a hospital or institutions.

Coefficient of Correlation : An index which expresses in degree of correspondence or relationship between two sets of measures.

Cognition : Means knowledge. In the psychological sense it is the process by which individuals acquire and make use of knowledge.

Cognitive Behaviour Therapy : This therapy is designed to help an individual to understand and deal with the irrational beliefs and values that cause distress.

Cognitive Dissonance : It is an unpleasant state that occurs when individuals observe inconsistency between their attitudes on between their overt behaviours.

Colour Blindness : A visual defect that makes it impossible or difficult for a person to distinguish between colours that most people can see. In the most common form of colour blindness there is a confusion of red and green.

Colour Constancy : It is the tendency to see familiar objects as of the same colour regardless of illumination and apparent changes in hue.

Collective Unconscious : It is formulated by Carl Jung. According to him a portion of the unconscious, a pool of ancient images which come forth in ghost stories, dreams, literature, art, religious beliefs inherited and common to all human kind. It is in contrast to personal unconscious.

Community Psychology : It is a branch of psychology developed out of the dissatisfaction of the services rendered by clinical psychology and psychotherapeutic techniques. It focuses its attention on the individuals adaption to community structure of which he is a member. It adapts preventive measures and tries to resolve community problems. According to community psychology, the community is the cause of all individual problems and it also can serve as the therapeutic factor and resolve the problems.

Complimentary Colours: When colours are mixed it results in producing an achromatic colour.

Concept: It is a representation of the common properties of objects, events or ideas that are otherwise distinguishable.

Concept Formation: Acquiring a concept by the process of grouping the objects, events or ideas, all of them share common attributes or properties. For example, the Vigotsky concept formation test.

Concurrent Validity: Validity of a test is established by comparing the scores of a test with those obtained on another test of well established validity.

Conditioned Response: In the classical conditioning theory, in the experimental situation a response is made to a particular stimulus. For example the experimental dog is made to respond to the ringing of the bell. On hearing the bell sound the dog expects that food is coming. Thus the dog's response is conditioned to the bell sound.

Conditioned Reflex: A reflex initially set off by a sensory stimulus A, and later aroused by the stimulus B, which has been associated with the stimulus A.

Cone: Retinal receptor cell in the eye which responds in terms of colour and gives vision acquity.

Conflict: In facing two or more incompatible motives an organism is put into a dilemma as to which is to be chosen.

Conjunctive Concept: A concept for which all examples have in common two or more attributes.

Conscience: Conscience is an internal recognition of standards of right and wrong, good and bad, moral and immoral by which the individual judges his own conduct.

Conscious: Conscious is an affective term, which is denoting or relating to a part of human mind that makes aware of a person's self-environment and mental activity to certain extent determines his choice of action. It is only a small part of the mind.

Consciousness: It is a noun form, has a long histroy. Different definitions and explanations are given. According William James consciousness is the stream of thought or subjective life. According to Bertrand Russell consciousness has two different meanings. One is, we are said to he conscious of something and the other is, consciousness is regarded as a quality of mental occurrence, not having any relation to other things. One definition is that consciousness is awareness of one's behaviour, its courses and effects.

Consensus: The extent of actions shown by one person are also shown by others.

Contour: Outline or boundary of an object.

Contrast: Significant difference in stimulation, as between silence and noise, dark and light.

Convergent Thinking : A form of thinking in which there is a search for a single answer to a question. For example, 2 + 2 = ? in contrast to divergent thinking in which there are many solutions.

Corpus Collosum: Fiber system connecting the homotopic areas of the two hemispheres. Split-brain patterns are those where corpus collosum has been severed.

Cortex: External layer, synonymous with neocortex.

Cranial Nerves: A set of 12 pairs of nerves conveying sensory and motor signals to and from the head.

Critinism: Retarded mental and physical development resulting from thyroid insufficiency.

Criterion: In testing, the job or other measure of performance that the test is supposed to predict.

Culture: Social organization, including language, customs, traditions, attitudes which characterizes a group of people or individuals.

(D)

Dark Adaptation: Adaptation of the eye to low levels of illumination.

Deference Mechanisms: Adjustments that enable a person to avoid facing a painful fact or an unpleasant situation.

Delusion: A false belief, a common symptom of paranoid schizophrenia.

Dendrite: Tree like process of the receiving end of the neurons.

Dependent Variable: A variable that changes as a result of a change in the independent variable.

Deviation Score: It is the difference between an obtained score and the mean, indicated by the letter-$\overline{X}$

Diencephalon: The region of the brain that includes the hypothalamus, thalamus and epithalamus. It is the posterior part of the forebrain from which the parts develop.

Difference Limen or Threshold: The smallest difference that can be perceived between two stimuli.

Discrimination: In learning, a differential response- reaction to a positive stimulus and not to a negative stimulus.

Divergent Thinking: Thinking which goes off in many directions in search of solutions to a problem, contrast with convergent thinking.

Dizygotic Twins: Twins who develop from two fertilized ova. Fraternal twins contrast with monozygotic twins, who develop from one fertilized ovum.

Doppler Effect: The rise in pitch of a sound when it approaches us and its lowering as it moves away.

Down Syndrome: It as a form of mental retardation with noticeable physical signs, like slanted eyes, thick tongue and lips, straight hair etc.

Dream: Dream, generally, is a mental activity. It is the form of an imagined series of events occurring during certain phases of sleep. According to Freud dream is a sublimated form of our repressed unfulfilled sex desires living in the inconscious mind.

Drive: It is a basic tendency to activity. It is also a condition within the organism which predisposes the organism to some action, for example, hunger, thirst, sexual feeling.

DSM-Ill: The present revision of the Diagnostic Statistical Manual of mental disorders published by the American Psychatric Association, 1980.

Dual-coding Theory: The theory that materials in memory are represented in two systems, a system of images and a verbal system.

Duplicity Theory of Vision: The theory of vision which has the essential fact that vision consists of two separate mechanisms.

Dyscelculla: A learning disability of which the major symptom is difficulty with calculations.

Dyslexia: A learning disability of which the major symptom is disability in reading.

(E)

Eardrum: The membrane at the entrance to the middle ear that is set into vibration by sound waves striking it.

Ectomorphy: One of the body types identified by Sheldon. The long thin angular highly muscled person is called like this.

EEG : Electroencephalogram.

Effectors: An organ that can produce a response, a muscle or gland.

Efferent Neuron: A neuron which conducts impulses away from the central nervous system towards an effector. It is in contrast with afferent neuron.

Ego: Concept mainly used in psychoanalytic theory. It is a major part of the personality that behaves relatively postponing pleasures of the id and directing the energies of the id into socially acceptable channels.

Eidetic Imagery: A clear vivid identical image of an object or event evolved after it has been experienced. Eidetic imagery is often but inaccurately referred to as photographic memory.

Elective Complex: In psychoanalysis the female child's erotic attachment to her father and antagonism towards her mother at about the age of four during the phallic stage. These Id impulses are usually repressed in some way by the fear of punishment.

Electroconvulsive Shock Therapy: It is in therapeutic technique often used with severely depressed patients, convulsions are produced by passing an electric current through the brain.

Electroencephalogram: It is an instrument records the brain waves. The record is with written of the spontaneous electrical activity of the brain, obtained by attaching electrodes to the scalp and greatly amplifying the voltage changes, commonly referred to as the record of brain waves.

Embryo: In human species this referred to as the conceptual period from second week to eight weak following the conception.

Emotion: It is a stirred up state of an organism. An excited state of mind based on the physiological departure from homeostasis.

Empathy: The feeling of a person into the feeling of another person, in the work of art or into the activity.

Empirical Criterion Method: It is a method for developing personality tests that selects the items that people who differ in the trait to be tested differently.

Encoding: Putting malerials into a form (code) that the memory system can handle.

Endocrine System: This refers to the ductless glands of the endocrine system of internal secretion. The hormones secreted by the endocrine glands regulate the bodily functions and they are important for shaping personality traits. They secrete the hormones directly into the blood streams.

Endomorph: One of the body types identified by Sheldon which is characterized by round or large body and short arms and legs.

Epilepsy: A condition characterized by recurrent seizures of various types associated with a disturbance of consciousness.

Episodic Memory: Memory of a particular incident or a particular time or place.

Estrogen: One of the several female sex hormones, secreted by the overies, they stimulate the maturing of the females sex organs and the development of her secondary sex characteristics.

Ethology: A branch of zoology concerned with the behaviour of animals particularly in their natural habitation and in this sense allied to psychology.

Etiology: It refers to all the causative factors of diseases and abnormal conditions.

Eugenics: It is the study and application of the principles to improve the genetic qualities of a population.

Evolution: The development of organic species through natural selection.

Evolutionary Theory of Smelling: The theory that advocates that smelling has evolved because the response has adaptive value.

Excitation: A general term for positive influence favouring the occurrence of a response, contrast to inhibition.

Exocrine Gland: A gland that secretes the products through ducts to the site of action, contrasted with endocrine glands.

Experimental Method: The method in which conditions are controlled and the sources of error are eliminated and the effects of the independent variables upon the dependent variable are measured.

Extinction: It is a state in which there is a gradual weekening of a conditioned response by repeated stimulations without reinforcements.

Extrovert: It is a personality type in which interests are directed towards the outer world and social life rather than towards oneself.

Extrasensory Perception (ESP): It is a process of perception developed by J.B. Rhine. It is a perceptual experience consisting of perception not mediated by sense organ stimulation.

(F)

Facial Nerve: A cranial nerve that innervates facial musulature and some sensory receptors.

Factor Analysis: It is a statistical technique by means of which common factors are identified from the correlations among a large number of tests, ratings and other measures.

Fallopian Tube: The structure between the ovum and uterus where fertilization takes place in humans. It is also called oviduct.

Fantasy: It is daydreaming, fanciful and uncontrolled imagination, free imagination.

Fat : A large complex of carbohydrate that provides long term energy storage.

Fechner's Law: S = K log 1 + A. The perceived intensity of a stimulus (S) varies with logarithm of physical intensity (1), A is an intercept-constant.

Feeblemined: Mentally deficit persons whose intelligence quotients are for below the normal people.

Feeling-tone: The feeling of pleasantness or unpleasantness of an experience.

Fetus: In human species the conceptions from the eighth week following conception to birth.

Figure-Ground Relationship: In the visual field the tendency of a pattern to stand out sharply definitely against an extended background.

Fixation: Stereotyped response developed as a consequence of conflict. In psychoanalysis arrested development of the libido in some particular stage of psychosexual development through excessive gratification or deprivation.

Flashbulb Memory: An immediate and vivid recollection of a notable event often a stressful one.

Fluid Intelligence: The term applied by Raymond Cattell to explain the cognitive ability that is not influenced by prior learning. It is contrast to crystallized intelligence.

Forebrain: The uppermost part of the brain which shows its most advanced development in the higher vertibrates. It includes hypothalamus and related structures.

Fovea: The region in the retina area that is closely packed with cones. The region is stimulated by the image of an object which the eye looks directly.

Frame of Reference: The background of reference against which a person's judgments are made.

Fraternal Twins: Two children in the fetus developed from two egg-cells.

Free Association: In psychoanalysis the technique of making the patient to report all thoughts as they come to his conscious mind.

Free Morpheme: A morpheme which has meaning when it stands alone, a word for example, is in contrast with bound morphene.

Frequency Distribution: A distribution in which scores are arranged from highest to lowest with the intervals of fixed size.

Frequency Polygon: Graph or figure obtained by plotting a frequency distribution. Frequencies are plotted as points above the midpoints of the intervals laid off along the horizontal axis.

Frontal Lobe: Of the four lobes this is the biggest one in the cerebral area. It consists of motor projection areas and the speech controlling areas. This is responsible for reasoning and problem solving.

Frustration: Any state of affair which prevents a person from obtaining a desired goal, the condition of being thwarted in some purpose.

Frustration-aggression Hypothesis: It is stated that all aggressive behaviour stem from the feeling of frustration, but not all frustration lead to aggression.

Fugue: It is a neurotic behavioural symptom in which the individual makes a flight from home and starts a new life elsewhere.

(G)

Galvanic Skin Response (GSR): A lowering of the resistance of the skin to the passage of electrornic current, normally occurs in emotional states.

Ganglion: A group of nerve cell bundles lying outside the brain and spinal cord forming a sort of nerve centre.

Ganglion Cell: In the retina, one of the nerve cell, fibres form the optic nerve.

Gene: The basic unit of the transmission of heredity attributes an amount of DNA located in a chromosome.

Genius: Persons of unparalleled achievements, formerly a person with an IQ of exceeding 150 to 180, but more recent discussions impose an additional criterion of supreme accomplishment contrast with giftedness.

Genotype: The inherited characteristics of an organism also a totality of genes possessed by it: contrast with phonotype.

Germ Cell: It a reproductive cell, a sperm or an egg cell which units with other cell to form a new individual of species.

Gestalt Psychology: It is a school of psychology developed in Germany which emphasises the wholeness of behaviour, placing more importance on perception. According to this school the whole dominates and gives meaning to the parts.

Giftedness: Persons having extraordinary level of intelligence are said to be gifted persons.

Glial Cells: Neurally inactive cells of the brain and spinal cord which perform supportive nutritive functions.

Gradient Texture or Texture Gradient: The texture of the object provides certain cues to our perceptual process. The nearer objects appear rough, irregular and full of dusts, while the distant objects appear smooth, regular and clean, for example houses, roads, mountains, lawns have these characteristics.

Graphic Rating Scale: A scale in which the author makes his judgements by marking somewhere along the line, one end of which describes the high, the other end low degree of the trairt.

Group Dynamics: This term is used to describe the group behaviour in which there is constant mutual give and take among the individuals in a group.

Group Therapy: It is a counselling technique in which groups of patients discuss their personal difficulties under the direction of a therapist.

Gray Matter: Parts of the Central Nervous System (CNS) containing many cell bodies that are dominated by cell bodies which are devoid of myclin, contrast to white matter.

Gyrus: It is a ridged or raised portion of a convoluted brain surface contrast to sulcus.

(H)

Habit: A stimulus-response bond that is acquired automatically and often gradually through occurrence of stimulus-response reinforcement contingencies.

Habituation: A form of non-associative learning characterized by a reduction in response strength following repeated presentation of the stimulus.

Hair Cell: A hearing receptor to the organ of corti., so called because the cell contains hair like projections.

Hallucination: Generally it is a false perception. A sensory perception without any known physical basis.

Halo Effect: In making judgements of people there is a tendency to rat an individual high or low on the traits, because he or she is known as to be high or low on one or few traits.

Health Psychology : The application of psychology to the promotion and maintenance of health, the prevention and treatment of illness and the identification of causes of what makes people healthy and what makes them illness.

Heredity : It is the biological transmission of traits from parents and grand parents to the off springs by way of germ plasm.

Heroin : It is a white crystalline powder derived from morphine a habit forming drug.

Hetrosexual Behaviour: It is male-female sexual behaviour. Interest in and attachment to a member of the opposite sex.

Heuristics: The basic principles or rules that allow as to make social judgements early and rapidly in social situations. For this the techniques the mind use to sort out, analyse also use information. It is a mental short - ents.

Hierarchial Needs: Maslow's concept of the sequential emergence of motives: physiological, safety, and security, love and belongingness, self-esteem, and self-activatization.

Hindbrain: The lowest portion of the brainstem just above and continuous with the spinal cord and just below and continuous with the midbrain. It consists of medulla oblongata, pons and cerebellum.

Hippocampus: A portion of the cerebral hemispheres found curled in the basal medial part of the temporal lobe that is thought to be important for learning and memory.

Histogram: It is a frequency graph drawn to the scores of the test in which the distribution is represented by rectangular, which vary in height with the frequency of the scores.

Hormone: A chemical substance secreted by the endocrine glands.

Hue: The aspect of colour determined by the length of the light wave and giving it its name.

Humanism: It is a feeling of devotion to and concern for human welfare, and mode of thought or action or way of life in which human interests, values and dignity predominate.

Humanistic Psychology: It is a system of psychology in which the view of human functioning is emphasised. The uniqueness of the individuals, their subjective experience of the world and their strivings and the potential for growth are stressed. It is often identified as a third force in psychology in contrast to psychoanalytic and behavioural theories.

Hyperactivity: It is a type of disorder of children who cannot remain quiet and attentive and also cannot apply themselves in situations which require their concentration.

Hypnosis: Hypnosis is an artificially induced state of semi-consciousness, characterized by an increased suggestibility to the words of the hypnotist used.

Hypochondriasis: A type of disorder in which the individual misinterprets physical signs and sensations as abnormal and is preoccupied with fears of having a serious disease.

Hypothalamus: It refers to a group of nucli in the forebrain, a part of the diencephalon, lying ventral to the thalamus. This controls many emotional and motivational processes.

Hysteria: An abnormal behavioural pattern characterized by dissociation and often accompanied by fits, palsies or functional disorders.

(I)

Iconic Memory : A very brief type of memory that stores the sensory impressions of a scene.

ID : In psychoanalysis it is the part of the personality that is concerned with immediate gratification of the basic bodily needs and urges.

Identical Twins : Two children developed from a single fertilized Ovum. These two children are always of the same sex, and closely alike in appearance.

Idiot : A person with the lowest amount of mental capacity. His IQ will be less than 25.

Idiot Savant : A person with mental capacity falls between mentally retarded and normal persons. The idiot savant has some special ability with which he can do things which normal persons cannot do.

Illusion : It is mistaking one thing for another. A physical stimulus is mistaken for a real object. A piece of rope may be mistaken for a snake.

Imbecile : It is a type of mental retardation or feeble mindedness whose. IQ falls between 25 and 40, it is just one step about the idiot.

Imitation : It is copying what is seen in another person. It may be some event, scene which attracts a person to do the same thing.

Immediate Memory Span : A number of items (numbers and words) reproduced immediately after a short exposure for learning.

Image : It is an antificial imitation or statue, effigy, constellation or figure. It is the presentation of the external form of a person or a thing in art. Psychologically it is the picture of something in the mind.

Imagery : Imagery is the plural form of image and it is also stated as imageries. Imagery means images in groups.

Imagination : It is the act of forming a mental picture which is not actually present to the senses. It is a mental image or idea. It is the result of mental images.

Incentive : Any external condition or goal which impels a person or an organism to action or to achieve the goal.

Incus : Generally it is the anvil, the middle bone of three bones at the middle ear that transmits sound variations from the eardrum to the cochlea.

Independent Variable : In an experiment any variable which serves as a basis for making a prediction or in verifying the hypothesis the independent variable plays an important role.

Induced Suggestibility : It is one of the important aspects of hypnosis. The hypnotised persons seem to be much more open to suggestion from the hypnotist than they would be in making state.

Inferential Statistics : The statistics which are used to generalize the results from a representative sample to a larger population, contrast to descriptive statistics.

Inferior Colliculus : The inner most pair of pulges in the tectum of the midbrain, a lower coordination centre for hearing.

Inferiority Complex : It is the term used by Adler to describe an individual's failure to fulfil his ambitions and by means of pervasive feeling of the personal inadequacy is resolved.

Information Processing Theory : In the study of memory this technique is applied for information processing, it is a type of communication system.

Inhibition : Generally it refers to any interference with response. It is stopping or restricting temporarily some activity.

Insight : It is getting sudden solution to a problem. Also looking the self into the origion of motive and conflict.

Integration : The process of bringing together and analysis of the parts into a whole.

Instinct : It is an unleaned behaviour pattern, biologically based, goal directed behaviour which is specific to species. For example, nestbuilding in birds.

Intellectualization : A defense mechanism whereby a person tries to gain detachment from emotionally threatening situation by dealing with it in abstract intellectual terms.

Intelligence : The mental ability to profit from experience and environment.

Intelligence Quotient (IQ) : It is the ratio determined by mental age by a specific test devided by the chronological ages.

Interest : It is an attribute with which one attends to anything his feeling accompanying.

Intervening Variable : An event inferred to occur between stimulus and response, thus accounting to one response rather than another to the same stimulus. The intervening variable may be inferred without further specification and may become an object of investigation. It is interferring in the experimental process and may affect the accuracy of the result.

Interview : It is a conversation between an interviewer and the subject with a purpose of obtaining some definite information, either for the subject's interest or for information gathering.

Intrinsic Motivation : A motivation in which the action and the results served by the action are organically or inherently related.

Introspection : One of the methods used in psychological studies. A form of reporting the mental content subjectively, the conscious events and experience.

Introvert : A person who reveals a strong tendency to find satisfaction in an inner life of thought and fantasy. It is contrast to extrovert.

Involutional Malencholia : According to DSM-III, it is an affective disorder among women who are going through menopause and among men whose sexual adequacy is diminising. The symptoms are degenerative and sometimes regressive and peranoid.

Iris : A smooth muscle structure whose constructions open or close the pupil to regulate the amount of light entering the eye, its pigment determines the colour of the eyes.

Ishihara's test : A test for defects in hue sensitivity (colour blindness) that depends on numerals that are perceived by the subject as outlined by an array of coloured dots that differ in hue and brightness.

(J)

James-Lange Theory : It is the earliest classical theory of emotion named on the two persons who independently evolved it. The theory stresses that the stimulus first leads to bodily responses; and then the awareness of these responses constitutes the experience of emotion.

Just Noticeabale Difference (JND) : A clearly noticeable physical change in a stimulus; a measure of the difference threshold. It is the smallest difference between two stimuli that can be detected.

(K)

Kinesthesis : The sensory impressions which come from the joins, muscles, and tendons and provide information about the positions and movements of the parts of the body.

Kymograph : A paper covered drum that rotates at a controlled speed. A pen connected to a physiological instrument reconds the physiological events on the kymograph as a moving line.

(L)

Latent Dream Content : According to Freud the unfulfilled desires are supressed and sent to the unconscious mind and they try to come out in some form in dream. But could not come out in original form. They come out in disguished form which freud called them as latent content.

Latent Learning : The learning experience that is not recognized untill the opportunity is given to use it.

Law of Effect : This was formulated by Thorndike. According to him we learn responses that could reward and unlearn responses that do not reward.

Learning : A fairly permanent change occurs in the behaviour of the organism as a result of experience.

Learning Curve : Graph showing the proficiency in the learning situation to practice. What is plotted is performance and learning is the assumed state underlying it.

Learning Disability : It is a failure to develop a normal capacity to learn, usually of some specific skill.

Lesion : It is an abnormal change in a particular part of the body tissue through injury or disease.

Level of Aspiration : It is a level of accomplishment which a person sets as a goal. Achieving the goal is perceived as success and the inability to achieve it is a failure.

Libido : Freud in his psychoanalysis used this concept. It refers to general sexuality which provides energy for all types of behaviour.

Lie Detector : It is a device used to detect guilt, which depends upon the physiological changes in the body like pulse, blood pressure, heart rate, sweeting and so on.

Limbic System : It is a structure in and around the core of the old forebrain that likely integrates motivational and emotional patterns such as arousal, sleep and feeding.

Linguistics : It is the study of human speech in its various aspects of language such as phonetics, phonology, morphology, syntax and semntics.

Long-term memory (LTM) : It is a memory system of long duration, following after short-term and intermediate memory.

Longitudinal Study : Research study conducted on individuals continuously for a long time.

Lysergic Acid Dicthylamide (LSD) : It is a hallucinogenic drug which can induce vivid perceptual experiences, expansiveness of the mind, hallucination and disorganized thinking.

(M)

Macrocephaly : An abnormal condition in human beings which is characterised by large head, causes mental retardation.

Mania : An intense emotional state in which the individual is elated, talkative, distractible, has flights of ideas and delusions and engages in purposeless activities.

Manic Depresive Psychosis : A type of mental illness in which the individual has extreme swings of mood from elation to depression and back again.

Manifest Dream Content : According to Freud's Psychoanalytic theory the unfulfilled wishes that lay supressed in unconscious mind try to come out in disguised form during sleep in dreams. The disguised form of the dream content is called manifest content.

Marijuna : The herb plant, hallucinatic in characteristic which enhances the mood and sense.

Maturation : It is a state of growth and development and the state of reaching fullfledged condition of the organism.

Maze : It is an instrument used in learning experiments. It consists of runways with correct pathways and blind alloys.

Mean : It is an average measure of scores; a measure of central tendency. It is computed by summing up the scores and dividing it by the number of cases.

Median : In a frequency distribution where the middle point or fifty per cent of the scores fall.

Meditation : Meditation is a form of mental state in which some kind of altered state of consciousness occurs. It is practised in several ways. Mainly it is intended to control the mind. It slows down the rudly rushing thoughts and sensations.

Memory : It is a store house of all the learned materials and experiences in human beings.

Memory Drum : An instrument with electrical device used to present verbal materials in rote learning experiments.

Memory Image : It is the result of recalling the past experience or the fact of the past which has sensory character. Thus memory image is the recalled fact or experience which is having the realistic sensory character as if we are actually seeing or hearing or smelling an object. The memory image essentially should have the real sensory experience.

Memory Span : It is the amount of material which a person can grasp in the trail without having to retain it for more than a few seconds. That is how much materials a person can remember when they are presented for a short period of intervals.

Memory Trace : The change in the nervous system that persists between the time that something is learned and the time that it is recalled.

Mental Age (MA): A unit of scale proposed by Binet for determining the level of intelligence of a person if an intelligence test is administered properly. Generally the score on an intelligence test for individual is expressed in terms of years and months. The individual's score is divided by the actual age of the person and the result is his mental age.

Mental Retardation : It is a condition of sub-average at the intelligence level. The person is deficit of adjustment in behaviour, which is manifesting first in the childhood period. There are four levels of mental retardation according to American Association of Mental Deficiency. They are mild retardation, moderate retardation, severe retardation and profound retardation. It is also indicated by the terms, mental deficiency and feeblemindedness.

Mental Set : It is the readiness of an organism to respond to in certain way.

Mesencephalon : It is the middle brain, which is involved in vision and hearing. One of the three primary embryonic ventricles.

Mesomorphy : It is one of the types of shelton's body type. The person of this type is vigorous, powerfully boned and muscled type of body build. This is in contrast to ectomorphy and endomorphy.

Metencephalon : Anterior portion of the rhompencephalon composed of the cerebellum and the pons.

Microcephaly : A person of abnormally small head and brain are caused by a disease trauma, a condition that results in mental retardation.

Midbrain : It is the middle section of the brain and it is continuous with and just above the hind brain and just below the forebrain. It contains tectum, tigmemdum and cerebral pedunclues.

Middle Ear : It is the air filled space between the ear drum and the inner ear. It contains the three auditory bones, malleus, incus and stapes.

Minesota Multiphasic Personality Inventory (MMPI) : It a personality inventory design to assess personality traits. It consists of 550 items evaluating personality by psychiatric categories, masculinity, femininity and social introversion.

Mitosis : It is the process of cell division in which the chromosomes duplicate, then join in pairs within the cell nucleus. The pairs split with one member of each duplicated pair going to one of the two daughter cells which have been formed.

Mode : The most often occurring score in a frequency distribution.

Mongolism : A type of hereditary feeblemindedness caused by an excess chromosome in the germinal material.

Monocular Cue : Any cue to depth or distance that depends on sensations from only one eye.

Morphine : A powerful, addictive narcotic drug.

Moron : A mentally deficient person just below the normal and just above the inbecile.

Motivation : A sets or drives within the organism which impels him to action.

Motive : A need or aspiration or ambition which initiates for action.

Moon Illusion : A visual illusion in which the moon appears larger, the closer it is to the horizon.

Moral Behaviour : Behaviour that is taken to be ethical in any culture.

Mnemonic Device : It is an artificial technique for helping the individual to remember the items to be learned.

Morpheme : The smallest meaningful unit in the structure of a language, whether a word or affix, or man.

Multiple Personality : It is an extreme form of dissociation in which the individual's personality is split into separate personalities often alternating with each other. The memories of one of the split-off personalities are not accessible to others.

Myelencephalon : It is developed from the bottom bulging of the neural tube of the developing embryonic structure. It is evolved from rhobencephalon. It is the spinal brain myclencephalon includes the pons, medulla oblongata and the fourth ventricle.

Myolin Sheath : It is an unusual and abrupt change in gene structure and hence in the bodily characteristic governs cellular impulses to travel faster.

Myopia: It is an eye condition in which the lens is too thick or the eyeball too long in an anterior to posterior dimension. The person cannot focus the image of distant objects in the retina.

(N)

Narcissism : In Freud's psychoanalysis the erotic feelings are associated with the individual's own body or self. It is nothing but self-love, ego centrism.

Narcotic : It is a drug which in moderate doses blunts the senses, relieves pains, and produce prolonged sleep, but in heavy doses it causes stupour coma, the person becomes addictive when used constantly.

Nativism-empiricism Issue : The nativistics, Kant and Descartes state that we are born with the ability to perceive world. In contrast Berkeley and Locke state that we learn our ways of perceiving through experience with objects in the world about us.

Natural Selection : It refers to Darwin's theory of natural selection. The proposition is that the inheritance of structures is determined by their usefulness to the species to the struggle for servival. Individuals of the species who can best meet the environmental demands survive and reproduce.

Nature-Nurture Interaction : It is the interaction between heredity and environment on the development of the individual traits and behaviour.

Neocortex : It is the part of the cerebral cortex that is unique to mammals.

Neonate : A new born infant.

Nerve : A bundle of axons in the periphreal system, generally with protective and supportive sheath.

Nerve Impulse : It is an electrochemical excitation in the walls of a nerve cell that allows the cell body and dendrities to communicate with the terminal buttons through axon.

Nerve tract : It is a collection of axons in the central system having a common origin, termination and function.

Neuron : Nerve cell, the active element of the nervous system.

Neurosis : It is the disorder of the nerves. But recent description is that it is the disorder in which anxiety is believed to be the most important factor. Hence it includes, phobias, obsessions, compulsions, anxiety attacks, hysteric and hypochondria.

Nonsense Syllables : It is an artificial word which is consisting of one vowel and two consonents, for example, REQ, TOZ.

Norm : It refers to a standard idea defined or an average score of some defined group.

Normal Curve : It is a statistical concept referring to a symmetrical bellshaped distribution.

Nucleus : The area within a cell that contains the genetic material.

Null-hypothesis : It is negatively stated hypothesis. There is no true difference between the experimental group and the control group after they have been treated differently in an experiment.

(O)

Object Constancy : The perception of objects is relatively unchanged, inspite of variations in placement, illumination distance and so on.

Occipital Lobe : The hindmost part of the cerebral hemisphere mainly concerned with visual functions.

Oedipus Complex : In Freudian psychoanalysis this refers to the erotic attachment of the little boy aged four years, in the phallic stage for his mother and the desire to eliminate the father as a rivel. These impulses normally repressed in some way through fear of punishment and as he grows older.

Operant Conditioning : It is learning theory formulated by skinner. According to this, in learning activity the responses which do something or operate on the environment are strengthened through positive or negative reinforcement. It is also known as instrumental coil conditioning.

Optic Nerve : It is one of the pairs of cranial nerves made up of axons of ganglion cells of the retina and connecting through relay centres of the thalamus with the visual centres of the brain.

Oral Stage : In Freudian terms, the oral stage is the psychosexual development in which the infant acquires its most desired pleasures from the activities of the mouth like chawing, suaking, biting and so on.

Organ of Corti : It is a structure located in the basilar membrane of the spiraling cochlea of the inner ear. It contains the hair cells, the receptor cells for hearing.

Orthopsetic Principle : It is Werner's principle that development proceeds from a global undifferentiated response towards one of increasing differentiation and hierarchical organization of the elements constituting it.

Ossicles : This consist of three bones in the middle ear that transmit movements of the eardrum to the fluid part of the inner ear.

Outgroup : A group of which one is not a member, an expelled member.

Oval Window : The membrane which covered is opening to the fluid filled inner ear, sound vibrations are carried by the ossicles to the oval window which communicates with scala vestibuli.

Overies : Female gonads or reproductive organ.

Overtone : In a complex tone frequencies showing resonance of part of the vibrating body to simple multiples of the fundamental frequency. In a complex tone any of the tones may be of higher pitch.

Ova : Female reproductive cells.

Ovulation : Release of an ovum by the follicles of the overies.

Ovum : Female egg cell.

Oxygen Debt : the amount of oxygen necessary to restore a tissue to equilibrium after its metabolic need for oxygen has exceeded the available supply. Hence, a debt occurs.

(P)

Pain : It is an uncomfortable sensation caused in the body. Any stimulation of free nerve endings that gives rise to pain sensation.

Paired Associates : It is a learning method in which items are learned in pairs, one item serves as the stimulus and the other serves as a response to be given by the learner.

Panic Attack : It is a sudden unexpectable fear and intense apprehense alongwith physiological symptoms, such as palpitations, chest pains, chocking and smoothening, sensations, dizziness, faintness, sweating and shaking. A sense of impending doom, fear of dying, going grazy or doing something uncontrolled. These are often found to be parts of a panic attack.

Pancreas : It is the gland that lies between the stomach and small intestine and contains the endocrine alpha and beta cells that secrete glucogen and insulin.

Paradoxical Intention : A variation of flooting in which the client is urged not meanly to express fears, but to confront them in exaggerated form.

Paradoxical Cold : It refers to inadequate stimulation of cold receptors by an intense warm stimulus.

Paradoxical Warm : It refers to inadequate stimulation of warm receptors by an intense cold stimulus.

Parasympathetic Division : It is a division of the Autonomic Nervous System (ANS), an afferent nerve supply that innervates the viscera and discretely stimulates digestive functions as well as forming part of certain reflexes of the smooth muscles.

Parasympathetic Ganglia : This refers to the collections of call bodies of the postganglionic PNS nerve fibres. In ganglia lie near the smooth muscles and glands which the postganglionic fibres innervate.

Parasympathetic Nervous System : It is a division of autonomic nervous system mainly concerned with protecting and conserving the body resources, preserving normal functions, and maintaining a calm emotional state. It functions in contrast with sympathetic nervous system.

Parasis : It is a consequence of syphilis in which organic damage to the brain is caused by the micro organisms which results in profound mental and neurological symptoms.

Paranoia : It is the general term for delusions of persecution, grandiosity, or both, found in several pathological conditions, can be produced by some larger doses of drugs.

Parieter Lobe : It is a major division of either cerebral hemisphere lying between the frontal and occipital lobes and is above the temporal lobe.

Parkinson's Disease : It is a presenile dementia, characterized by uncontrollable and severe muscle tremours such as a stiff gait, a mask like expressionless face and withdrawal.

Paranoid Personality : This individual is always expecting that he would be mistreated by others. He becomes suspicious, secretive, jealous and argumentative. He/she will not accept blames and appears cold and unemotional.

Passive-aggressive Personality : Those who are with passive-aggressive personality indirectly resist the demands of others by such exasperating behaviour as being late, missing appointments, not replying to calls and making foolish mistakes.

Pathology : The anatomical, physiological and psychological deviations of a disease or disorder of these abnormalities.

Penis : External genitals of the males, exclusive of the gonads and scratum.

Perception : A conscious event initiated by sensations, but including memory classification and integration of the input from several sense organs.

Perilymph : The fluid of the labrinths ear that surrounds it and it also contains in the scala vestibuli and scalatympani of the cochlea.

Peripheral Nervous System : The part of the nervous system that connects receptors with the central nervous system and the central nervous system to glands and muscles.

Person Perception : Inferences made about the disposition, motives and intentions of a person, also called impression formation.

Personal Unconscious : According to Carl Jung the part of the unconscious that contains specific experiences of a person. It is in contrast to collective unconscious.

Personalism : According to the attribution theory the tendency of the individual is to disregard information that indicates that his disposition is the cause of bad outcomes.

Personality : Personality is the unique integration of desirable permanent characteristics or traits that make the individual different from other individuals and at the same time determines how others respond to him.

Personality Dimension : The personality traits which vary from one extreme to the other extreme, For example, excitment-calm, love-hetred, etc.

Personality Inventory: A list of statements about personal characteristics, the measures give insight into an individual's personality when referred to norms obtained from groups having known characteristics.

Personality Structure: The personality traits of a person are organized in him in such a way that they make their appearance in behaviour.

Personality Test: A test that could assess the general disposition and adjustment of a person.

Petitmal Epilepsy: A form of epilepsy in which there are momentary loss of consciousness.

Phallic Stage: In Freud's psychoanalysis it is referred as a stage of psychosexual development from third to fifth year, during this period children become highly interested in their sexual organs and develop and resolve their oedipus and Electra complexes.

Phenomenology: In psychological studies the theoretical point of view advocating the study of subjective awareness and experience. It is also referring to naive reports of experiences made by individuals having no experience.

Phonotype: In genetics we speak about phonotype and genotype. Phonotype refers to the characteristics which appear in a living organism, depending upon the dominance of the genes and the interaction between the genotype and environment. This is in contrast with the genotype.

Phi Phenomenon: The gestalt psychologists formulated this theory of apparant movement of a single light, seen when it lights at a short distance apart is flarshed alternatively in rapid successions.

Phobia : It is an irrational and excessive fear of a specific object, activity, or situation which is not dangerous.

Phoneme : It is the smallest unit of sound that shows a difference in the meaning of a word of a language.

Phonetic Symbolism: The meaning conveyed by the sound of an utterence.

Phonetics: It is a branch of linguistics which analyses and classifies spoken sounds as they are produced by the organs of speech and received by the a ear.

Phonology: The study of language in terms of speech sounds.

Physiological Zero Point: The temperature which is experienced as neither warm nor cold.

Physiological Motive: It is the organic need which is needed for food or water and other essential materials.

Pitch: It is the psychological attribute of hearing which depends for the most part on the frequency of the sound waves.

Pituitary Gland: It is one of the glands of the endocrine system located at the base the brain. It is said to the master gland having control over all other glands.

Place Theory: It is the Helmholtz's theory. According to this theory, a particular place on the basilar membrane and the reasoning fibres, these are stimulated by a particular frequency of sound waves.

Pleasure Principle: In Freudian theory, the principle that the individual impulsively seeks to satisfy the id's urges either directly or in fantasy. It is in contrast with the reality priniciple.

Pons: It is the upper part of the hindbrain. The ventral surface enlargement on the interior part of the hindbrain formed by crossing tracts from higher parts of the brain on route to the cerebellum.

Prejudice: It is prejudgment, expressing hatred mildly, generally, it is a hostile attritude forward objects, persons, places etc. It is an unverified false belief.

Prestige: A feeling of having high opinion about oneself, a feeling that he/she is more important than others in some situations.

Projection: It is a type of defense mechanism in which the person attributes this in which the person attributes his own undersirable traits to some person or thing that he disliked.

Projective Test: A kind of personality test, generally unstructured figure without providing any clue. The person is expected while answering the test to reveal his own ideas, wishes and needs.

Propaganda: Widespread promotion of a particular item or idea, and influence people for a positive attitude in the desired direction by the propagandist.

Psychograph: A profile of personality traits.

Prevalence: Epidemiologically, the percentage of population that is facing a given disorder at some particular time or period.

Prevention, Primary: A community based health term indicating the procedures designed to prevent the development of disorder in some vulnerable community.

Prevention, Secondary: A community mental health term for the early diagnosis of disorder in an effort to reduce its duration and the impact.

Prevention Tertiary: A community mental health technique applying the efforts to reduce the difficulties and the severities of the problems.

Proactive Inhibition: A learning process in which the earlier learning interfers with later learning and inhibits the recalling.

Prosencephalon: Prosencephalon is the forebrain emerged from the top portion of the bulging in the neural tube. The fully developed of this part become the forebrain. This is the more anterior part (end-point) of the embryonic brain. This subsequently develops into two divisions as telencephalon and diencephaless.

Psychology: A systematic study human behaviour right from the time of birth till the time of death.

Psychosis: A severe type of mental disorder.

Psychoanalysis: A therapeutic technique developed by Sigmund Freud and his followers.

Psychoneurosis: It is a behaviour disorder in which anxiety always plays an important role. The symptoms are hysteria, phobias and other neuroses.

Psychopathology : A field of study which makes a scientific study of mental disorders. Psychological functioning and disordered behaviour are the main subject matter of this field.

Psychophysics: The procedure applied in the study of the relationship between physical stimuation and the corresponding psychological reactions.

Psychosomatic Disorder: A physical disorder which is aggravated by emotional stress and other psychological factors. It is also called as psychophysiological disorders.

Psychosomatic Medicine: The branch of medical science which is concerned with psychosomatic disorder.

Psychotherapy: It is a branch of therapeutic technique for the treatment of mental and emotional disorders by psychological methods, such as direction, suggestion, reconstruction, persuasion and counselling.

Puberty: It is a stage of transition from sexual immaturity to maturity.

Punishment: The use of any painful discomforting stimulus in order to decrease the probable undesirable behaviour.

Pupil: The opening part of the eye by the iris through which light enters the eyes.

Purkin' Shift : The change in apparant brightness of varies wavelengths with the change from daylight (cone) to night vision.

Putrid: One of the seven primary qualities of smell according to the chemical theory, a quality determined by negative changes on the molecule smalled. It is a foul small as in hydrogan sulplite.

Pyknik: One of the types of body build which Kretchmer had formulated. Pyknik persons are soft, round and overweight.

(R)

Race : A large division of human kind distinguished by some common ancestry and by greater frequency of a collection of inherited bodily traits than is observed in other racial groups.

Radial Fibres : Smooth muscle fibres that radiate from an opening such as pupil, their contraction increases the size of the opening.

Random Sampling : Samples collected selectively from a population without any bias and purely by choice.

Range : In an experimental data, the scores may be from lowest to highest levels. The difference between these two ends is called range.

Rating Scale : In making judgements a marking device is used, in which a rater can make his judgments about himself or about others.

Rationalization : It is a defense mechanism in which the person interprets his own idea or behaviour in which the real motive is concealed in favour of an acceptable motive.

Rapid Eye Movement (REM) : The eye movement which occurs during sleep normally indicate dreaming. The movements of the eye balls generate electrical activity which is recorded by the attaching electrodes near the eye.

Reaction formation : It is a defense mechanism in which an unconscious and an unacceptable impulse or feeling that could cause anxiety is converted into its opposite so that it can become conscious and it can be expressed.

Reaction Time : It is the interval time between the onset of stimulus and the beginning of the individual's response.

Recall : It is the process of bringing up the learned materials that are stored in the memory to the conscious level for making use of them.

Receptors : The cell groups which receive stimulations from outside.

Recessive gene : The gene which will not express itself in a trait unless paired with another recessive gene.

Reciprocal Inhibition : It is Wolpe's theory. According to this it is weakening anxiety to certain stimuli by providing a means of responding that inhibits the anxiety.

Recognition : It is identifying the response to an object previously learned or seen or heard. It takes place after the process of recall.

Reflex : It is a rapid and unlearned response to some sensory stimulus.

Reflex Area : A pathway which extends from a receptor to an effector and is followed by a neural impulse to produce a reflex.

Regression : In an extreme anxiety state the person in his thought process returns to an earlier stage in which he was happy without much worry or anxiety to get rid of his present frustration.

Reinforcement : In Skinner's operant conditioning, increasing the probabillity that a response will recur either by presenting a contingent positive event or by removing a negative one.

Reliability : It is a statistical measure, lost of self-consistency of a test, generally expressed by coefficient of correlation representing the relationship between two sets of measurement of the same thing.

Remembering : It is recalling something learned of an earlier time and making use of it.

Resonance : It is a vibrating response of an object, for example, the sound of a piano wire to a frequency pressed on it.

Response : An activity, a movement, or an answer or a glandular reaction resulting from some stimulation. It is a reaction to a stimulus.

Reticular Activating System (RAS) : It is a system of ill defined nerve pathways and connections within the brain stem, lying outside the well defined nerve pathways and it is an important arousal mechanism.

Reticular Formation : It is an extension part of the brain stem, extending from the medulla through thalamus, which is involved in arousal.

Retina : It is a part of the eye, sensitive to light, containing the rods and cones.

Retrieval : The process of calling back the stored material in the memory.

Retroactive Inhibition : The interfernece of recently learned material in recalling the material learned earlier.

Rhinencephalon (Smell Brain) : It is that part of the forebrain that contains olfactory tube and it is concerned with sense of smell. It is recently discovered part of the brain.

Rhobencephalon (Hind Brain) : This part of the brain is located behind the midbrain and also in the region of cerebellum, medulla oblongata, the pons and the fourth ventricle. It is involved in maintaining the balance of the organism and it has two divisions as (a) metencephalon and (b) myclencephalon.

Ribonucleic Acid (RNA) : A nucleic acid that implements information found in deoxyribonucleic acid (DNA). The two forms of RNA are the transfer RNA and the messenger RNA.

Rodopsin : It is a light sensitive substance containing the rods and cones located in the rear side of the eye.

Rods : One of the light sensitive receptor cells in the retina area that are most active at low levels of light.

Round Window : A membrane separating the cochlear duct from the middle ear cavity.

Rorschach Test : It is a projective test used for assessing personality traits. The person is asked to interprete freely Ink-blots on ten cards, shown one at a time.

Rote Learning : It is learning of verbal sequences with little attention to meaning.

Role Playing : A method of teaching principles. It is generally acting out the behaviour and the individual behaviour is expected of a certain role in a hypothatical and in staged situation.

(S)

Saccule : It is a fluid filled sac of the vertibule of the inner ear.

Sadism : It is a pathological motive that insticates a person to inflict pain upon another person, mainly on the opposite sex.

Sample : A small part of quantity of something intended as representative of whole as specimen of the population or large group. It is a statistical term.

Sampling Error : The variation in a distribution of scores or of statistics derived from them to be attributed to the fact that measurements are made on a variable sample from a large population. Thus the sampling errors, persist even though all measurements are made accurate.

Saturation : The colour dimensions that describle the purity; if highly, saturated it appears to be pure hue and free from gray, but if it is of low saturation it appears to have great deal of gray mixed with it.

Scapegoating : A form of displaced aggression in which an innoccent but helpless victim is blamed and punished as the source of the scapegoaters frustration.

Schizophrenia : It is a functional psychotic disorder in which there is a lack of harmony or split between aspects of personality functioning; specifically between emotion and behaviour. The symptoms may be autism, hallucinations and delusions.

Secondary Reinforcer : A stimulus that has become reinforcing factor through earlier association with a reinforcing stimulus, which we call as secondary reinforcer.

Secondary Sex Characteristics : During the course of development the physical features distinguish the mature male from the mature female, apart from the reproductive organs in man, the deeper voice of the male and the growth of the beard are the indicators. In female the appearance of pubic hairs and the development of breast are the indicators.

Selective Breeding : It is a method of studying genetic influences by mating animals that display certain traits. The subjects with specified characteristics are mated so that the transmission of these characteristics to the offspring can be studied.

Self-actualization : The highest order of motive in Maslow's hierarchy of motives. A person's basic tendency is towards the maximum actualization of his potentials.

Self-concept : The composite of ideas, feelings and attitudes an individual has about himself.

Self-consciousness : It is a form of highest self-awareness when an individual is specially concerned about the reactions of others to him.

Self-hypnosis : It is a process of making auto-suggestion the person need not go to the hypnotist.

Self-perception : A person's awareness of himself, which differs. from self-consciousness because it may take the form of obective self-appraisal.

Semantic Differential : It is a method developed by Osgood for using rating scales and factor analysis in studying the connotative meanings of words.

Semantic Memory : It is a memory that is necessary in the use of language.

Semiscircular Canals : In the inner ear there are three curved tubular fluid filled canals, perpendicular to one another and registering three planes. They sense the turning of the head and its coming to rest.

Sensory Memory : It is the memory of some sensations which exist only for a very short period. The retention of sights and sounds as sense impressions lasting only for a second.

Sensory Motor Stage : Piaget had developed a theory of intelligence specifically for the childhood period. The period extending from birth to two years, during which period the child knows the world through sensory perceptions and motor actions, but finally comes to realize that objects exist independently of his perceptions.

Sensory Neuron : The neuron which is having contact with receptors, and carried messages to the central nervous system, also known as afferent neurons.

Shape Constancy: It is a tendency to perceive objects as they are irrespective of their shape at different positions. The object seen from different angles which distract the rational images to be perceived as having a constant shape.

Short-term Memory : The memory which is remembered and retained for about fifteen to thirty seconds only. It is in contrast to long-term memory.

Signal Detection Theory : It is a theory of the sensory and judgement processes involved in psychophysical determinations. The theory states that when people report whether a signal is present or absent they decide what they experience is whether it is the background stimulation or an actual signal.

Size Constancy : It is the tendency to see a familiar object as of its actual size irrespective of its distance from us.

Skewed Distribution : A frequency distribution that is not symmetrical or off the centre as compared with normal curve.

Skill : It refers to the ability to be profited by the individual by training or practice.

Smooth Muscles : The muscles which are under the control of the autonomic nervous system, also called involuntary muscles.

Social Attitudes : Attitudes held in common with the members of others groups.

Social Class : People in a society are grouped and classified on the basis of income occupation, family standing, living area and the type of life lived by them.

Social Intelligence : It is the ability of an individual to get along pleasantly with others.

Socialization : The new born child as he grows older and older and learns to accommodate himself with others in the family and with outside members and as and when he moves in the society adapts all the living habits of the culture and becomes a fullfledged citizen of the group by means of socializing himself.

Social Psychology : That branch of psychology which seeks to understand the nature and causes of individual behaviour in social situations.

Sociogram : A diagram representing interpersonal relations between members of a group in action. Sociogram represents the structure of the group relationship.

Sociopathic Personality : The personality disorder of the guilt and anxiety free individual where impulsive and other destructive behaviour often brings this offender into conflict with the society.

Somatotherapy : It is the treatment of personality maladjustment or mental illness by drugs, electric shocks and other methods directly affecting bodily processes.

Somatic Motor Fibres : The axons which carry commands from the central nervous system to the muscles attached to bones at efferent neurons.

Somatic Nervous System : All the peripheral nerve structures are not in the automatic systems; the sensory fibres sending the sensory receptors and the motor fibres serving selected fibres.

Somatic Therapy : The treatment of mental and emotional disorders by physical methods acting on bodily processes.

Spectrum : It is the range of physical stimuli to which a receptor responds. Most of them are used in connection with the full range lengths of light waves.

Spinal Cord : The part of.the nervous system enclosed in the vertebral column.

Spinal Nerves : The thirty-one pairs of peripheral nerves extending from spaces between the vertebrae of the spinal column to various parts of the body.

Spinal Reflex : It is a complex reflex circuit which passes through the spinal cord, but not directly through a higher centre.

Split Brain : The brain in which the two hemispheres are no longer joined together, because the corpus collosum cuts it and separates their functionings.

Split-half Method : It is a technique by which the reliability of a test is established by correlating the two halves of a test scores, i.e., odd items one half and even items the other half.

Spontaneous Recovery : Reappearance of an estinguished conditioned response after a lapse of time.

Standard Deviation (SD) : It is a measure of variability in a frequency distribution. It is obtained by the square root of the mean of the squares of the amount by which each case departs from the mean.

Standard Error : The standard deviation of the sampling distribution of a mean and of certain other derived statistics. It can be interpreted as any other standard deviation.

Stanford-Binet Tests : An inelligence test originally designed and developed by Alfred-Binet has been revised in Stanford University by Terman and Merrill and named it as this.

Stapes : It is one of the three bones of the middle ear that transmits sound vibrations from the eardrum to the inner ear stirup-shaped as attached to the oval window.

Stereotype : Expectations and opinions about large groups of people, allowing the persons holding them to disregard the individuality. A stereotype is overtly concentrated and resistance to disconfirming the evidences. Generally, it is a fired set of beliefs and ideas held by the members of a group.

Stimulus : It refers to any source of energy which affects the sense organs which could arouse a response in the organism.

Storage : It is putting the materials into memory, some of them are accounted as short-term memory and some long-term memory depending upon the type of memory.

Stress : Any stimulus internal or external that disturbes the healthy and dynamic equilibrium of the system of the body and mind.

Stress Tolerance : The individual's ability to withstand stress without disability.

Stretch Reflex : The reflex contraction of a muscle to stretch of that muscle.

Striated Muscles : The skeletel muscles under the control of central nervous system. It is often called as voluntary muscles.

Sublimation : It is a defence mechanism in which the person substitutes the socially unapproved behaviour pattern with socially acceptable forms.

Super Ego : It is in Freud's personality theory the super-ego plays very crucial roles. This part of personality is consisting of parental and social standards internalized by the ego-ideal or aspiration.

Surface Trait : It is a personality component formulated by Raymond Cattell. It is one of the clusters of manifest traits that seem to belong together in expressing an underlying source trait.

Survival of the Fittest : It refers to the evolutionary theory formulated by Darwin. According to this, the heightened probability is that the best-adapted (fittest) variants of a spacies will survive and reproduce.

Sympathetic Nervous System : It is the division of the human nervous system mainly concerned with emotional excitment. It prepares the organism for emergency and in general acts in opposition to the parasympathetic nervous system.

Synapse : It is supposed to be the juncture of two neurons. The juncture involves proximity, very close, but not touching, connection takes place between the axons terminals of one neuron and the dendrite or cell body of another.

Syndrome : It is a pattern of symptoms characteristic of a disease or disorder.

Synesthesis : A stimulus registration by both of its own and by another sensory system.

Systematic Desensitization : It is a form of behaviour therapy devised by Wolpe which is based on the principle of counter conditioning. A person in a state of deep relaxation imagines threatening objects in a hierarchy from least to the most feared state.

(T)

Taste Bud : A cluster of 50 to 150 cells that detects tastes. Taste buds are found in papillac on the tongue.

Taxonomy : It is the technique of classification of organisms.

Tactum : It is the roof of the midbrain consisting of inferior and superior colli.

Tachistoscope : An instrument for determining the span of attention. It is a wooden board having a brief small window through which a list of words exposed one at a time for a few seconds and the subject is expected to notice them and repeat or write down the same later on.

Telencephalon : It is the frontal division of the prosencephalon (Forebrain). It has several important structures of the cephalon (brain). The telencephalon consists of cerebral hemispheres, cortex, basal ganglia, olfactory branches and limbic system. It is involved in higher complex activities.

Telepathy : It is a type of extra-sensory perception in which what is perceived depends upon thought transference from one person to another.

Tegmentum : It is the medical portion of the midbrain.

Temperament : A type of emotional characteristic like mood or tone.

Temporal Lobe : A portion of the cerebral hemisphere lying in front of the occipital lobe. It has the auditory projection areas. It is divided into large number of areas, mostly association areas. Several autonomic functions like respiration, change in blood pressures are also affected by electrical stimulation of some portions of temporal lobe.

Test : It is like an instrument, used to test some of the psychological processes. It may also consists of a group of items carefully selected and designed to measured some performance.

Test-retest Method : It is one of the methods of establishing reliability of a test. It is administered two times on the same population to the same persons.

Thalamus : It is one of the most influential area of the fore brain and cerebral cortex. It lies just above the pons and mid brain. The thalamus is a complex relay and control centre for incoming information. Even other information that travel from one part of the cortex to another pass through thalamus. The thalamus functions like the pons.

Themetic Apperception Test (TAT) : It is a projective test developed by Murray for assessing personality. The test consists of ambiguous pictures from which the individual is required to make out stories.

Timbre : The quality of sounds which equals us to tall one kind of sound from another when they are all of the same type of frequency.

Thinking : It is an implicit behaviour taking place inwardly or an inner behaviour involving symbols for ideas and objects.

Tip-of-the Tongue Phenomenon : The temporary or sudden inability to call forth a word or name that the speaker is quite certain that he/she has known.

Trait : It is an enduring attribute of an individual which is manifested in his behaviour in a variety of situations.

Trait theory : A theory of personality that stresses traits as enduring and permanently persisting aspects the personality.

Tranquillizer : The drug that is manufactured to reduce anxiety and manifestations of psychosis.

Transcedental Meditation : It is a form of meditation practised by some who follow the Maharishi's TM. The meditative state is induced by repecting a particular sound or phase called mantra over and over again. It is also used as a therapeutic technique. Harbert Benson an American cardiologist applied this technique to his patients and redressed their problems.

Transfer of Training : It is the process transfering the skill learned and practised in one area to another area.

Trial and Error Learning : In complex operation or in instrumental learning the organism selects the correct response in a specific situation by eleminating erring trails and retaining correct ones.

Trichromatic Theory : It is the theory of colour vision formulated by Young-Helmholtz which assumes that the three receptors are accounted for the elementary processes.

Trisomy 21 : It is a condition of having three chromosomes in the fertilization process instead of two or the number 21 chromosome (XXY), the cause of Down Syndrome.

Turner Syndrome : It is a disorder prevailing among females in which the absence of an X chromosome keeps the stature short and the breasts incompletely developed and brings sterility.

Type Theory : This theory states that human subjects can profitably be classified into a small number of categories or types. In each case or type there are characteristics in common that set its members most from other categories or types.

Type : A group of persons who are presumed to be having certain physical or mental characteristics in common which may serve to identify them from other groups.

Type-A Behaviour : This is with regards to the behaviours of the impatients, hard-driving aggressive individuals who are susceptible to heart diseases.

Type-B Behaviour : In this category the patient relaxed aggressive reactions of same one who does not have Type-A behaviour and who is not as susceptible to heart disease as the other.

T-Test : It is a statistical measure applied in treating the test scores to determine the level of difference between and among the groups. It indicates the level of relationship between two measures.

(U)

Unconditioned Response: It is a response given by an unconditioned stimulus without any practice or trail. The organism spontaneouly makes its response to the natural stimulus.

Unconditioned Stimulus: A stimulus that produces a consistant response in an organism as soon as the stimulus is exposed.

Unconscious: It is a state of the mind not having any awareness of things happening around the organism.

Unconscious Motives : Activities carried not by the organism without any awareness.

Unpleasantness: It is unacceptable feeling tone causing displeasure. This is opposite to pleasantness.

Unipolar Neuron: A nerve cell with both axon and dendrite arising from a single process of the cell body.

Urethra : The output tube leading from the bladder to the vagina (female) or penis (male).

Uterus: It is that part of the female reproductive apparatus in which the embryo develops.

(V)

Vagina: Muscular tube that is penetrated in the sexual intercourse and forms a passage between both the uterus and the bladder and the external overice.

Vagus Nerve: It is tenth cranial nerve that contains many fibres of the autonomic nervous system.

Validity: It is the predictive significance of a test. The extent to which a test measures what it is intended to measure.

Variable: In psychological measurement there are certain factors which determine the extenisity of their influences. The effect of independent variable is seen in the dependent variable.

Veins: The blood vessels that return blood from tissues to the heart.

Ventricle: A large hallow area inside the brain.

Vertibra: One of the individual joint bones of the vertibral column.

Valance, Positive or Negative: This term is proposed by Kurt Lewin for the positive (attractive) or (unattractive) negative properties of an object which indicate the movement of an organism towards or away from it. The positive makes it to move towards and the negative makes it to move away from it.

Verbal Intelligence Quotient : The intelligence test scores obtained from the tests of language, reasoning, vocabulary, general knowledge and comprehension are stated as verbal, Intelligence Quotients.

Verbal Test: Any test material involves the used of language is verbal test.

Vertical-horizontal Illusion: It is a tendency to see the vertical lines longer than the horizontal lines though they are of the same length.

Vestibule: The inner ear spreads right back into the skull. It is made up of three parts like the semicircular canals, the vestibule and the cochlea. The vestibule is just a receiving end transmitting centre.

Visceral Motor Fibres : The axons which carry orders from the central nervous system in the internal organs of the body.

Visual Acuity: It is the ability of the eye to distinguish much of the details.

Visual Cliff : A device to study the infant's depth perception.

Visual Projection Area: The area of the cerebral cortex that is the terminal point of more pathways for vision.

Vitamin: Any of the group of organic compounds whose accessary actions are necessary to the chemical reactions that support for healthy life.

Vitamin-D: The vitamin that regulates the calcium and phosphorus levels of the body.

Vineland Adaptive Behaviour Scale (VABS): It is an instrument used for assessing how many age-oriented socially adaptive behaviour a child is engaged in.

Vitreous Humour: The transperant jelly-like substance of pupil in our visual field making us to believe that they have influence.

Volley Theory: It is a theory of neural action in audition, that holds that squadrons of neurons fire out of phase with other squadrons in order to increase the frequencies with which people can respond.

(W)

Wachsler Adult Intelligence Scale (WAIS): It is an individual adult intelligence scale developed by Wachsler, which provides for verbal, performance and composite IQ scores.

Warm Spot: A small part on the skin surface that is sensitive to warm or hot stimuli.

Wave Amplitude: It is the intensity of electromagnetic energy.

Wavelength: This is the frequency of electromagnetic energy.

Weber's Law: ($\Delta I/I = k$) Where I is the magnitude of a reference stimulus and ΔI is the amount of the change required to produce a just noticeable difference.

Wechsler Intelligence Scale for Children (WSSC): It is the test developed by Wechsler for children to measure intelligence.

White Matter: The areas of the central nervous system, specially the cortex area is made up of the myelinated axons of neurons. It is in contrast to gray matter.

White Noise: The noise composed of sound waves of all frequencies.

Withdrawal: It is a defense mechanisms against the sress in which the person is strongly listless and indifferent and avoids all contacts with the world.

Word-association Test: It is a test used for personality assessment and diagnose in which a person gives the first word that comes to her/him.

(X)

X-axis (abscissa): The horizontal axis on a graph, generally is the independent variable.

X-Chromosome: The X-Chromosome may be paired with another X-chromosome. This determines that the child will be a female. If a X-chromosome is combined with a Y-chromosome then the child will be a male. It is the Y-chromosome that determine the sex of the child.

XYY Symptom: It is an abnormal combination of chromosomes in which a male has an extra Y sex chromosome; generally associated with unnatural aggressiveness. The persons will be tall, thin, men have criminal tendencies, have low IQs, but in certain specific area the persons may show extraordinary abilities.

(Y)

Y-axis (Ordinate): It is the vertical axis on a graph, generally, represents the dependent variable.

Y-chromosome: The chromosome that combined with an X-chromosome determines the male sex.

Young Helmbltz Theory: It is a theory of colour vision that determines three types of receptors in the retina area mainly representing red, green and blue respectively.

(Z)

Z- Score : It is a statistical measure which expresses in numerical value the distance of a particular score above or below the mean (M) in the units of standard deviation

$$Z = \frac{(X - \overline{X})}{S}$$

Zaigarnik Effect : The tendency of the individuals to remember interrupted or unfinished works better than the completed works.

Zygote :. A fertilized egg cell (ovum) formed when the male sperm cell and the female ovum united.

BIBLIOGRAPHY

Abel, T.M., Kinder, E.F. *The Subnormal Adolescent Girl.* New York : Columbia University Press, 1942.

Adinarayan, S.P. Before and after Independence: A Study of Racial and Communal Attitudes in India. The *British Journal of Psychology.* 1953, 44, 3, 108-115.

Adler, A. *Individual Psychology.* Littlefield : Adams & Co., Peterson, N.J. : 1959.

Adler, A. *Understanding Human Nature.* Greenwitch Conm : Fawcet, 1927.

Adorno, T.W., Frenkel-Brunwik, E., Lavinson, D. and **Sanford, N.** *Authoritarian Personality.* New York : Harper and Brothers, 1950.

Ainsworth, M.D., Bell, S.S.M., and **Slaton, D.J.,** Individual differences in the development of some attachment behaviour, *Merriall-Palmer Quarterly,* 1973, *10*, 123-143.

Allport, F.H. *Social Psychology.* Boston : Houghton Mifflin, 1928.

Allport, G.W. *Personality : Psychological Interpretation,* New York, Henry Holt, 1937.

Allport, G.W. and **Vernon, P.E.** *Allport-Vernon Study of Values,* New York Houghton Mifflin, 1931.

Allport, G.W. "Attitudes", *Handbook of Social Psychology.* Edited by Carl Murchison, Worcester, Massachusetts: Clark University Press, 1935.

Allport, G.W. Eidetic Imagery, *Brit. J. Psychol.* 1924, *15*, 99 - 120.

Allport G.W. *The Nature of Prejudice.* Boston : Beacon, 1954.

Alschuler, R.H. and **L.W. Hettwick,** *Painting and Personality.* Chicago University of Chicago Press, 2 Vol. 1947.

Amerback, *et al., Psychological Stress Brings About Negative Physical Changes,* 1981.

Anand, B.K., and **Brobacle, J.R.** Hypothalamic control of food intake, *Yale Journal of Biology and Medicine.* 1951, *24*, 123, 140.

Anastasi, Anne, *Psychological Testing* (2nd ed.), New York : Macmillan, 1961.

Anderson, J.R. and **Bower, G.H.** Recognition and retrieval process in free recall, *Psychol, Review,* 1972, *79*, 97 - 123.

Andreasen, N.E., Flaum, M. Swayze, V.O., O'Leary, D.S., Alliger, R., Cohan, G., Earhardt J., and **Yuth, W.T.,** Intelligence and brian structure in normal individuals, *American Journal of Psychiatry, 150*, 130-134, 1993.

Ardrey, R. *The Territorial Imperative,* New York, Atheneuns, 1960.

Arnold, M.B. *Emotion and Personality,* New York : Columbia University Press, 1960.

Aronfread, J. *Conduct and Consience,* New York: Academic Press, 1970.

Aronson, E. *The Social Animal.* Sanfrancisco: Freeman, 1972.

Atkinson, R.C., and **Raugh, M.R.** An application of mnemonic keyword method to the acquisition of a Russian Vocabulary, *Journal of Experimental Psychology.* 1975.

Ack, A.J., *Cognitive therapy and emotional disorders,* New York : International University Press, 1976.

Baker, R.W. A continued investigation on susceptibility to distraction in academically underachieving and achieving male college students. *Journal of educational psychology,* 1965, *56,* 254 - 258.

Bain, A. *Mental Science,* New York : D. Appletion & Co., 1968.

Balikov, S.M. and **Glesner, J.I.** Body and Brain Volume, 1948. *The Human brain: In Figures and Tables,* New York: Basic Books, 1948.

Balint, R., Seelanahamung des "Schavens" optische Ataxic reumaliche storung des Aufmerksemeit Monatshtra, *Neural, 25,* 51-81, 1909.

Bardier, *Les functions digestives,* Paris, O Doin, 1911.

Barker, R.G., Dembo, T., and **Lewin, K.** Frustration and regression, *An experiment with young children,* University of Iowa Studies on *Child Welfare, 10,* No. 386, 1941.

Baron, R.A., Effects of negative air icons on interpersonal attraction: Evidence for intensification, *Journal of Personality and Social Psychology,* 1987

Baron, R.A., and **Byrne, D.** *Social Psychology,* (5th ed.), New Delhi: Prentice-Hall of India Private Limited, 1998.

Barr, H.L., Langs, R.J., Holt, R.R., Goldberger, L, and **Klein, G.S.L.S.D.** *Personality and experience,* New York: Wiley, 1972.

Barron, F. The Psychology of Imagination, *Scientific American, 199,* 150-160, 1958.

Barron, F. *Creativity and Psychological Health,* New York : Van Nostrand Reinehart, 1983.

Baumeister, R.F. Choking under pressure : Self-consciousness and paradoxical effects of incentives on skilled performance. *Jorunal of Personality and Social Psychology,* 1984, 46, 410-420.

Bayley, N. Some increasing parent-child similarities during the growth of children, *Journal of Educational Psychology,* 1954, *45,* 19.

Bayley, Nancy, Developmental problems of the mentally retarded child, In I. Philips (Ed)., *Prevention and treatment of mental retardation,* New York : Basic Book, 1966.

Beach, T.A. The descent, of instinct, *Psychological Review,* 1955, *62,* 401 - 410.

Bell, C. Idea of new anatomy of the brain, In **W. Dennis,** *Readings in the history of psychology,* New York : Appleton Century Crofts, 1948, Reprinted of Bell, 1911.

Bern, D.J. Self-perception theory: In Berkoweitz (ed.), *Advances in Experimental Social Psychology,* Vol. 6, New York: Academic Press, 1972.

Berlyne D.E. *Conflict, arousal and curiosity,* New York : McGraw-Hill, 1960.

Beson, A.S. *Sensory Functions and limitations* of the vestibular system, In **R. Warren** and **A.H. Wertheiga** (eds), *Perception and control of self-motion.* Lawrence Erlbaum Associates, Hilsdates, 1990.

Bettill, R.B. Theory & Fluid and crystalized Intelligence — a critical experiment, *Journal of Educational Psychology,* 1963, *54,* 1 - 22.

Binet, A.S. and **Simon, T.** *The Development of intelligence in children,* Baltimore : Williams and Wilkins, 1916.

Blinkov, S.M. and **Glesner, J.I.,** *The Human Brain in Figures and Tables,* New York : Basic Books, 1968.

Blum, J.E., Jaru, L.P., and **Clark, E.T.** Rate of change on selective test of intelligence: A twenty year longitudinal study. *Journal of Gerentology,* 1970, 25, 171-176.

Bogardus, G.S. Measuring Social Distance, *The Journal of Applied Sociology,* 1925, 9 299-308.

Bogen, J.E., Schultz, D.H., and **Vogel, F.J.L.** Completeness of cailostomy shown by magnetic resonance imaging in the long term. *Archieves of Neurology,* 1988, *45*, 1203-1205.

Bonica, J.J. Pain research and therapy: Past and current status and future needs. In 2, Nig and J.J. Bonica (eds). *Pain discomfort and humanitarian care.* New York: Elsevir, 1987.

Boring, E., and **Holway, A.** The moon illusion and the angle of Regard, *Amer, J Psychol.,* 1940, *53*, 109 - 116.

Boring, F.G., A short historical perspective. In D.I. Mostofsky (Ed.) Attention : Contemporary and Analysis. New York : Appleton Century Crofts, 1970.

Bousfield, W.A., The occurrence of clustering in the recall of randomly arranged associates, *Journal of Gen. Psychology, 49*, 229-240, 1953.

Bower, G.H. and **Hilgard, E.R.,** *Theories of Hearings* (5th ed.) New Delhi, Prentice-Hall of India, 1986.

Bowlby, J. *Attachment and Love.* Vol. 1. New York : Basic Books, 1969.

Brewin, G.R. Attributions for industrial accidents: their relationship to rehabilitation outcome. *Journal of Social and Clinical Psychology,* 1984, 2, 156-165.

Broadbent, D.E., Effects of noise on behaviour. In **C.M. Harris** (Ed)., *Hand Book of Noise,* New York : McGraw-Hill 1957.

Brobeck, J.R. Food intake as a Mechanism of Temperature Regulation, Yale, J. *Biol, Med.* 1947 - 1948, *20*, 545 - 552.

Broca, P., *Bull, Soc Naot.* 2nd ed., 186.

Brodmann, K. From physiologic des. Gehirms, In Die ellgermeine, *Chirrsgic des., Vol. II* Strugart, Enke, 1914.

Bromley, D.B. *The psychology of human aging,* Baltimore: Penguin Books,1966.

Brophy, I.N. The luxury of anti-negro prejudice. *Public Opinion Quarterly,* 1946, *10*, 456-466.

Brown, R. *Words and things,* New York : Free Press, 1958.

Brown, T. *Lectures on philosophy of the human kind* 4, Vol. 16th ed. Edinburgh : William Taint (1846) 1920.

Brown, R.I., The effect of visual distraction on perception in subjects of subnormal intelligence, *British Journal of Social and Clinical Psychology,* 1964, *3*, 20-28.

Brownell, K.D. Behavioural Medicine, In **C.M. Franks, G.T. Wilson P.C., Kendell** and **K.D. Brownell** (eds.), *Annual review of behaviour therapy and practice.* (Vol.8), New York : Geulfond, 1982.

Bryan, T.W. and **Bryan, H.H.,** *Understanding learning disabilities,* (2nd ed.,) Sherman Oaks, Califf: Allied Publishing Company, 1976.

Buhler, C. and **Allen, M.** *Introduction to Humanistic Psychology,* Monetary, California Books/Cole Publishing Company, 1972.

Burt, C. The genetic determination of differences in intelligence : A study of monozygotic twins reared together and apart, *British Journal of Psychology,* 1966, *57*, 153 - 157.

Burt, C. The evidence for the concept of intelligence, *British Journal of Educational Psychology.* 158-177, 1955.

Cameron, N., and **Magaret, A.,** *Behaviour Pathology.* Bosten: Houghton Mifflin, 1951.

Cameron, N. Paranoid, Condition and Paranoia, In. 5. Aniati (Ed.), *American Handbook of Psychiatry.* New York : Basic Books, 1951.

Campbell, D.T. Social attitude and the acquired behavioural dispositions in Koch, *Psychology, a study of science,* Vol. 6, New York : Mcgraw Hill, 1963.

Candland, D.K. *Psychology : The Experimental Approach,* New York, McGraw-Hill Book Company, 1968.

Cannon, W.B. The James, Large theory of emotion : A critical examination and an alternative theory, *American Journal of Psychology,* 1927, *39*, 106 - 124.

Cannon, W.B. *Bodily changes in Pain, Hunger, Feer and Rage,* New York Appleton : 22nd, ed., 1929.

Caplan, G. *Principles of preventive psychiatry,* New York : Basic Book, 1964.

Carpenter, J., and **Eisenery, P.** Some relations between family background and personality, Jl. *psychol,* 1938, *6,* 115 - 156.

Cates, J. Psychology's man power, Report on the National Registrar of Scientific and Technical Personnel, *American Psychologists,* 1970, *25*, 254 - 268.

Carlson, A.J. *'The control of hunger in health and disease,* Chicago, University of Chicago Press, 1916.

Cattell, J. Mak. Mental testing and measurements, *Mind,* 1890, *15*, 373 - 381.

Cattell, R.B. *The scientific analysis of personality,* Chicago : Aldine, 1963.

Cesell, A. *Wolf Child and Human Child.* New York : Harper, 1941.

Chein, L., Gerand, D.L., Lee, H.S., and **Rosenfield, E.** *The Road to H.* New York. Basic Books, 1964.

Chen, W.K.C. The influence of oral propaganda material upon students' attitudes. *Arch. Psy.* 1933, *23*, 150-160.

Cherry, E.C. Some components on the recognition of speech with one and two ears, *Journal of Accoustical Society of America,* 1954, *26*, 554 - 559.

Child, I.L. *Humanistic Psychology and Research Tradition : their several virtues,* New York: John Wiley and Sons, *INC.*, 1973.

Chomsky, N. *Syntactive Structures,* The Hague : Mouton, 1957.

Chudhuri, H. *Philosophy of Meditation,* New York: Philadelphia Library, 1965.

Ciminero, R. Behavioural assessment: An overview., In A.R. Ciminero, K.S. Colhoun and H.E. Adams (Eds.), *Handbook of Behavioural Assessment,* New York : Wiley Interscience, 1977.

Clore, G.I., and **Byrne, D.** A reinforcement-effect model of attraction, In, T.L. Huston (ed) *Foundations of interpersonal attraction (pp.* 143-10).

Cohan, D.B., *Sleep and Dreeming: Origin, Natural and Functions,* Oxford : Pergamon, 1979.

Cohan, S., and **Milgrom, S.**, The experience of living in cities, *Science, 167*, 1461-1468, 1970.

Cohan, S. Environmental load and allocation of attention. In A. Bakam and E. Valins (Eds.), *Advances in Environmental Research,* Hilldete, N.J. Lawrence Erlbanum, 1977.

Cohn, T.S., and **Lee, W.** Attitude change and interpersonal attraction. *The Journal of Social Psychology,* 1961, *55*, 207-211.

Coleman, J.C. *Abnormal Psychology and Modem life* (5th ed.), Bombay, D.B. Taraporewala Sons & Co., Pvt. Ltd., 1976.

Collin's *English Dictionary,* London : Collins, 1984.

Conway, M., and **Ross, M.** Getting what you want by revisiting what you had. *Journal of Personality and Social Psychology,* 1984, *47*, 738-768.

Cooper, R.M. and **Zubek, J.P.B.** Effects of enriched and restricted early environment in the learning ability of bright and dull rats, *Canadian Journal of Psychology,* 1958, *12*, 109 - 164.

Constantini, E., and **Craik, K.H.** Personality and Politicians, California part leaders, 1960-1976. *Journal of Personality and Social Psychology,* 1980, *38*, 641-661.

Cowen, E.L. An overview and directions for future work, In E.L. Cowers., E.A., Gardner and M. Zax (eds), *Emergent Approach to Mental Health Problems,* New York : Appleton Century Crofts, 1967.

Cowen, E.L. The effectiveness of secondary prevention programmes using non-professionals in the school setting proceedings of 76, *Annual Convention of the American Psychological Association,* 1968, *2*, 705 - 706.

Cowen, E.L. Mothers, in the classroom, *Psychology today,* 1969.

Cowen, E.L. Broadening community mental health practicurn training, *Professional Psychology,* 1971, *2*, 159 - 168.

Cowen, E.L. *President's message,* Divisions of community psychology letter, 1975, 1-2.

Cumming, E., and **Henry, W.E.** *Growing Old: The Process of Disengagement,* New York: Basic Books, 1962.

Darwin, C. *On the origin of spacies by means of natural selection.*

Darwin, C. *The descent of man,* 1851.

Darwin, C. *Expression of Emotions in Men and Animals,* London : J. Murray, 1972.

Darwin, C. *Zoomomisa,* London : 1810, 3, 122.

Das, J.P. *Attention and Orienting Response in Retardete,* Lecture delivered in the Department of Psychology, Delhi University, 1975, August 29, 1975.

Datel, W.E., and **Generalti, J.A.** Reliability of Research Interpretation, *Jl. Projective Techniques,* 1958, *19*, 322 - 391.

Davidson, J.M. Camergo, C.A., and **Smith, E.R.**, Effects of Androgen on Sexual Behaviour in Hypogonedual Men, *Journal of Clinical Endocrinology and Metabolism,* 1979, *48*, 955 - 985.

Davison, G.C., and **Neale., J.M.**, *Abnormal Psychology* (5th Edition)., New York : John Wiley & Sons, 1990.

Deikman, A.J. Experimental Mediyayion, *Journal of Nervous and Mental Diseases,* 1963, *73*, 136-329.

Dement, W. An essay on dream : The role of physiology in understanding their nature, In, *New Directioning in Psychology-II,* New York : Holt Rinehart or Winston, 1965.

Dement, W., Wolpert, E.A., The relation of eye movements body motility and external stimuli of dream contents, *J Exp. Psychol.,* 1958, *55,* 543 - 553.

De Paul, B.M., Scone, J.L., and **Lassiter, C.D.** Deceiving and detecting deceit, In B.R. Schlenker (ed), *The Self and the Life.* New York: McGraw-Hill, 1985.

Dethier, V.G. and **Steller, E.,** *Animal behaviour* (2nd ed.) Englewood Cliffs, N.J. Prentice-Hall, 1964.

Diagnostic and **Statistical Manual of Mental Disorders,** First Edition 1952, Second Edition, 1960, Third edition 1980, Revised edition 1982, Washington : SC *American Psychiatry Association,* (APA).

Dockeray, F.C. *Psychology,* New York: Prentice-Hall, 1942.

Doob, L.W., *Social Psychology,* New York: Henry Holt &, Co., 1940.

Dovidio, J.H., Evans, N., and **Tylor, R.B.** Racial stereotype: The contents of their cognitive representations. *Journal of Experimental Social Psychology,* 1986, 22, 22-37.

Duck, S. Social and Personal Relationships. In **B.R. Miller** and **M.L. Knapp** (Eds.) *The Handbook of Interpersonal Communication,* Beverly Hills: C.A. Sage, 1985.

Duffy, E. *Activation and Behaviour,* New York : John Wiley & Sons, 1962.

Dugdala, R. *The Jukes : A study of crime, pauperism, disease, and heredity,* New York : G.P. Putram's Nons, *29,* 1910, (1877).

Dunlap, K. Are emotions teleslogical constructs? *American Journal of Psychology, 44*:146-162, 1932.

Duval, S., and **Wicklund, R.A.,** *A theory of objective selfawareness.* New York: Academic Press, 1972.

Easton, D.M., *Mechanisms of body functions:* New Delhi: Prentice Hall of India Private Limited, 1978.

Ebbinghaus, H. *Memory : A contribution to experimental psychology,* (N.A. Ranger & C.E. Bussenious Trans) New York : Columbia, 1913.

Edwards, A.J., and **Scannell, D.P.** *Educational Psychology,* Bombay Allied Publishers Private Limited, 1975.

Edwards, D.L. Political Frames of Reference as a Factor Influencing Recognition, *P. Abnormal Socio. Psychol.* 1941, *36,* 34 - 50.

Ehrlich, H.J., *The social psychology of prejudice,* New York : John Wiley and Sons, 1973.

Eiduson, B.T., Eiduson, S., and **Geller, A.,** Biochemistry, Genetic and Nature-nurture Problem, *Amer., JL., Psychiatry,* 1967, *119,* 342 - 550.

Einstein, A. *Why War?* Letter to Professor Freud, International Institute of Intellectual Corporation, League of Nations, 1933.

Ellis, N.R. The stimulus trace and Behavioural inadequacy. In Ellis, N.R. (Ed) *Mand Book of Mental Deficiency,* New York : McGraw Hill, 1963.

Ellis, A. *Reason and Emotion in Psychotherapy.* New York: Lyle Stuart, 1962.

Ellis, A. *The basic clinical theory of medical emotive therapy.* New York: Springer, 1977.

Epstein, W. *Variables of perceptual learning,* New York : McGraw Hill, 1967.

Epistein, W and **Shortz, P.** *Psychology in Progress,* New York : Holt Rinehart and Winston, *Inc.*, 1971.

Epistein, A.N., and **Fitgsimons, J.J.** and **Rolls, A.J.** Drinking induced by injection of angiotension into the brain of the rat, *Journal of Physiology* (London) 1970, *210,* 457 - 474.

Epstein, L.H., Brck, S., Figaeron, J., Farkan, G.H., Kardin, A.E., Daneme, D. and **Becker, D.** The effects of point economy and present management on Urine glucose and metabolic control in children with insulin dependent diabetics. *Journal of Applied Behaviour Analysis,* 1981, *14,* 365-376.

Erikson, Erik, H. Insight and Responsibility of Rorschach Interpretation, *JL. Projective Technique,* 1955, *19,* 372 - 381.

Erikson, Erik, H. Youth, Fidelity and Diversity, In Erik H. Erikson (Ed) *The Challenge of Youth,* New York, Double Day, 1963.

Erikson, E.H. *Childhood and Society,* New York: Nortion, 1965.

Erlanmeyes-Kimling, L., and **Jarvik, J.L.F.S.** Genetic and Intelligence a Review, *Science,* 1902, *42,* 1477 - 1479.

Eysenck, H.J. *The Biological Basis of Personality,* Springfield ill Charles C. Thomas, 1967.

Eysenck, H.J. and **Stanely Rachman,** *The cause and cures of Neurosis,* London : Routledge & Kegan Paul, 1965.

Eysenck., H.J., Wakefield, A., and **Friedman, A.F.,** Diagnosis and clinical assessment, The DSM-111, In M.R., Rosenweig and L.W. Porter (Eds.), *Annual Review of Psychology, Vol.* 34, Polo Alto., Calif : Annual Review., 1983.

Farah, M.J. *Visual Agnosia,* Cambridge Mass, M.D.T., Press, 1990.

Fechner, G.T. Elements of psychophysical (H.S. Hangfield Trans), In Brand (Ed), *The classical psychophysics,* Boston : Hougton Mifflin, 1912.

Fendrick, P. The influence of music distraction upon reading efficiency, *Journal of Educational Research, 31,* 267-271, 1937.

Fenigstein, A. Self-consciousness and overperception self as target. *Journal of Personality and Social Psychology,* 1984, *47,* 860-870.

Festinger, I., Schachter, S., and **Beck, E.** *Social Pressures on Informal Groups:* A study of a lousing community, New York: Harper Brother Ltd., 1950.

Finger, S. *Origins of Consiousness: A History of Explorations into Brain Functions.* New York: Oxford University Press, 1994.

Fisher, C.M. and **Adams, RV.** Transient Global Amnesia, *Transaction of American Neurological Association,* 1958, *89,* 100.

Fisher, S. The distraction effect and information processing complexity, perception, *2,* 79-89, 1973.

Fleisher, R.A., and **Chartkoff, J.M.** Effects of dominance and sex on leader selections in dyadic work groups. *Journal of Personality and Social Psychology,* 1986, *50,* 94-99.

Flourens, P. *Psychologic Compare,* 1864.

Folkee, V.S. Forming relationship and matching hypotheses, *Personality and Social Psychology Bulletin,* 1982, *8,* 631-636.

Fordyee, W.E., Borckway, J.A., Bergman, J.A., and **Spengler, D.** Acute back pain: A control group comparison of behavioural *versus* transitional methods. *Journal of Behavioural Medicine.* 1986, *9*, 127-148.

Franks, C. Alcohol, Alcoholism and Conditioning, *Journal of Mental Science, 104,* 14-33, 1958.

Freeman, W. and **Watt, J.W.** The thalamic projection to frontal Lobe, *Res. Rubl. Ass. Res. Sen, Ment.* 1948, *27*, 200, 209.

Freud, S. *New Introductory Lectures on Psycho Analysis.* New York: Norton, 1933.

Freud, S. Repression, Alcoholism and Conditioning, *Journal of Mental Science, 104,* 14-33, 1958.

Freud, S. *Interpretation of Dreams* (1899 - 1900) Standard Edition of complete works of Sigmund Freud, London : Hogarth Press, 1955.

Freud, S. Why War? Letter to Professor Einstein, In J. Strachey (Ed.), *Collected Papers of Sigmund Freud* Vol. 5 New York : Basic Book, 1959.

Freud, S. *General Interpretation to Psychoanalysis,* New York : Liveright, 1920.

Freud, S. *Psychopathology of Everyday life,* New York, Mentor Book New American Library World Literature, Columbia, 1913.

Freud, S. (1933) *New Introductory Lectures,* In Vol. XXII of the Standard Edition, London : Hogarth Press, 1964.

Fritsah, G., and **Hitgig, E.** On the electrical excitablity of the cerebrum. In G. Von Bomin Ced. *The cerebral cortex,* Springfield, Ill: Charles C Dhemas, 1960.

Fromm, Erich, *Escape from Freedom,* New York : Avon Books, 1941.

Fromm Erich, *The Resolution of Hope Towards, Humanised Technology,* New York : Holt Rinebant and Winston, 1968.

Fromm Erich, *Man for His Himself,* Greenwich, Conn : Fawcetee Book, 1947.

Fromm Erich, *The Sane Society,* Greenwich, Conn : Fawcetee Books, 1955.

Fromm Erich, *The heart of man, its genius for good and evil,* New York, Haper & Row, 1984.

Fuller, J.L. Experimental deprivation and better behaviour, *Science,* 1967, *158*, 164, 1652.

Galenter, E. Contemporary Psychophysics, In **R. Brown (Ed) Galenter, E. Hess, E.H.** and **G. Mandler,** *New Directions, In psychology,* New York : Holt, 1962.

Gall, F.J. Anatomic et physiologic, due system nerveuen *General et du ceruean en particular, 4 Vol.* Paris : D. Hankel, 1811 - 1819.

Gallagher, J.J. A comparison of brain injured and non-injured mentally retarded children on several psychological variables. *Monograph research children development,* 2, 5, 65, 1957.

Gallagher, M., Nagahara, A.H., and **Berwell, H.D.** Cognition and hippocampal systems in againg: Animal models in, **J.I. Mcgaugh, N.R. Weinberger** and **G. Lynch** (Eds.) *Brain and Memory : Modulations and Mediations of Neuroplasticity,* pp. 103-126, New York: Oxford University Press, 1995.

Galton, F. Psychomatric Experiments, *Brain,* 1879.

Galton, F. *Inquiring into Human Faculty and its Development,* London Macmillan, 1883.

Galton, F. *Hereditary Genius : An enquiry into its laws and consequences,* London: Macmillan, 1869.

Ganguli, H.C. and **Rao, M.N.** Industrial noise as a health problem, *Journal of Indian Medical Association, 1-24,* 1952.

Garrett, H.E. *General Psychology,* 2nd ed, New Delhi, Eurasia Publishing House Pvt. Ltd., 1975.

Gates, A.I. Recitation as a Factor in Memorising, *Arch, Psychol.,* 1917, No. 40.

Gauber, A.E., Terrell, G. and **Werthimer, M.** (Eds)., *Contemporary approaches to Creativity Thinking,* New York : Atherton, 1962.

Gazzoniza, M.S. Nature's mind: The biological roots of thinking, emotion, sexuality, language and intelligence. New York: Basic Books, 1972.

Geen, R.C., and **Dennerstein, E.** (eds.), *Aggression : Theoretial and Conceptual Reviews.* New York: Academic Press, 1993.

Geier, A. First approach to the study of leadership in small groups, *Journal of Communication,* 1969, *17,* 316-323.

Geissler, I.R. The Measurement of Attention, *American Journal of Psychology, 20,* 473-529, 1909.

Gellagher, J.A. *Analysis of Research in the Education of Gifted Children,* State of illusion, Office of the Superintendent of Public Illustration, 1960.

Geortzel, V. and **Geortzel, M.G.** *Cradles of Eminence,* Boston: Little Brown, 1962.

Gerard, E.R.W. What is memory? *Scientific American,* 1953, 190, 118 - 126.

Geschwind, N., and **Lavitsky, W.** Human brain left-right asymmetrics in temporal lobe region, *Science* 1968, *161,* 186-187.

Gesell, A. *The mental growth of the pre-school children,* New York: Macmillan, 1925.

Gesell, A. and **Thompson, H.** Learning and maturation in identical infant twins, In **R.G. Barran, J.S. Kourin,** and **K.F. Wight,** *Child Behaviour and Development,* New York : McGraw-Hill, 1993.

Getzels, J.W. and **Jackson, P.W.** *Creativity and Intelligence,* New York, *Wiley,* 1962.

Gibson, J.J. The visual perception of objective motion and subjective movement, Psychological Review, *61,* 304-314, 1954.

Gilbert, D., and **Janes, E.E.** *Journal of Personality and Social Psychology:* 1986, 50.

Gill, M.M., and **Brmman, M.** *Hypnosis and Related States,* New York: International University Press, 1959.

Goddard, H.A. *The Kallikak family : A study in the heredity of feeble mindedness,* New York : Macmillan, 1913.

Goldfield, M.R. Towards delineation of the therapeutic change principles. *American Psychologists,* 1980, *35,* 991-999.

Goldman, G. Clinical perspectives. In A. Simon and L.T. Epistein (eds). *Ageing in Modern Society,* Washington, D.C.: American Psychiatric Association, 1968.

Goldstein, M.J., Rodnick., E.H., and **Judd, L.K.,** *Schizoid Therapy, News Week.,* June, 7, 1971, 77 - 78.

Gondole, J.C. The enhancements of creativity through long and short-term exercise programme, *Journal of Social Behaviour and Personality,* 1985, *1,* 77-82.

Goodenough, F.L. and **Harris, D.B.** Studies in the psychology of children's drowing II, 1928 - 1949, *Psychological Bulletin,* 1958, *47*, 309-433.

Goodenough, F.L. *Measurement of Intelligence by Drawing,* New York, Hardcourt Brace, World, 1926.

Goodenough, F.C. A New approach to a measurement of the intelligence in young children, *J. Genet. Psychol.* 1926, *33*, 185 - 211.

Gottesmans, J.I. Hereditability of Personality, *Psychological Monograph,* 1963, *77*, 1-21.

Gottlinch, G. Prenatal behaviour of birds, *Quarterly Review of Biology,* 1968, *48*, 148 - 174.

Gottsman, L., and **Shields, J.,** Schizophrenia in twins 16 years consecutive admissions to a psychiatric clinic, *British Jl. Psychiatry,* 1966, *112*, 809 - 818.

Gough, H.G. *Manual forth California Psychological Inventory* (Revised)., Pole Alto., Colif: Consultings Psychological Press, 1960.

Gould, S.J.L., *The Mismeasure of Man,* New York; Norton, 1981.

Gray and **Wedderburn, A.A.I.,** Grouping strategies with simultaneous stimuli, *Quarterly Journal of Experimental Psychology, 12*, 180-184, 1960.

Green, R. Age-intelligence relationship between ages sixteen and sixty four: A rising trend. *Developmental Psychology,* 1969, *1*, 618-627.

Greenberg, J., Pyszcymaki, T., and **Solomon, S.** The Self-serving attributional bias: Beyond self-presentation, *Journal of Experimental Social Psychology,* 1982, *18*, 56-67.

Greenfield., J.C., and **Wolfson, J.M.,** Microphobia vera, *Arch-neuro psychiatry,* 1935, *33*, 1296 - 1316.

Greiano, W., Brothers, T., and **Berscheid, E.** Heights and Attraction: Do men and women see eye to eye? *Journal of Personality,* 1978, *46*, 128-145.

Grossman, S.P. *A Textbook of Physiological Psychology,* New York, John Wiley & Sons, 1967.

Gulford, J.P. Traits of Creativity, in H.H. Anderson (ED) *Creativity and its Cultivation,* New York : Harper, 1959.

Guilford, J.P., and **Fruchter, B.** *Fundamental Statistics in Psychology and Education,* New York: McGraw-Hill Book Company, 1987.

Guilford, J.P. Three faces of intellect, *American Psychologist,* 1959, *24*, 469-479.

Guilford, J.P. Factorial Angles to Psychology, *Psychological Review,* 1961, *68*, 1-20.

Guthrie, E.R. Association by contiguity, In Koch (ed) *Psychology : A Study of Science,* Vol. 2, New York : McGraw-Hill, 1958.

Haber, R.N. Eidetic Images, *Scientific American, April,* 1969, 36 - 44.

Haldane, J.B.S. *The interaction of nature and nurture.* Anna!s of Eugenics, *13*, 197-205, 1946

Haley, J., Direct study of child-parent interaction: Observation of the family of the schizophrenic, *American Journal of Orthopsychiat,* 1960, 460-417.

Hall, G.S., and **Van da Castle, R.L.** *The content analysis of dreams.* New York: Appleton-Century Crofts, 1961.

Hall, C.S. The inheritances of emotionality, *Sigma XI Quart,* 1938, *26*, 17-27. Temperament: a survey of animal studies, *Psychol. Bull.,* 1942, *38*, 909 -943.

Hall, N.S. The first 500 days of a child's life, *child study monthly,* 1996, 97, *2*, 330 - 342.

Hallersberg, E.F. The Horn Hallensberg test and adjustment to reality, *American J. Orthopsychiet,* 1945, *15*, 690-710.

Harrigan, J.A., and **Rosenchal, R.** Physicians' head and body positions as determinants of perceived rapport. *Journal of Applied Social Psychology,* 1983, *13*, 496-509.

Hassor, L., Margaret, A., and **Camern, N.** The Production of language disorganization through personalized distraction, *Journal of Applied Psychology, 33*, 89-299, 1952.

Hatfield, E., and **Webster, G.W.** *A newlook at love. Readings.* M.A. Addison Wesley, 1981.

Hatfield, E., and **Sprecher,** S. *Mirror-Mirror the importance of look in everyday* life - Albeny. New York: State University of New York, 1986.

Hathaway, S.R., and **McKinley, J.C.** *The Minnesota Multiphasic Personality Inventory,* Minneapolies: University of Minnesota Press, 1943.

Havighurst, R.J. *Human Development and Education,* New York: Longmans Green, 1953.

Havighurst, R.J., Niengartein, and **Tobin, B.F.** Disengagement and partners of aging. In **B.L. Negartein** (ed.) *Middle age and aging,* Chicago: University of Chicago Press, 1968.

Hebb, D.O. *The organization of behaviour,* New York, Wiley, 1949.

Heider, F. *The psychology of interpersonal relations.* New York: Wiley, 1981.

Helmholtz, H. Von. *Handbuch der physiogischen optick,* Liepzig : Voss 1st ed. 1867.

Helmholtz, H. Von. *Helmholtz's te tretise on physiological optics* (MO J.P.C. Sruhu, Trans, &, Ed.) Vol. 1 and 2, New York : Dover, 1962.

Helmholtz, H. Von. (1846) *Physiological optics,* Optical Society of America, 1924.

Helpage India, Directory of Oldage Homes in India, New Delhi; 1995.

Helson, H. Adaptation level theory, in S. Koch (ed) *Psychology : A study of science,* study-1, conceptual and systematic, Vol.1, sensory, perceptional and physiological, New York : McGraw-Hill, 1958.

Helson, H. *Adaptation, level theory,* New York : Harper & Row, 1964.

Hering, E. (1907), *Grundzuge der Letre von Lichtsinna,* Berlin Springer, 1920.

Nerman, F.L. The effects of noise on certain psychological process, *Arachialogical psychology,* No. *147*, 1933.

Hersch, C. From Mental Health to social action, clinical psychology in Historic and perspective, *American psychologists,* 1968, *24*, 909 - 916.

Hess, E.H., Effects of drugs on importing behaviour. In Uhr, L., and Miller, J.G.(eds.), *Drugs and behaviour.*

Hess, E.H. Imprinting, *Science,* 1959, *170*, 133 - 141.

Hess, E.H. Space perception in the chick, *Scientific American, July,* 1959, 71-80.

Hikbig, E. Uber de electrische, Errgbarkeit des Grosshings, *Arch. Anat. Physiol,* 1890.

Hilgard, E.R. Atkinson, R.C. and **Atkinson, ROB.,** *Introduction to psychology,* 6th ed., New Delhi, Oxford and IBH Publishing Co., 1975.

Hilgard, E.R. The Nature of Conditioned Response, the case for and against stimulus substitution, *Psychol. Rev.* 1936, *43*, 366 - 385.

Hilgard, E.R. The human hypnosis with some comments on alteration strategem paradigms. *Americal Psychiatrist,* 1973 (6), *28*, 973-982.

Hilgard, E.R., and **Bower, G.H.** *Theories of learning,* Englewood, N.J.: Prentice-Hall, 1975.

Hilgard, E.R., *Hypnotic suggestibility,* New York: Harcourt, Brace. Jonovich, 1965.

Hillsdeale, N. *Perception and control of self-emotion resources for ecological psychology,* Lawrence Erbaum Association, 1950.

Hirschberg, G., and **Gilliland, A.R.** Parent child relationships on attitudes, *The Journal of Abnormal Psychology,* 1942, *37*, 125-130.

Hobbes, T. *Leviatham,* The English Works of Thomas Hobbes, 4 Vols. London : Bohn, 1840 (originally published in 1651).

Hobbs, N. Group centred therapy, In R. Rogers, *Client Centerd Therapy,* Boston: Houghton Mifflin, 1951.

Hochberg, J.E. *Perception,* Englewood Cliffs, N.J. Prentice-Hall, 1964.

Hochlery, F.H. and **M. Leemay,** Arterioraphic correlates of handedness, *Neurology, 25,* 218-222, 1975.

Hofatactter, A.R. The changing composition of intelligence: A study in technique, *Journal of Genetic Psychology, 85,* 159-164, 1954.

Hollander, E.P. *Leadership dynamics: Practical guide to effective relationship.* New York: Free Press, 1978.

Holohan, C.G. Environmental effects on outdoor social behaviour in low income urban neighbourhoods. A naturalistic investigation, *Journal of Applied Social Psychology,* 6, 48-63, 1976.

Holzberg, J.D. Reliability examined, In M.A. Rickrosovsiantina *Rorschach Psychology,* New York : Wiley, 1960.

Horney, Karen, *Self-analysis,* New York : Norton, *196*, 1942.

Horowitz, E.L. The development of attitudes towards the Negro, *Arch.,* Psy., 1936, No. 194.

Hovey, H.B. *Effects of general distraction on the higher thought processes, American Journal of Psychology,* 1928, *40*, 585 - 591.

Hovland, C.I. Experimental studies in Rate-learning theory, *J Exp. Psychol,* 1938, *23,* 172 - 190.

Hull, C.L. *Principles of behaviour,* New York : Appleton, Century Crofts, 1943.

Hull, J.C. A self-awarness of the causes and effects of alcohol consumption, *Journal of Abnormal Psychology,* 1981, *90*, 586-600.

Inban, F.E. *Lie deduction and criminal interrogation,* Boltimore : Williamas and Wilkins, 1942.

Irwin, G.S. Infant Speech. *Scientific America,* 1949, *81,* 22-24.

Iscoe, I., Bloom, B., and **Spielberger, C.D.** (Ed.) *Community Psychology in Transition,* Washington, D.C. : Henisphar, 1977.

Isen, A.M. Towards understanding the role of effect in organism, In **R.S. Wyer,** Jr., and **T.K. Srull** (eds.), *Handbook of Social Cognition,* Vol.3. Hillsdale, N.J.: Erllbaum, 1984.

Jack, A.M. An experimental study of conduct behaviour in pre-school children, *Univ. IOWE, Stud-Clinical Welfare,* 1934, *9*, No. 3, pt. 1., 1-65.

James, W. *The Principle of Psychology,* New York : Holt, 1890-2 Vol.

Jay, S.M., Ozolines, C.H., and **Caldwell, S.** Assessment of Children's diseases during painful medical procedures, *Health Psychology,* 1983, *2*, 133-147.

Jeannered, M. *The Neural and Behavioural Organization of Goal Directed Movements,* Oxford: The Clarendon Press, 1988.

Jenkin, R.I., The varieties of children's behavioural problems and family dynamics, *Amer., Jl., Psychiatry,* 1968, *124*, 1440-1445.

Jenkins, J.G. and **Dallanback, K.M.** Oblives since during sleep and waking, *Amer. J. Psychol.,* 1924, *35*, 605-612.

Jenson, A.R. How much can we boost I.Q. and scholastic achievement? *Harvard Educational Review,* 1969, *39*, 1-123.

Johnson, J.T., Jammert, J.B. III and **Pettigrew, T.E.** Casual and dispositional inference: Evidence of incommitment judgements. *Journal of Experimental Social Psychology,* 1984, 20, 567-585.

Jones, E.E., and **Davis, K.E.** From acts to disposition: The attribution process in person perception. In **Berkowietz** (ed)., *Advance in Experimental Social Psychology,* Vol. 2, New York: Academic Press, 1965.

Jung, C.G. Psychological Types, London : Routledge, 1923.

Jung, C.G. *Analytical psychology, Its theory and practice,* New York, Pantheon Press, 1927, Random House, 1968.

Kaads, B. Brain mechanism related to aggressive behaviour in **C.D. Clemente** and **D.B. Lindslay** (eds.) *Aggression and defense,* Barkelay : University of California, 1962.

Kahnaman, D., Attention and efforts, New Jersey : Prentice Hill, 1973.

Kallmann, F.J., *Heredity in health and mental disorder : Principles of psychiatric genetics in the light of comparative twin studies,* New York : W.W. Norton, 1953.

Kanfer, F.H., and **Busemeyer, J.R.** The use of problem solving and decision making in behaviour therapy. *Clinical Psychologycal Review*, 1982, *2*, 239-266.

Kant, O., Clinical Investigation of Simple Schizophrenia., *Psychiatric Quarterly,* 1948, Vol. *22*, 141.

Kantovitz, B.H. *Environmental Psychology : Behaviour in the Physical World,* Philadelphia: W.B. Sonders, 1977.

Katz, K.D., and **Kahn, R.I.** *The Social Psychology of Organizations,* New York: Wiley, 1966.

Kautuvich, K.V. An attempt at associative reflex theory in alcoholism. *Psychological Abstract,* 1930, *4*, 493.

Keister, M.E. The behaviour of young children in failure, *Univ. IOWA, Stud., Child Welfare,* 1938, *14*, 29-82.

Kelly, G.A. *The psychology of persnal constructs,* New York: Norton, 1955.

Kelley, K., Byrne, D., Pzzybla, D.P.S., Eberly, C.C., Eberly, B.W., Greudlinger, V., Wan, C.K. and **Grosky, J.** Chronic self-destructiveness: Conceptualization, *Motivation and Emotion.* 1985, *9*, 139-151.

Kendler, H.H. *Basic Psychology,* London : H. Methuen & Co., Ltd., 1963.

Kendlar, H.H. and **Kendler, T.S.,** Selective attention *versus* meditation, *Psychological Bulletin, 66,* 282 - 288, 1966.

Kerry, R.J. Phobic of outer space, J. *Mental Science,* 1960, *106,* 1383-1387.

Kettog, H.H. Attraction in social interaction. In **E.E. Jones** *et al.* (eds.) *Attribution: Perceiving the causes of behaviour.* Morriaton, N.J. Beneral Learning Press, 1972.

Kheitman, N. Pattern of dreaming, *Scientific American,* No. 1960, *203,* 82-83, Freasan &, Co.,

Kiesler, C.A., Collins, B.E., and **Miller, N.** *Attitude Change: Critical analysis of theoretical approaches,* New York: Wiley, 1969.

Kimble, G.A. Garmezy, N. and **Zigler, E.** *Principles Psychology,* New Delhi : Wiley, Eastern Limited, 1985.

Kinsely, A.C., Pomeroy, W.O. Mortin, C.E. and **Gebbard, PH.** *Sexual behaviour in the human female,* Philadelpha, Sounders, 1953.

Kirscht, J.P. Preventive health behaviour: A review of research issues. *Health Psychology,* 1983, *2,* 277-301.

Klabanoff, L.B., Parental attributes of mothers of schizophremics, brain-injured and retarded and normal children, *American Journal of Orthopsychiatry,* 1959, *29,* 445-454, 480.

Klaf F.S. and **Davis., C.A.,** Homosexuality and Paranoid Schizophrenia : A survey of 100 cases and controls, *Amer. Jl. Psychiatry,* 1960, *116,* 1170-1075.

Klein, M. *The Psychoanalysis of Children,* London: Hogarth, 1932.

Klienberg, O. *Negro Intelligence and Selective Migration,* New York : Academic Press, 1965.

Klinberg, O. *Social Psychology,* New York : Rinehart and Winston, 1950.

Klopifer, B. and **Kellay, D. Meg.** *The Rorschach Techniques,* Yonken, World Books Co., 1946.

Kluver, H. and **B.C. Bucy.** Preliminary analysis of the temporal lobes in monkeys. *Archieves of Neurology and Psychiatry,* 1939.

Koffka, K. *Principles of Gestalt Psychology,* London : Routledge and Kegan Paul, 1950.

Koffka, K. *The growth of the mind,* (Trans, D.M. Ogdon), London, Kegan Paul, Trench, Trubener, 1924.

Kohler, W. *Gestalt Psychology,* New York, Liveright, 1929.

Kohler, W. *The mentality of apes* (W.E. Winter Trans) London : Kegan Paual, Trench, Taubner, & Co., (1925-1927).

Kolb., D.L.C. *Modern clinical psychology* (9th ed)., Pulladelphiq: Sounders, 1977.

Kolb, B. and **Whishaw, I.Q.** *Fundamentals of Human Neuropsychology* (4th ed), New York, W.A. Freeman and Company, 1996.

Korgman, W.M. *The physical growth of children: An Appraisal of studies;* 1950-1955, Monograph, *Soci. Res. Child Development* 1956, *20,* (No. 1).

Kovelman, J.A., and **A.B. Scheibel,** A neurohistological correlate of schzopherenic, *Biological Psychiatry,* 1984, *19,* 1601.

Kraepalin, E., *One Hundred Years of Psychiatry,* New York : The Citadel Press, 1926.

Krantz, D.S., and **Glass, D.C.** Personality: behavioural patterns and Physical illness: Conceptual and theoretical issues. In **W.D. Cemetry** (ed.), *Handbook of behavioural Science*, New York: Guilford, 1984.

Krech, D., and **Crutchfield, R.S.**, *Elements of Psychology*, Alfred A. Knopf, 1959.

Krech, D., and **Crutchfield, R.S.** *Theory and Problems of Social Psychology*, McGraw-Hill Book Company, 1944.

Kretschmer, *Physique and Character*, London : Kegan Paul, 9, 1925.

Kubis, J.F. cited in **Smith, B.M.** The polygraph, In **Atkinson, R.C.** (Ed) *Contemporary Psychology*, San Francisco, Freeman, 1962.

Kubler, Ross, *E. Death and dying*. New York: Macmillan, 1969.

Kugal, R.B., and **Parsons, M.H.** *Children of depreivation: changing course of familial retardation*. Washinton, D.C. Department of Health, Education and Welfare Children's Bureau, 1967.

Kulpe, *Outlines of Psychology*, 1895.

Lackaley, A., **Ortiz, Y.**, and **Hepbor, C.** Social categorization and discreminatory behaviour: Extinguishing the minimal intergroup Discremination effect. *Journal of Personality and Social Psychology*, 1980, *19*, 773-783.

Ladd-Franklin, C. On theories of light sensation, *Mind* 1893, *2*, 473-489.

Lane, H.L. **Cataria, A.C.** and **Stevens, S.S.** Voice level : Autophonic Scale, Perceived Loudness and effect of side tone. *Jl. Accoust, Soc., Amer*, 1961, *33*, 160-167.

Lapiere, R.T. and **Farnsworth, P.R.** *Social Psychology*, New York: McGraw-Hill Book Company, 1949.

Lashley, K.S. Cerebral function in learning, *Psychobiology*, 1920, *2*, 55-136.

Lashley, K.S. Studies of cerebral function in Learning IV : Vicarious function after the destruction of the visual area, *Amer. J. Physiol*, 1922, *89*, 44-71.

Lashley, K.S. *Brain Mechanism and Intelligence*, Chicago : University of Chicago Press, 1929.

Lashley, K.S. Integrative function of the cerebral Cortex, *Physio, Rev.*, 1933, *13*, 1-42.

Lashley, K.S. In search of engram. *Symposia of the Society for Experimental Biology*, 1950, *4*, 454-482.

Lasswell, H.D., *Psychopathalogy and Politics*, Chicago: Chicago University Press, 1930.

Lavy, S.M., **Herberman, R.B.**, **Walurish, A.**, **Schlier, B.**, and **Afppman, M.** Prognostic risk: Measurement in primary breast cancers by behavioural and immunological parameters, *Health Psychology*, 1985, *4*, 99-113.

Lazarus, A.A. Behaviour therapy and beyond, New York: McGraw-Hill, 1971.

Lebon, G. The *Crowd*, London : Unwin, 1925.

Lemert, E.M. Paranoic and the dynamies exclusion, *Sociometry*, 1962, 25, 2-25.

Lenner, W.G., **Gibbs, E.L.**, and **Gibbs, F.A.** Inheritance of cerebral dysnythmic and epilepsy, *Archieves of Neurological Pshychology*, 1940, *44*, 115.

Lerner, I.M. *Heredity, evolution and society*, Sanfrancisco, W.A. Freeman and Co., 1968.

Leukel, F. *Introduction to Physiological Psychology*, Saint Louis: The C.V. Mosby Company, 1968.

Levitt, E.E. Research in psychotherapy, with children, In **A.E. Bergin,** and **S.L. Garfied** (eds), *Handbook of Psychotherapy and behaviour change: An empirical analysis,* New York: Wiley, 1971.

Levitt, M. Contingent feedback, familiarization and infant affect: How a stranger becomes a friend? *Developmental Psychology,* 1980, *16*, 426-482.

Lewin, K. *Field theory of social science,* New York: Harper and Row, 1951.

Lewin, K. Group decision and social change, In *Social Psychology* (ed), **Maccoby, Newcomb,** and **Heartly,** New York: Henry Holt and Co., 1958.

Libert, R.M., Paulos, R.M., and **Marmor, G.S.,** *Developmental Psychology,* 2nd ed., New Delhi: Prentice-Hall of India Pvt., Ltd., 1970.

Lindgren, H.C., An *Introduction to Social Psychology,* (2nd. ed), New Delhi: Wiley Eastern Private Limited, 1974.

Littel, I. August - 17, Sad country tunes may turn your brown eyes red, Albeny, *Times Union, 8-1-1981.*

Loftus, E.F., and **Loftus, G.R.,** On the performance of stored information in the human brain, *American Psychologist,* 1980, *35*, 409-420.

Lorenz, K. *King Solomon's ring,* New York: Croswell, 1952.

Leronz, K. The conception of instinctive behaviour, (1937), In **C.H. Schiller** (ed.), *Instinctive behaviour,* New York, International University Press, 1957.

Lorenz, K. *On aggression,* New York: Harcourt, Brace World, 1966.

Lubin, W.H., and **Goodnow, J.J.** Impaired performance with acute sleep loss, *Psychological Monograph,* 1959, 73.

Luh, C.W. The condition of retention, *Psychological Monograph,* 1922, *31*, Nos. 142, No.3.

Luria, A.R. *The working brain,* Hermoodwork: England, Penguin, 1973.

McClearn, G.E. The inheritance of behaviour. In **L. Postman** (Ed.), *Psychology in the making,* New York: Knopf, 1963.

McChenhaum, D.H., and **Asurnow, J.** Cognitive behavioural modification and matacognitive development: Implications for the classroom. In **B.C. Kendall,** and **S.D. Hollon,** (eds.), *Cognitive Behavioural Interventions: Theory, Research and Procedures.* New York : Academic Press, 1979.

McClelland, D.C., and **Boyatxis, R.E.** Leadership motive pattern and long term success in management, *Journal of Applied Psychology,* 1982, *62*, 737-743.

McClelland, D.C., and **Atkinson, J.W.** The projective expression of needs, The effect of different intensities of the hunger drive on perception, *Jl. Psychol.* 1948, *25*, 205-222.

McGaugh, J.L. A multi trace view of memory storage processes, In **D. Bovet,** (ed)., *Attuali Orientameds' dalla ricerca Sall' apprendmend to memoria, Academic Nazionale dei linces,* Rome, 1968.

McDougall, W. *Social Psychology,* (8th. ed.), London: Metheun, 1908.

McDowell, A.A., Brown, W.I., and **McTee, A.C.** Sex as a factor in spatial delayed response performance by rheus monkeys, *Journal of Comparative Physiological Psychology,* 1900, *33*, 429-432.

Machover, K. *Personality Projection in the Drawing of the Human Figure,* Springfield : Thomas, 1949.

Mackinnon, D.W. The nature and nurture of creative talent, *American Psychologists,* 1962, *17*, 484-495.

Maddi, S.R. Humanistic Psychology : Allport and Murray, In **J. Wolpman** and **R. Heina** (Eds), *Concept of Personality,* Chicago : Aldine, 1963.

Magendie, F. *Laconsurles function at les miladies da systeme nervous,* Pairs : 1839, 2 Vol.

Magendie, F. *Lahabeuch Der Physiologic,* Tubingen : Ostrander, 1826.

Malzack, R. Neuropsychological basis of pain measurement, *Advances in pain research* 323-341, 1994.

Mander, G., and **Kessen, W.** *Language of psychology,* New York : Wiley, 1959.

Marowitsch, N.S. Transient Globel Amnesia, *Neuroscience and Biobehavioural Review,* 1982, *7*, 35-43.

Marshall, J.F., Richardson, J.S. and **Teitelbann, P.,** Nigrostriatal Bundle damage and the Lateral Hypothalamic Syndrome, *Journal of comparative and physiological psychology,* 1974, *87*, 820-830.

Maslow, A.M. *Psychology of Science - A Reconnaissance,* Chicago Henry Regnery, 1966.

Maslow, A.M. *Motivation and Personality,* (2nd ed) New York : Harper and Row Publishers, 1970.

Maslow, A.M., Emotional blocks to creativity, R. *Individ., Psychol,* 1958, *14*, 51-56.

Maslow, A.M. Self-Actualization and beyond, In **Bugental, J.F.T.** (Ed)., *Challenges of Humanistic Psychology,* New York : McGraw-Hill, 1967.

Masters, W.H. and **Johnson, V.E.** *Human Sexual Responses,* Boston Little Brown, 1966.

Masling, I, The influence of situational and Interpersonal variables in projective testing, *Psychological Bulletin,* 1960, *57*, 65-85.

Mazzucch, A.G., Moretti, P. Cafpra and **M. Parama,** Neuropsychological functions in the following up of transient of Global amensia, *Brain,* 1980, *103*, 161-178.

Mednick, S.A. The associative basis of the creative process, *Phychological Review,* 1962, *69,* 220-252.

Meed, M. *Culture and Commitments,* New York : Natural History Press, Doubleday, 1970.

Meenes, M and **Morton, M.A.,** Characteristics of eidetic phenomenon *J. Gen.Psychol.,* 1936, *14*, 374-391.

Mendel, G. Experiments in plant-hybridization, 1865, Reprited in **J.A. Peters** (ed) *Classic papers in genetic,* Englewood Cliffs N.J. Prentice-Hall Inc., 1859.

Meyer, H.H., Walker, W.B., and **Liwin, G.H.** Motive patterns and risk preference associated with entrepreneurship, *Journal of Abnormal and Social Psychology, 63,* 570-574, 1961.

Micheal, D. and **Treisman, M.** Motion Sciences, *Evolutionary Hypothesis, 197*, 493-495, 1977.

Miller, D.T., and **Ross, M.** Self-serving Biases in the attribution causality: Facts and fictions, *Psychological Bulletin,* 1975, *82*, 813-825.

Millon, T. On the nature of clinical health psychology. In **T. Millon, C., Green** and **R. Meagher,** (eds), *Handbook of Clinical Health Psychology*, New York: Norton Press, 1982.

Milner, P.M. *Physiological Psychology,* New York, Holt Rinetart and Winston, 1970.

Moray, N. *Attention, Selective Processes in Vision, and Hearing,* London Hutchison, Informational Ltd., 1969.

Moray, N. and **O'Brain, J.** Signal detection theory applied to selective listening, *Journal of Acoustical Society of America,* 765-772, 1967.

Morgan, C.T. *Physiological Psychology* (3rd ed) New York : McGraw-Hill, 1965.

Morgan, J.B., The Overcoming of distraction and other resistances, *Archcological Psychology,* 1916, *5*, No. 35.

Morgan, C.T., and **King, R.A.,** *Introduction to Psychology,* (4th ed.), New Delhi : Tata-McGraw-Hill.

Morgan, C.T., and **Murray, H.** Method for investigating fantasies : the thematic appreception test, *Acta Neursol Psychiat,* 1933, *34,* 289-306.

Morgan, C.L. *Introduction to Comparative Psychology,* New York Scribener's, 1894.

Morgan, A.H., Johnson, D.L., and **Hilgard, E.R.** The stability of Hypnotic suggestibility: A longitudinal study, *International Journal of Clinical and Experimental Hypnosis,* 1974.

Morgan, A.H. The Heritability of hypnotic suggestibility of twins. *Journal of Abnormal Psychology,* 1973, *82*, 56-61.

Morre, R.A., and **Selzer, M.L.** Male homosexuality, paranoia and the schizophrenia, *Amer. Jl. Psychiatry,* 1963, *119*, 143-747.

Morris, C.G. *Psychology : An Introduction,* Englewood Cliffs Prentice, Hall, Inc. 1973.

Morries, D. The *Naked ape,* New York, McGraw-Hill, 1968.

Moruzzi, G and **Megoun, H.W.** Brain stem reticular formation and activation of the EEG. *Electrosencophelography and clinical neurophysiology,* 1949, *1*, 455-473.

Moscovitel, M. Do 'Pets' have long or short ears? Mental imagery and neuroimagining, *Trends in Neuroscience,* 1984, *17*, 292-294.

Moskowitz., M.J. and **Orgel, A.R.** *General Psychology,* Boston Honghton Mifflin Company, 1969. '

Muller, J. *Hanbuch des physiologic das menschen,* Coblenz Holscher, 1844, 398.

Muller, J. *Elements of physiology,* Vol. 1, (W. Bely, Trans) London : Taylor and Welton, 1838.

Munn, N.L. *Introduction to psychology,* Boston : Houghton Mifflin, 1962.

Murphy, G. *An Introduction to psychology,* New York : Harper and Row Publishers, 1951.

Murray, H.A. *Exploration in personality,* New York : Oxford University Press, 1938.

Mussen, P. Some personality and social factors related to change in children's attitudes toward Negroes. *Journal of Abnormal and Social Psychology,* 1950, *45*, 423-441.

Myrdal, G. *An American Dilemma,* New York : Harper, 1944.

Navran, I., Communication and adjustment in marriage, *Family Processes,* 1967, *612*, 173-182.

Nelson, P.C., and **Phares, E.J.**, Anxiety discrepancies between need value and expectancy and internal and external control, *Psychological Review, 28*, 663-666, 1975.

Newcomb, T.M. *Personality and Social Change,* New York: Dryden, 1943.

Newcomb, T.M. *Social Psychology,* London: Tavistack Publications, Limited, 1952.

Newman, H.H. Freeman, F.N. and **Holzinger, K.T.** *Twins : A study of heredity and environment,* Chicago : University of Chicago Press, 1937.

Newman, E.B. Forgetting of meaningful material during sleep and waking, *Amer, J. Psychol.,* 1939, *52*, 65-71.

Nezu, A.M. Efficacy of a social problem involving theraphy approach for unipolar depression. *Journal of Counselling and Clinical Psychology,* 1986, *54*, 156-202.

Nihira, K., Foster, R., Sekheas, M., and **Laland, H.,** *AAMD Adaptive behaviour scale,* Washinton, D.C.; American Association Mental Difficiency, 1975.

Norman, W.T. Towards an adequate toxonomy of personality attributes, replicated factor structure in peer nomination personality rating. *Journal of abnormal and social psychology,* 1963, *66*, 574-583.

Ornstein, B.E. *The Psychology of Consciousness,* San Francisco: Freeman, 1972.

Page, M.L. The modification of ascendant behaviour in pre-school children, *Univ., Dowa, Stud., Child Welfare,* 1936, *12,* No. 3., 1-69.

Paivio, A. *Imagery and Verbal Processes,* New York, Holt Rinehart and Winston, 1971.

Panksepp, J. Towards a general psychological theory of emotion, *Behavioural and Brain Science,* 1982, *5*, 407-467.

Papez, J.W. A proposed mechanism of emotion, *Archieves of Neurology and Psychiatry,* 1937, *38*, 704-744.

Paramesh, C.R. *Creativity and Personality,* Madras : Janata Book House, 1972.

Paul, G.L. *Insight versus desensitization in psychotherapy,* Stanford Cliffs: Stanford University Press, 1966.

Paul, G.L. and **Shannon, D.T.** Treatment of anxiety through systematic desensitization therapy groups, *Journal of Abnormal Psychology,* 1966, *71*, 124-135.

Paul. G.L., and **Lentz, R.J.** *Psychological treatment of chronicalmental patients: Milieu versus social leaming programmes,* Cambridge, Mass: Harvard University Press, 1977.

Pavlov, I.P. *The work of the digestive glands,* London : Griffin, 1902.

Pavlov, I.P. *Conditioned reflex, An Investigation of the physiological activity of the cerebral cortex,* London : Oxford, 1927.

Pavlov, I.P. *Lectures in conditioned reflexes* (W.A. Gautt, Trans) New York: International, 1928-A.

Pavlov, I.P. *Lectures on conditioned reflexes,* New York, Liveright, 1928-B.

Pavlov, I.P. The reply of physiologist to psychologists, *psychological review,* 1932, *39*, 91-98.

Penfield, W., and **Jesper, H.,** *Epilepsy and the fundamental Anatomy of the Human Brain,* Boston: Little Brown, 1954.

Penfield, W., and **Roberts, L.,** *Speech and brain mechanism,* Princeton W.J.: Princeton University Press, 1959.

Perls, F.S. *Gestral Therapy: Verbatim,* Moth Utah : Real People Press, 1969.

Peterson, L. and **Peterson, M.J.** Short term retention of individual verbal items, *J. Exp. Psychol,* 1959, *58*, 193-198.

Piaget, J. *The language and thought of the child,* New York : Harcourt, Brace and World, 1926.

Piaget, J. The Genetic approach to the psychology of thought, *Journal of education psychology,* 1961, *52*, 275-281.

Posner, M.D., and **Boies., S.F.**, Component of attention, *Psychological Review, 78,* 391-408, 1976.

Posner, M.K. and **Peterson, S.E.**, The attention system of the brain, *Annual Review of Neuroscience,* 1990, *13*, 25-42.

Postman, L. and **Egan, J.P.** *Experimental psychology, An introduction,* Luthiana : Kalyani Publishers, 1949.

Poulton, E.C. Two-Channel listening, *Journal* of *experimental psychology,* 1953, *46*, 91-96.

Premack, D. Toward empirical behaviour laws: Positive reinforcement, *Psychological Review,* 1959, *66*, 219-233.

Pretty, G.H., and **Seligman, G.** Affect and over justification of facts, *Journal of Personality and Social Psychology,* 1984, 46, 1241-1253.

Proshanaky, H.M. Environmental psychology, and the real world, *American psychologists, 31,* 303-310(6), 1976.

Proshanaky, H.M. Ittalson, W.H., and **Rivlin L.G.** (Eds)., *Environmental Psychology:* Man and his physical setting, New York : Holt Rinehart and Winston, 1970.

Pyle, W.R. and Synder, J.C. The most economical unit for committing to memory, *J. edu. psychol,* 190, 2, 133-142.

Rajamanickam, M. *Concept Achievement Test,* Agra : Rakhi Prashan Publications, 2002

Randell, Jr. J.H. *Philosophy : An Introduction,* New York Baners of Noble, Inc. 1960.

Rappaport, J. *Community Psychology, Values, Research and Action,* New York : Holt Rinehart and Winston, 1977.

Rhine, J.B. *Extrasensory Perception,* Boston : Brace Hemprier, 1964.

Rodnick, E.H., The psychopathology of development : Investigating the etiology of schizophrenia, *Amer., Jl., Orthopsychiatry,* 1968, *38,* 784-798.

Roessler, R., and **N. Greenfield** (Eds)., *Physiological Correlates of psychological disorders,* The University or Wisconsin Press, 1962.

Rogers, C.R. *Counselling and psychotherapy,* Boston: Houghton Mifflin, 1942.

Rogers, S.R. *On becoming a person,* Boston : Houghtn Mifflin, 1961.

Rogers, C. *Client-centred therapy,* Boston: Houghton Mufflin, 1951.

Rogers, C.R. *Carl Rogers on encounter groups,* New York: Harper and Row, 1970.

Rolando, L. *Saggio Sapra la veva.otutture del cerevelloe sapra le junzioni del sisteme nervosa,* Turin, 1809.

Roland, P.E. *Brain and activation,* New York, John Wiley, 1993.

Roland, P.E. and **B. Gulyas,** Visual imagery and visual representation, *Trends in Neuroscience,* 1944, *110*, 281-287.

Romanes, G.J. *Animal Intelligence,* New York : D. Appleton, 1888.

Rorschach, H. *Psychodiagnostics,* New York : Cruna and Stratton, 1941.

Rose, A.M. *The Negro in America,* Boston, Mass : Beacon Press, 1956.

Rosenthal, R. Experimental outcome: Orientation and the results of the psychological experiments. *Psychological Bulletin,* 1964, 61, 605-612.

Rosenzweig, S. *Picture Frustration Study, Adult Fonn,* copyright, 1948.

Rosenzweig, M.R., Liaman, A.L. and **Breadlove, S.M.** *Biological psychology,* Sunderland, Massechusetts : Sinauer Associates, Inc., 1996.

Rosenzweig., M.R., Bennett, E.L., Colombo, A.J., Lee, D.W., and **P.A., Serrans,** Short-term, Intermediate and Long-term memory, *Behauioural Brain Research,* 1993.

Roslansky, J.D. (ed.), *Creativety,* London: North Holland Publishing Company, 1970.

Rosow, I. *Social interaction of the aged,* New York: The Free Press, 1967.

Ross, L. The intuitive psychologist and his short commings: Distortion in the attribution processes. In **Borkoweitz** (ed.), *Advances in Experimental Social Psychology,* No. 10, New York: Academic Press, 1977.

Rotter, D.L. Patient question asking in physician-client interaction, *Health Psychology,* 1983, *3,* 496-509.

Ruch, F.L. *Psychology and Life,* (5th ed.), Chicago: Scott Foreman and Company, 1958.

Rush, A.J., Back A.T., Kavaes, M., Weisseaherger, J., and **Hollon, S.D.** Comparison of the effects of cognitive therapy on helplessness and self-concept, *American Journal of Psychiatry,* 1962, *139,* 862-866.

Russell, B. *An Outline of Philosophy,* (6th ed.), London: George Allen and Unwin Ltd., 1951.

Salzman, L. An evaluation of shock therapy, *Amer. Jl. Psychiatry, 147,* 119-179.

Sandler, J. Aversion methods. In **H. Kanfer.,** and **S.P. Goldstein** (eds), Helping people change : A *Textbook of Methods* (3rd ed.), Elimsford, New York: Pergaman, 1986.

Sarason, I. *Personality: An objective approach.* New York: Wiley, 1966.

Sargent, S.S., and **Williamson, R.S.** *Social Psychology,* (2nd ed.), New York: The Ronald Press, 1958.

Scarle, L.V. The organization of hereditary maze brightness and maze dullness, *Genetic Psychology Monographs,* 1949, *39,* 279-325.

Schachter, S. *Emotions, obesity and crime,* New York: Academic Press, 1971.

Schachter, S., and **Sriger, J.E.,** Cognitive, Social and Physiological determinants of emotional state, *Psychol. Rev.* 1962, *62,* 329-389.

Schacter, D.L. Memory Amensia and frontal lobe dysfunction, *Psychobiology,* 1987, *15,* 21-36.

Schaffer, H.R., and **Emerson, P.F.** Patterns of responses in early human development. *Journal of Child Psychology and Psychiatry,* 1964, *5,* 1-13.

Schaffer, M.F., and **Carver, C.S.,** Optimism, coping and health: Assessment and implications of generalized outcome experiences, *Health Psychology,* 1985, *4,* 219-247.

Schwatz, B., The effect in Philadelphia of Pennsylvania increases penalities for rape and attempted rape. *Jl. Crim. Law. Criminal Police Science,* 1968, 51(4), 501-515.

Scoria, W.B., and **Milnew,** Loss of recent memory after bilateral hippocampus lesions, *Journal of Neurology, Annual Review of Neurosciences,* 1982, *5*, 242-273.

Scsville, W.B., and **Milnec, B.** Loss of recent memory after bilateral hippocampus lesions, *Journal of Neurology, Neurosurgery and Psychiatry,* 1957, *20*, 11-21.

Sears, R.R. Experimental studies of projection, attribution of traits. *Journal of Social Psychology,* 1936, *7*, 151-163.

Semmes, J., and **Jurner, B.** Effects of cortial laisons on somatosensory task, *Journal of Investigation on Dermatology,* 1977, *60*, 181-189.

Shanas, E. The aged report on health problems, In **Simon** and **I.D. Epistein** (eds.), *Aging and Modern Society,* Washington, D.C.: American Psychiatry Society, 1968.

Semmes, J. Hemispheric specialization: A possible clue to mechanism, *Neuropsychologia, 6*, 11-26, 1968.

Shannon, C.E. A mathematical theory of communication, Bull., *Syst, Tech. J.* 1948, *27*, 379-423, 623-656.

Shaw, M.E. *Group dynamics: The Psychology of small group behaviour,* New York: McGraw-Hill, 1971.

Sherif, M. *The psychology of social norms,* New York : Harper and Brothers, 1936.

Sherif, M., and **Cantril, H.** *The Psychology of Ego-Involvements.* New York: John Wiley and Sons, 1958.

Sheilds, J. *Monozygotic twins brought up apart and brought up together,* London: Oxford University Press, 1962.

Shelton, W.H. and **Stevens, S.S.** *The varieties of temperament,* New York : Harper, 1942.

Shils, Authoritarianism: Right and Left in **R. Chrestic, M. Johada Pers** (Ed.), *Studies in the scope of and Methods of the Authoritarian Personality,* Glencoe, Ill. Free Press, 1954.

Singer, Jl. and **Antrobus, J.S.** Eye movements during fantasies Imaginary and Suppressing fantasies, *Arch., Gan. Psychiat,* 1965, *12*, 71-76.

Sinnot, E.W., Dunn, L.C. and **Dobzhansky, T.** *Principles of Genetics,* New York: McGraw-Hill, 1958.

Skeels, H.M. The study of the effects of differential stimulation on mentally retarted children. A follow-up report, *American Journal of Mental Deficiences,* 1942, *46*, 340-300.

Skinner, B.F. *The behaviour of organism: An experimental analysis,* New York : Appleton - Century Crofts, 1938.

Skinner, B.F. Drive and Reflex Strength-II, *Joumal of General psychology,* 1932, *6*, 38-48.

Skinner, B.F., Extinction with and without subgoal reinforcemental, *Jl. Comol. Psychol.,* 1938, *26*, 121-133.

Smith, K.R., Intermittant loud noise and mental performance, *Science, 114*, 132-133, 1951.

Smith, R.K. and **Noble C.E.** Effects of a memonic Technique applied to verbal learning and memory, *Perceptual and Motor Skills,* 1969, *21*, 123-134.

Smith, K.V. and **Smith, M.F.** *Cybemetic principles of learning and Educational Design,* New York : Holt Rinehart and Winston, 1986.

Smith, R.T. A comparison of social environmental factors in monozygotic and dizygotic twins testing on assumption. In **Vendenberg** (Ed) *Methods and goals in human behaviour genetics,* New York, Academic Press, 1965.

Smith S.M., Brown, H.O., Thomas, J.E.P., and **Goodman, L.S.** The look of cerebral effects, *Anesthesiology, 8,* 1-14, 1947.

Smith, D., King, M., and **Hoebal, B.G.,** Lateral hypothalamic control of killing, Evidence for a cholinoceptive mechanism, *Science, 167,* 900-1991, 1970.

Smuts, Jan. C. *Holism and Evolution,* New York : The Viking Press, Inc., 1961.

Sparrow, S.S. Balls, D.A., and **Clechotti, D.V.,** *Vimaland and Adaptive Behaviour Scale,* Circule Pines, Minn: American Guidance Service, 1984.

Spearman, C. General Intelligence, objectively determined and measured, *Amer, J. Psychol,* 1904, *15,* 201-293.

Speer, G., The mental development of children of normal and feebleminded mother, 19th year book, *Net.,* Soc. *Stud. Educ., Part II,* 1940.

Sperry, R.W. Lateral specialization in the surgically separated hemispheres. In. **F.O. Schimitt,** and **F.C. Wordern** (eds), *Neuroscience: 3rd studies programme.* Cambridge: MIT Press, 1974.

Speery, R., Stamm, J., and **Minor, N.** Relearning tests for interocular tranfer following divisions of optic chisma corpus callosurn in cats, *Journal of Comparative Physiological Psychology,* 1956, *40,* 529-533.

Splading, D.A. The first experimental behaviourist, *Journal of general psychology,* 1962, *67,* 299-307.

Spielberger, S.D. (Ed) *Community Psychology : Perspective in Training and Research,* New York : Appleton, Century Crofts, 1970.

Spiken, C.C. and **McCandles, B.R.** The concept of intelligence in the philosophy of science, *psychol, Rev.* 1954, *61,* 255-260.

Spitz, B.R. Motherless infants, *Child Development,* 1949, *20,* 145-155.

Stagner, R. and **Moskowit, T.F.** *Psychology,* New York : McGraw-Hill Book Company, Inc., 1952.

Stampfl, T.C., and **Lewis, D.J.** Essentials of impulsive theory; A learning theory based on psychodynamic behavioural therapy, *Journal of Abnormal Psychology,* 1967, *72,* 496-503.

Stemberg, R.J., Conway, B.E., Ketron, J.L., and **Bernstein, M.,** People's conceptions of intelligence, *Journal of personality and Social psychology,* Attitudes and Social Cognition, *41,* 37-56, 1981.

Stephens, J.J., Astrup, C. and **Mangrum, J.C.,** Progress in Schizophrenia, *Archives of general psychiatry, 16,* 693-98, 1967.

Stevens, S.S. Mathematics, Measurement and Psychophysics, In **S.S. Stevens** (Ed.), *Handbook of Experimental Psychology,* New York : Wiley, 1951.

Stevens, S.S. On Psychophysical Law. *Psychological Review,* 1957, *64,* 153-181.

Stevens, S.S. *Psychophysics* (Ed.) by Geraldine Stevens, New York : John Wiley of Sons, 1975.

Stevens, S.S. On the brightness of lights and loudness of sounds, *Science,* 1953, *118*, 576 (Abstracts).

Stiller, B. Die pashologic des appettits, *Arch-Verdan,* 1915, *21*, 23.

Stone, P.J., Bales, R.F. and **Namenwirth, J.** The General Enquirer: A computer system for content analysis and retrival based on the sentence as a unit of information, *Behavioural Science, 7,* 1-15, 1962.

Stone, A.A., and **Neele, J.M.** Effects of severe daily events on mood. *Journal of Personality and Social Psychology,* 1984, *46*, 137-144.

Storr, A. *Human Aggression,* New York, Athananm, 1968.

Stunkard, A. Obesity and denial of hunger, *Psychometric Medicine,* 1959, *21*, 281-299.

Strange, J.R., *Abnormal psychology,* Bombay : Tata McGraw-Hill Publishing Company Ltd., 1965.

Strasman, H.D., Thalen, M.B., and **Sahein, E.H.** A prisoner of war syndrome: Apathy as a reaction to severe stress. *American Journal of Psychiatry,* 1956, *112*, 998-1003.

Sullivan, H.S. *The Interpersonal Therapy of Psychiatry,* New York: Norton, 1953.

Suntherland, N.S., *Breakdown,* New York : Stein and Day, 1976.

Super, D.S., Breach, W.F., Jr. and **Shoy, J.B.,** The effect of distraction on test results, *Joumal of Educational Psychology, 38*, 373-378, 1947.

Swis, J., and **Flatcher, B.** The relative efficacy of avoident and non-avoident coping strategies: Ametaanalysis. *Health Psychology,* 1985, *4*, 249-288.

Sydow, Vons, G. and **Rinne,** A very unequal identical twins, *Acta Paediatricia,* 1958, *47*, 163-171.

Tajhl, H., and **Cook, F.I.** The mass media and judgement of risk: Distinguishing impact on personal and social level judgements. *Journal of Personality and Social Psychology,* 1984, *47*, 693-708.

Tallard, G.A. *The pathology of memory,* New York, Academic Press, 1969.

Taylor, H., and **Cook, F.I.** The mass media and judgement of risk: Distinguishing impact on personal and social level judgements. *Journal of Personality and Social Psychology,* 1984, *47*, 693-708.

Taylor, C.W., and **Holland, J.W.** Development and application of tests of creativity, *Review of Educational Research,* 1961, *33*, 91-92.

Teitelbaum P., and **Stellar, E.** Recovery from the failure to act produced by hypothalamic lesions, *Science,* 1954, *120*, 894-895.

Terhune, W.B., The phobic syndrome — its nature and treatment, Jl., Ark Med., Science, 1961, *58*, 230-236.

Terman, L.M. *The measurement of Intelligence,* Boston : Houghton Mifflin, 1916.

Terman, L.M., and **Merrill, Maud, A.** *Stanford-Binet Intelligence Scale,* Boston: Houghton Mifflin, 1960.

Thamas, W.I., and **Znaicki, F.,** *The Polish Peasant in Europeand and America,* Boston: Bodger, 1918.

Thomas, D.W. and **Mayer, J.** The search for the secret of fat, *Psychology today,* 1973, 7, 74-79.

Thompson, W.R. The Inheritance and Development of Intelligence, *Publication of Association for research in Nervous and mental diseases,* 1954, *33*, 209-331.

Thorndike, E.L. *Educational Psychology* (2. Vols), New York : Teacher's College, Columbia University, 1913.

Thorndike, E.L. *The Psychology of arithmetic,* New York : Macmillan, 1922.

Thorndike, E.L., Animal intelligence : An experimental study of the associative processes in Animals, *Psychological Monography,* 1898, No. 8.

Thorndike, E.L., The measurement of creativity, *Teachers* College Record, 1963, *64*, 422-2, 424.

Thorndike, R.L. and **Hagan, E.** *Measurement and Evaluation in Psychology and Education,* (3rd ed.) New Delhi: Wiley Eastern Private Limited, 1970.

Thurstone, L.L. Primary Mental Abilities, *Psychometric Memory graph-1,* 1938.

Tinbergen, N. *The study of instinct,* Oxford, England : Oxford University Press, 1951.

Titchener, E.B. *Experimental Psychology,* New York : Macmillan, 1902.
To-day's Health, 1967, *45*, 82-84.

Tolman, E.E. *Purposive behaviours in animals and men,* New York Appleton-Century-Crofts, 1932.

Torrence, E.P. *Torrence Test of Creativity Thinking : Mannal and scoring guide,* Prenceton, N.J. : Personal Press, 1966.

Trabasso, T.R. and **Bower, G.H.,** *Attention in learning theory and Research,* 1968, New York : John Wiley, 1970.

Treisman, Motion Sickness : Evolutionary Hypothesis, *Science,* 1977, 197, 493-495.

Trevers, Robert, M.M., *Man's Information System:* A Primer for Media Specialists and Educational Technologists - Seranton : Chandler Publishing Co., 1970.

Tryon, R.C. Genetic differences in maze learning in rats, National society for the study of education, *The thirty nine year book.* Blomington, Illinois : Public School Publishing Co., 1940.

Tulving, E. Episodic Semantic Memory, In **E. Tulving** and **W. Donaldson** (Eds.) *Organization of memory,* New York : Academic Press, 1972.

Turk, T.C., Rudy, T.E., and **Salovey, P.** Health protection: Attitudes and behaviour of LPNS, teachers and college students. *Health Psychology,* 1984, *3*, 189-210.

Underwood, B.J. The representativeness of rate verbal learning In **A.N. Melton** (Ed.) *categories of human learning,* New York : Academic Press, 1964, United States Department of Health, 1961.

Valins, S. The perception and labelling of bodily changes as determinants of emotional behaviour. In **P. Back** (Ed.), *Physiological correlates of emotions,* New York : Academic Press, 1970, 229-243.

Vendenberg, S.G. *(Ed.) Methods and Goals in Human Behaviour Genetics*, New York : Academic Press, 1965.

Vernon, P.E., *The structure of human abilities* (Rev. ed.), London, Methuen, 1960.

Vetter, B.W., and **Green, M.** Personality and group factors in the atheists. *The Journal of Abnormal and Social Psychology,* 1934, No. *169*.

Vigotsky, L.S. *Thought and language,* Translated from the Russion and published by Hamsman and G. Vankan, Cambridge Mass : M.T. Press, 1962.

Voeglin, W.I. The treatment of alcoholism by utilizing a conditioned reflex. *American Journal of Medical Science,* 1940, *198*, 402-420.

Wachner, T.S. Interpretation of Spontaneous drawing and paintings, *Genetic Psychol Monogr,* 1946, *33*, 3-70.

Wald, G. and **Jackson, B.** Activity and nutritional deprination, *Proc. Nat., Acad. Sciences,* 1944, *9*, 255-263.

Wallach, M.A., and **Kogan, N.** *Modes of thinking in young children : A study of the creativity intelligence distinction,* New York : Holt Rinehart and Winston, 1985.

Wallach, H. The Perception of Motion, *Scientific American,* July, 1959, 56-69.

Warner, R.B., and **Sugarman, D.B.,** Attribution of personality based physical appearance, speech and handwriting, *Journal of Personality and Social Psychology,* 1986, *50*, 792-799.

Watson, J.B. *Behaviourism,* (2nd ed.) Chicago : The Chicago University Press, 1930.

Watson, J.B. *Behaviourism : An introduction to comparative psychology,* New York : Holt Rinehart and Winston, 1914.

Watson, J.B. *Behaviourism,* New York : Norton, 1924.

Weber, E.H. *Arch. Anat. Physiol,* 1847.

Wechsler, D. Intelligence defined and underined : A relative appraisal, *American Psychology,* 1975, *30*, 135-139.

Wechsler, D. *The measurement and appraisal of adult intelligence* (4th ed.) Baltimore, Williams and Wilkins, 1958.

Weinstein, N.D. Why it would not happen in me: Perceptions of risk favours and susceptibility, *Health Psychology,* 1984, *3*, 401-457.

Weissberg, R.P., Gesten, E.I., Rapkin, H.D., Cowen, E.I., Davidson, E., De Apodaca, M.H., and **McKin, H.J.** Evaluation of a social problem solving training programme for suburban and inner city third grade children, *Journal of Consulting and Clinical Psychology,* 1981, *49*, 251-261.

Wells, J.K., Howard, G.S., Nowlin, W.F., and **Vargas, M.J.,** Presurgical anxiety and postsurgical pain and adjustment: Effects a stress inoculation procedure. *Journal of Consulting and Clinical Psychology,* 1986, *34*, 830-835.

Wenger, M.A., Jones, F.N. and **Jones, M.A.,** *Physiological Psychology,* New York : Holt, 1956.

Werthimer, M. Undersuchlungan zer Lehra Vander, *Gestalt, Psychol,* Forsch, 1923.

Werthimer, M. Experimentable Studies uner des Sahen Van Bewegung, *Z. Psychol,* 1912, *61*, 161-265.

Werthimer, M. Psychomotor coordination of auditory and visual space of birth, *Science,* 1961, *134*, 1962.

White, A.B., Fichtenbaum, L., and **Dollard, J.A.,** A content measure of changes attributable to psychotherapy, *American Jl., Ortho psychiatry, 16*, 41-49, 1966.

Whitty, C.W. and **O.Z. Zanwill** (ed) *Arnnesia,* London : Butterworth, 1966.

Wickens, D.D. and **Meyer, D.R.** *Psychology,* New York : Holt Rinehart and Winston, 1961.

Wiggins, J.S., Wiggins, N. and **Conger, J.C.,** Correlates of hetrosexual somantic preferences, *Journal of Personality and Social Psychology,* 1968, *10,* 82-90.

Wilke, W.H., An Experimental comparison of the speech, radio and printed page as propaganda devices. *Arch. Psychol.* 1934, No. *169.*

Wilson, G.T. Chemical aversion conditioning as a treatment for alcoholics: A Reanalysis. *Behaviour Research and Therapy,* 1987, *25,* 513-515.

Wolf, W. *Endocrinology in Modern Practice,* (2nd ed) Philadelphia Sounders, 1939.

Wolpe, J. *The practice of behavior therapy,* New York: Pergamon Press, 1969.

Wolpert, E.A., Studies in psychophysiology of dreams : II : An electromyographic study of dreaming, *Arch, Gen. Psychiat,* 1960, *2,* 231-241.

Woodworth, R.S. and **Marquis, D.G.** *Psychology,* London : Metheun, 1952.

Wright, P.H. Self referent motivation and the intrinsic quality of friendship. *Journal of Personal and Social Psychology,* 1984, *10,* 115-130.

Writtenborn, J.R., Factorial equation for tests of attention psychometrika, *8,* 19-55, 1963.

Wurdt, W. *Physiologische Psychologic,* Leipzig : Engelmann, 1902.

Yamamota, N. Experimental Scoring manual for Minnestoa tests of creative thinking and writing, Kent, Ohio, *Bureau of Educational Research,* Kent State University, 1904.

Young, P.T. *Emotion in man and animal.* New York : Wiley, 1943.

Young, M. N. and **Gibson, W. B.** *How to develop an Exceptional Memory,* Philadelphia, Chilton, 1962.

Young, K. *Social Attitudes,* New York: Holt, 1931.

Zax, M. and **Specter, G.A.,** *An introduction to community psychology,* New York : Wiley, 1974.

Zigler, E., Abelson, W.K.D., and **Seitz., V.** Motivational factors in the performance of economically disadvantaged children on the peadody picture vocabulary Test, *Child Development, 44,* 294-303, 1973.

Zimmerman, F.T., Burgemeister, D.B. and **T.J. Putman,** Effects of Glutamic Acid on mental functioning in children and in adolescents *Arch. Neurol. Psychiat,* 1946, *56,* 429-506.

INDEX